THIRD EDITION

BIOLOGY
THE ESSENTIALS

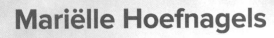

Mariëlle Hoefnagels

THE UNIVERSITY OF OKLAHOMA

MEDIA CONTRIBUTIONS BY

Matthew S. Taylor

McGraw Hill Education

BIOLOGY: THE ESSENTIALS, THIRD EDITION

Published by McGraw-Hill Education, 2 Penn Plaza, New York, NY 10121. Copyright © 2019 by McGraw-Hill Education. All rights reserved. Printed in the United States of America. Previous editions © 2016 and 2013. No part of this publication may be reproduced or distributed in any form or by any means, or stored in a database or retrieval system, without the prior written consent of McGraw-Hill Education, including, but not limited to, in any network or other electronic storage or transmission, or broadcast for distance learning.

Some ancillaries, including electronic and print components, may not be available to customers outside the United States.

This book is printed on acid-free paper.

2 3 4 5 6 7 8 9 LMN 21 20 19 18

ISBN 978-1-259-82491-3
MHID 1-259-82491-8

Executive Portfolio Manager: *Michelle Vogler*
Senior Product Developer: *Anne Winch*
Marketing Manager: *Britney Ross*
Senior Content Project Manager: *Vicki Krug*
Lead Content Project Manager: *Christina Nelson*
Senior Buyer: *Laura Fuller*
Senior Designer: *Tara McDermott*
Senior Content Licensing Specialist: *Lori Hancock*
Cover Image: *©Nadiia Zamedianska/Shutterstock*
Compositor: *MPS Limited*

All credits appearing on page or at the end of the book are considered to be an extension of the copyright page.

Library of Congress Cataloging-in-Publication Data

Hoefnagels, Mariëlle, author.
  Biology: the essentials / Mariëlle Hoefnagels, The University of
  Oklahoma; media contributions by Matthew S. Taylor.
  Third edition. | New York, NY: McGraw-Hill Education, 2018.
  LCCN 2017034401 | ISBN 9781259824913 (alk. paper)
  LCSH: Biology—Study and teaching (Higher)
  LCC QH315.H634 2018 | DDC 570.76—dc23 LC record available
  at https://lccn.loc.gov/2017034401

The Internet addresses listed in the text were accurate at the time of publication. The inclusion of a website does not indicate an endorsement by the authors or McGraw-Hill Education, and McGraw-Hill Education does not guarantee the accuracy of the information presented at these sites.

mheducation.com/highered

Brief Contents

UNIT 1 Science, Chemistry, and Cells

1 The Scientific Study of Life 2
2 The Chemistry of Life 20
3 Cells 48
4 The Energy of Life 68
5 Photosynthesis 84
6 Respiration and Fermentation 98

UNIT 2 DNA, Inheritance, and Biotechnology

7 DNA Structure and Gene Function 112
8 DNA Replication, Binary Fission, and Mitosis 138
9 Sexual Reproduction and Meiosis 154
10 Patterns of Inheritance 170
11 DNA Technology 196

UNIT 3 Evolution and Diversity

12 Forces of Evolutionary Change 216
13 Evidence of Evolution 238
14 Speciation and Extinction 254
15 Evolution and Diversity of Microbial Life 272
16 Evolution and Diversity of Plants 300
17 Evolution and Diversity of Animals 318

UNIT 4 Ecology

18 Populations 356
19 Communities and Ecosystems 372
20 Preserving Biodiversity 402

UNIT 5 Plant Anatomy and Physiology

21 Plant Form and Function 420
22 Reproduction and Development of Flowering Plants 442

UNIT 6 Animal Anatomy and Physiology

23 Animal Tissues and Organ Systems 460
24 The Nervous System and the Senses 476
25 The Endocrine System 502
26 The Skeletal and Muscular Systems 516
27 The Circulatory and Respiratory Systems 534
28 The Digestive and Urinary Systems 556
29 The Immune System 580
30 Animal Reproduction and Development 598

About the Author

©Davenport Photos

Mariëlle Hoefnagels is a professor in the Department of Biology and the Department of Microbiology and Plant Biology at the University of Oklahoma, where she teaches courses in introductory biology, mycology, and science writing. She has received the University of Oklahoma General Education Teaching Award and the Longmire Prize (the Teaching Scholars Award from the College of Arts and Sciences). She has also been awarded honorary memberships in several student honor societies.

Dr. Hoefnagels received her BS in environmental science from the University of California at Riverside, her MS in soil science from North Carolina State University, and her PhD in plant pathology from Oregon State University. Her dissertation work focused on the use of bacterial biological control agents to reduce the spread of fungal pathogens on seeds. In addition to authoring *Biology: The Essentials* and *Biology: Concepts and Investigations,* her recent publications have focused on creating investigative teaching laboratories and integrating technology into introductory biology classes. She also maintains a blog on teaching nonmajors biology, and she frequently gives presentations on study skills and related topics to student groups across campus.

DEDICATION

To my students

Mariëlle Hoefnagels

An Introduction for Students Using This Textbook

I have been teaching nonmajors biology at the University of Oklahoma since 1997 and over that time have encountered many students who fear science in general and biology in particular. The complexity, abstractions, and unfamiliar terms can be overwhelming, and some students believe they can't do well because they're just not "into science." In writing this book, I have focused on students and what you need to be successful in a nonmajors biology class.

In my experience, a big part of the problem is that many students just don't have the right study skills—they focus too much on superficial learning such as memorizing definitions, but they don't immediately grasp the power of *understanding* the material. I've created the following features to help you make the transition from memorizing to understanding.

- **Concept Maps** A new *Survey the Landscape* concept map at the start of each chapter illustrates how the pieces of the entire unit fit together. Each chapter ends with a *Pull It Together* concept map that makes connections between key terms within the chapter. Using these concept maps together will help you understand how the major topics covered throughout the book relate to one another.
- **Learn How to Learn** Each chapter in this book contains a tip that focuses on study skills that build understanding. Don't try to implement them all at once; choose one that appeals to you and add more as you determine what works best for you.
- **What's the Point?** This brief introduction helps explain the importance of the chapter topic. A companion feature is *What's the Point? Applied,* which appears near the end of each chapter and builds on the chapter's content by explaining a wide-ranging topic that is relevant to your life.

- **Summary Illustrations** Created specifically for the summary, these figures tie together the material in a visual way to help you learn relationships among the topics in the chapter. See if you can explain the relationships in your own words, then go back to review any sections you have trouble explaining.
- **Progress Bars** The bars found at the bottom of most pages should help you keep in mind where you are in the chapter's big picture.
- **Why We Care** These boxes reinforce the applications of specific topics to the real world.
- **Burning Question** In this feature, I answer questions from students who are either in my classes or who have written to me with a "burning question" of their own.
- **Miniglossaries** Most chapters have one or more miniglossaries, brief lists of key terms that help you define and distinguish between interrelated ideas. You can use the miniglossaries to create flashcards, concept maps, and other study aids.
- **Scientific Literacy** These new thought questions at the end of each chapter will help you practice thinking like a scientist about relevant social, political, or ethical issues.
- **Connect®** The content in this textbook is integrated with a wide variety of digital tools available in Connect that will help you learn the connections and relationships that are critical to understanding how biology really works.

Although developing study skills is a major step on the pathway to success, a student's mindset is important too. If you believe that you can develop your talents for biology—even if it takes some hard work—then you set the stage for a successful semester. Anyone can be a "science person."

I hope that you enjoy this text and find that the study tips and tools help you develop an understanding of biology.

Mariëlle Hoefnagels

Burning Question 23.1

How does the body react to food poisoning?

In a healthy body, the organ systems operate so seamlessly that you are unlikely to notice them. But suppose you eat food that is tainted with bacteria, viruses, mold, or other contaminants. Within a few hours to a few days, your body's reactions are hard to miss.

First, receptors in the digestive system signal the brain that toxins are in the gut. The brain responds by triggering vomiting, which ejects partially digested food from the stomach. Meanwhile, water moves from the circulatory system into the intestines. This fluid contributes to diarrhea that flushes toxins out of the body.

Vomiting and diarrhea dehydrate the body. In response, endocrine glands release a hormone called ADH (antidiuretic hormone) into the blood. ADH travels in blood vessels and binds to receptors in the kidneys. The kidney's cells respond by saving water, returning it to the blood instead of eliminating it in urine.

Some forms of food poisoning are accompanied by fever, shivers, and fatigue. These are responses of the immune system as it fights invaders. A feverish body is inhospitable to some bacteria. The shivers—a rapid series of muscle contractions—help raise the body's temperature. And what about fatigue? The body uses a lot of energy to maintain a fever and to produce immune cells. Ordinarily, we eat to replenish our reserves. But if food and liquids won't stay down, the digestive system cannot absorb nutrients or water. Both nutrient depletion and dehydration contribute to low energy.

©Alix Minde/PhotoAlto/Getty Images RF

Submit your burning question to marielle.hoefnagels@mheducation.com

(fire): ©Ingram Publishing/SuperStock RF

Nervous system controls the skeletal and muscular systems, which balance and move the body.

Food and drinks previously entered the digestive system.

Respiratory system absorbs O_2 and gives off CO_2.

Integumentary system gives off excess heat produced by active muscles.

Endocrine system helps regulate heart rate, metabolic rate, and body fluid composition.

Circulatory system transports O_2, water, food molecules, hormones, and metabolic wastes.

Wastes accumulate in the digestive and urinary systems for elimination later.

Lymphatic system collects and transports plasma leaking out of blood vessels.

Gametes develop in reproductive system.

Immune system protects the body from infection if injury occurs.

Figure 23.12 Organ System Functions: A Summary.
Photo: ©James Woodson/Getty Images RF

Author's Guide to Using This Textbook

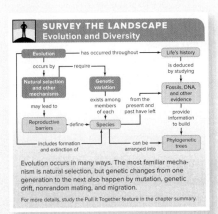

©Davenport Photos

This guide lists key chapter features and describes some of the ways that I use them in my own classes.

Learn How to Learn
Focus on Understanding, Not Memorizing

When you are learning the language of biology, be sure to concentrate on how each new term fits with the others. Are you studying multiple components of a complex system? Different steps in a process? The levels of a hierarchy? As you study, always make sure you understand how each part relates to the whole. For example, you might jot down brief summaries in the margins of your notes, or you could use lists of boldfaced terms in a chapter to make your own concept map.

Learn How to Learn **study tips help students develop their study skills.**

Each chapter has one *Learn How to Learn* study tip, and a complete list is in Appendix F. I present a *Study Minute* in class each week, with examples of how to use these study tips.

SURVEY THE LANDSCAPE
Evolution and Diversity

Evolution occurs in many ways. The most familiar mechanism is natural selection, but genetic changes from one generation to the next also happen by mutation, genetic drift, nonrandom mating, and migration.

For more details, study the Pull It Together feature in the chapter summary.

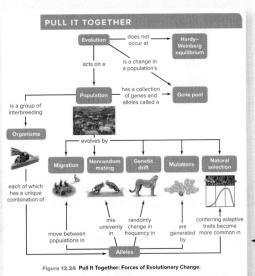

Figure 12.24 **Pull It Together: Forces of Evolutionary Change.**

Concept maps help students see the big picture.

New *Survey the Landscape* concept maps at the start of each chapter illustrate how the pieces of the entire unit fit together. These new figures integrate with the existing *Pull It Together* concept maps in the chapter summary.

After spending class time discussing the key points in constructing concept maps, I have my students draw concept maps of their own.

The *Learning Outline* introduces the chapter's main headings and helps students keep the big picture in mind.

Each heading is a complete sentence that summarizes the most important idea of the section. Students can also flip to the end of the chapter before starting to read; the chapter summary and *Pull It Together* concept map can serve as a review or provide a preview of what's to come.

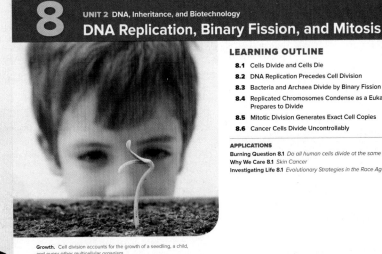

8
UNIT 2 DNA, Inheritance, and Biotechnology
DNA Replication, Binary Fission, and Mitosis

LEARNING OUTLINE

8.1 Cells Divide and Cells Die
8.2 DNA Replication Precedes Cell Division
8.3 Bacteria and Archaea Divide by Binary Fission
8.4 Replicated Chromosomes Condense as a Eukaryotic Cell Prepares to Divide
8.5 Mitotic Division Generates Exact Cell Copies
8.6 Cancer Cells Divide Uncontrollably

APPLICATIONS
Burning Question 8.1 *Do all human cells divide at the same rate?*
Why We Care 8.1 *Skin Cancer*
Investigating Life 8.1 *Evolutionary Strategies in the Race Against Cancer*

Growth. Cell division accounts for the growth of a seedling, a child, and every other multicellular organism.
©PhotoAlto/Getty Images RF

Investigating Life 12.1 | Bacterial Evolution Goes "Hog Wild" on the Farm

Although infectious diseases were once the leading cause of human death, antibiotics had made many bacteria-caused diseases manageable by the mid-1900s. Since that time, bacteria have become resistant not only to the original penicillin but also to the many manufactured antibiotics that followed it. Now antibiotic-resistant bacteria are common, creating new obstacles for physicians treating infectious disease.

Medical practices contribute to the rise of antibiotic-resistant bacteria, but so do farms. Antibiotics promote rapid animal growth when added to the food of cattle, chickens, swine, and other livestock. This practice comes at a cost to public health. The animals' manure contains not only antibiotics but also bacteria that are resistant to the drugs. These microbes swap genes with their neighbors (see figure 8.7). Farms have therefore become breeding grounds for antibiotic-resistant bacteria.

To learn more about this problem, researchers from China and the United States collected manure from three Chinese pig farms where antibiotics are used. Control manure came from pigs that had never been exposed to the drugs. When the team tallied the number of resistance genes in bacterial DNA extracted from each sample, they found that manure from antibiotic-treated animals had many more resistance genes than did control manure (figure 12.A).

But farmers often compost pig manure and then spread it on their croplands. Do the genes persist under those conditions? To find out, the researchers collected samples from compost piles and from the soil in nearby fields. DNA analysis revealed that compost and soil from farms using antibiotics had more diverse resistance genes than did soil from a forest.

These results have serious implications, and not just for farm workers. Since composted animal manure is spread over fields, crops may become contaminated with antibiotic-resistant bacteria. Meat from treated livestock may also harbor resistance genes. When we eat the crops or the meat, bacteria in our intestines may take up the resistance genes.

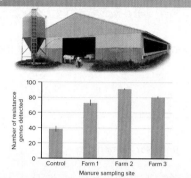

Figure 12.A Antibiotic Resistance on the Farm. The diversity of antibiotic-resistance genes was significantly higher in the manure of antibiotic-fed pigs than in that of untreated animals. Error bars represent the standard error of the mean (see appendix B).

Changes in government policy and consumer awareness may soon decrease the use of antibiotics on farms. As demand for meat from antibiotic-free animals grows, farmers will have an economic incentive to find alternatives to the drugs. The need for change is urgent because some antibiotics may become useless if current practices continue. Evolution never stops, but a thorough understanding of natural selection and bacteria can help us slow the rise of antibiotic resistance.

Source: Zhu, Yong-Guan, and seven coauthors, including James M. Tiedje. 2013. Diverse and abundant antibiotic resistance genes in Chinese swine farms. *Proceedings of the National Academy of Sciences*, vol. 110, pages 3435–3440.

Investigating Life **boxes focus on what introductory science students need:** an understanding of the process of science, an ability to interpret data, and an awareness of how scientific research contributes to our understanding of evolution.

Each box describes a real experiment focusing on an evolutionary topic related to the chapter's content. The studies touch on concepts found in other units; you can encourage students to draw a concept map illustrating the relationships between ideas. You might also use the case as a basis for discussion of the nature of science.

Assignable Connect activities contain questions focused on the process of science, data interpretation, and how the study contributed to our understanding of evolution.

What's the Point? and *What's the Point? Applied* **boxes help relate chapter topics to life outside the classroom.**

These boxes can be used as a starting point for traditional lecture or as the basis for class discussion.

What's the **Point?** ▼

©Comstock Images RF

"I wish I had your metabolism!" Perhaps you have overheard a calorie-counting friend make a similar comment to someone who stays slim on a diet of fattening foods.

In that context, the word *metabolism* means how fast a person burns food. But biochemists define metabolism as all of the chemical reactions that build and break down molecules within any cell. How are these two meanings related?

Interlocking networks of metabolic reactions supply the energy that every cell needs to stay alive. In humans, teams of metabolizing cells perform specialized functions such as digestion, muscle movement, hormone production, and countless other activities. It all takes a reliable energy supply—food, which each of us "burns" at a different rate.

This chapter describes the fundamentals of metabolism, including how cells organize, regulate, and fuel the chemical reactions that sustain life.

What's the **Point?** ▼ APPLIED

Metabolism describes all the chemical reactions in a cell. Because our cells always lose energy as heat, they require constant energy input to continue fueling their reactions. So the familiar definition of metabolism—how fast a person burns calories in food—relates to the rate at which cellular reactions are occurring. What can you do to make your cells use the energy in food more quickly?

Exercise speeds up the body's energy metabolism in several ways. Immediately after exercise, cells work to rebuild ATP and other energy reserves, so caloric demands are high. Also, body temperature remains elevated for hours after exercise, speeding chemical reactions and contributing to increased metabolism. Regular exercise also increases the size of muscle cells, which require more energy than fat cells even when at rest. Exercise also increases the abundance of enzymes and other proteins that regulate energy metabolism. For example, proteins that

©Corbis RF

transport fatty acids into cells become more numerous after one to two weeks of exercise, providing cells with easier access to energy.

Caffeine may also accelerate metabolism. Although caffeine contains zero calories, many people can attest to the "energy boost" that it provides. Caffeine increases the release of fatty acids into the blood and raises the heart rate, giving cells quick access to energy reserves. However, studies have shown that getting too little sleep (a side effect of excess caffeine) disturbs normal metabolism.

Finally, metabolism slows down when the body receives too few calories. Hormones then signal the body that it is entering a starved state. In response, cells begin to conserve energy via several mechanisms. One way to keep your metabolism high is therefore to maintain your blood sugar level by eating multiple small, healthy meals throughout the day.

Burning Questions **cover topics that students wonder about.**

Every chapter in the book answers one or more *Burning Questions,* encouraging readers to ask questions of their own. I ask my students to write down a Burning Question on the first day of class. I answer all of them during the semester, whenever a relevant topic comes up in class.

Burning Question 5.1

Why do leaves change colors in the fall?

Most leaves are green throughout a plant's growing season, although there are exceptions; some ornamental plants, for example, have yellow or purple foliage. The familiar green color comes from chlorophyll *a,* the most abundant pigment in photosynthetic plant parts.

But the leaf also has other photosynthetic pigments. Carotenoids contribute brilliant yellow, orange, and red hues. Purple pigments, such as anthocyanins, are not photosynthetically active, but they do protect leaves from damage by ultraviolet radiation.

Carotenoids are less abundant than chlorophyll, so they usually remain invisible to the naked eye during the growing season. As winter approaches, however, deciduous plants prepare to shed their leaves. Anthocyanins accumulate while chlorophyll degrades, and the now "unmasked" accessory pigments reveal their colors for a short

time as a spectacular autumn display. These pigments soon disappear as well, and the dead leaves turn brown and fall to the ground.

Spring brings a flush of fresh green leaves. The energy to produce the foliage comes from glucose the plant produced during the last growing season and stored as starch. The new leaves make food throughout the spring and summer, so the tree can grow—both above the ground and below—and produce fruits and seeds.

As the days grow shorter and cooler in autumn, the cycle will continue, and the colorful pigments will again participate in one of nature's great disappearing acts.

Submit your burning question to marielle.hoefnagels@mheducation.com

(leaves): ©Carlos E. Santa Maria/Shutterstock RF

The chapter summary highlights key points and terminology from the chapter.

Chapter summary illustrations help students see the big picture.

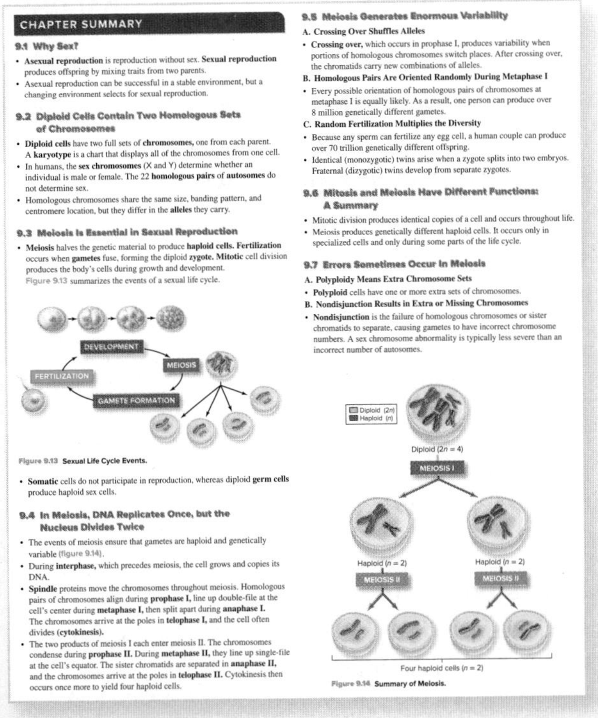

CHAPTER SUMMARY

9.1 Why Sex?
- **Asexual reproduction** is reproduction without sex. **Sexual reproduction** produces offspring by mixing traits from two parents.
- Asexual reproduction can be successful in a stable environment, but a changing environment selects for sexual reproduction.

9.2 Diploid Cells Contain Two Homologous Sets of Chromosomes
- **Diploid cells** have two full sets of **chromosomes**, one from each parent. A **karyotype** is a chart that displays all of the chromosomes from one cell.
- In humans, the **sex chromosomes** (X and Y) determine whether an individual is male or female. The 22 **homologous pairs** of **autosomes** do not determine sex.
- Homologous chromosomes share the same size, banding pattern, and centromere location, but they differ in the **alleles** they carry.

9.3 Meiosis Is Essential in Sexual Reproduction
- **Meiosis** halves the genetic material to produce **haploid cells**. **Fertilization** occurs when **gametes** fuse, forming the diploid **zygote**. **Mitotic** cell division produces the body's cells during growth and development. Figure 9.13 summarizes the events of a sexual life cycle.

Figure 9.13 Sexual Life Cycle Events.

- **Somatic** cells do not participate in reproduction, whereas diploid **germ cells** produce haploid sex cells.

9.4 In Meiosis, DNA Replicates Once, but the Nucleus Divides Twice
- The events of meiosis ensure that gametes are haploid and genetically variable (figure 9.14).
- During **interphase**, which precedes meiosis, the cell grows and copies its DNA.
- **Spindle** proteins move the chromosomes throughout meiosis. Homologous pairs of chromosomes align during **prophase I**, line up double-file at the cell's center during **metaphase I**, then split apart during **anaphase I**. The chromosomes arrive at the poles in **telophase I**, and the cell often divides (**cytokinesis**).
- The two products of meiosis I each enter meiosis II. The chromosomes condense during **prophase II**. During **metaphase II**, they line up single-file at the cell's equator. The sister chromatids are separated in **anaphase II**, and the chromosomes arrive at the poles in **telophase II**. Cytokinesis then occurs once more to yield four haploid cells.

9.5 Meiosis Generates Enormous Variability
A. Crossing Over Shuffles Alleles
- **Crossing over**, which occurs in prophase I, produces variability when portions of homologous chromosomes switch places. After crossing over, the chromatids carry new combinations of alleles.

B. Homologous Pairs Are Oriented Randomly During Metaphase I
- Every possible orientation of homologous pairs of chromosomes at metaphase I is equally likely. As a result, one person can produce over 8 million genetically different gametes.

C. Random Fertilization Multiplies the Diversity
- Because any sperm can fertilize any egg cell, a human couple can produce over 70 trillion genetically different offspring.
- Identical (monozygotic) twins arise when a zygote splits into two embryos. Fraternal (dizygotic) twins develop from separate zygotes.

9.6 Mitosis and Meiosis Have Different Functions: A Summary
- Mitotic division produces identical copies of a cell and occurs throughout life.
- Meiosis produces genetically different haploid cells. It occurs only in specialized cells and only during some parts of the life cycle.

9.7 Errors Sometimes Occur in Meiosis
A. Polyploidy Means Extra Chromosome Sets
- **Polyploid** cells have one or more extra sets of chromosomes.

B. Nondisjunction Results in Extra or Missing Chromosomes
- **Nondisjunction** is the failure of homologous chromosomes or sister chromatids to separate, causing gametes to have incorrect chromosome numbers. A sex chromosome abnormality is typically less severe than an incorrect number of autosomes.

Figure 9.14 Summary of Meiosis.

SCIENTIFIC LITERACY

Review Burning Question 10.1, which describes the inheritance pattern of the metabolic disease called PKU. Today, genetic testing for many disorders is relatively easy and inexpensive. Do prospective parents have an obligation to determine how likely they are to conceive a child with a genetic disorder? What are some possible drawbacks of learning more about one's own genetics? What are some possible advantages to oneself and to society?

New *Scientific Literacy* questions help students understand where biology intersects with ethics, politics, and social issues.

Write It Out **and** ***Mastering Concepts*** **questions are useful for student review or as short in-class writing assignments.**

I compile them into a list of *Guided Reading Questions* that help students focus on material I cover in class. I also use them as discussion questions in Action Centers, where students can come for additional help with course material.

7.5 Mastering Concepts

1. Which steps in gene expression require energy?
2. Why do cells regulate which genes are expressed?
3. How does a repressor protein help regulate the expression of a bacterial operon?
4. Explain how epigenetic modifications change the likelihood of transcription.
5. What is the role of transcription factors in gene expression?

Figure It Out **questions reinforce chapter concepts and typically have numeric answers (supporting student math skills).**

Students can work on these in small groups, in class, or in Action Centers. Most could easily be used as clicker questions as well.

Figure It Out

Compare the number of molecules of ATP generated from 100 glucose molecules undergoing aerobic respiration versus fermentation.

Answer: 3600 (theoretical yield) for aerobic respiration; 200 for fermentation.

 **connect®**

McGraw-Hill Connect® is a highly reliable, easy-to-use homework and learning management solution that utilizes learning science and award-winning adaptive tools to improve student results.

Homework and Adaptive Learning

- Connect's assignments help students contextualize what they've learned through application, so they can better understand the material and think critically.

- Connect will create a personalized study path customized to individual student needs through SmartBook®.

- SmartBook helps students study more efficiently by delivering an interactive reading experience through adaptive highlighting and review.

Over **7 billion questions** have been answered, making McGraw-Hill Education products more intelligent, reliable, and precise.

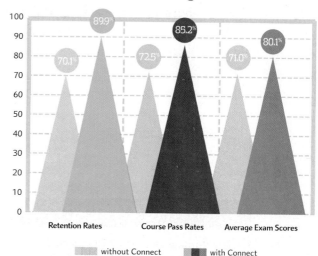

Connect's Impact on Retention Rates, Pass Rates, and Average Exam Scores

Chart data points: 70.1%, 89.9%, 72.5%, 85.2%, 71.0%, 80.1%

Categories: Retention Rates, Course Pass Rates, Average Exam Scores

Legend: without Connect, with Connect

Using **Connect** improves retention rates by **19.8%**, passing rates by **12.7%**, and exam scores by **9.1%**.

73% of instructors who use **Connect** require it; instructor satisfaction **increases** by 28% when **Connect** is required.

Quality Content and Learning Resources

- Connect content is authored by the world's best subject matter experts, and is available to your class through a simple and intuitive interface.

- The Connect eBook makes it easy for students to access their reading material on smartphones and tablets. They can study on the go and don't need internet access to use the eBook as a reference, with full functionality.

- Multimedia content such as videos, simulations, and games drive student engagement and critical thinking skills.

Robust Analytics and Reporting

- Connect Insight® generates easy-to-read reports on individual students, the class as a whole, and on specific assignments.

- The Connect Insight dashboard delivers data on performance, study behavior, and effort. Instructors can quickly identify students who struggle and focus on material that the class has yet to master.

- Connect automatically grades assignments and quizzes, providing easy-to-read reports on individual and class performance.

©Hero Images/Getty Images

Impact on Final Course Grade Distribution

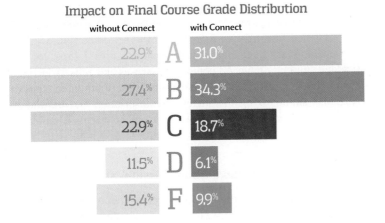

without Connect		with Connect
22.9%	A	31.0%
27.4%	B	34.3%
22.9%	C	18.7%
11.5%	D	6.1%
15.4%	F	9.9%

More students earn **As** and **Bs** when they use **Connect**.

Trusted Service and Support

- Connect integrates with your LMS to provide single sign-on and automatic syncing of grades. Integration with Blackboard®, D2L®, and Canvas also provides automatic syncing of the course calendar and assignment-level linking.

- Connect offers comprehensive service, support, and training throughout every phase of your implementation.

- If you're looking for some guidance on how to use Connect, or want to learn tips and tricks from super users, you can find tutorials as you work. Our Digital Faculty Consultants and Student Ambassadors offer insight into how to achieve the results you want with Connect.

Changes by Chapter

Chapter 1 (The Scientific Study of Life):

Developed new miniglossary comparing sexual and asexual reproduction; revised figure 1.12 to include experimental design.

Chapter 2 (The Chemistry of Life):

Clarified definitions in miniglossary of matter; added periodic table entry and definitions to figure 2.4; developed new table 2.3 to summarize water's characteristics; improved illustration of cellulose in figure 2.19 to show hydrogen bonds; omitted vitamin D as an example of a steroid; updated nutrition label in *What's the Point? Applied* to conform with latest FDA guidelines; wrote new *Investigating Life* on defensive chemicals in ants; simplified and improved summary figures for clarity.

Chapter 3 (Cells):

Clarified functions of free-floating and membrane-bound ribosomes. Added the following ebook-specific learning tools: table summarizing cell junctions; table summarizing the structures in eukaryotic cells.

Chapter 4 (The Energy of Life):

Explained how kinetic energy relates to an object's temperature; made small changes to section 4.1 to clarify the passage on energy transformations; clarified definition of negative feedback; improved illustrations of plant cells in figure 4.17. Added the following ebook-specific learning tool: table showing types of energy.

Chapter 5 (Photosynthesis):

Expanded miniglossary of leaf anatomy; revised caption of figure 5.5 to clarify components of photosystems (based on SmartBook user data); improved description of electron transport chain in the light reactions; clarified passage on C_4 pathway; wrote new *Investigating Life* on solar-powered salamanders. Added the following ebook-specific learning tool: table summarizing photosynthetic pigments.

Chapter 6 (Respiration and Fermentation):

Changed chapter title to complement "Photosynthesis" chapter title; revised caption of figure 6.2 to include the role of electron carriers (based on SmartBook user data); clarified in several places throughout the chapter that *proton* is synonymous with *hydrogen ion (H⁺)*; improved figure 6.9 to show how nitrogen from amino acids becomes a metabolic waste (based on SmartBook user data). Added the following ebook-specific learning tools: table showing where respiration occurs in prokaryotes and eukaryotes; table comparing respiration and photosynthesis.

Chapter 7 (DNA Structure and Gene Function):

Omitted the implication that transcription is a stage of protein synthesis (i.e., proteins are produced only in translation); added new miniglossary to help students understand the relationships between nucleotides, genes, chromosomes, and genomes (based on SmartBook user data); clarified that each cell contains many different tRNA molecules; improved figure 7.8 by zooming in on the codon/anticodon interaction (based on SmartBook user data); added photo of translation to complement the translation art in figure 7.9; expanded coverage of epigenetics, both in the main narrative and in *Burning Question 7.1*; added Ebola and Zika viruses to table 7.2, which lists viruses that infect humans; improved viral replication figure 7.18 to show receptors on the entire cell surface; wrote new subsection within section 7.8 explaining how influenza causes symptoms; improved and expanded miniglossary of viruses; clarified *Investigating Life* section and reworked figure 7.A to include evolutionary tree; improved summary figures 7.25, 7.26, and 7.28; added summary table 7.3 comparing viruses and cells. Added the following ebook-specific learning tools: table describing three types of RNA; tables summarizing the stages of transcription and translation (based on SmartBook user data); table summarizing regulated points in protein production.

Chapter 8 (DNA Replication, Binary Fission, and Mitosis):

Improved definitions in miniglossary of cell division; used the words *align* and *line up* consistently (in referring to chromosome movements) to conform with changes in chapter 9; modified *Burning Question 8.1* to include cancer cells; briefly mentioned newer cancer therapies (such as immunotherapy); wrote new *Investigating Life* essay that explains how evolutionary principles can be used in planning chemotherapy; added miniglossary of cell division terms to chapter summary.

Chapter 9 (Sexual Reproduction and Meiosis):

Used the words *align, line up,* and *orient* consistently (when referring to chromosome movements); explicitly listed in the narrative three mechanisms that generate genetic variability and added new *Figure It Out* problem in section 9.5 (based on SmartBook user data); defined recombinant and parental chromatids to improve consistency with chapter 10 and added both terms to a miniglossary; revised figure 9.15 (*Pull It Together*) to improve the connections among the terms.

Chapter 10 (Patterns of Inheritance):

Clarified some headings and subheadings to better reflect chapter content; changed alleles for yellow and green seeds from *G* and *g* to *Y* and y in art and narrative; clarified cookbook analogy by relating it back to chapter 7; added an explanation for why certain alleles are recessive; clarified that cells with incorrect chromosome numbers may not have exactly two alleles per gene; reworked *Burning Question 10.1* to focus more on the warning label; improved miniglossary of genetic terms by adding some terms and revising others; improved illustration of test cross (figure 10.8); clarified explanation of the product rule; more clearly distinguished recombinant chromatids from recombinant offspring (based on SmartBook user data); clarified explanation of ABO blood group system; improved explanations of pleiotropy and many gene/one phenotype situations; connected environmental effects on gene expression to epigenetics; reworked figure 10.25 to add the frequency of each possible skin color phenotype; updated *Investigating Life* essay to include two-toxin strategy for slowing the evolution of Bt-resistant insects; added new summary figure 10.26 to show the connection between mutations and Punnett squares; improved summary figure 10.27 to better illustrate the light bulb analogy. Added the following ebook-specific learning tools: new figure depicting the P, F1, and F2 generations (based on SmartBook user data); new summary figure showing a timeline that includes the main genetics-related events described in the chapter.

Chapter 11 (DNA Technology):

Expanded passage on ethical issues related to transgenic organisms; added content on high-throughput DNA sequencing methods; updated data on DNA exonerations; added content on cloning in plants, including a new illustration (figure 11.10); added new subsection to section 11.4 on CRISPR-Cas9, including a new illustration (figure 11.15).

Chapter 12 (Forces of Evolutionary Change):

Improved figure 12.8 to show the connection between natural selection and DNA; added table listing misconceptions about evolution and showing how a biologist would address each (based on SmartBook user data); added new *Burning Question* about whether there is a "pinnacle of evolution"; modified figure 12.13 to make the Hardy–Weinberg equations more prominent; modified figure 12.14 to include three phenotypes for directional selection; clarified the distinction between the bottleneck effect and natural selection; wrote new *Investigating Life* on antibiotic-resistant bacteria from livestock; reworked figure 12.24 (*Pull It Together*) to make it more informative.

Chapter 13 (Evidence of Evolution):

Added the proposed Anthropocene epoch to geologic timescale (figure 13.2); improved figure 13.15 for clarity and to add a lemur example. Added the following ebook-specific learning tools: miniglossary of estimating a fossil's age; miniglossary of comparative anatomy.

Chapter 14 (Speciation and Extinction):

Made small changes to several evolutionary trees to ensure consistent use of the word *ancestor*; added terms to miniglossary of speciation and extinction; revised *Why We Care 14.1* to add new illustration and information about why extinctions are important; wrote new *Burning Question 14.2* ("Did rabbits come from frogs?"); clarified the relationship between genus and species (based on SmartBook user data); wrote a new *Investigating Life* essay on plant "protection rackets." Added the following ebook-specific learning tools: miniglossary of reproductive barriers; new figure showing multiple ways to depict the same evolutionary relationships.

Chapter 15 (Evolution and Diversity of Microbial Life):

Made small changes to several evolutionary trees to ensure consistent use of the word *ancestor*; clarified that the outer membrane is considered part of the cell wall in bacteria; added miniglossary of prokaryote anatomy and revised miniglossary of prokaryote diversity; reworked figure 15.12 to clarify aerobic and anaerobic habitats; foreshadowed in section 15.2C that proteobacteria and cyanobacteria are related to the bacteria participating in endosymbiosis, then returned to that idea in section 15.3A and in figure 15.19; clarified explanation of nitrogen fixation; referred specifically to human microbiota; wrote new *Burning Question 15.2* about areas on Earth without life; added new figure 15.21 to illustrate the evolution of multicellularity; clarified basidiomycete life cycle in figure 15.35; added illustration (figure 15.37) showing fungi in everyday life; based on heat map data, clarified the differences between arbuscular mycorrhizae and ectomycorrhizae and between endophytes and mycorrhizae; revised figure 15.40 to show resources exchanged between the partners in a lichen. Added the following ebook-specific learning tools: miniglossary of types of algae; table of plasmodial and cellular slime mold life cycles; miniglossary of types of protozoa; miniglossary of fungal anatomy; miniglossary of fungal partnerships.

Chapter 16 (Evolution and Diversity of Plants):

In section 16.1's narrative, clarified relationship between zygote and sporophyte (based on SmartBook user data); reworked figure 16.10 to clarify that fern gametophytes do not self-fertilize; in section 16.4, clarified that ovules develop into seeds in narrative and corresponding art. Added the following ebook-specific learning tool: table listing key plant-adaptations.

Chapter 17 (Evolution and Diversity of Animals):

Clarified arrows depicting gastrulation in figure 17.5; added new miniglossary of arthropod diversity; modified figure 17.30 to better highlight the three groups of primates; clarified that *Ardipithecus* species are extinct and mentioned the recently discovered *H. naledi* fossils; added evolutionary tree to figure 17.C (*Investigating Life*).

Chapter 18 (Populations):

Updated demographic data for the world population in art and narrative; improved explanation of the demographic transition and added new illustration (figure 18.14); updated information on China's one-child policy. Added the following ebook-specific learning tool: miniglossary of population growth.

Chapter 19 (Communities and Ecosystems):

Made small corrections to convection cells in figure 19.4; added new figure 19.14 to illustrate mutualism and commensalism; updated data about mercury in tuna; added new *Burning Question 19.2*, comparing bottled water with tap water; clarified the meaning of the word *eutrophication*; wrote new *Investigating Life* essay on monarch butterfly migration.

Chapter 20 (Preserving Biodiversity):

Added the term *Anthropocene* and a new illustration (figure 20.2) illustrating where human impacts on the biosphere are most intense; mentioned the acronym HIPPO at the start of the chapter; added landfills as a source of water pollution; improved narrative and figure 20.9 explaining acid deposition; updated narrative and improved figure 20.10 explaining greenhouse effect; added graph to figure 20.12 showing decline in the extent of Arctic sea ice; added advice for people who fish to *Burning Question 20.5*; clarified figure 20.19 (*Pull It Together*). Added the following ebook-specific learning tools: table listing consequences of global climate change; miniglossary of pollution.

Chapter 21 (Plant Form and Function):

Added art of shoot apical meristem to figure 21.15; clarified that *axillary bud* and *lateral bud* are synonymous; wrote new *Burning Question 21.2* about controlled burns; clarified that hormones are present in xylem sap; added photo of a wilted plant (figure 21.21) and a corresponding description of why plants wilt when soil is too dry.

Chapter 22 (Reproduction and Development of Flowering Plants):

Clarified passage on flower structure; added miniglossary of the angiosperm life cycle (based on SmartBook user data); clarified passage on coevolution between flowers and pollinators; clarified role of cotyledons in eudicots and monocots; added photo of coconut to figure 22.9 to show a water-dispersed fruit; added new *Why We Care 22.1* on "talking plants"; annotated figure 22.15 to show how photoperiod affects flowering time.

Chapter 23 (Animal Tissues and Organ Systems):

Modified art for simple columnar epithelium in figure 23.2 to better match the accompanying photo; added new *Burning Question 23.1* on the body's reaction to food poisoning; clarified narrative and figure 23.8 to identify the stimulus, sensor, control center, and effector(s); added miniglossary of negative feedback; clarified definition of ectotherm. Added the following ebook-specific learning tools: miniglossary of animal anatomy and physiology; miniglossary of animal tissues; miniglossary of temperature homeostasis.

Chapter 24 (The Nervous System and the Senses):

Added new miniglossary of neuron anatomy; clarified definitions of *membrane potential* and *resting potential* (based on SmartBook user data); clarified why the inside of a resting neuron has a net negative charge; labeled the voltage meters in figures 24.4 and 24.5 to clarify their function; added context to figure 24.13 illustrating the blood–brain barrier; added information about concussions to section 24.6; wrote new *Burning Question 24.2* explaining whether we use 10% of our brain; improved figure 24.19 by showing context for the olfactory bulb and olfactory epithelium; added new miniglossary of vision; clarified in figure 24.24 that the overlying membrane in the cochlea does not consist of cells; expanded description of cochlear implants. Added the following ebook-specific learning tools: miniglossary of membrane potentials; miniglossary of smell and taste; miniglossary of hearing.

Chapter 25 (The Endocrine System):

Added paragraph about negative feedback loops to section 25.1; clarified that internal hormone receptors may be in the cytosol or in the nucleus and elaborated that steroid hormones may either stimulate or inhibit protein production (based on SmartBook user data); completed descriptions of effects of ADH and oxytocin in figure 25.4; reworked *Burning Question 25.1* to focus on endocrine disruptors; adjusted labels in figure 25.7 to add the role of the hypothalamus as a sensor; adjusted labels in figure 25.9 to add the role of the pancreas as a sensor; reworked figure 25.11 showing the correlation between obesity and diabetes; reworked the *What's the Point? Applied* box to focus on chronic stress. Added the following ebook-specific learning tool: summary table of hormones and their functions.

Chapter 26 (The Skeletal and Muscular Systems):

Clarified illustration of scoliosis (figure 26.3); revised figures in section 26.4 for clarity and improved page layout; improved description of the sarcomere and of the cross bridges in the sliding filament model; added a paragraph about sports balms to *Burning Question 26.2*; added miniglossary of the muscular system to the chapter summary. Added the following ebook-specific learning tool: table outlining the steps of muscle contraction.

Chapter 27 (The Circulatory and Respiratory Systems):

Improved consistency between ABO blood type passage in section 27.1 and related material in section 10.6; clarified the roles of the pulmonary and systemic circulation, especially with regard to O_2 and CO_2 (based on SmartBook user data); wrote new *Burning Question 27.3* on extreme exercise; added terms to the miniglossary of circulation; clarified blood pressure monitors in figure 27.13. Added the following ebook-specific learning tools: miniglossary of the heartbeat; miniglossary of breathing.

Chapter 28 (The Digestive and Urinary Systems):

Improved consistency in the use of *ions* and *salts* throughout the chapter (based on SmartBook user data); added information about how a high-fiber diet lowers cholesterol and helps regulate blood sugar; updated figure 28.4 to reflect new nutrition label regulations; added *Burning Question 28.1* about fad diets; clarified that the stomach does not absorb the proteins it begins to digest (based on SmartBook user data); clarified illustration of the large intestine (figure 28.19).

Chapter 29 (The Immune System):

Improved explanation of lymph; clarified narrative, figure 29.7, and figure 29.10 to show clonal selection for both T cells and B cells; added paragraph about cancer immunotherapy; reworked figure 29.13 illustrating the effects of immunodeficiencies; clarified that mast cells and basophils participate in allergies; added new *Burning Question 29.2* about tick-transmitted meat allergies; added narrative about "retraining" the immune system in children with peanut allergies.

Chapter 30 (Animal Reproduction and Development):

Clarified description of external fertilization; improved explanation of how oocytes enter uterine tubes; changed *sexually transmitted diseases* to *sexually transmitted infections* to recognize that not all infections lead to visible disease symptoms; added a labeled sperm cell to figure 30.12 to remind students where the acrosome is (based on SmartBook user data); clarified two descriptions in table 30.4; improved the explanation and illustration (figure 30.15) of the placenta's structure and function; added labels to clarify the stages of childbirth in figure 30.18; added new summary figure 30.20 to illustrate the paths of sperm and egg cells. Added the following ebook-specific learning tools: miniglossary of embryonic support structures; new summary table showing a timeline of human development (based on SmartBook user data).

Acknowledgments

It takes an army of people to make a textbook, and while I don't work with everyone directly, I greatly appreciate the contributions of each person who makes it possible.

Matt Taylor continues to be my right-hand man, participating in every stage of book development; in addition, he has seamlessly integrated the book's approach into our digital assets. His hard work, expertise, and eye for detail have improved every chapter in large and small ways. In addition, Sarah Greenwood has scrutinized every illustration, contributing a valuable student perspective to this book.

I appreciate the help of my colleagues at the University of Oklahoma, including Dr. Doug Gaffin, Dr. Ben Holt, Dr. Heather Ketchum, Dr. Cameron Siler, Dr. Doug Mock, and Lynn Nichols. Helpful colleagues from other institutions include Dr. Tamar Goulet, who has provided insightful comments on LearnSmart prompts.

My team at McGraw-Hill is wonderful. Thank you to Managing Director Thomas Timp and Executive Portfolio Manager Michelle Vogler, who help us navigate the ever-changing terrain in the publishing world. Product Developer Anne Winch continues to amaze us with her insights and sense of humor. Marketing Manager Britney Ross and Market Development Manager Beth Theisen are skillful and enthusiastic marketers. Emily Tietz continues to provide excellent service in photo selections. I also appreciate Program Manager Angie Fitzpatrick and Content Project Manager Vicki Krug for capably steering the book through production. Also among the talented folks at McGraw-Hill are Lead Digital Product Analyst Eric Weber, Content Licensing Specialist Lori Hancock, Designer Tara McDermott, and Assessment Content Project Manager Christina Nelson. Thanks to all of you for all you do.

MPS produced the art and composed the beautiful page layouts. I appreciate their artistic talent and creative ideas for integrating the narrative with the illustrations.

My family and friends continue to encourage me. Thank you to my parents, my sister, and my in-laws for their pride and continued support. I also thank my friends Kelly Damphousse, Ben and Angie Holt, Michael Markham and Kristi Isacksen, Karen and Bruce Renfroe, Ingo and Andrea Schlupp, Clarke and Robin Stroud, Matt Taylor and Elise Knowlton, Mark Walvoord, and Michael Windelspecht. Smudge and Snorkels occasionally keep me company in the office as well. Finally, my husband, Doug Gaffin, is always there for me, helping in countless large and small ways. I could not do this work without him.

Content Reviewers

Nicole Ashpole
University of Mississippi School of Pharmacy
Eddie Chang
Imperial Valley College
Ray Emmett
Daytona State College
Michele Engel
University of California Bakersfield

M. Cameron Harmon
Fayetteville Technical Community College
Manjushri Kishore
Heartland Community College
Jocelyn Krebs
University of Alaska, Anchorage
Catarina Mata
Borough of Manhattan Community College

Julie Posey
Columbus State Community College
Randal Snyder
SUNY Buffalo State
Robert Stark
California State University, Bakersfield
Ellen Young
College of San Mateo

Focus Group Participants

Nancy Buschhaus
University of Tennessee at Martin
Jocelyn Cash
Central Piedmont Community College
Matthew Cox
Central Carolina Technical College
Christina Fieber
Horry Georgetown Technical College

Michele B. Garrett
Guilford Technical Community College
Bridgette Kirkpatrick
Collin College
Elizabeth A. Mays
Illinois Central College
Reid L. Morehouse
Ivy Tech Community College

Caroline Odewumi
Florida A&M University
Tanya Smutka
Inver Hills Community College
Pamela Thinesen
Century College
Martin Zahn
Thomas Nelson Community College

Detailed Contents

Brief Contents iii | About the Author iv | An Introduction for Students Using This Textbook vi | Author's Guide to Using This Textbook vii | McGraw-Hill Connect® x | Changes by Chapter xii | Acknowledgments xvi

UNIT 1 Science, Chemistry, and Cells

1 The Scientific Study of Life 2

©IM_photo/Shutterstock RF

1.1 What Is Life? 3
A. Life Is Organized 5
B. Life Requires Energy 5
C. Life Maintains Internal Constancy 6
D. Life Reproduces, Grows, and Develops 6
E. Life Evolves 7

1.2 The Tree of Life Includes Three Main Branches 9

1.3 Scientists Study the Natural World 10
A. The Scientific Method Has Multiple Interrelated Parts 10
B. An Experimental Design Is a Careful Plan 12
C. Theories Are Comprehensive Explanations 13
D. Scientific Inquiry Has Limitations 14
E. Biology Continues to Advance 16

Burning Question 1.1 *Are viruses alive? 8*
Why We Care 1.1 *It's Hard to Know What's Bad for You 15*
Burning Question 1.2 *Why am I here? 16*
Investigating Life 1.1 *The Orchid and the Moth 16*

2 The Chemistry of Life 20

©ML Harris/Getty Images

2.1 Atoms Make Up All Matter 21
A. Elements Are Fundamental Types of Matter 21
B. Atoms Are Particles of Elements 22
C. Isotopes Have Different Numbers of Neutrons 23

2.2 Chemical Bonds Link Atoms 24
A. Electrons Determine Bonding 25
B. In an Ionic Bond, One Atom Transfers Electrons to Another Atom 25
C. In a Covalent Bond, Atoms Share Electrons 26
D. Partial Charges on Polar Molecules Create Hydrogen Bonds 28

2.3 Water Is Essential to Life 29
A. Water Is Cohesive and Adhesive 29
B. Many Substances Dissolve in Water 29
C. Water Regulates Temperature 30
D. Water Expands As It Freezes 30
E. Water Participates in Life's Chemical Reactions 31

2.4 Cells Have an Optimum pH 32

2.5 Cells Contain Four Major Types of Organic Molecules 33
A. Large Organic Molecules Are Composed of Smaller Subunits 33
B. Carbohydrates Include Simple Sugars and Polysaccharides 34
C. Proteins Are Complex and Highly Versatile 36
D. Nucleic Acids Store and Transmit Genetic Information 38
E. Lipids Are Hydrophobic and Energy-Rich 40

Why We Care 2.1 *Acids and Bases in Everyday Life 34*
Burning Question 2.1 *What does it mean when food is "organic" or "natural"? 35*
Why We Care 2.2 *Sugar Substitutes and Fake Fats 42*
Burning Question 2.2 *What is junk food? 43*
Investigating Life 2.1 *Chemical Warfare on a Tiny Battlefield 44*

3 Cells 48

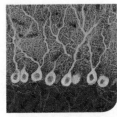
©Thomas Deerinck, NCMIR/ Science Source

3.1 Cells Are the Units of Life 49
A. Simple Lenses Revealed the First Glimpses of Cells 49
B. Microscopes Magnify Cell Structures 49
C. All Cells Have Features in Common 51

3.2 Different Cell Types Characterize Life's Three Domains 52
A. Domains Bacteria and Archaea Contain Prokaryotic Organisms 52
B. Domain Eukarya Contains Organisms with Complex Cells 53

3.3 A Membrane Separates Each Cell from Its Surroundings 54

3.4 Eukaryotic Organelles Divide Labor 56
  A. The Nucleus, Endoplasmic Reticulum, and Golgi Interact to Secrete Substances 57
  B. Lysosomes, Vacuoles, and Peroxisomes Are Cellular Digestion Centers 59
  C. Mitochondria Extract Energy from Nutrients 60
  D. Photosynthesis Occurs in Chloroplasts 60

3.5 The Cytoskeleton Supports Eukaryotic Cells 62

3.6 Cells Stick Together and Communicate with One Another 64

Burning Question 3.1 *Is it possible to make an artificial cell from scratch? 55*
Why We Care 3.1 *Most of Your Cells Are Not Your Own 57*
Investigating Life 3.1 *Bacterial Magnets 65*

4 The Energy of Life 68

©maxpro/Shutterstock RF

4.1 All Cells Capture and Use Energy 69
  A. Energy Allows Cells to Do Life's Work 69
  B. Life Requires Energy Transformations 70

4.2 Networks of Chemical Reactions Sustain Life 71
  A. Chemical Reactions Absorb or Release Energy 71
  B. Linked Oxidation and Reduction Reactions Form Electron Transport Chains 71

4.3 ATP Is Cellular Energy Currency 72
  A. Energy in ATP Is Critical to the Life of a Cell 72
  B. ATP Represents Short-Term Energy Storage 73

4.4 Enzymes Speed Reactions 74
  A. Enzymes Bring Reactants Together 74
  B. Many Factors Affect Enzyme Activity 74

4.5 Membrane Transport May Release Energy or Cost Energy 75
  A. Passive Transport Does Not Require Energy Input 76
  B. Active Transport Requires Energy Input 78
  C. Endocytosis and Exocytosis Use Vesicles to Transport Substances 79

Why We Care 4.1 *Enzymes Are Everywhere 74*
Burning Question 4.1 *Do hand sanitizers work? 75*
Investigating Life 4.1 *Does Natural Selection Maintain Cystic Fibrosis? 80*

5 Photosynthesis 84

© Rodrigo A. Torres/ Glowimages

5.1 Life Depends on Photosynthesis 85

5.2 Photosynthetic Pigments Capture Sunlight 86

5.3 Chloroplasts Are the Sites of Photosynthesis 87

5.4 Photosynthesis Occurs in Two Stages 88

5.5 The Light Reactions Begin Photosynthesis 89
  A. Light Striking Photosystem II Provides the Energy to Produce ATP 90
  B. Electrons from Photosystem I Reduce $NADP^+$ to NADPH 91

5.6 The Carbon Reactions Produce Carbohydrates 92

5.7 C_3, C_4, and CAM Plants Use Different Carbon Fixation Pathways 93

Burning Question 5.1 *Why do leaves change colors in the fall? 89*
Burning Question 5.2 *Does air have mass? 91*
Why We Care 5.1 *Weed Killers 93*
Investigating Life 5.1 *Salamanders Snack on Sugars from Solar Cells 95*

6 Respiration and Fermentation 98

©Three Images/Lifesize/ Getty Images RF

6.1 Cells Use Energy in Food to Make ATP 99

6.2 Cellular Respiration Includes Three Main Processes 100

6.3 In Eukaryotic Cells, Mitochondria Produce Most ATP 101

6.4 Glycolysis Breaks Down Glucose to Pyruvate 102

6.5 Aerobic Respiration Yields Much More ATP than Glycolysis Alone 103
  A. Pyruvate Is Oxidized to Acetyl CoA 103
  B. The Krebs Cycle Produces ATP and High-Energy Electron Carriers 104
  C. The Electron Transport Chain Drives ATP Formation 105

6.6 How Many ATPs Can One Glucose Molecule Yield? 106

6.7 Other Food Molecules Enter the Energy-Extracting Pathways 107

6.8 Fermenters Acquire ATP Only in Glycolysis 108

Why We Care 6.1 *Some Poisons Inhibit Respiration 103*
Burning Question 6.1 *How do diet pills work? 105*
Burning Question 6.2 *What happens during hibernation? 107*
Investigating Life 6.1 *Hot Plants Offer Heat Rewards 109*

UNIT 2 DNA, Inheritance, and Biotechnology

7 DNA Structure and Gene Function 112

©G. Murti/Science Source

7.1 DNA Is a Double Helix 113

7.2 DNA Stores Genetic Information: An Overview 115

7.3 Transcription Uses a DNA Template to Build RNA 116

7.4 Translation Builds the Protein 118
  A. Translation Requires mRNA, tRNA, and Ribosomes 118
  B. Translation Occurs in Three Steps 119
  C. Proteins Must Fold Correctly after Translation 120

7.5 Cells Regulate Gene Expression 121
  A. Operons Are Groups of Bacterial Genes That Share One Promoter 121
  B. Eukaryotic Organisms Use Many Regulatory Methods 121

7.6 Mutations Change DNA 123
  A. Mutations Range from Silent to Devastating 123
  B. What Causes Mutations? 124
  C. Mutations Are Important for Many Reasons 125

7.7 Viruses Are Genes Wrapped in a Protein Coat 126
  A. Viruses Are Smaller and Simpler than Cells 126
  B. Viral Replication Occurs in Five Stages 127

7.8 Viruses Infect All Cell Types 128
  A. Bacteriophages May Kill Cells Immediately or "Hide" in a Cell 128
  B. Animal Viruses May Cause Immediate Cell Death 129
  C. Some Animal Viruses Linger for Years 129
  D. Viruses Cause Diseases in Plants 131

7.9 Drugs and Vaccines Help Fight Viral Infections 131

7.10 Viroids and Prions Are Other Noncellular Infectious Agents 132

Why We Care 7.1 *Poisons That Block Protein Production* 120
Burning Question 7.1 *Is there a gay gene?* 125
Burning Question 7.2 *Why do we get sick when the weather turns cold?* 128
Investigating Life 7.1 *Clues to the Origin of Language* 133

8 DNA Replication, Binary Fission, and Mitosis 138

©PhotoAlto/Getty Images RF

8.1 Cells Divide and Cells Die 139
  A. Sexual Life Cycles Include Mitosis, Meiosis, and Fertilization 139
  B. Cell Death Is Part of Life 140

8.2 DNA Replication Precedes Cell Division 141

8.3 Bacteria and Archaea Divide by Binary Fission 142

8.4 Replicated Chromosomes Condense as a Eukaryotic Cell Prepares to Divide 143

8.5 Mitotic Division Generates Exact Cell Copies 145
  A. DNA Is Copied During Interphase 145
  B. Chromosomes Divide During Mitosis 146
  C. The Cytoplasm Splits in Cytokinesis 148

8.6 Cancer Cells Divide Uncontrollably 148
  A. Chemical Signals Regulate Cell Division 148
  B. Cancer Cells Are Malignant 149
  C. Cancer Treatments Remove or Kill Abnormal Cells 149
  D. Genes and Environment Both Can Increase Cancer Risk 150

Burning Question 8.1 *Do all human cells divide at the same rate?* 145
Why We Care 8.1 *Skin Cancer* 150
Investigating Life 8.1 *Evolutionary Strategies in the Race Against Cancer* 150

9 Sexual Reproduction and Meiosis 154

©IT Stock/age fotostock RF

9.1 Why Sex? 155

9.2 Diploid Cells Contain Two Homologous Sets of Chromosomes 156

9.3 Meiosis Is Essential in Sexual Reproduction 157

9.4 In Meiosis, DNA Replicates Once, but the Nucleus Divides Twice 158

9.5 Meiosis Generates Enormous Variability 160
  A. Crossing Over Shuffles Alleles 160
  B. Homologous Pairs Are Oriented Randomly During Metaphase I 161
  C. Random Fertilization Multiplies the Diversity 162

9.6 Mitosis and Meiosis Have Different Functions: A Summary 162

9.7 Errors Sometimes Occur in Meiosis 164
  A. Polyploidy Means Extra Chromosome Sets 164
  B. Nondisjunction Results in Extra or Missing Chromosomes 164

Burning Question 9.1 *If mules are sterile, then how are they produced? 160*
Why We Care 9.1 *Multiple Births 164*
Investigating Life 9.1 *Evolving Germs Select for Sex in Worms 167*

10 Patterns of Inheritance 170

10.1 Chromosomes Are Packets of Genetic Information: A Review 171

10.2 Mendel's Experiments Uncovered Basic Laws of Inheritance 172
  A. Dominant Alleles Appear to Mask Recessive Alleles 173

©Rick Gomez/Corbis/Getty Images

  B. For Each Gene, a Cell's Two Alleles May Be Identical or Different 174

10.3 The Two Alleles of a Gene End Up in Different Gametes 175
  A. The Simplest Punnett Squares Track the Inheritance of One Gene 175
  B. Meiosis Explains Mendel's Law of Segregation 176

10.4 Genes on Different Chromosomes Are Inherited Independently 178
  A. Tracking Two-Gene Inheritance May Require Large Punnett Squares 178
  B. Meiosis Explains Mendel's Law of Independent Assortment 178
  C. The Product Rule Is a Useful Shortcut 178

10.5 Genes on the Same Chromosome May Be Inherited Together 180
  A. Genes on the Same Chromosome Are Linked 180
  B. Studies of Linked Genes Have Yielded Chromosome Maps 181

10.6 Inheritance Patterns Are Rarely Simple 182
  A. Incomplete Dominance and Codominance Add Phenotype Classes 182
  B. Relating Genotype to Phenotype May Be Difficult 182

10.7 Sex-Linked Genes Have Unique Inheritance Patterns 184
  A. X-Linked Recessive Disorders Affect More Males than Females 184

  B. X Inactivation Prevents "Double Dosing" of Proteins 184

10.8 Pedigrees Show Modes of Inheritance 186

10.9 Most Traits Are Influenced by the Environment and Multiple Genes 188
  A. The Environment Can Alter the Phenotype 188
  B. Polygenic Traits Depend on More than One Gene 189

Burning Question 10.1 *Why does diet soda have a warning label? 174*
Burning Question 10.2 *Is male baldness really from the female side of the family? 186*
Why We Care 10.1 *The Origin of Obesity 188*
Investigating Life 10.1 *Heredity and the Hungry Hordes 190*

11 DNA Technology 196

11.1 DNA Technology Is Changing the World 197

11.2 DNA Technology's Tools Apply to Individual Genes or Entire Genomes 198
  A. Transgenic Organisms Contain DNA from Multiple Sources 198

©Tony Gutierrez/AP Images

  B. DNA Sequencing Reveals the Order of Bases 200
  C. PCR Replicates DNA in a Test Tube 202
  D. DNA Profiling Detects Genetic Differences 202

11.3 Stem Cells and Cloning Add New Ways to Copy Cells and Organisms 205
  A. Stem Cells Divide to Form Multiple Cell Types 205
  B. Cloning Produces Identical Copies of an Organism 206

11.4 Many Medical Tests and Procedures Use DNA Technology 208
  A. DNA Probes Detect Specific Sequences 208
  B. Preimplantation Genetic Diagnosis Can Screen Embryos for Some Diseases 208
  C. Genetic Testing Can Detect Existing Diseases 209
  D. Gene Therapy Uses DNA to Treat Disease 209
  E. CRISPR-Cas9 Cuts and Edits Specific Genes 210
  F. Medical Uses of DNA Technology Raise Many Ethical Issues 211

Burning Question 11.1 *Is selective breeding the same as genetic engineering? 198*
Burning Question 11.2 *What are the uses of DNA testing? 204*
Why We Care 11.1 *Gene Doping 211*
Investigating Life 11.1 *Weeds Get a Boost from Their Transgenic Cousins 212*

UNIT 3 Evolution and Diversity

12 Forces of Evolutionary Change 216

©Steven Hunt/Stone/Getty Images

12.1 Evolution Acts on Populations 217

12.2 Evolutionary Thought Has Evolved for Centuries 218
  A. Many Explanations Have Been Proposed for Life's Diversity 218
  B. Charles Darwin's Voyage Provided a Wealth of Evidence 219
  C. *On the Origin of Species* Proposed Natural Selection as an Evolutionary Mechanism 220
  D. Evolutionary Theory Continues to Expand 222

12.3 Natural Selection Molds Evolution 223
  A. Adaptations Enhance Reproductive Success 223
  B. Natural Selection Eliminates Poorly Adapted Phenotypes 224
  C. Natural Selection Does Not Have a Goal 224
  D. What Does "Survival of the Fittest" Mean? 225

12.4 Evolution Is Inevitable in Real Populations 226
  A. At Hardy–Weinberg Equilibrium, Allele Frequencies Do Not Change 226
  B. In Reality, Allele Frequencies Always Change 227

12.5 Natural Selection Can Shape Populations in Many Ways 228

12.6 Sexual Selection Directly Influences Reproductive Success 230

12.7 Evolution Occurs in Several Additional Ways 231
  A. Mutation Fuels Evolution 231
  B. Genetic Drift Occurs by Chance 231
  C. Nonrandom Mating Concentrates Alleles Locally 233
  D. Migration Moves Alleles Between Populations 233

Why We Care 12.1 *Dogs Are Products of Artificial Selection 220*
Burning Question 12.1 *Is there such a thing as a "pinnacle of evolution"? 226*
Why We Care 12.2 *The Unending War with Bacteria 229*
Investigating Life 12.1 *Bacterial Evolution Goes "Hog Wild" on the Farm 234*

13 Evidence of Evolution 238

©Martin Shields/ Alamy Stock Photo

13.1 Clues to Evolution Lie in the Earth, Body Structures, and Molecules 239

13.2 Fossils Record Evolution 241
  A. The Fossil Record Is Often Incomplete 241
  B. The Age of a Fossil Can Be Estimated in Two Ways 242

13.3 Biogeography Considers Species' Geographical Locations 243
  A. The Theory of Plate Tectonics Explains Earth's Shifting Continents 243
  B. Species Distributions Reveal Evolutionary Events 244

13.4 Anatomical Comparisons May Reveal Common Descent 245
  A. Homologous Structures Have a Shared Evolutionary Origin 245
  B. Vestigial Structures Have Lost Their Functions 245
  C. Convergent Evolution Produces Superficial Similarities 246

13.5 Embryonic Development Patterns Provide Evolutionary Clues 246

13.6 Molecules Reveal Relatedness 248
  A. Comparing DNA and Protein Sequences May Reveal Close Relationships 248
  B. Molecular Clocks Help Assign Dates to Evolutionary Events 249

Burning Question 13.1 *Does the fossil record include transitional forms? 241*
Why We Care 13.1 *An Evolutionary View of the Hiccups 248*
Investigating Life 13.1 *Evolving Backwards 250*

14 Speciation and Extinction 254

©Kike Calvo/National Geographic/Getty Images

14.1 What Is a Species? 255
  A. Linnaeus Classified Life Based on Appearance 255
  B. Species Can Be Defined Based on the Potential to Interbreed 255

14.2 Reproductive Barriers Cause Species to Diverge 256
  A. Prezygotic Barriers Prevent Fertilization 258
  B. Postzygotic Barriers Prevent Development of a Fertile Offspring 258

14.3 Spatial Patterns Define Two Types of Speciation 259
  A. Allopatric Speciation Reflects a Geographical Barrier 259
  B. Sympatric Speciation Occurs in a Shared Habitat 260
  C. Determining the Type of Speciation May Be Difficult 261

14.4 Speciation May Be Gradual or May Occur in Bursts 262

14.5 Extinction Marks the End of the Line 263

14.6 Biological Classification Systems Are Based on Common Descent 265
  A. The Taxonomic Hierarchy Organizes Species into Groups 265
  B. A Cladistics Approach Is Based on Shared Derived Traits 265

C. Cladograms Depict Hypothesized
Evolutionary Relationships 266

D. Many Traditional Groups Are Not Clades 267

Burning Question 14.1 *Can people watch evolution and speciation in action?* 258

Why We Care 14.1 *Recent Species Extinctions* 264

Burning Question 14.2 *Did rabbits come from frogs?* 267

Investigating Life 14.1 *Plant Protection Rackets May Stimulate Speciation* 268

15 Evolution and Diversity of Microbial Life 272

©Europics/Newscom

15.1 Life's Origin Remains Mysterious 273
A. The First Organic Molecules May Have Formed in a Chemical "Soup" 274
B. Clays May Have Helped Monomers Form Polymers 275
C. Membranes Enclosed the Molecules 276
D. Early Life Changed Earth Forever 276

15.2 Prokaryotes Are a Biological Success Story 277
A. What Is a Prokaryote? 277
B. Prokaryote Classification Traditionally Relies on Cell Structure and Metabolism 278
C. Prokaryotes Include Two Domains with Enormous Diversity 280
D. Bacteria and Archaea Are Essential to All Life 281

15.3 Eukaryotic Cells and Multicellularity Arose More Than a Billion Years Ago 284
A. Endosymbiosis Explains the Origin of Mitochondria and Chloroplasts 284
B. Multicellularity May Also Have Its Origin in Cooperation 286

15.4 Protists Are the Simplest Eukaryotes 287
A. What Is a Protist? 287
B. Algae Are Photosynthetic Protists 287
C. Some Heterotrophic Protists Were Once Classified as Fungi 289
D. Protozoa Are Diverse Heterotrophic Protists 290

15.5 Fungi Are Essential Decomposers 292
A. What Is a Fungus? 292
B. Fungal Classification Is Based on Reproductive Structures 293
C. Fungi Interact with Other Organisms 294

Burning Question 15.1 *Does new life spring from simple molecules now, as it did in the past?* 276

Why We Care 15.1 *Antibiotics and Other Germ Killers* 282

Burning Question 15.2 *Are there areas on Earth where no life exists?* 284

Burning Question 15.3 *Why and how do algae form?* 285

Why We Care 15.2 *Preventing Mold* 296

Investigating Life 15.1 *Shining a Spotlight on Danger* 297

16 Evolution and Diversity of Plants 300

©Joe Scherschel/National Geographic/Getty Images

16.1 Plants Have Changed the World 301
A. Green Algae Are the Closest Relatives of Plants 301
B. Plants Are Adapted to Life on Land 303

16.2 Bryophytes Are the Simplest Plants 306

16.3 Seedless Vascular Plants Have Xylem and Phloem but No Seeds 308

16.4 Gymnosperms Are "Naked Seed" Plants 310

16.5 Angiosperms Produce Seeds in Fruits 312

Burning Question 16.1 *Do all plants live on land?* 303

Burning Question 16.2 *What are biofuels?* 304

Why We Care 16.1 *Gluten and Human Health* 313

Investigating Life 16.1 *Genetic Messages from Ancient Ecosystems* 314

17 Evolution and Diversity of Animals 318

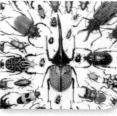
©Imagemore Co, Ltd./ Getty Images RF

17.1 Animals Live Nearly Everywhere 319
A. What Is an Animal? 319
B. Animal Life Began in the Water 319
C. Animal Features Reflect Shared Ancestry 320
D. Biologists Also Consider Additional Characteristics 322

17.2 Sponges Are Simple Animals That Lack Differentiated Tissues 323

17.3 Cnidarians Are Radially Symmetrical, Aquatic Animals 324

17.4 Flatworms Have Bilateral Symmetry and Incomplete Digestive Tracts 325

17.5 Mollusks Are Soft, Unsegmented Animals 326

17.6 Annelids Are Segmented Worms 327

17.7 Nematodes Are Unsegmented, Cylindrical Worms 328

17.8 Arthropods Have Exoskeletons and Jointed Appendages 329
A. Arthropods Have Complex Organ Systems 329
B. Arthropods Are the Most Diverse Animals 330

17.9 Echinoderm Adults Have Five-Part, Radial Symmetry 334

17.10 Most Chordates Are Vertebrates 335

17.11 Chordate Diversity Extends from Water to Land to Sky 337
   A. Tunicates and Lancelets Are Invertebrate Chordates 337
   B. Hagfishes and Lampreys Have a Cranium but Lack Jaws 338
   C. Fishes Are Aquatic Vertebrates with Jaws, Gills, and Fins 338
   D. Amphibians Live on Land and in Water 340
   E. Reptiles Were the First Vertebrates to Thrive on Dry Land 340
   F. Mammals Are Warm, Furry Milk-Drinkers 342

17.12 Fossils and DNA Tell the Human Evolution Story 343
   A. Humans Are Primates 343
   B. Anatomical and Molecular Evidence Documents Primate Relationships 344
   C. Human Evolution Is Partially Recorded in Fossils 345
   D. Environmental Changes Have Spurred Human Evolution 347
   E. Migration and Culture Have Changed *Homo sapiens* 347

Burning Question 17.1 *Are there really only nine kinds of animals?* 331
Why We Care 17.1 *Your Tiny Companions* 333
Burning Question 17.2 *Did humans and dinosaurs ever coexist?* 347
Investigating Life 17.1 *Discovering the "Fishapod"* 349

UNIT 4 Ecology

18 Populations 356

©Donal Husni/NurPhoto via Getty Images

18.1 Ecology Is the Study of Interactions 357

18.2 A Population's Size and Density Change Over Time 358

18.3 Births and Deaths Help Determine Population Size 359

18.4 Natural Selection Influences Life Histories 361
   A. Organisms Balance Reproduction Against Other Requirements 361
   B. Opportunistic and Equilibrium Life Histories Reflect the Trade-Off Between Quantity and Quality 362

18.5 Population Growth May Be Exponential or Logistic 363
   A. Growth Is Exponential When Resources Are Unlimited 363
   B. Population Growth Eventually Slows 364
   C. Many Conditions Limit Population Size 364

18.6 The Human Population Continues to Grow 366
   A. Birth and Death Rates Vary Worldwide 366
   B. The Ecological Footprint Is an Estimate of Resource Use 368

Burning Question 18.1 *How do biologists count animals in the open ocean?* 361
Why We Care 18.1 *Controlling Animal Pests* 365
Investigating Life 18.1 *A Toxic Compromise* 369

19 Communities and Ecosystems 372

©Pedro Ladeira/SambaPhoto/ Getty Images

19.1 Organisms Interact Within Communities and Ecosystems 373

19.2 Earth Has Diverse Climates 374

19.3 Biomes Are Ecosystems with Distinctive Communities of Life 376
   A. The Physical Environment Dictates Where Each Species Can Live 376
   B. Terrestrial Biomes Range from the Lush Tropics to the Frozen Poles 377
   C. Aquatic Biomes Include Fresh Water and the Oceans 381

19.4 Community Interactions Occur Within Each Biome 381
   A. Many Species Compete for the Same Resources 382
   B. Symbiotic Interactions Can Benefit or Harm a Species 383
   C. Herbivory and Predation Link Species in Feeding Relationships 383
   D. Closely Interacting Species May Coevolve 384
   E. A Keystone Species Has a Pivotal Role in the Community 385

19.5 Succession Is a Gradual Change in a Community 386

19.6 Ecosystems Require Continuous Energy Input 388
   A. Food Webs Depict the Transfer of Energy and Atoms 388
   B. Heat Energy Leaves Each Food Web 390
   C. Harmful Chemicals May Accumulate in the Highest Trophic Levels 391

19.7 Chemicals Cycle Within Ecosystems 392
   A. Water Circulates Between the Land and the Atmosphere 393
   B. Autotrophs Obtain Carbon as CO_2 394
   C. The Nitrogen Cycle Relies on Bacteria 395
   D. The Phosphorus Cycle Begins with the Weathering of Rocks 397
   E. Excess Nitrogen and Phosphorus Cause Problems in Water 397

Burning Question 19.1 *Why is there a "tree line" above which trees won't grow?* 377
Why We Care 19.1 *What Happens After You Flush* 388
Why We Care 19.2 *Mercury on the Wing* 390
Burning Question 19.2 *Is bottled water safer than tap water?* 393
Why We Care 19.3 *The Nitrogen Cycle in Your Fish Tank* 396
Investigating Life 19.1 *Winged Migrants Sidestep Parasites* 398

20 Preserving Biodiversity 402

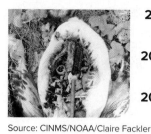

Source: CINMS/NOAA/Claire Fackler

20.1 Earth's Biodiversity Is Dwindling 403

20.2 Many Human Activities Destroy Habitats 404

20.3 Pollution Degrades Habitats 406
- A. Water Pollution Threatens Aquatic Life 406
- B. Air Pollution Causes Many Types of Damage 407

20.4 Global Climate Change Alters and Shifts Habitats 409
- A. Greenhouse Gases Warm Earth's Surface 409
- B. Global Climate Change Has Severe Consequences 411

20.5 Exotic Invaders and Overexploitation Devastate Many Species 412
- A. Invasive Species Displace Native Organisms 412
- B. Overexploitation Can Drive Species to Extinction 413

20.6 Some Biodiversity May Be Recoverable 414
- A. Protecting and Restoring Habitat Saves Many Species at Once 414
- B. Some Conservation Tools Target Individual Species 414
- C. Conserving Biodiversity Involves Scientists and Ordinary Citizens 415

Burning Question 20.1 *What are the best ways to reverse habitat destruction? 406*

Burning Question 20.2 *How can people reduce their contribution to water pollution? 408*

Burning Question 20.3 *What does the ozone hole have to do with global climate change? 409*

Burning Question 20.4 *How can small lifestyle changes reduce air pollution and global climate change? 411*

Burning Question 20.5 *How can people help slow the spread of invasive species? 412*

Burning Question 20.6 *Can everyday buying decisions help protect overharvested species? 413*

Why We Care 20.1 *Environmental Legislation 415*

Investigating Life 20.1 *Up, Up, and Away 416*

UNIT 5 Plant Anatomy and Physiology

21 Plant Form and Function 420

©Bob Gibbons/Alamy Stock Photo

21.1 Vegetative Plant Parts Include Stems, Leaves, and Roots 421

21.2 Soil and Air Provide Water and Nutrients 422
- A. Plants Require 16 Essential Elements 422
- B. Leaves and Roots Absorb Essential Elements 423

21.3 Plant Cells Build Tissues 424
- A. Plants Have Several Cell Types 424
- B. Plant Cells Form Three Main Tissue Systems 426

21.4 Tissues Build Stems, Leaves, and Roots 427
- A. Stems Support Leaves 427
- B. Leaves Are the Primary Organs of Photosynthesis 427
- C. Roots Absorb Water and Minerals, and Anchor the Plant 429

21.5 Plants Have Flexible Growth Patterns, Thanks to Meristems 430
- A. Plants Grow by Adding New Modules 430
- B. Plant Growth Occurs at Meristems 431
- C. In Primary Growth, Apical Meristems Lengthen Stems and Roots 431
- D. In Secondary Growth, Lateral Meristems Thicken Stems and Roots 432

21.6 Vascular Tissue Transports Water, Minerals, and Sugar 434
- A. Water and Minerals Are Pulled Up to Leaves in Xylem 434
- B. Sugars Are Pushed to Nonphotosynthetic Cells in Phloem 436
- C. Parasitic Plants Tap into Another Plant's Vascular Tissue 437

Burning Question 21.1 *What's the difference between fruits and vegetables? 422*

Why We Care 21.1 *Boost Plant Growth with Fertilizer 424*

Burning Question 21.2 *What are controlled burns? 433*

Burning Question 21.3 *Where does maple syrup come from? 437*

Investigating Life 21.1 *An Army of Tiny Watchdogs 438*

22 Reproduction and Development of Flowering Plants 442

©Gay Bumgarner/Alamy Stock Photo

22.1 Angiosperms Reproduce Sexually and Asexually 443

22.2 The Angiosperm Life Cycle Includes Flowers, Fruits, and Seeds 444
- A. Flowers Are Reproductive Organs 445
- B. The Pollen Grain and Embryo Sac Are Gametophytes 445
- C. Pollination Brings Pollen to the Stigma 446
- D. Double Fertilization Yields Zygote and Endosperm 446
- E. A Seed Is an Embryo and Its Food Supply Inside a Seed Coat 447
- F. The Fruit Develops from the Ovary 448
- G. Fruits Protect and Disperse Seeds 449

22.3 Plant Growth Begins with Seed Germination 450

22.4 Hormones Regulate Plant Growth and Development 451
  A. Auxins and Cytokinins Are Essential for Plant Growth 452
  B. Gibberellins, Ethylene, and Abscisic Acid Influence Plant Development in Many Ways 453

22.5 Light Is a Powerful Influence on Plant Life 454

22.6 Plants Respond to Gravity and Touch 455

Burning Question 22.1 *How can a fruit be seedless?* 451
Why We Care 22.1 *Talking Plants* 452
Investigating Life 22.1 *A Red Hot Chili Pepper Paradox* 456

UNIT 6 Animal Anatomy and Physiology

23 Animal Tissues and Organ Systems 460

©James King-Holmes/Science Source

23.1 Specialized Cells Build Animal Bodies 461

23.2 Animals Consist of Four Tissue Types 462
  A. Epithelial Tissue Covers Surfaces 462
  B. Most Connective Tissues Bind Other Tissues Together 464
  C. Muscle Tissue Provides Movement 464
  D. Nervous Tissue Forms a Rapid Communication Network 465

23.3 Organ Systems Are Interconnected 466
  A. The Nervous and Endocrine Systems Coordinate Communication 466
  B. The Skeletal and Muscular Systems Support and Move the Body 466
  C. The Digestive, Circulatory, and Respiratory Systems Work Together to Acquire Energy 467
  D. The Urinary, Integumentary, Immune, and Lymphatic Systems Protect the Body 467
  E. The Reproductive System Produces the Next Generation 468

23.4 Organ System Interactions Promote Homeostasis 468

23.5 Animals Regulate Body Temperature 469

Why We Care 23.1 *Two Faces of Plastic Surgery* 465
Burning Question 23.1 *How does the body react to food poisoning?* 466
Burning Question 23.2 *Can biologists build artificial organs?* 471
Investigating Life 23.1 *Sniffing Out the Origin of Feathers* 472

24 The Nervous System and the Senses 476

©Cary Wolinsky/Getty Images

24.1 The Nervous System Forms a Rapid Communication Network 477

24.2 Neurons Are the Functional Units of a Nervous System 478
  A. A Typical Neuron Consists of a Cell Body, Dendrites, and an Axon 478
  B. The Nervous System Includes Three Classes of Neurons 478

24.3 Action Potentials Convey Messages 479
  A. A Neuron at Rest Has a Negative Charge 480
  B. A Neuron's Membrane Potential Reverses During an Action Potential 480
  C. The Myelin Sheath Speeds Communication 482

24.4 Neurotransmitters Pass the Message from Cell to Cell 482

24.5 The Peripheral Nervous System Consists of Nerve Cells Outside the Central Nervous System 484

24.6 The Central Nervous System Consists of the Spinal Cord and Brain 486
  A. The Spinal Cord Transmits Information Between Body and Brain 486
  B. The Brain Is Divided into Several Regions 486
  C. Many Brain Regions Participate in Memory 488
  D. Damage to the Central Nervous System Can Be Devastating 489

24.7 The Senses Connect the Nervous System with the Outside World 490
  A. Sensory Receptors Respond to Stimuli by Generating Action Potentials 490
  B. Continuous Stimulation May Cause Sensory Adaptation 491

24.8 The General Senses Detect Touch, Temperature, and Pain 491

24.9 The Senses of Smell and Taste Detect Chemicals 492

24.10 Vision Depends on Light-Sensitive Cells 494

24.11 The Sense of Hearing Begins in the Ears 496

Burning Question 24.1 *Do neurons communicate at the speed of light?* 480
Why We Care 24.1 *Drugs and Neurotransmitters* 484
Burning Question 24.2 *Do I really use only 10% of my brain?* 488
Burning Question 24.3 *Do humans have pheromones?* 493
Why We Care 24.2 *Correcting Vision* 495
Burning Question 24.4 *What is an ear infection?* 496
Investigating Life 24.1 *Scorpion Stings Don't Faze Grasshopper Mice* 498

25 The Endocrine System 502

©Purestock/SuperStock RF

25.1 The Endocrine System Uses Hormones to Communicate 503

25.2 Hormones Stimulate Responses in Target Cells 504
   A. Water-Soluble Hormones Trigger Second Messenger Systems 504
   B. Lipid-Soluble Hormones Directly Alter Gene Expression 505

25.3 The Hypothalamus and Pituitary Gland Oversee Endocrine Control 505
   A. The Posterior Pituitary Stores and Releases Two Hormones 507
   B. The Anterior Pituitary Produces and Secretes Six Hormones 507

25.4 Hormones from Many Glands Regulate Metabolism 508
   A. The Thyroid Gland Sets the Metabolic Pace 508
   B. The Parathyroid Glands Control Calcium Level 509
   C. The Adrenal Glands Coordinate the Body's Stress Responses 509
   D. The Pancreas Regulates Blood Glucose 510
   E. The Pineal Gland Secretes Melatonin 511

25.5 Hormones from the Ovaries and Testes Control Reproduction 512

Burning Question 25.1 *What are endocrine disruptors? 508*
Why We Care 25.1 *Anabolic Steroids in Sports 512*
Investigating Life 25.1 *Addicted to Affection 513*

26 The Skeletal and Muscular Systems 516

©Jeff J Mitchell/Getty Images

26.1 Skeletons Take Many Forms 517

26.2 The Vertebrate Skeleton Features a Central Backbone 518

26.3 Bones Provide Support, Protect Internal Organs, and Supply Calcium 519
   A. Bones Consist Mostly of Bone Tissue and Cartilage 519
   B. Bone Meets Bone at a Joint 521
   C. Bones Are Constantly Built and Degraded 521
   D. Bones Help Regulate Calcium Homeostasis 522

26.4 Muscle Movement Requires Contractile Proteins and ATP 522
   A. Actin and Myosin Filaments Fill Muscle Cells 524
   B. Sliding Filaments Are the Basis of Muscle Cell Contraction 524
   C. Motor Neurons Stimulate Muscle Contraction 526

26.5 Muscle Cells Generate ATP in Multiple Ways 527

26.6 Muscle Fiber Types Influence Athletic Performance 528

Why We Care 26.1 *Bony Evidence of Murder, Illness, and Evolution 519*
Burning Question 26.1 *Is creatine a useful dietary supplement? 527*
Burning Question 26.2 *Why does heat soothe sore muscles and joints? 529*
Investigating Life 26.1 *Did a Myosin Gene Mutation Make Humans Brainier? 530*

27 The Circulatory and Respiratory Systems 534

©Philippe Plailly/Science Source

27.1 Blood Plays a Central Role in Maintaining Homeostasis 535
   A. Plasma Carries Many Dissolved Substances 536
   B. Red Blood Cells Transport Oxygen 536
   C. White Blood Cells Fight Infection 536
   D. Blood Clotting Requires Platelets and Plasma Proteins 537

27.2 Animal Circulatory Systems Range from Simple to Complex 538

27.3 Blood Circulates Through the Heart and Blood Vessels 539

27.4 The Human Heart Is a Muscular Pump 540
   A. The Heart Has Four Chambers 540
   B. The Right and Left Halves of the Heart Deliver Blood Along Different Paths 540
   C. Cardiac Muscle Cells Produce the Heartbeat 541
   D. Exercise Strengthens the Heart 542

27.5 Blood Vessels Form the Circulation Pathway 542
   A. Arteries, Capillaries, and Veins Have Different Structures 543
   B. Blood Pressure and Velocity Differ Among Vessel Types 544

27.6 The Human Respiratory System Delivers Air to the Lungs 545
   A. The Nose, Pharynx, and Larynx Form the Upper Respiratory Tract 546
   B. The Lower Respiratory Tract Consists of the Trachea and Lungs 548

27.7 Breathing Requires Pressure Changes in the Lungs 550

27.8 Red Blood Cells Carry Most Oxygen and Carbon Dioxide 551

Burning Question 27.1 *What is the difference between donating whole blood and donating plasma? 537*
Burning Question 27.2 *What causes bruises? 539*
Burning Question 27.3 *If some exercise is good, is more exercise better? 542*
Why We Care 27.1 *Unhealthy Circulatory and Respiratory Systems 549*
Investigating Life 27.1 *In (Extremely) Cold Blood 552*

28 The Digestive and Urinary Systems 556

©Ingram Publishing RF

28.1 Animals Maintain Nutrient, Water, and Ion Balance 557

28.2 Digestive Systems Derive Energy and Raw Materials from Food 558

28.3 A Varied Diet Is Essential to Good Health 559

28.4 Body Weight Reflects Food Intake and Activity Level 560
  A. Body Mass Index Can Identify Weight Problems 560
  B. Starvation: Too Few Calories to Meet the Body's Needs 561
  C. Obesity: More Calories Than the Body Needs 561

28.5 Most Animals Have a Specialized Digestive Tract 562
  A. Acquiring Nutrients Requires Several Steps 562
  B. Digestive Tracts May Be Incomplete or Complete 563
  C. Diet Influences Digestive Tract Structure 564

28.6 The Human Digestive System Consists of Several Organs 565
  A. Muscles Underlie the Digestive Tract 566
  B. Digestion Begins in the Mouth 566
  C. The Stomach Stores, Digests, and Churns Food 566
  D. The Small Intestine Digests and Absorbs Nutrients 567
  E. The Large Intestine Completes Nutrient and Water Absorption 569

28.7 Animals Eliminate Nitrogenous Wastes and Regulate Water and Ions 570

28.8 The Urinary System Produces, Stores, and Eliminates Urine 572

28.9 Nephrons Remove Wastes and Adjust the Composition of Blood 573
  A. Nephrons Interact Closely with Blood Vessels 573
  B. Urine Formation Includes Filtration, Reabsorption, and Secretion 574
  C. Hormones Regulate Kidney Function 574

Burning Question 28.1 *Which diets lead to the most weight loss?* 561
Burning Question 28.2 *What is lactose intolerance?* 568
Why We Care 28.1 *The Unhealthy Digestive System* 570
Why We Care 28.2 *Urinary Incontinence* 572
Burning Question 28.3 *What can urine reveal about health and diet?* 573
Why We Care 28.3 *Kidney Failure, Dialysis, and Transplants* 575
Investigating Life 28.1 *The Cost of a Sweet Tooth* 576

29 The Immune System 580

Source: CDC/James Gathany

29.1 Many Cells, Tissues, and Organs Defend the Body 581
  A. White Blood Cells Play Major Roles in the Immune System 581
  B. The Lymphatic System Produces and Transports Many Immune System Cells 582
  C. The Immune System Has Two Main Subdivisions 582

29.2 Innate Defenses Are Nonspecific and Act Early 583
  A. External Barriers Form the First Line of Defense 583
  B. Internal Innate Defenses Destroy Invaders 584

29.3 Adaptive Immunity Defends Against Specific Pathogens 586
  A. Helper T Cells Play a Central Role in Adaptive Immunity 586
  B. Cytotoxic T Cells Provide Cell-Mediated Immunity 587
  C. B Cells Direct the Humoral Immune Response 588
  D. The Secondary Immune Response Is Stronger Than the Primary Response 590

29.4 Vaccines Jump-Start Immunity 590

29.5 Several Disorders Affect the Immune System 592
  A. Autoimmune Disorders Are Devastating and Mysterious 592
  B. Immunodeficiencies Lead to Opportunistic Infections 592
  C. Allergies Misdirect the Immune Response 593

Why We Care 29.1 *Severe Burns* 584
Why We Care 29.2 *Protecting a Fetus from Immune Attack* 587
Burning Question 29.1 *Why do we need multiple doses of some vaccines?* 591
Burning Question 29.2 *Can people be allergic to meat?* 592
Investigating Life 29.1 *The Hidden Cost of Hygiene* 594

30 Animal Reproduction and Development 598

©UIG via Getty Images

30.1 Animal Development Begins with Reproduction 599
  A. Reproduction Is Asexual or Sexual 599
  B. Development Is Indirect or Direct 600

30.2 Males Produce Sperm Cells 601
  A. Male Reproductive Organs Are Inside and Outside the Body 601
  B. Spermatogenesis Yields Sperm Cells 602
  C. Hormones Influence Male Reproductive Function 603

30.3 Females Produce Egg Cells 604
  A. Female Reproductive Organs Are Inside the Body 604
  B. Oogenesis Yields Egg Cells 605
  C. Hormones Influence Female Reproductive Function 606
  D. Hormonal Fluctuations Can Cause Discomfort 607

**30.4 Reproductive Health Considers Contraception and
Disease 608**

30.5 The Human Infant Begins Life as a Zygote 611
  A. Fertilization Initiates Pregnancy 611
  B. The Preembryonic Stage Ends
     When Implantation Is Complete 612
  C. Organs Take Shape During the Embryonic Stage 614
  D. Organ Systems Become Functional in the Fetal Stage 615
  E. Muscle Contractions in the Uterus Drive Childbirth 616

Burning Question 30.1 *When can conception occur? 608*
Why We Care 30.1 *Substances That Cause Birth Defects 613*
Investigating Life 30.1 *Playing "Dress Up" on the Reef 617*

Appendix A   Answers to Multiple Choice Questions A-1
Appendix B   Brief Guide to Statistical Significance A-2
Appendix C   Units of Measure A-5
Appendix D   Periodic Table of the Elements A-6
Appendix E   Amino Acid Structures A-7
Appendix F   Learn How to Learn A-8

Glossary G-1
Index I-1

THIRD EDITION

BIOLOGY

THE ESSENTIALS

LEARNING OUTLINE

1.1 What Is Life?

1.2 The Tree of Life Includes Three Main Branches

1.3 Scientists Study the Natural World

APPLICATIONS

Burning Question 1.1 *Are viruses alive?*
Why We Care 1.1 *It's Hard to Know What's Bad for You*
Burning Question 1.2 *Why am I here?*
Investigating Life 1.1 *The Orchid and the Moth*

Biology Is Everywhere. Central Park is an oasis of green in New York City, but life thrives in the city's streets and buildings too.
©IM_photo/Shutterstock RF

Learn How to Learn
Real Learning Takes Time

You got good at basketball, running, dancing, art, music, or video games by putting in lots of practice. Likewise, you will need to commit time to your biology course if you hope to do well. To get started, look for the Learn How to Learn tip in each chapter of this textbook. Each hint is designed to help you use your study time productively. With practice, you'll discover that all concepts in biology are connected. The Survey the Landscape figure in every chapter highlights each chapter's place in the "landscape" of the entire unit. Use it, along with the more detailed Pull It Together concept map in the chapter summary, to see how each chapter's content fits into the unit's big picture.

SURVEY THE LANDSCAPE
Science, Chemistry, and Cells

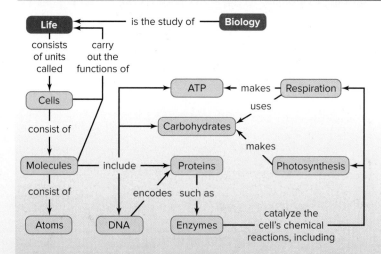

Organisms from all three branches of life share a unique combination of characteristics. Biologists are scientists who use evidence to test hypotheses about life.

For more details, study the Pull It Together feature in the chapter summary.

©Jeff Gynane/Getty Images RF

Imagine a biologist. If you are like many people, you may have pictured someone in a lab coat, carefully recording a mouse's reaction to some new drug. But this view of biology as something that happens only in a laboratory is much too limited. Indeed, we need not even leave home to study biology. Life is in parks, backyards, and the strips between streets and sidewalks. It's also in office buildings and restaurants, not only because we are alive but also because countless microorganisms live everywhere, smaller than the eye can see. The food you have eaten today was (until recently, anyway) alive. Biology really is everywhere.

Biology is frequently in the news, in the form of stories about fossils, weight loss, cancer, genetics, climate change, and the environment. Topics such as these enjoy frequent media coverage because this is an exciting time to study biology. Not only is the field changing rapidly, but its new discoveries and applications might change your life. DNA technology has brought us genetically engineered bacteria that can manufacture pharmaceutical drugs—and genetically engineered corn plants that produce their own pesticides. One day, physicians may routinely cure inherited diseases by supplementing faulty DNA with a functional "patch."

This book will bring you a taste of modern biology and help you make sense of the science-related news you see every day. Chapter 1 begins your journey by introducing the scope of biology and explaining how science teaches us what we know about life.

1.1 What Is Life?

Welcome to biology, the scientific study of life. The second half of this chapter explores the meaning of the term *scientific,* but first we will consider the question, "What is life?" We all have an intuitive sense of what life is. If we see a rabbit on a rock, we know that the rabbit is alive and the rock is not. But it is difficult to state just what makes the rabbit alive. Likewise, in the instant after an individual dies, we may wonder what invisible essence has transformed the living into the dead.

One way to define life is to list its basic components. The **cell** is the basic unit of life; every **organism,** or living individual, consists of one or more cells. Every cell has an outer membrane that separates it from its surroundings. This membrane encloses the water and other chemicals that carry out the cell's functions. One of those biochemicals, deoxyribonucleic acid (DNA), is the informational molecule of life (figure 1.1). Cells use genetic instructions—as encoded in DNA—to produce proteins, which enable cells to carry out their functions in tissues, organs, and organ systems.

A list of life's biochemicals, however, provides an unsatisfying definition of life. After all, placing DNA, water, proteins, and a membrane in a test tube does not create life. And a crushed insect still contains all of the biochemicals that it had immediately before it died.

Figure 1.1 Informational Molecule of Life. All cells contain DNA, a series of "recipes" for proteins that each cell can make.
©SMC Images/The Image Bank/Getty Images

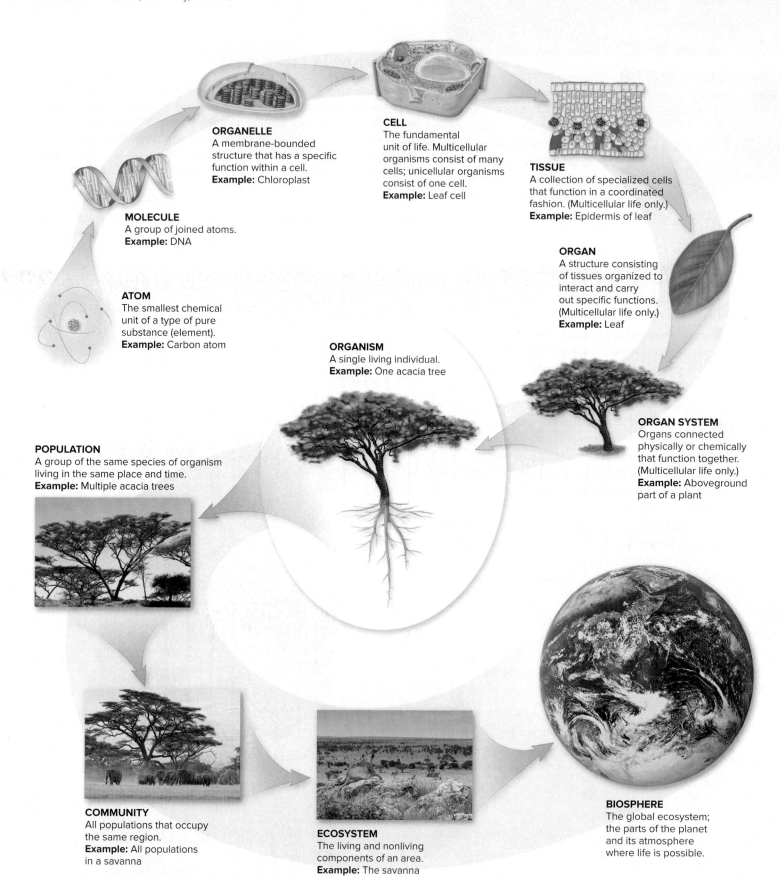

ORGANELLE
A membrane-bounded structure that has a specific function within a cell.
Example: Chloroplast

CELL
The fundamental unit of life. Multicellular organisms consist of many cells; unicellular organisms consist of one cell.
Example: Leaf cell

TISSUE
A collection of specialized cells that function in a coordinated fashion. (Multicellular life only.)
Example: Epidermis of leaf

MOLECULE
A group of joined atoms.
Example: DNA

ORGAN
A structure consisting of tissues organized to interact and carry out specific functions. (Multicellular life only.)
Example: Leaf

ATOM
The smallest chemical unit of a type of pure substance (element).
Example: Carbon atom

ORGANISM
A single living individual.
Example: One acacia tree

ORGAN SYSTEM
Organs connected physically or chemically that function together. (Multicellular life only.)
Example: Aboveground part of a plant

POPULATION
A group of the same species of organism living in the same place and time.
Example: Multiple acacia trees

COMMUNITY
All populations that occupy the same region.
Example: All populations in a savanna

ECOSYSTEM
The living and nonliving components of an area.
Example: The savanna

BIOSPHERE
The global ecosystem; the parts of the planet and its atmosphere where life is possible.

Figure 1.2 Life's Organizational Hierarchy. This diagram applies life's organizational hierarchy to a multicellular organism (an acacia tree). Green arrows represent the hierarchy up to the level of the organism; blue arrows represent levels that include multiple organisms. Photos: (population): ©Gregory G. Dimijian, M.D./Science Source; (community): ©Daryl Balfour/Gallo Images/Getty Images; (ecosystem): ©Bas Vermolen/Getty Images; (biosphere): ©StockTrek/Getty Images

What Is Life?   The Tree of Life Includes Three Main Branches   Scientists Study the Natural World

In the absence of a concise definition, scientists have settled on five qualities that, in combination, constitute life. Table 1.1 summarizes them, and the rest of section 1.1 describes each one in more detail. An organism is a collection of structures that function together and exhibit all of these qualities (see Burning Question 1.1). Note, however, that each trait in table 1.1 may also occur in nonliving objects. A rock crystal is highly organized, but it is not alive. A fork placed in a pot of boiling water absorbs heat energy and passes it to the hand that grabs it, but this does not make the fork alive. A fire can "reproduce" and grow, but it lacks most of the other characteristics of life. It is the *combination* of these five characteristics that makes life unique.

A. Life Is Organized

Just as the city where you live belongs to a county, state, and nation, living matter also consists of parts organized in a hierarchical pattern (figure 1.2). At the smallest scale, all living structures are composed of particles called **atoms,** which bond together to form **molecules.** These molecules can form **organelles,** which are compartments that carry out specialized functions in cells (note that not all cells contain organelles). Many organisms consist of single cells. In multicellular organisms such as the tree illustrated in figure 1.2, however, the cells are organized into specialized **tissues** that make up **organs.** Multiple organs are linked into an individual's **organ systems.**

We have now reached the level of the organism, which may consist of just one cell or of many cells organized into tissues, organs, and organ systems. Organization in the living world extends beyond the level of the individual organism as well. A **population** includes members of the same species occupying the same place at the same time. A **community** includes the populations of different species in a region, and an **ecosystem** includes both the living and nonliving components of an area. Finally, the **biosphere** consists of all parts of the planet that can support life.

Biological organization is apparent in all life. Humans, eels, and evergreens, although outwardly very different, are all organized into specialized cells, tissues, organs, and organ systems. Single-celled bacteria, although less complex than animals or plants, still contain DNA, proteins, and other molecules that interact in highly organized ways.

An organism, however, is more than a collection of successively smaller parts. **Emergent properties** are new functions that arise from interactions among a system's components, much as flour, sugar, butter, and chocolate can become brownies—something not evident from the parts themselves. Figure 1.3 shows another example of emergent properties: the thoughts and memories produced by interactions among the neurons in a person's brain. For an emergent property, the whole is greater than the sum of the parts.

Emergent properties explain why structural organization is closely tied to function. Disrupt a structure, and its function ceases. Brain damage, for instance, disturbs the interactions between brain cells and can interfere with memory, coordination, and other brain functions. Likewise, if a function is interrupted, the corresponding structure eventually breaks down, much as unused muscles begin to waste away. Biological function and form are interdependent.

B. Life Requires Energy

Inside each cell, countless chemical reactions sustain life. These reactions, collectively called metabolism, allow organisms to acquire and use energy and nutrients to build new structures, repair old ones, and reproduce.

TABLE 1.1	Characteristics of Life: A Summary
Characteristic	**Example**
Organization	Atoms make up molecules, which make up cells, which make up tissues, and so on.
Energy use	A kitten uses the energy from its mother's milk to fuel its own growth.
Maintenance of internal constancy (homeostasis)	Your kidneys regulate your body's water balance by adjusting the concentration of your urine.
Reproduction, growth, and development	An acorn germinates, develops into an oak seedling, and, at maturity, reproduces sexually to produce its own acorns.
Evolution	Increasing numbers of bacteria survive treatment with antibiotic drugs.

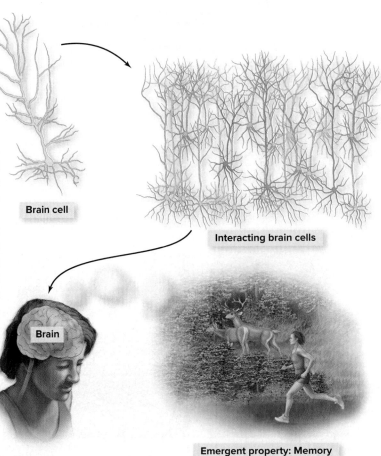

Brain cell

Interacting brain cells

Brain

Emergent property: Memory

Figure 1.3 An Emergent Property—From Cells to Memories. Highly branched cells interact to form a complex network in the brain. Memories, consciousness, and other qualities of the mind emerge only when these cells interact in a certain way.

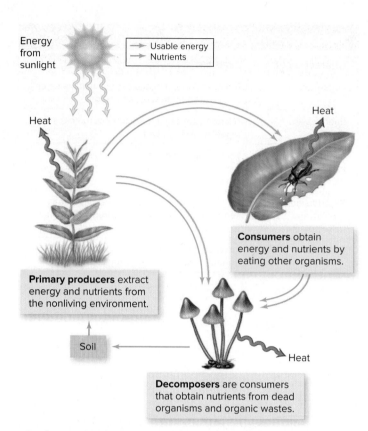

Energy from sunlight

→ Usable energy
→ Nutrients

Heat

Heat

Consumers obtain energy and nutrients by eating other organisms.

Primary producers extract energy and nutrients from the nonliving environment.

Soil

Heat

Decomposers are consumers that obtain nutrients from dead organisms and organic wastes.

Figure 1.4 **Life Is Connected.** All organisms extract energy and nutrients from the nonliving environment or from other organisms. Decomposers recycle nutrients back to the nonliving environment. At every stage along the way, heat is lost to the surroundings.

Biologists divide organisms into broad categories, based on their source of energy and raw materials (figure 1.4). **Primary producers,** also called *autotrophs,* make their own food by extracting energy and nutrients from nonliving sources. The most familiar primary producers are the plants and microbes that capture light energy from the sun, but some bacteria can derive chemical energy from rocks. **Consumers,** in contrast, obtain energy and nutrients by eating other organisms, living or dead; consumers are also called *heterotrophs* (*hetero-* means "other"). You are a consumer, relying on energy and atoms from food to stay alive. **Decomposers** are heterotrophs that absorb energy and nutrients from wastes or dead organisms. These organisms, which include fungi and some bacteria, recycle nutrients to the nonliving environment.

Within an ecosystem, organisms are linked into elaborate food webs, beginning with primary producers and continuing through several levels of consumers (including decomposers). But energy transfers are never 100% efficient; some energy is always lost to the surroundings in the form of heat (see figure 1.4). Because no organism can use it as an energy source, heat represents a permanent loss from the cycle of life. All ecosystems therefore depend on a continuous stream of energy from an outside source, usually the sun. ⓘ *food webs,* section 19.6A

C. Life Maintains Internal Constancy

The conditions inside cells must remain within a constant range, even if the surrounding environment changes. For example, a cell must maintain a certain temperature; it will die if it becomes too hot or too cold. The cell must also take in nutrients, excrete wastes, and regulate its many chemical reactions to prevent a shortage or surplus of essential substances. **Homeostasis** is this state of internal constancy, or equilibrium.

Because cells maintain homeostasis by counteracting changes as they occur, organisms must be able to sense and react to stimuli. To illustrate this idea, consider the mechanisms that help maintain your internal temperature at about 37°C (figure 1.5). When you go outside on a cold day, you may begin to shiver; heat from these muscle movements warms the body. In severe cold, your lips and fingertips may turn blue as your circulatory system sends blood away from your body's surface. Conversely, on a hot day, sweat evaporating from your skin helps cool your body.

D. Life Reproduces, Grows, and Develops

Organisms reproduce, making other individuals that are similar to themselves (figure 1.6). Reproduction transmits DNA from generation to generation; this genetic information defines the inherited characteristics of the offspring.

Reproduction occurs in two basic ways: asexually and sexually. In **asexual reproduction,** genetic information comes from only one parent, and all offspring are virtually identical. One-celled organisms such as bacteria reproduce asexually by doubling and then dividing the contents of the cell. Many multicellular organisms also reproduce asexually. A strawberry plant, for instance, produces "runners" that sprout leaves and roots, forming new plants that are identical to the parent. Fungi produce countless asexual spores, visible as the green, white, or black powder on moldy bread or cheese. Some animals, including sponges, reproduce asexually when a fragment of the parent animal detaches and develops into a new individual.

In **sexual reproduction,** genetic material from two parents unites to form an offspring, which has a new combination of inherited traits. By mixing genes at each generation, sexual reproduction results in tremendous diversity in a

a.

b.

Figure 1.5 **Temperature Homeostasis.** (a) Shivering and (b) sweating are responses that maintain body temperature within an optimal range.

(a): ©Design Pics/Kristy-Anne Glubish RF; (b): ©John Rowley/Getty Images RF

a.   b.

Figure 1.6 **Asexual and Sexual Reproduction.** (a) Identical plantlets develop along the runners of a wild strawberry plant. (b) Two swans protect their offspring, the products of sexual reproduction.

(a): ©Dorling Kindersley/Getty Images; (b): ©Jadranko Markoc/flickr/Getty Images RF

Miniglossary | Reproduction	
Asexual reproduction	Only one parent passes genetic information to offspring; produces genetically identical offspring (except for mutations); adaptive in unchanging environments
Sexual reproduction	Genetic material from two parents combines to form offspring; produces genetically variable offspring; adaptive in changing environments

population. Genetic diversity, in turn, enhances the chance that some individuals will survive even if conditions change. Sexual reproduction is therefore a very successful strategy, especially in an environment where conditions change frequently; it is extremely common among plants, animals, and fungi.

If each offspring is to reproduce, it must grow and develop to adulthood. Each young swan in figure 1.6, for example, started as a single fertilized egg cell. That cell divided over and over, developing into an embryo. Continued cell division and specialization yielded the newly hatched swans, which will eventually mature into adults that can also reproduce—just like their parents.

E. Life Evolves

One of the most intriguing questions in biology is how organisms become so well-suited to their environments. A beaver's enormous front teeth, which never stop growing, are ideal for gnawing wood. Tubular flowers have exactly the right shapes for the beaks of their hummingbird pollinators. Some organisms have color patterns that enable them to fade into the background (figure 1.7).

These examples, and countless others, illustrate adaptations. An **adaptation** is an inherited characteristic or behavior that enables an organism to survive and reproduce successfully in its environment.

Where do these adaptive traits come from? The answer lies in natural selection. The simplest way to think of natural selection is to consider two facts. First, populations produce many more offspring than will survive to reproduce; these organisms must compete for limited resources such as food and habitat. A single mature oak tree may produce thousands of acorns in one season, but only a few are likely to germinate, develop, and reproduce. The rest die. Second, no organism is exactly the same as any other. Genetic mutations—changes in an organism's DNA sequence—generate variability in all organisms, even those that reproduce asexually.

Of all the offspring in a population, which ones will outcompete the others and live long enough to reproduce? The answer is those with the best adaptations to the current environment; conversely, the poorest competitors are most likely to die before reproducing. A good definition of **natural selection,** then, is the enhanced reproductive success of certain individuals from a population based on inherited characteristics.

Figure 1.7 **Hiding in Plain Sight.** This pygmy seahorse is barely visible in its coral habitat, thanks to its unique body shape, skin color, and texture.

©Mark Webster Wwwphoteccouk/Getty Images

Burning Question 1.1

Are viruses alive?

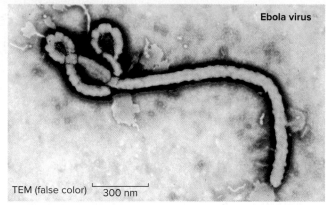

Ebola virus

TEM (false color) |— 300 nm —|

Source: CDC/Frederick Murphy

Many people combine viruses and bacteria into the category of "germs." This terminology makes sense because both viruses and bacteria are microscopic and can cause disease. But they are not the same thing.

A bacterium is a cell, complete with a membrane, cytoplasm, DNA, and proteins. Viruses, on the other hand, are not cells. Instead, the simplest virus consists of a protein shell surrounding a small amount of genetic material. Other viruses have more complex features, but no virus has the structure or functions of a cell.

Most biologists do not consider a virus to be alive because it does not metabolize, respond to stimuli, or reproduce on its own. Instead, a virus must enter a living host cell to manufacture more of itself.

Nevertheless, viruses do have some features in common with life, including evolution. Each time a virus replicates inside a host cell, random mutations occur in its genetic information. The resulting variability among the new viruses is subject to natural selection. That is, some variants are better than others at infecting and replicating in host cells. Many mutant viruses die out, but others pass their successful gene versions to the next generation. Over time, natural selection shapes the genetic composition of each viral population.

Submit your burning question to
marielle.hoefnagels@mheducation.com

Figure 1.8 shows one example of natural selection. The illustration shows a population of bacteria in which a mutation has occurred in one cell. If antibiotics are present, the drug kills most of the unmutated cells. The mutated cell, however, is unaffected and can reproduce. After many generations of exposure to the drug, antibiotic-resistant cells are common.

The same principle applies to all populations. In general, individuals with the best combinations of genes survive and reproduce, while those with less suitable characteristics fail to do so. Over many generations, individuals with adaptive traits make up most or all of the population.

But the environment is constantly changing. Continents shift, sea levels rise and fall, climates warm and cool. What happens to a population when the selective forces that drive natural selection change? Only some organisms survive: those with the "best" traits in the *new* environment. Features that may once have been rare become more common as the reproductive success of individuals with those traits improves. Notice, however, that this outcome depends on variability within the population. If no individual can reproduce in the new environment, the species may go extinct.

Natural selection is one mechanism of **evolution,** which is a change in the genetic makeup of a population over multiple generations. Although evolution can also occur in other ways, natural selection is the mechanism that selects for adaptations. Charles Darwin became famous in the 1860s after he published a book describing the theory of evolution by natural selection; another naturalist, Alfred Russel Wallace, independently developed the same idea at around the same time.

Evolution is the single most powerful idea in biology. As unit 3 describes in detail, the similarities among existing organisms strongly suggest that all species descend from a common ancestor. Evolution has molded the life that has populated the planet since the first cells formed almost 4 billion years ago, and it continues to act today.

1.1 Mastering Concepts

1. List life's organizational hierarchy from smallest to largest, starting with atoms and ending with the biosphere.

2. The bacteria in figure 1.8 reproduce asexually, yet they are evolving. What is their source of genetic variation?

Figure 1.8 **Natural Selection.** *Staphylococcus aureus* (commonly called "staph") is a bacterium that causes skin infections. A bacterium undergoes a random genetic mutation that (by chance) makes the cell resistant to an antibiotic. The presence of the antibiotic increases the reproductive success of the resistant cell and its offspring. After many generations, nearly all of the bacteria in the population are antibiotic-resistant. Conversely, if antibiotics are absent, the antibiotic-resistance trait remains rare.

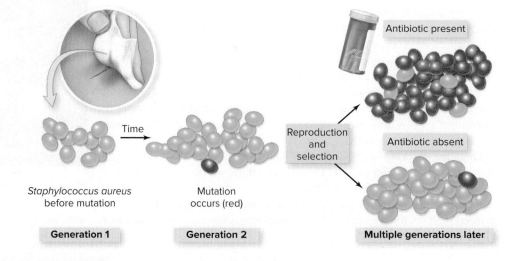

Antibiotic present

Time

Reproduction and selection

Antibiotic absent

Staphylococcus aureus before mutation

Mutation occurs (red)

Generation 1

Generation 2

Multiple generations later

1.2 The Tree of Life Includes Three Main Branches

Biologists have been studying life for centuries, documenting the existence of everything from bacteria to blue whales. An enduring problem has been how to organize the ever-growing list of known organisms into meaningful categories. **Taxonomy** is the science of naming and classifying organisms.

The basic unit of classification is the **species,** which designates a distinctive "type" of organism. Closely related species are grouped into the same **genus.** Together, the genus and a specific descriptor denote the unique, two-word scientific name of each species. A human, for example, is *Homo sapiens.* (Note that scientific names are always italicized and that the genus—but not the specific descriptor—is capitalized.) Scientific names help taxonomists and other biologists communicate with one another.

But taxonomy involves more than simply naming species. Taxonomists also strive to classify organisms according to what we know about evolutionary relationships; that is, how recently one type of organism shared an ancestor with another type. The more recently they diverged from a shared ancestor, the more closely related we presume the two types of organisms to be (figure 1.9). Researchers infer these relationships by comparing anatomical, behavioral, cellular, genetic, and biochemical characteristics.

Genetic evidence suggests that all species fall into one of three **domains,** the broadest (most inclusive) taxonomic category. Figure 1.10 depicts the three domains: **Bacteria, Archaea,** and **Eukarya.** The species in domains Bacteria and Archaea are superficially similar to one another; all are prokaryotes, meaning that their DNA is free in the cell and not confined to an organelle called a nucleus. Major differences in DNA sequences separate these two domains from each other. The third domain, Eukarya, contains all species of eukaryotes, which are unicellular or multicellular organisms whose cells contain a nucleus.
ⓘ *prokaryotes and eukaryotes*, section 3.2

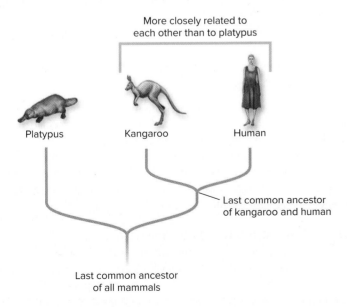

Figure 1.9 Simple Evolutionary Tree. The common ancestor of kangaroos and humans lived more recently than did the common ancestor that both groups share with a platypus. This diagram depicts one tiny twig in the overall tree of life.

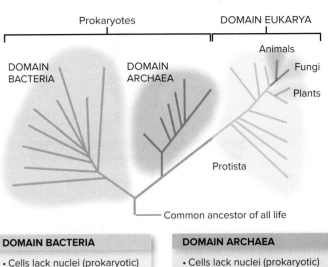

DOMAIN BACTERIA
- Cells lack nuclei (prokaryotic)
- Most are unicellular

TEM (false color) 1 μm

DOMAIN ARCHAEA
- Cells lack nuclei (prokaryotic)
- Most are unicellular

SEM (false color) 1 μm

DOMAIN EUKARYA
- Cells contain nuclei (eukaryotic)
- Unicellular or multicellular

Protista (multiple kingdoms)
- Unicellular or multicellular
- Autotrophs or heterotrophs

Kingdom Animalia
- Multicellular
- Heterotrophs (by ingestion)

LM 200 μm

Kingdom Fungi
- Most are multicellular
- Heterotrophs (by external digestion)

Kingdom Plantae
- Multicellular
- Autotrophs

Figure 1.10 Life's Diversity. The three domains of life arose from a hypothetical common ancestor, shown at the base of the evolutionary tree.

Photos: (Bacteria): ©Heather Davies/SPL/Getty Images RF; (Archaea): ©Eye of Science/ Science Source; (Protista): ©Melba/age fotostock; (Animalia): USDA/ARS/Scott Bauer; (Fungi): ©Corbis RF; (Plantae): USDA/Keith Weller

	Taxonomic group	Humans are in:
Most inclusive	Domain	Eukarya
	Kingdom	Animalia
	Phylum	Chordata
	Class	Mammalia
	Order	Primates
	Family	Hominidae
	Genus	*Homo*
Least inclusive	Species	*Homo sapiens*

Figure 1.11 The Taxonomic Hierarchy. The eight major levels in the taxonomic hierarchy range from domain to species. This example applies the hierarchy to our own species, *Homo sapiens*.

The species in each domain are further subdivided into **kingdoms;** figure 1.10 shows the kingdoms within domain Eukarya. Three of these kingdoms—Animalia, Fungi, and Plantae—are familiar to most people. Within each one, organisms share the same general strategy for acquiring energy. The plant kingdom contains autotrophs, whereas fungi and animals are consumers that differ in the details of how they obtain food. But the fourth group of eukaryotes, the Protista, contains a huge collection of unrelated species. Protista is a convenient but artificial "none of the above" category for the many species of eukaryotes that are not plants, fungi, or animals.

Each kingdom is likewise divided into multiple phyla, which are in turn divided into still smaller groups. Figure 1.11 depicts the complete taxonomic hierarchy, which is described in more detail in section 14.6.

1.2 Mastering Concepts

1. If the human and kangaroo in figure 1.9 switched places, would the evolutionary tree show different relationships among the three animals from what it shows now? Why or why not?
2. How are domains related to kingdoms?
3. List and describe the four main groups of eukaryotes.

1.3 Scientists Study the Natural World

The idea of biology as a "rapidly changing field" may seem strange if you think of science as a collection of facts. After all, the parts of a frog are the same now as they were 50 or 100 years ago. But memorizing frog anatomy is not the same as thinking scientifically. Scientists use evidence to answer questions about the natural world. If you compare a bullfrog to a rattlesnake, for instance, can you determine how the frog can live in water and on land, whereas the snake survives in the desert? Understanding anatomy simply gives you the vocabulary you need to ask these and other interesting questions about life.

A. The Scientific Method Has Multiple Interrelated Parts

Scientific knowledge arises from application of the **scientific method,** which is a general way of using evidence to answer questions and test ideas (figure 1.12). Although this diagram may give the impression that science is a tedious, step-by-step process, that is not at all true. Instead, science combines thinking, detective work, collaborating with other scientists, learning from mistakes, and noticing connections. The resulting insights have taught us everything we know about the natural world.

Observations and Questions The scientific method begins with observations and questions. The observations may rely on what we can see, hear, touch, taste, or smell, or they may be based on existing knowledge and experimental results. Often, a great leap forward happens when one person makes connections between previously unrelated observations. Charles Darwin, for example, developed the idea of natural selection by combining his understanding of Earth's long history with his detailed observations of

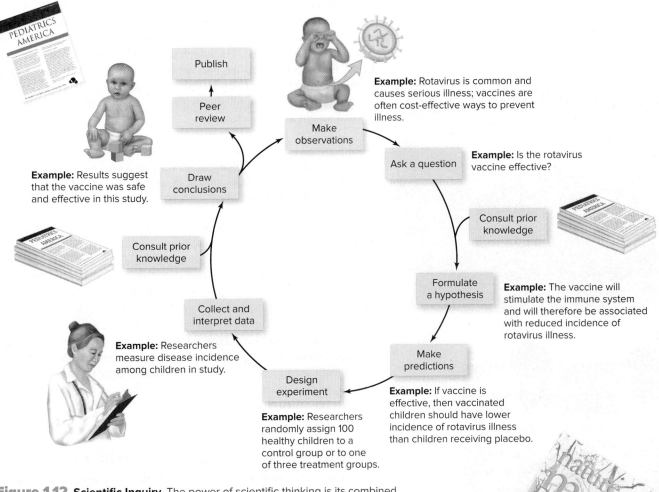

Publish

Peer review

Make observations

Example: Rotavirus is common and causes serious illness; vaccines are often cost-effective ways to prevent illness.

Example: Results suggest that the vaccine was safe and effective in this study.

Draw conclusions

Ask a question

Example: Is the rotavirus vaccine effective?

Consult prior knowledge

Consult prior knowledge

Formulate a hypothesis

Example: The vaccine will stimulate the immune system and will therefore be associated with reduced incidence of rotavirus illness.

Collect and interpret data

Example: Researchers measure disease incidence among children in study.

Make predictions

Design experiment

Example: If vaccine is effective, then vaccinated children should have lower incidence of rotavirus illness than children receiving placebo.

Example: Researchers randomly assign 100 healthy children to a control group or to one of three treatment groups.

Figure 1.12 Scientific Inquiry. The power of scientific thinking is its combined emphasis on logic and creativity. This figure is simplified; in reality, teams of scientists work on multiple "steps" simultaneously. In addition, note that data may come from observations of the natural world or from experimentation; this example uses an experiment.

organisms. Another great advance occurred decades later, when biologists realized that mutations in DNA generate the variation that Darwin saw but could not explain.

Hypothesis and Prediction A **hypothesis** is a tentative explanation for one or more observations. The hypothesis is the essential "unit" of scientific inquiry. To be useful, the hypothesis must be testable—that is, there must be a way to collect data that can support or reject it. Interestingly, a hypothesis cannot be *proven* true, because future discoveries may contradict today's results. Nevertheless, a hypothesis becomes widely accepted when multiple lines of evidence support it, no credible data refute it, and plausible alternative hypotheses have been rejected.

A hypothesis is a general statement that should lead to specific **predictions.** Often, a prediction is written as an if-then statement. As a simple example, suppose you hypothesize that your lawn mower stopped working because it ran out of gas. A reasonable prediction would be, "If I put fuel into the tank, then my lawn mower should start."

a.

b.

Figure 1.13 Different Types of Science. (a) Scientists track the number of migratory birds that visit a wildlife refuge each year—an example of discovery science. (b) Controlled experiments can help food scientists objectively compare techniques for roasting or brewing coffee.

(a): U.S. Fish & Wildlife Service/J&K Hollingsworth; (b): ©Corbis/age fotostock RF

Data Collection Investigators draw conclusions based on data (figure 1.13). The data may come from careful observations of the natural world, an approach called discovery science. The National Audubon Society's annual Christmas Bird Count is a case in point. For more than a century, thousands of "citizen scientists" have documented the ups and downs of hundreds of bird species nationwide. Another way to gather data is to carry out an experiment to test a hypothesis under controlled conditions (section 1.3B explores experimental design in more detail).

Discovery and experimentation work hand in hand. Consider the well-known connection between cigarettes and lung cancer. In the late 1940s, scientists showed that smokers are far more likely than nonsmokers to develop cancer. Since that time, countless laboratory experiments have revealed how the chemicals in tobacco damage living cells.

Analysis and Peer Review After collecting and interpreting data, investigators decide whether the evidence supports or falsifies the hypothesis. Often, the most interesting results are those that are unexpected, because they provide new observations that force scientists to rethink their hypotheses; figure 1.12 shows this feedback loop. Science advances as new information arises and explanations continue to improve.

Once a scientist has enough evidence to support or reject a hypothesis, he or she may write a paper and submit it for publication in a scientific journal. The journal's editors send the paper to anonymous reviewers who are knowledgeable about the research topic. In a process called **peer review,** these scientists independently evaluate the validity of the methods, data, and conclusions. Overall, peer review ensures that journal articles—the tangible products of the global scientific conversation—are of high quality.

B. An Experimental Design Is a Careful Plan

Scientists test many hypotheses with the help of experiments (table 1.2). An **experiment** is an investigation carried out in controlled conditions. This section considers a real study that tested the hypothesis that a new vaccine protects against a deadly virus. The virus, called *rotavirus,* causes severe diarrhea and takes the lives of hundreds of thousands of young children each year. An effective, inexpensive vaccine would prevent many childhood deaths.

Sample Size One of the most important decisions that an investigator makes in designing an experiment is the **sample size,** which is the number of individuals assigned to each treatment. The sample size in the rotavirus study was approximately 100 infants per treatment. In general, the larger the sample size, the more credible the results of a study.

Variables A systematic consideration of variables is also important in experimental design. A **variable** is a changeable element of an experiment, and there are several types. The **independent variable** is the factor that an investigator directly manipulates to determine whether it causes another variable to change. In the rotavirus study, the independent variable was the dose of the vaccine. The **dependent variable** is any response that might *depend on* the value of the independent variable, such as the number of children with rotavirus-related illness during the study period.

A **standardized variable** is anything that the investigator holds constant for all subjects in the experiment, ensuring the best chance of detecting the effect of the independent variable. Because rotavirus infection is most common among very young children, the test of the new vaccine included only infants younger than 12 weeks. Furthermore, vaccines work best in people with healthy immune

TABLE 1.2   **Experimental Design: A Summary**

Component	Definition	Example
Sample size	The number of subjects in a treatment group or control group	100 infants
Variables		
Independent variable	A variable that an investigator manipulates to determine whether it influences the dependent variable	Dose of vaccine
Dependent variable	A variable that an investigator measures to determine whether it is affected by the independent variable	Number of children with illness caused by rotavirus
Standardized variable	Any variable that an investigator intentionally holds constant for all subjects in an experiment, including the control group	Age and health of children in study
Control	Basis for comparison to treatment group(s); control subjects may remain untreated or receive a placebo	Placebo lacking active ingredient in vaccine

systems, so the study excluded infants who were ill or had weak immunity. Age and health were therefore among the study's standardized variables.

Controls   Well-designed experiments compare one or more groups undergoing treatment to a group of "normal" (untreated) individuals. The experimental **control** is the untreated group, and it is important because it provides a basis for comparison in measuring the effect of the independent variable. Ideally, the only difference between the control and any other experimental group is the one factor being tested.

Experimental controls may take several forms. Sometimes, the control group simply receives a "zero" value for the independent variable. If a gardener wants to test a new fertilizer in her garden, she may give some plants a lot of fertilizer, others only a little, and still others—the control plants—none. In medical research, a control group might receive a **placebo,** an inert substance that resembles the treatment given to the experimental group. The control infants in the rotavirus study received a placebo that contained all components of the vaccine except the active ingredient.

Statistical Analysis   Once an experiment is complete, the investigator compiles the data and decides whether the results support the hypothesis. Look at the results in figure 1.14. Did the vaccine prevent illness, or do the data simply reflect random chance? The researchers concluded that the vaccine was effective, but only after applying a statistical analysis.

Researchers may use many different statistical tests, depending on the type of data. All such tests consider both variation and sample size to estimate the probability that the results arose purely by chance. If this probability is low, then the results are considered **statistically significant.** Appendix B shows how scientists use error bars and other notation to illustrate statistical significance in graphs.

C. Theories Are Comprehensive Explanations

Outside of science, the word *theory* is often used to describe an opinion or a hunch. For instance, immediately after a plane crash, experts offer their "theories" about the cause of the disaster. These tentative explanations are really untested hypotheses.

In science, the word *theory* has a distinct meaning. Like a hypothesis, a **theory** is an explanation for a natural phenomenon, but a theory is typically broader in scope than a hypothesis. For example, the germ theory—the idea that some microorganisms cause human disease—is the foundation for medical

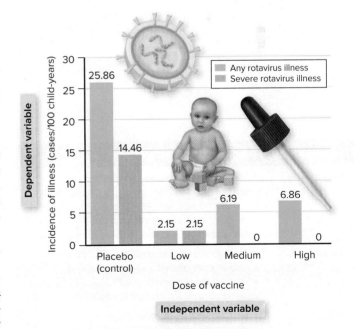

Figure 1.14   Vaccine Test. In this test of a new vaccine, the independent variable was the vaccine dose; control infants received a placebo. The statistical analysis (not shown) suggests that all vaccine doses prevented rotavirus illness when compared with the placebo.

Nectar tubes

Figure 1.15 **Prediction Confirmed.** When Charles Darwin saw this Madagascar orchid, he predicted that its pollinator would have long, thin mouthparts that could reach the bottom of the elongated nectar tubes. He was right; the unknown pollinator turned out to be a moth with an extraordinarily long tongue.

©Kjell Sandved/Alamy Stock Photo

Miniglossary | Scientific Knowledge

Hypothesis	A tentative, falsifiable explanation for one or more observations. If tests support a hypothesis, it may be incorporated into broader theories. Outside of science, *hypothesis* is used interchangeably with *theory*.
Theory	A falsifiable, comprehensive explanation for a natural phenomenon, typically backed with many lines of evidence. Nonscientists often use the term as a synonym for *opinion* or *hunch*.
Fact	A repeatable observation that everyone can agree on. Nonscientists often view facts as the only reliable presentation of reality, but a collection of facts by itself does not explain anything. Theories provide these explanations.

microbiology. Individual hypotheses relating to the germ theory are much narrower, such as the suggestion that rotavirus causes illness. Not all theories are as "large" as the germ theory, but they generally encompass multiple hypotheses. Note also that the germ theory does not imply that *all* microbes make us sick or that all illnesses have microbial causes. But it does explain many types of disease.

A second difference between a hypothesis and a theory is acceptance and evidence. A hypothesis is tentative, whereas theories reflect broader agreement. This is not to imply that theories are not testable; in fact, the opposite is true. Every scientific theory is *potentially* falsifiable, meaning that a particular set of observations could prove the theory wrong. The germ theory remains widely accepted because many observations support it and no reliable tests have disproved it. The same is true for the theory of evolution and many other scientific theories.

Another quality of a scientific theory is its predictive power. A good theory not only ties together many existing observations, but also suggests predictions about phenomena that have yet to be observed. Both Charles Darwin and naturalist Alfred Russel Wallace used the theory of evolution by natural selection to predict the existence of a moth that could pollinate orchid flowers with unusually long nectar tubes (figure 1.15). Decades later, scientists discovered the long-tongued insect (see Investigating Life 1.1). A theory weakens if subsequent observations do not support its predictions.

What is the relationship between facts and a theory? One definition of the word *fact* is "a repeatable observation that everyone can agree on." It is a fact that a dropped pencil falls toward the ground; no reasonable person disagrees with that statement. Gravity is a *fact;* gravitational *theory* explains the forces that cause pencils and other objects to fall.

Biologists also consider biological evolution to be a fact. Yet the phrase "theory of evolution" persists, because evolution is both a fact *and* a theory. Both terms apply equally well. The evidence for genetic change over time is so persuasive and comes from so many different fields of study that to deny the existence of evolution is unrealistic. Nevertheless, biologists do not understand everything about how evolution works. Many questions about life's history remain, but the debates swirl around *how,* not *whether,* evolution occurs.

D. Scientific Inquiry Has Limitations

Scientific inquiry is neither foolproof nor always easy to implement. One problem is that experimental evidence may lead to multiple interpretations, and even the most carefully designed experiment can fail to provide a definitive answer (see Why We Care 1.1). Consider the observation that animals fed large doses of vitamin E live longer than similar animals that do not ingest the vitamin. So, does vitamin E slow aging? Possibly, but excess vitamin E also causes weight loss, and other research has connected weight loss with longevity. Does vitamin E extend life, or does weight loss? The experiment alone does not distinguish between these possibilities.

Another limitation is that researchers may misinterpret observations or experimental results. For example, centuries ago, scientists sterilized a bottle of broth, corked the bottle shut, and observed bacteria in the broth a few days later. They concluded that life arose directly from the broth. The correct explanation, however, was that the cork did not keep airborne bacteria out. Although scientists may make mistakes in the short term, science is self-correcting in the long run because it remains open to new data and new interpretations.

A related problem is that the scientific community may be slow to accept new evidence that suggests unexpected conclusions. Every investigator should

Why We Care 1.1 | It's Hard to Know What's Bad for You

You have probably heard reports that a food previously considered healthy is actually bad for you, or vice versa. These conflicting reports may tempt you to mistakenly conclude that scientific studies are no better than guesswork.

Instead, the problem lies in the fact that some questions are extremely hard to answer. Take, for example, the artificial sweetener saccharin (see Why We Care 2.2). In 1977, the U.S. Food and Drug Administration (FDA) proposed a ban on saccharin, based on a handful of studies suggesting that the sweetener caused bladder cancer in rats. Congress opted to require warning labels on products containing saccharin. In 1991, the FDA withdrew its proposed ban, and in 1998, saccharin was rated as "not classifiable as to its carcinogenicity to humans." Two years later, legislation removed the warning label requirement.

©Vitalii Hulai/
Shutterstock RF

This tangled history raises an important issue: Why can't science reply yes or no to the seemingly simple question of whether saccharin is bad for you? To understand the answer, consider one of the studies that prompted the FDA to propose the ban on saccharin in the first place. Researchers divided 200 rats into two groups. The control animals ate standard rodent chow, whereas the experimental group got the same food supplemented with saccharin. At reproductive maturity the animals were bred, and the researchers fed the offspring the same dose of saccharin throughout their lives as well. To measure the incidence of cancer, they counted the tumors in both generations of rats for 24 months or until the rats died, whichever came first. Figure 1.A summarizes the results.

At first glance, the conclusion seems inescapable: Saccharin causes cancer in male lab rats. But closer study reveals several hidden complexities that make the data hard to interpret. First, the dose of saccharin was huge: 5% of the rats' diets, for life. The equivalent dose in humans would require drinking hundreds of cans of saccharin-sweetened soda every day. In addition, the experimental rats weighed much less than the control rats by the end of the study, suggesting that high doses of the sweetener are toxic. Rather than causing cancer directly, the saccharin may have simply made the animals more susceptible to disease. Moreover, follow-up studies using other animals were inconclusive.

Perhaps the scientists should have studied the saccharin–cancer connection in humans instead. Unfortunately, however, documenting a link between any food and cancer in people is extremely difficult. One strategy might be to measure the incidence of cancer in saccharin users versus nonusers. But with so many other possible causes of cancer—smoking, poor diet, exposure to job-related chemicals, genetic predisposition—it is difficult to separate out just the effects of saccharin.

So what are we to make of the mixed news reports? It is hard to know, but one thing is certain: No matter what the headlines say, one study, especially a small one, cannot reveal the whole story.

Figure 1.A The Saccharin Scare. The studies examining the link between saccharin and bladder cancer in rats are summarized in these graphs. Sample sizes ranged from 36 to 49 rats per treatment.

Source: Data adapted from Office of Technology Assessment Report, October 1977, *Cancer Testing Technology and Saccharin*, page 52.

Male rats

% with tumors — Parents: 19; Offspring: 27, 3 / 0

Female rats

Legend: Saccharin-fed, Controls

% with tumors — Parents: 0, 0; Offspring: 4, 0

try to keep an open mind about observations, not allowing biases or expectations to cloud interpretation of the results. But it is human nature to be cautious in accepting an observation that does not fit what we think we know. The careful demonstration that life does not arise from broth surprised many people who believed that mice sprang from moldy grain and that flies came from rotted beef. More recently, it took many years to set aside the common belief that stress caused stomach ulcers. Today, we know that a bacterium causes most ulcers.

Although science is a powerful tool for answering questions about the natural world, it cannot answer questions of beauty, morality, ethics, or the meaning of life (see Burning Question 1.2). Nor can we directly study some phenomena that occurred long ago and left little physical evidence. Consider the many experiments that have attempted to re-create the chemical reactions that might have produced life on early Earth. Although the experiments produce interesting results and reveal ways that these early events may have occurred, we cannot know if they accurately reflect conditions at the dawn of life.

©Stockbyte/Getty Images RF

Burning Question 1.2

Why am I here?

The Burning Questions featured in each chapter of this book came from students. On the first day of class, I always ask students to turn in a "Burning Question"— anything they have always wondered about biology. I answer most of the questions as the relevant topics come up during the semester.

Why not answer *all* of the questions? It is because at least one student often asks something like, "Why am I here?" or "What is the meaning of life?" Such puzzles have fascinated humans throughout the ages, but they are among the many questions that we cannot approach scientifically. Biology can explain how you developed after a sperm from your father fertilized an egg cell from your mother. But no one can develop a testable hypothesis about life's meaning or the purpose of human existence. Science must remain silent on such questions.

Instead, other ways of knowing must satisfy our curiosity about "why." Philosophers, for example, can help us see how others have considered these questions. Religion may also provide the meaning that many people seek. Part of the value of higher education is to help you acquire the tools you need to find your own life's purpose.

Submit your burning question to
marielle.hoefnagels@mheducation.com

©Getty Images/
Photodisc RF

E. Biology Continues to Advance

Science is just one of many ways to investigate the world, but its strength is its openness to new information. Theories change to accommodate new knowledge. The history of science is full of long-established ideas that changed as we learned more about nature, often thanks to new technology. People thought that Earth was flat and at the center of the universe before inventions and data analysis revealed otherwise. Similarly, biologists thought all organisms were plants or animals until microscopes unveiled a world of life invisible to our eyes.

Technology is the practical application of scientific knowledge. Science and technology are therefore intimately related. For example, thanks to centuries of scientific inquiry, we understand many of the differences between humans and bacteria. We can exploit these differences to invent new antibiotic drugs that kill germs without harming our own bodies. These antibiotics, in turn, can be useful tools that help biologists learn even more about bacterial cells. The new scientific discoveries spawn new technologies, and so on.

Biology is changing rapidly because technology has expanded our ability to spy on living cells, compare DNA sequences, track wildlife, and make many other types of observations. Scientists can now answer questions about the natural world that previous generations could never have imagined.

1.3 Mastering Concepts

1. Identify the elements of the experiment summarized in Why We Care 1.1.
2. What is a statistically significant result?
3. What is the difference between a hypothesis and a theory, and why are some theories regarded as facts?
4. What are some limitations of scientific inquiry?
5. Compare and contrast *science* and *technology*.

Investigating Life 1.1 | The Orchid and the Moth

Each chapter of this book has a box that examines how biologists use systematic, scientific observations to solve a different evolutionary puzzle from life's long history. This first installment of "Investigating Life" revisits the story of the orchid plant pictured in figure 1.15.

In a book published in 1862, Charles Darwin wrote that certain orchids in Madagascar have nectar tubes "eleven and a half inches long, with only the lower inch and a half filled with very sweet nectar." He speculated that the moths that reach the nectar must have tongues "capable of extension to a length of between ten and eleven inches." However, the pollinator had not yet been discovered.

Alfred Russel Wallace picked up the story in a book published in 1895, summarizing how natural selection could explain this unusual orchid. He said that only the moths with the longest tongues would be able to reach the nectar reward in the flower. At the same time, orchids with the deepest nectar tubes would be pollinated most easily, because the moths would rub against the flower's reproductive parts while reaching for the nectar. Therefore, moths with long tongues and orchids with deep nectar tubes "each confer on the other an advantage in the battle of life."

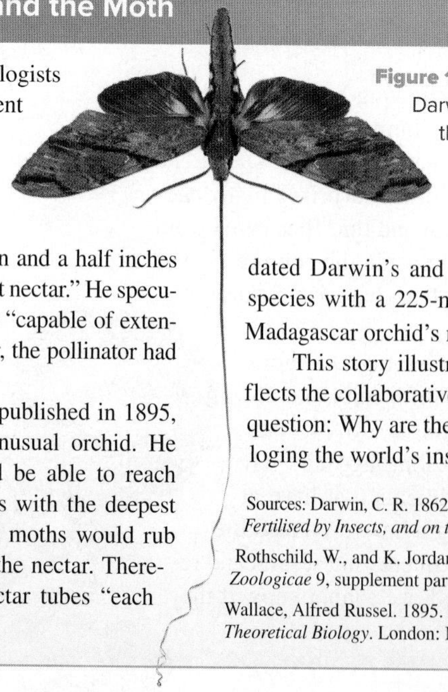

Figure 1.B Found at Last. More than 40 years after Charles Darwin predicted its existence, scientists finally discovered the sphinx moth that pollinates the Madagascar orchid.

©The Natural History Museum/Alamy Stock Photo

A taxonomic publication from 1903 finally validated Darwin's and Wallace's predictions. The authors described a moth species with a 225-millimeter (8-inch) tongue, well-suited for sipping the Madagascar orchid's nectar (figure 1.B).

This story illustrates how theories lead to testable predictions and reflects the collaborative nature of science. Darwin and Wallace asked a simple question: Why are these nectar tubes so long? Decades later, biologists cataloging the world's insect species finally solved the puzzle.

Sources: Darwin, C. R. 1862. *On the Various Contrivances by Which British and Foreign Orchids Are Fertilised by Insects, and on the Good Effects of Intercrossing*. London: John Murray, pages 197–198.

Rothschild, W., and K. Jordan. 1903. A revision of the lepidopterous family Sphingidae. *Novitates Zoologicae* 9, supplement part 1, page 32.

Wallace, Alfred Russel. 1895. *Natural Selection and Tropical Nature: Essays on Descriptive and Theoretical Biology*. London: MacMillan and Co., pages 146–148.

What's the **Point?** ▼ APPLIED

This chapter explained how biologists define, classify, and answer questions about life. Biologists explore life at every scale—from the molecules that form life to the biosphere that contains it. Although each biological discipline has its own research objectives, they all share a reliance on hypothesis-testing, evidence, and technology.

Some biologists work in a lab, where they may use chemistry to learn more about diabetes or Alzheimer disease. They may use microscopes to test hypotheses about how cells work, leading to new insights about cancer. Some lab biologists are developing life-saving artificial tissues and organs.

Other biologists work outdoors, monitoring populations and communities in an effort to learn why some species become endangered or go extinct. They may also collaborate with geologists and climatologists to conduct large-scale experiments that reveal how global climate change affects life from the tropics to the poles.

Some biological questions, like those related to disease, might seem more relevant to humans than others. But like all organisms, humans are made of cells and form populations and communities in the global ecosystem. Studying life's many characteristics, forms, and interactions—as biologists do every day—teaches us more about ourselves.

©Aurora Photos/Alamy Stock Photo RF

CHAPTER SUMMARY

1.1 What Is Life?

- A combination of characteristics distinguishes life: organization, energy use, internal constancy, reproduction and development, and evolution.

A. Life Is Organized (figure 1.16)

- **Atoms** form **molecules.** These molecules form the **organelles** inside many **cells.** An **organism** consists of one or more cells. In most multicellular organisms, cells form **tissues** and then **organs** and **organ systems.**
- Whether unicellular or multicellular, multiple individuals of the same species make up **populations;** multiple populations form **communities.** **Ecosystems** include living communities plus their nonliving environment. The **biosphere** is composed of all of the world's ecosystems.
- **Emergent properties** arise from interactions among the parts of an organism.

B. Life Requires Energy

- Life requires energy to maintain its organization and functions. **Primary producers** make their own food, using energy and nutrients from the nonliving environment. **Consumers** eat other organisms, living or dead. **Decomposers** recycle nutrients to the nonliving environment.
- Because of heat losses, all ecosystems require constant energy input from an outside source, usually the sun.

C. Life Maintains Internal Constancy

- Organisms must maintain **homeostasis,** an internal state of constancy in changing environmental conditions.

D. Life Reproduces, Grows, and Develops

- Organisms reproduce asexually, sexually, or both. **Asexual reproduction** yields virtually identical copies of one parent, whereas **sexual reproduction** generates tremendous genetic diversity by combining and scrambling DNA from two parents.

E. Life Evolves

- In **natural selection,** environmental conditions select for organisms with inherited traits (**adaptations**) that increase the chance of survival and reproduction.
- **Evolution** through natural selection explains how common ancestry unites all species, producing diverse organisms with many similarities.

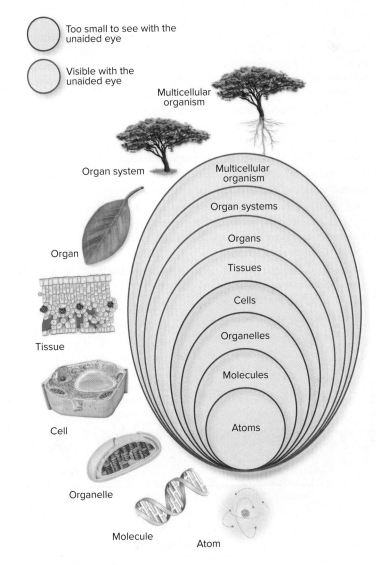

Too small to see with the unaided eye

Visible with the unaided eye

Multicellular organism

Organ system

Organ

Tissue

Cell

Organelle

Molecule

Atom

Multicellular organism

Organ systems

Organs

Tissues

Cells

Organelles

Molecules

Atoms

Figure 1.16 Life Is Organized: A Summary.

1.2 The Tree of Life Includes Three Main Branches

- **Taxonomy** is the science of classification. Biologists classify types of organisms, or **species,** according to probable evolutionary relationships. A **genus,** for example, consists of closely related species.
- The two broadest taxonomic levels are domain and kingdom.
- The three **domains** of life are **Archaea, Bacteria,** and **Eukarya.** Within each domain, mode of nutrition and other features distinguish the **kingdoms.**

1.3 Scientists Study the Natural World

A. The Scientific Method Has Multiple Interrelated Parts

- Scientific inquiry, which uses the **scientific method,** is a way of using evidence to evaluate ideas about the natural world.
- A scientist makes observations, raises questions, and uses reason to construct a testable explanation, or **hypothesis.** Specific **predictions** follow from a scientific hypothesis.
- After collecting data and making conclusions based on the evidence, the investigator may seek to publish scientific results. **Peer review** ensures that published studies meet high standards for quality.

B. An Experimental Design Is a Careful Plan

- An **experiment** is a test of a hypothesis carried out in controlled conditions.
- The larger the **sample size,** the more credible the results of an experiment.
- **Variables** are changeable elements in an experiment. The **independent variable** is the factor that the investigator manipulates. The **dependent variable** is what the investigator measures to determine the outcome of the experiment. **Standardized variables** are held constant for all subjects, including the **control** group (often a set of subjects receiving no treatment or a **placebo**).
- Figure 1.17 shows a typical arrangement of a graph summarizing experimental results. **Statistically significant** results are unlikely to be due to chance.

C. Theories Are Comprehensive Explanations

- A **theory** is more widely accepted and broader in scope than a hypothesis.
- The acceptance of scientific ideas may change as new evidence accumulates.

D. Scientific Inquiry Has Limitations

- The scientific method does not always yield a complete explanation, or it may produce ambiguous results. Science cannot answer all questions—only those for which it is possible to develop testable hypotheses.

E. Biology Continues to Advance

- **Technology** is the practical application of scientific knowledge. Advances in science lead to new technologies, and vice versa.

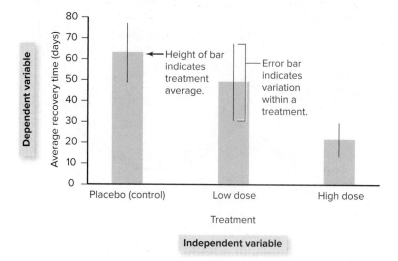

Figure 1.17 **Experimental Results: An Example.**

MULTIPLE CHOICE QUESTIONS

1. Which of the following is smaller than an organelle?
   a. An organ   c. A cell
   b. A molecule   d. A tissue

2. All of the following are characteristics of life EXCEPT
   a. evolution.   c. homeostasis.
   b. reproduction.   d. multicellularity.

3. The concentration of salts in blood remains relatively steady, regardless of a person's diet. This situation best illustrates
   a. homeostasis.
   b. life's organizational hierarchy.
   c. autotrophy.
   d. evolution.

4. Because plants extract nutrients from soil and use sunlight as an energy source, they are considered to be
   a. autotrophs.   c. heterotrophs.
   b. consumers.   d. decomposers.

5. Evolution through natural selection will occur most rapidly for populations of plants that
   a. are already well adapted to the environment.
   b. live in an unchanging environment.
   c. are in the same genus.
   d. reproduce sexually and live in an unstable environment.

6. What is the correct way to write the scientific name for humans?
   a. Homo sapiens   c. *Homo Sapiens*
   b. *Homo sapiens*   d. homo sapiens

7. In an experiment to test the effect of temperature on the rate of bacterial reproduction, temperature would be the
   a. standardized variable.
   b. independent variable.
   c. dependent variable.
   d. control variable.

8. A scientist has just observed a new phenomenon and wonders how it happens. What is the next step in his or her discovery of the answer?
   a. Observe
   b. Hypothesize
   c. Experiment
   d. Peer review

9. Can a theory be proven wrong?
   a. No, theories are exactly the same as facts.
   b. No, because there is no good way to test a theory.
   c. Yes, a new observation or interpretation of data could disprove a theory.
   d. Yes, theories are exactly the same as hypotheses.

10. Which of the following statements is false?
    a. Emergent properties are functions that arise from the interactions between an organism's parts.
    b. Two of the three domains contain prokaryotic organisms.
    c. When medical researchers test the effectiveness of a new drug, they typically compare the new drug to a placebo.
    d. For a scientific study to be considered valid, the researchers must conduct experiments.

Answers to Multiple Choice questions are in appendix A.

WRITE IT OUT

1. Describe each of the five characteristics of life, and list several nonliving things that possess at least two of these characteristics.
2. Imagine two related species of unicellular protists living together in a pond. Write the organizational hierarchy of this ecosystem, starting with "atom" and ending with "ecosystem." Give an example of a structure at each level.
3. Draw and explain the relationships among primary producers, consumers, and decomposers.
4. Describe the main differences between asexual and sexual reproduction. Should changing or unchanging conditions favor each type of reproduction?
5. Why is a cell, and not an atom or a molecule, considered the basic unit of life?
6. Think of an analogy that will help you remember the differences between populations, communities, and ecosystems.
7. Other than the brownie example given in the text, name an example of emergent properties from everyday life.
8. Explain why populations of organisms are typically well adapted to their environment.
9. How are the members of the three domains similar? How are they different?
10. Give two examples of questions that cannot be answered using the scientific method. Explain your reason for choosing each example.
11. If you dissect and label the parts of an earthworm, are you "doing science"? Why or why not? Give an example of a testable hypothesis that could result from dissecting organisms.
12. List each step of the scientific method and explain why it is important.
13. Design an experiment to test the following commonly held belief: "Eating chocolate causes zits." Include sample size, independent variable, dependent variable, the most important variables to standardize, and an experimental control.

SCIENTIFIC LITERACY

Review Why We Care 1.1, which describes conflicting data about the healthfulness of artificial sweeteners. Think of some other foods that some people consider healthy and others consider unhealthy. Pick one, then search the Internet for one website defending the food and for another website condemning the food as unhealthy. What evidence suggests the food is healthy or unhealthy? Which website is more convincing? Why?

Answers to Mastering Concepts, Write It Out, Scientific Literacy, and Pull It Together questions can be found in the Connect ebook.
connect.mheducation.com

PULL IT TOGETHER

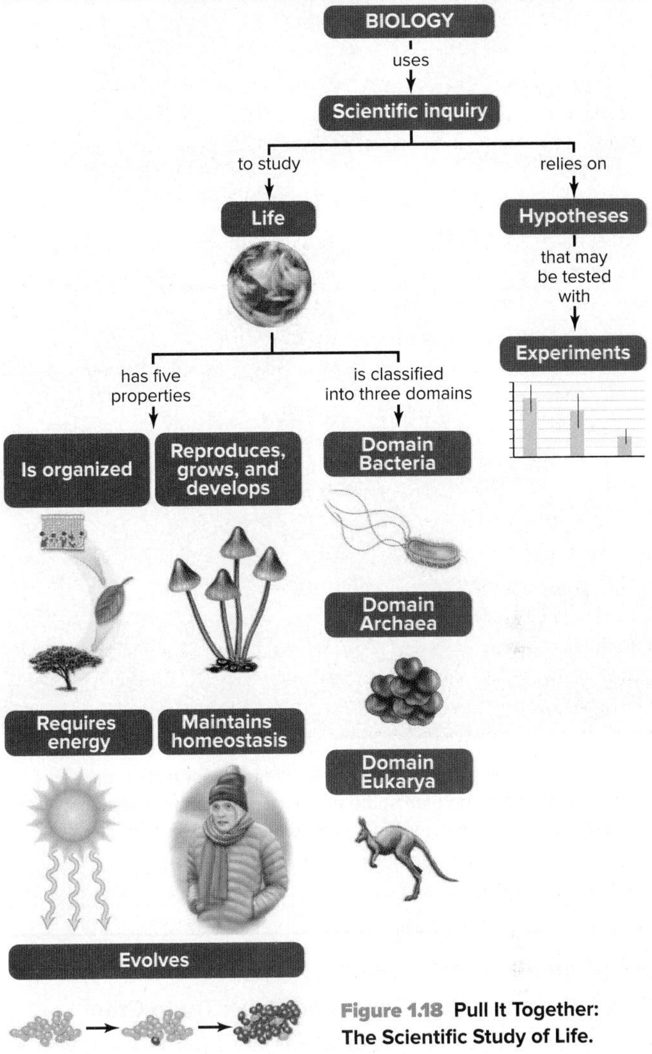

Figure 1.18 Pull It Together: The Scientific Study of Life.

Refer to figure 1.18 and the chapter content to answer the following questions.

1. What are the elements of a controlled experiment?
2. What is the relationship between natural selection and evolution?
3. Review the Survey the Landscape figure in the chapter introduction, paying special attention to the units of life and their components. Connect these structures to the Pull It Together concept map.

Design element: Burning Question (fire background): ©Ingram Publishing/Super Stock

LEARNING OUTLINE

2.1 Atoms Make Up All Matter

2.2 Chemical Bonds Link Atoms

2.3 Water Is Essential to Life

2.4 Cells Have an Optimum pH

2.5 Cells Contain Four Major Types of Organic Molecules

APPLICATIONS

Why We Care 2.1 *Acids and Bases in Everyday Life*

Burning Question 2.1 *What does it mean when food is "organic" or "natural"?*

Why We Care 2.2 *Sugar Substitutes and Fake Fats*

Burning Question 2.2 *What is junk food?*

Investigating Life 2.1 *Chemical Warfare on a Tiny Battlefield*

Life Is Chemistry. Soil, water, and air provide the elements that make up plants. When we eat the plants, the elements they contain become part of our own bodies. This gardener is digging up a fresh crop of organic red potatoes.

©ML Harris/Getty Images RF

Learn How to Learn
Organize Your Time, and Don't Try to Cram

Get a calendar and study the syllabus for every class you are taking. Write each due date in your calendar. Include homework assignments, quizzes, and exams, and add new dates as you learn them. Then, block out time well before each due date to work on each task. Success comes much more easily if you take a steady pace instead of waiting until the last minute.

SURVEY THE LANDSCAPE
Science, Chemistry, and Cells

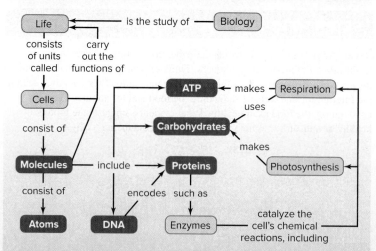

All life is composed of chemical substances, including not only water but also DNA, proteins, and other organic molecules. Each of these substances is composed of atoms.

For more details, study the Pull It Together feature in the chapter summary.

If you are like many people, you may not feel as though chemistry has much to do with your life. But it does. The easiest connection to make is with your food. Look at any nutrition label from a food package. The chemicals listed on the label are a subset of the same ones that your own body is made of. You really are what you eat!

©Photodisc/Getty Images RF

There are many other examples of everyday chemistry as well. Chemistry allows technicians to test urine for everything from drugs to sugar to the hormones that indicate pregnancy. Chemical reactions account for the damaging effects of acid rain and air pollution. The chemicals we add to swimming pools prevent the growth of algae and the spread of disease. The flea medicines we apply to our dogs and cats are chemicals that disrupt the insect life cycle. And we water and fertilize our plants because of chemistry. The list is literally endless.

Life is made of chemicals, and chemical reactions sustain the life of every individual. Understanding biology is impossible without an introduction to chemistry. This chapter describes how tiny particles called atoms come together to form the molecules of life.

2.1 Atoms Make Up All Matter

If you have ever touched a plant in a restaurant to see if it's fake, you know that we all have an intuitive sense of what life is made of. Most living leaves feel moist and pliable; a fake one is dry and stiff. But what does chemistry tell us about the composition of life?

Your desk, your body, your sandwich, your dog, a swimming pool, a plastic plant—indeed, all objects in the universe, including life on Earth—are composed of matter and energy. **Matter** is any material that takes up space, such as organisms, rocks, the oceans, and gases in the atmosphere. This chapter and the next concentrate on the building blocks that make up living matter. Physicists define **energy,** on the other hand, as the ability to do work. In this context, *work* means moving matter. Heat, light, and chemical bonds are all forms of energy; chapters 4, 5, and 6 discuss the energy of life in detail.

A. Elements Are Fundamental Types of Matter

The matter that makes up every object in the universe consists of one or more elements. A chemical **element** is a pure substance that cannot be broken down by chemical means into other substances. Examples of elements include oxygen (O), carbon (C), nitrogen (N), sodium (Na), and hydrogen (H).

Scientists had already noticed patterns in the chemical behavior of the elements by the mid-1800s, and several had proposed schemes for organizing the elements into categories. Nineteenth-century Russian chemist Dmitry Mendeleyev invented the chart that we still use today. This chart, called the **periodic table,** arranges the elements in such a way that their chemical properties repeat in each vertical column. Figure 2.1 illustrates an abbreviated periodic table, emphasizing the elements that make up organisms.

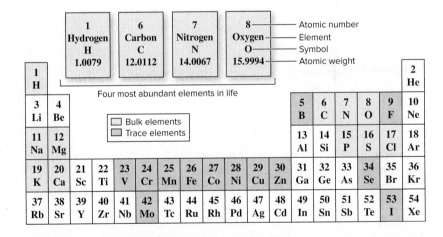

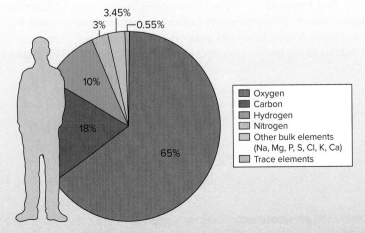

Figure 2.1 Chemical Elements. This abbreviated periodic table shows the first 54 elements, each with a unique atomic number and symbol; a complete periodic table appears in appendix D. The pie chart shows the distribution (by weight) of the elements that compose the human body.

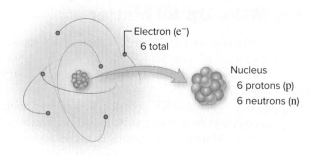

Figure 2.2 **Atom Anatomy.** The nucleus at the center of an atom is made of protons (*green*) and neutrons (*tan*). A cloud of electrons surrounds the nucleus. This example has six protons, so it is a carbon atom.

TABLE 2.1	Types of Particles in an Atom		
Particle	**Charge**	**Mass**	**Location**
Electron	Negative (–)	~0	Surrounding nucleus
Neutron	None	1	Nucleus
Proton	Positive (+)	1	Nucleus

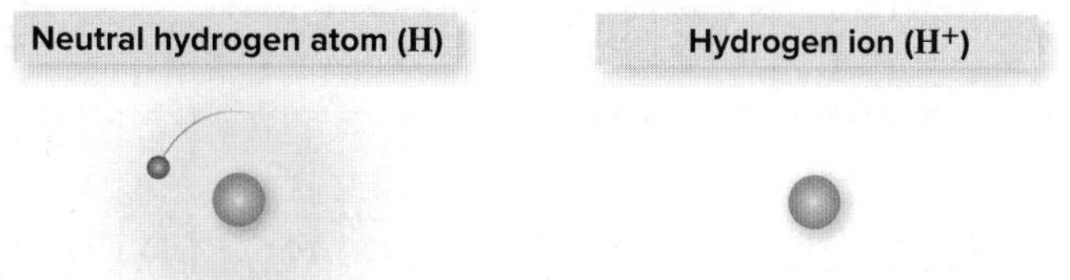

Neutral hydrogen atom (H)

Hydrogen ion (H⁺)

Particle	Charge		Particle	Charge
1 proton (p)	+1		1 proton (p)	+1
0 neutrons (n)	0		0 neutrons (n)	0
1 electron (e⁻)	–1		0 electrons (e⁻)	0
Net charge on atom	0		Net charge on atom	+1

Figure 2.3 **Two Forms of Hydrogen.** A neutral (uncharged) atom of hydrogen consists of one proton and one electron, whose charges balance one another. A hydrogen ion (H⁺) has a net positive charge because it has lost its electron.

About 25 elements are essential to life. Of these, the **bulk elements** are required in the largest amounts because they make up the vast majority of every living cell. The four most abundant bulk elements in life are carbon, hydrogen, oxygen, and nitrogen; these elements account for 96% of the human body. About 3.5% of the body consists of other bulk elements, including phosphorus (P), sulfur (S), sodium (Na), magnesium (Mg), potassium (K), and calcium (Ca). **Trace elements,** such as iron (Fe) and zinc (Zn), are required in small amounts; they total only about 0.5% of the human body.

A person whose diet is deficient in any essential element can become ill or die. The thyroid gland, for example, requires the trace element iodine (I). If the diet does not supply enough iodine, the thyroid may become enlarged, forming a growth called a goiter in the neck. Similarly, blood cells require iron to carry oxygen to the body's tissues. An iron-poor diet can cause anemia, which is a decline in the number of red blood cells.

B. Atoms Are Particles of Elements

An **atom** is the smallest possible "piece" of an element that retains the characteristics of the element. An atom is composed of three types of particles (figure 2.2 and table 2.1). **Protons,** which carry a positive charge, and **neutrons,** which are uncharged, together form a central **nucleus.** Negatively charged **electrons** surround the nucleus. Compared with a proton or neutron, an electron is essentially weightless.

For simplicity, most illustrations of atoms show the electrons closely hugging the nucleus. In reality, however, if the nucleus of a hydrogen atom were the size of a small marble, the entire atom would have a diameter slightly longer than a football field! Thus, most of an atom's mass is concentrated in the nucleus, while the electron cloud occupies virtually all of its volume.

How can this electron cloud, which is mostly empty space, account for the solid "feel" of the objects in our world? The fact that the electrons are in constant motion helps explain this paradox. A good analogy is a ceiling fan. When the fan is not spinning, it is easy to move your hand between two blades. But when the fan is on, the rotating blades essentially form a solid disk.

Each element has a unique **atomic number,** the number of protons in the nucleus. Hydrogen, the simplest type of atom, has an atomic number of 1. In contrast, an atom of uranium has 92 protons. Elements are arranged sequentially in the periodic table by atomic number, which appears above each element's symbol (see figure 2.1).

When the number of protons equals the number of electrons, the atom is electrically neutral; that is, it has no net charge. An **ion** is an atom (or group of atoms) that has gained or lost electrons and therefore has a net negative or positive charge. One common positively charged ion is hydrogen, H⁺ (figure 2.3); others include sodium (Na⁺), and potassium (K⁺). Negatively charged ions include hydroxide (OH⁻) and chloride (Cl⁻). Ions participate in many biological processes, including the transmission of messages in the nervous system. They also form ionic bonds, discussed in section 2.2. ⓘ *action potential,* section 24.3C

C. Isotopes Have Different Numbers of Neutrons

An atom's **mass number** is the total number of protons and neutrons in its nucleus. Because neutrons and protons have the same mass (see table 2.1), subtracting the atomic number from the mass number yields the number of neutrons in an atom.

All atoms of an element have the same number of protons but not necessarily the same number of neutrons. An **isotope** is any of these different forms of a single element (figure 2.4). For example, carbon has three isotopes, designated ^{12}C (six neutrons), ^{13}C (seven neutrons), and ^{14}C (eight neutrons). The superscript denotes the mass number of each isotope.

Figure It Out

The most abundant isotope of iron (Fe) has a mass number of 56. If Fe has an atomic number of 26, how many neutrons are in each atom of ^{56}Fe?

Answer: 30.

Often one isotope of an element is very abundant, and others are rare. For example, about 99% of carbon isotopes are ^{12}C, and only 1% are ^{13}C or ^{14}C. An element's **atomic weight**, which is the average mass of all atoms of an element, is typically close to the mass number of the most abundant isotope (see figure 2.4b).

Most elements have both stable (nonradioactive) isotopes and **radioactive isotopes,** which emit energy as rays or particles when they break down into more stable forms. Every radioactive isotope has a characteristic half-life, which is the time it takes for half of the atoms in a sample to emit radiation, or "decay" to a different, more stable form. Scientists have determined the half-life of each radioactive isotope experimentally. Depending on the isotope, the half-life might range from a fraction of a second to millions or even billions of years. Physicists use large samples of isotopes and precise measurements to calculate the longest half-lives.

Radioactive isotopes have many uses in medicine and science, ranging from detecting broken bones to determining the ages of fossils. But the same properties that make radioactive isotopes useful can also make them dangerous. Exposure to excessive radiation can lead to radiation sickness, and radiation-induced mutations of a cell's DNA can cause cancer (see chapter 8). The lead-containing "bib" that a dentist places on your chest during mouth X-rays protects you from radiation. ⓘ *radiometric dating*, section 13.2B

2.1 Mastering Concepts

1. Which four chemical elements do organisms require in the largest amounts?
2. Where in an atom are protons, neutrons, and electrons located?
3. What does an element's atomic number indicate?
4. What is the relationship between an atom's mass number and an element's atomic weight?
5. How are the isotopes of an element different from one another?

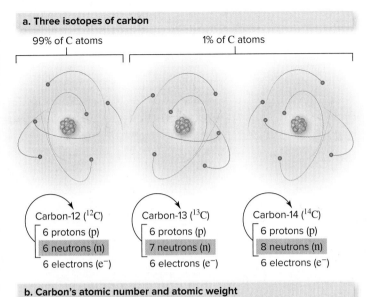

a. Three isotopes of carbon

99% of C atoms   1% of C atoms

Carbon-12 (^{12}C)	Carbon-13 (^{13}C)	Carbon-14 (^{14}C)
6 protons (p)	6 protons (p)	6 protons (p)
6 neutrons (n)	7 neutrons (n)	8 neutrons (n)
6 electrons (e⁻)	6 electrons (e⁻)	6 electrons (e⁻)

b. Carbon's atomic number and atomic weight

6 ——— Atomic number (number of protons in C)
Carbon ——— Element
C ——— Symbol
12.0112 ——— Atomic weight (average mass of all C atoms)

Figure 2.4 Three Isotopes of Carbon. (a) Carbon's atomic number is 6, so its nucleus always contains six protons. These three carbon isotopes, however, have different numbers of neutrons. (b) Thanks to the 1% of carbon atoms that have more than six neutrons, the atomic weight of carbon is just over 12.

Miniglossary	Matter
Element	A pure substance that cannot be broken down by chemical means into other substances; each element has a characteristic atomic number
Atom	The smallest unit of an element that retains the characteristics of that element
Atomic number	The number of protons in an atom's nucleus
Mass number	The number of protons plus the number of neutrons in an atom's nucleus
Isotope	Any of the different forms of the same element; each isotope is characterized by a different number of neutrons
Atomic weight	The average mass of all atoms of an element

2.2 Chemical Bonds Link Atoms

Like all organisms, you are composed mostly of carbon, hydrogen, oxygen, and nitrogen atoms. But the arrangement of these atoms is not random. Instead, your atoms are organized into molecules (see figure 1.2). A **molecule** is two or more chemically joined atoms.

Some molecules, such as the gases hydrogen (H_2), oxygen (O_2), and nitrogen (N_2), consist of two atoms of the same element. More often, however, the elements in a molecule are different. A **compound** is a molecule composed of two or more different elements. Carbon monoxide (CO), for example, is a compound consisting of one carbon and one oxygen atom. Likewise, water (H_2O) is made of two atoms of hydrogen and one of oxygen. Many large biological compounds, including DNA and proteins, consist of tens of thousands of atoms.

A compound's characteristics can differ strikingly from those of its separate elements. Consider table salt, sodium chloride. Sodium (Na) is a silvery, highly reactive solid metal, whereas chlorine is a yellow, corrosive gas. But when equal numbers of these two atoms combine, the resulting compound forms the familiar white salt crystals that we sprinkle on food—an excellent example of an emergent property. Another example is methane, the main component of natural gas. Its components are carbon, a black sooty solid, and hydrogen, a light, combustible gas. ⓘ *emergent properties,* section 1.1A

Scientists describe molecules by writing the symbols of their constituent elements and indicating the number of atoms of each element as subscripts. For example, methane is written CH_4, which denotes a molecule with one carbon atom and four hydrogen atoms. This representation of the atoms in a compound is termed a *molecular formula.* Table salt's formula is NaCl, that of water is H_2O, and that of the gas carbon dioxide is CO_2.

What forces hold together the atoms that make up each of these molecules? To understand the answer, we must first learn more about how electrons are arranged around the nucleus.

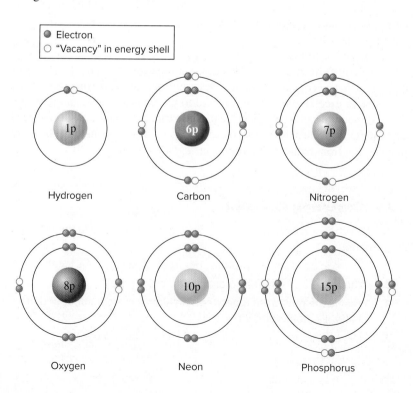

- ● Electron
- ○ "Vacancy" in energy shell

1p — Hydrogen

6p — Carbon

7p — Nitrogen

8p — Oxygen

10p — Neon

15p — Phosphorus

Figure 2.5 **Energy Shells.** Shown here are models of atoms from six different elements. Each ring represents one energy shell, and each pair of electrons represents one orbital.

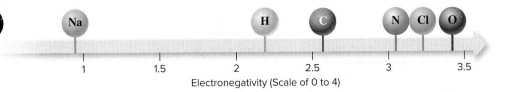

Figure 2.6 Unequal Attraction. Atoms vary widely in their electronegativity, which is the ability to attract electrons.

Electronegativity (Scale of 0 to 4)

A. Electrons Determine Bonding

Electrons occupy distinct energetic regions around the nucleus. They are constantly in motion, so it is impossible to determine the exact location of any electron at any given moment. Instead, chemists use the term **orbitals** to describe the most likely location for an electron relative to its nucleus. Each orbital can hold up to two electrons. Consequently, the more electrons in an atom, the more orbitals they occupy.

We can envision any atom's electrons as occupying a series of concentric **energy shells**, each having a higher energy level than the one inside it (figure 2.5). The number of orbitals in each shell determines the number of electrons the shell can hold. The lowest energy shell, for example, contains just one orbital and thus holds up to two electrons. The next two shells each contain four orbitals and therefore hold as many as eight electrons each.

Electrons occupy the lowest energy level available to them, starting with the innermost one. As each energy shell fills, any additional electrons must reside in higher energy shells. For example, hydrogen has only one electron in the lowest energy orbital, and helium has two. Carbon has six electrons; two occupy the lowest energy orbital, and four are in the next energy shell. Oxygen, with eight electrons total, has two electrons in the lowest energy orbital and six at the next higher energy level.

An atom's **valence shell** is its outermost occupied energy shell. Atoms are most stable when their valence shells are full. The gases helium (He) and neon (Ne), for example, are inert—that is, they are chemically unreactive. Because their outermost shells are full, they exist in nature without combining with other atoms.

For most atoms, however, the valence shell is only partially filled. Such an atom will become most stable if its valence-shell "vacancies" fill. As you will soon see, atoms may donate, steal, or share electrons to arrive at exactly the right number. The exact "strategy" that an atom uses depends in part on its **electronegativity,** which measures the atom's ability to attract electrons on a scale of 0 to 4 (figure 2.6). Oxygen, for example, has high electronegativity compared to sodium. Elements with high electronegativity tend to strip electrons away from those with lower values. Elements with moderate electronegativity often share electrons.

Whether electrons are stolen or shared, the transfer of electrons from one atom to another creates a **chemical bond,** an attractive force that holds atoms together. The remainder of this section describes three types of chemical bonds that are important in biology (table 2.2).

B. In an Ionic Bond, One Atom Transfers Electrons to Another Atom

Sometimes, two atoms have such different electronegativities that one actually takes one or more of its partner's electrons. Recall that an atom is most stable if its valence shell is full. The most electronegative atoms, such as chlorine (Cl), are usually those whose valence shells have only one "vacancy." Likewise, sodium (Na) and other weakly electronegative atoms have only one electron in the outermost shell. Neither chlorine

TABLE 2.2 Chemical Bonds: A Summary

Type	Chemical Basis	Strength	Example
Ionic bond	One atom donates one or more electrons to another atom, forming oppositely charged ions that attract each other.	Strong but breaks easily in water	Sodium chloride (NaCl)
Covalent bond	Two atoms share pairs of electrons.	Strong	O–H bond within water molecule
Hydrogen bond	An atom with a partial negative charge attracts an atom with a partial positive charge. Hydrogen bonds form between adjacent molecules or between different parts of a large molecule.	Weak	Attraction between adjacent water molecules

Figure 2.7 **Table Salt, an Ionically Bonded Molecule.** (a) A sodium atom (Na) can donate its "spare" electron to a chlorine atom (Cl), which has seven electrons in its outermost shell. After the transfer, the valence shells of both atoms are full. The resulting ions (Na^+ and Cl^-) form the compound sodium chloride, NaCl. (b) Na^+ and Cl^- ions occur in a repeating pattern that produces salt crystals.

- Electron
- "Vacancy" in energy shell

11p	17p		11p (+)	17p (−)
Na	+	Cl	→	NaCl

a.

Na^+ Cl^-

b.

nor sodium would benefit from sharing. Instead, sodium is most stable if it simply releases its extra electron to chlorine, which needs this "scrap" electron to complete its own valence shell (figure 2.7).

An ion is an atom that has lost or gained electrons. The atom that has lost electrons—such as the sodium atom in figure 2.7—is an ion carrying a positive charge. Conversely, the one that has gained electrons—chlorine in this case—acquires a negative charge. An **ionic bond** results from the electrical attraction between two ions with opposite charges. In general, such bonds form between an atom whose outermost shell is almost empty and one whose valence shell is nearly full.

The ions in figure 2.7 have bonded ionically to form NaCl. In NaCl, the most stable configuration of Na^+ and Cl^- is a three-dimensional crystal. Ionic bonds in crystals are strong, as demonstrated by the stability of the salt in your shaker. Those same crystals, however, dissolve when you stir them into water. As described in section 2.3, water molecules pull ionic bonds apart.

C. In a Covalent Bond, Atoms Share Electrons

So far, we have seen ionic bonds in which one highly electronegative atom fills its outermost shell by taking one or more electrons from another atom. However, it is also possible for two atoms to fill their outermost shells by pooling their resources. In a **covalent bond,** two atoms share electrons. The shared electrons travel around both nuclei, strongly connecting the atoms together. Most of the bonds in biological molecules are covalent.

Figure 2.8 shows three examples of molecules made of atoms joined by covalent bonds. The top row, for example, illustrates methane, CH_4. A carbon atom has six electrons, two of which occupy its innermost shell. That leaves four electrons in its valence shell, which has a capacity of eight. Carbon therefore requires four more electrons to fill its outermost shell. A carbon atom can attain the stable eight-electron configuration by sharing electrons with four hydrogen atoms, each of which has one electron in its only shell. Similarly, the middle row of figure 2.8 shows how oxygen and hydrogen atoms combine to produce water (H_2O), and the bottom row shows how two oxygen atoms form oxygen gas, O_2.

Covalent bonds are usually depicted as lines between the interacting atoms, with each line representing one bond (figure 2.9). Each single bond contains two electrons, one from each atom. Atoms can also share two pairs of electrons,

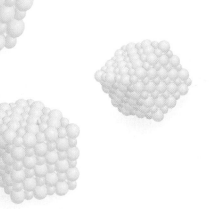

Figure It Out

Use the information in figures 2.1 and 2.5 to predict the number of covalent bonds that nitrogen (N) forms.

Answer: 3.

Name	Molecular formula	Reaction			Structural formula

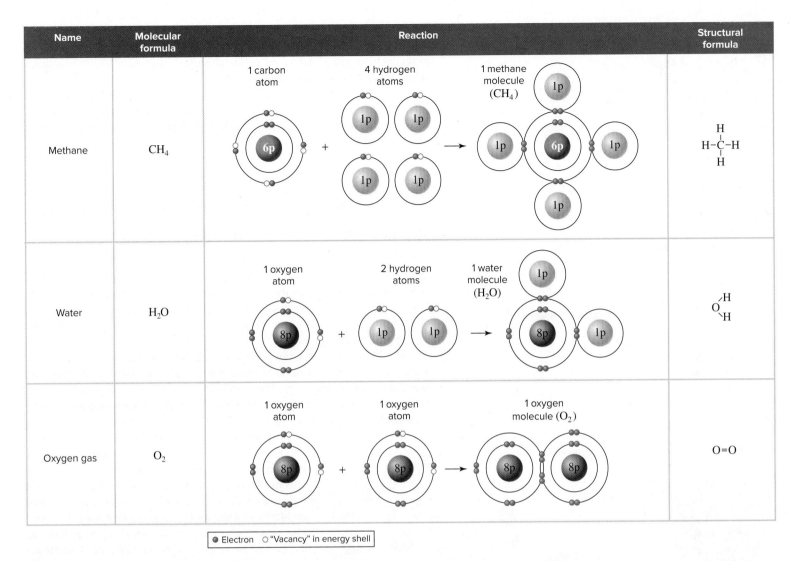

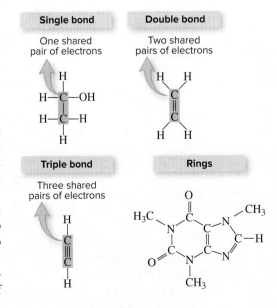

● Electron   ○ "Vacancy" in energy shell

Figure 2.8 Atoms Share Electrons in Covalent Bonds. Methane (CH_4), water (H_2O), and oxygen gas (O_2) all consist of atoms that complete their outermost shells by sharing electrons with their neighbors, forming covalent bonds.

forming a double covalent bond. The O_2 molecule in figure 2.8, for example, has one double bond. The greater the number of shared electrons, the stronger the bond. A triple covalent bond (three shared pairs of electrons) is therefore extremely strong. Multiple elements, joined primarily with single and double covalent bonds, form the large chains and rings that characterize the molecules of life (see the ring structure in figure 2.9).

Covalent bonding means "sharing," but the partnership is not necessarily equal. A **polar covalent bond** is a lopsided union in which one nucleus exerts a much stronger pull on the shared electrons than does the other nucleus. Polar bonds form whenever a highly electronegative atom such as oxygen shares electrons—unequally—with a less electronegative partner such as carbon or hydrogen.

Polar covalent bonds are critical to biology. As described in section 2.2D, they are responsible for hydrogen bonds, which in turn help define not only the unique properties of water (see section 2.3) but also the shapes of DNA and proteins (section 2.5).

In contrast, in a **nonpolar covalent bond,** two atoms exert approximately equal pull on their shared electrons. A bond between two atoms of

Figure 2.9

Versatile Carbon. Each carbon atom in a molecule shares four pairs of electrons, forming combinations of single, double, and triple covalent bonds. In some molecules, carbon combines with other elements to form rings.

Figure It Out

The electronegativity of potassium (K) is 0.82. Using the scale in figure 2.6, what type of bond should form between K and Cl?

Answer: Ionic.

Figure It Out

If C is in a covalent bond with N, will the partial charge on N be positive or negative? (*Hint*: Use figure 2.6.)

Answer: N will have a partial negative charge.

the same element, such as a carbon–carbon bond, is nonpolar; after all, a bond between two identical atoms must be electrically balanced. H_2, N_2, and O_2 are all nonpolar molecules. Carbon and hydrogen atoms have similar electronegativity. A carbon–hydrogen bond is therefore also nonpolar.

Ionic bonds, polar covalent bonds, and nonpolar covalent bonds represent points along a continuum. If one atom is so electronegative that it rips electrons from another atom's valence shell, an ionic bond forms. If one atom tugs at shared electrons much more than the other, a covalent bond is polar. And two atoms of similar electronegativity share electrons equally in nonpolar covalent bonds. Notice that the bond type depends on the *difference* in electronegativity, so the same element can participate in different types of bonds. Oxygen, for example, forms nonpolar bonds with itself (as in O_2) and polar bonds with hydrogen (as in H_2O).

D. Partial Charges on Polar Molecules Create Hydrogen Bonds

When a covalent bond is polar, the negatively charged electrons spend more time around the nucleus of the more electronegative atom than around its partner. The "electron-hogging" atom therefore has a partial negative charge (written as "δ^-"), and the less electronegative partner has an electron "deficit" and a partial positive charge (δ^+).

In a **hydrogen bond,** opposite partial charges on *adjacent molecules*—or within a single large molecule—attract each other. The name comes from the fact that the atom with the partial positive charge is always hydrogen. The atom with the partial negative charge, on the other hand, is a highly electronegative atom such as oxygen or nitrogen.

Water provides the simplest illustration of hydrogen bonds (figure 2.10). Each water molecule has a "boomerang" shape. Moreover, the two O–H bonds in water are polar, with the nucleus of the oxygen atom attracting the shared electrons more strongly than do the hydrogen nuclei. Each hydrogen atom in a water molecule therefore has a partial positive charge, which attracts the partial negative charge of the oxygen atom on an adjacent molecule. This attraction is the hydrogen bond. The partial charges on O and H, plus the bent shape, cause water molecules to stick to one another and to some other substances. (This slight stickiness is another example of an emergent property, because it arises from interactions between O and H.)

Hydrogen bonds are relatively weak compared with ionic and covalent bonds. In one second, the hydrogen bonds between one water molecule and its nearest neighbors break and re-form some 500 billion times. Even though hydrogen bonds are weak, they account for many of water's unusual characteristics—the subject of section 2.3. In addition, the collective strength of multiple hydrogen bonds helps stabilize some large molecules, including proteins and DNA (see section 2.5).

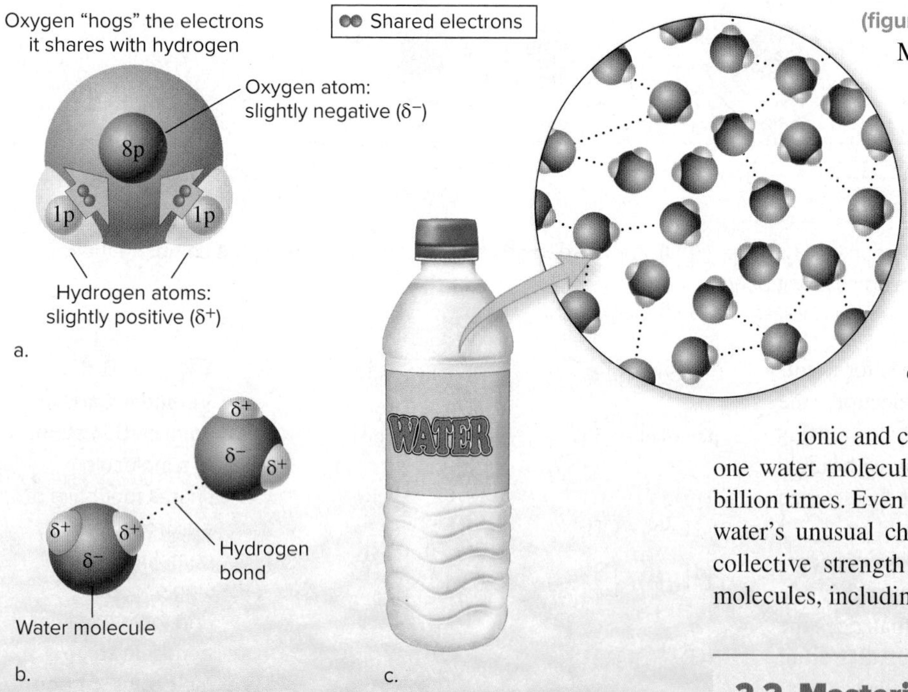

Oxygen "hogs" the electrons it shares with hydrogen

Shared electrons

Oxygen atom: slightly negative (δ^-)

8p

1p   1p

Hydrogen atoms: slightly positive (δ^+)

a.

δ^+

δ^- δ^+

δ^+ δ^+

δ^-

Hydrogen bond

Water molecule

b.

WATER

c.

Figure 2.10 Hydrogen Bonds in Water. (a) An oxygen atom attracts electrons more strongly than do the hydrogen atoms in a water molecule. The O atom therefore bears a partial negative charge (δ^-), and the H atoms carry partial positive charges (δ^+). (b) The hydrogen bond is the attraction between partial charges on adjacent molecules. (c) In liquid water, many molecules stick to one another with hydrogen bonds.

2.2 Mastering Concepts

1. How are atoms, molecules, and compounds related?
2. How does the number of valence electrons determine an atom's tendency to form bonds?
3. Explain how electronegativity differences between atoms result in each type of chemical bond.

2.3 Water Is Essential to Life

Although water may seem to be a rather ordinary fluid, it is anything but. The tiny, three-atom water molecule has extraordinary properties that make it essential to all organisms, which explains why the search for life on other planets begins with the search for water. Indeed, life on Earth began in water, and for at least the first 3 billion years of life's history on Earth, all life was aquatic (see chapter 15). It was not until some 475 million years ago, when plants and fungi colonized land, that life could survive without being surrounded by water. Even now, terrestrial organisms cannot live without it. This section explains some of the properties that make water central to biology. Table 2.3 summarizes water's properties.

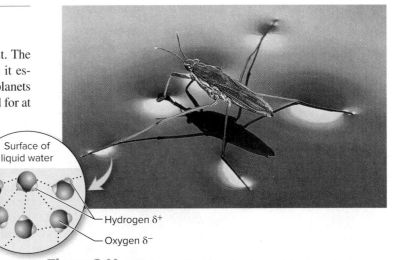

Surface of liquid water

Hydrogen δ+
Oxygen δ−

Figure 2.11 **Running on Water.** A lightweight body and water-repellent legs allow this water strider to "skate" across a pond without breaking the water's surface tension.

Photo: ©Herman Eisenbeiss/Science Source

A. Water Is Cohesive and Adhesive

Hydrogen bonds contribute to a property of water called **cohesion**—the tendency of water molecules to stick together. Without cohesion, water would evaporate instantly from most locations on Earth's surface. Cohesion also contributes to the observation that you can sometimes fill a glass so full that water is above the rim, yet it doesn't flow over the side unless disturbed.

This tendency of a liquid to hold together at its surface is called surface tension, and not all liquids exhibit it. Water has high surface tension because it is cohesive. At the boundary between water and air, the water molecules form hydrogen bonds with neighbors to their sides and below them in the liquid. These bonds tend to hold the surface molecules together, creating a thin "skin" that is strong enough to support small animals without breaking through (figure 2.11).

A related property of water is **adhesion,** the tendency to form hydrogen bonds with substances other than water. For example, when water soaks into a paper towel, it is adhering to the molecules that make up the paper.

Both adhesion and cohesion are at work when water seemingly defies gravity as it moves from a plant's roots to its highest leaves (figure 2.12). This movement depends upon cohesion of water within the plant's internal conducting tubes. Water entering roots is drawn up through these tubes as water molecules evaporate from leaf cells. Adhesion to the walls of the conducting tubes also helps lift water to the topmost leaves of trees. ⓘ *transpiration,* section 21.6A

B. Many Substances Dissolve in Water

Another reason that water is vital to life is that it can dissolve a wide variety of chemicals. To illustrate this process, picture the slow disappearance of table salt as it dissolves in water. Although the salt crystals seem to vanish, the sodium and chloride ions remain. Water

1 Water evaporates through pores in leaves.

2 Evaporating molecules pull water up stem.

3 Water molecules are pulled into roots.

Figure 2.12 **Defying Gravity.** Thanks to hydrogen bonds, water evaporating from the leaves of a palm tree is replaced by water pulled up from the soil and through the tree's trunk.

Photo: ©Getty Images/flickr RF

(water background): ©Getty Images/flickr RF

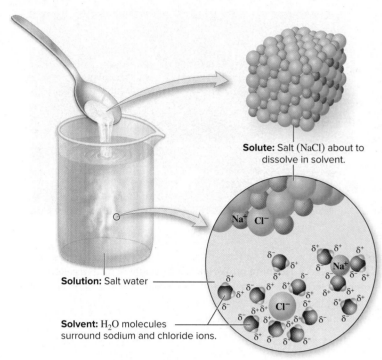

Solute: Salt (NaCl) about to dissolve in solvent.

Na⁺ Cl⁻

Na⁺

Cl⁻

Solution: Salt water

Solvent: H₂O molecules surround sodium and chloride ions.

Figure 2.13 Dissolving Salt. As salt crystals dissolve, polar water molecules surround each sodium and chloride ion.

molecules surround each ion individually, separating them from one another (figure 2.13).

In this example, water is a **solvent:** a chemical in which other substances, called **solutes,** dissolve. A **solution** consists of one or more solutes dissolved in a liquid solvent. In a so-called aqueous solution, water is the solvent. But not all solutions are aqueous. According to the rule "Like dissolves like," polar solvents such as water dissolve polar molecules; similarly, nonpolar solvents dissolve nonpolar substances.

Scientists divide chemicals into two categories, based on their affinity for water. **Hydrophilic** substances are either polar or charged, so they readily dissolve in water (the term literally means "water-loving"). Examples include sugar, salt, and ions. Electrolytes are ions in the body's fluids, and the salty taste of sweat illustrates water's ability to dissolve them. Sports drinks replace not only water but also sodium, potassium, magnesium, and calcium ions that are lost in perspiration during vigorous exercise.

Not every substance, however, is water-soluble. Nonpolar molecules are called **hydrophobic** ("water-fearing") because they do not dissolve in, or form hydrogen bonds with, water. Butter and oil are hydrophobic because they are made mostly of carbon and hydrogen, which form nonpolar bonds with each other. This is why water alone will not remove grease from hands, dishes, or clothes. Detergents contain molecules that attract both water and fats, so they can dislodge greasy substances and carry the mess down the drain with the wastewater.

C. Water Regulates Temperature

Another unusual property of water is its ability to resist temperature changes. When molecules absorb energy, they move faster. Water's hydrogen bonds tend to counteract this molecular movement; as a result, more heat is needed to raise water's temperature than is required for most other liquids, including alcohols. Because an organism's fluids are aqueous solutions, the same effect holds: An organism may encounter considerable heat before its body temperature becomes dangerously high. Likewise, the body cools slowly in cold temperatures.

At a global scale, water's resistance to temperature change explains why coastal climates tend to be mild. People living along the California coast have good weather year-round because the Pacific Ocean's steady temperature helps keep winters warm and summers cool. Far away from the ocean, in the central United States, winters are much colder and summers are much hotter. These differences in local climate contribute to the unique ecosystems that occur in each region. (Chapter 19 describes climate in more detail.)

Hydrogen bonds also mean that a lot of heat is required to evaporate water. **Evaporation** is the conversion of a liquid into a vapor. When sweat evaporates from skin, individual water molecules break away from the liquid droplet and float into the atmosphere. Surface molecules must absorb energy to escape, and when they do, heat energy is removed from those that remain, drawing heat out of the body—an important part of the mechanism that regulates body temperature (figure 2.14).

D. Water Expands As It Freezes

Water's unusual tendency to expand upon freezing also affects life. In liquid water, hydrogen bonds are constantly forming and breaking, and the water molecules are relatively close together. But in an ice crystal, the hydrogen bonds are stable, and the molecules are "locked" into roughly hexagonal

Figure 2.14 Cooling Off. A fan speeds the evaporation of sweat. As water molecules evaporate from the skin's surface, they carry heat energy with them. The result: a cool sensation.

shapes. Therefore, the less-dense ice floats on the surface of the denser liquid water below (figure 2.15).

This characteristic benefits aquatic organisms. When the air temperature drops, a small amount of water freezes at the pond's surface. This solid cap of ice retains heat in the water below. If ice were to become denser upon freezing, it would sink to the bottom. The lake would then gradually turn to ice from the bottom up, entrapping the organisms that live there.

The formation of ice crystals inside cells, however, can be deadly. The expansion of ice inside a frozen cell can rupture the delicate outer membrane, killing the cell. How, then, do organisms survive in extremely cold weather? Mammals have thick layers of insulating fur and fat that help their bodies stay warm. Icefishes that live in the cold waters surrounding Antarctica have a different adaptation: They produce antifreeze chemicals that prevent their cells from freezing solid.

E. Water Participates in Life's Chemical Reactions

Life exists because of thousands of simultaneous chemical reactions. In a **chemical reaction,** two or more molecules "swap" their atoms to yield different molecules; that is, some chemical bonds break and new ones form. Chemists depict these reactions as equations with the **reactants,** or starting materials, to the left of an arrow; the **products,** or results of the reaction, are listed to the right.

Consider what happens when the methane in natural gas burns inside a heater, gas oven, or stove:

$$CH_4 + 2O_2 \longrightarrow CO_2 + 2H_2O$$

methane + oxygen $\longrightarrow$ carbon dioxide + water

In words, this equation says that one methane molecule combines with two oxygen molecules to produce a carbon dioxide molecule and two molecules of water. The bonds of the methane and oxygen molecules have broken, and new bonds have formed in the products.

Note that each side of the equation shows the same number of atoms of each element: that is, one carbon, four hydrogens, and four oxygens. Atoms are neither created nor destroyed in a chemical reaction; rather, they are simply rearranged.

Nearly all of life's chemical reactions occur in the watery solution that fills and bathes cells. Moreover, water is either a reactant in or a product of many of these reactions. In photosynthesis, for example, plants use the sun's energy to assemble food out of just two reactants: carbon dioxide and water (see chapter 5). Section 2.5 describes two other water-related reactions, hydrolysis and dehydration synthesis, that are vital to life.

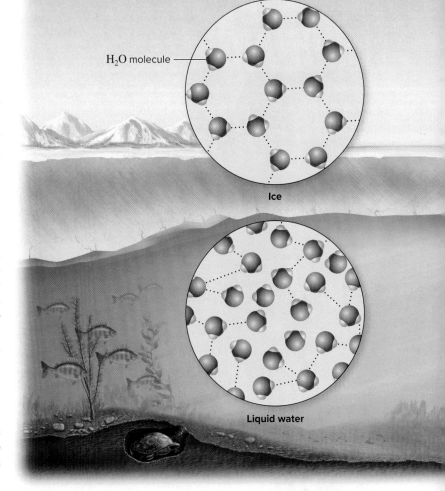

H_2O molecule

Ice

Liquid water

Figure 2.15 Ice Floats. The water molecules in ice form hexagons stabilized by hydrogen bonds. Ice is therefore less dense than—and floats on top of—the liquid water in this lake.

2.3 Mastering Concepts

1. How are cohesion and adhesion important to life?
2. Distinguish between a solute and a solvent and between a hydrophilic and a hydrophobic molecule.
3. How does water help an organism regulate its body temperature?
4. How does the density difference between ice and liquid water affect life?
5. What happens in a chemical reaction?
6. How does water participate in the chemistry of life?

TABLE 2.3	**Characteristics of Water: A Summary**
Characteristic	**Example**
Cohesion and adhesion	Water moves from a plant's roots to its leaves.
Dissolves substances	Salt dissolves in water.
Regulates temperature	Coastal climates are more mild than inland climates.
Expands as it freezes	Ice floats on a lake's surface.
Chemical reactant and product	Photosynthesis reactions require water molecules.

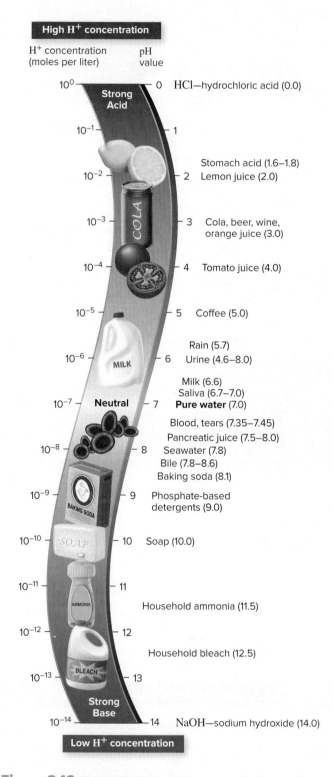

High H⁺ concentration

H⁺ concentration (moles per liter)	pH value	
10^0	0	HCl—hydrochloric acid (0.0)
		Strong Acid
10^{-1}	1	
10^{-2}	2	Stomach acid (1.6–1.8) Lemon juice (2.0)
10^{-3}	3	Cola, beer, wine, orange juice (3.0)
10^{-4}	4	Tomato juice (4.0)
10^{-5}	5	Coffee (5.0)
10^{-6}	6	Rain (5.7) Urine (4.6–8.0)
10^{-7}	7	Milk (6.6) Saliva (6.7–7.0) **Pure water (7.0)** **Neutral**
10^{-8}	8	Blood, tears (7.35–7.45) Pancreatic juice (7.5–8.0) Seawater (7.8) Bile (7.8–8.6) Baking soda (8.1)
10^{-9}	9	Phosphate-based detergents (9.0)
10^{-10}	10	Soap (10.0)
10^{-11}	11	
10^{-12}	12	Household ammonia (11.5)
10^{-13}	13	Household bleach (12.5)
10^{-14}	14	**Strong Base** NaOH—sodium hydroxide (14.0)

Low H⁺ concentration

Figure 2.16 **The pH Scale.** A neutral solution has a pH of 7. The higher the concentration of hydrogen ions (H⁺), the more acidic the solution (pH < 7). The lower the H⁺ concentration, the more basic (alkaline) the solution (pH > 7).

2.4 Cells Have an Optimum pH

One of the most important substances dissolved in water is one of the simplest: H⁺ ions. Each H⁺ is a hydrogen atom stripped of its electron; in other words, it is simply a proton (see figure 2.3). But its simplicity belies its enormous effects on living systems. Too much or too little H⁺ can ruin the shapes of critical molecules inside cells, rendering them nonfunctional.

One source of H⁺ is pure water. At any time, about one in a million water molecules spontaneously breaks into two pieces, producing one hydrogen ion (H⁺) and one hydroxide ion (OH⁻):

$$H_2O \longrightarrow H^+ + OH^-$$

In pure water, the number of hydrogen ions exactly equals the number of hydroxide ions. A **neutral** solution likewise contains as much H⁺ as it does OH⁻.

Some substances, however, alter this balance. An **acid** is a chemical that adds H⁺ to a solution, making the concentration of H⁺ ions exceed the concentration of OH⁻ ions. Examples include hydrochloric acid (HCl) and sour foods such as vinegar and lemon juice (see Why We Care 2.1). Adding acid to pure water releases H⁺ ions into the solution:

$$HCl \longrightarrow H^+ + Cl^-$$

A **base** is the opposite of an acid: It makes the concentration of OH⁻ ions exceed the concentration of H⁺ ions. Bases work in one of two ways. They come apart to directly add OH⁻ ions to the solution, or they absorb H⁺ ions. Either way, the result is the same: The balance between H⁺ and OH⁻ shifts toward OH⁻. Two common household bases are baking soda and sodium hydroxide (NaOH), an ingredient in oven and drain cleaners. When NaOH dissolves in water, it releases OH⁻ into solution:

$$NaOH \longrightarrow Na^+ + OH^-$$

If mixed together, acids and bases neutralize each other. The acid releases protons, while the base either absorbs the H⁺ or releases OH⁻.

Scientists use the **pH scale** to measure how acidic or basic a solution is. The pH scale ranges from 0 to 14, with 7 representing a neutral solution such as pure water (figure 2.16). An acidic solution has a pH lower than 7, whereas an **alkaline,** or basic, solution has a pH greater than 7. Thus, 0 represents a strongly acidic solution and 14 represents an extremely basic one.

Each unit on the pH scale represents a 10-fold change in H⁺ concentration. A solution with a pH of 4 is therefore 10 times more acidic than one with a pH of 5, and it is 100 times more acidic than one with a pH of 6.

All species have characteristic pH requirements. Straying too far from the normal pH can be deadly, yet organisms frequently encounter conditions that could alter their internal pH. They can maintain homeostasis because of **buffers,** pairs of weak acids and bases that resist pH changes.

2.4 Mastering Concepts

1. How do acids and bases affect a solution's H⁺ concentration?
2. How do the values of 0, 7, and 14 relate to the pH scale?
3. How do organisms maintain their pH within certain limits?

2.5 Cells Contain Four Major Types of Organic Molecules

Organisms are composed mostly of water and **organic molecules,** chemical compounds that contain both carbon and hydrogen. (Burning Question 2.1 explains the use of the term *organic* in describing food.) As you will see later in this unit, plants and other autotrophs can produce all the organic molecules they require, whereas heterotrophs—including humans—must obtain their organic building blocks from food.

Life uses a tremendous variety of organic compounds. Organic molecules consisting almost entirely of carbon and hydrogen are called hydrocarbons; methane (CH_4) is the simplest example. Because a carbon atom forms four covalent bonds, however, this element can assemble into much more complex molecules, including long chains, intricate branches, and rings (see figure 2.9). Many organic compounds also include other essential elements, such as oxygen, nitrogen, phosphorus, or sulfur. A peek ahead at the molecules illustrated in this section reveals the diversity of shapes and sizes of organic molecules. Without carbon's versatility, organic chemistry—and life—would be impossible.

All organisms, from bacteria to plants to people, consist largely of the same four types of organic molecules: carbohydrates, proteins, nucleic acids, and lipids. This unity in life's chemistry is powerful evidence that all species inherited the same basic chemical structures and processes from a common ancestor.

Name	Structure	Formula
Hydroxyl group	—O—H	—OH
Carboxyl group	—C(=O)(O—H)	—COOH
Amino group	—N(H)(H)	$-NH_2$
Phosphate group	—O—P(=O)(O⁻)—O⁻	$-PO_4^{-2}$

Figure 2.17 Molecular Connectors. Each of these chemically reactive groups of atoms occurs in one or more types of organic molecules. Look through the illustrations in the rest of section 2.5 to find examples of each type.

A. Large Organic Molecules Are Composed of Smaller Subunits

Proteins, nucleic acids, and some carbohydrates all share a property in common with one another: They are **polymers,** which are chains of small molecular subunits called **monomers.** A polymer is made of monomers that are linked together, just as a train is made of individual railcars.

Railcars include two connectors, which enable one to hook to another in a long train. Similarly, organic molecules have small groups of atoms that serve the same coupling function. Figure 2.17 shows four of the most common examples: hydroxyl, carboxyl, amino, and phosphate groups. As you study this section, you will see that these distinctive groups of atoms participate in the reactions that create life's large organic molecules.

Cells use a chemical reaction called **dehydration synthesis** (also called a condensation reaction) to link monomers into polymers (figure 2.18a). In this reaction, a protein called an enzyme removes an –OH (hydroxyl group) from one molecule and a hydrogen atom from another, forming H_2O and a new covalent bond between the two smaller components. (The term *dehydration* means that water is lost.) By repeating this reaction many times, cells can build extremely large polymers consisting of thousands of monomers. ⓘ *enzymes,* section 4.4

The reverse reaction, called **hydrolysis,** breaks the covalent bonds that link monomers

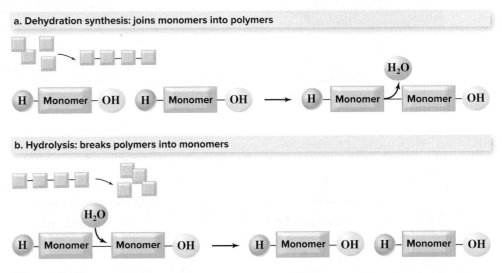

Figure 2.18 Opposite Reactions. (a) In dehydration synthesis, water is removed and a new covalent bond forms between two monomers. (b) In hydrolysis, water breaks the bond between monomers.

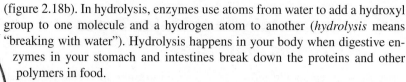
(figure 2.18b). In hydrolysis, enzymes use atoms from water to add a hydroxyl group to one molecule and a hydrogen atom to another (*hydrolysis* means "breaking with water"). Hydrolysis happens in your body when digestive enzymes in your stomach and intestines break down the proteins and other polymers in food.

Table 2.4 reviews the characteristics of the four major types of organic molecules in life. The rest of this section takes a closer look at each one.

B. Carbohydrates Include Simple Sugars and Polysaccharides

Although they range from sweet to starchy, foods such as candy, sugary fruits, cereal, potatoes, pasta, and bread all share a common characteristic. They are rich in **carbohydrates,** organic molecules that consist of carbon, hydrogen, and oxygen, often in the proportion 1:2:1.

Carbohydrates are the simplest of the four main types of organic compounds, mostly because just a few monomers account for the most common types in cells. The two main groups of carbohydrates are simple sugars and complex carbohydrates.

Sugars (Simple Carbohydrates) The smallest carbohydrates, the **monosaccharides,** usually contain five or six carbon atoms (figure 2.19a). A **disaccharide** ("two sugars") is two monosaccharides joined by dehydration synthesis. Figure 2.19b shows how sucrose (table sugar) forms when a molecule of glucose bonds to a molecule of fructose. Lactose, or milk sugar, is also a disaccharide.

Together, the sweet-tasting monosaccharides and disaccharides are called sugars, or simple carbohydrates. Their function in cells is to provide a ready source of energy, which is released when their bonds are broken (see chapter 6). Sugarcane sap and sugar beet roots contain abundant sucrose, which the plants use to fuel growth. The disaccharide maltose provides energy in sprouting seeds; beer brewers also use it to promote fermentation.

Short chains of monosaccharides on cell surfaces are important in immunity. For example, a person's blood type—A, B, AB, or O—refers to the

Carbohydrates (starch); lipids

Proteins; lipids

Carbohydrates (cellulose)

©Ingram Publishing/Alamy Stock Photo RF

TABLE 2.4 The Macromolecules of Life: A Summary

Type of Molecule	Chemical Structure	Function(s)
Carbohydrates		
Simple sugars	Monosaccharides and disaccharides	Provide quick energy
Complex carbohydrates (cellulose, chitin, starch, glycogen)	Polysaccharides (polymers of monosaccharides)	Support cells and organisms (cellulose, chitin); store energy (starch, glycogen)
Proteins	Polymers of amino acids	Carry out nearly all the work of the cell
Nucleic acids (DNA, RNA)	Polymers of nucleotides	Store and use genetic information, and transmit it to the next generation
Lipids		
Triglycerides (fats)	Glycerol + 3 fatty acids	Store energy
Phospholipids	Glycerol + 2 fatty acids + phosphate group (see chapter 3)	Form major part of biological membranes
Steroids	Four fused rings, mostly of C and H	Stabilize animal membranes; sex hormones

a. Monosaccharides

HOCH₂ ... OH

Ribose
C₅H₁₀O₅

CH₂OH

Glucose
C₆H₁₂O₆

HOCH₂ ... H

CH₂OH

Fructose
C₆H₁₂O₆

b. Disaccharide formation and breakdown

Glucose
C₆H₁₂O₆

Fructose
C₆H₁₂O₆

OH HO

H₂O
Dehydration
synthesis

Hydrolysis

H₂O

O

Sucrose
C₁₂H₂₂O₁₁

c. Polysaccharides

Cellulose

SEM (false color) 50 nm

Starch

LM 10 μm

Glycogen

TEM (false color) 1 μm

Figure 2.19 Carbohydrates. (a) Monosaccharides such as ribose, glucose, and fructose each consist of a single ring. (b) Disaccharides form by dehydration synthesis. (c) Polysaccharides such as cellulose, starch, and glycogen are long chains of glucose monomers.

Photos: (c, cellulose): ©BioPhoto Associates/Science Source; (c, starch): ©Dr. Keith Wheeler/Science Source; (c, glycogen): ©Marshall Sklar/SPL/Science Source

combination of carbohydrates attached to the surface of his or her red blood cells. A transfusion of the "wrong" blood type can trigger a harmful immune reaction. ⓘ *blood type,* section 27.1B

Complex Carbohydrates Chains of monosaccharides are collectively called complex carbohydrates. **Polysaccharides** ("many sugars") are huge molecules consisting of hundreds or thousands of monosaccharide monomers (figure 2.19c). The most common polysaccharides are cellulose, chitin, starch, and glycogen. All are long chains of glucose, but they differ from one another by the orientation of the bonds that link the monomers.

Cellulose forms part of plant cell walls. Although it is the most common organic compound in nature, humans cannot digest it. Yet cellulose is an important component of the human diet, making up much of what nutrition labels refer to as "fiber." A high-fiber diet reduces the risk of colon cancer. No one knows exactly why fiber has this effect. One possible explanation is that fiber eases the movement of food through the digestive tract, so it may shorten the length of time that harmful chemicals linger within the intestines. Cotton fibers, wood, and paper consist largely of cellulose. ⓘ *plant cell wall,* section 3.6

Chitin is the second most common polysaccharide in nature. The cell walls of fungi contain chitin, as do the flexible exoskeletons of insects, spiders, and crustaceans. Like cellulose, chitin also supports cells. It resembles a glucose polymer, except that it also contains nitrogen atoms. Because chitin is tough, flexible, and biodegradable, it is used in the manufacture of surgical thread.

Starch and glycogen have similar structures and functions. Both act as storage molecules that readily break down into their glucose monomers when cells need a burst of energy. Most plants store starch. Potatoes, rice, and wheat are all starchy, high-energy staples in the human diet. On the other hand, glycogen occurs in animal and fungal cells. In humans, for example, skeletal muscles and the liver store energy as glycogen.

Burning Question 2.1

What does it mean when food is "organic" or "natural"?

©Keith Brofsky/UpperCut Images/
Getty Images RF

The word *organic* has multiple meanings. To a chemist, an organic compound contains carbon and hydrogen. Chemically, all food is therefore organic. To a farmer or consumer, however, organic foods are produced according to a defined set of standards.

The U.S. Department of Agriculture (USDA) certifies crops as organically grown if the farmer did not apply pesticides (with few exceptions), petroleum-based fertilizers, or sewage sludge. Organically raised cows, pigs, and chickens cannot receive growth hormones or antibiotics, and they must have access to the outdoors and eat organic food. In addition, food labeled "organic" cannot be genetically engineered or treated with ionizing radiation.

A natural food may or may not be organic. The term *natural* refers to the way in which foods are processed, not how they are grown. Standards for what constitutes a natural food are fuzzy. The USDA specifies that meat and poultry labeled as natural cannot contain artificial ingredients or added color, but no such standards exist for other foods.

**Submit your burning question to
marielle.hoefnagels@mheducation.com**

a. Amino acids

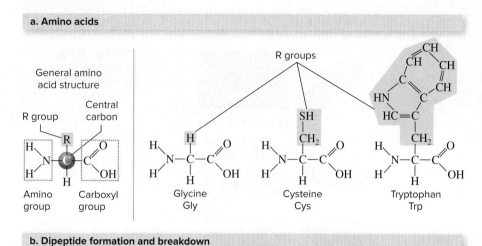

General amino acid structure

R group   Central carbon

Amino group   Carboxyl group

R groups

Glycine Gly

Cysteine Cys

Tryptophan Trp

b. Dipeptide formation and breakdown

Dehydration synthesis

Hydrolysis

H_2O

H_2O

Peptide bond

Amino acid   Amino acid   Dipeptide

Figure 2.20 **Amino Acids.** (a) An amino acid is composed of an amino group, a carboxyl group, and one of 20 R groups attached to a central carbon atom. Three examples appear here. (b) A peptide bond forms by dehydration synthesis, joining two amino acids together. Hydrolysis breaks the peptide bond.

©Comstock/Jupiter Images RF

C. Proteins Are Complex and Highly Versatile

Proteins do more jobs in the cell than any other type of biological molecule. Cells produce thousands of kinds of proteins, which control all the activities of life; illness or death can result if even one is missing or faulty. To name one example, the protein insulin controls the amount of sugar in the blood. The failure to produce insulin leads to one form of diabetes, an illness that can be deadly. (i) *diabetes*, section 25.4D

Amino Acid Structure and Bonding

A **protein** is a chain of monomers called **amino acids.** Each amino acid has a central carbon atom bonded to four other atoms or groups of atoms (figure 2.20a). One is a hydrogen atom; another is a carboxyl group; a third is an **amino group,** a nitrogen atom single-bonded to two hydrogen atoms ($-NH_2$); and the fourth is a side chain, or **R group,** which can be any of 20 chemical groups.

Organisms use 20 types of amino acids; figure 2.20a shows three of them. (Appendix E includes a complete set of amino acid structures.) The R groups distinguish the amino acids from one another, and they have diverse chemical structures. An R group may be as simple as the lone hydrogen atom in glycine or as complex as the two rings of tryptophan. Some R groups are acidic or basic; some are strongly hydrophilic or hydrophobic.

Just as the 26 letters in our alphabet combine to form a nearly infinite number of words in many languages, mixing and matching the 20 amino acids gives rise to an endless diversity of unique proteins. This variety means that proteins have a seemingly limitless array of structures and functions.

The dehydration synthesis reaction connects amino acids to each other; a **peptide bond** is the resulting covalent bond that links each amino acid to its neighbor (figure 2.20b). Two linked amino acids form a dipeptide; three form a tripeptide. Long chains of amino acids are **polypeptides.** A polypeptide is called a protein once it folds into its functional shape; a protein may consist of one or more polypeptide chains.

Where do the amino acids in your own proteins come from? Humans can synthesize most of them from scratch. However, eight amino acids are considered "essential" because they must come from protein-rich foods such as meat, fish, dairy products, beans, and tofu. Digestive enzymes catalyze the hydrolysis reactions that release amino acids from proteins in food. The body then uses these monomers to build its own polypeptides.

Protein Folding
Unlike polysaccharides, most proteins do not exist as long chains inside cells. Instead, the polypeptide chain folds into a unique three-dimensional

structure determined by the order and kinds of amino acids. Biologists describe the conformation of a protein at four levels (figure 2.21):

- **Primary structure:** The amino acid sequence of a polypeptide chain. This sequence determines all subsequent structural levels.

- **Secondary structure:** A "substructure" with a defined shape, resulting from hydrogen bonds between parts of the polypeptide. These interactions fold the chain of amino acids into coils, sheets, and loops. Each protein can have multiple areas of secondary structure.

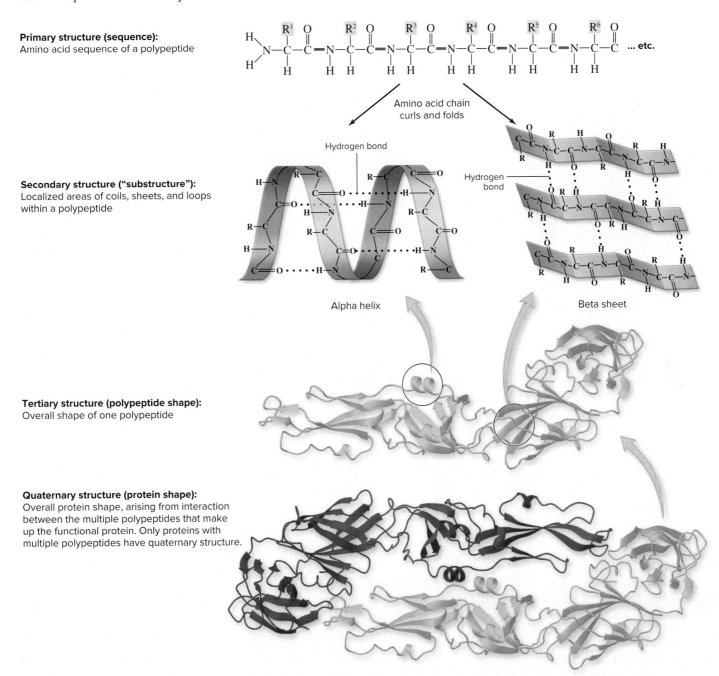

Primary structure (sequence):
Amino acid sequence of a polypeptide

Amino acid chain curls and folds

Secondary structure ("substructure"):
Localized areas of coils, sheets, and loops within a polypeptide

Alpha helix

Beta sheet

Tertiary structure (polypeptide shape):
Overall shape of one polypeptide

Quaternary structure (protein shape):
Overall protein shape, arising from interaction between the multiple polypeptides that make up the functional protein. Only proteins with multiple polypeptides have quaternary structure.

Figure 2.21 Four Levels of Protein Structure. The amino acid sequence of a polypeptide forms the primary structure, while hydrogen bonds create secondary structures such as a helix or sheet. The tertiary structure is the overall three-dimensional shape of a protein. The interaction of multiple polypeptides forms the protein's quaternary structure.

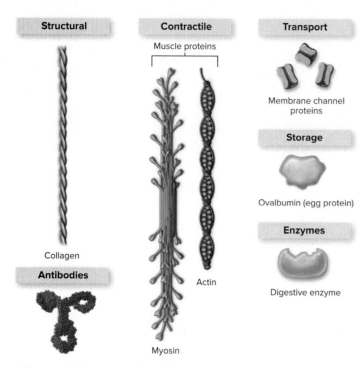

Collagen

Antibodies

Contractile

Muscle proteins

Actin

Myosin

Transport

Membrane channel proteins

Storage

Ovalbumin (egg protein)

Enzymes

Digestive enzyme

Structural

Figure 2.22 **Protein Diversity.** The function of a protein is a direct consequence of its shape. Shown here are a few of the thousands of types of known proteins.

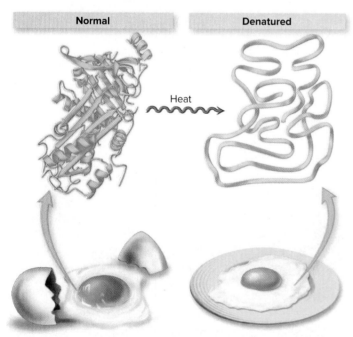

Normal

Denatured

Heat

Figure 2.23 **Denatured Proteins.** The proteins in a raw egg white are clear and fluid. As the egg is cooked, however, heat causes the proteins to denature. The protein chains unravel and refold at random, ruining their overall shape. The new arrangement causes the proteins to become rubbery and white.

- **Tertiary structure:** The overall shape of a polypeptide, arising primarily through interactions between R groups and water. Inside a cell, water molecules surround each polypeptide. The hydrophobic R groups move away from water toward the protein's interior. In addition, hydrogen bonds and ionic bonds form between the peptide backbone and some R groups. Covalent bonds between sulfur atoms in some R groups further stabilize the structure. These disulfide bridges are abundant in structural proteins such as keratin, which forms hair, scales, beaks, feathers, wool, and hooves.

- **Quaternary structure:** The shape arising from interactions between multiple polypeptide subunits of the same protein. The protein in figure 2.21 consists of two polypeptides; similarly, the oxygen-toting blood protein hemoglobin is composed of four polypeptide chains. Only proteins consisting of multiple polypeptides have quaternary structure.

As detailed in chapter 7, an organism's genetic code specifies the amino acid sequence of each protein. A genetic mutation may therefore change a protein's primary structure. The protein's secondary, tertiary, and quaternary structures all depend upon the primary structure. Genetic mutations are often harmful because they result in misfolded, nonfunctional proteins.

Denaturation: Loss of Function It is impossible to overstate the importance of a protein's shape in determining its function. Examine figure 2.22, which illustrates the major categories of protein function: structural support, contraction, transport, storage, enzymes, and antibodies. Notice the great variety of protein shapes, reflecting their different jobs in the cell. A digestive enzyme, for example, has a groove that holds a food molecule in just the right way to break the nutrient apart. Muscle proteins form long, aligned fibers that slide past one another, shortening their length to contract the muscle. Membrane channels have pores that admit some molecules but not others into a cell.

Proteins are therefore vulnerable to conditions that alter their shapes. Heat, excessive salt, or the wrong pH can **denature** a protein, changing its shape so that it can no longer carry out its function. As an example, consider what happens to an egg as it cooks (figure 2.23). Heat disrupts the hydrogen bonds that maintain protein shape. The proteins unfold, then clump and refold randomly as the once-clear egg protein turns solid white. Typically, a denatured protein will not "renature." That is, there is no way to uncook an egg.

Humans prevent microbes from spoiling food by denaturing proteins. These proteins are not necessarily those in the food itself. Rather, when we heat foods or preserve them in salt or vinegar, we are denaturing *microbial* proteins. Without functional proteins, the microbes die, and the food's shelf life is extended.

D. Nucleic Acids Store and Transmit Genetic Information

How does a cell "know" which amino acids to string together to form a particular protein? The answer is that each protein's primary structure is encoded in the sequence of a **nucleic acid,** a polymer consisting of monomers called nucleotides. Cells contain two types of nucleic acids: **DNA (deoxyribonucleic acid)** and **RNA (ribonucleic acid).**

Each **nucleotide** monomer consists of three components (figure 2.24a). At the center is a five-carbon sugar—ribose in RNA and deoxyribose in DNA. Attached to one of the sugar's carbon atoms is at least one phosphate group (PO_4). Attached to the opposite side of the sugar is a **nitrogenous base:** adenine (A), guanine (G), thymine (T), cytosine (C), or uracil (U). DNA contains A, C, G, and T, whereas RNA contains A, C, G, and U.

Dehydration synthesis links nucleotides together (figure 2.24b). In this reaction, a covalent bond forms between the sugar of one nucleotide and the phosphate group of its neighbor.

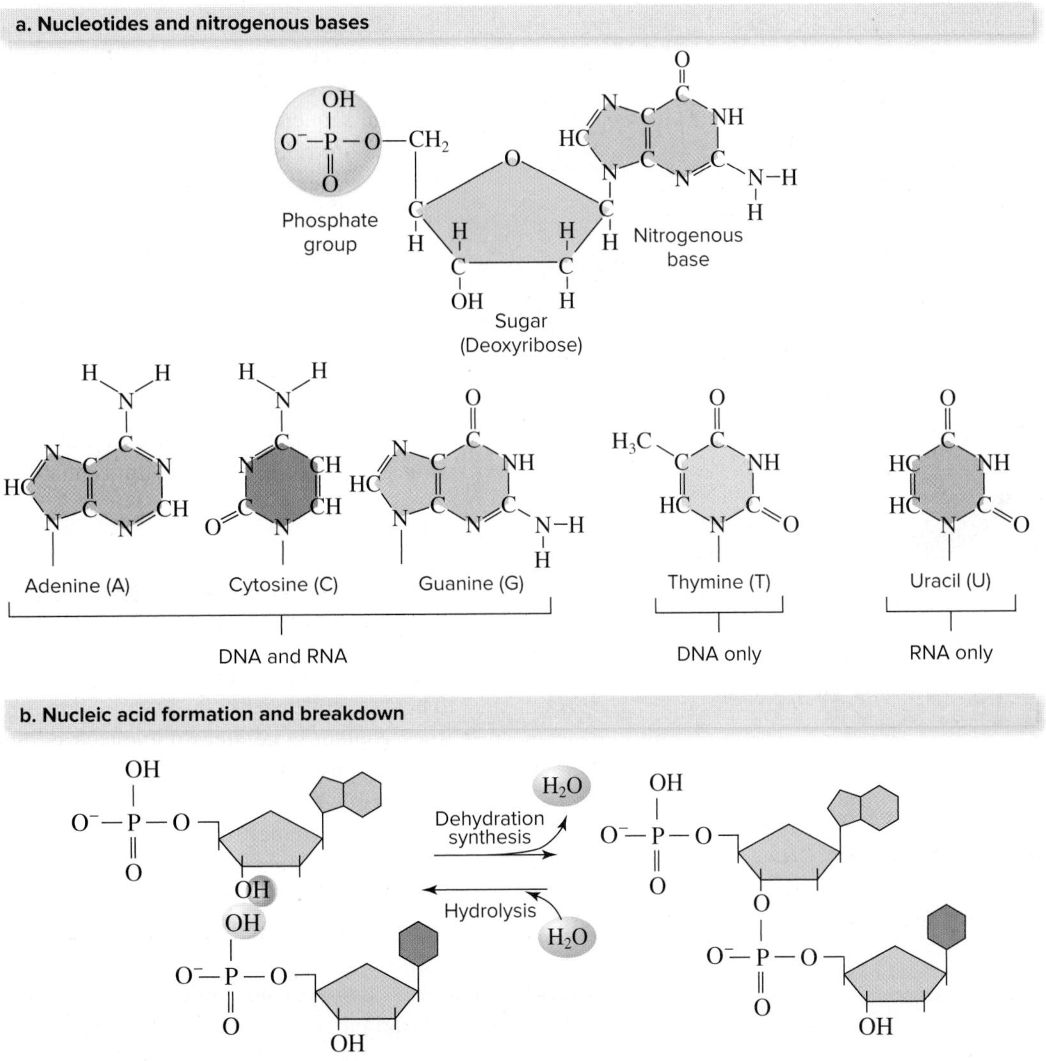

a. Nucleotides and nitrogenous bases

b. Nucleic acid formation and breakdown

Figure 2.24   Nucleotides. (a) A nucleotide consists of a sugar, one or more phosphate groups, and one of several nitrogenous bases. In DNA, the sugar is deoxyribose, whereas RNA nucleotides contain ribose. In addition, the base thymine appears only in DNA; uracil is only in RNA. (b) Dehydration synthesis joins two nucleotides together, and hydrolysis breaks them apart.

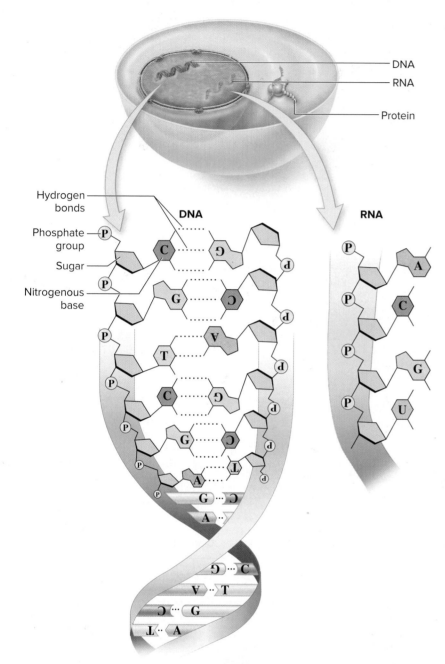

Figure 2.25 Nucleic Acids: DNA and RNA. DNA consists of two strands of nucleotides entwined to form a double-helix shape held together by hydrogen bonds (dotted lines). RNA is usually single-stranded. Both molecules participate in the production of proteins in cells.

A DNA polymer is a double helix that resembles a spiral staircase. Alternating sugars and phosphates form the rails of the staircase, and nitrogenous bases form the rungs (figure 2.25). Hydrogen bonds between the bases hold the two strands of nucleotides together: A with T, C with G. The two strands are therefore complementary to, or "opposites" of each other. Because of complementary base pairing, one strand of DNA contains the information for the other, providing a mechanism for the molecule to replicate. ⓘ *DNA replication*, section 8.2

DNA's main function is to store genetic information; its sequence of nucleotides "tells" a cell which amino acids to string together to form each protein. (This process, which is summarized at the top of figure 2.25, is described in detail in chapter 7.) Every organism inherits DNA from its parents (or parent, in the case of asexual reproduction). Slight changes in DNA from generation to generation, coupled with natural selection, account for many of the evolutionary changes that have occurred throughout life's history. As a result, DNA and protein sequences reveal important information about how species are related to one another. ⓘ *molecular evidence for evolution*, section 13.6

Unlike DNA, RNA is typically single-stranded (see figure 2.25). One function of RNA is to enable cells to use the protein-encoding information in DNA. In addition, a modified RNA nucleotide, adenosine triphosphate (ATP), carries the energy that cells use in many biological functions. ⓘ *ATP*, section 4.3

If DNA encodes only protein, where do the rest of the molecules in cells come from? The answer relates to the diverse functions of proteins: Some of them synthesize the carbohydrates, nucleic acids, and lipids that are essential to a cell's function.

E. Lipids Are Hydrophobic and Energy-Rich

Lipids are organic compounds with one property in common: They do not dissolve in water. They are hydrophobic because they contain large areas dominated by nonpolar carbon–carbon and carbon–hydrogen bonds (see section 2.2C). Moreover, lipids are not polymers consisting of long chains of monomers. Instead, they have extremely diverse chemical structures.

This section describes two groups of lipids: triglycerides and steroids. Another important group, phospholipids, forms the majority of cell membranes; section 3.3 describes them.

Triglycerides A **triglyceride** (more commonly known as a fat) consists of three long hydrocarbon chains called **fatty acids** bonded to **glycerol,** a three-carbon molecule that forms the triglyceride's backbone. Although triglycerides do not consist of long strings of similar monomers, cells nevertheless use dehydration synthesis to produce them (figure 2.26). Enzymes link three fatty acids to one glycerol molecule, yielding three water molecules per triglyceride.

Many dieters try to avoid fats. Red meat, butter, margarine, oil, cream, cheese, lard, fried foods, and chocolate are all examples of high-fat foods. (A high fat content characterizes many unhealthy foods; see Burning

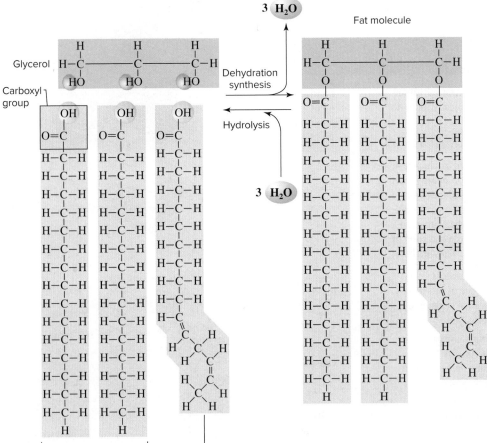

Figure 2.26 **Fat Molecule.** A triglyceride, or fat molecule, consists of three fatty acids bonded to glycerol. Saturated fatty acid chains contain only single carbon–carbon bonds. In unsaturated fatty acids, one or more double bonds bend the chains.

Question 2.2.) Nutrition labels divide these fats into two groups: saturated and unsaturated. The degree of saturation is a measure of a fatty acid's hydrogen content. A **saturated fatty acid** contains all the hydrogens it possibly can. That is, single bonds connect all the carbons, and each carbon has two hydrogens (see the straight chains in figure 2.26). Animal fats are saturated and tend to be solid; bacon fat and butter are two examples. Most nutritionists recommend a diet low in saturated fats, which tend to clog arteries and cause heart disease.

An **unsaturated fatty acid** has at least one double bond between carbon atoms (see the bent fatty acid in figure 2.26). A polyunsaturated fat has many such double bonds. These fats have an oily (liquid) consistency at room temperature. Olive oil, for example, is an unsaturated fat. These fats are healthier than are their saturated counterparts.

Why are saturated fats like butter solid at room temperature, while unsaturated olive oil is a liquid? The answer relates to the shapes of the fatty acid "tails" (figure 2.27a,b). When the tails are straight, as they are in butter, the fat molecules form tight, dense stacks. When double bonds cause kinks in the tails, as in olive oil, the molecules cannot pack tightly together. The fat molecules in butter are like flat pieces of paper in neat piles. The molecules in olive oil are more like crumpled paper.

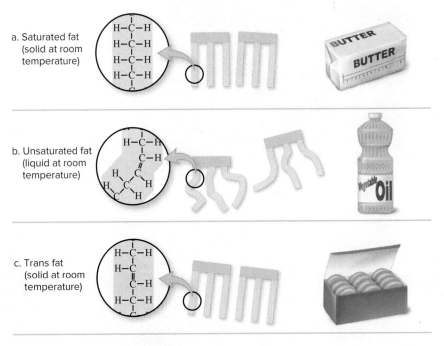

a. Saturated fat (solid at room temperature)

b. Unsaturated fat (liquid at room temperature)

c. Trans fat (solid at room temperature)

Figure 2.27 **Solid or Liquid?** (a) The straight fatty acid chains of saturated fats stack neatly, easily forming solids. (b) Double bonds in liquid unsaturated fats create kinks that prevent the fat molecules from packing tightly. (c) The fatty acids in a trans fat contain double bonds yet remain straight, so the fat remains a solid.

Why We Care 2.2 | Sugar Substitutes and Fake Fats

Many weight-conscious people turn to artificial sweeteners and fat substitutes to cut calories while still enjoying their favorite foods. Chemically, how do these sugar and fat replacements compare with the real thing?

Artificial Sweeteners

Table sugar delivers about 4 Calories per gram. (As described in chapter 4, a nutritional Calorie—with a capital C—is a measure of energy that represents 1000 calories.) Consuming artificial sweeteners reduces calorie intake in either of two ways. Some of the additives are truly calorie-free. Others contain calories but taste hundreds of times sweeter than sugar, so a tiny amount of artificial sweetener achieves the same effect as a teaspoon of sugar. Figure 2.A shows a few popular artificial sweeteners. They include:

- **Sucralose** (sold as Splenda): This sweetener is a close relative of sucrose, except that three chlorine (Cl) atoms replace three of sucrose's hydroxyl groups. Sucralose is about 600 times sweeter than sugar, and the body digests little if any of it, so it is virtually calorie-free.

- **Aspartame** (sold as NutraSweet and Equal): Surprisingly, aspartame's chemical structure does not resemble sugar. Instead, it consists of two amino acids, phenylalanine and aspartic acid. Aspartame delivers about 4 Calories per gram, but it is about 200 times sweeter than sugar.

- **Saccharin** (sold as Sweet'n Low and Sugar Twin): This sweetener, which has only 1/32 of a Calorie per gram, consists of a double-ring structure with nitrogen and sulfur. (Saccharin's eventful history as a food additive is the topic of Why We Care 1.1.)

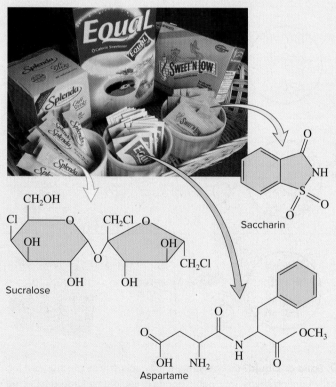

Figure 2.A   Three Artificial Sweeteners.

Photo: ©McGraw-Hill Education/Jill Braaten, photographer

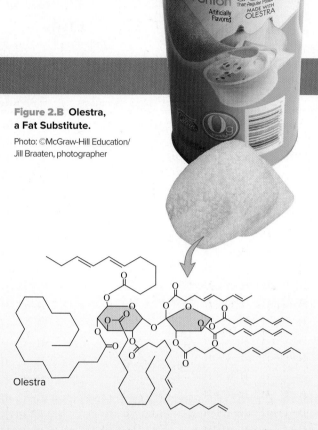

Figure 2.B   Olestra, a Fat Substitute.

Photo: ©McGraw-Hill Education/Jill Braaten, photographer

Olestra

Fat Substitutes

Because fat is so energy-rich (about 9 Calories per gram), cutting fat is a quick way to trim calories from the diet. It is important to remember, however, that some dietary fat is essential for good health. Fat aids in the absorption of some vitamins and provides fatty acids that human bodies cannot produce. Fats also lend foods taste and consistency.

Fat substitutes are chemically diverse. The most common ones are based on carbohydrates, proteins, or even fats, and a careful reading of nutrition labels will reveal their presence in many processed foods.

- **Carbohydrate-based fat substitutes:** Modified food starches, dextrins, guar gum, pectin, and cellulose gels are all derived from polysaccharides, and they all mimic fat's "mouth feel" by absorbing water to form a gel. Depending on whether they are indigestible (cellulose) or digestible (starches), these fat substitutes deliver 0 to 4 Calories per gram. They cannot be used to fry foods.

- **Protein-based fat substitutes:** These food additives are derived from egg whites or whey (the watery part of milk). When ground into "microparticles," these proteins mimic fat's texture as they slide by each other in the mouth. Protein-based fat substitutes deliver about 4 Calories per gram, and they cannot be used in frying.

- **Fat-based fat substitute:** Olestra (marketed as Olean) is a hybrid molecule that combines a central sucrose molecule with six to eight fatty acids (figure 2.B). Its chief advantage is that it tastes and behaves like fat—even for frying. Olestra is currently approved only for savory snacks such as chips. It is indigestible and calorie-free, but some people have expressed concern that olestra removes fat-soluble vitamins as it passes through the digestive tract. Others have publicized its reputed laxative properties. Most people, however, do not experience problems after eating small quantities of olestra.

Sugar and fat substitutes can be useful for people who cannot—or do not wish to—eat much of the real thing. But nutritionists warn that these food additives should not take the place of a healthy diet and moderate eating habits.

Food chemists have discovered how to turn vegetable oils into solid fats such as margarine, shortening, and peanut butter. A technique called partial hydrogenation adds hydrogen to the oil to solidify it—in essence, partially saturating a formerly unsaturated fat. One byproduct of this process is **trans fats,** which are unsaturated fats whose fatty acid tails are straight, not kinked (figure 2.27c). Trans fats are common in fast foods, fried foods, and many snack products, and they raise the risk of heart disease even more than saturated fats. A healthy diet should be as low as possible in trans fats.

Despite their unhealthful reputation, fats and oils are vital to life. Fat is an excellent energy source, providing more than twice as much energy as equal weights of carbohydrate or protein. Animals must have dietary fat for growth; this requirement explains why human milk is rich in lipids, which fuel the brain's rapid growth during the first 2 years of life. Fats also slow digestion, and they are required for the use of some vitamins and minerals. Nutrition experts therefore recommend that people eat nuts and oily fish such as salmon, which contain polyunsaturated omega-3 fatty acids and other "good fats."

Fat-storing cells aggregate as adipose tissue in animals. The adipose tissue that forms most of the fat in human adults is important because it cushions organs and helps to retain body heat. Excess body fat, however, is associated with diabetes, heart disease, and an elevated risk of cancer.

As different as carbohydrates, proteins, and lipids are, food chemists have discovered ways to use all three substances to make artificial sweeteners and fat substitutes. Why We Care 2.2 describes how they do it.

Steroids Steroids are lipids that have four interconnected carbon rings. Cortisone is an example of a steroid, as is cholesterol (figure 2.28). Cholesterol is a key part of animal cell membranes. In addition, animal cells use cholesterol as a starting material to make other lipids, including the sex hormones testosterone and estrogen. ⓘ *steroid hormones*, section 25.2B

Although cholesterol is essential, an unhealthy diet can easily contribute to cholesterol levels that are too high, increasing the risk of cardiovascular disease. Because saturated fats stimulate the liver to produce more cholesterol, it is important to limit dietary intake of both saturated fats and cholesterol.

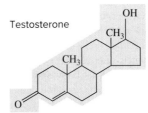

Figure 2.28 Steroids. All steroid molecules consist of four interconnected rings. Cholesterol and testosterone are two variations on this theme.

Burning Question 2.2

What is junk food?

Your favorite potato chips contain carbohydrates, protein, and fat—three of the four main types of organic molecules in cells. If you must obtain these molecules from your diet, why are chips considered a junk food?

In general, junk foods such as chips and candy are high in fat or sugar (or both) but low in protein and complex carbohydrates. They also typically have few vitamins and minerals. Junk foods therefore are high in calories but deliver little nutritional value.

Many junk foods also contain chemical additives. One common ingredient in packaged cookies, pies, and other baked goods is partially hydrogenated vegetable oil, a type of chemically processed fat. Partial hydrogenation causes fats to remain solid at room temperature; it also produces trans fats, which have been linked to several diseases (see section 2.5E). Some junk foods also contain artificial colors, flavor enhancers, artificial flavors, and preservatives that make food look or taste more appealing without adding nutritional value. One example is monosodium glutamate (MSG). This chemical consists of an amino acid and a sodium atom connected by an ionic bond. It enhances the flavor of many packaged snacks and fast foods, imparting a savory taste. Moreover, preservatives such as BHA and BHT increase the shelf life of many junk foods. These chemicals prevent oxygen from interacting with fat, so it takes longer for the food to become stale.

Potato chips, pizza, fries, candy bars, snack cakes, and other junk foods are hard to resist because they tap into our desire to eat sweet, salty, and fatty foods. These snacks are tasty, appealing, easily available, and often cheap. But for a more nutritious diet, reach for whole grains, fresh fruits, and vegetables instead. ⓘ *healthful diet*, section 28.3

Submit your burning question to
marielle.hoefnagels@mheducation.com

(chicken): ©Burke/Triolo/Brand X Pictures RF

2.5 Mastering Concepts

1. Distinguish between hydrolysis and dehydration synthesis.
2. Compare and contrast the structures of polysaccharides, proteins, and nucleic acids.
3. What is the significance of a protein's shape, and how can that shape be destroyed?
4. What are some differences between RNA and DNA?
5. What are the components of a triglyceride?
6. Explain why fats are essential to a healthful diet.
7. List an example of a carbohydrate, protein, nucleic acid, and lipid, and name the function of each.

Investigating Life 2.1 | Chemical Warfare on a Tiny Battlefield

Chemical warfare may be illegal among humans, but in other species, it is extremely common. Many types of bacteria, plants, fungi, and animals produce toxic chemicals that ward off competitors, herbivores, and predators. Fire ants, for example, produce a potent venom that is deadly to most other insects.

Fire ants are native to South America, but they have invaded much of the southeastern United States. Recently, however, another invasive South American ant species—the tawny crazy ant—has been competing with fire ants. How do these newcomers survive a fire ant's killer sting?

Tawny crazy ants have venom of their own, but their success does not stem from a superior chemical. Rather, crazy ants use their venom in a unique defensive behavior. Immediately after contact with fire ant venom, a crazy ant bends its abdomen underneath its body and squirts its own venom on its mouthparts. It then runs its front legs through its mouthparts and grooms itself, smearing its venom over its entire body. Somehow this unusual behavior protects the crazy ant.

To find out how, researchers set up a lab experiment using two groups of crazy ants. In one group, the venom gland in each ant's abdomen was plugged with nail polish. The other crazy ants retained open glands; these ants formed a control group. The scientists then placed two fire ants in a vial with either a treated or a control crazy ant. Once fire ant venom touched the crazy ant, the researchers removed the fire ants and monitored the crazy ant for 8 hours. Fewer than half of the crazy ants (48%) with plugged venom glands withstood the attack, whereas almost all (98%) of the control ants survived (figure 2.C). The team considered this strong evidence that crazy ant venom detoxifies fire ant venom.

Further experiments suggested that one component of crazy ant venom, formic acid, is the "secret weapon" that protects against fire ant venom. By

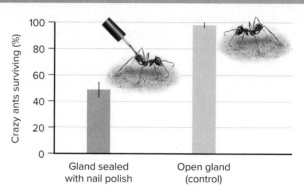

Figure 2.C Venom Beats Venom. Shortly after contact with a fire ant's venom, many of the 44 crazy ants with sealed venom glands died, whereas most of the 43 crazy ants with open venom glands survived. (Error bars represent standard errors; see appendix B.)

reducing the pH, formic acid may deactivate enzymes in the fire ant venom, indirectly protecting crazy ants from the poison.

Defensive chemicals enable tawny crazy ants to invade their rivals' nests and steal their food with little risk. This advantage has allowed crazy ants to displace fire ants in parts of their newly adopted country. It's chemical warfare, and crazy ants have the antidote to the fire ants' best weapons.

Source: LeBrun, Edward G., Nathan T. Jones, and Lawrence E. Gilbert. 2014. Chemical warfare among invaders: A detoxification interaction facilitates an ant invasion. *Science*, vol. 343, pages 1014–1017.

What's the **Point?** ▼ APPLIED

Life—like all matter in the universe—is composed of atoms. Every day we eat countless atoms, many of which become incorporated into the molecules that make up our cells and tissues. Some of these chemicals are useful to our bodies, whereas others may be harmful in excess quantities. Fortunately, nutrition labels list the chemicals that make up our foods.

Figure 2.D lists the nutrition facts for one slice of a hypothetical frozen pizza. The label indicates the mass and percent daily value for many of the chemicals in each slice. Notice that the pizza contains elements such as sodium, calcium, iron, and potassium, which are highlighted in blue. One slice contains 40% of the calcium that you need each day. Highlighted in yellow are three of the four major groups of macromolecules: lipids (fats and cholesterol), carbohydrates, and proteins. The subtypes of fats and carbohydrates are also listed. Almost half of the fat molecules in this food are saturated, which means that they are solid at room temperature. Since

Figure 2.D Nutrition Label. The label lists the complete nutritional breakdown for each serving, including Calories, various organic molecules (*yellow*), and minerals (*blue*).

few of the carbohydrate molecules are fiber or simple sugars, most of them are probably starch. Finally, the label also lists other important organic molecules called vitamins.

Not included in this simplified label is a list of ingredients, which often includes several other chemicals that are not necessary in our diet. Preservatives help processed foods stay fresh longer, and artificial colors and flavors make them look and taste appealing. Some people question whether eating these added chemicals is healthy (see Burning Question 2.2). Understanding chemistry therefore helps us to make informed decisions about the foods that we eat.

Nutrition Facts

6 servings per container

Serving Size	**1 slice**

Amount Per Serving	
Calories	**350**

	% Daily Value
Total Fat 10g	**15%**
Saturated Fat 4g	**20%**
Trans Fat 0g	
Cholesterol 30mg	**10%**
Sodium 400mg	**17%**
Total Carbohydrate 44g	**15%**
Dietary Fiber 2g	**8%**
Total Sugars 3g	
Includes 1g Added Sugars	**2%**
Protein 21g	

Vitamin D 0 mcg	0%
Calcium 520mg	40%
Iron 3mg	15%
Potassium 160 mg	4%

CHAPTER SUMMARY

2.1 Atoms Make Up All Matter

- All substances contain matter and **energy,** the ability to do work.
- All **matter** can be broken down into pure substances called **elements.**

A. Elements Are Fundamental Types of Matter

- **Bulk elements** are essential to life in large quantities. The most abundant are C, H, O, and N. **Trace elements** are required in smaller amounts.

B. Atoms Are Particles of Elements

- An **atom** is the smallest unit of an element. Positively charged **protons** and neutral **neutrons** form the **nucleus.** The negatively charged, much smaller **electrons** surround the nucleus.
- Elements are organized in the **periodic table** according to **atomic number** (the number of protons).
- An **ion** is an atom that gains or loses electrons.

C. Isotopes Have Different Numbers of Neutrons

- **Isotopes** of an element differ by the number of neutrons. A **radioactive isotope** is unstable.
- An element's **atomic weight** reflects the average **mass number** of all isotopes, weighted by the proportions in which they naturally occur.

2.2 Chemical Bonds Link Atoms (figure 2.29)

- A **molecule** is two or more atoms joined together; if they are of different elements, the molecule is called a **compound.**

A. Electrons Determine Bonding

- Electrons move constantly; they are most likely to occur in volumes of space called **orbitals.** Orbitals are grouped into **energy shells.**
- An atom's tendency to fill its **valence shell** with electrons drives it to form **chemical bonds** with other atoms.
- The more **electronegative** an atom, the more strongly it attracts electrons.

B. In an Ionic Bond, One Atom Transfers Electrons to Another Atom

- An **ionic bond** is an attraction between two oppositely charged ions, which form when one highly electronegative atom strips one or more electrons from another atom.

C. In a Covalent Bond, Atoms Share Electrons

- **Covalent bonds** form between atoms that can fill their valence shells by sharing one or more pairs of electrons.
- Atoms in a **nonpolar covalent bond** share electrons equally. If one atom is more electronegative than the other, a **polar covalent bond** forms.

D. Partial Charges on Polar Molecules Create Hydrogen Bonds

- **Hydrogen bonds** result from the attraction between opposite partial charges on adjacent molecules or between oppositely charged parts of a large molecule.

2.3 Water Is Essential to Life

A. Water Is Cohesive and Adhesive

- Water molecules stick to each other **(cohesion)** and to other substances **(adhesion).**

B. Many Substances Dissolve in Water

- A **solution** consists of a **solute** dissolved in a **solvent.**
- Water dissolves **hydrophilic** (polar and charged) substances but not **hydrophobic** (nonpolar) substances.

C. Water Regulates Temperature

- Water helps regulate temperature in organisms because it resists both temperature change and **evaporation.**
- Large bodies of water help keep coastal climates mild.

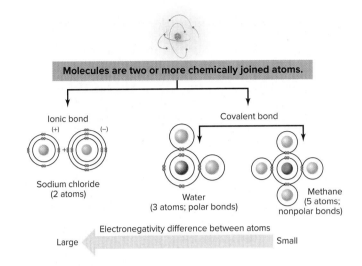

Molecules are two or more chemically joined atoms.

Ionic bond

(+)   (−)

Sodium chloride
(2 atoms)

Covalent bond

Water
(3 atoms; polar bonds)

Methane
(5 atoms;
nonpolar bonds)

Electronegativity difference between atoms

Large   Small

Figure 2.29 Atoms and Molecules.

D. Water Expands as It Freezes

- Ice floats because it is less dense than liquid water.

E. Water Participates in Life's Chemical Reactions

- In a **chemical reaction,** the **products** are different from the **reactants.**
- Most biochemical reactions occur in a watery solution.

2.4 Cells Have an Optimum pH

- In pure water, the concentrations of H^+ and OH^- are equal, so the solution is **neutral.** An **acid** adds H^+ to a solution, and a **base** adds OH^- or removes H^+.
- The **pH scale** measures H^+ concentration. Pure water has a pH of 7, acidic solutions have a pH below 7, and an **alkaline** solution has a pH between 7 and 14.
- **Buffers** consist of weak acid–base pairs that maintain the optimal pH ranges of body fluids by releasing or consuming H^+.

2.5 Cells Contain Four Major Types of Organic Molecules

A. Large Organic Molecules Are Composed of Smaller Subunits

- Many **organic molecules** consist of small subunits called **monomers,** which link together to form **polymers. Dehydration synthesis** is the chemical reaction that joins monomers together, releasing a water molecule (figure 2.30).
- The **hydrolysis** reaction uses water to break polymers into monomers.

B. Carbohydrates Include Simple Sugars and Polysaccharides

- **Carbohydrates** consist of carbon, hydrogen, and oxygen in the proportions 1:2:1.
- **Monosaccharides** are single-ring sugars such as glucose. Two bonded monosaccharides form a **disaccharide.** These simple sugars provide quick energy.
- **Polysaccharides** are complex carbohydrates consisting of hundreds of monosaccharides. They provide support and store energy.

C. Proteins Are Complex and Highly Versatile

- **Proteins** consist of **amino acids,** each composed of a central carbon atom bonded to a hydrogen atom, a carboxyl group, an **amino group,** and a variable **R group.** Amino acids join into **polypeptides** by forming **peptide bonds** through dehydration synthesis.

Dehydration synthesis builds large molecules from small subunits.

Cells build triglycerides out of glycerol + fatty acids.

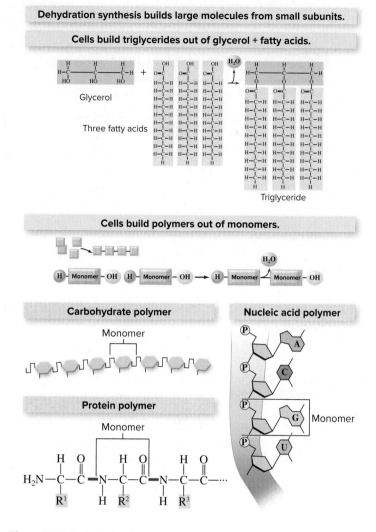

Glycerol

Three fatty acids

Triglyceride

Cells build polymers out of monomers.

H — Monomer — OH   H — Monomer — OH   →   H — Monomer — Monomer — OH

Carbohydrate polymer

Monomer

Protein polymer

Monomer

Nucleic acid polymer

Monomer

Figure 2.30 Dehydration Synthesis.

- A protein's three-dimensional shape is vital to its function. A **denatured** protein has a ruined shape.
- Proteins have a great variety of functions, participating in all the work of the cell.

D. Nucleic Acids Store and Transmit Genetic Information

- **Nucleic acids,** including **DNA** and **RNA,** are polymers consisting of **nucleotides.** Each nucleotide consists of a phosphate group, a sugar, and a **nitrogenous base.**
- DNA carries genetic information and transmits it from generation to generation. RNA helps use the DNA's information to make proteins.

E. Lipids Are Hydrophobic and Energy-Rich

- **Lipids** are diverse hydrophobic compounds consisting mainly of carbon and hydrogen.
- **Triglycerides** (fats) consist of **glycerol** and three **fatty acids,** which may be **saturated** (straight chains with no double bonds) or **unsaturated** (typically kinked chains with at least one double bond). Many processed foods contain unhealthy **trans fats,** unsaturated fats with straight chains.
- Fats store energy, slow digestion, cushion organs, and preserve body heat.
- **Steroids,** including cholesterol and sex hormones, are lipids consisting of four fused rings.

MULTIPLE CHOICE QUESTIONS

1. A neutral hydrogen atom with a mass number of 1 has _____ neutrons, _____ protons, and _____ electrons.
   a. 0; 1; 1
   b. 1; 1; 1
   c. 0; 1; 0
   d. 1; 0; 0

2. How many neutrons does a carbon-14 atom have in its nucleus?
   a. 6
   b. 7
   c. 8
   d. 14

3. An ionic bond forms when
   a. an electrical attraction occurs between two atoms of different charge.
   b. a nonpolar attraction is formed between two atoms.
   c. a valence electron is shared between two atoms.
   d. two atoms have similar attraction for electrons.

4. A covalent bond forms when
   a. electrons are present in a valence shell.
   b. a valence electron is removed from one atom and added to another.
   c. a pair of valence electrons is shared between two atoms.
   d. one atom attracts electrons more strongly than another atom.

5. Water dissolves salts because it
   a. is hydrophobic, and salts are also hydrophobic.
   b. forms covalent bonds with the atoms of the salt crystal.
   c. has partial positive and negative charges.
   d. evaporates quickly at room temperature.

6. A hydrogen bond is distinct from ionic and covalent bonds in that it
   a. is a weak attraction between two molecules.
   b. forms only between two hydrogen atoms.
   c. is considerably stronger than the other two types of bonds.
   d. occurs more commonly in lipids than in other types of molecules.

7. A hydrophilic substance is one that can
   a. form covalent bonds with hydrogen.
   b. dissolve in water.
   c. buffer a solution.
   d. mix with nonpolar solvents.

8. What type of chemical bond forms during a dehydration synthesis reaction?
   a. Covalent
   b. Hydrogen
   c. Ionic
   d. Polymer

9. A sugar is an example of a _____, whereas DNA is a _____ .
   a. protein; nucleic acid
   b. lipid; protein
   c. nucleic acid; lipid
   d. carbohydrate; nucleic acid

10. _____ are monomers that form polymers called _____.
   a. Nucleotides; nucleic acids
   b. Amino acids; nucleic acids
   c. Monoglycerides; triglycerides
   d. Carbohydrates; monosaccharides

Answers to Multiple Choice questions are in appendix A.

WRITE IT OUT

1. Describe how the number of protons, neutrons, and electrons in an atom affects its atomic number, mass number, and charge.

2. The vitamin biotin contains 10 atoms of carbon, 16 of hydrogen, 3 of oxygen, 2 of nitrogen, and 1 of sulfur. What is its molecular formula?

3. Consider the following atomic numbers: oxygen (O) = 8; fluorine (F) = 9; neon (Ne) = 10; magnesium (Mg) = 12. Draw the electron shells of each atom, and then predict how many bonds each atom should form.

4. Distinguish between nonpolar covalent bonds, polar covalent bonds, and ionic bonds.

5. If oxygen strongly attracts electrons, why is a covalent bond between two oxygen atoms considered nonpolar?

6. Can nonpolar molecules such as CH_4 participate in hydrogen bonds? Why or why not?

7. How does electronegativity explain whether a covalent bond is polar or nonpolar?

8. Give an example from everyday life of each of the following properties of water: cohesion, adhesion, ability to dissolve solutes, resistance to temperature change.

9. Draw from memory a diagram showing the interactions among a few water molecules.

10. Define *solute, solvent,* and *solution.*

11. How do hydrogen ions relate to the pH scale?

12. Sketch a monosaccharide, an amino acid, a nucleotide, a glycerol molecule, and a fatty acid. Then show how those smaller molecules form carbohydrates, proteins, nucleic acids, or fats.

13. How is an amino acid's R group analogous to a nucleotide's nitrogenous base?

14. Refer to figure 2.23 and explain why regulating body temperature is essential to survival.

15. Complete and explain the following analogy: a protein is to a knitted sweater as a denatured protein is to a _____.

16. A man on a very low-fat diet proclaims to his friend, "I'm going to get my cholesterol down to zero!" Why is achieving this goal impossible (and undesirable)?

17. You eat a sandwich made of starchy bread, ham, and cheese. What types of chemicals are in it?

18. Name three examples of emergent properties (see chapter 1) in chemistry.

SCIENTIFIC LITERACY

1. Review Burning Question 2.1, which describes the meaning of the words *organic* and *natural* in foods. Why might a manufacturer label its packaged foods as "natural"? Why might an organic food cost more than a conventional food? What steps might you take to decide if you should buy "natural," "organic," or conventional foods?

2. Review Why We Care 2.1, which describes acids and bases in foods. Some people claim that eating "acidifying" and "alkalizing" foods can alter the body's pH, affecting the risk of everything from cancer to osteoporosis. Find a website that supports these claims, and find another that disputes them. Which website seems more credible? Why?

PULL IT TOGETHER

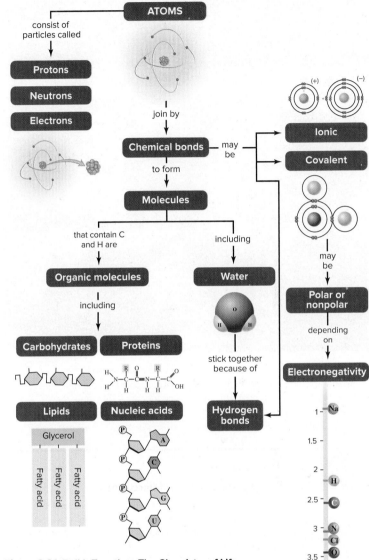

Figure 2.31 Pull It Together: The Chemistry of Life.

Refer to figure 2.31 and the chapter content to answer the following questions:

1. Compare the Survey the Landscape figure in the chapter introduction with the Pull It Together concept map. Are molecules alive? Can life exist without molecules? Why?

2. How do ions and isotopes fit into this concept map?

3. Besides water, which other molecules are essential to life?

4. Add *monomers, polymers, dehydration synthesis,* and *hydrolysis* to this concept map.

Design element: Burning Question (fire background): ©Ingram Publishing/Super Stock

3

Cells

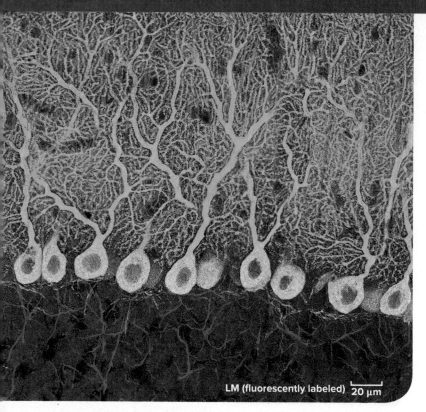

LM (fluorescently labeled) 20 μm

Brain Slice. Microscopes offer a detailed view of life's smallest components. The cells shown here, called neurons, occur in the brain. Each intricately branched neuron may form connections with more than 200,000 other cells.

©Thomas Deerinck, NCMIR/Science Source

LEARNING OUTLINE

3.1 Cells Are the Units of Life

3.2 Different Cell Types Characterize Life's Three Domains

3.3 A Membrane Separates Each Cell from Its Surroundings

3.4 Eukaryotic Organelles Divide Labor

3.5 The Cytoskeleton Supports Eukaryotic Cells

3.6 Cells Stick Together and Communicate with One Another

APPLICATIONS

Burning Question 3.1 *Is it possible to make an artificial cell from scratch?*
Why We Care 3.1 *Most of Your Cells Are Not Your Own*
Investigating Life 3.1 *Bacterial Magnets*

 Learn How to Learn
Interpreting Images from Microscopes

Any photo taken through a microscope should include information that can help you interpret what you see. First, read the caption and labels so you know what you are looking at—usually an organ, a tissue, or an individual cell. Then study the scale bar and estimate the size of the image. (For a review of metric units, consult appendix C.) Finally, check whether a light microscope (LM), scanning electron microscope (SEM), or transmission electron microscope (TEM) was used to create the image. Note that stains and false colors are often added to emphasize the most important features.

 SURVEY THE LANDSCAPE
Science, Chemistry, and Cells

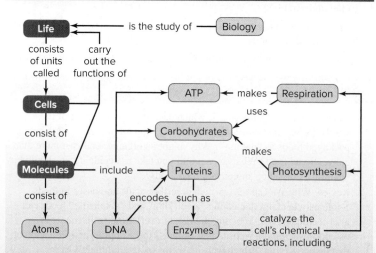

All life is organized into fundamental units called cells. The details of cell structure vary from organism to organism, but all cells share features that enable them to carry out life's functions.

For more details, study the Pull It Together feature in the chapter summary.

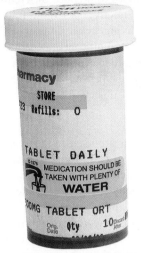

©Comstock/Alamy Stock Photo RF

"The cell is the fundamental unit of life." You have probably heard this line ever since you started learning about science in grade school. But who cares about tiny packets of life that you can only see with the help of a microscope?

To answer this question, consider first the amazing power of antibiotics. These drugs kill bacteria inside our bodies but leave our own cells alone. How? By inhibiting processes that occur only in bacterial cells. Conquering cancer is another compelling reason to study cell biology. In cancer, a person's own cells multiply out of control. Research revealing how cancer cells differ from normal cells has yielded spectacular new treatments that target these differences, producing fewer side effects than older treatments that destroy healthy cells, too.

Antibiotics, cancer treatments, and many other wonder drugs owe their success to generations of cell biologists who painstakingly documented the structures in and on cells—the subject of this chapter.

Figure 3.1 Ranges of Light and Electron Microscopes. Biologists use light microscopes and electron microscopes to view a world too small to see with the unaided eye, from large molecules to entire cells. This illustration uses the metric system to measure size (see appendix C).

3.1 Cells Are the Units of Life

A human, a plant, a mushroom, and a bacterium appear to have little in common other than being alive. However, on a microscopic level, these organisms share many similarities. For example, all organisms consist of one or more microscopic structures called **cells,** the smallest units of life that can function independently. Within cells, highly coordinated biochemical activities carry out the basic functions of life. This chapter introduces the cell, and the chapters that follow delve into the cellular events that make life possible.

A. Simple Lenses Revealed the First Glimpses of Cells

The study of cells began in 1660, when English physicist Robert Hooke melted strands of spun glass to create lenses. When he used a lens to look at cork, which is bark from a type of oak tree, it appeared to be divided into little boxes, left by cells that were once alive. Hooke called these units "cells" because they looked like the cubicles (Latin, *cellae*) where monks studied and prayed. Although Hooke did not realize the significance of his observation, he was the first person to see the outlines of cells. His discovery initiated a new field of science, now called cell biology.

Microscopes continued to improve into the nineteenth century, revealing details of the nucleus and other structures inside cells. In 1839, German biologists Mathias J. Schleiden and Theodor Schwann used their observations of many different plant and animal cells to formulate the **cell theory,** which originally had two main components: All organisms are made of one or more cells, and the cell is the fundamental unit of all life. German physiologist Rudolf Virchow added a third component in 1855, when he proposed that all cells come from preexisting cells (see Burning Question 3.1).

Like any scientific theory, the cell theory is *potentially* falsifiable—yet many lines of evidence support each of its components, making it one of the most powerful ideas in biology.

B. Microscopes Magnify Cell Structures

The unaided eye can see objects that are larger than about 0.2 mm (figure 3.1). Cells are typically smaller than this lower limit of human vision, so studying

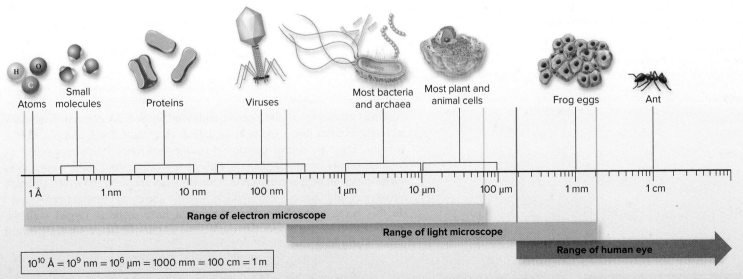

| Atoms | Small molecules | | Proteins | | Viruses | | Most bacteria and archaea | Most plant and animal cells | | Frog eggs | | Ant |

1 Å   1 nm   10 nm   100 nm   1 μm   10 μm   100 μm   1 mm   1 cm

Range of electron microscope

Range of light microscope

Range of human eye

10^{10} Å = 10^9 nm = 10^6 μm = 1000 mm = 100 cm = 1 m

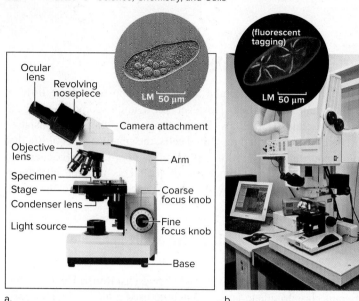

a.

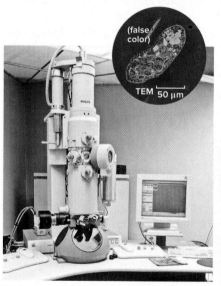

c.

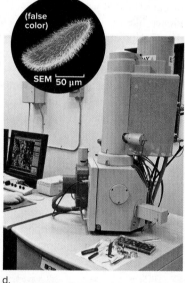

d.

Figure 3.2 **Light and Electron Microscopes: A Comparison.**
These photographs show four types of microscopes, along with sample images of a protist called *Paramecium*. (a) Compound light microscope. (b) Confocal microscope. (c) Transmission electron microscope. (d) Scanning electron microscope.

life at the cellular and molecular levels requires magnification. Cell biologists use a variety of microscopes to produce different types of images. As you will see, some microscopes show full-color structures and processes inside living cells. Others can greatly magnify cell structures, but with two significant drawbacks: They require that cells be killed, and they produce only black-and-white images. This section describes several types of microscopes; figure 3.1 provides a sense of the size of objects that each can reveal.

Light Microscopes Light microscopes are ideal for generating true-color views of living or preserved cells. Because light must pass through an object to reveal its internal features, however, the specimens must be transparent or thinly sliced to generate a good image.

Two types of light microscopes are the compound microscope and the confocal microscope (figure 3.2a, b). A compound scope uses two or more lenses to focus visible light through a specimen. The most powerful ones can magnify up to 1600 times and resolve objects that are 200 nanometers apart. A confocal microscope enhances resolution by focusing white or laser light through a lens to the object. The image then passes through a pinhole. The result is a scan of highly focused light on one tiny part of the specimen at a time. Computers can integrate multiple confocal images of specimens exposed to fluorescent dyes to produce spectacular three-dimensional peeks at living structures.

Transmission and Scanning Electron Microscopes
Light microscopes are useful, but their main disadvantage is that many cell structures are too small to see using light. Electron microscopes provide much greater magnification and resolution. Instead of using light, these microscopes use electrons.

A transmission electron microscope (TEM) sends a beam of electrons through a very thin slice of a specimen. The microscope translates differences in electron transmission into a high-resolution, two-dimensional image that shows the internal features of the object (figure 3.2c). TEMs can magnify up to 50 million times and resolve objects less than 1 angstrom (10^{-10} meters) apart.

A scanning electron microscope (SEM) scans a beam of electrons over a metal-coated, three-dimensional specimen. Its images have lower resolution than those of the TEM; in SEM, the maximum magnification is about 250,000 times, and the resolution limit is 1 to 5 nanometers. SEM's chief advantage is its ability to reveal textures on a specimen's external surface (figure 3.2d).

Both TEM and SEM provide much greater magnification and resolution than light microscopes. Nevertheless, they do have limitations. First, they are extremely expensive to build, operate, and maintain. Second, electron microscopy normally requires that a specimen be killed, chemically fixed, and placed in a vacuum. These treatments can distort natural structures. Light microscopy, in contrast, allows an investigator to view living organisms. Third, unlike light microscopes, all images from electron microscopes are black and white, although artists often add false color to highlight specific objects in electron micrographs. (In this book, each photo taken through a microscope is tagged with the magnification and the type of microscope; the addition of false color is also noted where applicable.)

C. All Cells Have Features in Common

Microscopes and other tools clearly reveal that although cells can appear very different, they all have some features that reflect their shared evolutionary history (see figure 1.10). For example, all cells contain DNA, the cell's genetic information. They also contain RNA, which participates in the production of proteins (see chapter 7). These proteins, in turn, are essential to life because they carry out all of the cell's work, from orchestrating reproduction to processing energy to regulating what enters and leaves the cell.

Since all cells require proteins, they also contain **ribosomes,** which are the structures that manufacture proteins. Each cell is also surrounded by a lipid-rich **cell membrane** (also called the plasma membrane) that forms a boundary between the cell and its environment (see section 3.3). The membrane encloses the **cytoplasm,** which includes all cell contents (except the nucleus, in cells that have one). The **cytosol** is the fluid portion of the cytoplasm.

One other feature common to nearly all cells is small size, typically less than 0.1 millimeter in diameter (see figure 3.1). Why so tiny? The answer is that nutrients, water, oxygen, carbon dioxide, and waste products enter or leave a cell through its surface. Each cell must have abundant surface area to accommodate these exchanges efficiently. As an object grows, however, its volume increases much more quickly than its surface area. Figure 3.3a illustrates this principle for a series of cubes, but the same applies to cells: Small size maximizes the ratio of surface area to volume.

Cells avoid surface area limitations in several ways. Nerve cells may be long (up to a meter or so), but they are also extremely slender, so the ratio of surface area to volume remains high. The flattened shape of a red blood cell maximizes its ability to carry oxygen, and the many microscopic extensions of an amoeba's membrane provide a large surface area for absorbing oxygen and capturing food (figure 3.3b). A transportation system that quickly circulates materials throughout the cell also helps.

The concept of surface area is everywhere in biology; many structures illustrate the principle that a large surface area maximizes contact with the environment. For example, a pine tree's pollen grains have extensions that enable them to float on air currents; root hairs have tremendous surface area for absorbing water; the broad, flat leaves of plants maximize exposure to light; a fish's feathery gills absorb oxygen from water; a jackrabbit's enormous ears help the animal lose excess body heat in the desert air—the list goes on and on. Conversely, low surface areas minimize the exchange of materials or heat with the environment. A hibernating animal, for example, conserves warmth by tucking its limbs close to its body; a cactus plant produces few if any leaves, reducing water loss in its dry habitat.

3.1 Mastering Concepts

1. Why are cells, not atoms, the basic units of life?
2. What are the three main components of the cell theory?
3. Rank the three main types of microscopes from lowest to highest potential magnification.
4. Which molecules and structures occur in all cells?
5. Describe adaptations that increase the ratio of surface area to volume in cells.

Size of cube		
1 cm	2 cm	3 cm

Surface area = height × width × number of sides		
1 cm × 1 cm × 6 = 6 cm²	2 cm × 2 cm × 6 = 24 cm²	3 cm × 3 cm × 6 = 54 cm²

Volume = height × width × length		
1 cm × 1 cm × 1 cm = 1 cm³	2 cm × 2 cm × 2 cm = 8 cm³	3 cm × 3 cm × 3 cm = 27 cm³

Ratio of surface area to volume		
6/1 = 6.0	24/8 = 3.0	54/27 = 2.0

a.

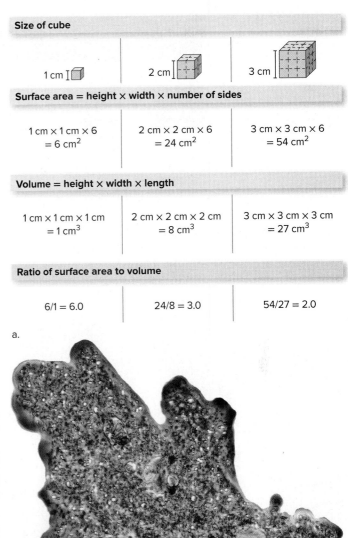

b.

Figure 3.3 Surface Area and Volume. (a) This simple example shows that larger objects have less surface area *relative to their volume* than do smaller objects with the same overall shape. (b) The membrane of this amoeba is highly folded, producing a large surface area relative to the cell's volume.

(b): ©Roland Birke/Photolibrary/Getty Images

Figure It Out

For a cube 5 centimeters on each side, calculate the ratio of surface area to volume.

Answer: 1.2.

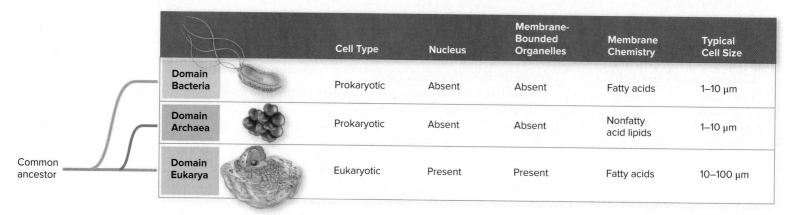

	Cell Type	Nucleus	Membrane-Bounded Organelles	Membrane Chemistry	Typical Cell Size
Domain Bacteria	Prokaryotic	Absent	Absent	Fatty acids	1–10 μm
Domain Archaea	Prokaryotic	Absent	Absent	Nonfatty acid lipids	1–10 μm
Domain Eukarya	Eukaryotic	Present	Present	Fatty acids	10–100 μm

Figure 3.4 **The Three Domains of Life.** Biologists distinguish domains Bacteria, Archaea, and Eukarya based on unique features of cell structure and biochemistry. The small evolutionary tree shows that archaea are the closest relatives of the eukaryotes.

Cytoplasm

Nucleoid (DNA)   Ribosomes   Cytosol

Flagellum

Capsule   Cell wall   Cell membrane

a.

SEM (false color)

2 μm

b.

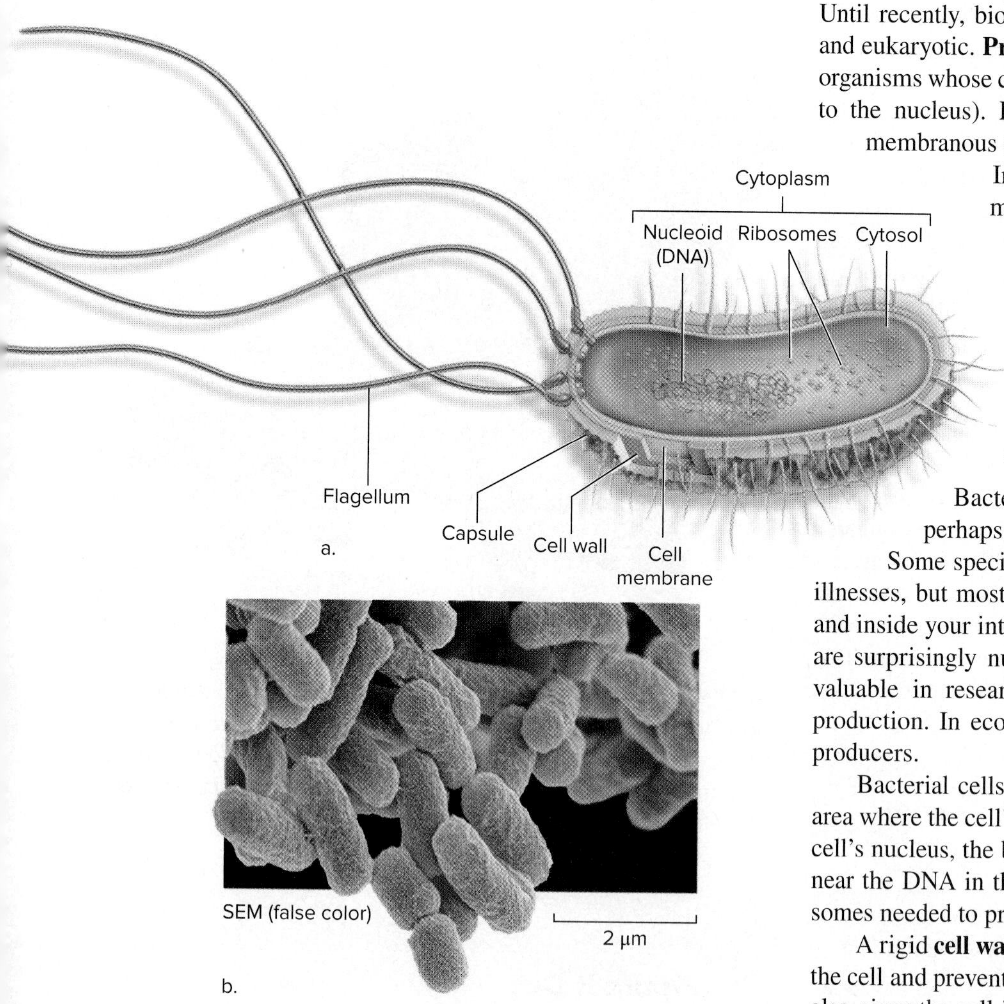

Figure 3.5 **Anatomy of a Bacterium.** (a) Bacterial cells are structurally simple. (b) Rod-shaped cells of *E. coli* inhabit human intestines.

(b): ©Steve Gschmeissner/SPL/Getty Images RF

3.2 Different Cell Types Characterize Life's Three Domains

Until recently, biologists organized life into just two categories: prokaryotic and eukaryotic. **Prokaryotes,** the simplest and most ancient forms of life, are organisms whose cells lack a nucleus (*pro* = before; *karyon* = kernel, referring to the nucleus). **Eukaryotes** have cells that contain a nucleus and other membranous organelles (*eu* = true).

In 1977, however, microbiologist Carl Woese studied key molecules in many cell types. He detected differences suggesting that prokaryotes actually include two forms of life that are distantly related to each other. Biologists subsequently divided life into three domains: **Bacteria, Archaea,** and **Eukarya** (figure 3.4).

A. Domains Bacteria and Archaea Contain Prokaryotic Organisms

Bacteria are the most abundant and diverse organisms on Earth, perhaps because they have existed longer than any other group. Some species, such as *Streptococcus* and *Escherichia coli,* can cause illnesses, but most are not harmful. In fact, the bacteria living on your skin and inside your intestinal tract are essential for good health. (These microbes are surprisingly numerous; see Why We Care 3.1.) Bacteria are also very valuable in research, food and beverage processing, and pharmaceutical production. In ecosystems, bacteria play critical roles as decomposers and producers.

Bacterial cells are structurally simple (figure 3.5a). The **nucleoid** is the area where the cell's circular DNA molecule congregates. Unlike a eukaryotic cell's nucleus, the bacterial nucleoid is not bounded by a membrane. Located near the DNA in the cytoplasm are the enzymes, RNA molecules, and ribosomes needed to produce the cell's proteins.

A rigid **cell wall** surrounds the cell membrane of most bacteria, protecting the cell and preventing it from bursting if it absorbs too much water. This wall also gives the cell its shape: usually rod-shaped (as in figure 3.5b), round, or spiral. Many antibiotic drugs, including penicillin, halt bacterial infection by interfering with the microorganism's ability to construct its protective cell wall. In some bacteria, polysaccharides on the cell wall form a capsule that adds protection or enables the cell to attach to surfaces.

Common ancestor

Many bacteria can swim in fluids. **Flagella** (singular: flagellum) are tail-like appendages that enable these cells to move. One or more flagella are anchored in the cell wall and underlying cell membrane. Bacterial flagella rotate like a propeller, moving the cell forward or backward.

Archaean cells resemble bacterial cells in some ways. Like bacteria, they are smaller than most eukaryotic cells, and they lack a nucleus and other organelles. Most have cell walls; flagella are also common. And like bacteria, most archaea are one-celled organisms. However, the resemblance to bacteria is only superficial. Archaea have their own domain because they build their cells out of biochemicals that are different from those in either bacteria or eukaryotes. Their ribosomes, however, share similarities with those of both bacteria and eukaryotes, and key DNA sequences suggest that archaea are actually the closest relatives of eukaryotes.

B. Domain Eukarya Contains Organisms with Complex Cells

An amazing diversity of organisms, ranging from microscopic protists to enormous whales, belong to domain Eukarya. Many eukaryotes, including some fungi and most protists, consist of only one cell. A few protists, most fungi, and all plants and animals are multicellular and easily visible with the unaided eye.

Despite their great differences in size and appearance, all eukaryotic organisms share many features on a cellular level. Figures 3.6 and 3.7 depict generalized animal and plant cells. Although both of the illustrated cells have many structures in common, there are some differences. Most notably, plant cells have chloroplasts and a cell wall, which animal cells lack.

One obvious feature that sets eukaryotic cells apart is their large size, typically 10 to 100 times greater than prokaryotic cells. The other main difference is that the cytoplasm of a eukaryotic cell is divided into **organelles** ("little organs"), compartments that carry out specialized functions. Examples include the nucleus, mitochondria, and chloroplasts. An elaborate system of internal membranes creates these compartments.

In general, organelles keep related biochemicals and structures close enough to make them function efficiently, without altering or harming other cellular contents. Compartmentalization also means that the cell maintains high concentrations of each biochemical only in certain organelles, not throughout the entire cell. The rest of this chapter describes the structure of the eukaryotic cell in greater detail.

3.2 Mastering Concepts

1. How do prokaryotic cells differ from eukaryotic cells?
2. Compare and contrast bacteria and archaea.
3. What is the relationship between cells and organelles?

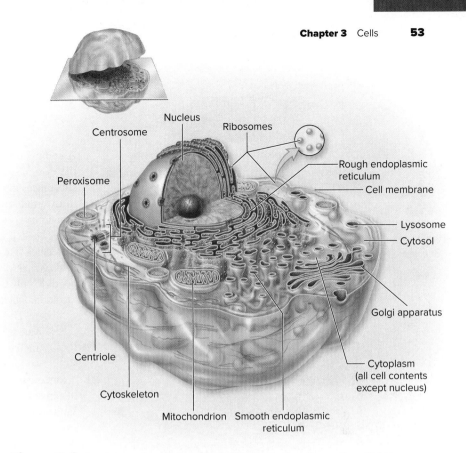

Figure 3.6 Anatomy of an Animal Cell. This illustration shows the relative sizes and locations of the components of an animal cell.

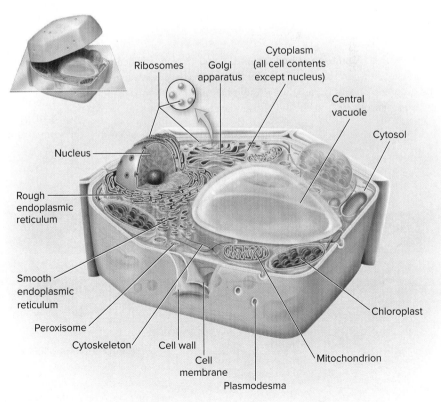

Figure 3.7 Anatomy of a Plant Cell. This generalized view illustrates key features of the plant cell. Note the cell wall, chloroplasts, and large central vacuole.

Phospholipid molecule

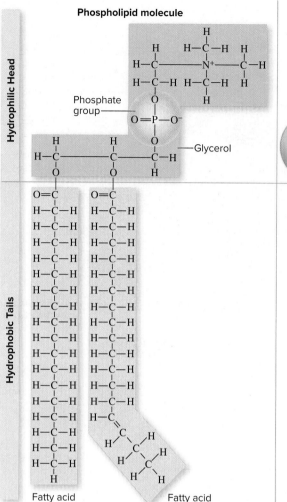

Figure 3.8 Membrane Phospholipid. A phospholipid molecule consists of a glycerol molecule attached to a hydrophilic phosphate "head" group and two hydrophobic fatty acid "tails." The right half of the illustration shows a simplified phospholipid structure.

3.3 A Membrane Separates Each Cell from Its Surroundings

A cell membrane is one feature common to all cells. The membrane separates the cytoplasm from the cell's surroundings. The cell's surface also transports substances into and out of the cell (see chapter 4), and it receives and responds to external stimuli. Inside a eukaryotic cell, internal membranes enclose the organelles.

The cell membrane is composed of phospholipids, which are organic molecules that resemble triglycerides (figure 3.8). In a triglyceride, three fatty acids attach to a three-carbon glycerol molecule. But in a **phospholipid,** glycerol bonds to only two fatty acids; the third carbon binds to a phosphate group attached to additional atoms. ⓘ *triglycerides,* section 2.5E

This chemical structure gives phospholipids unusual properties in water. The phosphate "head" end, with its polar covalent bonds, is attracted to water; that is, it is hydrophilic. The other end, consisting of two fatty acid "tails," is hydrophobic. In water, these molecules spontaneously arrange themselves into a **phospholipid bilayer:** a double layer of phospholipids (figure 3.9). In some ways, this bilayer resembles a cheese sandwich. The hydrophilic head groups (the "bread" of the sandwich) are exposed to the watery medium outside and inside the cell, whereas the hydrophobic tails face each other on the inside of the sandwich, like cheese between the bread slices. Unlike a sandwich, however, the bilayer forms a three-dimensional sphere, not a flat surface. ⓘ *hydrophilic and hydrophobic substances,* section 2.3B

Thanks to its hydrophobic middle portion, the phospholipid bilayer has selective permeability, meaning that some but not all substances can pass through it. Lipids and small, nonpolar molecules such as O_2 and CO_2 pass freely into and out of a cell. The fatty acid tails at the bilayer's interior, however, block ions and polar molecules such as glucose from passing through.

A cell membrane consists not only of a phospholipid bilayer but also of proteins and other molecules (figure 3.10). The membrane is often called a **fluid mosaic** because many of the molecules (the pieces of the "mosaic") drift laterally within the bilayer, a bit like pickpockets moving within a crowd of people. Steroid molecules maintain the membrane's fluidity as the temperature fluctuates. Both animal and plant cell membranes contain steroids; the cholesterol in animal membranes is the most familiar example. ⓘ *cholesterol,* section 2.5E

Whereas phospholipids and steroids provide the membrane's structure, proteins are especially important to its function. As you can see in figure 3.10, some of the proteins extend through the phospholipid bilayer, whereas others face only the inside or outside of the cell. Cells have multiple types of membrane proteins:

- **Transport proteins:** Transport proteins embedded in the phospholipid bilayer create passageways through which ions, glucose, and other polar substances pass into or out of the cell. Section 4.5 describes membrane transport in more detail.

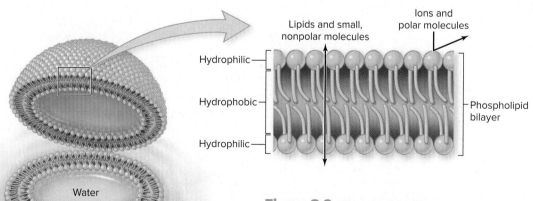

Figure 3.9 Phospholipid Bilayer. In water, phospholipids form a bilayer. The hydrophilic head groups are exposed to the water; the hydrophobic tails face each other, minimizing contact with water. This bilayer has selective permeability, allowing only some substances to pass freely.

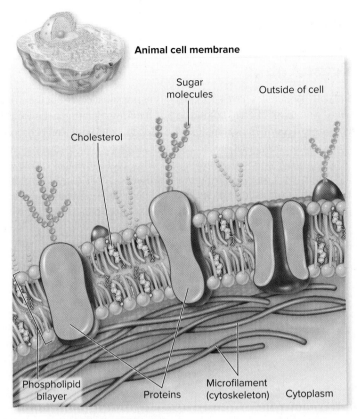

Animal cell membrane

Sugar molecules

Outside of cell

Cholesterol

Phospholipid bilayer

Proteins

Microfilament (cytoskeleton)

Cytoplasm

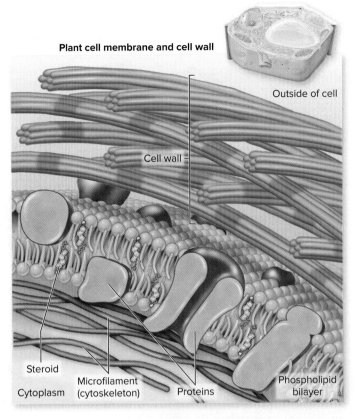

Plant cell membrane and cell wall

Outside of cell

Cell wall

Steroid

Cytoplasm

Microfilament (cytoskeleton)

Proteins

Phospholipid bilayer

Figure 3.10 **Anatomy of a Cell Membrane.** The cell membrane is a "fluid mosaic" of proteins embedded in a phospholipid bilayer. Steroid molecules, such as the cholesterol in animal cell membranes, add fluidity. The outer face of the animal cell membrane also features carbohydrate (sugar) molecules linked to proteins. A rigid wall of cellulose fibers surrounds each plant cell.

- **Enzymes:** These proteins facilitate chemical reactions that otherwise would proceed too slowly to sustain life. ⓘ *enzymes,* section 4.4
- **Recognition proteins:** Carbohydrates attached to cell surface proteins serve as "name tags" that help the body's immune system recognize its own cells.
- **Adhesion proteins:** These membrane proteins enable cells to stick to one another.
- **Receptor proteins:** Receptor proteins bind to molecules outside the cell and trigger a response inside the cell.

Understanding membrane proteins is a vital part of medicine, in part because at least half of all drugs bind to them. One example is omeprazole (Prilosec). This drug relieves heartburn by blocking some of the transport proteins that pump hydrogen ions into the stomach. Another is the antidepressant drug fluoxetine (Prozac), which prevents receptors on brain cell surfaces from absorbing a mood-altering biochemical called serotonin.

3.3 Mastering Concepts

1. Chemically, how is a phospholipid different from a triglyceride?
2. How does the chemical structure of phospholipids enable them to form a bilayer in water?
3. Where in the cell do phospholipid bilayers occur?
4. What are some functions of membrane proteins?

Burning Question 3.1

Is it possible to make an artificial cell from scratch?

We know exactly what cells are made of, from their DNA and RNA to the watery cytoplasm to the lipids and proteins that make up the membrane. Shouldn't we be able to make an artificial cell by combining those ingredients in a test tube?

So far, the answer is no. Making a cell from scratch is not as easy as mixing eggs, butter, flour, and sugar to make cookies. Although we know which chemicals are essential to life, we cannot simply blend them and wait for living cells to appear. That's because life is an emergent property of interacting molecules. These intricate relationships are extremely complex, and no one has ever controlled the participants with enough precision to craft a living cell.

If biologists ever do learn to make artificial cells, they could have practical uses. For example, with the right DNA, the cells could be coaxed to churn out biofuels, vaccines, and many other products.

Submit your burning question to
marielle.hoefnagels@mheducation.com

(petri dish): ©GlowImages/Alamy Stock Photo RF

3.4 Eukaryotic Organelles Divide Labor

In eukaryotic cells, organelles have specialized functions that carry out the work of the cell. If you think of a eukaryotic cell as a home, each organelle would be analogous to a room. For example, your kitchen, bathroom, and bedroom each hold unique items that suit the uses of those rooms. Likewise, each organelle has distinct sets of proteins and other molecules that fit the organelle's function. The "walls" of these cellular compartments are membranes, often intricately folded and studded with enzymes and other proteins. These folds provide tremendous surface area where many of the cell's chemical reactions occur.

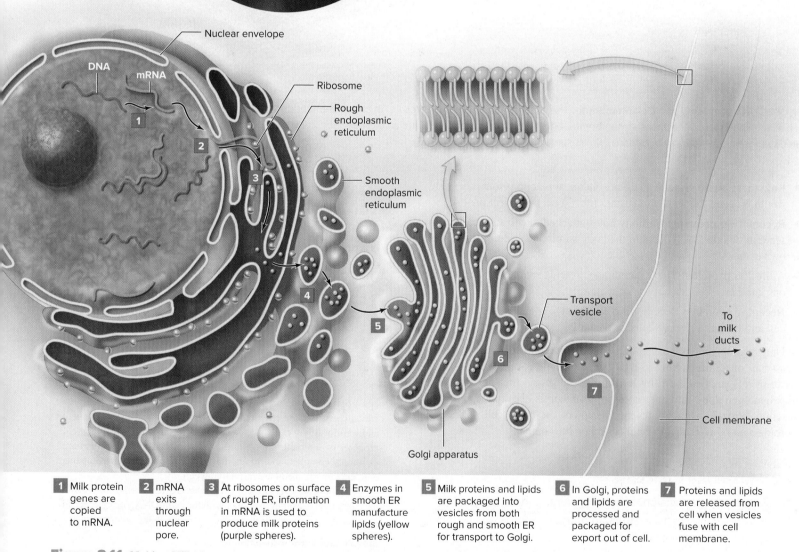

| 1 Milk protein genes are copied to mRNA. | 2 mRNA exits through nuclear pore. | 3 At ribosomes on surface of rough ER, information in mRNA is used to produce milk proteins (purple spheres). | 4 Enzymes in smooth ER manufacture lipids (yellow spheres). | 5 Milk proteins and lipids are packaged into vesicles from both rough and smooth ER for transport to Golgi. | 6 In Golgi, proteins and lipids are processed and packaged for export out of cell. | 7 Proteins and lipids are released from cell when vesicles fuse with cell membrane. |

Figure 3.11 **Making Milk.** Several types of organelles work together to produce and secrete milk from a cell in a mammary gland; the numbers (1) through (7) indicate each organelle's role. Note that membranes enclose each organelle and the entire cell.

Photo: ©Tim Flach/Getty Images

Many of the cell's internal membranes form a coordinated **endomembrane system,** which consists of several interacting organelles: the nuclear envelope, endoplasmic reticulum, Golgi apparatus, lysosomes, vacuoles, and cell membrane. As you will see, the organelles of the endomembrane system are connected by **vesicles,** small membranous spheres that transport materials inside the cell. These "bubbles" of membrane, which can pinch off from one organelle, travel within the cell, and fuse with another, are also considered part of the endomembrane system.

Interactions between the organelles of the endomembrane system enable cells to produce, package, and release complex mixtures of biochemicals. This section focuses on each step involved in the production and secretion of one such mixture: milk **(figure 3.11).**

A. The Nucleus, Endoplasmic Reticulum, and Golgi Interact to Secrete Substances

Special cells in the mammary glands of female mammals produce milk, which contains proteins, fats, carbohydrates, and water in a proportion ideal for the development of a newborn. Human milk is rich in lipids, which the rapidly growing baby's nervous system requires. (Cow's milk contains a higher proportion of protein, better suited to a calf's rapid muscle growth.) Milk also contains calcium, potassium, and antibodies that help jump-start the infant's immunity to disease.

The milk-producing cells of the mammary glands are dormant most of the time, but they undergo a burst of productivity shortly after the female gives birth. How do each cell's organelles work together to manufacture milk?

The Nucleus The process of milk production and secretion begins in the **nucleus** (see figure 3.11, step 1), the organelle that contains most of a eukaryotic cell's DNA. The function of DNA is to specify the "recipe" for every protein a cell can make (such as milk protein and enzymes required to synthesize carbohydrates and lipids). The cell copies the genes encoding these proteins into another nucleic acid, messenger RNA (mRNA).

The mRNA molecules exit the nucleus through **nuclear pores,** which are holes in the double-membrane **nuclear envelope** that separates the nucleus from the cytoplasm (figure 3.11, step 2, and **figure 3.12**). Nuclear pores are highly specialized channels composed of dozens of types of proteins. Traffic through the nuclear pores is busy, with millions of regulatory proteins entering and mRNA molecules leaving each minute.

Also inside the nucleus is the **nucleolus,** a dense spot that assembles the components of ribosomes. These ribosomal subunits leave the nucleus through the nuclear pores, and they come together in the cytoplasm to form complete ribosomes.

The Endoplasmic Reticulum and Golgi Apparatus The remainder of the cell, between the nucleus and the cell membrane, is the cytoplasm. The cytoplasm includes the cytosol, a watery mixture of ions, enzymes, RNA, and other dissolved substances. Organelles are also part of the cytoplasm, as are arrays of protein rods and tubules called the cytoskeleton (see section 3.5).

Why We Care 3.1 | Most of Your Cells Are Not Your Own

Many people are surprised to learn that nonhuman cells vastly outnumber the body's own cells. Microbiologists estimate that the number of bacteria living in and on a typical human is *10 times* the number of human cells! Although some of these bacteria can cause disease, most exist harmlessly on the skin and in the mouth and intestines. These inconspicuous guests also can help extract nutrients from food and prevent disease.

Red blood cells

So how many *human* cells make up a person's body? For adults, estimates range from about 10 trillion to 100 trillion. No one knows for sure, because counting living cells is very difficult. After all, the number of cells changes throughout life. A child's growth comes from cell division that adds new cells, not from the expansion of existing ones. Moreover, new cells arise as old cells die, so a "true" count is a moving target. Also, no one has found a good way to count them all. Cells come in so many different shapes and sizes that it is hard to extrapolate from a small sample to the whole body.

Photo: ©MedicalRF.com/Getty Images RF

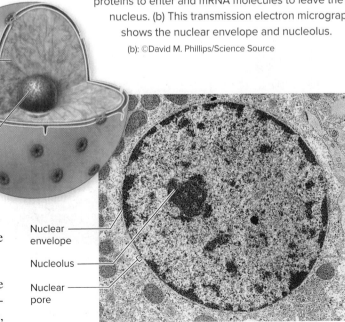

Figure 3.12 The Nucleus. (a) The nucleus contains DNA and is surrounded by two membrane layers, which make up the nuclear envelope. Large pores in the nuclear envelope allow proteins to enter and mRNA molecules to leave the nucleus. (b) This transmission electron micrograph shows the nuclear envelope and nucleolus.

(b): ©David M. Phillips/Science Source

Nuclear envelope

DNA

Nuclear pore

Nucleolus

a.

Nuclear envelope

Nucleolus

Nuclear pore

b.   TEM (false color)   2 μm

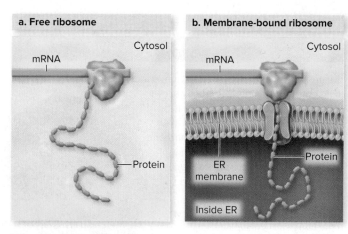

a. Free ribosome

Cytosol

mRNA

Protein

b. Membrane-bound ribosome

Cytosol

mRNA

Protein

ER membrane

Inside ER

Figure 3.13 Ribosomes. (a) Free ribosomes produce proteins used in the cell's cytosol. (b) Proteins produced by ribosomes attached to the rough ER's membrane are typically used in specialized organelles or in the cell membrane; they may also be secreted outside the cell.

Once in the cytoplasm, mRNA coming from the nucleus binds to a ribosome, which manufactures proteins (see figure 3.11, step 3). Free-floating ribosomes produce proteins that remain in the cell's cytosol (figure 3.13). But many proteins are destined for organelles, for the cell membrane, or for secretion (in milk, for example). In these cases, the entire complex of ribosome, mRNA, and partially made protein anchors to the membrane of the endoplasmic reticulum.

The **endoplasmic reticulum (ER)** is a network of sacs and tubules composed of membranes. This complex organelle originates at the nuclear envelope and winds throughout the cell (*endoplasmic* means "within the cytoplasm," and *reticulum* means "network"). Close to the nucleus, the membrane surface is studded with ribosomes making proteins that enter the inner compartment of the ER; these proteins are destined to be secreted from the cell. This section of the network is called the **rough endoplasmic reticulum** because the ribosomes give these membranes a roughened appearance (figure 3.14).

Adjacent to the rough ER, a section of the network called the **smooth endoplasmic reticulum** synthesizes lipids—such as those that will end up in the milk—and other membrane components (see figure 3.11, step 4, and figure 3.13). The smooth ER also houses enzymes that detoxify drugs and poisons. In muscle cells, a specialized type of smooth ER stores and delivers the calcium ions required for muscle contraction. ⓘ *muscle function,* section 26.4C

The lipids and proteins made by the ER exit the organelle in vesicles. A loaded transport vesicle pinches off from the tubular endings of the ER membrane (see figure 3.11, step 5) and takes its contents to the Golgi apparatus, the next stop in the production line. The **Golgi apparatus** is a stack of flat, membrane-enclosed sacs that functions as a processing center (figure 3.15). Proteins from the ER pass through the series of Golgi sacs, where they complete their intricate folding and become functional (see figure 3.11, step 6). Enzymes in the Golgi apparatus also manufacture and attach carbohydrates to proteins or lipids, forming the "name tags" recognized by the immune system (see section 3.3).

The Golgi apparatus sorts and packages materials into vesicles, which move toward the cell membrane. Some of the proteins received from the ER

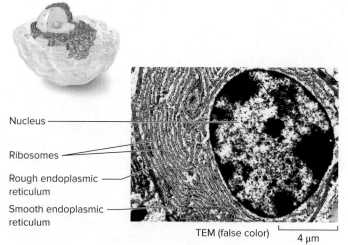

Nucleus

Ribosomes

Rough endoplasmic reticulum

Smooth endoplasmic reticulum

TEM (false color)   4 µm

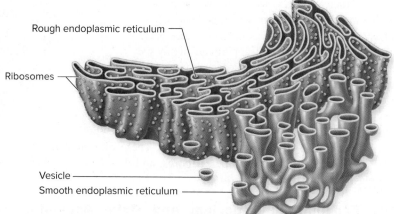

Rough endoplasmic reticulum

Ribosomes

Vesicle

Smooth endoplasmic reticulum

Figure 3.14 Rough and Smooth Endoplasmic Reticulum. The endoplasmic reticulum is a network of membranes extending from the nuclear envelope. Ribosomes dot the surface of the rough ER, giving it a "rough" appearance. The smooth ER is a series of interconnecting tubules and is the site for lipid production and other metabolic processes.

Photo: ©Prof. J. L. Kemeny/ISM/Phototake

will become membrane surface proteins; other substances (such as milk protein and fat) are packaged for secretion from the cell. In the production of milk, these vesicles fuse with the cell membrane and release the proteins outside the cell (see figure 3.11, step 7). The fat droplets stay suspended in the watery milk because they retain a layer of surrounding membrane when they leave the cell.

This entire process happens simultaneously in countless specialized cells lining the milk ducts of the mother's breast, beginning shortly after a baby's birth. When the infant suckles, hormones released in the mother's body stimulate muscles surrounding balls of these cells to contract, squeezing milk into the ducts that lead to the nipple.

B. Lysosomes, Vacuoles, and Peroxisomes Are Cellular Digestion Centers

Besides producing molecules for export, eukaryotic cells also break down molecules in specialized compartments. All of these "digestion center" organelles are sacs surrounded by a single membrane.

Lysosomes Lysosomes are organelles containing enzymes that dismantle and recycle food particles, captured bacteria, worn-out organelles, and debris (figure 3.16). They are so named because their enzymes lyse, or cut apart, their substrates.

The enzymes inside lysosomes originate in the rough ER. The Golgi apparatus detects these enzymes by recognizing a sugar attached to them, then packages them into vesicles that eventually become lysosomes. The lysosomes, in turn, fuse with transport vesicles carrying debris from outside or from within the cell. The lysosome's enzymes break down the large organic molecules into smaller subunits by hydrolysis, releasing them into the cytosol for the cell to use.

What keeps a lysosome from digesting the entire cell? The lysosome's membrane maintains the pH of the organelle's interior at about 4.8, much more acidic than the neutral pH of the rest of the cytoplasm. If one lysosome were to burst, the liberated enzymes would no longer be at their optimum pH, so they could not digest the rest of the cell. Nevertheless, a cell injured by extreme cold, heat, or another physical stress may initiate its own death by bursting all of its lysosomes at once. (i) *pH*, section 2.4; *cell death*, section 8.1B

Some cells have more lysosomes than others. White blood cells, for example, have many lysosomes because these cells engulf and dispose of debris and bacteria. Liver cells require many lysosomes to process cholesterol.

Malfunctioning lysosomes can cause illness. In Tay-Sachs disease, for example, a defective lysosomal enzyme allows a lipid to accumulate to toxic levels in nerve cells of the brain. The nervous system deteriorates, and an affected person eventually becomes unable to see, hear, or move. In the most severe forms of the illness, death usually occurs by age 5.

Vacuoles Most plant cells lack lysosomes, but they do have an organelle that serves a similar function. In mature plant cells, the large central **vacuole** contains a watery solution of enzymes that degrade and recycle molecules and organelles (see figure 3.7).

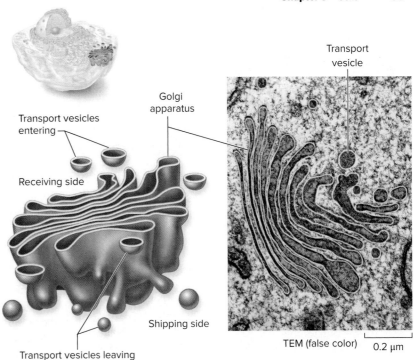

Figure 3.15 **The Golgi Apparatus.** This organelle consists of a series of flattened sacs, plus transport vesicles that deliver and remove materials. Proteins are sorted and processed as they move through the Golgi apparatus on their way to the cell surface or to a lysosome.

Photo: ©Biophoto Associates/Science Source

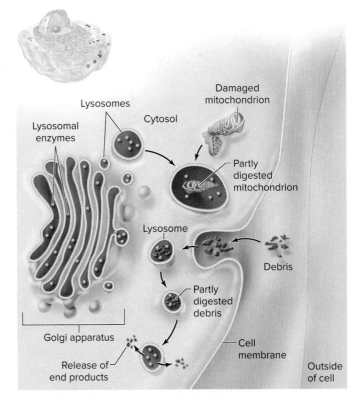

Figure 3.16 **Lysosomes.** Lysosomes contain enzymes that dismantle damaged organelles and other debris, then release the nutrients for the cell to use.

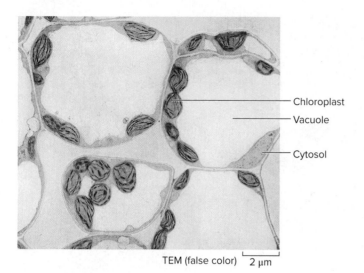

Chloroplast
Vacuole
Cytosol

TEM (false color) 2 μm

Figure 3.17 Vacuole. Much of the volume of a spinach leaf cell is occupied by the large central vacuole. The rest of each cell's contents (including numerous chloroplasts) is pushed to the edges of the cell.

©Biophoto Associates/Science Source

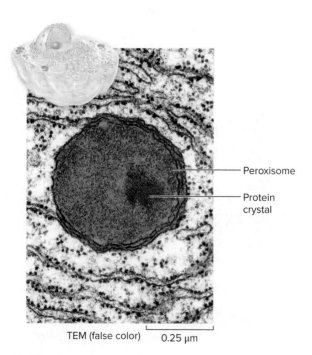

Peroxisome

Protein crystal

TEM (false color) 0.25 μm

Figure 3.18 Peroxisomes. Protein crystals give peroxisomes their characteristic appearance in an animal cell.

Photo: ©Don W. Fawcett/Science Source

The vacuole also has other roles. Most of the growth of a plant cell comes from an increase in the volume of its vacuole. In some plant cells, the vacuole occupies up to 90% of the cell's volume (figure 3.17). As the vacuole acquires water, it exerts pressure (called turgor pressure) against the cell membrane. Turgor pressure helps plants stay rigid and upright.

Besides water and enzymes, the vacuole also contains a variety of salts, sugars, and weak acids. Therefore, the pH of the vacuole's solution is usually somewhat acidic. In citrus fruits, the solution is very acidic, producing the tart taste of lemons and oranges. Water-soluble pigments also reside in the vacuole, producing blue, purple, and magenta colors in some leaves, flowers, and fruits.

Some protists have vacuoles, although their function is different from that in plants. The contractile vacuole in *Paramecium,* for example, pumps excess water out of the cell. In *Amoeba,* food vacuoles digest nutrients that the cell has engulfed.

Peroxisomes All eukaryotic cells contain **peroxisomes,** organelles that contain several types of enzymes that dispose of toxic substances. Although they resemble lysosomes in size and function, peroxisomes originate at the ER (not the Golgi) and contain different enzymes. In some peroxisomes, the concentration of enzymes reaches such high levels that the proteins condense into easily recognized crystals (figure 3.18).

Peroxisomes in liver and kidney cells help dismantle toxins from the blood. Peroxisomes also break down fatty acids and produce cholesterol and some other lipids. In a disease called adrenoleukodystrophy (ALD), a faulty peroxisomal enzyme causes fatty acids to accumulate to toxic levels in the brain, causing severe brain damage and eventually death.

C. Mitochondria Extract Energy from Nutrients

Growth, cell division, protein production, secretion, and many chemical reactions in the cytoplasm all require a steady supply of energy. **Mitochondria** (singular: mitochondrion) are organelles that use a process called cellular respiration to extract this needed energy from food (see chapter 6). With the exception of a few types of protists, all eukaryotic cells have mitochondria.

A mitochondrion has two membrane layers: an outer membrane and an intricately folded inner membrane that encloses the mitochondrial matrix (figure 3.19). Within the matrix is DNA that encodes proteins essential for mitochondrial function; ribosomes occupy the matrix as well. **Cristae** are the folds of the inner membrane. The cristae add tremendous surface area to the inner membrane, which houses the enzymes that catalyze the reactions of cellular respiration.

In most mammals, mitochondria are inherited from the female parent only. (This is because the mitochondria in a sperm cell degenerate after fertilization.) Mitochondrial DNA is therefore useful for tracking inheritance through female lines in a family. For the same reason, genetic mutations that cause defective mitochondria also pass only from mother to offspring. Mitochondrial illnesses are most serious when they affect the muscles or brain, because these energy-hungry organs depend on the functioning of many thousands of mitochondria in every cell.

D. Photosynthesis Occurs in Chloroplasts

Plants and many protists carry out photosynthesis, a process that uses energy from sunlight to produce glucose and other food molecules (see chapter 5). These nutrients sustain not only the photosynthetic organisms but also the consumers (including humans) that eat them.

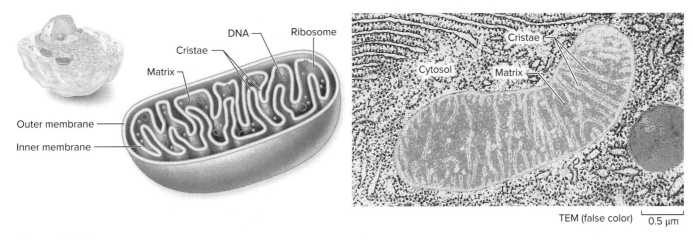

Figure 3.19 Mitochondria. Cellular respiration occurs inside mitochondria. Each mitochondrion contains a highly folded inner membrane where many of the reactions of cellular respiration occur.

Photo: ©Bill Longcore/Science Source

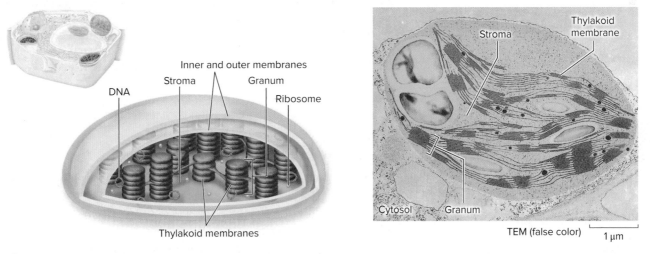

Figure 3.20 Chloroplasts. Photosynthesis occurs inside chloroplasts. Each chloroplast contains stacks of thylakoids that form the grana within the inner compartment, the stroma. Enzymes and light-harvesting pigments embedded in the membranes of the thylakoid convert the energy in sunlight to chemical energy.

Photo: ©Biophoto Associates/Science Source

The **chloroplast** (figure 3.20) is the site of photosynthesis in eukaryotes. Each chloroplast contains multiple membrane layers. Two outer membrane layers enclose an enzyme-rich fluid called the stroma. Within the stroma is a third membrane system folded into flattened sacs called thylakoids, which are stacked like pancakes to form structures called grana. Photosynthetic pigments such as chlorophyll are embedded in the thylakoid membranes.

A chloroplast is one representative of a larger category of plant organelles called plastids. Some plastids synthesize lipid-soluble red, orange, and yellow carotenoid pigments, such as those found in carrots and ripe tomatoes. Plastids that assemble starch molecules are important in cells specialized for food storage, such as those in potatoes and corn kernels. Interestingly, any plastid can convert into any other type. As a tomato ripens,

TABLE 3.1	Mitochondria and Chloroplasts: A Comparison	
	Mitochondria	**Chloroplasts**
Contain DNA and ribosomes?	Yes	Yes
Inner fluid	Matrix	Stroma
Internal membrane structures	Cristae	Thylakoids
Function	Produce ATP (cellular respiration)	Produce sugars (photosynthesis)
Location	Nearly all eukaryotic cells	Eukaryotic cells that carry out photosynthesis

for example, its green chloroplasts change into plastids that store red carotenoid pigments.

Like mitochondria, all plastids (including chloroplasts) contain DNA and ribosomes. The genetic material encodes proteins unique to plastid structure and function, including some of the enzymes required for photosynthesis.

Table 3.1 compares mitochondria and chloroplasts. The striking similarities between mitochondria and chloroplasts—both have their own DNA and ribosomes, and both are surrounded by double membranes—provide clues to the origin of eukaryotic cells, an event that occurred at least 1.5 billion years ago. According to the endosymbiosis theory, some ancient organism (or organisms) engulfed bacterial cells. Rather than digesting them as food, the host cells kept them on as partners: mitochondria and chloroplasts. The structures and genetic sequences of today's bacteria, mitochondria, and chloroplasts supply powerful evidence for this theory. (i) *endosymbiosis,* section 15.3A

Organelles divide a cell's work, just as the rooms in a house contain related items: Pots and dishes are in the kitchen, whereas blankets and pillows are in the bedroom. But highly specialized buildings also exist. A restaurant, for example, has an enormous kitchen and no bedrooms at all. Likewise, cells can also have specialized functions (figure 3.21). For example, a heart muscle cell is roughly cylindrical when compared with a neuron, which produces extensions that touch adjacent nerve cells. A leaf cell is packed with chloroplasts. The protective epidermis of an onion, on the other hand, is dry and tough; because it forms underground, it lacks chloroplasts. In each case, the mix of organelles inside each cell determines its functions. Keep these specialized structures and functions in mind as you study cell processes throughout this book.

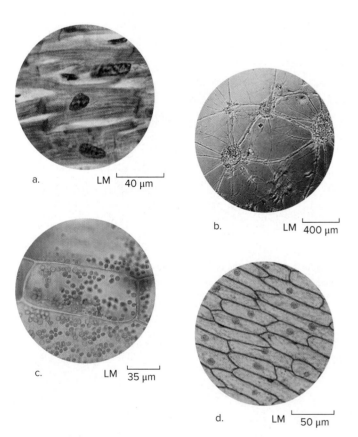

a.   LM |— 40 μm —|

b.   LM |— 400 μm —|

c.   LM |— 35 μm —|

d.   LM |— 50 μm —|

Figure 3.21 Specialized Cells. (a) Muscle cells in the heart look and behave differently from (b) the highly branched neurons that form the nervous system. (c) A plant's leaf cells contain chloroplasts, whereas (d) the cells making up the outer skin of an onion do not.

(a): ©Fuse/Getty Images RF; (b): ©Francois Paquet-Durand/Science Source; (c): ©Ed Reschke/Photolibrary/Getty Images; (d): ©Ted Kinsman/Science Source

3.4 Mastering Concepts

1. How do organelles help eukaryotic cells compensate for a small ratio of surface area to volume?
2. Which parts of a cell interact to produce and secrete a complex substance such as milk?
3. What is the function of the nucleus and its contents?
4. Which organelles are the cell's "recycling centers"?
5. Which organelle houses the reactions that extract chemical energy from nutrient molecules?
6. How are the functions of plastids essential to the life of a plant cell?
7. Describe how form fits function for three organelles.

3.5 The Cytoskeleton Supports Eukaryotic Cells

The cytosol of a eukaryotic cell contains a **cytoskeleton,** an intricate network of protein "tracks" and tubules. The cytoskeleton is a structural framework with many functions. It is a transportation system, and it provides the physical support necessary to maintain the cell's characteristic three-dimensional shape. It aids in cell division and helps connect cells to one another. The cytoskeleton also enables cells—or parts of a cell—to move.

The cytoskeleton includes three major components: microfilaments, intermediate filaments, and microtubules (figure 3.22). They are distinguished by protein type, diameter, and how they aggregate into larger structures. Other proteins connect these components to one another, creating an intricate meshwork.

The thinnest component of the cytoskeleton is the **microfilament,** a long rod composed of the protein **actin.** Each microfilament is only about 7 nanometers in diameter. Actin microfilament networks are part of nearly all eukaryotic cells. Muscle contraction, for example, relies on actin filaments and another protein, myosin. Microfilaments also provide strength for cells to survive stretching and compression, and they help to anchor one cell to another (see section 3.6). ⓘ *muscle movement,* section 26.4B

Intermediate filaments are so named because their 10-nanometer diameters are intermediate between those of microfilaments and microtubules. Unlike the other components of the cytoskeleton, which consist of a single protein type, intermediate filaments are made of a variety of proteins. They maintain a cell's shape by forming an internal scaffold in the cytosol and resisting mechanical stress. Intermediate filaments also help bind some cells together (see section 3.6).

A **microtubule** is composed of a protein called tubulin assembled into a hollow tube that is 23 nanometers in diameter. The cell can change the length of a microtubule rapidly by adding or removing tubulin molecules. Microtubules have many functions in eukaryotic cells. For example, they form a type of "trackway" along which substances move within a cell. Specialized motor proteins "walk" along the tracks toting an organelle, a vesicle, or other cargo. In addition, chapter 8 describes how microtubules split a cell's duplicated chromosomes apart during cell division.

In animal cells, structures called **centrosomes** organize the microtubules. (Plants typically lack centrosomes and assemble microtubules at sites scattered throughout the cell.) The centrosome contains two centrioles, which are visible in figure 3.6. The centrioles also indirectly produce the extensions that enable some cells to move: cilia and flagella (figure 3.23).

Cilia are short, numerous extensions resembling a fringe. Some protists, such as the *Paramecium* in figure 3.2, have thousands of cilia that enable the cells to "swim" in water. In the human respiratory tract, coordinated movement of cilia sets up a wave that propels particles up and out; other cilia can move an egg cell through the female reproductive tract. ⓘ *ciliates,* section 15.4D

Unlike cilia, flagella occur singly or in pairs, and a flagellum is much longer than a cilium. Flagella are more like tails, and their whiplike movement propels cells. Sperm cells in many species (including humans) have prominent flagella. A man whose sperm cells have defective flagella is infertile because the sperm are unable to swim to the egg cell.

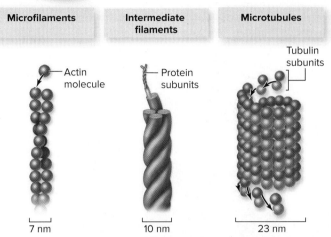

Microfilaments	Intermediate filaments	Microtubules

Actin molecule — 7 nm

Protein subunits — 10 nm

Tubulin subunits — 23 nm

Figure 3.22 **Proteins of the Cytoskeleton.** The cytoskeleton consists of three sizes of protein filaments, arranged in this figure from smallest to largest diameter.

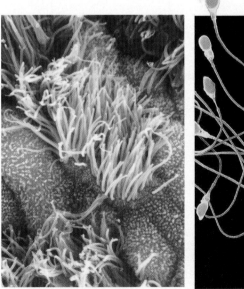
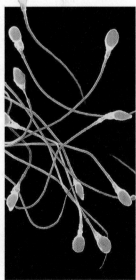

a.   SEM (false color) 4 μm   b. SEM (false color) 10 μm

Figure 3.23 **Cilia and Flagella.** (a) These cilia help eliminate dust and other foreign particles from the human respiratory tract. (b) The flagella on human sperm cells enable them to swim.

(a): ©D.W. Fawcett/Science Source; (b): ©Dr. Tony Brain/Science Source

3.5 Mastering Concepts

1. What are some functions of the cytoskeleton?
2. What are the main components of the cytoskeleton?
3. Why are cilia and flagella important?

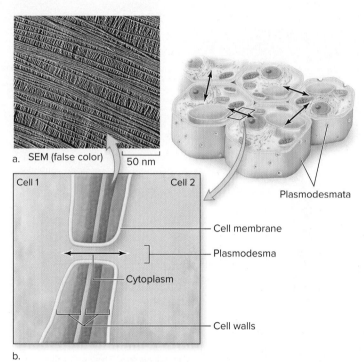

a.   SEM (false color)   50 nm

b.

Figure 3.24   The Plant Cell Wall. (a) Cellulose fibers make up the cell wall. (b) The walls of adjoining cells are composed of layers that each cell lays down. Plasmodesmata allow the exchange of materials between adjacent cells.

(a): ©BioPhoto Associates/Science Source

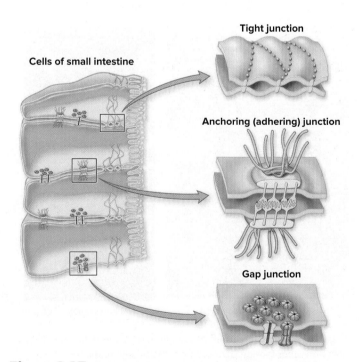

Figure 3.25   Animal Cell Connections. Tight junctions fuse neighboring cell membranes, anchoring junctions form "spot welds," and gap junctions allow small molecules to move from cell to cell.

3.6 Cells Stick Together and Communicate with One Another

So far, this chapter has described individual cells. But multicellular organisms, including plants and animals, are made of many cells that work together. How do these cells adhere to one another so that your body—or that of a plant—doesn't disintegrate in a heavy rain? Also, how do cells in direct contact with one another communicate to coordinate development and respond to the environment? This section describes how the cells of plant and animal tissues stick together and how neighboring cells share signals.

Cell walls surround the cell membranes of nearly all bacteria, archaea, fungi, algae, and plants. But *cell wall* is a misleading term: It is not just a barrier that outlines the cell. Cell walls impart shape, regulate cell volume, and prevent bursting when a cell takes in too much water.

Much of the plant cell wall consists of cellulose molecules aligned into countless crisscrossing fibers that impart great strength (figure 3.24a). Other polysaccharides glue adjacent cells together and add strength and flexibility. Each cell communicates with its neighbors through **plasmodesmata,** channels that connect adjacent plant cells (figure 3.24b). Plasmodesmata are essentially "tunnels" in the cell wall through which the cytoplasm of one plant cell can interact with that of another. ⓘ *cellulose,* section 2.5B

Animal cells lack cell walls. Instead, many animal cells secrete a complex extracellular matrix that holds them together and coordinates many aspects of cellular life. In these tissues, cells are not in direct contact with one another. In other tissues, however, the plasma membranes of adjacent cells directly connect to one another via several types of junctions (figure 3.25):

- A **tight junction** fuses animal cells together, forming an impermeable barrier between them. Proteins anchored in membranes connect to actin in the cytoskeleton and join cells into sheets, such as those lining the inside of the digestive tract. These connections allow the body to control where biochemicals move, since fluids cannot leak between the joined cells. For example, tight junctions prevent stomach acid from seeping into the tissues surrounding the stomach.

- An **anchoring (or adhering) junction** connects an animal cell to its neighbors or to the extracellular matrix, somewhat like a rivet. Proteins at each anchoring junction span the cell membrane and link to each cell's cytoskeleton. These junctions hold skin cells in place by anchoring them to one another and to the extracellular matrix.

- A **gap junction** is a protein channel that links the cytoplasm of adjacent animal cells, allowing exchange of ions, nutrients, and other small molecules. It is therefore analogous to plasmodesmata in plants. Gap junctions link heart muscle cells to one another, allowing groups of cells to contract together.

3.6 Mastering Concepts

1. What are the functions of a cell wall?
2. What is the chemical composition of a plant cell wall?
3. What are plasmodesmata?
4. What are the three types of junctions that link cells in animals?

Investigating Life 3.1 | Bacterial Magnets

Until recently, biologists thought that prokaryotic cells lacked any internal membranes. But microscopes revealed that some bacteria living in the ocean have small lipid bilayer spheres in their cytoplasm. Scientists found high concentrations of magnetic iron crystals within these membrane bubbles and aptly named them "magnetosomes."

When scientists found magnetosomes in ocean-dwelling bacteria, they already knew that Earth's magnetic field runs nearly vertically through the water. Experiments on the bacteria revealed that the magnetosomes align with magnetic field lines and that the microbes swim either against or with the field.

These studies showed how bacteria respond to magnetism, but they did not explain why orienting to magnetic fields is adaptive. A team of researchers aimed to answer this question. The observation that the bacteria do not always swim in the same direction along the magnetic field lines led them to hypothesize that another factor must influence bacterium movement. One clue was that these bacteria cannot survive if oxygen levels are too high or too low. So the investigators devised an experiment to test the hypothesis that magnetism and oxygen concentration jointly guided bacterial movement.

The scientists put the bacteria in a solution. They then drew the mixture into narrow glass tubes and sealed one end. When the team produced a magnetic field across one of the tubes, all of the bacteria turned toward the field. Some then swam forward, while others moved backward. They aggregated in a distinct band in the center of the tube, at their optimal oxygen concentration (figure 3.A). The scientists then switched the direction of the magnetic field. All of the bacteria turned 180 degrees, but none migrated out of the band in the center of the tube.

These results indicate that magnetic fields influence the direction that magnetosome-containing bacteria face, helping the cells find the shortest path to their preferred oxygen concentration. Decreasing the swimming distance saves energy for other cellular tasks, such as reproduction.

Scientists used powerful microscopes and clever experiments to reveal how some bacteria avoid getting lost at sea. Lipid-enclosed magnetosomes guide them like compasses through the deep unknown.

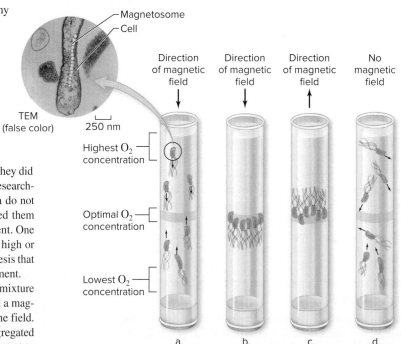

Figure 3.A Magnetic Orientation. (a) Bacteria with magnetosomes (*inset*) turn toward magnetic fields and (b) move in straight lines toward their optimal oxygen concentration. (c) Switching the direction of the magnetic field rotates the bacteria. (d) Without a magnetic field, bacteria move toward an optimal O_2 concentration but do not take a direct path. Small arrows in (a) and (d) indicate the direction of movement.

Photo: ©Dennis Kunkel/SPL/Science Source

Source: Frankel, Richard B., Dennis A. Bazylinski, Mark S. Johnson, and Barry L. Taylor. 1997. Magneto-aerotaxis in marine coccoid bacteria. *Biophysical Journal*, vol. 73, pages 994–1000.

What's the **Point?** ▼ APPLIED

All cells have features in common, reflecting their shared evolutionary history. These shared features include DNA, RNA, ribosomes, proteins, cytoplasm, and a cell membrane. At the same time, life's three domains differ in many ways, including DNA and RNA sequences, ribosome type, protein diversity, cytoplasm composition, and membrane structure. In addition to these differences, eukaryotic cells have variable numbers and types of organelles. Microscopes and cell chemistry techniques are vital tools that scientists use to discover the differences between cells.

Understanding the unique features of cells is useful in the fight against human disease. For example, researchers apply their understanding of cells in the ongoing development of antibiotics—medicines that kill bacteria. In the 1970s, scientists discovered that prokaryotic ribosomes are structurally different from those of eukaryotes. Some antibiotics

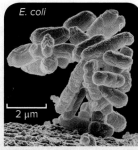

E. coli

2 μm

SEM (false color)

attack bacterial ribosomes, stopping protein synthesis in bacteria without shutting down vital functions in human cells. Other antibiotics prevent bacteria from building a cell wall (human cells lack a cell wall) or copying their DNA (human cells use different DNA replication proteins than do bacteria). Still, with antibiotic-resistant bacteria on the rise, biologists are racing to find new treatment options. Innovative approaches to drug discovery combine microscopy, chemistry, and math into a process that rapidly assesses if an antibiotic will work. Researchers will need any shortcut they can find to stay ahead in the race against antibiotic resistance.

Next time you take antibiotics, thank the scientists whose research made your treatment possible.

Photo: USDA-ARS/Eric Erbe/Chris Pooley

CHAPTER SUMMARY

3.1 Cells Are the Units of Life

A. Simple Lenses Revealed the First Glimpses of Cells

- **Cells** are the microscopic components of all organisms.
- The first person to see cells was Robert Hooke, who viewed cork with a crude lens in the late seventeenth century.
- The **cell theory** states that all life is composed of cells, that cells are the functional units of life, and that all cells come from preexisting cells.

B. Microscopes Magnify Cell Structures

- Light microscopes, transmission electron microscopes, and scanning electron microscopes are essential tools for viewing the parts of a cell.

C. All Cells Have Features in Common

- All cells have DNA, RNA, **ribosomes** that build proteins, and a **cell membrane** that is the interface between the cell and the outside environment (figure 3.26). This membrane encloses the **cytoplasm,** which includes a fluid portion called the **cytosol.**
- Complex cells also have specialized compartments called **organelles.**
- The surface area of a cell must be large relative to its volume.

3.2 Different Cell Types Characterize Life's Three Domains

- **Eukaryotic** cells have a nucleus and other organelles; **prokaryotic** cells lack these structures. Prokaryotic cells include bacteria and archaea.

A. Domains Bacteria and Archaea Contain Prokaryotic Organisms

- **Bacteria** are structurally simple, but they are abundant and diverse. Most have a **cell wall** and one or more **flagella.** DNA occurs in an area called the **nucleoid.**
- **Archaea** share some characteristics with bacteria and eukaryotes but also have unique structures and chemistry.

	Cell type	
	Prokaryotic	**Eukaryotic**
Nucleus	No	Yes
Membrane-bounded organelles	No	Yes
Typical size	1–10 μm	10–100 μm

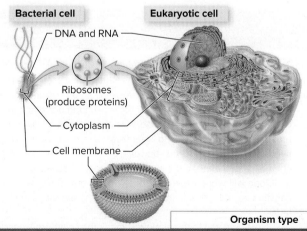

	Organism type	
	Prokaryotic	**Eukaryotic**
Number of cells in organism	Usually one	One or more
Domain(s)	Bacteria and Archaea	Eukarya

Figure 3.26 Cell Features: A Summary.

B. Domain Eukarya Contains Organisms with Complex Cells

- Domain **Eukarya** includes protists, plants, fungi, and animals. Most eukaryotic cells are larger than prokaryotic cells.

3.3 A Membrane Separates Each Cell from Its Surroundings

- A **phospholipid** consists of a phosphate group, a glycerol, and two fatty acids. A biological membrane consists of a **phospholipid bilayer** embedded with movable proteins and steroid molecules, forming a **fluid mosaic.**
- Membrane proteins carry out a variety of functions.

3.4 Eukaryotic Organelles Divide Labor

- The **endomembrane system** includes the nuclear envelope, endoplasmic reticulum, Golgi apparatus, lysosomes, vacuoles, cell membrane, and **vesicles** that transport materials within cells.

A. The Nucleus, Endoplasmic Reticulum, and Golgi Interact to Secrete Substances

- A eukaryotic cell houses DNA in a **nucleus. Nuclear pores** allow the exchange of materials through the two-layered **nuclear envelope;** assembly of the ribosome's subunits occurs in the **nucleolus.**
- The **smooth endoplasmic reticulum, rough endoplasmic reticulum,** and **Golgi apparatus** work together to synthesize, store, transport, and release molecules.

B. Lysosomes, Vacuoles, and Peroxisomes Are Cellular Digestion Centers

- A eukaryotic cell degrades wastes and digests nutrients in **lysosomes.**
- In plants, a watery **vacuole** degrades wastes, exerts turgor pressure, and stores acids and pigments.
- **Peroxisomes** help digest fatty acids and detoxify many substances.

C. Mitochondria Extract Energy from Nutrients

- **Mitochondria** house the reactions of cellular respiration. The **cristae** (folds) of the inner mitochondrial membrane add surface area.

D. Photosynthesis Occurs in Chloroplasts

- In the photosynthetic cells of plants and algae, **chloroplasts** use solar energy to make food.

3.5 The Cytoskeleton Supports Eukaryotic Cells

- The **cytoskeleton** is a network of protein rods and tubules that provides cells with form, support, and the ability to move.
- **Microfilaments,** the thinnest components of the cytoskeleton, are composed of the protein **actin.**
- **Intermediate filaments** consist of various proteins that strengthen the cytoskeleton.
- **Microtubules** are hollow tubes made of tubulin subunits. They form an internal trackway and include the fibers that separate chromosomes during cell division. **Centrosomes** organize microtubules in animal cells.
- **Cilia** are short, numerous extensions; flagella are less numerous but much longer. Both cilia and flagella aid in the movement of cells or materials.

3.6 Cells Stick Together and Communicate with One Another

- Most organisms other than animals have cell walls, which provide protection and shape. Plant cell walls consist of cellulose filaments connected by other molecules.
- **Plasmodesmata** are pores that extend through the cell walls of adjacent plant cells.
- Connections between animal cells include **tight junctions, anchoring junctions,** and **gap junctions.** Tight junctions create a seal between adjacent cells. Anchoring junctions are "spot welds" that secure cells in place. Gap junctions allow adjacent cells to exchange materials.

MULTIPLE CHOICE QUESTIONS

1. One property that distinguishes cells in domain Bacteria from those in domain Eukarya is the presence of
   a. a cell wall.   b. DNA.   c. flagella.   d. membranous organelles.

2. Which organelles are associated with the job of cellular digestion?
   a. Lysosomes and peroxisomes
   b. Golgi apparatus and vesicles
   c. Nucleus and nucleolus
   d. Smooth and rough endoplasmic reticulum

3. Within a single cell, which of the following is physically the smallest?
   a. Nuclear envelope   c. Phospholipid molecule
   b. Cell membrane   d. Mitochondrion

4. A human nerve cell that has an abnormal shape most likely has a defective
   a. cell wall.   b. cytoskeleton.   c. nucleus.   d. ribosome.

5. What type of cellular junction prevents stomach acid from leaking into the abdomen and digesting internal organs?
   a. Plasmodesmata   c. Tight junctions
   b. Anchoring junctions   d. Gap junctions

Answers to Multiple Choice questions are in appendix A.

WRITE IT OUT

1. How did microscopes contribute to the formulation of the cell theory?

2. List the features that all cells share, then name three structures or activities found in eukaryotic cells but not in bacteria or archaea.

3. If a eukaryotic cell is like a house, how is a prokaryotic cell like a one-room apartment?

4. Suppose you find a sample of cells at a crime scene. What criteria might you use to determine if the cells are from prokaryotes, plants, or animals?

5. Rank the following in order from smallest to largest: ant, prokaryotic cell, actin molecule, microtubule, nitrogen atom. What type of microscope (if any) would you need if you wanted to see each?

6. Which has a greater ratio of surface area to volume, a hippopotamus or a mouse? Which animal would lose heat faster in a cold environment and why?

7. List the chemicals that make up cell membranes.

8. Compare and contrast the phospholipid bilayer with two pieces of Velcro sticking to each other.

9. Imagine that you could engineer a cell that exchanges gases efficiently with the environment and quickly metabolizes sugars. Describe your cell's size and shape. What organelles would be abundant?

10. One way to understand cell function is to compare the parts of a cell to the parts of a factory. For example, the Golgi apparatus would be analogous to the factory's shipping department. How would the other cell parts fit into this analogy?

11. Choose an organelle in a human cell, and imagine that a disease causes that organelle to be faulty. How would the malfunctioning organelle affect the cell's function?

12. Why does a muscle cell contain many mitochondria? Why does a white blood cell contain many lysosomes?

13. List the components and functions of the cytoskeleton.

14. Describe how animal cells use junctions in different ways.

15. Chapter 1 explains emergent properties and describes the characteristics of life. Use this information to explain why life is an emergent property that appears at the level of the cell.

SCIENTIFIC LITERACY

Why We Care 3.1 explains that your body is home to trillions of bacteria, many them in the large intestine. Given this context, consider the effect of taking a probiotic capsule containing 500 million beneficial bacteria. Under what circumstances might the bacteria in the probiotic be likely to colonize the large intestine? How might you decide if you should consume a probiotic?

PULL IT TOGETHER

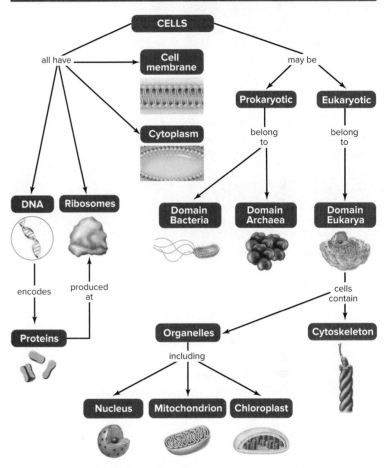

Figure 3.27 Pull It Together: Cells.

Refer to figure 3.27 and the chapter content to answer the following questions:

1. Review the Survey the Landscape figure in the chapter introduction, and then add *molecules, atoms, carbohydrates,* and *enzymes* to the Pull It Together concept map.

2. How might you connect the terms *proteins* and *cytoskeleton*?

3. Write the distinguishing features of domain Bacteria and domain Archaea on the concept map.

4. Add the main groups within domain Eukarya to this concept map.

5. Add *cytosol, chloroplast, lysosome,* and *vacuole* to this concept map.

Answers to Mastering Concepts, Write It Out, Scientific Literacy, and Pull It Together questions can be found in the Connect ebook.
connect.mheducation.com

4
The Energy of Life

Burning Calories. Riding a bicycle takes energy, which comes from metabolic reactions inside cells.

©maxpro/Shutterstock RF

LEARNING OUTLINE

4.1 All Cells Capture and Use Energy

4.2 Networks of Chemical Reactions Sustain Life

4.3 ATP Is Cellular Energy Currency

4.4 Enzymes Speed Reactions

4.5 Membrane Transport May Release Energy or Cost Energy

APPLICATIONS

Why We Care 4.1 *Enzymes Are Everywhere*
Burning Question 4.1 *Do hand sanitizers work?*
Investigating Life 4.1 *Does Natural Selection Maintain Cystic Fibrosis?*

Learn How to Learn
Focus on Understanding, Not Memorizing

When you are learning the language of biology, be sure to concentrate on how each new term fits with the others. Are you studying multiple components of a complex system? Different steps in a process? The levels of a hierarchy? As you study, always make sure you understand how each part relates to the whole. For example, you might jot down brief summaries in the margins of your notes, or you could use lists of boldfaced terms in a chapter to make your own concept map.

SURVEY THE LANDSCAPE
Science, Chemistry, and Cells

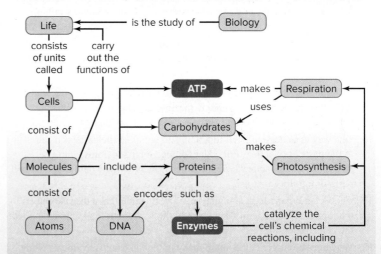

All cells require energy in the form of ATP to carry out their chemical reactions, acquire resources, and power their other activities. Enzymes are proteins that speed these reactions.

For more details, study the Pull It Together feature in the chapter summary.

©Comstock Images RF

"I wish I had your metabolism!" Perhaps you have overheard a calorie-counting friend make a similar comment to someone who stays slim on a diet of fattening foods.

In that context, the word *metabolism* means how fast a person burns food. But biochemists define metabolism as all of the chemical reactions that build and break down molecules within any cell. How are these two meanings related?

Interlocking networks of metabolic reactions supply the energy that every cell needs to stay alive. In humans, teams of metabolizing cells perform specialized functions such as digestion, muscle movement, hormone production, and countless other activities. It all takes a reliable energy supply—food, which each of us "burns" at a different rate.

This chapter describes the fundamentals of metabolism, including how cells organize, regulate, and fuel the chemical reactions that sustain life.

4.1 All Cells Capture and Use Energy

You're running late. You overslept, you have no time for breakfast, and you have a full morning of classes. You rummage through your cupboard and find something called an "energy bar"—just what you need to get through the morning. But what is energy?

A. Energy Allows Cells to Do Life's Work

Physicists define **energy** as the ability to do work—that is, to move matter. This idea, as abstract as it sounds, is fundamental to biology. Life depends on rearranging atoms and trafficking substances across membranes in precise ways. These intricate movements represent work, and they require energy.

Although it may seem strange to think of a "working" cell, all organisms do tremendous amounts of work on a microscopic scale. For example, a plant cell assembles glucose molecules into long cellulose fibers, moves ions across its membranes, and performs thousands of other tasks simultaneously. A gazelle grazes on a plant's tissues to acquire energy that will enable it to do its own cellular work. A crocodile eats that gazelle for the same reason.

The total amount of energy in any object is the sum of energy's two forms: potential and kinetic (figure 4.1). **Potential energy** is stored energy available to do work. A bicyclist at the top of a hill illustrates potential energy, as does a compressed spring. The covalent bonds of molecules, such as the ingredients in your energy bar, contain a form of potential energy called chemical energy. A concentration gradient is another form of potential energy (see section 4.5). ⓘ *covalent bonds,* section 2.2C

Kinetic energy is the energy of motion; any moving object possesses this form of energy. The bicyclist coasting down the hill in figure 4.1 demonstrates kinetic energy, as do molecules inside a cell. In fact, all of the chemical reactions that sustain life rely on collisions between moving molecules. The colder an object feels, the slower the movement of its atoms and molecules; this is why many cells die if conditions are too chilly.

Calories are units used to measure energy. One **calorie** (cal) is the amount of energy required to raise the temperature of 1 gram of water from 14.5°C to 15.5°C. The energy content of food, however, is usually measured in **kilocalories**

Figure 4.1 Potential and Kinetic Energy. (a) A bicyclist at the top of a hill has potential energy, which can be converted to kinetic energy as the cyclist coasts down. (b) A compressed spring has potential energy, which is released as kinetic energy when the spring is released. (c) Chemical energy is a form of potential energy, which is released when the molecule's bonds break.

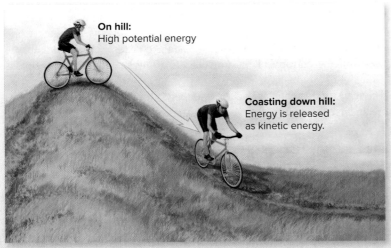

On hill:
High potential energy

Coasting down hill:
Energy is released as kinetic energy.

a.

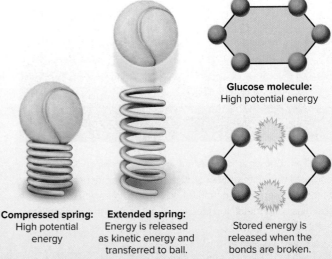

Glucose molecule:
High potential energy

Compressed spring:
High potential energy

Extended spring:
Energy is released as kinetic energy and transferred to ball.

Stored energy is released when the bonds are broken.

b.   c.

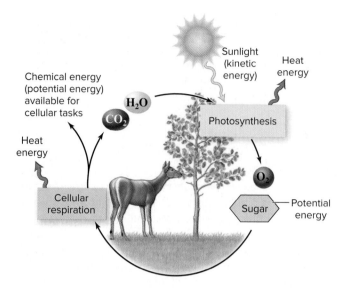

Figure 4.2 Energy Conversions. In photosynthesis, plants transform the kinetic energy in sunlight into potential energy stored in the chemical bonds of sugars and other organic molecules. Respiration, in turn, releases this potential energy. Heat energy is lost to the environment at every step along the way.

(kcal), each of which equals 1000 calories. (In nutrition, one food **Calorie**—with a capital C—is actually a kilocalorie.) A typical energy bar, for example, stores 240 kcal of potential energy in the chemical bonds of its carbohydrates, proteins, and fats. ⓘ *organic molecules,* section 2.5

B. Life Requires Energy Transformations

Physical laws describe the energy conversions vital for life, as well as those that occur in the nonliving world. They apply to all energy transformations—gasoline combustion in a car's engine, a burning chunk of wood, or a cell breaking down glucose. Two of these physical laws are especially relevant to the study of biology.

The first, called the law of energy conservation, states that energy cannot be created or destroyed, although it can be converted to other forms. This means that the total amount of energy in the universe does not change.

Living cells constantly convert energy from one form to another (figure 4.2). The most important energy transformations are photosynthesis and cellular respiration. In photosynthesis, plants and some microbes use carbon dioxide, water, and the kinetic energy in sunlight to produce sugars that are assembled into glucose and other carbohydrates. These molecules contain potential energy in their chemical bonds. During cellular respiration, the energy-rich glucose molecules change back to carbon dioxide and water, liberating the energy necessary to power life. Cells translate some of the potential energy in glucose into the kinetic energy of molecular motion and use that kinetic energy to do work.

A second physical law states that all energy transformations are inefficient because every reaction loses some energy to the surroundings as heat (see figure 4.2). If you eat your energy bar on the way to your first class, your cells can use the potential energy in its chemical bonds to make proteins, divide, or do other forms of work. But you will lose some energy as heat with every chemical reaction. This process is irreversible; cells cannot use energy that has been converted to heat.

Heat energy is disordered because it results from random molecular movements. Because heat is disordered and all energy eventually becomes heat, it follows that all energy transformations must head toward increasing disorder. **Entropy** is a measure of this randomness. In general, the more disordered a system is, the higher its entropy (figure 4.3).

Because organisms are highly organized, they may seem to defy the principle that entropy always increases. But organisms are not isolated from their surroundings. Instead, a constant stream of incoming energy and matter allows organisms to maintain their organization and stay alive, using the information in DNA. In other words, organisms can increase in complexity *as long as something else decreases in complexity by a greater amount.* Ultimately, life remains ordered and complex because the sun is constantly supplying energy to Earth. But the entropy of the universe as a whole, including the sun, is increasing.

The ideas in this chapter and the two that follow describe how organisms acquire and use the energy they need to sustain life.

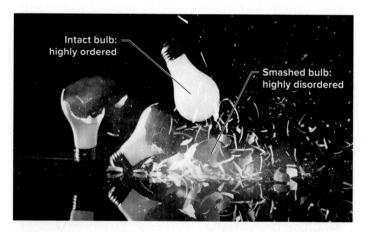

Figure 4.3 Entropy. In an instant, a highly organized lightbulb is transformed into broken glass and metal fragments. Entropy has irreversibly increased; no matter how many times you drop the smashed glass and metal, the pieces will not reorganize themselves into a lightbulb.

©Ryan McVay/Getty Images

4.1 Mastering Concepts

1. Describe how your body has potential and kinetic energy.

2. What are some energy conversions that occur in cells?

4.2 Networks of Chemical Reactions Sustain Life

The number of chemical reactions occurring in even the simplest cell is staggering. Thousands of reactants and products form interlocking pathways that resemble complicated road maps.

The word **metabolism** encompasses all of these chemical reactions in cells, including those that build new molecules and those that break down existing ones. Each reaction rearranges atoms into new compounds, and each reaction either absorbs or releases energy. Digesting your morning energy bar and using its carbohydrates to fuel muscle movement are part of your metabolism. Photosynthesis and respiration are part of the metabolism of the grass under your feet as you hurry to class.

A. Chemical Reactions Absorb or Release Energy

Biologists group metabolic reactions into two categories based on energy requirements: those that require energy input to proceed, and those that release energy (figure 4.4).

If a reaction requires an input of energy, the products contain more energy than the reactants. Reactions that build complex molecules from simpler components therefore typically require energy input. One example is photosynthesis. Glucose ($C_6H_{12}O_6$), the product of photosynthesis, contains more potential energy than carbon dioxide (CO_2) and water (H_2O), the reactants. The energy source that powers this reaction is sunlight.

In contrast, if a reaction releases energy, the products contain less energy than the reactants. Such reactions break large, complex molecules into their smaller, simpler components. Cellular respiration, the breakdown of glucose to carbon dioxide and water, is an example. The products, carbon dioxide and water, contain less energy than glucose.

What happens to the released energy? As we saw in section 4.1, some is lost to the environment as heat. But some of the energy can be used to do work. For example, the cell may use the energy to form bonds or to power reactions that require energy input. As we shall see, life's biochemistry is full of reactions that proceed only at the expense of energy released in other reactions.

B. Linked Oxidation and Reduction Reactions Form Electron Transport Chains

Electrons can carry energy. Most energy transformations in organisms occur in **oxidation–reduction ("redox") reactions,** which transfer energized electrons from one molecule to another. ⓘ *electrons,* section 2.1B

A redox reaction is similar to a person presenting a gift to a friend (figure 4.5). **Oxidation** means the loss of electrons—and a corresponding loss of energy— from a molecule, an atom, or an ion. In figure 4.5, the electron donor molecule being oxidized is analogous to the gift-giver. Conversely, **reduction** means a gain of electrons (and their energy); the electron acceptor being reduced is analogous to the woman receiving the package.

Oxidations and reductions occur simultaneously because electrons removed from one molecule during oxidation must join another molecule and

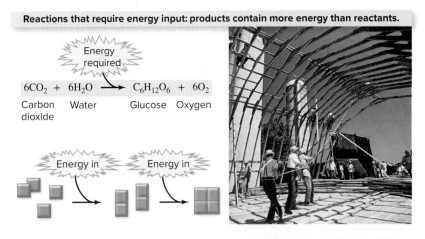

Reactions that require energy input: products contain more energy than reactants.

Energy required

$6CO_2$ + $6H_2O$ → $C_6H_{12}O_6$ + $6O_2$

Carbon dioxide · Water · Glucose · Oxygen

Energy in   Energy in

Reactions that release energy: reactants contain more energy than products.

Energy released

$6O_2$ + $C_6H_{12}O_6$ → $6CO_2$ + $6H_2O$

Oxygen · Glucose · Carbon dioxide · Water

Energy out   Energy out

Figure 4.4 Energy Required or Released. Some reactions, such as those that build complex molecules from small components, require an input of energy. This input is analogous to the energy used to build a barn out of boards and nails. Other reactions release energy, as when complex molecules (or old buildings) are dismantled.

Photos: (barn): ©Blair Seitz/Science Source; (demolition): ©ImageSource/Corbis RF

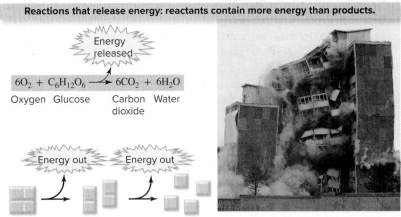

Electrons (●) are transferred from donor to acceptor.

Electron donor molecule

Oxidation

Oxidized molecule

Electron acceptor molecule

Reduction

Reduced molecule

Figure 4.5 Redox Reaction. An electron donor molecule loses electrons and is therefore being oxidized. The molecule that accepts the electrons is being reduced.

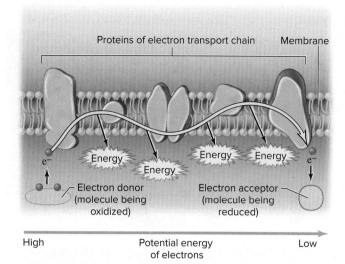

Figure 4.6 Electron Transport Chain. An electron donor molecule transfers an electron to the first protein in an electron transport chain. When the recipient passes the electron to its neighbor, energy is released. The electron continues along the chain, releasing energy at each step, until it reaches a final electron acceptor.

reduce it. That is, if one molecule is reduced (gains electrons), then another must be oxidized (loses electrons).

Groups of proteins that are electron-shuttling "specialists" often align in membranes. In an **electron transport chain,** each protein accepts an electron from the molecule before it and passes it to the next (figure 4.6), like a basketball team passing a ball from one player to another. As a result, each protein in the chain is first reduced and then oxidized. Small amounts of energy are released at each step, and the cell uses this energy in other reactions. As you will see in chapters 5 and 6, both photosynthesis and respiration rely on electron transport chains to harvest energy.

4.2 Mastering Concepts

1. What is metabolism on a cellular level?

2. Which reactions require energy input and which release energy?

3. Review figure 4.6. As electrons pass from the first to the second protein in the electron transport chain, which protein is oxidized and which is reduced? Explain your answer.

4.3 ATP Is Cellular Energy Currency

All cells contain a maze of interlocking chemical reactions, some releasing energy and others absorbing it. For example, digesting a snack releases energy. The covalent bonds of **ATP (adenosine triphosphate)** temporarily store much of the released energy. Cells then use the energy in ATP to power reactions that require energy input. In this way, cells indirectly use food energy to fuel muscle contractions and all other energy-requiring processes.

In eukaryotic cells, organelles called mitochondria produce most of a cell's ATP. As you will see in chapter 6, a mitochondrion uses the potential energy in the bonds of one glucose molecule to generate dozens of ATP molecules in cellular respiration. Not surprisingly, the most energy-hungry cells, such as those in the muscles and brain, also contain the most mitochondria.

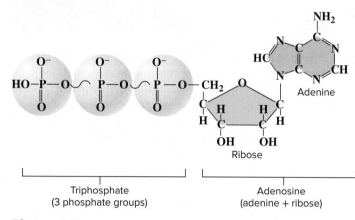

Figure 4.7 ATP's Chemical Structure. ATP is a nucleotide consisting of adenine, ribose, and three phosphate groups.

A. Energy in ATP Is Critical to the Life of a Cell

ATP is a type of nucleotide (figure 4.7). Its components are the nitrogen-containing base adenine, the five-carbon sugar ribose, and three phosphate groups (PO_4). The negative charges on neighboring phosphate groups repel one another, making the molecule unstable. It therefore releases energy when the covalent bonds between the phosphates break. ⓘ *nucleotide*s, section 2.5D

All cells depend on the potential energy in ATP to power their activities. When a cell requires energy for a chemical reaction, it "spends" ATP by removing the endmost phosphate group (figure 4.8). The products of this hydrolysis reaction are adenosine *di*phosphate (ADP, in which only two phosphate groups remain attached to ribose), the liberated phosphate group, and a burst of energy:

$$ATP + H_2O \longrightarrow ADP + \textcircled{P} + energy$$

In the reverse situation, energy can be temporarily stored by adding a phosphate to ADP, forming ATP and water:

$$ADP + \textcircled{P} + energy \longrightarrow ATP + H_2O$$

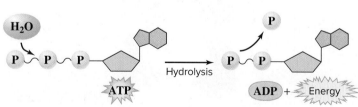

Figure 4.8 ATP Hydrolysis. Removing the endmost phosphate group of ATP yields ADP and a free phosphate group. The cell uses the released energy to do work.

The energy for this reaction comes from molecules that are broken down in other reactions, such as those in cellular respiration.

These reactions are fundamental to biology because ATP is the "go-between" that links reactions that require energy input with those that release energy. Coupled reactions, as their name implies, are simultaneous reactions in which one provides the energy that drives the other (figure 4.9). ATP hydrolysis is coupled to the reactions that require energy input, such as those that do work or synthesize new molecules. ① *hydrolysis*, section 2.5A

How does this coupling work? A cell uses ATP as an energy source by transferring its phosphate group to another molecule. This transfer may have either of two effects (figure 4.10). In one scenario, the presence of the phosphate may energize the target molecule, making it more likely to bond with other molecules. The other possible consequence is a change in the shape of the target molecule. For example, adding phosphate can force a protein to take a different shape; removing phosphate returns the protein to its original form. Changing the shape back and forth may allow the protein to shuttle substances across a membrane. ATP hydrolysis provides the energy.

ATP is sometimes described as energy "currency." Just as you can use money to purchase a variety of products, all cells use ATP in many chemical reactions to do different kinds of work. Besides transporting substances across cell membranes, other examples of jobs that require ATP include muscle contraction, moving chromosomes during cell division, and synthesizing the large molecules that make up cells.

ATP is also analogous to a fully charged rechargeable battery. A full battery represents a versatile source of potential energy that can provide power to many types of electronic devices. Although a dead battery is no longer useful as an energy source, you can recharge a spent battery to restore its utility. Likewise, the cell can use respiration to rebuild its pool of ATP.

B. ATP Represents Short-Term Energy Storage

Organisms require huge amounts of ATP. A typical human cell uses the equivalent of 2 billion ATP molecules a minute just to stay alive. Organisms recycle ATP at a furious pace, adding phosphate groups to ADP to reconstitute ATP, using the ATP to drive reactions, and turning over the entire supply every minute or so. If you ran out of ATP, you would die instantly.

Even though ATP is essential to life, cells do not stockpile it in large quantities. ATP's high-energy phosphate bonds make the molecule too unstable for long-term storage. Instead, cells store energy-rich molecules such as fats, starch, and glycogen. When ATP supplies run low, cells divert some of their lipid and carbohydrate reserves to the metabolic pathways of cellular respiration. This process soon produces additional ATP.

4.3 Mastering Concepts

1. How does ATP hydrolysis supply energy for cellular functions?

2. Describe the relationship between energy-requiring reactions, ATP hydrolysis, and cellular respiration.

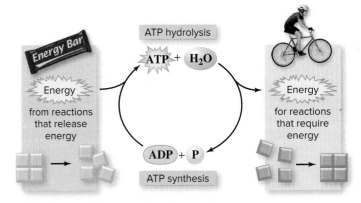

Figure 4.9 Coupled Reactions. Cells use ATP hydrolysis, a reaction that releases energy, to fuel reactions that require energy input. The cell regenerates ATP in other reactions, such as cellular respiration.

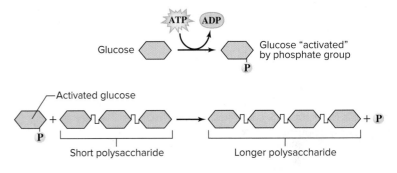

a. ATP energizes a target molecule, making it more likely to bond with other molecules.

E.g., ATP provides the energy to build large molecules out of small subunits.

ATP donates P group to glucose. Activated glucose then reacts with a short polysaccharide to build a longer polysaccharide.

b. ATP donates a phosphate group that changes the shape of the target molecule.

E.g., Phosphate group changes the shape of a membrane transport protein.

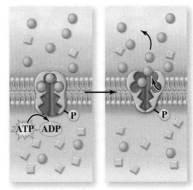

ATP donates P group to protein, allowing ions to move across the membrane.

Figure 4.10 ATP Use. When ATP donates a phosphate group to a molecule, the recipient may (a) be more likely to bond or (b) change its shape in a useful way.

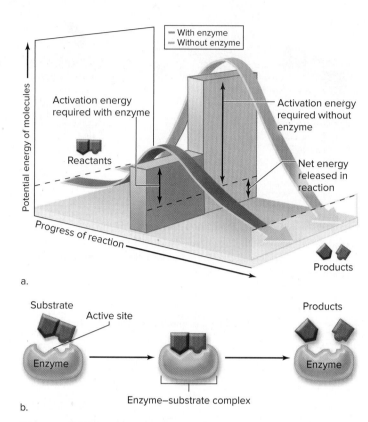

Figure 4.11 **How Enzymes Work.** (a) Enzymes lower the amount of energy required to start a reaction. The "walls" in this figure represent the activation energy for the same reaction, with and without an enzyme. (b) An enzyme's active site has a specific shape that binds to one or more substrates. After the reaction, the enzyme releases the products.

Why We Care 4.1 | Enzymes Are Everywhere

©Stockdisc/PunchStock RF

Enzymes are so critical to life that just one missing enzyme can have dramatic effects. Lactose intolerance is one example. People whose intestinal cells do not secrete an enzyme called lactase cannot digest milk sugar (lactose).

Phenylketonuria (PKU) is a much more serious disease. A PKU sufferer lacks an enzyme required to break down an amino acid called phenylalanine. When this amino acid accumulates in the bloodstream, it causes brain damage. People with PKU must avoid foods containing phenylalanine, including the artificial sweetener aspartame (NutraSweet).

Enzymes also have applications in cooking. Raw pineapple contains an enzyme that breaks down protein, which explains why you cannot put this fruit in gelatin. The enzymes will destroy the gelatin, which will not solidify. Some meat tenderizers contain the same enzyme, which breaks down muscle tissue and makes the meat easier to chew.

4.4 Enzymes Speed Reactions

Enzymes are among the most important of all biological molecules. An **enzyme** is an organic molecule that catalyzes (speeds up) a chemical reaction without being consumed. Most enzymes are proteins, although some are made of RNA.

Many of the cell's organelles, including mitochondria, chloroplasts, lysosomes, and peroxisomes, are specialized sacs of enzymes. Enzymes copy DNA, build proteins, digest food, recycle a cell's worn-out parts, and catalyze oxidation–reduction reactions, just to name a few of their jobs (see Why We Care 4.1). Without enzymes, all of these reactions would proceed far too slowly to support life. ⓘ *eukaryotic organelles,* section 3.4

A. Enzymes Bring Reactants Together

Enzymes speed reactions by lowering the **activation energy,** the amount of energy required to start a reaction (figure 4.11a). Even reactions that ultimately release energy require an initial "kick" to get started. The enzyme brings reactants (also called substrates) into contact with one another, so that less energy is required for the reaction to proceed. Just as it is easier to climb a small hill than a tall mountain, reactions occur more rapidly if the activation energy is low. Enzyme-catalyzed reactions therefore occur much faster—millions to billions of times faster—than they do in the absence of an enzyme.

Most enzymes can catalyze only one or a few chemical reactions. An enzyme that dismantles a fatty acid, for example, cannot break down the starch in your energy bar. The key to this specificity lies in the shape of the enzyme's **active site,** the region to which the substrates bind (figure 4.11b). The substrates fit like puzzle pieces into the active site. Once the reaction occurs, the enzyme releases the products. Note that the reaction does not consume or alter the enzyme. Instead, after the protein releases the products, its active site is empty and ready to pick up more substrate.

B. Many Factors Affect Enzyme Activity

The intricate network of metabolic pathways may seem chaotic, but in reality it is just the opposite. Cells precisely control the rates of their chemical reactions. If they did not, some vital compounds would always be in short supply, and others might accumulate to wasteful (or even toxic) levels.

One way to regulate a metabolic pathway is by **negative feedback** (also called feedback inhibition), in which a change triggers an action that reverses the change. For example, as a reaction's products accumulate, they inhibit the enzyme catalyzing the reaction; the reaction rate then slows or stops. But when the concentration of the reaction products falls, the block on the enzyme lifts, and the cell can once again carry out the reaction.

Negative feedback works in two general ways to prevent too much of a substance from accumulating (figure 4.12). In **noncompetitive inhibition,** product molecules bind to the enzyme at a location other than the active site in a way that alters the enzyme's shape so that it can no longer bind the substrate. Alternatively, in **competitive inhibition,** the product of a reaction binds to the enzyme's active site, preventing it from binding substrate. It is "competitive" because the product competes with the substrate to occupy the active site.

Enzymes are also very sensitive to conditions in the cell. If the pH or the salt concentration is too high or too low, an enzyme can become denatured and stop working. Temperature is also important. Enzyme action generally speeds up as the temperature climbs because reactants have more kinetic energy at

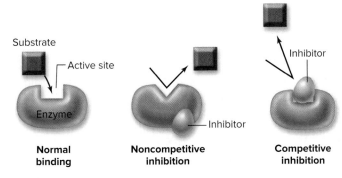

Figure 4.12 **Enzyme Inhibitors.** In noncompetitive inhibition, a substance binds to an enzyme at a location other than the active site, changing the enzyme's shape. A competitive inhibitor physically blocks an enzyme's active site.

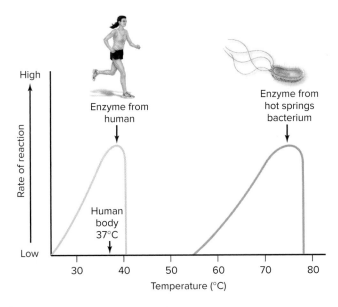

Figure 4.13 **Temperature Matters.** These graphs show how temperature affects the activity of enzymes from a human (*left*) and a bacterium that lives in hot springs (*right*). The microbes have heat-tolerant enzymes that function only at very high temperatures.

higher temperatures. Figure 4.13 illustrates this effect for two enzymes, one from humans and one from a bacterium that lives in very hot water. Note that the human enzyme works fastest at body temperature (37°C), but the bacterium's optimum temperature is much higher. If either enzyme gets too hot, however, it rapidly denatures and can no longer function; that is why reaction rates plummet above the temperature optimum. Hand sanitizers also owe their germ-killing effect partly to denatured enzymes (see Burning Question 4.1). ⓘ *denatured proteins*, section 2.5C

Pharmaceutical drugs can also inhibit enzyme function. Aspirin relieves pain by binding to an enzyme that cells use to produce pain-related molecules called prostaglandins. Likewise, some poisons are also enzyme inhibitors. For example, the active ingredient in the herbicide Roundup competitively inhibits an enzyme found in plant cells but not in animals.

4.4 Mastering Concepts

1. What do enzymes do in cells?
2. How does an enzyme lower a reaction's activation energy?
3. What is the role of negative feedback in enzyme regulation?
4. List three conditions that influence enzyme activity.

4.5 Membrane Transport May Release Energy or Cost Energy

The membrane surrounding each cell or organelle is a busy place. Like a well-used border crossing between two countries, raw materials enter and wastes exit in a continuous flow of traffic. How do membranes regulate this activity?

A biological membrane is a phospholipid bilayer studded with proteins (see section 3.3). This arrangement means that a membrane has **selective permeability.** That is, some substances pass freely through the bilayer, but others—such as the sugar from a digested energy bar—require help from proteins.

Thanks to the regulation of membrane transport, the interior of a cell is chemically different from the outside. Concentrations of some dissolved substances (solutes) are higher inside the cell than outside, and others are lower. Likewise, the inside of each organelle in a eukaryotic cell may be chemically quite different from the solution in the rest of the cell.

The term *gradient* describes any such difference between two neighboring regions. In a **concentration gradient,** a solute is more concentrated in one region

Burning Question 4.1

Do hand sanitizers work?

Bottles of alcohol-based hand sanitizers are everywhere: in handbags, in medical offices, at schools, and in shopping centers. They promise to kill bacteria and viruses, reducing the spread of disease. But do they really work?

Under ideal conditions, the answer is yes. For the sanitizer to be effective, it must contain at least 60% alcohol; check the label. And because alcohol evaporates so quickly, it is important to use enough of the product to kill germs. A dollop about the size of a dime is usually sufficient for the sanitizer to last through 30 seconds of constant hand-rubbing. Finally, the hands must be reasonably clean to begin with.

Alcohol-based sanitizers kill bacteria by disrupting cell membranes and by denaturing enzymes and other proteins. Influenza and other viruses with outer membrane "envelopes" are also vulnerable. But if your own cells have membranes and proteins too, why don't hand sanitizers damage the skin? The explanation is that the cells of your skin's outer layers cannot be killed—they are already dead.

**Submit your burning question to
marielle.hoefnagels@mheducation.com**

(sanitizer): ©McGraw-Hill Education/Richard Hutchings, photographer

TABLE 4.1   Movement Across Membranes: A Summary

Mechanism	Characteristics
Passive transport	Net movement is down concentration gradient; does not require energy input.
Simple diffusion	Substance moves across membrane without assistance from transport proteins.
	Area of low concentration ← Area of high concentration
Osmosis	Water diffuses across a selectively permeable membrane.
Facilitated diffusion	Substance moves across membrane with assistance from transport proteins.
Active transport	Net movement is against concentration gradient; requires transport protein and energy input, often from ATP.
	ATP → ADP + P
Transport using vesicles	Vesicle carries large particles into or out of a cell; requires energy input.
Endocytosis	Membrane engulfs incoming substance, enclosing it in a vesicle.
Exocytosis	Vesicle fuses with cell membrane, releasing substances outside of cell.

than in a neighboring region. For example, the images in table 4.1 all illustrate concentration gradients in which the solution on the right side of the membrane has a higher solute concentration than the solution on the left.

If a substance moves from an area where it is more concentrated to an area where it is less concentrated, it is said to be "moving down" or "following" its concentration gradient. As the solute moves, the gradient dissipates—that is, it disappears. Any concentration gradient will eventually dissipate *unless energy is expended to maintain it*. Why? Random molecular motion always increases the amount of disorder (entropy), and it costs energy to counter this tendency toward disorder. By the same token, however, an existing concentration gradient represents a form of stored potential energy. Cells therefore spend ATP to create some types of concentration differences, which they can later "cash in" to do work (see section 4.5B).

A. Passive Transport Does Not Require Energy Input

In **passive transport,** a substance moves across a membrane without the direct expenditure of energy. All forms of passive transport involve **diffusion,** the spontaneous movement of a substance from a region where it is more concentrated to a region where it is less concentrated. Because diffusion represents the dissipation of a chemical gradient—and the loss of potential energy—it does not require energy input.

For a familiar example of diffusion, picture what happens when you first place a tea bag in a cup of hot water: Near the tea bag, there are many more brown tea molecules than elsewhere in the cup (figure 4.14). Over time, however, the brownish color spreads to create a uniform brew.

Figure 4.14 Diffusion in a Cup.
The solute particles leaving a tea bag can move in any direction, with only a few paths leading back to the source. Eventually, the solutes are distributed uniformly throughout the cup.

Solvent
Solute

How do the tea molecules "know" which way to diffuse? The answer is, of course, that atoms and molecules know nothing. Diffusion occurs because all substances have kinetic energy; that is, they are in constant, random motion. To simplify the tea example, suppose each molecule can move randomly along one of 10 possible paths (in reality, the number of possible directions is infinite). Assume further that only one path leads back to the tea bag. Since 9 of the 10 possibilities point away from the tea bag, the tea molecules tend to spread out; that is, they move down their concentration gradient.

If diffusion continues long enough, the gradient disappears. Diffusion *appears* to stop at that point, but the molecules do not stop moving. Instead, they continue to travel randomly back and forth at the same rate, so at equilibrium the concentration remains equal throughout the solution.

Simple Diffusion: No Proteins Required

In a form of passive transport called **simple diffusion,** a substance moves down its concentration gradient without the use of a transport protein (see table 4.1). Substances may enter or leave cells by simple diffusion only if they can pass freely through the membrane. Lipids and small, nonpolar molecules such as oxygen (O_2) and carbon dioxide (CO_2), for example, diffuse easily across a biological membrane (see figure 3.9).

If gradients dissipate without energy input, how can a cell use simple diffusion to acquire essential substances or get rid of toxic wastes? The answer is that the cell maintains the gradients, either by continually consuming the substances as they diffuse in or by producing more of the substances that diffuse out. For example, mitochondria consume O_2 as soon as it diffuses into the cell, maintaining the O_2 gradient that drives diffusion. Respiration also produces CO_2, which diffuses out because its concentration always remains higher in the cell than outside.

Osmosis: Diffusion of Water Across a Selectively Permeable Membrane

Two solutions of different concentrations may be separated by a selectively permeable membrane through which water, but not solutes, can pass. In that case, water will diffuse down its own gradient toward the side with the high solute concentration. **Osmosis** is this simple diffusion of water across a selectively permeable membrane (see table 4.1 and figure 4.15).

A human red blood cell demonstrates the effects of osmosis (figure 4.16). The cell's interior is normally **isotonic** to the surrounding blood plasma, which

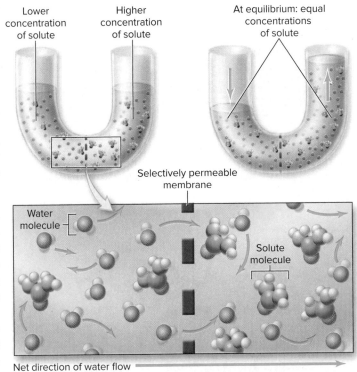

Figure 4.15 Osmosis. The selectively permeable membrane dividing this U-shaped tube permits water but not solutes to pass through. Water diffuses from the left side (low solute concentration) toward the right side (high solute concentration). At equilibrium, water flow is equal in both directions, and the solute concentrations will be equal on both sides of the membrane.

Figure It Out

A saltwater fish is accidentally placed into a tank containing fresh water. Why does the fish soon die?

Answer: Its cells absorb too much water, then burst.

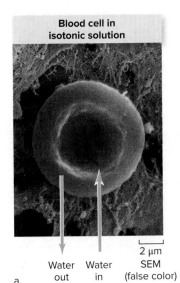

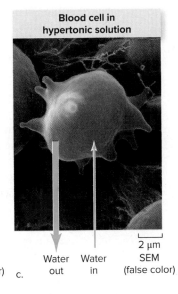

a. Water out / Water in / 2 μm SEM (false color)

b. Water out / Water in / 2 μm SEM (false color)

c. Water out / Water in / 2 μm SEM (false color)

Figure 4.16 Osmosis and Red Blood Cells. (a) A human red blood cell is isotonic to the surrounding plasma. Water enters and leaves the cell at the same rate, and the cell maintains its shape. (b) When the salt concentration of the plasma decreases, water flows into the cell faster than it leaves. The cell swells and may even burst. (c) In salty surroundings, the cell loses water and shrinks.

Photos: (a–c): ©David M. Phillips/Science Source

Membrane Transport May Release or Cost Energy

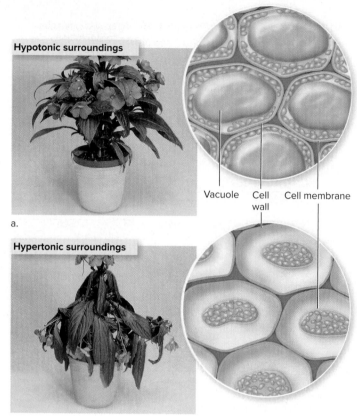

a.

b.

Figure 4.17 **Osmosis and Plant Turgor Pressure.** (a) The interior of a plant cell usually contains more concentrated solutes than its surroundings. Water enters the cell by osmosis, generating turgor pressure. (b) In a hypertonic environment, turgor pressure is low. The plant wilts.

Photos: (a, b): ©Nigel Cattlin/Science Source

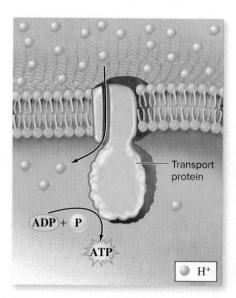

Figure 4.18 **Facilitated Diffusion.** H⁺ ions move across a membrane and down their concentration gradient through a transport protein. The protein uses potential energy in the concentration gradient to produce ATP.

means that the plasma's solute concentration is the same as the inside of the cell (*iso-* means "equal," and *tonicity* is the ability of a solution to cause water movement). Water therefore moves into and out of the cell at equal rates. In a **hypotonic** environment, the solute concentration is lower than it is inside the cell (*hypo-* means "under," as in hypodermic). Water therefore moves by osmosis into a blood cell placed into hypotonic surroundings; since animal cells lack a cell wall, the membrane may even burst. Conversely, **hypertonic** surroundings have a higher concentration of solutes than the cell's cytoplasm (*hyper-* means "over," as in hyperactive). In a hypertonic environment, a cell loses water, shrivels, and may die for lack of water.

Hypotonic and *hypertonic* are relative terms that can refer to the surrounding solution or to the solution inside the cell. The same solution might be hypertonic to one cell but hypotonic to another, depending on the solute concentrations inside the cells.

A plant's roots are often hypertonic to the soil, particularly after a heavy rain. Water rushes in, and the central vacuoles of the plant cells expand until the cell walls constrain their growth. **Turgor pressure** is the resulting force of water against the cell wall (figure 4.17). A limp, wilted piece of lettuce demonstrates the effect of lost turgor pressure. But the leaf becomes crisp again if placed in water, as individual cells expand like inflated balloons. Turgor pressure helps keep plants erect.

Facilitated Diffusion: Proteins Required Ions and polar molecules cannot freely cross the hydrophobic layer of a membrane; instead, transport proteins form channels that help these solutes cross. **Facilitated diffusion** is a form of passive transport in which a membrane protein assists the movement of a polar solute down its concentration gradient (see table 4.1). Facilitated diffusion releases energy because the solute moves from where it is more concentrated to where it is less concentrated.

Figure 4.18 shows an example of facilitated diffusion. Hydrogen ions (H⁺) are too hydrophilic to pass freely across a membrane, but a specialized protein forms a channel that allows the ions to pass through. This membrane protein is especially important because of its role in the cell: As the H⁺ gradient dissipates, the cell uses the released energy to generate ATP.

Membrane proteins can enhance osmosis, too. Although membranes are somewhat permeable to water, osmosis can be slow. The cells of many organisms, including bacteria, plants, and animals, use membrane proteins called aquaporins to increase the rate of water flow. Kidney cells control the amount of water that enters urine by changing the number of aquaporins in their membranes.

B. Active Transport Requires Energy Input

Both simple diffusion and facilitated diffusion dissipate an existing concentration gradient. Often, however, a cell needs to do the opposite: create and maintain a concentration gradient. A plant's root cell, for example, may need to absorb nutrients from soil water that is much more dilute than the cell's interior. In **active transport,** a cell uses a transport protein to move a substance *against* its concentration gradient—from where it is less concentrated to where it is more concentrated (see table 4.1). Because a gradient represents a form of potential energy, the cell must expend energy to create it; this energy often comes from ATP.

Cells must contain high concentrations of potassium (K⁺) and low concentrations of sodium (Na⁺) to perform many functions. In animals, for example, sodium and potassium ion gradients are essential for nerve and muscle function (see chapters 24 and 26). One active transport system in the

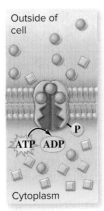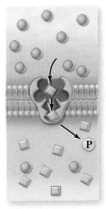

Outside of cell

○ Na⁺
□ K⁺

Cytoplasm

1 ATP binds to transport protein along with three Na⁺ from cytoplasm. ATP transfers phosphate to protein.

2 Phosphate changes the shape of the protein, moving Na⁺ across the membrane.

3 Two K⁺ from outside of cell bind to protein, causing phosphate release.

4 Release of phosphate changes the shape of the protein, moving K⁺ into the cytoplasm.

Figure 4.19 **Active Transport.** The sodium–potassium pump is a protein embedded in the cell membrane. It uses energy released in ATP hydrolysis to move sodium ions (Na⁺) out of the cell and potassium ions (K⁺) into the cell. The process costs energy because both types of ions are moving from where they are less concentrated to where they are more concentrated.

membranes of most animal cells is a protein called the **sodium–potassium pump** (figure 4.19), which uses ATP as an energy source to expel three Na⁺ for every two K⁺ it admits. Maintaining these ion gradients is costly: The million or more sodium–potassium pumps embedded in a cell's membrane use some 25% of the cell's ATP.

Concentration gradients are an important source of potential energy that cells can use to do work. For example, chapters 5 and 6 describe how cells establish concentration gradients of H⁺ during photosynthesis and respiration. A chloroplast or mitochondrion can control how and when the H⁺ gradient dissipates. As it does so, the organelle converts the potential energy stored in the gradient into another form of potential energy—that is, chemical energy in the bonds of ATP (see figure 4.18).

C. Endocytosis and Exocytosis Use Vesicles to Transport Substances

Most molecules dissolved in water are small, and they can cross cell membranes by simple diffusion, facilitated diffusion, or active transport. Large particles, however, must enter and leave cells with the help of a transport vesicle—a small sac that can pinch off of, or fuse with, a cell membrane.

In **endocytosis,** a cell membrane engulfs fluids and large molecules to bring them into the cell. When the cell membrane indents, a "bubble" of membrane closes in on itself. The resulting vesicle traps the incoming substance (figure 4.20). The formation and movement of this vesicle require energy.

The two main forms of endocytosis are pinocytosis and phagocytosis. In pinocytosis, the cell engulfs small amounts of fluids and dissolved substances. In **phagocytosis,** the cell captures and engulfs large particles, such as debris or even another cell. The vesicle then fuses with a lysosome, where hydrolytic enzymes dismantle the cargo. ⓘ *lysosomes,* section 3.4B

When biologists first viewed endocytosis in white blood cells in the 1930s, they thought a cell would gulp in anything at its surface. They now

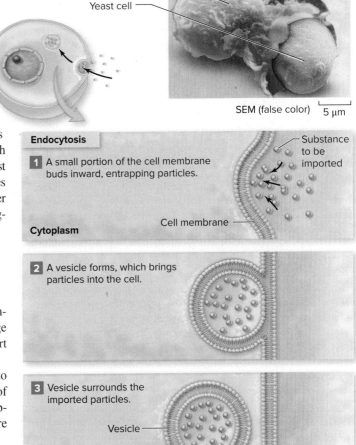

White blood cell
Yeast cell
SEM (false color) 5 μm

Endocytosis
Substance to be imported

1 A small portion of the cell membrane buds inward, entrapping particles.

Cell membrane
Cytoplasm

2 A vesicle forms, which brings particles into the cell.

3 Vesicle surrounds the imported particles.

Vesicle

Figure 4.20 **Endocytosis.** Large particles enter a cell by endocytosis. The inset (*top right*) shows a white blood cell engulfing a yeast cell by phagocytosis, a form of endocytosis.

Photo: ©Biology Media/Science Source

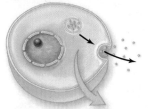

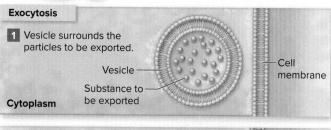

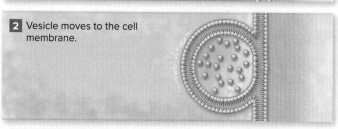

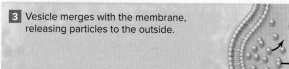

Figure 4.21 **Exocytosis.** Cells package substances to be secreted into vesicles, which fuse with the cell membrane to release the materials.

Exocytosis

1 Vesicle surrounds the particles to be exported.

Vesicle

Substance to be exported

Cell membrane

Cytoplasm

2 Vesicle moves to the cell membrane.

3 Vesicle merges with the membrane, releasing particles to the outside.

recognize a more selective form of the process. In receptor-mediated endocytosis, a receptor protein on a cell's surface binds a biochemical; the cell membrane then indents, drawing the substance into the cell. Liver cells use receptor-mediated endocytosis to absorb cholesterol-toting proteins from the bloodstream.

Exocytosis, the opposite of endocytosis, uses vesicles to transport fluids and large particles out of cells (figure 4.21). Inside a cell, the Golgi apparatus produces vesicles filled with substances to be secreted. The vesicle moves to the cell membrane and joins with it, releasing the substance outside the membrane. For example, the tip of a neuron releases neurotransmitters by exocytosis; these chemicals then stimulate or inhibit neural impulses in a neighboring cell. The secretion of milk into milk ducts, depicted in figure 3.11, is another example. As in endocytosis, moving the transport vesicle requires energy.

4.5 Mastering Concepts

1. What is diffusion?
2. What types of substances diffuse freely across a membrane?
3. What would happen to a plant cell in a *hypertonic* environment?
4. Why does it cost energy to maintain a concentration gradient?
5. Distinguish between simple diffusion, facilitated diffusion, and active transport.
6. How do exocytosis and endocytosis use vesicles to transport materials across cell membranes?

Investigating Life 4.1 | Does Natural Selection Maintain Cystic Fibrosis?

A single enzyme or membrane protein may seem too small to be very important—until you consider that a single faulty one can cause serious illness. Cystic fibrosis is one example. Each affected person lacks a membrane transport protein called CFTR.

The CFTR protein normally occurs in the membranes of cells that secrete watery fluids such as mucus. Its function is to move chloride ions (Cl^-) out of cells by active transport. As it does so, the solute concentration outside the cell increases, drawing water out by osmosis. CFTR therefore helps thin the mucus in the lungs and digestive tract, as well as other organs. As long as some CFTR proteins are functional, cystic fibrosis does not occur.

Patients with cystic fibrosis, however, have abnormal or missing CFTR proteins. The mucus remains thick, making breathing difficult and obstructing the movement of food and enzymes in the digestive tract. Cystic fibrosis may take patients' lives before they are old enough to reproduce.

With this background, scientists at the University of North Carolina asked why natural selection hasn't eliminated cystic fibrosis. A possible answer may lie with the CFTR proteins in the digestive tract. The bacteria that cause an intestinal disease called cholera produce a toxin that overstimulates CFTR. As a result, Cl^- and water pour from the lining of the small intestine and leave the body in watery diarrhea. The resulting dehydration can be deadly. The researchers hypothesized that the abnormal CFTR protein—the same one that causes cystic fibrosis—may actually help protect against cholera.

To test their hypothesis, the team bred three types of mice, differing only in the form of the CFTR proteins in their membranes: all abnormal, all normal, or a mix of both. All mice were then exposed to the cholera toxin. Sure enough, the fewer normal CFTR proteins in the membrane, the less fluid accumulated in the intestines (figure 4.A).

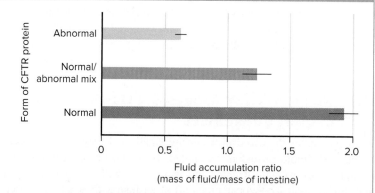

Figure 4.A **Cholera Toxin and CFTR.** Exposure to the cholera toxin causes fluid accumulation in the small intestines of mice. Mice with normal CFTR proteins lost the most fluid and were therefore the most susceptible to cholera. (Error bars represent standard errors; see appendix B.)

This study helps explain how natural selection might maintain cystic fibrosis in the human population. A person develops cystic fibrosis only if he or she has no functional CFTR proteins. Evolutionary biologists suggest that, in some areas of the world, cholera resistance may give people with some faulty CFTR proteins an advantage. From an evolutionary point of view, improved resistance to infectious disease apparently offsets losing some children to cystic fibrosis.

Source: Gabriel, Sherif E., K. N. Brigman, B. H. Koller, et al. Oct. 7, 1994. Cystic fibrosis heterozygote resistance to cholera toxin in the cystic fibrosis mouse model. *Science,* vol. 266, pages 107–109.

What's the **Point?** ▼ APPLIED

Metabolism describes all the chemical reactions in a cell. Because our cells always lose energy as heat, they require constant energy input to continue fueling their reactions. So the familiar definition of metabolism—how fast a person burns calories in food—relates to the rate at which cellular reactions are occurring. What can you do to make your cells use the energy in food more quickly?

Exercise speeds up the body's energy metabolism in several ways. Immediately after exercise, cells work to rebuild ATP and other energy reserves, so caloric demands are high. Also, body temperature remains elevated for hours after exercise, speeding chemical reactions and contributing to increased metabolism. Regular exercise also increases the size of muscle cells, which require more energy than fat cells even when at rest. Exercise also increases the abundance of enzymes and other proteins that regulate energy metabolism. For example, proteins that

©Corbis RF

transport fatty acids into cells become more numerous after one to two weeks of exercise, providing cells with easier access to energy.

Caffeine may also accelerate metabolism. Although caffeine contains zero calories, many people can attest to the "energy boost" that it provides. Caffeine increases the release of fatty acids into the blood and raises the heart rate, giving cells quick access to energy reserves. However, studies have shown that getting too little sleep (a side effect of excess caffeine) disturbs normal metabolism.

Finally, metabolism slows down when the body receives too few calories. Hormones then signal the body that it is entering a starved state. In response, cells begin to conserve energy via several mechanisms. One way to keep your metabolism high is therefore to maintain your blood sugar level by eating multiple small, healthy meals throughout the day.

CHAPTER SUMMARY

4.1 All Cells Capture and Use Energy

A. Energy Allows Cells to Do Life's Work

- **Energy** is the ability to do work. **Potential energy** is stored energy, and **kinetic energy** is action.
- Energy is measured in units called **calories.** One food **Calorie** is 1000 calories, or 1 **kilocalorie.**

B. Life Requires Energy Transformations

- Energy cannot be created or destroyed but only converted to other forms.
- Every reaction increases **entropy** (disorder) and loses heat energy to the environment.

4.2 Networks of Chemical Reactions Sustain Life

- **Metabolism** is the sum of the chemical reactions in a cell.

A. Chemical Reactions Absorb or Release Energy

- In reactions that require energy input, the products have more energy than the reactants. Conversely, in reactions that release energy, the products have less energy than the reactants.

B. Linked Oxidation and Reduction Reactions Form Electron Transport Chains

- Many energy transformations in organisms occur via **oxidation–reduction (redox) reactions. Oxidation** is the loss of electrons; **reduction** is the gain of electrons. Oxidation and reduction reactions occur simultaneously.
- In both photosynthesis and respiration, proteins shuttle electrons along **electron transport chains.**

4.3 ATP Is Cellular Energy Currency

A. Energy in ATP Is Critical to the Life of a Cell

- **ATP (adenosine triphosphate)** stores energy in its high-energy phosphate bonds. Cellular respiration generates ATP.
- Cells use the energy released in ATP hydrolysis to drive other reactions.

B. ATP Represents Short-Term Energy Storage

- ATP is too unstable for long-term storage. Instead, cells store energy as fats and carbohydrates.

4.4 Enzymes Speed Reactions

A. Enzymes Bring Reactants Together

- **Enzymes** are organic molecules (usually proteins) that speed biochemical reactions by lowering the **activation energy.**
- Substrate molecules fit into the enzyme's **active site.**

B. Many Factors Affect Enzyme Activity

- A reaction product may temporarily shut down its own synthesis whenever its levels rise. Such **negative feedback** may occur by **competitive inhibition** or **noncompetitive inhibition.**
- Enzymes have narrow ranges of conditions in which they function.

4.5 Membrane Transport May Release Energy or Cost Energy

- Membranes have **selective permeability,** which means they admit only some substances.

Miniglossary	Energy of Life
Metabolism	All of the chemical reactions inside a cell
Redox reaction	Chemical reaction involving the transfer of electrons
ATP	Molecule that transfers the energy required for many metabolic reactions
Coupled reaction	Two simultaneous reactions in which one (e.g., ATP hydrolysis) provides energy required by the other
Activation energy	Energy required to start a chemical reaction
Enzyme	Protein that speeds a chemical reaction by lowering the activation energy

- A **concentration gradient** is a difference in solute concentration between two neighboring regions, such as across a membrane. Gradients dissipate without energy input.

A. Passive Transport Does Not Require Energy Input

- All forms of **passive transport** involve **diffusion,** the dissipation of a chemical gradient by random molecular motion.
- In **simple diffusion,** a substance passes through a membrane along its concentration gradient without the aid of a transport protein.
- **Osmosis** is the simple diffusion of water across a selectively permeable membrane. Terms describing tonicity (**isotonic, hypotonic,** and **hypertonic**) predict whether cells will swell or shrink when the surroundings change. When plant cells lose too much water, the resulting loss of **turgor pressure** causes the plant to wilt.
- In **facilitated diffusion,** a membrane protein admits a substance along its concentration gradient without expending energy.

B. Active Transport Requires Energy Input

- In **active transport,** a carrier protein uses energy (ATP) to move a substance against its concentration gradient. For example, the **sodium–potassium pump** uses active transport to exchange sodium ions for potassium ions across an animal cell membrane.
- **Figure 4.22** shows how enzymes and ATP interact in a cell to generate energy for active transport and other activities that require energy input.

C. Endocytosis and Exocytosis Use Vesicles to Transport Substances

- In **endocytosis,** a cell engulfs liquids or large particles. Pinocytosis brings in fluids; **phagocytosis** brings in solid particles.
- In **exocytosis,** vesicles inside the cell carry substances to the cell membrane, where they fuse with the membrane and release the cargo to the outside of the cell.

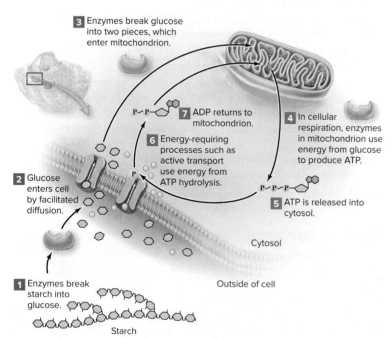

Processes in this figure that require energy input	Processes in this figure that release energy
ATP production: $ADP + P \rightarrow ATP + H_2O$ Active transport	Starch hydrolysis Facilitated diffusion Reactions of cellular respiration ATP hydrolysis: $ATP + H_2O \rightarrow ADP + P$

Figure 4.22 Enzymes, Energy, and ATP.

MULTIPLE CHOICE QUESTIONS

1. Which of the following is the best example of potential energy in a cell?
   a. Cell division
   b. A molecule of glucose
   c. Movement of a flagellum
   d. Exocytosis

2. How does ATP participate in coupled reactions?
   a. Hydrolysis of ATP fuels energy-requiring reactions.
   b. Synthesis of ATP fuels energy-requiring reactions.
   c. Hydrolysis of ADP fuels energy-releasing reactions.
   d. Synthesis of ADP fuels energy-releasing reactions.

3. How do proteins contribute to the function of an electron transport chain?
   a. They become oxidized and reduced.
   b. They undergo osmosis.
   c. They are involved in the hydrolysis of electrons.
   d. They consume electrons.

4. Where in a molecule of ATP is the stored energy that is used by the cell?
   a. Within the nitrogenous base, adenine
   b. Within the five-carbon ribose sugar
   c. In the covalent bonds between the phosphate groups
   d. In the bond between adenine and ribose

5. How does an enzyme affect the energy of a reaction?
   a. The enzyme lowers the activation energy.
   b. The enzyme lowers the net energy released.
   c. The enzyme raises the energy of the reactants.
   d. The enzyme raises the energy of the products.

6. Imagine that you place a drop of red food coloring in a glass of water. At first, the red molecules remain concentrated where you placed the drop, but the color eventually spreads throughout the water. This example illustrates
   a. hydrolysis.
   b. decreasing entropy.
   c. osmosis.
   d. diffusion.

7. A "naked egg" is a chicken egg surrounded only by a semipermeable membrane (the hard outer shell has been dissolved away during an extended soak in vinegar). Suppose you place a "naked egg" in pure water for 24 hours. What will happen to the egg?
   a. There will be no change.
   b. It will swell.
   c. It will exert turgor pressure against the cell wall.
   d. It will shrink.

8. A concentration gradient is an example of
   a. oxidation–reduction.
   b. potential energy.
   c. entropy.
   d. hydrolysis.

9. How does ATP relate to membrane transport?
   a. The movement of a substance down its concentration gradient through transport proteins requires the hydrolysis of ATP.
   b. The higher the concentration of ATP in the cell, the more permeable the membrane is to water and small, nonpolar molecules.
   c. Digestion produces a high concentration of ATP outside the cell, and the ATP enters the cell via facilitated diffusion.
   d. A cell uses the energy in ATP to transport substances against their concentration gradient.

Answers to Multiple Choice questions are in appendix A.

WRITE IT OUT

1. Some people claim that life's high degree of organization defies the physical law that says that entropy always increases. What makes this statement false?

2. List some examples of energy-requiring and energy-releasing reactions that have been introduced in previous chapters.

3. Why is ATP called the cell's "energy currency"?

4. In what ways is an enzyme's function similar to engineers digging a tunnel through a mountain rather than building a road over the peak?

5. Use what you know about enzymes to propose an explanation for why our bodies cannot digest the cellulose in dietary fiber. Also, why would a cell's fat-digesting enzymes not be able to digest an artificial fat such as Olestra (see chapter 2)?

6. Considering that enzymes are essential to all cells, including microbes, why might refrigeration and freezing help preserve food?

7. When a person eats a fatty diet, excess cholesterol accumulates in the bloodstream. Cells then temporarily stop producing cholesterol. What phenomenon described in the chapter does this control illustrate?

8. Diffusion is an efficient means of transport only over small distances. How does this relate to a cell's surface-area-to-volume ratio (chapter 3)?

9. Liver cells are packed with glucose. If the concentration of glucose in a liver cell is higher than in the surrounding fluid, what mechanism could the cell use to import even more glucose? Why would only this mode of transport work?

10. List three ways the content in this chapter relates to an organism's ability to maintain homeostasis.

11. Golden knifefish use an electric field to detect nearby objects and to communicate. Scientists know that ion channels in the membranes of the electric organ cells are largely responsible for generating the electric field. They also know that knifefish can quickly turn up or turn down the intensity of the electric field. What type of membrane transport might the knifefish use to quickly increase the number of ion channels in the cell membranes of its electric organ? What process would remove ion channels from the cell membranes?

SCIENTIFIC LITERACY

Review Burning Question 4.1, which explains how alcohol-based hand sanitizers disrupt the functions of enzymes and other biochemicals in living cells. Instead of alcohol, some hand sanitizers contain chemicals that kill germs by targeting specific structures inside bacterial cells. Why might these ingredients select for resistant bacteria, while alcohol-based ones do not? What factors might determine whether a person uses hand sanitizer rather than washing hands with soap and water (which wash away bacteria instead of killing them)? If you buy a hand sanitizer, what information should you look for before deciding which product to select?

PULL IT TOGETHER

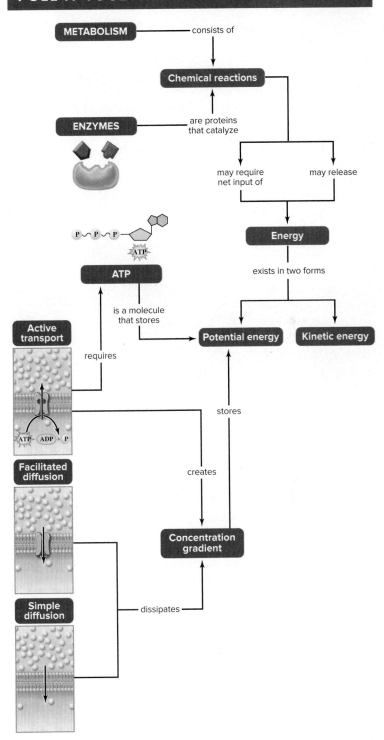

Figure 4.23 Pull It Together: The Energy of Life.

Refer to figure 4.23 and the chapter content to answer the following questions.

1. Review the Survey the Landscape figure in the chapter introduction, and then add *life, cells,* and *respiration* to the Pull It Together concept map.

2. What are some examples of potential energy and kinetic energy other than those included in the concept map?

3. Add *substrate, active site,* and *activation energy* to this concept map.

4. Where does passive transport fit in this concept map?

5

Photosynthesis

Food from Plants. A farmer walks through a rice terrace in China. Rice grains are a food staple for much of the world's population.

©Rodrigo A. Torres/Glowimages RF

LEARNING OUTLINE

5.1 Life Depends on Photosynthesis

5.2 Photosynthetic Pigments Capture Sunlight

5.3 Chloroplasts Are the Sites of Photosynthesis

5.4 Photosynthesis Occurs in Two Stages

5.5 The Light Reactions Begin Photosynthesis

5.6 The Carbon Reactions Produce Carbohydrates

5.7 C_3, C_4, and CAM Plants Use Different Carbon Fixation Pathways

APPLICATIONS

Burning Question 5.1 *Why do leaves change colors in the fall?*

Burning Question 5.2 *Does air have mass?*

Why We Care 5.1 *Weed Killers*

Investigating Life 5.1 *Salamanders Snack on Sugars from Solar Cells*

Learn How to Learn
A Quick Once-Over

Unless your instructor requires you to read your textbook in detail before class, try a quick preview. Start by reviewing the Survey the Landscape figure at the start of each chapter to see how the material fits with the rest of the unit, then read the chapter outline to identify the main ideas. It is also a good idea to look at the figures and the key terms in the narrative. Previewing a chapter should help you follow the lecture because you will already know the main ideas. In addition, note-taking will be easier if you recognize new vocabulary words from your quick once-over. Return to your book for an in-depth reading after class to help nail down the details.

SURVEY THE LANDSCAPE
Science, Chemistry, and Cells

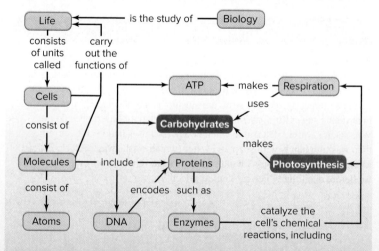

Photosynthetic cells use carbon dioxide, water, and sunlight to build carbohydrates. These molecules, which make up much of a plant's body, can also be used to store energy and generate ATP.

For more details, study the Pull It Together feature in the chapter summary.

©bo1982/E+/Getty Images RF

Most plants are easy to grow (compared with animals, anyway) because their needs are simple. Give a plant water, essential elements in soil, carbon dioxide, and light, and it will produce food and oxygen. These products build the plant's body and sustain its life. Meanwhile, animals and other consumers eat plants. A leafy foundation therefore supports Earth's ecosystems.

How can plants do so much with such simple raw materials? The answer lies in chloroplasts, microscopic solar panels inside each green cell. This chapter explains how chloroplasts use the sun's energy to conjure sugar out of thin air.

Photosynthesis

Carbon dioxide and water consumed

Glucose and oxygen produced

CO_2 + H_2O + light energy → $C_6H_{12}O_6$ + O_2

Leaf cell —

Chloroplasts —

TEM
15 μm (false color)

Figure 5.1 **Sugar from the Sun.** In photosynthesis, a plant produces carbohydrates and O_2 from simple starting materials: carbon dioxide, water, and sunlight.

Electron micrograph by Wm. P. Wergin, Courtesy Eldon H. Newcomb, University of Wisconsin-Madison

5.1 Life Depends on Photosynthesis

It is spring. A seed germinates underground, its tender roots and pale yellow stem extending rapidly in a race against time. For now, the seedling's sole energy source is food stored in the seed itself. If its shoot does not reach light before its reserves run out, the seedling will die. But if it makes it, the shoot turns green and unfurls leaves that spread and catch the light. The seedling begins to feed itself, and an independent new life begins.

The plant is an **autotroph** ("self feeder"), meaning it uses inorganic substances such as water and carbon dioxide (CO_2) to produce organic compounds. The opposite of an autotroph is a **heterotroph,** which is an organism that obtains carbon by consuming preexisting organic molecules. You are a heterotroph, and so are all other animals, all fungi, and many other microbes.

Autotrophs underlie every ecosystem on Earth. It is not surprising, therefore, that if asked to designate the most important metabolic pathway, most biologists would not hesitate to cite **photosynthesis:** the process by which plants, algae, and some bacteria harness solar energy and convert it into chemical energy.

Photosynthesis is a series of chemical reactions that use light energy to assemble CO_2 into glucose ($C_6H_{12}O_6$) and other carbohydrates (figure 5.1). The plant uses water in the process and releases oxygen gas (O_2) as a byproduct. The reactions of photosynthesis are summarized as follows:

$$6CO_2 + 6H_2O \xrightarrow{\text{light energy}} C_6H_{12}O_6 + 6O_2$$

This process provides not only food for the plant but also the energy, raw materials, and O_2 that support most heterotrophs (see figure 4.2). Animals, fungi, and other consumers eat the leaves, stems, roots, flowers, pollen, nectar, fruits, and seeds of the world's producers. Even the waste product of photosynthesis, O_2, is essential to much life on Earth.

Because humans live on land, we are most familiar with the contribution that plants make to Earth's terrestrial ecosystems. In fact, however, more than half of the world's photosynthesis occurs in the vast oceans, courtesy of countless algae and bacteria.

On land or in the water, Earth without photosynthesis would not be a living world for long. If the sky were blackened by a nuclear holocaust, cataclysmic volcanic eruption, or massive meteor impact, the light intensity reaching Earth's surface would decline to about a tenth of its normal level. Photosynthetic organisms would die as they depleted their energy reserves faster than they could manufacture more food. Animals that normally ate these autotrophs would go hungry, as would the animals that ate them. A year or even two might pass before enough life-giving light could penetrate the hazy atmosphere, but by then, it would be too late. The lethal chain reaction would already be well into motion, destroying food webs at their bases. No wonder biologists consider photosynthesis to be the most important metabolic process on Earth.

5.1 Mastering Concepts

1. How is an autotroph different from a heterotroph?
2. Describe photosynthesis in words and as a chemical reaction.
3. Why is photosynthesis essential to life?

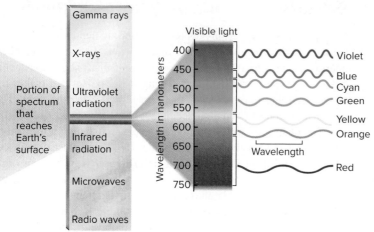

Figure 5.2 **The Electromagnetic Spectrum.** Sunlight reaching Earth consists of ultraviolet radiation, visible light, and infrared radiation, all of which are just a small part of a continuous spectrum of electromagnetic radiation. Photons with the shortest wavelengths carry the most energy.

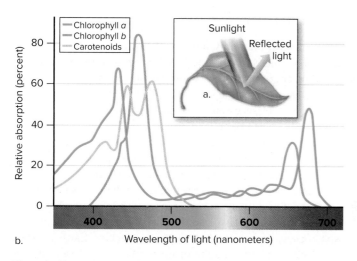

Figure 5.3 **Everything but Green.** (a) Overall, a leaf reflects green and yellow wavelengths of light, while absorbing the other wavelengths. (b) Each type of pigment absorbs some wavelengths of light and reflects others.

5.2 Photosynthetic Pigments Capture Sunlight

The sun releases energy in all directions, in the form of waves of electromagnetic radiation. After an 8-minute journey, about 2 billionths of the sun's total energy output reaches Earth's upper atmosphere. Of this, only about 1% is used for photosynthesis, yet this tiny fraction of the sun's power ultimately produces enough sugar to fill 600 billion Olympic-sized swimming pools each year. Light may seem insubstantial, but it is a powerful force on Earth.

Visible light is a small sliver of a much larger **electromagnetic spectrum,** the range of possible frequencies of radiation (figure 5.2). All electromagnetic radiation, including light, consists of **photons,** discrete packets of kinetic energy. A photon's **wavelength** is the distance it moves during a complete vibration. The shorter a photon's wavelength, the more energy it contains.

The sunlight that reaches Earth's surface consists of three main components: ultraviolet radiation, visible light, and infrared radiation. Of the three, ultraviolet radiation has the shortest wavelengths. Its high-energy photons damage DNA, causing sunburn and skin cancer. In the middle range is visible light, which provides the energy that powers photosynthesis; we perceive visible light of different wavelengths as distinct colors. Infrared radiation, with its longer wavelengths, contains too little energy per photon to be useful to organisms. Most of its energy is converted immediately to heat.

Plant cells contain several pigment molecules that capture light energy. The most abundant is **chlorophyll** *a,* a green photosynthetic pigment in plants, algae, and cyanobacteria. Photosynthetic organisms usually also have several types of **accessory pigments,** which are energy-capturing pigment molecules other than chlorophyll *a.* Chlorophyll *b* and carotenoids are accessory pigments in plants (see Burning Question 5.1).

The photosynthetic pigments have distinct colors because they absorb only some wavelengths of visible light, while transmitting or reflecting others (figure 5.3). Chlorophylls *a* and *b* absorb red and blue wavelengths; they appear green because they reflect green light. Carotenoids, on the other hand, reflect longer wavelengths of light, so they appear red, orange, or yellow. (Carrots, tomatoes, lobster shells, and the flesh of salmon all owe their distinctive colors to carotenoid pigments, which the animals must obtain from their diets.)

Only absorbed light is useful in photosynthesis. Accessory pigments absorb wavelengths that chlorophyll *a* cannot, so they extend the range of light wavelengths that a cell can harness. This is a little like the members of the same team on a quiz show, each contributing answers from a different area of expertise.

Figure It Out

Study figure 5.3. If you could expose plants to just one wavelength of light at a time, would a wavelength of 350 nm, 450 nm, or 600 nm produce the highest photosynthetic rate?

Answer: 450 nm.

5.2 Mastering Concepts

1. What are the main components of sunlight?

2. How does it benefit a photosynthetic organism to have multiple types of pigments?

5.3 Chloroplasts Are the Sites of Photosynthesis

In plants, leaves are the main organs of photosynthesis (figure 5.4). Their broad, flat surfaces expose abundant surface area to sunlight. But light is just one requirement for photosynthesis. Water is also essential; roots absorb this vital ingredient, which moves up stems and into the leaves. And plants also exchange CO_2 and O_2 with the atmosphere through **stomata** (singular: stoma), tiny openings in the epidermis of a leaf or stem. (The word *stoma* comes from the Greek word for "mouth.") ① *stomata*, section 21.3B

Most photosynthesis occurs in **mesophyll,** a collective term for the cells filling a leaf's interior (*meso-* means "middle," and *-phyll* means "leaf"). Leaf mesophyll cells contain abundant **chloroplasts,** the organelles of photosynthesis in plants and algae. Most photosynthetic cells contain 40 to 200 chloroplasts, which add up to about 500,000 per square millimeter of leaf—an impressive array of solar energy collectors.

Each chloroplast contains tremendous surface area for the reactions of photosynthesis. Two membranes enclose the **stroma,** the chloroplast's fluid inner region. This gelatinous fluid contains ribosomes, DNA, and enzymes. (Be careful not to confuse the *stroma* with a *stoma,* or leaf pore.) Suspended in the stroma are between 10 and 100 **grana** (singular: granum), each composed of a stack of 10 to 20 pancake-shaped thylakoids. Each **thylakoid,** in turn, consists of a membrane that is studded with photosynthetic pigments. The **thylakoid space** is the inner compartment enclosed by a thylakoid membrane.

Figure 5.4 **Leaf and Chloroplast Anatomy.** (a) The tissue inside a leaf is called mesophyll. (b) Each mesophyll cell contains multiple chloroplasts. (c) A chloroplast contains light-harvesting pigments, embedded in (d) the stacks of thylakoid membranes that make up each granum.

Photos: (leaves): ©Robert Glusic/Corbis RF; (mesophyll): Electron micrograph by Wm. P. Wergin, Courtesy Eldon H. Newcomb, University of Wisconsin-Madison

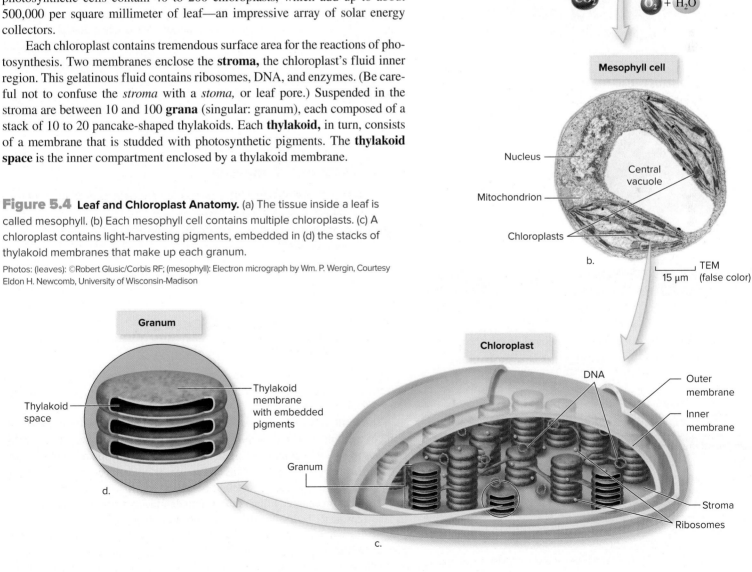

Leaf

Mesophyll cells

Stoma

CO_2 $O_2 + H_2O$

Mesophyll cell

Nucleus

Mitochondrion

Chloroplasts

Central vacuole

TEM 15 µm (false color)

b.

Granum

Thylakoid space

Thylakoid membrane with embedded pigments

Granum

d.

Chloroplast

DNA

Outer membrane

Inner membrane

Stroma

Ribosomes

c.

The Light Reactions Begin Photosynthesis   The Carbon Reactions Produce Carbohydrates C_3, C_4, and CAM Plants

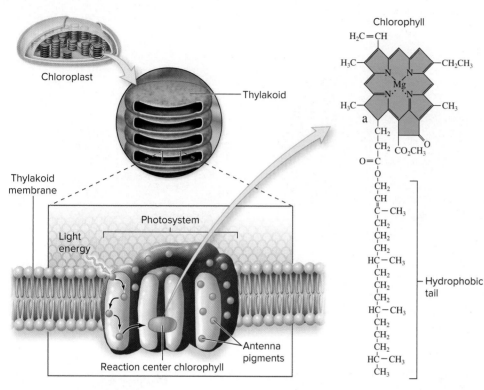

Chlorophyll

Figure 5.5 Photosystem. This diagram shows one of the many photosystems embedded in a typical thylakoid membrane. Each photosystem consists of proteins (*purple*) and pigments (*green*), including chlorophyll *a* and accessory pigments.

Anchored in the thylakoid membranes are many **photosystems,** which are clusters of pigments and proteins that participate in photosynthesis (figure 5.5). Each photosystem includes about 300 chlorophyll *a* molecules and 50 accessory pigments. The photosystem's **reaction center** includes a special pair of chlorophyll *a* molecules that actually use the energy in photosynthetic reactions. The other pigments of the photosystem make up the light-harvesting complex that surrounds the reaction center. These additional pigments are called **antenna pigments** because they capture photon energy and funnel it to the reaction center. If the array of pigments in a photosystem is like a quiz show team, then the reaction center is analogous to the one member who announces the team's answer to the show's moderator.

Each photosystem has a few hundred chlorophyll molecules, so why does only the reaction center chlorophyll actually participate in the photosynthetic reactions? A single chlorophyll *a* molecule can absorb only a small amount of light energy. But because many pigment molecules are arranged close together, each antenna pigment can quickly pass its energy to the reaction center, freeing the antenna to absorb other photons as they strike. Thus, the photosystem's organization greatly enhances the efficiency of photosynthesis.

5.3 Mastering Concepts

1. Describe the relationships among the chloroplast, stroma, grana, thylakoids, and photosystems.
2. How does the reaction center chlorophyll interact with the antenna pigments in a photosystem?

Miniglossary | Leaf Anatomy

Stoma	Opening through which a leaf exchanges gases with the atmosphere
Mesophyll cell	Leaf cell where photosynthesis occurs
Chloroplast	Organelle of photosynthesis in a plant cell
Granum	A stack of thylakoids suspended in a chloroplast
Thylakoid	A pancake-shaped compartment within a chloroplast; site of the light reactions
Thylakoid membrane	The lipid bilayer in which photosystems are embedded
Photosystem	Cluster of pigments and proteins that participate in the light reactions of photosynthesis
Thylakoid space	The compartment enclosed by the thylakoid membrane
Stroma	The fluid inside the chloroplast; site of the carbon reactions

5.4 Photosynthesis Occurs in Two Stages

Inside a chloroplast, photosynthesis occurs in two stages: the light reactions and the carbon reactions. Figure 5.6 summarizes the entire process, and sections 5.5 and 5.6 describe each part in greater detail.

The **light reactions** convert solar energy to chemical energy. (You can think of the light reactions as the "photo-" part of photosynthesis.) In the chloroplast's thylakoid membranes, pigment molecules in two linked photosystems capture kinetic energy from photons and store it as potential energy in the chemical bonds of two molecules: ATP and NADPH.

Recall from chapter 4 that **ATP** is a nucleotide that stores potential energy in the covalent bonds between its phosphate groups (see figure 4.7). ATP forms when a phosphate group is added to ADP. The other energy-rich product of the light reactions, **NADPH,** is a molecule that carries pairs of energized electrons. In photosynthesis, these electrons come from one of the two reaction center chlorophyll molecules. Once the light reactions are underway, chlorophyll, in turn, replaces its "lost" electrons by splitting water molecules, yielding O_2 as a waste product.

These two resources (energy and "loaded" electron carriers) set the stage for the second part of photosynthesis: the carbon reactions. In the **carbon reactions,** the chloroplast uses ATP, the high-energy electrons in NADPH, and CO_2 to produce sugar molecules. These reactions are the "-synthesis" part of photosynthesis. The ATP and NADPH come from the light reactions, and the CO_2 comes from the atmosphere (see Burning Question 5.2). Once inside the leaf, CO_2 diffuses into a mesophyll cell and across the chloroplast membrane into the stroma, where the carbon reactions occur.

Overall, photosynthesis is an oxidation–reduction (redox) process. "Oxidation" means that electrons are removed from an atom or molecule; "reduction" means electrons are added. As you will see, photosynthesis strips electrons from the oxygen atoms in H_2O (i.e., the oxygen atoms are oxidized). These electrons reduce the carbon in CO_2. Because oxygen atoms attract electrons more strongly than do carbon atoms (see chapter 2), moving electrons from oxygen to carbon requires energy. The energy source for this reaction is, of course, light. ⓘ *redox reactions,* section 4.2B

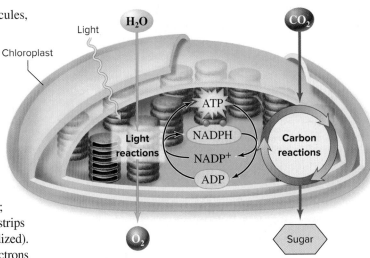

Figure 5.6 Overview of Photosynthesis. In the light reactions, pigment molecules capture light energy and transfer it to molecules of ATP and NADPH. The carbon reactions use this energy to build sugar molecules out of carbon dioxide.

5.4 Mastering Concepts

1. What happens in each of the two main stages of photosynthesis?
2. Where in the chloroplast does each stage occur?

5.5 The Light Reactions Begin Photosynthesis

A plant placed in a dark closet literally starves. Without light, the plant cannot generate ATP or NADPH. And without these critical sources of energy and electrons, the plant cannot produce sugars to feed itself. Once its stored reserves are gone, the plant dies. The plant's life thus depends on the light reactions of photosynthesis, which occur in the membranes of chloroplasts.

Burning Question 5.1

Why do leaves change colors in the fall?

Most leaves are green throughout a plant's growing season, although there are exceptions; some ornamental plants, for example, have yellow or purple foliage. The familiar green color comes from chlorophyll *a,* the most abundant pigment in photosynthetic plant parts.

But the leaf also has other photosynthetic pigments. Carotenoids contribute brilliant yellow, orange, and red hues. Purple pigments, such as anthocyanins, are not photosynthetically active, but they do protect leaves from damage by ultraviolet radiation.

Carotenoids are less abundant than chlorophyll, so they usually remain invisible to the naked eye during the growing season. As winter approaches, however, deciduous plants prepare to shed their leaves. Anthocyanins accumulate while chlorophyll degrades, and the now "unmasked" accessory pigments reveal their colors for a short

time as a spectacular autumn display. These pigments soon disappear as well, and the dead leaves turn brown and fall to the ground.

Spring brings a flush of fresh green leaves. The energy to produce the foliage comes from glucose the plant produced during the last growing season and stored as starch. The new leaves make food throughout the spring and summer, so the tree can grow—both above the ground and below—and produce fruits and seeds.

As the days grow shorter and cooler in autumn, the cycle will continue, and the colorful pigments will again participate in one of nature's great disappearing acts.

Submit your burning question to
marielle.hoefnagels@mheducation.com

(leaves): ©Carlos E. Santa Maria/Shutterstock RF

Miniglossary | Light Reactions

Light reactions	The steps of photosynthesis in which light energy is converted into chemical energy
Photosystem II	Group of pigments and proteins that uses light energy to energize electrons stripped from water
Photosystem I	Group of pigments and proteins that uses light energy to energize electrons received from photosystem II
Electron transport chain	Series of membrane proteins that shuttle electrons and use the released energy to create a proton gradient across a membrane
ATP synthase	Enzyme that uses the potential energy in a proton gradient to produce ATP

We have already seen that the pigments and proteins of the chloroplast's thylakoid membranes are organized into photosystems (see figure 5.5). More specifically, the thylakoid membranes contain two types of photosystems, dubbed "I" and "II." The entire sequence, from photosystem II through the enzyme that produces NADPH, is an electron transport chain.

Recall from chapter 4 that an **electron transport chain** is a group of proteins that shuttle electrons from carrier to carrier, releasing energy with each step. As you will see, the photosynthetic electron transport chain provides both the energy required for ATP synthesis and the electrons required for the production of NADPH.

Figure 5.7 depicts the arrangement of the photosystems and electron transport chain in the thylakoid membrane. Refer to this illustration as you work through the rest of this section.

A. Light Striking Photosystem II Provides the Energy to Produce ATP

Photosynthesis begins in the cluster of pigment molecules of photosystem II. These pigments absorb light and transfer the energy to a chlorophyll *a* reaction

Figure 5.7 The Light Reactions. (*1*) Chlorophyll molecules in photosystem II transfer light energy to electrons. (*2*) Electrons are stripped from water molecules, releasing O_2. (*3*) The energized electrons pass to photosystem I along a series of proteins. Each transfer releases energy that is used to pump protons (H^+) into the thylakoid space. (*4*) The resulting proton gradient is used to generate ATP. (*5*) In photosystem I, the electrons absorb more light energy and (*6*) are passed to $NADP^+$, generating the energy-rich NADPH. The inset (*upper right*) shows the light reactions in the context of the overall process of photosynthesis.

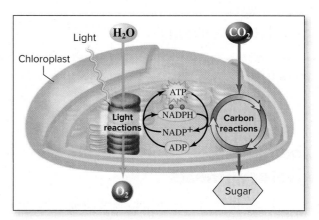

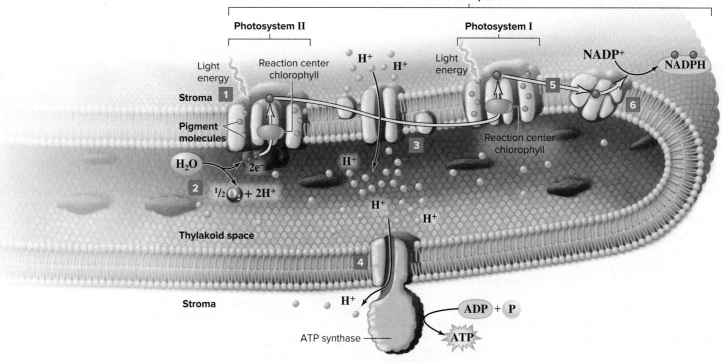

center, where it boosts two electrons to an orbital with a higher energy level. The "excited" electrons, now packed with potential energy, are ejected from the reaction center chlorophyll *a* molecule and begin their journey along the electron transport chain (figure 5.7, step 1). ⓘ *electron orbitals,* section 2.2A

How does the reaction center chlorophyll *a* molecule replace these two electrons? They come from water (H_2O), which donates two electrons when it splits into oxygen gas and two protons (H^+). Chlorophyll *a* picks up the electrons. The protons are released into the thylakoid space, and the O_2 is either used in the plant's respiration or released to the environment (step 2).

Meanwhile, the chloroplast uses the potential energy in the electrons to create a proton gradient (step 3). As the electrons pass along the electron transport chain, the energy they lose drives the active transport of protons from the stroma into the thylakoid space. The resulting proton (H^+) gradient across the thylakoid membrane represents a form of potential energy. ⓘ *active transport,* section 4.5B

ATP synthase is the enzyme complex that transforms the gradient's potential energy into chemical energy in the form of ATP (step 4). A channel in ATP synthase allows protons trapped inside the thylakoid space to return to the chloroplast's stroma. As the gradient dissipates, energy is released. The ATP synthase enzyme uses this energy to add phosphate to ADP, generating ATP. (As described in chapter 6, the same enzyme produces ATP in cellular respiration.)

This mechanism is similar to using a dam to produce electricity. Water accumulating behind a dam represents potential energy, like the proton gradient across a thylakoid membrane. To harness this potential energy, the dam's operators allow water to gush through a large pipe at the dam's base, turning blades that spin an electric generator. Likewise, ATP synthase generates ATP as it allows accumulated protons to pass from the thylakoid space into the stroma.

B. Electrons from Photosystem I Reduce NADP⁺ to NADPH

Photosystem I functions much as photosystem II does. Photon energy strikes energy-absorbing antenna pigment molecules, which pass the energy to the reaction center chlorophyll *a*. The reactive chlorophyll molecule boosts the electrons to a higher energy level and passes them along the chain (figure 5.7, step 5). The electrons from photosystem I are then replaced with electrons passing down from photosystem II.

At the end of the photosynthetic electron transport chain, the electrons energized at photosystem I reduce a molecule of NADP⁺ to NADPH (step 6). This NADPH is the electron carrier that will reduce carbon dioxide in the carbon reactions, while ATP will provide the energy. (Why We Care 5.1 explains how blocking the light reactions quickly leads to a plant's demise.)

Burning Question 5.2

Does air have mass?

"I feel as light as air." You may have heard a friend say this on a day when she's feeling optimistic, as if all of the burdens of life had been lifted from her shoulders. The phrase is correct in its assertion that air is extremely light. But air is not massless. The easiest way to demonstrate this idea is to walk or run into a strong headwind. Pushing against the moving air makes the journey extremely difficult.

Invisible gas molecules give air its mass. Each cubic centimeter of air at sea level on a cold day has some 30 trillion N_2, 8 trillion O_2, and 15 billion CO_2 molecules. We inhale these gases with each breath, but a plant exchanges them with the air through stomata in its leaves. The plant assembles the CO_2 into sugar molecules and polysaccharides, which represent the first step in the growth of new stems, leaves, and other organs. Much of the mass of these touchable, edible organs originates as invisible air molecules.

Submit your burning question to marielle.hoefnagels@mheducation.com

(girl): ©Corbis Super/Alamy Stock Photo RF

5.5 Mastering Concepts

1. How does light striking photosystem II lead to ATP production?
2. What is water's role in the light reactions?
3. What happens after light strikes photosystem I?
4. How are the electrons from photosystem I replaced?

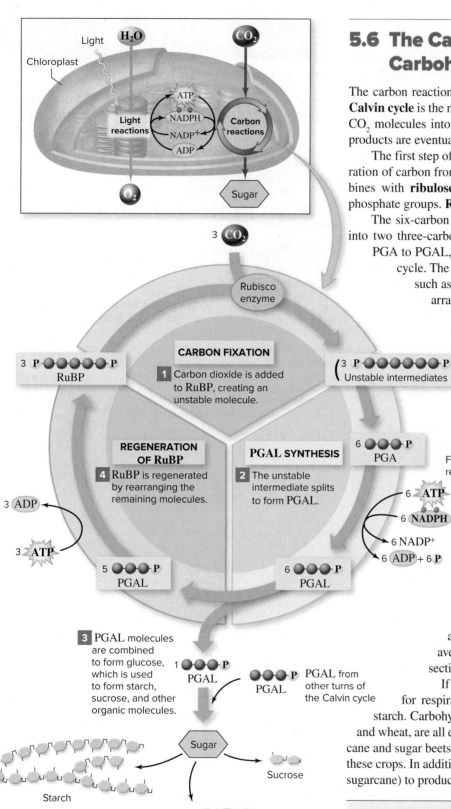

5.6 The Carbon Reactions Produce Carbohydrates

The carbon reactions, also called the Calvin cycle, occur in the stroma. The **Calvin cycle** is the metabolic pathway that uses NADPH and ATP to assemble CO_2 molecules into three-carbon carbohydrate molecules (figure 5.8). These products are eventually assembled into glucose and other sugars.

The first step of the Calvin cycle is **carbon fixation**—the initial incorporation of carbon from CO_2 into an organic compound. Specifically, CO_2 combines with **ribulose bisphosphate (RuBP),** a five-carbon sugar with two phosphate groups. **Rubisco** is the enzyme that catalyzes this first reaction.

The six-carbon product of the initial reaction immediately breaks down into two three-carbon molecules (PGA). Further steps in the cycle convert PGA to PGAL, which is the carbohydrate product that leaves the Calvin cycle. The cell can use PGAL to build larger carbohydrate molecules such as glucose and sucrose. Some of the PGAL, however, is rearranged to form additional RuBP, perpetuating the cycle.

Several fates await the carbohydrates produced in the carbon reactions. A plant's cells use about half of the sugar as fuel for their own cellular respiration, the metabolic pathway described in chapter 6. Roots, flowers, fruits, seeds, and other nonphotosynthetic plant parts could not grow without sugar shipments from green leaves and stems. Plants also combine glucose with other substances to manufacture additional compounds, including amino acids and a host of economically important products such as rubber, medicines, and spices.

Moreover, glucose molecules are the building blocks of the cellulose wall that surrounds every plant cell. Wood is mostly made of cellulose. The timber in the world's forests therefore stores enormous amounts of carbon. So do vast deposits of coal and other fossil fuels, which are the remains of plants and other organisms that lived long ago. Burning wood or fossil fuels releases this stored carbon into the atmosphere as CO_2. As the amount of CO_2 in the atmosphere has increased, Earth's average temperature has risen. ① *global climate change,* section 20.4

If a plant produces more glucose than it immediately needs for respiration or building cell walls, it may store the excess as starch. Carbohydrate-rich tubers and grains, such as potatoes, rice, corn, and wheat, are all energy-storing plant organs. Some plants, including sugarcane and sugar beets, store energy as sucrose instead. Table sugar comes from these crops. In addition, people use starch (from corn kernels) and sugar (from sugarcane) to produce biofuels such as ethanol.

Figure 5.8 The Carbon Reactions. ATP and NADPH from the light reactions power the Calvin cycle, which assembles CO_2 molecules into carbohydrates.

5.6 Mastering Concepts

1. What are the roles of CO_2, ATP, and NADPH in the Calvin cycle?
2. Use figure 5.8 to determine how many ATP molecules are used to produce a six-carbon glucose molecule.

5.7 C₃, C₄, and CAM Plants Use Different Carbon Fixation Pathways

The Calvin cycle is also known as the **C₃ pathway** because a three-carbon molecule, PGA, is the first stable compound in the pathway. Although all plants use the Calvin cycle, C₃ plants use *only* this pathway to fix carbon from CO_2. About 95% of plant species are C₃, including cereals and most trees.

C₃ photosynthesis is obviously a successful adaptation, but it does have a weakness: inefficiency. Photosynthesis has a theoretical efficiency rate of about 30% in ideal conditions, but a plant's efficiency in nature is typically as low as 0.1% to 3%.

How do plants waste so much solar energy? One contributing factor is a metabolic pathway called **photorespiration,** a series of reactions that begin when the rubisco enzyme adds O_2 instead of CO_2 to RuBP. The product of this reaction does not enter the Calvin cycle. The plant therefore loses CO_2 that it fixed in previous turns of the cycle, wasting both ATP and NADPH.

A plant with open stomata minimizes its photorespiration rate. This is because CO_2 and O_2 compete for rubisco's active site; when stomata are open, CO_2 from the atmosphere enters the leaf, and O_2 produced in the light reactions diffuses out. But when the weather heats up, plants face a trade-off. If the stomata remain open too long, a plant may lose water, wilt, and die. If the plant instead closes its stomata, CO_2 runs low, and O_2 builds up in the leaves. Under these conditions, photorespiration becomes much more likely, and photosynthetic efficiency plummets.

In hot, dry climates, plants that minimize photorespiration have a significant competitive advantage. One way to improve efficiency is to ensure that rubisco always encounters high CO_2 concentrations. The C₄ and CAM pathways are two adaptations that do just that.

C₄ plants physically separate the Calvin cycle from the O_2-rich air spaces in the leaf **(figure 5.9)**. The light reactions occur in mesophyll cells, as does a carbon-fixation reaction called the C₄ pathway. In the **C₄ pathway,** CO_2 combines with a three-carbon "ferry" molecule to form a four-carbon compound; rubisco is not involved. The four-carbon molecule then moves into adjacent **bundle-sheath cells** that surround the leaf veins. The CO_2 is liberated inside these cells, which contain rubisco. The Calvin cycle then fixes the carbon a second time. Meanwhile, at the cost of two ATP molecules, the three-carbon "ferry" returns to the mesophyll to pick up another CO_2.

About 1% of plants use the C₄ pathway. All are flowering plants growing in hot, sunny environments, including crabgrass and crop plants such as sugarcane and corn. C₄ plants are less abundant, however, in cooler, moister habitats. In those environments, the ATP cost of ferrying each CO_2 from a mesophyll cell to a bundle-sheath cell apparently exceeds the benefits of reduced photorespiration.

Another energy- and water-saving strategy is crassulacean acid metabolism (CAM). Plants that use the **CAM pathway** add a new twist: They open their stomata only at night, fix CO_2, then fix it again in the Calvin cycle during the day. Unlike in C₄ plants, both fixation reactions occur in the same cell.

A CAM plant's stomata open at night, when the temperature drops and humidity rises. CO_2 diffuses in. Mesophyll cells incorporate the CO_2 into a four-carbon compound, which they store in large vacuoles. The stomata close during the heat of the day, but the stored molecule moves from the vacuole to a chloroplast and releases its CO_2. The chloroplast then fixes the CO_2 in the Calvin cycle. The CAM pathway reduces photorespiration by generating high CO_2 concentrations inside chloroplasts.

Why We Care 5.1 | Weed Killers

One low-tech way to kill an unwanted plant is to deprive it of light. Gardeners who want to convert a lawn into a garden, for example, might kill the grass by covering it with layers of newspaper or cardboard for several weeks. The light reactions of photosynthesis cannot occur in the dark; the plants die.

©image100/Corbis RF

Many herbicides also stop the light reactions. For example, a weed killer called diuron blocks electron flow in photosystem II. Paraquat, noted for its use in destroying marijuana plants, diverts electrons from photosystem I. Either way, blocking electron flow prevents the production of ATP and NADPH. Without these critical products, photosynthesis cannot continue.

Other herbicides take a different approach. Accessory pigments called carotenoids protect plants from damage caused by free radicals. Triazole herbicides kill plants by blocking carotenoid synthesis. No longer protected from free-radical damage, the cell's organelles are destroyed.

Still other weed killers exploit pathways not directly related to photosynthesis. For instance, glyphosate (Roundup) inhibits an enzyme that plants require for amino acid synthesis. Another herbicide, 2,4-D, mimics a plant hormone called auxin (see chapter 22); no one knows exactly why the treated plant dies.

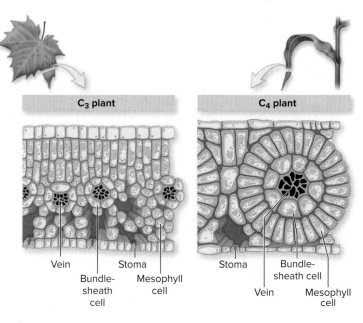

C₃ plant   C₄ plant

Vein   Stoma   Stoma   Bundle-sheath cell
Bundle-sheath cell   Mesophyll cell   Vein   Mesophyll cell

Figure 5.9 C₃ and C₄ Leaf Anatomy. In C₃ plants, the light reactions and the Calvin cycle occur in mesophyll cells. In C₄ plants, the light reactions occur in mesophyll, but the inner ring of bundle-sheath cells houses the Calvin cycle.

About 3% to 4% of plant species, including pineapple and cacti, use the CAM pathway. All CAM plants are adapted to dry habitats. In cool environments, however, CAM plants cannot compete with C_3 plants. Their stomata are open only at night, so CAM plants have much less carbon available to their cells for growth and reproduction.

Figure 5.10 compares and contrasts C_3, C_4, and CAM plants.

Figure 5.10 C_3, C_4, and **CAM Pathways Compared.** The C_4 and CAM pathways are adaptations that minimize photorespiration.

Photos: (sycamore): ©Tony Sweet/Digital Vision/Getty Images RF; (corn): ©Pixtal/age fotostock RF; (cactus): ©Lucky-photographer/Shutterstock RF

5.7 Mastering Concepts

1. Why is the Calvin cycle also called the C_3 pathway?
2. How does photorespiration counter photosynthesis?
3. Describe how a C_4 plant minimizes photorespiration.
4. How is the CAM pathway similar to C_4 metabolism, and how is it different?

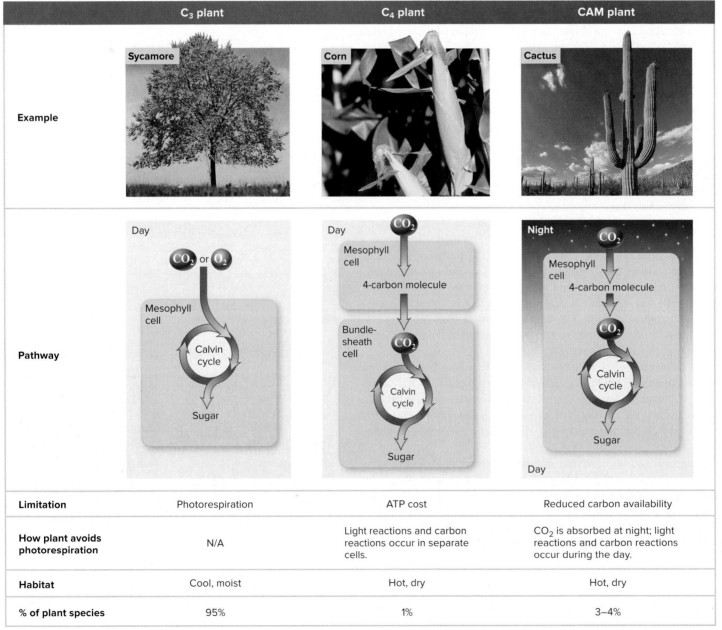

	C_3 plant	C_4 plant	CAM plant
Example	Sycamore	Corn	Cactus
Pathway			
Limitation	Photorespiration	ATP cost	Reduced carbon availability
How plant avoids photorespiration	N/A	Light reactions and carbon reactions occur in separate cells.	CO_2 is absorbed at night; light reactions and carbon reactions occur during the day.
Habitat	Cool, moist	Hot, dry	Hot, dry
% of plant species	95%	1%	3–4%

Investigating Life 5.1 | Salamanders Snack on Sugars from Solar Cells

Most animals have an indirect relationship with photosynthesis: Autotrophs make food, which animals eat. But spotted salamanders are unusual animals: Their eggs have live-in algae of their own.

Spotted salamanders are amphibians that mate in temporary ponds, where the females lay masses of fertilized eggs. Each egg contains a tiny embryo and is surrounded by a thick jelly layer. Then, something strange happens: Microscopic green algae somehow find each egg and enter the jelly layer. The photosynthetic algae use CO_2 from the water to produce sugars in the protective confines of their new homes.

Do salamander embryos receive any of the sugars that the algae produce? To find out, researchers incubated algae-infected salamander eggs in a solution containing CO_2 that was "tagged" with a radioactive isotope of carbon. After the incubation period, they measured the amount of radioactive carbon in each egg and embryo. Any radioactive carbon they detected in a salamander embryo could come from one of two sources. The carbon might simply diffuse in from the solution, without any help from the algae. Alternatively, the algae might use the tagged carbon to produce sugars in photosynthesis, then transfer some of the radioactive sugar to the embryo. ⓘ *diffusion*, section 4.5A

To differentiate between these two possibilities, the team incubated some eggs in the light, allowing both diffusion and photosynthesis to occur. A second set of eggs was incubated in total darkness. Photosynthesis is not possible in the dark, but diffusion continues. Subtracting the amount of radioactive carbon in dark-treated embryos from the amount in light-treated embryos should therefore reveal how much sugar moves from algae to embryo.

After the experiment was complete, the researchers discovered that eggs and embryos incubated in the light incorporated more radioactive carbon than did their dark-treated counterparts, a sure sign that the algae were sharing their carbon with their tiny hosts (figure 5.A). This sugar supplement can help a developing embryo survive.

What's in it for the algae? They probably benefit from the partnership as well. A developing salamander embryo releases CO_2 in respiration (see chap-

Sample	Average Net Radioactivity Difference (Light Minus Dark)	Average Hourly Change in Carbon	Source of Carbon Increase
Whole egg			
Embryo	12,041 dpm*	294.5 ng	Carbon fixation by algae in egg
Embryo alone	627 dpm*	15.4 ng	Transfer of sugars from algae to embryo

*dpm = disintegrations per minute, a measure of radioactivity

Figure 5.A Thanks for the Snack. Using a radioactive isotope of carbon, researchers measured the amount of carbon transferred from egg-dwelling green algae to salamander embryos.

Photo: ©George Grall/National Geographic Creative

ter 6). Perhaps this extra shot of CO_2 makes photosynthesis more efficient for the algae, completing the exchange of materials between two allies from different kingdoms of life.

Source: Graham, Erin R., Scott A. Fay, Adam Davey, and Robert W. Sanders. 2013. Intracapsular algae provide fixed carbon to developing embryos of the salamander *Ambystoma maculatum. Journal of Experimental Biology,* vol. 216, pages 452–459.

What's the **Point?** ▼ APPLIED

This chapter describes how autotrophs convert solar energy into chemical energy. Since people do not share this ability, it may seem hard to apply photosynthesis to our own lives. But most organisms, including humans, rely directly or indirectly on the sugars produced in photosynthesis. Even the most dedicated meat-lovers would starve without the plants that sustain their favorite foods.

It may seem surprising, then, that the first cells lacked the ability to capture sunlight. Once photosynthetic microbes arose some 3.5 billion years ago, however, their activities slowly began altering Earth's atmosphere. Until about 2.4 billion years ago, the atmosphere contained little O_2. But because O_2 is a waste product of photosynthesis, it gradually accumulated

©Pixtal/age fotostock RF

over the next couple of billion years as microbes used light energy to make sugars. Photosynthesis therefore allows the energy-harvesting pathways of aerobic cellular respiration to exist.

In addition, O_2 gas reacted with free oxygen atoms high in the atmosphere, producing ozone (O_3). As ozone accumulated, it blocked some of the sun's ultraviolet radiation. Lower levels of harmful radiation meant less genetic damage and allowed new varieties of life to arise. Today, the ozone layer continues to protect you: A thinner ozone layer in the southern hemisphere is associated with faster sunburns and a higher incidence of skin cancer than in the northern hemisphere. We can therefore thank plants for helping protect us from the sun, for the air we breathe, and for the foods we eat.

CHAPTER SUMMARY

5.1 Life Depends on Photosynthesis

- **Autotrophs** produce their own organic compounds from inorganic starting materials such as CO_2 and water. **Heterotrophs** rely on organic molecules produced by other organisms.
- **Photosynthesis** converts kinetic energy in light to potential energy in the bonds of carbohydrates, according to the following chemical equation:

$$6CO_2 + 6H_2O \xrightarrow{\text{light energy}} C_6H_{12}O_6 + 6O_2$$

- Plants, algae, and some bacteria are autotrophs. Food and oxygen produced in photosynthesis are critical to life in terrestrial and aquatic habitats.

5.2 Photosynthetic Pigments Capture Sunlight

- Visible light is a small part of the **electromagnetic spectrum.**
- **Photons** move in waves. The shorter the **wavelength,** the more kinetic energy per photon.
- **Chlorophyll** *a* is the primary photosynthetic pigment in plants. **Accessory pigments** absorb wavelengths of light that chlorophyll *a* cannot absorb.

5.3 Chloroplasts Are the Sites of Photosynthesis

- Plants exchange gases with the environment through pores called **stomata.**
- Leaf **mesophyll** cells contain abundant **chloroplasts.**
- A chloroplast contains a gelatinous fluid called the **stroma.** This fluid surrounds the **grana,** which are stacks of pancake-shaped **thylakoid** membranes. Photosynthetic pigments are embedded in the thylakoid membranes, which enclose the **thylakoid space.**
- A **photosystem** consists of proteins, **antenna pigments,** and a **reaction center.**

5.4 Photosynthesis Occurs in Two Stages

- The **light reactions** of photosynthesis produce **ATP** and **NADPH;** these molecules provide energy and electrons for the sugar-producing **carbon reactions** (figure 5.11).
- Photosynthesis is a redox reaction in which water is oxidized and CO_2 is reduced to glucose.

5.5 The Light Reactions Begin Photosynthesis

A. Light Striking Photosystem II Provides the Energy to Produce ATP

- Photosystem II captures light energy and sends electrons from reactive chlorophyll *a* along the **electron transport chain.**
- Electrons from chlorophyll are replaced with electrons from water. O_2 is the waste product.
- The energy released in the electron transport chain drives the active transport of protons (H^+) into the thylakoid space. The protons diffuse out through channels in **ATP synthase.** This movement powers the production of ATP.

B. Electrons from Photosystem I Reduce $NADP^+$ to NADPH

- Light striking photosystem I re-energizes the electrons, which pass to an enzyme that uses them to reduce $NADP^+$. The product of this reaction is NADPH.

5.6 The Carbon Reactions Produce Carbohydrates

- The carbon reactions use energy from ATP and electrons from NADPH in **carbon fixation** reactions that add CO_2 to organic compounds.
- In the **Calvin cycle, rubisco** catalyzes the reaction of CO_2 with **ribulose bisphosphate (RuBP)** to yield two molecules of PGA. These are converted to PGAL, the immediate product of photosynthesis. PGAL later becomes glucose and other carbohydrates.
- Plants use sugar produced in photosynthesis to generate ATP, grow, nourish nonphotosynthetic plant parts, and produce many biochemicals.

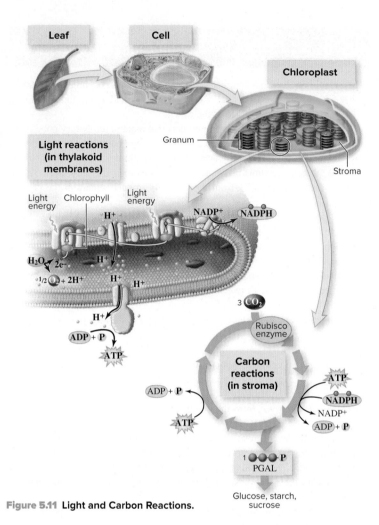

Figure 5.11 Light and Carbon Reactions.

5.7 C_3, C_4, and CAM Plants Use Different Carbon Fixation Pathways

- The Calvin cycle is also called the C_3 **pathway.** Most plant species are C_3 plants, which use only this pathway to fix carbon.
- **Photorespiration** wastes carbon and energy when rubisco reacts with O_2 instead of CO_2.
- The C_4 **pathway** reduces photorespiration by separating two carbon fixation reactions into different cells. In mesophyll cells, CO_2 is fixed as a four-carbon molecule, which moves to a **bundle-sheath cell** and liberates CO_2 to be fixed again in the Calvin cycle.
- In the **CAM pathway,** desert plants such as cacti open their stomata and take in CO_2 at night, storing the fixed carbon in vacuoles. During the day, they split off CO_2 and fix it in chloroplasts in the same cells.

MULTIPLE CHOICE QUESTIONS

1. Where does the energy come from to drive photosynthesis?
   a. A chloroplast   b. ATP   c. The sun   d. Glucose

2. Animals and other _____ rely on _____ that carry out photosynthesis.
   a. autotrophs; autotrophs   c. autotrophs; heterotrophs
   b. heterotrophs; heterotrophs   d. heterotrophs; autotrophs

3. Photosynthesis is essential to animal life because it provides
   a. CO_2 for respiration.   c. organic molecules.
   b. O_2.   d. Both b and c are correct.

4. Only high-energy light can penetrate the ocean and reach photosynthetic organisms in coral reefs. What color light would you predict these organisms use?
   a. Red   b. Yellow   c. Blue   d. Orange

5. The light reactions and the carbon reactions are connected by the participation of which molecules?
   a. ATP and ADP   c. Glucose
   b. NADPH and NADP   d. Both a and b are correct.

6. A plant that only opens its stomata at night is a
   a. C_2 plant.   b. C_3 plant.   c. C_4 plant.   d. CAM plant.

Answers to Multiple Choice questions are in appendix A.

WRITE IT OUT

1. Imagine that black ash from multiple volcanic eruptions made photosynthesis impossible anywhere on Earth for many years. What would be the consequence to plants? To animals? To microbes?

2. Would a plant grow better in a room painted blue or in a room painted green? Explain your answer.

3. In some ways, chlorophyll can be thought of as a solar panel like the ones used to generate electricity at power plants. Using terms such as *granum, photosystem, ATP,* and *chlorophyll,* explain this comparison.

4. Define these terms and arrange them from smallest to largest: *thylakoid membrane; photosystem; chloroplast; granum; reaction center.*

5. Determine whether each of the following molecules is involved in the light reactions, the carbon reactions, or both and explain how: O_2, CO_2, carbohydrates, chlorophyll *a,* photons, NADPH, ATP, H_2O.

6. In the early 1600s, a Flemish scientist monitored the weight of willow trees and their surrounding soil for 5 years. Because he had applied large amounts of water and the soil had lost little weight, he concluded (incorrectly) that plants grew solely by absorbing water. What is the actual source of the added biomass? Explain your answer.

7. In 1771, Joseph Priestley found that if he placed a mouse in an enclosed container with a lit candle, the mouse would die. But if he also added a plant to the container, the mouse could live. Explain this observation.

8. In 1941, biologists exposed photosynthesizing cells to water containing a heavy oxygen isotope, designated ^{18}O. The "labeled" isotope appeared in the O_2 gas released in photosynthesis, showing that the oxygen came from the water. Where would the ^{18}O have ended up if the researchers had used ^{18}O-labeled CO_2 instead of H_2O?

9. Explain how C_4 photosynthesis is based on a spatial arrangement of structures, whereas CAM photosynthesis is temporally based.

10. Listed here are three misconceptions about photosynthesis; explain why each statement is false. (a) Only plants are autotrophs. (b) Plants do not need cellular respiration because they carry out photosynthesis. (c) Chlorophyll is the only photosynthetic pigment.

11. Considering the requirements for photosynthesis, why can't most plants live on unbroken pavement or in bare sand?

SCIENTIFIC LITERACY

Review Why We Care 5.1, which describes how several chemical weed killers work. What potential consequences to yourself, to your family, to your neighbors, and to the environment might you consider before using chemical weed killers? Use the Internet to determine if your concerns are justified. Which websites did you use? Do you think those sites are convincing? Why or why not?

PULL IT TOGETHER

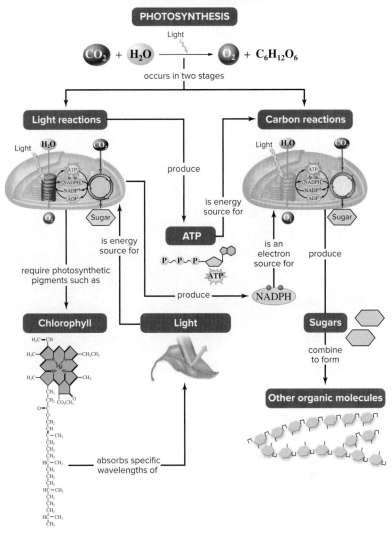

Figure 5.12 Pull It Together: Photosynthesis.

Refer to figure 5.12 and the chapter content to answer the following questions.

1. How would you incorporate the Calvin cycle, rubisco, C_3 plants, C_4 plants, and CAM plants into this concept map?

2. Where do humans and other heterotrophs fit into this concept map?

3. Build another small concept map showing the relationships among the terms *chloroplast, stroma, grana, thylakoid, photosystem,* and *chlorophyll.*

4. Review the Survey the Landscape figure in the chapter introduction, and then add *enzymes, cells, molecules,* and *respiration* to the Pull It Together concept map.

Respiration and Fermentation

LEARNING OUTLINE

6.1 Cells Use Energy in Food to Make ATP

6.2 Cellular Respiration Includes Three Main Processes

6.3 In Eukaryotic Cells, Mitochondria Produce Most ATP

6.4 Glycolysis Breaks Down Glucose to Pyruvate

6.5 Aerobic Respiration Yields Much More ATP than Glycolysis Alone

6.6 How Many ATPs Can One Glucose Molecule Yield?

6.7 Other Food Molecules Enter the Energy-Extracting Pathways

6.8 Fermenters Acquire ATP Only in Glycolysis

APPLICATIONS

Why We Care 6.1 *Some Poisons Inhibit Respiration*
Burning Question 6.1 *How do diet pills work?*
Burning Question 6.2 *What happens during hibernation?*
Investigating Life 6.1 *Hot Plants Offer Heat Rewards*

Precious Oxygen. This scuba diver could not survive underwater without a tank of compressed air. The gas mixture includes oxygen, which our cells need to release energy from food.

©Three Images/Lifesize/Getty Images RF

Learn How to Learn
Don't Skip the Figures

As you read the narrative in the text, pay attention to the figures; they are there to help you learn. Some figures summarize the narrative, making it easier for you to see the "big picture." Other illustrations show the parts of a structure or the steps in a process. Still others summarize a technique or help you classify information. Flip through this book and see if you can find examples of each type.

SURVEY THE LANDSCAPE
Science, Chemistry, and Cells

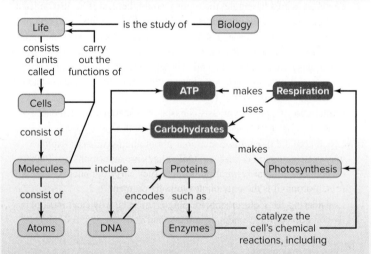

The reactions of aerobic respiration consume carbohydrates and oxygen gas. The overall function is to store energy in ATP, which powers cell activities. Meanwhile, the carbon is released as carbon dioxide, a gaseous waste.

For more details, study the Pull It Together feature in the chapter summary.

©pchoui/iStock/
Getty Images RF

Our need for oxygen is absolute; we lose consciousness after just a few minutes without it. We can extract it only from air, which explains why swimmers must emerge to breathe every few moments, and why scuba divers carry tanks filled with compressed air.

Life also demands a steady supply of food. Plants make their own food, but animals—like the bluebird pictured here—have to eat.

There is an intimate relationship between our twin requirements to eat and breathe. Both are essential for the production of ATP, the power-packed molecule that supplies energy for life's activities. This chapter describes how cells make those little ATP molecules that nothing can live without.

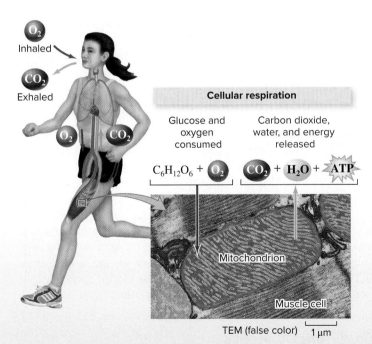

Cellular respiration

| Glucose and oxygen consumed | Carbon dioxide, water, and energy released |

$$C_6H_{12}O_6 + O_2 \longrightarrow CO_2 + H_2O + ATP$$

Mitochondrion

Muscle cell

TEM (false color) 1 μm

Figure 6.1 **Breathing and Cellular Respiration.** The athlete breathes in O_2, which is distributed to all cells. In mitochondria, the O_2 participates in the reactions of cellular respiration, which generate energy-rich ATP. CO_2, a metabolic waste, is exhaled.
Photo: ©Thomas Deerinck, NCMIR/SPL/Science Source

6.1 Cells Use Energy in Food to Make ATP

No cell can survive without **ATP**—adenosine triphosphate. Without this energy-toting molecule, you could not have developed from a fertilized egg into an adult. You could not breathe, chew, talk, circulate your blood, walk, or listen to music. Without ATP, a plant could not grow. A fungus could not produce mushrooms. A bacterial cell could not divide or move. Like a car without gasoline, a cell without ATP would simply die. ⓘ *ATP,* section 4.3

ATP is essential because it powers nearly every activity that requires energy input in the cell: synthesis of DNA, RNA, proteins, carbohydrates, and lipids; active transport across membranes; separation of duplicated chromosomes during cell division; muscle contraction; and many others. This constant need for ATP explains the need for a steady food supply. All organisms, from trees to whales to bacteria, use the potential energy stored in food to make ATP.

Where does the food come from in the first place? Chapter 5 explains the answer: In most ecosystems, plants and other autotrophs use photosynthesis to make organic molecules such as glucose ($C_6H_{12}O_6$) out of carbon dioxide (CO_2) and water (H_2O). Light supplies the energy. The carbohydrates produced in photosynthesis feed not only autotrophs but also all of the animals, fungi, and microbes that share the ecosystem (see figure 4.2).

All cells need ATP, but they don't all produce it in the same way. In **aerobic respiration,** a cell uses oxygen gas (O_2) and glucose to generate ATP. Plants, animals, and most microbes, especially those in O_2-rich environments, use aerobic respiration (the main subject of this chapter). Some cells, however, can generate ATP without using O_2. Section 6.8 describes fermentation, an O_2-free pathway that is most common in microorganisms.

The overall equation for aerobic respiration is essentially the reverse of photosynthesis:

$$\text{glucose} + \text{oxygen} \longrightarrow \text{carbon dioxide} + \text{water} + \text{ATP}$$
$$C_6H_{12}O_6 + 6O_2 \longrightarrow 6CO_2 + 6H_2O + 36\ \text{ATP}$$

This equation reveals that aerobic cellular respiration requires organisms to acquire O_2 and get rid of CO_2 (figure 6.1). In humans and many other animals, the circulatory system carries inhaled O_2 to cells. O_2 diffuses into the cell's mitochondria, the sites of respiration. Meanwhile, CO_2 diffuses out of the cells and into the bloodstream to be exhaled at the lungs.

Many people mistakenly believe that plants do not use cellular respiration because they are photosynthetic. In fact, plants use O_2 to respire about half of the glucose they produce. Why do plants have a reputation for producing O_2, if they also consume it? The reason is that plants incorporate much of the remaining glucose into cellulose, starch, and other stored organic molecules. Therefore, they absorb much more CO_2 in photosynthesis than they release in respiration, and they release much more O_2 than they consume.

The rest of this chapter describes how cells use the potential energy in food to generate ATP. Like photosynthesis, the journey entails several overlapping metabolic pathways and many different chemicals. But if we consider energy release in major stages, the logic emerges.

6.1 Mastering Concepts

1. Why do all organisms need ATP?
2. What is the overall equation for cellular respiration?
3. Why do plants carry out photosynthesis and respiration?

6.2 Cellular Respiration Includes Three Main Processes

The chemical reaction that generates ATP is straightforward: An enzyme tacks a phosphate group onto ADP, yielding ATP. As described in chapter 4, however, ATP synthesis requires an input of energy. The metabolic pathways of respiration harvest potential energy from food molecules and use it to make ATP. This section briefly introduces these pathways; later sections explain them in more detail.

Like photosynthesis, respiration is an oxidation–reduction reaction. The pathways of aerobic respiration oxidize (remove electrons from) glucose and reduce (add electrons to) O_2. Because of oxygen's strong attraction for electrons, this reaction is "easy," like riding a bike downhill. It therefore releases energy, which the cell traps in the bonds of ATP. ⓘ *redox reactions,* section 4.2B

This reaction does not happen all at once. If a cell released all the potential energy in glucose's chemical bonds in one uncontrolled step, the sudden release of heat would destroy the cell; in effect, it would act like a tiny bomb. Rather, the chemical bonds and atoms in glucose are rearranged one step at a time, releasing a tiny bit of energy with each transformation. Some of this energy is released as heat, but much of it is stored in the chemical bonds of ATP.

Biologists organize the intricate biochemical pathways of respiration into three main groups: glycolysis, the Krebs cycle, and electron transport (figure 6.2). In **glycolysis** (literally, "breaking sugar"), a six-carbon glucose molecule splits into two three-carbon molecules of **pyruvate.** This process harvests energy in two forms. First, some of the electrons from glucose are transferred to an electron carrier molecule called **NADH.** Second, glycolysis generates two molecules of ATP.

Additional reactions, including a "transition step" and the **Krebs cycle,** oxidize the pyruvate and release CO_2. Enzymes rearrange atoms and bonds in ways that transfer the pyruvate's potential energy and electrons to ATP, NADH, and another electron carrier molecule—**FADH$_2$.**

By the time the Krebs cycle is complete, the carbon atoms that made up the glucose are gone—liberated as CO_2. The cell has generated a few molecules of ATP, but most of the potential energy from glucose now lingers in the high-energy electron carriers, NADH and FADH$_2$. The cell uses them to generate more ATP.

The **electron transport chain** transfers energy-rich electrons from NADH and FADH$_2$ through a series of membrane proteins. As electrons pass from carrier to carrier in the electron transport chain, the energy is used to create a gradient of hydrogen ions. (Recall from chapter 2 that a hydrogen ion, H^+, is simply a hydrogen atom stripped of its electron—in other words, it is a proton. See figure 2.3.) The mitochondrion uses the potential energy stored in this proton gradient to generate ATP. An enzyme called **ATP synthase** forms a channel in the membrane, releasing the protons and using their potential energy to add phosphate to ADP. (As described in section 5.5, the same enzyme generates ATP in the light reactions of photosynthesis.) In the meantime, the "spent" electrons are transferred to O_2, generating water as a waste product.

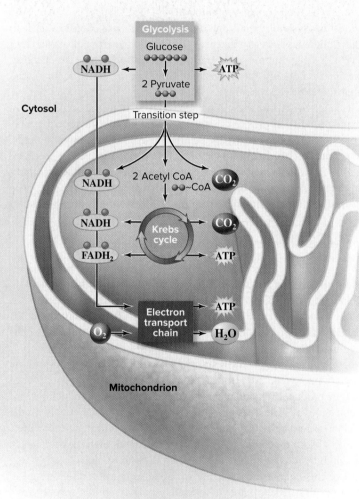

Figure 6.2 Overview of Aerobic Cellular Respiration.
The reactions of aerobic respiration harvest energy from glucose, producing ATP. First, glycolysis splits glucose. Then the transition step and the Krebs cycle complete glucose oxidation, releasing six molecules of CO_2. Throughout these reactions, electrons (*purple dots*) stripped from glucose attach to the electron carriers NADH and FADH$_2$, which tote them to the electron transport chain. Except for glycolysis, these reactions occur inside the mitochondria of eukaryotic cells.

6.2 Mastering Concepts

1. Why do the reactions of respiration occur step-by-step instead of all at once?

2. What occurs in the three stages of cellular respiration?

6.3 In Eukaryotic Cells, Mitochondria Produce Most ATP

Glycolysis always occurs in the cytosol, but the location of the other pathways in aerobic respiration depends on the cell type. In bacteria and archaea, the enzymes of the Krebs cycle are in the cytosol, and electron transport proteins are embedded in the cell membrane. The eukaryotic cells of protists, plants, fungi, and animals, however, contain **mitochondria,** specialized organelles that house the other reactions of cellular respiration (figure 6.3).

A mitochondrion is bounded by two membranes: an outer membrane and a highly folded inner membrane. **Cristae** are folds of the inner membrane. The **intermembrane compartment** is the area between the two membranes, and the mitochondrial **matrix** is the fluid enclosed within the inner membrane.

In a eukaryotic cell, the two pyruvate molecules produced in glycolysis cross both of the mitochondrial membranes and move into the matrix. Here, enzymes cleave pyruvate and carry out the Krebs cycle. Then, the electron carriers $FADH_2$ and NADH move to the inner mitochondrial membrane, which is studded with many copies of the electron transport proteins and ATP synthase. The inner membrane's cristae provide tremendous surface area on which the reactions of the electron transport chain occur.

Electron transport chains and ATP synthase also occur in the thylakoid membranes of chloroplasts, which generate ATP in the light reactions of photosynthesis (see chapter 5). Similar enzymes operate in respiring bacteria and archaea, making ATP synthase one of the most highly conserved proteins over evolutionary time.

Mitochondria and chloroplasts share another similarity, too: Both types of organelles contain DNA and ribosomes. Mitochondrial DNA encodes ATP synthase and most of the proteins of the electron transport chain. Not surprisingly, a person with abnormal versions of these genes may be very ill or even die. The worst mitochondrial diseases affect the muscular and nervous systems. Muscle and nerve cells are especially energy-hungry; each one may contain as many as 10,000 mitochondria. When their mitochondria fail, these cells cannot carry out their functions.

6.3 Mastering Concepts

1. What are the parts of a mitochondrion?
2. Which reactions occur in each part of a mitochondrion?

Miniglossary | Mitochondrion Anatomy

Mitochondrion	Organelle that houses many of the reactions of cellular respiration
Matrix	Fluid enclosed by the inner mitochondrial membrane; site of the transition step and Krebs cycle
Inner mitochondrial membrane	Site of the electron transport chain and ATP synthase enzyme
Intermembrane compartment	Space between the outer and inner membranes of a mitochondrion

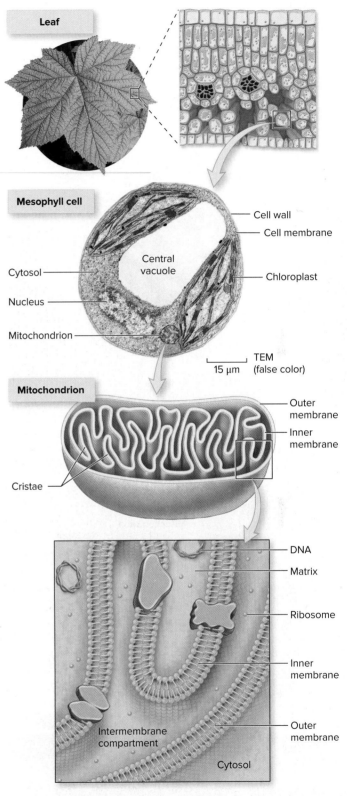

Figure 6.3 Anatomy of a Mitochondrion. Each mitochondrion includes two membranes. The inner membrane encloses fluid called the matrix, and the space between the inner and outer membranes is the intermembrane compartment.

Photos: (leaf): ©Robert Glusic/Corbis RF; electron micrograph by Wm. P. Wergin, Courtesy Eldon H. Newcomb, University of Wisconsin-Madison

6.4 Glycolysis Breaks Down Glucose to Pyruvate

Glycolysis is a more-or-less universal metabolic pathway that splits glucose into two three-carbon pyruvate molecules. The name of the pathway reflects its function: *glyco-* means "sugar," and *-lysis* means "to break."

The entire process of glycolysis requires many steps, all of which occur in the cell's cytosol. **Figure 6.4** shows a simplified version, with an emphasis on the major steps. Note that none of the steps requires O_2, so cells can use glycolysis in both oxygen-rich and oxygen-free environments.

The reactions of glycolysis are divided into two stages, the first of which is labeled "energy investment" in figure 6.4. In steps 1 and 2, the cell spends two

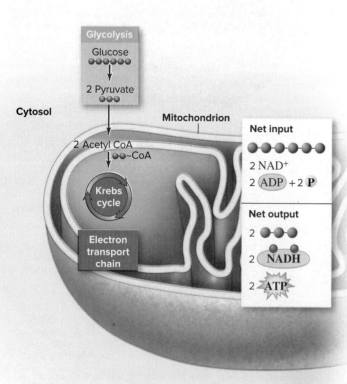

Figure 6.4 Glycolysis. Glycolysis splits glucose into two pyruvate molecules, producing a net yield of two ATPs and two NADHs. The illustration of the mitochondrion shows an overview of glycolysis; the rest of the figure shows simplified versions of the "energy investment" and "energy harvest" stages. (Each gray sphere represents a carbon atom.)

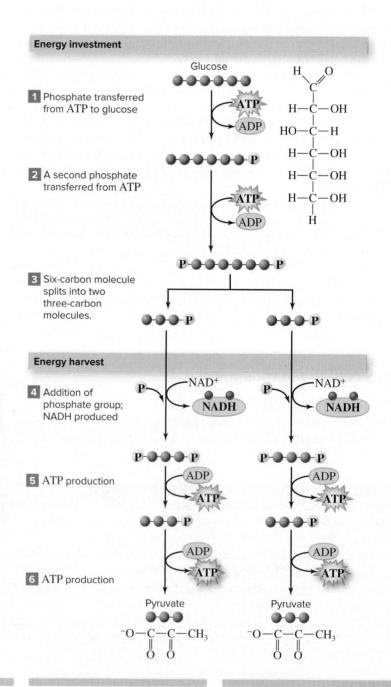

molecules of ATP to activate glucose, redistributing energy in the molecule and splitting it in half (step 3). Then, in the "energy harvest" stage, the cell generates a return on its initial investment, producing two molecules of NADH (step 4) plus four molecules of ATP (steps 5 and 6). Overall, the net gain is two NADHs and two ATPs per molecule of glucose.

Glucose contains considerable bond energy, but cells recover only a small portion of it as ATP and NADH during glycolysis. Most of the potential energy of the original glucose molecule remains in the two pyruvate molecules. Cells that carry out fermentation, such as yeasts that produce wine and beer, survive on this paltry ATP yield (see section 6.8). But as you will see, the pathways of aerobic respiration extract much more of that energy.

6.4 Mastering Concepts

1. Overall, what happens in glycolysis?
2. What is the net gain of ATP and NADH for each glucose molecule undergoing glycolysis?

6.5 Aerobic Respiration Yields Much More ATP than Glycolysis Alone

Overall, aerobic cellular respiration taps much of the potential energy remaining in the pyruvate molecules that emerge from the pathways of glycolysis. The Krebs cycle and electron transport chain are the key ATP-generating processes. This section explains what they do, and Why We Care 6.1 describes poisons that interfere with their work.

A. Pyruvate Is Oxidized to Acetyl CoA

After glycolysis, pyruvate moves into the mitochondrial matrix, where a preliminary "transition step" further oxidizes each pyruvate molecule (figure 6.5). First, a molecule of CO_2 is removed, and NAD^+ is reduced to NADH. The remaining two-carbon molecule, called an acetyl group, is transferred to a molecule called coenzyme A to form acetyl coenzyme A (abbreviated acetyl CoA). **Acetyl CoA** is the compound that enters the Krebs cycle.

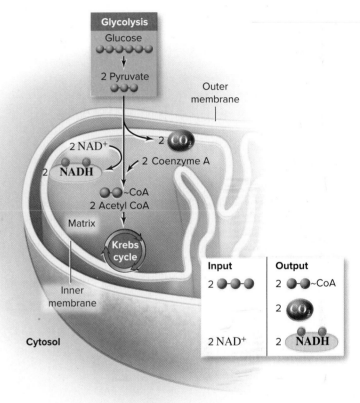

Figure 6.5 Transition Step. After pyruvate moves into a mitochondrion, it is oxidized to form a two-carbon acetyl group, CO_2, and NADH. The acetyl group joins with coenzyme A to form acetyl CoA, the molecule that enters the Krebs cycle.

Why We Care 6.1 | Some Poisons Inhibit Respiration

Many toxic chemicals kill by blocking one or more reactions in respiration. Here are a few examples:

Krebs cycle inhibitor

*   Arsenic binds to part of a molecule needed for the formation of acetyl CoA. It therefore blocks the Krebs cycle.

Electron transport inhibitors

*   Some mercury compounds stop an oxidation–reduction reaction early in the electron transport chain.
*   Cyanide blocks the final transfer of electrons to O_2. When proteins in the electron transport chain have no place to "dump" their electrons, the process grinds to a halt.

*   Carbon monoxide (CO) blocks electron transport at the same point as cyanide. This colorless, odorless gas is a byproduct of incomplete fuel combustion. CO from unvented heaters, stoves, and fireplaces can accumulate to deadly levels in homes. Car exhaust and cigarette smoke are other sources of CO.

Proton gradient inhibitor

*   A poison called DNP makes the inner mitochondrial membrane permeable to protons, blocking formation of the proton gradient necessary to drive ATP synthesis.

©Oliver Hoffmann/ Shutterstock

Miniglossary | Aerobic Respiration

Glycolysis	Series of reactions in which glucose splits into two pyruvate molecules; occurs in cytosol; produces ATP and NADH
Transition step	Series of reactions in which pyruvate is oxidized to acetyl CoA and CO_2; occurs in mitochondrial matrix; produces NADH
Krebs cycle	Series of reactions that oxidize acetyl CoA to CO_2; occurs in the mitochondrial matrix; produces ATP, NADH, and $FADH_2$
Mitochondrial electron transport chain	Series of electron carriers that accept electrons from NADH and $FADH_2$ and use the released energy to create a proton (H^+) gradient across a membrane
ATP synthase	Enzyme that uses the potential energy in a proton (H^+) gradient to produce ATP

Figure 6.6   Krebs Cycle. In the mitochondrial matrix, *(1)* acetyl CoA enters the Krebs cycle and *(2, 3)* is oxidized to two molecules of CO_2. *(4, 5, 6)* In the rest of the Krebs cycle, potential energy is trapped as ATP, NADH, and $FADH_2$.

B. The Krebs Cycle Produces ATP and High-Energy Electron Carriers

The Krebs cycle completes the oxidation of each acetyl group, releasing CO_2 (figure 6.6). The cycle begins when acetyl CoA sheds the coenzyme and combines with a four-carbon molecule (step 1). The resulting six-carbon molecule is called citrate; the Krebs cycle is therefore also known as the citric acid cycle.

The remaining steps in the Krebs cycle rearrange and oxidize citrate through several intermediates. Along the way, two carbon atoms are released as CO_2 (steps 2 and 3). In addition, some of the transformations transfer energy-rich electrons to NADH or $FADH_2$ (steps 2, 3, 5, and 6); others produce ATP (step 4). Eventually, the molecules in the Krebs cycle re-create the original four-carbon acceptor molecule. The cycle can now repeat.

Since one glucose molecule yields two acetyl CoA molecules, the Krebs cycle turns twice for each glucose. Thus, the combined net output to this point (glycolysis, transition step, and the Krebs cycle) is four ATP molecules, 10 NADH molecules, and two $FADH_2$ molecules. All six carbon atoms are gone, released as CO_2.

Besides continuing the breakdown of glucose, the Krebs cycle also has another function not directly related to respiration. The cell uses intermediate compounds formed in the Krebs cycle to manufacture other organic molecules, such as amino acids or fats. Section 6.7 explains that the reverse process also occurs; amino acids and fats can enter the Krebs cycle to generate energy from food sources other than carbohydrates.

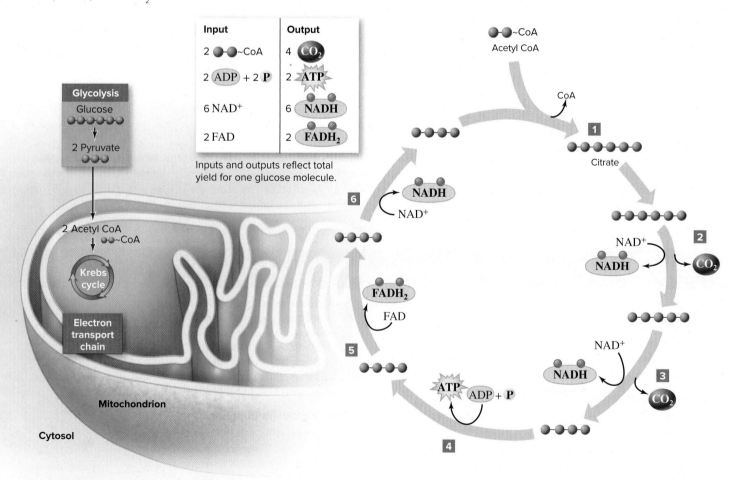

C. The Electron Transport Chain Drives ATP Formation

The products generated so far are CO_2, ATP, NADH, and $FADH_2$. The cell ejects the CO_2 as waste and uses ATP to fuel essential processes. But what becomes of the high-energy electron carriers (NADH and $FADH_2$)? They transfer their cargo to an electron transport chain in the inner mitochondrial membrane.

The electron transport chain harnesses the energy from these electrons in stages (figure 6.7). The first protein in the chain accepts electrons from NADH; $FADH_2$ donates its electrons to the second protein. The electrons then pass to the next protein in the chain, and the next, and so on. The final electron acceptor is O_2, which combines with H^+ to form water. Along the way, some of the proteins use energy from the electrons to pump H^+ from the matrix into the intermembrane compartment.

NADH and $FADH_2$ therefore deliver the energy that the electron transport chain uses to establish a proton (H^+) gradient across the inner mitochondrial membrane; this gradient represents a form of potential energy (see chapter 4). The mitochondrion harvests this energy as ATP in the final stage of cellular respiration, with the help of the ATP synthase enzyme. Protons move down their gradient through ATP synthase back into the matrix, and a phosphate group is linked to ADP. (Section 5.5A compared the function of ATP synthase to the turbines that generate electricity in a dam.) The ATP synthase enzyme therefore captures the potential energy of the proton gradient and saves it in a form the cell can use: ATP.

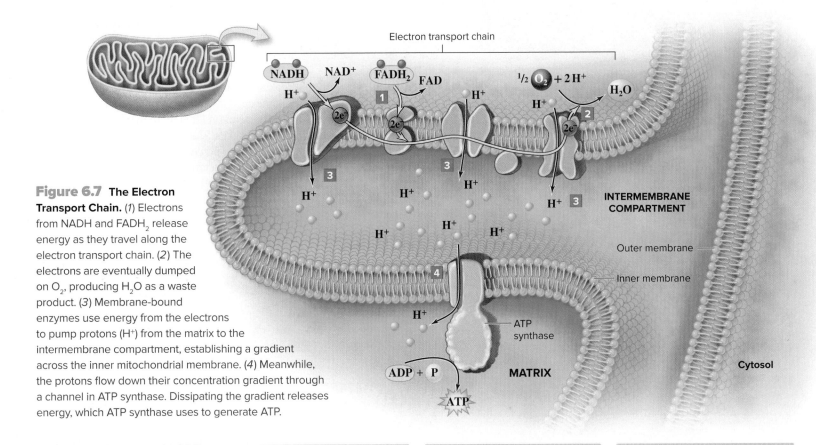

Figure 6.7 The Electron Transport Chain. (*1*) Electrons from NADH and $FADH_2$ release energy as they travel along the electron transport chain. (*2*) The electrons are eventually dumped on O_2, producing H_2O as a waste product. (*3*) Membrane-bound enzymes use energy from the electrons to pump protons (H^+) from the matrix to the intermembrane compartment, establishing a gradient across the inner mitochondrial membrane. (*4*) Meanwhile, the protons flow down their concentration gradient through a channel in ATP synthase. Dissipating the gradient releases energy, which ATP synthase uses to generate ATP.

6.6 How Many ATPs Can One Glucose Molecule Yield?

To estimate the yield of ATP produced from every glucose molecule that enters aerobic cellular respiration, we can add the maximum number of ATPs generated—directly and indirectly—in glycolysis, the transition step, the Krebs cycle, and the electron transport chain (figure 6.8).

Glycolysis yields two ATPs, as does the Krebs cycle (one ATP each from two turns of the cycle). In addition, each glucose yields two NADH molecules from glycolysis and two more from acetyl CoA production. Two turns of the Krebs cycle yield an additional six NADHs and two $FADH_2$s.

In theory, the ATP yield from electron transport is three ATPs per NADH and two ATPs per $FADH_2$. Electrons from the 10 NADHs from glycolysis, the transition step, and the Krebs cycle therefore yield up to 30 ATPs; electrons from the two $FADH_2$ molecules yield four more. Add the four ATPs from glycolysis and the Krebs cycle, and the total is 38 ATPs per glucose. However, NADH from glycolysis must be shuttled into the mitochondrion, usually at a cost of one ATP for each NADH. This reduces the net theoretical production of ATPs to 36.

In reality, some protons leak across the inner mitochondrial membrane on their own, and the cell spends some energy to move pyruvate and ADP into the matrix. These "expenses" lower the actual ATP yield to about 30 per glucose. The number of calories stored in 30 ATPs is about 32% of the total calories stored in the glucose bonds; the rest of the potential energy in glucose is lost to the environment as heat. This may seem wasteful, but for a biological process, it is reasonably efficient. To put this energy yield into perspective, an automobile uses only about 20% to 25% of the energy contained in gasoline's chemical bonds; the rest is released as heat.

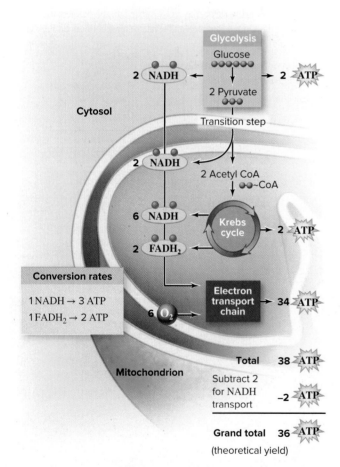

Figure 6.8 Energy Yield of Respiration. Breaking down glucose to carbon dioxide can theoretically yield as many as 36 ATPs, mostly from the electron transport chain.

6.7 Other Food Molecules Enter the Energy-Extracting Pathways

So far we have focused on the complete oxidation of glucose. But food also includes starch, proteins, and lipids that contribute calories to the diet. These molecules also enter the energy pathways (figure 6.9).

The digestion of starch from potatoes, wheat, rice, beans, and other carbohydrate-rich food begins in the mouth and continues in the small intestine. Enzymes snip the long starch chains into individual glucose monomers, which generate ATP as described in this chapter. Another polysaccharide, glycogen, follows essentially the same path as starch. ⓘ *carbohydrates,* section 2.5B

Proteins are digested into monomers called amino acids. The cell does not typically use these amino acids to produce ATP; instead, most of them are incorporated into new proteins. When an organism depletes its immediate carbohydrate supplies, however, cells may use amino acids as an energy source. First, nitrogen is stripped from the amino acid and excreted, often as urea. The remainder of each molecule enters the energy pathways as pyruvate, acetyl CoA, or an intermediate of the Krebs cycle, depending on the amino acid. ⓘ *amino acids,* section 2.5C

Meanwhile, enzymes in the small intestine digest fat molecules from food into glycerol and three fatty acids, which enter the bloodstream and move into the body's cells. (Burning Question 6.1 describes a diet pill that blocks this process.) Enzymes convert the glycerol to pyruvate, which then proceeds through the rest of cellular respiration as though it came directly from glucose. The fatty acids enter the mitochondria, where they are cut into many two-carbon pieces that become acetyl CoA. From here, the pathways continue as they would for glucose. ⓘ *lipids,* section 2.5E

Fats contain more calories per gram than any other food molecule; after all, a single fat molecule may yield dozens of two-carbon acetyl CoA groups for the Krebs cycle. Conversely, the body can also store excess energy from either carbohydrates or fat by doing the reverse: diverting acetyl CoA away from the Krebs cycle and using the two-carbon fragments to build fat molecules. These lipids are stored in fat tissue that the body can use for energy if food becomes scarce. Some hibernating animals use this metabolic strategy; see Burning Question 6.2.

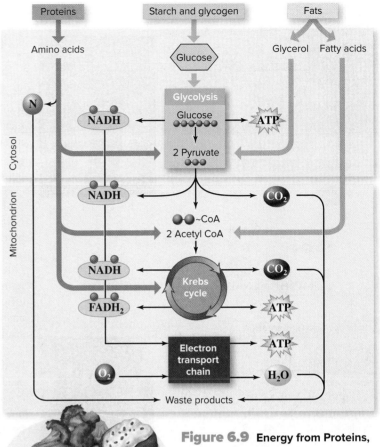

Figure 6.9 **Energy from Proteins, Polysaccharides, and Fats.** Although glucose is the primary source of energy, cells can also use proteins, starch, glycogen, and fats to generate ATP.

6.7 Mastering Concepts

1. At which points do digested polysaccharides, proteins, and fats enter the energy pathways?
2. How does the body store extra calories as fat?

Burning Question 6.2

©Dorling Kindersley/Getty Images RF

What happens during hibernation?

When winter arrives and snow blankets the ground, some mammals become dormant. Hibernation slows the metabolism of these animals, allowing them to conserve energy when food is scarce. In some hibernating squirrels, body temperature approaches freezing, and the breathing rate falls to as low as a couple of breaths per minute.

The reduced energy demand of hibernation does not mean the requirement for ATP disappears completely. Aerobic respiration must continue, albeit at a greatly reduced rate. Where does the energy come from? During winter, a hibernating bear's cells slowly extract energy from stored fat to maintain aerobic respiration. Other mammals, such as chipmunks, periodically break hibernation to snack on cached food.

Submit your burning question to
marielle.hoefnagels@mheducation.com

a. Alcoholic fermentation

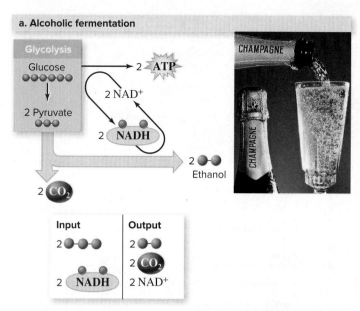

b. Lactic acid fermentation

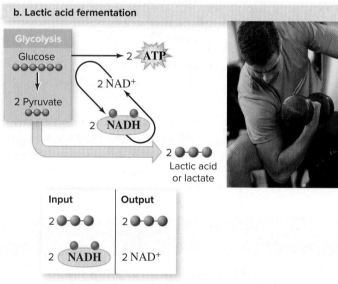

Figure 6.10 **Fermentation.** In fermentation, ATP comes only from glycolysis. (a) Yeasts produce ethanol and carbon dioxide by alcoholic fermentation; one possible product is champagne. (b) Lactic acid fermentation occurs in some bacteria and, occasionally, in mammalian muscle cells.

Photos: (a): ©Brand X Pictures/PunchStock RF; (b): ©Corbis RF

6.8 Fermenters Acquire ATP Only in Glycolysis

Most of the known organisms on Earth, including humans, use aerobic cellular respiration. Nevertheless, life thrives without O_2 in waterlogged soils, deep puncture wounds, sewage treatment plants, and your own digestive tract, to name just a few places. In the absence of O_2, the microbes in these habitats generate ATP using metabolic pathways that are anaerobic (meaning they do not use O_2). Fermentation is one such pathway.

In organisms that use fermentation, glycolysis still yields two ATPs, two NADHs, and two molecules of pyruvate per molecule of glucose. But the NADH does not donate its electrons to an electron transport chain, nor is the pyruvate further oxidized.

Instead, in **fermentation,** electrons from NADH reduce pyruvate. This process regenerates NAD⁺, which is essential for glycolysis to continue. But fermentation produces no additional ATP. This pathway is therefore far less efficient than respiration. Not surprisingly, fermentation is most common among microorganisms that live in sugar-rich environments where food is essentially unlimited.

Many of the microbes that live in human intestines make their entire living by fermentation. One example is a disease-causing protist (an amoeba) that causes a form of dysentery in humans. Others, including the bacterium *Escherichia coli,* use O_2 when it is available but switch to fermentation when it is not. Most multicellular organisms, however, require too much energy to rely on fermentation exclusively.

Of the many fermentation pathways that exist, one of the most familiar produces ethanol (an alcohol). In **alcoholic fermentation,** pyruvate is converted to ethanol and CO_2, while NADH is oxidized to produce NAD⁺ (figure 6.10a). Alcoholic fermentation produces wine from grapes, beer from barley, and cider from apples.

In **lactic acid fermentation,** a cell uses NADH to reduce pyruvate, but in this case, the products are NAD⁺ and lactic acid or its close relative, lactate (figure 6.10b). The bacterium *Lactobacillus,* for example, ferments the lactose in milk, producing the lactic acid that gives yogurt its sour taste.

Fermentation also occurs in human muscle cells. During vigorous exercise, muscles work so strenuously that they consume their available oxygen supply. In this "oxygen debt" condition, the muscle cells can acquire ATP only from glycolysis. The cells use lactic acid fermentation to generate NAD⁺ so that glycolysis can continue. Lactate concentrations therefore rise. After the exercise, when the circulatory system catches up with the muscles' demand for O_2, liver cells convert lactate back to pyruvate. Mitochondria then process the pyruvate as usual.

One common misconception about intense exercise is that lactic acid buildup causes the pH to drop in muscle cells, provoking soreness a day or two later. This idea, however, is a myth. Instead, microscopic tears in muscle tissue are now thought to be the culprit responsible for delayed muscle soreness.

Figure It Out

Compare the number of molecules of ATP generated from 100 glucose molecules undergoing aerobic respiration versus fermentation.

Answer: 3600 (theoretical yield) for aerobic respiration; 200 for fermentation.

6.8 Mastering Concepts

1. How many ATP molecules per glucose does fermentation produce?
2. What are two examples of fermentation pathways?

Investigating Life 6.1 | Hot Plants Offer Heat Rewards

Think of an organism that feels warm. Did you think of yourself? A puppy? Your cat? Chances are you did not picture a plant. Yet some plants, including one called *Philodendron solimoesense,* do warm their flowers to several degrees above ambient temperature (figure 6.A).

The flowers generate heat with a metabolic pathway that diverts electrons that enter the electron transport chain. NADH and $FADH_2$ still donate electrons to their usual acceptors, but the electrons next pass immediately to O_2 instead of traveling along the rest of the chain. The potential energy stored in the electrons is released as heat rather than helping the mitochondrion produce ATP.

Australian researchers wondered what the plant gains by heating its flowers. They did a simple set of experiments to find out whether warm flowers attract pollinators. First, they measured the temperature of *Philodendron* flowers and found that the central spike peaked at 40°C, about 15° above ambient temperature.

Next, the researchers turned their attention to beetles that pollinate the flowers. The team measured the amount of CO_2 produced by active and resting beetles at temperatures from 20°C to 35°C. (CO_2 production is an indirect measure of energy use.) Resting beetles emitted approximately the same amount of CO_2 at all temperatures, but active ones (such as those that would visit flowers) produced much less CO_2 at 30°C than they did at 20°C (see the graph in figure 6.A).

The results suggest that insects save energy by loitering on or near the flowers, energy that they can use to find food or lure mates. The hot flowers therefore enhance the reproductive success of both *Philodendron* and the beetles.

Source: Seymour, Roger S., Craig R. White, and Marc Gibernau. November 20, 2003. Heat reward for insect pollinators. *Nature,* vol. 426, pages 243–244.

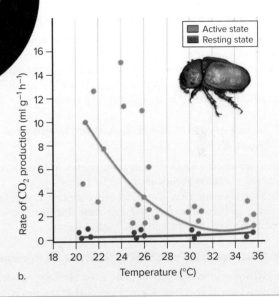

Figure 6.A **Energy Saver.** (a) The central spike of this *Philodendron solimoesense* flower generates heat. (b) The warm surroundings help active beetles—the flower's pollinators—save energy.

(a): ©Marc Gibernau, CNRS

a.

b.

What's the **Point?** ▼ APPLIED

Every living cell in your body requires ATP. As you have just learned, cells use oxygen and the energy in food molecules to produce ATP during aerobic respiration. Not surprisingly, we cannot last more than a few minutes without oxygen. On the other hand, people can survive for weeks without eating.

You might hear news stories about someone living for extended periods without food. In 2013, a woman was rescued after being buried for 16 days under a collapsed building with access only to air and some water. Many years earlier, a lean-bodied social activist from India, Mahatma Gandhi, voluntarily went hungry for 21 days before willingly ending his protest. Neither example approaches the limit to human survival without food, which is about 60 days (or less for people with little body fat).

©Stockbyte/PunchStock RF

How can humans survive for so long without nourishment? Part of the explanation is that metabolism slows to conserve energy, but the cell's reactions cannot stop entirely. Stored nutrients provide the rest of the answer. After only a few hours without food, cells begin to tap into stored glucose and other carbohydrates to produce ATP. These reserves last for about a day. Fat reserves are then metabolized, followed eventually by muscles, enzymes, antibodies, and other essential proteins. The body literally digests itself to maintain aerobic respiration until food again becomes available.

Unlike nutrients, cells cannot store O_2 in large amounts. The *What's the Point? Applied* in chapter 27 explains how some humans take oxygen debt to the limit.

CHAPTER SUMMARY

6.1 Cells Use Energy in Food to Make ATP

- Every cell requires **ATP** to power reactions that require energy input.
- **Aerobic respiration** is a biochemical pathway that produces ATP by extracting energy from glucose in the presence of oxygen.
- The overall reaction for cellular respiration is

$$C_6H_{12}O_6 + 6O_2 \longrightarrow 6CO_2 + 6H_2O + 36 \text{ ATP}$$

- In humans and many other animals, the respiratory system provides the oxygen that aerobic cellular respiration requires.
- Autotrophs such as plants also use aerobic respiration to generate ATP.

6.2 Cellular Respiration Includes Three Main Processes

- In respiration, electrons stripped from glucose are used to reduce O_2.
- Nearly all cells use **glycolysis** as the first step in harvesting energy from glucose. The **Krebs cycle** and an **electron transport chain** follow.
- The electron transport chain establishes a proton (H^+) gradient that powers the production of ATP by the enzyme **ATP synthase.**

6.3 In Eukaryotic Cells, Mitochondria Produce Most ATP

- In eukaryotes, the Krebs cycle and electron transport chain occur in organelles called **mitochondria.**
- Each mitochondrion has two membranes enclosing a central **matrix. Cristae** are the folds of the inner membrane. The space between the two membranes is the **intermembrane compartment.**

6.4 Glycolysis Breaks Down Glucose to Pyruvate

- In glycolysis, glucose is split into two molecules of **pyruvate.**
- The reactions of glycolysis also produce **NADH** and two ATPs.

6.5 Aerobic Respiration Yields Much More ATP than Glycolysis Alone

A. Pyruvate Is Oxidized to Acetyl CoA

- Pyruvate moves into the mitochondrial matrix, where it is broken down into **acetyl CoA** and CO_2. This "transition step" also produces NADH.

B. The Krebs Cycle Produces ATP and High-Energy Electron Carriers

- Acetyl CoA enters the Krebs cycle. This series of oxidation–reduction reactions occurs in the matrix and produces ATP, NADH, $FADH_2$, and CO_2.

C. The Electron Transport Chain Drives ATP Formation

- Energy-rich electrons from NADH and $FADH_2$ fuel an electron transport chain in the inner mitochondrial membrane. Electrons move along a series of proteins that release energy at each step. O_2 accepts the electrons at the end of the chain, producing water.
- Proteins in the electron transport chain pump H^+ from the matrix into the intermembrane compartment. As protons diffuse back into the matrix through ATP synthase, their potential energy drives ATP production.

6.6 How Many ATPs Can One Glucose Molecule Yield?

- In aerobic respiration, each glucose molecule theoretically yields 36 ATP molecules (figure 6.11). The actual yield is about 30 ATPs per glucose.

6.7 Other Food Molecules Enter the Energy-Extracting Pathways

- Polysaccharides are digested to glucose before undergoing cellular respiration. Amino acids enter the energy pathways as pyruvate, acetyl CoA, or an intermediate of the Krebs cycle. Fatty acids enter as acetyl CoA, and glycerol enters as pyruvate.

6.8 Fermenters Acquire ATP Only in Glycolysis

- **Fermentation** pathways oxidize NADH to NAD^+, which is recycled to glycolysis, but these pathways do not produce additional ATP. **Alcoholic fermentation** produces ethanol and CO_2, whereas **lactic acid fermentation** generates lactic acid or lactate as a waste.

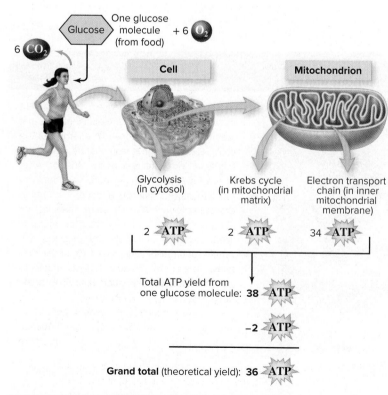

Figure 6.11 Theoretical ATP Yield for One Glucose Molecule.

MULTIPLE CHOICE QUESTIONS

1. Which of the following best describes aerobic respiration?
   a. The production of ATP from glucose in the presence of oxygen
   b. The production of pyruvate in the absence of oxygen
   c. The production of pyruvate using energy from the sun
   d. The production of ATP from glucose in the absence of oxygen

2. Which stage in cellular respiration produces the most ATP?
   a. Glycolysis   c. Krebs cycle
   b. Transition step   d. Electron transport

3. What is the role of ATP synthase?
   a. It uses ATP to make glucose.
   b. It uses a hydrogen ion gradient to make ATP.
   c. It uses ATP to make a hydrogen ion gradient.
   d. It synthesizes ATP directly from glucose.

4. Where in a eukaryotic cell does glycolysis occur?
   a. The cytosol
   b. The outer mitochondrial membrane
   c. The inner mitochondrial membrane
   d. The mitochondrial matrix

5. Which molecule has the greatest amount of potential energy?
   a. Pyruvate   b. Acetyl CoA   c. Glucose   d. CO_2

6. If a substance causes holes to form in the inner mitochondrial membrane, which process would be affected first?
   a. The donation of electrons to O_2
   b. Glycolysis
   c. The production of ATP by ATP synthase
   d. The formation of acetyl CoA

7. The CO_2 produced in respiration comes mainly from
   a. glycolysis.
   b. the Krebs cycle.
   c. the electron transport chain.
   d. All of the above processes generate CO_2.

8. Fats can be broken down into acetyl CoA for use in the Krebs cycle. Fats can also
   a. be built from excess acetyl CoA for energy storage.
   b. function as an electron carrier in the electron transport chain.
   c. be broken down directly into ATP.
   d. be broken down directly into NADH.

9. Why is it important to regenerate NAD^+ during fermentation?
   a. To help maintain the reactions of glycolysis
   b. So that it can transfer an electron to the electron transport chain
   c. To maintain the concentration of pyruvate in a cell
   d. To produce alcohol or lactic acid for the cell

Answers to Multiple Choice Questions are in appendix A.

WRITE IT OUT

1. *Respiration* contains the Latin word root *spiro,* which means "to breathe." Why is the process described in this chapter called cellular respiration?
2. All steps of cellular respiration are closely connected. Describe the problems that would occur if glycolysis, the transition step, the Krebs cycle, or the electron transport chain were not working.
3. How does aerobic respiration yield so much ATP from each glucose molecule, compared with glycolysis alone?
4. Health-food stores sell a product called "pyruvate plus," which supposedly boosts energy. Why is this product unnecessary? What would be a much less expensive substitute that would accomplish the same thing?
5. At what point does O_2 enter the energy pathways of aerobic respiration? What is the role of O_2? Why does respiration stop if a person cannot breathe? Why would a cell die if it could not make ATP?
6. How might a mitochondrion's double membrane make cellular respiration more efficient than if it had a single membrane?
7. Some types of beer are bottled with yeast. These beers are not carbonated at bottling, but if you open them a few weeks later, they will bubble. Explain the source of this carbonation.
8. A student runs 5 kilometers each afternoon at a slow, leisurely pace. One day, she runs 2 km as fast as she can. Afterward she is winded. She thought she was in great shape! What, in terms of energy metabolism, has she experienced?
9. Explain the fact that species as diverse as humans and yeasts use the same biochemical pathways to extract energy from nutrient molecules.
10. Compare the number of ATP molecules required to produce one glucose molecule in photosynthesis (see figure 5.8) with the number of ATP molecules generated per glucose in aerobic respiration (see figure 6.8). How do these numbers compare to the ATP yield from fermentation?

SCIENTIFIC LITERACY

Review Why We Care 6.1, which lists several poisons that inhibit aerobic respiration. One of the chemicals listed, arsenic, became the center of media attention after physician and TV personality Dr. Oz issued a warning about arsenic levels in apple juice. The FDA disagreed, saying that the arsenic levels in apple juices are safe. As a consumer, how can you decide which story to believe? Who might be more motivated to stretch the truth: Dr. Oz or the FDA? Why?

PULL IT TOGETHER

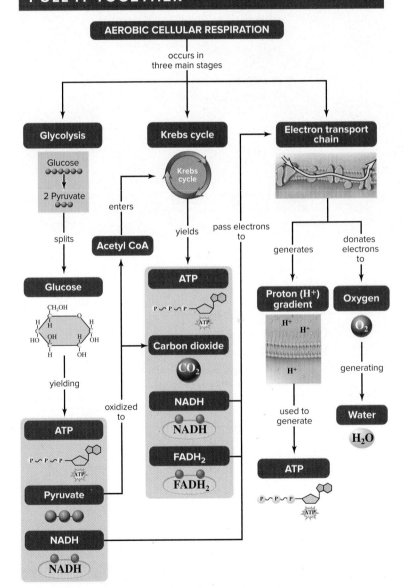

Figure 6.12 Pull It Together: Respiration and Fermentation.

Refer to figure 6.12 and the chapter content to answer the following questions.
1. Add the locations of each stage of respiration to this map.
2. How many ATP, NADH, CO_2, $FADH_2$, and H_2O molecules are produced at each stage of respiration?
3. What do cells do with the ATP they generate in respiration?
4. Review the Survey the Landscape figure in the chapter introduction. Explain the connection between respiration and photosynthesis.

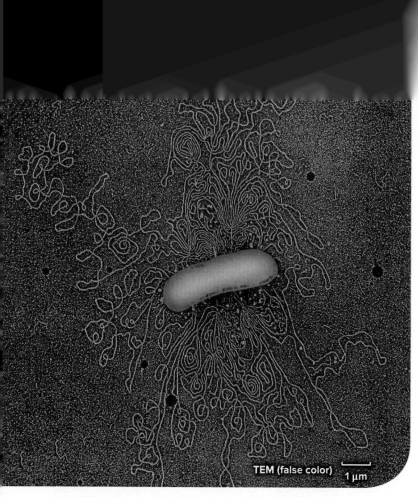

TEM (false color) 1 μm

Lots of DNA. Genetic material bursts from this bacterium, illustrating just how much DNA is packed into a single cell.
©G. Murti/Science Source

LEARNING OUTLINE

7.1 DNA Is a Double Helix

7.2 DNA Stores Genetic Information: An Overview

7.3 Transcription Uses a DNA Template to Build RNA

7.4 Translation Builds the Protein

7.5 Cells Regulate Gene Expression

7.6 Mutations Change DNA

7.7 Viruses Are Genes Wrapped in a Protein Coat

7.8 Viruses Infect All Cell Types

7.9 Drugs and Vaccines Help Fight Viral Infections

7.10 Viroids and Prions Are Other Noncellular Infectious Agents

---

APPLICATIONS

Why We Care 7.1 *Poisons That Block Protein Production*

Burning Question 7.1 *Is there a gay gene?*

Burning Question 7.2 *Why do we get sick when the weather turns cold?*

Investigating Life 7.1 *Clues to the Origin of Language*

Learn How to Learn
Explain It, Right or Wrong

As you work through any multiple choice question, such as those at the end of each chapter or on an exam, make sure you can explain why each correct choice is right. You can also test your understanding by taking the time to explain why each of the other choices is wrong.

SURVEY THE LANDSCAPE
DNA, Inheritance, and Biotechnology

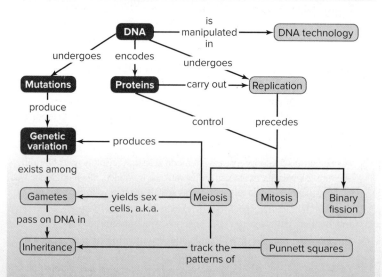

DNA is an information storage molecule; its main function is to carry the "recipes" for the proteins that do the cell's work. A cell will also express mutated DNA or genetic material received from viruses, producing a variety of proteins.

For more details, study the Pull It Together feature in the chapter summary.

What's the **Point?** ▼

The nucleic acid DNA is one of the most familiar molecules, the subject matter of movies and headlines. Criminal trials hinge on DNA evidence; the idea of human cloning raises questions about the role of DNA in determining who we are; and DNA-based discoveries are yielding new diagnostic tests, medical treatments, and vaccines.

©Creatista/Shutterstock RF

More important than DNA's role in society is its role in life itself. DNA acts as a sort of "library" that stores the information required for cells to function. Every species passes this vital molecule from generation to generation, with major or minor changes along the way. We begin this genetics unit with a look at DNA's structure and its role in the cell. Armed with this knowledge, toward the end of the chapter, you will learn how viruses make us sick by "spiking" our cells with their own genetic information.

7.1 DNA Is a Double Helix

Life depends on **DNA (deoxyribonucleic acid),** a molecule with a remarkable function: It stores the information that each cell needs to produce proteins. As described in section 2.5, proteins are chains of amino acids that fold into intricate, three-dimensional shapes.

Proteins are vital not only to the life of individual cells but also to defining many of an organism's overall traits. Some proteins do their jobs by interacting with other chemicals inside cells. Enzymes, for example, are proteins that speed up the cell's chemical reactions. These reactions break down food, build the cell's structures, place blood type molecules on red blood cell surfaces, produce the melanin pigment in your skin, and so on. Other proteins have jobs that do not relate directly to chemical reactions. Hemoglobin, for example, is a protein that carries oxygen in the bloodstream; other proteins determine the color of your eyes.

Because proteins have so many critical functions, the instructions in DNA make life possible. In fact, before a cell divides, it first makes an exact replica of its DNA. This process, described in chapter 8, copies all of the information that will enable the next generation of cells to live.

Given DNA's central role in life, it is amazing to consider that biologists only confirmed that DNA is the genetic material in the 1950s. Even then, nobody completely understood its chemical structure. They knew that **nucleotides** form DNA's building blocks, and they knew that each nucleotide included one type of base: adenine (A), cytosine (C), guanine (G), and thymine (T).

The breakthrough came in 1953. U.S. biochemist James Watson and English physicist Francis Crick used two lines of evidence to deduce DNA's structure. First, biochemist Erwin Chargaff had shown that the amount of guanine in a DNA molecule always equals the amount of cytosine, and the amount of adenine always equals the amount of thymine. Second, English physicist Maurice Wilkins and chemist Rosalind Franklin used a technique called X-ray diffraction to determine the three-dimensional shape of the molecule. The X-ray diffraction pattern revealed a regularly repeating structure of building blocks.

Watson and Crick combined these clues to build a ball-and-stick model of the DNA molecule. The now familiar double helix included equal amounts of G and C and of A and T, and it had the sleek symmetry revealed in the X-ray diffraction pattern (figure 7.1).

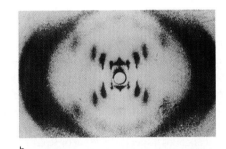

a.   b.

c.

d.

Figure 7.1 Discovery of DNA's Structure. (a) Rosalind Franklin produced (b) X-ray images that were crucial in the discovery of DNA's structure. (c) Maurice Wilkins, Francis Crick, and James Watson (*first, third,* and *fifth from the left*) shared the 1962 Nobel Prize in Physiology or Medicine for their discovery. Franklin had died in 1958, and by the rules of the award, she could not be included. (d) A model of the DNA double helix.

(a, b): ©Science Source/Science Source; (c): ©Bettmann/Corbis/Getty Images; (d): ©CNRI/Science Source

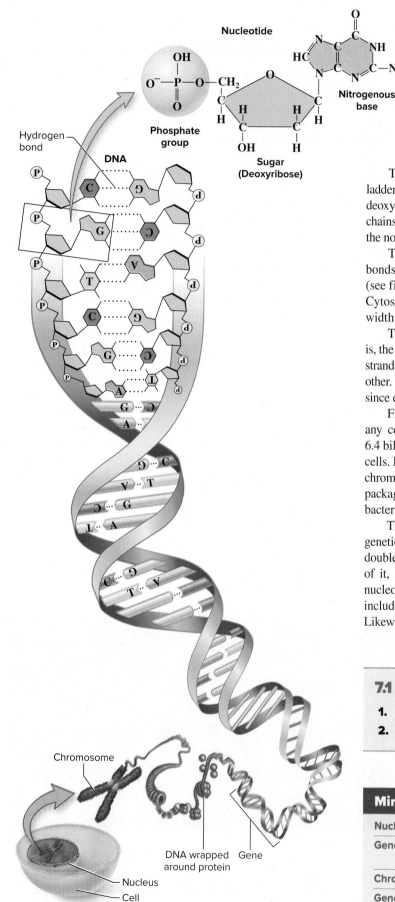

Nucleotide

OH

O⁻—P—O—CH₂

O

Phosphate group

Sugar (Deoxyribose)

Nitrogenous base

Hydrogen bond

DNA

Chromosome

DNA wrapped around protein   Gene

Nucleus

Cell

Figure 7.2 From Nucleotide to Chromosome. The DNA double helix consists of two strands of nucleotides, oriented in opposite directions. A eukaryotic cell's nucleus contains chromosomes, which consist of DNA wrapped around specialized proteins. A gene is a segment of DNA that encodes a protein or RNA molecule.

The DNA double helix resembles a twisted ladder. The twin rails of the ladder, also called the sugar–phosphate "backbones," are alternating units of deoxyribose and phosphate joined with covalent bonds (figure 7.2). The two chains are parallel to each other, but they are oriented in opposite directions, like the northbound and southbound lanes of a highway.

The ladder's rungs are A–T and C–G base pairs joined by hydrogen bonds. These base pairs arise from the chemical structures of the nucleotides (see figure 2.24). Adenine and guanine are bases with a double ring structure. Cytosine and thymine each have a single ring. Each A–T pair is the same width as a C–G pair because each includes a double- and a single-ringed base.

The two strands of a DNA molecule are **complementary** to each other; that is, the sequence of each strand determines the sequence of the other. An A on one strand means a T on the opposite strand, and a G on one strand means a C on the other. The two strands are therefore somewhat like a photograph and its negative, since each is sufficient to define the other.

Figure 7.2 shows a small amount of DNA. In reality, the amount of DNA in any cell is immense; in humans, for example, each nucleus contains some 6.4 billion base pairs. An organism's **genome** is all of the genetic material in its cells. In a eukaryotic cell, the majority of the genome is divided among multiple chromosomes housed inside the cell's nucleus; each **chromosome** is a discrete package of DNA coiled around proteins. On the other hand, the genome of a bacterial cell typically consists of one circular chromosome.

The chromosome at the bottom of figure 7.2 is tightly coiled; to use its genetic information, the cell must "unpack" the chromosome and expose the double helix. Once unpacked, much of the DNA has no known function. Some of it, however, encodes RNA and proteins. A **gene** is a sequence of DNA nucleotides that encodes a specific protein or RNA molecule; the human genome includes 20,000 to 25,000 genes scattered on its 23 pairs of chromosomes. Likewise, a bacterial chromosome is also divided into multiple genes.

7.1 Mastering Concepts

1. How did Watson and Crick decipher the structure of DNA?

2. Describe DNA's components and three-dimensional structure.

Miniglossary | Hierarchy of DNA Structure

Nucleotide	Building block of nucleic acids (DNA and RNA)
Gene	A sequence of DNA nucleotides that encodes a specific protein or RNA molecule
Chromosome	A DNA molecule wrapped around proteins
Genome	All of the genetic material in a cell

7.2 DNA Stores Genetic Information: An Overview

In the 1940s, biologists deduced that each gene somehow controls the production of one protein. In the next decade, Watson and Crick described this relationship between nucleic acids and proteins as a flow of information they called the "central dogma" (figure 7.3). First, in **transcription,** a cell "rewrites" a gene's DNA sequence to a complementary RNA molecule. Then, in **translation,** the information in RNA is used to assemble a different class of molecule: a protein (just as an interpreter translates one language into another).

A gene is therefore somewhat like a recipe in a cookbook. A recipe specifies the ingredients and instructions for assembling one dish, such as brownies. Likewise, a protein-encoding gene contains the instructions for assembling a polypeptide, amino acid by amino acid (the polypeptide subsequently folds to form the finished protein). A cookbook that contains many recipes is analogous to a chromosome, which is an array of genes. A person's entire collection of cookbooks, then, is analogous to a genome.

RNA (ribonucleic acid) is a multifunctional nucleic acid that differs from DNA in several ways (figure 7.4). First, its nucleotides contain the sugar ribose instead of deoxyribose. Second, RNA has the nitrogenous base uracil, which behaves like thymine; that is, uracil binds with adenine in complementary base pairs. Third, unlike DNA, RNA can be single-stranded (although it often folds into loops). Finally, RNA can catalyze chemical reactions, a role not known for DNA.

Three types of RNA (abbreviated mRNA, tRNA, and rRNA) interact to synthesize proteins. The function of each is explained later in this chapter, but a brief overview is helpful at this point. **Messenger RNA (mRNA)** carries the information that specifies a protein. The mRNA is divided into genetic "code words" called codons; a **codon** is a group of three consecutive mRNA bases that corresponds to one amino acid. **Transfer RNA (tRNA)** molecules are "connectors" that bind an mRNA codon at one end and a specific amino acid at the other. Their role is to carry each amino acid to the correct spot along the mRNA molecule. **Ribosomal RNA (rRNA)** combines with proteins to form a **ribosome,** the physical location where translation occurs.

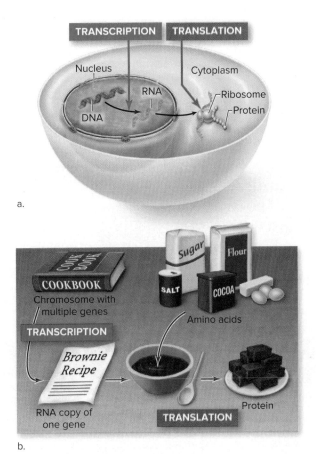

a.

b.

Figure 7.3 DNA to RNA to Protein. (a) The central dogma of biology states that information stored in DNA is copied to RNA (transcription), which is used to assemble proteins (translation). (b) DNA stores the information used to make proteins, just as a recipe stores the information needed to make brownies.

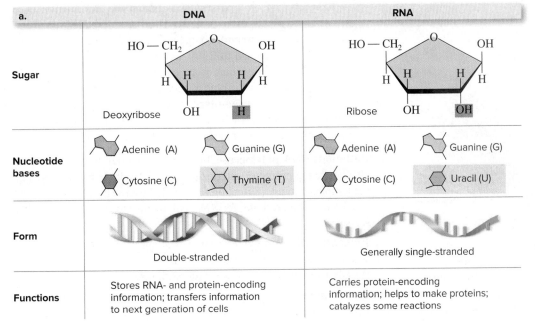

a.	DNA		RNA		b. Complementary base pairs
Sugar	HO—CH₂ O OH H H H H Deoxyribose OH H		HO—CH₂ O OH H H H H Ribose OH OH		**DNA** pairs with **RNA** Adenine (A) — Uracil (U) Cytosine (C) — Guanine (G) Guanine (G) — Cytosine (C) Thymine (T) — Adenine (A)
Nucleotide bases	Adenine (A) Guanine (G) Cytosine (C) Thymine (T)		Adenine (A) Guanine (G) Cytosine (C) Uracil (U)		**RNA** pairs with **RNA** Adenine (A) — Uracil (U) Cytosine (C) — Guanine (G) Guanine (G) — Cytosine (C) Uracil (U) — Adenine (A)
Form	Double-stranded		Generally single-stranded		
Functions	Stores RNA- and protein-encoding information; transfers information to next generation of cells		Carries protein-encoding information; helps to make proteins; catalyzes some reactions		

Figure 7.4 DNA and RNA Compared. (a) DNA and RNA differ in structure and function. (b) RNA contains uracil, not thymine. Like thymine, uracil pairs with adenine in complementary base pairs.

Mutations Change DNA   What Are Viruses?   Viruses Infect All Cell Types   Fighting Viral Infections   Viroids and Prions

To illustrate DNA's function with a concrete example, suppose a cell in a female mammal's breast is producing milk to feed an infant (see figure 3.11). One of the proteins in milk is albumin. Inside the nucleus of a milk-producing cell, an enzyme first transcribes the albumin gene's DNA sequence (albumin's "recipe") to a complementary sequence of RNA. After some modification, the RNA emerges from the nucleus and binds to a ribosome. There, in translation, amino acids are assembled in a specific order to produce the albumin protein.

Notice that the amino acid sequence in albumin is dictated by the sequence of nucleotides in the RNA molecule. The RNA, in turn, was transcribed from DNA. In this way, DNA provides the recipe for albumin and every other protein.

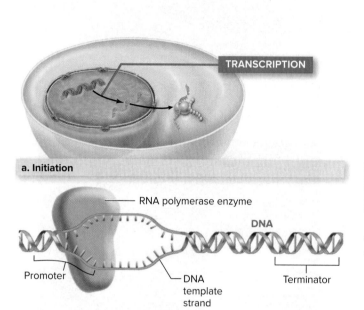

a. Initiation

RNA polymerase enzyme

DNA

Promoter

DNA template strand

Terminator

b. Elongation

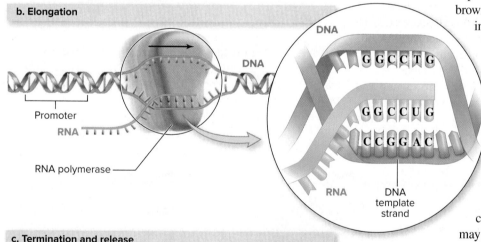

DNA

DNA

Promoter

RNA

RNA polymerase

GGCCTG

GGCCUG

CCGGAC

RNA

DNA template strand

c. Termination and release

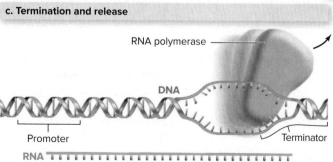

RNA polymerase

DNA

Promoter

Terminator

RNA

Figure 7.5 Transcription Builds mRNA. (a) During initiation, RNA polymerase binds to DNA. (b) During elongation, RNA polymerase adds nucleotides to the mRNA strand. (c) Termination occurs at the terminator sequence in the gene; RNA polymerase and mRNA detach from the DNA, which returns to its double-helix form.

7.2 Mastering Concepts

1. What is the relationship between a gene and a protein?
2. How do transcription and translation use genetic information?
3. What are the three types of RNA, and how does each contribute to protein synthesis?

7.3 Transcription Uses a DNA Template to Build RNA

Transcription produces an RNA copy of one gene. Recall from our brownie analogy that a gene is a recipe for a protein. According to this analogy, transcription is like opening a cookbook to a particular page and copying just the recipe for the dish you want to prepare. After the copy is made, the book can return safely to the shelf. Just as you would then use the instructions on the copy to make your meal, the cell uses the information in RNA—and not the DNA directly—to make each protein.

DNA is a double helix, but only one of the two strands contains the information encoding each protein. This strand, called the **template strand,** contains the DNA sequence that is actually copied to RNA. Depending on the gene, the template may be either of the two DNA strands. How does the cell "know" which strand to transcribe? The enzymes that carry out transcription recognize the **promoter,** a DNA sequence that not only signals a gene's start but also indicates which of the two strands is the template.

Transcription occurs in three stages (figure 7.5):

1. **Initiation: RNA polymerase,** the enzyme that builds an RNA chain, attaches to the promoter and unzips the DNA double helix, exposing the template strand. (Often, proteins called transcription factors must first bind to the promoter for RNA polymerase to attach to the DNA; see section 7.5.)

2. **Elongation:** RNA polymerase moves along the DNA template strand, adding RNA nucleotides that are complementary to exposed bases on the DNA template strand (see figure 7.5, inset).

3. **Termination:** A **terminator** sequence in DNA signals the end of the gene. Upon reaching the terminator, the RNA polymerase enzyme separates

from the DNA template and releases the newly produced RNA. The DNA molecule then resumes its usual double-helix shape.

As the RNA molecule forms, it curls into a three-dimensional shape dictated by complementary base pairing within the molecule. The finished RNA may function as mRNA, tRNA, or rRNA. The observation that the cell's DNA encodes all types of RNA—not just mRNA—has led to debate over the definition of the word *gene*. Originally, a gene was defined as any stretch of DNA that encodes one protein. More recently, however, the definition has expanded to include any DNA sequence that is transcribed to RNA. The phrase *gene expression* can therefore mean the production of either a functional RNA molecule or a protein.

In bacteria and archaea, ribosomes may begin translating mRNA to a protein before transcription is even complete. In eukaryotic cells, however, mRNA is usually altered before it leaves the nucleus to be translated (figure 7.6). A short sequence of modified nucleotides, called a cap, is added to one end of the mRNA molecule. At the opposite end, 100 to 200 adenines are added, forming a "poly A tail." Together, the cap and poly A tail help ensure that ribosomes attach to the correct end of the mRNA.

In archaea and in eukaryotic cells, only part of an mRNA molecule is translated into an amino acid sequence. Figure 7.6 shows that an mRNA molecule consists of alternating sequences called introns and exons. **Introns** are portions of the mRNA that are removed before translation. The remaining portions, called **exons,** are spliced together to form the mature mRNA that leaves the nucleus to be translated. (One tip for remembering this is that *ex*ons are the parts of the mRNA that are actually *ex*pressed or that *ex*it the nucleus.)

The amount of genetic material devoted to introns can be immense. The average exon is 100 to 300 nucleotides long, whereas the average intron is about 1000 nucleotides long. Some mature mRNA molecules consist of 70 or more spliced-together exons; the cell therefore simply discards much of the RNA produced in transcription.

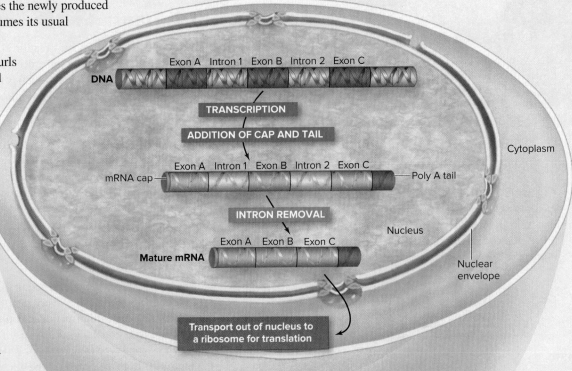

Figure 7.6 Processing mRNA. In eukaryotic cells, a nucleotide cap and poly A tail are added to mRNA, and introns are removed. Finally, the mature mRNA exits the nucleus.

7.3 Mastering Concepts

1. What happens during each stage of transcription?
2. Where in the cell does transcription occur?
3. What is the role of RNA polymerase in transcription?
4. What are the roles of the promoter and terminator sequences?
5. How is mRNA modified before it leaves the nucleus of a eukaryotic cell?

Miniglossary │ Gene Expression	
Transcription	Process in which a cell builds an mRNA copy of DNA
Translation	Process in which a cell builds a protein using the information in mRNA
Template strand	The DNA strand that is transcribed
Codon	A three-nucleotide mRNA sequence that encodes one amino acid or a "stop translation" signal
Genetic code	The "dictionary" that relates each codon with an amino acid or stop signal

Mutations Change DNA   What Are Viruses?   Viruses Infect All Cell Types   Fighting Viral Infections   Viroids and Prions

Figure It Out

Write the sequence of the mRNA molecule transcribed from the following DNA template sequence: T T A C A C T T G C A A C

Answer: AAUGUGAACGUUG.

Figure 7.7 **The Genetic Code.** In translation, mRNA codons are matched with amino acids as specified in this "dictionary" of the genetic code. Most mRNA codons correspond to an amino acid. Three stop codons, however, signal the ribosome to stop translating. (Appendix E shows the chemical structures of all 20 amino acids.)

7.4 Translation Builds the Protein

Transcription copies the information encoded in a DNA base sequence into the complementary language of mRNA. Once transcription is complete and mRNA has left the nucleus, the cell is ready to translate the mRNA "message" into a sequence of amino acids. If mRNA is like a copy of a recipe, then translation is like preparing the dish.

On paper, translating a molecule of mRNA is easy, thanks to the work of biologists who, in the 1960s, deciphered the **genetic code**—the set of "rules" by which a cell uses the codons in mRNA to assemble amino acids into a protein. Figure 7.7 shows the complete genetic code, the product of about a decade of research conducted in many laboratories. The rest of this section explains how cells implement the genetic code in protein synthesis.

A. Translation Requires mRNA, tRNA, and Ribosomes

Translation—the actual construction of the protein—requires three main types of participants. We have already met the first type: mRNA, the molecule that contains the genetic information encoding a protein. As illustrated in figure 7.7, each three-base codon in mRNA specifies one amino acid.

The second type of participant is tRNA. These "bilingual" molecules carry amino acids from the cytosol to the mRNA being translated. Each tRNA includes an **anticodon,** a three-base loop on tRNA that is complementary to one mRNA codon. The other end of each tRNA molecule carries the amino acid corresponding to that codon. For example, a tRNA with the anticodon sequence AAG always picks up the amino acid phenylalanine, which is encoded by the codon UUC (see figure 7.7).

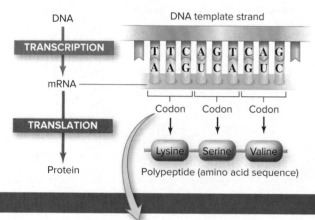

The Genetic Code

Second letter of codon

First letter of codon		U	C	A	G	Third letter of codon
U	UUU UUC	Phenylalanine (Phe; F)	UCU UCC UCA UCG — Serine (Ser; S)	UAU UAC — Tyrosine (Tyr; Y)	UGU UGC — Cysteine (Cys; C)	U C
	UUA UUG	Leucine (Leu; L)		UAA **Stop** UAG **Stop**	UGA **Stop** UGG Tryptophan (Trp; W)	A G
C	CUU CUC CUA CUG	Leucine (Leu; L)	CCU CCC CCA CCG — Proline (Pro; P)	CAU CAC — Histidine (His; H) CAA CAG — Glutamine (Gln; Q)	CGU CGC CGA CGG — Arginine (Arg; R)	U C A G
A	AUU AUC AUA — Isoleucine (Ile; I) AUG **Start** Methionine (Met; M)		ACU ACC ACA ACG — Threonine (Thr; T)	AAU AAC — Asparagine (Asn; N) AAA AAG — Lysine (Lys; K)	AGU AGC — Serine (Ser; S) AGA AGG — Arginine (Arg; R)	U C A G
G	GUU GUC GUA GUG — Valine (Val; V)		GCU GCC GCA GCG — Alanine (Ala; A)	GAU GAC — Aspartic acid (Asp; D) GAA GAG — Glutamic acid (Glu; E)	GGU GGC GGA GGG — Glycine (Gly; G)	U C A G

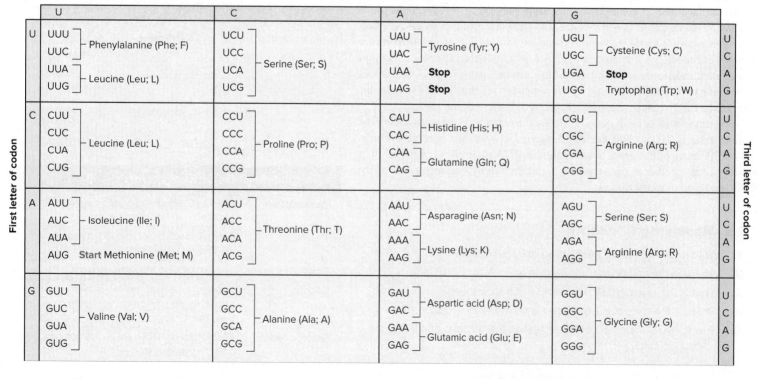

The remaining participant in translation is the ribosome. Ribosomes are the sites of translation. That is, in the recipe analogy in figure 7.3, a ribosome is the "bowl" where the ingredients come together (and tRNA molecules are helpers that carry those ingredients to the bowl). Each cell has many ribosomes, which may be free in the cytosol or attached to the rough endoplasmic reticulum (see figure 3.13). As you will soon see, each ribosome has one large and one small subunit that join at the start of translation.

B. Translation Occurs in Three Steps

The process of translation is divided into three stages, during which mRNA, tRNA molecules, and ribosomes come together, link amino acids into a chain, and then dissociate again (figure 7.8).

1. **Initiation:** One end of the mRNA molecule (the "cap" end) bonds with a small ribosomal subunit. The first mRNA codon to specify an amino acid is usually AUG, which attracts a tRNA that carries the amino acid methionine. A large ribosomal subunit attaches to the small subunit to complete initiation.

2. **Elongation:** A tRNA molecule carrying the second amino acid (glycine in figure 7.8) binds to the second codon, GGA in this case. The two amino

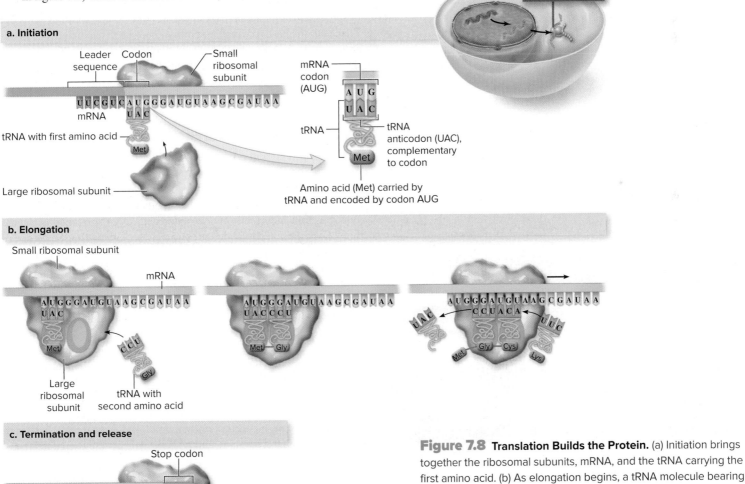

Figure 7.8 Translation Builds the Protein. (a) Initiation brings together the ribosomal subunits, mRNA, and the tRNA carrying the first amino acid. (b) As elongation begins, a tRNA molecule bearing the second amino acid binds to the second codon. The first amino acid forms a covalent bond with the second amino acid. Additional tRNAs bring subsequent amino acids encoded in the mRNA. (c) Termination occurs when a release factor protein binds to the stop codon. All components of the translation machine are released, along with the completed polypeptide.

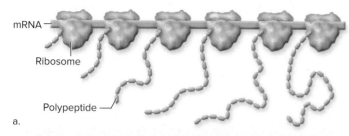

mRNA

Ribosome

Polypeptide

a.

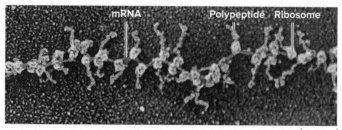

mRNA · Polypeptide Ribosome

b. · TEM (false color) ⊢50 nm⊣

Figure 7.9 **Efficient Translation.** (a) Multiple ribosomes can simultaneously translate one mRNA. (b) This micrograph shows about 30 ribosomes producing proteins from the same mRNA.

(b): ©Dr. Elena Kiseleva/SPL/Science Source

Figure It Out

How many different three-codon sequences encode the amino acid sequence Phe-Val-Ala? *Hint:* Refer to figure 7.7.

Answer: 2 (Phe) × 4 (Val) × 4 (Ala) = 32.

Why We Care 7.1 | Poisons That Block Protein Production

We learned in chapter 6 that some poisons kill because they interfere with respiration. Here we list a few poisons that inhibit protein synthesis. A cell that cannot make proteins quickly dies.

Death cap mushroom

- **Amanatin:** This toxin occurs in a mushroom called the "death cap." Amanatin inhibits RNA polymerase, making transcription impossible.

- **Diphtheria toxin:** Certain bacteria secrete a toxin that causes the respiratory illness diphtheria. The toxin inhibits a protein that helps add amino acids to a polypeptide chain during translation.

- **Antibiotics:** Antibiotics that bind to bacterial ribosomes include clindamycin, chloramphenicol, tetracyclines, and gentamicin. When its ribosomes are disrupted, a bacterium cannot make proteins, and it dies.

Photo: ©Barrie Watts/Oxford Scientific/Getty Images

acids, methionine and glycine, align, and a covalent bond forms between them. With that bond in place, the ribosome releases the first tRNA. Next, the ribosome moves down the mRNA by one codon. A third tRNA enters, carrying its amino acid. This third amino acid aligns with the other two and forms a covalent bond to the second amino acid in the growing chain, and the tRNA attached to glycine is released. In this way, the polypeptide grows one amino acid at a time, as tRNAs continue to deliver their cargo.

3. **Termination:** Elongation halts at a "stop" codon (UGA, UAG, or UAA). No tRNA molecules correspond to these stop codons. Instead, proteins called release factors bind to the stop codon, prompting the participants of translation to separate from one another. The ribosome releases the last tRNA, the ribosomal subunits separate and are recycled, and the new polypeptide is released.

Protein synthesis can be very speedy. A cell in the human immune system, for example, can manufacture 2000 identical antibody proteins per second, helping the body respond quickly to infections. How can protein synthesis occur fast enough to meet all of a cell's needs? One way the cell maximizes efficiency is by producing multiple copies of each mRNA; moreover, many ribosomes may simultaneously translate one mRNA molecule **(figure 7.9)**. These ribosomes zip along the mRNA, incorporating some 15 amino acids per second. Thanks to this fast-moving "assembly line," a cell can make many copies of a protein from the same mRNA. ⓘ *antibodies,* section 29.3C

On the other hand, blocking protein synthesis can quickly kill a cell. Why We Care 7.1 describes a few poisons that make transcription or translation impossible.

C. Proteins Must Fold Correctly after Translation

The newly synthesized protein cannot do its job until it folds into its final shape (see figure 2.21). Some regions of the amino acid chain attract or repel other parts, contorting the polypeptide's overall shape. Enzymes catalyze the formation of chemical bonds, and "chaperone" proteins stabilize partially folded regions.

Proteins can fold incorrectly if the underlying DNA sequence is altered (see section 7.6) because the encoded protein may have the wrong sequence of amino acids. In some forms of a serious genetic illness called cystic fibrosis, for example, a membrane protein does not fold correctly into its final form. Alzheimer disease is associated with a protein that forms an abnormal mass in brain cells because of improper folding. "Mad cow disease" and similar conditions in sheep and humans are caused by abnormal clumps of misfolded proteins called prions in nerve cells (see section 7.10).

In addition to folding, some proteins must be altered in other ways before they become functional. For example, insulin, which is 51 amino acids long, is initially translated as the 80-amino-acid polypeptide proinsulin. Enzymes cut proinsulin to form insulin. A different type of modification occurs when polypeptides join to form larger protein molecules. The oxygen-carrying blood protein hemoglobin, for example, consists of four polypeptide chains (two alpha and two beta) encoded by separate genes.

7.4 Mastering Concepts

1. What happens in each stage of translation?
2. Where in the cell does translation occur?
3. How are polypeptides modified after translation?

7.5 Cells Regulate Gene Expression

Producing proteins costs tremendous amounts of energy. For example, an *E. coli* cell spends 90% of its ATP on gene expression. Transcription and translation require energy, as does the synthesis of nucleotides, enzymes, ribosomal proteins, and other molecules that participate in these processes. Removing the introns and making other modifications to the mRNA require still more energy. ⓘ *ATP,* section 4.3

Cells constantly produce essential proteins, such as the enzymes involved in respiration. Considering the high cost of making a protein, however, it makes sense that cells save energy by not producing unneeded proteins. This section describes some of the mechanisms that regulate gene expression.

A. Operons Are Groups of Bacterial Genes That Share One Promoter

Many bacteria, including *E. coli,* live in animal intestines; their food sources change from hour to hour. To maximize efficiency, the bacteria should produce enzymes that degrade only the foods that are actually available. For example, *E. coli* requires certain enzymes to absorb and degrade the sugar lactose. How does the bacterium "know" to produce these enzymes only when lactose is present?

The answer relates to the way that genes are organized in *E. coli* and other bacteria. An **operon** is a group of related genes plus a promoter and an operator that control the transcription of the entire group at once. The promoter, as described earlier, is the site to which RNA polymerase attaches to begin transcription. The **operator** is a DNA sequence located between the promoter and the protein-encoding regions. If a protein called a **repressor** binds to the operator, it prevents the transcription of the genes.

Figure 7.10 shows one example: *E. coli's* **lac operon,** which consists of three genes that encode lactose-degrading enzymes, plus a promoter and operator. To understand how the *lac* operon works, first imagine an *E. coli* cell in an environment lacking lactose. Expressing the three genes would be a waste of energy. The repressor protein therefore binds to the operator, preventing RNA polymerase from transcribing the genes. The genes are effectively "off." But when lactose is present, the sugar attaches to the repressor, changing its shape so that it detaches from the DNA. RNA polymerase is now free to transcribe the genes. After translation, the resulting enzymes confer a (temporary) new trait: the ability to absorb and degrade the lactose.

B. Eukaryotic Organisms Use Many Regulatory Methods

In multicellular eukaryotes, the control of gene expression is more complex than in bacteria because different cell types express different combinations of genes. A cell in an animal embryo, for example, must express the proteins that dictate the formation of body parts in the correct places. A skin cell in an adult would not need those proteins but would need others, such as those required to deposit pigments into the skin.

To understand this idea, consider how a chef uses the same set of cookbooks to prepare recipes for breakfast, lunch, and dinner. The cook does not prepare all possible dishes in all cookbooks simultaneously; rather, he or she selects only the recipes required for each meal.

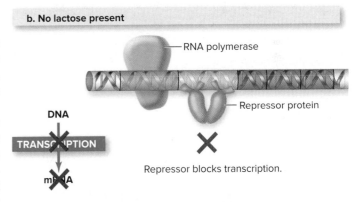

a. The *lac* operon

Bacterial cell

DNA   Promoter   Operator   Genes encoding enzymes that break down lactose   1   2   3

b. No lactose present

RNA polymerase

DNA

TRANSCRIPTION

mRNA

Repressor protein

Repressor blocks transcription.

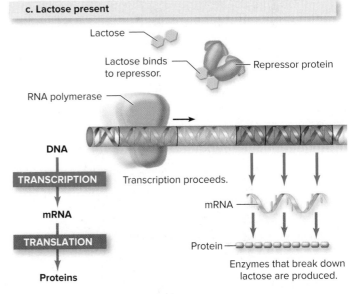

c. Lactose present

Lactose

Lactose binds to repressor.

Repressor protein

RNA polymerase

DNA

TRANSCRIPTION   Transcription proceeds.

mRNA

TRANSLATION

Proteins

mRNA

Protein

Enzymes that break down lactose are produced.

Figure 7.10 **The *Lac* Operon.** (a) An operon consists of a promoter, an operator, and a group of related genes. (b) In the absence of lactose, a repressor protein binds to the operator and prevents transcription. The enzymes are not produced. (c) If lactose is present, it binds to the repressor, which subsequently releases the operator. Transcription proceeds.

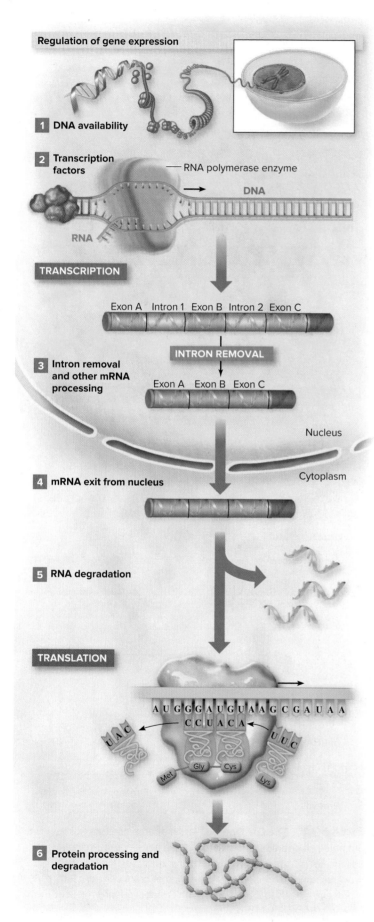

Regulation of gene expression

1 DNA availability

2 Transcription factors — RNA polymerase enzyme

DNA

RNA

TRANSCRIPTION

Exon A   Intron 1   Exon B   Intron 2   Exon C

INTRON REMOVAL

3 Intron removal and other mRNA processing

Exon A   Exon B   Exon C

Nucleus

4 mRNA exit from nucleus

Cytoplasm

5 RNA degradation

TRANSLATION

A U G G G A U G U A A G C G A U A A
C C U A C A

UAC   UUC

Met   Gly   Cys   Lys

6 Protein processing and degradation

The rest of this section describes a few of the ways in which eukaryotic cells control whether each gene is "on" or "off." **Figure 7.11** illustrates where each mechanism fits into the overall process of gene expression.

DNA Availability A chromosome's DNA must be unwound for its genes to be expressed (figure 7.11, step 1). Enzymes in a cell can modify stretches of DNA by adding and removing regulatory "tags" consisting of methyl groups ($-CH_3$). Tagged regions of a chromosome are folded in a way that makes the DNA unavailable; removing the tags means the genes may be expressed. Likewise, the chromosome's proteins can also be chemically modified in a way that either exposes DNA to be transcribed or tucks it away.

The study of these and other modifications is part of a field of research called **epigenetics.** The definition of this term is evolving, but biologists agree that it concerns changes in gene expression that do not involve changes to the DNA sequence itself. Abnormalities in epigenetic mechanisms can lead to birth defects, diabetes, and other serious illnesses.

The placement of epigenetic markers can persist after a cell divides. However, most of the tags are wiped away when sperm fertilizes egg, starting each new generation with a virtually clean slate. After conception, the tags on our chromosomes change throughout life in response to diet, exercise, stress, exposure to toxins, and other environmental factors. These changes help explain why identical twins—who have matching DNA—are not exactly the same.

Transcription Factors In eukaryotic cells, groups of proteins called **transcription factors** bind DNA at specific sequences that regulate transcription (figure 7.11, step 2). RNA polymerase cannot bind to a promoter or initiate transcription of a gene in the absence of transcription factors. The transcription factors form a pocket for RNA polymerase, activating transcription.

Figure 7.12 shows how transcription factors prepare a promoter to receive RNA polymerase. The first transcription factor to bind is attracted to a part of the promoter called the TATA box. The TATA binding protein attracts other transcription factors. Finally, RNA polymerase joins the complex, binding just in front of the start of the gene sequence. With RNA polymerase in place, transcription can begin.

Defects in transcription factors underlie some diseases. Cancer, for example, is a family of illnesses in which cells divide out of control. Proteins provide the signals that normally regulate cell division. Since transcription factors help turn genes on and off, defective transcription factors can cause these fine-tuned signals to be disrupted. Cancer develops because the cells are unable to stop dividing. ⓘ *cancer,* section 8.6

mRNA Processing One gene can encode multiple proteins if different combinations of exons are included in the final mRNA (figure 7.11, step 3). For example, researchers know of a gene in fruit flies that can theoretically be spliced into more than 38,000 different configurations!

mRNA Exit from Nucleus For a protein to be produced, mRNA must leave the nucleus and attach to a ribosome (figure 7.11, step 4). If the mRNA fails to leave, the gene is silenced.

mRNA Degradation Not all mRNA molecules are equally stable. Some are rapidly destroyed, perhaps before they can be translated, whereas others persist long enough to be translated many times (figure 7.11, step 5).

Figure 7.11 Regulating Gene Expression. Eukaryotic cells have many ways to control whether each gene is turned on or off.

Protein Processing and Degradation Some proteins must be altered before they become functional (figure 7.11, step 6). Producing insulin, for example, requires a precursor protein to be cut in two places. If these modifications fail to occur, the insulin protein cannot function.

In addition, to do its job, a protein must move from the ribosome to where the cell needs it. For example, a protein secreted in milk must be escorted to the Golgi apparatus and be packaged for export (see figure 3.11). A gene is effectively silenced if its product never moves to the correct destination. Finally, like RNA, not all proteins are equally stable. Some are degraded shortly after they form, whereas others persist longer.

A human cell may express hundreds to thousands of genes at once. Unraveling the complex regulatory mechanisms that control the expression of each gene is an enormous challenge. Biologists now have the technology to begin navigating this regulatory maze. The work has just begun, but the payoff will be a much better understanding of cell biology, along with many new medical applications. The same research may also help scientists understand how external influences on gene expression contribute to complex traits, such as homosexuality; see Burning Question 7.1.

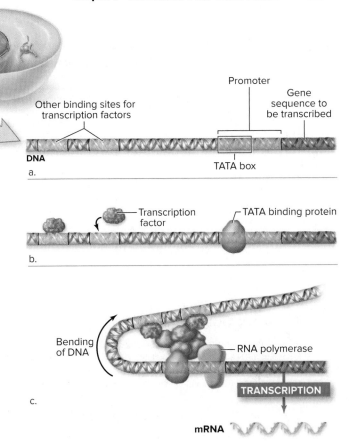

7.5 Mastering Concepts

1. Which steps in gene expression require energy?
2. Why do cells regulate which genes are expressed?
3. How does a repressor protein help regulate the expression of a bacterial operon?
4. Explain how epigenetic modifications change the likelihood of transcription.
5. What is the role of transcription factors in gene expression?

Figure 7.12 How Transcription Factors Work. (a) Promoters and other binding sites regulate transcription. (b) Transcription factors, including the TATA binding protein, bind to these sites. (c) DNA bends, bringing the transcription factors together. The RNA polymerase enzyme can now initiate transcription.

7.6 Mutations Change DNA

A **mutation** is any change in a cell's DNA sequence. The change may occur in a gene or in a regulatory region such as a promoter. Many people think that mutations are always harmful, perhaps because some of them cause such dramatic changes (figure 7.13). Although some mutations do cause illness, they also provide the variation that makes life interesting (and makes evolution possible).

To continue the cookbook analogy introduced earlier, a mutation in a gene is similar to an error in a recipe. A small typographical error might be barely noticeable. A minor substitution of one ingredient for another might hurt (or improve) the flavor. But serious errors such as missing ingredients or truncated instructions are likely to ruin the dish.

A. Mutations Range from Silent to Devastating

A mutation may change one or a few base pairs or affect large portions of a chromosome. Some are detectable only by using DNA sequencing techniques, while others may be lethal. The rest of this section describes the major types of mutations in detail. ⓘ *DNA sequencing, section 11.2B*

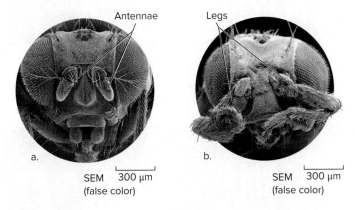

Figure 7.13 Legs on the Head. (a) Normally, a fruit fly has two small antennae between its eyes. (b) This fly has legs growing where antennae should be. It has a mutation in a gene that affects development.

(a): ©Andrew Syred/SPL/Science Source; (b): ©Eye of Science/Science Source

TABLE 7.1 Types of Mutations: A Sentence Analogy

Original sentence	THE ONE BIG FLY HAD ONE RED EYE
Substitution mutation	TH**Q** ONE BIG FLY HAD ONE RED EYE
Deletion of three nucleotides	THE ONE BIG HAD ONE RED EYE
Insertion of three nucleotides	THE ONE BIG **WET** FLY HAD ONE RED EYE
Insertion of one nucleotide (frameshift)	THE ONE **Q**BI GFL YHA DON ERE DEY

Figure It Out

Suppose that a substitution mutation replaces the first "A" in the following mRNA sequence with a "U":

A A A G C A G U A C U A

How many amino acids will be in the polypeptide chain?

Answer: Zero.

Table 7.1 illustrates some of the major types of mutations, using sentences composed of three-letter words. A **substitution mutation** is the replacement of one DNA base with another. Such a mutation is "silent" if the mutated gene encodes the same protein as the original gene version. Mutations can be silent because more than one codon encodes most amino acids.

Often, however, a substitution mutation changes a base triplet so that it specifies a different amino acid. The substituted amino acid may drastically alter the protein's shape, changing its function. Sickle cell disease results from this type of mutation in a gene encoding hemoglobin (figure 7.14).

In other cases, a base triplet specifying an amino acid changes into one that encodes a "stop" codon. This shortens the protein product, which can profoundly influence the organism. At least one of the mutations that give rise to cystic fibrosis, for example, shortens a membrane protein from its normal 1480 amino acids to only 493. The faulty protein cannot function.

An **insertion mutation** adds one or more nucleotides to a gene; a **deletion mutation** removes nucleotides. Either type of mutation may be a **frameshift mutation,** in which nucleotides are added or deleted by a number other than a multiple of three. Because triplets of DNA bases specify amino acids, such an addition or deletion disrupts the codon reading frame. Frameshift mutations are therefore likely to alter the sequence of amino acids or cause premature stop codons. Either way, a frameshift usually devastates a protein's function. Some mutations that cause cystic fibrosis, for example, reflect the addition or deletion of just one or two nucleotides.

Even if a small insertion or deletion does not shift the reading frame, the effect might still be significant if the change drastically alters the protein's shape. The most common mutation that causes severe cystic fibrosis deletes a single group of three nucleotides. The resulting protein lacks just one amino acid, but it cannot function.

Some mutations affect extensive regions of DNA. For example, a large part of a chromosome may be deleted, duplicated, or inverted. Any of these events can bring together DNA segments that were not previously joined.

B. What Causes Mutations?

Some mutations occur spontaneously—that is, without outside causes. A spontaneous substitution mutation usually originates as a DNA replication error, but replication errors can also cause insertions and deletions. Mutations may also occur during meiosis, a type of cell division required for sexual reproduction (see chapter 9). ① *DNA replication,* section 8.2

Exposure to harmful chemicals or radiation may also damage DNA. A **mutagen** is any external agent that induces mutations. Examples include the ultraviolet radiation in sunlight, X-rays, radioactive fallout from atomic bomb tests and nuclear accidents, and chemicals in tobacco and in

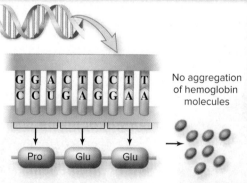

a. Normal red blood cells

G G A	C T C	C T T
C C U	G A G	G A A

Pro — Glu — Glu

No aggregation of hemoglobin molecules

SEM 6 μm (false color)

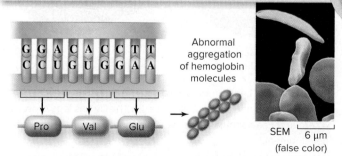

b. Sickled red blood cells

G G A	C A C	C T T
C C U	G U G	G A A

Pro — Val — Glu

Abnormal aggregation of hemoglobin molecules

SEM 6 μm (false color)

Figure 7.14 Sickle Cell Mutation. Sickle cell anemia usually results from a mutation in a hemoglobin gene. (a) Normal hemoglobin molecules do not aggregate, enabling the red blood cell to assume a rounded shape. (b) A substitution mutation causes hemoglobin molecules to clump (aggregate) into long, curved rods that deform the red blood cell.

Photos: (a): ©Micro Discovery/Corbis Documentary/Getty Images; (b): ©Dr. Gopal Murti/SPL/Science Source

the environment (such as pollution in soil, air, or water). The more contact a person has with mutagens, the higher the risk for cancer. Coating skin with sunscreen, wearing a lead "bib" during dental X-rays, and avoiding tobacco all lower cancer risk by reducing exposure to mutagenic chemicals and radiation.

C. Mutations Are Important for Many Reasons

One reason that mutations are important is that they produce genetic variability. They are the raw material of evolution because they generate new **alleles,** or variants of genes. Some of these new alleles are "neutral" and have no effect on an organism's fitness. Your reproductive success, for example, does not ordinarily depend on your eye color or shoe size. As unit 3 explains, however, variation has important evolutionary consequences. In every species, individuals with some allele combinations reproduce more successfully than others. In natural selection, the environment "edits out" the less favorable allele combinations.

The importance of mutations in evolution became clear with the discovery of **homeotic genes,** which encode transcription factors that are expressed during the development of an embryo. Specifically, homeotic genes control the formation of an organism's body parts. The flies in figure 7.13 show what happens when homeotic genes are mutated. Having parts in the wrong places is, of course, usually harmful. But studies of many species reveal that mutations in homeotic genes have profoundly influenced animal evolution (see section 13.5). Limb modifications such as arms, wings, and flippers trace their origins to homeotic mutations.

Mutations sometimes enhance an organism's reproductive success. Consider, for example, the antibiotic drugs that kill bacteria by targeting membrane proteins, enzymes, and other structures. Random mutations in bacterial DNA sometimes encode new versions of these targeted proteins. Such a mutation may give a bacterium a new trait—antibiotic resistance—that the cell is likely to pass on to its descendants whenever antibiotics are present. The medical consequences are immense. Antibiotic-resistant bacteria have become more and more common, and many people now die of bacterial infections that once were easily treated with antibiotics.

Mutations can also be enormously useful in science and agriculture. Geneticists frequently induce mutations to learn how genes normally function. For example, biologists discovered how genes control flower formation by studying mutant plants in which flower parts form in the wrong places. Plant breeders also induce mutations to develop new varieties of many crop species (figure 7.15). Some kinds of grapefruits, rice, cotton, oats, lettuce, begonias, and many other plants owe their existence to breeders who treated cells with radiation and then selected mutated individuals with interesting new traits. Chapter 11 describes new biotechnology tools that manipulate DNA directly.

a.   b.   c.

Figure 7.15 Useful Mutants. (a) Rio Red grapefruits and several types of (b) rice and (c) cotton are among the many plant varieties that have been developed by using radiation to induce mutations.

(a): ©Nigel Cattlin/Alamy Stock Photo; (b): ©Pallava Bagla/Corbis via Getty Images; (c): ©Scott Olson/Getty Images

Burning Question 7.1

Is there a gay gene?

Research linking human behavior to individual genes is extremely difficult for several reasons. First, genes encode proteins, not behaviors, so the question of a "gay gene" is somewhat misleading. Second, to establish a clear link to DNA, a researcher must be able to define and measure a behavior. This in itself is difficult because people disagree about what it means to be homosexual. Third, multiple genes are likely to be involved. Fourth, an individual who possesses an allele associated with a trait will not necessarily express the allele; many genes in each cell remain "off" at any given time. To complicate matters, the environment contributes mightily to gene expression.

Nevertheless, research has yielded some evidence of a biological component to homosexuality, at least in males. For example, a male homosexual's identical twin is much more likely to also be homosexual than is a nonidentical twin, indicating a strong genetic contribution. In addition, the more older brothers a male has, the more likely he is to be homosexual. This "birth order" effect occurs only for siblings with the same biological mother; having older stepbrothers does not increase the chance that a male is homosexual. That means that events before birth, not social interactions with brothers, are apparently responsible for the effect.

Scientists are now exploring the role of epigenetics in sexual orientation. In 2015, for example, researchers found five epigenetic markers that are more common in homosexual men than in heterosexuals. However, no one knows whether (or how) these epigenetic markers affect sexual orientation.

So is there a gay gene? The answer remains elusive. But we can say without a doubt that both the environment and genetics play important roles.

Submit your burning question to marielle.hoefnagels@mheducation.com

7.6 Mastering Concepts

1. What are the types of mutations, and how does each alter the encoded protein?
2. What causes mutations?
3. In what ways are mutations important?

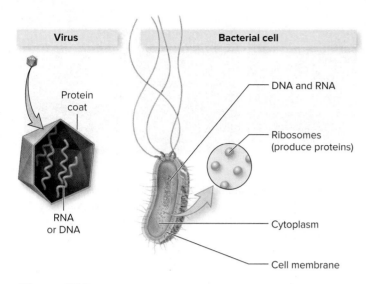

| Virus | Bacterial cell |

Figure 7.16 Viruses and Bacteria Compared. The simplest viruses consist of a protein coat surrounding either RNA or DNA. Viruses are much smaller than, and lack many of the features of, living cells.

TABLE 7.2	**Some Viruses That Infect Humans**
Genetic Material	**Virus (Disease)**
DNA	Poxvirus (smallpox)
	Herpesviruses (oral and genital herpes; chickenpox)
	Epstein–Barr virus (mononucleosis, Burkitt lymphoma)
	Papillomaviruses (warts, cervical cancer)
	Hepatitis B virus
RNA	Human immunodeficiency virus (AIDS)
	Poliovirus
	Influenza viruses
	Measles virus
	Mumps virus
	Rabies virus
	Rhinovirus (common cold)
	West Nile virus
	Hepatitis A and C viruses
	Ebola virus
	Zika virus

7.7 Viruses Are Genes Wrapped in a Protein Coat

So far, this chapter has explained the structure and function of DNA and RNA in cells. As we have seen, genes act as "recipes" for the cell's proteins. The rest of this chapter describes a logical extension of this idea: If a cell receives new genes, it can produce a new set of proteins. In this case, the genes come from viruses.

Because bacteria and viruses are microscopic and cause disease, many people mistakenly lump them together as "germs." Viruses, however, are not bacteria. In fact, they are not even cells. A **virus** is a small, infectious agent that is simply genetic information enclosed in a protein coat (figure 7.16). The 2000 or so known species of viruses therefore straddle the boundary between the chemical and the biological. Even though viruses are extremely simple, they cause a long list of diseases that range from the merely inconvenient to the deadly. Smallpox, influenza, the common cold, rabies, polio, chickenpox, warts, mononucleosis, and AIDS are just a few examples.

A. Viruses Are Smaller and Simpler than Cells

A typical virus is much smaller than a cell. At about 10 μm (microns) in diameter, an average human cell is perhaps one tenth the diameter of a human hair. A bacterium is about one tenth again as small, at about 1 μm (1000 nm) long. The average virus, with a diameter of about 80 nm, is more than 12 times smaller than a bacterium.

Viruses are simple structures that lack many of the characteristics of cells. A virus does not have a nucleus, organelles, ribosomes, a cell membrane, or even cytoplasm. Only a few types of viruses contain enzymes. All viruses, however, have the following two features in common:

- **Genetic information.** All viruses contain genetic material (either RNA or DNA) that carries the "recipes" for their proteins (table 7.2).
- **Protein coat.** A **protein coat** surrounds the viral genetic material. The protein coat's shape determines a virus's overall form (figure 7.17). Many viruses are spherical or icosahedral (a 20-faced shape built of triangular sections). Others are rod-shaped, oval, or filamentous.

Some viruses have other features as well. For example, some have an **envelope,** a lipid- and protein-rich outer layer derived from the host cell's

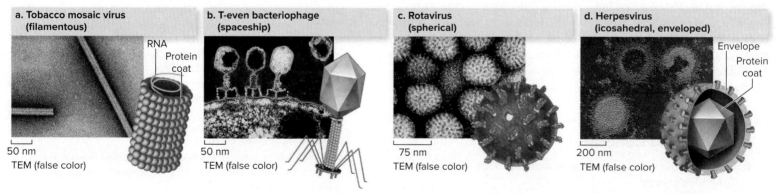

a. Tobacco mosaic virus (filamentous)
50 nm
TEM (false color)

b. T-even bacteriophage (spaceship)
50 nm
TEM (false color)

c. Rotavirus (spherical)
75 nm
TEM (false color)

d. Herpesvirus (icosahedral, enveloped)
200 nm
TEM (false color)

Figure 7.17 Virus Variety. (a) Tobacco mosaic viruses cause disease in plants. (b) T-even viruses infect bacteria. (c) Rotavirus causes severe diarrhea in young children. (d) Herpesviruses induce cold sores and rashes.

Photos: (a): ©Mary Martin/Science Source; (b): ©Eye of Science/Science Source; (c): Source: CDC/Dr. Erskine Palmer & Byron Skinner; (d): ©G. Murti/Science Source

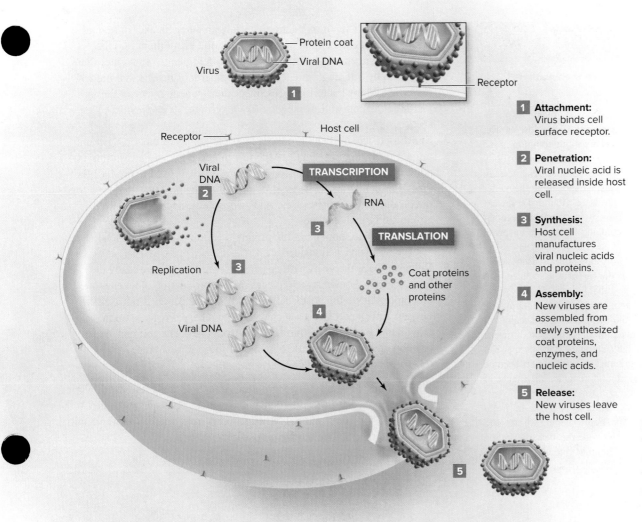

Protein coat

Viral DNA

Virus

Receptor

Receptor

Host cell

Viral DNA

TRANSCRIPTION

RNA

Replication

TRANSLATION

Coat proteins and other proteins

Viral DNA

1 Attachment:
Virus binds cell surface receptor.

2 Penetration:
Viral nucleic acid is released inside host cell.

3 Synthesis:
Host cell manufactures viral nucleic acids and proteins.

4 Assembly:
New viruses are assembled from newly synthesized coat proteins, enzymes, and nucleic acids.

5 Release:
New viruses leave the host cell.

Figure 7.18 **Viral Replication.** The five basic steps of viral replication apply to any virus, whether the host cell is prokaryotic or eukaryotic.

membrane. Proteins embedded in the envelope help the virus invade a new host cell. One example of an enveloped virus is the human immunodeficiency virus (HIV), which causes acquired immunodeficiency syndrome (AIDS). The influenza virus also has an envelope. ⓘ *cell membrane,* section 3.3.

Despite having relatively few components, a virus's overall structure can be quite intricate and complex. For example, **bacteriophages** (sometimes simply called "phages") are viruses that infect bacteria. Some phages have parts that resemble tails, legs, and spikes; they look like the spacecrafts once used to land on the moon (see figure 7.17b).

B. Viral Replication Occurs in Five Stages

The production of new viruses is very different from cell division. When a cell divides, it doubles all of its components and splits in two. Virus production, on the other hand, more closely resembles the assembly of cars in a factory.

A cell infected with one virus may produce and release hundreds or thousands of new viral particles. Whatever the host species or cell type, the same basic processes occur during a viral infection (figure 7.18):

1. **Attachment:** A virus attaches to a host cell by adhering to a receptor molecule on the cell's surface. Generally, the virus can attach only to a cell within which it can reproduce. HIV cannot infect skin cells, for example, because its receptors occur only on certain white blood cells.

Burning Question 7.2

Why do we get sick when the weather turns cold?

Scientists have proposed many explanations for the observation that colds, the flu, and other illnesses are more common in the winter. For example, we know that low body temperature slows the immune response. In addition, decreased sun exposure in the winter may lead to lower levels of vitamin D, further reducing immune function. Being crowded together indoors during the winter also allows cold and flu viruses to spread easily from host to host.

However, these hypotheses don't fully explain flu season, which lasts from November to March. Why is the flu virus associated with the winter months? Some studies suggest that the virus is most stable at low temperatures; the amount of moisture in the air also plays a key role. Dry winter air dehydrates the respiratory tract, making it more susceptible to viral infection. In addition, each cough or sneeze releases respiratory droplets laden with countless viruses that can stay suspended in dry air for hours. In moist air, the droplets become heavy with water and soon fall to the ground, reducing the chance of the viruses finding a new host.

Conventional wisdom says that venturing outside in frigid weather without a coat causes colds. But simply keeping warm won't protect you from respiratory viruses. Instead, the best prevention strategies are to drink plenty of water, humidify your house, wash your hands frequently, and—in the case of the flu—endure an annual flu shot.

Submit your burning question to
marielle.hoefnagels@mheducation.com

2. **Penetration:** The viral genetic material can enter the cell in several ways. Animal cells engulf virus particles and bring them into the cytoplasm via endocytosis. Viruses that infect plants often enter their host cells by hitching a ride on the mouthparts of insects that munch on leaves. Many bacteriophages inject their genetic material through a hole in the cell wall, somewhat like a syringe. ⓘ *endocytosis,* section 4.5C

3. **Synthesis:** The host cell produces multiple copies of the viral genome; mutations during this stage are the raw material for viral evolution. In addition, the information encoded in viral DNA is used to produce the virus's proteins. The host cell provides all of the resources required for the production of new viruses: ATP, ribosomes, nucleotides, tRNA, amino acids, and enzymes.

4. **Assembly:** The subunits of the protein coat join, and then genetic information is packed inside.

5. **Release:** Once the virus particles are assembled, they are ready to leave the cell. Some bacteriophages induce production of an enzyme that breaks down the host's cell wall, killing the cell as it releases the viruses. Enveloped viruses such as HIV and herpesviruses, on the other hand, bud from the host cell by exocytosis. The cell may die as enveloped viruses carry off segments of the cell membrane. ⓘ *exocytosis,* section 4.5C

The amount of time between initial infection and cell death varies. Bacteriophages need as little as half an hour to infect a cell and replicate. At the other extreme, for some animal viruses, years may elapse between initial attachment and the final burst of viral particles.

7.7 Mastering Concepts

1. How are viruses similar to and different from cells?
2. What features do all viruses share?
3. Describe the five steps in viral replication.
4. What is the source of energy and raw materials for the synthesis of viruses in a host cell?

7.8 Viruses Infect All Cell Types

Following attachment to the host cell and penetration of the viral genetic material, viruses may or may not immediately cause cell death.

A. Bacteriophages May Kill Cells Immediately or "Hide" in a Cell

Bacteriophages, the viruses that infect bacteria, have two viral replication strategies: the lytic and lysogenic pathways (figure 7.19). In a **lytic infection,** a virus enters a bacterium, immediately replicates, and causes the host cell to burst (lyse) as it releases a flood of new viruses (see figure 7.19a). The newly released viruses infect other bacteria, repeating the process until all of the cells are dead.

In a **lysogenic infection,** the genetic material of a virus is replicated along with the bacterial chromosome, but the cell lives and reproduces as usual (see figure 7.19b). Many lysogenic viruses use enzymes to cut the host cell DNA and join their own DNA with the host's. A **prophage** is the DNA of a lysogenic bacteriophage that is inserted into the host chromosome. Other lysogenic viruses maintain their DNA apart from the chromosome. Either way, however, when the cell divides, the viral genes replicate, too. The virus gets a "free ride," infecting new cells without actually triggering viral production.

During a lysogenic stage, the viral DNA does not damage the host cell. Only a few viral proteins are produced, most functioning as a "switch" that determines whether the virus should become lytic. At some signal, such as stress from DNA damage or cell starvation, these viral proteins trigger a lytic infection cycle that kills the cell and releases new viruses that infect other cells. The next generation of viruses may enter a lytic or lysogenic replication cycle, depending on the condition of the host cells.

B. Animal Viruses May Cause Immediate Cell Death

A person can acquire a viral infection by inhaling the respiratory droplets of an ill person or by ingesting food or water contaminated with viruses. Blood transfusions, sexual contact, and the use of contaminated needles can also spread viruses.

Once a viral infection is established, the death of infected cells produces a wide range of symptoms that reflect the types of host cells destroyed. Consider, for example, the flu. Influenza viruses infect cells lining the human airway. As cells produce and release new viruses, the infection spreads rapidly in the lungs, throat, and nose. The dead and damaged cells cause the respiratory symptoms of influenza, including cough and sore throat.

What about the other symptoms, such as fever and body aches? These classic signs of influenza result from the immune system's response to the viral infection. Cells of the immune system release signaling molecules that induce fever; a high body temperature speeds other immune responses. The signaling molecules also trigger inflammation, which causes body aches and fatigue. These immune reactions usually defeat the virus, but in the meantime, they can also make a person rather miserable! Chapter 29 describes the human immune system in more detail.

C. Some Animal Viruses Linger for Years

In a **latent infection,** viral genetic information inside an animal cell lies dormant; even as the infected host cell divides, new viruses are not produced. However, the virus may be reactivated later. A latent virus in an animal cell is therefore similar to a lysogenic phage in a bacterial cell.

Many people harbor latent infections of the herpes simplex virus type I, which causes cold sores on the lips. After initial infection, the viral DNA remains in skin cells indefinitely. When a cell becomes stressed or damaged, new viruses are assembled and leave the cell to infect other cells. Cold sores, which reflect the localized death of these cells, periodically recur at the site of the original infection.

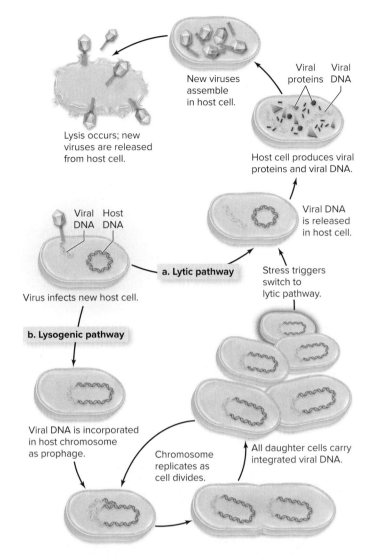

Lysis occurs; new viruses are released from host cell.

New viruses assemble in host cell.

Viral proteins   Viral DNA

Host cell produces viral proteins and viral DNA.

Viral DNA is released in host cell.

Viral DNA Host DNA

a. Lytic pathway

Virus infects new host cell.

Stress triggers switch to lytic pathway.

b. Lysogenic pathway

Viral DNA is incorporated in host chromosome as prophage.

Chromosome replicates as cell divides.

All daughter cells carry integrated viral DNA.

Figure 7.19 Lysis and Lysogeny. (a) In the lytic pathway, the host cell bursts (lyses) when new virus particles leave the cell. (b) In lysogeny, viral DNA replicates along with the cell, but new viruses are not produced. Stress in the host cell may trigger a lysogenic virus to switch to the lytic pathway.

Miniglossary | Viruses

Virus	An infectious agent consisting of genetic information enclosed in a protein coat
Bacteriophage	A virus that infects bacteria
Lytic infection	An infection by a bacteriophage that causes the host bacterium to burst immediately after assembling new viruses
Lysogenic infection	An infection by a bacteriophage in which the viral genetic material replicates along with the host genome, but infected cells do not immediately produce new viruses; the infection may become lytic at any time
Latent infection	In animals, a viral infection in which viral genetic information replicates along with the host genome, but infected cells do not immediately produce new viruses; when the infection is reactivated, infected cells begin producing viruses

HIV is another virus that can remain latent inside a human cell (figure 7.20). HIV belongs to a family of viruses called retroviruses, all of which have an RNA genome. The virus infects helper T cells, which are part of the immune system. Once inside the cell, HIV's reverse transcriptase enzyme transcribes the viral RNA to DNA. The DNA then inserts itself into the host cell's DNA. Cells with active infections express the viral genes, producing and releasing many new HIV particles. These viruses go on to infect other T cells, so the number of helper T cells gradually declines. Eventually, the loss of T cells leaves the body defenseless against infections or cancer. AIDS is the result. ⓘ *T cells,* section 29.3A

In some of the body's T cells, however, HIV's genetic information remains dormant, forming a latent reservoir of infected but inactive cells. These latent infections are especially hard to treat with anti-HIV drugs. Finding ways to coax the virus out of latency would greatly improve the efficacy of anti-HIV drugs; this is one of many active areas of HIV research.

Because latent viruses persist by signaling their host cells to divide continuously, some cause cancer. A latent infection by some strains of human pap-

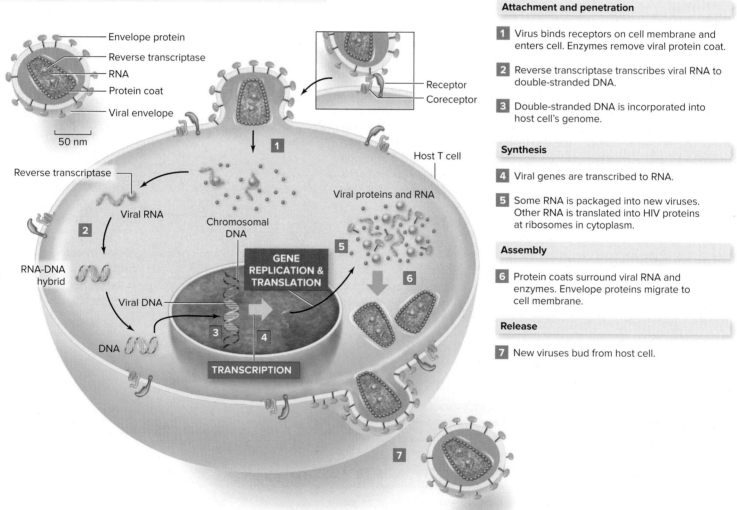

Attachment and penetration

1 Virus binds receptors on cell membrane and enters cell. Enzymes remove viral protein coat.

2 Reverse transcriptase transcribes viral RNA to double-stranded DNA.

3 Double-stranded DNA is incorporated into host cell's genome.

Synthesis

4 Viral genes are transcribed to RNA.

5 Some RNA is packaged into new viruses. Other RNA is translated into HIV proteins at ribosomes in cytoplasm.

Assembly

6 Protein coats surround viral RNA and enzymes. Envelope proteins migrate to cell membrane.

Release

7 New viruses bud from host cell.

Figure 7.20 HIV Replication. HIV's genetic material (RNA) is transcribed to DNA, which integrates into the host T cell's chromosome. The production of viruses eventually kills the cell, damaging the person's immune system.

illomavirus can cause cervical cancer. Epstein–Barr virus is another example. More than 80% of the human population carries this virus, which infects B cells of the immune system. A person who is initially exposed to the virus may develop mononucleosis ("mono"). The virus later maintains a latent infection in B cells. In a few people, especially those with weakened immune systems, the virus eventually causes a form of cancer called Burkitt lymphoma.

D. Viruses Cause Diseases in Plants

Like all organisms, plants can have viral infections (figure 7.21). The first virus ever discovered was tobacco mosaic virus, which affects not only tobacco but also tomatoes, peppers, and more than 120 other plant species.

To infect a plant cell, a virus must penetrate waxy outer leaf layers and thick cell walls. Most viral infections spread when plant-eating insects such as leafhoppers and aphids move virus-infested fluid from plant to plant on their mouthparts. ① *plant cell wall,* section 3.6

Once inside a plant, viruses multiply at the initial site of infection. The killed plant cells often appear as small dead spots on the leaves. Over time, the viruses spread from cell to cell through plasmodesmata (bridges of cytoplasm between plant cells). They can also move throughout a plant by entering the vascular tissues that distribute sap. Depending on the location and extent of the viral infection, symptoms may include blotchy, mottled leaves or abnormal growth. A few symptoms, such as the streaking of some flower petals, appear beautiful to us.

Although plants do not have the same forms of immunity as do animals, they can fight off viral infections. For example, virus-infected cells may "commit suicide" before the infection has a chance to spread to neighboring cells. Alternatively, a plant cell may destroy the mRNA transcribed from viral genes. Since the viral mRNA is never translated into proteins, this defense prevents the assembly of new viruses.

a.   b.

Figure 7.21 **Sick Plants.** (a) Cucumber mosaic virus causes a characteristic mottling (spotting) of squash leaves. (b) A virus has also caused the streaking on the petals of these tulips.

(a): ©Nigel Cattlin/Science Source; (b): ©Anna Yu/Getty Images RF

7.8 Mastering Concepts

1. Compare and contrast lysogenic and lytic viral infections.
2. Which flu symptoms are caused directly by the virus, and which are caused by the human immune response?
3. What is a latent animal virus?
4. Describe how HIV replicates in host cells.
5. What are some symptoms of a viral infection in plants?

7.9 Drugs and Vaccines Help Fight Viral Infections

Halting a viral infection is a challenge, in part because viruses invade living cells. Some antiviral drugs interfere with enzymes or other proteins that are unique to viruses, but overall, researchers have developed few medicines that inhibit viruses without killing infected host cells. Many viral diseases therefore remain incurable.

Vaccination remains our most potent weapon against many viral diseases (figure 7.22). A **vaccine** "teaches" the immune system to recognize one or more molecular components of a virus without actually exposing the person to

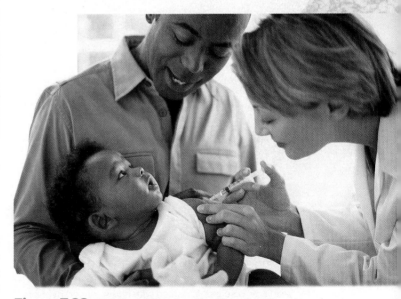

Figure 7.22 **Childhood Vaccination.** Countless lives have been saved by vaccines that medical researchers have developed against a suite of deadly diseases caused by viruses.

©Science Photo Library/Getty Images RF

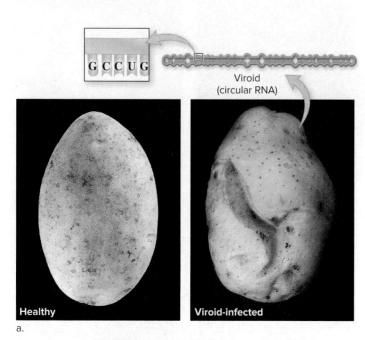

Viroid
(circular RNA)

GCCUG

Healthy

Viroid-infected

a.

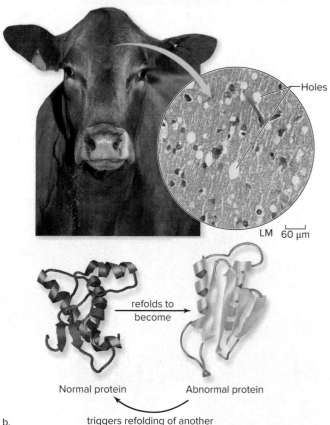

Holes

LM 60 μm

refolds to
become

Normal protein   Abnormal protein

b.   triggers refolding of another

Figure 7.23 **Viroids and Prions.** (a) Potatoes affected by the potato spindle tuber viroid often have prominent cracks. (b) A brain affected by "mad cow disease" is riddled with holes caused by abnormal prion proteins.

Photos: (healthy potato): ©SerAlexVi/iStock/360/Getty Images RF; (infected potato): ©Nigel Cattlin/Alamy Stock Photo; (cow): ©Pixtal/age fotostock RF; (brain tissue): ©Ralph Eagle Jr./Science Source

the disease. Some vaccines confer immunity for years, whereas others must be repeated annually. The influenza vaccine is an example of the latter. Flu viruses mutate rapidly, so this year's vaccine is likely to be ineffective against next year's strains. (To learn more about flu season, see Burning Question 7.2.) ⓘ *vaccines,* section 29.4

Thanks to successful global vaccination programs, smallpox has vanished from human populations, and polio is nearly defeated. Childhood vaccinations have greatly reduced the incidence of measles, mumps, and many other potentially serious illnesses. Unfortunately, researchers have been unable to develop vaccines against many deadly viruses, including HIV.

Antiviral drug development is complicated by the genetic variability of many viruses. Consider the common cold. Many different cold viruses exist, and their genomes mutate rapidly. As a result, a different virus strain is responsible every time you get the sniffles. Developing drugs that work against all of these variations has so far proved impossible. Even if a drug inactivated 99.99% of cold-causing viruses, the remaining 0.01% would be resistant. These viruses would replicate, and natural selection would rapidly render the drug ineffective.

The antibiotic drugs that kill bacteria never work against viruses, which lack the structures targeted by antibiotics. Physicians sometimes prescribe the drugs for viral infections anyway. This needless exposure to antibiotics selects for drug-resistance mutations in harmless bacteria, which can later share their genes with disease-causing microbes.

Nevertheless, a viral infection can sometimes trigger bacterial growth. For example, patients sometimes develop sinus infections as a complication of influenza or the common cold. Physicians may prescribe antibiotics to treat these secondary bacterial infections, but the drugs will not affect the underlying virus.

7.9 Mastering Concepts

1. How are viral infections treated and prevented?
2. Explain why antiviral drugs are difficult to develop.

7.10 Viroids and Prions Are Other Noncellular Infectious Agents

The idea that something as simple as a virus can cause devastating illness may seem amazing. Yet some infectious agents are even simpler than viruses.

A **viroid** is a highly wound circle of RNA that lacks a protein coat; it is simply naked RNA that can infect a plant cell. Although viroid RNA does not encode protein, it can nevertheless cause severe disease in many important crop plants (figure 7.23a). Apparently the viroid's RNA interferes with the plant's ability to produce one or more essential proteins.

Another type of infectious agent is a **prion,** which stands for "*pro*teinaceous *in*fectious particle." A prion protein (PrP for short) is a normal membrane protein that can exist in multiple three-dimensional shapes, at least one of which is abnormal and can cause disease (figure 7.23b). Upon contact with an abnormal form of PrP, a normal prion protein switches to the abnormal PrP configuration. The change triggers another round of protein refolding, and so on. As a result of this chain reaction, masses of abnormal prion proteins accumulate inside cells. ⓘ *protein folding,* section 2.5C

The misshapen prion proteins cause brain cells to die. The brain eventually becomes riddled with holes, like a sponge. Mad cow disease is one example. A cow may acquire this disease by ingesting infected cattle. Because of mad cow disease, governments now ban the practice of feeding cattle the processed remains of other cattle.

7.10 Mastering Concepts

1. How are viroids and prions different from viruses?

2. How do viroids and prions cause disease?

Investigating Life 7.1 | Clues to the Origin of Language

As you chat with your friends and study for your classes, you may take language for granted. Although communication is not unique to humans, a complex spoken language does set us apart from other organisms. Every human society has language; without it, people could not transmit information from one generation to the next, so culture could not develop. Its importance to human evolutionary history is therefore incomparable. But how and when did such a crucial adaptation arise?

Scientists have pinpointed a gene, called *FOXP2*, that has many functions related to language. People who inherit a mutated version of the gene have difficulty controlling the muscles of the mouth and face, leading to pronunciation problems. They also have relatively low intelligence and trouble applying simple grammatical rules. *FOXP2* encodes a transcription factor, a protein that binds to DNA and controls the expression of other genes. This gene is not solely responsible for language acquisition. But the fact that it encodes a transcription factor explains how it can simultaneously affect both muscle control and brain structure.

To learn more about the evolution of language, scientists compared the sequences of the 715 amino acids that make up the FOXP2 protein in humans, several other primates, and mice (figure 7.A). In the 70 million or so years since the mouse and primate lineages split, the FOXP2 protein has seldom changed. A mutation in the mouse *FOXP2* gene changed one amino acid; a different amino acid changed in orangutans. Yet, after humans split from chimpanzees—an event that occurred just 5 or 6 million years ago—the FOXP2 protein changed twice. Initially, the new, human-specific *FOXP2* allele would have been rare, as are all mutations. Today, however, nearly everyone has the same allele of *FOXP2*.

In a follow-up study, the researchers used ancient DNA to learn more about when the beneficial allele arose. They found that the Neandertal version of *FOXP2* had the same mutations as the version of the gene in modern humans. Both human-specific mutations had therefore already occurred by 300,000 to 400,000 years ago, the time when modern humans and Neandertals last shared a common ancestor.

The study of *FOXP2* is important because it helps us understand a critical period in human history. The new, human-specific *FOXP2* version evidently conferred such improved language skills that individuals with the allele produced more offspring than those with any other version. By natural selection, the new allele quickly became fixed in the human population. Without

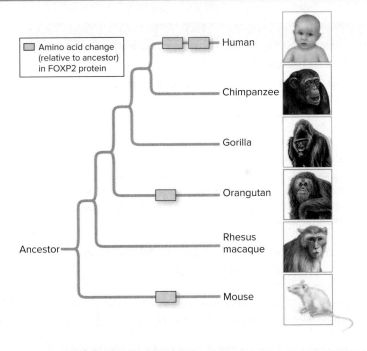

those events, human communication and culture (including everything you chat about with your friends) might never have happened. ⓘ *human evolution,* section 17.12.

Figure 7.A FOXP2 Protein Compared. This evolutionary tree shows how the 715 amino acids of the FOXP2 protein differ in mice and various primates. Each blue box represents a difference of 1 amino acid.

Sources: Enard, Wolfgang, Molly Przeworski, Simon E. Fisher, and five coauthors, including Svante Pääbo. August 22, 2002. Molecular evolution of *FOXP2*, a gene involved in speech and language. *Nature*, vol. 418, pages 869–872.

Krause, Johannes, and 12 coauthors, including Wolfgang Enard and Svante Pääbo. November 6, 2007. The derived *FOXP2* variant of modern humans was shared with Neandertals. *Current Biology,* vol. 17, pages 1908–1912.

What's the Point? ▼ APPLIED

This chapter described how cells use information stored in DNA to make proteins. As you have seen, transcription builds an mRNA molecule that is translated at the ribosomes. The process of protein synthesis is so essential to life that some variation of it is found in every known organism. This central feature of life may lead you to wonder, "Why are proteins so important?"

Proteins are the blue-collar workers of the cell. They are cellular security guards, controlling what crosses each membrane. They are postal carriers, delivering and receiving chemical messages. They are bus drivers, moving materials within the cell. Proteins called enzymes speed up the cell's chemical reactions; other proteins form scaffolds that physically support the cell. With these and many other roles for proteins, it is no surprise that protein synthesis occurs in almost every living cell in the human body.

DNA changes—mutations—can mean that proteins are misshapen and unable to function. The effects may range from

minor differences in appearance to serious illnesses to death. If a "security guard" protein is unable to do its job, traffic into or out of the cell may be blocked. Cystic fibrosis traces its origins to one such mutated membrane protein. If an enzyme is not working, a critical chemical reaction will not proceed; lactose intolerance is an example. If a "scaffold" protein does not function, tissues may not develop correctly, causing conditions such as Marfan syndrome.

Understanding the connection between DNA and proteins is critical to learning why cells—and entire organisms—look and function as they do. Later in this unit, you will learn how genes pass from generation to generation, helping you to understand family resemblances and why everything from baldness to cancer seems to run in families.

Photo: ©Brand X Pictures RF

CHAPTER SUMMARY

7.1 DNA Is a Double Helix

- Watson and Crick combined many clues to propose the double-helix structure of DNA.
- **DNA** is made of building blocks called **nucleotides.** The rungs of the DNA "ladder" consist of **complementary** base pairs (A with T, and C with G). Hydrogen bonds hold the two strands together.
- An organism's **genome** includes all of its genetic material. In eukaryotic cells, the genome is divided among multiple **chromosomes.**
- A **gene** is a sequence of DNA that is transcribed to RNA, typically encoding a protein (figure 7.24). Proteins, in turn, interact with other chemicals and perform other functions that are critical to life.

7.2 DNA Stores Genetic Information: An Overview

- Three types of **RNA** (**mRNA, rRNA,** and **tRNA**) participate in gene expression (figure 7.25).
- To produce a protein, a cell **transcribes** a gene's information to mRNA, which is **translated** into a sequence of amino acids.

7.3 Transcription Uses a DNA Template to Build RNA

- Transcription consists of three stages: initiation, elongation, and termination.
- The process begins when the **RNA polymerase** enzyme binds to a **promoter** sequence on the DNA **template strand.** RNA polymerase then builds an RNA molecule. Transcription ends when RNA polymerase reaches a **terminator** sequence in the DNA.
- After transcription, the cell adds a cap and poly A tail to mRNA. **Introns** are cut out of RNA, and the remaining **exons** are spliced together. The finished mRNA molecule then leaves the nucleus.

7.4 Translation Builds the Protein

- Each group of three consecutive mRNA bases is a **codon** that specifies one amino acid (or signals translation to stop).
- The correspondence between codons and amino acids is the **genetic code.**

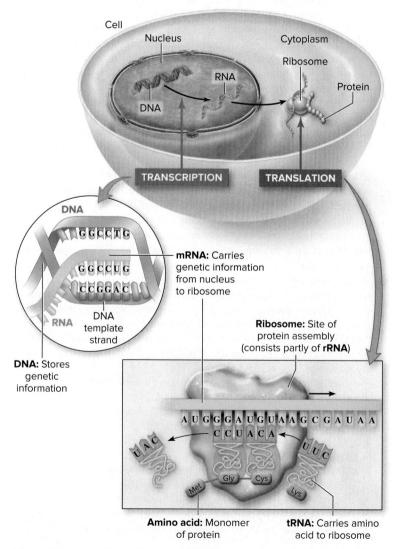

Figure 7.24 Protein Production: A Summary.

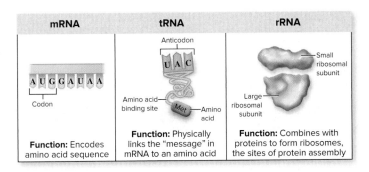

Figure 7.25 Three Types of RNA.

A. Translation Requires mRNA, tRNA, and Ribosomes

- mRNA carries a protein-encoding gene's information. rRNA associates with proteins to form **ribosomes,** which support and help catalyze protein synthesis.
- Each type of tRNA has an end with an **anticodon** sequence complementary to one mRNA codon; the other end of the tRNA carries the corresponding amino acid.

B. Translation Occurs in Three Steps

- The three stages of translation are initiation, elongation, and termination.
- In initiation, mRNA joins with a small ribosomal subunit and a tRNA carrying an amino acid. A large ribosomal subunit then joins the small one.
- In the elongation stage, additional tRNA molecules carrying amino acids bind to subsequent mRNA codons. The ribosome moves along the mRNA as the chain grows.
- Termination occurs when the ribosome reaches a stop codon. The ribosome is released, and the new polypeptide breaks free.

C. Proteins Must Fold Correctly after Translation

- A protein does not function until it has folded into a specific shape. A polypeptide may also be shortened or combined with others to form the finished protein.

7.5 Cells Regulate Gene Expression

- Protein synthesis requires substantial energy input.

A. Operons Are Groups of Bacterial Genes That Share One Promoter

- In bacteria, **operons** coordinate the expression of grouped genes whose encoded proteins participate in the same metabolic pathway. *E. coli*'s *lac* operon is a well-studied example. Transcription does not occur if a **repressor** protein binds to the **operator** sequence of the DNA.

B. Eukaryotic Organisms Use Many Regulatory Methods

- **Epigenetic** mechanisms—chemical modifications that alter the activity of regions of DNA—play important roles in regulating gene expression.
- Proteins called **transcription factors** bind to DNA and regulate which genes a cell transcribes.
- Other regulatory mechanisms include inactivating regions of a chromosome; alternative splicing; controls over mRNA stability and translation; and controls over protein folding and movement.

7.6 Mutations Change DNA

- A **mutation** adds, deletes, or changes nucleotides in a DNA sequence (figure 7.26).

A. Mutations Range from Silent to Devastating

- A **substitution mutation** replaces one base with another. The resulting mRNA may encode the wrong amino acid or substitute a stop codon for an amino acid–encoding codon. Substitution mutations can also be "silent."

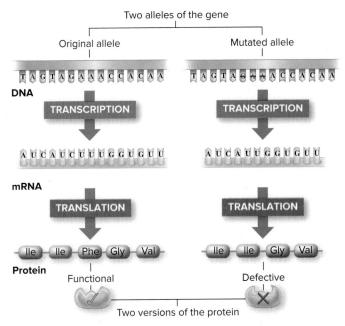

Figure 7.26 Mutations Generate New Alleles.

- **Insertion mutations** and **deletion mutations** add or remove nucleotides. In a **frameshift mutation,** inserting or deleting nucleotides disrupts the reading frame and changes the sequence of the encoded protein. Insertions or deletions in groups of three nucleotides do not alter the reading frame.

B. What Causes Mutations?

- A gene can mutate spontaneously, especially during DNA replication. **Mutagens,** such as chemicals or radiation, also induce some mutations.

C. Mutations Are Important for Many Reasons

- Mutations generate new **alleles,** which are the raw material for evolution.
- If a mutation occurs in a **homeotic gene,** an animal's body parts may form in the wrong places. Such mutations have revealed much about the genes that control development.
- Induced mutations help scientists deduce gene function and help plant breeders produce fruits and flowers with useful new traits.

7.7 Viruses Are Genes Wrapped in a Protein Coat

A. Viruses Are Smaller and Simpler than Cells

- A **virus** is a nucleic acid (DNA or RNA) in a **protein coat.** An **envelope** derived from the host cell's membrane surrounds some viruses.
- A virus must infect a living cell to reproduce; table 7.3 compares cells and viruses.
- Many viruses, including some **bacteriophages,** have relatively complex structures.

B. Viral Replication Occurs in Five Stages

- The stages of viral replication within a host cell are attachment, penetration, synthesis, assembly, and release (see figure 7.27).

7.8 Viruses Infect All Cell Types

A. Bacteriophages May Kill Cells Immediately or "Hide" in a Cell

- In a **lytic infection,** new viruses are immediately assembled and released.
- In a **lysogenic infection,** the virus's nucleic acid replicates along with that of a dividing cell without causing symptoms. The viral DNA may integrate as a **prophage** into the host chromosome.

TABLE 7.3   Cells and Viruses Compared

Feature	Cells	Viruses
Size	Typically 1–100 μm	Typically ~80 nm
Genetic material	DNA	DNA or RNA
Protein coat	No	Yes
Cell membrane	Yes	No
Viral envelope	No	Some viruses
Nucleus/membrane-bounded organelles	Eukaryotes only	No
Cytoplasm	Yes	No
Ribosomes	Yes	No
Enzymes	Yes	Some viruses
Metabolism	Yes	No
Independent replication	Yes	No

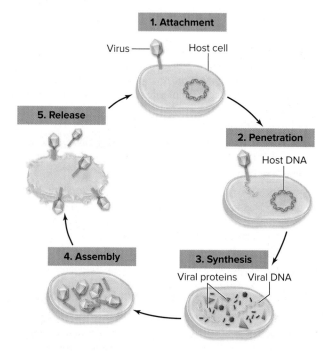

Figure 7.27 Viral Replication: A Summary.

B. Animal Viruses May Cause Immediate Cell Death

• The effects of a virus depend on the cell types it infects. Viruses cause disease by killing infected cells and by stimulating immune responses.

C. Some Animal Viruses Linger for Years

• HIV and some other viruses can cause **latent infections,** which are dormant inside animal cells.
• Some latent viruses are associated with cancer.

D. Viruses Cause Diseases in Plants

• Viruses infect plant cells, then spread via plasmodesmata.

7.9 Drugs and Vaccines Help Fight Viral Infections

• Antiviral drugs and **vaccines** combat some viral infections.
• Antibiotics, drugs that kill bacteria, are ineffective against viruses.

7.10 Viroids and Prions Are Other Noncellular Infectious Agents

• **Viroids** are naked RNA molecules that infect plant cells.
• A **prion** protein can take multiple shapes, at least one of which can cause diseases of the nervous system.

MULTIPLE CHOICE QUESTIONS

1. If one strand of DNA has the sequence ATTGTCC, then the sequence of the complementary strand will be
   a. TAACAGG.   c. ACCTCGG.
   b. CGGAGTT.   d. CCTGTTA.

2. Transcription copies a _____ to a complementary _____ molecule.
   a. chromosome; DNA   c. gene; RNA
   b. genome; RNA   d. DNA sequence; ribosome

3. Choose the DNA sequence from which this mRNA sequence was transcribed: AUACGAUUA.
   a. TATGCTAAT   c. UAUCGUAAU
   b. UTUGCUTTU   d. CTCAGCTTC

4. What is the job of tRNA during translation?
   a. It carries amino acids to the mRNA.
   b. It triggers the formation of a covalent bond between amino acids.
   c. It binds to the small ribosomal subunit.
   d. It triggers the termination of the protein.

5. Certain portions of the mRNA transcribed from the tropomyosin gene can act as either introns or exons. As a result,
   a. one gene may encode many different possible proteins.
   b. each codon may encode many different amino acids.
   c. an amino acid may correspond to many different codons.
   d. a single protein may determine many different traits.

6. If adenine in the tenth nucleotide position of a gene mutates to cytosine, then the _____ amino acid in the protein encoded by this gene could change.
   a. first   c. tenth
   b. fourth   d. thirtieth

7. At which stage in viral replication does the genetic information enter the host cell?
   a. Penetration   c. Assembly
   b. Synthesis   d. Release

8. What is a prion?
   a. A highly wound circle of RNA
   b. A virus that has not yet acquired its envelope
   c. A protein that can alter the shape of a second protein
   d. The protein associated with a latent virus

Answers to Multiple Choice questions are in appendix A.

WRITE IT OUT

1. Describe the three-dimensional structure of DNA.

2. What is the function of DNA?

3. Write the complementary DNA sequence of each of the following base sequences:
   a. A G G C A T A C C T G A G T C
   b. G T T T A A T G C C C T A C A
   c. A A C A C T A C C G A T T C A

4. Arrange the following objects in order from smallest to largest: nucleotide, nitrogenous base, gene, nucleus, cell, codon, chromosome.

5. List the three major types of RNA and their functions.

6. Some people compare DNA to a blueprint stored in the office of a construction company. Explain how this analogy would extend to transcription and translation.

7. Where in a eukaryotic cell do transcription and translation occur?

8. Write the sequence of the mRNA molecule transcribed from the following template DNA sequence:

   G G A A T A C G T C T A G C T A G C A

9. How many codons are in the mRNA molecule that you wrote for question 8?

10. If the sequence of a template strand of DNA is AAAGCAGTACTA, what would be the corresponding amino acid sequence?

11. The roundworm *C. elegans* has 556 cells when it hatches. Each cell contains the entire genome but expresses only a subset of the genes. Therefore, the cells "specialize" in particular functions. List all of the ways that a roundworm cell might silence the unneeded genes.

12. If a protein is 1259 amino acids long, what is the minimum size of the gene that encodes the protein? Why might the gene be longer than the minimum?

13. If a gene is like a cake recipe, then a mutated gene is like a cake recipe containing an error. List the major types of mutations, and describe an analogous error in a cake recipe.

14. A protein-encoding region of a gene has the following DNA sequence:

    T T T C A T C A G G A T G C A A C A

    Determine how each of the following mutations alters the amino acid sequence:

    a. Substitution of an A for the T in the first position
    b. Substitution of a G for the C in the seventeenth position
    c. Insertion of a T between the fourth and fifth DNA bases
    d. Insertion of a GTA between the twelfth and thirteenth DNA bases
    e. Deletion of the first DNA nucleotide

15. How might the effects of a mutation in a gene's promoter differ from the effects of a similar mutation in the gene's protein-encoding region? How might the effects differ if the mutation occurs in an intron versus an exon? Does the type of mutation affect your answer?

16. The amount of melanin in the skin is controlled by genes, yet melanin is not a protein. How can this be?

17. Describe the basic parts of a virus and how each contributes to viral replication.

18. Your biology lab instructor gives you a petri dish of agar covered with visible colonies. Your lab partner says the colonies are viruses, but you disagree. How do you know the colonies are bacteria?

19. Imagine a hybrid virus with the protein coat of virus X and the DNA of virus Y. Will a host cell infected with this hybrid virus produce virus X, virus Y, a mix of X and Y viruses, or hybrid viruses? Explain your answer.

20. Why do antibiotics such as penicillin kill bacteria but leave viruses unharmed?

21. Search the Internet for information about the injectable flu vaccine (a "flu shot"). Why is the flu shot administered annually when many other vaccines last for years? Is it possible for a flu shot to cause influenza?

Answers to Mastering Concepts, Write It Out, Scientific Literacy, and Pull It Together questions can be found in the Connect ebook.
connect.mheducation.com

Design element: Burning Question (fire background): ©Ingram Publishing/Super Stock

SCIENTIFIC LITERACY

Review Burning Question 7.2, which explains why you might be likely to become sick in the winter. Scientists studying the rhinovirus (which infects the nose and air passages) found that it multiplies more quickly when human cells are below normal body temperature because the immune system functions less efficiently. What might you do to keep your airways warm while outside in cold weather? Why might frequent hand-washing be even more important?

PULL IT TOGETHER

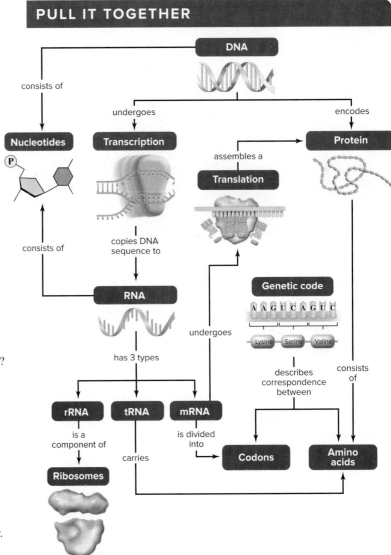

Figure 7.28 Pull It Together: DNA Structure and Gene Function.

Refer to figure 7.28 and the chapter content to answer the following questions.

1. Why is protein production essential to cell function?

2. Where do promoters, terminators, stop codons, transcription factors, RNA polymerase, and ribosomes fit into this concept map?

3. How would viruses fit into this concept map?

4. Review the Survey the Landscape figure in the chapter introduction, and then explain why a mutation in DNA sometimes causes protein function to change.

LEARNING OUTLINE

8.1 Cells Divide and Cells Die

8.2 DNA Replication Precedes Cell Division

8.3 Bacteria and Archaea Divide by Binary Fission

8.4 Replicated Chromosomes Condense as a Eukaryotic Cell Prepares to Divide

8.5 Mitotic Division Generates Exact Cell Copies

8.6 Cancer Cells Divide Uncontrollably

APPLICATIONS

Burning Question 8.1 *Do all human cells divide at the same rate?*
Why We Care 8.1 *Skin Cancer*
Investigating Life 8.1 *Evolutionary Strategies in the Race Against Cancer*

Growth. Cell division accounts for the growth of a seedling, a child, and every other multicellular organism.
©PhotoAlto/Getty Images RF

Learn How to Learn
Write It Out—Really!

Get out a pen and a piece of scratch paper, and answer the open-ended "Write It Out" questions at the end of each chapter. This tip applies even if the exams in your class are multiple choice. Putting pen to paper (as opposed to just saying the answer in your head) forces you to organize your thoughts and helps you discover the difference between what you know and what you only THINK you know.

SURVEY THE LANDSCAPE
DNA, Inheritance, and Biotechnology

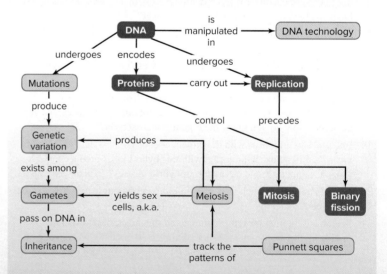

Prokaryotic cells divide by binary fission; eukaryotic cells use mitotic or meiotic division. Proteins regulate the entire process and copy all of the cell's DNA shortly before the split occurs.

For more details, study the Pull It Together feature in the chapter summary.

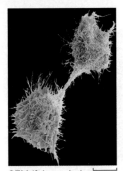

SEM (false color) ⊢—⊣ 5 μm

©Steve Gschmeissner/SPL/
Getty Images RF

Your cells are too small to see without a microscope, so it is hard to appreciate just how many you lose as you sleep, work, and play. Each minute, for example, you shed tens of thousands of dead skin cells. If you did not have a way to replace these building blocks, your body would literally wear away. Instead, cells in your deep skin layers divide and replace the ones you lose. Each new cell lives an average of about 35 days, so you will gradually replace your entire skin in the next month or so—without even noticing!

This chapter describes how and when cells divide to reproduce, grow, or repair injuries. Cell division is under tight control. As you will see, an organism that fails to keep cell division in check risks deadly consequences: cancer.

8.1 Cells Divide and Cells Die

Without a doubt, DNA is an amazing molecule. In chapter 7, you learned about DNA's main function: to specify the "recipes" for all of the proteins in a cell. The production of proteins is essential to life, so it stands to reason that every cell needs a complete set of DNA instructions. Where does this DNA come from? The scientists who discovered the structure of DNA surmised the answer: A cell can copy its own DNA. This process of DNA replication must occur before a cell's nucleus and cytoplasm can split in two. ⓘ *eukaryotic cell structure,* section 3.2B

Cell division produces a continuous supply of replacement cells everywhere in your body. But cell division has other functions as well. No living organism can reproduce without cell division, and the growth and development of a multicellular organism also require the production of new cells.

This chapter explores the opposing forces of cell division and cell death and considers what happens if either process goes wrong. We begin by exploring cell division's role in reproduction, growth, and development.

A. Sexual Life Cycles Include Mitosis, Meiosis, and Fertilization

Organisms must reproduce—generate other individuals like themselves—for a species to persist. For a single-celled organism, the most straightforward (and ancient) method is **asexual reproduction,** in which one cell replicates its genetic material and splits into two. Except for the occasional mutation, asexual reproduction generates genetically identical offspring. Bacteria and archaea, for example, reproduce asexually via a simple type of cell division called binary fission (see section 8.3). Many protists and multicellular eukaryotes also reproduce asexually.

Sexual reproduction, in contrast, is the production of offspring whose genetic makeup comes from two parents. Each parent contributes a sex cell, and the fusion of these cells signals the start of the next generation. Because sexual reproduction mixes up and recombines traits, the offspring are genetically different from each other and their parents.

Figure 8.1 illustrates how two types of cell division, meiosis and mitosis, interact in a sexual life cycle. **Meiosis** is a specialized process that gives rise to nuclei that are genetically different from one another (see chapter 9). In humans and many other species, these nuclei are packaged into **gametes:** sperm cells (produced by males) and egg cells (produced by females). The variation among gametes explains why siblings generally look different from one another (except for identical twins).

Fertilization is the union of the sperm and the egg cell, producing a zygote (the first cell of the new offspring). Immediately after fertilization, the other type of cell division—mitotic—takes over. **Mitosis** divides a eukaryotic cell's genetic information into two identical nuclei.

Each of the trillions of cells in your body retains the genetic information that was present in the fertilized egg. Inspired by the astonishing precision with which this occurs, geneticist Herman J. Müller wrote in 1947:

In a sense we contain ourselves, wrapped up within ourselves, trillions of times repeated.

MITOSIS

MITOSIS

MEIOSIS   MEIOSIS

Gametes
(sperm and egg cells)

Zygote
(fertilized egg)

FERTILIZATION

Figure 8.1 **Sexual Reproduction.** In the life cycle of humans and many other organisms, adults produce gametes by meiosis. Fertilization unites sperm and egg, forming a zygote. Mitotic cell division accounts for the growth of the new offspring.

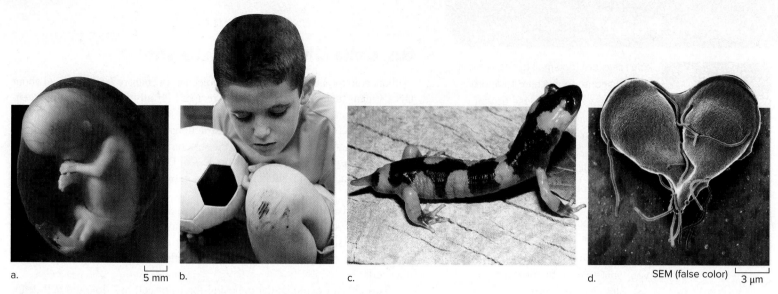

a.   5 mm   b.   c.   d.   SEM (false color) 3 μm

Figure 8.2 Functions of Mitotic Cell Division. Cells divide mitotically as a eukaryotic organism (a) grows and (b) repairs damaged tissues. (c) Some species can regenerate lost body parts; this salamander is regrowing its tail. (d) Other organisms, such as this protist, reproduce asexually by mitotic cell division. (a): ©Brand X Pictures/PunchStock RF; (b): ©Image Source/Getty Images RF; (c): ©Joseph T. Collins/Science Source; (d): Source: CDC

This quotation eloquently expresses the powerful idea that every cell in the body results from countless rounds of cell division, each time forming two genetically identical cells from one.

Mitotic cell division explains how you grew from a single cell into an adult, how you repair damage after an injury, and how you replace the cells that you lose every day (figure 8.2). Likewise, mitotic cell division accounts for the growth and development of plants, mushrooms, and other multicellular eukaryotes and for asexual reproduction in protists and many other eukaryotes.

B. Cell Death Is Part of Life

The development of a multicellular organism requires more than just cell division. Cells also die in predictable ways, carving distinctive structures. **Apoptosis,** also called "programmed cell death," is a normal part of development. Like cell division, apoptosis is a precise, tightly regulated sequence of events.

During early development, both cell division and apoptosis shape new structures. For example, the feet of both ducks and chickens start out as webbed paddles when the birds are embryos (figure 8.3). The webs of tissue remain in the duck's foot throughout life. In the chicken, however, individual toes form as cells between the digits die. Likewise, cells in the tail of a tadpole die as the young frog develops into an adult.

Throughout an animal's life, cell division and cell death are in balance, so tissue neither overgrows nor shrinks. Cell division compensates for the death of skin and blood cells, a little like adding new snow (cell division) to a snowman that is melting (apoptosis). Both cell division and apoptosis also help protect the organism. For example, cells divide to heal a scraped knee; apoptosis peels away sunburnt skin cells that might otherwise become cancerous (see section 8.6).

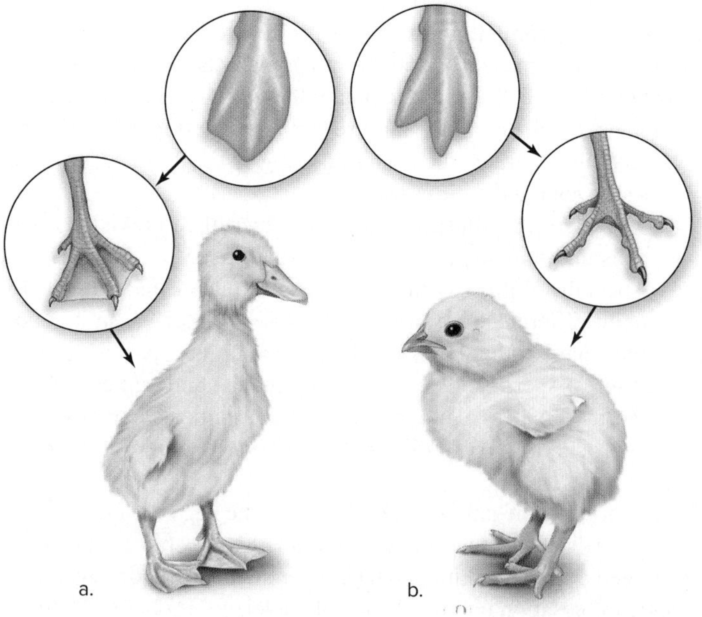

Figure 8.3 Apoptosis Carves Toes. The feet of embryonic birds have webbing between the digits. (a) A duck's foot retains this webbing. (b) In a developing chicken foot, the toes take shape when the cells between the digits die.

a.   b.

8.1 Mastering Concepts

1. Explain the roles of mitotic cell division, meiosis, and fertilization in the human life cycle.

2. Why are both cell division and apoptosis necessary for the development of an organism?

8.2 DNA Replication Precedes Cell Division

Before any cell divides—by binary fission, mitotically, or meiotically—it must first duplicate its entire **genome,** which consists of all of the cell's genetic material. The genome may consist of one or more **chromosomes,** individual molecules of DNA with their associated proteins. Chapter 7 describes the cell's genome as a set of "cookbooks" (chromosomes), each containing "recipes" (genes) that encode proteins. In DNA replication, the cell copies all of this information, letter by letter. Without a full set of instructions, a new cell may die.

Recall from figure 7.2 that DNA is a double-stranded nucleic acid. Each strand of the double helix is composed of nucleotides. Hydrogen bonds between the nitrogenous bases of complementary nucleotides hold the two strands together. That is, the base adenine (A) pairs with thymine (T), whereas cytosine (C) forms complementary base pairs with guanine (G).

When Watson and Crick reported DNA's chemical structure, they understood that they had uncovered the key to DNA replication. They envisioned DNA unwinding, exposing unpaired bases that would attract their complements, and neatly knitting two double helices from one. This route to replication, which turned out to be essentially correct, is called semiconservative because each DNA double helix ends up with one complete strand from the original molecule (figure 8.4).

DNA does not, however, replicate by itself. Instead, an army of enzymes copies DNA just before a cell divides. Enzymes called helicases unwind and "unzip" the DNA molecule. **DNA polymerase** is the enzyme that adds new DNA nucleotides that are complementary to the bases on each exposed strand. As the new DNA strands grow, hydrogen bonds form between the complementary bases, and two DNA molecules form in the place of one. ⓘ *enzymes,* section 4.4

Replication enzymes work simultaneously at hundreds of points, called origins of replication, on each DNA molecule (figure 8.5). This arrangement is similar to the way that hurried office workers might split a lengthy report into short pieces and then divide the sections among many copy machines operating at the same time. Thanks to this division of labor, copying the billions of DNA nucleotides in a human cell takes only 8 to 10 hours. Replication proceeds in both directions at once from each origin of replication; **ligase** enzymes form covalent bonds between adjacent DNA segments.

DNA replication is incredibly accurate. DNA polymerase "proofreads" as it goes, discarding mismatched nucleotides and inserting correct ones. After proofreading, DNA polymerase has an error rate of only about 1 in a billion nucleotides. Other repair enzymes help ensure the accuracy of DNA replication by cutting out and replacing incorrect nucleotides.

Nevertheless, mistakes occasionally remain. The result is a **mutation,** which is any change in a cell's DNA sequence. To extend the cooking analogy, a mutation is similar to a mistake in one of the recipes in a cookbook. Section 7.6 describes the many ways that a mutation can affect the life of a cell.

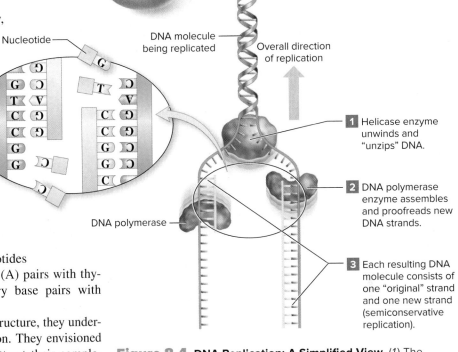

1 Helicase enzyme unwinds and "unzips" DNA.

2 DNA polymerase enzyme assembles and proofreads new DNA strands.

3 Each resulting DNA molecule consists of one "original" strand and one new strand (semiconservative replication).

Figure 8.4 DNA Replication: A Simplified View. (*1*) The helicase enzyme unwinds and separates the two DNA strands. (*2*) The DNA polymerase enzyme assembles new DNA strands, using nucleotides that are complementary to each exposed strand. (*3*) The process ends with two identical double-stranded DNA molecules.

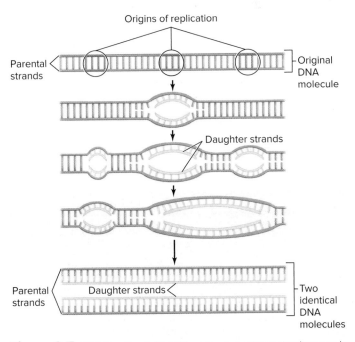

Figure 8.5 Dividing the Job. DNA replication occurs simultaneously at many points along a chromosome.

Figure It Out

Write the complementary strand for the following DNA sequence: TCAATACCGATTAT.

Answer: AGTTATGGCTAATA.

Overall, DNA replication requires a great deal of energy because a large, organized nucleic acid contains much more potential energy than do many individual nucleotides. Energy from ATP is required to synthesize nucleotides and to create the covalent bonds that join them together in the new strands of DNA. Many of the enzymes that participate in DNA replication, including helicase and ligase, also require energy to catalyze their reactions. ⓘ *ATP*, section 4.3

8.2 Mastering Concepts

1. Why does DNA replicate?
2. What are the steps of DNA replication?
3. Why do enzymes work at multiple origins of replication?
4. What happens if DNA polymerase makes an error?

8.3 Bacteria and Archaea Divide by Binary Fission

Like all organisms, bacteria and archaea transmit DNA from generation to generation as they reproduce. In prokaryotes, reproduction occurs by **binary fission,** an asexual process that replicates DNA and distributes it (along with other cell parts) into two daughter cells (figure 8.6). ⓘ *prokaryotic cells, section 3.2A*

Miniglossary	Cell Division
Mitosis	Divides a eukaryotic cell's genetic information into two identical nuclei
Meiosis	Divides a eukaryotic cell's genetic information into genetically unique nuclei
Binary fission	Replicates and divides a prokaryotic cell's DNA into two daughter cells
Cytokinesis	Distributes the cytoplasm into two daughter cells following division of a cell's chromosomes

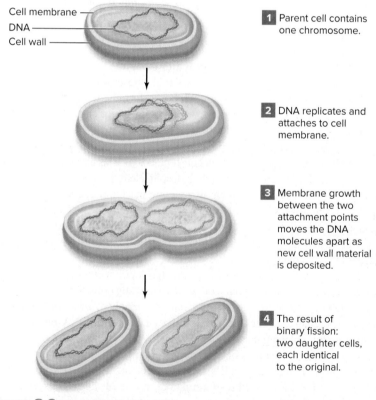

Figure 8.6 Binary Fission. A dividing prokaryotic cell replicates its DNA, grows, and then splits into two identical daughter cells.

Each prokaryotic cell contains one circular chromosome. As the cell prepares to divide, its DNA replicates. The chromosome and its duplicate attach to the inner surface of the cell. The cell membrane grows between the two DNA molecules, separating them. Then the cell pinches in half to form two daughter cells from the original one.

In optimal conditions, some bacterial cells can divide every 20 minutes. Those few microbes that remain after you brush your teeth therefore easily repopulate your mouth as you sleep. Their metabolic activities produce the notoriously foul-smelling "morning breath."

Asexual organisms generally have few ways to generate genetic diversity. For example, in eukaryotes that reproduce asexually, random mutations in DNA account for most or all of the genetic variation. Bacteria and archaea undergo mutations as well, but they can also acquire new genetic material from other sources (figure 8.7). For example, one cell may transfer a copy of some of its DNA to another cell through an appendage called a sex pilus. Alternatively, a cell may absorb stray bits of DNA that are released after another cell dies. Gene transfer among bacterial cells has profound implications in many fields, including medicine. As bacteria continue to swap antibiotic-resistance genes, for example, many once-curable diseases are becoming impossible to treat (see Why We Care 12.2).

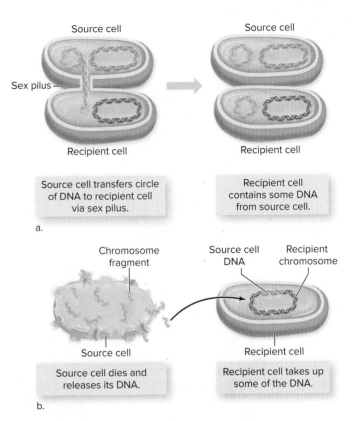

Figure 8.7 **Getting New Genes.** Prokaryotes can get additional DNA (a) via a sex pilus or (b) by scooping up genes released by dying cells.

8.3 Mastering Concepts

1. What are the events of binary fission?
2. What are the sources of genetic variation in bacteria and archaea?

8.4 Replicated Chromosomes Condense as a Eukaryotic Cell Prepares to Divide

Binary fission is relatively uncomplicated because the genetic material in prokaryotic cells typically consists of a single circular DNA molecule, so cell division is relatively simple. In a eukaryotic cell, however, distributing DNA into daughter cells is more complex because the genetic information consists of multiple chromosomes inside a nucleus.

Each species has a characteristic number of chromosomes. A mosquito's cell has 6 chromosomes; grasshoppers, rice plants, and pine trees all have 24; humans have 46; dogs and chickens have 78; a carp has 104.

With so much genetic information, a eukaryotic cell must balance two needs. On one hand, the cell must have access to the information in its DNA. On the other hand, a dividing cell must package its DNA into a portable form that can easily move into the two daughter cells (figure 8.8). DNA packing is therefore comparable to winding a very long piece of yarn into a compact ball. Just as a ball of yarn occupies less space and is more portable than a tangled pile of loose yarn, condensed DNA is easier for the dividing cell to manage than is an unwound chromosome.

To learn how cells maintain this balance, we must look closely at a chromosome's structure. Eukaryotic chromosomes consist of **chromatin,** which is a collective term for all of the cell's DNA and its associated proteins. These proteins include the many enzymes that help replicate the DNA and transcribe it to RNA

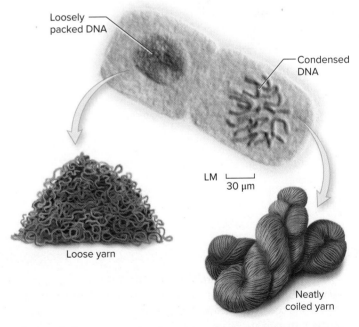

Figure 8.8 **Two Views of DNA.** Loosely packed DNA is available for replication and transcription. Before a cell divides, however, the DNA condenses into compact chromosomes. These two forms of DNA are comparable to loose and coiled yarn.

Photo: ©Clouds Hill Imaging Ltd./Getty Images

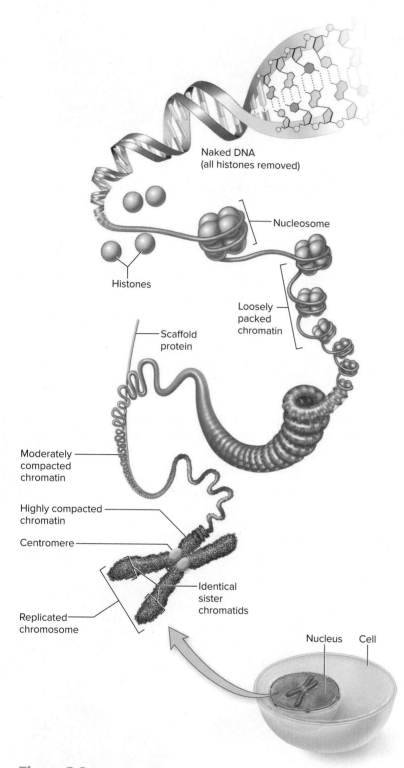

Figure 8.9 **Parts of a Chromosome.** After DNA replication, the chromatin condenses into a more compact form by winding around histones and scaffold proteins. Once condensed, the chromosome appears X-shaped because two identical chromatids are attached side-by-side at the centromere.

(see chapter 7). Others serve as scaffolds around which DNA entwines, helping to pack the DNA efficiently inside the cell.

Stretched end to end, the DNA in one human cell would form a thread some 2 meters long. If the DNA bases of all 46 human chromosomes were typed as A, C, T, and G, the several billion letters would fill 4000 books of 500 pages each! How can a cell only 100 microns in diameter contain so much material?

The explanation is that chromatin is organized into **nucleosomes,** each consisting of a stretch of DNA wrapped around eight proteins (histones). A continuous thread of DNA connects nucleosomes like beads on a string (figure 8.9). When the cell is not dividing, chromatin is barely visible because the nucleosomes are loosely packed together. The cell can therefore access the information in the DNA to produce the proteins that it needs. DNA replication in preparation for cell division also requires that the cell's DNA be unwound.

The chromosome's appearance changes shortly after DNA replication. The nucleosomes gradually fold into progressively larger structures, until the chromosome takes on its familiar, compact shape. Once condensed, a chromosome has readily identifiable parts (see figure 8.9). Two **chromatids** make up the replicated chromosome. Because these paired chromatids have identical DNA sequences, they are called "sister chromatids." The **centromere** is a small section of DNA and associated proteins that attaches the sister chromatids to each other. It often appears as a constriction in a replicated chromosome. As a cell's genetic material divides, the centromere splits, and the sister chromatids move apart (see section 8.5). At that point, each chromatid becomes an individual chromosome.

8.4 Mastering Concepts

1. What is the relationship between DNA, histones, and chromatin?

2. Sketch and label the main parts of a duplicated chromosome.

Miniglossary	Chromosomes
Chromatin	DNA and its associated proteins
Chromosome	A single continuous molecule of DNA wrapped around protein. Eukaryotic cells contain multiple linear chromosomes, whereas bacterial cells typically have one circular chromosome.
Chromatid	One of two identical attached copies that make up a replicated chromosome
Centromere	A small region of a chromosome where sister chromatids attach to each other

8.5 Mitotic Division Generates Exact Cell Copies

Suppose you scrape your leg while sliding into second base during a softball game. At first, the wound bleeds, but the blood soon clots and forms a scab. Underneath the dried crust, cells of the immune system clear away trapped dirt and dead cells. At the same time, undamaged skin cells bordering the wound begin to divide repeatedly, producing fresh, new daughter cells that eventually fill the damaged area (see Burning Question 8.1).

Those actively dividing skin cells illustrate the **cell cycle,** which describes the events that occur in one complete round of cell division. Biologists divide the cell cycle into stages (figure 8.10). **Interphase** is the interval between successive cell divisions; protein synthesis, DNA replication, and many other events occur during interphase. Next is **mitosis,** during which the contents of the nucleus divide. In **cytokinesis,** the cell splits into two daughter cells. After cytokinesis is complete, each daughter cell enters interphase, and the cell cycle begins anew.

Mitotic cell division occurs some 300 million times per minute in your body, replacing cells lost to abrasion or cell death. In each case, the products of cell division are two daughter cells, each receiving complete, identical genetic instructions, plus the molecules and organelles they need for their own metabolism.

A. DNA Is Copied During Interphase

Biologists once mistakenly described interphase as a time when the cell is at rest. The chromatin is unwound and therefore barely visible, so the cell appears inactive. However, interphase is actually a very active time. The cell produces proteins and carries out its functions, from photosynthesis to muscle contraction to insulin production to bone formation. DNA replication also occurs during this stage.

Interphase is divided into "gap" phases (designated G_1 and G_2), separated by a "synthesis" (S) phase. During G_1 **phase,** the cell grows, carries out its basic functions, and produces the new organelles and other components it will require if it divides.

During **S phase,** enzymes replicate the cell's genetic material and repair damaged DNA (see section 8.2). As S phase begins, each chromosome includes one DNA molecule. By the end of S phase, each chromosome consists of two attached sister chromatids.

In an animal cell, another event that occurs during S phase is the duplication of the centrosome. **Centrosomes** are structures that organize the mitotic **spindle,** a set of microtubule proteins that coordinate the movements of the chromosomes during mitosis. Each centrosome includes proteins enclosing a pair of barrel-shaped centrioles. Most plant cells lack centrosomes; they organize their spindle fibers throughout the cell. ⓘ *microtubules,* section 3.5

In G_2 **phase,** the cell continues to grow but also prepares to divide, producing the proteins that will help coordinate mitosis. The DNA winds more tightly around its associated proteins, and this event signals the start of mitosis. Interphase has ended.

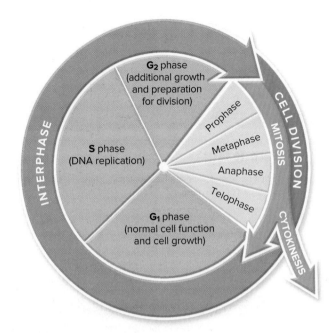

Figure 8.10 **The Cell Cycle.** Interphase includes gap phases (G_1 and G_2), when the cell grows and some organelles duplicate. During the synthesis (S) phase of interphase, DNA replicates. Mitosis divides the replicated genetic material between two nuclei. Cytokinesis then splits the cytoplasm in half, producing two identical daughter cells.

Burning Question 8.1

Do all human cells divide at the same rate?

The short answer to this question is no. Cells divide at different rates depending on several factors, including the type of cell. As mentioned in this chapter's "What's the Point?" box, skin cells divide rapidly. Other tissues that are subject to abrasion, such as cells in the mouth, also have a high division rate. On the other hand, some cells hardly divide at all. Mature brain cells called neurons rarely or never divide, although the glial cells that surround neurons do. Brain cancer, if it develops, often occurs when glial cells fail to control cell division.

Age also affects the rate of cell division. In general, cells of a younger person divide more quickly than those of an older individual. Of course, as a person is growing, cell division must be faster in the ends of bones and other areas, compared to a person who is already fully grown. And cancer—which is characterized by an abnormally high rate of cell division—becomes more likely as a person ages (see section 8.6).

Submit your burning question to
marielle.hoefnagels@mheducation.com

(cancer cells): Source: National Institutes of Health (NIH)/USHSS

Figure It Out

A cell that has completed interphase contains ___ times as much DNA as a cell at the start of interphase.

Answer: Two.

B. Chromosomes Divide During Mitosis

Overall, mitosis separates the genetic material that replicated during S phase. Biologists divide mitosis into four main stages (figure 8.11); note, however, that the process does not actually stop as each stage ends.

During **prophase,** DNA coils very tightly, shortening and thickening the chromosomes (see figure 8.9). As the chromosomes condense, they become visible when stained and viewed under a microscope. For now, the chromosomes remain randomly arranged in the nucleus.

Also during prophase, the two centrosomes migrate toward opposite ends of the cell, and the spindle begins to form. The nucleolus—the darkened area in the nucleus—disappears. The nuclear envelope breaks into small pieces, as does the surrounding endoplasmic reticulum. The spindle fibers are now free to attach to the chromosomes.

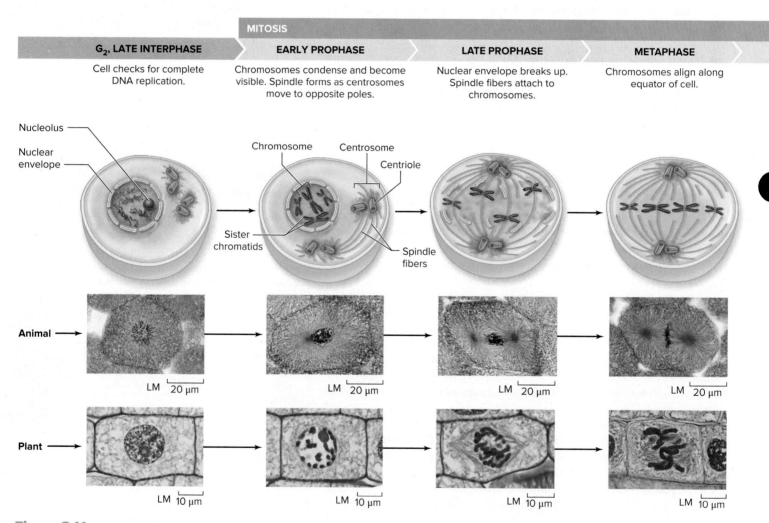

MITOSIS

G₂, LATE INTERPHASE	EARLY PROPHASE	LATE PROPHASE	METAPHASE
Cell checks for complete DNA replication.	Chromosomes condense and become visible. Spindle forms as centrosomes move to opposite poles.	Nuclear envelope breaks up. Spindle fibers attach to chromosomes.	Chromosomes align along equator of cell.

Nucleolus
Nuclear envelope
Chromosome
Centrosome
Centriole
Sister chromatids
Spindle fibers

Animal
LM 20 μm   LM 20 μm   LM 20 μm   LM 20 μm

Plant
LM 10 μm   LM 10 μm   LM 10 μm   LM 10 μm

Figure 8.11 **Stages of Mitosis.** Mitotic cell division includes similar stages in all eukaryotes, including animals and plants. Notice that the cell entering mitosis has four chromosomes (two red and two blue), as does each of the two resulting daughter cells.

Photos: (all animal): ©Ed Reschke/Photolibrary/Getty Images; (all plant): ©Ed Reschke

As **metaphase** begins, the spindle lines up the chromosomes along the center, or equator, of the cell. This arrangement ensures that each cell will receive one copy of each chromosome.

In **anaphase,** the centromeres split and some spindle fibers shorten as they pull the sister chromatids (now chromosomes) toward opposite poles of the cell. At the same time, other microtubules in the spindle lengthen in a way that moves the poles farther apart, stretching the dividing cell.

Telophase, the final stage of mitosis, essentially reverses the events of prophase. The spindle disassembles, and the chromosomes begin to unwind. In addition, a nuclear envelope and nucleolus form at each end of the stretched-out cell. As telophase ends, the division of the genetic material is complete, and the cell contains two nuclei—but not for long.

Figure It Out

1. A mosquito cell in interphase has six chromosomes. How many sister chromatids does the same cell have during metaphase?

2. A human cell in early prophase has 46 chromosomes. How many chromosomes does one of the daughter cells have immediately after mitosis and cytokinesis?

Answer 1: 12. **Answer 2:** 46.

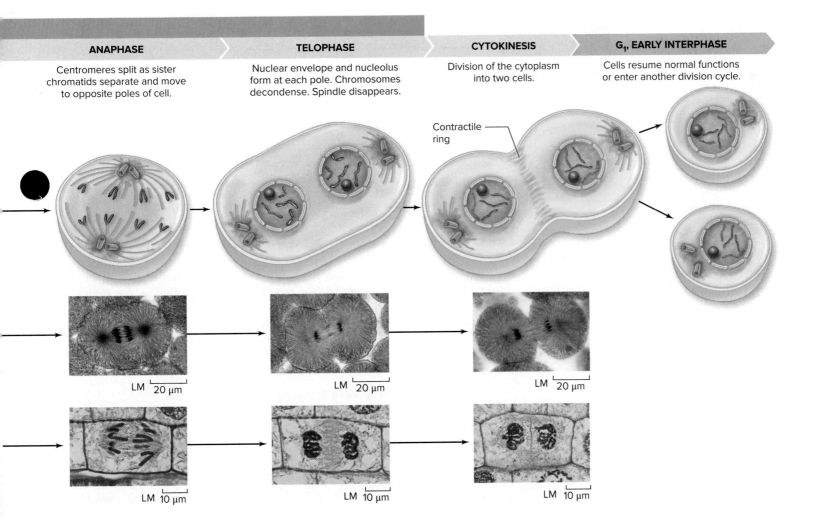

ANAPHASE

Centromeres split as sister chromatids separate and move to opposite poles of cell.

TELOPHASE

Nuclear envelope and nucleolus form at each pole. Chromosomes decondense. Spindle disappears.

CYTOKINESIS

Division of the cytoplasm into two cells.

G₁, EARLY INTERPHASE

Cells resume normal functions or enter another division cycle.

Contractile ring

LM 20 μm

LM 20 μm

LM 20 μm

LM 10 μm

LM 10 μm

LM 10 μm

Mitotic Division Generates Exact Cell Copies   Cancer Cells Divide Uncontrollably

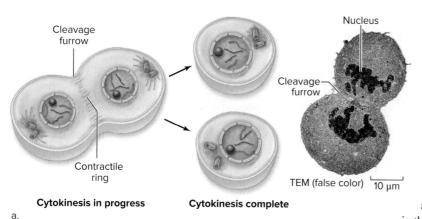

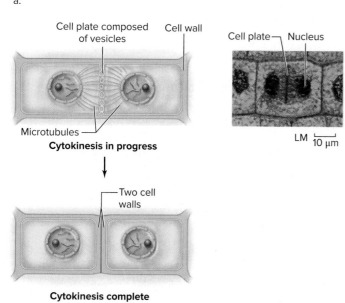

Figure 8.12 Cytokinesis. (a) In an animal cell, the first sign of cytokinesis is an indentation called a cleavage furrow, formed by a ring of protein filaments. (b) In plant cells, the cell plate is the first stage in the formation of a new cell wall.

Photos: (a): ©Dr. Gopal Murti/Science Source; (b): ©Wood/Custom Medical Stock Photo/Getty Images

C. The Cytoplasm Splits in Cytokinesis

In cytokinesis, the cytoplasm and the two nuclei are distributed into the two forming daughter cells, which then physically separate. The process differs somewhat between animal and plant cells (figure 8.12).

In an animal cell, the first sign of cytokinesis is the **cleavage furrow,** a slight indentation around the middle of the dividing cell. A ring of proteins beneath the cell membrane contracts like a drawstring, separating the daughter cells.

Unlike animal cells, plant cells are surrounded by cell walls. A dividing plant cell must therefore construct a new wall that separates the two daughter cells. The first sign of cell wall construction is the **cell plate,** a structure that appears at the midline of the dividing plant cell. The cell plate grows and consolidates as vesicles from the Golgi apparatus deliver cellulose, other polysaccharides, and proteins. The resulting layer of cellulose fibers embedded in surrounding material makes a strong, rigid wall that gives a plant cell its shape. ⓘ *cell wall,* section 3.6

> ### 8.5 Mastering Concepts
>
> **1.** What are the three main events of the cell cycle?
>
> **2.** What happens during interphase?
>
> **3.** Suppose a centromere does not split during anaphase; describe the chromosomes in the daughter cells.
>
> **4.** What happens during each stage of mitosis?
>
> **5.** Distinguish between mitosis and cytokinesis.

8.6 Cancer Cells Divide Uncontrollably

Some cells divide more or less constantly. The cells at a plant's root tips, for example, may divide throughout the growing season, exploring soil for water and nutrients. Likewise, stem cells in bone marrow constantly produce new blood cells. On the other hand, skin cells quit dividing once a scrape has healed; mature brain cells rarely divide. How do any of these cells "know" what to do?

A. Chemical Signals Regulate Cell Division

Cells divide in response to a variety of chemical signals, many of which originate outside the cell. Proteins that stimulate cell division bind to receptors on a receiving cell's membrane, and then a cascade of chemical reactions inside the cell initiates division. At a wound site, for example, a protein called epidermal growth factor stimulates our cells to produce new skin underneath a scab; in plants, protein signals induce the formation of abnormal growths called galls.

In addition, several internal "checkpoints" ensure that a cell does not enter one stage of the cell cycle until the previous stage is complete. Some of the checkpoints screen for damaged DNA. If the genetic material is damaged beyond repair, a signaling protein may trigger apoptosis. A cell that fails to pass a checkpoint correctly will not progress to the next stage. These checkpoints are therefore somewhat like the guards that check passports at border crossings, denying entry to travelers without proper documentation.

Precise timing of the many chemical signals that regulate the cell cycle is essential. Too little cell division, and an injury may go unrepaired; too much, and an abnormal growth forms. Understanding these signals has helped reveal how diseases such as cancer arise.

B. Cancer Cells Are Malignant

What happens when the body loses control over cell division and apoptosis? Sometimes, a **tumor**—an abnormal mass of tissue—forms. Biologists classify tumors into two groups (figure 8.13). **Benign tumors** are usually slow-growing and harmless, unless they become large enough to disrupt nearby tissues or organs. A tough capsule surrounding the tumor prevents it from invading nearby tissues or spreading to other parts of the body. Warts and moles are examples of benign tumors of the skin.

In contrast, a **malignant tumor** invades adjacent tissue. Because it lacks a surrounding capsule, a malignant tumor is likely to **metastasize,** meaning that its cells can break away from the original mass and travel in the bloodstream or lymphatic system to colonize other areas of the body. **Cancer** is a class of diseases characterized by malignant cells.

Solid tumors of the breast, lung, skin, and other major organs are the most familiar forms of cancer. But cells in the blood-forming tissues of the bone marrow can also divide out of control. Leukemia is a group of cancers characterized by the excessive production of the wrong kinds of blood cells.

Whatever its form, cancer begins when a single cell accumulates genetic mutations that cause it to break through its cell cycle controls. Each cancerous cell passes its loss of cell cycle control to its daughter cells. With enough nutrients and space, cancer cells can divide uncontrollably and eternally. As they do so, they may crush vital organs, block the body's passageways, and divert nutrients from other body cells.

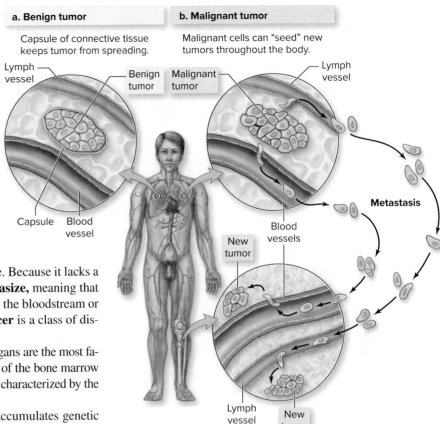

a. Benign tumor
Capsule of connective tissue keeps tumor from spreading.
Lymph vessel
Benign tumor
Malignant tumor
Capsule   Blood vessel

b. Malignant tumor
Malignant cells can "seed" new tumors throughout the body.
Lymph vessel
Metastasis
Blood vessels
New tumor
Lymph vessel   New tumor

Figure 8.13 Benign and Malignant Tumors. (a) A capsule of connective tissue prevents a benign tumor from invading adjacent tissues. (b) A malignant tumor lacks a capsule and therefore can spread throughout the body in blood and lymph.

C. Cancer Treatments Remove or Kill Abnormal Cells

Traditional cancer treatments include surgical tumor removal, drugs (chemotherapy), and radiation. Chemotherapy drugs, usually delivered intravenously, are intended to stop cancer cells anywhere in the body from dividing. Radiation therapy uses directed streams of energy from radioactive isotopes to kill tumor cells in limited areas. ⓘ *isotopes,* section 2.1C

Chemotherapy and radiation are relatively "blunt tools" that target rapidly dividing cells, whether cancerous or not. Examples of cells that divide frequently include those in the bone marrow, digestive tract, and hair follicles. The death of these cells accounts for the most notorious side effects of cancer treatment: fatigue, a weakened immune system, nausea, and hair loss.

Newer cancer treatments take specific aim at cancer cells without the devastating side effects. These include not only targeted drugs but also techniques that stimulate the immune system to attack the cancer cells.

The success of any cancer treatment depends on many factors, including the type of cancer and the stage in which it is detected. Surgery can cure cancers that have not spread. Once cancer metastasizes, however, it becomes difficult to locate and treat all of the tumors. Moreover, DNA replication errors introduce mutations in rapidly dividing cancer cells. Treatments that shrank the original tumor may have no effect on this new, changed growth (see Investigating Life 8.1).

TABLE 8.1	**Reducing Cancer Risk**
Risk Factor	**Ways to Reduce Risk**
Unhealthy diet	Eat less saturated fat; eat more fruits and vegetables
Obesity	Maintain healthy body weight; get regular, vigorous exercise
Tobacco use	Quit smoking or chewing tobacco, or never start
Ultraviolet radiation	Avoid UV radiation from sunlight and tanning beds
Cancer-causing viruses	Use condoms to avoid sexually transmitted diseases (e.g., human papillomavirus, which is associated with cervical cancer)
Detection time	Use self tests and medical exams for early detection
Family history	Cannot be avoided

Why We Care 8.1 | Skin Cancer

Cancer has many forms, some inherited and others caused by radiation or harmful chemicals. Exposure to ultraviolet radiation from the sun or from tanning beds, for example, increases the risk of skin cancer because UV radiation damages DNA. If mutations occur in genes encoding proteins that control the pace of cell division, cells may begin dividing out of control, forming a malignant tumor on the skin.

How might a person determine whether a mole, sore, or growth on the skin is cancerous? The abnormal skin may vary widely in appearance, and only a physician can tell for sure. Nevertheless, most skin cancers have a few features in common. "ABCD" is a shortcut for remembering these four characteristics:

©Stockbyte/Getty Images RF

Asymmetry: One half of the area looks different from the other.

Borders: The borders of the growth are irregular, not smooth.

Color: The color varies within a patch of skin, from tan to dark brown to black. Other colors, including red, may also appear.

Diameter: The diameter of a cancerous area is usually greater than 6 mm, which is about equal to the size of a pencil eraser.

D. Genes and Environment Both Can Increase Cancer Risk

Proteins control both the cell cycle and apoptosis. Genes encode proteins, so genetic mutations (changes in genes) play a key role in causing cancer. So far, researchers know of hundreds of genetic mutations that contribute to cancer. Where do the cancer-causing mutations come from? Sometimes, a person inherits mutated versions of the genes from one or both parents. The parent may also have had cancer, or the mutations may have arisen spontaneously in sperm- or egg-producing cells. Often, however, people develop cancer after exposure to harmful chemicals, radiation, and viruses, all of which may alter their genes (see Why We Care 8.1). Poor diet and exercise habits, sun exposure, and tobacco use also raise cancer risks (see table 8.1).

8.6 Mastering Concepts

1. What prevents normal cells from dividing when they are not supposed to?
2. List and describe traditional and newer cancer treatments.
3. What is the relationship between genetic mutations and cancer?

Investigating Life 8.1 | Evolutionary Strategies in the Race Against Cancer

Many cancer patients face a frustrating cycle. Their tumors shrink after grueling chemotherapy treatments, sparking dreams of a full recovery. But success is fleeting. The cancer soon returns, harder to treat than ever.

This sad reality underscores a huge challenge in the fight against cancer: natural selection. Every tumor contains a mix of drug-resistant and susceptible cells that compete for space and nutrients. In the absence of drugs, the resistant cells are at a disadvantage because they spend considerable energy producing the proteins that maintain their resistance. However, standard chemotherapy treatment—a rapid-fire barrage of drugs at doses so strong that the patient can barely tolerate them—kills susceptible cells and changes the balance of power. Freed from competition, drug-resistant cells divide rapidly.

At Florida's Moffitt Cancer Center, Robert Gatenby uses this idea to refine cancer treatment. Gatenby hypothesizes that maintaining some susceptible cells should help to *stabilize* tumor growth by keeping drug-resistant cells in check. Gatenby outlined two "adaptive therapies" that could do just that. In one treatment plan, the drug doses are constant but relatively low, and applications are skipped if tumors are stable or shrinking. In the other, the dose starts high but declines as tumors shrink.

To test these plans, Gatenby's research team obtained lab mice that had been "seeded" with human breast cancer cells. All of the mice received the same chemotherapy drug, paclitaxel, but the treatment plans differed. One group of mice received the "standard therapy," five injections of the highest tolerable dose over 2.5 weeks. The other mice received one of the two adaptive therapies.

Mice receiving the standard therapy had the fastest-growing tumors (figure 8.A). In mice that periodically skipped doses, tumors grew somewhat more slowly, but after 4 months, the two groups fared equally poorly. The clear winners were the mice that received declining drug doses; their tumors remained small, and most mice were eventually weaned off the drug.

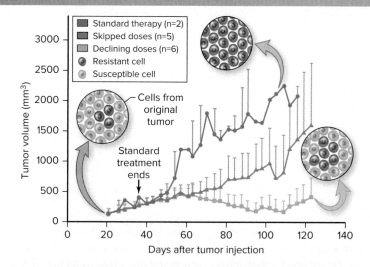

Figure 8.A Adaptive Therapy. Tumors contain a mix of susceptible and drug-resistant cells. In mice receiving standard therapy, tumors grew quickly after treatment stopped. Periodically skipping doses delayed but did not prevent this outcome. Declining doses, however, were associated with stable tumor size, presumably because susceptible cells survived and kept resistant ones in check. (Error bars represent one standard error; see appendix B.)

Adaptive therapy is now being tested on people with prostate cancer. Time will tell whether this approach—based on the principles of natural selection and competition—will keep resistant cells at bay.

Source: Enriquez-Navas, Pedro M., and 10 coauthors, including Robert A. Gatenby. 2016. Exploiting evolutionary principles to prolong tumor control in preclinical models of breast cancer. *Science Translational Medicine*, vol. 8, issue 327. 327ra24. [doi:10.1126/scitranslmed.aad7842]

What's the **Point?** ▼ APPLIED

Lavinia Warren (0.81 m tall), circa 1860

During cell division, one cell becomes two. The next round of division yields four cells, and the following round produces eight. Each new cell grows as it produces proteins and takes in water and other nutrients.

A person's adult height reflects cell division during childhood and adolescence, especially in the ends of the bones. If the thousands of cells at each end of a bone are all actively dividing, then the bone tissue quickly expands, and the person gets taller.

Some people are much, much taller than average. This condition, called gigantism, occurs when bone cells receive too many cell division signals. Over many years, even a small boost in the cell division rate may produce bones that are much longer than normal. The tallest person known to have lived, Robert Wadlow, had gigantism (see figure 25.5). He never stopped growing, and he was 2.72 meters (nearly 9 feet) tall when he died.

At the other extreme, slow cell division produces a type of dwarfism. In this disorder, bone cells do not receive enough signals to divide during childhood. One person with this condition is Chandra Bahadur Dangi of Nepal. At only about 0.55 m (1 foot, 9 inches) tall, he is the shortest known adult. Another example is Lavinia Warren, pictured here at about age 20; she was a famous performer in the mid-1800s.

Dwarfism has other causes as well. The most common form of dwarfism, called achondroplasia, reflects a problem with the cells of the limb bones. Even though the chemicals signaling cell division are present, the bone cells fail to divide, and the arms and legs remain short.

The natural variation in human height shines the spotlight on a larger message: Tissue growth and repair rely on cell division. As you review how cells divide, think back on this essay and remember that faulty signals can produce surprisingly large—or small—effects.

Photo: Source: Library of Congress Prints and Photographs Division [LC-DIG-cwpbh-02976]

CHAPTER SUMMARY

8.1 Cells Divide and Cells Die

A. Sexual Life Cycles Include Mitosis, Meiosis, and Fertilization

- **Asexual reproduction** generates virtually identical copies of an organism.
- In **sexual reproduction,** two parents produce genetically variable **gametes** by **meiosis. Fertilization** (the joining of two gametes) produces a zygote.
- **Mitotic** cell division produces identical eukaryotic cells used in growth, tissue repair, and asexual reproduction.

B. Cell Death Is Part of Life

- **Apoptosis** is programmed cell death. Both cell division and cell death occur throughout the normal development and life of a multicellular organism.

8.2 DNA Replication Precedes Cell Division

- A dividing cell must first duplicate its **genome,** which may consist of one or more **chromosomes.**
- In DNA replication, helicase enzymes unwind and unzip the DNA molecule. **DNA polymerase** adds DNA nucleotides. Replication occurs simultaneously at many locations along the DNA; **ligase** forms covalent bonds between adjacent segments of the newly created DNA strands.
- Enzymes repair damaged DNA and fix mistakes made during replication.
- **Mutations** are changes in a cell's DNA sequence.

8.3 Bacteria and Archaea Divide by Binary Fission

- During **binary fission,** DNA first replicates, then the two chromosomes attach to the cell membrane. Cell growth between the attachment points separates the chromosomes into two identical daughter cells.
- Table 8.2 compares binary fission and mitotic cell division.

8.4 Replicated Chromosomes Condense as a Eukaryotic Cell Prepares to Divide

- A chromosome consists of **chromatin** (DNA plus protein). In eukaryotic cells, chromatin is organized into **nucleosomes.**

TABLE 8.2 Cell Division in Prokaryotes and Eukaryotes

	Prokaryotes	Eukaryotes
Preceded by DNA replication?	Yes	Yes
DNA condenses?	No	Yes
Chromosome(s) attach(es) to cell membrane?	Yes	No
Requires mitotic spindle?	No	Yes
Cytoplasm splits?	Yes	Yes

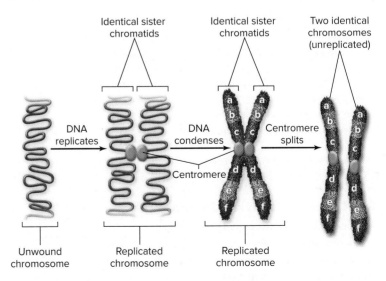

Figure 8.14 Chromosomes and Chromatids Compared.

- A replicated chromosome consists of two identical sister **chromatids** attached at a section of DNA called a **centromere** (figure 8.14).

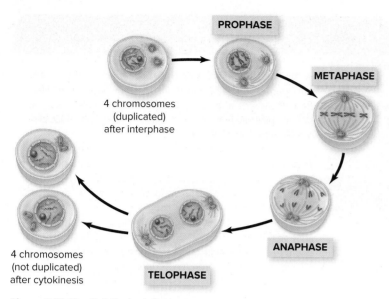

PROPHASE

METAPHASE

4 chromosomes
(duplicated)
after interphase

ANAPHASE

4 chromosomes
(not duplicated)
after cytokinesis

TELOPHASE

Figure 8.15 **The Cell Cycle: A Summary.**

8.5 Mitotic Division Generates Exact Cell Copies

- The **cell cycle** is a sequence of events in which a cell is preparing to divide (interphase), dividing its DNA (**mitosis**), and dividing its cytoplasm (cytokinesis). Table 8.3 summarizes some of the vocabulary related to chromosomes and the cell cycle; figure 8.15 illustrates the stages of mitosis.

A. DNA Is Copied During Interphase

- **Interphase** includes gap periods (G_1 **phase** and G_2 **phase**) when the cell grows and produces molecules required for cell function and division. DNA replicates during the synthesis period (**S phase**).
- In animal cells, the **centrosome** duplicates during interphase; the two centrosomes organize the microtubule proteins that form the mitotic **spindle.**

B. Chromosomes Divide During Mitosis

- In **prophase,** the chromosomes condense and the spindle forms; in addition, the nucleolus disassembles, the nuclear envelope breaks up, and spindle fibers attach to the chromosomes.
- In **metaphase,** replicated chromosomes line up along the cell's equator.
- In **anaphase,** the sister chromatids of each replicated chromosome separate; each sister chromatid is now a chromosome in its own right.
- In **telophase,** the chromosomes arrive at each end of the cell, the spindle breaks down, and nuclear envelopes form.

C. The Cytoplasm Splits in Cytokinesis

- **Cytokinesis** is the physical separation of the two daughter cells.
- In animal cells, a contractile ring forms a **cleavage furrow,** dividing the cell in two. Plant cells divide as a **cell plate** forms at the midline of a dividing cell.

8.6 Cancer Cells Divide Uncontrollably

A. Chemical Signals Regulate Cell Division

- External molecular signals normally stimulate cell division.
- Cell division may pause or halt at multiple checkpoints during the cell cycle.

B. Cancer Cells Are Malignant

- **Tumors** can result from excess cell division. A **benign tumor** does not spread, but a **malignant tumor** invades nearby tissues and **metastasizes** if it reaches the bloodstream or lymph.
- **Cancer** is a family of diseases characterized by malignant cells.

TABLE 8.3 Mitosis Terminology: A Summary

Term	Definition
Interphase	Stage of the cell cycle in which chromosomes replicate and the cell grows
Mitosis	Division of a cell's chromosomes into two identical nuclei
Prophase	Stage of mitosis when chromosomes condense, the spindle forms, and the nuclear envelope breaks up (*pro-* = before)
Metaphase	Stage of mitosis when chromosomes line up along the center of the cell (*meta-* = middle)
Anaphase	Stage of mitosis when the spindle pulls sister chromatids toward opposite poles of the cell
Telophase	Stage of mitosis when chromosomes arrive at opposite poles and nuclear envelopes form (*telo-* = end)
Cytokinesis	Distribution of cytoplasm to daughter cells following division of a cell's chromosomes
Cleavage furrow	Indentation in cell membrane of an animal cell undergoing cytokinesis
Cell plate	Material that forms the beginnings of the cell wall in a plant cell undergoing cytokinesis
Centrosome	Structure that organizes the microtubules that make up the spindle in animal cells
Spindle	Array of microtubule proteins that move chromosomes during mitosis and meiosis

C. Cancer Treatments Remove or Kill Abnormal Cells

- Surgery, chemotherapy, and radiation are common cancer treatments. Success depends on the type of cancer and whether it has spread.

D. Genes and Environment Both Can Increase Cancer Risk

- Cancer can result from mutations in genes encoding the proteins that normally regulate the cell cycle. These mutations may be inherited or induced by environmental triggers.

MULTIPLE CHOICE QUESTIONS

1. A DNA molecule is placed in a test tube containing fluorescently tagged nucleotides. DNA replication is induced. After replication,
   a. only one DNA molecule would have two fluorescent strands.
   b. both strands of each DNA molecule would be half-fluorescent.
   c. each DNA molecule would have one fluorescent strand.
   d. both DNA molecules would be completely fluorescent.

2. A chromosome is made of
   a. DNA.
   b. histones.
   c. chromatin.
   d. All of the above are correct.

3. Which of the following best explains why binary fission can occur without a spindle like that found in mitotic cells?
   a. The cell is small, so there is less material to divide.
   b. There is only one chromosome, and it attaches to the membrane.
   c. The prokaryotic DNA does not need to replicate.
   d. The DNA is transferred through the sex pilus.

4. If you were to look at a sample of actively dividing leaf cells, in what stage of the cell cycle would you find most of the cells?
   a. Interphase
   b. Prophase
   c. Metaphase
   d. Telophase

5. Which stage of the cell cycle occurs immediately *after* the stage in which the chromosomes become visible?
   a. Interphase
   b. Metaphase
   c. Anaphase
   d. Cytokinesis

6. What would happen to an animal cell if interphase and mitosis occurred in the absence of cytokinesis?
   a. The number of nuclei in the cell would increase over time.
   b. The amount of DNA in the cell would decrease over time.
   c. The cell would enter S phase.
   d. The cell would become smaller over time.

7. Why are cell cycle control checkpoints so important?
   a. Because they determine how quickly a cell's DNA gets copied
   b. Because they ensure that metaphase always follows anaphase
   c. Because they help prevent cells with damaged DNA from dividing
   d. Because they ensure that mitosis occurs continuously in all body cells

Answers to Multiple Choice questions are in appendix A.

WRITE IT OUT

1. Explain how cell division and cell death work together to form a functional multicellular organism.

2. Use words and diagrams to describe how DNA is copied in a cell.

3. If a cell contains all the genetic material it needs to synthesize protein, why must the DNA also replicate?

4. Sketch and describe the events that occur when a bacterial cell divides.

5. List the ways that binary fission is similar to and different from mitosis.

6. Seconds after thoroughly washing your hands, you grab a doorknob with your right hand. One million bacteria are transferred from the doorknob to your hand. You immediately put on latex gloves for a lab dissection; the gloves provide a warm, moist environment that is ideal for bacterial growth. If the bacteria divide once every 40 minutes, approximately how many bacteria will be on your right hand in 2 hours?

7. Obtain a rubber band and twist it as many times as you can. What happens to the overall shape of the rubber band? How is this similar to what happens to chromosomes as a cell prepares to divide? How is it different?

8. Label the arrows connecting the chromosome images in figure 8.14 with the phase of the cell cycle in which each event occurs.

9. Which contains more DNA: a cell in G_1 or a cell in G_2 phase?

10. If you draw on your skin with a permanent marker, the markings will fade in a couple of days. What does this simple demonstration reveal about cell division in your skin? What can you infer about tattoos?

11. Why do chemotherapy and radiation sometimes kill hair follicle cells along with cancer cells, while leaving many other cells unaffected?

12. A protein called p53 promotes the expression of genes encoding DNA repair enzymes. Badly damaged DNA prompts p53 to trigger apoptosis, and the cell dies. Why might mutations in the gene encoding p53 be associated with a high risk for cancer?

Answers to Mastering Concepts, Write It Out, Scientific Literacy, and Pull It Together questions can be found in the Connect ebook.
connect.mheducation.com

Design element: Burning Question (fire background): ©Ingram Publishing/Super Stock

SCIENTIFIC LITERACY

Suppose you read a news report of an individual who was exposed to a certain chemical early in life and subsequently developed cancer. Would such a story convince you to avoid exposure to that chemical? Would you be convinced if you read two similar stories? Search the Internet to learn how scientists determine whether a particular chemical causes cancer in humans. Why might it be difficult to conclude with certainty that a chemical is absolutely safe?

PULL IT TOGETHER

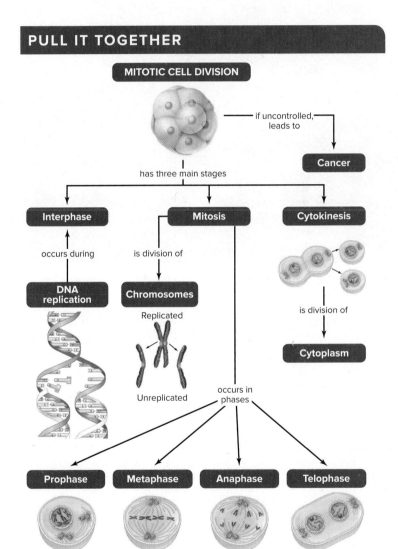

Figure 8.16 **Pull It Together: DNA Replication, Binary Fission, and Mitosis.**

Refer to figure 8.16 and the chapter content to answer the following questions.

1. Add *DNA polymerase, nucleotides,* and *complementary base pairing* to this concept map.

2. Add *cell growth* and *tissue repair* to this concept map.

3. Review the Survey the Landscape figure in the chapter introduction and then connect *DNA* and *proteins* to the Pull It Together concept map in at least two ways each.

LEARNING OUTLINE

9.1 Why Sex?

9.2 Diploid Cells Contain Two Homologous Sets of Chromosomes

9.3 Meiosis Is Essential in Sexual Reproduction

9.4 In Meiosis, DNA Replicates Once, but the Nucleus Divides Twice

9.5 Meiosis Generates Enormous Variability

9.6 Mitosis and Meiosis Have Different Functions: A Summary

9.7 Errors Sometimes Occur in Meiosis

APPLICATIONS

Burning Question 9.1 *If mules are sterile, then how are they produced?*
Why We Care 9.1 *Multiple Births*
Investigating Life 9.1 *Evolving Germs Select for Sex in Worms*

Pollinator. As a bumblebee searches for nectar in a sunflower, pollen covers its body. Inside these tiny yellow particles are the sunflower's sperm nuclei. The bee collects nectar from many neighboring plants as well, and some of the sperm-toting pollen grains stick to the female parts of other flowers. Insects and other pollinators are therefore crucial to sexual reproduction in many flowering plants.

©IT Stock/age fotostock RF

Learn How to Learn
Don't Neglect the Boxes

You may be tempted to skip the boxed readings in a chapter because they're not "required." Read them anyway. The contents should help you remember and visualize the material you are trying to learn. And who knows? You may even find them interesting.

SURVEY THE LANDSCAPE
DNA, Inheritance, and Biotechnology

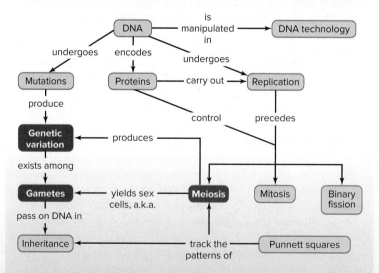

How can the same two parents produce offspring that look so different from one another? Meiosis explains the answer. This specialized process generates the genetically variable nuclei that are packaged into gametes, the cells used in sexual reproduction.

For more details, study the Pull It Together feature in the chapter summary.

Humans reproduce sexually, as do the pet dogs, cats, birds, gerbils, and fish that share our lives. But did you know that most plants also use sexual reproduction, as does the mold that grows on your stale bread? Sex occurs in almost all species of multicellular life.

©rSnapshotPhotos/
Shutterstock RF

Sexual reproduction requires the production of sperm and egg cells. We already know, from chapter 8, how cells use mitosis to make virtually identical copies of themselves. This chapter explains meiosis, which resembles mitosis in some ways. But whereas mitosis makes identical copies, meiosis does something different: It generates sperm and egg cells, each of which is genetically unique. These specialized cells lie at the heart of sexual reproduction.

9.1 Why Sex?

Humans are so familiar with our way of reproducing that it can be hard to remember that there is any other way to make offspring. In fact, however, reproduction occurs in two main forms: asexual and sexual (figure 9.1). In **asexual reproduction,** an organism simply copies its DNA and splits the contents of one cell into two (see sections 8.3 and 8.5). Some DNA may mutate during DNA replication, but the offspring are virtually identical. Examples of asexual organisms include bacteria, archaea, and single-celled eukaryotes such as the amoeba in figure 9.1a. Many plants, fungi, and other multicellular organisms also reproduce asexually. ⓘ *mutations,* section 7.6; *DNA replication,* section 8.2

Sexual reproduction, in contrast, requires two parents. The male parent contributes sperm cells, one of which fertilizes a female's egg cell to begin the next generation. Later in this chapter, you will learn that each time the male produces sperm, he scrambles the genetic information that he inherited from his own parents. A similar process occurs as the female produces eggs. The resulting variation among sex cells ensures that the offspring from two parents are genetically different from one another.

Attracting mates takes a lot of energy, as does producing and dispersing sperm and egg cells. Yet the persistence of sexual reproduction over billions of years and in many diverse species attests to its success. Why does such a costly method of reproducing persist, and why is asexual reproduction comparatively rare?

Although no one knows the full answer to this question, many studies point to the benefit of genetic diversity in a changing environment. The mass production of identical offspring makes sense in habitats that never change, but conditions rarely remain constant in the real world. Temperatures rise and fall, prey species disappear, and new parasites emerge (see Investigating Life 9.1). Genetic variability increases the chance that at least some individuals will have a combination of traits that allows them to survive and reproduce, even if some poorly suited individuals die. Asexual reproduction typically cannot create or maintain this genetic diversity, but sexual reproduction can.

Figure 9.1 Asexual and Sexual Reproduction. (a) A single-celled amoeba reproduces asexually, generating identical offspring by splitting in two. This process takes about 20 minutes. (b) These puppies were conceived sexually. They look different because each inherited a unique combination of alleles from its parents.

(a): (all): ©Biophoto Associates/Science Source; (b): ©Eric Isselee/123RF

9.1 Mastering Concepts

1. How do asexual and sexual reproduction differ?
2. What are the advantages of asexual reproduction?
3. Why does sexual reproduction persist even though it requires more energy than asexual reproduction?

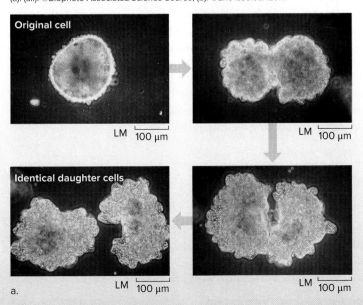

Original cell

Identical daughter cells

LM 100 µm   LM 100 µm

LM 100 µm   LM 100 µm

a.

b.

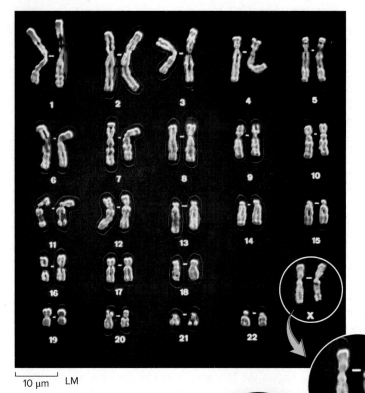

10 μm LM

Figure 9.2 Human Karyotype.
A diploid human cell contains 23 pairs of chromosomes. Autosomes are numbered 1 through 22; the insets show sex chromosomes for a female (XX) and a male (XY).

(all): ©CNRI/Science Source

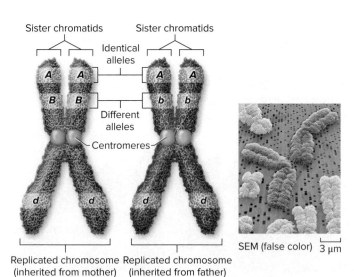

SEM (false color) 3 μm

Replicated chromosome (inherited from mother)   Replicated chromosome (inherited from father)

Homologous pair of chromosomes

Figure 9.3 Homologous Pair. These homologous chromosomes carry alleles for three genes. (Capital and lowercase letters represent different alleles.)

Photo: ©Andrew Syred/Science Source

9.2 Diploid Cells Contain Two Homologous Sets of Chromosomes

Before exploring sexual reproduction further, a quick look at a cell's chromosomes is in order. Recall from chapters 7 and 8 that a **chromosome** is a single molecule of DNA and its associated proteins.

A sexually reproducing organism consists mostly of **diploid cells** (abbreviated 2n), which contain two full sets of chromosomes; one set is inherited from each parent. Each diploid human cell, for example, contains 46 chromosomes (figure 9.2). The photo in figure 9.2 illustrates a **karyotype,** a size-ordered chart of all the chromosomes in a cell. Notice that the 46 chromosomes are arranged in 23 pairs; the mother and the father each contributed one member of each pair.

Of the 23 chromosome pairs in a human cell, 22 pairs consist of **autosomes**—chromosomes that are the same for both sexes. The remaining pair is made up of the two **sex chromosomes,** which determine whether an individual is female or male. Females have two X chromosomes, whereas males have one X and one Y chromosome.

The two members of most chromosome pairs are homologous to each other. In a **homologous pair,** the two chromosomes look alike and have the same sequence of genes. (The word *homologous* means "having the same basic structure.") The physical similarities between any two homologous chromosomes are evident in figure 9.2: They share the same size, centromere position, and pattern of light- and dark-staining bands. In addition, the two members of a homologous pair of chromosomes carry the same sequence of genes. For example, chromosome 21 includes 367 genes, always in the same order.

Homologous chromosomes, however, are not identical—after all, nobody has two identical parents! Instead, the two homologs differ in the combination of **alleles,** or versions, of the genes they carry (figure 9.3). As described in chapter 7, each allele of a gene encodes a different version of the same protein. A chromosome typically carries exactly one allele of each gene, so a person inherits one allele per gene from each parent. Depending on the parents' chromosomes, the two alleles may be identical or different. Overall, however, the members of each homologous pair of chromosomes are at least slightly different from each other.

If you think of a gene as a "recipe" for a protein (see figure 7.3), then a chromosome is like a cookbook—that is, a collection of recipes. Inheriting a set of chromosomes from each parent is like acquiring two complete sets of cookbooks, each containing slightly different recipes for the same foods.

Unlike the autosome pairs, however, the X and Y chromosomes are not homologous to each other. X is much larger than Y, and its genes are completely different. Nevertheless, in males, the sex chromosomes behave as homologous chromosomes during meiosis.

9.2 Mastering Concepts

1. What are autosomes and sex chromosomes?
2. Draw a karyotype for a cell with a diploid number of 8.
3. How are the members of a homologous pair similar and different?

9.3 Meiosis Is Essential in Sexual Reproduction

Sexual reproduction poses a practical problem: maintaining the correct chromosome number. We have already seen that most cells in the human body contain 46 chromosomes. If a baby arises from the union of a sperm and egg, then why does the child not have 92 chromosomes per cell (46 from each parent)? And shouldn't cells in the next generation have 184 chromosomes?

In fact, the chromosome number does not double with each generation. The explanation is that sperm cells and egg cells are not diploid. Rather, they are **haploid cells** (abbreviated *n*); that is, they contain only one full set of genetic information instead of the two sets that characterize diploid cells.

These haploid cells, called **gametes,** are sex cells that combine to form a new offspring. **Fertilization** merges the gametes from two parents, creating a new cell: the diploid **zygote,** which is the first cell of the new organism (figure 9.4). The zygote has two full sets of chromosomes, one set from each parent. In most species, the zygote begins dividing mitotically shortly after fertilization.

Thus, the life of a sexually reproducing, multicellular organism requires two ways to package DNA into new cells. **Mitosis,** described in chapter 8, divides a eukaryotic cell's chromosomes into two identical daughter nuclei. Mitotic cell division produces the cells needed for growth, development, and tissue repair. **Meiosis,** the subject of this chapter, forms genetically variable nuclei, each containing half as many chromosomes as the organism's diploid cells.

Only **germ cells** can undergo meiosis. In humans and other animals, these specialized diploid cells occur only in the ovaries and testes. The rest of the body's diploid cells, called **somatic cells,** do not participate directly in reproduction. Muscle cells and neurons are examples of somatic cells.

To make sense of this, consider your own life (figure 9.5). It began when a small, swimming sperm cell carrying 23 chromosomes from your father wriggled toward your mother's comparatively enormous egg cell, also containing 23 chromosomes. You were conceived when the sperm fertilized the egg cell. At that moment, you were a one-celled zygote, with 46 chromosomes. That first cell then began dividing, generating identical copies of itself to form an embryo, then a fetus, an infant, a child, and eventually an adult. Once you reached reproductive maturity, diploid cells in your testes or ovaries produced haploid gametes of your own, perpetuating the cycle.

The human life cycle is of course most familiar to us, and many animals reproduce in essentially the same way. Gametes are the only haploid cells in our life cycle; all other cells are diploid. Sexual reproduction, however, can take many other forms as well. In some organisms, including plants, both the haploid and the diploid stages are multicellular. Chapters 16, 17, 22, and 30 describe the life cycles of plants and animals in more detail.

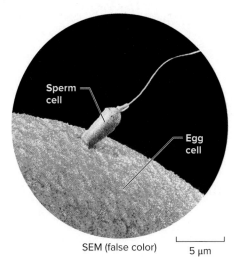

Figure 9.4
Human Gametes.
Note the size difference between the human sperm and egg cells.
©Francis Leroy, Biocosmos/Science Source

Sperm cell

Egg cell

SEM (false color)   5 μm

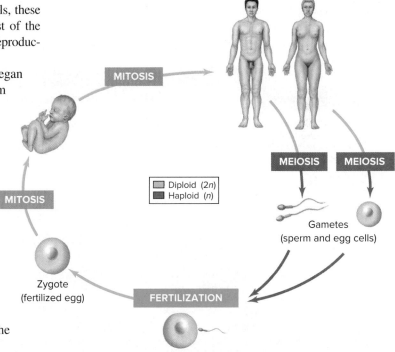

Diploid (2*n*)
Haploid (*n*)

MITOSIS

MEIOSIS   MEIOSIS

Gametes (sperm and egg cells)

MITOSIS

Zygote (fertilized egg)

FERTILIZATION

Figure 9.5 **The Human Life Cycle.** Adults produce haploid gametes—sperm and egg cells—by meiosis. Fertilization produces the diploid zygote, and then mitotic cell division enables the zygote to grow and develop into an adult.

9.3 Mastering Concepts

1. How do haploid and diploid nuclei differ?
2. What are the roles of meiosis, gamete formation, and fertilization in sexual life cycles?
3. What is a zygote?
4. What is the difference between somatic cells and germ cells?

9.4 In Meiosis, DNA Replicates Once, but the Nucleus Divides Twice

The mitotic cell cycle described in chapter 8 includes three main parts: interphase, mitosis, and cytokinesis. During interphase, the cell grows, synthesizes molecules, carries out its functions, and replicates its DNA. During mitosis and cytokinesis, the nucleus and cytoplasm split into two. Mitotic cell division creates identical copies by replicating a cell's DNA once and then dividing once.

Meiosis is closely related to mitosis. For example, the **interphase** that comes before meiosis is similar to interphase in the mitotic cell cycle. The cell grows, DNA replicates, and the cell produces the enzymes and other proteins necessary to divide the cell. Afterward, each of the cell's chromosomes consists of two identical sister chromatids attached at a centromere. Finally, late in interphase, chromatin begins to condense, and the cell produces the microtubule proteins that will become the spindle. The names of the meiotic phases are also similar to those in mitosis.

Despite these similarities, meiosis has two unique outcomes. First, meiosis includes two divisions (meiosis I and meiosis II) that distribute the DNA from one specialized diploid cell into four haploid nuclei. Second, meiosis shuffles genetic information, setting the stage for each haploid nucleus to receive a unique mixture of alleles.

Figure 9.6 illustrates the stages of meiosis. During **prophase I** (that is, prophase of meiosis I), the replicated chromosomes condense. A **spindle** begins to form from microtubules assembled at the centrosomes, spindle attachment points grow on each centromere, and the nuclear envelope breaks up.

The events described so far resemble those of prophase of mitosis, but something unique happens during prophase I of meiosis: Each chromosome lines up

Figure 9.6 **The Stages of Meiosis.** A diploid nucleus (containing four chromosomes) gives rise to four genetically different haploid nuclei, each containing two chromosomes.

Photos: (all): ©Ed Reschke/Photolibrary/Getty Images

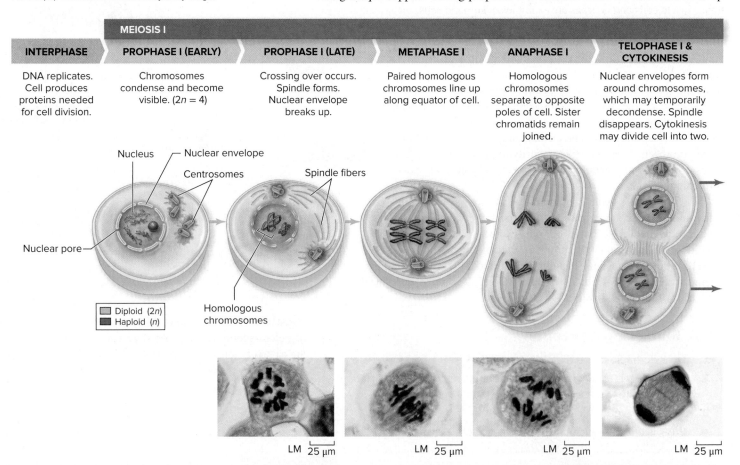

MEIOSIS I					
INTERPHASE	**PROPHASE I (EARLY)**	**PROPHASE I (LATE)**	**METAPHASE I**	**ANAPHASE I**	**TELOPHASE I & CYTOKINESIS**
DNA replicates. Cell produces proteins needed for cell division.	Chromosomes condense and become visible. (2n = 4)	Crossing over occurs. Spindle forms. Nuclear envelope breaks up.	Paired homologous chromosomes line up along equator of cell.	Homologous chromosomes separate to opposite poles of cell. Sister chromatids remain joined.	Nuclear envelopes form around chromosomes, which may temporarily decondense. Spindle disappears. Cytokinesis may divide cell into two.

Nucleus — Nuclear envelope

Centrosomes

Spindle fibers

Nuclear pore

Diploid (2n)
Haploid (n)

Homologous chromosomes

LM ⊢ 25 μm   LM ⊢ 25 μm   LM ⊢ 25 μm   LM ⊢ 25 μm

Why Sex?   Diploid Cells Have Two Sets of Chromosomes   Meiosis Is Essential in Sexual Reproduction   DNA Replicates Once; Nucleus Divides Twice

next to its homolog. Section 9.5 describes how this arrangement allows for an allele-shuffling mechanism called crossing over. (Mules are typically sterile because their germ cells cannot complete this stage; see Burning Question 9.1.)

In **metaphase I,** the spindle arranges the paired homologs down the center of the cell. Each member of a homologous pair attaches to a spindle fiber stretching to one pole. The stage is therefore set for the homologous pairs to separate in **anaphase I,** and the chromosomes complete their movement to opposite poles in **telophase I. Cytokinesis** typically occurs after telophase I, splitting the original cell into two. ⓘ *cytokinesis,* section 8.5C

A second interphase precedes meiosis II in many species. During this time, the chromosomes unfold into very thin threads. The cell produces proteins, but the DNA does not replicate a second time.

Meiosis II strongly resembles mitosis. The process begins with **prophase II,** when the chromosomes again condense and become visible. In **metaphase II,** the spindle arranges the chromosomes along the center of each cell. In **anaphase II,** the centromeres split, and the separated sister chromatids move to opposite poles. In **telophase II,** nuclear envelopes form around the separated sets of chromosomes. Cytokinesis then separates the nuclei into individual cells. The overall result: One diploid cell has divided into four haploid cells.

Figure It Out

A cell that is entering prophase I contains __ times as much DNA as one daughter cell at the end of meiosis.

Answer: Four.

Figure It Out

A mosquito cell that is about to begin meiosis has six chromosomes. How many chromosomes does each daughter cell have during metaphase II?

Answer: Three.

9.4 Mastering Concepts

1. What happens during interphase?

2. How do the events of meiosis I, meiosis II, and cytokinesis produce four haploid cells from one diploid cell?

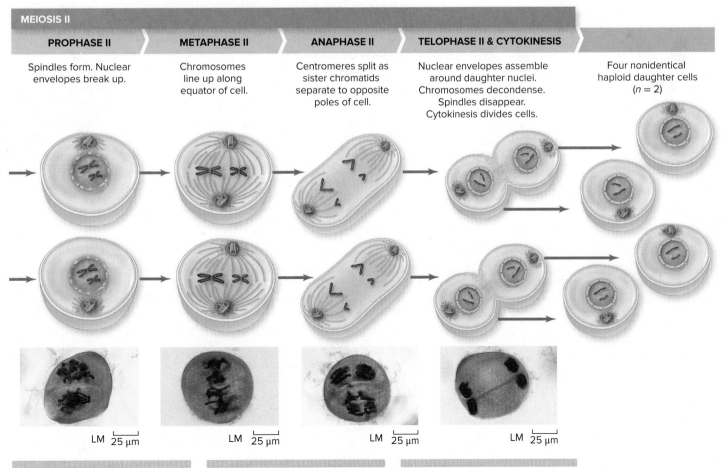

MEIOSIS II

PROPHASE II	METAPHASE II	ANAPHASE II	TELOPHASE II & CYTOKINESIS	
Spindles form. Nuclear envelopes break up.	Chromosomes line up along equator of cell.	Centromeres split as sister chromatids separate to opposite poles of cell.	Nuclear envelopes assemble around daughter nuclei. Chromosomes decondense. Spindles disappear. Cytokinesis divides cells.	Four nonidentical haploid daughter cells (*n* = 2)

LM ⌐25 µm⌐   LM ⌐25 µm⌐   LM ⌐25 µm⌐   LM ⌐25 µm⌐

Meiosis Generates Enormous Variability   Mitosis and Meiosis: A Summary   Errors Sometimes Occur in Meiosis

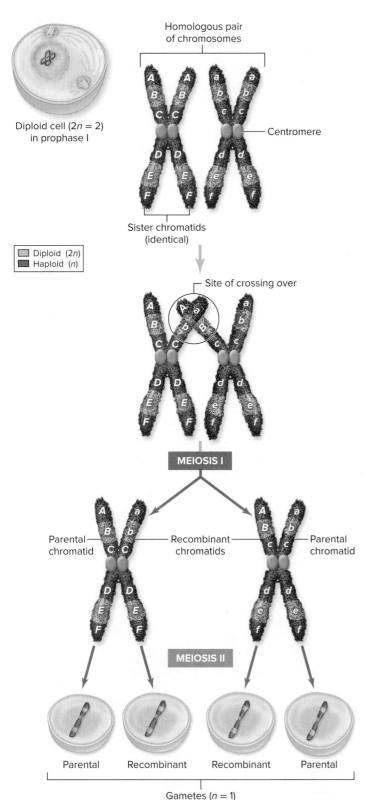

Diploid cell (2*n* = 2)
in prophase I

Homologous pair
of chromosomes

Centromere

Sister chromatids
(identical)

Diploid (2*n*)
Haploid (*n*)

Site of crossing over

MEIOSIS I

Parental
chromatid

Recombinant
chromatids

Parental
chromatid

MEIOSIS II

Parental   Recombinant   Recombinant   Parental

Gametes (*n* = 1)

Figure 9.7 Crossing Over. In crossing over, portions of homologous chromosomes swap places. This process generates genetic diversity because each of the resulting chromatids has a unique combination of alleles.

9.5 Meiosis Generates Enormous Variability

By creating new combinations of alleles, meiosis generates astounding genetic variety among the offspring from just two parents. This section describes three mechanisms that account for this diversity: crossing over, random orientation of chromosomes during metaphase I, and random fertilization.

A. Crossing Over Shuffles Alleles

Crossing over is a process in which two homologous chromosomes exchange genetic material (figure 9.7). During prophase I, the homologs align themselves precisely, gene by gene. The chromosomes are attached at a few points along their lengths, where the homologs exchange chromosomal material.

As an example, consider what takes place in your own ovaries or testes. You inherited one member of each homologous pair from your mother; the other came from your father. Crossing over means that pieces of these homologous chromosomes physically change places during meiosis.

Suppose, for instance, that one chromosome carries the genes that dictate hair color, eye color, and finger length. Perhaps the version you inherited from your father has the alleles that specify blond hair, blue eyes, and short fingers. The homolog from your mother is different; its alleles dictate black hair, brown eyes, and long fingers. Now, suppose that crossing over occurs between

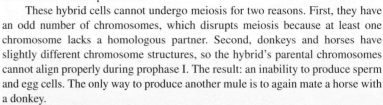

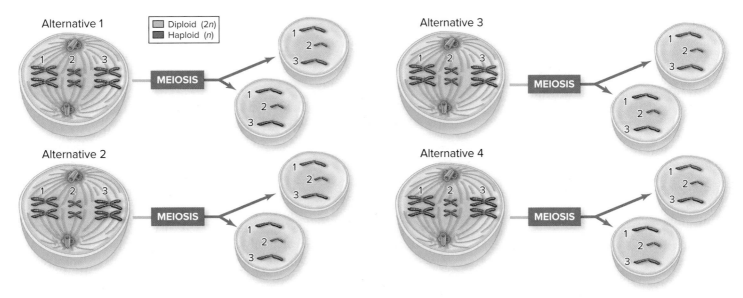

Figure 9.8 Random Orientation. A germ cell containing three homologous pairs of chromosomes can generate eight genetically different gametes. Note that this number does not include the effects of crossing over.

the homologous chromosomes. Afterward, one chromatid might carry alleles for blond hair, brown eyes, and long fingers; another would specify black hair, blue eyes, and short fingers. These two chromatids are termed "recombinant" because they combine alleles from your two parents. The two chromatids that did not participate in crossing over, however, would remain unchanged and are termed "parental." Note that although all of the alleles in your ovaries or testes came from your parents, half of the chromatids—the recombinant ones—now contain new allele *combinations*.

The result of crossing over is four unique chromatids in place of two pairs of identical chromatids. Each chromatid will end up in a separate haploid cell. Thus, crossing over ensures that each haploid cell will be genetically different from the others.

B. Homologous Pairs Are Oriented Randomly During Metaphase I

Figure 9.8 reveals a second way that meiosis creates genetic variability. At metaphase I, pairs of homologous chromosomes line up at the cell's center. Examine the orientation of the chromosomes in the cell labeled "Alternative 1." All of the blue chromosomes are on top, whereas the red homologs are on the bottom. In anaphase I, the chromosomes separate, and the resulting nuclei contain either all blue or all red chromosomes.

The next time a cell in the same individual undergoes meiosis, the orientation of the chromosomes may be the same, or it may not be. The arrangement of chromosomes at metaphase I is random, and all four alternatives shown in figure 9.8 are equally probable. Most of the time, gametes will inherit a mix of genetic material from both parents.

The number of possible arrangements is related to the chromosome number. For two pairs of homologs, each resulting gamete may have any of four (2^2) unique chromosome configurations. For three pairs, as shown in

Figure It Out

A chromosome carries alleles D, e, and F; its homolog carries alleles d, E, and F. If crossing over occurs once between genes D and E, what alleles would the resulting recombinant chromatids carry?

Answer: One would carry d, e, and F; the other would carry D, E, and F.

Miniglossary	Variability in Meiosis
Homologous pair	Two chromosomes that have the same gene sequence but may have different alleles of those genes
Crossing over	The exchange of genetic material between homologous chromosomes, producing variability among the four chromatids in the homologous pair; occurs during prophase I
Parental chromatid	Chromatid that does not participate in crossing over and therefore retains its original allele combination
Recombinant chromatid	Chromatid carrying a new allele combination after crossing over
Random orientation	The random orientation of homologous pairs of chromosomes during metaphase I; the many possible combinations mean that each round of meiosis produces daughter cells with different allele combinations

Monozygotic (identical) twins

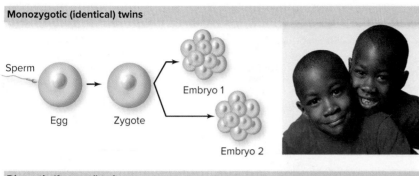

Dizygotic (fraternal) twins

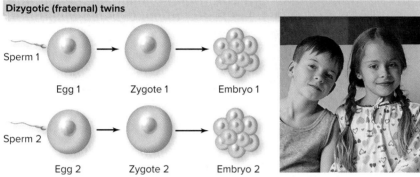

Figure 9.9 **Two Ways to Make Twins.** Monozygotic twins are genetically identical because they come from the same zygote. Dizygotic (fraternal) twins are no more alike than nontwin siblings because they start as two different zygotes.

Photos: (identical): ©Barbara Penoyar/Getty Images RF; (fraternal): ©Image Source Black/Getty Images RF

figure 9.8, eight (2^3) unique configurations can occur in the gametes. Extending this formula to humans, with 23 chromosome pairs, each gamete contains one of 8,388,608 (2^{23}) possible chromosome combinations—all equally likely.

C. Random Fertilization Multiplies the Diversity

We have already seen that every diploid cell undergoing meiosis is likely to produce haploid nuclei with different combinations of chromosomes. Furthermore, it takes two to reproduce. In one mating, any of a woman's 8,388,608 possible egg cells can combine with any of the 8,388,608 possible sperm cells of a partner. One couple could therefore theoretically create more than 70 trillion ($8,388,608^2$) genetically unique individuals! And this enormous number is an underestimate, because it does not take into account the additional variation from crossing over.

With so much potential variability, the chance of two parents producing genetically identical children seems exceedingly small. How do the parents of identical twins defy the odds? The answer is that identical twins result from just one fertilization event. The resulting zygote or embryo splits in half, creating separate, identical babies (figure 9.9). Identical twins are called *monozygotic* because they derive from one zygote. When the embryos fail to separate completely, the twins remain conjoined, or physically attached to one another.

In contrast, nonidentical (fraternal) twins occur when two sperm cells fertilize two separate egg cells. The twins are therefore called *dizygotic*. Triplets and higher order multiple births occur when three or more babies develop at the same time (see Why We Care 9.1).

9.5 Mastering Concepts

1. How does crossing over shuffle alleles?
2. Explain how to arrive at the estimate that one human couple can produce over 70 trillion unique offspring.
3. How are identical twins different from fraternal twins?

9.6 Mitosis and Meiosis Have Different Functions: A Summary

Mitosis and meiosis are both mechanisms that divide a eukaryotic cell's genetic material, and they have many events in common. However, the two processes also differ in many ways (figure 9.10):

- Mitosis occurs in somatic cells throughout the life cycle, whereas meiosis occurs only in germ cells and only at some stages of life.
- Homologous chromosomes align with each other during meiosis but not mitosis. This alignment allows for crossing over, which also occurs only in meiosis.
- Following mitosis, cytokinesis occurs once for every DNA replication event. The product of mitotic division is therefore two daughter cells. In

Figure It Out

A dog has 39 pairs of chromosomes. Considering only the orientation of homologous chromosomes during metaphase I, how many genetically different puppies are possible from the mating of two dogs?

Answer: One dog: 5.5×10^{11} possible gametes; two dogs: 3×10^{23} possible puppies.

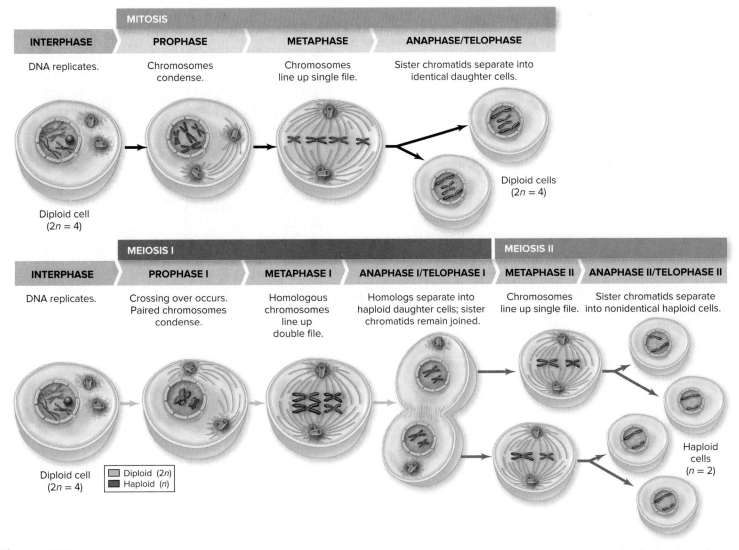

Figure 9.10 Mitosis and Meiosis Compared. Mitotic division adds and replaces identical cells, whereas meiosis produces haploid nuclei with new genetic combinations. (Some stages are omitted for clarity.)

meiosis, cytokinesis occurs twice, although the DNA has replicated only once. One cell therefore yields four daughter cells.

- After mitosis, the chromosome number in the two daughter cells is the same as in the parent cell. Depending on the species, either haploid or diploid cells can divide mitotically. In contrast, only diploid cells divide by meiosis, producing four haploid daughter cells.

- Mitotic division yields identical daughter cells for growth, repair, and asexual reproduction. Meiotic division generates genetically variable daughter cells used in sexual reproduction. Table 9.1 compares asexual and sexual reproduction.

9.6 Mastering Concepts

1. In what ways are mitosis and meiosis similar?
2. In what ways are mitosis and meiosis different?

TABLE 9.1 Asexual and Sexual Reproduction Compared

	Asexual	Sexual
Requires two parents?	No	Yes
Produces identical offspring?	Yes (except for mutations)	No
Mutations occur?	Yes	Yes
Homologous chromosomes pair up?	No	Yes
Crossing over occurs?	No	Yes
Adaptive in changing environment?	No	Yes

Meiosis Generates Enormous Variability   Mitosis and Meiosis: A Summary   Errors Sometimes Occur in Meiosis

Why We Care 9.1 | Multiple Births

Triplets, quadruplets, and higher order multiple births have become more common since the 1980s. How do they arise?

Triplets come about in several ways. The least common route is for a single embryo to split and develop into three genetically identical babies (monozygotic triplets). Alternatively, if three sperm fertilize three separate egg cells, the triplets will be fraternal (trizygotic). Most commonly, however, an embryo splits and forms two identical babies, and a separate embryo develops into an additional, nonidentical baby. Higher order multiples likewise usually include combinations of identical and fraternal siblings. Identical quadruplets are exceedingly rare, occurring perhaps once in 11 million deliveries. Monozygotic quintuplets are even more unusual, with only one set ever known to have been born.

Two trends account for the rising incidence of multiple births. First, older women are more likely than younger women to have multiple births, and childbearing among older women has become more common. Second, couples have increasingly sought treatment for infertility. Some fertility drugs stimulate a woman to release more than one egg cell. If sperm fertilize all of them, a multiple birth could result. Another infertility therapy is *in vitro* fertilization, in which sperm fertilize egg cells harvested from a woman's ovaries in the lab.

©Nancy R. Cohen/Getty Images RF

One or more embryos judged most likely to result in a live birth are then implanted into the woman's uterus. Multiple births often result.

9.7 Errors Sometimes Occur in Meiosis

Considering the number of separate events that take place in meiosis, it is not surprising that things occasionally take a wrong turn. The result can be gametes with extra or missing chromosomes. Even small chromosomal abnormalities can have devastating effects on health.

A. Polyploidy Means Extra Chromosome Sets

An error in meiosis, such as the failure of the spindle to form properly, can produce a **polyploid** cell with one or more complete sets of extra chromosomes (*polyploid* means "many sets"). For example, if a sperm with the normal 23 chromosomes fertilizes an abnormal egg cell with two full sets (46), the resulting zygote will have three copies of each chromosome (69 total), a type of polyploidy called triploidy. Most human polyploids fail to live past the very early stages of development.

Polyploidy is an important force in plant evolution. In contrast to humans, about 30% of flowering plant species tolerate polyploidy well, and many crop plants are polyploids. The durum wheat in pasta is tetraploid (it has four sets of seven chromosomes), and the wheat species in bread is a hexaploid, with six sets of seven chromosomes.

B. Nondisjunction Results in Extra or Missing Chromosomes

Some gametes have just one extra or missing chromosome. The cause of the abnormality is an error called **nondisjunction,** which occurs when chromosomes fail to separate at either anaphase I or anaphase II (figure 9.11). The result is a sperm or egg cell with two copies of a particular chromosome or none at all. When such a gamete fuses with another at fertilization, the resulting zygote has either 45 or 47 chromosomes instead of the normal 46.

a. Nondisjunction in meiosis I

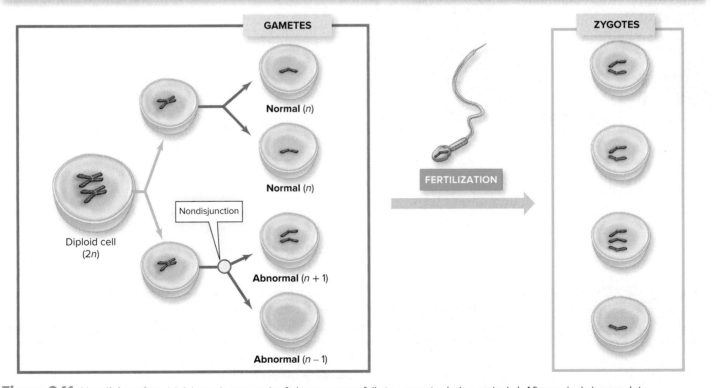

Figure 9.11 Nondisjunction. (a) A homologous pair of chromosomes fails to separate during meiosis I. After meiosis is complete, two gametes have two copies of the chromosome and two gametes lack the chromosome. (b) Sister chromatids fail to separate during meiosis II. One gamete therefore has an extra chromosome, and one is missing the chromosome. The other two gametes are unaffected. (All chromosomes other than the ones undergoing nondisjunction are omitted for clarity.)

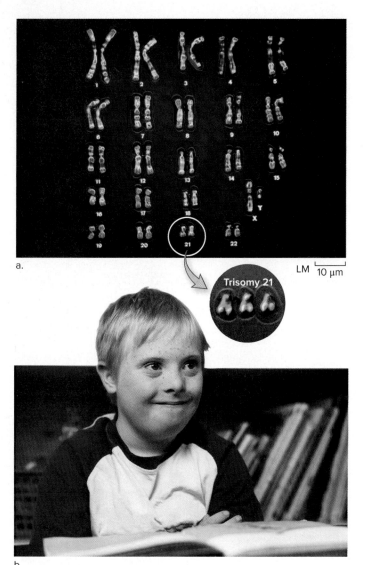

a.

Trisomy 21

LM |—| 10 μm

b.

Figure 9.12 **Trisomy 21.** (a) A normal human karyotype reveals 46 chromosomes, in 23 pairs. (b) A child who inherits three copies of chromosome 21 has Down syndrome.

(a, both): ©CNRI/Science Source; (b): ©George Doyle/Stockbyte/Getty Images RF

Most embryos with incorrect chromosome numbers cease developing before birth; they account for about half of all spontaneous abortions (miscarriages) that occur early in a pregnancy. Extra genetic material, however, causes fewer problems than missing material. This is why most children with the wrong number of chromosomes have an extra one—a trisomy—rather than a missing one. The rest of this section describes some syndromes that occur in humans who inherit two many or too few chromosomes.

Extra Autosomes: Trisomy 21, 18, or 13 A person with trisomy 21, the most common cause of Down syndrome, has three copies of chromosome 21 (figure 9.12). An affected person has distinctive facial features and a unique pattern of hand creases. Intelligence varies greatly; some children have profound mental impairment, whereas others learn well. Many affected children die before their first birthdays, often because of congenital heart defects. People with Down syndrome also have an above-average risk for leukemia and Alzheimer disease.

The probability of giving birth to a child with trisomy 21 increases dramatically as a woman ages. For women younger than 30, the chances of conceiving a child with the syndrome are 1 in 3000. For a woman of 48, the incidence jumps to 1 in 9. An increased likelihood of nondisjunction in older females apparently accounts for this age association, but no one knows for sure why older women might have problems completing meiosis.

Trisomy 21 is the most common autosomal trisomy, but that is only because the fetus is most likely to remain viable. Trisomies 18 and 13 are the next most common, but few infants with these genetic abnormalities survive. Trisomies undoubtedly occur with other chromosomes, but the embryos fail to develop at all.

Extra or Missing Sex Chromosomes: XXX, XXY, XYY, and XO Nondisjunction can produce a gamete that contains two X or Y chromosomes instead of only one. Fertilization then produces a zygote with too many sex chromosomes: XXX, XXY, or XYY (table 9.2). A gamete may also lack a sex chromosome altogether. If one gamete contains an X chromosome and the other gamete has neither X nor Y, the resulting zygote is XO. Interestingly, medical researchers have never reported a person with one Y and no X chromosome. When a zygote lacks an X chromosome, so much genetic material is missing that it probably cannot sustain more than a few cell divisions.

TABLE 9.2   Sex Chromosome Abnormalities

Chromosomes	Name of Condition	Approximate Incidence
XXX	Triplo-X	1 in 1500 females
XXY	Klinefelter or XXY syndrome	1 in 750 males
XYY	Jacobs or XYY syndrome	1 in 1000 males
XO	Turner syndrome	1 in 2000 females

9.7 Mastering Concepts

1. What is polyploidy?
2. How can nondisjunction during meiosis lead to gametes with extra or missing chromosomes?
3. Draw a diagram to show how nondisjunction of all chromosomes during meiosis I in one parent could lead to polyploid offspring. Use $2n = 6$ for the starting cells; assume the other parent's gamete contributes the normal number of chromosomes.
4. Explain why few children are born with extra autosomes (with the exception of trisomy 21).
5. Why do sex chromosome abnormalities include XO but not OY?

Investigating Life 9.1 | Evolving Germs Select for Sex in Worms

Sexual reproduction is a hassle. Why spend energy to attract mates when you could just make identical copies of yourself and ensure that your genome makes it to the next generation?

Surprisingly, the answer to the puzzle of sexual reproduction may lie in disease. That's because tiny, fast-breeding parasites typically evolve more quickly than their hosts. Perhaps sexually reproducing hosts, which generate new allele combinations in each generation, have the best chance to survive in an environment full of constantly changing parasites.

A team of Indiana University researchers tested this hypothesis by studying disease-causing bacteria and their host, a microscopic roundworm. These worms can be hermaphrodites, meaning that the same individual can generate both male and female sex cells. A hermaphrodite worm can reproduce sexually with a male, producing varied offspring; it can also self-reproduce, in which case the offspring are identical to the parent. Which reproductive strategy is most effective when parasites are present? (i) *roundworms,* section 17.7

The research team studied a roundworm population containing both sexually reproducing and self-reproducing individuals. One group of worms was allowed to reproduce in petri dishes containing living bacteria that evolved with the hosts and therefore got better at infecting the worms in every generation. Another worm group reproduced in dishes containing dead bacteria.

The researchers documented the rate of sexual reproduction in each group over 30 worm generations (figure 9.A). At the start of the experiment, sexual reproduction was rare in both groups. But the rate of sexual reproduction increased within a few generations for the population exposed to living bacteria. As each generation of bacteria got better at infecting their hosts, the worms became more likely to shuffle the genetic deck and produce a greater proportion of offspring that were "unfamiliar" to the bacteria. On the other hand, worms exposed to dead bacteria continued to self-reproduce.

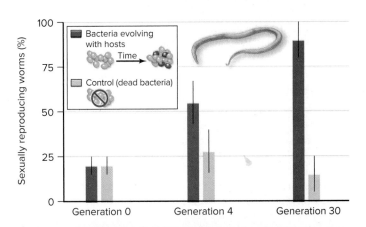

Figure 9.A **Shift to Sexual Reproduction.** Over 30 worm generations, live bacteria selected for sexual reproduction in their hosts, whereas dead bacteria did not. (Error bars represent two standard errors; see appendix B.)

In another experiment, a population consisting entirely of self-reproducing worms was exposed to the disease-causing bacteria. After producing identical offspring for only 20 generations, the population was extinct.

Both results reinforce the same conclusion. Sex costs time and energy, but the alternative—easy reproduction of a doomed allele combination—may be even costlier. Sexual reproduction produces variability and boosts the chance of reproductive success in a rapidly changing pool of parasites.

Source: Morran, Levi, Olivia Schmidt, Ian Gelarden, Raymond Parrish II, and Curtis Lively. July 8, 2011. Running with the Red Queen: Host-parasite coevolution selects for biparental sex. *Science,* vol. 333, pages 216–218.

What's the **Point?** ▼ APPLIED

Meiosis explains how a diploid cell produces four genetically different haploid nuclei. In humans, meiosis produces egg cells in females and sperm in males. When egg and sperm unite at fertilization, a zygote forms—the first cell of the next generation. Mitosis follows, and each daughter cell contains the same genetic material as the zygote. Nine months and countless mitotic events later, a single child is born—that is, unless something unusual occurs.

Sometimes, an early embryo splits just a few hours or days after fertilization. If the split is complete, then each cell line develops independently, resulting in identical twins. However, on occasion the split is incomplete, and conjoined twins develop. The twins may be equal in size, but in other cases, only one twin completes development. The undeveloped twin is considered parasitic because it relies entirely on its sibling for survival. In 2013, doctors in China removed a 20-centimeter-long parasitic twin growing inside the abdomen of a 2-year-old boy. The undeveloped twin was using nutrients from the boy and putting pressure on his internal organs.

Conjoined twins

Rarely, one embryo completely absorbs its fraternal twin very early in development. The surviving twin continues development with two different cell lines, each with a unique set of DNA. An individual with two distinct cell lines is called a chimera.

Human chimeras explain some baffling medical mysteries. In the late 1990s, a woman named Karen needed a kidney transplant. Her DNA was compared with that of her adult children to see if either of them could be a donor. The results were unexpected: Neither appeared to be Karen's child! Two years later, doctors discovered that Karen's ovaries had different DNA from the rest of her body. Apparently, when she was still a ball of cells in her mother's womb, Karen absorbed her nonidentical female twin, the cells of which multiplied and developed into her ovaries.

Photo: ©Justin Sullivan/Getty Images

CHAPTER SUMMARY

9.1 Why Sex?

- **Asexual reproduction** is reproduction without sex. **Sexual reproduction** produces offspring by mixing traits from two parents.
- Asexual reproduction can be successful in a stable environment, but a changing environment selects for sexual reproduction.

9.2 Diploid Cells Contain Two Homologous Sets of Chromosomes

- **Diploid cells** have two full sets of **chromosomes,** one from each parent. A **karyotype** is a chart that displays all of the chromosomes from one cell.
- In humans, the **sex chromosomes** (X and Y) determine whether an individual is male or female. The 22 **homologous pairs** of **autosomes** do not determine sex.
- Homologous chromosomes share the same size, banding pattern, and centromere location, but they differ in the **alleles** they carry.

9.3 Meiosis Is Essential in Sexual Reproduction

- **Meiosis** halves the genetic material to produce **haploid cells. Fertilization** occurs when **gametes** fuse, forming the diploid **zygote. Mitotic** cell division produces the body's cells during growth and development. Figure 9.13 summarizes the events of a sexual life cycle.

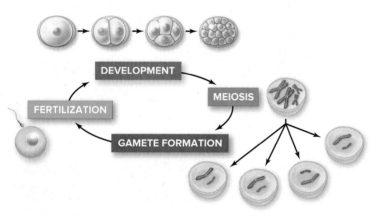

Figure 9.13 Sexual Life Cycle Events.

- **Somatic** cells do not participate in reproduction, whereas diploid **germ cells** produce haploid sex cells.

9.4 In Meiosis, DNA Replicates Once, but the Nucleus Divides Twice

- The events of meiosis ensure that gametes are haploid and genetically variable (figure 9.14).
- During **interphase,** which precedes meiosis, the cell grows and copies its DNA.
- **Spindle** proteins move the chromosomes throughout meiosis. Homologous pairs of chromosomes align during **prophase I,** line up double-file at the cell's center during **metaphase I,** then split apart during **anaphase I.** The chromosomes arrive at the poles in **telophase I,** and the cell often divides (**cytokinesis**).
- The two products of meiosis I each enter meiosis II. The chromosomes condense during **prophase II.** During **metaphase II,** they line up single-file at the cell's equator. The sister chromatids are separated in **anaphase II,** and the chromosomes arrive at the poles in **telophase II.** Cytokinesis then occurs once more to yield four haploid cells.

9.5 Meiosis Generates Enormous Variability

A. Crossing Over Shuffles Alleles

- **Crossing over,** which occurs in prophase I, produces variability when portions of homologous chromosomes switch places. After crossing over, the chromatids carry new combinations of alleles.

B. Homologous Pairs Are Oriented Randomly During Metaphase I

- Every possible orientation of homologous pairs of chromosomes at metaphase I is equally likely. As a result, one person can produce over 8 million genetically different gametes.

C. Random Fertilization Multiplies the Diversity

- Because any sperm can fertilize any egg cell, a human couple can produce over 70 trillion genetically different offspring.
- Identical (monozygotic) twins arise when a zygote splits into two embryos. Fraternal (dizygotic) twins develop from separate zygotes.

9.6 Mitosis and Meiosis Have Different Functions: A Summary

- Mitotic division produces identical copies of a cell and occurs throughout life.
- Meiosis produces genetically different haploid cells. It occurs only in specialized cells and only during some parts of the life cycle.

9.7 Errors Sometimes Occur In Meiosis

A. Polyploidy Means Extra Chromosome Sets

- **Polyploid** cells have one or more extra sets of chromosomes.

B. Nondisjunction Results in Extra or Missing Chromosomes

- **Nondisjunction** is the failure of homologous chromosomes or sister chromatids to separate, causing gametes to have incorrect chromosome numbers. A sex chromosome abnormality is typically less severe than an incorrect number of autosomes.

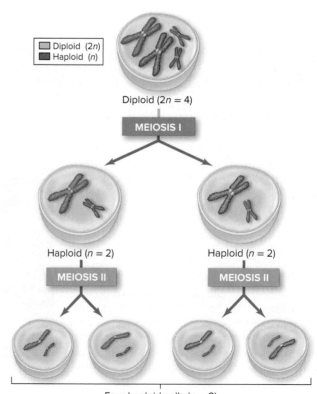

Figure 9.14 Summary of Meiosis.

MULTIPLE CHOICE QUESTIONS

1. Compared to other forms of reproduction, the unique feature of sex is
   a. the ability of a cell to divide.
   b. the production of offspring.
   c. the ability to generate new genetic combinations.
   d. Each of the above is unique to sexual reproduction.

2. Meiosis explains why
   a. you inherited half of your DNA from each of your parents.
   b. the sister chromatids in a chromosome are identical to each other.
   c. each of your somatic cells contains the same DNA.
   d. zygotes contain half as much DNA as somatic cells.

3. Which of the following is *not* a mechanism that contributes to diversity?
   a. Random fertilization
   b. Cytokinesis
   c. Crossing over
   d. Chromosome alignment during metaphase I

4. What event occurs soon after a cell completes meiosis I?
   a. Homologous chromosomes pair up.
   b. Homologous chromosomes move apart from each other.
   c. DNA is replicated for the second time.
   d. Cytokinesis divides the cytoplasm.

5. How many chromatids are in a human cell during metaphase I?
   a. 23   b. 46   c. 92   d. 184

6. Which of the following best describes what happens when fraternal twins are conceived?
   a. One sperm cell fertilizes one egg cell.
   b. One sperm cell fertilizes two egg cells.
   c. Two sperm cells fertilize one egg cell.
   d. Two sperm cells fertilize two egg cells.

7. Nondisjunction results from an error at which stage of meiosis?
   a. Prophase II   c. Anaphase I or II
   b. Metaphase I or II   d. Telophase I

Answers to Multiple Choice questions are in appendix A.

WRITE IT OUT

1. Explain why evolution often selects traits that promote genetic diversity.
2. Most cells in a sexually reproducing organism have two sets of chromosomes. Explain this observation and describe its significance to meiosis.
3. Sketch the relationships among mitosis, meiosis, and fertilization in a sexual life cycle.
4. What is the difference between haploid and diploid cells? Are your skin cells haploid or diploid? What about gametes?
5. In some animals, females can reproduce by themselves—that is, without males. In this process, called parthenogenesis, the young develop from unfertilized eggs. Use the Internet to find a species that uses parthenogenesis. How does parthenogenesis work in that species? What are the differences between meiosis in that species and the events of "typical" meiosis?
6. How are mitosis and meiosis different?
7. Draw all possible metaphase I chromosomal arrangements for a cell with a diploid number of 8.
8. Is it possible for a boy–girl pair of twins to be genetically identical? Why or why not?
9. List some examples of chromosomal abnormalities, and explain how each relates to an error in meiosis.

SCIENTIFIC LITERACY

Some large companies in the United States cover the cost for female employees who choose to freeze some of their eggs, enabling the women to postpone having children. The frozen eggs can be retrieved years later, then fertilized and implanted into the woman's uterus. From a biological perspective, why might freezing eggs be an attractive option for a young woman who wishes to postpone motherhood, and why might the pregnancy rate decline if an older woman's eggs are frozen? Use section 9.7 and the Internet to guide your response. In your opinion, does a company's offer to pay for this procedure create a work environment that is more inviting to women? Why or why not?

PULL IT TOGETHER

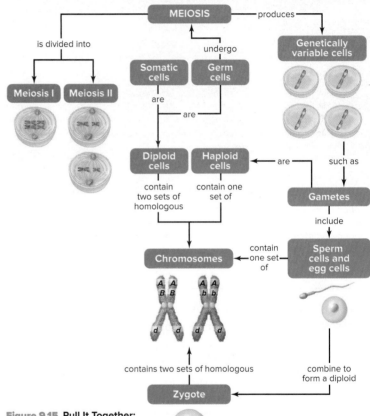

Figure 9.15 Pull It Together: Sexual Reproduction and Meiosis.

Refer to figure 9.15 and the chapter content to answer the following questions.

1. Fit the following terms into this concept map: *chromatid, centromere, nondisjunction, fertilization,* and *mitosis.*
2. Create a separate concept map that includes these terms: *crossing over, gamete, autosome,* and *homologous pair.* You may add other terms to the map as well.
3. Where do the members of each pair of homologous chromosomes in a diploid cell come from?
4. Review section 9.5 and the Survey the Landscape figure in the chapter introduction. What two processes in meiosis I generate genetic variation among gametes? What other process produces genetic variation?

Answers to Mastering Concepts, Write It Out, Scientific Literacy, and Pull It Together questions can be found in the Connect ebook.
connect.mheducation.com

Design element: Burning Question (fire background): ©Ingram Publishing/Super Stock

Human Diversity. Everyone inherits a unique combination of DNA sequences from his or her parents. Combined with environmental influences, these genetic sequences determine not only our appearance but many other traits as well.
©Rick Gomez/Corbis/Getty Images

LEARNING OUTLINE

10.1 Chromosomes Are Packets of Genetic Information: A Review

10.2 Mendel's Experiments Uncovered Basic Laws of Inheritance

10.3 The Two Alleles of a Gene End Up in Different Gametes

10.4 Genes on Different Chromosomes Are Inherited Independently

10.5 Genes on the Same Chromosome May Be Inherited Together

10.6 Inheritance Patterns Are Rarely Simple

10.7 Sex-Linked Genes Have Unique Inheritance Patterns

10.8 Pedigrees Show Modes of Inheritance

10.9 Most Traits Are Influenced by the Environment and Multiple Genes

APPLICATIONS

Burning Question 10.1 *Why does diet soda have a warning label?*
Burning Question 10.2 *Is male baldness really from the female side of the family?*
Why We Care 10.1 *The Origin of Obesity*
Investigating Life 10.1 *Heredity and the Hungry Hordes*

Learn How to Learn
Be a Good Problem Solver

This chapter is about the principles of inheritance, and it contains many practice problems. Need help? The **How to Solve a Genetics Problem** guide at the end of this chapter shows a systematic, step-by-step approach to solving three of the most common types of problems. Keep using the guide until you feel comfortable solving any problem type.

SURVEY THE LANDSCAPE
DNA, Inheritance, and Biotechnology

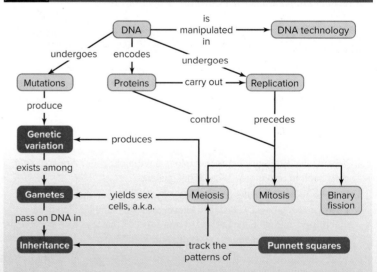

When gametes from two parents combine to produce offspring, what traits might the offspring inherit and express? Punnett squares help organize the possibilities.

For more details, study the Pull It Together feature in the chapter summary.

What's the **Point?** ▼

Interest in heredity is probably as old as humankind itself. People throughout time have wondered at their similarities, and they have used their intuition about inheritance to select for superior varieties of everything from poodles to ornamental flowers. But the *systematic* study of inheritance began with a nineteenth-century Austrian monk named Gregor Mendel; subsequent efforts to learn the basic principles of genetics and inheritance continued well into the twentieth century.

©Rodrusoleg/iStockphoto/
Getty Images RF

Today, genetics and DNA are familiar to nearly everyone, and the entire set of genetic instructions to build a person—the human genome—has been deciphered. Even so, when a family meets with a genetic counselor or physician to learn about an inherited illness, they encounter the same principles of heredity that Mendel derived in his early experiments with pea plants. Our look at genetics begins the traditional way, with Mendel, but we can appreciate his genius in light of what we now know about DNA.

10.1 Chromosomes Are Packets of Genetic Information: A Review

A healthy young couple, both with family histories of cystic fibrosis, visits a genetic counselor before deciding whether to have children. The counselor suggests genetic tests, which reveal that both the man and the woman are carriers of cystic fibrosis. The counselor tells the couple that each of their future children has a 25% chance of inheriting this serious illness. How does the counselor arrive at that one-in-four chance? This chapter will explain the answer. ⓘ *genetic testing,* section 11.4C

First, however, recall from chapter 7 that cells contain DNA, a molecule that encodes all of the information needed to sustain life. Human DNA includes about 25,000 genes. A **gene** is a portion of DNA whose sequence of nucleotides (A, C, G, and T) encodes a protein; the organism's proteins, in turn, help determine many of its characteristics. When a gene's nucleotide sequence mutates, the encoded protein—and the corresponding trait—may also change. Each gene in a population can therefore exist as one or more **alleles,** or alternative forms, each arising from a different mutation.

The DNA in the nucleus of a eukaryotic cell is divided among multiple **chromosomes,** which are long strands of DNA associated with proteins. Recall that a **diploid cell** contains two sets of chromosomes, with one set inherited from each parent. The human genome consists of 46 chromosomes, arranged in 23 pairs **(figure 10.1a)**. Of these, 22 pairs are **autosomes,**

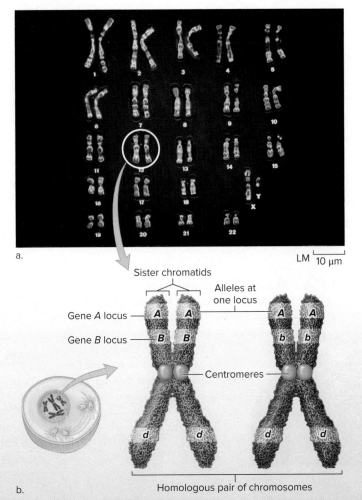

LM ⊢ 10 μm

a.

Sister chromatids

Alleles at one locus

Gene *A* locus — *A* *A*   *A* *A*

Gene *B* locus — *B* *B*   *b* *b*

Centromeres

d   *d*   *d*   *d*

b.

Homologous pair of chromosomes

Figure 10.1 Homologous Chromosomes. (a) A human diploid cell contains 23 pairs of chromosomes. (b) Each chromosome has one allele for each gene. For the chromosome pair in this figure, both alleles for gene *A* are identical; the same is true for gene *D*. The chromosomes carry different alleles for gene *B*.

(a): ©CNRI/Science Source

Trait	Dominant allele	Recessive allele
Seed color	Yellow (Y)	Green (y)
Seed form	Round (R)	Wrinkled (r)
Pod color	Green (G)	Yellow (g)
Pod form	Inflated (V)	Constricted (v)
Flower color	Purple (P)	White (p)
Flower position	Axial (A)	Terminal (a)
Stem length	Tall (L)	Short (l)

Figure 10.2 **Mendel and His Peas.**
Gregor Mendel was an Austrian monk whose studies of pea plants revealed the basic principles of inheritance. Mendel's experiments helped him deduce the inheritance patterns of many pea plant characteristics. (Section 10.2 explains the terms *dominant* and *recessive*.)

Photo: ©James King-Holmes/Science Source

which are the chromosomes that are the same for both sexes. The single pair of **sex chromosomes** determines a person's sex: A female has two X chromosomes, whereas a male has one X and one Y.

With the exception of X and Y, the chromosome pairs are homologous (figure 10.1b). As described in chapter 9, the two members of a **homologous pair** of chromosomes look alike and have the same sequence of genes in the same positions. (A gene's *locus* is its physical place on the chromosome.) But the two homologs may or may not carry the same alleles. Since each homolog comes from a different parent, each person inherits two alleles for each gene in the human genome.

Figure 7.3 illustrates an analogy that may help clarify the relationships among these terms. Each chromosome is like a cookbook; the human genome is a "library" that consists of 46 such volumes, arranged in 23 pairs of similar books. The entire cookbook library includes about 25,000 recipes, each analogous to one gene.

The two alleles for each gene, then, are comparable to two of the many ways to prepare brownies; some recipes include nuts, for example, whereas others use different types of chocolate. The two "brownie recipes" in a cell may be exactly the same as, slightly different from, or very different from each other. Furthermore, with the exception of identical twins, everyone inherits a unique combination of alleles for all of the genes in the human genome. For example, you may have black hair and brown eyes, whereas your best friend has brown hair and green eyes. The two of you do not look alike because you have different alleles for the hair and eye color genes.

Another important idea to review from chapter 9 is the role of meiosis and fertilization in a sexual life cycle (see figure 9.5). **Meiosis** is a specialized form of cell division that occurs in diploid germ cells and gives rise to **haploid cells,** each containing just one set of chromosomes. In humans, these haploid cells are **gametes**—sperm or egg cells. **Fertilization** unites the gametes from two parents, producing the first cell of the next generation. Gametes are the cells that convey chromosomes from one generation to the next, so they play a critical part in the study of inheritance.

No one can examine a gamete and say for sure which allele it carries for every gene. As we shall see in this chapter, however, for some traits, we can use knowledge of a person's characteristics and family history to say that a gamete has a 100% chance, 50% chance, or 0% chance of carrying a specific allele. With this information for both parents, it is simple to calculate the probability that a child will inherit the allele.

10.1 Mastering Concepts

1. How are chromosomes, DNA, genes, and alleles related?
2. How do meiosis, fertilization, diploid cells, and haploid cells interact in a sexual life cycle?

10.2 Mendel's Experiments Uncovered Basic Laws of Inheritance

Of all the people who have studied inheritance, one nineteenth-century investigator, Gregor Mendel, made the most lasting impression on what would become the science of genetics (figure 10.2). Mendel was born in 1822 and spent his early childhood in a small village in what is now the Czech Republic, where he learned early how to tend fruit trees. After finishing school ahead of

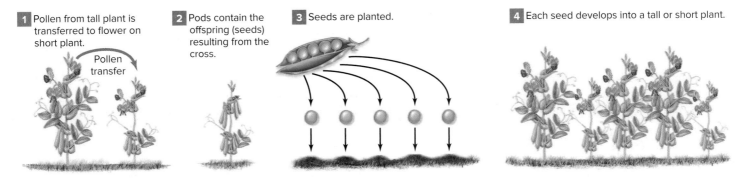

1 Pollen from tall plant is transferred to flower on short plant.

Pollen transfer

2 Pods contain the offspring (seeds) resulting from the cross.

3 Seeds are planted.

4 Each seed develops into a tall or short plant.

Figure 10.3 Breeding Peas. Gregor Mendel used this technique to set up carefully designed crosses of pea plants so he could observe the appearance of traits in the next generation.

schedule, Mendel became a priest at a monastery, where he could teach and do research in natural science.

The young man eagerly learned how to artificially pollinate crop plants to control their breeding. The monastery sent him to earn a college degree at the University of Vienna, where courses in the sciences and statistics fueled his interest in plant breeding. Mendel began to think about experiments to address a compelling question that had confounded other plant breeders: Why did some traits disappear, only to reappear a generation later?

The pea plant was Mendel's choice for studying heredity. Pea plants are easy to grow, develop quickly, and produce many offspring. Also, peas have many traits that appear in two easily distinguishable forms (see figure 10.2). For example, seeds may be round or wrinkled, yellow or green. Pods may be inflated or constricted. Stems may be tall or short.

Pea plants also have another advantage for studies of inheritance: It is easy to control which plants mate with which (figure 10.3). An investigator can take pollen from the male flower parts of one plant and apply it to the female part of the same plant (self-fertilization) or another plant (cross-fertilization). The resulting offspring are seeds that develop inside pods; each pea represents a genetically unique offspring, analogous to you and your siblings. Traits such as seed color or seed shape are evident right away; for other characteristics, such as plant height or flower color, the investigator must sow the seeds and observe each plant that develops.

A. Dominant Alleles Appear to Mask Recessive Alleles

From 1857 to 1863, Mendel crossed and cataloged some 24,034 plants through several generations. He observed consistent ratios of traits in the offspring and deduced that the plants transmitted distinct units, or "elementen" (now called genes).

Mendel's first experiments dealt with single traits that have two expressions, such as yellow or green seed colors. He noted that some plants were always **true-breeding;** that is, self-fertilization always produced offspring resembling the parent plant. Plants derived from green seeds, for example, always produced green seeds when self-fertilized. But crosses involving plants grown from yellow seeds were more variable. Sometimes these plants were true-breeding, but others were **hybrids:** Their offspring were mixed, including both yellow and green peas (figure 10.4).

Cross-fertilization experiments yielded other intriguing results. For example, Mendel crossed plants derived from green seeds with plants grown from yellow seeds. Sometimes, the pods contained only yellow seeds; the green trait

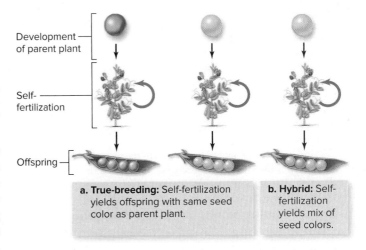

Development of parent plant

Self-fertilization

Offspring

a. True-breeding: Self-fertilization yields offspring with same seed color as parent plant.

b. Hybrid: Self-fertilization yields mix of seed colors.

Figure 10.4 True-Breeding and Hybrid Plants. (a) Pea plants derived from green seeds were always true-breeding. Some plants grown from yellow seeds were true-breeding, but others (b) were hybrids.

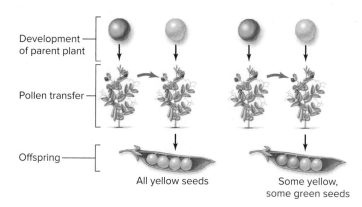

Development of parent plant

Pollen transfer

Offspring

All yellow seeds

Some yellow, some green seeds

Figure 10.5 **Yellow Is Dominant.** When Mendel crossed a plant derived from a green seed with a plant grown from a yellow seed, the offspring could be all yellow, or they could be a mix of green and yellow peas. Crosses such as these led Mendel to conclude that yellow seed color is dominant over green.

seemed to vanish, although it could reappear in the next generation. Other times, the pods contained green and yellow seeds (figure 10.5). Mendel noticed a similar mode of inheritance when he studied other pea plant characteristics: One trait seemed to obscure the other. Mendel called the masking trait *dominant;* the trait being masked was called *recessive.* The yellow-seed trait, for example, is dominant over the green trait.

Although Mendel referred to *traits* as dominant or recessive, modern biologists reserve these terms for *alleles.* A **dominant** allele encodes a protein that exerts its effects whenever it is present; a **recessive** allele encodes a protein whose effect is masked if a dominant allele is also present. When a gene has only two alleles, it is common to symbolize the dominant allele with a capital letter (such as *Y* for yellow) and the recessive allele with the corresponding lowercase letter (*y* for green).

The "dominance" of an allele may seem to imply that it "dominates" in the population as a whole. The most common allele, however, is not always dominant. In humans, the allele that causes a form of dwarfism called achondroplasia is dominant, but it is very rare—as is the dominant allele that causes Huntington disease. Conversely, blue eyes are the norm in people of northern European origin, but the alleles that produce this eye color are recessive.

The term *dominant* may also conjure images of a bully that forces a weak, recessive allele into submission. After all, the recessive allele seems to hide when a dominant allele is present, emerging from its hiding place only if the dominant allele is absent. How does the recessive allele "know" what to do? In fact, alleles cannot hide, emerge, or know anything. A recessive allele remains a part of the cell's DNA, regardless of the presence of a dominant allele. It only seems to hide because it typically encodes a nonfunctional protein. If a dominant allele is also present, the organism usually has enough of the functional protein to maintain its normal appearance (although section 10.6 describes some exceptions). It is only when both alleles are recessive that the lack of the functional protein becomes noticeable. Burning Question 10.1 describes a health-related consequence of a nonfunctional protein encoded by a recessive allele.

Burning Question 10.1

Why does diet soda have a warning label?

Foods containing the artificial sweetener aspartame carry a warning label that says "Contains phenylalanine." Since other sugar substitutes lack similar words of caution, aspartame must pose a unique threat. What is it?

A peek at aspartame's biochemistry reveals the answer. Aspartame contains an amino acid called phenylalanine. In most people, an enzyme converts phenylalanine into another amino acid. A mutated allele of the gene encoding this enzyme, however, results in the production of an abnormal, nonfunctional protein. People who have just one copy of the recessive allele are healthy because the cell has enough of the normal protein, thanks to the dominant allele. The person is healthy, so the recessive allele seems to "vanish."

Individuals who inherit two copies of the recessive allele, however, have a disorder called phenylketonuria (abbreviated PKU). These people cannot produce the normal enzyme. Phenylalanine accumulates to toxic levels, causing intellectual disability and other problems. Avoiding foods containing the artificial sweetener aspartame helps minimize the effects of PKU—hence the warning.

Submit your burning question to
marielle.hoefnagels@mheducation.com

(label): ©David Tietz/Editorial Image, LLC

B. For Each Gene, a Cell's Two Alleles May Be Identical or Different

Mendel chose traits encoded by genes with only two alleles, but some genes have hundreds of forms. Regardless of the number of possibilities, however, a diploid cell can have only two alleles for each gene. After all, each diploid individual has inherited one set of chromosomes from each parent, and each chromosome carries only one allele per gene. (Errors in meiosis, however, may result in cells with extra chromosome copies; see section 9.7.)

For a given gene, a diploid cell's two alleles may be identical or different. The **genotype** expresses the genetic makeup of an individual, and it is written as a pair of letters representing the alleles (figure 10.6). An individual that is **homozygous** for a gene has two identical alleles, meaning that both parents contributed the same gene version. If both of the alleles are dominant, the individual's genotype is homozygous dominant (written as *YY,* for example). If both alleles are recessive, the individual is homozygous recessive (*yy*). An individual with a **heterozygous** genotype, on the other hand, has two different alleles for the gene (*Yy*). That is, the two parents each contributed different genetic information.

The organism's genotype is distinct from its **phenotype,** or observable characteristics (see figure 10.6). Seed color, flower color, and stem length are examples of pea plant phenotypes that Mendel studied. Your own phenotype includes not only your height, eye color, shoe size, number of fingers and toes, skin color, and hair texture but also other traits that are not readily visible, such as your blood type or the specific shape of your hemoglobin proteins (see figure 7.14). As described in section 10.9, most phenotypes result from a complex interaction between genes and environment. Mendel, however, chose traits controlled exclusively by genes.

Mendel's observation that some but not all yellow-seeded pea plants were true-breeding arises from the two possible genotypes for the yellow phenotype (homozygous dominant and heterozygous). All homozygous plants are true-breeding because all of their gametes contain the same allele. Heterozygous plants, however, are not true-breeding because they may pass on either the dominant or the recessive allele. These plants are hybrids.

Today, biologists use additional terms to describe organisms. A **wild-type** allele, genotype, or phenotype is the most common form or expression of a gene in a population. Wild-type fruit flies, for example, have two antennae and one pair of wings. A **mutant** allele, genotype, or phenotype is a variant that arises when a gene undergoes a mutation. Mutant phenotypes for fruit flies include having multiple pairs of wings and having legs instead of antennae growing out of the head (see figure 7.13).

10.2 Mastering Concepts

1. Why did Gregor Mendel choose pea plants as his experimental organism?
2. Distinguish between dominant and recessive; heterozygous and homozygous; phenotype and genotype; wild-type and mutant.

10.3 The Two Alleles of a Gene End Up in Different Gametes

Mendel kept careful tallies of the offspring from countless crosses, which required a systematic accounting of multiple generations of plants. Biologists still use Mendel's system of standardized names to keep track of inheritance patterns. The purebred **P generation** (for "parental") is the first set of individuals being mated; the F_1 **generation,** or first filial generation, is the offspring from the P generation (*filial* derives from the Latin word for "son"). The F_2 **generation** is the offspring of the F_1 plants, and so on. Although these terms are applicable only to lab crosses, they are analogous to human family relationships. If you consider your grandparents the P generation, your parents are the F_1 generation, and you and your siblings are the F_2 generation.

A. The Simplest Punnett Squares Track the Inheritance of One Gene

Mendel began by deducing the rules of inheritance for single genes. He began with a P generation consisting of true-breeding plants derived from yellow seeds (YY) and true-breeding plants derived from green seeds (yy). The F_1 offspring produced in this cross were yellow seeds (genotype Yy). The green trait therefore seemed to disappear in the F_1 generation.

Genotype	Phenotype
Homozygous dominant (YY)	Yellow
Heterozygous (Yy)	Yellow
Homozygous recessive (yy)	Green

Figure 10.6 **Genotypes and Phenotypes Compared.** A pea's genotype for the "seed color" gene consists of the two alleles that the seed inherited from its parents. Its phenotype is its outward appearance: yellow or green.

Miniglossary | Genetic Terms

Chromosomes and genes

Chromosome	A continuous molecule of DNA plus associated proteins
Gene	A sequence of DNA that encodes a protein
Locus	The physical location of a gene on a chromosome
Allele	One of the two or more variants of a gene

Dominant and recessive

Dominant allele	An allele that is expressed if present in the genotype
Recessive allele	An allele whose expression is masked by a dominant allele

Genotypes and phenotypes

Genotype	An individual's allele combination for a particular gene
Homozygous	Possessing identical alleles of one gene
Heterozygous	Possessing different alleles of one gene
Phenotype	An observable characteristic
Wild-type	The most common allele, genotype, or phenotype in a population
Mutant	An allele, genotype, or phenotype resulting from a change (mutation) in a gene
True-breeding	Homozygous; self-fertilization produces offspring identical to self for a given trait
Hybrid	Heterozygous; self-fertilization produces offspring with mixed genotypes and phenotypes

Generations

P generation	True-breeding (homozygous) parents
F_1 generation	Offspring of P generation
F_2 generation	Offspring of F_1 generation

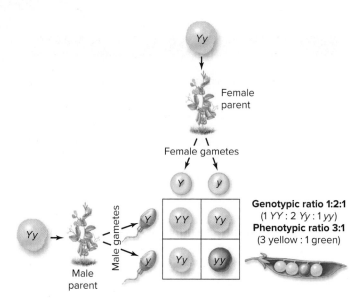

Figure 10.7 Punnett Square. This diagram depicts Mendel's monohybrid cross of two heterozygous plants grown from yellow seeds (*Yy*). The four compartments within the Punnett square contain the genotypes and phenotypes of all possible offspring.

Genotypic ratio 1:2:1
(1 *YY* : 2 *Yy* : 1 *yy*)
Phenotypic ratio 3:1
(3 yellow : 1 green)

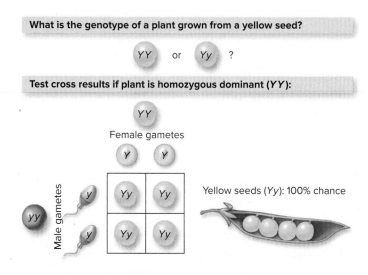

What is the genotype of a plant grown from a yellow seed?

YY   or   *Yy*   ?

Test cross results if plant is homozygous dominant (*YY*):

Yellow seeds (*Yy*): 100% chance

Test cross results if plant is heterozygous (*Yy*):

Yellow seeds (*Yy*): 50% chance
Green seeds (*yy*): 50% chance

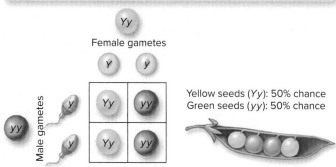

Figure 10.8 Test Cross. A test cross with a homozygous recessive (*yy*) plant reveals whether a pea plant grown from a yellow seed is homozygous dominant (*YY*) or heterozygous (*Yy*).

Next, he used the F₁ plants to set up a **monohybrid cross:** a mating between two individuals that are both heterozygous for the same gene (figure 10.7). The resulting F₂ generation had both yellow and green phenotypes, in a ratio of 3:1; that is, for every three yellow seeds, he observed one green seed.

A **Punnett square** is a diagram that uses the genotypes of two parents to reveal which allele combinations their offspring may inherit. The Punnett square in figure 10.7, for example, shows how the green phenotype reappeared in the F₂ generation. Both parents are heterozygous (*Yy*) for the seed color gene. Each therefore produces some gametes carrying the *Y* allele and some gametes carrying *y*. All three possible genotypes may therefore appear in the F₂ generation, in the ratio 1 *YY*: 2 *Yy*: 1 *yy*. The corresponding phenotypic ratio is three yellow seeds to one green seed, or 3:1. Mendel saw similar results for all seven traits that he studied.

Note that all Punnett squares, including the one in figure 10.7, show the *probabilities* that apply to each offspring. That is, if two pea plants produce exactly four offspring, there will not necessarily be exactly one with genotype *YY*, two with *Yy*, and one with *yy*. Similarly, the chance of tossing a fair coin and seeing "heads" is 50%, but two tosses will not necessarily yield one head and one tail. If you toss the coin 1000 times, however, you will likely approach the expected 1:1 ratio of heads to tails. Pea plants are ideal for genetics studies in part because they produce many offspring in each generation.

Mendel could tally the number of plants with each phenotype, but he also needed to keep track of each genotype. He knew that green seeds were always homozygous recessive (*yy*). But what was the genotype of each yellow seed, *YY* or *Yy?* He had no way to tell just by looking, so he set up breeding experiments called test crosses to distinguish between the two possibilities. A **test cross** is a mating between an individual of unknown genotype and a homozygous recessive individual (figure 10.8). If a plant grown from a yellow seed was crossed with a *yy* plant and the offspring were all yellow seeds, Mendel knew the unknown genotype was *YY;* if the cross produced seeds of both colors, he knew it must be *Yy*.

Figure It Out

Holstein cattle suffer from the condition citrullinemia, in which homozygous recessive calves die within a week of birth because they cannot break down ammonia that is produced when amino acids are metabolized. If a cow that is heterozygous for the citrullinemia gene is inseminated by a bull that is homozygous dominant, what is the probability that a calf inherits citrullinemia?

Answer: 0%.

B. Meiosis Explains Mendel's Law of Segregation

All of Mendel's breeding experiments and calculations added up to a brilliant description of basic genetic principles. Without any knowledge of chromosomes or genes, Mendel used his data to conclude that genes occur in alternative versions (which we now call alleles). He further determined that each individual inherits two alleles for each gene and that these alleles

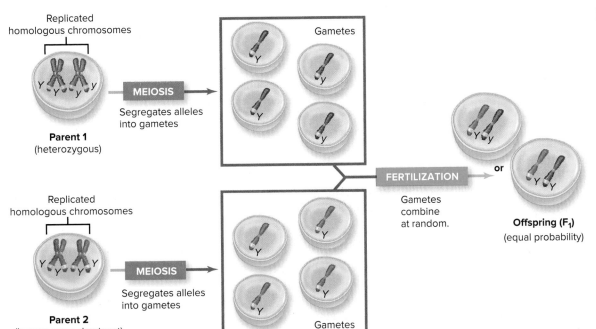

Replicated homologous chromosomes

MEIOSIS

Segregates alleles into gametes

Parent 1
(heterozygous)

Gametes

FERTILIZATION   or

Gametes combine at random.

Offspring (F₁)
(equal probability)

Replicated homologous chromosomes

MEIOSIS

Segregates alleles into gametes

Parent 2
(homozygous dominant)

Gametes

Figure 10.9 Mendel's Law of Segregation. During meiosis, the four chromatids in a homologous pair of chromosomes segregate from each other and are packaged into separate gametes. At fertilization, gametes combine at random to form the next generation.

may be the same or different. Finally, he deduced his **law of segregation,** which states that the two alleles of each gene are packaged into separate gametes; that is, they "segregate," or move apart from each other, during gamete formation.

Mendel's law of segregation makes perfect sense in light of what we now know about reproduction. During meiosis I, homologous pairs of chromosomes separate and move to opposite poles of the cell. After a plant of genotype Yy undergoes meiosis, half of the gametes carry Y and half carry y (figure 10.9). A YY plant, on the other hand, can produce only Y gametes. When gametes from the two plants meet at fertilization, they combine at random. About 50% of the time, both gametes carry Y; the other 50% of the time, one contributes Y and the other, y.

This principle of inheritance applies to all diploid species, including humans. Return for a moment to the couple and their genetic counselor introduced in section 10.1. Cystic fibrosis arises when a person has two recessive alleles for a particular gene on chromosome 7. Genetic testing revealed that the man and the woman are both carriers. In genetic terms, this means that although neither has the disease, both are heterozygous for the gene that causes cystic fibrosis. Just as in Mendel's monohybrid crosses, each of their children has a 25% chance of inheriting two recessive alleles (figure 10.10).

Mother: healthy carrier
Female gametes

	F	**f**
F	**FF** Healthy non-carrier	**Ff** Healthy carrier
f	**Ff** Healthy carrier	**ff** Affected

Father: healthy carrier
Male gametes

Healthy noncarrier (*FF*): 25% chance
Healthy carrier (*Ff*): 50% chance
Affected (*ff*): 25% chance

Figure 10.10 Inheritance of Cystic Fibrosis. This Punnett square shows the possible results of a mating of two carriers of cystic fibrosis. The siblings in the photo have cystic fibrosis; the masks deliver medicine to treat their lung problems.

Photo: ©ZUMA Wire Service/Alamy Stock Photo

10.3 Mastering Concepts

1. What is a monohybrid cross, and what are the genotypic and phenotypic ratios expected in the offspring of the cross?
2. How are Punnett squares helpful in following the inheritance of single genes?
3. What is a test cross, and why is it useful?
4. How do Punnett squares reflect the events of meiosis?

Miniglossary | Tracking inheritance

Punnett square	Diagram showing how the alleles in two parents' gametes might combine at fertilization
Monohybrid cross	Mating between two individuals that are both heterozygous for one gene
Dihybrid cross	Mating between two individuals that are both heterozygous for two genes
Test cross	Mating between an individual of unknown genotype and a homozygous recessive individual

Complex Inheritance Patterns   Inheritance of Sex-Linked Genes   Pedigrees Show Modes of Inheritance   Environment and Genes Affect Traits

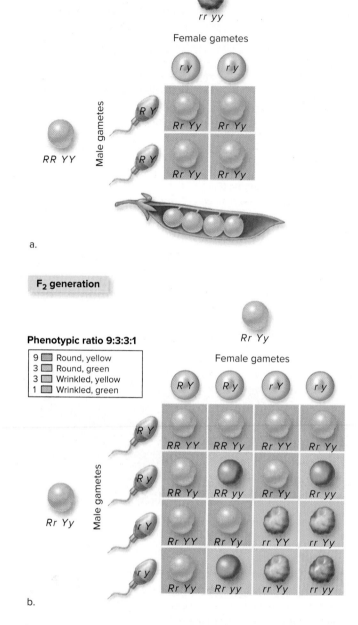

F₁ generation

Female gametes

Male gametes

a.

F₂ generation

Phenotypic ratio 9:3:3:1

9	Round, yellow
3	Round, green
3	Wrinkled, yellow
1	Wrinkled, green

Female gametes

Male gametes

b.

Figure 10.11 Generating a Dihybrid Cross. (a) In the parental generation, one parent is homozygous recessive for two genes; the other is homozygous dominant. The F₁ generation is therefore heterozygous for both genes. (b) A dihybrid cross is a mating between two plants from the F₁ generation. Phenotypes occur in a distinctive ratio in the resulting F₂ generation.

10.4 Genes on Different Chromosomes Are Inherited Independently

Mendel's law of segregation arose from his studies of the inheritance of single traits. He next asked himself whether the same law would apply if he followed two characters at the same time. Mendel therefore began another set of breeding experiments in which he simultaneously examined the inheritance of two characteristics: pea shape and pea color.

A. Tracking Two-Gene Inheritance May Require Large Punnett Squares

A pea's shape may be round or wrinkled (determined by the R gene, with the dominant allele specifying round shape; see figure 10.2). At the same time, its color may be yellow or green (determined by the Y gene, with the dominant allele specifying yellow).

As he did before, Mendel began with a P generation consisting of true-breeding parents (figure 10.11a). He crossed plants grown from wrinkled, green seeds with plants derived from round, yellow seeds. All F₁ offspring were heterozygous for both genes ($Rr\ Yy$) and therefore had round, yellow seeds.

Next, Mendel crossed the F₁ plants with each other (figure 10.11b). A **dihybrid cross** is a mating between two individuals that are each heterozygous for the same two genes. Each $Rr\ Yy$ individual in the F₁ generation produced equal numbers of gametes of four different types: $R\ Y$, $R\ y$, $r\ Y$, and $r\ y$. After Mendel completed the crosses, he found four phenotypes in the F₂ generation, reflecting all possible combinations of seed shape and color. The Punnett square predicts that the four phenotypes will occur in a ratio of 9:3:3:1. That is, nine of 16 offspring should be round, yellow seeds; three should be round, green seeds; three should be wrinkled, yellow seeds; and just one should be a wrinkled, green seed. This prediction almost exactly matches Mendel's results.

B. Meiosis Explains Mendel's Law of Independent Assortment

Based on the results of the dihybrid cross, Mendel proposed a second law of inheritance. The **law of independent assortment** states that during gamete formation, the alleles for one gene do not influence the alleles for another gene. That is, alleles Y and y are randomly packaged into gametes, independent of alleles R and r. With this second set of experiments, Mendel had again inferred a principle of inheritance based on meiosis (figure 10.12).

Interestingly, Mendel found some trait combinations for which a dihybrid cross did not yield the expected phenotypic ratio. Mendel could not explain this result. No one could, until Thomas Hunt Morgan's work led to the chromosomal theory of inheritance. As you will see in section 10.5, the law of independent assortment does not apply to genes that are close together on the same chromosome.

C. The Product Rule Is a Useful Shortcut

Punnett squares become cumbersome when analyzing more than two genes. A Punnett square for three genes has 64 boxes; for four genes, 256 boxes. Fortunately, there is a shortcut that relies on the rules of probability on which

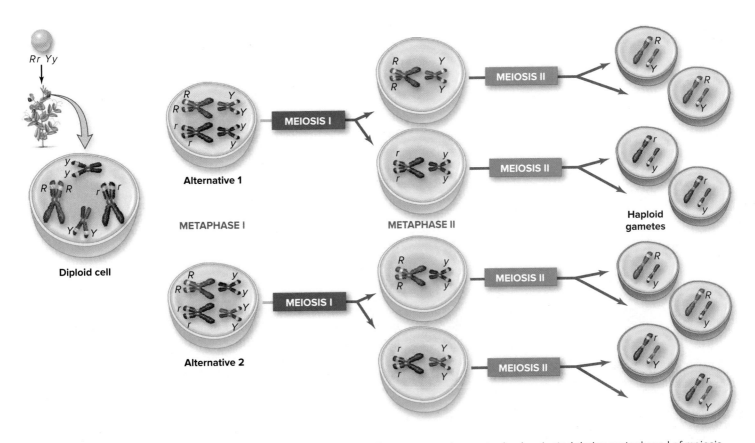

Figure 10.12 Mendel's Law of Independent Assortment. Homologous chromosome pairs are randomly oriented during metaphase I of meiosis. The exact allele combination in a gamete depends on which chromosomes happen to be packaged together. An individual of genotype *Rr Yy* therefore produces approximately equal numbers of four types of gametes: *R Y, r y, R y,* and *r Y.*

Punnett squares are based. According to the **product rule,** the probability that multiple independent events will occur simultaneously (for example, an offspring inheriting specific alleles for two genes) can be calculated by multiplying the chances of each event occurring alone.

As an example, the product rule can predict the chance of obtaining wrinkled, green seeds (*rr yy*) from dihybrid (*Rr Yy*) parents. The probability that multiple *Rr* plants will produce *rr* offspring is 25%, or ¼, and the chance of two *Yy* plants producing a *yy* individual is ¼. According to the product rule, the chance of dihybrid parents (*Rr Yy*) producing homozygous recessive (*rr yy*) offspring is therefore ¼ multiplied by ¼, or ¹⁄₁₆. Now consult the 16-box Punnett square for Mendel's dihybrid cross (see figure 10.11). As expected, only one of the 16 boxes contains *rr yy.* Figure 10.13 applies the product rule to three traits.

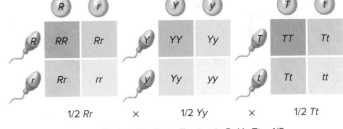

1/2 *Rr* × 1/2 *Yy* × 1/2 *Tt*

Probability that offspring is *Rr Yy Tt* = 1/8

Figure 10.13 The Product Rule. What is the chance that two parents that are heterozygous for three genes (*Rr Yy Tt*) will give rise to an offspring with that same genotype? To find out, multiply the individual probabilities for each gene.

10.4 Mastering Concepts

1. What is a dihybrid cross, and what is the phenotypic ratio expected in the offspring of the cross?

2. How does the law of independent assortment reflect the events of meiosis?

3. How can the product rule be used to predict the results of crosses in which multiple genes are studied simultaneously?

Figure It Out

A man and a woman each have dark eyes, dark hair, and freckles. The genes encoding these traits are on separate chromosomes. The woman is heterozygous for each of these genes, but the man is homozygous. Assume that the *B* allele confers dark eyes, *H* confers dark hair, and *F* confers freckles. Use the product rule to determine the probability that a child will share the genotype of each parent.

Answer: ½ × ½ × ½ = ⅛ chance for each parent.

Punnett square (assumes genes assort independently)

Body color:
Gray (*A*) or black (*a*)
Wings:
Normal (*B*) or vestigial (*b*)

aa bb

a b | *a b* | *a b* | *a b*

A B
| Aa Bb | Aa Bb | Aa Bb | Aa Bb |

A b
| Aa bb | Aa bb | Aa bb | Aa bb |

a B
| aa Bb | aa Bb | aa Bb | aa Bb |

Aa Bb

a b
| aa bb | aa bb | aa bb | aa bb |

Expected genotypic ratio if genes assort independently = 1:1:1:1
1 *Aa Bb*   1 *Aa bb*   1 *aa Bb*   1 *aa bb*

Actual genotypic ratio ≈ 5:1:1:5
5 *Aa Bb*   1 *Aa bb*   1 *aa Bb*   5 *aa bb*

Figure 10.14 Linked Genes. The Punnett square shows that if genes *A* and *B* are on separate chromosomes, the cross should yield approximately equal numbers of offspring of each genotype. However, the actual genotypic ratio is skewed toward two of the four offspring classes. This result reveals that genes *A* and *B* do not assort independently; that is, they are linked.

10.5 Genes on the Same Chromosome May Be Inherited Together

Biologists did not appreciate the significance of Gregor Mendel's findings until 1900, when three botanists working independently each rediscovered the principles of inheritance. They eventually found the paper that Mendel had published in 1866, and other scientists demonstrated Mendel's ratios again and again in several species. At about the same time, advances in microscopy were allowing scientists to observe and describe chromosomes for the first time.

It soon became apparent that what Mendel called "elementen" (later renamed "genes") have much in common with chromosomes. Both genes and chromosomes, for example, come in pairs. In addition, alleles of a gene are packaged into separate gametes, as are the members of a homologous pair of chromosomes. Finally, both genes and chromosomes are inherited in random combinations.

As biologists cataloged traits and the chromosomes that transmit them in several species, they realized that the number of traits far exceeds the number of chromosomes. Fruit flies, for example, have only four pairs of chromosomes, but they have dozens of different bristle patterns, body colors, eye colors, wing shapes, and other characteristics. How might a few chromosomes control so many traits? The answer: Each chromosome carries many genes.

A. Genes on the Same Chromosome Are Linked

Linked genes are carried on the same chromosome; they are therefore inherited together. Unlike genes on different chromosomes, they do not assort independently during meiosis. The seven traits that Mendel followed in his pea plants all happened to be transmitted on separate chromosomes. Had the same chromosome carried these genes, Mendel's dihybrid crosses would have generated markedly different results.

The inheritance pattern of linked genes was first noticed in the early 1900s when Thomas Hunt Morgan at Columbia University studied the inheritance of pairs of traits in fruit flies. The data began to indicate four **linkage groups,** collections of genes that tended to be inherited together. Within each linkage group, dihybrid crosses did not produce the proportions of offspring that Mendel's law of independent assortment predicts (figure 10.14). Because the number of linkage groups was the same as the number of homologous pairs of chromosomes, scientists eventually realized that each linkage group was simply a set of genes transmitted on the same chromosome.

Nevertheless, the researchers did sometimes see offspring with trait combinations not seen in either parent. How could this occur? The answer turned out to involve yet another event in gamete formation: **crossing over,** an exchange of genetic material between homologous chromosomes during prophase I of meiosis (see figure 9.7). After crossing over, no two chromatids in a homologous pair of chromosomes are identical (figure 10.15).

Crossing over is a random process, and it might or might not occur between two linked genes. For any pair of genes on the same chromosome, then, most gametes receive a **parental chromatid,** which retains the allele combinations from one parent. But when crossing over happens between two genes, some of the gametes receive a **recombinant chromatid** carrying a mix of maternal and paternal alleles. A recombinant offspring is one that inherits a recombinant chromatid.

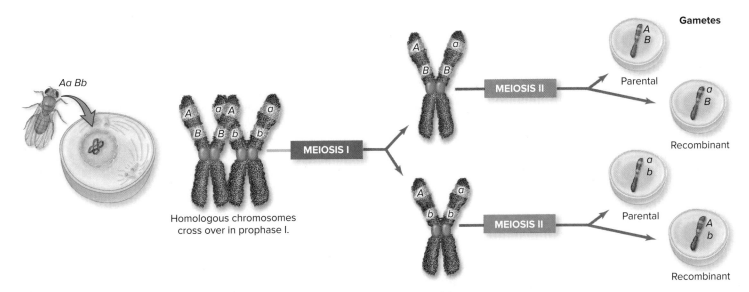

Figure 10.15 Crossing Over. Linkage between two alleles is interrupted if crossing over occurs at a point between the two genes. As a result, some gametes contain recombinant arrangements of the alleles.

B. Studies of Linked Genes Have Yielded Chromosome Maps

Morgan wondered why some crosses produced a higher proportion of recombinant offspring than others. Might the differences reflect the physical relationships between the genes on the chromosome? Alfred Sturtevant, Morgan's undergraduate assistant, explored this idea. In 1911, Sturtevant proposed that the farther apart two alleles are on the same chromosome, the more likely crossing over is to separate them—simply because more space separates the genes (figure 10.16a).

Sturtevant's idea became the basis for mapping genes on chromosomes. By determining the percentage of recombinant offspring, investigators can infer how far apart the genes are on one chromosome. Crossing over frequently separates alleles on opposite ends of the same chromosome, so recombinant offspring occur frequently. In contrast, a crossover would rarely separate alleles lying very close together on the chromosome, and the proportion of recombinant offspring would be small. Geneticists use this correlation between crossover frequency and the distance between genes to construct **linkage maps,** which are diagrams of gene order and spacing on chromosomes.

In 1913, Sturtevant published the first genetic linkage map, depicting the order of five genes on the X chromosome of the fruit fly (figure 10.16b). Researchers then rapidly mapped genes on all four fruit fly chromosomes. Linkage maps for the human chromosomes followed over the next half century. These linkage maps provided the rough drafts to which DNA sequence information was added later in the century. Today, entire genomes are routinely sequenced using powerful computers. ⓘ *DNA sequencing,* section 11.2B

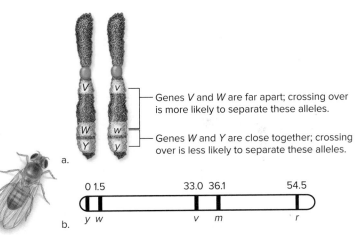

Genes *V* and *W* are far apart; crossing over is more likely to separate these alleles.

Genes *W* and *Y* are close together; crossing over is less likely to separate these alleles.

a.

0 1.5   33.0 36.1   54.5

b. *y w*   *v   m*   *r*

Figure 10.16 Breaking Linkage. (a) Crossing over is more likely to separate the alleles of genes *V* and *W* than to separate the alleles of genes *W* and *Y*. (b) A linkage map of a fruit fly chromosome, showing the locations of five genes. The numbers represent crossover frequencies relative to the leftmost gene, *y*.

10.5 Mastering Concepts

1. How do patterns of inheritance differ for unlinked versus linked pairs of genes?
2. What is the difference between recombinant and parental chromatids, and how do they arise?
3. How do biologists use crossover frequencies to map genes on chromosomes?

Miniglossary	Gene Linkage
Linked genes	Genes carried on the same chromosome
Parental chromatid	A chromatid carrying the alleles of only one parent
Recombinant chromatid	A chromatid carrying a mix of alleles from both parents
Recombinant offspring	An individual that inherits a recombinant chromatid

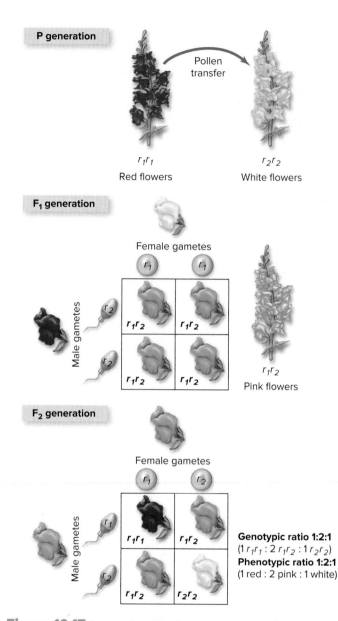

Figure 10.17 **Incomplete Dominance.** In snapdragons, a cross between a plant with red flowers (r_1r_1) and a plant with white flowers (r_2r_2) produces heterozygous plants with pink flowers (r_1r_2). The red and white phenotypes reappear in the F_2 generation.

Genotypic ratio 1:2:1
($1\,r_1r_1 : 2\,r_1r_2 : 1\,r_2r_2$)
Phenotypic ratio 1:2:1
(1 red : 2 pink : 1 white)

Figure It Out

A woman with type AB blood has children with a man who has type O blood. What is the chance that they have a child with type A blood? Type B? AB? O?

10.6 Inheritance Patterns Are Rarely Simple

Offspring traits frequently do not occur in the proportions that Punnett squares predict. Sections 10.6 and 10.7 describe some situations that may produce phenotypic ratios other than those Mendel observed.

A. Incomplete Dominance and Codominance Add Phenotype Classes

For the traits Mendel studied, one allele is completely dominant and the other is completely recessive. The phenotype of a heterozygote is therefore identical to that of a homozygous dominant individual. For many genes, however, heterozygous offspring do not share the phenotype of either parent. As you will see, biologists apply unique notation to alleles for these traits. Designating the alleles with capital and lowercase letters does not work because neither allele is necessarily dominant over the other.

When a gene shows **incomplete dominance,** the heterozygote has a phenotype that is intermediate between those of the two homozygotes. For example, a red-flowered snapdragon plant of genotype r_1r_1 crossed with a white-flowered r_2r_2 plant gives rise to an r_1r_2 plant with pink flowers (figure 10.17). The single copy of allele r_1 in the pink heterozygote directs less pigment production than the two copies in a red-flowered r_1r_1 plant. Although the red color seems to be "diluted" in the all-pink F_1 generation, the Punnett square shows that crossing two of these pink plants can yield F_2 offspring with red, white, or pink flowers.

In **codominance,** two different alleles are fully expressed in the phenotype. For example, the ABO blood typing system is important in determining whose blood a person can receive in a transfusion. A person's ABO blood type is determined by the *I* gene (figure 10.18). Alleles I^A and I^B encode enzymes that transfer either an "A" or a "B" molecule onto the surfaces of red blood cells. Allele i is recessive, so a person with genotype ii has type O blood with neither molecule A nor molecule B. A person whose blood cells express only enzyme A (genotype I^AI^A or I^Ai) has type A blood; likewise, someone with only enzyme B (genotype I^BI^B or I^Bi) has type B blood. But genotype I^AI^B yields type AB blood. The I^A and I^B alleles are codominant because both are equally expressed when both are present. ⓘ *ABO blood groups,* section 27.1B

What is the difference between the recessive i allele and the codominant I^A and I^B alleles? The i allele encodes a nonfunctional protein, whereas alleles I^A and I^B code for functional proteins. People with genotype I^AI^B therefore have molecule A and molecule B on the surfaces of their red blood cells.

The ABO blood type example also illustrates what can happen when one gene has three or more possible alleles: The number of phenotypes increases. In this case, two codominant alleles (I^A and I^B) and one recessive allele (i) produce six genotypes and four phenotypes.

B. Relating Genotype to Phenotype May Be Difficult

Some conditions are especially difficult to trace through families. For example, one gene may influence multiple phenotypes; conversely, mutations in multiple genes may produce the same phenotype. Although the basic rules of inheritance apply to each gene, the patterns of phenotypes that appear in grandparents, parents, and siblings may be hard to interpret.

In **pleiotropy,** a single gene affects multiple, seemingly unrelated phenotypes. Pleiotropy arises when one protein is important in different biochemical pathways or affects more than one body part or process. The conditions they cause can be difficult to trace through families because individuals with different subsets of symptoms may appear to have different illnesses. A collection of disorders called Marfan syndrome offers one example of pleiotropy (figure 10.19). In Marfan syndrome, a single faulty connective tissue protein affects many organ systems, with symptoms including long limbs, spindly fingers, a caved-in chest, a weakened aorta, and lens dislocation. ⓘ *connective tissue,* section 23.2B

Another situation that can complicate the interpretation of inheritance patterns is that mutations in multiple genes sometimes produce the same phenotype. For example, blood clot formation requires 11 chemical reactions. A different gene encodes each enzyme in the pathway, and clotting disorders may result from mutations in any of these genes. The phenotypes are the same, but the genotypes differ.

One gene's product may also influence the expression of another gene. As a familiar example, consider male pattern baldness and the "widow's peak" hairline; the receding hairline associated with male pattern baldness hides the effects of the "widow's peak" allele. Likewise, gene interactions account for some ABO blood type inconsistencies between parents and their children. Rarely, a child whose blood type tests as O has parents whose genotypes indicate that type O offspring are not possible. The explanation traces to a protein that physically links the A and B molecules to the cell surface. If a gene called *H* is mutated, then that protein is nonfunctional, and A and B cannot attach. A person with the extremely rare genotype *hh* therefore has blood that always tests as type O, even though he or she may have a genotype indicating type A, B, or AB blood.

Genotypes	Phenotypes		
	Surface molecules	ABO blood type	
$I^A I^A$ $I^A i$	Only A	Type A	
$I^B I^B$ $I^B i$	Only B	Type B	
$I^A I^B$	Both A and B	Type AB	
ii	None	Type O	

Figure 10.18 Codominance. The I^A and I^B alleles of the *I* gene are codominant, meaning that both are fully expressed in a heterozygote. Allele *i* is recessive.

10.6 Mastering Concepts

1. How do incomplete dominance and codominance increase the number of phenotypes observed in a population?

2. What is pleiotropy?

3. How can the same phenotype stem from many different genotypes?

4. This section explains two ways that a person can have blood that tests as type O. Explain how each relates to specific alleles of genes *I* and *H*.

Miniglossary | Relating Genotype to Phenotype

Complete dominance	One allele masks the expression of another in a heterozygote.
Incomplete dominance	Two alleles confer an intermediate phenotype in a heterozygote.
Codominance	Two alleles are both fully expressed in a heterozygote.
Pleiotropy	One gene affects multiple, seemingly unrelated phenotypes.

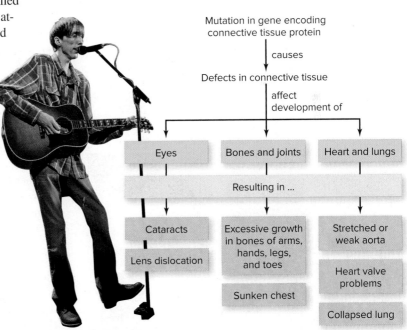

Figure 10.19 Pleiotropy. In Marfan syndrome, a single mutated gene encodes a defective connective tissue protein. The resulting effects on the body are widespread. Singer Bradford Cox was born with Marfan syndrome.

Photo: ©Roger Kisby/Getty Images

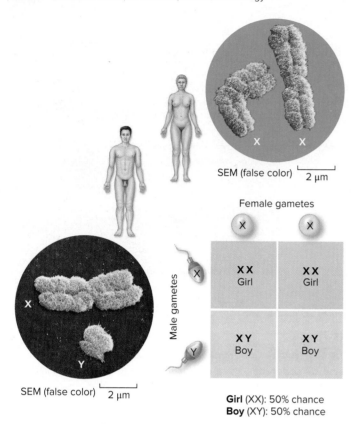

SEM (false color) |—— 2 μm ——|

Female gametes

	X	X
X	**X X** Girl	**X X** Girl
Y	**X Y** Boy	**X Y** Boy

Male gametes

SEM (false color) |—— 2 μm ——|

Girl (XX): 50% chance
Boy (XY): 50% chance

Figure 10.20 Inheritance of Sex. In humans, each egg contains 23 chromosomes, one of which is a single X chromosome. A sperm cell's 23 chromosomes include either an X or a Y chromosome. If a Y-bearing sperm cell fertilizes an egg, the baby will be a male (XY). If an X-bearing sperm cell fertilizes an egg, the baby will be a female (XX).

Photos: (both): ©Andrew Syred/Science

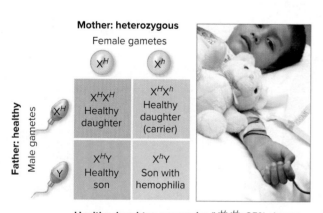

Mother: heterozygous

Female gametes

	X^H	X^h
X^H	$X^H X^H$ Healthy daughter	$X^H X^h$ Healthy daughter (carrier)
Y	$X^H Y$ Healthy son	$X^h Y$ Son with hemophilia

Father: healthy / Male gametes

Healthy daughter, noncarrier ($X^H X^H$): 25% chance
Healthy daughter, carrier ($X^H X^h$): 25% chance
Healthy son ($X^H Y$): 25% chance
Affected son ($X^h Y$): 25% chance

Figure 10.21 Inheritance of Hemophilia. In a cross between a heterozygous woman (a "carrier" of hemophilia A) and a healthy male, the chance of having a son with hemophilia is 25%. The boy in the photo is receiving a blood transfusion.

Photo: ©Mauro Fermariello/Science Source

10.7 Sex-Linked Genes Have Unique Inheritance Patterns

Huntington disease, cystic fibrosis, and other diseases that are caused by genes on autosomes affect both sexes equally. A few conditions, including red–green color blindness and hemophilia, however, occur much more frequently in males than females. Phenotypes that affect one sex more than the other are **sex-linked;** that is, the alleles controlling them are on the X or Y chromosome.

In many species, including humans, the sexes have equal numbers of autosomes but differ in the types of sex chromosomes they have. Females have two X chromosomes, whereas males have one X and one much smaller Y chromosome (figure 10.20). Because the mother can pass on only the X chromosome, the sex chromosome carried by the sperm (X or Y) determines the sex of the baby.

The Y chromosome plays the largest role in human sex determination. All human embryos start with rudimentary female structures (see figure 30.16). An embryo having a working copy of a particular gene on the Y chromosome develops into a male. This sex-determining gene encodes a protein that switches on other genes that direct the undeveloped testes to secrete the male sex hormone testosterone. Cascades of other gene activities promote the development of male sex organs while the embryonic female structures break down.

Despite this critical role, the Y chromosome carries fewer than 100 genes. Scientists therefore know of very few Y-linked disorders; most involve defects in sperm production. The human X chromosome, on the other hand, carries more than 1000 protein-encoding genes, most of which have nothing to do with sex determination. Most human sex-linked traits are therefore **X-linked:** They are controlled by genes on the X chromosome.

A. X-Linked Recessive Disorders Affect More Males than Females

Recessive alleles cause most X-linked disorders in humans, although a few are associated with dominant alleles (table 10.1). A human female inherits an X chromosome from both parents; a male inherits his X chromosome from his mother (see figure 10.20). A female therefore exhibits an X-linked recessive disorder only if she inherits the recessive allele from both parents. A male, in contrast, expresses every allele on his X chromosome, whether dominant or recessive.

Figure 10.21 shows the inheritance of hemophilia A, a disorder with an X-linked recessive mode of inheritance. In hemophilia, a protein called a clotting factor is missing or defective. Blood therefore clots slowly, and bleeding is excessive. The heterozygous female parent in the Punnett square does not exhibit symptoms because her dominant allele encodes a functional blood-clotting protein. When she has children with a normal male, however, each son has a 50% chance of being affected, and each daughter has a 50% chance of being a carrier.

B. X Inactivation Prevents "Double Dosing" of Proteins

Relative to males, female mammals have a "double dose" of every gene on the X chromosome. Cells balance this inequality by **X inactivation,** in which a cell shuts off all but one X chromosome in each cell. This process happens early in the embryonic development of a mammal.

TABLE 10.1 Some X-Linked Disorders in Humans

Disorder	Genetic Explanation	Characteristics
X-linked recessive inheritance		
Duchenne muscular dystrophy	Mutant allele for gene encoding dystrophin	Rapid muscle degeneration early in life
Fragile X syndrome	Unstable region of X chromosome has unusually high number of CCG repeats	Most common form of inherited intellectual disability
Hemophilia A	Mutant allele for gene encoding blood clotting protein (factor VIII)	Uncontrolled bleeding, easy bruising
Red–green color blindness	Mutant alleles for genes encoding receptors for red or green (or both) wavelengths of light	Reduced ability to distinguish between red and green
Rett syndrome	Mutant allele for DNA-binding protein expressed in nerve cells	Multiple severe developmental problems; occurs almost exclusively in females; affected male fetuses rarely survive to birth
X-linked dominant inheritance		
Extra hairiness (congenital generalized hypertrichosis; some forms)	Mechanism unknown	Many more hair follicles than normal
Hypophosphatemic rickets (some forms)	Mutant allele for gene involved in phosphorus absorption	Defective bones caused by low blood phosphorus
Retinitis pigmentosa (some forms)	Mutant allele for cell-signaling protein; mechanism unknown	Partial blindness caused by defects in retina

Which X chromosome becomes inactivated—the one inherited from the father or the one from the mother—is a random event. As a result, a female expresses the paternal X chromosome alleles in some cells and the maternal alleles in others. Moreover, when a cell with an inactivated X chromosome divides mitotically, all of the daughter cells have the same X chromosome inactivated. Because the inactivation occurs early in development, females have patches of tissue that differ in their expression of X-linked alleles. Figure 10.22 shows how inactivation of an X-linked coat color gene causes the distinctive orange and black fur patterns of calico and tortoiseshell cats, which are always female (except for rare XXY males).

X chromosome inactivation also explains another interesting observation: X-linked dominant disorders are typically less severe in females than in males. Thanks to X chromosome inactivation, a female who is heterozygous for an X-linked gene will express a dominant disease-causing allele in only some of her cells. As a result, she experiences less severe symptoms than an affected male, who expresses the dominant allele in every cell.

An X-linked neurological disorder called Rett syndrome, for example, may be mild or severe in a girl, depending on how many of her cells express the Rett allele. Male offspring who inherit the Rett allele typically die before birth. Nearly all people with Rett syndrome are therefore female.

Figure It Out

What is the chance of a daughter expressing a recessive X-linked disease if her mother is a symptomless carrier and her father has the disease?

Answer: 50%.

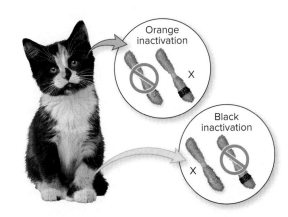

Figure 10.22 X Inactivation. In cats, the X chromosome carries a coat color gene with alleles for black or orange coloration. Calico cats are heterozygous for this gene; one of the two X chromosomes is inactivated in each colored patch. (A different gene accounts for the white background.) Can you explain why calico cats are almost always female?

Photo: ©Siede Preis/Getty Images RF

10.7 Mastering Concepts

1. What determines a person's sex?
2. Why do males and females express recessive X-linked alleles differently?
3. Why does X inactivation occur in female mammals?

Burning Question 10.2

Is male baldness really from the female side of the family?

Male pattern baldness is the distinctive hair loss that many men (and some women) experience as they enter their 20s, 30s, and 40s. It develops when hormones called androgens are present in high concentrations. Testosterone and dihydrotestosterone (DHT) are androgens; they bind to receptors in hair follicle cells, then interact with the DNA to stop growth of the hair follicle. The gene encoding the androgen receptor resides on the X chromosome, which males receive from their mothers.

Why don't women suffer from baldness as often as males? The amount of testosterone is the deciding factor. The higher the concentration of testosterone, the stronger the influence of the "baldness allele." Men typically have more of this sex hormone than women—hence the name, *male* pattern baldness.

The X-linked androgen receptor gene, however, does not tell the full story of male pattern baldness. A region on chromosome 20—an autosome—is also associated with a genetic predisposition for the condition. Either parent can pass this autosomal allele to a child. Multiple genes therefore control baldness (and many other conditions, as described in section 10.9).

Submit your burning question to
marielle.hoefnagels@mheducation.com

(man): ©Image Source Pink/Getty Images RF

10.8 Pedigrees Show Modes of Inheritance

Although Gregor Mendel did not study human genetics, our species nevertheless has some "Mendelian traits": those determined by single genes with alleles that are either dominant or recessive. Several thousand phenotypes fit these criteria, and most of the corresponding genes are on autosomes. Because both sexes have two copies of each autosome, genes on those chromosomes affect both sexes equally.

Genes on autosomes exhibit two modes of inheritance: autosomal dominant and autosomal recessive (table 10.2). An **autosomal dominant** disorder is expressed whether a person inherits one or two copies of the disease-causing allele. It therefore typically appears in every generation. Because the allele is dominant and located on an autosome, one or both of the affected individual's parents must also have the disorder (unless the disease-causing allele arose by a new mutation).

Inheriting an **autosomal recessive** disorder requires that a person receive the disease-causing allele from both parents. Each parent must therefore carry at least one copy of the allele, either because they are homozygous recessive (and therefore have the disease) or because they are heterozygous. A person who is heterozygous is called a *carrier* because he or she is unaffected by the disorder but still has a 50% chance of passing the disease-causing allele to the next generation. If both parents are carriers, autosomal recessive conditions may seem to disappear in one generation, only to reappear in the next.

Pedigree charts depicting family relationships and phenotypes are useful for determining a disorder's mode of inheritance (figure 10.23). In a pedigree chart, squares indicate males, and circles denote females. Colored shapes indicate individuals with the disorder, and half-filled shapes represent known carriers. Horizontal lines connect parents. Siblings connect to their parents by vertical lines and to each other by an elevated horizontal line.

TABLE 10.2   Some Autosomal Dominant and Autosomal Recessive Disorders in Humans

Disorder	Genetic Explanation	Characteristics
Autosomal dominant inheritance		
Achondroplasia	Mutant allele on chromosome 4 causes deficiency of receptor protein for growth factor.	Dwarfism with short limbs, but normal-size head and trunk
Familial hypercholesterolemia	Mutant allele on chromosome 2 encodes faulty cholesterol-binding protein.	High cholesterol, heart disease
Huntington disease	Mutant allele on chromosome 4 encodes protein that misfolds and forms clumps in brain cells.	Progressive uncontrollable movements and personality changes, beginning in middle age
Marfan syndrome	Mutant allele on chromosome 15 causes connective tissue disorder.	Long limbs, sunken chest, lens dislocation, spindly fingers, weakened aorta
Autosomal recessive inheritance		
Albinism	Mutant allele on chromosome 11 encodes faulty protein required for pigment production.	Lack of pigmentation in skin, hair, and eyes
Cystic fibrosis	Mutant allele on chromosome 7 encodes faulty chloride channel protein.	Lung infections and congestion, poor fat digestion, infertility, poor weight gain, salty sweat
Phenylketonuria (PKU)	Mutant allele on chromosome 12 causes enzyme deficiency in pathway that breaks down the amino acid phenylalanine.	Intellectual disability caused by buildup of metabolic byproducts
Tay-Sachs disease	Mutant allele on chromosome 15 causes deficiency of lysosome enzyme.	Nervous system degeneration caused by buildup of byproducts

a. Achondroplasia (autosomal dominant)

©Roger Bacon/Reuters/Alamy Stock Photo

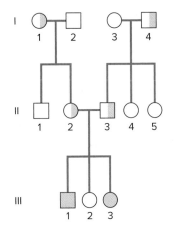

b. Albinism (autosomal recessive)

©Eric Lafforgue/Alamy Stock Photo

c. Red–green color blindness (X-linked recessive)

©BSIP/Science Source

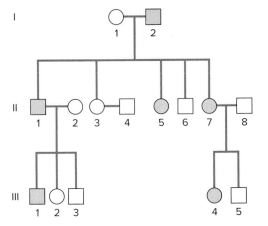

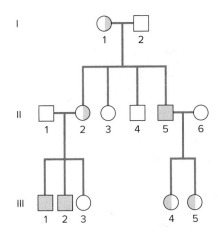

	Normal	Carrier	Affected
Female	○	◐	●
Male	□	◧	■

Figure 10.23 Pedigrees. (a) A pedigree for an autosomal dominant trait; affected individuals are homozygous dominant or heterozygous. (b) A pedigree for an autosomal recessive trait; affected individuals are homozygous recessive. (c) A pedigree for an X-linked recessive trait. Males are affected if they inherit the allele on their single X chromosome; females are typically carriers.

Figure 10.23 shows typical pedigrees for genes carried on autosomes and the X chromosome. In studying the differences among these three pedigrees, notice especially the patterns of the colored shapes. For example, affected individuals appear in every generation for autosomal dominant traits, whereas recessive conditions often skip generations. Carriers appear only in pedigrees depicting the inheritance of recessive traits. Also, note that males are most commonly affected by X-linked recessive disorders. Burning Question 10.2 describes a common example of a sex-influenced condition—male pattern baldness.

10.8 Mastering Concepts

1. How are pedigrees helpful in determining a disorder's mode of inheritance?

2. For each of the pedigrees in figure 10.23, determine the genotype of individual #1 in the first row.

Miniglossary | Modes of Inheritance

Autosomal dominant trait	Condition caused by dominant allele on non-sex chromosome; expressed if an individual inherits one or two copies of the allele
Autosomal recessive trait	Condition caused by recessive allele on non-sex chromosome; expressed only if an individual inherits two copies of the allele
X-linked trait	Condition caused by allele on the X chromosome
Pedigree	Chart used to determine a disorder's mode of inheritance

10.9 Most Traits Are Influenced by the Environment and Multiple Genes

Mendel's data were clear enough for him to infer principles of inheritance because he observed characteristics determined by single genes with two easily distinguished alleles (see figure 10.2). Moreover, the traits he selected are unaffected by environmental conditions. A genetic counselor can likewise be confident in telling two cystic fibrosis carriers that each of their children has a 25% chance of getting the disease. But the counselor cannot calculate the probability that the child will be an alcoholic, have depression, be a genius, or wear size 9 shoes. The reason is that multiple genes and the environment control these and most other traits.

A. The Environment Can Alter the Phenotype

The environment can profoundly affect gene expression; that is, a gene may be active in one circumstance but inactive in another. As a simple example, temperature influences the quantity of pigment molecules in the fur of some animals. Siamese cats and Himalayan rabbits have light-colored bodies but dark ears, noses, paws, and tails, thanks to differences in gene expression between warm and cool body parts (figure 10.24). ⓘ *epigenetics,* section 7.5B

In humans, fetal alcohol syndrome is an example of the effect of environment on phenotype: Prenatal exposure to alcohol can cause a baby to develop facial abnormalities and cognitive impairment. Likewise, personal circumstances ranging from hormone levels to childhood experiences to diet influence a person's susceptibility to depression, alcoholism, and type 2 diabetes.

These three diseases have a genetic component as well, but sorting out the relative contributions of "nature" and "nurture" is difficult. Studies of twins are often helpful, as are careful observations of everything from family composition to brain structures. Burning Question 7.1, for example, explains some strategies that researchers use to explore the genetic connection to homosexuality. Discovering the exact mechanism by which genes interact with external stimuli is an active area of research.

Figure 10.24 Temperature and Fur Color. Siamese cats have a mutation in a gene encoding an enzyme required for pigment production. The enzyme is active only at the relatively cool temperatures of the paws, ears, snout, and tail; these areas are darkly pigmented. At higher temperatures, the enzyme is inactive. Pigment production is therefore reduced where the skin is warmest.
©Photodisc/Getty Images RF

Why We Care 10.1 | The Origin of Obesity

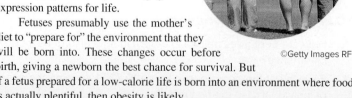

At first glance, the cause of obesity seems simple: If a person eats more calories than he or she expends, the body stores the excess calories as fat. As fat accumulates, body weight climbs. According to this view, a person's genes are irrelevant to his or her body weight. In reality, however, obesity reflects the combined action of genes and the environment.

Several genes are associated with obesity. One example is the gene that encodes leptin, a hormone that helps curb appetite. Individuals who inherit mutant alleles for this gene never feel full, leading to overeating and obesity.

The environment can also influence the expression of the genes that a person inherits. For example, scientists have found that mothers who ingest low amounts of carbohydrates—sugars and starches—give birth to children who are especially likely to become obese later in life. Evidence suggests that epigenetic modifications occur while a developing child is still in the womb, permanently altering gene expression patterns for life.

Fetuses presumably use the mother's diet to "prepare for" the environment that they will be born into. These changes occur before birth, giving a newborn the best chance for survival. But if a fetus prepared for a low-calorie life is born into an environment where food is actually plentiful, then obesity is likely.

©Getty Images RF

Even human diseases with simple, single-gene inheritance patterns can have an environmental component. Cystic fibrosis, for example, is a single-gene disorder. Because cystic fibrosis patients are so susceptible to infection, however, the course of the illness depends on which infectious agents a person encounters.

B. Polygenic Traits Depend on More than One Gene

Unlike cystic fibrosis, most inherited traits are **polygenic,** meaning that the phenotype reflects the activities of more than one gene. Eye color is a familiar example of a polygenic trait. Multiple enzymes, encoded by multiple genes, influence the production and distribution of pigments in the eye's iris. Eye color is among the few traits unaffected by external conditions.

To complicate matters further, the environment also affects the expression of both single-gene and polygenic traits (see Why We Care 10.1). For example, in plants, polygenic traits typically include flower color, the density of leaf pores (stomata), and crop yield. But genes alone do not determine these phenotypes. Soil pH can affect flower color, CO_2 concentration can influence the number of stomata per square centimeter, and nutrient and water availability helps determine crop production. ⓘ *stomata,* section 21.3B

When the frequencies of all the phenotypes associated with a polygenic trait are plotted on a graph, they form a characteristic bell-shaped curve. Human height, for example, ranges from very short to very tall, with most people somewhere in the middle. Height is a product of genetics, childhood nutrition, and health care. Figure 10.25 shows a continuum of gene expression for skin color, a trait that is affected by genes and exposure to sunlight. Body weight and intelligence are other traits that are both polygenic and influenced by the environment.

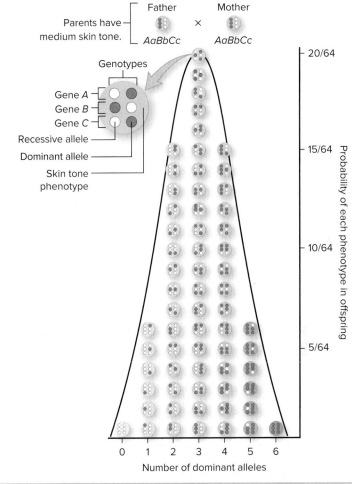

a.

b.

Figure 10.25 **Variation in Human Skin Color.** (a) Multiple genes interact to determine the quantity of pigment in skin cells. This example illustrates three genes: *A*, *B*, and *C*. Two parents who are heterozygous for all three genes may have children with light skin, dark skin, or any shade in between. In this example, the probabilities of the seven potential phenotypes form a bell curve, with medium skin tones being more common than very light or very dark skin tones. (b) The girls in the photo, while not from the same family, represent a sample of the variation in human skin color.

10.9 Mastering Concepts

1. How can the environment affect a phenotype?
2. What is polygenic inheritance, and how is it different from codominance?

Investigating Life 10.1 | Heredity and the Hungry Hordes

Hungry insects and other animals can devastate crops that we grow for food and fiber. The pink bollworm is especially voracious. The adults of this species are moths that lay eggs on cotton bolls. When the eggs hatch, the pink caterpillars tunnel into the boll and eat the seeds, damaging the cotton fibers.

One tool that keeps bollworms and other caterpillars at bay is a chemical called "Bt toxin," named after the bacterium in which the toxin was discovered. In the 1990s, scientists genetically modified cotton plants to produce Bt toxin in all of their cells. The Bt toxin binds to receptor molecules in a bollworm's digestive tract, leaving the insect unable to digest food. Nibbling on any plant part therefore spells death to the caterpillar. (i) *transgenic plants,* section 11.2A

Bt plants, however, pose a problem: They kill most of the susceptible caterpillars, leaving the resistant individuals to produce the next generation. To combat this selective pressure, farmers growing Bt crops surround each field with a buffer strip planted with a conventional (non-Bt) variety of the same crop.

The buffer strategy slows the evolution of Bt resistance because the Bt-resistance alleles are recessive. Suppose that a homozygous recessive, resistant moth (*rr*) emerging from the Bt crop encounters a susceptible, heterozygous mate from the buffer strip (*RR* or *Rr*). Their heterozygous offspring are susceptible; they will die if they eat the Bt plants (figure 10.A). A ready pool of susceptible mates from the buffer strip should therefore keep the recessive allele rare.

But some countries might not require buffer strips, and a resistant variety of bollworms could quickly spread worldwide. Anticipating this problem, biotech companies have developed plants with a second Bt toxin that binds to a different receptor in the insect's gut. This strategy should greatly delay the emergence of resistant bollworms because random mutations would have to occur in two genes—encoding the two receptor proteins—to confer resistance to both toxins. So far, researchers have concluded that the refuge strategy and new transgenic cotton varieties have put the brakes on the evolution of Bt resistance.

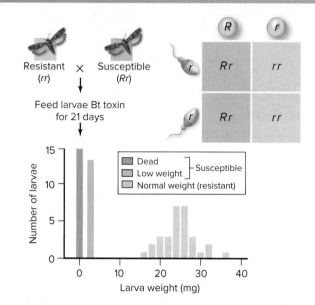

Figure 10.A **Recessive Resistance.** When a heterozygous, susceptible insect is bred with a resistant mate, half the offspring should be resistant and half should be susceptible. Tests with Bt toxin show that this is indeed the case: Half of the larvae thrived in the presence of Bt toxin, while the susceptible ones died or were very small.

Sources: Morin, Shai, Robert W. Biggs, Mark S. Sisterson, and 10 other authors (including Bruce E. Tabashnik). April 29, 2003. Three cadherin alleles associated with resistance to *Bacillus thuringiensis* in pink bollworm. *Proceedings of the National Academy of Sciences*, vol. 100, pages 5004–5009.

Tabashnik, Bruce E., and five other authors. July 21, 2009. Asymmetrical cross-resistance between *Bacillus thuringiensis* toxins Cry1Ac and Cry2Ab in pink bollworm. *Proceedings of the National Academy of Sciences*, vol. 106, pages 11889–11894.

What's the **Point?** ▼ APPLIED

Armed with the basic rules of inheritance, anyone can make predictions about the genotypes and phenotypes of offspring. Understanding inheritance is especially important to people who have decided to start a family.

A pregnant woman may delight in seeing her unborn child, thanks to ultrasound technology. But the same scans that reveal the size, position, and sex of the fetus can also show possible abnormalities. For many conditions, a physician cannot be sure of the diagnosis without ordering a test of the fetus's chromosomes.

How is it possible to see chromosomes hidden inside the cells of a fetus, which is itself tucked into the mother's uterus? A technician begins by obtaining a small amount of the fluid surrounding the developing fetus. Fetal cells in the fluid can then be used to prepare a photograph of the

©JGI/Jamie Grill/Blend Images LLC RF

fetus's chromosomes. This image may reveal extra chromosomes, missing chromosomes, or the movement of genetic material from one chromosome to another.

The technician can also extract DNA from the fetal cells to get a closer look at the genes on the chromosomes. Suppose, for example, that the child's father has hemophilia, and the mother has a family history of the same disease. Since hemophilia is an X-linked recessive disorder, they calculate that the child's chance of inheriting the disease may be as high as 50%. The mother therefore requests a test of the fetus's DNA for alleles associated with hemophilia. The results of the test may set her mind at ease. On the other hand, if she learns that the child has inherited the disease, she may consult a counselor who can advise her on the challenges the family will face in the years to come.

CHAPTER SUMMARY

10.1 Chromosomes Are Packets of Genetic Information: A Review

- **Genes** encode proteins; mutations create new **alleles** (figure 10.26a).
- A **chromosome** is a continuous molecule of DNA with associated proteins. In humans, a **diploid cell** contains 22 **homologous pairs** of **autosomes** and one pair of **sex chromosomes.**
- **Meiosis** gives rise to **haploid cells. Fertilization** unites haploid **gametes** and restores the diploid number.

10.2 Mendel's Experiments Uncovered Basic Laws of Inheritance

- Gregor Mendel studied inheritance in pea plants because they are easy to breed, develop quickly, and produce abundant offspring.

A. Dominant Alleles Appear to Mask Recessive Alleles

- An individual is **true-breeding** for a trait if self-fertilization yields only offspring that resemble the parent. A **hybrid** individual produces a mix of offspring when self-fertilized.
- A **dominant** allele is always expressed if it is present; a **recessive** allele is masked by a dominant allele. "Dominant" does not necessarily mean "most common."

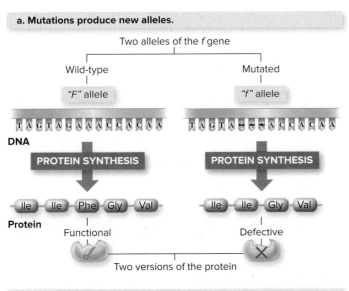

a. Mutations produce new alleles.

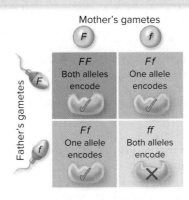

b. A Punnett square tracks the inheritance of these alleles among all possible offspring.

Figure 10.26 From Mutation to Punnett Square.

B. For Each Gene, a Cell's Two Alleles May Be Identical or Different

- An individual's **genotype** may be **heterozygous** (two different alleles for a gene) or **homozygous** (both alleles are the same).
- A **phenotype** is any observable characteristic of an organism.
- A **wild-type** allele is the most common in a population. A change in a gene is a mutation and may result in a **mutant** phenotype.

10.3 The Two Alleles of a Gene End Up in Different Gametes

- In genetic crosses, the purebred parental generation is designated **P**; the next generation is F_1; and the next is F_2.

A. The Simplest Punnett Squares Track the Inheritance of One Gene

- A **monohybrid cross** is a mating between two individuals that are heterozygous for the same gene.
- **Punnett squares** are useful for predicting the allele combinations that the offspring of two parents might inherit (figure 10.26b).
- A **test cross** is one way to reveal the unknown genotype of an individual with a dominant phenotype; the individual is crossed with a homozygous recessive mate.

B. Meiosis Explains Mendel's Law of Segregation

- Mendel's **law of segregation** states that the two alleles of the same gene separate into different gametes. Each individual therefore receives one allele of each gene from each parent.

10.4 Genes on Different Chromosomes Are Inherited Independently

A. Tracking Two-Gene Inheritance May Require Large Punnett Squares

- A **dihybrid cross** is a mating between individuals that are heterozygous for two genes.

B. Meiosis Explains Mendel's Law of Independent Assortment

- According to Mendel's **law of independent assortment,** the inheritance of one gene does not affect the inheritance of another gene on a different chromosome. Independent assortment occurs because homologous chromosomes are oriented randomly during metaphase I of meiosis.

C. The Product Rule Is a Useful Shortcut

- The **product rule** is an alternative to Punnett squares for following the inheritance of two or more traits at a time.

10.5 Genes on the Same Chromosome May Be Inherited Together

A. Genes on the Same Chromosome Are Linked

- **Linked genes** are located on the same chromosome. **Linkage groups** are collections of linked genes that are often inherited together.
- The farther apart two genes are on a chromosome, the more likely **crossing over** is to separate their alleles. If crossing over occurs between two alleles, some offspring will inherit **recombinant chromatids;** otherwise, offspring will inherit **parental chromatids.**

B. Studies of Linked Genes Have Yielded Chromosome Maps

- Breeding studies reveal the crossover frequencies used to create **linkage maps**—diagrams that show the order of genes on a chromosome.

10.6 Inheritance Patterns Are Rarely Simple

A. Incomplete Dominance and Codominance Add Phenotype Classes

- Heterozygotes for alleles with **incomplete dominance** have phenotypes intermediate between those of the two homozygotes. **Codominant** alleles are both expressed in a heterozygote. Figure 10.27 uses a lightbulb analogy to compare dominance relationships.

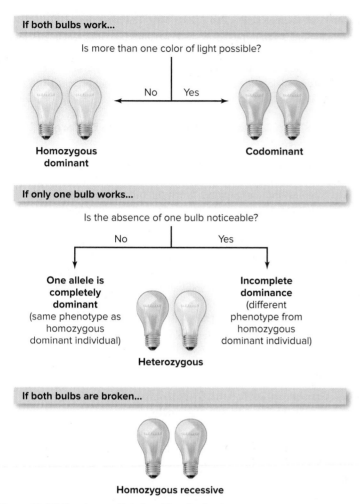

If both bulbs work...

Is more than one color of light possible?

No | Yes

Homozygous dominant

Codominant

If only one bulb works...

Is the absence of one bulb noticeable?

No | Yes

One allele is completely dominant (same phenotype as homozygous dominant individual)

Incomplete dominance (different phenotype from homozygous dominant individual)

Heterozygous

If both bulbs are broken...

Homozygous recessive

Figure 10.27 Dominance Relationships: An Analogy.

B. Relating Genotype to Phenotype May Be Difficult
- A **pleiotropic** gene affects multiple phenotypes.
- When multiple proteins participate in a biochemical pathway, mutations in genes encoding any of the proteins can produce the same phenotype.
- The activity of one gene may mask the effect of another.

10.7 Sex-Linked Genes Have Unique Inheritance Patterns
- Genes controlling **sex-linked** traits are located on the X or Y chromosome.
- Scientists know of many more **X-linked** disorders (encoded by genes on the X chromosome) than Y-linked disorders.

A. X-Linked Recessive Disorders Affect More Males than Females
- A male expresses every allele on his one X chromosome. A female has two X chromosomes, so she may inherit a dominant allele that masks the expression of a recessive one.

B. X Inactivation Prevents "Double Dosing" of Proteins
- **X inactivation** randomly shuts off all but one X chromosome in each cell.

10.8 Pedigrees Show Modes of Inheritance
- An **autosomal dominant** disorder can be inherited from one affected parent. An **autosomal recessive** disorder must be inherited from both parents. X-linked recessive disorders affect mostly males.
- **Pedigrees** trace phenotypes in families and reveal the mode of inheritance.

10.9 Most Traits Are Influenced by the Environment and Multiple Genes

A. The Environment Can Alter the Phenotype
- Most traits have environmental as well as genetic influences.

B. Polygenic Traits Depend on More than One Gene
- A **polygenic trait** varies continuously in its expression.

MULTIPLE CHOICE QUESTIONS

1. In the list of four terms below, which term is the *second* most inclusive?
   a. Genome   b. Allele   c. Chromosome   d. Gene

2. If a plant that is homozygous dominant for a gene is crossed with a plant that is homozygous recessive for the same gene, what is the probability that the offspring will have the same genotype as one of the parents?
   a. 100%   b. 50%   c. 25%   d. 0%

3. Which of the following is a possible gamete for an individual with the genotype *PP rr?*
   a. *P P*   b. *P r*   c. *p r*   d. *r r*

4. Use the product rule to determine the chance of obtaining an offspring with the genotype *Rr Yy* from a dihybrid cross between parents with the genotype *Rr Yy.*
   a. ½   b. ¼   c. ⅛   d. ¹⁄₁₆

5. Refer to the linkage map in figure 10.16b. A crossover event is most likely to occur between which pair of genes?
   a. *w* and *v*   c. *y* and *w*
   b. *y* and *r*   d. *v* and *m*

6. Suppose a woman is a symptomless carrier of a recessive X-linked disease. She is expecting a daughter. If her husband has the disease, what is the chance that the girl will have the disease?
   a. 100%   b. 50%   c. 25%   d. 0%

Answers to Multiple Choice questions are in appendix A.

WRITE IT OUT

1. Select one gene mentioned in the chapter, then explain the link between an organism's genotype (for that gene) and its corresponding phenotype. Make sure to use the term *protein* in your answer.

2. List three genes (mentioned in this chapter or not) that do not affect physical appearance. Do these genes contribute to an organism's phenotype?

3. Create a flow chart with connections between the terms *P generation, F_1 generation,* and *F_2 generation.*

4. A woman with fair skin, blond hair, and blue eyes gives birth to fraternal twins; the father has dark brown skin, dark hair, and brown eyes. One twin has blond hair, brown eyes, and light skin, and the other has dark hair, brown eyes, and dark skin. What Mendelian law does this real-life case illustrate?

5. If two different but linked genes are located very far apart on a chromosome, how may the inheritance pattern create the appearance of independent assortment?

6. Explain how each of the following appears to disrupt Mendelian ratios: incomplete dominance, codominance, pleiotropy. In each case, what is happening at the level of the protein?

7. In the ABO blood type system, the enzymes that link molecules A and B to red blood cells each consist of 354 amino acids; their sequences differ by four amino acids. The allele encoding the *i* allele has a frameshift mutation; it therefore encodes a nonfunctional protein. Combine the concepts of chapters 7 and 10 to diagram how blood cells acquire their ABO phenotype, starting with transcription of the *I^A*, *I^B*, or *i* allele and ending with the surface of the red blood cells.

8. Calico cats have large patches of orange and black fur; in tortoiseshell cats, the patches are smaller. In which type of cat does X inactivation occur earlier in development? How do you know?

9. Explain the following "equation":

Genotype + Environment = Phenotype

SCIENTIFIC LITERACY

Review Burning Question 10.1, which describes the inheritance pattern of the metabolic disease called PKU. Today, genetic testing for many disorders is relatively easy and inexpensive. Do prospective parents have an obligation to determine how likely they are to conceive a child with a genetic disorder? What are some possible drawbacks of learning more about one's own genetics? What are some possible advantages to oneself and to society?

GENETICS PROBLEMS

See the *How to Solve a Genetics Problem* section at the end of this chapter for step-by-step guidance.

1. In rose bushes, red flowers (*FF* or *Ff*) are dominant to white flowers (*ff*). A true-breeding red rose is crossed with a white rose; two flowers of the F_1 generation are subsequently crossed. What will be the most common genotype of the F_2 generation?

2. In Mexican hairless dogs, a dominant allele confers hairlessness. However, inheriting two dominant alleles is lethal; the fetus dies before birth. Suppose two dogs that are heterozygous for the hair allele mate. Predict the genotypic and phenotypic ratios of the puppies that are born.

3. A species of ornamental fish comes in two colors; red is dominant and gray is recessive. Emily mates her red fish with a gray fish. If 50 of the 100 babies are red, what is the genotype of Emily's fish?

4. Two lizards have green skin and large dewlaps (genotype *Gg Dd*). If they mate and 32 offspring are born, how many of the offspring are expected to be homozygous recessive for both genes? (Assume that the traits assort independently.)

5. A fern with genotype *AA Bb Cc dd Ee* mates with another fern with genotype *aa Bb CC Dd ee*. What proportion of the offspring will be heterozygous for all genes? (Assume the genes assort independently.) *Hint:* Use the product rule.

6. In Fraggles, males are genotype XY and females are XX. Silly, a male Fraggle, has a rare X-linked recessive disorder that makes him walk backwards. He mates with Lilly, who is a carrier for the disorder. What proportion of their male offspring will walk backwards?

Answers to Mastering Concepts, Write It Out, Scientific Literacy, and Pull It Together questions can be found in the Connect ebook. **connect.mheducation.com**

Design element: Burning Question (fire background): ©Ingram Publishing/Super Stock

PULL IT TOGETHER

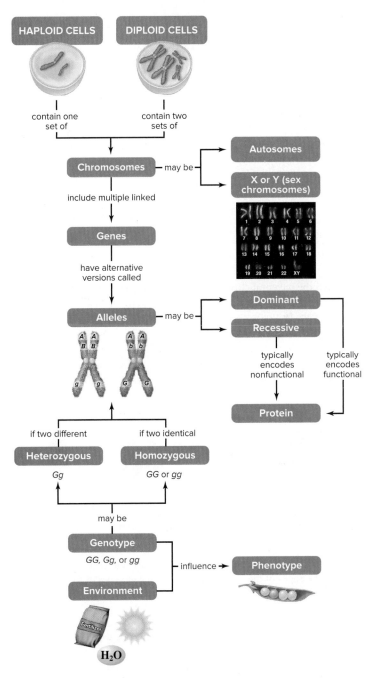

Figure 10.28 Pull It Together: Patterns of Inheritance.

Refer to figure 10.28 and the chapter content to answer the following questions.

1. Compare the Survey the Landscape figure in the chapter introduction with the Pull It Together concept map. Explain the connections between *alleles, genetic variation, gametes, Punnett squares,* and *inheritance*.

2. Analyze the concept map and then explain the effects of a mutated allele. Use *allele, dominant, recessive, genotype,* and *phenotype* in your answer.

3. Add *meiosis, gametes, mutations, incomplete dominance, codominance,* and *pleiotropy* to this concept map.

HOW TO SOLVE A GENETICS PROBLEM

One Gene

Sample problem: Phenylketonuria (PKU) is an autosomal recessive disorder. If a man with PKU marries a woman who is a symptomless carrier, what is the probability that their first child will be born with PKU?

1. **Write a key.** Pick ONE letter to represent the gene in your problem. Use the capital form of your letter to symbolize the dominant allele; use the lowercase letter to symbolize the recessive allele.

   Sample: The dominant allele is *K;* the recessive allele is *k.*

2. **Summarize the problem's information.** Make a table listing the phenotypes and genotypes of both parents.

   Sample:

	Male	**Female**
Phenotype	Has PKU	No PKU (carrier)
Genotype	*kk*	*Kk*

3. **Sketch the parental chromosomes and gametes.** Use the genotypes in your table to draw the alleles onto chromosomes. Then draw short arrows to show the homologous chromosomes moving into separate gametes for each parent.

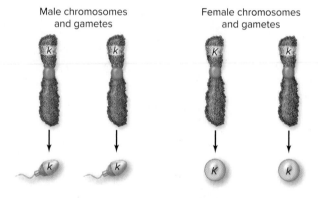

Male chromosomes and gametes   Female chromosomes and gametes

4. **Make a Punnett square.** Arrange the gametes you sketched in step 3 along the edges of the square, and fill in the genotypes of the offspring.

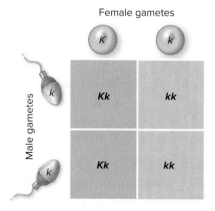

Female gametes

	K	*k*
k	**Kk**	**kk**
k	**Kk**	**kk**

Male gametes

5. **Calculate the genotypic ratio.** Count the number of squares that contain each offspring genotype.

   Sample: 2 *Kk* : 2 *kk*

6. **Calculate the phenotypic ratio.** Count the number of squares that contain each offspring phenotype.

   Sample: 2 PKU carriers : 2 PKU sufferers

7. **Calculate the probability of each phenotype.** Divide each number in step 6 by the total number of squares (4) and multiply by 100.

   Sample: 50% probability that a child will be a carrier; 50% probability that a child will have PKU

Two Genes (Punnett Square)

Sample problem: A student collects pollen (male sex cells) from a pea plant that is homozygous recessive for the genes controlling seed form and seed color. She uses the pollen to fertilize a plant that is heterozygous for both genes. What is the probability that an offspring plant has the same genotype and phenotype as the male parent? Assume the genes are not linked.

1. **Write a key.** Pick ONE letter to represent each of the genes in your problem. Use the capital form of your letter to symbolize the dominant allele; use the lowercase letter to symbolize the recessive allele.

   Sample: For seed form, the dominant allele (round) is *R;* the recessive allele (wrinkled) is *r;* for seed color, the dominant allele (yellow) is *Y;* the recessive allele (green) is *y.*

2. **Summarize the problem's information.** Make a table listing the phenotypes and genotypes of both parents.

   Sample:

	Male	**Female**
Phenotype	Wrinkled, green	Round, yellow
Genotype	*rr yy*	*Rr Yy*

3. **Sketch the parental chromosomes and gametes.** Use the genotypes in your table to draw the alleles onto two sets of chromosomes, one for each parent. The law of independent assortment means that you need to draw all possible configurations. So redraw the chromosomes, this time switching the order of the alleles in one pair. Then draw short arrows to show the chromosomes separating, and sketch the four possible gametes for each parent. (For homozygous parents, multiple gametes will have the same genotype.)

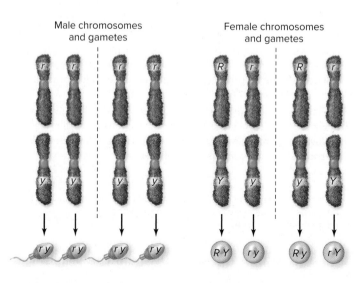

Male chromosomes and gametes   Female chromosomes and gametes

4. **Make a Punnett square.** Arrange the gametes you sketched in step 3 along the edges of the square, and fill in the genotypes of the offspring.

Female gametes

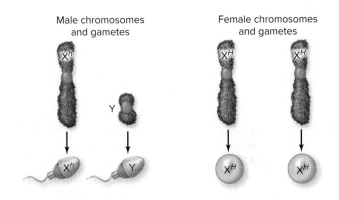

5. **Calculate the genotypic ratio.** Count the number of squares that contain each offspring genotype.

   Sample: 4 *Rr Yy* : 4 *rr yy*: 4 *Rr yy* : 4 *rr Yy*

6. **Calculate the phenotypic ratio.** Count the number of squares that correspond to each possible phenotype combination.

   Sample: 4 round, yellow : 4 wrinkled, green : 4 round, green : 4 wrinkled, yellow

7. **Calculate the probability of each phenotype.** Divide each number in step 6 by the total number of squares (16) and multiply by 100.

   Sample: 25% probability that an offspring has the same genotype and phenotype as the male parent.

Two Genes (Product Rule)

The product rule is a simpler way to solve a multi-gene problem and eliminates the need for a large Punnett square. To use the product rule in the previous sample problem, first calculate the probability that the parents (*rr* × *Rr*) produce an offspring with genotype *rr* (½, or 50%). Then calculate the chance that *yy* × *Yy* parents produce a *yy* offspring (½, or 50%). Multiply the two probabilities to calculate the probability that both events occur simultaneously: ½ × ½ = ¼, or 25%. See section 10.4 for more on the product rule.

X-Linked Gene

Sample problem: Hemophilia is caused by an X-linked recessive allele. If a man who has hemophilia marries a healthy woman who is not a carrier, what is the chance that their child will have hemophilia?

1. **Write a key.** Pick ONE letter to represent the gene in your problem. Use the capital form of your letter to symbolize the dominant allele; use the lowercase letter to symbolize the recessive allele.

   Sample: The dominant allele is *H;* the recessive allele is *h*. Because these alleles are on the X chromosome, inheritance will differ between males and females. It is therefore best to designate the chromosomes and alleles together as X^H and X^h.

2. **Summarize the problem's information.** Make a table listing the phenotypes and genotypes of both parents.

   Sample:

	Male	**Female**
Phenotype	Has hemophilia	Healthy
Genotype	$X^h Y$	$X^H X^H$

3. **Sketch the parental chromosomes and gametes.** Use the genotypes in your table to draw the alleles onto chromosomes. Then draw short arrows to show the chromosomes moving into separate gametes for each parent.

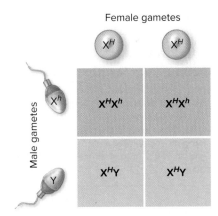

4. **Make a Punnett square.** Arrange the gametes you sketched in step 3 along the edges of the square; fill in the genotypes of the offspring.

5. **Calculate the genotypic ratio.** Count the number of squares that contain each offspring genotype.

   Sample: 2 $X^H X^h$: 2 $X^H Y$

6. **Calculate the phenotypic ratio.** Count the number of squares that correspond to each possible phenotype.

   Sample: 2 female carriers : 2 healthy males

7. **Calculate the probability of each phenotype.** Divide each number in step 6 by the total number of squares (4) and multiply by 100.

   Sample: 50% probability that a child will be a female carrier; 50% probability that a child will be a healthy male. No child, male or female, will have hemophilia.

LEARNING OUTLINE

11.1 DNA Technology Is Changing the World

11.2 DNA Technology's Tools Apply to Individual Genes or Entire Genomes

11.3 Stem Cells and Cloning Add New Ways to Copy Cells and Organisms

11.4 Many Medical Tests and Procedures Use DNA Technology

APPLICATIONS

Burning Question 11.1 *Is selective breeding the same as genetic engineering?*

Burning Question 11.2 *What are the uses of DNA testing?*

Why We Care 11.1 *Gene Doping*

Investigating Life 11.1 *Weeds Get a Boost from Their Transgenic Cousins*

Wrongfully Convicted. Johnny Pinchback (*center*) is freed after spending 27 years in prison for a sexual assault that he did not commit. DNA testing finally exonerated him.

©Tony Gutierrez/AP Images

Learn How to Learn

Write Your Own Test Questions

Have you ever tried putting yourself in your instructor's place by writing your own multiple-choice test questions? It's a great way to pull the pieces of a chapter together. The easiest questions to write are based on definitions and vocabulary, but those will not always be the most useful. Try to think of questions that integrate multiple ideas or that apply the concepts in a chapter. Write 10 questions, and then let a classmate answer them. You'll probably both learn something new.

SURVEY THE LANDSCAPE
DNA, Inheritance, and Biotechnology

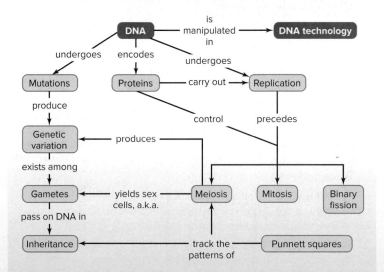

Biologists can manipulate DNA in many ways, from determining its nucleotide sequence to mixing genes from multiple species. As a result, we can now accomplish feats that were unimaginable a generation ago.

For more details, study the Pull It Together feature in the chapter summary.

Thanks to more than a century of scientific research, we now understand DNA's structure and function, and we know how DNA passes from generation to generation. We can also manipulate DNA in many ways: We can copy it, figure out its sequence of nucleotides, switch it on and off, and search for specific pieces of it inside a living cell. We can also cut and paste it, transferring genes from one species to another.

©Edward Kinsman/ Science Source

The application of these DNA technology tools can produce outcomes that some might consider frivolous, such as a glow-in-the-dark pet fish. This animal contains a gene, originally isolated from a sea anemone, that encodes a fluorescent protein. But other applications of DNA technology are anything but silly. As you will learn in this chapter, the ability to manipulate DNA can help us solve crimes, save lives, and learn more about our place in the tree of life, among many other worthy goals.

11.1 DNA Technology Is Changing the World

The title of this section is not an exaggeration: DNA-based technologies have affected nearly every imaginable facet of society. **DNA technology** is a broad term that usually means the manipulation of DNA for some practical purpose. This chapter describes some of the ways in which DNA technology has become a powerful tool in research, medicine, agriculture, criminal justice, and many other fields.

DNA technology became possible only after biologists learned the structure and function of DNA. Recall from chapter 7 that the DNA double helix is composed of nucleotides and that the function of DNA is to provide the "recipes" for the cell's proteins (figure 11.1). As described in chapter 8, enzymes copy these recipes as a cell prepares to divide, so that nearly every cell in a multicellular organism carries the same DNA sequence.

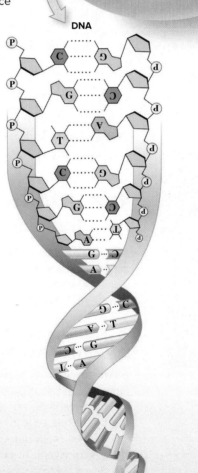

Figure 11.1 DNA Structure and Function. The DNA double helix specifies which proteins a cell can make. Thanks to the tools of DNA technology, biologists can add new DNA to a cell, determine DNA's sequence, copy it, or use it to identify an individual.

Burning Question 11.1

Is selective breeding the same as genetic engineering?

Simply put, the answer to this question is no. Selective breeding, also called artificial selection, yields new varieties of plants and animals by selecting for or against traits that already occur in a population. For example, suppose that researchers want to develop a new variety of carrots lacking orange pigments. They would allow those rare plants with pale carrots to breed only among themselves. Over many generations, the result would be a line of white carrots. If, instead, only plants with darkly pigmented roots breed, the offspring might include purple or red carrots. Breeders used these selective breeding strategies to develop a rainbow of carrot colors (figure 11.A). ⓘ *artificial selection,* section 12.2C

Introducing new DNA—genetic engineering—is a totally different way to develop new plant and animal varieties. We have already seen, for instance, that Bt corn plants contain genes that were originally isolated from bacteria (see Investigating Life 10.1). Likewise, some transgenic bacteria produce insulin and other human proteins, thanks to our ability to transfer DNA from one species to another.

A third technique for developing new varieties of plants and animals is random mutagenesis. Researchers use chemicals or radiation to induce genetic mutations in an organism's DNA. These mutations may cause damage, but they sometimes also bring about interesting new characteristics such as sweeter fruits or higher yields (see figure 7.15). This technique falls somewhere in the middle of the spectrum between selective breeding and transgenic technology. It does not rely on preexisting mutations, as does selective breeding, but it is much less controlled than transgenic technology. ⓘ *mutations,* section 7.6

Figure 11.A Selectively bred carrots.
Source: USDA/Stephen Ausmus

Submit your burning question to marielle.hoefnagels@mheducation.com

We have also learned that each person inherits a unique DNA sequence from his or her parents. As a result, with the exception of identical twins, each person is genetically different. Yet chapter 10 showed how we can trace the inheritance of particular alleles from child to parent to grandparent, and so on. Using the same logic, it is easy to see the power of DNA as a tool for tracing evolutionary history. Throughout the billions of years of life's history, descent with modification has produced countless unique species. Analyzing the differences in their DNA can reveal their relationships with one another.

DNA technology applies these facts (and many more) to open entirely new ways to learn about life's history, to prevent and relieve human suffering, to protect the environment, and to enforce the law. As you will see, many of the more familiar applications of DNA technology are in medicine. For example, technicians can test a person's DNA for many alleles associated with inherited illnesses, marking a huge advance in disease screening and diagnosis. Stem cells often make headlines as well, especially because the ability to manipulate gene expression in these cells may offer treatments for diseases that currently have no cure.

Many people also know that genetically modified organisms (often abbreviated GMOs) have made their way into the human food supply, mostly in the form of herbicide- and insect-resistant crop plants. Yet another familiar use of DNA technology is DNA profiling, which can help solve crimes and match parents with their offspring.

As helpful as DNA technology can be, the ability to manipulate DNA also carries both risks and ethical questions. This chapter describes not only some of the tools and applications but also some of the downsides of DNA technology.

11.1 Mastering Concepts

1. What is DNA technology?
2. In what fields is DNA technology useful?

11.2 DNA Technology's Tools Apply to Individual Genes or Entire Genomes

Some applications of DNA technology require moving a gene from one cell to another; others require comparisons among multiple genomes. This section explores a few of the tools that biologists use to manipulate everything from short stretches of DNA to the entire genetic makeup of a cell.

A. Transgenic Organisms Contain DNA from Multiple Sources

As we saw in chapter 7, virtually all species use the same genetic code. It therefore makes sense that one type of organism can express a gene from another species, even if the two are distantly related. Biologists take advantage of this fact by coaxing cells to take up **recombinant DNA,** which is genetic material that has been spliced together from multiple sources (typically, different species). A **transgenic organism** is an individual that receives recombinant DNA.

Scientists first accomplished this feat of "genetic engineering" in *E. coli* bacteria in the 1970s, but many microbes, plants, and animals have since been genetically modified. When cells containing the recombinant DNA divide, all of their daughter cells also harbor the new genes. These transgenic organisms express their new genes just as they do their own, producing the desired protein along with all of the others that they normally make. (Note that new varieties of animals and plants may also come from selective breeding. These organisms, however, are not transgenic, as described in Burning Question 11.1.)

Transgenic Bacteria and Yeasts How do scientists produce a transgenic organism? The first step is to obtain DNA from a source cell—usually a bacterium, a plant, or an animal (figure 11.2, step 1). The next step is to insert this source DNA into a **plasmid,** a small circle of double-stranded DNA that is separate from the cell's chromosome (step 2).

The next step is to construct a recombinant plasmid (step 3). To generate DNA fragments that can be spliced together, researchers use **restriction enzymes,** which are proteins that cut double-stranded DNA at a specific base sequence. Some restriction enzymes generate single-stranded ends that "stick" to each other by complementary base pairing. When plasmid and donor DNA is cut with the same restriction enzyme and the fragments are mixed, the single-stranded sticky ends of some plasmids form base pairs with those of the donor DNA.

Next, the researchers move the recombinant plasmid into a recipient cell (step 4). Zapping a bacterial cell with electricity opens temporary holes that admit naked DNA. Alternatively, "gene guns" shoot DNA-coated pellets directly into cells. DNA can also be packaged inside a fatty membrane "bubble" that fuses with the recipient cell's membrane, or it can be hitched to a virus that subsequently infects the recipient cell. ⓘ *viruses,* section 7.7

In the pharmaceutical industry, transgenic bacteria produce dozens of drugs, including human insulin to treat diabetes (see figure 15.17), blood-clotting factors to treat hemophilia, immune system biochemicals, and fertility hormones. Other genetically modified bacteria produce the amino acid phenylalanine, which is part of the artificial sweetener aspartame (see Why We Care 2.2). Still others degrade petroleum, pesticides, and other soil pollutants.

Single-celled fungi (yeasts) can also be genetically modified. For example, transgenic yeast cells produce a milk-curdling enzyme called chymosin used by many U.S. cheese producers. The baking, brewing, and wine industries, which rely on yeasts for fermentation, may increasingly use transgenic yeast cells in the future. ⓘ *fermentation,* section 6.8

Transgenic Plants One tool for introducing new genes into plant cells is a bacterium called *Agrobacterium tumefaciens.* In nature, these bacteria enter the plant at a wound and inject a plasmid into the host's cells. The plasmid normally encodes proteins that stimulate the infected plant cells to divide rapidly, producing a tumorlike gall where the bacteria live. (The name of the plasmid, Ti, stands for "tumor inducing.")

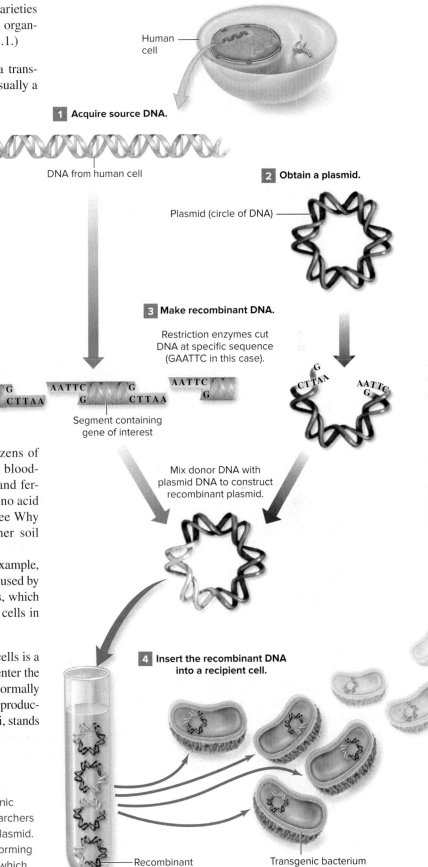

1 Acquire source DNA.

DNA from human cell

2 Obtain a plasmid.

Plasmid (circle of DNA)

3 Make recombinant DNA.

Restriction enzymes cut DNA at specific sequence (GAATTC in this case).

Segment containing gene of interest

Mix donor DNA with plasmid DNA to construct recombinant plasmid.

4 Insert the recombinant DNA into a recipient cell.

Recombinant plasmid

Transgenic bacterium containing human DNA

Figure 11.2 Transgenic Bacteria. The first steps in building a transgenic bacterium are to (*1*) isolate source DNA and (*2*) obtain a plasmid. (*3*) Researchers use the same restriction enzyme to cut DNA from the donor cell and the plasmid. When the pieces are mixed, the "sticky ends" of the DNA fragments join, forming recombinant plasmids. (*4*) The plasmids are delivered into bacterial cells, which pass on the recombinant DNA to their descendants as they reproduce.

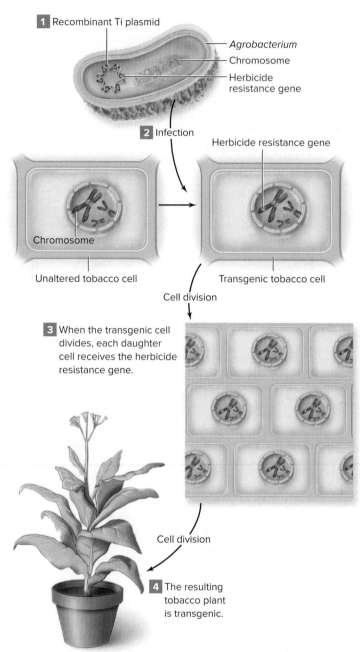

Figure 11.3 **Transgenic Plant.** (*1*) This genetically modified *Agrobacterium* cell contains a recombinant Ti plasmid encoding a gene that confers herbicide resistance. (*2*) The bacterium infects a tobacco plant cell, inserting the Ti plasmid into the plant cell's DNA. (*3*) The transgenic plant cells can be grown into (*4*) tobacco plants that express the herbicide resistance gene in every cell.

Scientists can replace some of the Ti plasmid's own genes with other DNA, such as a gene encoding a protein that confers herbicide resistance (figure 11.3, step 1). They allow the transgenic *Agrobacterium* to inject these recombinant plasmids into plant cells (step 2). All plants that grow from the infected cells should express the new herbicide-resistance gene (steps 3 and 4). The farmer who plants the crop can therefore spray the field with herbicides, killing weeds without harming the genetically modified plants.

Biologists have used a similar technique to produce corn and cotton varieties that produce their own insecticides (see Investigating Life 10.1). The insect-killing protein originated in a bacterium called *Bacillus thuringiensis,* abbreviated Bt. Any insect that nibbles on a plant expressing the Bt protein dies. These genetically modified Bt crops save farmers time and money because they greatly reduce the need for sprayed insecticides.

Besides tolerating herbicides or producing insecticides, transgenic crop plants may also resist viral infections, survive harsh environmental conditions, or contain nutrients that they otherwise wouldn't. Transgenic potato plants, for example, may someday be used to produce vaccines, and tobacco plants enhanced with bacterial enzymes may help degrade leftover explosives at contaminated military installations.

Transgenic Animals So far, we have described the use of DNA technology to genetically modify bacteria, yeasts, and plants. Biologists use a different technique to produce transgenic mice and other animals. Typically, they pack recombinant DNA into viruses that can infect a gamete or fertilized egg. As the transgenic animal develops, it carries the foreign genes in every cell.

Transgenic animals have many applications. A transgenic mouse "model" for a human gene can reveal how a disease begins, enabling researchers to develop drugs that treat the disease in its early stages. Transgenic farm animals can secrete human proteins in their milk or semen, yielding abundant supplies of otherwise rare substances that are useful as drugs. Transgenic salmon have been engineered for rapid growth, so the fish can reach a marketable size faster than their wild relatives. On a more whimsical note, a glow-in-the-dark fish was the first genetically modified house pet (see the photo in this chapter's What's the Point box).

Ethical Issues Although transgenic organisms have many practical uses, some people question whether their benefits outweigh their potential dangers. Some object to the "unnatural" practice of combining genes from organisms that would never breed in nature. Others fear that ecological disaster could result if genetically modified organisms spread their new genes to wild species (see Investigating Life 11.1).

Still others worry that unfamiliar protein combinations in transgenic crops could be unsafe to eat. A 2016 analysis of more than 1000 studies suggested that genetically modified organisms are probably not increasing the incidence of disease or food allergies. This debate will likely continue, however, as new varieties of transgenic crops hit the market.

Finally, genetically modified seeds may be expensive, reflecting the high cost of developing and testing the plants. The farmers who stand to gain the most from transgenic plants are often unable to afford them.

B. DNA Sequencing Reveals the Order of Bases

Scientists often want to know the nucleotide sequences of genes, chromosomes, or entire genomes. Researchers can use DNA sequence information to predict protein sequences, as described in chapter 7, or they can compare DNA sequences among species to determine evolutionary relationships (see section 13.6). How do investigators get the DNA sequence information they need?

Figure 11.4 illustrates how first-generation DNA sequencing instruments work. Step 1 in the figure shows the components of the reaction mixture. The DNA polymerase enzyme generates a series of DNA fragments that are complementary to the DNA being sequenced. Short, single-stranded pieces of DNA called primers are required by DNA polymerase to begin replication. Also included are normal nucleotides, supplemented with low concentrations of specially modified "terminator" nucleotides tagged with fluorescent labels. Each time DNA polymerase incorporates one of these modified nucleotides instead of a normal one, the new DNA chain stops growing. ⓘ *DNA replication,* section 8.2

Step 2 in figure 11.4 shows the products of the replication reactions: a group of fragments that differ in length from one another by one end base. Once a collection of such pieces is generated, a technique called **electrophoresis** separates the fragments by size (step 3). The researcher can deduce the sequence by "reading" the fragments from smallest to largest (step 4). The data appear as a sequential readout of the wavelengths of the fluorescence from the labels.

The most famous application of DNA sequencing technology has been the Human Genome Project. This worldwide effort was aimed at sequencing all 3.2 billion base pairs of the human genome. The sequence, which was completed in 2003, revealed unexpected complexities. Although our genome includes approximately 25,000 protein-encoding genes, our cells can produce some 400,000 different proteins. Furthermore, only about 1.5% of the human genome sequence actually encodes protein.

How can so few genes specify so many proteins? Part of the answer lies in introns. By removing different combinations of introns from an mRNA molecule, a cell can produce several proteins from one gene—a departure from the old idea that each gene encodes exactly one protein. So far, no one understands exactly how a cell "decides" which introns to remove. ⓘ *introns,* section 7.3

And what is the function of the 98.5% of our genome that does not encode proteins? Some of it consists of regulatory sequences that control gene expression. In addition, much of our DNA is transcribed to rRNA and tRNA. Chromosomes also contain many pseudogenes: noncoding DNA sequences that are very similar to protein-encoding genes and that are transcribed, but whose mRNA is not translated into protein. Pseudogenes may be remnants of old genes that once functioned in our nonhuman ancestors; eventually, they mutated too far from the normal sequence to encode a working protein. ⓘ *regulation of gene expression,* section 7.5; *mutations,* section 7.6

The human genome is also riddled with highly repetitive sequences that have no known function. The most abundant types of repeats are transposons, DNA sequences that can "jump" within the genome. Transposons make up about 45% of human DNA. The genome also contains many tandem repeats (or "satellite DNAs"). These sequences consist of one or more bases repeated many times without interruption, such as CACACA or ATTCGATTCG. The exact number of repeats varies from person to person. As described in section 11.2D, DNA profiling technology measures variation in these areas.

Researchers are comparing the human genome to the DNA sequences of dozens of other species, from bacteria and archaea to protists, fungi, plants, and other animals. The similarities and differences have yielded insights into the sequences that unite all life and those that make each species unique.

Such whole genome comparisons are now faster and cheaper than ever. The cost of sequencing a human genome, for example, has plunged from $100 million in 2001 to about $1000 today. This massive price drop occurred because high-throughput sequencing machines have replaced slower, first-generation technology.

A high-throughput sequencing machine carries out thousands of reactions at once. In general, the process starts by shattering a cell's DNA into many

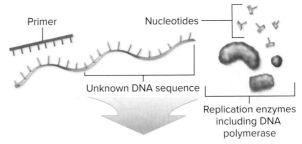

1 Each solution contains the unknown DNA sequence, replication enzymes, primers, normal nucleotides (A, C, T, and G), and a small amount of one type of labeled "terminator" nucleotide (A*, C*, T*, or G*).

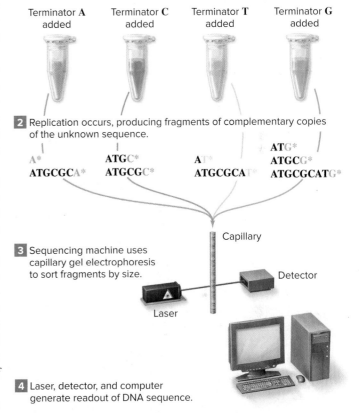

2 Replication occurs, producing fragments of complementary copies of the unknown sequence.

3 Sequencing machine uses capillary gel electrophoresis to sort fragments by size.

4 Laser, detector, and computer generate readout of DNA sequence.

Figure 11.4 First-Generation DNA Sequencing. The DNA polymerase enzyme makes complementary copies of an unknown DNA sequence. But the copies are terminated early, thanks to chemically modified "terminator" nucleotides. Sorting the fragments by size reveals the sequence.

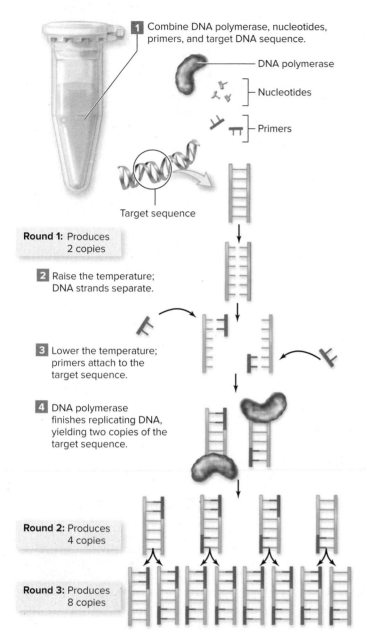

1 Combine DNA polymerase, nucleotides, primers, and target DNA sequence.

— DNA polymerase

— Nucleotides

— Primers

Target sequence

Round 1: Produces 2 copies

2 Raise the temperature; DNA strands separate.

3 Lower the temperature; primers attach to the target sequence.

4 DNA polymerase finishes replicating DNA, yielding two copies of the target sequence.

Round 2: Produces 4 copies

Round 3: Produces 8 copies

Figure 11.5 Polymerase Chain Reaction. Inside a PCR reaction tube, DNA polymerase uses primers and plenty of nucleotides to produce millions of copies of a target DNA sequence.

Figure It Out

Suppose a researcher needs a million copies of a viral gene. She decides to use PCR on a sample of fluid containing one copy of the gene. If one round of PCR takes 2 minutes, how long will it take the researcher to obtain her million-fold amplification?

Answer: Producing 1 million copies would take 20 rounds, or 40 minutes.

overlapping pieces, each about 500 base pairs long. Many copies of each fragment are produced; these copies form clusters that are immobilized on a specialized plate. The technician adds primers, DNA polymerase, and one type of nucleotide to the entire plate at once. As DNA polymerase adds a complementary nucleotide to a fragment, the reaction generates a signal that a computer can detect. This cycle is repeated hundreds of times for each plate. When finished, the computer assembles the overlapping sequences of the entire collection to determine the finished DNA sequence.

Future machines will be even more efficient, sequencing single long DNA molecules. Costs will therefore likely continue to drop, further accelerating the already dizzying pace at which researchers generate genomic data.

C. PCR Replicates DNA in a Test Tube

The **polymerase chain reaction (PCR)** rapidly produces millions of copies of a selected DNA sequence in a test tube. Thanks to this extremely powerful and useful tool, trace amounts of DNA extracted from a single hair follicle or a few skin cells left at a crime scene can yield enough genetic material to reveal a person's unique DNA profile.

PCR borrows heavily from a cell's DNA copying machinery. As illustrated in step 1 of figure 11.5, a PCR reaction tube includes the target DNA sequence to be replicated, DNA polymerase enzymes, a supply of the four types of DNA nucleotides, and two types of short, laboratory-made primers that are complementary to opposite ends of the target sequence.

PCR occurs in an automated device called a thermal cycler, which controls key temperature changes. The reaction begins when heat separates the two strands of the target DNA (figure 11.5, step 2). Next, the temperature is lowered, and the short primers attach to the separated target strands by complementary base pairing (step 3). DNA polymerase adds nucleotides to the primers and builds sequences complementary to the target sequence (step 4). The new strands then act as templates in the next round of replication, which is initiated immediately by raising the temperature to separate the strands once more. The number of pieces of DNA doubles with every round of PCR.

Since its invention in the 1980s, PCR has found an enormous variety of applications. Forensic scientists often work with DNA samples that are too tiny to analyze. With PCR, however, they can make thousands or millions of copies of a particular sequence. Once amplified, the DNA can easily be examined to help establish family relationships, identify human remains, convict criminals, and exonerate the falsely accused. When used to amplify the nucleic acids of microorganisms, viruses, and other parasites, PCR is important in agriculture, veterinary medicine, environmental science, and human health care. In genetics, PCR is both a crucial basic research tool and a way to identify known disease-causing genes in a cell's genome. Evolutionary biologists use PCR to amplify DNA from long-dead plants and animals. The list goes on and on (see Burning Question 11.2).

PCR's greatest weakness, ironically, is its extreme sensitivity. A blood sample contaminated by leftover DNA from a previous PCR run or by a stray eyelash dropped from the person running the reaction can yield a false result.

D. DNA Profiling Detects Genetic Differences

On average, each person's DNA sequence differs from that of a nonrelative by just one nucleotide out of 1000. Finding these small differences by sequencing and comparing entire genomes would be time-consuming, costly, and impractical.

Instead, **DNA profiling** uses just the most variable parts of the genome to detect genetic differences between individuals.

The most common approach to DNA profiling is to examine **short tandem repeats (STRs),** which are sequences of a few nucleotides that are repeated in noncoding regions of DNA. People within a population have different numbers of these repeats. Figure 11.6 shows an STR site on chromosome 5 for three men. The first man has seven copies of the STR on the chromosome that he inherited from one of his parents; he has nine copies on his other chromosome. His genotype for this STR is therefore "7, 9." The other two men have different genotypes.

To generate a DNA profile, a technician extracts DNA from a person's cells and uses PCR to amplify the DNA at each of 13 STR sites, leaving the rest of the DNA alone (figure 11.6, step 1). A fluorescent label is incorporated into the DNA at the STR sites during the PCR reaction. The technician can then use electrophoresis and a fluorescence imaging system to determine the number of repeats at each site (steps 2 and 3).

Statistical analysis plays a large role in DNA profiling. For example, suppose that DNA extracted from a hair found on a murder victim's body matches DNA from a suspect at all 13 STR sites (figure 11.7a). What is the probability that the matching DNA patterns come from the same person—the suspect—rather than from two individuals who happen to share the same DNA sequences? To find out, investigators consult databases that compile the frequency of each STR variant in the population. The statistical analysis suggests that the probability that any two unrelated individuals have the same pattern at all 13 STR markers is one in 250 trillion.

Conversely, a suspect can use dissimilar DNA profiles as evidence of his or her innocence. Since 1989, DNA analysis of stored evidence has proved the innocence of more than 300 people serving time in prison for violent crimes they did not commit (see the chapter-opening photo and figure 11.7b).

In addition to STRs in nuclear DNA, analysis of mitochondrial DNA is also sometimes useful. Mitochondrial DNA is typically only about 16,500 base pairs long, far shorter than the billions of nucleotides in nuclear DNA. But because each cell contains multiple mitochondria, each of which contains many DNA molecules, mitochondria can often yield useful information even when nuclear DNA is badly degraded. Investigators extract mitochondrial DNA from hair follicles, bones, and teeth, then use PCR to amplify the variable regions for sequencing. (i) *mitochondria,* section 3.4C

Because everyone inherits mitochondria only from his or her mother, this technique cannot distinguish between siblings. It is very useful, however,

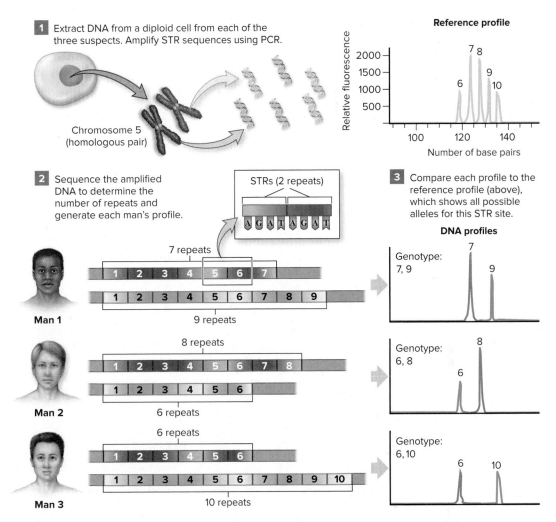

1 Extract DNA from a diploid cell from each of the three suspects. Amplify STR sequences using PCR.

Chromosome 5 (homologous pair)

2 Sequence the amplified DNA to determine the number of repeats and generate each man's profile.

STRs (2 repeats)

AGATAGAT

7 repeats

Man 1

9 repeats

8 repeats

Man 2

6 repeats

6 repeats

Man 3

10 repeats

Reference profile

Relative fluorescence

3 Compare each profile to the reference profile (above), which shows all possible alleles for this STR site.

DNA profiles

Genotype: 7, 9

Genotype: 6, 8

Genotype: 6, 10

Figure 11.6 DNA Profiling. The human genome contains regions of short tandem repeats (STRs) that are genetically variable. DNA profiling techniques detect differences in the number of repeats at multiple STRs; this figure illustrates one STR site. (*1*) DNA extracted from cells of each man is amplified using PCR. (*2*) Sequencing the DNA and (*3*) comparing to reference profiles reveals each man's pattern at the STR site. Figure 11.7 shows complete profiles for all three men.

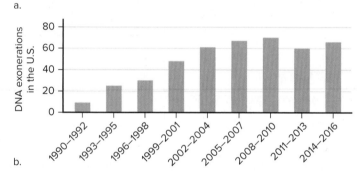

STR Locus		Genotypes			
Chromosome (Locus name)	Number of possible repeats	DNA from crime scene	Man 1	Man 2	Man 3
2 (TPOX)	4–16	10, 11	9, 10	10, 11	8, 8
3 (D3S1358)	8–20	17, 19	19, 19	17, 19	15, 16
4 (FGA)	13–51	20, 24	22, 24	20, 24	21, 23
5 (D5S818)	**6–18**	**6, 8**	**7, 9**	**6, 8**	**6, 10**
5 (CSF1PO)	5–16	7, 10	10, 11	7, 10	6, 12
7 (D7S820)	5–16	13, 14	8, 12	13, 14	9, 10
8 (D8S1179)	7–20	11, 12	13, 14	11, 12	13, 13
11 (TH01)	3–14	7, 7	9, 12	7, 7	8, 9
12 (vWA)	10–25	15, 12	17, 19	15, 12	18, 19
13 (D13S317)	5–17	10, 10	14, 14	10, 10	9, 13
16 (D16S539)	4–16	9, 13	10, 11	9, 13	9, 9
18 (D18S51)	7–40	15, 21	20, 21	15, 21	14, 18
21 (D21S11)	12–41	30, 32	16, 18	30, 32	29, 30

a.

b.

Figure 11.7 **Guilt and Innocence.** (a) DNA evidence collected at a crime scene is compared to the DNA of three suspects. Man 2's profile matches the evidence from the crime scene. Note that the STR site highlighted in red is featured in figure 11.6. (b) In the United States, DNA evidence has exonerated more than 300 prisoners who were serving time for crimes they did not commit. (Data are from the National Registry of Exonerations.)

for verifying the relationship between woman and child. For example, children who were kidnapped during infancy can be matched to their biological mothers or grandmothers. The study of human evolution also has benefited from mitochondrial DNA analysis, which has revealed the genetic relationships among subpopulations from around the world. ⓘ *human evolution, section 17.12*

11.2 Mastering Concepts

1. What are some uses for transgenic organisms?
2. What are the steps in producing a transgenic organism?
3. How do tagged nucleotides participate in first-generation and high-throughput methods of DNA sequencing?
4. What is the function of the 98.5% of the human genome that does not encode protein?
5. How does PCR work, and why is it useful?
6. How are short tandem repeats used in DNA profiling?
7. Why do investigators sometimes analyze mitochondrial DNA instead of nuclear DNA?

Burning Question 11.2

What are the uses of DNA testing?

DNA testing includes DNA profiling and other methods of analyzing genetic material, often from a tiny DNA sample that has been amplified by PCR. DNA testing has so many practical applications that it is impossible to explore them all in a few paragraphs. Listed here, however, are a few uses that are not already described in the chapter.

In criminal justice, investigators sometimes use DNA profiling to generate leads when they have no suspects. They compare DNA collected from a crime scene with a database of DNA profiles. A partial match to someone whose DNA profile is on file can lead investigators to a close relative who committed the crime.

Many people also use DNA to learn more about their families. DNA testing can verify that a man is the father of a child or determine whether newborn twins are identical or fraternal. And genealogical DNA testing can use hundreds of thousands of variable DNA sequences in the human genome to paint a portrait of a family's ethnic history.

Food safety increasingly relies on DNA testing as well. In 2013, for example, DNA testing revealed that some products labeled as 100% beef in Europe actually contained ground horse meat or pork. Researchers can also use DNA to verify other claims on food labels, notably the absence of potentially dangerous allergens such as peanuts.

Submit your burning question to marielle.hoefnagels@mheducation.com

Identical or fraternal?

(twin boys): ©Blend Images/Alamy Stock Photo RF

11.3 Stem Cells and Cloning Add New Ways to Copy Cells and Organisms

The public debate over stem cells and cloning combines science, philosophy, religion, and politics in ways that few other modern issues do (figure 11.8). What is the biology behind the headlines?

A. Stem Cells Divide to Form Multiple Cell Types

A human develops from a single fertilized egg into an embryo and then a fetus—and eventually into an infant, a child, and an adult—thanks to mitotic cell division. As development continues, more and more cells become permanently specialized into muscle, skin, liver, brain, and other cell types. All contain the same DNA, but some genes become irreversibly "turned off" in specialized cells (see figure 7.11). Once committed to a fate, a mature cell rarely reverts to another type.

Animal development therefore relies on stem cells. In general, a **stem cell** is any undifferentiated cell that can give rise to specialized cell types. When a stem cell divides mitotically to yield two daughter cells, one remains a stem cell, able to divide again. The other specializes.

Animals have two general categories of stem cells: embryonic and adult (figure 11.9). **Embryonic stem cells** give rise to all cell types in the body (including adult stem cells) and are therefore called "totipotent"; *toti-* comes from the Latin word for "entire." **Adult stem cells** are more differentiated and produce a limited subset of cell types. For example, stem cells in the skin replace cells lost through wear and tear, and stem cells in the bone marrow produce all of the cell types that make up blood. Adult stem cells are "pluripotent"; *pluri-* means "many" in Latin.

Stem cells are important in biological and medical research. With the correct combination of chemical signals, medical researchers should theoretically

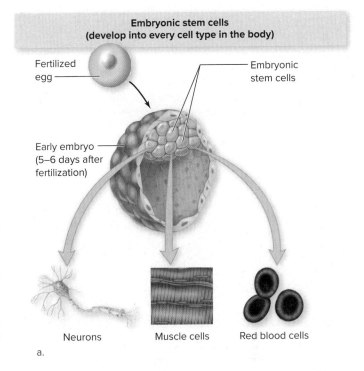

Embryonic stem cells
(develop into every cell type in the body)

Fertilized egg

Embryonic stem cells

Early embryo
(5–6 days after fertilization)

Neurons   Muscle cells   Red blood cells

a.

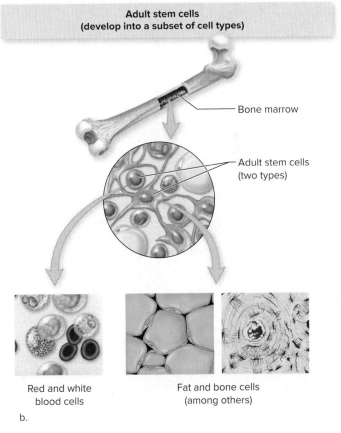

Adult stem cells
(develop into a subset of cell types)

Bone marrow

Adult stem cells
(two types)

Red and white blood cells

Fat and bone cells (among others)

b.

Figure 11.9 **Stem Cells.** (a) Human embryonic stem cells are derived from a ball of cells that forms several days after fertilization. These stem cells give rise to all of the body's cell types. (b) The adult body also contains stem cells, but they may not have the potential to develop into as many different cell types as do embryonic stem cells.

a.   b.

Figure 11.8 **Stem Cell Controversy.** Debates over stem cells often pit (a) people such as actor Michael J. Fox who advocate the use of embryonic stem cells in medicine against (b) people who have moral objections.

(a): ©Congressional Quarterly/Getty Images; (b): ©Getty Images

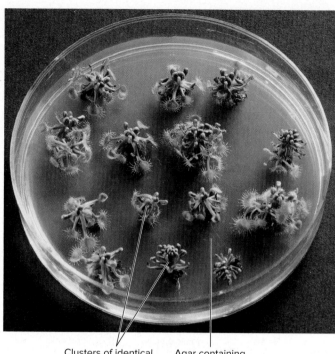

Clusters of identical plantlets | Agar containing nutrients and hormones

a.

b.

Figure 11.10 Cloning in a Dish. (a) When plant tissue is cultured with the correct combination of hormones and nutrients, it gives rise to genetically identical plantlets. This petri dish contains clusters of carnivorous sundew plantlets. (b) Plantlets are transferred to jars providing hormones, nutrients, and room to grow.

(a): ©Rosenfeld Images Ltd/Science Source; (b): ©Philippe Psaila/Science Source

be able to coax stem cells to divide in the laboratory and produce blood cells, neurons, or any other cell type. Many people believe that stem cells hold special promise as treatments for neurological disorders such as Parkinson disease and spinal cord injuries, since neurons ordinarily do not divide to replace injured or diseased tissue. Stem cell therapies may also conquer blood cancers, eye problems, diabetes, heart disease, and many other illnesses that are currently incurable.

The practical benefits would extend beyond treating illness. Currently, pharmaceutical companies test new drugs primarily on whole organisms, such as mice and rats. The ability to test on just kidney or brain cells, for example, would allow researchers to better predict the likely side effects of a new drug. It might also reduce the need for laboratory animals.

Both embryonic and adult stem cells have advantages and disadvantages for medical use. Embryonic stem cells are extremely versatile, but a patient's immune system would probably reject tissues derived from another individual's cells. In addition, research on embryonic stem cells is controversial because of their origin. In fertility clinics, technicians fertilize eggs *in vitro*, and only a few of the resulting embryos are ever implanted into a woman's uterus. Researchers destroy some of the "spare" embryos at about 5 days old to harvest the stem cells. (The other embryos are either stored for possible later implantation or discarded.) Many people consider it unethical to use human embryos in research, even if those embryos would otherwise have been thrown away.

Biologists are also investigating adult stem cells in hair follicles, bone marrow, the lining of the small intestine, and other locations in the body. A patient's immune system would not reject tissues derived from his or her own adult stem cells. These stem cells are less abundant than embryonic stem cells, however, and they usually give rise to only some cell types.

New techniques may eliminate some of these drawbacks. Researchers have discovered how to induce adult cells to behave like embryonic stem cells. This technique could allow differentiated cells taken from an adult to be turned into stem cells, which could then be coaxed to develop into any other cell type. Time will tell how useful these "induced pluripotent stem cells" will be or whether they will match the medical potential of embryonic stem cells.

B. Cloning Produces Identical Copies of an Organism

Imagine being able to grow a new individual, genetically identical to yourself, from a bit of skin or the root of a hair. Although humans cannot reproduce in this way, many organisms do the equivalent. They develop parts of themselves into genetically identical individuals—clones—that then detach and live independently.

Cloning simply means asexual reproduction. In its simplest form, asexual reproduction consists of the division of a single cell. In bacteria, archaea, and single-celled eukaryotes such as *Amoeba*, the cell's DNA replicates, and then the cell splits into two identical, individual organisms (see chapter 8). Although the details of cell division differ between prokaryotes and eukaryotes, the result is the same: One individual becomes two.

Most plants, fungi, and animals reproduce sexually, but at least some organisms in each kingdom also use asexual reproduction. This strategy is especially common in plants and fungi. Asexual reproduction is much less common in animals, but sponges, coral animals, hydra, and jellyfishes all can "bud" new individuals that break away from the parent.

Plants are especially easy to clone (figure 11.10). Commercial plant growers use a technique called tissue culture to clone carnivorous plants and many other species in petri dishes or small jars. But anyone can clone a favorite houseplant by placing a portion of a stem in water, allowing roots to develop, and transplanting the young plant into a pot of soil.

Mammals do not naturally clone themselves. In 1996, however, researcher Ian Wilmut and his colleagues in Scotland used a new procedure to produce Dolly the sheep, the first clone of an adult mammal. The researchers used a cloning technique called **somatic cell nuclear transfer.** A *somatic cell* is any cell that makes up an animal's body, other than stem cells or gamete-forming cells. To clone an animal, a nucleus is moved from a DNA donor's somatic cell into an egg cell without a nucleus. This egg, now containing a full set of DNA, develops into a clone of the original donor. ⓘ *somatic cells,* section 9.3

Figure 11.11 illustrates the cloning technique in more detail. First, the researchers obtain the nucleus from a DNA donor's somatic cell (steps 1 and 2). They then obtain an egg cell from another animal (usually of the same species) and remove its nucleus (steps 3 and 4). Transferring the "donor" nucleus to the denucleated egg cell (step 5) forms a cell that divides mitotically to form an embryo (step 6). The researchers implant the embryo in a surrogate mother's uterus (step 7). The embryo then develops into a newborn clone of the adult DNA donor (step 8).

Since Dolly's birth, researchers have used somatic cell nuclear transfer to clone other mammals as well, including dogs, cats, mice, bulls, and a champion horse that had been castrated. Cloning may even help rescue endangered species or recover extinct species.

Many people wonder whether humans can and should be cloned. Reproductive cloning could help infertile couples to have children. Scientists could also use cloned human embryos as a source of stem cells, which could be used to grow "customized" artificial organs that the patient's immune system would not reject. This application of cloning is called therapeutic cloning.

Despite the potential benefits, however, human cloning carries unresolved ethical questions. For example, most clones die early in development. Even the tiny percentage of clones that make it to birth often have abnormalities. In addition, therapeutic cloning still requires the destruction of an embryo to harvest the stem cells. As we have already seen, many people question the practice of making human embryos only to destroy them. Finally, both reproductive and therapeutic cloning require unfertilized human eggs. The removal of eggs from a woman's ovaries is costly and poses medical risks.

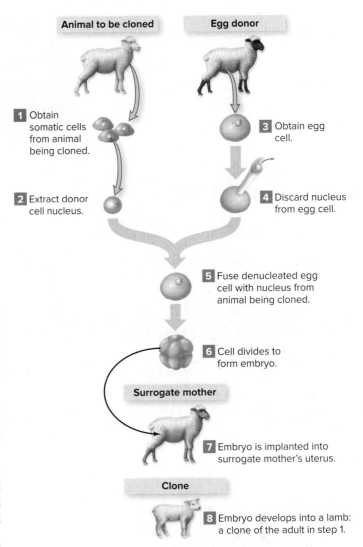

1 Obtain somatic cells from animal being cloned.

2 Extract donor cell nucleus.

3 Obtain egg cell.

4 Discard nucleus from egg cell.

5 Fuse denucleated egg cell with nucleus from animal being cloned.

6 Cell divides to form embryo.

Surrogate mother

7 Embryo is implanted into surrogate mother's uterus.

Clone

8 Embryo develops into a lamb: a clone of the adult in step 1.

Animal to be cloned

Egg donor

Figure 11.11 Reproductive Cloning. The first steps in somatic cell nuclear transfer are to (1) obtain a somatic cell from the animal to be cloned and (2) extract the nucleus from that cell. Then, (3) the researcher obtains an egg cell from another animal and (4) discards its nucleus. (5) The donor nucleus is placed into the egg cell. (6) This cell develops into an embryo, which (7) is implanted into the uterus of a surrogate mother. (8) The clone is born after gestation is complete.

11.3 Mastering Concepts

1. Describe the differences among embryonic, adult, and induced pluripotent stem cells.
2. What are the potential medical benefits of stem cells?
3. Summarize the steps scientists use to clone an adult mammal.
4. Why is the technique used to clone mammals called somatic cell nuclear transfer?

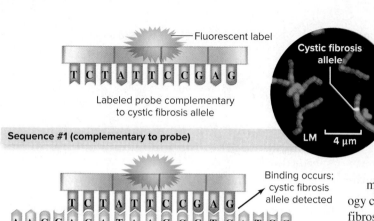

—Fluorescent label

T C T A T T C C G A G

Labeled probe complementary
to cystic fibrosis allele

Sequence #1 (complementary to probe)

T C T A T T C C G A G
A A G C A G A T A A G G C T C A T C G

Binding occurs;
cystic fibrosis
allele detected

Sequence #2 (not complementary to probe)

T C T A T T C C G A G
T A T G C A T C G A T T A T G A T A C

No binding;
cystic fibrosis
allele not
detected

Cystic fibrosis
allele

LM 4 μm

Figure 11.12 **DNA Probe.** A DNA probe is a labeled piece of
single-stranded DNA that binds to a complementary target sequence.
In this case, the probe reveals the presence of a cystic fibrosis allele.
Photo: ©Patrick Landmann/Science Source

11.4 Many Medical Tests and Procedures Use DNA Technology

The list of human illnesses is long. Worldwide, the top causes of death include heart disease, stroke, cancer, and infection (see table 18.2); all of these diseases have environmental and genetic components. But some ailments, including hemophilia, Tay-Sachs disease, sickle cell disease, and dozens of others, are entirely caused by mutated alleles of single genes. This section describes how DNA technology can help prevent, detect, and treat genetic diseases. Although we use cystic fibrosis as an example, the same techniques are applicable (at least in theory) to any illness associated with a single gene.

A. DNA Probes Detect Specific Sequences

The ability to detect the alleles that cause cystic fibrosis and other genetic illnesses is crucial to the medical applications of DNA technology. At first glance, however, all DNA looks alike: a sequence of A, C, G, and T. With billions of nucleotides in a single cell, how can biologists search through an entire genome to find just the piece they need to "see"?

The answer is a **DNA probe,** a single-stranded sequence of nucleotides that is used to detect a complementary DNA sequence (figure 11.12). A typical probe is a short, synthetic strand of DNA that is labeled with either a radioactive isotope or a fluorescent tag. For example, a researcher can construct a probe that is complementary to part of an allele known to be associated with cystic fibrosis. A region of the DNA to be tested is separated into single strands and immobilized on a solid surface. If the DNA contains a nucleotide sequence that is complementary to the probe, then the probe binds to that region. The radioactivity or wavelength emitted by the probe reveals the presence of the cystic fibrosis allele.

B. Preimplantation Genetic Diagnosis Can Screen Embryos for Some Diseases

Imagine a young couple that wants a child. Both of the prospective parents know they are symptomless carriers of cystic fibrosis. Can DNA technology help the couple ensure that their baby is free of the disease? Although no one can guarantee a cystic fibrosis–free baby, a technique called **preimplantation genetic diagnosis (PGD)** can screen embryos for the disease-causing allele and therefore greatly reduce the odds of having an affected child.

The process begins with *in vitro* fertilization (literally, fertilization "in glass"), in which the man's sperm fertilize several of the woman's eggs in a laboratory dish. The resulting zygotes develop into embryos, each consisting of eight genetically identical cells. A technician then selects an embryo for PGD. He or she removes one cell from the embryo (figure 11.13); the loss of this cell will not affect the embryo's subsequent development. DNA extracted from that single cell undergoes PCR, amplifying the region of DNA where the cystic fibrosis gene is located. A DNA probe specific for one or more cystic fibrosis alleles can then determine whether the embryo's cells contain the disease-causing DNA sequences.

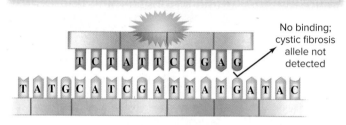

a.

LM 25 μm

Figure 11.13

**Preimplantation Genetic
Diagnosis.** (a) A single cell
from a human embryo is
removed for testing. (b) PGD
was used to help ensure
that this child was free of a
potentially fatal disease.

(a): ©RAJAU/PHANIE/Science Source,
Inc.; (b): ©Reuters/Alamy Stock Photo

b.

If the allele is detected, the embryo can be discarded, and others can be tested. Any embryo that lacks the disease-causing allele is a good candidate to be placed into the woman's body. If the embryo implants into the uterus and develops into a baby, the child is very likely to be born without cystic fibrosis.

There is a small chance, however, that the baby may be born with the disease despite PGD. Human error is one possible explanation. By amplifying DNA sequences that occur in just one or two copies from a single cell, PGD pushes PCR to its limits. As we have already seen, PCR is extremely sensitive to contamination; stray DNA that is accidentally amplified can lead to a false result. A second explanation relates to the fact that researchers have identified hundreds of mutations that can cause cystic fibrosis. PGD tests for the most common disease-causing alleles, but the baby may have inherited rare variants that the test cannot detect.

C. Genetic Testing Can Detect Existing Diseases

The same genetic tests used in PGD are also useful for testing fetuses, newborns, older children, and adults for disease-causing alleles. Instead of searching the DNA from an embryonic cell, however, the tests detect the alleles in DNA from cells taken from blood, saliva, or body tissues.

For example, newborns are routinely screened for a genetic disorder called phenylketonuria (PKU). Cells from unborn children can also be tested for disease-causing alleles; the parents can use the information to decide whether to terminate the pregnancy or to prepare for life with a special needs child.

Genetic testing has many applications in adults as well. People who suspect they may be heterozygous carriers of cystic fibrosis might choose to be tested for the disease-causing allele before deciding whether to have children. Likewise, a woman with a family history of breast cancer might be tested for damage to a gene called *BRCA1,* which is strongly associated with susceptibility to that disease. A positive test for the mutant allele might prompt the woman to have her breasts surgically removed to prevent the cancer from ever arising. And patients who already have breast cancer often have DNA from their tumors screened for genes encoding estrogen receptors. The results can indicate which treatments might be most promising.

D. Gene Therapy Uses DNA to Treat Disease

Cystic fibrosis and most other genetic illnesses currently have no cure, but **gene therapy** may someday provide new treatment options by adding healthy DNA to a person's cells. The new DNA supplements the function of a faulty gene (figure 11.14).

The gene therapy strategy illustrated in figure 11.14 shares some similarities with producing transgenic organisms (see section 11.2A), in that new DNA is introduced into existing cells. But the two techniques are also different in key ways. First, in gene therapy, the healthy gene introduced into a cell is from humans, not another species. Second, a typical transgenic organism can theoretically pass the foreign genes to the next generation, whereas a gene therapy patient would only receive new genes in the cell type that needs correction. Other cell types, including the germline cells that produce sperm and egg cells, would be left alone.

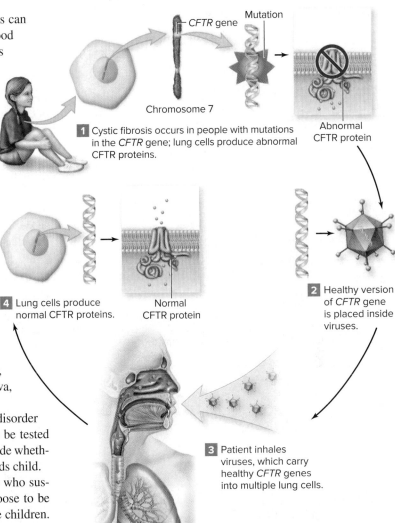

CFTR gene

Mutation

Chromosome 7

1 Cystic fibrosis occurs in people with mutations in the *CFTR* gene; lung cells produce abnormal CFTR proteins.

Abnormal CFTR protein

2 Healthy version of *CFTR* gene is placed inside viruses.

3 Patient inhales viruses, which carry healthy *CFTR* genes into multiple lung cells.

4 Lung cells produce normal CFTR proteins.

Normal CFTR protein

Figure 11.14 Gene Therapy. The overall goal of gene therapy is to supplement a faulty gene with a normal, healthy version. In this example, a genetically modified virus delivers a healthy *CFTR* gene to the lungs of a person with cystic fibrosis.

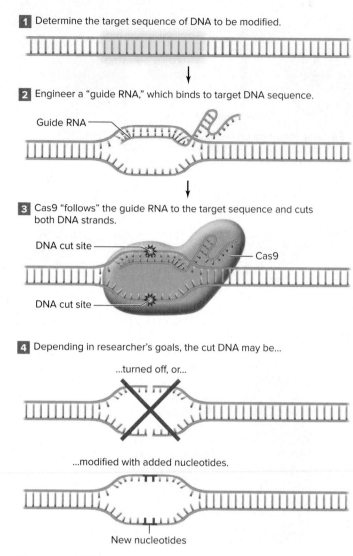

1 Determine the target sequence of DNA to be modified.

2 Engineer a "guide RNA," which binds to target DNA sequence.

Guide RNA

3 Cas9 "follows" the guide RNA to the target sequence and cuts both DNA strands.

DNA cut site

Cas9

DNA cut site

4 Depending in researcher's goals, the cut DNA may be...

...turned off, or...

...modified with added nucleotides.

New nucleotides

Figure 11.15 Gene Editor. (*1*) To use CRISPR-Cas9, the researcher first determines the sequence of the DNA to be edited. (*2*) A short "guide RNA" molecule, complementary to the target DNA, is added to the cell. (*3*) Cas9 binds to the guide RNA and cuts the DNA. (*4*) The researcher can pair CRISPR-Cas9 with other tools that complete the editing process.

Gene therapy is challenging for several reasons. The new gene must be delivered directly to only those cells that express the faulty allele. Viruses may be ideal for carrying DNA into target cells because they typically infect only a limited range of cells. But for gene therapy to be safe, the viruses must not trigger an immune reaction, and the new DNA must not induce mutations that cause cancer. In addition, the gene therapy patient must express the new genes long enough for his or her health to improve.

Gene therapy trials in humans pose significant risks. In 1999, for example, 18-year-old Jesse Gelsinger received a massive infusion of viruses carrying a gene to correct an inborn error of metabolism. He died within days from an overwhelming immune system reaction. Gelsinger's death led to stricter rules for conducting experiments. Nevertheless, gene therapy research and clinical trials continue, with promising results for diseases including cystic fibrosis, sickle cell disease, and some forms of inherited blindness and immune disorders.

E. CRISPR-Cas9 Cuts and Edits Specific Genes

For most of the last 10,000 years, people who wanted to develop a new crop variety had only one choice: the slow, limited process of selective breeding (artificial selection). Around 1930, plant breeders began to bombard huge numbers of seeds with radiation or chemicals to induce mutations in the DNA. They would then plant the seeds, wait for the plants to grow, and scrutinize each one, all in hopes of identifying a handful that might have new and useful traits (see Burning Question 11.1). ⓘ *artificial selection,* section 12.2C

Now, however, a tool called CRISPR-Cas9 promises to revolutionize this process. Rather than inducing random mutations throughout an entire genome, CRISPR-Cas9 allows researchers to edit specific genes. Moreover, it applies to all kinds of organisms—not just plants.

How does CRISPR-Cas9 work? In its simplest form, the tool has two parts (figure 11.15). One component is Cas9, an enzyme that cuts double-stranded DNA. The other is a short piece of "guide RNA" that binds to complementary DNA and shows Cas9 exactly where to cut. After the cut is complete, the next steps depend on the researcher's objectives. Paired with other DNA technology tools, CRISPR-Cas9 can turn a gene off or add new DNA to it. Because researchers can design the guide RNA to bind to and cut any short sequence of DNA, CRISPR-Cas9 allows for unprecedented precision in genome editing.

CRISPR-Cas9 may someday help medical researchers fix the broken genes that trigger cancer and other illnesses. Or they may improve the body's ability to fight cancer. Technicians might extract a cancer patient's immune system cells and alter the DNA so the cells attack molecules that are unique to the patient's tumors. Once returned to the patient's body, the modified cells would be primed to attack the cancer.

Although not yet routinely used in human medicine, CRISPR-Cas9 has enormous potential in many other fields of research as well. In one study, researchers used CRISPR-Cas9 to snip genes out of the mosquito genome so the insects could no longer transmit malaria. Other scientists are working to craft disease-resistant livestock, chickens lacking the proteins that induce egg allergies, and honeybees that obsessively clean their hives to prevent the spread of parasites. Still others propose using the technology to alter the DNA of today's pigeons in ways that would, in effect, "bring back" the extinct passenger pigeon. Plants could be altered to boost yield, improve disease resistance, or survive in drought-stricken fields.

So far, the main limitation is that the guide RNA is so short that it sometimes binds to DNA that is similar but not identical to the target sequence. Cas9 therefore sometimes cuts DNA in the wrong part of the genome, but researchers are working to find ways to overcome this obstacle.

F. Medical Uses of DNA Technology Raise Many Ethical Issues

The use of DNA technology in medicine can prevent or reduce human suffering in many ways: by improving the chance of having healthy children, by detecting diseases early if they do occur, and by offering the prospect of new treatments for illnesses that currently have no cure.

But these techniques also present ethical dilemmas. A thorough treatment of ethics is beyond the scope of this book, but the rest of this section offers a small sampling of some questions that accompany the use of DNA technology in medicine.

In vitro fertilization and preimplantation genetic diagnosis, for example, are costly. Should these techniques be available only to the wealthy? And consider the diagnosis of a genetic disease in an unborn child. A woman who is pregnant with a fetus that carries a genetic abnormality may decide to end the pregnancy rather than carrying the child to term. Does the morality of her decision depend on the severity of the illness? In other words, should we reserve fetal screening for life-threatening illnesses, or is it morally permissible to use it for milder conditions as well? What about using genetic tests to select for or against embryos with traits that do not affect health at all, such as sex or eye color?

Genetic testing in older children and adults may also lead to sticky questions. For example, a genetic test that reveals a high risk for cancer may be beneficial if it leads to lifestyle changes that promote a longer, healthier life. On the other hand, genetic testing can detect alleles associated with diseases for which effective treatments are not yet available. The test results may therefore lead to depression or anxiety without improving the chance of treatment or a cure.

Gene therapy also comes with its share of dilemmas. This new form of treatment currently carries so many risks that its use is extremely limited. Once the technology is perfected, however, how should it be used? Only for debilitating diseases, or for less serious conditions as well? Is it right to use the techniques of gene therapy to enhance a person's appearance or athletic performance (see Why We Care 11.1)? What about using DNA technology to alter the DNA in a person's germline so that future generations contain the new gene? The answer to this question is not trivial; tinkering with germline DNA could affect our future evolution in unforeseen ways.

Why We Care 11.1 | Gene Doping

©Corbis RF

If we can use genes to cure diseases, it must also be possible to use DNA to make a healthy person even "better." For example, it should be possible to inject genes that make an athlete stronger, faster, or better able to withstand the physical stress of competition.

"Gene doping" is the use of DNA to enhance the function of a healthy person. The techniques would be essentially the same as those used in gene therapy: New genes would be introduced into existing cells, and the proteins encoded by those genes would change the cells' function. The difference is that rather than curing a disease, the goal of gene doping is to give an athlete a competitive edge. An introduced gene might induce the growth of extra muscle, for example. Alternatively, an endurance athlete might use gene doping to boost the production of erythropoietin (EPO), a protein that stimulates red blood cell formation.

In some ways, gene doping might seem like a more attractive option than using anabolic steroids or other performance-enhancing drugs. After all, the altered cells are simply making more of the same proteins that they can already produce. But this view ignores the immense challenges and risks that gene doping entails.

Both gene doping and gene therapy share the same difficulties. One problem is getting the genes into the right cells without triggering an immune reaction; another is activating the DNA once it is in the target cells. The dangers are the same, too. For example, if a newly introduced gene inserts itself into the wrong part of a chromosome, it might cause a mutation that triggers cancer. The gene might also have unwanted side effects. Too much EPO can be life-threatening if it causes the blood to become too thick to flow in arteries and veins. And if something does go wrong, no one knows how to reverse the gene doping process; getting the introduced gene back out is impossible.

For now, technical difficulties and risks have kept gene doping from developing into a practical option for athletes seeking an edge. As the process improves, however, gene doping may become common. That prospect has led many sports organizations to simultaneously ban the practice and seek improved detection methods.

11.4 Mastering Concepts

1. Explain how and why a researcher might use a DNA probe.
2. Compare and contrast preimplantation genetic diagnosis and genetic testing.
3. What is gene therapy?
4. Describe how CRISPR-Cas9 targets a specific gene for editing.
5. What are some examples of ethical questions raised by the medical use of DNA technology?

Investigating Life 11.1 | Weeds Get a Boost from Their Transgenic Cousins

Among the most widely grown transgenic plants are those that resist glyphosate, the active ingredient in the herbicide called Roundup (see figure 11.3). Farmers who use these transgenic crops can then spray their fields with the herbicide, killing weeds but leaving the crops unharmed.

Glyphosate typically kills plants by inhibiting an enzyme that plants require for survival. Some transgenic crops contain a "booster" gene that enables the plant to produce 20 times more of this critical enzyme than normal plants. The extra dose of the enzyme allows the plant to survive high levels of glyphosate treatment, giving crops a major advantage over their weedy competitors.

On the other hand, cultivated plants and their weedy relatives often cross-pollinate, a process that could lead to transgenic weeds. A group of Chinese researchers wanted to learn what would happen if transgenic cultivated rice cross-pollinated with weedy (noncultivated) rice. The researchers transferred pollen from transgenic rice to weedy rice, producing two types of hybrid plants: weedy rice expressing the booster gene and weedy rice not expressing the gene (the control group). They then grew both types of hybrid rice in an herbicide-free environment—conditions the weeds would encounter outside of a farm—and measured the ability of the plants to reproduce.

Their results were clear: Even in the absence of herbicides, weeds with the booster gene had a striking reproductive advantage. They had a higher seed germination rate and produced more seeds than did weeds without the beneficial gene (figure 11.B). Plant growth also increased in weeds with the booster gene.

The implications of these findings are unknown, but they could be serious. Climate change is driving researchers to develop new transgenic crops that resist drought and heat. If the new genes move to other plants, then weeds are likely to become even "weedier." These new plant varieties, which might be difficult to control, could prove to be formidable competitors for the crops that sustain our food supply.

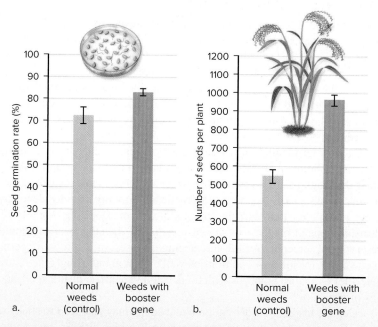

Figure 11.B Getting Weedier. (a) Weedy rice with the glyphosate booster gene had a higher seed germination rate and (b) produced more seeds than similar plants without the gene.

Source: Wang, Wei and 9 coauthors. 2014. A novel 5-enolpyruvoylshikimate-3-phosphate (EPSP) synthase transgene for glyphosate resistance stimulates growth and fecundity in weedy rice (*Oryza sativa*) without herbicide. *New Phytologist*, vol. 202, pages 679–688.

What's the **Point?** ▼ APPLIED

Scientists use DNA technology to make transgenic organisms, determine evolutionary relationships, detect genetic differences between individuals, and treat or prevent diseases, among many other applications. Here we will travel back in time to the mid-1980s for a glimpse at the first criminal case that investigators solved using DNA profiling.

In the central England village of Narborough, a young woman was raped and murdered. Police collected evidence from the scene and determined the killer's blood type. However, the list of possible suspects was long, and the case went cold. Three years later, the murderer struck again. This time, police identified a 17-year-old suspect. During interrogation, the man admitted to killing the second victim, but he denied involvement in the first crime.

Meanwhile, University of Leicester professor Alec Jeffreys worked to perfect a genetic comparison technique called DNA profiling. A couple of months after the second murder, the

investigators asked Jeffreys to apply his new technique to the evidence found at the crime scenes. Prosecutors hoped to prove that their suspect had killed both women.

As expected, DNA evidence collected from the first victim matched DNA found on the second victim, indicating that both women were killed by the same man. But the DNA did not match the suspect in police custody. Instead, it matched that of another man: Colin Pitchfork. The innocent teen bound for jail was exonerated, and the true killer was put behind bars.

In the three decades since this pioneer case, DNA profiling has led to the conviction of many criminals. It has also allowed over 300 innocent prisoners to go free (see figure 11.7b). DNA has proved to be a powerful tool in law and many other disciplines; scientists are likely to unravel new applications of the double helix in the future.

Photo: ©Hill Street Studios/Matthew Palmer/Blend Images LLC RF

CHAPTER SUMMARY

11.1 DNA Technology Is Changing the World

- Many disciplines benefit from **DNA technology,** the practical application of knowledge about DNA. Figure 11.16 summarizes some of the tools and techniques of DNA technology.

11.2 DNA Technology's Tools Apply to Individual Genes or Entire Genomes

A. Transgenic Organisms Contain DNA from Multiple Sources

- **Transgenic organisms** are important in industry, research, and agriculture.

Technology or Tool	Definition
Restriction enzyme	Protein that cuts double-stranded DNA at a specific base sequence
Recombinant DNA	Genetic material that has been cut with restriction enzymes and spliced with DNA from other sources
Transgenic organism	An individual with recombinant DNA 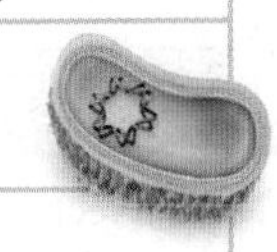
DNA sequencing	Determines the nucleotide sequence of DNA fragments
PCR (polymerase chain reaction)	Amplifies DNA in a test tube using the cell's replication machinery
DNA profiling	Uses DNA sequencing and PCR to detect genetic differences among individuals
Stem cells 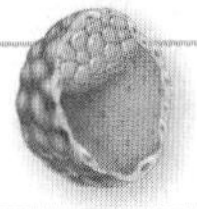	Cells found in embryos and some adult tissues that can give rise to other cell types
Cloning	Makes an identical copy of an organism 
Somatic cell nuclear transfer	A type of cloning that combines a nucleus taken from one individual's body cell with a denucleated egg cell from another individual to produce the first cell of a new organism
DNA probe	A single-stranded sequence of DNA, labeled with a radioactive isotope or fluorescent tag, used to detect the presence of a known sequence of nucleotides
Preimplantation genetic diagnosis	Uses PCR and DNA probes to detect genetic diseases in embryos that might later be implanted in a woman's uterus
Genetic testing	Uses PCR and DNA probes to detect genetic diseases in fetuses, newborns, older children, and adults
Gene therapy 	Employs viruses to insert healthy genes into cells
CRISPR-Cas9	Uses enzymes and "guide RNA" to delete, add, or alter specific genes

Figure 11.16 Miniglossary of DNA Technology.

- **Restriction enzymes** and **plasmids** are tools that help researchers construct **recombinant DNA** and introduce it to recipient cells.

B. DNA Sequencing Reveals the Order of Bases

- In first-generation DNA sequencing machines, DNA polymerase incorporates modified nucleotides into a copy of DNA, generating DNA fragments of various lengths. Using **electrophoresis** to sort the fragments by size reveals the DNA sequence.
- Only 1.5% of the 3.2 billion base pairs of the human genome encode protein. The remaining 98.5% of the human genome encodes rRNA, tRNA, regulatory sequences, pseudogenes, transposons, and other repeats.
- High-throughput machines have dramatically reduced the cost of DNA sequencing.

C. PCR Replicates DNA in a Test Tube

- In the **polymerase chain reaction (PCR),** DNA separates into two strands, and DNA polymerase adds complementary nucleotides to each strand. Repeated cycles of heating and cooling allow for rapid amplification of the target DNA sequence.
- PCR finds many applications in research, forensics, medicine, agriculture, and other fields.

D. DNA Profiling Detects Genetic Differences

- Individuals vary genetically in single bases and **short tandem repeats (STRs). DNA profiling** detects these differences.
- Investigators can use known frequencies of alleles in the population to calculate the probability that two DNA samples match purely by chance.
- Analysis of mitochondrial DNA can verify maternal relationships.

11.3 Stem Cells and Cloning Add New Ways to Copy Cells and Organisms

A. Stem Cells Divide to Form Multiple Cell Types

- **Stem cells** give rise to specialized cell types. **Embryonic stem cells** generate all cells in the body; **adult stem cells** produce only a limited subset of cell types.
- Induced pluripotent stem cells are adult cells that are converted to stem cells. They may eliminate some ethical issues associated with embryonic stem cells.

B. Cloning Produces Identical Copies of an Organism

- Researchers use a technique called **somatic cell nuclear transfer** to clone adult mammals.
- Human reproductive and therapeutic cloning have potential medical applications, but they also involve ethical dilemmas.

11.4 Many Medical Tests and Procedures Use DNA Technology

A. DNA Probes Detect Specific Sequences

- A **DNA probe** is a single-stranded fragment of DNA that is labeled. The probe binds to any complementary DNA, revealing its location.

B. Preimplantation Genetic Diagnosis Can Screen Embryos for Some Diseases

- In **preimplantation genetic diagnosis (PGD),** a human embryo can be tested for a variety of diseases before being implanted into a woman's uterus.

C. Genetic Testing Can Detect Existing Diseases

- With the help of DNA probes, genetic material extracted from cells of a fetus, a child, or an adult can be tested for disease-causing alleles.

D. Gene Therapy Uses DNA to Treat Disease

- **Gene therapy** places a functional gene into cells that are expressing a faulty gene.

E. CRISPR-Cas9 Cuts and Edits Specific Genes
- When combined with other tools, CRISPR-Cas9 allows researchers to silence targeted genes or modify them to encode new traits.

F. Medical Uses of DNA Technology Raise Many Ethical Issues
- Because of its risks, high expense, and potential to alter human life, DNA technology raises a number of ethical questions.

MULTIPLE CHOICE QUESTIONS

1. If a restriction enzyme cuts between G and A whenever it encounters the sequence GAATTC, how many fragments will be produced when the enzyme digests DNA with the following sequence?
   TGAGAATTCAACTGAATTCAAATTCGAATTCTTAGC
   a. Two
   b. Three
   c. Four
   d. Five

2. Which of the following is *not* a reason that scientists make transgenic organisms?
   a. To increase the global diversity of organisms
   b. To produce human drugs
   c. To increase the human food supply
   d. To help plants tolerate harsh environments

3. The function of electrophoresis is to
   a. break a long DNA sequence into fragments.
   b. incorporate nucleotides into a DNA template.
   c. sort DNA fragments by size.
   d. cause DNA fragments to bind together.

4. Why is PCR useful?
   a. Because it replicates all the DNA in a cell
   b. Because it can produce large amounts of DNA from small amounts
   c. Because it produces a heat-tolerant DNA polymerase
   d. Because it occurs in an automated device

5. Suppose an investigator at the scene of a murder finds a pair of gloves. Back at the lab, she discovers small fragments of skin and hair on the gloves. What fast test might she do with this evidence to help solve the case?
   a. Remove the nucleus of a skin cell and place it in an egg cell.
   b. Extract DNA and determine genotypes at multiple STR sites.
   c. Extract DNA and sequence the genome.
   d. Use DNA probes to search for specific gene sequences.

6. What is an induced pluripotent stem cell?
   a. A cell from which the nucleus has been removed
   b. A cell extracted from an early embryo
   c. A specially treated somatic cell that can develop into any cell type
   d. A specially treated embryonic stem cell that develops into one specialized cell type

7. Dolly the sheep was the first clone of an adult mammal. Dolly was genetically identical to the sheep that
   a. gave birth to her.
   b. donated the egg that developed into her.
   c. donated her chromosomes.
   d. Dolly was equally similar to all of these sheep.

8. A DNA probe with sequence TCAGGCTTCAG would bind most strongly to which of the following DNA fragments?
   a. AGTCCGAAGTC
   b. TCAGGCTTCAG
   c. GACTTCGGACT
   d. UGAGGCUUGAG

9. Preimplantation genetic diagnosis would be least useful in detecting a ___ disease-causing allele.
   a. dominant
   b. recessive
   c. common
   d. rare

10. What is the role of a virus in gene therapy?
    a. It causes the disease that gene therapy is aiming to cure.
    b. It carries the healthy DNA into the patient's cells.
    c. It carries the faulty DNA out of the patient's cells.
    d. It reveals which cells carry the DNA causing the disease.

Answers to Multiple Choice questions are in appendix A.

WRITE IT OUT

1. What techniques might researchers use to produce transgenic bacteria that make human growth hormone (a drug used to treat extremely short stature)?

2. Transgenic crops often require fewer herbicides and insecticides than conventional crops. In that respect, they could be considered environmentally friendly. Use the Internet to research the question of why some environmental groups oppose transgenic technology.

3. Describe why sorting DNA fragments by size is useful in first-generation DNA sequencing methods.

4. Explain how the ingredients in a PCR reaction tube replicate DNA.

5. In a 2013 investigation, researchers discovered that meat packaged as "100% beef" in Europe actually contained traces of horse meat and pork. What DNA technology techniques might the researchers have used to uncover the truth about the origin of the meat?

6. Why are entire genomes not used for DNA profiling?

7. What would be the advantage of sequencing an entire chromosome from a DNA sample collected at a crime scene? What would be the disadvantage or limitations of this approach?

8. Mature neurons in the brain do not replicate. Why are stem cells an intriguing solution to patients who suffer from traumatic brain injury?

9. Unneeded genes in an adult animal cell are permanently inactivated, making it impossible for most specialized cells to turn into any other cell type. How does this arrangement save energy inside a cell? Why does the ability to clone an adult mammal depend on techniques for reactivating these "dormant" genes?

10. Scientists are interested in cloning an extinct animal called the gastric brooding frog. This strange frog swallows its eggs and broods its young within its stomach. So far, scientists have successfully used cloning to make an embryo of the frog, but they have yet to raise one to maturity. What steps might the scientists have used to clone this extinct species? Why was it important for scientists to determine before the experiments that the great barred frog is a close relative of the gastric brooding frog?

11. Why We Care 11.1 describes some potential applications of gene doping. What are some examples of ethical issues that gene doping presents? What might the prospect of gene doping mean for the future of sports?

12. Describe gene therapy, and explain the ethical issues that gene therapy presents.

13. Use the Internet to research an application of CRISPR-Cas9. For your chosen application, describe how scientists have used or plan to use the gene editing tool.

14. If a cell's genome is analogous to a cookbook and a gene is analogous to a recipe, what is an analogy for preimplantation genetic diagnosis? For gene testing? For gene therapy?

SCIENTIFIC LITERACY

Review Burning Question 11.1, which describes the difference between selective breeding and genetic engineering. Why do you think that selectively bred plants are generally acceptable to consumers, whereas many people avoid genetically engineered foods? Some scientists are suggesting "rewilding" crops—that is, genetically modifying crop plants to restore genes that were bred out many generations ago. The genes, which would come from ancient plant varieties that had not undergone selective breeding, might improve drought tolerance or confer other useful traits. Do you think the resulting plants, which would technically be genetically modified organisms, would be more acceptable to consumers than plants receiving genes from a bacterium? Why or why not?

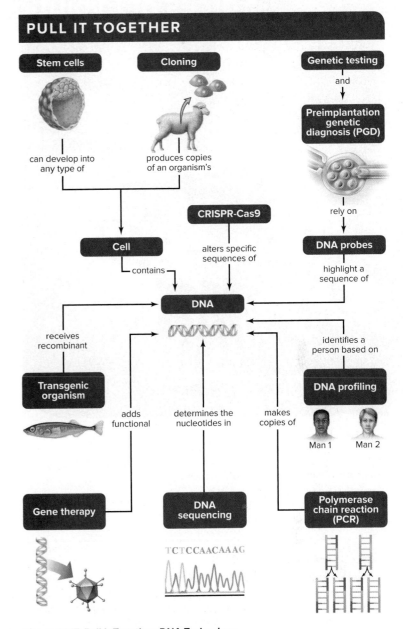

PULL IT TOGETHER

Figure 11.17 Pull It Together: DNA Technology.

Refer to figure 11.17 and the chapter content to answer the following questions.

1. Review the Survey the Landscape figure in the chapter introduction. Given DNA's role in the cell, why do the basic tools of DNA technology summarized in figure 11.17 have applications in such diverse fields of study?

2. How does *PCR* relate to *DNA profiling* and *preimplantation genetic diagnosis?*

3. Add the terms *restriction enzyme, plasmid, virus, DNA polymerase,* and *short tandem repeat* to this concept map.

4. How is a patient who receives gene therapy similar to and different from a transgenic organism?

Answers to Mastering Concepts, Write It Out, Scientific Literacy, and Pull It Together questions can be found in the Connect ebook.
connect.mheducation.com

LEARNING OUTLINE

12.1 Evolution Acts on Populations

12.2 Evolutionary Thought Has Evolved for Centuries

12.3 Natural Selection Molds Evolution

12.4 Evolution Is Inevitable in Real Populations

12.5 Natural Selection Can Shape Populations in Many Ways

12.6 Sexual Selection Directly Influences Reproductive Success

12.7 Evolution Occurs in Several Additional Ways

APPLICATIONS

Why We Care 12.1 *Dogs Are Products of Artificial Selection*
Burning Question 12.1 *Is there such a thing as a "pinnacle of evolution"?*
Why We Care 12.2 *The Unending War with Bacteria*
Investigating Life 12.1 *Bacterial Evolution Goes "Hog Wild" on the Farm*

Try to Eat Me Now. When threatened, a porcupinefish swallows air, puffing into a round shape that is too big for many predators to swallow. Sharp spines add a further deterrent. Why do porcupinefish have these signature traits? Natural selection provides the answer.

©Steven Hunt/Stone/Getty Images

Learn How to Learn
Take the Best Possible Notes

Some students take notes only on what they consider "important" during a lecture. Others write down words but not diagrams. Both strategies risk losing vital information and connections between ideas that could help in later learning. Instead, write down all you can during lecture, including examples, analogies, and sketches of the diagrams the instructor uses. It will be much easier to study later if you have a complete picture of what happened in every class.

SURVEY THE LANDSCAPE
Evolution and Diversity

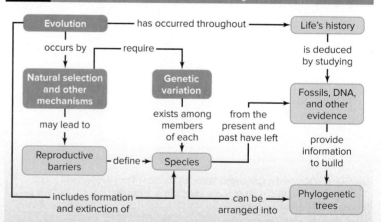

Evolution occurs in many ways. The most familiar mechanism is natural selection, but genetic changes from one generation to the next also happen by mutation, genetic drift, nonrandom mating, and migration.

For more details, study the Pull It Together feature in the chapter summary.

What's the **Point?** ▼

As the chapters in this unit will repeatedly demonstrate, evolution occurs everywhere and is obvious in many ways. Evolution explains the porcupinefish's defenses, the unusual shape of the bleeding heart flower, and countless features of other species from microbes to humans.

Evolution serves as an enormously compelling conceptual framework for many observations about life. In fact, geneticist Theodosius Dobzhansky gave this title to a much-quoted article he wrote: "Nothing in Biology Makes Sense Except in the Light of Evolution." This chapter explains how this powerful process works.

Evolution does not, however, answer one question that fascinates many people: How did life begin in the first place? Because little evidence remains from life's ancient origin, this question is difficult to answer scientifically; chapter 15 describes some of what we do know.

©C Squared Studios/
Getty Images RF

12.1 Evolution Acts on Populations

Scientific reasoning has profoundly changed thinking about the origin of species. Just 250 years ago, no one knew Earth's age. A century later, scientists learned that Earth is millions of years old (or older), but many believed that a creator made all species in their present form. Today's scientists, relying on a wide range of evidence, accept evolution as the explanation for life's diversity.

But what *is* evolution? A simple definition of **evolution** is descent with modification. "Descent" implies inheritance; "modification" refers to changes in traits from generation to generation. For example, we see evolution at work in the lions, tigers, and leopards that all descended from one ancestral cat species.

Evolution has another, more specific, definition as well. Recall from chapter 7 that a gene is a DNA sequence that encodes a protein; in part, an organism's proteins determine its traits. Moreover, each gene can have multiple versions, or alleles. We have also seen that a **population** consists of interbreeding members of the same species (see figure 1.2). Biologists say that evolution occurs in a population when some alleles become more common, and others less common, from one generation to the next. A more precise definition of evolution, then, is genetic change in a population over multiple generations.

According to this definition, evolution is detectable by examining a population's **gene pool**—its entire collection of genes and their alleles. Evolution is a change in **allele frequencies;** an allele's frequency is calculated as the number of copies of that allele, divided by the total number of alleles in the population. Suppose, for example, that a gene has two possible alleles, *A* and *a*. In a population of 100 diploid individuals, the gene has 200 alleles. If 160 of those alleles are *a*, then the frequency of *a* is 160/200, or 0.8. In the next generation, *a* may become either more or less common. Because an individual's alleles do not change, *evolution occurs in populations, not in individuals.*

The allele frequencies for each gene determine the characteristics of a population (figure 12.1). Many people in Sweden, for example, have alleles conferring blond hair and blue eyes; a population of Asians would contain many more alleles specifying darker hair and eyes. If Swedes migrate to Asia and interbreed with the locals (or vice versa), regional allele frequencies change.

Figure 12.1 **Same Genes, Different Alleles.** Human populations originating in different regions of the world have unique allele frequencies. Blond hair and blue eyes are typical of people from northern European countries, whereas people originating on the Asian continent usually have darker coloration.
(woman): ©Stockdisc/PunchStock RF; (man): ©Red Chopsticks/Getty Images RF

Miniglossary	Populations and Evolution
Population	Interbreeding members of the same species
Gene pool	A population's entire collection of genes and alleles
Allele frequency	The number of copies of an allele, divided by the total number of alleles for the same gene in the population
Microevolution	Small, generation-by-generation changes to a population's gene pool
Macroevolution	Large-scale evolutionary events, such as the appearance of new species

Some people use the term *microevolution* to refer to the small, generation-by-generation changes occurring in every population or species. This chapter describes the most common ways that such changes occur. Over long periods, these same processes give rise to what is sometimes called *macroevolution*, which includes the appearance of new species. Chapter 14 explains this process in more detail, and chapters 15 through 17 explore the resulting diversity of life.

12.1 Mastering Concepts

1. What are two ways to define evolution?
2. Why can't evolution act on individuals?

12.2 Evolutionary Thought Has Evolved for Centuries

Although Charles Darwin typically receives credit for developing the theory of evolution, people began pondering life's diversity well before his birth. This section offers a brief glimpse into the history of evolutionary thought.

A. Many Explanations Have Been Proposed for Life's Diversity

People have tried to explain the diversity of life for a very long time (figure 12.2). In ancient Greece, Aristotle recognized that all organisms are related in a hierarchy of simple to complex forms, but he believed that all members of a species were created identical to one another. This idea influenced scientific thinking for nearly 2000 years.

Several other ideas were also considered fundamental principles of science well into the 1800s. Among them was the concept of a "special creation," the sudden appearance of organisms on Earth. People believed that this creative event was planned and purposeful, that species were fixed and unchangeable, and that Earth was relatively young.

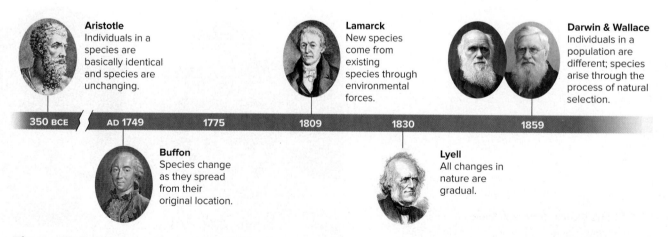

Aristotle Individuals in a species are basically identical and species are unchanging.

Lamarck New species come from existing species through environmental forces.

Darwin & Wallace Individuals in a population are different; species arise through the process of natural selection.

350 BCE   AD 1749   1775   1809   1830   1859

Buffon Species change as they spread from their original location.

Lyell All changes in nature are gradual.

Figure 12.2 Early Evolutionary Thought. Many scientists made significant contributions over the years, developing the foundation that Charles Darwin and Alfred Russel Wallace used to describe natural selection as the mechanism for evolution.

Evolution Acts on Populations   Evolutionary Thought Evolved for Centuries   Natural Selection Molds Evolution   Evolution Is Inevitable in Real Populations

Scientists struggled to reconcile these beliefs with compelling evidence that species could in fact change. Fossils, which had been discovered at least as early as 500 BCE, were at first thought to be oddly shaped crystals or faulty attempts at life that arose spontaneously in rocks. But by the mid-1700s, the increasingly obvious connection between living organisms and fossils argued against these ideas.

Scientists used religious stories to explain the existence of fossils without denying the role of a creator. Yet some of the fossils depicted organisms not seen before. Because people believed that species created by God could not become extinct, these fossils presented a paradox. The conflict between ideology and observation widened as geologists discovered that different rock layers revealed different groups of fossilized organisms, many of them now extinct (figure 12.3).

Newer rock layers

Older rock layers

Figure 12.3 **Earth's History Revealed in Rocks.** Sedimentary rock layers, such as these in the Grand Canyon, often contain fossil evidence of organisms that lived (and died) when the layer was formed. Layers near the bottom are older than those on top.
©Terry Moore/Stocktrek Images/Getty Images RF

In 1749, French naturalist Georges-Louis Buffon became one of the first to openly suggest that closely related species arose from a common ancestor and were changing—a radical idea at the time. By moving the discussion into the public arena, he made possible a new consideration of evolution and its causes from a scientific point of view. Still, no one had proposed how species might change.

Then, in 1809, French taxonomist Jean Baptiste de Lamarck proposed the first scientifically testable evolutionary theory. He reasoned that organisms that used one part of their body repeatedly would increase their abilities, very much like weight lifters developing strong arms. Conversely, disuse would weaken an organ until it disappeared. Lamarck surmised (incorrectly) that these changes would pass to future generations.

With these new theories and ideas, people were beginning to accept the concept of evolution but did not yet understand how it could result in the formation of new species. Ultimately, Charles Darwin (1809–1882) became the first to solve this puzzle.

B. Charles Darwin's Voyage Provided a Wealth of Evidence

Young Charles Darwin attended Cambridge University in England and, at the urging of his family, completed studies to enter the clergy. Meanwhile, he also followed his own interests. He joined geological field trips and met several eminent geology professors.

Eventually, Darwin was offered a position as a collector and captain's companion aboard the HMS *Beagle*. Before the ship set sail for its 5-year voyage in 1831 (figure 12.4), the botany professor who had arranged Darwin's position gave the young man the first volume of Charles Lyell's *Principles of Geology*. Darwin picked up the second and third volumes in South America. By the time he had finished reading these works, Darwin was an avid proponent of Lyell's idea that natural processes are slow and steady and that Earth is much older than previously thought—perhaps millions or hundreds of millions of years old.

Darwin recorded his observations as the ship journeyed around the coast of South America. He noted forces that uplifted new land, such as

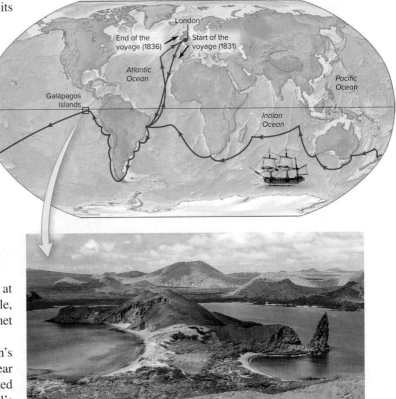

Figure 12.4 **The Voyage of the *Beagle*.** Darwin observed life and geology throughout the world during the journey of the HMS *Beagle*. Many of Darwin's ideas about natural selection and evolution had their origins in the observations he made on the Galápagos Islands.
Photo: ©DC_Colombia/iStock/Getty Images RF

Why We Care 12.1 | Dogs Are Products of Artificial Selection

People have been breeding dogs for thousands of years, beginning with domesticated wolves. Dog fanciers now recognize hundreds of breeds, each the product of artificial selection for a different trait that originally occurred as natural genetic variation. Bloodhounds, for example, are selected for their keen sense of smell. Border collies herd livestock (or anything else that moves), and the sleek greyhound is bred for speed.

Behind the carefully bred traits, however, lurk small gene pools and extensive inbreeding, which may harm the health of purebred show animals (table 12.A). Dog breeders can select for desired characteristics, but they can't always avoid the hereditary health problems associated with each breed.

To take one familiar example, dachshunds have extended torsos, short legs, and a bold demeanor. These traits make dachshunds excellent hunters of badgers and other animals that live in dens. Unfortunately, these dogs often suffer from painful degeneration of the discs between the vertebrae; skin, thyroid, and eye problems are also common in the breed.

TABLE 12.A Purebred Plights

Breed	Health Problem(s)
Cocker spaniel	Nervousness, ear infections, hernias, kidney problems
Collie	Blindness, bald spots, seizures
Dalmatian	Deafness
German shepherd	Hip dysplasia
Golden retriever	Lymphatic cancer, muscular dystrophy, skin allergies, hip dysplasia, absence of one testicle
Great Dane	Heart failure, bone cancer
Labrador retriever	Dwarfism, blindness
Shar-pei	Skin disorders

©Liliya Kulianionak/Shutterstock RF

earthquakes and volcanoes, and the constant erosion that wore it down. He marveled at forest plant fossils interspersed with sea sediments and at shell fossils in a mountain cave. Darwin tried to reconstruct the past from contemporary observations and wondered how each fossil had arrived where he found it.

In the fourth year of the voyage, the HMS *Beagle* spent 5 weeks in the Galápagos Islands, off the coast of Ecuador. The notes and samples Darwin brought back would form the seed of his theory of evolution by natural selection.

C. *On the Origin of Species* Proposed Natural Selection as an Evolutionary Mechanism

Toward the end of the voyage, Darwin began to assimilate all he had seen and recorded. Pondering the great variety of organisms in South America and their relationships to fossils and geology, he began to think that these were clues to how new species originate.

Descent with Modification Darwin returned to England in 1836, and by 1837 he had begun assembling his notes in earnest. In March 1837, Darwin consulted ornithologist (bird expert) John Gould about the finches and other birds that the *Beagle* brought back from the Galápagos Islands. Gould could tell from bill structures that some of the finches ate small seeds, whereas others ate large seeds, fruits, or insects. In all, he described 13 distinct types of finch, each different from the birds on the mainland yet sharing some features.

Darwin thought that all of the finch species on the Galápagos had probably descended from a single ancestral type of finch that had flown to the islands and, finding a relatively unoccupied new habitat, flourished. Over the next few million years, the finch population gradually branched in several directions. Different groups ate insects, fruits, and seeds of different sizes, depending on the resources each island offered. Darwin also noted changes in other species, including Galápagos tortoises. He coined the phrase "descent with modification" to describe gradual changes from an ancestral type.

Malthus's Ideas on Populations In September 1838, Darwin read a work that helped him further understand the diversity of finches on the Galápagos Islands. Economist and theologian Thomas Malthus's *Essay on the Principle of Population,* written 40 years earlier, stated that food availability, disease, and war limit the size of a human population. Wouldn't other organisms face similar limitations? If so, then individuals that could not obtain essential resources would die. ⓘ *human population, section 18.6*

The insight Malthus provided was that individual members of a population were not all the same, as Aristotle had taught. Instead, individuals better able to obtain resources were more likely to survive and reproduce. This would explain the observation that more individuals are produced in a generation than survive; they do not all obtain enough vital resources to live. Over time, environmental challenges would eliminate the more poorly equipped variants, and gradually, the population would change.

The Concept of Natural Selection Darwin used the term *natural selection* to describe "this preservation of favourable variations and the rejection of injurious variations." Biologists later modified the definition to add modern genetics terminology. We now say that **natural selection** occurs when individuals with certain genotypes—those that are best suited to the environment—have greater reproductive success than other individuals.

Darwin got the idea of natural selection from thinking about artificial selection (also called selective breeding). In **artificial selection,** a human chooses one or a few desired traits, such as milk production or leaf size, and then allows only the individuals that best express those qualities to reproduce (figure 12.5). Artificial selection is responsible not only for agriculturally important varieties of animals and plants but also for the many breeds of domesticated cats and dogs (see Why We Care 12.1). Darwin himself raised pigeons and developed several new breeds by artificial selection.

How did natural selection apply to the diversity of finch species on the Galápagos? Long ago, some finches flew from the mainland to one island. Eventually, that island population outgrew the supply of small seeds, and birds that could eat nothing else starved. But finches that could eat other things, perhaps because of an inherited difference in bill structure, survived and reproduced. Since their food was plentiful, these once-unusual birds gradually came to make up more of the population.

Darwin further realized that he could extend this idea to multiple islands, each of which had a slightly different habitat and therefore selected for different varieties of finches. A new species might have arisen when a population adapted to so many new conditions that its members could no longer breed with the original group (see chapter 14). In a similar way, new species have evolved throughout the history of life as populations have adapted to different resources. All species are therefore ultimately united by common ancestry.

Publication of *On the Origin of Species* Darwin continued to work on his ideas until 1858, when he received a manuscript from British naturalist Alfred Russel Wallace. Wallace had observed the diverse insects, birds, and mammals of South America and southeast Asia, and his manuscript independently proposed that natural selection was the driving force of evolution.

Both Darwin's and Wallace's papers were presented at a scientific conference later that year. In 1859, Darwin finally published the 490-page *On the Origin of Species by Means of Natural Selection, or Preservation of Favoured Races in the Struggle for Life.* It would form the underpinning of modern life science.

Table 12.1 summarizes Darwin's main arguments in support of natural selection. He began with three observations. First, individuals in a species are different from one another, and at least some of this variation is heritable. Second, essential resources such as food and space are limited in every habitat. Third, in every population, more offspring are born than can survive. These observations led Darwin to infer that organisms engage in a "struggle for existence"—that is, they must compete for scarce resources. He also inferred that those individuals with the most adaptive traits would be most likely to "win" the competition, reproduce, and pass those favorable traits to the next generation. Darwin's final inference was that over many generations, natural selection could change a population or even give rise to new species. Darwin's own observations of ants, pigeons, orchids, and many other organisms provided abundant support for his ideas.

Some members of the scientific community happily embraced Darwin's efforts. Upon reading *On the Origin of Species,* his friend Thomas Henry Huxley remarked, "How stupid of me not to have thought of that." Others, however, were less appreciative. People in some religious denominations perceived a clash with their beliefs that all life arose from separate special creations, that species did not change, and that nature was harmonious and purposeful. Perhaps most disturbing to many people was the idea that humans were just one more species competing for resources.

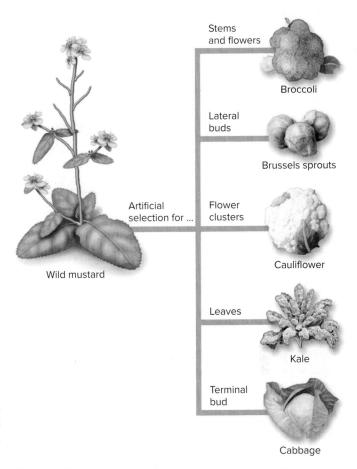

Figure 12.5 Artificial Selection. By selecting for different traits, plant breeders used one type of wild mustard to create all five of these vegetable varieties.

TABLE 12.1	**The Logic of Natural Selection: A Summary**

Observations of Nature

1. **Genetic variation:** Within a species, no two individuals are exactly alike. Some of this variation is heritable.
2. **Limited resources:** Every habitat contains limited supplies of the resources required for survival.
3. **Overproduction of offspring:** More individuals are born than survive to reproduce.

Inferences from Observations

1. **Struggle for existence:** Individuals compete for the limited resources that enable them to survive.
2. **Unequal reproductive success (natural selection):** The inherited characteristics of some individuals make them more likely to obtain resources, survive, and reproduce.
3. **Descent with modification:** Over many generations, a population's characteristics can change by natural selection, even giving rise to new species.

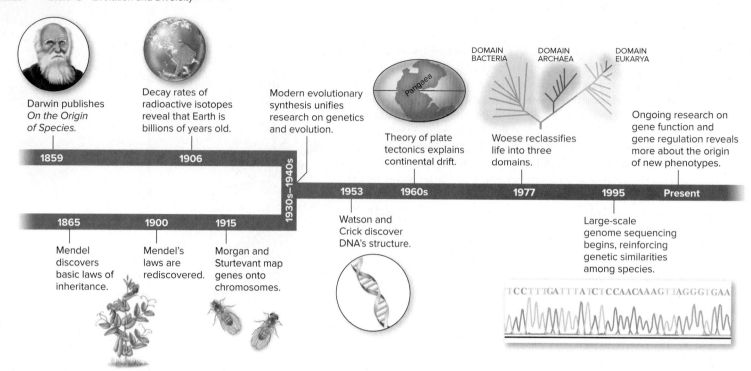

Figure 12.6 Evolutionary Theory Since Darwin. Charles Darwin and Gregor Mendel laid a foundation for evolutionary theory, but thousands of scientists since that time have added to our understanding of evolution. This figure shows a few examples of their work.

©IT Stock/PunchStock RF

D. Evolutionary Theory Continues to Expand

Although Charles Darwin's arguments were fundamentally sound, he could not explain all that he saw. For instance, he did not understand the source of variation within a population, nor did he know how heritable traits were passed from generation to generation. Ironically, Austrian monk Gregor Mendel was solving the puzzle of inheritance at the same time that Darwin was pondering natural selection (see chapter 10). Mendel's work, however, remained obscure until after Darwin's death.

Since Darwin's time, scientists have learned much more about genes, chromosomes, and the origin and inheritance of genetic variation (figure 12.6). In the 1930s, scientists finally recognized the connection between natural selection and genetics. They unified these ideas into the **modern evolutionary synthesis,** which suggests that genetic mutations create heritable variation and that this variation is the raw material upon which natural selection acts.

After the discovery of DNA's structure in the 1950s, the picture became even clearer. We now know that mutations are changes in DNA sequence (see chapter 7) and that mutations occur at random in all organisms. Sexual reproduction amplifies this variability by shuffling and reshuffling parental alleles to produce genetically different offspring (see chapter 9). Today, overwhelming evidence supports the theory of evolution by natural selection (see chapter 13); even so, some misconceptions about evolution persist (table 12.2).

TABLE 12.2 Selected Misconceptions About Evolutionary Theory

Misconception	Biologist's Explanation
Biological evolution explains the origin of life.	Biological evolution did not begin until life existed.
Evolution is a random process.	Some mechanisms of evolution, such as mutations, do occur randomly. Natural selection, however, is nonrandom because the environment selects against poorly adapted individuals.
In a changing environment, all individuals in a population simultaneously develop beneficial adaptations.	Adaptations become "fixed" in a population over multiple generations, as individuals with beneficial adaptations are most likely to survive, reproduce, and pass their genes to the next generation.

12.2 Mastering Concepts

1. How might artificial selection and natural selection produce the same result? Which process would be faster? Why?

2. What did Darwin observe that led him to develop his ideas about the origin of species?

3. What is the modern evolutionary synthesis?

12.3 Natural Selection Molds Evolution

Natural selection is the most famous, and often the most important, mechanism of evolution. This section explains the basic requirements for natural selection to occur. The rest of this chapter describes natural selection and several other forces of evolutionary change in more detail.

A. Adaptations Enhance Reproductive Success

As Darwin knew, organisms of the same species are different from one another, and every population produces more individuals than resources can support (figure 12.7). Some members of any population will not survive to reproduce. A struggle for existence is therefore inevitable.

The variation inherent in each species means that some individuals in each population are better than others at obtaining nutrients and water, avoiding predators, tolerating temperature changes, attracting mates, or reproducing. The heritable traits conferring these advantages are **adaptations**—features that provide a selective advantage because they improve an organism's ability to survive and reproduce.

The word *adaptation* can be confusing because it has multiple meanings. For example, a student might say, "I have adapted well to college life," but short-term changes in an individual do not constitute evolution. Adaptations in the evolutionary sense include only those structures, behaviors, or physiological processes that are heritable and that contribute to reproductive success.

In any population, individuals with the best adaptations are most likely to reproduce and pass their advantage to their offspring. Because of this "differential reproductive success," a population changes over time, with the best available adaptations to the existing environment becoming more common with each generation (figure 12.8).

Natural selection requires preexisting genetic diversity. Ultimately, this diversity arises largely by chance. Nevertheless, it is important to realize that natural selection itself is not a random process. Instead, it selectively eliminates most of the individuals that are least able to compete for resources or cope with the prevailing environment.

a.

b.

Figure 12.7 **Requirements for Natural Selection.** (a) This group of people illustrates genetic variation within the human population. (b) Dandelions produce many offspring, but few survive.

(a): ©Punchstock/Sockbyte RF; (b): ©Angelo Cavalli/Getty Images RF

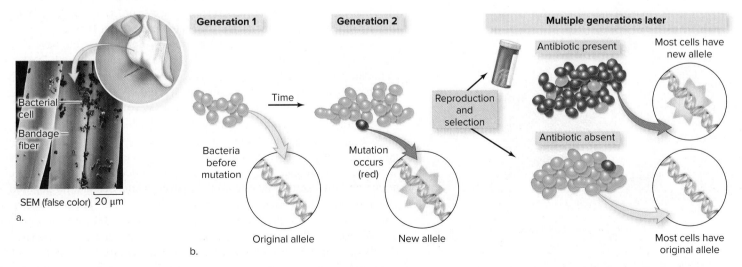

Figure 12.8 **Natural Selection.** (a) *Staphylococcus aureus* is a bacterium that causes skin infections. (b) By chance, a cell undergoes a random genetic mutation. The population is then exposed to an antibiotic. The drug kills most of the unmutated cells, but the mutated cell is unaffected and can reproduce. After many generations of exposure to the antibiotic, the mutation is common.

Photo: (a): ©Paul Gunning/Science Source

Three Modes of Natural Selection   Sexual Selection Influences Reproductive Success   Evolution Occurs in Several Additional Ways

Figure 12.9 **Struggle to Survive.** Newly hatched green sea turtles scurry toward the water. The tiny hatchlings are defenseless against shorebirds, crabs, and other predators; fewer than 1% will survive to maturity. On average, those with the most adaptive combination of traits have the greatest chance of survival.
©Image Source/PunchStock RF

B. Natural Selection Eliminates Poorly Adapted Phenotypes

Recall from chapter 10 that an individual's phenotype is its observable properties, most of which arise from a combination of environmental influences and the action of multiple genes. By "weeding out" individuals with poorly adapted phenotypes, natural selection indirectly changes allele frequencies in the population (figure 12.9). The scientific literature contains countless examples, some of which are described in the Investigating Life boxes of this book. For example, disease-causing bacteria select for sexual reproduction in nematode worms (Investigating Life 9.1); genetically modified cotton plants producing the Bt toxin eliminate Bt-susceptible moths from the population (Investigating Life 10.1); and antibiotics select for drug-resistant bacteria on farms (Investigating Life 12.1).

Environmental conditions constantly change, so evolution never really stops. After all, the phenotype that is "best" depends entirely on the time and place; a trait that is adaptive in one set of circumstances may become a liability in another. Some orchids, for example, produce flowers that are pollinated by only one or a few species of wasp (figure 12.10). The orchids release chemicals that mimic a female wasp's pheromones. When a male wasp visits the flower, seeking a mate, pollen from the orchid sticks to its body. The insect later deposits the pollen on another orchid. As long as wasps are present, this exclusive relationship benefits the plants; they do not waste energy by attracting animals that are unlikely to visit another flower of the same species. But if the wasps went extinct, the orchids could no longer reproduce. In that case, having an exclusive relationship with one or a few pollinator species would doom the orchids.

C. Natural Selection Does Not Have a Goal

Because most species have become more complex over life's long evolutionary history, many people erroneously believe that natural selection leads to ever more "perfect" organisms or that evolution works toward some long-term goal. Explanations that use the words *need* or *in order to* typically reflect this misconception. For example, a person might say, "The orchids started producing pheromones because they needed to attract wasps" or "The orchids make pheromones in order to trick the male wasps into visiting their flowers."

Both explanations are incorrect because evolution does not have a goal. How could it? No known mechanism allows the environment to tell DNA how to mutate and generate the alleles needed to confront future conditions. Nor does natural selection strive for perfection; if it did, the vast majority of species in life's history would still exist. Instead, most are extinct.

Several factors combine to prevent natural selection from producing all of the traits that a species might find useful. First, every genome has limited potential, imposed by its evolutionary history. The structure of the human skeleton, for example, will not allow for the sudden appearance of wheels, no matter how useful they might be on paved roads. Second, no population contains every allele needed to confront every possible change in the environment. If the right alleles aren't available at the right time, an environmental change may wipe out a species (figure 12.11). Third, disasters such as floods and volcanic eruptions can indiscriminately eliminate the best allele combinations, simply by chance (see section 12.7). And finally, some harmful genetic traits are out of natural selection's reach; one example is Huntington disease, a severe genetic illness that typically appears after reproductive age.

Figure 12.10 **Evolving Together.** The intimate relationship between this orchid and its pollinator (a wasp) is efficient for the plant. But if the insect goes extinct, the orchid pays the price, too.
©Tim Gainey/Alamy Stock Photo

a.

2 cm

b.

Figure 12.11 Extinction. (a) Sea scorpions once thrived worldwide but became extinct some 250 million years ago. (b) These petrified trees in the Arizona desert are from a plant family that is now extinct in the northern hemisphere, thanks to continental drift.

(a): ©Francois Gohier/Science Source; (b): ©B.A.E. Inc./Alamy Stock Photo RF

In contrast to natural selection, artificial selection does have a goal: Humans select for specific, desired traits. However, we affect other species in so many ways that the distinction between artificial selection and human-influenced natural selection can be confusing. In general, if humans alter the environment but do not select which individuals breed, the term *natural selection* applies.

D. What Does "Survival of the Fittest" Mean?

Natural selection is often called "the survival of the fittest," but this phrase is not entirely accurate or complete. In everyday language, the "fittest" individual is the one in the best physical shape: the strongest, fastest, or biggest. Physical fitness, however, is not the key to natural selection (although it may play a part).

Rather, in an evolutionary sense, **fitness** refers to an organism's genetic contribution to the next generation. A large, quick, burly elk scores zero on the evolutionary fitness scale if poor eyesight makes it vulnerable to an early death in the jaws of a wolf. On the other hand, a mayfly that dies in the act of producing thousands of offspring is highly fit.

Because successful reproduction is the only way for an organism to perpetuate its genes, fitness depends on the ability to survive just long enough to reproduce. Plants that grow, flower, produce seeds, and die within just a few weeks may have fitness equal to a redwood tree that lives for centuries (figure 12.12).

These examples illustrate an important point: *By itself, survival is not enough.* Paradoxically, natural selection promotes any trait that increases fitness, even if the trait virtually guarantees an individual's death. For example, a male praying mantis may not resist if the female begins to eat his head during copulation. The male's passive behavior is adaptive because the extra food the female obtains in this way will enhance the chance of survival for their young. Likewise, a male Australian redback spider somersaults his abdomen into the jaws of his mate during copulation. Unlike the mantis, his body is too small to offer the female a nutritional benefit. Instead, the male's suicidal behavior prolongs copulation, so that he sires the most possible offspring. As in the case of the mantis, the male spider does not survive—but his alleles will.

a.

b.

Figure 12.12 Fitness Is Reproductive Success. One key to fitness is living long enough to reproduce. (a) For a redwood, that may take centuries. (b) For an annual plant, it may take just a few weeks.

(a): ©Comstock/Jupiter Images RF; (b): ©Christopher PB/Shutterstock RF

Burning Question 12.1

Is there such a thing as a "pinnacle of evolution"?

One common illustration of human evolution depicts a slouching chimpanzee that transforms by several stages into a caveman and then into an upright modern human. The misleading implications are that chimps evolved into humans and that the process was goal-oriented and directional—in other words, that humans are the "pinnacle of evolution."

But viewing evolutionary processes as a ladder, with humans at the top and every other species evolving toward that "goal," is a mistake. Instead, evolution has produced a tree of life, with successful organisms at every branch tip. We occupy one of those tips.

We might therefore reframe the question to ask whether any species could colonize every habitat on Earth. Would humans fit the bill? No. We can survive in a great variety of locations, but we can make only brief excursions into water and onto the highest mountaintops. What about other species? Again, the answer is no. To take an extreme example, a trout's adaptations to a cold mountain stream are worthless in parched desert sands. Even ubiquitous pests such as cheat grass, dandelions, cockroaches, and rats can't withstand conditions that are too hot, too cold, too dry, or too wet. The variety of habitats on Earth—ocean, freshwater, tundra, prairie, desert, forest—is too great for one species to be able to live everywhere.

©Christopher Kimmel/Aurora Open/Getty Images RF

Submit your burning question to
marielle.hoefnagels@mheducation.com

Fitness includes not only the total number of offspring produced but also the proportion that reach reproductive age. Some organisms, including humans, have few offspring but invest large amounts of energy in each one. Insects and many other species produce thousands of young but invest minimally in each. The optimal balance between "quality" and "quantity" may vary greatly, even among individuals within a population. Section 18.4 further describes this evolutionary trade-off.

Many adaptations contribute to an organism's overall fitness. Being able to overcome poor weather conditions, combat parasites and other disease-causing organisms, evade predators, and compete for resources all enhance an organism's chance of reaching reproductive age. At that point, the ability to attract mates (or pollinators, in the case of many flowering plants) affects the number of offspring an organism produces.

Some people wonder whether humans (or any other species with a knack for outcompeting its rivals) could be considered a "pinnacle of evolution." Burning Question 12.1 discusses this possibility.

12.3 Mastering Concepts

1. What is an adaptation, and how do adaptations become more common within a population?
2. What is the role of genetic variation in natural selection?
3. How can natural selection favor different phenotypes at different times?
4. What is "fitness" in the context of evolution?

12.4 Evolution Is Inevitable in Real Populations

Shifting allele frequencies in populations are the small steps of change that collectively drive evolution. Given the large number of genes in any organism and the many factors that can alter allele frequencies (including but not limited to natural selection), evolution is not only possible but unavoidable. This section explains why.

A. At Hardy–Weinberg Equilibrium, Allele Frequencies Do Not Change

The study of population genetics relies on the intimate relationship between allele frequencies and **genotype frequencies.** Each genotype's frequency is the number of individuals with that genotype, divided by the total size of the population. For example, if 64 of the 100 individuals in a population are homozygous recessive, then the frequency of that genotype is 64/100, or 0.64.

Hardy–Weinberg equilibrium is the highly unlikely situation in which allele frequencies and genotype frequencies do not change from one generation to the next. It occurs only in populations that meet the following assumptions: (1) natural selection does not occur; (2) mutations do not occur, so no new alleles arise; (3) the population is infinitely large, or at least large enough to eliminate random changes in allele frequencies; (4) individuals mate at random; and (5) individuals do not migrate into or out of the population.

Figure 12.13 illustrates the equations that represent the relationship between allele frequencies and genotype frequencies. The figure shows a hypothetical population of ferrets that meets all of the assumptions of Hardy–Weinberg equilibrium. In the parent generation, the frequencies of alleles *D* and *d* are 0.6 and 0.4, respectively; the corresponding genotype frequencies are 0.36 for *DD*, 0.48 for *Dd*, and 0.16 for *dd*. The ferrets mate at random, as illustrated in the Punnett square. In the next generation, allele and genotype frequencies remain the same. Evolution therefore has not occurred. Conversely, if the allele and genotype frequencies had changed from one generation to the next, we would have concluded that the population had evolved.

Besides providing a framework for determining whether evolution has occurred, these equations are useful because they allow us to infer characteristics of a population based on limited information. One application is the use of known allele frequencies to estimate genotype frequencies in a population. DNA profiling, for example, relies on population databases that contain the known frequencies of each allele at 13 sites in the human genome. Forensic analysts can use this information to calculate the probability that two people share the same genotype across all 13 sites. ⓘ *DNA profiling, section 11.2D*

B. In Reality, Allele Frequencies Always Change

As illustrated in figure 12.13, a population at Hardy–Weinberg equilibrium does not evolve. A real population, however, violates some or all of the assumptions of Hardy–Weinberg equilibrium. Allele frequencies change when natural selection, mutations, genetic drift, nonrandom mating, or migration occurs.

All of these events are common. If you think about the human population, for example, natural selection reduces the frequency of alleles that cause deadly genetic illnesses in childhood. Moreover, genetic mutations can and do happen. The large size of the human population minimizes the chance of random changes in allele frequencies, but they occasionally occur (see section 12.7B). Finally, we do not choose our mates at random, and migration alters allele frequencies as we move and mix.

These forces act on populations of other species as well, so allele frequencies always change over multiple generations. In other words, evolution is inevitable. Even though its assumptions do not apply to real populations, the concept of Hardy–Weinberg equilibrium does serve as a basis of comparison to reveal when evolution is occurring. Additional studies can then reveal which mechanism of evolution is acting on the population. Sections 12.5 through 12.7 describe these mechanisms of evolution in more detail.

Allele frequencies: $p + q = 1$

Definition/equation	Example
p = frequency of dominant allele	p = frequency of D (dark fur) = 0.6
q = frequency of recessive allele	q = frequency of d (tan fur) = 0.4
$p + q = 1$	$0.6 + 0.4 = 1$

Genotype frequencies: $p^2 + 2pq + q^2 = 1$

Definition/equation	Example
p^2 = frequency of DD genotype	$0.6 \times 0.6 = 0.36$
$2pq$ = frequency of Dd genotype	$2 \times 0.6 \times 0.4 = 0.48$
q^2 = frequency of dd genotype	$0.4 \times 0.4 = 0.16$
$p^2 + 2pq + q^2 = 1$	$(0.6)^2 + (2 \times 0.6 \times 0.4) + (0.4)^2 = 1$

Reproduction (random mating)

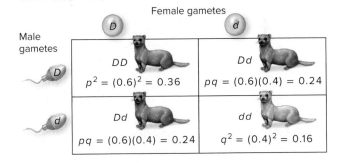

Female gametes — D — d

Male gametes

	D	d
D	DD — $p^2 = (0.6)^2 = 0.36$	Dd — $pq = (0.6)(0.4) = 0.24$
d	Dd — $pq = (0.6)(0.4) = 0.24$	dd — $q^2 = (0.4)^2 = 0.16$

Figure 12.13 **No Evolution.** At Hardy–Weinberg equilibrium, allele frequencies remain constant from one generation to the next; evolution does not occur.

Figure It Out

Assume that one in 3000 Caucasian babies in the United States is born with cystic fibrosis, a disease caused by a recessive allele. The value of q^2 is therefore $1/3000 = 0.0003$; q is the square root of 0.0003, or 0.018. Use this information to estimate the frequency of heterozygotes (symptomless carriers) in the American Caucasian population.

Answer: If $q = 0.018$, then $p = 0.982$; the frequency of heterozygotes is $2 \times 0.982 \times 0.018 = 0.035$, or 3.5%.

12.4 Mastering Concepts

1. What five conditions are required for Hardy–Weinberg equilibrium?

2. Why doesn't Hardy–Weinberg equilibrium occur in real populations?

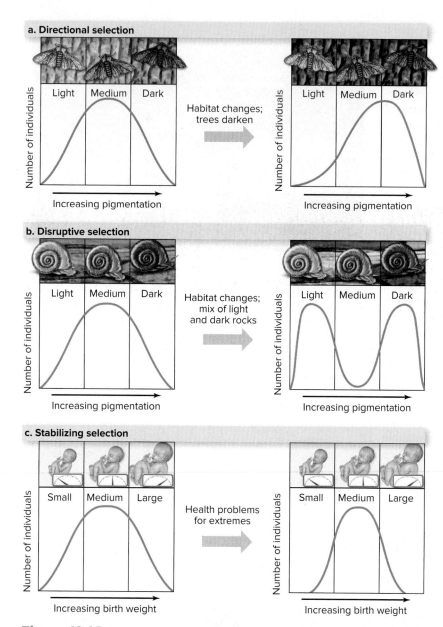

Figure 12.14 **Types of Natural Selection.** (a) Directional selection results from selection against one extreme phenotype. (b) In disruptive selection, two extreme phenotypes each confer a selective advantage over the intermediate phenotype. (c) Stabilizing selection maintains an intermediate expression of a trait by selecting against extreme variants.

12.5 Natural Selection Can Shape Populations in Many Ways

Of all the mechanisms by which a population can evolve, natural selection is probably the most important. Natural selection changes the genetic makeup of a population by favoring the alleles that contribute to reproductive success and selecting against those that do not.

Natural selection, however, does not eliminate alleles directly. Instead, individuals with the "best" phenotypes are most likely to pass their alleles to the next generation; those with poorly suited phenotypes are less likely to survive long enough to reproduce. Three modes of natural selection—directional, disruptive, and stabilizing—are distinguished by their effects on the phenotypes in a population (figure 12.14).

In **directional selection,** one extreme phenotype is fittest, and the environment selects against the others. A change in tree trunk color from light to dark, for example, may select for dark-winged moths and against light-winged individuals. The rise of antibiotic resistance among bacteria also reflects directional selection, as described in Why We Care 12.2; so does the increase in pesticide-resistant insects (see Investigating Life 10.1). The fittest phenotype may initially be rare, but its frequency increases over multiple generations as the environment changes—for example, after exposure to the antibiotic or insecticide.

In **disruptive selection,** two or more extreme phenotypes are fitter than the intermediate phenotype. Consider, for example, a population of marine snails that live among brown rocks encrusted with white barnacles. The white snails near the barnacles are camouflaged, and the dark brown ones on the bare rocks likewise blend in. The snails that are neither white nor dark brown are most often seen and eaten by predatory shorebirds.

A third form of natural selection, called **stabilizing selection,** occurs when extreme phenotypes are less fit than the optimal intermediate phenotype. Human birth weight illustrates this tendency to stabilize. Very small or very large newborns are less likely to survive than are babies of intermediate weight. By eliminating all but the individuals with the optimal phenotype, stabilizing selection tends to reduce the variation in a population. It is therefore most common in stable, unchanging environments.

These three models of natural selection might seem to suggest that for each trait, only one or a few beneficial alleles ought to persist in the population. The harmful alleles should gradually become less common until they disappear, while the others become "fixed" in the population. For some genes, however, multiple alleles persist indefinitely in the population at more or less constant frequencies—even harmful alleles. This situation seems contrary to natural selection; how can it occur?

One circumstance that can maintain harmful alleles occurs when an individual with two different alleles for a gene (a heterozygote) has greater fitness than those whose two alleles are identical (homozygotes). Heterozygotes can maintain a harmful recessive allele in a population, even if homozygous recessive individuals have greatly reduced fitness.

The best documented example in humans is sickle cell disease. The disease-causing allele encodes an abnormal form of hemoglobin. The abnormal hemoglobin proteins do not fold properly; instead, they form chains that bend a red blood cell into a characteristic sickle shape (see figure 7.14). In a person who is homozygous recessive for the sickle cell allele, all the red blood cells are affected. Symptoms include anemia, joint pain, a swollen spleen, and frequent, severe infections; the person may not live long enough to reproduce. On the other hand, a person who is heterozygous for the sickle cell allele is only mildly affected. Some of his or her red blood cells may take abnormal shapes, but the resulting mild anemia is not usually harmful.

Still, if the sickle cell allele causes such severe problems in homozygous recessive individuals, why hasn't natural selection eliminated it from the population? The answer is that heterozygotes also have a reproductive edge over people who are homozygous for the *normal* hemoglobin allele. Specifically, heterozygotes are resistant to a severe infectious disease, malaria. When a mosquito carrying a protist called *Plasmodium* feeds on a human with normal hemoglobin—a homozygous dominant person—the parasite enters the red blood cells. Eventually, infected blood cells burst, and the parasite travels throughout the body. The resulting bouts of fever, chills, fatigue, and nausea are severe; if damaged blood cells block blood vessels, the patient may die of organ failure. But the sickled red blood cells of an infected carrier—a heterozygote—halt the parasite's spread. ⓘ *malaria*, section 15.4D

Not surprisingly, the sickle cell allele's frequency is highest in parts of the world where malaria is most common (figure 12.15). In these areas, the heterozygotes remain healthiest; they are resistant to malaria, but they are not seriously ill from sickle cell disease. They therefore have more children (on average) than do people who are homozygous for either allele. Unfortunately, two carriers have a 25% chance of producing a child who is homozygous recessive (see figure 12.15b). These children pay the evolutionary price for the genetic protection against malaria.

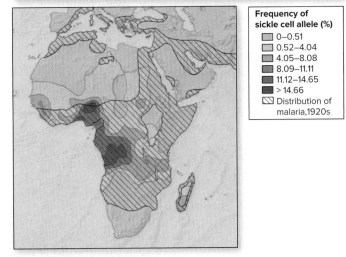

a. Distribution of sickle cell allele and malaria in humans

Frequency of sickle cell allele (%)
- 0–0.51
- 0.52–4.04
- 4.05–8.08
- 8.09–11.11
- 11.12–14.65
- > 14.66
- Distribution of malaria, 1920s

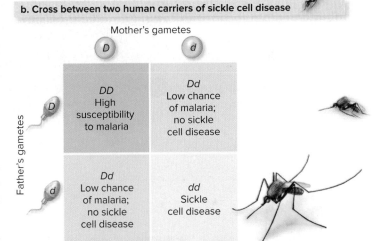

b. Cross between two human carriers of sickle cell disease

Mother's gametes

Father's gametes

	D	d
D	DD High susceptibility to malaria	Dd Low chance of malaria; no sickle cell disease
d	Dd Low chance of malaria; no sickle cell disease	dd Sickle cell disease

Figure 12.15 Heterozygote Advantage. (a) The sickle cell allele is common where malaria is prevalent. (b) In regions where malaria thrives, heterozygotes for the sickle cell allele are most likely to survive long enough to reproduce. However, two heterozygotes have a 25% chance of producing a child with sickle cell anemia.

12.5 Mastering Concepts

1. Distinguish among directional, disruptive, and stabilizing selection.
2. How can natural selection maintain harmful alleles in a population?

Why We Care 12.2 | The Unending War with Bacteria

Antibiotics revolutionized medical care in the twentieth century and enabled people to survive many once-deadly bacterial infections. Unfortunately, the miracle of antibiotics is under threat, and many infections that once were easily treated are re-emerging as killers. Ironically, the overuse and misuse of antibiotics are partly responsible for the problem. Physicians overprescribe antibiotics, patients fail to take them as directed, and farmers feed them to livestock even in the absence of infection.

By saturating the environment with antibiotics, we have profoundly affected the evolution of bacteria (see Investigating Life 12.1). Antibiotics kill susceptible bacteria and leave the resistant ones alone. The survivors multiply, producing a new generation of antibiotic-resistant bacteria. This is an example of natural selection in action.

Today, many infections are caused by so-called "superbugs" that resist all antibiotics but one. Worse, some laboratory strains are resistant to all antibiotics. Researchers fear that the discovery of new antibiotics will not keep up with the rate at which resistant strains evolve and spread around the world.

Everyone can help fight antibiotic-resistant bacteria. Physicians should prescribe antibiotics only when needed, and patients should take the whole prescription as directed. Frequent hand-washing with soap and water also helps prevent the spread of disease.

a.

12.6 Sexual Selection Directly Influences Reproductive Success

In many vertebrate species, the sexes look alike. The difference between a male and a female house cat, for example, is not immediately obvious. In some species, however, natural selection can maintain a **sexual dimorphism,** which is a difference in appearance between males and females. One sex may be much larger or more colorful than the other, or one sex may have distinctive structures such as horns or antlers.

Some of these sexually dimorphic features may seem to violate natural selection. For example, female cardinals are brown and inconspicuous, but their male counterparts have vivid red feathers that make them much more visible to predators. Similarly, the extravagant tail of a peacock is brightly colored and makes flying difficult. How can natural selection allow for traits that apparently reduce survival?

The answer is that a special form of natural selection is at work. **Sexual selection** is a type of natural selection resulting from variation in the ability to obtain mates. If female cardinals prefer bright red males, then showy plumage directly increases a male's chance of reproducing. Because the brightest males get the most opportunities to mate, alleles that confer red plumage are common in the population.

Figure 12.16 Fighting to Mate. (a) Two bighorn rams butt heads in Montana. (b) Male ox beetles fight during the mating season.

(a): ©mlharing/Getty Images RF;
(b): ©James H. Robinson/Science Source

b.

Sexual selection has two forms. In one type, the members of one sex compete among themselves for access to the opposite sex (figure 12.16). Mate choice plays no part in deciding the winner. Male bighorn sheep, for example, use their horns to battle for the right to mate with multiple females. The strongest rams are therefore the most likely to pass on their alleles. In the other type of sexual selection, the members of one sex (usually female) choose their mates from among multiple individuals of the opposite sex (figure 12.17).

Why do males usually show the greatest effects of sexual selection? In most (but not all) vertebrate species, females spend more time and energy rearing offspring than do males. Because of this high investment in reproduction, females tend to be selective about their mates. Males are typically less choosy and must compete for access to females.

The evolutionary origin of the males' elaborate ornaments remains an open question. One possibility is that long tail feathers and bright colors are costly to produce and maintain; they are therefore indirect advertisements of good health or disease resistance. Likewise, the ability to win fights with competing males could also be an indicator of good genes. A female who chooses a high-quality male will increase not only his fitness but also her own.

a.   b.

Figure 12.17 Attracting Females. (a) Male weaver birds build nests, then females select mates based on nest quality. (b) The male bird-of-paradise displays bright plumes that attract females.

(a): ©James Warwick/Getty Images; (b): ©TeeJe/Moment/Getty Images RF

12.6 Mastering Concepts

1. How does sexual selection promote traits that decrease survival?
2. Describe two ways in which competition for access to mates can lead to sexual selection.

12.7 Evolution Occurs in Several Additional Ways

Natural selection is responsible for adaptations that enhance survival and reproduction, but it is not the only mechanism of evolution. This section describes four more ways that a population can evolve: mutation, genetic drift, nonrandom mating, and migration. All occur frequently, and each can, by itself, disrupt Hardy–Weinberg equilibrium. The changes in allele frequencies that constitute evolution therefore occur nearly all the time.

A. Mutation Fuels Evolution

A change in an organism's DNA sequence introduces a new allele to a population. The new trait may be harmful, neutral, or beneficial, depending on how the mutation affects the sequence of the encoded protein.

Mutations are the raw material for evolution because genes contribute to phenotypes, and natural selection acts on phenotypes. For example, random mutations in bacterial DNA may change the shapes of key proteins in the cell's ribosomes or cell wall. Some of these mutations may confer a new phenotype—resistance to an antibiotic, for example. If exposure to antibiotics selects for that phenotype, then the mutations will pass to the next generation. ⓘ *mutations*, section 7.6

As we saw in section 12.3C, one common misconception about evolution is that a mutation produces a novel adaptation precisely when a population "needs" it to confront a new environmental challenge. For example, many people mistakenly believe that antibiotics *create* resistance; that is, that resistance arises in bacteria *in response* to exposure to the drugs.

In reality, genes do not "know" when to mutate. The only way antibiotic resistance arises is if some bacteria happen to have a mutation that confers resistance, *before* exposure to the drug. The drug creates a situation in which these variants can flourish. That trait will then become more common within the population by natural selection. If no bacteria start out resistant, the drug kills the entire population.

A mutation affects evolution only if subsequent generations can inherit it. In asexually reproducing organisms such as bacteria, each mutated cell gives rise to mutant offspring (if the mutation does not prevent reproduction). In a multicellular organism, however, a mutation can pass to the next generation only if it arises in a germ cell (i.e., one that will give rise to gametes; see chapter 9). For example, a cigarette smoker with lung cancer will not pass any smoking-induced mutations to her children because her egg cells will not contain the altered DNA.

B. Genetic Drift Occurs by Chance

Genetic drift is a change in allele frequencies that occurs purely by chance. Unlike mutation, which increases diversity, genetic drift tends to eliminate alleles from a population.

All forms of genetic drift are rooted in sampling error, which occurs when a sample does not match the larger group from which it is drawn (figure 12.18). Such sampling errors are most likely to affect small populations. Suppose, for example, that one allele of a gene occurs at a very low frequency in a population. If, by chance, none of the individuals carrying the rare allele happens to reproduce, that variant will disappear from the population. Even if some do reproduce, the allele still might not pass to the next generation. After all, the events of meiosis ensure that each allele has only a 50% chance of passing to

Miniglossary | Mechanisms of Evolution

Natural selection	Differential reproductive success based on inherited traits
Mutation	A change in an organism's DNA sequence
Genetic drift	Change in allele frequencies that occurs purely by chance; includes the founder and bottleneck effects
Nonrandom mating	Choosing mates based on location, physical traits, or other factors
Migration	Movement of individuals or alleles into or out of a population

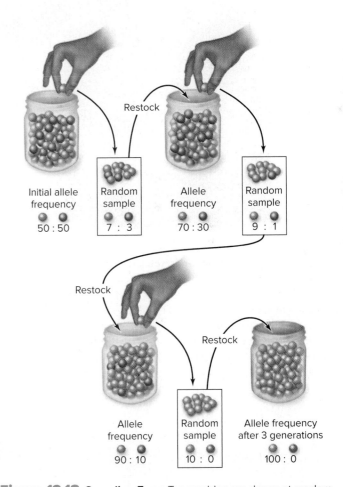

Figure 12.18 Sampling Error. Ten marbles are drawn at random from the first jar (*upper left*). Even though 50% of the marbles in the jar are blue, the random sample contains only 30% blue marbles. In the next "generation," 30% of the marbles are blue, but the random sample contains just one blue marble. The third sample contains no blue marbles and therefore eliminates the blue "allele" from the population, purely by chance.

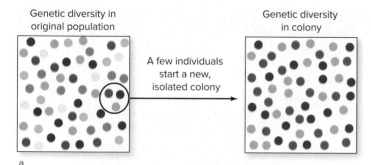

Genetic diversity in
original population

A few individuals
start a new,
isolated colony

Genetic diversity
in colony

a.

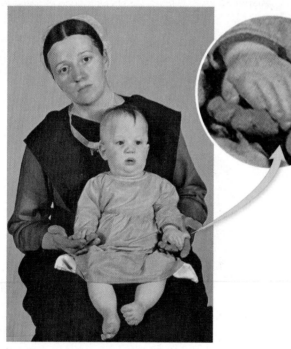

b.

Figure 12.19 **The Founder Effect.** (a) A few individuals leave
their original population and begin a new colony. The new population
has an altered allele frequency—and less genetic diversity—
compared to the original population. (b) Ellis–van Creveld syndrome
is characterized by dwarfism, extra fingers, and other symptoms. This
recessive disorder occurs in 7% of the Old Order Amish population of
Lancaster County, Pennsylvania, but is extremely rare elsewhere.

(b): ©Dr. Victor A. McKusick/Johns Hopkins Hospital

each offspring. A rare allele can therefore vanish from a population—not
because it reduces fitness but simply by chance.

The Founder Effect   One cause of genetic drift is the **founder effect,**
which occurs when a small group of individuals leaves its home population
and establishes a new, isolated settlement. The small group's random "allele
sample" may not represent the allele frequencies of the original population.
Some traits that were rare in the original population may therefore be more
frequent in the new population. Likewise, other traits will be less common or
may even disappear.

The Amish people of Pennsylvania provide a famous example of the
founder effect. About 200 followers of the Amish denomination immigrated to North America from Switzerland in the 1700s. One couple,
who immigrated in 1744, happened to carry the recessive allele associated with Ellis–van Creveld syndrome (figure 12.19). This allele
is extremely rare in the population at large. Intermarriage among
the Amish, however, has kept the disease's incidence high in this
subgroup more than two centuries after the immigrants arrived.

The Bottleneck Effect   Genetic drift also may result from a
population **bottleneck,** which occurs when a population's size drops
rapidly over a short period. The bottleneck randomly eliminates many
alleles that were present in the larger ancestral population (figure 12.20).
Even if the few remaining individuals mate and restore the population's numbers, the loss of genetic diversity is permanent. Note that the loss of alleles is
random; the bottleneck effect is therefore different from natural selection,
which weeds out alleles that reduce fitness.

Cheetahs are currently undergoing a population bottleneck. Until 10,000
years ago, these cats were common in many areas. Today, just two isolated
populations live in South and East Africa, numbering only a few thousand
animals. Inbreeding has made the South African cheetahs so genetically alike
that even unrelated animals can accept skin grafts from each other. Researchers
attribute this genetic uniformity to at least two bottlenecks: one that occurred
at the end of the most recent ice age, when habitats changed drastically, and
another when humans slaughtered many cheetahs during the 1800s.

North American species have undergone severe bottlenecks as well. Bison,
for example, nearly went extinct because of overhunting in the 1800s. And habitat
loss has caused the population of greater prairie chickens to plummet from about

Genetic diversity in
original population

Genetic diversity after
bottleneck event

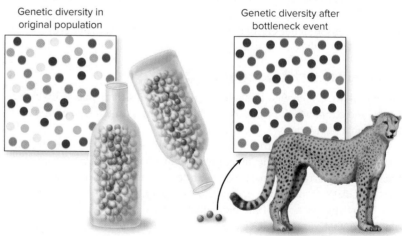

Figure 12.20 **The Bottleneck Effect.** A drastic decline in a
population's size—a population bottleneck—eliminates many alleles
at random. Even if the survivors rebuild the population, their genetic
diversity is greatly reduced relative to the ancestral population.

Original cheetah population
contains 25 different alleles
of a particular gene.

Cheetah population
is drastically reduced.

Repopulation occurs.
Only three different
alleles remain.

100 million in 1900 to several hundred today. The loss of genetic diversity in cheetahs, bison, and prairie chickens invites disaster: A single change in the environment might doom them all.

C. Nonrandom Mating Concentrates Alleles Locally

In a population with completely random mating, every individual has an equal chance of mating with any other member of the population. Wind-pollinated plants illustrate random mating, as do sponges and other aquatic animals that cannot move; these animals simply release their sperm into the water.

In most species, however, mating is not random. As we have already seen, many animals exhibit some preference in mate choice, including sexual selection (figure 12.21). Many other factors also influence mating, including geographical restrictions and physical access to the opposite sex. In humans, mate selection and culture are intertwined; many people choose local partners with a level of education, income, and religious beliefs similar to their own.

The practice of artificial selection is another way to reduce random mating. As described in section 12.2C, humans select those animals or plants that have a desired trait; they then allow only those "superior" individuals to mate. The result is a wide variety of subpopulations (such as different breeds of dogs) that humans maintain by selective breeding.

D. Migration Moves Alleles Between Populations

Migration is the movement of individuals into or out of a population (figure 12.22). The migrating brown rabbit in the figure will add new alleles to the population of black rabbits, increasing the local genetic diversity in its new home. But "migration" does not require the movement of entire individuals. Wind can carry a plant's pollen for miles, spreading one individual's alleles to a new population.

Over time, sustained migration has reduced the genetic differences between human populations. For example, isolated European populations once had unique allele frequencies for many genetic diseases. Geographical barriers, such as mountain ranges and large bodies of water, historically restricted migration and kept the gene pools separate. Highways, trains, and airplanes, however, have eliminated physical barriers to migration. Eventually, migration should make these regional differences disappear.

Figure 12.21 **Nonrandom Mating.** If mating among toads were random, all individuals would have an equal chance of reproducing. Instead, a male mates only if his song attracts a willing female. This male toad inflates his throat pouch and generates a call.

©Creatas/PunchStock RF

Figure 12.22 **Migration.** A migrating animal transfers its alleles to another population.

12.7 Mastering Concepts

1. How do mutations affect an organism's phenotype?
2. South China tigers once had two color patterns (orange/black and blue/gray), which did not affect an individual's reproductive success. Over many decades, the tiger population has drastically declined, and all blue/gray individuals have disappeared. What evolutionary process eliminated this color pattern?
3. What is the difference between the founder effect and a population bottleneck?
4. How do nonrandom mating and migration result in evolutionary change?

Investigating Life 12.1 | Bacterial Evolution Goes "Hog Wild" on the Farm

Although infectious diseases were once the leading cause of human death, antibiotics had made many bacteria-caused diseases manageable by the mid-1900s. Since that time, bacteria have become resistant not only to the original penicillin but also to the many manufactured antibiotics that followed it. Now antibiotic-resistant bacteria are common, creating new obstacles for physicians treating infectious disease.

Medical practices contribute to the rise of antibiotic-resistant bacteria, but so do farms. Antibiotics promote rapid animal growth when added to the food of cattle, chickens, swine, and other livestock. This practice comes at a cost to public health. The animals' manure contains not only antibiotics but also bacteria that are resistant to the drugs. These microbes swap genes with their neighbors (see figure 8.7). Farms have therefore become breeding grounds for antibiotic-resistant bacteria.

To learn more about this problem, researchers from China and the United States collected manure from three Chinese pig farms where antibiotics are used. Control manure came from pigs that had never been exposed to the drugs. When the team tallied the number of resistance genes in bacterial DNA extracted from each sample, they found that manure from antibiotic-treated animals had many more resistance genes than did control manure (figure 12.A).

But farmers often compost pig manure and then spread it on their croplands. Do the genes persist under those conditions? To find out, the researchers collected samples from compost piles and from the soil in nearby fields. DNA analysis revealed that compost and soil from farms using antibiotics had more diverse resistance genes than did soil from a forest.

These results have serious implications, and not just for farm workers. Since composted animal manure is spread over fields, crops may become contaminated with antibiotic-resistant bacteria. Meat from treated livestock may also harbor resistance genes. When we eat the crops or the meat, bacteria in our intestines may take up the resistance genes.

Figure 12.A Antibiotic Resistance on the Farm. The diversity of antibiotic-resistance genes was significantly higher in the manure of antibiotic-fed pigs than in that of untreated animals. Error bars represent the standard error of the mean (see appendix B).

Changes in government policy and consumer awareness may soon decrease the use of antibiotics on farms. As demand for meat from antibiotic-free animals grows, farmers will have an economic incentive to find alternatives to the drugs. The need for change is urgent because some antibiotics may become useless if current practices continue. Evolution never stops, but a thorough understanding of natural selection and bacteria can help us slow the rise of antibiotic resistance.

Source: Zhu, Yong-Guan, and seven coauthors, including James M. Tiedje. 2013. Diverse and abundant antibiotic resistance genes in Chinese swine farms. *Proceedings of the National Academy of Sciences*, vol. 110, pages 3435–3440.

What's the **Point?** ▼ APPLIED

Evolution occurs in every population; bacteria are no exception. The evolutionary changes in microbes have important implications in human medicine. Why We Care 12.2 and Investigating Life 12.1 highlight one problem: In the hundred years since the discovery of antibiotics, many strains of resistant bacteria have appeared. Health professionals can minimize this problem by prescribing antibiotics only when needed and by educating the public about the proper use of the drugs.

Viruses also evolve. The surprising and sometimes deadly appearance of new influenza strains such as H1N1 is evidence of how viruses change over time. Rapid mutations in viral DNA provide enormous variation that fuels natural selection. Understanding viral evolution allows researchers to stay a step ahead, producing vaccines that protect against the new strains of the flu that arise each year. Unfortunately, scientists have not been able to develop vaccines against all viral infections. The rapid

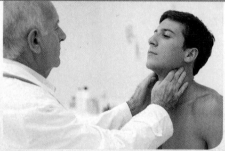

©PhotoAlto sas/Alamy Stock Photo RF

evolution of HIV, for example, has made this virus a moving target; an effective anti-HIV vaccine has yet to be invented. A better understanding of how the virus evolves and evades the immune system might lead to better treatments against HIV.

Studying evolution has led to innovative ideas in the treatment of disease. For example, the evolution of chemotherapy-resistant cancer cells has inspired researchers to develop new drug delivery strategies (see Investigating Life 8.1). Also, studying intestinal worms has led to new theories about the origin of allergies. As explained in Investigating Life 29.1, these findings may make parasites a viable option for treating an overactive immune system.

In the future, biologists hope to apply evolutionary theory to answer questions about everything from aging to mental illness. A thorough understanding of this powerful process will undoubtedly lead to a healthier future.

CHAPTER SUMMARY

12.1 Evolution Acts on Populations

- Biological **evolution** is descent with modification. One way to detect evolution is to look for a shift in the **gene pool** of a **population; allele frequencies** change from one generation to the next when evolution occurs.
- A small-scale genetic change within a species is called microevolution, and it occurs by several mechanisms (figure 12.23). Over the long term, microevolutionary changes also explain macroevolutionary events such as the emergence of new species.

12.2 Evolutionary Thought Has Evolved for Centuries

A. Many Explanations Have Been Proposed for Life's Diversity

- Early attempts to explain life's diversity relied on belief in a creator.
- Geology laid the groundwork for evolutionary thought. Lower rock layers are older than those above, suggesting an evolutionary sequence for fossils within them.
- Lamarck proposed a testable mechanism of evolution, but it was erroneously based on use and disuse of traits acquired during an organism's life.

B. Charles Darwin's Voyage Provided a Wealth of Evidence

- During the voyage of the HMS *Beagle,* Darwin observed the distribution of organisms in diverse habitats and their relationships to geological formations. After much thought and consideration of input from other scientists, he developed his theory of the origin of species by means of natural selection.

C. *On the Origin of Species* Proposed Natural Selection as an Evolutionary Mechanism

- **Natural selection** is based on multiple observations: Individuals vary for inherited traits; many more offspring are born than survive; and life is a struggle to acquire limited resources. The environment eliminates poorly adapted individuals, so only those with the best adaptations reproduce.
- **Artificial selection** is based on similar requirements, except that a human breeder decides exactly which individuals reproduce.
- *On the Origin of Species* offered abundant evidence for descent with modification.

D. Evolutionary Theory Continues to Expand

- The **modern evolutionary synthesis** unifies ideas about DNA, mutations, inheritance, and natural selection.

12.3 Natural Selection Molds Evolution

A. Adaptations Enhance Reproductive Success

- Individuals with the best **adaptations** to the current environment are most likely to leave offspring. Therefore, their alleles become more common in the population over time.
- Natural selection requires variation, which arises ultimately from random mutations.

B. Natural Selection Eliminates Poorly Adapted Phenotypes

- Natural selection weeds out some phenotypes, causing changes in allele frequencies over multiple generations.

C. Natural Selection Does Not Have a Goal

- Natural selection does not work toward an objective, nor can it achieve perfectly adapted organisms.

D. What Does "Survival of the Fittest" Mean?

- Organisms with the highest evolutionary **fitness** are the ones that have the greatest reproductive success. Many traits contribute to an organism's fitness.

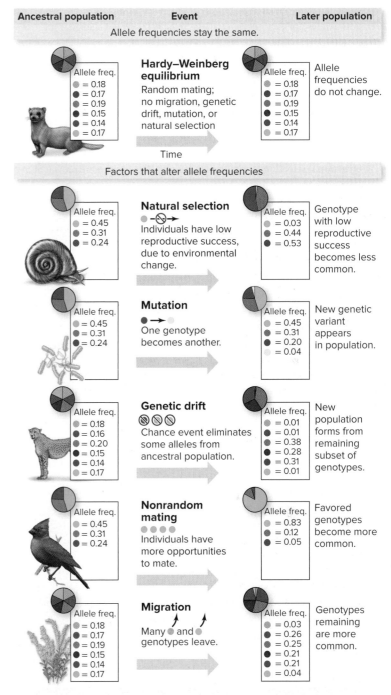

Figure 12.23 Mechanisms of Evolution: A Summary. At Hardy–Weinberg equilibrium, allele frequencies remain unchanged. Each mechanism of evolution changes allele frequencies. Different colored dots represent the alleles for a particular gene. The left image of each pair shows the allele frequencies in the ancestral population; the right image shows the frequencies after evolutionary change has occurred.

12.4 Evolution Is Inevitable in Real Populations

- Calculations of allele frequencies and **genotype frequencies** allow biologists to detect whether evolution has occurred.

A. At Hardy–Weinberg Equilibrium, Allele Frequencies Do Not Change

- It is possible to calculate the frequencies of genotypes and phenotypes in a population by inserting allele frequencies into the equation $p^2 + 2pq + q^2 = 1$.

- If a population meets all assumptions of **Hardy–Weinberg equilibrium,** evolution does not occur because allele frequencies do not change from generation to generation.

B. In Reality, Allele Frequencies Always Change

- The conditions for Hardy–Weinberg equilibrium do not occur together in natural populations, suggesting that allele frequencies always change from one generation to the next.

12.5 Natural Selection Can Shape Populations in Many Ways

- In **directional selection,** one extreme phenotype becomes more prevalent in a population.
- In **disruptive selection,** two or more extreme phenotypes survive at the expense of intermediate forms.
- In **stabilizing selection,** an intermediate phenotype has an advantage over individuals with extreme phenotypes.
- Harmful recessive alleles may persist in populations where heterozygous carriers have a reproductive advantage over homozygotes.

12.6 Sexual Selection Directly Influences Reproductive Success

- **Sexual dimorphisms** differentiate the sexes. They result from **sexual selection,** a form of natural selection in which inherited traits—even those that seem nonadaptive—make an individual more likely to mate.
- The competition that promotes sexual selection may or may not involve mate choice by members of the opposite sex.

12.7 Evolution Occurs in Several Additional Ways

A. Mutation Fuels Evolution

- Mutation alters allele frequencies by changing one allele into another, sometimes providing new phenotypes for natural selection to act on.

B. Genetic Drift Occurs by Chance

- In **genetic drift,** allele frequencies change purely by chance events, especially in small populations. The **founder effect** and population **bottlenecks** are forms of genetic drift.

C. Nonrandom Mating Concentrates Alleles Locally

- Nonrandom mating causes some alleles to concentrate in subpopulations.

D. Migration Moves Alleles Between Populations

- As individuals migrate, they can carry new alleles into existing populations, altering allele frequencies.

MULTIPLE CHOICE QUESTIONS

1. Microevolution applies to changes that occur
   a. only within small populations of organisms.
   b. in small genes.
   c. in the allele frequencies of a population or species.
   d. in small cells such as bacteria.

2. Biological evolution describes how ____ change from one generation to the next.
   a. individuals
   b. allele frequencies
   c. phenotype frequencies
   d. communities

3. Suppose a population of 200 individuals has a gene with two alleles (G and g). If 100 alleles in the population are g, then the frequency of the g allele in the population is ____.
   a. 0.1
   b. 0.25
   c. 0.5
   d. 1.0

4. What is the most accurate way to explain the relationship between antibiotics and antibiotic-resistant bacteria?
   a. Exposing bacteria to antibiotics causes DNA mutations that confer antibiotic resistance.
   b. A population of bacteria evolves antibiotic resistance alleles in order to survive exposure to antibiotics.
   c. Bacteria with alleles that confer antibiotic resistance are most likely to survive when exposed to antibiotics.
   d. Antibiotics bind to alleles that confer antibiotic susceptibility, causing changes that make the gene pool more resistant.

5. Suppose that after an environmental change, foxes with shorter legs than average are most likely to survive and reproduce. What type of selection will act on this population in the coming generations?
   a. Directional selection
   b. Stabilizing selection
   c. Disruptive selection
   d. Normalizing selection

6. Huntington disease is caused by a rare, lethal, dominant allele. Symptoms typically begin in midlife. Which answer best explains why natural selection has not eliminated the Huntington allele?
   a. Females prefer males with the Huntington allele.
   b. The disease does not interfere with the ability to reproduce.
   c. New mutations generate the allele.
   d. Natural selection can eliminate only common alleles.

7. A population of 100 sea stars is in Hardy–Weinberg equilibrium. The trait for long arms is completely dominant to the trait for short arms. In this population, 40% of the alleles for this trait are dominant, and 60% are recessive. What percent of the sea stars in this population are heterozygous?
   a. 64%
   b. 52%
   c. 48%
   d. 26%

8. Sexual selection
   a. typically occurs after mating has already taken place.
   b. is one type of artificial selection.
   c. may result in sexual dimorphism.
   d. explains why sexual reproduction occurs in so many species.

9. Darwin observed that different types of organisms were found on either side of a geographic barrier. The barrier was preventing
   a. migration.
   b. genetic drift.
   c. sexual selection.
   d. mutation.

10. Which of the following processes is nonrandom?
    a. A population bottleneck
    b. Natural selection
    c. The founder effect
    d. Mutation

Answers to Multiple Choice questions are in appendix A.

WRITE IT OUT

1. List and describe five mechanisms of evolution.

2. How did the work of other scientists influence Charles Darwin's thinking?

3. Explain how understanding evolution is important to medicine, agriculture, and maintaining the diversity of organisms on Earth.

4. Write a paragraph that describes the connections among the following terms: *gene, nucleotide, allele, phenotype, population, genetic variation, natural selection,* and *evolution.*

5. Jellyfish Lake, located on the Pacific island of Palau, is home to millions of jellyfish. Many years ago, sea levels dropped and the jellyfish were trapped in the basin. The lake houses no predators, and the jellyfish's sting has weakened. Jellyfish Lake is now a popular tourist attraction where snorkelers can swim among the jellyfish. Explain how Jellyfish Lake is evidence for evolution.

6. Influenza and smallpox are diseases caused by different types of viruses. Scientists must produce a new influenza vaccine each year, whereas the smallpox vaccine eradicated the disease. Explain these results from an evolutionary perspective.

7. Explain how harmful recessive alleles can persist in populations, even though they prevent homozygous individuals from reproducing.

8. Fraggles are mythical, mouselike creatures that live underground beneath a large vegetable garden. Of the 100 Fraggles in this population, 84 have green fur and 16 have gray fur. A dominant allele *F* confers green fur, and a recessive allele *f* confers gray fur. Assuming Hardy–Weinberg equilibrium is operating, answer the following questions. (a) What is the frequency of the gray allele *f*? (b) What is the frequency of the green allele *F*? (c) How many Fraggles are heterozygotes (*Ff*)? (d) How many Fraggles are homozygous recessive (*ff*)? (e) How many Fraggles are homozygous dominant (*FF*)?

9. Describe the competing selective forces acting on peacock tails. Together, do these selective forces produce disruptive, directional, or stabilizing selection?

10. Some researchers suggest that a giraffe's long neck results from competition for foliage with other types of animals; others say it is the product of sexual selection. How might each mechanism explain how a long-necked species evolved from an ancestral population with short necks? How does each explanation compare to how Lamarck might have explained it?

SCIENTIFIC LITERACY

Burning Question 12.1 explains why an organism cannot be adapted to all environments. Even humans cannot live in all parts of the world. However, technology allows us to live in habitats that would otherwise be deadly. For example, the cold temperatures, low oxygen level, and lack of food on a high mountain would soon kill an unclothed man. However, the same man could survive indefinitely in the same location if he had a warm house and clothing, breathed from an oxygen tank, and grew food in a sophisticated greenhouse. Are human technologies the products of natural selection? Is a human who lives in a harsh environment "adapted" to that environment, even if he must use technology to survive? What areas of the world might be especially difficult to inhabit, even with the use of current or potential future technologies? Why?

PULL IT TOGETHER

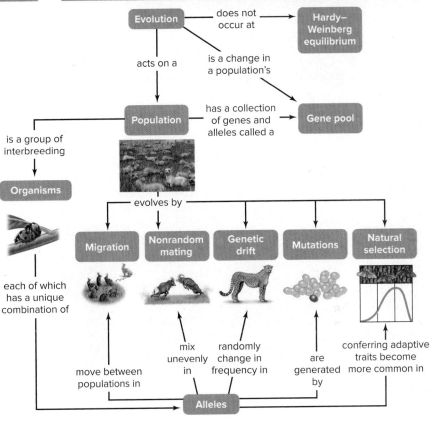

Figure 12.24 Pull It Together: Forces of Evolutionary Change.

Refer to figure 12.24 and the chapter content to answer the following questions.

1. Review the Survey the Landscape figure in the chapter introduction. When has evolution occurred in life's history? How do scientists know that evolution has occurred in the past?

2. Describe situations in which the five mechanisms of evolution shown in the concept map would occur.

3. Add the terms *genotype, phenotype, allele frequencies, founder effect, bottleneck effect,* and *sexual selection* to this concept map.

Answers to Mastering Concepts, Write It Out, Scientific Literacy, and Pull It Together questions can be found in the Connect ebook.
connect.mheducation.com

Design element: Burning Question (fire background): ©Ingram Publishing/Super Stock

Evidence of Evolution

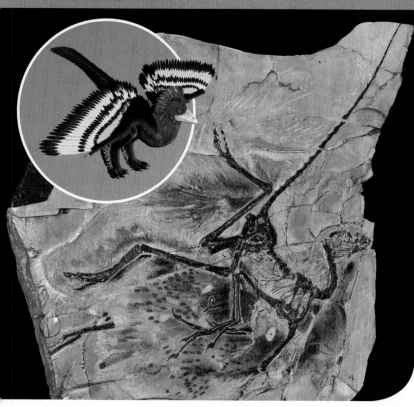

Feathery Dinosaur. The dinosaur *Anchiornis huxleyi* lived about 160 million years ago in China. It had feathers, revealing a close evolutionary relationship to birds. The model (inset) shows what the animal might have looked like.

(fossil): ©Martin Shields/Alamy Stock Photo; (inset): ©National Geographic Creative/Alamy·Stock Photo

LEARNING OUTLINE

13.1 Clues to Evolution Lie in the Earth, Body Structures, and Molecules

13.2 Fossils Record Evolution

13.3 Biogeography Considers Species' Geographical Locations

13.4 Anatomical Comparisons May Reveal Common Descent

13.5 Embryonic Development Patterns Provide Evolutionary Clues

13.6 Molecules Reveal Relatedness

APPLICATIONS

Burning Question 13.1 *Does the fossil record include transitional forms?*
Why We Care 13.1 *An Evolutionary View of the Hiccups*
Investigating Life 13.1 *Evolving Backwards*

Learn How to Learn
Why Rewrite Your Notes?

Your notes are your record of what happened in class, so why should you rewrite them after a lecture is over? One excellent reason is that the abbreviations and shorthand that make perfect sense while you take notes will become increasingly mysterious as the days or weeks go by. Rewriting the information in complete sentences not only reinforces learning but also makes your notes much easier to study before an exam.

SURVEY THE LANDSCAPE
Evolution and Diversity

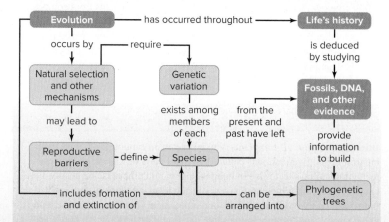

Life has a rich evolutionary history. Traces of evolution are in fossils, geography, anatomy, patterns of embryonic development, and the sequences of life's molecules.

For more details, study the Pull It Together feature in the chapter summary.

©McGraw-Hill Education/Carlyn Iverson

Most of life's history happened before humans came along, so can evolution actually be tested? Although no experiment can re-create the conditions that led to today's diversity of life, evolution is testable. In fact, its validity has been verified repeatedly over the past 150 years.

A mountain of evidence supports the idea of common descent. Extinct organisms have left traces of their existence, both as fossils and as the genetic legacy that all organisms have inherited. The distribution of life on Earth offers other clues, as does the study of everything from anatomical structures to protein sequences. Laboratory experiments and field observations of natural populations likewise support the case for evolution and suggest likely mechanisms for evolutionary change. No other scientifically testable hypothesis explains and unifies all of these observations as well as common descent.

13.1 Clues to Evolution Lie in the Earth, Body Structures, and Molecules

The millions of species alive today did not just pop into existence all at once; they are the result of continuing evolutionary change that started with organisms living billions of years in the past. Many types of clues enable us to hypothesize about how modern species evolved from extinct ancestors and to understand the relationships among organisms that live today.

Comparisons among living organisms provided early evidence of the evolutionary relationships among species. Abundant additional data came from **paleontology,** the study of fossil remains or other clues to past life (figure 13.1). The discovery of many new types of fossils in the early 1800s created the climate that allowed Charles Darwin to make his tremendous breakthrough. As people recognized that fossils must represent snapshots from the history of life, scientists developed theories to explain that history. The geographical locations of fossils and modern species provided additional clues. ⓘ *Darwin,* section 12.2B

In Darwin's time, some scientists suspected that Earth was hundreds of millions of years old, but no one knew the exact age. We now know that Earth's history is about 4.6 billion years long, a duration that most people find nearly unimaginable. Scientists describe the events along life's long evolutionary path in the context of the **geologic timescale,** which divides Earth's history into a series of eons and eras defined by major geological or biological events such as mass extinctions (figure 13.2). ⓘ *mass extinctions,* section 14.5

Fossils and biogeographical studies provided the original evidence for evolution, revealing when species most likely diverged from common ancestors in the context of other events happening on Earth. Comparisons of embryonic

Figure 13.1 **A Gallery of Fossils.** Clockwise from left: A horsetail plant that lived hundreds of millions of years ago; petrified wood from Arizona; a dinosaur egg from at least 100 million years ago; fossilized feces of a turtle; extinct arthropods called trilobites; a *Triceratops* skull; a *Ginkgo* leaf; an exceptionally well-preserved fish.

(horsetail and *Ginkgo*): ©Biophoto Associates/Science Source; (wood): ©Bill Florence/Shutterstock RF; (egg): ©Millard H. Sharp/Science Source; (feces): ©Sinclair Stammers/Science Source; (trilobites): ©Siede Preis/Getty Images RF; (fish): ©Alan Morgan RF; (*Triceratops*): ©Francois Gohier/Science Source

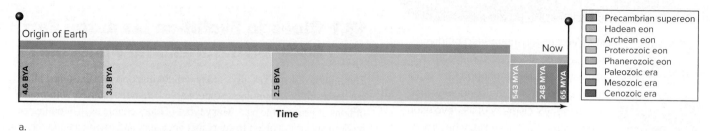

a.

Eon	Era	Period	Epoch	MYA	Important events
Phanerozoic eon	Cenozoic era ("Age of Mammals")	Quaternary	Holocene*		Human civilization
				0.01	
			Pleistocene		*Homo sapiens*, large mammals; ice ages
				1.8	
		Neogene	Pliocene		Early humans; modern whales
				5.3	
			Miocene		First great apes; other mammals continue to diversify; modern birds; expansion of grasslands
				23.8	
		Paleogene	Oligocene		Elephants, horses; grasses
				33.7	
			Eocene		Mammals and flowering plants continue to diversify; first whales
				54.8	
			Paleocene		First primates; mammals, birds, and pollinating insects diversify
				65	
	Mesozoic era ("Age of Reptiles")	Cretaceous			Widespread dinosaurs until extinction at end of Cretaceous; flowering plants diversify; present-day continents form
				144	
		Jurassic			First birds; cycads and ferns abundant; giant reptiles on land and in water; first flowering plants
				206	
		Triassic			First dinosaurs; first mammals; therapsids and thecodonts; forests of conifers and cycads
				248	
	Paleozoic era ("Age of Amphibians" / "Age of Fishes")	Permian			First conifers; fewer amphibians, more reptiles; cotylosaurs and pelycosaurs; Pangaea supercontinent forms
				290	
		Carboniferous			First reptiles; ferns abundant; amphibians diversify; first winged insects
				354	
		Devonian			First bony fishes, corals, crinoids; first amphibians; first seed plants; arthropods diversify
				417	
		Silurian			First vascular plants and terrestrial invertebrates; first fish with jaws
				443	
		Ordovician			Algae, invertebrates, jawless fishes; first land plants
				490	
		Cambrian			"Explosion" of sponges, worms, jellyfish, "small shelly fossils"; first fossils of most modern animal phyla; trilobites
				543	
Precambrian supereon	Proterozoic eon				O_2 from photosynthesis accumulates in atmosphere; first eukaryotes; first multicellular organisms; Ediacaran organisms
				2500	
	Archean eon				Life starts; first bacteria and archaea
				3800	
	Hadean eon				Earth forms
				4600	

*Some geologists propose ending the Holocene and adding a new epoch, the Anthropocene, to acknowledge human impacts on Earth.

b.

Figure 13.2   **The Geologic Timescale.** (a) Scientists divide Earth's 4.6-billion-year history into four eons. The three earliest eons are combined into the Precambrian supereon, which lasted more than 4 billion years. The most recent eon, the Phanerozoic, started 543 MYA. (b) Fossil evidence paints a detailed portrait of life's history during the Phanerozoic, which includes three eras (Paleozoic, Mesozoic, and Cenozoic). Red lines indicate the five largest mass extinction events of the Phanerozoic eon. (BYA = billion years ago; MYA = million years ago)

development and anatomical structures provided additional data. An entirely new type of evidence emerged in the 1960s and 1970s, when scientists began analyzing the sequences of DNA, proteins, and other biological molecules. Since then, the explosion of molecular data has revealed in unprecedented detail how species are related to one another.

Chapter 12 explained how natural selection and other processes drive evolutionary changes, both in the past and today. This chapter examines the different approaches to studying evolution in both living and extinct species.

13.1 Mastering Concepts

1. What is the geologic timescale?
2. What types of information provide the clues that scientists use in investigating evolutionary relationships?

13.2 Fossils Record Evolution

A **fossil** is any evidence of an organism from more than 10,000 years ago (the end of the Pleistocene epoch). Fossils come in all sizes, documenting the evolutionary history of everything from microorganisms to dinosaurs to humans. These remains, the oldest of which formed more than 3 billion years ago, give us our only direct evidence of organisms that preceded human history. They occur all over the world and represent all major groups of organisms, revealing much about the geological past. For example, the abundant remains of extinct marine animals called ammonites in Oklahoma indicate that a vast, shallow ocean once submerged what is now the central United States **(figure 13.3)**.

Fossils do more than simply provide a collection of ancient remains of plants, animals, and microbes. They also allow researchers to test predictions about evolution. Investigating Life 17.1 describes an excellent example: the discovery of *Tiktaalik,* an extinct animal with characteristics of both fishes and amphibians. Based on many lines of evidence showing the close relationship between these two groups, biologists had long predicted the existence of such an animal. *Tiktaalik* finally provided direct fossil evidence of the connection.

A. The Fossil Record Is Often Incomplete

Researchers sometimes assemble groups of fossils that reveal, step-by-step, the evolution of one species into another (see Burning Question 13.1). For example, biologists have found many fossils revealing intermediate stages in the evolution of whales and dolphins from land animals (see figure 13.9). Usually, however, the fossil record is incomplete, meaning that some of the features marking the transition from one group to another are not recorded in fossils.

Several explanations account for this partial history. First, the vast majority of organisms never leave a fossil trace. Soft-bodied organisms, for example, are much less likely to be preserved than are those with teeth, bones, or shells. Organisms that decompose or are eaten after death, rather than being buried in sediments, are also unlikely to fossilize. Second, erosion or the movements of Earth's continental plates have destroyed many fossils that did form. Third, scientists are unlikely to ever discover the many fossils that must be buried deep in the Earth or submerged under water.

Burning Question 13.1

Does the fossil record include transitional forms?

For some lineages, the answer to this question is a definitive "yes." In general, the fossil record is most complete for groups that evolved recently and that have bones, teeth, or other hard parts that fossilize readily. We therefore have excellent evidence for the transition from fishes to amphibians, from reptiles to mammals, and from terrestrial mammals to whales.

In addition, the chapter-opening photo shows one of many fossils that help biologists understand how today's birds evolved from feathered dinosaurs. A trove of fossils from many extinct reptiles reveal that birds and certain dinosaurs share uncanny similarities in bone structure, posture, feathers, and many other features. ⓘ *reptiles,* section 17.11E

Undoubtedly, the fossil record does contain many gaps. Perhaps the most famous example dates back several hundred million years to an event called the "Cambrian explosion." Despite its name, this phrase refers not to a cataclysmic event but rather to a time during which many new groups of animals first appeared in the ocean. Fossils from this period of rapid evolutionary change are scarce.

It is easy to imagine why the fossil record will never be complete. The remains of extinct organisms may never have fossilized in the first place, the fossils may have been destroyed, or paleontologists simply haven't found them yet. Each new find, however, fills another gap in the jigsaw puzzle of life's history.

Submit your burning question to
marielle.hoefnagels@mheducation.com

Figure 13.3 **Big Change.** Ammonite fossils such as these are common in land-locked Oklahoma, indicating that what is now the central United States was once covered by an ocean.
©Jean-Claude Carton/Photoshot

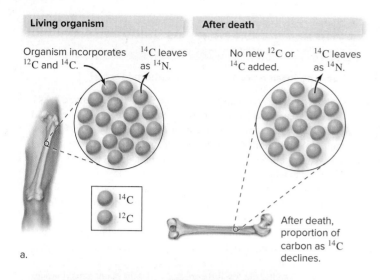

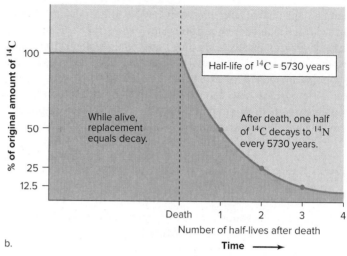

c.

Figure 13.4 **Carbon-14 Dating.** (a) Living organisms accumulate radioactive carbon-14 (^{14}C) by photosynthesis or eating other organisms. During life, ^{14}C is replaced as fast as it decays to nitrogen-14 (^{14}N). After death, no new ^{14}C enters the body, so the proportion of ^{14}C to ^{12}C declines. (b) During one half-life, 50% of the remaining radioactive atoms in a sample decay. (c) Measuring the proportion of ^{14}C to ^{12}C allows scientists to determine how long ago a fossilized organism—such as this woolly mammoth—died.

(c): ©Ethan Miller/Getty Images

B. The Age of a Fossil Can Be Estimated in Two Ways

Scientists use two general approaches to estimate when a fossilized organism lived: relative dating and absolute dating.

Relative Dating Relative dating places a fossil into a sequence of events without assigning it a specific age. It is usually based on the presumption that lower rock strata are older than higher layers (see figure 12.3). The farther down a fossil is, therefore, the longer ago the organism lived—a little like a memo at the bottom of a stack of papers being older than a sheet near the top. Relative dating therefore places fossils in order from "oldest" to "most recent."

Absolute Dating and Radioactive Decay Researchers use **absolute dating** to assign an age to a fossil by testing either the fossil itself or the sediments above and below the fossil. Either way, the dates usually are expressed in relation to the present. For example, scientists studying the dinosaur in the chapter-opening photo showed that this animal lived about 160 million years ago (MYA).

Radiometric dating is a type of absolute dating that uses radioactive isotopes as a "clock." Recall from chapter 2 that each isotope of an element has a different number of neutrons. Some isotopes are naturally unstable, which causes them to emit radiation as they radioactively decay. Each radioactive isotope decays at a characteristic and unchangeable rate, called its half-life. The **half-life** is the time it takes for half of the atoms in a sample of a radioactive substance to decay. If an isotope's half-life is 1 year, for example, 50% of the radioactive atoms in a sample will have decayed in a year. In another year, half of the remaining radioactive atoms will decay, leaving 25%, and so on. If we measure the amount of a radioactive isotope in a sample, we can use the isotope's known half-life to deduce when the fossil formed. ⓘ *isotopes,* section 2.1C

One radioactive isotope often used to assign dates to fossils is carbon-14 (^{14}C; figure 13.4). Carbon-14 has a half-life of 5730 years; it decays to the more stable nitrogen-14 (^{14}N). Organisms accumulate ^{14}C during photosynthesis or by eating organic matter. One in every trillion carbon atoms present in living tissue is ^{14}C; most of the rest are ^{12}C, a nonradioactive (stable) isotope. When an organism dies, however, its intake of carbon, including ^{14}C, stops. As the body's ^{14}C decays without being replenished, the ratio of ^{14}C to

¹²C decreases. This ratio is then used to determine when death occurred, up to about 40,000 years ago.

For example, radioactive carbon dating determined the age of fossils of vultures that once lived in the Grand Canyon. The birds' remains have about one-fourth the ¹⁴C-to-¹²C ratio of a living organism. Therefore, about two half-lives, or about 11,460 years, passed since the animals died. It took 5730 years for half of the ¹⁴C to decay, and another 5730 years for half of what was left to decay to ¹⁴N.

Another widely used radioactive isotope, potassium-40 (⁴⁰K), decays to argon-40 (⁴⁰Ar) with a half-life of 1.3 billion years. It is valuable in dating rocks that are about 300,000 years old or older.

One limitation of ¹⁴C and potassium–argon dating is that they leave a gap, resulting from the different half-lives of the radioactive isotopes. To cover the missing years, researchers use isotopes with intermediate half-lives or turn to other techniques.

13.2 Mastering Concepts

1. Why is the fossil record useful, even if it doesn't represent every type of organism that ever lived?
2. Distinguish between relative and absolute dating of fossils.
3. How does radiometric dating work?

13.3 Biogeography Considers Species' Geographical Locations

Geographical barriers such as mountains and oceans greatly influence the origin of species (see chapter 14). It is therefore not surprising that the studies of geography and biology overlap in one field, **biogeography,** the study of the distribution of species across the planet.

A. The Theory of Plate Tectonics Explains Earth's Shifting Continents

Earth's geological history has been extremely eventful. Fossils tell the story of ancient seafloors rising all the way to Earth's "ceiling": the Himalayan Mountains. Littering the mountains of Nepal are countless fossilized ammonites. How did fossils of marine animals end up more than 3600 meters above sea level?

The answer is that Earth's continents are in motion, an idea called "continental drift" **(figure 13.5).** According to the theory of **plate tectonics,** Earth's surface consists of several rigid layers, called tectonic plates, that move in response to forces acting deep within the planet. In some places where plates come together, one plate dives beneath another, forming a deep trench. In other areas, mountain ranges form as the plates become wrinkled and distorted. Long ago, the plate that carries the Indian subcontinent moved slowly north and collided with the Eurasian plate. The mighty Himalayas—once an ancient seafloor—rose at the boundary, lifting the marine fossils toward the sky.

Meanwhile, at areas where plates move apart, molten rock seeps to Earth's surface and forms new plate material at the seafloor. As a result, oceans now separate continents that were once joined. This slow-motion dance of the continental plates has dramatically affected life's history as oceans shifted, land bridges formed and disappeared, and mountain ranges emerged.

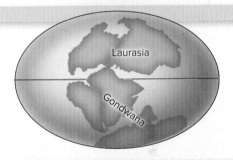

Figure 13.5 A Changing World. The locations of Earth's continents have changed with time, due to shifting tectonic plates.

Embryonic Development Patterns   Molecules Reveal Relatedness

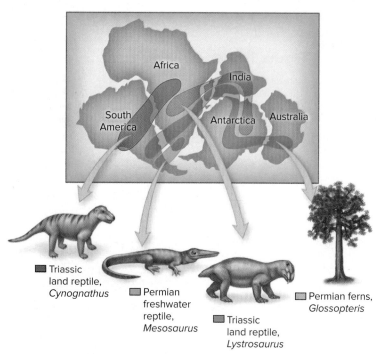

Figure 13.6 **Growing Apart.** Continental drift explains the modern-day distributions of fossils representing life in the southern hemisphere at the time of Pangaea.

It may seem hard to imagine that Earth's continents have not always been located where they are now. But a wealth of evidence, including the distribution of some key fossils, indicates that the continents were once united (figure 13.6). Deep-sea probes that measure seafloor spreading, along with the locations of the world's earthquake-prone and volcanic "hot spots," reveal that the continents continue to move today.

B. Species Distributions Reveal Evolutionary Events

Biogeographical studies have shed light on past evolutionary events. The rest of this section describes two examples.

The Rise and Fall of the Marsupials Marsupials are pouched mammals such as kangaroos, koalas, and sugar gliders. The newborns are tiny, hairless, blind, and helpless. As soon as they are born, they crawl along the mother's fur to tiny, milk-secreting nipples inside her pouch.

Marsupials were once the most widespread land mammals on Earth. By about 110 MYA, however, a second group, the placental mammals, had evolved. The young of placental mammals develop within the female's body, nourished in the uterus by the placenta. Baby placental mammals are born more fully developed than are marsupials, giving them a better chance of survival after birth. Because of this reproductive advantage, placental mammals soon displaced marsupials on most continents, including North America. ⓘ *mammals,* section 17.11F

Nevertheless, fossil evidence suggests that marsupials were diverse and abundant in South America until about 1 or 2 MYA, long after their counterparts on most other continents had disappeared. The reason is that water separated South America from North America until about 3 MYA. But sediments eroding from both continents eventually created a new land bridge that permitted migration between North and South America. The resulting invasion of placental mammals spelled extinction for most South American marsupials.

Australia's marsupials remained isolated from competition with placental mammals for much longer; Australia separated from the other continents about 60 or 70 MYA (see figure 13.5). In fact, Australia remains unique in that most of its native mammals are still marsupials.

Wallace's Line Biogeography figured prominently in the early history of evolutionary thought. Alfred Russel Wallace, the British naturalist who independently discovered natural selection along with Charles Darwin, had noticed unique assemblages of birds and mammals on either side of an imaginary line in the Malay Archipelago (figure 13.7). The explanation for what came to be called "Wallace's line" turned out to be a deepwater trench that separated the islands, even as sea levels rose and fell. The watery barrier prevented the migration of most species, so evolution produced a unique variety of organisms on each side of Wallace's line. ⓘ *Wallace,* section 12.2C

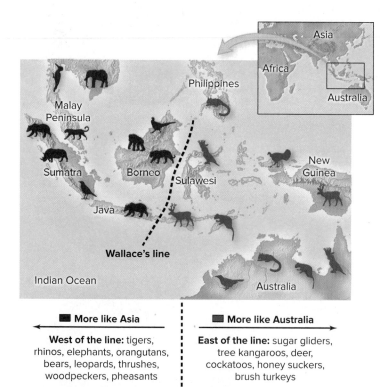

Figure 13.7 **Wallace's Line.** As Alfred Russel Wallace traveled around the Malay Archipelago, he noticed distinct patterns of animal life on either side of an imaginary boundary, which eventually came to be called Wallace's line.

13.3 Mastering Concepts

1. How have the positions of Earth's continents changed over the past 200 million years?

2. How do biogeographical observations help biologists interpret evolutionary history?

13.4 Anatomical Comparisons May Reveal Common Descent

Many clues to the past come from the present. As unit 1 explains, all life is made of cells, and eukaryotic cells are very similar in the structure and function of their membranes and organelles. On a molecular scale, cells share many similarities in their enzymes, signaling proteins, and metabolic pathways. Unit 2 describes another set of common features: the relationship between DNA and proteins, and the mechanisms of inheritance. We now turn to the whole-body scale, where comparisons of anatomy and physiology reveal still more commonalities among modern species.

A. Homologous Structures Have a Shared Evolutionary Origin

Two structures are termed **homologous** if the similarities between them reflect common ancestry. Homologous genes, chromosomes, anatomical structures, or other features are similar in their configuration, position, or developmental path.

The organization of the vertebrate skeleton illustrates homology. All vertebrate skeletons support the body, are made of the same materials, and consist of many of the same parts. Amphibians, birds and other reptiles, and mammals typically have four limbs, and the numbers and positions of the bones that make up those appendages are strikingly similar (figure 13.8). The simplest explanation is that modern vertebrates descended from a common ancestor that originated this skeletal organization. Each group gradually modified the skeleton as species adapted to different environments.

Note that homologous structures share a common evolutionary origin, but they may not have the same function. The middle ear bones of mammals, for example, originated as bones that supported the jaws of primitive fishes, and they still exist as such in some vertebrates. These bones are homologous and reveal our shared ancestry with fishes. Likewise, the forelimbs pictured in figure 13.8 have varying functions.

Homology is a powerful tool for discovering evolutionary relationships. For example, as described in section 13.6, a newly sequenced gene or genome can be compared with homologous genes from other species to infer how closely related any two species are. Similarly, fossilized structures are often compared with homologous parts in known species. Comparative studies can also provide clues to the origin of human features, including the hiccups (see Why We Care 13.1).

B. Vestigial Structures Have Lost Their Functions

As environmental changes select against some structures, others persist even if they are not used. A **vestigial** structure has no apparent function in one species, yet it is homologous to a functional organ in another. Figure 13.9 shows three examples of animals with vestigial structures: a mole, a boa constrictor, and a whale. A mole has vestigial eyes (and greatly reduced ears); this animal spends most of its life underground. Boa constrictors and pythons lack external limbs, as do all snakes, but their skeletons reveal the bones of tiny hindlimbs. (See section 13.5 and Investigating Life 13.1 for more on snake evolution.) Whales also have vestigial hindlimbs, retained from vertebrate ancestors that used legs to walk on land. Their aquatic habitat, meanwhile, selected for forelimbs modified into flippers.

Figure 13.8 **Homologous Limbs.** Although all of these forelimbs have different functions, the underlying bones are similar.

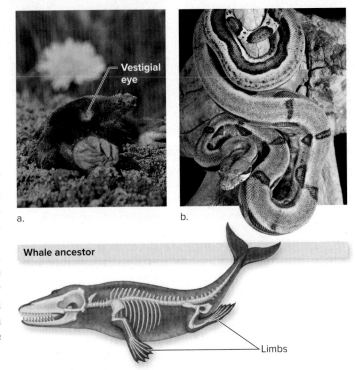

a. b.

Whale ancestor

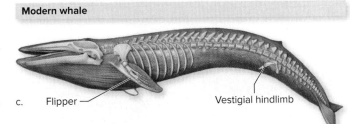

Limbs

Modern whale

c. Flipper Vestigial hindlimb

Figure 13.9 **Vestigial Structures.** (a) A mole's tiny vestigial eyes are covered with skin and fur. (b) Boa constrictors and some other snakes have vestigial hindlimbs. (c) Whales descended from mammals with four limbs; some modern whales retain a vestigial pelvis and hindlimbs.

Embryonic Development Patterns   Molecules Reveal Relatedness

Salamander

Crayfish

a. Cave animals

Cactus

Euphorbia

b. Desert plants

Figure 13.10 Convergent Evolution. (a) The blind salamander (*left*) and the cave crayfish (*right*), both from Florida, lack eyes and pigment. (b) Cacti from the desert in Mexico (*left*) have adaptations similar to those of *Euphorbia* plants from the Namib Desert in Africa (*right*).

(a, both): ©Danté Fenolio/Science Source; (b, cactus): ©DLILLC/Corbis RF; (b, *Euphorbia*): ©Natphotos/Getty Images RF

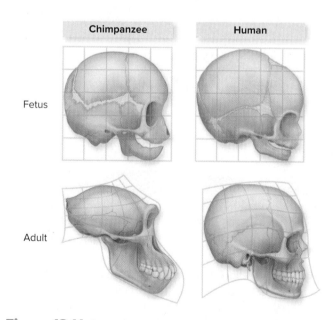

Chimpanzee

Human

Fetus

Adult

Figure 13.11 Same Parts, Different Proportions. The fetuses of humans and chimpanzees have very similar skulls, but the parts grow at different rates as they mature into adults.

Humans have several vestigial organs. The tiny muscles that make hairs stand on end helped our furry ancestors conserve heat or show aggression; in us, they apparently serve only as the basis of goose bumps. Human embryos have tails, which usually disintegrate long before birth; in other vertebrates, tails persist into adulthood. Above our ears, a trio of muscles (which most of us can't use) help other mammals move their ears in a way that improves hearing. Each vestigial structure links us to other animals that still use these features.

C. Convergent Evolution Produces Superficial Similarities

Some body parts that appear superficially similar in structure and function are not homologous. Rather, they are **analogous,** meaning that the structures evolved independently. Flight, for example, evolved independently in birds and in insects. The bird's wing is a modification of vertebrate limb bones, whereas the insect's wing is an outgrowth of the exoskeleton that covers its body. The wings have the same function—flight—and enhance fitness in the face of similar environmental challenges. The differences in structure, however, indicate they do not have a common developmental pathway. They are analogous, not homologous.

Analogous structures are often the product of **convergent evolution,** which produces similar adaptations in organisms that do not share the same evolutionary lineage. The absence of pigmentation and eyes in cave animals provides a compelling example of convergent evolution, as does the similar appearance of unrelated desert plants from Mexico and Africa (figure 13.10).

Likewise, the similarities between sharks and dolphins illustrate the power of selective forces in shaping organisms. A shark is a fish, whereas a dolphin is a mammal that evolved from terrestrial ancestors. The two animals are not closely related; their last common ancestor lived hundreds of millions of years ago. Nevertheless, their marine habitat and predatory lifestyle have selected for many shared adaptations, including the streamlined body and the shape and locations of the fins or flippers.

13.4 Mastering Concepts

1. What can homologous structures reveal about evolution?
2. What is a vestigial structure? What are some examples of vestigial structures in humans and other animals?
3. What is convergent evolution?

13.5 Embryonic Development Patterns Provide Evolutionary Clues

Because related organisms share many physical traits, they must also share the processes that produce those traits. Developmental biologists study how the adult body takes shape from its single-celled beginning. Careful comparisons of developing body parts can be enlightening. As just one example, figure 13.11 shows how skulls that look alike as fetuses can develop in different ways, depending on how each part grows in proportion to the others.

Developmental biologists have also photographed embryos and fetuses of a variety of vertebrate species (figure 13.12). The images reveal homologies in the overall structure of the embryonic bodies and in specific features such as the tail.

Chick	Mouse	Human

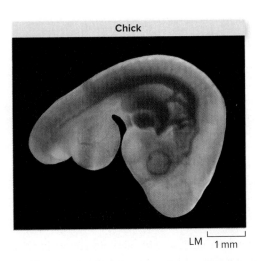
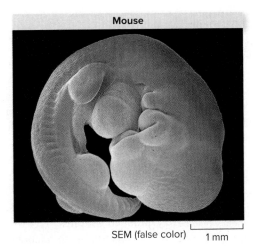
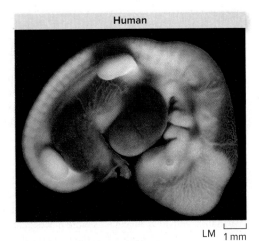

LM ⌐1 mm SEM (false color) ⌐1 mm LM ⌐1 mm

Figure 13.12 Embryo Resemblances. Vertebrate embryos appear alike early in development. As development continues, parts grow at different rates in different species, and the embryos begin to look less similar.

(chick): ©Oxford Scientific/Getty Images; (mouse): ©Steve Gschmeissner/Science Source; (human): ©Science Source

More recently, the discovery of genes that contribute to development has spawned the field of evolutionary developmental biology (or "evo-devo" for short). Recall from chapter 7 that a gene is a region of DNA that encodes a protein. Some genes encode proteins that dictate how an organism will develop. One goal of evo-devo research is to identify these genes and determine how mutations can give rise to new body forms.

A basic question in developmental biology, for example, is how a clump of identical cells transforms into a body with a distinct head, tail, segments, and limbs. One way to learn more about this process is to study **homeotic genes,** which encode proteins that regulate development. Mutations in homeotic genes lead to organisms with structures in abnormal or unusual places—see, for example, the fruit fly with legs growing out of its head in figure 7.13.

Homeotic genes occur in all animal phyla studied to date, as well as in plants and fungi, and they provide important clues about development. Consider, for example, a pair of homeotic genes that influence limb formation in vertebrates (figure 13.13). In the chick embryo in figure 13.13a, the gene labeled *A* prompts wing development, whereas the gene labeled *B* stimulates formation of the legs. No limbs occur where both genes are expressed, such as in the midsection of the body. Now compare the chick with the pattern of gene expression in the python (figure 13.13b). The same two genes are expressed along most of the snake's body; as a result, the animal never develops forelimbs at all. (Another gene prevents the development of the python's vestigial hindlimbs.)

These examples only scratch the surface of the types of information about evolution that biologists can learn by studying development. The relatively new evo-devo field is sure to yield many more insights into evolution.

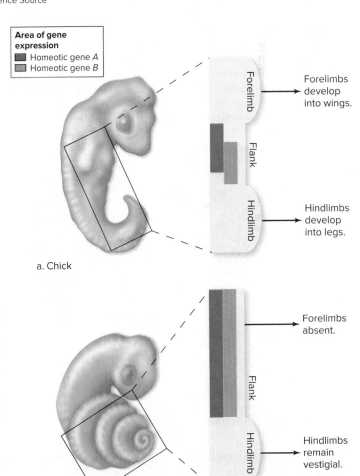

Area of gene expression
- ■ Homeotic gene *A*
- ■ Homeotic gene *B*

Forelimb — Flank — Hindlimb

Forelimbs develop into wings.

Hindlimbs develop into legs.

a. Chick

Flank — Hindlimb

Forelimbs absent.

Hindlimbs remain vestigial.

b. Python

Figure 13.13 Homeotic Genes. (a) Two homeotic genes are expressed unequally along the length of a chicken embryo. Where both genes are expressed, no limbs form. (b) The same pair of homeotic genes prevents the development of forelimbs in a python embryo.

13.5 Mastering Concepts

1. How does the study of embryonic development reveal clues to a shared evolutionary history?

2. Why are evolutionary biologists interested in how genes influence development?

Embryonic Development Patterns   Molecules Reveal Relatedness

Why We Care 13.1 | An Evolutionary View of the Hiccups

It happens to everyone: We eat or drink too fast, or we laugh too hard, and then the hiccups begin. A hiccup is an involuntary muscle spasm that causes a person to inhale sharply. At the same time, a flap of tissue called the epiglottis blocks the airway to the lungs, producing the classic "hic" sound. This quick intake of air doesn't prevent or solve any known problem. So why do we get the hiccups at all?

Hiccups result from irritation of one or both of the phrenic nerves that trigger contraction of the diaphragm, which controls breathing. The great distance between the origin of these nerves (in the neck) and the diaphragm (below the lungs) offers many opportunities for irritation and, thus, hiccups. A more practical arrangement would be for the nerves that control breathing to emerge from the spinal cord nearer the diaphragm. But we inherited these nerves from our fishy ancestors, whose gills are near where the breathing-control nerves emerge (figure 13.A).

A second clue to the origin of hiccups comes from a close examination of young amphibians. Tadpoles have both lungs and gills, so they can gulp air or extract oxygen from water. In the latter case, they pump mouthfuls of water across their gills; at the same time, the glottis closes to keep water out of the lungs. The tadpole's breathing action almost exactly matches what happens in a human hiccup.

Taken together, these two lines of anatomical evidence suggest that the hiccup is an accident of evolution—a remnant of our shared evolutionary history with the vertebrates that came before us.

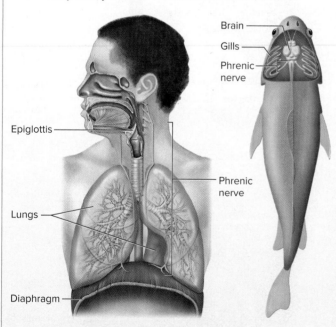

Figure 13.A Hiccup Origins. Humans have long phrenic nerves compared to those of a fish, providing ample opportunities for irritation—and the hiccups.

13.6 Molecules Reveal Relatedness

The evidence for evolution described so far in this chapter is compelling, but it is only the beginning. Since the 1970s, biologists have compiled a wealth of additional data by comparing the molecules inside the cells of diverse organisms. The results have not only confirmed many previous studies but have also added unprecedented detail to our ability to detect and measure the pace of evolutionary change.

The molecules that are most useful to evolutionary biologists are nucleic acids (DNA and RNA) and proteins. As described in chapter 2, nucleic acids are long chains of subunits called nucleotides, whereas proteins are composed of amino acids. Cells use the information in nucleic acids to produce proteins.

This intimate relationship between nucleic acids and proteins is itself a powerful argument for common ancestry. All species use the same genetic code in making proteins, and cells use the same 20 amino acids. The fact that biologists can move DNA among species to create transgenic organisms is a practical reminder of the universal genetic code. ⓘ *transgenic organisms,* section 11.2A

A. Comparing DNA and Protein Sequences May Reveal Close Relationships

To study molecular evolution, biologists compare nucleotide and amino acid sequences among species. It is highly unlikely that two unrelated species would evolve precisely the same DNA and protein sequences by chance. It is more likely that the similarities were inherited from a common ancestor and that differences arose by mutation after the species diverged.

DNA The ability to rapidly sequence DNA has led to an explosion of information. DNA differences can be assessed for just a few bases, for one gene, for families of genes with related structures or functions, or for whole genomes. Biologists routinely locate a gene in one organism, then scan huge databases to study homologous genes in other species. ⓘ *DNA sequencing,* section 11.2B

The recent explosion of DNA sequences has also helped explain how evolution works. We now know that cells may add new functions by acquiring DNA from other organisms and by duplicating genes. And we have already seen that studying gene expression can reveal differences among closely related species (see figure 13.13). The list of applications is endless, ranging from pinpointing the origin of diseases to monitoring the evolution of pesticide-resistant insects (see Investigating Life 10.1).

Proteins Like DNA, homologous protein sequences also often support fossil and anatomical evidence of evolutionary relationships. One study, for instance, found seven of 20 proteins to be identical in humans and chimps, our closest relatives. Many other proteins have only minor sequence differences from one species to another. The keratin of sheep's wool, for example, is virtually identical to that of human hair. The similarity reflects the shared evolutionary history of all mammals.

To study even broader groups of organisms, biologists use proteins that are present in all species. One example is cytochrome *c,* which is part of the electron transport chain in mitochondria (see section 6.5C). Figure 13.14 shows that the more closely related two species are, the more alike is their cytochrome *c* amino acid sequence.

B. Molecular Clocks Help Assign Dates to Evolutionary Events

A biological molecule can act as a "clock." A **molecular clock** uses DNA sequences to estimate the time when multiple organisms diverged from a common ancestor. In the example shown in figure 13.15, biologists used the known mutation rate for a gene, plus the number of DNA sequence differences between species 1 and species 2, to estimate that the two species last shared an ancestor about 50 million years ago.

DNA sequences helped estimate the time since humans and chimpanzees diverged. Many human and chimp genes differ in 4% to 6% of their nucleotides, and substitutions occur at an estimated rate of 1% per 1 million years. Therefore, the two species diverged about 4 million to 6 million years ago.

Molecular clock studies, however, are not quite as straightforward as glancing at a wristwatch. DNA replication errors occur in different regions of a chromosome at different rates. In addition, if too much time has passed since two species diverged, the same site may have undergone multiple changes, which would be impossible to detect. Researchers must account for these limitations when interpreting a molecular clock.

13.6 Mastering Concepts

1. How does analysis of DNA and proteins support other evidence for evolution?
2. How can molecular clocks help determine when two species diverged from a common ancestor?

Cytochrome c Evolution	
Organism	**Number of amino acid differences from humans**
Chimpanzee	0
Rhesus monkey	1
Rabbit	9
Cow	10
Pigeon	12
Bullfrog	18
Fruit fly	25
Yeast	40

Figure 13.14 Cytochrome c Comparison. The more recent the shared ancestor with humans, the fewer the differences in the amino acid sequence for the respiratory protein cytochrome c.

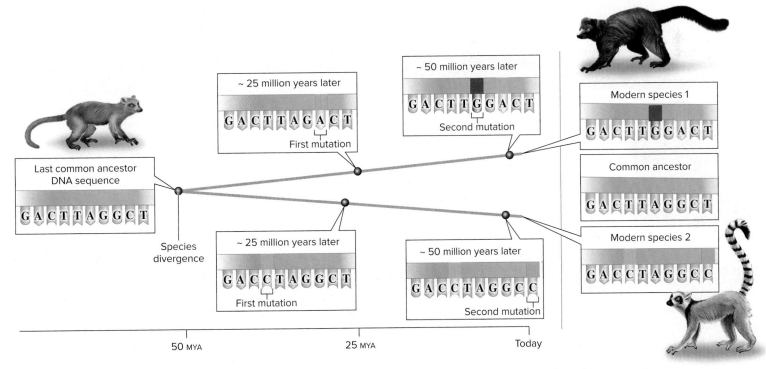

Figure 13.15 Molecular Clock. If a sequence of DNA accumulates mutations at a regular rate, then the number of sequence changes can act as a "clock" that tells how much time has passed since two species last shared a common ancestor. In this hypothetical example of a lemur gene, mutations occur about once every 25 million years, suggesting that the two lemur species diverged about 50 MYA.

Embryonic Development Patterns   Molecules Reveal Relatedness

Investigating Life 13.1 | Evolving Backwards

Evolution sometimes seems to run in reverse, such as when species lose features that their ancestors had. Snakes provide a notable example. These animals lack the limbs that characterize most other vertebrate species.

Biologists know quite a bit about the evolution of snakes. Anatomical information, including vestigial hind legs in some snakes, clearly indicates that snakes evolved from lizards with four legs. Additional fossil evidence shows that the forelimbs disappeared before the hindlimbs.

Although some events in snake evolution are clear, for a long time no one knew the answer to a puzzle: *Where* in the world did snakes lose their limbs, on land or in the ocean? Fossils have offered a breakthrough in the debate over snake origins.

Paleontologists from Argentina and Brazil uncovered a critical clue when they reported finding three fossilized snakes in the Patagonia region of Argentina. The snakes, which they named *Najash rionegrina*, lived about 90 MYA. *Najash* was the first snake ever found to have not only functional legs and a pelvis but also a sacrum—a bone connecting the pelvis to the spine. The sacrum is important because lizards and other terrestrial vertebrates have the same bone. *Najash* is therefore more primitive than any snake ever found, including fossils of marine snakes (figure 13.B). Moreover, both *Najash*'s features and the rock where it was found suggest that it was terrestrial.

Taken together, these two pieces of evidence seem to settle the matter: Snakes originated on land. These animals apparently lost their limbs as they adapted to a burrowing lifestyle. Since that time, snakes have diversified into thousands of species. The existence of so many snake species provides evidence that in evolution, going backwards can sometimes be a good thing.

Source: Apesteguía, Sebastián, and Hussam Zaher. April 20, 2006. A Cretaceous terrestrial snake with robust hindlimbs and a sacrum. *Nature*, vol. 440, pages 1037–1040.

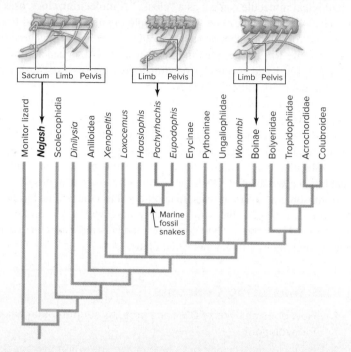

Figure 13.B *Najash,* **the Fossil Snake.** *Najash* was one of the most primitive snakes, whereas all three types of extinct marine snakes with legs arose later. This evidence suggests that snakes evolved on land and later colonized water.

What's the **Point?** ▼ APPLIED

©Fuse/Getty Images RF

Perhaps no scientific issue is more tantalizing and entangled with philosophy than the question of what makes us human. One way to look for answers is to apply DNA analysis techniques to study the evolutionary relationship between humans and chimpanzees, our closest living relatives.

Large, international teams of scientists have sequenced the 3 billion or so DNA nucleotides that make up the chimpanzee and human genomes. The complete DNA sequences for both species reveal exactly how much we have in common: The coding regions (the sequences that specify proteins) are 99% alike. Scientists must still scrutinize both genomes to identify the 25,000 or so coding regions, and they must learn which alleles confer which traits. The result will be an unprecedented view of human (and chimpanzee) biology and evolution.

One tantalizing possibility is that we may soon learn precisely which genes define humans. By comparing the chimp and human genomes, scientists can search for regions that have changed since humans and chimps last shared a common ancestor. These are the sequences that define humans.

Comparing our genome with the chimp genome may also teach us what accounts for our uniquely human features. Considering how genetically similar humans and chimps are, our external appearances are strikingly different. Perhaps humans and chimpanzees express the same proteins but at different times, or perhaps mutations in human DNA have made some genes stop working. The chimp and human genomes will help biologists test these hypotheses.

Studying DNA sequences cannot fully explain what makes us human. Whatever the source of our humanity, however, it is revealed partly in our ability to make reasoned decisions and in our compassion for others. The journal article announcing the chimpanzee genome sequence advocates the protection of chimps in the wild, where these animals are endangered. The paper ends with this statement: "We hope that elaborating how few differences separate our species will broaden recognition of our duty to these extraordinary primates that stand as our siblings in the family of life."

CHAPTER SUMMARY

13.1 Clues to Evolution Lie in the Earth, Body Structures, and Molecules

- The **geologic timescale** divides Earth's history into segments defined by major events such as mass extinctions.
- Evidence for evolutionary relationships comes from **paleontology** (the study of past life), from biogeography, and from comparisons of the physical and biochemical characteristics of species (figure 13.16).

13.2 Fossils Record Evolution

- **Fossils** are the remains of ancient organisms.

A. The Fossil Record Is Often Incomplete

- Many organisms that lived in the past did not leave fossil evidence.

B. The Age of a Fossil Can Be Estimated in Two Ways

- The position of a fossil in the context of others provides a **relative date**.
- **Radiometric dating** uses radioactive isotopes to estimate the **absolute date** when an organism lived (figure 13.17). The length of an isotope's **half-life** determines whether it is useful for ancient or recent objects.

13.3 Biogeography Considers Species' Geographical Locations

- **Biogeography** is the study of the distribution of species on Earth.

A. The Theory of Plate Tectonics Explains Earth's Shifting Continents

- The **plate tectonics** theory indicates that forces deep inside Earth have moved the continents throughout much of life's history, creating and eliminating geographical barriers.

B. Species Distributions Reveal Evolutionary Events

- Biogeography provides insight into large- and small-scale evolutionary events.

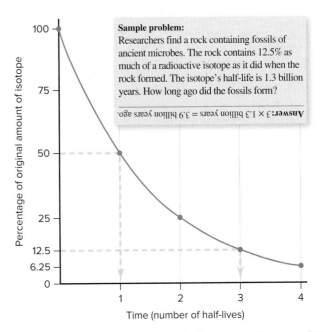

Sample problem:
Researchers find a rock containing fossils of ancient microbes. The rock contains 12.5% as much of a radioactive isotope as it did when the rock formed. The isotope's half-life is 1.3 billion years. How long ago did the fossils form?

Answer: 3 × 1.3 billion years = 3.9 billion years ago.

Figure 13.17 Using Radioactive Isotopes in Radiometric Dating: An Example.

13.4 Anatomical Comparisons May Reveal Common Descent

A. Homologous Structures Have a Shared Evolutionary Origin

- **Homologous** anatomical structures and molecules have similarities that indicate they were inherited from a shared ancestor, although they may differ in function.

B. Vestigial Structures Have Lost Their Functions

- **Vestigial** structures have no function in an organism but are homologous to functioning structures in related species.

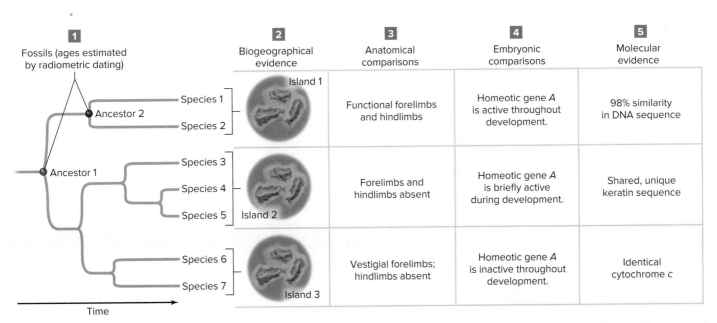

Figure 13.16 Using Evidence to Build an Evolutionary Tree. This illustration shows how five lines of evidence help researchers understand the relationships among seven hypothetical species on three islands.

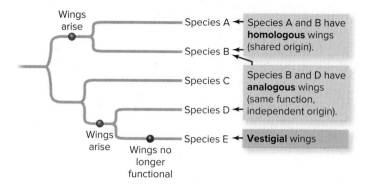

Figure 13.18 Homologous, Analogous, and Vestigial Structures Compared.

C. Convergent Evolution Produces Superficial Similarities

- **Analogous** structures are similar in function but do not reflect shared ancestry. **Convergent evolution** can produce analogous structures.
- Figure 13.18 illustrates the evolutionary origins of homologous, analogous, and vestigial structures.

13.5 Embryonic Development Patterns Provide Evolutionary Clues

- Evolutionary developmental biology combines the study of development with the study of DNA sequences. Many genes, including **homeotic genes,** influence the development of an organism's body parts; mutations in homeotic genes therefore may lead to new phenotypes.

13.6 Molecules Reveal Relatedness

A. Comparing DNA and Protein Sequences May Reveal Close Relationships

- Similarities in molecular sequences are unlikely to occur by chance; descent from a shared ancestor is more likely.
- DNA sequence comparisons provide an indication of the relationships among species, as can the amino acid sequences of proteins.

B. Molecular Clocks Help Assign Dates to Evolutionary Events

- A **molecular clock** compares DNA sequences to estimate the time when two species diverged from a common ancestor.

MULTIPLE CHOICE QUESTIONS

1. Why is the fossil record incomplete?
   a. Because many organisms with soft body parts never fossilize
   b. Because many fossils are destroyed by geological events
   c. Because many fossils are buried where no one will ever find them
   d. All of the above are correct.

2. Fossils found in deeper layers of the Earth generally have _____ ^{14}C than fossils found in the upper layers.
   a. more
   b. less
   c. the same amount of
   d. different types of

3. You discover that a 24,000-year-old fossil has one-fourth the concentration of a radioactive isotope compared to a living organism. What is the half-life of this isotope?
   a. 3000 years
   b. 6000 years
   c. 8000 years
   d. 12,000 years

4. The study of biogeography is most concerned with the
   a. correct placement of species on the evolutionary tree.
   b. precise rock layer in which a fossil is found.
   c. current and past distribution of species on Earth.
   d. predicted locations of future extinction "hot spots."

5. Octopuses and cuttlefish are mollusks that have a single-lens eye. Their common ancestor also had a single-lens eye. What phrase describes the relationship between the octopus eye and the cuttlefish eye?
   a. Homologous structures
   b. Vestigial structures
   c. Analogous structures
   d. Convergent structures

6. Ground beetles have useless hindwings. In related species of beetles, the hindwings function in flight. What term describes ground beetles' hindwings?
   a. Homeotic
   b. Vestigial
   c. Analogous
   d. Fossilized

7. Scorpions occupy every continent except Antarctica, and all scorpions fluoresce under ultraviolet light. What do these observations most likely suggest about the origin of scorpion fluorescence?
   a. The common ancestor of these scorpions fluoresced.
   b. Scorpion fluorescence evolved independently on each continent.
   c. Scorpion fluorescence is a vestigial characteristic.
   d. Scorpion fluorescence evolved recently.

8. How does the activity of a homeotic gene relate to evolution?
   a. Homeotic genes serve as markers for convergent evolution.
   b. Organisms with similar homeotic genes have the same vestigial structures.
   c. Mutations in homeotic genes can produce new body plans.
   d. The presence of homeotic genes helps to identify a fossil as that of an animal.

9. As described in chapter 7, a genetic mutation does not necessarily change the amino acid sequence of a protein. For a given gene, which molecule would you expect to change the most over evolutionary time?
   a. The DNA sequence should change more than the protein sequence.
   b. The protein sequence should change more than the DNA sequence.
   c. The two should have exactly the same number of changes.
   d. The answer depends on the gene.

10. Which of the following would be most useful for comparing ALL known groups of organisms?
    a. DNA encoding ribosomal RNA
    b. DNA encoding the keratin protein
    c. Mitochondrial DNA
    d. Y chromosome DNA

Answers to Multiple Choice questions are in appendix A.

WRITE IT OUT

1. Explain the significance of the geologic timescale in the context of evolution.

2. When considering events in the context of geologic time, an event that seems to have occurred "a long time ago" may actually be recent relative to the age of the Earth. Suppose all the events since the origin of the Earth (4.6 BYA) were sped up into an hour-long video. How many seconds from the end of the video would dinosaurs go extinct (65 MYA)?

3. Why are transitional fossils especially useful for understanding evolutionary relationships?

4. The bubonic plague swept through western Europe in 1348. Suppose researchers use ^{14}C dating on the skeletons of suspected plague victims, and they discover that about 50% of the original amount of ^{14}C remains in the bones. Are the bones the remains of plague victims?

5. Index fossils represent organisms that were widespread but lived during relatively short periods of time. How are index fossils useful in relative dating?

6. How do molecular sequences provide different information than relative and absolute dating?

7. Suppose that collaborating research teams found fossils of the same extinct species in eastern South America and western Africa. What can the researchers conclude about the age of these fossils without using absolute or relative dating techniques?

8. How did the discovery of Wallace's line demonstrate the predictive power of evolution?

9. Why is it important for evolutionary biologists to be able to distinguish between homologous and analogous anatomical structures?

10. Suppose that plants in the San Francisco Bay area and in southern Chile share a common seed dispersal method. Scientists determine that the evolutionary divergence of these plants happened long before this seed dispersal method arose in each plant. What term relates the seed dispersal method of the San Francisco Bay plant to that of the southern Chile plant? Explain your answer.

11. Many species look similar as embryos. What causes them to appear different as adults? Why does the study of development give insights into evolutionary relationships?

12. Give examples of how the field of evolutionary developmental biology uses evidence from anatomical comparisons and from homeotic genes.

13. Some genes are more alike between human and chimp than other genes are from person to person. Does this mean that chimps are humans or that humans with different alleles are different species? What other explanation fits the facts?

14. Search the Internet to learn about the evidence suggesting that birds and dinosaurs are closely related. If scientists could extract DNA from dinosaur fossils, how could they use the sequences to learn more about the origin of birds?

15. Why are molecular clocks useful, and what are their limitations?

16. Evolutionary biologists often try to assign an approximate date when two organisms last shared a common ancestor. Why do you think that molecular evidence often yields an earlier date than fossil evidence?

SCIENTIFIC LITERACY

1. Genetic anthropology combines the study of DNA with physical evidence such as fossils to reveal the history of the human species. Use the Internet to research the goals and methods of the Genographic Project, HapMap, or the Human Genome Diversity Project. What are the benefits of these projects? What ethical issues arise from this type of research?

2. Review Burning Question 13.1, which explains why the fossil record is incomplete for the evolutionary transition between some groups of organisms. Which is more logical: To assume that a transitional form existed even though its fossils have not been found, or to assume that the transitional form did not exist since we have not seen the fossils? Explain your answer.

Answers to Mastering Concepts, Write It Out, Scientific Literacy, and Pull It Together questions can be found in the Connect ebook.
connect.mheducation.com

Design element: Burning Question (fire background): ©Ingram Publishing/Super Stock

PULL IT TOGETHER

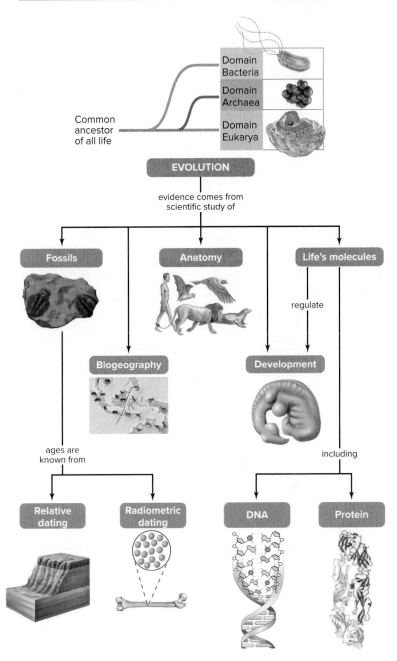

Figure 13.19 Pull It Together: Evidence of Evolution.

Refer to figure 13.19 and the chapter content to answer the following questions.

1. Review the Survey the Landscape figure in the chapter introduction. What diagrams do scientists use to visualize evolutionary relationships? Add this term to the concept map.

2. Write a phrase to connect *fossils* and *biogeography* and a separate phrase to connect *DNA* and *anatomy*.

3. Add the following terms to this concept map: *homologous structures, vestigial structures, homeotic genes,* and *molecular clock.*

4. Provide an example of each line of evidence for evolution.

LEARNING OUTLINE

14.1 What Is a Species?

14.2 Reproductive Barriers Cause Species to Diverge

14.3 Spatial Patterns Define Two Types of Speciation

14.4 Speciation May Be Gradual or May Occur in Bursts

14.5 Extinction Marks the End of the Line

14.6 Biological Classification Systems Are Based on Common Descent

APPLICATIONS

Burning Question 14.1 *Can people watch evolution and speciation in action?*

Why We Care 14.1 *Recent Species Extinctions*

Burning Question 14.2 *Did rabbits come from frogs?*

Investigating Life 14.1 *Plant Protection Rackets May Stimulate Speciation*

Endangered Primate. Brazil's golden lion tamarin is one of many species in danger of extinction, thanks to the destruction of its forest habitat.
©Kike Calvo/National Geographic/Getty Images

Learn How to Learn
Take Notes on Your Reading

Taking notes as you read should help you not only to retain information but also to identify what you don't understand. Before you take notes, skim through the assigned pages once; otherwise, you may have trouble distinguishing between main points and minor details. Then read them again. This time, pause after each section and write the most important ideas in your own words. What if you can't remember or don't understand well enough to summarize the passage? Read it again, and if that doesn't work, ask for help with whatever isn't clear.

SURVEY THE LANDSCAPE
Evolution and Diversity

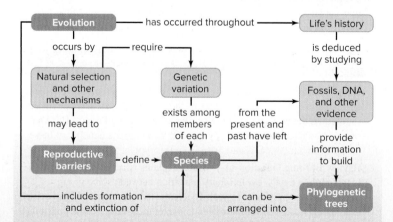

Evolutionary processes sometimes lead to reproductive barriers within a population, and the isolated groups may become unique species. Meanwhile, existing species may go extinct. Phylogenetic trees track these large-scale evolutionary changes.

For more details, study the Pull It Together feature in the chapter summary.

Over billions of years, many new species have appeared on Earth, but they did not suddenly pop into existence. Instead, populations slowly became isolated into subgroups that could not interbreed. This chapter describes how small evolutionary changes can lead to the development of entirely new species.

©Getty Images RF

Yet most species that have ever lived are now extinct. Many species have vanished following global disasters. Most dinosaurs died out 65 million years ago, though some survived and evolved into today's birds. More recently, the explosive growth of the human population has doomed countless species, with unknown consequences to ecosystems on land and in water.

The opposing, ongoing processes of species formation and extinction have likely been a part of evolution since life began. This chapter explains how these processes occur.

14.1 What Is a Species?

Throughout the history of life, the types of organisms have changed. New species have appeared, and others have gone extinct. The term **macroevolution** describes these large, complex changes in life's panorama. Macroevolutionary events tend to span very long periods, whereas the microevolutionary processes described in chapter 12 happen so rapidly that we can sometimes observe them over just a few years (see Burning Question 14.1). Nevertheless, the many small changes that accumulate in a population by microevolution eventually lead to large-scale macroevolution.

Evolution has produced an obvious diversity of life. A bacterium, for example, is clearly distinct from a tree or a bird (figure 14.1). At the same time, some organisms are more closely related than others; the bird in figure 14.1 is more similar to a chicken than it is to a tree. To make sense of these observations, biologists recognize the importance of grouping similar individuals into **species**—that is, distinct types of organisms. This task requires agreement on what the word *species* means. Perhaps surprisingly, the definition has changed over time and is still the topic of vigorous debate among biologists.

A. Linnaeus Classified Life Based on Appearance

Swedish botanist Carolus Linnaeus (1707–1778) was not the first to ponder what constitutes a species, but his contributions last to this day. Linnaeus defined species as "all examples of creatures that were alike in minute detail of body structure."

Linnaeus also devised a hierarchical system for classifying species, as described in section 14.6. His classifications organized life's diversity and helped scientists communicate with one another. His system did not, however, consider the role of evolutionary relationships. Linnaeus thought that each species was created separately and could not change. Therefore, species could not appear or disappear, nor were they related to one another.

Charles Darwin (1809–1882) finally connected species diversity to evolution. He predicted that classifications would come to resemble genealogies, or extended "family trees." As the theory of evolution by natural selection became widely accepted, scientists no longer viewed classifications merely as ways to organize life. They considered them to be hypotheses about life's evolutionary history.

B. Species Can Be Defined Based on the Potential to Interbreed

In the 1940s, biologists incorporated reproduction and genetics into the question of what constitutes a species. According to the **biological species concept,** a species is a population, or group of populations, whose members can interbreed and produce fertile offspring. **Speciation,** the formation of new species, occurs when some members of a population can no longer successfully interbreed with the rest of the group.

Figure 14.1 Distinctive Species. Bacteria, a tree, and a bird are about as dissimilar as three types of organisms can be.

SEM (false color) |⎯⎯| 3 μm

(bacteria): ©S. Lowry/University Ulster/Getty Images; (tree): ©Sieboldianus/E+/Getty Images RF; (bird): ©Erich Kuchling/Westend61/Getty Images RF

Figure 14.2 How Many Species? Linnaeus would have categorized these butterflies based on their physical appearance. The biological species definition, however, provides an objective rule for determining whether each group really is a separate species.
©IT Stock Free/Alamy Stock Photo RF

How might this happen? A new species can form if a population somehow becomes divided. Recall from chapter 12 that a population's **gene pool** is its entire collection of genes and their alleles. An intact, interbreeding population shares a common gene pool. After a population splits in two, however, micro-evolutionary changes such as mutations, natural selection, and genetic drift can lead to genetic divergence between the groups. With the accumulation of enough differences in their separate gene pools, the two groups can no longer produce fertile offspring even if they come into contact once again. In this way, microevolution becomes macroevolution.

The biological species definition does not rely on physical appearance, so it is much less subjective than Linnaeus's observations. Under the system of Linnaeus, it would be impossible to determine whether two similar-looking butterflies belong to different species (figure 14.2). Using the biological species concept, however, we can say that they belong to one species if the two groups can produce fertile offspring together.

Nevertheless, the biological species concept raises several difficulties. First, it cannot apply to asexually reproducing organisms, such as bacteria, archaea, and many fungi and protists. Second, it is impossible to apply the biological species definition to extinct organisms. Third, some types of organisms have the *potential* to interbreed in captivity, but they do not do so in nature. Fourth, reproductive isolation is not always absolute. Closely related species of plants, for example, may occasionally produce fertile offspring together, even though their gene pools mostly remain separate.

As a result, the biological species concept does not provide a perfect way to determine the "boundaries" of each species. DNA sequence analysis has helped to fill in some of these gaps. Biologists working with bacteria and archaea, for example, use a stretch of DNA that encodes ribosomal RNA to define species. If the DNA sequences of two specimens are more than 97% identical, they are considered to be the same species. These genetic sequences, however, still present some ambiguity because they cannot reveal whether genetically similar organisms currently share a gene pool. ⓘ *DNA sequencing,* section 11.2B

Despite these difficulties, reproductive isolation is the most common criterion used to define species. The rest of this chapter therefore uses the biological species concept to describe how speciation occurs.

14.1 Mastering Concepts

1. How are macroevolution and microevolution related?
2. How does the biological species concept differ from Linnaeus's definition of the term *species*?
3. What are some of the challenges in defining species?

Miniglossary	Macroevolution
Biological species	A group of organisms that can potentially interbreed and produce fertile offspring
Speciation	The formation of new species
Gene pool	A population's entire collection of genes and alleles
Reproductive barrier	Mechanism that prevents groups of organisms from sharing a gene pool
Extinction	The death of all the individuals of a species

14.2 Reproductive Barriers Cause Species to Diverge

In keeping with the biological species concept, a new species forms when one portion of a population can no longer breed and produce fertile offspring with the rest of the population. That is, the separate groups no longer share a gene pool, and each begins to follow its own, independent evolutionary path.

One portion of a population can become reproductively isolated in many ways because successful reproduction requires so many complex events. Any interruption in courtship, fertilization, embryo formation, or offspring development can be a reproductive barrier.

Biologists divide the many mechanisms of reproductive isolation into two broad groups: prezygotic and postzygotic. Prezygotic reproductive barriers prevent the formation of a zygote, or fertilized egg; collectively, they are the most common way for one gene pool to become isolated from another. Postzygotic reproductive barriers, which act after fertilization, reduce the fitness of a hybrid offspring. (A **hybrid,** in this context, is the offspring of individuals from two different species.) Figure 14.3 summarizes the reproductive barriers; the rest of this section describes them in detail.

Barrier	Description	Example	Illustration	
PREZYGOTIC REPRODUCTIVE ISOLATION				
Habitat isolation	Different environments	Ladybugs feed on different plants.		
Temporal isolation	Active or fertile at different times	Field crickets mature at different rates.		
Behavioral isolation	Different courtship activities	Frog mating calls differ.		
Mechanical isolation	Mating organs or pollinators incompatible	Sage species use different pollinators.		
Gametic isolation	Gametes cannot unite.	Sea urchin gametes are incompatible.		
POSTZYGOTIC REPRODUCTIVE ISOLATION				
Hybrid inviability	Hybrid offspring fail to reach maturity.	Hybrid eucalyptus seeds and seedlings are not viable.		
Hybrid infertility (sterility)	Hybrid offspring unable to reproduce	Lion-tiger cross (liger) is infertile.		
Hybrid breakdown	Second-generation hybrid offspring have reduced fitness.	Offspring of hybrid mosquitoes have abnormal genitalia.		

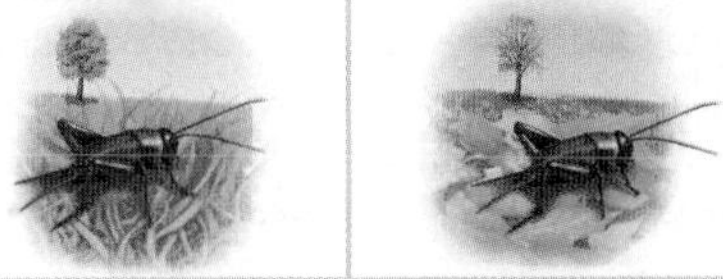

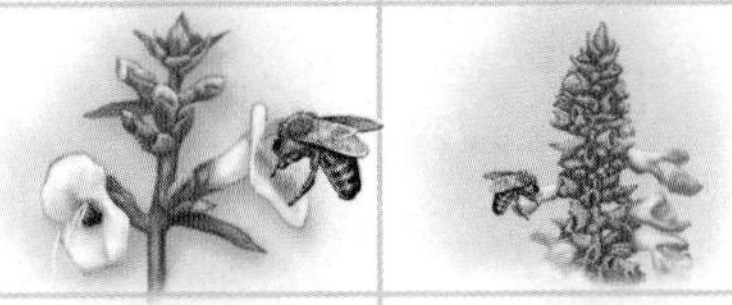

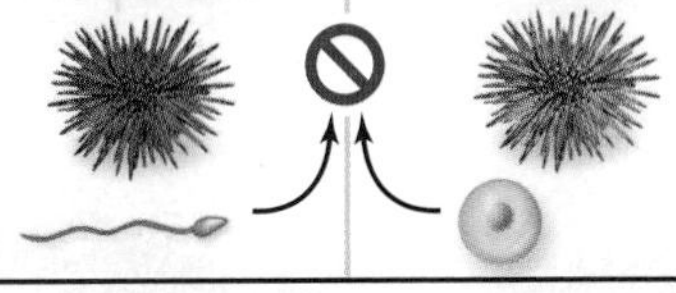

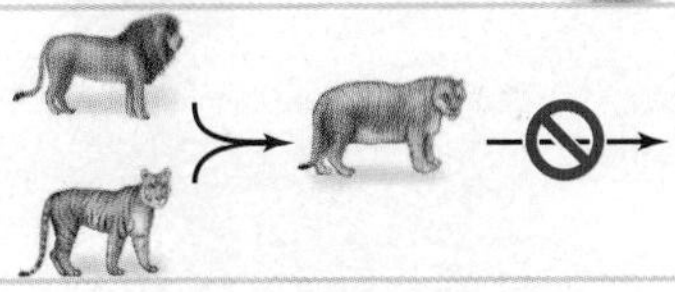

Figure 14.3 Reproductive Barriers. Prezygotic and postzygotic reproductive barriers can prevent two related species from producing fertile offspring.

Burning Question 14.1

Can people watch evolution and speciation in action?

Microevolution is ongoing in every species, so it is not surprising that biologists have documented many instances of evolution "in action." Medicine provides the most familiar context. One example is the discovery that HIV evolved from a virus that occurs in chimpanzees; another is the rise of antibiotic resistance among the bacteria that cause staph infections and tuberculosis.

The use of pesticides has provided ample examples as well. Many people know that populations of DDT-resistant insects skyrocketed shortly after people began using DDT to kill mosquito larvae. Likewise, Investigating Life 10.1 describes the selection for moth larvae that are resistant to Bt, an insecticidal protein.

Speciation is observable as well. In laboratory studies, flies reared for multiple generations on one food source became reproductively isolated from their counterparts reared on a different diet. Outside the lab, mosquito populations that have been isolated for more than 100 years in the tunnels of the London Underground can no longer breed with their aboveground counterparts—a sure sign that a new species has formed.

Submit your burning question to
marielle.hoefnagels@mheducation.com

(mosquito): Source: USDA

Mule

©Antonio Esparraga Godoy/Alamy Stock Photo

A. Prezygotic Barriers Prevent Fertilization

Mechanisms of **prezygotic reproductive isolation** affect the ability of two species to combine gametes and form a zygote. As you can see in figure 14.3, two of the prezygotic barriers keep members of two different species from ever encountering each other. The populations may live in different places (habitat isolation) or be active at different times of the day or year (temporal isolation).

Even if individuals from two species do encounter each other, prezygotic barriers may still prevent breeding. The two species may use such different mating rituals that they are not attracted to each other (behavioral isolation). For example, mate selection in many birds is based on intricate courtship dances. Any variation in the ritual from one group to another could prevent them from mating.

Prezygotic barriers may apply even if the members of two species attempt to mate. In animals, the male and female parts of two related species may not match; in plants, the insect that pollinates one species may be unable to fit into the flower of another. Both scenarios illustrate a reproductive barrier called mechanical isolation, which prevents male and female gametes from meeting even if mating does occur.

The final prezygotic barrier is gametic isolation. If a sperm cannot fertilize an egg cell, then no reproduction will occur. For example, many marine organisms, such as sea urchins, simply release sperm and egg cells into the water. These gametes display unique surface molecules that enable an egg to recognize sperm of the same species. In the absence of a "match," fertilization will not occur, and the gene pools will remain separate.

B. Postzygotic Barriers Prevent Development of a Fertile Offspring

Individuals of two different species may produce a hybrid zygote. Even then, **postzygotic reproductive isolation** may keep the species separate by selecting against the hybrid offspring (see figure 14.3).

One type of postzygotic reproductive barrier is hybrid inviability. In this case, a hybrid embryo dies before reaching reproductive maturity, typically because the genes of its parents are incompatible. Alternatively, the hybrid offspring may develop to adulthood but be unable to produce offspring of its own (hybrid infertility). The most familiar example of this postzygotic barrier is a mule, which is the hybrid offspring of a female horse and a male donkey. Meiosis does not occur in the mule's germ cells because the two parents contribute different numbers of chromosomes. As a result, the mule cannot produce gametes. Similarly, a liger is the hybrid offspring of a male lion and a female tiger. Like mules, ligers are usually sterile. ⓘ *homologous chromosomes,* section 9.2

Some species produce hybrid offspring that are fertile. When the hybrids reproduce, however, their offspring may have abnormalities that reduce their fitness. Some second-generation hybrid offspring of two mosquito species, for example, have abnormal genitalia that make mating difficult. This last type of postzygotic reproductive barrier is called hybrid breakdown.

14.2 Mastering Concepts

1. How do reproductive barriers lead to speciation?
2. Write a real or fictitious example other than those listed in figure 14.3 of each type of reproductive barrier.

14.3 Spatial Patterns Define Two Types of Speciation

Reproductive barriers keep related species apart, but how do these barriers arise in the first place? More specifically, how can two populations of the same species evolve along different pathways, eventually yielding two species?

The most obvious way is to physically separate the populations so that they do not exchange genes. Eventually, the genetic differences between the populations would give rise to one or more reproductive barriers. Yet speciation can also occur when two populations have physical contact with each other. Biologists recognize these different circumstances by dividing the geographical setting of speciation into two categories: allopatric and sympatric (figure 14.4).

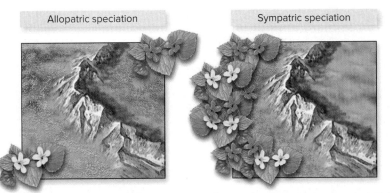

Allopatric speciation | Sympatric speciation

No contact between populations | Continuous contact between populations

Figure 14.4 **Speciation and Geography.** Allopatric and sympatric speciation are distinguished based on whether populations are separated by a physical barrier or mingle within a shared area.

A. Allopatric Speciation Reflects a Geographical Barrier

In **allopatric speciation,** a new species forms when a geographical barrier physically separates a population into two groups that cannot interbreed (*allo-* means "other," and *patria* means "fatherland"). The barrier may be a river, desert, glacier, mountain range, large body of water, dam, farm, or city. Rising sea levels may also trap populations on isolated islands.

If the separate parts of a population cannot contact each other, migration between them stops. Meanwhile, mutations and the other forces of microevolution continue to alter allele frequencies in each group. The result may be one or more reproductive barriers. When the descendants of the original two populations can no longer interbreed, one species has branched into two.

Island groups offer ideal opportunities for allopatric speciation. For example, 11 subspecies of tortoise occupy the Galápagos islands (figure 14.5).

Figure 14.5 **Allopatric Speciation in Tortoises.** Descendants of the first tortoises to arrive on the Galápagos have colonized most of the islands, evolving into many subspecies. Unique habitats have selected for different sets of adaptations in the tortoises.

Photos: (domed): ©Millard H. Sharp/Science Source; (saddleback): ©Nancy Nehring/iStock/360i/Getty Images RF

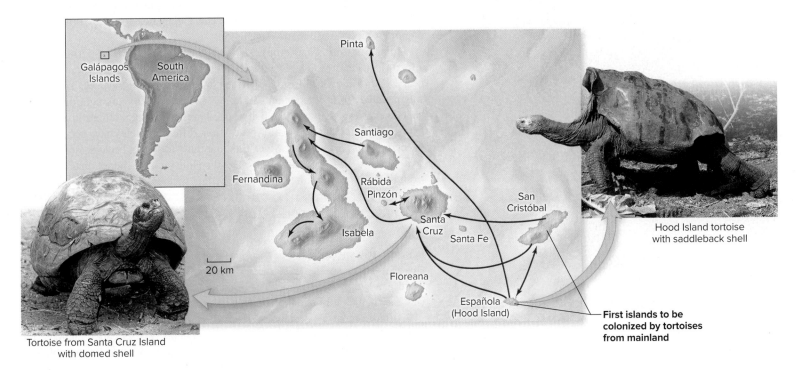

Galápagos Islands / South America

Pinta

Santiago

Fernandina

Rábida
Pinzón

Isabela

Santa Cruz

Santa Fe

San Cristóbal

20 km

Floreana

Española (Hood Island)

First islands to be colonized by tortoises from mainland

Hood Island tortoise with saddleback shell

Tortoise from Santa Cruz Island with domed shell

Figure 14.6 Allopatric Speciation in Pupfish. The pupfish species that lives in Devil's Hole cannot breed with pupfish from nearby springs.

Photos: (Devil's Hole): ©Robyn Beck/AFP/Getty Images; (both fish): ©Stone Nature Photography/Alamy Stock Photo

According to DNA analysis, a few newcomers from the South American mainland first colonized either San Cristóbal or Española (Hood Island) a couple of million years ago. Tortoises soon floated on ocean currents to nearby islands, where they encountered new habitats that selected for different adaptations, especially in shell shape. Dry islands with sparse vegetation selected for notched shells that enable the tortoises to reach for higher food sources. The Hood Island tortoise in figure 14.5 illustrates this characteristic "saddleback" shape. On islands with lush, low-growing vegetation, the tortoises have domed shells; the tortoise from Santa Cruz Island is an example.

Although many of the subspecies look noticeably different from one another, the tortoises can interbreed. They are not yet separate species, but the genetic similarities among tortoises on each island suggest that migration from island to island was historically rare. The Galápagos tortoises illustrate an ongoing process of allopatric speciation.

In addition to island groups, isolated springs also offer opportunities for allopatric speciation. The Devil's Hole pupfish, which inhabits a warm spring near Death Valley, California, provides one example (figure 14.6). The spring was isolated from other bodies of water about 50,000 years ago, preventing genetic exchange between the fish trapped in the spring and those in the original population. Generation after generation, different alleles have accumulated in each pupfish population. Since the time that the spring became isolated, the gene pool has shifted enough that a Devil's Hole pupfish cannot mate with fish from nearby springs. It has become a distinct species.

B. Sympatric Speciation Occurs in a Shared Habitat

In **sympatric speciation,** a new species arises while living in the same physical area as its parent species (*sym-* means "together"). Among evolutionary biologists, the idea of sympatric speciation can be controversial. After all, how can a new species arise in the midst of an existing population?

Often sympatric speciation reflects the fact that a habitat that appears uniform actually consists of many microenvironments. Fishes called cichlids, for example, have diversified within African lakes. Figure 14.7 shows two cichlids in

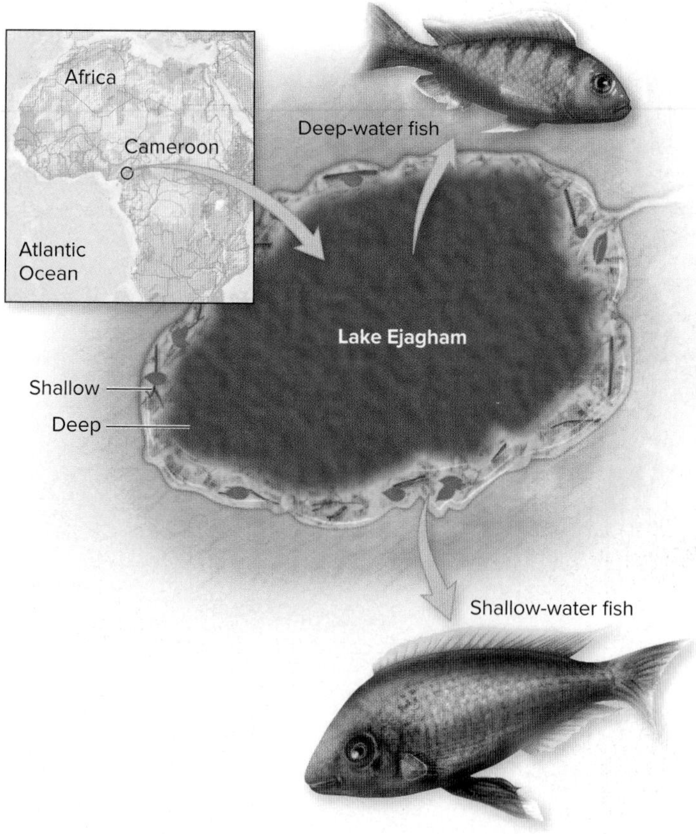

Figure 14.7 Sympatric Speciation. Cichlids in the deepest waters of Cameroon's Lake Ejagham have smaller bodies than do shallow-water fish. Both varieties belong to the same species, at least for now.

Cameroon's tiny Lake Ejagham. This 18-meter-deep lake has distinct ecological zones. Its bottom is muddy near the center, whereas leaves and twigs cover the sandy bottom near the shore. The two types of fish belong to the same species, but the larger ones consume insects near the shore, whereas the smaller ones eat tiny floating prey in the deeper waters. The fish breed where they eat, so the two forms typically remain isolated. As genetic differences between the subpopulations continue to accumulate, sympatric speciation may occur.

In plants, sympatric speciation may occur when cells become **polyploid,** meaning they acquire extra sets of chromosomes. Nearly half of all flowering plant species are natural polyploids, as are about 95% of ferns. Moreover, many major crops, including wheat, corn, sugarcane, potatoes, and coffee, are derived from polyploid plants. ⓘ *polyploid cell,* section 9.7A

Polyploidy sometimes arises when gametes from two different species fuse. Cotton plants provide an example (figure 14.8). An Old World species of wild cotton has 26 large chromosomes, whereas one from Central and South America has 26 small chromosomes. The two species interbred, forming a diploid hybrid with 26 chromosomes (13 large and 13 small). This hybrid was sterile. But eventually the chromosome number doubled. The resulting cotton plant is a fertile polyploid with 52 chromosomes (26 large and 26 small); this new polyploid species formed sympatrically in the midst of its ancestors. Farmers around the world cultivate this species to harvest cotton for cloth.

C. Determining the Type of Speciation May Be Difficult

Biologists sometimes debate whether a speciation event is allopatric or sympatric. One reason for the disagreement is that the definitions represent points along a continuum, from complete reproductive isolation to continuous intermingling. Another difficulty is that we may not be able to detect the barriers that are important to other species. For example, a researcher might perceive a patch of forest to be uniform and conclude that speciation events occurring there are sympatric. But to a tiny insect, the distance between the forest floor and the treetops may represent an insurmountable barrier. In that case, speciation would be considered allopatric.

The problem of perspective also leads to debate over the size of the geographical barrier needed to separate two populations, which depends on the distance over which a species can spread its gametes. A plant with windblown pollen or a fungus producing lightweight spores encounters few barriers to gene exchange; pollen and spores can travel thousands of miles in the upper atmosphere. On the other hand, a desert pupfish cannot migrate out of its aquatic habitat, so the isolation of its pool instantly creates a geographical barrier. The same circumstance would not deter gene exchange in species that walk or fly between pools.

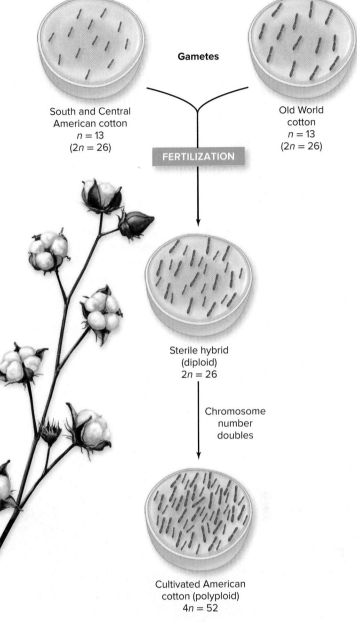

Gametes

South and Central American cotton
$n = 13$
$(2n = 26)$

Old World cotton
$n = 13$
$(2n = 26)$

FERTILIZATION

Sterile hybrid (diploid)
$2n = 26$

Chromosome number doubles

Cultivated American cotton (polyploid)
$4n = 52$

Figure 14.8 Polyploid Cotton. Cultivated American cotton is a polyploid species derived from Old World and New World ancestors.

14.3 Mastering Concepts

1. Distinguish between allopatric and sympatric speciation, and provide examples of each.
2. How can polyploidy contribute to sympatric speciation?
3. Why is it sometimes difficult to determine whether speciation is allopatric or sympatric?

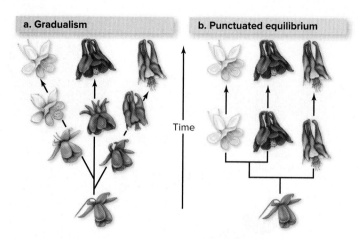

a. Gradualism

b. Punctuated equilibrium

Time

Figure 14.9 **Evolution—Both Gradual and in Bursts.**
(a) In gradualism, species arise in small, incremental steps.
(b) Punctuated equilibrium produces the same result, except that the new species arise in rapid bursts followed by periods of little change.

14.4 Speciation May Be Gradual or May Occur in Bursts

Darwin envisioned one species gradually transforming into another through a series of intermediate stages. The pace as he saw it was slow, although not necessarily constant. This idea, which became known as **gradualism,** held that evolution proceeds in small, incremental changes (figure 14.9a).

If the gradualism model is correct, then "slow and steady" evolutionary change should be evident in the fossil record. Microscopic protists such as foraminiferans and diatoms, for example, have evolved gradually. Vast populations of these asexual organisms span the oceans, leaving a rich fossil record in sediments. Since isolated populations rarely form in this uniform environment, it is unsurprising that speciation has been gradual. ⓘ *protists,* section 15.4

Much of the fossil record, however, supports a different pattern. In **punctuated equilibrium,** relatively brief bursts of rapid speciation interrupt long periods of little change (see figure 14.9b). Fossils of diverse animals such as bryozoans, mollusks, and mammals all reveal many examples of rapid evolution followed by periods of stability.

Evolution seems to happen "suddenly" in punctuated equilibrium, with few transitional fossils documenting the evolution of one species into another. What accounts for the missing transitional forms? One explanation is that the fossil record is incomplete, for reasons explored in chapter 13. Another is that the predicted "missing links" may have been too rare to leave many fossils. After all, periods of rapid speciation would mean that few examples of any given transitional form ever existed, so we are unlikely to find their remains.

The punctuated equilibrium model fits well with the concept of allopatric speciation. Consider the isolated population of desert pupfish that became genetically distinct from its ancestral population (see figure 14.6). If the climate changed and the spring containing the new species rejoined its "old" spring, the fossil record might show that a new fish species suddenly appeared with its ancestors—after all, 50,000 years is a blink of an eye in geologic time. Afterward, unless the environment changed again, a period of stability would ensue.

A rapid bout of speciation may also occur when some members of a population inherit a key adaptation that gives them an advantage. For example, after flowering plants appeared 144 million years ago, they diversified rapidly into the hundreds of thousands of species that now inhabit Earth. The new adaptation—the flower—apparently unleashed an entirely new set of options for reproduction, prompting rapid diversification. ⓘ *flowering plants,* section 16.5

Rapid speciation may also occur when some members of a population inherit adaptations that enable them to survive a major environmental change. After the poorly suited organisms perish, the survivors diversify as they exploit the new resources in the changed environment. Mammals, for example, underwent an enormous burst of speciation when dinosaur extinctions opened up many new habitats about 65 million years ago (figure 14.10).

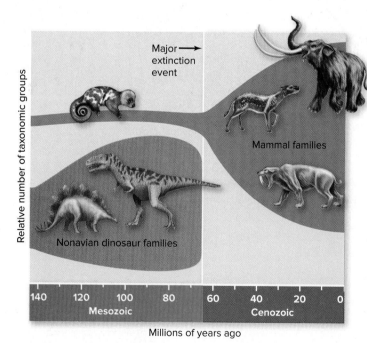

Relative number of taxonomic groups

Major → extinction event

Mammal families

Nonavian dinosaur families

| 140 | 120 | 100 | 80 | 60 | 40 | 20 | 0 |

Mesozoic   Cenozoic

Millions of years ago

Figure 14.10 **Speciation Following a Mass Extinction.**
Many ecological niches were vacated when nonavian ("non-bird") dinosaurs went extinct about 65 million years ago. Mammals diversified and flourished in the aftermath.

14.4 Mastering Concepts

1. Describe the theories of gradualism and punctuated equilibrium.
2. How can the fossil record support both gradualism and punctuated equilibrium?

14.5 Extinction Marks the End of the Line

An **extinction** occurs when all members of a species have died. The change that wipes out a species may be habitat loss, new predators, or new diseases. Extinction may also be a matter of bad luck: Sometimes no individual of a species survives a volcanic eruption or asteroid impact.

No matter what the external trigger of an extinction, the root cause is always the same: Species die out if evolution fails to meet the pace of environmental change. Any species will eventually vanish if its gene pool does not contain the "right" alleles necessary to sustain the population; genetic diversity is therefore essential in a changing environment.

Biologists distinguish between two different types of extinction events. The **background extinction rate** results from the steady, gradual loss of species due to normal evolutionary processes. Paleontologists have used the fossil record to calculate that the background rate is roughly 0.1 to 1.0 extinctions per year per million species. Most extinctions overall have occurred as part of this more-or-less constant background rate.

Earth has also witnessed several periods of **mass extinctions,** when a great number of species disappeared in a relatively short time. The geologic timescale in figure 13.2 shows five major mass extinction events over the past 500 million years (red lines indicate mass extinctions). These events have had a great influence on Earth's history because they have periodically opened vast new habitats for surviving species to diversify (see figure 14.10).

Paleontologists study clues in Earth's sediments to understand the events that lead to mass extinctions. For example, the **impact theory** suggests that meteorites or comets have crashed to Earth, producing huge debris clouds that blocked sunlight and triggered extinctions in a deadly chain reaction. Without sunlight, plants died; animals likewise perished without food and shelter. A meteor impact 65 million years ago apparently doomed nearly all of the dinosaurs; evidence includes thin layers of earth that are rich in iridium, an element rare on Earth but common in meteorites (figure 14.11).

Movements of Earth's crust may also explain some mass extinctions. The crust is divided into many pieces, called tectonic plates. During Earth's history, these plates have drifted apart and come back together. Climates changed as continents moved toward or away from the poles, and colliding continents caused shallow coastal areas packed with life to disappear. Mountain ranges grew, destroying some habitats and creating new ones. ⓘ *plate tectonics,* section 13.3A

Many biologists warn that we are in the midst of a sixth mass extinction—this one caused by human actions. Ecologists estimate that the extinction rate is now about 20 to 200 extinctions per million species per year. Habitat loss and habitat fragmentation, pollution, introduced species, and overharvesting combine to imperil many species (see chapter 20). Why We Care 14.1 lists a few of the many vertebrate species that have recently become extinct, but the problem extends throughout all kingdoms of life. The loss of so many species is likely to severely disrupt the ecosystems we rely on.

Figure 14.11 Impact Theory Evidence. This distinctive layer of rock (inset) marks the cataclysmic end of the Cretaceous period, about 65 million years ago.

Photo: ©Francois Gohier/Science Source

14.5 Mastering Concepts

1. What factors can cause or hasten extinction?
2. Compare mass extinctions to the background extinction rate.
3. How have humans influenced extinctions?

Miniglossary | Speciation and Extinction

Allopatric speciation	A new species arises when a geographical barrier physically separates a population into two groups that cannot interbreed
Sympatric speciation	A new species arises while living in the same physical area as its parent species
Gradualism	The hypothesis that evolution occurs in small, incremental changes
Punctuated equilibrium	The hypothesis that evolution occurs in periods of rapid change following intervals of little change
Background extinction rate	The gradual loss of species due to normal evolutionary processes
Mass extinction	Rapid loss of species in a relatively short period

Why We Care 14.1 | Recent Species Extinctions

Species extinctions have occurred throughout life's long history. They continue today, often accelerated by human activities (see chapter 20 and figure 14.A). Overharvesting contributes to species extinctions, as does habitat loss to agriculture, urbanization, damming, and pollution. Introduced plants and animals can deplete native species by competing with or preying on them.

The International Union for Conservation of Nature keeps track of plant and animal extinctions. Table 14.A lists a few species of vertebrate animals that have disappeared during the past few centuries. This list is far from complete; many more species of animals (both vertebrate and invertebrate) and plants have become extinct during the same time. Countless others are threatened or endangered, meaning that they are at risk for extinction.

Why does it matter if human activities drive species extinct? One reason is that future generations are deprived of the opportunity to enjoy their existence. More importantly, all strands in the web of life are interconnected, and we rely on other species to maintain ecosystem function. Even humans cannot escape this basic ecological principle.

Figure 14.A Extinct. The Pyrenean ibex has been extinct since 2000. Attempts to clone the species using preserved DNA have so far been unsuccessful.

©Flickr Open/Getty Images RF

TABLE 14.A   Recent Vertebrate Extinctions

Name	Cause of Extinction	Former Location
Fishes		
Chinese paddlefish (*Psephurus gladius*)	Habitat destruction	China
Las Vegas dace (*Rhinichthys deaconi*)	Habitat destruction	North America
Amphibians		
Palestinian painted frog (*Discoglossus nigriventer*)	Habitat destruction	Israel
Southern day frog (*Taudactylus diurnus*)	Undetermined	Australia
Reptiles		
Yunnan box turtle (*Cuora yunnanensis*)	Habitat destruction, overharvesting	China
Martinique lizard (*Leiocephalus herminieri*)	Undetermined	Martinique
Birds		
Dodo (*Raphus cucullatus*)	Habitat destruction, overharvesting	Mauritius
Moa (*Megalapteryx diderius*)	Overharvesting	New Zealand
Laysan honeycreeper (*Himatione sanguinea*)	Habitat destruction	Hawaii
Black mamo (*Drepanis funerea*)	Habitat destruction, introduced predators	Hawaii
Passenger pigeon (*Ectopistes migratorius*)	Overharvesting	North America
Great auk (*Alca impennis*)	Overharvesting	North Atlantic
Mammals		
Quagga (*Equus quagga quagga*)	Overharvesting	South Africa
Steller's sea cow (*Hydrodamalis gigas*)	Overharvesting	Bering Sea
Javan tiger (*Panthera tigris sondaica*)	Habitat destruction, overharvesting	Indonesia
Caspian tiger (*Panthera tigris virgata*)	Habitat destruction, overharvesting	Central Asia
Yangtze River dolphin (*Lipotes vexillifer*)	Habitat destruction, overharvesting	China
Pyrenean ibex (*Capra pyrenaica pyrenaica*)	Overharvesting, competition with livestock	Spain, Portugal, France

14.6 Biological Classification Systems Are Based on Common Descent

Darwin proposed that evolution occurs in a branched fashion, with each species giving rise to other species as populations occupy and adapt to new habitats. As described in chapter 13, ample evidence has shown him to be correct.

The goal of modern classification systems is to reflect this shared evolutionary history. **Systematics,** the study of classification, therefore incorporates two interrelated specialties: taxonomy and phylogenetics. **Taxonomy** is the science of describing, naming, and classifying species; **phylogenetics** is the study of evolutionary relationships among species. This section describes how biologists apply the evidence for evolution to the monumental task of organizing life's diversity into groups.

A. The Taxonomic Hierarchy Organizes Species into Groups

Carolus Linnaeus, the biologist introduced at the start of this chapter, made a lasting contribution to systematics. He was the first investigator to give every species a two-word name. The first word refers to the genus (plural: genera), and the second word designates the species. The scientific name for humans, for example, is *Homo sapiens*. Linnaeus also grouped similar genera into a nested hierarchy of orders, classes, and kingdoms.

Although scientists now use additional categories, Linnaeus's idea is the basis of the taxonomic hierarchy used today (figure 14.12). The three domains—Archaea, Bacteria, and Eukarya—are the most inclusive levels. Each domain is divided into kingdoms, which in turn are divided into phyla, then classes, orders, families, genera, and species. A **taxon** (plural: taxa) is a group at any rank; that is, domain Eukarya is a taxon, as is the order Liliales and the species *Aloe vera*.

The more features two organisms have in common, the more taxonomic levels they share (figure 14.13). A human, a squid, and a fly are all members of the animal kingdom, but their many differences place them in separate phyla. A human, rat, and pig are more closely related—all belong to the same kingdom, phylum, and class (Mammalia). A human, an orangutan, and a chimpanzee are even more closely related, sharing the same kingdom, phylum, class, order, and family (order Primates, family Hominidae). Figure 1.11 shows the full classification for humans.

B. A Cladistics Approach Is Based on Shared Derived Traits

Biologists illustrate life's diversity in the form of phylogenetic (evolutionary) trees, which depict relationships based on descent from shared ancestors. Multiple lines of evidence are used to construct these trees (see figure 13.16). Anatomical features of fossils and existing organisms are useful, as are behaviors, physiological adaptations, and molecular sequences.

In the past, systematists constructed phylogenetic tree diagrams by comparing as many characteristics as possible among species. Those organisms with the most characteristics in common would be neighbors on the tree's branches. Basing a tree entirely on similarities, however, can be misleading. As just one example, many types of cave animals are eyeless and lack pigments (see figure 13.10). But these resemblances do not mean that the species that occupy caves are closely related to one another; instead, they are the product of

Taxonomic group	*Aloe vera* classification	Number of species
Domain	Eukarya	Several million
Kingdom	Plantae	~375,000
Phylum	Anthophyta	~235,000
Class	Liliopsida	~65,000
Order	Liliales	~1200
Family	Asphodelaceae	785
Genus	*Aloe*	500
Species	*Aloe vera*	1

Figure 14.12 Taxonomic Hierarchy. Life is divided into domains, then kingdoms, then numerous smaller categories. This diagram shows the complete classification for the plant *Aloe vera*.

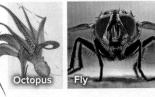

a.   b.

Figure 14.13 More Similarities, More Shared Levels. (a) An octopus and a fly are both eukaryotes classified in the animal kingdom, but they do not share other taxonomic levels. (b) A chimpanzee and an orangutan share enough similarities to be classified in the same family.

(octopus): ©Alex Bramwell/Moment/Getty Images; (fly): ©Kimberly Hosey/Getty Images RF; (chimp): ©Alexander Rieber/EyeEm/Getty Images RF; (orangutan): ©MedioImages/SuperStock RF

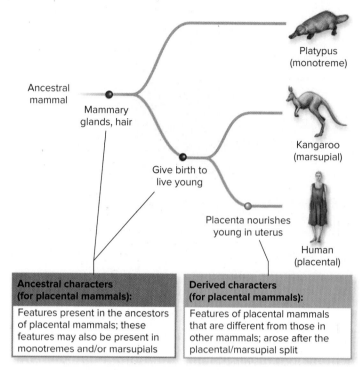

Ancestral mammal

Mammary glands, hair

Give birth to live young

Placenta nourishes young in uterus

Platypus (monotreme)

Kangaroo (marsupial)

Human (placental)

Ancestral characters (for placental mammals):	Derived characters (for placental mammals):
Features present in the ancestors of placental mammals; these features may also be present in monotremes and/or marsupials	Features of placental mammals that are different from those in other mammals; arose after the placental/marsupial split

Figure 14.14   Ancestral and Derived Characters. The placenta is a derived character that was not present in the common ancestor that placental mammals share with marsupials and monotremes.

Figure It Out

How many clades are represented in the phylogenetic tree in figure 14.15?

Answer: 13.

Miniglossary	Biological Classification
Ancestral characters	Features present in the common ancestor of a clade
Clade	Group of organisms consisting of a common ancestor and all of its descendants
Cladistics	Phylogenetic system that groups organisms by characters that best indicate shared ancestry
Cladogram	Phylogenetic tree built on shared derived characters
Derived characters	Features of an organism that are different from those found in a clade's ancestors
Outgroup	Comparator organism outside the group being studied; useful for identifying ancestral traits
Systematics	The combined study of taxonomy and evolutionary relationships among organisms

convergent evolution. If the goal of a classification system is to group related organisms together, then attending only to similarities might lead to an incorrect classification. ⓘ *convergent evolution,* section 13.4C

A cladistics approach solves this problem. Widely adopted beginning in the 1990s, **cladistics** is a phylogenetic system that defines groups by distinguishing between ancestral and derived characters. **Ancestral characters** are inherited attributes that resemble those of the ancestor of a group; an organism with **derived characters** has features that are different from those found in the group's ancestor (figure 14.14).

In making a diagram such as figure 14.14, how do researchers know which characters are ancestral and which are derived? They choose an **outgroup** consisting of comparator organisms that are not part of the group being studied. For example, in a cladistic analysis of mammals that give birth to live young, an appropriate outgroup might be monotremes. Features that are present in all mammals, such as mammary glands and hair, are assumed to be ancestral. For placental mammals, derived features include the placenta and other characteristics that do not appear in monotremes or marsupials.

C. Cladograms Depict Hypothesized Evolutionary Relationships

The result of a cladistics analysis is a **cladogram,** a treelike diagram built using shared derived characters (figure 14.15). The basic "unit" of a cladogram is a **clade,** a group of organisms consisting of a common ancestor and all of its descendants. A clade is therefore a group of species united by a single evolutionary pathway. For example, modern birds form a clade because they all descended from the same group of reptiles. A clade may contain any number of species, as long as all of its members share an ancestor that organisms outside the clade do not share. ⓘ *reptiles,* section 17.11E

All cladograms have features in common. The tips of the branches represent taxa. Existing species, such as birds and turtles, are at the tips of longer branches in figure 14.15; the nonavian ("nonbird") dinosaurs are extinct and therefore occupy a shorter branch. Each node in a cladogram indicates where two groups arose from a common ancestor (see Burning Question 14.2). A branching pattern of lines therefore represents populations that diverge genetically, splitting off to form a new species. The branching pattern also implies the passage of time, as indicated by the arrow at the bottom of figure 14.15.

The emphasis in a cladogram is not physical similarities but rather historical relationships. To emphasize this point, imagine a lizard, a crocodile, and a chicken. Which resembles the lizard more closely: the crocodile or the chicken? Clearly, the most *similar* animals are the lizard and the crocodile. But these resemblances are only superficial. The shared derived characters tell a more complete story of evolutionary history. Based on the evidence, crocodiles are more closely related to birds than they are to lizards.

A common mistake in interpreting cladograms is to incorrectly assume that a taxon must be closely related to both groups that appear next to it on the tree. In figure 14.15, for example, mammals are adjacent to both turtles and amphibians. Does this mean that rabbits are as closely related to frogs as they are to turtles? To find out, look at the amount of time that has passed since mammals last shared a common ancestor with each group. Because the common ancestor of mammals and turtles existed more recently, these groups are more closely related than are mammals and amphibians.

All phylogenetic trees are based on limited and sometimes ambiguous information. They are therefore not peeks into the past but rather tools that

Figure 14.15 **Reading a Cladogram.** Each clade consists of a common ancestor and all of its descendants. The more recently any two groups shared a common ancestor, the more closely related they are.

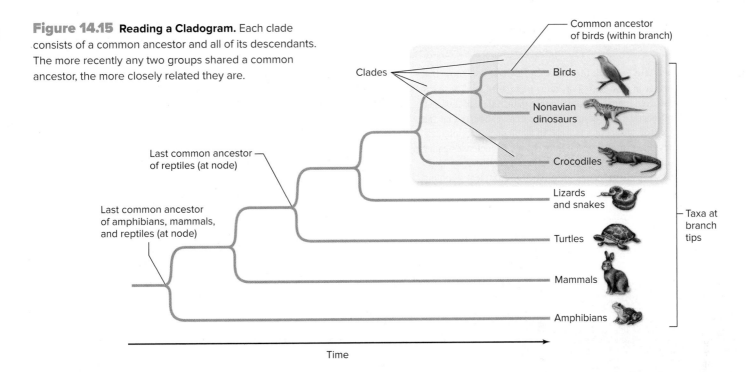

researchers can use to construct hypotheses about the relationships among different types of organisms. These investigators can then add other approaches to test the hypotheses.

D. Many Traditional Groups Are Not Clades

Contemporary scientists using a cladistics approach typically assign names only to clades; incomplete clades or groups that combine portions of multiple clades are not named. Many familiar groups of species, however, are not clades.

For example, according to the traditional Linnaean classification system, class Reptilia includes turtles, lizards, snakes, crocodiles, and the extinct dinosaurs, but it excludes birds. The cladogram in figure 14.15, however, places birds in the same clade with the reptiles based on their many shared derived characteristics. Most biologists therefore now consider birds to be reptiles, so they make a distinction between birds and nonavian reptiles.

Nor does the kingdom Protista form a clade. Protists include mostly single-celled eukaryotes that do not fit into any of the three eukaryotic kingdoms (plants, fungi, and animals). Yet all three of these groups share a common eukaryotic ancestor with the protists. Biologists are currently struggling to divide kingdom Protista into clades, an immense task (see chapter 15).

As yet another example, a group consisting of endothermic (formerly called "warm-blooded") animals includes only birds and mammals. This group is not a clade because it excludes the most recent common ancestor of birds and mammals, which was an ectotherm (formerly called "cold-blooded").

14.6 Mastering Concepts

1. Describe the taxonomic hierarchy.
2. What are the strengths of a cladistics approach over a more traditional approach to phylogeny?
3. Distinguish between ancestral and derived characters.

Burning Question 14.2

Did rabbits come from frogs?

Suppose you examine figure 14.15 and conclude that today's reptiles and mammals—including rabbits—evolved from frogs. Is your interpretation correct?

The answer is no. To understand why, remember that shared ancestors are at nodes (branching points) buried within the tree. The descendants of those ancestors are at the tips. The first node in figure 14.15 represents the common ancestor of *all* species on the tree, including today's frogs, reptiles, and mammals. A species at one branch tip—say, a frog—cannot be an ancestor of a rabbit or any other species at a different branch tip.

If rabbits did not *come from* frogs, what is the evolutionary relationship between these species? Figure 14.15 shows the answer: Frogs and rabbits are modern descendants of the same shared ancestor near the root of the tree. Although we do not know exactly what that animal looked like, we know that each lineage has been evolving independently for many, many generations. As a result, frogs and rabbits do not look very much alike—but the features they do share are clues to the past.

Submit your burning question to
marielle.hoefnagels@mheducation.com

Investigating Life 14.1 | Plant Protection Rackets May Stimulate Speciation

Although most people are familiar with the nectar in flowers, few know that some plants also store nectar in their leaves or stems (figure 14.B). These extrafloral nectaries have evolved independently in more than 100 plant families.

Why are extrafloral nectaries so common? One explanation is that the plants are involved in a sort of "protection racket." In a human protection racket, a business pays money to a criminal group in exchange for defense against harm from third parties. Likewise, plants with extrafloral nectaries may "pay" sugary nectar to ants and wasps in exchange for protection from leaf-gobbling herbivores. More leaf area, in turn, means more energy for reproduction.

Thanks to this fitness boost, ancient plants that recruited insect defenders may have been able to exploit areas that were unavailable to their unprotected ancestors. These new niches would likely have had unique selective pressures, driving evolutionary changes that could have led to speciation. If so, then plant

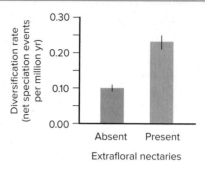

Figure 14.C **Plant Diversification.** Plant families with extrafloral nectaries have higher diversification rates than plant families without them.

lineages with extrafloral nectaries should have higher speciation rates than lineages without these adaptations.

To test this prediction, researchers studied fossils and consulted previously published data to construct a detailed evolutionary tree depicting most plant families. Then they used the age and size of each family to calculate its diversification rate: A relatively young plant family with many species has a high diversification rate, whereas an old family with just a few species has a low rate.

When they compared the rates, they found that plant lineages with members containing extrafloral nectaries have diversified much more quickly than lineages without the sugary rewards (figure 14.C). Note that extrafloral nectaries do not directly cause speciation or decrease extinctions. Rather, the analysis supported the team's prediction: Having insect defenders may allow plants to occupy new habitats, which may eventually lead to speciation.

Plants have forged many ecological interactions with animals, but few are as intriguing as the "protection racket" explored in this study. Amazingly, this mutually beneficial interaction—a criminal act in human society—seems to have spawned the evolution of many new plant species.

Figure 14.B **Extrafloral Nectary.** An ant drinks nectar from an extrafloral nectary on a leaf.
©Dr. Morley Read/Stockbyte/Getty Images RF

Source: Weber, Marjorie G., and Anurag A. Agrawal. 2014. Defense mutualisms enhance plant diversification. *Proceedings of the National Academy of Sciences*, vol. 111, pages 16442–16447.

What's the **Point?** ▼ APPLIED

Speciation and extinction are the large-scale, macroevolutionary changes in species diversity. Most of life's history unfolded before humans evolved, but since our arrival, we have rapidly shaped both processes.

Human actions sometimes initiate reproductive barriers, the first step toward speciation. For example, in the last half-century, some populations of blackcap warblers have begun migrating to the United Kingdom. Before backyard bird feeders became popular, this location did not have enough food to support the birds. Warblers that migrate to the United Kingdom are more likely to interbreed with one another than with birds that migrate farther south. Human-influenced migration patterns are therefore creating a reproductive barrier in warblers.

The growing human population has also caused many extinctions. Why We Care 14.1 lists some species exterminated by habitat destruction and overharvesting. Pollution and climate change threaten other species, such as tropical amphibians. Also, humans move species around the world, with potentially disastrous consequences. European starlings introduced to the

European starling

United States in the late 1800s compete with bluebirds and woodpeckers for nesting space, causing these native bird populations to crash.

Some conservationists predict that up to three in 10 species will become extinct in the next half-century, exceeding the background extinction rate by a factor of more than 1000. We need other organisms for food, medicine, energy, clean air, and clean water, among other uses. However, ecosystem recovery took millions of years following Earth's previous mass extinctions. Therefore, many people are trying to slow or even reverse today's species extinctions.

One effort is an attempt to recover extinct species using cloning techniques. For example, researchers in the Lazarus Project successfully produced embryos of an extinct frog. Although none of the embryos has survived, the project's limited success suggests that "de-extinction" may be possible. But cloning alone cannot keep up with the pace of extinctions in the wild. Chapter 20 explains other strategies for battling the biodiversity crisis. ⓘ *cloning*, section 11.3B

Photo: Source: U.S. Fish & Wildlife Service/Dave Menke

CHAPTER SUMMARY

14.1 What Is a Species?

- **Macroevolution** refers to large-scale changes in life's diversity, including extinctions and the appearance of new **species.**

A. Linnaeus Classified Life Based on Appearance

- Carolus Linnaeus grouped similar-looking organisms and invented a classification system for life. He assumed, however, that species do not change; Darwin's work added an evolutionary context to classification.

B. Species Can Be Defined Based on the Potential to Interbreed

- The **biological species concept** defines species based on reproductive isolation.
- **Speciation** is the formation of a new species, which occurs when a population's **gene pool** is divided and each part takes its own evolutionary course.

14.2 Reproductive Barriers Cause Species to Diverge

- Figure 14.16 summarizes the main types of reproductive barriers.

A. Prezygotic Barriers Prevent Fertilization

- **Prezygotic reproductive isolation** occurs before or during fertilization. It includes obstacles to mating such as space, time, and behavior; mechanical mismatches between male and female reproductive organs; and molecular mismatches between gametes.

B. Postzygotic Barriers Prevent Development of a Fertile Offspring

- **Postzygotic reproductive isolation** results in **hybrid** offspring that die early in development, are infertile, or produce a second generation of offspring with abnormalities.

14.3 Spatial Patterns Define Two Types of Speciation

A. Allopatric Speciation Reflects a Geographical Barrier

- **Allopatric speciation** occurs when a geographical barrier separates a population. The two populations then diverge genetically to the point that their members can no longer produce fertile offspring together.

B. Sympatric Speciation Occurs in a Shared Habitat

- **Sympatric speciation** enables populations that occupy the same area to diverge. It may occur when a **polyploid** organism (with one or more extra chromosome sets) is reproductively isolated from its parent.

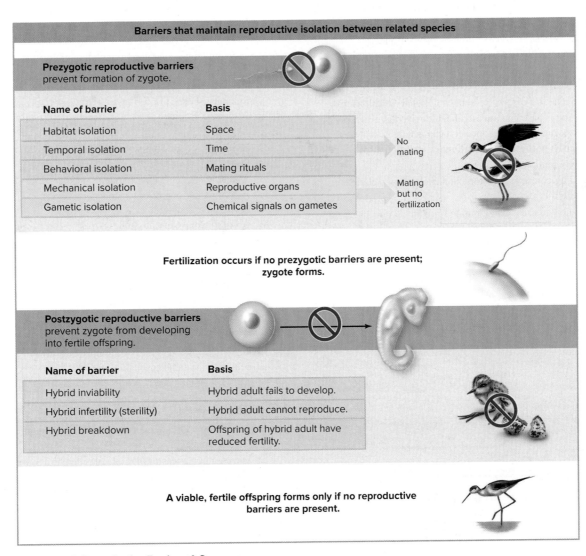

Figure 14.16 Reproductive Barriers: A Summary.

C. Determining the Type of Speciation May Be Difficult

- The distinction between allopatric and sympatric speciation is not always straightforward, partly because it is difficult to define the size and significance of a geographical barrier.

14.4 Speciation May Be Gradual or May Occur in Bursts

- Evolutionary change occurs at many rates, from slow and steady **gradualism** to the periodic bursts that characterize **punctuated equilibrium.**

14.5 Extinction Marks the End of the Line

- **Extinction** is the disappearance of a species.
- The **background extinction rate** reflects steady, ongoing losses of species.
- Historically, **mass extinctions** have resulted from global changes such as continental drift. The **impact theory** suggests that a meteorite or comet can also cause mass extinctions (such as the one that occurred 65 million years ago).
- Human activities are increasing the extinction rate.

14.6 Biological Classification Systems Are Based on Common Descent

- The study of **systematics** includes **taxonomy** (the science of classification) and **phylogenetics** (the study of species relationships).

A. The Taxonomic Hierarchy Organizes Species into Groups

- Biologists use a taxonomic hierarchy to classify life's diversity, with **taxa** ranging from domain to species.

B. A Cladistics Approach Is Based on Shared Derived Traits

- Biologists use phylogenetic trees to represent evolutionary relationships.
- **Cladistics** defines groups based on **ancestral** and **derived characters.** An **outgroup** helps researchers detect ancestral characters.

C. Cladograms Depict Hypothesized Evolutionary Relationships

- A **cladogram** shows evolutionary relationships as a branching hierarchy with nodes representing common ancestors. In a cladogram, a **clade** consists of an ancestor plus all of its descendants.
- **Figure 14.17** shows how to use a cladogram to find the common ancestor shared by two groups.

D. Many Traditional Groups Are Not Clades

- Group names such as "protist" and "algae" remain in common usage, but they do not reflect evolutionary relationships.

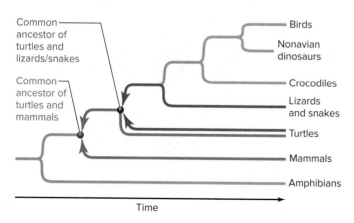

Figure 14.17 Finding Common Ancestors in a Cladogram.

MULTIPLE CHOICE QUESTIONS

1. Macroevolution is distinct from microevolution in that macroevolution
   a. results in changes to life's diversity.
   b. results in changes to the DNA of organisms.
   c. affects larger organisms.
   d. can be observed.

2. The biological species concept defines species based on
   a. external appearance.
   b. the number of adaptations to the same habitat.
   c. ability to interbreed.
   d. DNA and protein sequences.

3. A mule is the offspring of a male donkey and a female horse. Mules are unable to produce offspring. What reproductive barrier separates horses and donkeys?
   a. Mechanical isolation   c. Hybrid inviability
   b. Gametic isolation   d. Hybrid infertility

4. How can infertility occur in a hybrid whose parents have different numbers of chromosomes?
   a. The difference prevents mitotic cell division.
   b. The cells of the hybrid cannot grow, so the embryo dies.
   c. Meiosis is blocked, so gametes cannot form.
   d. Mitosis is altered, so the gametes are not viable.

5. A mountain range separates a population of gorillas. After many generations, the gorillas on different sides of the mountain range cannot produce viable, fertile offspring. What has happened?
   a. Sympatric speciation   c. Postzygotic speciation
   b. Allopatric speciation   d. Cladistic speciation

6. Why might speciation occur at an unusually rapid pace?
   a. A new phenotype enables organisms to exploit the environment in new ways.
   b. A rapidly changing environment selects for new phenotypes.
   c. A dominant group of organisms goes extinct, paving the way for the evolution of new species.
   d. All of the above are possibilities.

7. Why is a species with a small population more likely than one with a large population to undergo an extinction?
   a. Because small populations are exposed to greater environmental changes
   b. Because genetic diversity is likely to be lower in a small population
   c. Because the mutation rate is too low
   d. Because they take too long to produce offspring

8. Flying animals have diverse evolutionary histories. They therefore
   a. belong to the same species.
   b. share DNA and protein sequences with a common ancestor.
   c. do not form a single clade.
   d. share a common developmental pathway.

9. Refer to figure 14.17. Which of the following common ancestors existed the longest time ago?
   a. The common ancestor of crocodiles and birds
   b. The common ancestor of amphibians and turtles
   c. The common ancestor of lizards and mammals
   d. The common ancestor of nonavian dinosaurs and birds

10. Which of the groups in figure 1.10 represents a clade?
    a. The prokaryotes   c. Kingdom Protista
    b. Domain Eukarya   d. None of the above is a clade.

Answers to Multiple Choice questions are in appendix A.

WRITE IT OUT

1. How has the meaning of the term *species* changed since the time of Linnaeus?

2. What type of reproductive barrier applies to each of these scenarios?

   a. Humans introduced apple trees to North America in the 1800s. Insects called hawthorn flies, which feed and mate on hawthorn plants, quickly discovered the new fruits. Some flies preferred the taste of apples to their native host plants. Because these flies mate where they eat, this difference in food preference quickly led to a reproductive barrier.

   b. Water buffalo and cattle can mate with each other, but the embryos die early in development.

   c. Scientists try to mate two species of dragonfly that inhabit the same pond at the same time of day. However, females never allow males of the other species to mate with them.

   d. One species of reed warbler is active in the upper parts of the tree canopy while another species of reed warbler is active in the lower canopy. Both species are active during the day.

   e. Scientists mate two parrots from different populations to see if speciation has occurred. The parrots mate over and over again, but the male's sperm never fertilizes the female's egg.

3. If the apple-feeding flies from question 2a form a different species from their hawthorn-feeding relatives, which type of speciation has occurred?

4. Polyploidy is a common mechanism of speciation in plants. When the chromosome number of a plant doubles, the offspring may not be able to breed with the parent plant. Alternatively, gametes of related species may fuse, producing polyploid offspring unable to reproduce with either parent. How are these two mechanisms of speciation similar? How might you draw each speciation event on an evolutionary tree?

5. How does natural selection predict a gradualistic mode of evolution? Does the presence of fossils that are consistent with punctuated equilibrium mean that natural selection does not occur?

6. What information would you need to determine the background extinction rate hundreds of millions of years ago? How might you determine the current extinction rate?

7. Examine the cladogram in figure 14.17 and answer the following questions:

   a. Which taxon is most closely related to birds?

   b. Are lizards more closely related to crocodiles or turtles?

   c. Which taxon is ancestral to all the others?

   d. Redraw the tree so that birds are next to mammals without changing the evolutionary relationships among any of the taxa.

8. Figure 16.3 summarizes the hypothesized evolutionary relationships among living plants. Do the gymnosperms form a clade? Explain your answer.

9. Figure 17.3 shows a phylogenetic tree for animals. How many clades are depicted in the figure?

SCIENTIFIC LITERACY

Review Why We Care 14.1. Why do species become extinct? Choose a species that has recently become extinct, and describe some possible evolutionary consequences to other species that interacted with that species before its extinction. Should humans only be concerned with saving organisms with which we directly interact?

PULL IT TOGETHER

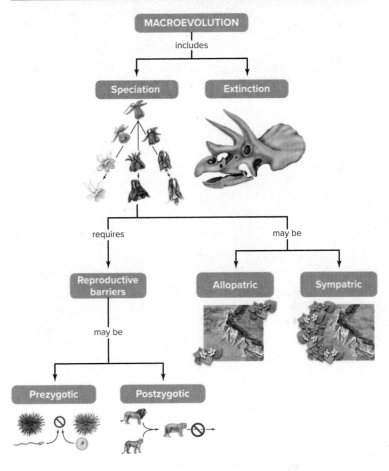

Figure 14.18 Pull It Together: Speciation and Extinction.

Refer to figure 14.18 and the chapter content to answer the following questions.

1. Review the Survey the Landscape figure in the chapter introduction and then add *phylogenetic trees, species, fossils, DNA,* and *anatomical structures* to figure 14.18.

2. Add *fertilization* and *offspring* to the concept map.

3. Add *gradualism* and *punctuated equilibrium* to this concept map.

Answers to Mastering Concepts, Write It Out, Scientific Literacy, and Pull It Together questions can be found in the Connect ebook.
connect.mheducation.com

15

Evolution and Diversity of Microbial Life

LEARNING OUTLINE

15.1 Life's Origin Remains Mysterious

15.2 Prokaryotes Are a Biological Success Story

15.3 Eukaryotic Cells and Multicellularity Arose More Than a Billion Years Ago

15.4 Protists Are the Simplest Eukaryotes

15.5 Fungi Are Essential Decomposers

APPLICATIONS

Burning Question 15.1 *Does new life spring from simple molecules now, as it did in the past?*

Why We Care 15.1 *Antibiotics and Other Germ Killers*

Burning Question 15.2 *Are there areas on Earth where no life exists?*

Burning Question 15.3 *Why and how do algae form?*

Why We Care 15.2 *Preventing Mold*

Investigating Life 15.1 *Shining a Spotlight on Danger*

Blue Glow of Algae. Microscopic algae emit neon blue light at Gudong Beach in eastern China. Investigating Life 15.1 explores why these beautiful microbes glow.

©Europics/Newscom

Learn How to Learn
Use All Your Resources

Online quizzes, animations, and other resources can help you learn biology. As you study, take regular breaks from reading and explore the digital resources associated with your book. For example, practice questions can help you learn the material and test your understanding. Also, check for animations that take you through complex processes one step at a time. Sometimes the motion of an animation can help you understand what's happening more easily than studying a static image.

SURVEY THE LANDSCAPE
Evolution and Diversity

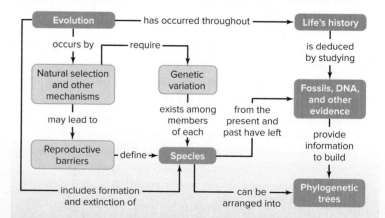

Primitive microscopic cells—the first life on Earth—arose billions of years ago. They eventually diversified into bacteria and archaea. Later, eukaryotic microbes arose and evolved into today's protists and fungi. These microscopic organisms are essential to all other life, playing critical roles at scales from individual health to the functioning of the entire biosphere.

For more details, study the Pull It Together feature in the chapter summary.

©McGraw-Hill Education/
John Thoeming

When it comes to biodiversity, most people think about plants and animals. But the microbes that inhabit Earth are just as important.

We can't see most microbes with the unaided eye, yet they profoundly influence human life. Most obviously, some types of microbes cause deadly illnesses, not only in humans but also in other animals and in plants. Most people also know that microbes play a role in the production of cheese, beer, wine, bread, and other household items.

Microbes also have played most of the starring roles in the history of life on Earth; plants, animals, and even fungi are newcomers compared with the bacteria and archaea. Even today, ecosystems would grind to a halt without the photosynthesis and decomposition services that microbes provide.

This chapter introduces the tiny world of microbiology. It combines the bacteria, archaea, protists, and fungi based on just one shared feature: Some or all of their members are microscopic. As you will see, these distantly related organisms are as diverse as they are vital to life on Earth.

15.1 Life's Origin Remains Mysterious

Reconstructing life's start is like reading all the chapters of a novel except the first. A reader can get some idea of the events and setting of the opening chapter from clues throughout the novel. Similarly, scattered clues from life through the ages reflect events that may have led to the origin of life.

Scientists describe the origin and history of life in the context of the **geologic timescale,** which divides time into eons and eras defined by major geological or biological events. Figure 15.1 shows a simplified version of the geologic timescale; see figure 13.2 for a more complete version.

The study of life's origin begins with astronomy and geology. Earth and the solar system's other planets formed about 4.6 BYA (billion years ago) as solid matter condensed out of a vast expanse of dust and gas swirling around the early sun. The red-hot ball that became Earth cooled enough to form a crust by about 4.2 to 4.1 BYA, when the surface temperature ranged from 500°C to 1000°C and atmospheric pressure was 10 times what it is now.

The geological evidence paints a chaotic picture of this Hadean eon, including volcanic eruptions, earthquakes, and ultraviolet radiation. Analysis of craters on other objects in the solar system suggests that comets, meteorites, and possibly asteroids bombarded Earth's surface during its first 500 million to 600 million years. These impacts repeatedly boiled off the seas and vaporized rocks to carve the features of the fledgling world.

Still, organic molecules could probably interact in protected pockets of the environment. At some point, an entity arose that could survive, thrive, reproduce, and diversify. The clues from geology and paleontology suggest that from 4.2 to 3.85 BYA, simple cells (or their precursors) arose.

Unfortunately, direct evidence of the first life is likely gone because most of Earth's initial crust has been destroyed. Erosion tears rocks and minerals into particles, only to be built up again into sediments that become heated and compressed. Seafloor is dragged into Earth's interior at deep-sea trenches, where it is melted and recycled. Therefore, although Earth formed 4.6 BYA, the oldest rocks that remain today date to only about 3.85 BYA. ⓘ *plate tectonics, section 13.3A*

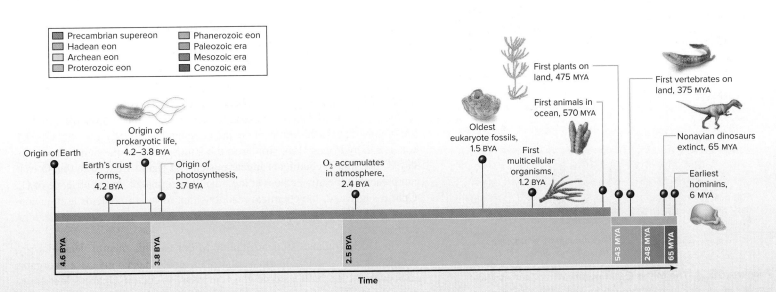

Figure 15.1 Highlights in Life's History. In this simplified geologic timescale, the size of each eon and era is proportional to its length in years. (BYA = billion years ago; MYA = million years ago)

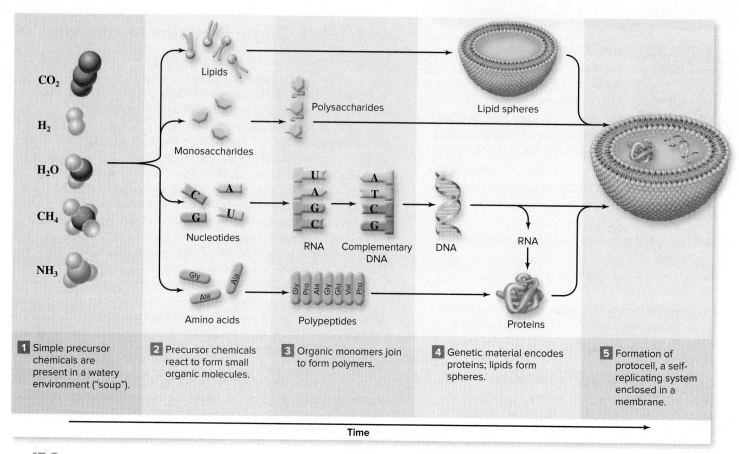

1. Simple precursor chemicals are present in a watery environment ("soup").

2. Precursor chemicals react to form small organic molecules.

3. Organic monomers join to form polymers.

4. Genetic material encodes proteins; lipids form spheres.

5. Formation of protocell, a self-replicating system enclosed in a membrane.

Time

Figure 15.2 **Pathway to a Cell.** The steps leading to the origin of life on Earth may have started with the formation of organic molecules from simple precursors. However it originated, the first cell would have contained self-replicating molecules enclosed in a phospholipid bilayer membrane.

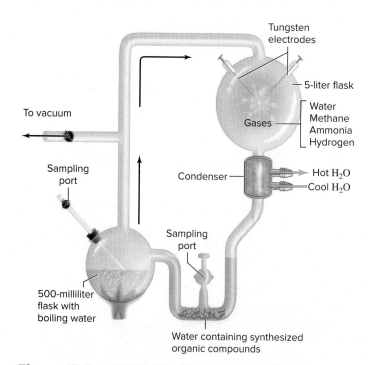

Figure 15.3 **The Miller Experiment.** When Stanley Miller passed an electrical spark through heated gases, the mixture generated amino acids and other organic molecules.

This section describes some of the major steps in the chemical evolution that eventually led to the first cell; figure 15.2 summarizes one possible version of the process. (Today, however, new life is unlikely to originate from nonliving matter; Burning Question 15.1 explains why.)

A. The First Organic Molecules May Have Formed in a Chemical "Soup"

Early Earth was different from today's planet. The atmosphere's current gas mixture includes nitrogen (N_2), oxygen (O_2), carbon dioxide (CO_2), and water vapor (H_2O). What might it have been like 4 BYA?

Russian chemist Alex I. Oparin hypothesized in his 1938 book, *The Origin of Life,* that Earth's atmosphere included methane (CH_4), ammonia (NH_3), water, and hydrogen (H_2), similar to the atmospheres of the outer planets today. These simple chemicals appear in step 1 of figure 15.2. Note that O_2 was not present in the atmosphere at the time life originated. In the absence of O_2, Oparin suggested, the chemical reactions that form amino acids and nucleotides could have occurred (see figure 15.2, step 2). ⓘ *organic molecules,* section 2.5

In 1953, graduate student Stanley Miller and his mentor, Harold Urey, decided to test whether Oparin's atmosphere could indeed give rise to organic molecules. Miller built a sterile glass enclosure to contain Oparin's four gases, through which he passed sparks to simulate lightning (figure 15.3). He condensed the gases in a narrow tube; from there, the liquid passed into a flask.

Boiling the fluid caused gases to evaporate back into the synthetic "atmosphere," completing the loop. After a few failures and adjustments, Miller saw the condensed liquid turn yellowish, then varying shades of red, pink, and yellow-brown. Chemical analysis revealed a variety of amino acids, some found in life.

The Miller experiment went down in history as the first attempt to re-create chemical conditions on Earth before life arose. Miller and many others later extended his results by altering conditions or using different starting materials. For example, methane and ammonia could form hydrogen cyanide (HCN), which produced amino acids in the presence of ultraviolet light and water. "Soups" that included phosphates yielded nucleotides, including the biological energy molecule ATP. Other experiments produced carbohydrates or lipids like those in biological membranes. ⓘ *ATP,* section 4.3

The experiment has survived criticisms that Earth's early atmosphere actually contained CO_2, a gas not present in Miller's original setup. Organic molecules still form, even with an adjusted gas mixture.

B. Clays May Have Helped Monomers Form Polymers

Once the organic building blocks (monomers) of macromolecules were present, they had to have linked into chains (polymers). This process, depicted in step 3 of figure 15.2, may have happened on hot clays or other minerals that provided ample, dry surfaces. ⓘ *monomers and polymers,* section 2.5

Clays may have played an important role in early organic chemistry, for at least three reasons. First, the sheetlike minerals in clays can form templates on which chemical building blocks could have linked to build larger molecules. Second, iron pyrite and some other minerals in clay can release electrons, providing energy to form chemical bonds. Third, these minerals may also have acted as catalysts to speed chemical reactions.

For example, the first RNA molecules could have formed on clay surfaces (figure 15.4). Not only do the positive charges on clay's surface attract and hold negatively charged RNA nucleotides, but clays also promote the formation of the covalent bonds that link the nucleotides into chains. They even attract other nucleotides to form a complementary strand. About 4 BYA, clays might have been fringed with an ever-increasing variety of growing polymers. Some of these might have become the macromolecules that would eventually build cells.

The origin of RNA is important because life requires an informational molecule. That molecule may have been RNA, or something like it, because RNA is a versatile molecule. It stores genetic information and uses it to make proteins. RNA can also catalyze chemical reactions and duplicate on its own. As Stanley Miller summed it up, "The origin of life is the origin of evolution, which requires replication, mutation, and selection. Replication is the hard part. Once a genetic material could replicate, life would have just taken off."

Perhaps pieces of RNA on clay surfaces formed, accumulated, grew longer, became more complex in sequence, and changed as replication errors led to mutations. Some molecules would have been more stable than others, leading to an early form of natural selection. The term **RNA world** describes how self-replicating RNA may have been an essential precursor to life on Earth.

At some point, RNA might have begun encoding proteins, just short chains of amino acids at first. An RNA molecule may eventually have grown long enough to encode the enzyme reverse transcriptase, which copies RNA to DNA. With DNA, the chemical blueprints of life found a much more stable home. Protein enzymes eventually took over some of the functions of catalytic RNAs. Step 4 in figure 15.2 shows this stage in life's origin.

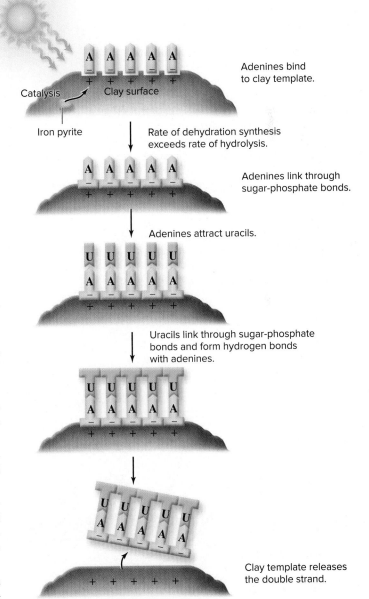

Adenines bind to clay template.

Catalysis — Clay surface

Iron pyrite

Rate of dehydration synthesis exceeds rate of hydrolysis.

Adenines link through sugar-phosphate bonds.

Adenines attract uracils.

Uracils link through sugar-phosphate bonds and form hydrogen bonds with adenines.

Clay template releases the double strand.

Figure 15.4 A Possible Role for Clay. Chains of nucleotides may have formed on clay templates. In this hypothesized scenario, iron pyrite ("fool's gold") was the catalyst for polymer formation, and sunlight provided the energy.

Fungi Are Essential Decomposers

C. Membranes Enclosed the Molecules

Meanwhile, lipids would have been entering the picture. Under the right temperature and pH conditions, phospholipids could have formed membrane-like structures, some of which left evidence in ancient sediments. Laboratory experiments show that pieces of membrane can indeed grow on structural supports and break free, forming a bubble. ⓘ *phospholipids,* section 3.3

Perhaps an ancient membrane bubble enclosed a collection of organic molecules to form a cell-like assemblage, or protocell (see figure 15.2, step 5). These hypothetical, ancient aggregates of RNA, DNA, proteins, and lipids were simple precursors of cells. The capacity of nucleic acids to mutate may have enabled protocells to become increasingly self-sufficient, giving rise eventually to the reaction pathways of metabolism.

D. Early Life Changed Earth Forever

Several types of early cells probably prevailed for millions of years, competing for resources and sharing genetic material. Eventually, however, a type of cell arose that was the last shared ancestor of all life on Earth today.

The first cells lived in the absence of O_2 and probably used organic molecules as a source of both carbon and energy. Another source of carbon, however, was the CO_2 in the atmosphere. Photosynthetic bacteria and archaea eventually evolved that could use light for energy and atmospheric CO_2 as a carbon source (see chapter 5). These microbes no longer relied on organic compounds in their surroundings for food.

Photosynthesis probably originated in aquatic bacteria that used hydrogen sulfide (H_2S) instead of water as an electron donor. These first photosynthetic microorganisms would have released sulfur, rather than O_2, into the environment. Eventually, changes in pigment molecules enabled some of the microorganisms to use H_2O instead of H_2S as an electron donor. Cells using this new form of photosynthesis released O_2 as a waste product. The O_2 would have bubbled out of the water and into the air, changing the composition of Earth's atmosphere (figure 15.5).

The evolution of photosynthesis forever altered life on Earth. Photosynthetic organisms formed the base of new food chains. In addition, natural selection began to favor aerobic organisms that could use O_2 in metabolism, while anaerobic species would persist in pockets of the environment away from O_2. Ozone (O_3) also formed from O_2 high in the atmosphere, blocking the sun's damaging ultraviolet radiation. The overall result was an explosion of new life that eventually gave rise to today's microbes, plants, fungi, and animals.

Although most people associate photosynthesis only with plants, microbes living in water or along shorelines were the only photosynthetic organisms for most of life's history. Plants did not even colonize dry land until about 475 million years ago. Even today, about half of the O_2 generated in photosynthesis comes from aquatic organisms, mostly living in the oceans.

The remaining sections in this chapter offer a taste of the diversity in the microbial world, beginning with a tour of prokaryotic cells. A sampling of protists comes next, and the chapter ends by describing the fungi.

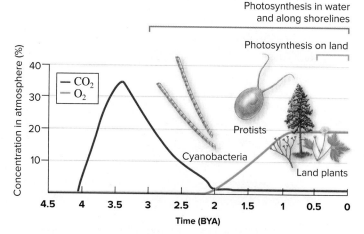

Figure 15.5 Photosynthesis and the Atmosphere. Billions of years ago, photosynthetic microbes began to pump O_2 into the water and atmosphere, permanently altering Earth's habitats.

15.1 Mastering Concepts

1. Describe the conditions on Earth before life began.
2. What can we learn from simulations of early Earth?
3. Why is RNA likely to have been pivotal in life's beginnings?
4. How did early life change Earth?

15.2 Prokaryotes Are a Biological Success Story

The microscopic world of life may be invisible to the naked eye, but its importance is immense. The earliest known fossils closely resemble today's bacteria, suggesting that the first cells were prokaryotic. As we have already seen, ancient photosynthetic microbes changed Earth's chemistry by contributing O_2 to the atmosphere. Along the road of evolution, bacteria probably gave rise to the chloroplasts and mitochondria of eukaryotic cells (see section 15.3).

The reign of the prokaryotes continues today. Virtually no place on Earth is free of bacteria and archaea; their cells live within rocks and ice, high in the atmosphere, far below the ocean's surface, in thermal vents, nuclear reactors, hot springs, animal intestines, plant roots, and practically everywhere else. Many species prefer hot, cold, acidic, alkaline, or salty habitats that humans consider "extreme." This section describes the diversity and importance of prokaryotic life.

A. What Is a Prokaryote?

A **prokaryote** is a single-celled organism that lacks a nucleus and membrane-bounded organelles. In contrast to a prokaryote, a **eukaryotic cell** has a nucleus and other membrane-bounded organelles, such as mitochondria and chloroplasts. At about 1 to 10 μm long, a typical prokaryotic cell is 10 to 100 times smaller than most eukaryotic cells (figure 15.6).

DNA sequences and other lines of evidence suggest the existence of two prokaryotic domains: **Bacteria** and **Archaea** (figure 15.7). Microbiologists have probably discovered just a tiny fraction of prokaryotic life on Earth. Soil, water, and even the human body teem with microbes that have yet to be named and described. The total number of species in the two prokaryotic domains may be anywhere between 100,000 and 10,000,000; no one knows.

Before embarking on a tour of prokaryote biology and ecology, it is worth noting that the term *prokaryote* has become controversial among microbiologists. The reason is that the word falsely implies a close evolutionary relationship between bacteria and archaea, despite DNA evidence indicating that

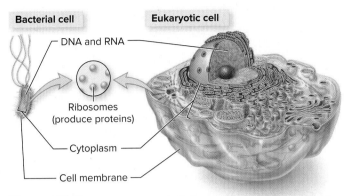

Figure 15.6 Prokaryotes and Eukaryotes Compared. All cells have nucleic acids, ribosomes, cytoplasm, and a cell membrane. Prokaryotic cells, however, are typically smaller and lack membrane-bounded organelles.

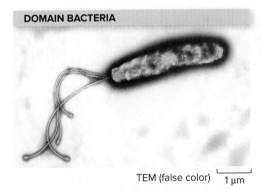

DOMAIN BACTERIA

TEM (false color) 1 μm

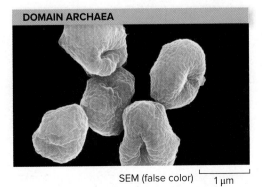

DOMAIN ARCHAEA

SEM (false color) 1 μm

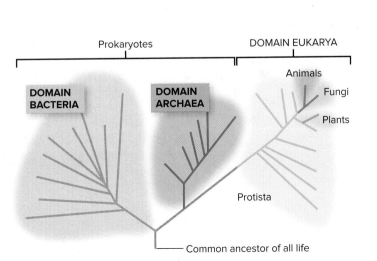

Figure 15.7 Diversity of Prokaryotic Life. Domains Bacteria and Archaea form two of the three main branches of life.

Photos: (Bacteria): ©Heather Davies/SPL/Getty Images RF; (Archaea): ©Eye of Science/Science Source

Fungi Are Essential Decomposers

Similarities Between Bacteria and Archaea

- Prokaryotic cells (no nucleus or other membrane-bounded organelles)
- Size ~1–10 μm
- Circular chromosome
- Predominantly unicellular
- Some can fix nitrogen or grow at temperatures above 80°C

Features Unique to Bacteria

- Cell wall typically composed of peptidoglycan
- Membrane based on fatty acids
- Some use chlorophyll in photosynthesis
- Cannot generate methane
- Sensitive to streptomycin
- Genes do not contain introns

Features Unique to Archaea

- Cell wall composed of molecules other than peptidoglycan
- Membrane based on nonfatty acid lipids
- Do not use chlorophyll
- Some generate methane
- Insensitive to streptomycin
- Genes may contain introns

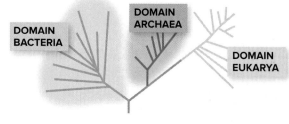

DOMAIN BACTERIA

DOMAIN ARCHAEA

DOMAIN EUKARYA

Figure 15.8 **Bacteria and Archaea Compared.** Bacteria and archaea share several structural and metabolic similarities. However, the many features unique to each group support their separation into two domains.

archaea are actually more closely related to eukaryotes. Nevertheless, many biologists continue to use the term as a handy shortcut for describing all cells that lack nuclei.

B. Prokaryote Classification Traditionally Relies on Cell Structure and Metabolism

Bacteria and archaea look similar under the microscope, although they differ in some details of their chemical composition and cell processes (figure 15.8). How do microbiologists classify the diversity of life within these two domains, given the tiny cell sizes and scarcity of distinctive internal structures?

The answer to that question has evolved over time. For hundreds of years, biologists classified microbes based on close scrutiny of their cells and metabolism. More recent studies of DNA sequences have revealed that the traditional classification criteria almost certainly group together organisms that are only distantly related to one another. The "old" criteria remain useful, however, because they are based on characteristics that are relatively easy to observe using a microscope and well-defined laboratory tests.

Viewing cells with a microscope is an essential step in identifying bacteria and archaea. Light microscopes (and sometimes electron microscopes) reveal the internal and external features unique to each species. Figure 15.9 illustrates a typical bacterial cell; in reading through this section, remember that a given cell may have some or all of the structures pictured.

Internal Cell Structures   Like the cells of other organisms, all bacteria and archaea are bounded by a cell membrane that encloses cytoplasm, DNA, and ribosomes. A prokaryotic cell's DNA typically consists of one circular chromosome. The **nucleoid** is the region where this DNA is located, along with some RNA and a few proteins. Unlike the nucleus of a eukaryotic cell, a membranous envelope does not surround the nucleoid.

The cells of many bacteria and archaea also contain one or more **plasmids,** circles of DNA apart from the chromosome. The genes on a plasmid may encode the proteins necessary to copy the plasmid and transfer it to another cell. Other genes may provide the ability to resist an antibiotic or toxin, cause disease, or alter the cell's metabolism. Recombinant DNA technology uses plasmids to ferry genes from one kind of cell to another. ⓘ *transgenic organisms,* section 11.2A

Ribosomes are structures that use the information in RNA to assemble proteins (see chapter 7). Bacterial, archaean, and eukaryotic ribosomes all make proteins in essentially the same way, but they are structurally different from one another. Some antibiotics, such as streptomycin, kill bacteria without harming eukaryotic host cells by exploiting this difference. Why We Care 15.1 describes more examples of how antibiotics work.

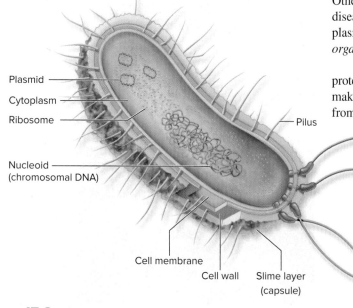

Plasmid

Cytoplasm

Ribosome

Nucleoid (chromosomal DNA)

Pilus

Flagellum

Cell membrane

Cell wall

Slime layer (capsule)

Figure 15.9 **Prokaryotic Cell.** This diagram shows the internal and external structures that are typical of a prokaryotic cell.

External Cell Structures The **cell wall** is a rigid barrier that surrounds the cells of most bacteria and archaea. In most species of bacteria, the cell wall contains **peptidoglycan,** a complex polysaccharide that does not occur in the cell walls of archaea. The wall gives the cell its shape (figure 15.10). Three of the most common forms are **coccus** (spherical), **bacillus** (rod-shaped), and **spirillum** (spiral- or corkscrew-shaped). In addition, the arrangement of the cells in pairs, clusters (*staphylo-*), or chains (*strepto-*) is sometimes important in classification. The disease-causing bacterium *Staphylococcus,* for example, forms grapelike clusters of spherical cells.

In some bacteria, the cell wall includes a protective outer membrane consisting of lipid, polysaccharide, and protein. The outer membrane causes the toxic effects of many disease-causing bacteria, a subject we return to in section 15.2C. Parts of this outer layer trigger a strong immune response, including fever and inflammation that help the body eliminate the bacteria (see chapter 29).

Many prokaryotic cells have other distinctive structures outside the cell wall (see figure 15.9). A slime layer (or capsule) is a sticky layer of proteins or polysaccharides that may surround the cell wall. The slime layer has many functions, including attachment to surfaces, resistance to drying, and protection from immune system cells.

Some cells have **pili** (singular: pilus), which are short, hairlike projections made of protein. Attachment pili enable cells to adhere to objects. The bacterium that causes cholera, for example, uses pili to attach to a human's intestinal wall. Other projections, called sex pili, aid in the transfer of DNA from cell to cell.

Not all prokaryotes can move, but many can do so. For example, cells may move toward or away from an external stimulus such as food, toxins, oxygen, or light. Cells that can move have a **flagellum,** which is a whiplike extension that rotates like a propeller. The bacterium in figure 15.9 has multiple flagella; other cells have one or a few. (Some eukaryotic cells also have flagella, but they are not homologous to those on bacterial or archaean cells.)

Endospores Some types of bacteria produce **endospores,** which are dormant, thick-walled structures that can survive harsh conditions (figure 15.11). The endospore wall surrounds DNA and a small amount of cytoplasm. An endospore can withstand boiling, drying, ultraviolet radiation, and disinfectants. Once environmental conditions improve, the endospore germinates and develops into a normal cell.

One spore-forming soil bacterium is *Clostridium botulinum.* Food canning processes typically include a high-pressure heat sterilization treatment to destroy endospores of this species. If any endospores survive, they may germinate inside the can, producing cells that thrive in the absence of oxygen. The cells produce a toxin that causes botulism, a severe (and sometimes deadly) form of food poisoning. Green beans, corn, and other vegetables that are improperly home-canned are the most frequent sources of food-borne botulism.

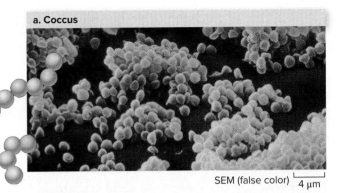

a. Coccus

SEM (false color)   4 μm

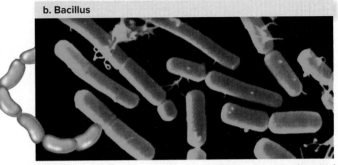

b. Bacillus

SEM (false color)   10 μm

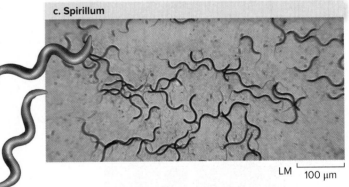

c. Spirillum

LM   100 μm

Figure 15.10 Cell Shapes. (a) Cocci are spherical. (b) A bacillus is rod-shaped. (c) A spirillum is spiral-shaped. Photos: (a): ©Biology Pics/Science Source; (b): ©SciMAT/Science Source; (c): ©Ed Reschke/Photolibrary/Getty Images

Miniglossary	Prokaryote Anatomy
Nucleoid	The region of a prokaryotic cell where chromosomal DNA is located
Cell wall	A rigid barrier surrounding the cell membrane
Pilus	A short, hairlike projection used in bacterial attachment or DNA transfer
Flagellum	A whiplike projection that enables a cell to move
Endospore	A thick-walled structure that protects bacterial DNA from harsh conditions

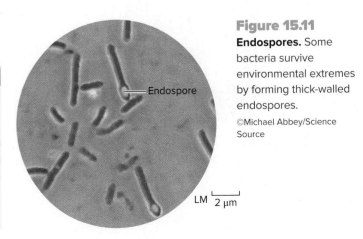

Figure 15.11
Endospores. Some bacteria survive environmental extremes by forming thick-walled endospores.
©Michael Abbey/Science Source

Endospore

LM   2 μm

Fungi Are Essential Decomposers

Miniglossary | Prokaryote Diversity

Prokaryote	A single-celled organism that lacks a nucleus and membrane-bounded organelles; classified in either domain Bacteria or domain Archaea
Bacteria and Archaea	Two domains of prokaryotic organisms distinguished from each other based on molecular evidence and other criteria
Autotroph	An organism that acquires carbon from inorganic sources, typically CO_2
Heterotroph	An organism that acquires carbon by consuming other organisms
Phototroph	An organism that derives energy from the sun
Chemotroph	An organism that derives energy from chemicals
Aerobe	An organism that uses O_2 to produce ATP
Anaerobe	An organism that can produce ATP in the absence of O_2

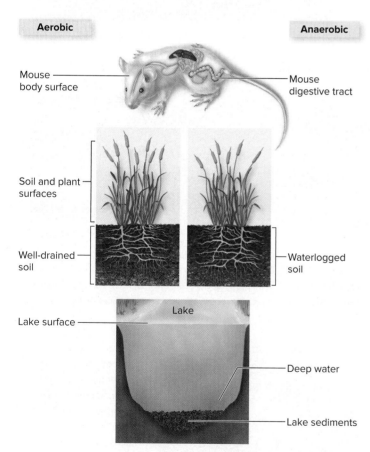

Aerobic   **Anaerobic**

Mouse body surface

Mouse digestive tract

Soil and plant surfaces

Well-drained soil

Waterlogged soil

Lake

Lake surface

Deep water

Lake sediments

Figure 15.12 **Habitats with and without Oxygen.** Aerobic locations house obligate aerobes and facultative anaerobes; anaerobic habitats are home to obligate and facultative anaerobes.

Another spore-forming bacterium is *Bacillus anthracis.* This organism, ordinarily found in soil, can cause a deadly disease called anthrax when inhaled. Cultures of *B. anthracis* can be dried to induce endospore formation, then ground into a fine powder that remains infectious for decades. This property makes anthrax a potential biological weapon.

Metabolic Diversity   Over billions of years, bacteria and archaea have developed a tremendous variety of chemical reactions that allow them to metabolize everything from organic matter to metal. One way to group microorganisms is to examine some of these key metabolic pathways.

The methods by which organisms acquire carbon and energy form one basis for classification. **Autotrophs,** for example, are "self-feeders"; they assemble their own organic molecules using inorganic carbon sources such as carbon dioxide (CO_2). Plants and algae are the most familiar autotrophs. **Heterotrophs,** on the other hand, are "other feeders," acquiring carbon by consuming organic molecules produced by other organisms. *Escherichia coli,* a notorious intestinal bacterium, is a heterotroph. The organism's energy sources are also important. **Phototrophs** derive energy from the sun; **chemotrophs** oxidize inorganic or organic chemicals. ⓘ *redox reactions,* section 4.2B

By combining these terms, a biologist can describe how a microbe fits into the environment. Plants and cyanobacteria, for example, are photoautotrophs; they use sunlight (*photo-*) for energy and CO_2 (*auto-*) for carbon, as described in chapter 5. Many disease-causing bacteria are chemoheterotrophs because they use organic molecules from their hosts as sources of both carbon and energy. Animals and fungi are also chemoheterotrophs.

In addition, oxygen requirements are often important in classification. **Obligate aerobes** require O_2 for generating ATP in cellular respiration (see chapter 6). For **obligate anaerobes,** O_2 is toxic, and they live in habitats that lack it. *Clostridium tetani,* the bacterium that causes tetanus when it infects a deep puncture wound, is one example. **Facultative anaerobes,** which include the intestinal microbes *E. coli* and *Salmonella,* can live either with or without O_2.

Figure 15.12 shows habitats with varying O_2 availability. Oxygen-rich areas include a mouse's skin, a plant's leaves, and the surface of a lake. Anaerobic habitats include the animal's digestive tract, deep water, and the lake's sediments. Note that soil may contain abundant O_2 or be anaerobic, depending on whether it is waterlogged or well-drained.

C. Prokaryotes Include Two Domains with Enormous Diversity

For many decades, the tendency to lump together all prokaryotic organisms hid much of the diversity in the microbial world. We now know of so many species of bacteria and archaea in so many habitats that it would take many books to describe them all—and many more species remain undiscovered. This section contains a small sampling of this extraordinary diversity; Burning Question 15.2 explores some of the few locations that are microbe-free.

Domain Bacteria   Scientists have identified 23 phyla within domain Bacteria, but the evolutionary relationships among them remain unclear. Figure 15.13 illustrates two examples.

Proteobacteria exemplify the overall diversity within the domain. Some carry out photosynthesis. Others play important roles in nitrogen or sulfur cycling, whereas still others form a medically important group that includes enteric bacteria and vibrios. The bacterium that causes stomach ulcers in humans is a proteobacterium, as are the intestinal bacteria *E. coli* and *Salmonella*.

Cyanobacteria form another lineage of bacteria. Billions of years ago, these autotrophs were the first to produce O_2 in photosynthesis. Cyanobacteria remain important in ecosystems: They make up some of the photosynthetic plankton at the base of aquatic food chains, and they form symbiotic relationships with fungi on land (see section 15.5C). Proteobacteria and cyanobacteria are notable for another reason as well: Their ancestors gave rise to the mitochondria and chloroplasts in eukaryotic cells (see section 15.3A).

Spirochaetes are spiral-shaped organisms. One example is the bacterium that can cause Lyme disease when transmitted to humans in a tick's bite. Another spirochaete causes the sexually transmitted disease syphilis.

Many other bacteria are medically important as well. As we have already seen, *Bacillus anthracis* causes anthrax; another endospore-former, *Clostridium tetani,* causes tetanus. Other disease-causing bacteria include *Staphylococcus* and *Streptococcus*. But bacteria can also be our allies in medicine. For example, actinobacteria are filamentous, soil-dwelling microbes that produce infection-fighting antibiotics such as streptomycin.

Domain Archaea Archaea are often collectively described as "extremophiles" because scientists originally found them in places that lacked oxygen or that were extremely hot, acidic, or salty (figure 15.14). At first, the organisms were informally grouped by habitat. The thermophiles, for example, live in habitats such as boiling hot springs and hydrothermal vents, whereas the halophiles prefer salt concentrations of up to 30%. Acidophiles tolerate habitats with a pH as low as 1.0—low enough to dissolve metal! Methanogens live in stagnant waters and the anaerobic intestines of many animals, releasing huge quantities of methane gas. As more archaea are discovered in moderate environments such as soil or the open ocean, however, formal classification is becoming more important. This process is ongoing; in fact, scientists do not yet agree on how many phyla of archaea exist, let alone the number of species.

The importance of archaea in ecosystems is slowly becoming clearer as scientists decipher more about their roles in global carbon, nitrogen, and sulfur cycles. Many live in ocean waters and sediments, a hard-to-explore habitat in which the role of archaea is especially poorly understood. Their immense numbers, however, suggest that archaea are critical players in ocean ecology.

D. Bacteria and Archaea Are Essential to All Life

Many people think of microbes as harmful "germs" that cause disease. Indeed, most of the familiar examples of bacteria listed in the previous section are pathogens (disease-causing organisms). Although some bacteria do make people sick, most microbes do not harm us at all. This section describes some of the many ways that bacteria and archaea affect our lives.

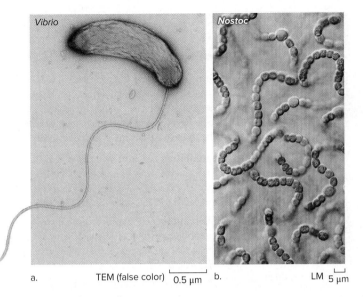

a.   TEM (false color) 0.5 μm   b.   LM 5 μm

Figure 15.13 Two Types of Bacteria. (a) *Vibrio cholerae,* a proteobacterium. (b) Filaments of *Nostoc,* a cyanobacterium.

(a): ©Dr. Gopal Murti/Science Source; (b): ©McGraw-Hill Education/Don Rubbelke

SEM
(false color)   1 μm

Figure 15.14 Extremophiles. Archaea such as *Sulfolobus* thrive in boiling mud pools. This is the Námafjall hot springs area in Iceland.

(hot spring): ©Dr. Mariëlle Hoefnagels; (inset): ©Eye of Science/Science Source

Fungi Are Essential Decomposers

Why We Care 15.1 | Antibiotics and Other Germ Killers

©Keith Brofsky/Getty Images RF

When a person develops a bacterial infection, a physician may prescribe antibiotics. These drugs typically inhibit structures and functions present in bacterial, but not human, cells. Some mechanisms of action include:

- **Inhibiting cell wall synthesis:** Penicillin is an antibiotic that interferes with cell wall formation. A bacterium that cannot make a rigid cell wall will burst and die.

- **Disrupting cell membranes:** Polymyxin antibiotics exploit differences between bacterial and eukaryotic cell membranes. (i) *membranes,* section 3.3

- **Inhibiting transcription:** Rifamycin antibiotics prevent RNA synthesis in bacteria by binding to a bacterial form of RNA polymerase. (i) *transcription,* section 7.3

- **Inhibiting protein assembly:** The antibiotics streptomycin, chloramphenicol, and erythromycin bind to bacterial ribosomes, killing bacteria without harming us. (i) *ribosomes,* section 7.4A

- **Inhibiting enzymes:** Theoretically, antibiotics could block any bacterial metabolic pathway that does not occur in host cells. Sulfanilamide, for example, interferes with a bacterial enzyme that participates in an essential chain of chemical reactions. (i) *enzymes,* section 4.4

Vital Links in Ecosystems Although it may seem hard to believe that one-celled organisms can be essential, the truth is that all other species would die without bacteria and archaea. For example, microbes play crucial roles in the global carbon cycle. They decompose organic matter in soil and water, releasing CO_2. Other microorganisms absorb CO_2 in photosynthesis. And all kinds of microbes, both heterotrophs and autotrophs, are eaten by countless organisms in every imaginable habitat. Chapter 19 explains these community interactions in more detail. (i) *carbon cycle,* section 19.7B

Another essential process is **nitrogen fixation,** the chemical reactions in which prokaryotes convert atmospheric nitrogen gas (N_2) into forms that plants and other organisms can absorb, such as ammonium (NH_4^+). Nitrogen is a component of protein, DNA, and many other organic molecules. The only organisms that can use N_2 directly (by fixing nitrogen) are a few species of bacteria and archaea. Ultimately, most of Earth's nitrogen would be locked in the atmosphere if not for nitrogen-fixers. The nitrogen cycle—and therefore all life—would eventually cease without these crucial microbes. (i) *nitrogen cycle,* section 19.7C

Some nitrogen-fixing bacteria live in soil or water. Others, such as those in the genus *Rhizobium,* induce the formation of nodules in the roots of clover and other plants in the legume family (figure 15.15). Inside the nodules, *Rhizobium* cells share the nitrogen that they fix with their hosts; in exchange, the bacteria receive nutrients and protection.

Beneficial and Pathogenic Microbes No matter how much soap you use or how hard you scrub, you are a habitat for microorganisms. A menagerie of microbes lives on your skin and in your mouth, urogenital tract, upper respiratory tract, and large intestine (see section 28.6E). Collectively called the human microbiota, many of these microscopic companions are beneficial because they help crowd out disease-causing bacteria.

Most people never notice these invisible residents unless something disrupts their personal microbial community. Suppose, for example, that your cat scratches your leg and the wound becomes infected. If you take antibiotics to fight the infection, the drug will probably also kill off some of the normal microbes in your body. As they die, harmful ones can take their place. The resulting microbial imbalance in the intestines or genital tract causes unpleasant side effects such as diarrhea or a vaginal yeast infection. These problems subside in time as the normal microbes divide and restore their populations.

Although most bacteria in and on the human body are harmless, some cause disease. (So far, no archaea are linked to human illnesses.) To cause an infection, bacteria must first enter the body. Animal bites transmit some bacteria, as can sexual activity. A person can also inhale air containing respiratory droplets from a sick coworker or ingest bacteria in contaminated food or water. Bacteria can also enter the body through open wounds.

Once inside the host, pili or slime capsules attach the pathogens to host cells. As the invaders multiply, disease may develop. Some symptoms result from damage caused by the bacteria themselves. The cells may produce enzymes that break down host tissues, for example, or they may release toxins that harm the host's circulatory, digestive, or nervous system. These toxins may help the pathogens invade the host, acquire nutrients, escape the immune system, or spread to new hosts. (i) *enzymes,* section 4.4

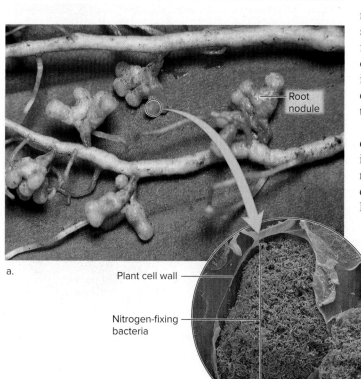

a.

Root nodule

Plant cell wall

Nitrogen-fixing bacteria

SEM (false color) ⊢——⊣ 10 μm

b.

Figure 15.15 **Nitrogen-Fixing Bacteria.** *Rhizobium* bacteria infect these sweet clover roots, producing root nodules where nitrogen fixation occurs. The inset shows a cross section of a nodule, revealing the bacteria inside one root cell.

(a): ©Science Source; (b): ©Andrew Syred/Science Source

Microbiologists divide bacterial toxins into two categories: exotoxins and endotoxins. Exotoxins are toxic proteins that diffuse out of a bacterial cell. *Staphylococcus aureus* and *Clostridium botulinum* are two examples of bacteria that produce exotoxins; *S. aureus* causes toxic shock syndrome and infections of the skin and sinuses, and *C. botulinum* causes botulism.

Rather than diffusing out of a bacterial cell, endotoxins are part of the bacterial outer membrane and are released only when the cell dies. Consider, for example, *E. coli,* a normal inhabitant of animal intestines. Sometimes, cattle droppings containing *E. coli* contaminate water, milk, raw fruits and vegetables, hamburger, and other foods (figure 15.16). Most cells of *E. coli* are harmless, but one particularly nasty variety multiplies inside the body. Its endotoxin can cause belly pain, bloody diarrhea, and, in some cases, life-threatening kidney failure. Outbreaks of foodborne *E. coli* have led to widely publicized recalls of everything from raw spinach to ground beef to unpasteurized apple juice. Raw eggs and other foods contaminated with animal feces also may contain *Salmonella,* a close relative of *E. coli.*

E. coli and *Salmonella* are two examples of microbes that thrive in foods that have been improperly refrigerated or inadequately cooked. The toxins they produce in the food—not infection with the bacteria themselves—produce the vomiting and diarrhea associated with food poisoning.

Human Uses of Prokaryotes Humans have exploited the metabolic talents of microbes for centuries, long before we could see their cells under a microscope (figure 15.17). For instance, many foods are the products of bacterial metabolism. Vinegar, sauerkraut, sourdough bread, pickles, olives, yogurt, and cheese are just a few examples; organic acids released in fermentation produce the tart flavors of these foods. ⓘ *fermentation,* section 6.8

Microbes also have many industrial applications, many of which are related to the burgeoning field of biotechnology (see chapter 11). Vats of fermenting bacteria can produce enormous quantities of vitamin B_{12} and of useful

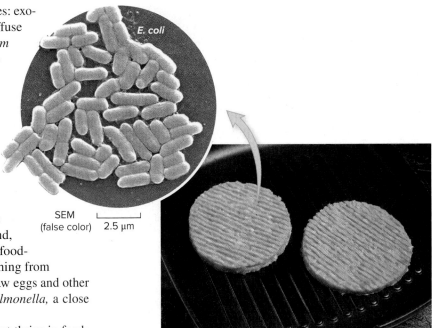

SEM
(false color)   2.5 µm

Figure 15.16 **Cook with Care.** Undercooked hamburger meat is a common source of *E. coli.*
(meat): ©Ingram Publishing/SuperStock RF; (bacteria): Source: CDC

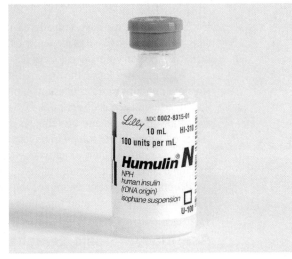

a.   b.   c.

Figure 15.17 **Bacteria at Work.** (a) Bacteria that ferment milk participate in the manufacture of cheddar cheese. (b) Transgenic bacteria produce many drugs, including human insulin. (c) Raw sewage is sprayed on a trickling filter at a municipal wastewater treatment plant. Bacterial biofilms on the filter degrade the organic matter in the sewage. (a): ©Joe Munroe/Science Source; (b): ©Eric Carr/Alamy Stock Photo; (c): ©Jonathan A. Meyers/Science Source

Fungi Are Essential Decomposers

Burning Question 15.2

Are there areas on Earth where no life exists?

Sand, bare rock, and polar ice may seem devoid of life, but they are not. Scientists using microscopes and molecular tools have discovered microbes living in some of the hottest, coldest, wettest, driest, saltiest, highest, most radioactive, and most pressurized places on the planet—including places where no other organism can survive.

There are a few places, however, that humans keep artificially microbe-free for the sake of our own health. For example, people in many professions use autoclaves, radiation, and filters to sterilize everything from surgical tools to medicines and bandages to processed foods. Sterilization kills microbes that could otherwise cause infections, food poisoning, or other illnesses.

Our own bodies are home to many, many microbes, both inside and out, yet we maintain many germ-free fluids and tissues; they include the sinuses, muscles, brain and spinal cord, ovaries and testes, blood, urine in kidneys and the bladder, and semen before it enters the urethra. These areas are among the few places where microbes do not ordinarily live; if a bacterial infection does occur, the resulting illness can be deadly.

Submit your burning question to
marielle.hoefnagels@mheducation.com

chemicals such as ethanol and acetone. Transgenic bacteria mass-produce human proteins, including insulin and blood-clotting factors. In addition, heat-tolerant enzymes isolated from bacteria can degrade proteins and fats in hot water, boosting the cleaning power of detergents used in laundry machines and dishwashers. ⓘ *transgenic bacteria,* section 11.2A

Water and waste treatment also use bacteria and archaea. Sewage treatment plants in most communities, for example, rely on slimy biofilms consisting of countless microbes that degrade organic wastes. And a technique called bioremediation uses microorganisms to metabolize and detoxify pollutants such as petroleum and mercury. Given the right conditions, microbes can consume huge quantities of organic wastes. For example, oil- and gas-munching bacteria consumed much of the oil spilled in the Gulf of Mexico after the disastrous *Deepwater Horizon* well blowout.

15.2 Mastering Concepts

1. What are two domains that contain prokaryotes?

2. Without looking at figure 15.9, sketch the features of a typical prokaryotic cell. What are the functions of each structure?

3. What terms do microbiologists use to describe carbon sources, energy sources, and oxygen requirements?

4. In what ways are bacteria and archaea essential to eukaryotic life?

5. What adaptations enable pathogenic bacteria to enter the body and cause disease?

15.3 Eukaryotic Cells and Multicellularity Arose More Than a Billion Years Ago

Until this point, we have considered the origin and diversity of prokaryotic cells. According to fossil evidence, the first cells in domain **Eukarya** descended from their prokaryotic ancestors at least 1.9 to 1.4 BYA.

We may never know the origin of the membranes that make up the nuclear envelope, endoplasmic reticulum, Golgi apparatus, and other membranes in the eukaryotic cell (see section 3.4). The membranes of these organelles consist of phospholipids and proteins, as does the cell membrane. Perhaps an ancient cell's membrane folded in on itself, eventually pinching off inside the cell to form a complex internal network (figure 15.18). Unfortunately, we can only speculate about that aspect of eukaryotic cell evolution. Some details, however, are becoming clear. For example, the endosymbiont theory may explain the origin of two types of membrane-bounded organelles.

A. Endosymbiosis Explains the Origin of Mitochondria and Chloroplasts

The **endosymbiont theory** proposes that mitochondria and chloroplasts originated as free-living proteobacteria and cyanobacteria that began living

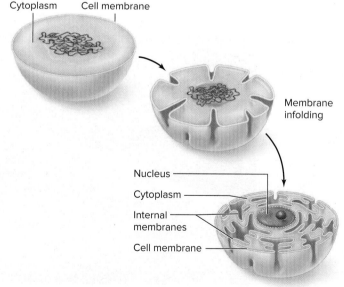

Cytoplasm   Cell membrane

Membrane infolding

Nucleus
Cytoplasm
Internal membranes
Cell membrane

Figure 15.18 Membrane Infolding. A highly folded cell membrane may have formed an internal membrane network as a possible step in the origin of eukaryotic cells.

inside other cells (figure 15.19). The term *endosymbiont* derives from *endo-*, meaning "inside," and *symbiont*, meaning "to live together." Mitochondria must have come first, because virtually all eukaryotes have these organelles. Chloroplasts came later, in the lineages that eventually gave rise to photosynthetic protists and plants. ⓘ *mitochondria*, section 3.4C; *chloroplasts*, section 3.4D

After the ancient endosymbiosis events, many genes moved from the DNA of the organelles to the nuclei of the host cells. These genetic changes made the captured microorganisms unable to live on their own outside their hosts. Over time, they came to depend on one another for survival. The result of this biological interdependency, according to the endosymbiont theory, is the compartmentalized cells of modern eukaryotes.

The evidence supporting the idea that mitochondria and chloroplasts originated as independent organisms includes:

- similarities in size, shape, and membrane structure between the organelles and some types of bacteria;
- the double membrane surrounding mitochondria and chloroplasts, a presumed relic of the original engulfing event;
- the observation that mitochondria and chloroplasts are not assembled in cells but instead divide, as do bacterial cells;
- the similarity between the photosynthetic pigments in chloroplasts and those in cyanobacteria;
- the observation that mitochondria and chloroplasts contain DNA, RNA, and ribosomes, which are similar to those in bacterial cells; and
- DNA sequence analysis, which shows a close relationship between mitochondria and proteobacteria, and between chloroplasts and cyanobacteria. ⓘ *DNA sequencing*, section 11.2B

Burning Question 15.3

Why and how do algae form?

Algae are common aquatic protists, but they are often inconspicuous. Sometimes, however, their populations grow so large that they seem to take over; ponds and poorly maintained swimming pools can turn bright green with algae. This population explosion, also called an algal bloom, occurs whenever nutrients and sunlight are abundant.

Algal blooms are normal in some ecosystems, such as in many ponds. A bloom where water is normally clear, however, usually indicates that nutrients from sewage, fertilizer, or animal waste are polluting the waterway. The use of lawn fertilizers, for example, is a common cause of algal blooms in ponds in residential settings. ⓘ *eutrophication*, section 19.7E

Submit your burning question to
marielle.hoefnagels@mheducation.com

(algal bloom): ©Michael Marten/Science Source

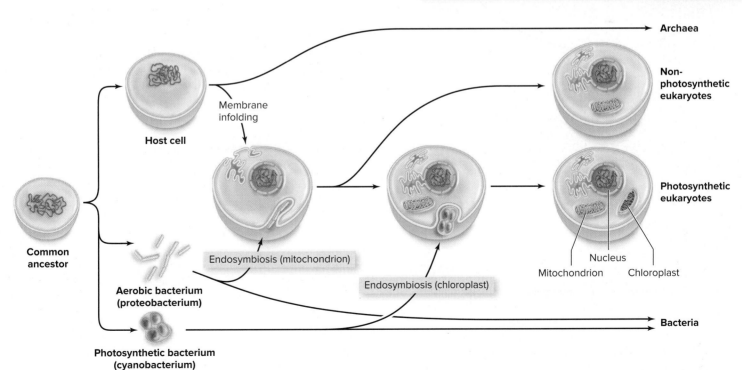

Figure 15.19 The Endosymbiont Theory. Mitochondria and chloroplasts may have originated when ancient host cells engulfed bacterial cells.

Archaea

Non-photosynthetic eukaryotes

Photosynthetic eukaryotes

Bacteria

Membrane infolding

Host cell

Common ancestor

Aerobic bacterium (proteobacterium)

Endosymbiosis (mitochondrion)

Endosymbiosis (chloroplast)

Photosynthetic bacterium (cyanobacterium)

Nucleus

Mitochondrion   Chloroplast

Fungi Are Essential Decomposers

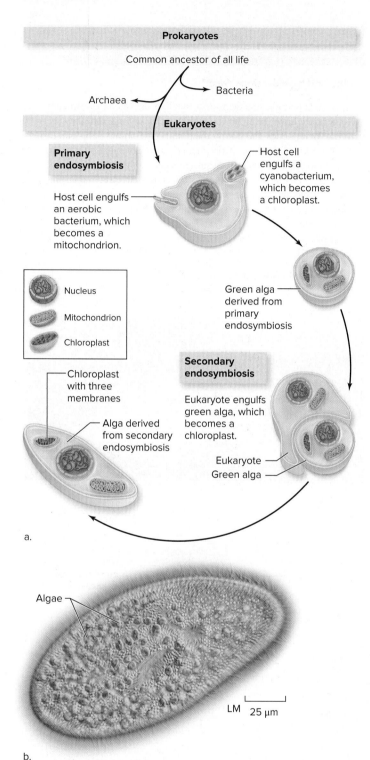

Prokaryotes

Common ancestor of all life

Archaea

Bacteria

Eukaryotes

Primary endosymbiosis

Host cell engulfs an aerobic bacterium, which becomes a mitochondrion.

Host cell engulfs a cyanobacterium, which becomes a chloroplast.

Nucleus

Mitochondrion

Chloroplast

Green alga derived from primary endosymbiosis

Chloroplast with three membranes

Alga derived from secondary endosymbiosis

Secondary endosymbiosis

Eukaryote engulfs green alga, which becomes a chloroplast.

Eukaryote

Green alga

a.

Algae

LM 25 μm

b.

Figure 15.20 Primary and Secondary Endosymbiosis Compared. (a) In primary endosymbiosis, a host cell engulfs a bacterium. Some chloroplasts, including those in green algae, originated in a primary endosymbiosis event. Other chloroplasts originated by secondary endosymbiosis, in which a host cell engulfed a eukaryote. (b) Eukaryotic algae live inside this protist, *Paramecium bursaria*. This partnership, and other similar ones, provide evidence for the endosymbiont theory.

(b): ©Michael Abbey/Science Source

Endosymbiosis has been a potent force in eukaryote evolution. In fact, the chloroplasts of some types of photosynthetic protists apparently derive from a secondary endosymbiosis—that is, from a eukaryotic cell engulfing another eukaryote (figure 15.20). In these species, three or four membranes surround the chloroplasts; some of their cells even retain remnants of the engulfed cell's nucleus. No one knows exactly how many times, or when, these endosymbiosis events happened. Biologists are just beginning to unravel the events surrounding the origins of all the different types of chloroplasts.

B. Multicellularity May Also Have Its Origin in Cooperation

Another critical step leading to the evolution of plants, fungi, and animals was the origin of multicellularity, which occurred about 1.2 BYA. No one knows how eukaryotes came to adopt a multicellular lifestyle. The fossil record is essentially silent on the transition, mostly because the first multicellular organisms lacked hard parts that fossilize readily. We do know, however, that multicellularity arose independently in multiple lineages. After all, genetic evidence clearly suggests that plants, fungi, and animals arose from different lineages of multicellular protists.

We also know that some multicellular organisms consist of cells that bear an uncanny resemblance to one-celled protists; figure 15.21 shows an example. How might the transition to multicellularity have occurred? Perhaps many individual cells came together, joined, and took on specialized tasks to form a multicellular organism. The life cycle of modern-day protists called cellular slime molds illustrates this possibility (see figure 15.29). Alternatively, a single-celled organism may have divided, and the daughter cells may have remained stuck together rather than separating. After many rounds of cell division, these cells may have begun expressing different subsets of their DNA. The result would have been a multicellular organism with specialized cells—similar to the way in which modern animals and plants develop from a single fertilized egg cell.

No matter how it happened, the origin of multicellularity ushered in the possibility of specialized cells, which allowed for new features such as an upright orientation. The resulting explosion in the variety of body sizes and forms introduced new evolutionary possibilities and opened new habitats for other organisms.

15.3 Mastering Concepts

1. What is the evidence that mitochondria and chloroplasts descend from cells?

2. List a logical sequence of evolutionary events that starts with a prokaryote and ends with a multicellular eukaryote.

Chlamydomonas

resembles individual cells in

Volvox

Figure 15.21 From One to Many. A single-celled green alga called *Chlamydomonas* resembles individual cells in its close relative, the many-celled *Volvox*.

15.4 Protists Are the Simplest Eukaryotes

We now embark on a tour of the protists, the simplest eukaryotes. As you will see, the metabolic diversity among protists means they have an astonishingly wide variety of functions and roles in human life.

A. What Is a Protist?

Until recently, biologists recognized four eukaryotic kingdoms: Protista, Plantae, Fungi, and Animalia. The plants, fungi, and animals are distinguished based on their characteristics. Loosely defined, plants are multicellular eukaryotes that carry out photosynthesis; fungi are mostly multicellular eukaryotes that obtain food by external digestion; and animals are multicellular eukaryotes that obtain food by ingestion.

Kingdom Protista, in contrast, was defined by *exclusion*. An organism was designated a **protist** if it was a eukaryote that did not fit the description of a plant, a fungus, or an animal. Kingdom Protista was, in effect, a convenient but artificial "none of the above" category (figure 15.22). Not surprisingly, the nearly 100,000 named species of protists are extremely diverse, displaying great variety in size, nutrition, locomotion, reproduction, and cell surfaces.

Biologists have traditionally grouped the protists based on the organisms that they resemble: the plantlike algae, funguslike water molds and slime molds, and animal-like protozoa. Modern systematists, however, group organisms based on evolutionary relationships. DNA sequences provide the most objective measure of relatedness. Based on these new molecular data, the former kingdom Protista has shattered into dozens of groups whose relationships to one another remain uncertain. ⓘ *systematics,* section 14.6

Because the classification of protists is in transition and many of the new groupings are not universally accepted, this chapter uses the traditional approach to classification. Protistan classification will continue to evolve as research reveals new molecular sequences, but it will likely remain a work in progress for years to come.

B. Algae Are Photosynthetic Protists

Most people probably think of algae as pond scum, but **algae** is a general term that refers to any photosynthetic protist that lives in water. (Although the cyanobacteria were traditionally called "blue-green algae," most biologists now reserve the term *algae* for eukaryotes.) The cells of algae contain chloroplasts that house yellow, gold, brown, red, or green photosynthetic pigments.

Algae produce much of the O$_2$ in Earth's atmosphere and form part of the plankton that supports food webs in oceans, lakes, rivers, and ponds. In addition, algae living among the threads of fungi on rocks and tree bark form lichens, which play a crucial role in building soil from bare rock (see section 15.5C). This section describes some major types of algae. ⓘ *food webs,* section 19.6A

Dinoflagellates The marine protists known as **dinoflagellates** have two flagella of different lengths. One of the flagella propels the cell with a whirling motion (the Greek *dinein* means "to whirl"); the other mainly acts as a rudder. In addition, many dinoflagellates have cell walls that consist of overlapping cellulose plates (figure 15.23).

A red tide is a sudden population explosion, or "bloom," of dinoflagellates that turn the water red, orange, or brown. Some of these algae produce toxins,

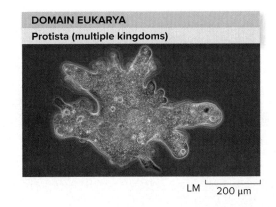

DOMAIN EUKARYA
Protista (multiple kingdoms)

LM ⊢ 200 μm

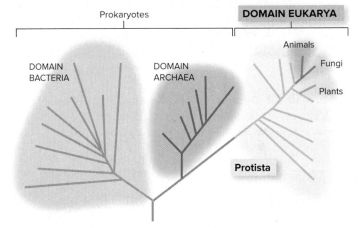

Prokaryotes

DOMAIN EUKARYA

DOMAIN BACTERIA

DOMAIN ARCHAEA

Animals

Fungi

Plants

Protista

Figure 15.22 The Protists. Many separate lineages are informally classified as protists; each may eventually be considered its own kingdom.

Photo: ©Melba/age fotostock RF

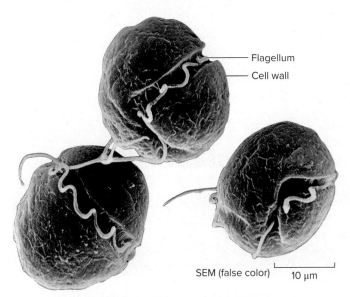

Flagellum

Cell wall

SEM (false color) ⊢ 10 μm

Figure 15.23 Dinoflagellates. Note the two flagella and the cellulose plates that make up the cell walls.

©David M. Phillips/Science Source

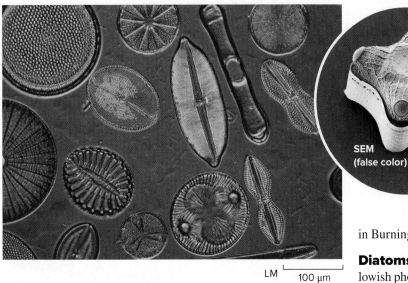

which can make red tides deadly. A person who eats clams, oysters, scallops, or mussels tainted with dinoflagellate toxins may develop paralytic shellfish poisoning. In this context, the recommendation to "never eat shellfish in a month without an R" makes sense because toxic algae blooms are most frequent in May, June, July, and August. In reality, however, most commercially harvested shellfish are tested for toxins and are therefore safe to eat year-round.

A red tide is one type of harmful algal bloom, an overgrowth of algae that release toxins or harm ecosystems in other ways. Usually these blooms occur in response to a boost in the nutrient content of the water, as described in Burning Question 15.3.

SEM (false color) 50 μm

LM 100 μm

Figure 15.24 Diatoms. The "glass houses" (silica cell walls) of these photosynthetic protists exhibit a dazzling variety of forms.
©Jan Hinsch/Science Source; (inset): ©Steve Gschmeissner/Science Source

Diatoms and Brown Algae The diatoms and brown algae contain a yellowish photosynthetic pigment that gives these organisms a golden, olive green, or brown color. **Diatoms** are unicellular algae with ornate, two-part silica cell walls that fit together like a shoebox and its lid (figure 15.24). These protists occupy just about every moist habitat on Earth, from damp soil to fish tanks.

Although diatoms occur nearly everywhere, most live in oceans. Their tiny photosynthetic cells can reach huge densities, removing CO_2 from the atmosphere and providing food for zooplankton.

Over millions of years, the glassy shells of diatoms have accumulated as thick deposits on the ocean floor. The abrasive shells mined from these deposits are used in swimming pool filters, polishes, toothpaste, and many other products. Diatoms also impart the reflective quality of paints used in roadway signs and license plates.

The **brown algae** are the most complex and largest protists. These multicellular algae live in marine habitats all over the world. The kelps, which are the largest of the brown algae, produce enormous underwater forests that provide food and habitat for many animals (figure 15.25).

Figure 15.25 Giant Kelp. These brown algae form huge underwater forests near coastlines. Each individual may be dozens of meters long. ©Ralph A. Clevenger/Getty Images

Figure 15.26 A Red Alga. The blades of this alga, called dulse, grow to 50 cm. People on the northern Atlantic coast consider it a healthy snack.
©Andrew J. Martinez/Science Source

Humans consume several species of kelp, especially in east Asian cuisine. Algin, a chemical extracted from the cell walls of brown algae, is used as an emulsifying, thickening, and stabilizing agent in products including ice cream, candies, chocolate, salad dressings, sauces, soft drinks, beer, cough syrup, toothpaste, cosmetics, polishes, latex paint, and paper.

Red Algae Most **red algae** are relatively large (figure 15.26), although some are microscopic. These marine organisms can live in water exceeding 200 meters in depth, thanks to reddish and bluish photosynthetic pigments that absorb wavelengths of light that chlorophyll *a* cannot capture.

Humans use red algae in many ways. Agar, for example, is a polysaccharide in the cell walls of some species. This jellylike substance is used as a culture medium for microorganisms in petri dishes; agar is sometimes also used as a gel in canned meats and as a thickener in ice cream and yogurt. Another useful product is carrageenan, a polysaccharide that emulsifies fats in chocolate bars and stabilizes paints, cosmetics, and creamy foods.

Green Algae The **green algae** are the protists that share the most similarities with plants. Their habitats and body forms are diverse; they may be unicellular, filamentous, colonial, or multicellular (figure 15.27). The multicellular species may have rootlike and stemlike parts, but these structures are far less specialized than the true roots, stems, and leaves of plants.

One well-studied green alga is *Chlamydomonas*, a unicellular organism that reproduces asexually and sexually; scientists study these algae to learn about the evolution of sex. A classroom favorite is the colonial green alga, *Volvox*. Hundreds to thousands of *Volvox* cells form hollow balls; the cells move their flagella in coordinated waves to propel the sphere. New colonies remain within the parental ball of cells until they burst free. *Volvox* and *Chlamydomonas* have been important in studies of the evolution of multicellularity (see figure 15.21).

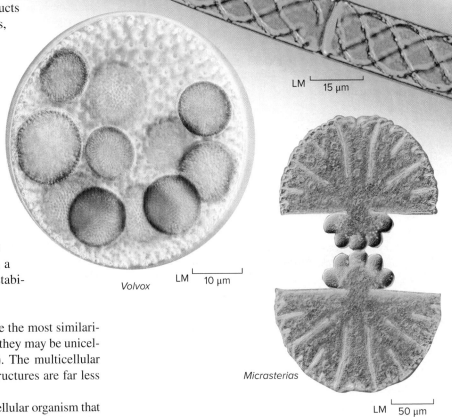

Spirogyra

LM 15 μm

Volvox LM 10 μm

Micrasterias

LM 50 μm

Figure 15.27 Green Algae. Green algae have a variety of body forms, from solitary microscopic cells to complex multicellular forms. (*Spirogyra*): ©Nuridsany et Perennou/Science Source; (*Volvox*): ©Biology Pics/Science Source; (*Micrasterias*): ©M. I. Walker/Science Source

C. Some Heterotrophic Protists Were Once Classified as Fungi

Water molds and slime molds are protists that resemble fungi in some ways: They are heterotrophic, and some produce filamentous feeding structures similar to those in fungi. Nevertheless, molecular evidence clearly indicates that neither group is closely related to fungi.

Water Molds The **water molds** are decomposers or parasites of plants and animals in moist environments. Like fungi, these protists produce filaments that secrete digestive enzymes into their surroundings and absorb the nutrients. Swimming spores help them disperse in water and wet soil. Unlike fungi, however, the cell walls of water molds contain cellulose.

The best-known water molds are those that ruin crops, causing such diseases as downy mildew of grapes and lettuce. The water mold *Phytophthora infestans*, which means "plant destroyer," causes late blight of potatoes (figure 15.28). This disease triggered the devastating Irish potato famine from 1845 to 1847, during which more than a million people starved and millions more emigrated from Ireland.

Figure 15.28 Water Mold. *Phytophthora infestans* is a water mold that causes late blight of potatoes. This disease was responsible for the Irish potato famine in the mid-1840s. ©W.E. Fry, Plant Pathology, Cornell University

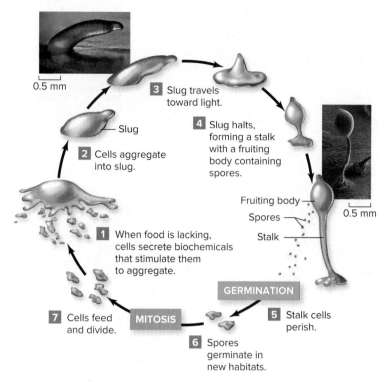

`0.5 mm`

3 Slug travels toward light.

4 Slug halts, forming a stalk with a fruiting body containing spores.

— Slug

2 Cells aggregate into slug.

Fruiting body

Spores

`0.5 mm`

Stalk

1 When food is lacking, cells secrete biochemicals that stimulate them to aggregate.

GERMINATION

7 Cells feed and divide.

MITOSIS

5 Stalk cells perish.

6 Spores germinate in new habitats.

Figure 15.29 Life Cycle of a Cellular Slime Mold. (*1*) Starvation stimulates cells to (*2*) aggregate into a multicellular "slug," which (*3*) crawls to a new habitat and (*4*) forms a fruiting body that releases spores. (*5*) Stalk cells die, but (*6*) spores develop into (*7*) amoeboid cells that consume bacteria. Only asexual reproduction is shown.

Photos: (slug): ©Carolina Biological Supply Company/Medical Images; (fruiting body): ©David Scharf/Science Source

Slime Molds Slime molds live in damp habitats such as forest floors, engulfing bacteria and other microbes on decaying vegetation. Their life cycles are extremely unusual: They can exist either as single cells or as large masses that behave as one multicellular organism.

The feeding stage of a **plasmodial slime mold** consists of a huge cell: a mass of thousands of nuclei enclosed by one cell membrane. This conspicuous, slimy, bright yellow or orange mass may be up to 25 cm in diameter. It migrates along the forest floor, engulfing its food. In times of drought or starvation, the gigantic cell halts and forms fruiting bodies, which produce thick-walled reproductive cells called spores. When favorable conditions return, the spores germinate and form new cells that resume feeding.

In contrast, a **cellular slime mold** consists of individual cells that retain their membranes throughout the life cycle. The cells exist as haploid feeding amoebae. When food becomes scarce, the amoebae secrete chemical signals that stimulate the neighboring cells to aggregate into a sluglike structure (figure 15.29). The "slug" moves toward light, stops, and forms a stalk topped by a fruiting body that produces spores. The cells of the stalk perish, but the spores survive; wind, water, or animals carry them to new habitats. The spores then germinate, and the cycle begins anew.

D. Protozoa Are Diverse Heterotrophic Protists

Finding a list of characteristics that unite the diverse **protozoa** is difficult. Most are unicellular, and the vast majority are heterotrophs, but several autotrophic species exist. Some can swim, but others cannot. Some are free-living, but others are parasites. Most are asexual, but sexual reproduction occurs in many species.

This section describes four groups of distantly related protozoa that are defined by locomotion and morphology. New molecular techniques are redefining the protozoa, but until the newer system of classification is better defined and more widely accepted, these four groups remain practical for general biology, education, and medicine.

Flagellated Protozoa The **flagellated protozoa** are unicellular organisms with one or more flagella (figure 15.30). Most are free-living in fresh water, the ocean, and soil, but a few parasitic species harm humans.

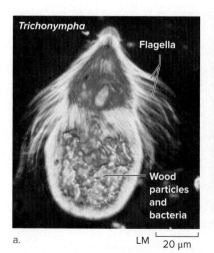

Trichonympha

Flagella

Wood particles and bacteria

a.

LM `20 μm`

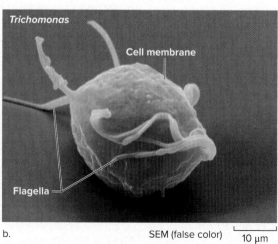

Trichomonas

Cell membrane

Flagella

b.

SEM (false color) `10 μm`

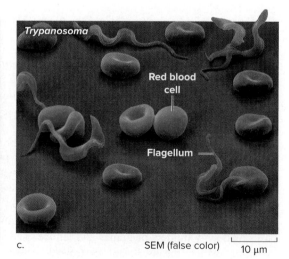

Trypanosoma

Red blood cell

Flagellum

c.

SEM (false color) `10 μm`

Figure 15.30 Flagellated Protozoa. (a) *Trichonympha* is a protist that lives in termites. Note the fringe of flagella. (b) *Trichomonas vaginalis* causes the sexually transmitted disease trichomoniasis. (c) A blood smear from a patient with African sleeping sickness reveals *Trypanosoma brucei* (*purple*) among the blood cells (*red*). Each trypanosome features a single flagellum. (a): ©Eric V. Grave/Science Source; (b, c): ©Eye of Science/Science Source

For example, *Trichomonas vaginalis* resides in the urogenital tracts of both men and women. It is sexually transmitted and causes a form of vaginitis in females. *Giardia* is a flagellate that causes "hiker's diarrhea," or giardiasis. People ingest the cysts of the organism in contaminated water. Another group of disease-causing flagellates are the trypanosomes, whip-shaped parasites that invade the bloodstream and brain. Insects transmit trypanosomes to humans, causing illnesses such as African sleeping sickness and Chagas disease.

Amoeboid Protozoa The **amoeboid protozoa** produce cytoplasmic extensions called pseudopodia (Latin, meaning "false feet"), which are important in locomotion and food capture. The most studied species is *Amoeba proteus,* a common freshwater microbe that engulfs bacteria, algae, and other protists in its pseudopodia (see figure 15.22).

The **foraminiferans,** or forams, are an ancient group of mostly marine amoeboid protozoa. They have complex shells made of durable minerals (figure 15.31). Their populations are immense: About one-third of the ocean floor is made of the shells of marine forams. Paleontologists studying extinct forams have learned which species correlate with oil and gas deposits. The shells are also useful in dating rock strata.

a. LM ⊢200 μm⊣   b. LM ⊢500 μm⊣

Figure 15.31 Foraminiferans and Kin. (a) Thin threads of cytoplasm extend from the calcium-rich shell of this foram. (b) These protists, called radiolarians, are close relatives of the forams. Their silica shells come in a wide variety of shapes and sizes.

(a): ©Peter Parks/Oceans-Image/Photoshot/Newscom; (b): ©Eric V. Grave/Science Source

Ciliates The **ciliates** are complex, unicellular protists characterized by abundant hairlike cilia (figure 15.32). The cilia have multiple functions. Waves of moving cilia propel the organism through the water. Cilia also sweep bacteria, algae, and other ciliates into the cell's oral groove. ⓘ *cilia,* section 3.5

Most ciliates are free-living, motile cells such as *Paramecium.* Nearly one third of ciliates are symbiotic, living in the bodies of crustaceans, mollusks, and vertebrates. Some inhabit the stomachs of cattle, where they house bacteria that break down the cellulose in grass. Others are parasites. The ciliate *Ichthyophthirius multifilis,* for example, is familiar to aquarium owners as the cause of a freshwater fish disease called "ich."

Apicomplexans The **apicomplexans** are nonmotile, spore-forming, internal parasites of animals. The name *apicomplexa* comes from the apical complex, a cluster of microtubules and organelles at one end of the cell. This structure, visible only with an electron microscope, apparently helps the parasite attach to and invade host cells.

Apicomplexans include several organisms that cause illness. *Cryptosporidium* is a genus containing several species that cause diarrhea, cramps, fever, vomiting, and dehydration. Tough-walled "crypto" cysts can survive for days in the chlorinated water of public pools, a common source of this illness. Another example is *Toxoplasma gondii,* a protist that infects cats and other mammals. A person who handles feces from infected cats can accidentally ingest *Toxoplasma* cysts. The infection can pass to a fetus, which is why pregnant women should avoid cat litter boxes.

Malaria is another example of an illness caused by an apicomplexan. Four species of *Plasmodium* cause mosquito-borne malaria in humans. Despite decades of research, malaria continues to be the world's most significant infectious disease. No effective vaccine exists, and *Plasmodium*

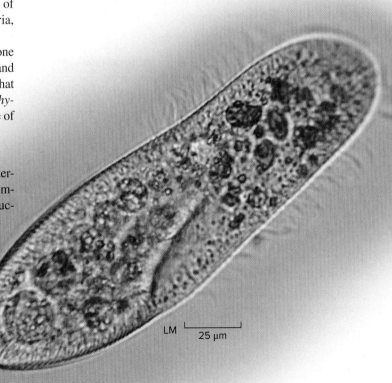

LM ⊢25 μm⊣

Figure 15.32 A Ciliate. Numerous hairlike cilia on the cell's exterior give ciliates their name. This is *Paramecium.*

©Nancy Nehring/E+/Getty Images RF

continues to develop resistance to drugs that were effective in the past. Malaria prevention efforts therefore focus on repelling and killing mosquitoes.

Not everyone is equally susceptible to malaria. People with one copy of the recessive sickle cell allele are much less likely to contract malaria than are people with two dominant alleles. In areas of the world where malaria is endemic, human populations have a relatively high incidence of the sickle cell allele (see figure 12.15). In malaria-free areas, the sickle cell allele is much rarer. This pattern illustrates the selective force that malaria exerts on the human population. ⓘ *sickle cell mutation,* section 7.6A

15.4 Mastering Concepts

1. What features define the protists?
2. Describe examples illustrating why protists are important.
3. Compare and contrast the algae, slime molds, water molds, and protozoa.

15.5 Fungi Are Essential Decomposers

The members of kingdom **Fungi** live nearly everywhere—in soil, in and on plants and animals, in water, even in animal dung. Microscopic fungi infect the cells of protists, while massive fungi extend enormous distances. For example, a single underground fungus extends over nearly 9 million square meters in an Oregon forest. Mycologists (biologists who study fungi) have identified about 80,000 species of fungi, but 1.5 million or so are thought to exist.

A. What Is a Fungus?

Fungi are more closely related to animals than to plants (figure 15.33). This finding may surprise those who notice the superficial similarities between plants and fungi. Unlike plants, however, fungi cannot carry out photosynthesis. Moreover, fungi share many chemical and metabolic features with animals.

Fungi have a unique combination of characteristics:

- The cells of fungi are eukaryotic.
- Fungi are heterotrophs, as are animals, but these two groups acquire food in different ways. Animals ingest their food and digest it internally; fungi secrete enzymes that break down organic matter outside their bodies. The fungus then absorbs the nutrients.
- Fungal cell walls are composed primarily of the modified carbohydrate chitin. This tough, flexible molecule also forms the exoskeletons of some animals. ⓘ *carbohydrates,* section 2.5B
- The storage carbohydrate of fungi is glycogen, the same as for animals (see figure 2.19).
- Most fungi are multicellular, although **yeasts** are unicellular.

The body of a fungus is much more extensive than just a mushroom or the visible fuzz on a moldy piece of food (figure 15.34). Instead, fungi usually consist of an enormous number of **hyphae** (singular: hypha), which are microscopic, threadlike filaments. Hyphae branch rapidly within a food source, growing and absorbing nutrients at their tips. A **mycelium** is a mass of aggregated hyphae that may form visible strands in soil or decaying wood.

While the feeding hyphae remain hidden in the food, the reproductive structures emerge at the surface. Most fungi produce abundant **spores,** which

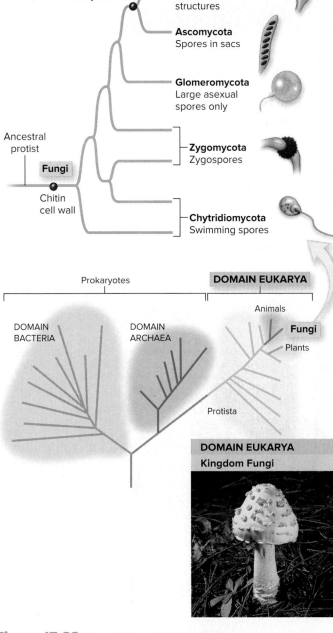

Figure 15.33 Fungal Diversity. Kingdom Fungi contains five phyla, distinguished mainly on the basis of spore type.

Photo: ©Corbis RF

are microscopic reproductive cells (see Why We Care 15.2). Spores can be asexually or sexually produced. Either way, spores that land in a suitable habitat can germinate and give rise to hyphae, starting a new colony.

B. Fungal Classification Is Based on Reproductive Structures

Mycologists classify fungi into five phyla based on the presence and types of sexual structures (see figure 15.33). The **chytridiomycetes** (phylum Chytridiomycota), or chytrids, produce gametes and asexual spores with flagella. **Zygomycetes** (phylum Zygomycota) produce thick-walled sexual spores called zygospores. **Glomeromycetes** (phylum Glomeromycota) do not reproduce sexually at all; instead, they have large, distinctive asexual spores, the largest of which are visible with the unaided eye.

The remaining two phyla, the **ascomycetes** (phylum Ascomycota) and the **basidiomycetes** (phylum Basidiomycota), are the most complex fungi. Their hyphae aggregate to form a **fruiting body,** a large, specialized, sexual spore-producing organ such as a mushroom, morel, puffball, or truffle. The two phyla differ, however, in the ways they produce sexual spores. Ascomycetes produce sexual spores in characteristic sacs, and basidiomycetes release sexual spores from club-shaped structures.

Figure 15.35 shows how the fruiting body fits into the basidiomycete life cycle. In step 1 of the life cycle, the fusion of two haploid hyphae creates a

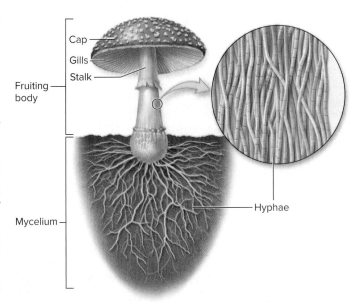

Figure 15.34 The Fungal Body. A mushroom arises from hyphae penetrating the fungus's food source. The mushroom itself is composed of hyphae that are tightly aligned to form a solid structure.

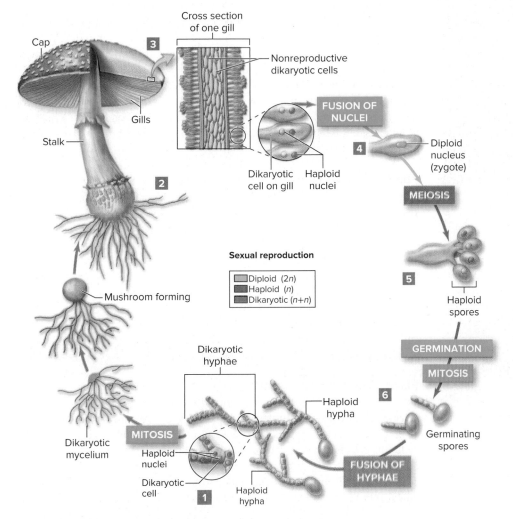

Figure 15.35 Basidiomycete Reproduction. (*1*) Hyphae unite and form a dikaryotic cell with two nuclei. (*2*) This dikaryotic fungus grows and forms a mushroom. (*3*) Club-shaped cells form on gills. (*4*) Two nuclei fuse, forming a diploid zygote that undergoes meiosis. (*5*) The resulting haploid spores fall from the mature mushroom. (*6*) The spores germinate and new haploid hyphae grow.

Fungi Are Essential Decomposers

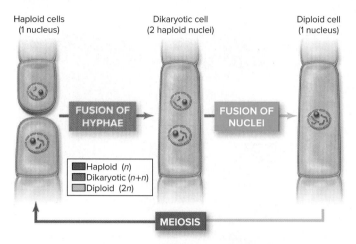

Haploid cells (1 nucleus)   Dikaryotic cell (2 haploid nuclei)   Diploid cell (1 nucleus)

FUSION OF HYPHAE

FUSION OF NUCLEI

■ Haploid (*n*)
■ Dikaryotic (*n+n*)
□ Diploid (2*n*)

MEIOSIS

Figure 15.36 Haploid, Dikaryotic, or Diploid?
In basidiomycetes, dikaryotic cells form when haploid cells fuse but the two nuclei remain separate. Once the nuclei fuse, the cell is diploid.

a.

b.   c.

Figure 15.37 Fungi in Everyday Life. (a) Mold colonies thrive in damp walls after a flood. (b) A bag of edible mushrooms. (c) Roquefort cheese.

dikaryotic mycelium. The term *dikaryotic* means "two nuclei," and it refers to the fact that the nuclei from the two fused hyphae remain separate (figure 15.36). This mycelium typically grows unseen within its food source. When environmental conditions are favorable, however, one or more mushrooms emerge (figure 15.35, step 2).

Step 3 of the life cycle shows that numerous dikaryotic, club-shaped cells line the mushroom's gills. Inside each one, the haploid nuclei fuse, giving rise to a diploid zygote (step 4). The zygote immediately undergoes meiosis, yielding four haploid nuclei (step 5). Each nucleus migrates into a spore, which germinates after dispersal by wind or water (step 6). A new colony begins to grow.

Sometimes, a circle of mushrooms emerges from the ground all at once. The growth pattern of the underground mycelium explains this phenomenon. Hyphae extend outward in all directions from a colony's center. Mushrooms poke up at the margins of the mycelium, creating the "fairy rings" of folklore.

C. Fungi Interact with Other Organisms

Many people know that fungi can cause disease and turn foods moldy, so these organisms have a rather unsavory reputation. But this reputation is undeserved. Although some fungi do harm people, many others are beneficial. Here we profile a few interesting examples of fungi that affect our world.

Decomposers Many fungi secrete enzymes that break down dead plants and animals, releasing inorganic nutrients and recycling them to plants. These decomposers are, in a sense, the garbage processors of the planet.

In forests, fungi that decay wood play a vital role in Earth's carbon cycle. Unfortunately, their talent for degrading the cellulose and lignin in fallen logs also makes them serious pests in another context: They cause dry rot in wooden wall studs and other building materials.

Likewise, most molds become obvious only when they are decaying our own foods and possessions. The general term *mold* includes many types of fungi. For example, a zygomycete forms a black fuzz on bread, fruits, and vegetables, but many of the common molds that ravage flood-damaged homes are ascomycetes (figure 15.37a).

Food and Medicine Commercial mushroom growers cultivate many edible fungi (figure 15.37b). The white "button mushroom" is most familiar, but shiitakes, oyster mushrooms, and other basidiomycete species are available as well. Ascomycetes called truffles and morels also are prized for their delicious flavors.

Fungi also find their way into the human diet in less obvious ways. For example, a species of the ascomycete *Penicillium* lends its sharp flavors to Roquefort cheese (figure 15.37c). Fermentation by yeasts is essential for baking bread, brewing beer, and making wine. Another ascomycete ferments soybean pulp to produce soy sauce. ⓘ *fermentation,* section 6.8

Other species of *Penicillium* are famous for secreting the antibiotic penicillin. Cyclosporine, a drug that suppresses the immune systems of organ transplant recipients, comes from an ascomycete, too.

Plant and Animal Health Fungi cause a wide variety of diseases in plants and animals (figure 15.38). Figure 15.38a, for example, shows an ascomycete called a powdery mildew fungus, which causes a common disease of garden plants. Ascomycetes also cause most other fungal diseases of plants, including Dutch elm disease and chestnut blight.

Animals also suffer from fungal diseases. The dead spider in figure 15.38b is infected with an ascomycete called *Cordyceps.* Eerie spikes release the fungal spores, which may go on to infect another victim. A fungal infection is also responsible for a disease called "white nose syndrome" in hibernating bats (figure 15.38c). The fungus, which thrives in cold cave temperatures, coats the bat's muzzle and wings. Because bats roost close together, fungal spores spread easily from host to host. For reasons that remain unclear, affected bats are unlikely to survive hibernation.

A few fungi threaten human health by causing mild to fatal infections, allergic reactions, or poisonings. For example, fungi can infect our skin, hair, and nails, causing ringworm, athlete's foot, and other irritating diseases. Some fungi, such as *Candida albicans,* inhabit the mucous membranes of the mouth, intestines, and vagina. Normally these fungi are harmless, but they can cause yeast infections under some conditions. Mold spores can trigger allergies. Moreover, some fungi are deadly poisonous, and others are hallucinogenic.

Endophytes and Mycorrhizae

Not all fungi that colonize plant tissues are harmful. **Endophytes,** for example, are hyphae that live between the cells of a plant's tissues without triggering disease symptoms (*endo-* means inside, and *-phyte* means plant). Every known plant, from mosses to angiosperms, harbors endophytes. The ubiquity of endophytes has led some researchers to comment that "all plants are part fungi." Endophytes can help defend plants against disease.

Mycorrhizae (literally, "fungus-roots") are specialized associations between fungi and living roots (figure 15.39). Fungal hyphae extend into the soil, absorbing water and minerals that the fungus shares with the plant; in return, the fungus gains carbohydrates that the plant produces in photosynthesis. Some of the oldest known plant fossils show evidence of mycorrhizae, indicating that plants and fungi moved onto land together hundreds of millions of years ago.

Glomeromycetes form the most common types of mycorrhizae. The fungi pierce the host plant's root cells and produce highly branched arbuscules through which the partners exchange materials (figure 15.39a). Basidiomycetes and ascomycetes form a different type of mycorrhiza. In ectomycorrhizae, the fungal hyphae wrap around root cells and reach into the surrounding soil (figure 15.39b); the hyphae do not, however, penetrate the root cells as they exchange materials with the plant.

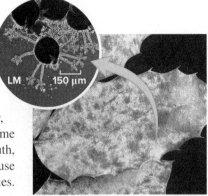

a. Powdery mildew b. Spider with *Cordyceps*

c. Bat with white nose syndrome

Figure 15.38 Pathogenic Fungi. (a) Powdery mildew fungi produce tiny reproductive structures (inset) on leaves. (b) *Cordyceps* fungi killed this spider. (c) The fungus that causes white nose syndrome is visible on this bat's muzzle. The inset shows the fungal spores (*purple*).

(a, leaf): ©Nigel Cattlin/Alamy Stock Photo; (a, inset): ©Scenics & Science/Alamy Stock Photo; (b): ©Morley Read/Alamy Stock Photo; (c, bat): Source: Greg Turner/Pennsylvania Game Commission/USGS; (c, inset): ©Deborah J. Springer

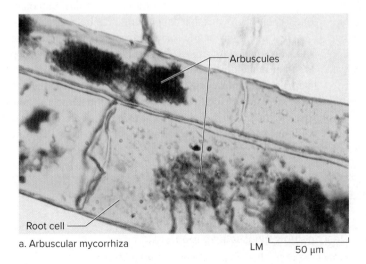
a. Arbuscular mycorrhiza

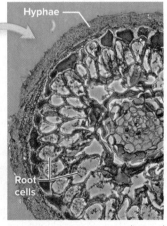
b. Ectomycorrhiza

Figure 15.39 Mycorrhizae. (a) The arbuscules of a glomeromycete occupy these root cells. (b) The creamy white root tips of a pine tree are colonized by an ectomycorrhizal fungus. A cross-section of one root tip reveals a sheath of hyphae.

(a): ©Biophoto Associates/Science Source; (b, root tips): R Henrik Nilsson, Erik Kristiansson, Martin Ryberg and Karl-Henrik Larsson, "Approaching the taxonomic affiliation of unidentified sequences in public databases – an example from the mycorrhizal fungi," *BMC Bioinformatics* 2005, 6:178, doi:10.1186/1471-2105-6-178; (b, cross section): ©Biology Pics/Science Source

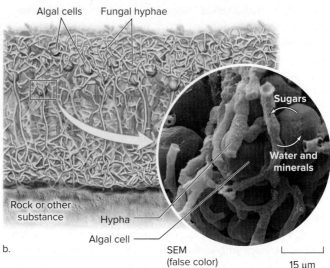

Algal cells   Fungal hyphae

a.

b.

Rock or other substance

Hypha

Algal cell

Sugars

Water and minerals

SEM (false color)

15 μm

Figure 15.40 **Anatomy of a Lichen.** (a) This rock is home to an assortment of lichens. (b) A cross section of a lichen reveals fungal hyphae wrapping tightly around their photosynthetic "partner" cells.

Photos: (a): ©seraficus/Getty Images RF; (b, inset): ©Eye of Science/Science Source

Many edible fungi depend on their ectomycorrhizal relationships with live tree roots; these mushrooms are difficult to cultivate commercially. The popularity and high price of these wild delicacies have lured many mushroom pickers into the woods. Scientists are debating whether the wild mushroom trade will harm populations of fungi—and the trees that rely on them—in the long term.

Lichens A **lichen** forms when a fungus, either an ascomycete or a basidiomycete, harbors green algae or cyanobacteria among its hyphae (figure 15.40). Each member provides resources to the relationship. The photosynthetic partner contributes sugars; the fungus absorbs water and minerals from the surroundings.

Lichens are sometimes called "dual organisms" because the two species—the fungus and its photosynthetic partner—appear to be one individual when viewed with the unaided eye. The body forms can vary widely. Many lichens are colorful, flattened crusts, but others form upright structures that resemble mosses or miniature shrubs. Still others are long, scraggly growths that dangle from tree branches.

Just about any stable surface, from tree bark to boulders to soil, can support lichens. They survive dehydration by suspending their metabolism, only to revive when moisture returns. Polluted habitats, however, are hostile to lichens. Lichens absorb toxins but cannot excrete them. Toxin buildup hampers photosynthesis, and the lichen dies. Disappearance of native lichens is a sign that pollution is disturbing the environment; scientists therefore use lichens to monitor air quality.

15.5 Mastering Concepts

1. What combination of characteristics defines fungi?
2. Describe how fungi acquire food.
3. How do scientists classify the five phyla of fungi?
4. How do fungi benefit humans?
5. Compare and contrast endophytes, mycorrhizae, and lichens.

Investigating Life 15.1 | Shining a Spotlight on Danger

The explosion of blue-green light depicted in the chapter-opening photo is a beautiful sight that is not limited to eastern China. Protists capable of bioluminescence (the production of light by an organism) are common throughout the world's oceans.

Why has evolution selected some algae to release light when they are disturbed? One clue is that herbivores called copepods graze on bioluminescent dinoflagellates at night. The light produced by the dinoflagellates might attract fish that eat the copepods (figure 15.A). If so, then copepods grazing on bioluminescent dinoflagellates should face a higher risk of predation than do copepods given only nonbioluminescent dinoflagellates to eat.

Researchers tested their hypothesis by placing copepods and a stickleback fish in large jars in the laboratory. Half of the jars received bioluminescent dinoflagellates, whereas the other jars contained dinoflagellates that would not emit light. The experimenters then darkened the room, allowing copepods to graze on dinoflagellates and sticklebacks to prey on copepods. After a few hours, they counted the remaining copepods.

The results aligned with the predictions: Stickleback fish eat more copepods when bioluminescent dinoflagellates are present (see the graph in figure 15.A). This outcome suggests that a protist can avoid being eaten by increasing the threat of predation on its grazers; the copepod benefits more from fleeing than from continuing to graze. Since light-emitting dinoflagellates are less likely to be a copepod's dinner, natural selection maintains bioluminescence in these algae.

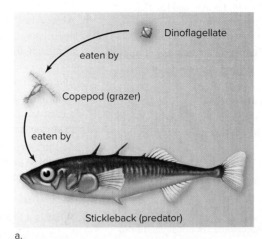
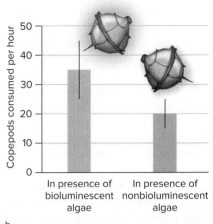

a.

b.

Figure 15.A **Dangerous Light.** (a) A simplified experimental food chain. (b) Sticklebacks ate more copepods when light-emitting algae were present than they did when only nonbioluminescent algae were nearby. (Error bars indicate standard errors; see appendix B.)

If you ever visit a coastline where bioluminescent algae thrive, drop a pebble in the water. You'll see sparkles of light from the protists you disturbed—a beacon to predators that a grazer may be nearby. And if you look closely, you might also see a fish looking for a copepod dinner.

Source: Abrahams, Mark V. and Linda D. Townsend. 1993. Bioluminescence in dinoflagellates: a test of the burglar alarm hypothesis. *Ecology,* vol. 74, pages 258–260.

What's the **Point?** ▼ APPLIED

This chapter offers a taste of the microbial world, providing a sampling of the diversity of life that is smaller than the eye can see. Many people collectively call microorganisms "germs." This is somewhat of a misnomer, as the majority of microbes are either harmless or beneficial. However, some prokaryotes, protists, and fungi do cause disease, and understanding these organisms is important to human health.

Bacteria can be dangerous when they work together. Biofilms are complex communities in which bacterial cells communicate, protect one another, and even form structures with specialized functions. Persistent biofilms that form in catheters, on teeth, and in the lungs and sinuses are much more resistant to antibiotic treatment than are individual cells. Microbiologists are therefore learning to disrupt biofilm formation by silencing the signals that bacteria use to communicate.

One example of a dangerous protist is *Pfiesteria,* a dinoflagellate that lives in water off the east coast of the United States. *Pfiesteria* can produce extremely potent toxins that kill fish and accumulate in shellfish, making them poisonous. In humans, *Pfiesteria* toxins can cause rashes, open sores,

fatigue, erratic heartbeat, breathing difficulty, personality changes, and extreme memory loss. However, the effects of the toxin subside without treatment.

Some of the smallest fungi can be harmful to humans, causing skin infections such as athlete's foot or yeast infections of the mouth or vagina. *Aspergillus flavus* colonizes grains and nuts; when consumed, this fungus's potent toxin causes severe liver damage. Picking and eating wild mushrooms can also be extremely risky. For example, the "death angel" mushroom quietly destroys the liver, kidney, and other organs. By the time symptoms become pronounced, the damage is irreversible, and an emergency organ transplant is the only hope for survival.

The more scientists learn about microorganisms, the more we can fight human diseases. But we can also appreciate the amazing capabilities of Earth's simplest organisms. Prokaryotes, protists, and fungi benefit humans in more ways than they harm us; microbes are certainly much more than just "germs."

Dental plaque (a biofilm)

Bacteria

Bacterial secretions

SEM (false color) 5 μm

Photo: ©Steve Gschmeissner/SPL/Getty Images RF

CHAPTER SUMMARY

15.1 Life's Origin Remains Mysterious

- Earth and the rest of the solar system formed about 4.6 BYA, and life first left evidence on Earth by about 3.7 BYA. The **geologic timescale** describes these and many other events in life's history.

A. The First Organic Molecules May Have Formed in a Chemical "Soup"

- On early Earth, simple precursor chemicals may have combined to form life's organic building blocks.

B. Clays May Have Helped Monomers Form Polymers

- Amino acids and nucleotides may have formed proteins and nucleic acids on hot clay surfaces. The **RNA world** theory proposes that RNA preceded the formation of the first cells.

C. Membranes Enclosed the Molecules

- Phospholipid bubbles may have enclosed proteins and nucleic acids, forming cell precursors (protocells).

D. Early Life Changed Earth Forever

- Early organisms permanently changed the physical and chemical conditions in which life continued to evolve.

15.2 Prokaryotes Are a Biological Success Story

A. What Is a Prokaryote?

- **Prokaryotes** (domains **Bacteria** and **Archaea**) have cells that lack nuclei.

B. Prokaryote Classification Traditionally Relies on Cell Structure and Metabolism

- Prokaryotic cells contain DNA and **ribosomes,** and they are surrounded by a cell membrane. The chromosome is in an area called the **nucleoid. Plasmids** are circles of DNA apart from the chromosome.
- Most prokaryotes have a **cell wall** made of **peptidoglycan.** The wall gives the cell its shape: a spherical **coccus,** rod-shaped **bacillus,** or spiral-shaped **spirillum.**
- External cell structures may include a carbohydrate-rich slime layer (or capsule), short projections called **pili,** and **flagella** that provide movement.
- Some bacteria survive harsh conditions by forming protective **endospores.**
- Prokaryotes may acquire carbon from inorganic sources (**autotrophs**) or organic sources (**heterotrophs**); energy sources include light (**phototrophs**) or chemicals (**chemotrophs**). Cells may require oxygen (**obligate aerobes**), live with or without oxygen (**facultative anaerobes**), or die in the presence of oxygen (**obligate anaerobes**).

C. Prokaryotes Include Two Domains with Enormous Diversity

- Domain Bacteria includes proteobacteria, cyanobacteria, spirochaetes, and actinobacteria; domain Archaea contains methanogens, halophiles, thermophiles, and many organisms that thrive in moderate conditions.

D. Bacteria and Archaea Are Essential to All Life

- All life depends on the bacteria and archaea that recycle organic matter and **fix nitrogen.** Humans harbor and use many beneficial microbes.

15.3 Eukaryotic Cells and Multicellularity Arose More Than a Billion Years Ago

- The internal membranes of **eukaryotic cells** (domain **Eukarya**) may have formed when the cell membrane folded in on itself repeatedly.

A. Endosymbiosis Explains the Origin of Mitochondria and Chloroplasts

- The **endosymbiont theory** proposes that chloroplasts and mitochondria originated as free-living bacteria that were engulfed by larger host cells.

B. Multicellularity May Also Have Its Origin in Cooperation

- Multicellularity evolved multiple times in life's history.

15.4 Protists Are the Simplest Eukaryotes

A. What Is a Protist?

- **Protists** are eukaryotes that are not plants, fungi, or animals. Classification of protists is changing, thanks to molecular sequence data.

B. Algae Are Photosynthetic Protists

- Photosynthesis by **algae** supports food webs and releases oxygen. The many lineages of algae include **dinoflagellates, diatoms, brown algae, red algae,** and **green algae.**

C. Some Heterotrophic Protists Were Once Classified as Fungi

- **Plasmodial slime molds, cellular slime molds,** and **water molds** are funguslike in some ways, but none is closely related to the fungi.

D. Protozoa Are Diverse Heterotrophic Protists

- Most **protozoa** are heterotrophs, and most have motile cells. Groups include **flagellated protozoa, amoeboid protozoa** (including amoebae and the **foraminiferans**), **ciliates,** and **apicomplexans.**

15.5 Fungi Are Essential Decomposers

A. What Is a Fungus?

- **Fungi** are heterotrophs that produce chitin cell walls and glycogen.
- A fungal body typically includes a **mycelium** built of threads called **hyphae,** which may form a **fruiting body. Yeasts** are single-celled fungi.
- Fungi reproduce using asexual and sexual **spores.**

B. Fungal Classification Is Based on Reproductive Structures

- Five main groups of fungi are **chytridiomycetes, zygomycetes, glomeromycetes, ascomycetes,** and **basidiomycetes.** Each produces distinctive types of spores.

C. Fungi Interact with Other Organisms

- Many foods and antibiotics derive from fungi. Fungi have essential roles in ecosystems as decomposers. Most plant pathogens are fungi.
- Many fungal species form partnerships with photosynthetic organisms. **Endophytes** occupy plant tissues without triggering disease symptoms; **mycorrhizae** are specialized associations between fungi and roots; a **lichen** consists of a fungus that harbors cyanobacteria or green algae among its hyphae.
- Figure 15.41 shows some familiar habitats for a tiny selection of the diversity of bacteria, protists, and fungi.

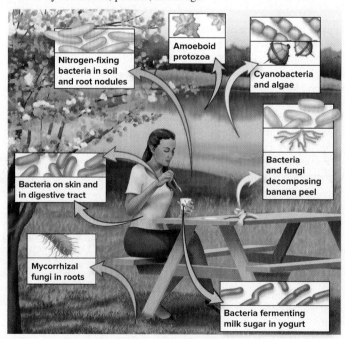

Nitrogen-fixing bacteria in soil and root nodules

Amoeboid protozoa

Cyanobacteria and algae

Bacteria and fungi decomposing banana peel

Bacteria on skin and in digestive tract

Mycorrhizal fungi in roots

Bacteria fermenting milk sugar in yogurt

Figure 15.41 Microbes in a Selection of Habitats.

MULTIPLE CHOICE QUESTIONS

1. Which of the following must be true for natural selection to occur in an "RNA world"?
   a. RNA molecules must turn into DNA molecules.
   b. RNA molecules must undergo mutations.
   c. RNA molecules must replicate.
   d. Both b and c are correct.

2. Which type of organism uses inorganic chemicals as a source of both energy and carbon?
   a. A photoautotroph   c. A chemoautotroph
   b. A photoheterotroph   d. A chemoheterotroph

3. Some protist lineages arose from secondary endosymbiosis. How many membranes would surround the chloroplasts of these organisms?
   a. 0   b. 1   c. 2   d. 3 or more

4. Why is the classification of protists based on DNA sequences useful?
   a. Because only protists have DNA
   b. Because it has confirmed the traditional categories of protists
   c. Because it has revealed evolutionary relationships, even among organisms that look different
   d. All of the above are correct.

5. Fungi are traditionally classified based on their _____.
   a. habitats   c. spore types
   b. dikaryotic cells   d. metabolism

Answers to Multiple Choice questions are in appendix A.

WRITE IT OUT

1. If you were developing a new "broad-spectrum" antibiotic to kill a wide variety of bacteria, which cell structures and pathways would you target? Which of those targets also occur in eukaryotic cells, and why is that important?

2. The amoeba *Pelomyxa palustris* is a single-celled eukaryote with no mitochondria, but it contains symbiotic bacteria that can live in the presence of oxygen. How does this observation support the endosymbiont theory?

3. Why might overwatering your plants make them more susceptible to infection by some kinds of plant-decomposing protists?

4. Review figure 15.33. Are fungi more closely related to animals or to plants? What characteristics do fungi share with plants? What characteristics do fungi share with animals?

5. Hyphae are highly branched structures. How does their extensive surface area contribute to their functions?

SCIENTIFIC LITERACY

Review Why We Care 15.2 and Burning Question 2.2, which describe chemical preservatives in foods. Search the Internet for two websites: one that argues that a widely used preservative called BHT is unhealthy and another that argues it is safe to eat. Which website is more believable, and what is the most persuasive argument on that site?

Answers to Mastering Concepts, Write It Out, Scientific Literacy, and Pull It Together questions can be found in the Connect ebook.
connect.mheducation.com

PULL IT TOGETHER

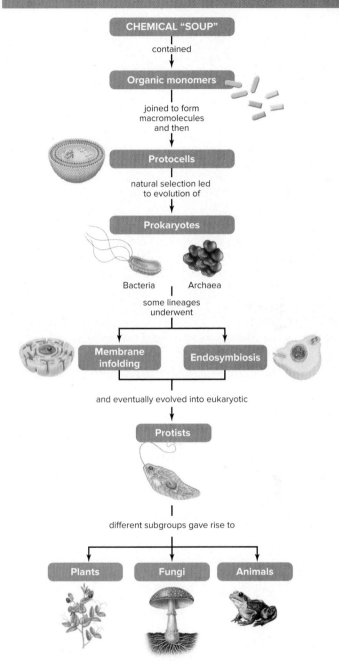

Figure 15.42 Pull It Together: Evolution and Diversity of Microbial Life.

Refer to figure 15.42 and the chapter content to answer the following questions.

1. Review the Survey the Landscape figure in the chapter introduction, then explain the evidence that supports the following events in figure 15.42: the origin of protocells from simple chemicals; the origin of prokaryotes; endosymbiosis; and the evolution of plants, fungi, and animals.

2. Add labeled arrows to this concept map that depict the relationships that connect (a) leaves and endophytes; (b) roots and mycorrhizal fungi; and (c) the fungal and photosynthetic partners in a lichen.

Cork. This man is prying the outer bark off a cork oak tree in France. The bark will be cleaned and used to make wine bottle stoppers. Over the next decade, the tree will regrow its cork, which can then be harvested again.

©Joe Scherschel/National Geographic/Getty Images

LEARNING OUTLINE

16.1 Plants Have Changed the World

16.2 Bryophytes Are the Simplest Plants

16.3 Seedless Vascular Plants Have Xylem and Phloem but No Seeds

16.4 Gymnosperms Are "Naked Seed" Plants

16.5 Angiosperms Produce Seeds in Fruits

APPLICATIONS

Burning Question 16.1 *Do all plants live on land?*
Burning Question 16.2 *What are biofuels?*
Why We Care 16.1 *Gluten and Human Health*
Investigating Life 16.1 *Genetic Messages from Ancient Ecosystems*

Learn How to Learn
Think While You Search the Internet

Your class assignments may require you to use the Internet. But the Internet is full of misinformation, so you must evaluate every site you visit. Collaborative sites such as Wikipedia may be unreliable because anyone can change any article. For other sites, ask the following questions: Are you looking at someone's personal page? Is there an educational, governmental, nonprofit, or commercial sponsor? Is the author reputable? Does the page contain facts or opinions? Are the facts backed up with documentation? Taking the time to find the answers to these questions will help ensure that the sites you use are credible.

SURVEY THE LANDSCAPE
Evolution and Diversity

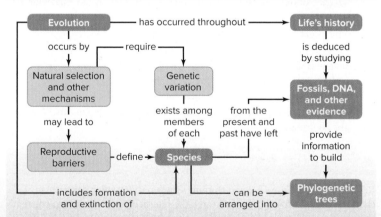

Nearly everywhere on land, plants produce the carbohydrates and oxygen that sustain entire food webs. The ancestors of plants lived in water, but natural selection gradually selected for adaptations that enabled them to dominate terrestrial ecosystems.

For more details, study the Pull It Together feature in the chapter summary.

©Pixtal/age fotostock RF

Most people know that plants are essential to animal life. Vegetation provides the food and habitat that support nearly every ecosystem on land, and photosynthesis produces oxygen.

Yet plants serve us in many unexpected ways as well. Peat moss is a common component of potting mix, we make wine bottle stoppers out of the thick bark of the cork oak tree, and cotton provides the fibers that make up T-shirts and towels. In the landscape and indoors, plants provide a beautiful variety of forms and textures. On a more utilitarian level, the roots of plants bind soil particles, preventing erosion and water pollution. Some plants even clean the soil by absorbing toxic wastes.

Plants are also useful in science. To name just one example, plants carry out many of the same chemical processes that occur in your own cells, including respiration. Plants, however, are much easier to grow and study in the laboratory than are animals. Biologists have therefore studied everything from inheritance to immunity to Alzheimer disease by observing the cells of plants.

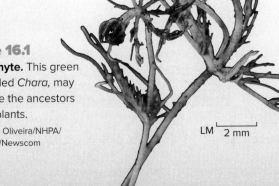

Figure 16.1

Charophyte. This green alga, called *Chara,* may resemble the ancestors of land plants.

©Paulo de Oliveira/NHPA/ Photoshot/Newscom

LM   2 mm

16.1 Plants Have Changed the World

If you glance at your surroundings in almost any outdoor setting, the plants are the first things you see. Grasses, trees, shrubs, ferns, or mosses exist nearly everywhere, at least on land. Members of kingdom **Plantae** dominate habitats from moist bogs to parched deserts. They are so familiar that it is difficult to imagine a world without plants.

Plants are autotrophs: Like the cyanobacteria and algae described in chapter 15, they use sunlight as an energy source to assemble CO_2 and H_2O into sugars (see chapter 5). The sugars, in turn, provide the energy and raw materials that maintain and build a plant's body.

Moreover, the chemical reactions of photosynthesis release oxygen gas, O_2, as a waste product. The evolution of photosynthesis billions of years ago set into motion a complex series of changes that would profoundly affect both the nonliving and the living worlds. The explosion of photosynthetic activity altered the atmosphere, gradually lowering CO_2 levels and raising O_2 content (see figure 15.5). Animals and other organisms that use aerobic respiration need this gas; in addition, O_2 in the atmosphere helps form the ozone layer that protects life from the sun's harmful ultraviolet radiation.

At first, cyanobacteria and algae were the only organisms to release O_2 in photosynthesis. But hundreds of millions of years ago, plants emerged from water and transformed the terrestrial landscape. As plants gradually expanded from the water's edge to the world's driest habitats, they formed the bases of intricate food webs, providing diverse habitats for many types of animals, fungi, and microbes.

Of course, plants remain essential to life today. Herbivores consume living leaves, stems, roots, seeds, and fruits. Dead leaves accumulating on the soil surface feed countless soil microorganisms, insects, and worms. When washed into streams and rivers, this leaf litter supports a spectacular assortment of fishes and other aquatic animals.

From a human perspective, farms and forests provide the foods we eat, the paper we read, the lumber we use to build our homes, many of the clothes we wear, and some of the fuel we burn. The list goes on and on. It is amazing to think that plants do so much with such modest raw materials: sunlight, water, minerals, and CO_2.

A. Green Algae Are the Closest Relatives of Plants

All plants, from mosses to maple trees, are multicellular organisms with eukaryotic cells. With the exception of a few parasitic species, plants are autotrophs. A careful reading of section 15.4B, however, will reveal that some algae have the same combination of traits. Which of the many lineages of algae gave rise to plants?

The answer is that green algae apparently share the most recent common ancestor with plants. About 475 million years ago (MYA), or perhaps earlier, one group of green algae related to today's **charophytes** likely gave rise to plants (figure 16.1). Evidence for this evolutionary connection includes chemical and structural similarities. For example, the chloroplasts of plants and green algae contain the same photosynthetic pigments. In addition, like green algae, plants have cellulose-rich cell walls and use starch as a nutrient reserve. Similar DNA sequences offer additional evidence of a close relationship. ⓘ *green algae,* section 15.4B; *polysaccharides,* section 2.5B

Nevertheless, the body forms of algae are quite different from those of plants, in part because water presents selective forces that

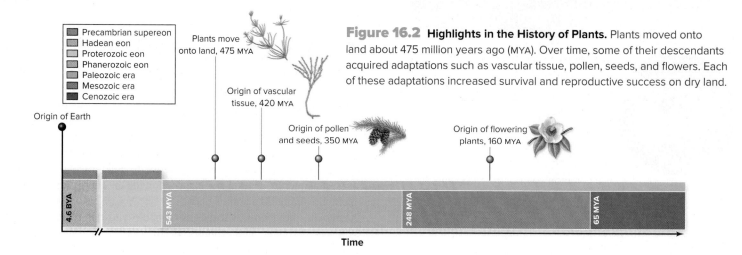

Figure 16.2 Highlights in the History of Plants. Plants moved onto land about 475 million years ago (MYA). Over time, some of their descendants acquired adaptations such as vascular tissue, pollen, seeds, and flowers. Each of these adaptations increased survival and reproductive success on dry land.

are far different from those in the terrestrial landscape. Consider the aquatic habitat. Light, water, minerals, and dissolved gases surround the whole body of a submerged green alga, and the buoyancy of water provides physical support. In sexual reproduction, an alga simply releases swimming gametes into the water.

On land, the water and minerals are in the soil, and only the aboveground part of the plant is exposed to light. Air provides much less physical support than

Figure 16.3 Plant Diversity. Biologists classify plants according to the presence or absence of vascular tissue, seeds, and flowers and fruits.

Photo: Source: USDA/Keith Weller

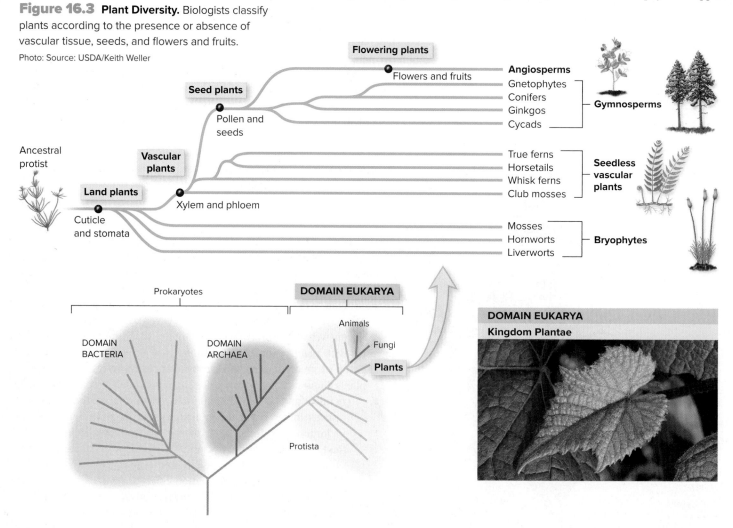

does water, and it dries out the plant's stem and leaves. Furthermore, the dispersal of gametes for sexual reproduction becomes more complicated on dry land.

These conditions have selected for unique adaptations in the body forms and reproductive strategies of plants (figure 16.2). (Note, however, that some plants spend their entire lives in water; see Burning Question 16.1.) As described in the next section, biologists use some of these features to organize land plants into four main groups (figure 16.3): the bryophytes, seedless vascular plants, gymnosperms, and angiosperms.

B. Plants Are Adapted to Life on Land

Figure 16.4 illustrates many of the adaptations that enable plants to produce food, grow upright, retain moisture, survive, and reproduce on land. Refer to this figure often as you read the rest of this section.

Obtaining Resources To carry out photosynthesis, plants need light, CO_2, water, and minerals. Aboveground stems typically support multiple leaves; the extensive surface area of the leaves maximizes exposure to sunlight and CO_2. Below the ground surface, highly branched root systems not only absorb water and minerals but also anchor the plant in the soil.

A plant that dries out will not survive. One water-conserving adaptation is the **cuticle,** a waxy coating that minimizes water loss from the aerial parts of a plant. Dry habitats such as deserts select for extra-thick cuticles; plants in moist habitats typically have thin cuticles.

Burning Question 16.1

Do all plants live on land?

Most plant species live on land, but thousands of them inhabit freshwater or brackish water habitats. Their lifestyles vary widely. Some float freely in the water; others are rooted to the muddy bottom. For some species of aquatic plants, the leaves are entirely submerged, and only the flowers peek above the water surface. One example is elodea, a plant commonly encountered in biology labs. In other species, the leaves float, as in water lilies. Still other aquatic plants, such as cattails, have leaves that emerge from the water.

Aquatic plants have adaptations suited to their watery homes. Submerged plants are surrounded by water, not air, so their leaves may lack stomata and have poorly developed xylem. The roots, if they form at all, are thin and delicate. Floating leaves, on the other hand, have stomata only on their upper surfaces, and the tissues contain large air chambers that make the plant buoyant.

Note that many aquatic organisms that are commonly called "plants" are actually algae. Chapter 15 describes the kelps, red algae, and green algae in detail.

Submit your burning question to
marielle.hoefnagels@mheducation.com

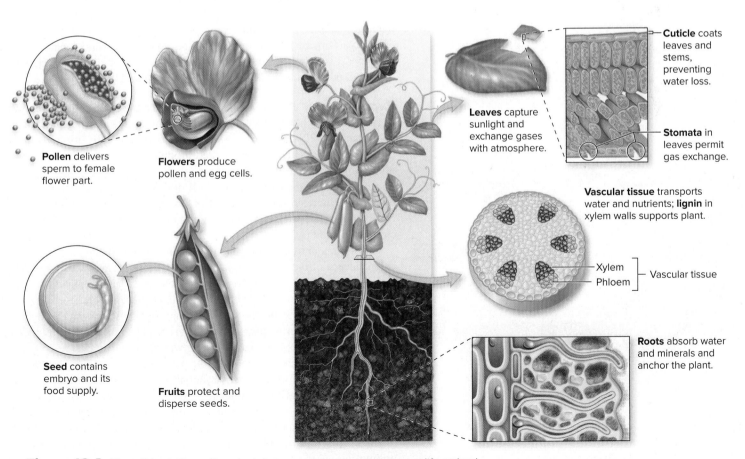

Pollen delivers sperm to female flower part.

Flowers produce pollen and egg cells.

Seed contains embryo and its food supply.

Fruits protect and disperse seeds.

Leaves capture sunlight and exchange gases with atmosphere.

Cuticle coats leaves and stems, preventing water loss.

Stomata in leaves permit gas exchange.

Vascular tissue transports water and nutrients; lignin in xylem walls supports plant.

Xylem
Phloem
Vascular tissue

Roots absorb water and minerals and anchor the plant.

Figure 16.4 Plant Adaptations. Pea plants have many features that support life on land.

Burning Question 16.2

What are biofuels?

Biofuels are plant-based substitutes for fossil fuels. These alternative fuels can help decrease reliance on foreign petroleum and should help reduce CO_2 emissions and the associated problem of global climate change. ⓘ *global climate change,* section 20.4

Two types of biofuels are biodiesel and ethanol. As the name suggests, biodiesel is a diesel fuel substitute. Currently, most biodiesel comes from oil extracted from crushed soybeans or canola seeds. Ethanol, the other main biofuel, is a gasoline substitute. Corn kernels and sugarcane are the main sources of ethanol.

To avoid driving up the price of food crops, researchers are looking for economical, nonfood sources of biofuels. The inedible stems of corn or of prairie grasses such as switchgrass would be ideal; bacterial and fungal enzymes easily break the cellulose in the plant cell walls into simple sugars to use in ethanol production. One problem, however, is that the stems also contain lignin, a complex molecule that interferes with cellulose extraction. So far, the heat and acid treatment needed to eliminate the lignin is too costly and inefficient to make cellulose-derived ethanol economical.

Biofuels are promising, but it is important to realize that they are not exactly "carbon-neutral." Most biofuel crops require fertilizers and pesticides—both of which come from fossil fuels and cause additional environmental problems of their own.

Submit your burning question to
marielle.hoefnagels@mheducation.com

(gas pump): ©NithidPhoto/Getty Images RF

The waxy cuticle is impermeable not only to water but also to gases such as CO_2 and O_2. Plants exchange these gases with the atmosphere through **stomata,** which are pores in the epidermis of stems and leaves. Two guard cells surround each stoma and control whether the pore is open or closed. Water also escapes from the plant's tissues through open stomata. Plants close their stomata in dry weather, minimizing water loss. ⓘ *plant epidermis,* section 21.3B

Internal Transportation and Support The division of labor in a plant poses a problem: Roots need the food produced at the leaves, whereas leaves and stems need water and minerals from soil. In the simplest plants, the bryophytes, cell-to-cell diffusion meets these needs. Other plants have **vascular tissue,** a collection of tubes that transport sugar, water, and minerals throughout the plant. Fossil evidence suggests that the earliest species of vascular plants originated more than 420 million years ago (see figure 13.2).

The two types of vascular tissue are xylem and phloem. **Xylem** (pronounced "zy-lem") conducts water and dissolved minerals from the roots to the leaves. **Phloem** (pronounced "flow-um") transports sugars produced in photosynthesis to the roots and other nongreen parts of the plant. This internal transportation system has supported the evolution of specialized roots, stems, and leaves, many of which have adaptations that enable plants to exploit extremely dry habitats. (Chapter 21 describes these tissues in detail.)

In addition, xylem is rich in **lignin,** a complex polymer that strengthens cell walls. The additional support from lignin means that vascular plants can grow tall and form branches. This increase in size was adaptive because taller plants have the edge over their shorter neighbors in the competition for sunlight. Larger plants, including trees, also triggered evolutionary changes in other organisms by providing new habitats and more diverse food sources for arthropods, vertebrates, and other land animals. (Lignin also poses challenges in biofuel production, as described in Burning Question 16.2.)

Reproduction Plants and green algae have a life cycle called **alternation of generations,** in which a multicellular diploid stage alternates with a multicellular haploid stage (figure 16.5). In the **sporophyte** (diploid) generation, some cells undergo meiosis and produce haploid spores; these spores divide mitotically to form the gametophyte. The haploid **gametophyte,** in turn, produces gametes by mitotic cell division; these sex cells fuse at fertilization. The resulting zygote is the first cell of the next sporophyte generation, and the cycle starts anew. (Read more about mitosis and meiosis in chapters 8 and 9.)

A prominent trend among land plants is a change in the relative sizes and independence of the gametophyte and sporophyte generations (figure 16.6). In a bryophyte such as a moss, for example, the green gametophyte is the most prominent generation, and the brown sporophyte depends on it for nutrition. In more complex plants, the sporophyte is photosynthetic and much larger than the gametophyte. Ferns, pines, and flowering plants have gametophytes that range from microscopic to barely visible with the unaided eye. Keep this evolutionary trend in mind as you study the plant life cycles in this chapter.

Plant reproduction has other variations as well. The sperm cells of mosses and ferns swim in a film of water to reach an egg, limiting the distance over which gametes can spread. Gymnosperms and angiosperms can reproduce over far greater distances, thanks to pollen (see figure 16.4). **Pollen** consists of the male gametophytes of seed plants; each pollen grain produces sperm. In **pollination,** wind or animals deliver pollen to female plant parts, eliminating the need for moisture in sexual reproduction.

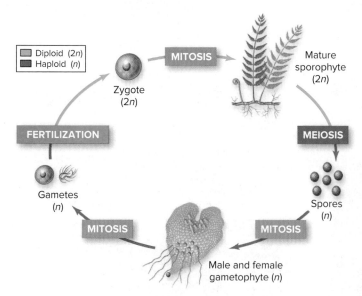

Figure 16.5 Alternation of Generations. Plants have multicellular haploid (gametophyte) and diploid (sporophyte) generations. Note the relationships among the sporophyte, spores, gametophyte, and gametes.

	Bryophytes	Seedless Vascular Plants	Gymnosperms	Angiosperms
GAMETOPHYTE (haploid generation)	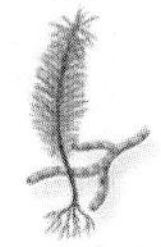	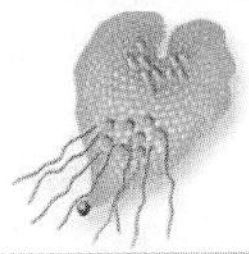	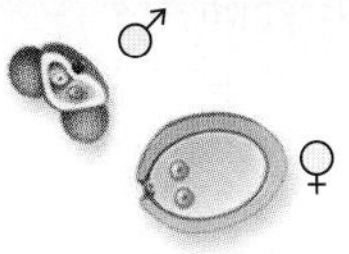	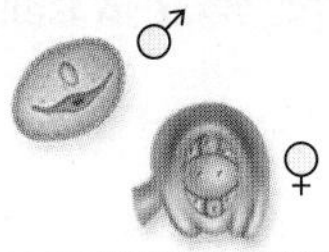
Size relative to sporophyte?	Varies	Small	Microscopic	Microscopic
Depends on sporophyte for nutrition?	No	No	Yes	Yes
SPOROPHYTE (diploid generation)				
Size relative to gametophyte?	Varies	Large	Large	Large
Depends on gametophyte for nutrition?	Yes	No	No	No

Figure 16.6 Changes in the Generations. As plants became more complex, the gametophyte generation was reduced to just a few cells that depend on the sporophyte for nutrition.

Gymnosperms and angiosperms also share another reproductive adaptation: seeds. A **seed** is a dormant plant embryo packaged with a food supply; a tough outer coat keeps the seed's interior from drying out. The food supply sustains the young plant between the time the seed germinates and when the seedling begins photosynthesis.

The origin of pollen and seeds, which occurred more than 350 million years ago, was a significant event in the evolution of plants. The gametes and spores of mosses and ferns—the seedless plants—take little energy to produce, but they are short-lived and tend to remain close to the parent plant. Gymnosperms and angiosperms, in contrast, use pollen and seeds to disperse over great distances, even in dry conditions. Moreover, seeds can remain dormant for years, germinating when conditions are favorable. These adaptations give gymnosperms and angiosperms a competitive edge in many habitats.

Two additional reproductive adaptations occur only in the angiosperms: flowers and fruits. **Flowers** are reproductive structures that produce pollen and egg cells. After fertilization, parts of the flower develop into a **fruit** that contains the seeds. Flowers and fruits help angiosperms protect and disperse both their pollen and their offspring. These adaptations, which arose about 160 million years ago, are spectacularly successful; angiosperms far outnumber all other plants, both in numbers and in species diversity.

16.1 Mastering Concepts

1. How have plants changed the landscape, and how are they vital to life today?
2. What evidence suggests that plants evolved from green algae?
3. Suppose a plant has a mutation that prevents it from closing its stomata. What would be the consequence?
4. Describe the reproductive adaptations of plants.
5. What features differentiate the four major groups of plants?

Miniglossary | Plant Reproduction

Alternation of generations	Life cycle featuring multicellular diploid and haploid stages
Sporophyte	Diploid stage of the plant life cycle, during which some cells undergo meiosis and produce haploid spores
Spores	Haploid cells that develop into the gametophyte generation
Gametophyte	Haploid stage of the plant life cycle, during which some cells undergo mitosis and produce haploid gametes
Gametes	Sperm and egg cells
Pollen grain	Male gametophyte of a seed plant; carried by wind or animals, eliminating the need for moisture in reproduction
Seed	A plant embryo and its food supply, packaged inside a seed coat; produced by gymnosperms and angiosperms
Flower	Angiosperm structure that produces pollen and egg cells; the site of fertilization
Fruit	Seed-containing structure that is unique to angiosperms; develops from flower parts after fertilization

Angiosperms Produce Seeds in Fruits

Figure 16.7 **A Gallery of Bryophytes.** (a) The umbrella-shaped portions of these liverwort gametophytes produce sperm or egg cells. (b) The tapered hornlike structures of a hornwort are sporophytes, below which the flat gametophytes are visible. (c) Short sporophytes topped with dark capsules peek above the gametophytes of sphagnum moss.

(a): ©Dr. Jeremy Burgess/Science Source; (b): ©Steven P. Lynch RF; (c): ©McGraw-Hill Education/Steven P. Lynch

Liverwort

a.

Hornwort

b.

16.2 Bryophytes Are the Simplest Plants

Bryophytes are seedless plants that lack vascular tissue (and its associated lignin). Because they lack physical support, bryophytes are typically small, compact plants. Their small size means that each cell can absorb minerals and water directly from its surroundings. Materials move from cell to cell within the plant by diffusion and osmosis, not within specialized transport tissues.

Although bryophytes lack true leaves and roots, many have structures that are superficially similar to these organs. For example, photosynthesis occurs at flattened leaflike areas. In addition, hairlike extensions called rhizoids cover a bryophyte's lower surface, anchoring the plant to its substrate. Unlike true roots, rhizoids cannot tap distant sources of water. Many species are therefore restricted to moist, shady habitats that are unlikely to dry out. Others tolerate periods of drought by entering dormancy until moisture returns.

Biologists classify the 24,000 or so species of bryophytes into three phyla (figure 16.7):

- **Liverworts** (phylum Marchantiophyta) have various gametophyte forms, from flat to upright and "leafy." The liverworts may be the bryophytes most closely related to ancestral land plants.

- **Hornworts** (phylum Anthocerotophyta) are named for their sporophytes, which are shaped like tapered horns.

- **Mosses** (phylum Bryophyta) are the closest living relatives to the vascular plants. The gametophytes resemble short "stems" with many "leaves." The brown or green sporophyte looks nothing like the gametophyte.

Figure 16.8 shows the sexual life cycle of a moss. The sporophyte is a stalk attached to the leafy gametophyte. At the tip of the stalk, specialized cells inside a sporangium undergo meiosis and produce haploid spores. After the spores are released, they germinate, giving rise to new haploid gametophytes. Gametes form by mitosis in separate sperm- and egg-producing structures on the gametophyte. Sperm swim to the egg cell in a film of water that coats the plants. Sexual reproduction therefore requires water, another factor that limits these plants to moist areas. The sporophyte generation begins at fertilization, with the formation of the diploid zygote. This cell divides mitotically, producing the sporophyte's stalk.

Moss

c.

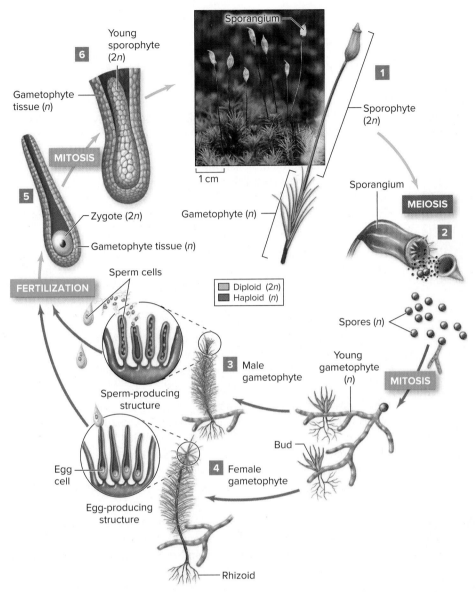

Figure 16.8 Life Cycle of a Moss. (*1*) In the sporophyte, cells in sporangia undergo meiosis, which yields haploid spores (*2*) that develop into male and female gametophytes. (*3*) Male gametophytes produce sperm that swim in a film of water to (*4*) the female gametophytes, which produce egg cells. (*5*) Gametes join and form a zygote, which (*6*) develops into a new sporophyte.

Photo: ©Ed Reschke

Bryophytes play important roles in ecosystems. For example, mosses can survive on bare rock or in a very thin layer of soil. As their tissues die, they contribute organic matter, helping build soil that larger plants subsequently colonize. In forest canopies, bryophytes living on tree limbs help build an organic soil that sustains entire communities of tree-dwelling organisms.

Mosses also find many human uses. Houseplant lovers recognize peat moss as an ingredient in potting mix. Peat comes from partially decomposed sphagnum moss harvested from enormous bogs. The dried moss is unusually spongy, absorbing 20 times its weight in water. When mixed with soil, peat slowly releases water to plant roots. People also burn peat as cooking fuel or to generate electricity.

16.2 Mastering Concepts

1. Describe the three main groups of bryophytes.
2. Name two reasons mosses usually live in moist, shady habitats.
3. How do bryophytes reproduce?

©Steven P. Lynch RF

Angiosperms Produce Seeds in Fruits

Club moss

a.

Spike moss

b.

Whisk fern

c.

Horsetail

d.

True fern

e.

Figure 16.9 **A Gallery of Seedless Vascular Plants.**
(a) The club moss produces upright stems. (b) The spike moss has scalelike foliage. (c) Small, round, spore-producing structures are visible on the highly branched stems of this whisk fern. (d) The stems of a horsetail plant produce spores at their tips. (e) This narrow beech fern has the fronds typical of a true fern.

(a): ©imageBROKER/Alamy Stock Photo RF; (b): ©Geoff Bryant/Science Source;
(c): ©Dr. Mariëlle Hoefnagels; (d): ©Ed Reschke; (e): ©Rod Planck/Science Source

16.3 Seedless Vascular Plants Have Xylem and Phloem but No Seeds

The 12,000 or so species of **seedless vascular plants** have xylem and phloem but do not produce seeds. Unlike the bryophytes, the seedless vascular plants typically have true roots, stems, and leaves. In many species, the leaves and roots arise from underground stems called rhizomes (not to be confused with the rhizoids of bryophytes). Rhizomes sometimes also store carbohydrates that provide energy for the growth of new leaves and roots.

The seedless vascular plants include two phyla divided into four main lineages (figure 16.9):

- **Club mosses** (phylum Lycopodiophyta) are small plants with simple leaves that resemble scales or needles. The name reflects their club-shaped reproductive structures. Their close relatives are the spike mosses. Collectively, club mosses and spike mosses are sometimes called lycopods.

- **Whisk ferns** (phylum Pteridophyta) are simple plants that have rhizomes but not roots. Most species have no obvious leaves. Their name comes from their highly branched stems, which resemble whisk brooms.

- **Horsetails** (phylum Pteridophyta) grow along streams or at the borders of forests. These plants produce branched rhizomes, which give rise to green aerial stems bearing spores at their tips. Horsetails are also called scouring rushes because their stems and leaves contain abrasive silica particles. Native Americans used horsetails to polish bows and arrows, and early colonists and pioneers used them to scrub pots and pans.

- **True ferns** (phylum Pteridophyta) make up the largest group of seedless vascular plants, with about 11,000 species. The fronds, or leaves, of ferns are their most obvious feature; some species are popular as ornamental plants. Ferns were especially widespread and abundant during the Carboniferous period, when their huge fronds dominated warm, moist forests. Their remains form most coal deposits.

Figure 16.10 illustrates the life cycle of a fern. The sporophyte produces haploid spores by meiosis in collections of sporangia on the underside of each frond. Once shed, the spores germinate and develop into tiny, heart-shaped gametophytes that produce gametes by mitotic cell division. The swimming sperm require a film of water to reach the egg cell. The gametes fuse, forming a zygote. This diploid cell divides mitotically and forms the sporophyte, which quickly dwarfs the gametophyte.

Many seedless vascular plants live in shady, moist habitats. Like bryophytes, these plants cannot reproduce sexually in the absence of moisture. Most live on land, where their roots and rhizomes help stabilize soil and prevent erosion. But not all species are terrestrial. The tiny fern *Azolla* lives in water, where its leaves house cyanobacteria that fix nitrogen. In Asia, rice farmers cultivate *Azolla* to help fertilize their crops. ① *nitrogen fixation,* section 15.2D

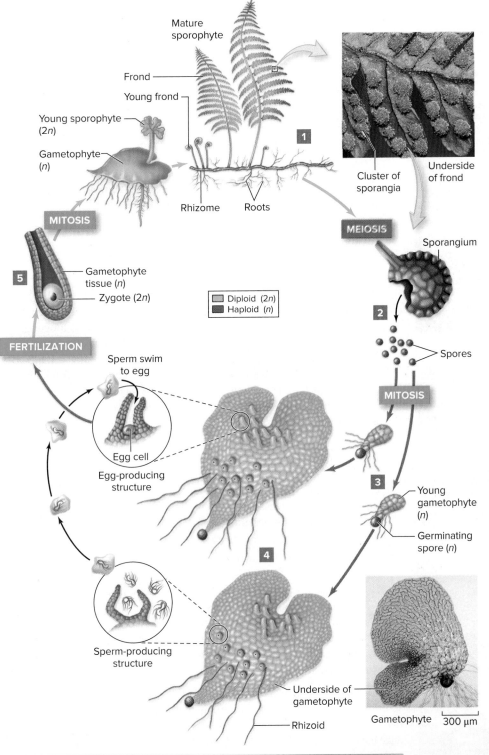

Mature sporophyte

Frond

Young frond

Young sporophyte (2*n*)

Gametophyte (*n*)

MITOSIS

Rhizome Roots

5

Gametophyte tissue (*n*)

Zygote (2*n*)

Diploid (2*n*)
Haploid (*n*)

FERTILIZATION

Sperm swim to egg

Egg cell

Egg-producing structure

Sperm-producing structure

Rhizoid

Cluster of sporangia

Underside of frond

MEIOSIS

Sporangium

2

Spores

MITOSIS

3

Young gametophyte (*n*)

Germinating spore (*n*)

4

Underside of gametophyte

Gametophyte 300 μm

Figure 16.10 **Life Cycle of a True Fern.**
(*1*) Sporangia on the sporophyte's fronds house cells that (*2*) produce spores by meiosis. (*3*) Haploid spores develop into gametophytes, which (*4*) produce egg cells and sperm cells. Sperm swim in water to reach eggs. (*5*) These gametes join and produce a zygote, which develops into the sporophyte.

Photos: (sporangia): ©Ed Reschke/Photolibrary/Getty Images; (gametophyte): ©Les Hickok and Thomas Warne, C-Fern

16.3 Mastering Concepts

1. Describe the four groups of seedless vascular plants.
2. How do seedless vascular plants reproduce?
3. How are seedless vascular plants similar to and different from bryophytes?

©A. Aleksandravicius/Getty Images RF

Figure 16.11 **A Gallery of Gymnosperms.** (a) Cycads are seed plants with cones that form in the center of a crown of large leaves. A seed cone is shown here. (b) Ginkgo leaves turn yellow in the fall. The lower photo shows the fleshy seed. (c) This pinyon pine is a conifer. The seed cone has woody scales. (d) *Ephedra* is a gnetophyte with cones that resemble tiny flowers.

(a, cycad): ©Alena Brozova/Alamy Stock Photo RF; (a, cycad cone): ©Pat Pendarvis; (b, ginkgo tree): ©Light of Peace/Flickr/Getty Images RF; (b, ginkgo seed): ©G. R. "Dick" Roberts/Natural Sciences Image Library; (c, conifer tree): ©Jack Dykinga/Nature Picture Library/Getty Images; (c, conifer cone): ©Ed Reschke/Photolibrary/Getty Images; (d, gnetophyte): ©Jeff Foott/Discovery Channel Images/Getty Images; (d, gnetophyte cones): ©Steven P. Lynch/McGraw-Hill Education

16.4 Gymnosperms Are "Naked Seed" Plants

The first seed plants were gymnosperms. The term **gymnosperm** derives from the Greek words *gymnos,* meaning "naked," and *sperma,* meaning "seed." The seeds of these plants are "naked" because they are not enclosed in fruits.

Living gymnosperms are remarkably diverse in reproductive structures and leaf types. The sporophytes of most gymnosperms are woody trees or shrubs, although a few species are more vinelike. Leaf shapes range from tiny reduced scales to needles, flat blades, and large fernlike leaves. The 800 or so species of gymnosperms group into four phyla (figure 16.11):

- **Cycads** (phylum Cycadophyta) live primarily in tropical and subtropical regions. They have palmlike leaves, and they produce large cones. Cycads dominated Mesozoic era landscapes. Today, cycads are popular ornamental plants, but many species are near extinction in the wild because of slow growth, low reproductive rates, and shrinking habitats.

- The **ginkgo** (phylum Ginkgophyta), also called the maidenhair tree, has distinctive, fan-shaped leaves. Only one species exists. It no longer grows wild in nature, but it is a popular cultivated tree. Ginkgos have male and female organs on separate plants; landscapers avoid planting female ginkgo trees because the fleshy seeds produce a foul odor. Although clinical trials have not conclusively supported their medical benefits, some people believe that extracts of ginkgo leaves may improve memory and concentration.

- **Conifers** (phylum Pinophyta) such as pine trees are by far the most familiar gymnosperms. These plants often have needlelike or scalelike leaves, and they produce egg cells and pollen in cones. Conifers are commonly called "evergreens" because most retain their leaves all year, unlike deciduous trees. This term is somewhat misleading, however, because conifers do shed their needles. They just do it a few needles at a time, turning over their entire needle supply every few years.

- **Gnetophytes** (phylum Gnetophyta) include some bizarre seed plants. Some details of their life history suggest a close relationship with the flowering plants, but molecular evidence places these puzzling plants with the conifers. One example is *Welwitschia,* a slow-growing desert plant with a single pair of large, strap-shaped leaves that persist throughout the life of the plant. *Ephedra* is also a gnetophyte. This plant was once used in weight-loss remedies, but its use in dietary supplements was banned after a series of ephedra-related deaths.

Pine trees illustrate the gymnosperm life cycle (figure 16.12). The mature sporophyte produces **cones,** the organs that bear the reproductive structures. Each female cone scale bears two ovules on its upper surface; the **ovules** produce the female reproductive cells (and eventually develop into seeds). Through meiosis, each ovule produces four haploid megaspores, only one of which develops into a female gametophyte. Over many months, the female gametophyte gives rise to two to six egg cells. At the same time, male cones bear sporangia on thin, delicate scales. Through meiosis, these sporangia produce microspores, which eventually become windblown pollen grains (male gametophytes). Pollination occurs when pollen grains settle between the scales of female cones and adhere to a sticky secretion.

The pollen grain germinates, giving rise to a pollen tube that grows through the ovule toward an egg cell. Two haploid sperm nuclei develop inside the pollen

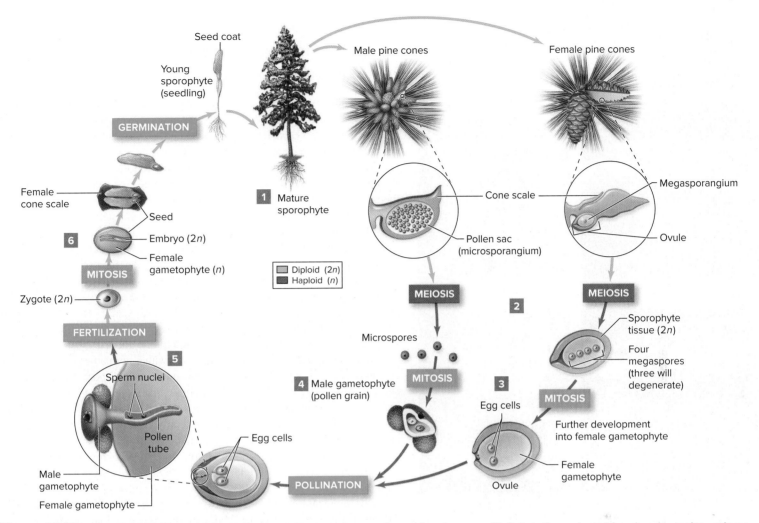

Figure 16.12 Life Cycle of a Pine. (*1*) The mature sporophyte produces male and female cones. (*2*) Cells in the male and female cone scales undergo meiosis, producing spores. (*3*) Female cone scales have two ovules (only one is shown), each of which yields four spores. One spore eventually develops into an egg-producing gametophyte. (*4*) The male cones produce spores that develop into pollen grains, the male gametophytes. (*5*) A pollen grain delivers a sperm nucleus to an egg cell via a pollen tube. The fertilized egg (zygote) will become the embryo. (*6*) The embryo is packaged inside a seed, which will eventually germinate and yield a pine seedling.

tube; one sperm nucleus fertilizes the haploid egg cell, and the other disintegrates. The resulting zygote is the first cell of the sporophyte generation. The whole process is so slow that fertilization occurs about 15 months after pollination.

Within the ovule, the haploid tissue of the female gametophyte nourishes the developing diploid embryo, which soon becomes dormant. Meanwhile, the ovule develops a tough, protective seed coat. The seed may remain in the cone for another year. Eventually, however, the seed is shed and dispersed by wind or animals. If conditions are favorable, the seed germinates, giving rise to a new tree that can begin the cycle again.

16.4 Mastering Concepts

1. What are the characteristics of gymnosperms?
2. What are the four groups of gymnosperms?
3. What is the role of cones in conifer reproduction?
4. What happens during and after pollination in gymnosperms?

©Westend61/Alamy Stock Photo RF

Angiosperms Produce Seeds in Fruits

Cattail

a.

Fruit

Flower

Protective bract

Banana

b.

Passion flower

c.

Pampas grass

d.

Figure 16.13 A Gallery of Angiosperms. The angiosperms exhibit an astonishing variety of flowers and fruits. (a) The brown cylindrical "tails" of a cattail plant consist of tiny brown flowers. (b) Tubular, pale yellow banana flowers form in clusters; developing fruits are also visible in this photo. (c) Passion vines are tropical plants with very showy flowers. (d) Pampas grass produces large plumes of wind-pollinated flowers.

(a): ©Hans Reinhard/OKAPIA/Science Source; (b): ©Igor Prahin/Getty Images RF; (c): ©Pat Pendarvis; (d): ©Southern Stock/The Image Bank/Getty Images

©Burke/Triolo Productions/ Getty Images RF

16.5 Angiosperms Produce Seeds in Fruits

The **angiosperms,** or flowering plants (phylum Magnoliophyta), make up 95% of all modern plant species. At least 250,000 angiosperm species exist. Examples include apple trees, corn, roses, petunias, lilies, grasses, and many other familiar plants, including those we grow for our own food.

Many people think of flowers only as decorations for the human habitat. Although many ornamental plants are the products of selective breeding for their spectacular blooms, not all flowers are showy, sweet-smelling beauties (figure 16.13). The flowers of wind-pollinated plants such as grasses and oak trees, for example, are typically plain and easily overlooked.

Biologists are still working to sort out the evolutionary relationships among the angiosperms. The two largest clades, the eudicots and the monocots, together account for about 97% of all flowering plants. The **eudicots** have two cotyledons (the first leaf structures to arise in the embryo), and their pollen grains feature three or more pores. About 175,000 species exist, representing about two-thirds of all angiosperms. The diverse eudicots include roses, daisies, sunflowers, oaks, tomatoes, beans, and many others.

Most other angiosperms are **monocots,** which are named for their single cotyledon; in addition, their pollen grains have just one pore. (Monocots and eudicots also differ by other characteristics, further described in chapter 21.) Examples of the 70,000 species of monocots are orchids, lilies, grasses, bananas, and ginger. The grasses include not only lawn plants but also sugarcane and grains such as rice, wheat, barley, and corn.

The angiosperm life cycle is similar to that of gymnosperms in some ways (figure 16.14). For example, the sporophyte is the only conspicuous generation, and both types of plants produce pollen and seeds. Yet the life cycles differ in important ways. Most obviously, the reproductive organs in angiosperms are flowers, not cones. Another difference is that an angiosperm's ovules develop into seeds inside the flower's ovary. The ovary develops into the fruit, which helps to protect and disperse the seeds. By comparison, a gymnosperm's seeds are produced "naked" on the female cone's scales (see figure 16.12).

Pollination triggers one other unique feature of the angiosperm life cycle. In **double fertilization,** two sperm nuclei enter the female gametophyte. One sperm nucleus fertilizes the egg, producing the zygote that will develop into the embryo. The other sperm nucleus fertilizes a pair of nuclei in the female gametophyte's central cell. The resulting triploid nucleus develops into the **endosperm,** a tissue that supplies nutrients to the germinating seedling. The embryo and endosperm, together with a seed coat, make up the seed; one or more seeds develop inside each fruit.

Endosperm tissue often contains energy-rich starch or oils. For example, the endosperm of wheat and other grains is starchy; bakers grind these seeds into flour to make bread and other baked goods. Coconuts and castor seeds are two sources of useful oils derived from endosperm. Endosperm also contains proteins, such as the gluten in wheat. (Why We Care 16.1 describes gluten.)

Wind and animals play key roles in angiosperm reproduction. Grasses and maples are common examples of wind-pollinated plants, but many other species rely on animal "couriers" that unwittingly carry pollen from flower to flower. Of course, animals do not pollinate plants as an act of charity; they usually visit flowers in search of food. Large petals, bright colors, alluring scents, and sweet rewards such as nectar attract pollinators.

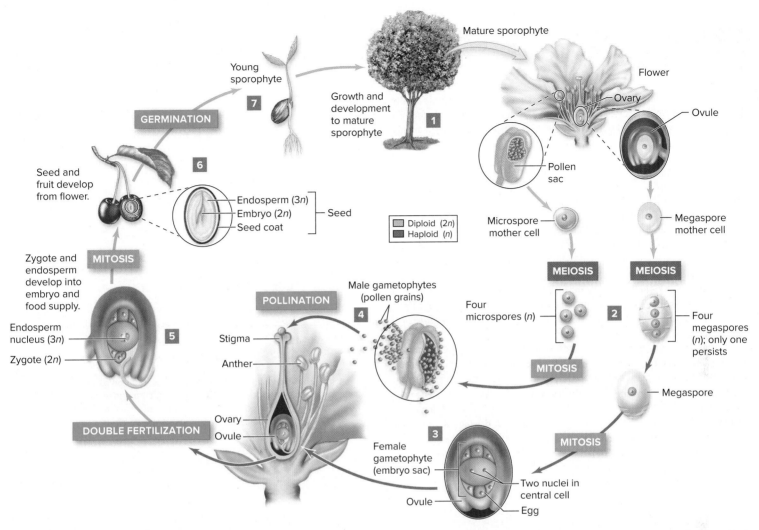

Figure 16.14 Life Cycle of an Angiosperm. (*1*) The mature sporophyte produces flowers. (*2*) Cells in a flower's pollen sac and ovary undergo meiosis, producing spores. (*3*) Of the four spores produced in each ovule, one develops into the female gametophyte, which includes one egg cell and two nuclei in a central cell. (*4*) Spores produced in the pollen sac develop into pollen grains, the male gametophytes. (*5*) A pollen grain delivers two sperm nuclei. One fertilizes the egg and the other fertilizes the nuclei in the central cell, forming a triploid cell that develops into the endosperm. (*6*) Each ovule develops into a seed; the fruit develops from the ovary wall. (*7*) Seed germination reveals the young sporophyte.

Like pollination, seed dispersal also usually involves wind or animals. Some fruits, such as those of dandelions and maples, have "parachutes" or "wings" that promote wind dispersal. Others, however, spread only with the help of animals. Some have burrs that cling to animal fur. Others are sweet and fleshy, attracting animals that eat the fruits and later spit out or discard the seeds in feces. ⓘ *types of fruits,* section 22.2F

16.5 Mastering Concepts

1. What are the two largest groups of angiosperms?
2. In what ways are the life cycles of angiosperms similar to and different from those of conifers?
3. What is the relationship between flowers and fruits?
4. Describe two ways that animals participate in angiosperm reproduction and dispersal.

Why We Care 16.1 | Gluten and Human Health

Gluten is a mixture of proteins produced in the endosperm of wheat and other grains. During seed germination, the proteins (and other materials in the endosperm) nourish the angiosperm embryo.

People with gluten sensitivity must take care to avoid wheat and other gluten-rich grains. Reactions to gluten vary widely. In people with moderate sensitivity, eating gluten causes bloating and abdominal discomfort. If gluten sensitivity is severe (celiac disease), gluten triggers inflammation in the small intestine. The result may be painful digestive problems, vitamin deficiencies, and many other symptoms. People with a gluten allergy may have a rapid whole-body reaction to ingested gluten, leading to hives, swelling, headache, or even anaphylactic shock.

Fortunately for people with any of these problems, gluten-free foods are increasingly common at many grocery stores.

Investigating Life 16.1 | Genetic Messages from Ancient Ecosystems

When an organism dies, its DNA usually degrades rapidly. But in some special cases, the genetic material remains intact indefinitely. Freezing is one way to preserve DNA. An ideal source of diverse ancient DNA is therefore a landscape that once teemed with life but that has since become permanently frozen.

One such example is the land bridge, called Beringia, that once connected present-day northeastern Siberia to Alaska. Long ago, giant mammals such as mammoth, bear, bison, and large cats roamed the grassy Beringian landscape. But climates shift, and much of Beringia is now permanently frozen land in Siberia, Alaska, and the Yukon.

Can DNA preserved in frozen Siberian soil reveal which plants supported ecosystems hundreds of thousands—or even millions—of years ago? To find out, Danish researchers drove metal cylinders deep into Siberian permafrost and removed long, thin rods (called "cores") of ice, soil, and organic material. The deepest holes yield the oldest deposits because new sediments accumulate over old ones. Radiometric dating, pollen analysis, and other techniques helped them estimate the age of each layer of material in the cores. ⓘ *radiometric dating*, section 13.2B

Gene fragments extracted from the sediments allowed the researchers to reconstruct the ancient ecosystem. Herbs (grasses and other nonwoody plants) dominated the Beringian landscape 300,000 or so years ago but lost ground to shrubs over time (figure 16.A). The most dramatic decline of grasses occurred in the past 10,000 years, a time that coincided with the extinction of the mammoth and bison. Did one event cause the other? Or did the end of the last ice age cause both? What role did increasing human populations play? These questions remain unanswered for now, but ice cores collected in years to come may reveal more hidden messages from the past.

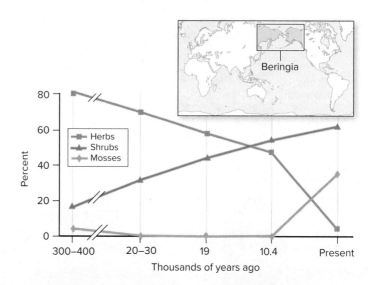

Figure 16.A **Changing Community.** DNA isolated from sediment samples reveals changes in the Siberian plant community over the past 400,000 years. Grasses and other herbs have become much less common, whereas shrubs and mosses have become more common.

Source: Willerslev, E., A. J. Hansen, J. Binladen, et al. May 2, 2003. Diverse plant and animal genetic records from Holocene and Pleistocene sediments. *Science*, vol. 300, pages 791–795.

What's the Point? ▼ APPLIED

This chapter introduced the amazing diversity in the plant kingdom. Humans have practical uses for many organisms in each of the main plant groups. Peat moss, for example, is used in potting mix and as a fuel source. The abrasive silica particles in horsetails make them useful scrubbing agents. Many conifers give us wood that we use in construction, and flowering plants may be beautiful, edible, or both. In addition, some chemicals within plants improve human health. Willow trees contain salicylic acid (an ingredient in aspirin), and opium poppies have pain-relieving chemicals such as morphine. For millennia, people have used the leaves and flowers of marijuana plants for religious, medicinal, and recreational purposes.

Herbal remedies such as echinacea, ginkgo, and St. John's wort feature plant chemicals as well. Unlike drugs, which are scrutinized by the U.S. Food and Drug Administration (FDA), testing herbal remedies is almost exclusively the responsibility of the manufacturer. With some exceptions, the manufacturer is typically not required to show that the product is safe or effective before it is distributed. However, the FDA intervenes if a dietary supplement is later found to be unsafe. In 2013, a diet aid called OxyElite Pro was recalled after the FDA

©marilyna/iStock/Getty Images RF

confirmed that many cases of liver failure were tied to its use. The FDA also steps in if manufacturers make false claims on their labels. Herbal remedies claiming to treat diabetes have recently come under scrutiny for labels that make unsubstantiated promises. The danger is real. People who rely on natural remedies may delay proper treatment, boosting the chance of serious complications. As with OxyElite Pro, many consumers who used the bogus diabetes treatments suffered real harm before the FDA intervened.

Some supplements have been on the market long enough that they are likely to be safe. St. John's wort, for example, is used so widely that the most common side effects and drug interactions are well known. Evidence suggests that extracts from this flowering plant may help treat mild depression; its effectiveness in treating other ailments, ranging from alcoholism to irritable bowel syndrome, requires much more study. Side effects are typically mild.

Humans need plants to keep us warm, safe, clean, fed, and healthy. Plants also support terrestrial food webs. Preserving plant diversity is therefore vital as the human population continues to grow.

CHAPTER SUMMARY

16.1 Plants Have Changed the World

- Members of kingdom **Plantae** provide food and habitat for other organisms, remove CO_2 from the atmosphere, and produce O_2. Humans rely on plants for food, lumber, clothing, paper, and many other resources.

A. Green Algae Are the Closest Relatives of Plants

- The ancestor of land plants may have resembled green algae called **charophytes.** Plants emerged onto land about 475 million years ago.
- Like many green algae, plants are multicellular, eukaryotic autotrophs that have cellulose cell walls and use starch as a carbohydrate reserve. Green algae and plants also use the same photosynthetic pigments. Unlike green algae, most plants live on land.

B. Plants Are Adapted to Life on Land

- Adaptations that enable plants to obtain and conserve resources include roots, leaves, a waterproof **cuticle,** and **stomata.**
- **Vascular tissue** is a transportation system inside many plants. **Xylem** transports water and minerals; **phloem** carries sugars. **Lignin** strengthens xylem cell walls, providing physical support.
- Figure 16.15 illustrates the **alternation of generations** in the plant life cycle. The diploid **sporophyte** produces haploid spores by meiosis; the haploid **gametophyte** produces haploid gametes by mitosis. Fertilization restores the diploid number.

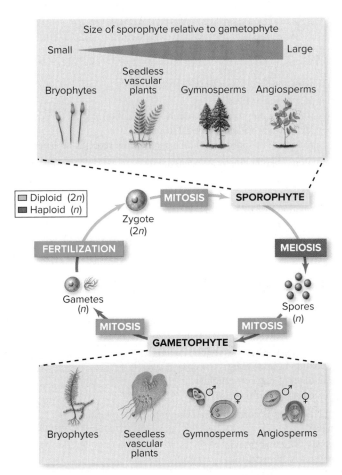

Figure 16.15 Alternation of Generations: A Summary.

- In the simplest plants, the gametophyte generation is most prominent; in more complex plants, the sporophyte dominates (see figure 16.15).
- In gymnosperms and angiosperms, reproductive adaptations include **pollen** and **seeds. Pollination** delivers sperm to egg. The resulting zygote develops into an embryo, which is packaged along with a food supply into a seed. Angiosperms also produce **flowers** and **fruits.**
- Plants are classified by the presence or absence of vascular tissue, seeds, flowers, and fruits.

16.2 Bryophytes Are the Simplest Plants

- **Bryophytes** are small plants lacking vascular tissue, leaves, roots, and stems. The three groups of bryophytes are **liverworts, hornworts,** and **mosses.**
- In bryophytes, the gametophyte stage is dominant. Sperm require water to swim to egg cells.

16.3 Seedless Vascular Plants Have Xylem and Phloem but No Seeds

- **Seedless vascular plants** have vascular tissue but lack seeds. This group includes **club mosses, whisk ferns, horsetails,** and **true ferns.**
- The diploid sporophyte generation is the most obvious stage of a fern life cycle, but the haploid gametophyte forms a tiny separate plant.
- In the sexual life cycle of ferns, collections of sporangia appear on the undersides of fronds. Meiosis occurs in the sporangia and yields haploid spores, which germinate in soil and develop into gametophytes. The gametophytes produce egg cells and swimming sperm.

16.4 Gymnosperms Are "Naked Seed" Plants

- **Gymnosperms** are vascular plants with seeds that are not enclosed in fruits. The four groups of gymnosperms are **cycads, ginkgos, conifers,** and **gnetophytes.**
- In pines (a type of conifer), **cones** house the reproductive structures. Male cones release pollen, and female cones produce egg cells inside **ovules.** Pollen germination yields a pollen tube, through which a sperm nucleus travels to an egg cell. After fertilization, the resulting embryo remains dormant in a seed until germination.
- In conifers, the sperm do not require water to swim to the egg cell. Instead, most gymnosperms rely on wind to spread pollen.

16.5 Angiosperms Produce Seeds in Fruits

- **Angiosperms** are vascular plants that produce flowers and fruits.
- The two largest clades of angiosperms are **eudicots** and **monocots.**
- Flowers produce pollen and egg cells. Wind or animals typically carry angiosperm pollen. In **double fertilization,** two sperm nuclei enter the female gametophyte. One fertilizes the egg cell, and the resulting zygote develops into the embryo. A second sperm nucleus fertilizes the gametophyte's central cell, producing a triploid cell that develops into the seed's **endosperm.**
- After pollination the flower develops into a fruit, which protects the developing seeds. The fruit also aids in dispersal, usually by wind or animals.
- Figure 16.16 summarizes plant diversity.

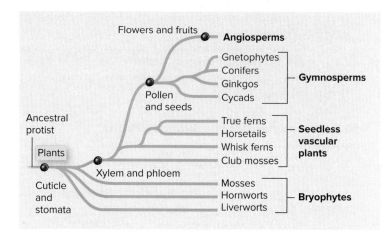

■ **Angiosperms (flowering plants)**
- \>250,000 species
- Monocots, eudicots
- Independent sporophyte
- Pollen and egg cells develop in flowers
- Usually pollinated by wind or animals
- Seeds develop inside fruits

■ **Pines and other gymnosperms**
- ~830 species
- Cycads, ginkgos, conifers, gnetophytes
- Independent sporophyte
- Pollen and seeds usually develop on cone scales
- Usually wind-pollinated

■ **Ferns and other seedless vascular plants**
- ~12,000 species
- Club mosses, whisk ferns, horsetails, true ferns
- Independent sporophyte

■ **Mosses and other bryophytes (nonvascular plants)**
- ~24,000 species
- Liverworts, hornworts, mosses
- Sporophyte depends on gametophyte for nutrition

Group	Swimming Sperm	Vascular Tissue	Pollen	Seeds	Flowers	Fruits
Bryophytes	Yes	No	No	No	No	No
Seedless vascular plants	Yes	Yes	No	No	No	No
Gymnosperms	No	Yes	Yes	Yes	No	No
Angiosperms	No	Yes	Yes	Yes	Yes	Yes

Figure 16.16 Plant Diversity: A Summary.

MULTIPLE CHOICE QUESTIONS

1. Which of the following is NOT a property common to both land plants and green algae?
   a. Photosynthesis
   b. Starch as a form of stored energy
   c. Cellulose cell walls
   d. The presence of a cuticle and stomata

2. In the alternation of generations in plants, the gametophyte is _____ and produces gametes by _____.
   a. haploid; mitosis   c. haploid; meiosis
   b. diploid; mitosis   d. diploid; meiosis

3. What conditions did plants face when they moved to land?
   a. They had less physical support from air than from water.
   b. Drying out became more likely.
   c. Water was not always available for gamete dispersal.
   d. All of the above are correct.

4. Which of the following is present in all land plants?
   a. Windblown sperm
   b. A seed coat that prevents embryos from drying out
   c. Vessels that transport water and nutrients throughout the plant
   d. A waxy cuticle that reduces water loss

5. How does the presence of vascular tissue (xylem and phloem) affect a plant?
   a. It reduces the plant's dependence on a moist environment.
   b. It allows specialization of roots, leaves, and stems.
   c. It allows for the growth of larger plants.
   d. All of the above are correct.

6. Why do many ferns require a shady, moist habitat?
   a. They lack vascular tissue.
   b. They use swimming sperm for sexual reproduction.
   c. They lack a water-retaining cuticle layer.
   d. The production of spores within the sporangium requires moisture.

7. Reproduction in a pine tree is associated with
   a. male and female flowers.
   b. windblown pollen.
   c. the formation of fruits.
   d. vascular tissue.

8. Which type of plant produces seeds enclosed in fruits?
   a. Bryophyte
   b. Gymnosperm
   c. Angiosperm
   d. True fern

9. In comparing the life cycle of an angiosperm to that of a human, pollination is analogous to ___ and the seed is analogous to the ___.
   a. childbirth; growth of the child
   b. sexual intercourse; baby
   c. production of egg cells; mother
   d. production of sperm; uterus

10. What plant group is correctly matched with an adaptation of its members?
    a. Cycads: Vascular tissue
    b. Liverworts: True leaves
    c. Whisk ferns: Seeds
    d. Conifers: Fruits

Answers to Multiple Choice questions are in appendix A.

WRITE IT OUT

1. What characteristics do all land plants have in common?

2. Review the alternation of generations. If you isolated all of the gametes that one gametophyte produced and analyzed the DNA, would you see variation among the gametes?

3. List the characteristics that distinguish the four major groups of plants, then provide an example of a plant within each group.

4. Give at least two explanations for the observation that bryophytes are much smaller than most vascular plants. How can increased height be adaptive? In what circumstances is small size adaptive?

5. Your friend John is admiring what he calls "little flowers" on a moss. How would you correct his statement? In what way might those structures be similar to flowers?

6. A fern plant can produce as many as 50 million spores a year. How are these spores similar to and different from seeds? In a fern population that is neither shrinking nor growing, approximately what proportion of these spores is likely to survive long enough to reproduce? What factors might determine whether an individual spore produces a new fern plant?

7. How do the adaptations of gymnosperms and angiosperms enable them to live in drier habitats than bryophytes and seedless vascular plants?

8. How do angiosperms differ from gymnosperms? How are the two groups of plants similar?

9. The immature fruit of the opium poppy produces many chemicals that affect animal nervous systems. In what way might these chemicals benefit the plant?

10. In a sentence or two, either support or refute the following statement: The pollen grains of angiosperms are homologous to the spores of bryophytes.

11. The spurge-laurel is a species of shrub. It produces berries that, if consumed, will cause internal bleeding and death in humans. On the other hand, some birds can eat the berries and remain unharmed. Why might it be advantageous for the plant to prevent mammals but not birds from eating its fruits?

12. Compare and contrast the life cycles of the four groups of plants. How does each group represent a variation on the theme of alternation of generations?

13. Suppose you and a friend are hiking and you see an unfamiliar plant. What observations would you make in trying to determine which type of plant it is?

SCIENTIFIC LITERACY

Review Burning Question 16.2 and then use the Internet to research the production and use of biofuels in your country. In your opinion, what should be the role of government in promoting the production and use of biofuels? What should be the role of private industry? Of scientists? Do citizens have an ethical responsibility to use biofuels rather than fossil fuels? Why or why not?

Answers to Mastering Concepts, Write It Out, Scientific Literacy, and Pull It Together questions can be found in the Connect ebook.
connect.mheducation.com

Design element: Burning Question (fire background): ©Ingram Publishing/Super Stock

PULL IT TOGETHER

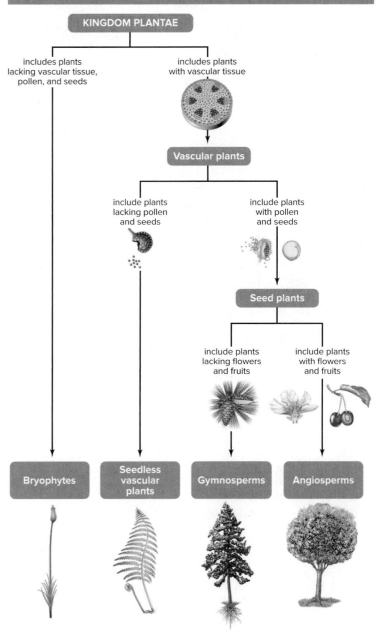

Figure 16.17 Pull It Together: Evolution and Diversity of Plants.

Refer to figure 16.17 and the chapter content to answer the following questions.

1. Review the relationship between natural selection and speciation in the Survey the Landscape figure in the chapter introduction. What environmental conditions selected for the main adaptations in each group of plants shown in figure 16.17?

2. Circle each plant group that produces spores.

3. How do bryophytes and seedless vascular plants reproduce if they lack pollen and seeds?

4. Describe the relationship between pollen and seeds.

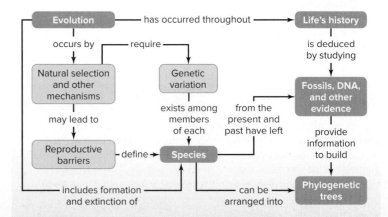

Colorful Beetles. This photo shows a few representatives of the amazingly diverse beetles. Biologists have described over 350,000 species of beetles, more than any other type of animal.

©Imagemore Co, Ltd./Getty Images RF

LEARNING OUTLINE

17.1 Animals Live Nearly Everywhere

17.2 Sponges Are Simple Animals That Lack Differentiated Tissues

17.3 Cnidarians Are Radially Symmetrical, Aquatic Animals

17.4 Flatworms Have Bilateral Symmetry and Incomplete Digestive Tracts

17.5 Mollusks Are Soft, Unsegmented Animals

17.6 Annelids Are Segmented Worms

17.7 Nematodes Are Unsegmented, Cylindrical Worms

17.8 Arthropods Have Exoskeletons and Jointed Appendages

17.9 Echinoderm Adults Have Five-Part, Radial Symmetry

17.10 Most Chordates Are Vertebrates

17.11 Chordate Diversity Extends from Water to Land to Sky

17.12 Fossils and DNA Tell the Human Evolution Story

APPLICATIONS

Burning Question 17.1 *Are there really only nine kinds of animals?*
Why We Care 17.1 *Your Tiny Companions*
Burning Question 17.2 *Did humans and dinosaurs ever coexist?*
Investigating Life 17.1 *Discovering the "Fishapod"*

Learn How to Learn
Flashcard Excellence

While making flashcards, you may be tempted to focus on definitions. For example, after reading this chapter, you might make a flashcard with "amnion" on one side and "membrane surrounding an embryo" on the other. This description is correct, but it won't help you understand the bigger picture. Instead, your flashcards should include realistic questions that cover both the big picture and the small details. Try making flashcards that pose a question, such as "Which animals are amniotes?" or "How are amniotes adapted for life on land?" Write the full answer on the other side, then practice writing the answers on scratch paper until you are sure you have them right.

SURVEY THE LANDSCAPE
Evolution and Diversity

Animal life began in the oceans, with a diverse array of simple body forms. Muscular and nervous systems gradually developed, expanding the range of lifestyles. Once plants invaded land, they provided new sources of food and shelter and acted as a selective force that sparked the evolution of diverse terrestrial animals.

For more details, study the Pull It Together feature in the chapter summary.

©Digital Vision/PunchStock RF

Think of any animal. There's a good chance that the example that popped into your head was a mammal such as a dog, cat, horse, or cow. Although it makes sense that we think first of our most familiar companions, the mammals represent only a tiny subset of organisms in kingdom Animalia.

Biologists have described about 1,300,000 animal species, and their diversity is astonishing. More than 1 million of the described animal species are insects. Only about 57,000 animal species are vertebrates, and most of those are fishes. Mammals make up a paltry 5800 or so species.

Why are animals important? Pets provide companionship, whereas other animals provide food in the form of milk, cheese, meat, and eggs. Animals also play important roles in ecosystems. They graze on vegetation, scavenge dead organic matter, till the earth, control the populations of other animals, pollinate flowers, and carry seeds to new habitats.

It is almost impossible to describe the beauty, diversity, and importance of animals in just one textbook chapter. The pages that follow provide a small sampling of the fascinating organisms that make up kingdom Animalia.

17.1 Animals Live Nearly Everywhere

The diversity of animals is astonishing. Members of kingdom **Animalia** live in us, on us, and around us. They are extremely diverse in size, habitat, body form, and intelligence. Whales are immense; roundworms can be microscopic. Bighorn sheep scale mountaintops; crabs scuttle on the deep ocean floor. Earthworms are squishy; clams surround themselves in heavy armor. Sponges are witless; humans, chimps, and dolphins are clever.

This chapter explores some of this amazing variety, describing nine of the 37 phyla of animals in detail; Burning Question 17.1 highlights three others. Well over 1 million species of animals are **invertebrates** (animals without backbones). **Vertebrates** (animals with backbones), such as mammals and birds, are much less diverse.

A. What Is an Animal?

Animals are diverse, but their shared evolutionary history means that all have some features in common. First, they are multicellular organisms with eukaryotic cells lacking cell walls. Second, all animals are heterotrophs, obtaining both carbon and energy from organic compounds produced by other organisms. Most animals ingest their food, break it down in a digestive tract, absorb the nutrients, and eliminate the indigestible wastes. ⓘ *animal cell,* section 3.2B

Third, animal development is unlike that of any other type of organism. After fertilization, the diploid zygote (the first cell of the new organism) divides rapidly. The early animal embryo begins as a solid ball of cells that quickly hollows out to form a **blastula,** a sphere of cells surrounding a fluid-filled cavity. No other organisms besides animals go through a blastula stage of development.

Fourth, animal cells secrete and bind to a nonliving substance called the extracellular matrix. This complex mixture of proteins and other substances enables some cells to move, others to assemble into sheets, and yet others to embed in supportive surroundings, such as bone or shell.

B. Animal Life Began in the Water

All of today's animals have their origins in aquatic ancestors (figure 17.1). The first animals, which arose about 570 million years ago (MYA), may have been related to aquatic protists called choanoflagellates. Although no one knows exactly what the first animal looked like, the Ediacaran organisms that thrived

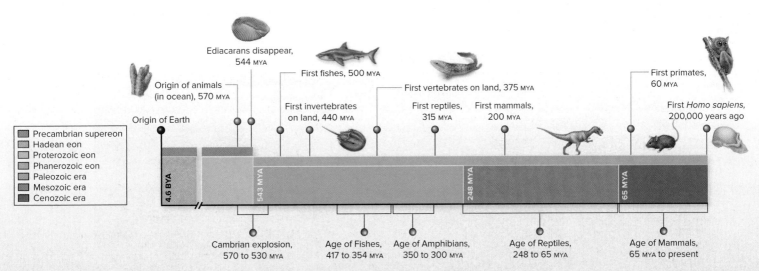

Figure 17.1 Highlights in the History of Animals. Animal life started in the oceans about 570 million years ago (MYA). About 440 MYA, some animals began to move from water onto land. Today, animals are abundant both on land and in water.

a.   b.

1 cm   1 cm

Figure 17.2 **Ancient Animals.** (a) *Dickinsonia* was an Ediacaran organism. (b) *Marrella* is one of many strange animals whose fossils have been discovered in the Burgess Shale.

(a): ©De Agostini Picture Library/Getty Images; (b): ©O. Louis Mazzatenta/National Geographic Creative

during the Precambrian left some clues. These puzzling organisms had shapes resembling disks, tubes, or fronds, but little is known about their lives. They vanished from the fossil record about 544 MYA and left no known modern descendants (figure 17.2a).

Animal life diversified spectacularly during the Cambrian period, which ended about 490 MYA. In the 40-million-year long "Cambrian explosion," most of today's phyla of animals originated in the Cambrian seas. Sponges, jellyfishes, arthropods, mollusks, and many types of worms all arose during this time. The Burgess Shale from the Canadian province of British Columbia preserves a glimpse of life from this time (figure 17.2b).

Aquatic animals were already diverse by the time plants and fungi moved onto land about 475 MYA. Arthropods, vertebrates, and other animals soon followed, diversifying further as they adapted to new food sources and habitats.

C. Animal Features Reflect Shared Ancestry

Figure 17.3 compiles the nine animal phyla described in this chapter into a phylogenetic tree. As you will see, the members of each phylum share similarities because they evolved from a common ancestor with those features. Moreover, the phyla are themselves grouped based on shared features of their appearance, physiology, embryonic development, and DNA. As you read this section, refer back to the tree to recall the positions of the branching points.

Figure 17.3 **Animal Diversity.** Biologists classify animals based on shared ancestry, as revealed by body form, developmental characteristics, and DNA sequences. This simplified evolutionary tree includes only nine of the 37 animal phyla.

Photo: Source: USDA/ARS/Scott Bauer

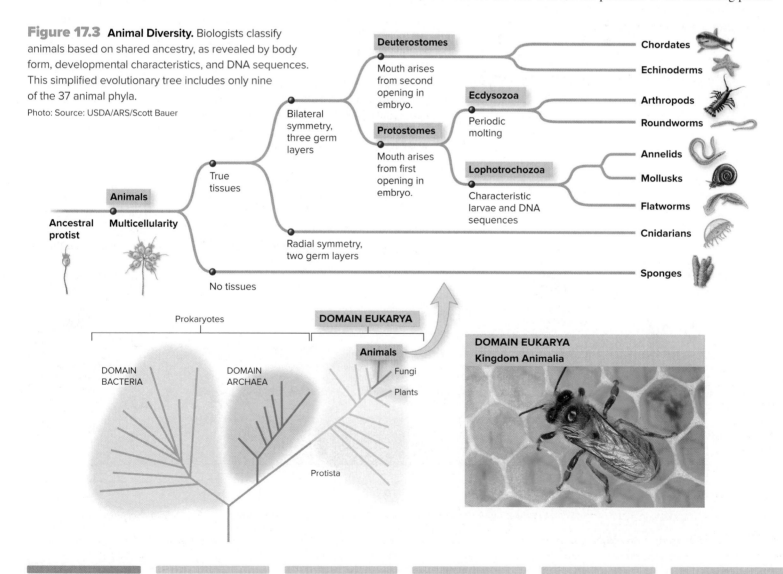

Cell and Tissue Organization The first major branching point separates animals based on whether their bodies contain true tissues. The simplest animals, the sponges, have several specialized cell types, but the cells do not interact to provide specific functions as they would in a true tissue. The other clade contains animals with true tissues. In most of these animals, multiple tissue types interact to form organs, which work together to circulate and distribute blood, dispose of wastes, and carry out other functions.

Body Symmetry and Cephalization Body symmetry is another major criterion used in animal classification (figure 17.4). Many sponges are asymmetrical; that is, they lack symmetry. Other sponges, jellyfishes, adult sea stars, and their close relatives have **radial symmetry,** a body form in which multiple similar parts are arranged around a central axis. Most animals, however, have **bilateral symmetry,** in which only one plane can divide the animal into mirror images. Bilaterally symmetrical animals such as crayfish have head (anterior) and tail (posterior) ends, and they typically move through their environment head first. Bilateral symmetry is correlated with **cephalization,** the tendency to concentrate sensory organs and a brain at an animal's head. These adaptations, in turn, mean a greater ability to evaluate and respond to environmental stimuli.

Embryonic Development: Two or Three Germ Layers Early embryos give other clues to evolutionary relationships (figure 17.5). In animals with true tissues, the blastula folds in on itself to generate the **gastrula,** a cup-shaped structure composed of two or three layers of tissue (called "primary germ layers"). **Ectoderm** is the outer tissue layer, and **endoderm** is the inner layer. Jellyfishes and their relatives have only those two layers. The gastrulas of all other animals with true tissues have **mesoderm,** a third germ layer that forms between the ectoderm and endoderm.

The gastrula's germ layers eventually give rise to all of the body's tissues and organs. Ectoderm develops into the skin and nervous system, whereas endoderm becomes the digestive tract and the organs derived from it. Mesoderm gives rise to the muscles, the bones, the circulatory system, and many other specialized structures. Overall, animals with three germ layers have much greater variety in body forms and functions than do animals with two germ layers.

Embryonic Development: Protostomes and Deuterostomes After an embryo has folded into a gastrula, the inner cell layer fuses with the opposite side of the embryo, forming a tube with two openings. This cylinder of endoderm will develop into the animal's digestive tract, with one opening becoming the mouth and the other becoming the anus. But which end is which?

In most **protostomes,** the gastrula's first indentation develops into the mouth, and the anus develops from the second opening. (*Protostome* literally means "mouth first.") In **deuterostomes,** the first indentation becomes the anus, and the mouth develops from the second opening. (*Deuterostome* means "mouth second.") Echinoderms and chordates are deuterostomes. We now know that some animals classified as protostomes do not conform to the "mouth first" pattern. Nevertheless, DNA sequences support their close relationship to other animals in the protostome clade.

As you can see from figure 17.3, protostomes are further divided into two main groups: ecdysozoans and lophotrochozoans. These groups are largely defined by their DNA sequences rather than by a combination of easily observable characteristics. Ecdysozoans, however, do share a visible feature (molting).

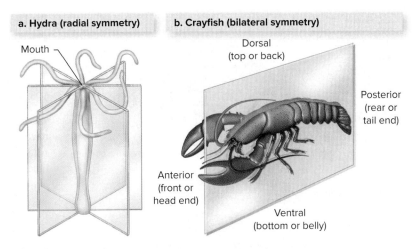

a. Hydra (radial symmetry)
b. Crayfish (bilateral symmetry)

Figure 17.4 Types of Symmetry. (a) A hydra has radial symmetry. (b) A crayfish has bilateral symmetry. Animals with bilateral symmetry have a front (anterior) and rear (posterior) end, and a dorsal (top or back) and ventral (bottom or belly) side.

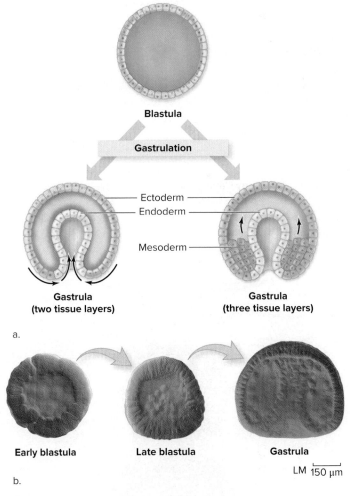

Figure 17.5 Two or Three Primary Germ Layers. (a) A fluid-filled ball of cells called a blastula folds in on itself and forms the two- or three-layered gastrula. (b) A sea star's blastula and gastrula.

(b, all): ©Herve Conge/Medical Images

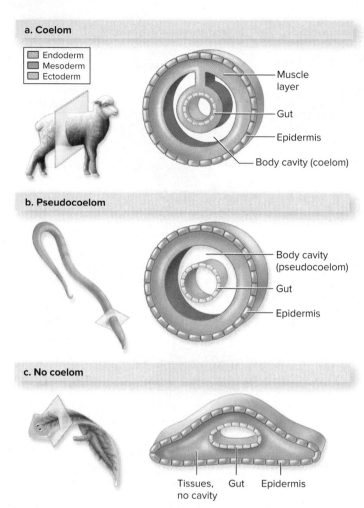

a. Coelom

- ☐ Endoderm
- ☐ Mesoderm
- ☐ Ectoderm

Muscle layer
Gut
Epidermis
Body cavity (coelom)

b. Pseudocoelom

Body cavity (pseudocoelom)
Gut
Epidermis

c. No coelom

Tissues, no cavity
Gut
Epidermis

Figure 17.6 Body Cavities. (a) Like many other animals, a sheep has a coelom. (b) A roundworm has a pseudocoelom. (c) A flatworm lacks a coelom. Note that these drawings are abstractions. In the sheep, for example, the internal organs grow into the coelom, greatly distorting the cavity's shape.

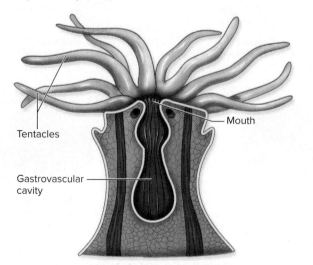

Tentacles
Mouth
Gastrovascular cavity

Figure 17.7 Incomplete Digestive Tract. An animal with a gastrovascular cavity, such as this sea anemone, takes in food and ejects undigested wastes through the same opening (its mouth).

D. Biologists Also Consider Additional Characteristics

Other adaptations have also been important milestones in animal evolution. This section describes some other characteristics that you will encounter as you learn about the animal phyla in this chapter.

Body Cavity (Coelom) A bilaterally symmetrical animal may or may not have a coelom (figure 17.6). The **coelom** (pronounced SEA-loam) is a fluid-filled body cavity that forms completely within the mesoderm. Animals that have a coelom include earthworms, snails, insects, sea stars, and chordates. In contrast, roundworms have a **pseudocoelom** ("false coelom"), a cavity that is lined partly with mesoderm and partly with endoderm. Flatworms lack a coelom, although evidence suggests their ancestors may have had body cavities.

The coelom's chief advantage is flexibility. As internal organs such as the heart, lungs, liver, and intestines develop, they push into the coelom. The fluid of the coelom cushions the organs, protects them, and enables them to shift as the animal bends and moves.

In many animals, the coelom or pseudocoelom serves as a hydrostatic skeleton that provides support and movement. In a **hydrostatic skeleton,** muscles push against a constrained fluid. An earthworm, for example, burrows through soil by alternately contracting and relaxing muscles surrounding its coelom. Note that jellyfishes, flatworms, and other invertebrates also have hydrostatic skeletons, even though they lack a coelom or pseudocoelom. Instead, their muscles push against fluid in the digestive tract or between body cells.

Digestive Tract A sponge lacks a digestive tract; instead, the animal has pores through which water enters and leaves the body. In other animals, the digestive tract may be incomplete or complete. Cnidarians and flatworms have an **incomplete digestive tract,** in which the mouth both takes in food and ejects wastes (figure 17.7). In these animals, digestion occurs in the **gastrovascular cavity,** which secretes digestive enzymes and distributes nutrients to all parts of the animal's body.

In humans and other animals with a **complete digestive tract,** food passes in one direction from mouth to anus (see chapter 28). A complete digestive tract allows the animal to process food stepwise. For example, cells near the mouth can secrete digestive enzymes into the tract, "downstream" cells can absorb nutrients, and those near the anus can help eject wastes.

Segmentation Segmentation is the division of an animal body into repeated parts. In centipedes, millipedes, and earthworms, the segments are clearly visible. Insects and vertebrates also have segmented bodies, although the subdivisions may be less obvious. Segmentation adds to the body's flexibility, and it enormously increases the potential for the development of specialized body parts. Antennae can form on an insect's head, for example, while wings or legs sprout from other segments.

17.1 Mastering Concepts

1. What characteristics do all animals share?
2. When and in what habitat did animals likely originate?
3. What features were used to build the animal phylogenetic tree?
4. What are the events of animal early embryonic development?
5. What are the two main types of digestive tracts?
6. What advantages does segmentation confer?

17.2 Sponges Are Simple Animals That Lack Differentiated Tissues

The **sponges** belong to phylum Porifera, which means "pore-bearers"—an apt description of these simple animals (figure 17.8). Unlike in other animals, a sponge's cells do not interact to form tissues. Their structural simplicity means that sponges bear little resemblance to the rest of the animal kingdom.

Habitat: Aquatic. Most are marine, although some live in fresh water.

Body Structure: A sponge's body is either radially symmetrical or asymmetrical. It is also hollow, and its body wall is riddled with pores. Examine the body wall in figure 17.8. Several types of cells are embedded in a jellylike matrix. Lining the inner surface of the body wall is an inner layer of flagellated "collar cells." (These cells strongly resemble choanoflagellates, the protists that may be the closest relatives to animals.) As the flagella on the collar cells wave, water moves into the sponge through the pores. Amoebocytes are cells that help digest food, distribute nutrients to other cells, and secrete skeletal components.

Feeding: Sponges are filter feeders. The water current produced by the collar cells carries not only oxygen but also bacteria and microscopic particles of organic matter—the sponge's food. (Contractile cells in the body wall can close the pores when the water contains too much sediment.) The collar cells trap and partially digest the food and pass it to amoebocytes. Water and wastes exit the sponge's central cavity through a large hole at the top.

Support and Movement: Protein fibers and sharp slivers (spicules) of silica or calcium carbonate provide support. Although some sponges can move very slowly, these animals are generally considered sessile, meaning they remain anchored to their substrate.

Reproduction: The sponge's porous body wall not only participates in feeding but also produces gametes. Sponges are hermaphrodites, which means the same individual makes both sperm and egg cells. The sperm are released into the water, but the animal retains its eggs. Meanwhile, sperm from nearby sponges enter its body through the pores. After fertilization, the zygote develops into a blastula, which is released and drifts briefly before settling into a new habitat. Some sponges also reproduce asexually by budding or fragmentation.

Defense: Spicules and toxic chemicals help sponges deter predators.

Effects on Humans: Some people use natural sponges in bathing. Also, the chemicals that protect sponges from predators may yield useful anticancer and antimicrobial drugs. Collecting sponges, however, can harm ecosystems.

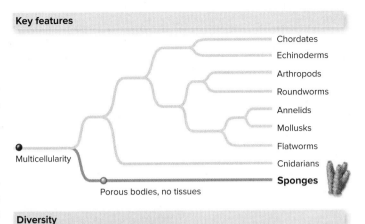

Key features

Diversity

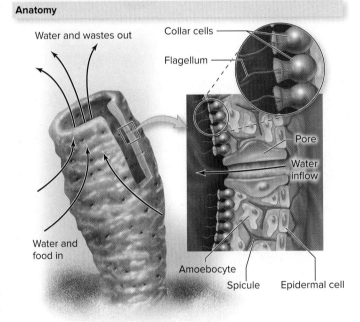

Anatomy

Figure 17.8 Sponges (Phylum Porifera).

Photos: (green sponge): ©Getty Images RF; (red sponge): ©Laurence F Tapper/YAY Micro/age fotostock RF

17.2 Mastering Concepts

1. What characteristics distinguish the sponges? (*Hint*: See figure 17.8.)
2. How is a sponge's body adapted to its aquatic habitat and sessile life?
3. Explain how the arrangement of cells in a sponge is adaptive to its feeding strategy.
4. In what ways are sponges important?

Key features

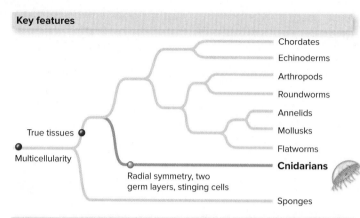

True tissues
Multicellularity

Chordates
Echinoderms
Arthropods
Roundworms
Annelids
Mollusks
Flatworms
Cnidarians

Radial symmetry, two
germ layers, stinging cells

Sponges

Diversity

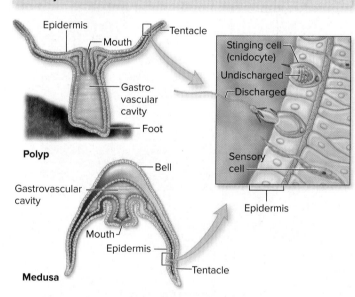

Jellyfish   Hydra   Coral reef   Sea anemones

Coral animal

Anatomy

Epidermis
Mouth
Tentacle
Gastro-vascular cavity
Foot

Polyp

Bell
Gastrovascular cavity
Mouth
Epidermis
Tentacle

Medusa

Stinging cell (cnidocyte)
Undischarged
Discharged
Sensory cell
Epidermis

Figure 17.9 Cnidarians (Phylum Cnidaria).

Photos: (jellyfish): ©Kevin Schafer/Alamy Stock Photo RF; (hydra): ©Ted Kinsman/Science Source; (coral reef): ©Comstock Images/PictureQuest RF; (coral animal): ©Leslie Newman & Andrew Flowers/Science Source; (anemones): ©Russell Illig/Getty Images RF

17.3 Cnidarians Are Radially Symmetrical, Aquatic Animals

Phylum Cnidaria (pronounced nigh-DARE-ee-ah) takes its name from the Greek word for "nettle," a stinging plant. The **cnidarians** all share the ability to sting predators and prey (figure 17.9).

Habitat: Aquatic (mostly marine, though some live in fresh water).

Body Structure: Cnidarians are radially symmetrical. One end of the body has an opening, the mouth, which is surrounded by a ring of tentacles. In a sessile cnidarian, called a **polyp,** a stalk holds the tentacles upward. In a **medusa,** the tentacles dangle downward from a free-swimming bell. In both body forms, the mouth leads to the dead-end gastrovascular cavity.

Diversity: Corals and sea anemones belong to a clade of cnidarians that exist exclusively as sessile polyps. A second clade contains hydras and the jellyfishes.

Feeding: Cnidarians are carnivores. Tentacles surrounding the mouth house cells called **cnidocytes,** which act as tiny harpoons that either inject venom or entangle the prey. The tentacles sense, grab, and sting passing prey, then stuff the meal into the gastrovascular cavity. Cells lining the digestive tract secrete enzymes that digest the food. After absorbing the nutrients, the animal ejects indigestible matter through the mouth.

Support and Movement: In all cnidarians, the two-layered body wall acts as a hydrostatic skeleton. Although a cnidarian's tissues do not form organs such as a brain or muscles, these animals can nevertheless make coordinated movements as they swim or capture prey. In the epidermis, groups of linked neurons called nerve nets coordinate the contraction of specialized cells. In this way, a jellyfish can force water out of its bell to propel itself through water. The same mechanism enables a sea anemone to stuff food into its gastrovascular cavity.

Reproduction: Cnidarians reproduce sexually and asexually.

Defense: Stinging cnidocytes are the main defense against predators.

Effects on Humans: Huge swarms of jellyfish are becoming increasingly common, presenting a nuisance for tourist destinations and fisheries. Jellyfish stings may cause skin irritation or cramps; a few species have toxins that can become lethal on contact. On the positive side, coral animals secrete calcium carbonate exoskeletons that have accumulated over many generations to build magnificent coral reefs. These unique ecosystems house many commercially important species of fishes and other animals, and they protect coastlines from erosion. As they build their calcium carbonate reefs, corals play an important role in the carbon cycle. A molecule originally isolated from corals (but now produced in the laboratory) is being developed into a sunscreen for human use. ⓘ *coral reefs,* section 19.3C

17.3 Mastering Concepts

1. What features do all cnidarians share?
2. Compare and contrast a polyp and a medusa.
3. How do cnidarians feed, move, and reproduce?
4. In what ways are cnidarians important?

17.4 Flatworms Have Bilateral Symmetry and Incomplete Digestive Tracts

Phylum Platyhelminthes includes the **flatworms.** (*Platy* means "flat," like a plate; *helminth* means "worm.") Some of these animals are surprisingly beautiful, whereas others look downright scary (figure 17.10).

Habitat: Free-living (usually aquatic) or parasitic on other animals.

Body Structure: Flatworms are bilaterally symmetrical animals that lack a coelom. Thanks to their flattened bodies, each cell is close to the body surface and can exchange materials with the environment.

Diversity: This phylum includes **free-living flatworms** (such as marine flatworms and planarians), **flukes,** and **tapeworms.**

Feeding: Free-living flatworms usually are predators or scavengers. The mouth opens into a muscular, tubelike pharynx at the body's midpoint; this structure delivers food to the highly branched gut, and it also ejects undigested food. In contrast, a parasitic tapeworm lacks a mouth and digestive system entirely. Instead, its hooks or suckers attach to the host's intestine, and the worm absorbs the host's already-digested food through its body wall. Other parasites, such as flukes, do have a digestive system.

Circulation and Respiration: Flatworms lack specialized circulatory and respiratory systems. CO_2 and O_2 simply diffuse through the body wall.

Excretion: Specialized structures maintain internal water balance and excrete nitrogenous wastes through pores on the body surface.

Nervous System: The flatworm nervous system can sense stimuli and coordinate movements. The planarian in figure 17.10, for example, has a ladderlike arrangement of nerve cords running the length of its body. The head end features a simple brain and sensory structures that detect touch, chemicals, and light.

Support and Movement: Flatworms have a hydrostatic skeleton and may creep or swim by contracting muscles in a rolling motion.

Reproduction: Many flatworms reproduce asexually. Free-living species, for example, may simply pinch in half and regenerate the missing parts. Sexual reproduction is also common. After mating, tapeworms release fertilized eggs that hatch once inside a new host. The larvae then mature in the host's body.

Defense: A tough outer layer protects parasitic flatworms against the host's digestive and immune systems; free-living forms secrete a protective mucus.

Effects on Humans: Worldwide, infections with flukes and tapeworms affect hundreds of millions of people and countless domesticated and wild animals.

Key features

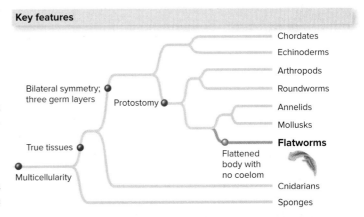

Diversity

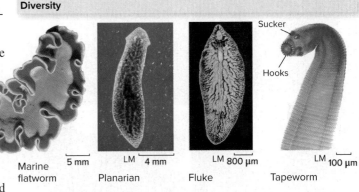

Marine flatworm — 5 mm
Planarian — LM 4 mm
Fluke — LM 800 μm
Tapeworm — LM 100 μm

Anatomy

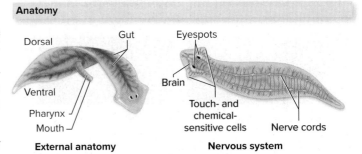

External anatomy — Dorsal, Gut, Ventral, Pharynx, Mouth

Nervous system — Eyespots, Brain, Touch- and chemical-sensitive cells, Nerve cords

Figure 17.10 Flatworms (Phylum Platyhelminthes).

Photos: (marine flatworm): ©Leslie Newman & Andrew Flowers/Science Source; (planarian): ©NHPA/M. I. Walker RF; (fluke): ©Volker Steger/Science Source; (tapeworm): ©Biophoto Associates/Science Source

17.4 Mastering Concepts

1. What features do all flatworms share?

2. How does the body shape of a flatworm enhance gas exchange?

3. How do flatworms eat, move, and reproduce?

4. In what ways are flatworms important?

Key features

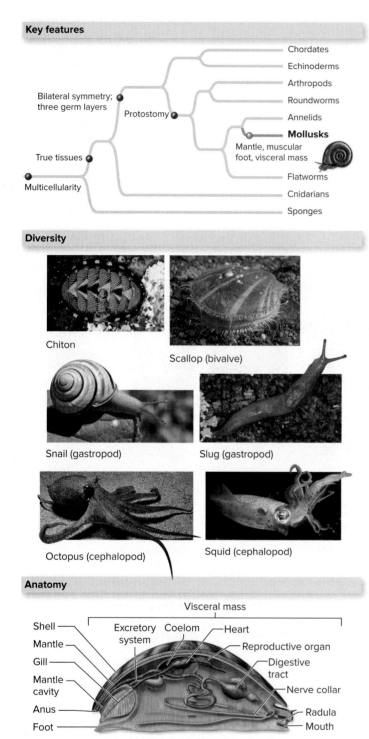

Diversity

Chiton

Scallop (bivalve)

Snail (gastropod)

Slug (gastropod)

Octopus (cephalopod)

Squid (cephalopod)

Anatomy

Visceral mass

Shell
Excretory system
Mantle
Gill
Mantle cavity
Anus
Foot

Coelom
Heart
Reproductive organ
Digestive tract
Nerve collar
Radula
Mouth

Figure 17.11 **Mollusks (Phylum Mollusca).**

Photos: (chiton) ©Kjell B. Sandved/Science Source; (scallop) ©Andrew J. Martinez/Science Source; (snail) ©Ivan Marjanovic/Shutterstock RF; (slug) ©McGraw-Hill Education/Steven P. Lynch; (octopus) ©Mark Conlin/Alamy Stock Photo; (squid) ©Comstock Images/PictureQuest RF

17.5 Mollusks Are Soft, Unsegmented Animals

Mollusks include many familiar animals on land, in fresh water, and in the ocean (figure 17.11). The word *mollusk* comes from the Latin word for "soft," reflecting the fleshy bodies in this phylum.

Habitat: Terrestrial, marine, and freshwater.

Body Structure: The **mantle** is a fold of tissue that secretes a shell in most species. A muscular **foot** provides movement, and an area called the **visceral mass** contains the complete digestive tract, along with the circulatory, excretory, and reproductive organs. Many mollusks have a **radula,** a tonguelike strap with teeth made of chitin (a tough polysaccharide). They use the radula to scrape food into their mouths.

Diversity: **Chitons** are marine animals with eight flat shells that overlap like shingles. **Bivalves,** such as clams and scallops, have two-part, hinged shells. **Gastropods** ("stomach-foot") include snails and slugs. The **cephalopods** ("head-foot") are marine mollusks such as octopuses and squids.

Feeding: Chitons scrape algae off rocks; bivalves filter food particles out of water; most gastropods are herbivores; and cephalopods are predators.

Circulatory System: Most mollusks have an open circulatory system in which blood flows throughout the body cavity. Cephalopods, however, have a closed circulatory system, with blood confined to vessels.

Respiratory System: Aquatic mollusks have gills; terrestrial snails and slugs have a lung derived from a space called the mantle cavity.

Excretory System: An excretory organ filters blood and produces urine.

Nervous System: The nervous system varies from simple and ladderlike to complex and cephalized. An octopus's nervous system includes a brain, a highly developed visual system, and an excellent sense of touch.

Support and Movement: All mollusks have a hydrostatic skeleton, and most have a supportive internal or external shell.

Reproduction: In sexual reproduction, fertilization may be external (in bivalves) or internal (in gastropods and cephalopods).

Defense: The hard shells of bivalves and snails protect against many predators. Squids and octopuses lack external shells, but they are speedy swimmers. They can also change their color and shape to match their background. When alarmed, they can squirt "ink" that cloaks their swift escape.

Effects on Humans: We harvest pearls from oysters, and we eat clams, mussels, oysters, snails, squids, and octopuses. However, bivalves can become poisonous if they accumulate pollutants or toxins. Snails and slugs are voracious consumers of garden plants, and invasive zebra mussels have disrupted aquatic ecosystems in the central United States. ⓘ *invasive species,* section 20.5A

17.5 Mastering Concepts

1. What adaptations do mollusks have in common?
2. How do mollusks feed, move, and protect themselves?
3. In what ways are mollusks important?

17.6 Annelids Are Segmented Worms

Earthworms and other segmented worms are **annelids.** The name of the phylum, Annelida, derives from the Latin word *annulus* ("little ring"), a reference to the segmented bodies of these animals (figure 17.12).

Habitat: Terrestrial, freshwater, and marine.

Body Structure: The most obvious characteristic of annelids is segmentation. They also have a complete digestive tract. A coelom separates the gut from the body wall and acts as a hydrostatic skeleton.

Diversity: Biologists recognize two main classes of annelids. One class contains the **leeches** and the **earthworms.** The other class, the **polychaetes,** contains the marine segmented worms.

Feeding: Earthworms ingest soil, digest the organic matter, and eliminate the indigestible particles. Some leeches suck blood from vertebrates, but most eat small prey such as arthropods, snails, or other annelids. Polychaetes are often filter feeders or deposit feeders, although some are predators.

Circulatory System: All except leeches have a closed circulatory system.

Respiratory System: Some polychaetes have feathery gills, but earthworms and leeches lack a specialized respiratory system. Because these animals exchange gases by diffusion through the body wall, they must remain moist.

Excretory System: The excretory system draws in fluid from the coelom, returns some ions and other substances to the blood, and discharges the waste-laden fluid outside the body through a pore.

Nervous System: The simple "brain" consists of a mass of nerve cells at the head end. These cells connect to a ventral nerve cord, with lateral nerves running through each segment. Together, the nerves stimulate contraction of the muscles in the body wall.

Support and Movement: Circular and longitudinal muscles push against the coelom as the worm crawls, burrows, or swims. Leeches crawl, inchworm-style, by using the suckers at each end of the body. Polychaetes have paddlelike structures that help them walk, swim, and dig.

Reproduction: Leeches and earthworms are hermaphrodites, so each individual has the reproductive organs of both sexes. In contrast, polychaetes have separate sexes.

Defense: Many annelids avoid predation by burrowing underground or in sediments. Polychaetes called tubeworms construct stationary tubes of chitin into which they can retract, and some also have powerful jaws.

Effects on Humans: Earthworms aerate and fertilize soil. Worm farms raise earthworms for sale as fishing bait or as soil conditioners. A blood-thinning chemical from leeches can stimulate circulation in surgically reattached digits and ears, and physicians sometimes apply leeches to remove excess blood that accumulates after damage to the nervous system.

17.6 Mastering Concepts

1. What features do all annelids share?
2. List examples of animals in each of the two classes of annelids.
3. How do annelids feed, exchange gases, and move?

Key features

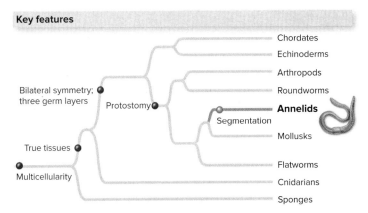

Diversity

Earthworm   Leech   Polychaete

Anatomy

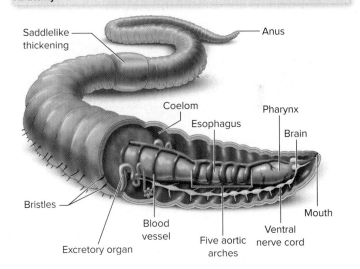

Figure 17.12 Annelids (Phylum Annelida).

Photos: (earthworm): ©David Chapman/Alamy Stock Photo; (leech): ©Edward Kinsman/Science Source; (polychaete): ©L. Newman & A. Flowers/Science Source

Key features

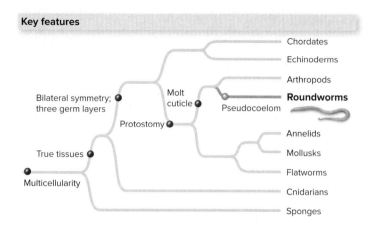

Diversity

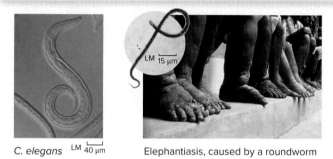

C. elegans LM |—| 40 μm

Elephantiasis, caused by a roundworm

Anatomy

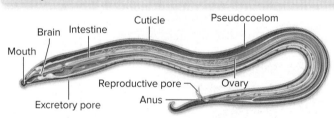

Figure 17.13 **Roundworms (Phylum Nematoda).**

Photos: (*C. elegans*): ©Sinclair Stammers/Science Source; (elephantiasis): ©R. Umesh Chandran, TDR, WHO/Science Source; (inset): Source: CDC/Dr. Mae Melvin

17.7 Nematodes Are Unsegmented, Cylindrical Worms

Most **roundworms** (phylum Nematoda) are barely visible to the unaided eye, but they are extremely abundant in every habitat (figure 17.13).

Habitat: Some nematodes parasitize plants or animals, but most are free-living in soil or in the sediments of aquatic ecosystems.

Body Structure: Nematodes are unsegmented worms with tapered ends and a complete digestive tract. An external layer of tissue secretes a tough cuticle, which is periodically molted.

Diversity: No one knows how many species of nematodes exist. According to one estimate, 80,000 species have been discovered, but hundreds of thousands more may remain undescribed.

Feeding: Free-living nematodes eat insect larvae, fungi, bacteria, or plants. Animal parasites may suck blood or consume food in the host's intestines; plant parasites use spearlike mouthparts to pierce cells and suck out the contents.

Circulation and Respiration: Nematodes lack specialized circulatory or respiratory organs. Instead, fluid in the pseudocoelom distributes nutrients, O_2, and CO_2 throughout the body.

Excretory System: Specialized cells maintain salt balance and eliminate nitrogenous wastes through an excretory pore.

Nervous System: Nematodes have a brain, which is connected to two nerve cords that run along the length of the body. Bristles and other sensory structures on the body surface enable the worms to detect touch and chemicals.

Support and Movement: The pseudocoelom acts as a hydrostatic skeleton. Nematodes are limited to back-and-forth, thrashing motions because only longitudinal (lengthwise) muscles act on the pseudocoelom. As a result, a nematode can neither crawl nor lift its body above its substrate.

Reproduction: Most species have separate sexes. Females produce huge numbers of tough, hard-to-kill eggs that survive drying and damaging chemicals.

Defense: The thick cuticle protects free-living and parasitic nematodes from the environment and from a host's immune system. Many nematodes can survive extreme heat, cold, or drying by entering a state of suspended animation.

Effects on Humans: The most familiar nematodes are parasites such as pinworms, hookworms, heartworms, and the *Trichinella* worms that are transmitted by eating undercooked pork. Nematodes also profoundly affect agriculture. Some are plant pathogens that cause diseases in and spread viruses among important food crops; others aid farmers by attacking insect pests. Finally, biologists use the nematode *Caenorhabditis elegans* in scientific research.

17.7 Mastering Concepts

1. What features do all roundworms share?
2. Contrast the bodies of roundworms, flatworms, and annelids.
3. How do nematodes feed and move?
4. How are roundworms important?

17.8 Arthropods Have Exoskeletons and Jointed Appendages

If diversity and sheer numbers are the measure of biological success, then the phylum Arthropoda certainly is the most successful group of animals. More than one million species of **arthropods** have been recorded already, and biologists speculate that this number could double. The number of individuals is also astonishing: According to some estimates, arthropods outnumber humans by more than 200 million to one.

Arthropoda means "jointed foot," a reference to the most distinctive feature of this phylum: their jointed appendages (figure 17.14). These appendages include not only feet and legs but also mouthparts, wings, antennae, copulatory organs, ornaments, and weapons.

In addition, all arthropods have an **exoskeleton,** a rigid outer covering that protects and supports the body. The arthropod exoskeleton is made mostly of chitin, protein, and (sometimes) calcium salts. Thin, flexible areas create moveable joints between body segments and within appendages. Although the exoskeleton is lightweight, it does have a drawback—to grow, an animal must molt and secrete a bigger one, leaving the animal vulnerable while its new exoskeleton is still soft. Molting is one feature that arthropods have in common with nematodes, the other group sharing their branch of the evolutionary tree.

A. Arthropods Have Complex Organ Systems

Habitat: Terrestrial, freshwater, and marine.

Body Structure: In addition to the distinctive exoskeleton and jointed appendages, arthropod bodies are segmented. Unlike in an annelid, an arthropod's segments do not all function alike. Instead, in many arthropods, the segments group into three major body regions: the head, thorax, and abdomen. Within each region, segments develop specialized functions such as walking or flying.

Feeding: Depending on the species, arthropods can eat almost anything, including dead organic matter, plant parts, and other animals.

Circulatory System: Arthropods have open circulatory systems. A heart propels the circulating fluid freely around the animal's organs.

Respiratory System: In most land arthropods, the body wall is perforated with holes that open into a series of branching tubes called tracheae, which transport oxygen and carbon dioxide to and from tissues. In contrast, aquatic arthropods have gills, and spiders and scorpions have stacked folds of tissue called book lungs.

Excretory System: Insects, spiders, and other terrestrial arthropods have specialized organs that collect and remove nitrogenous wastes while reabsorbing water. These organs deposit dry, nitrogen-rich waste into the posterior end of the digestive tract. The animal ejects the waste, together with undigested food, through its anus.

Nervous System: Thanks to a nervous system with a brain and ventral nerve cords, many arthropods are active, fast, and sensitive to their environment. Consider, for example, the speed with which a fly can avoid a swatter. Arthropod eyes, bristles, antennae, and other sensory systems can detect light, sound, touch, vibrations, air currents, and chemical signals. All of these clues help arthropods find food, identify mates, and escape predation.

Key features

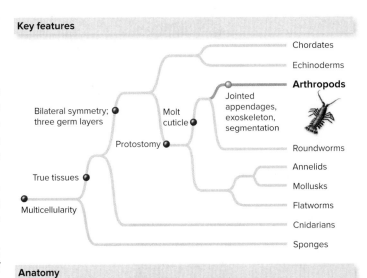

Anatomy

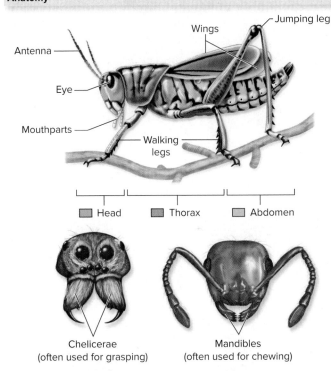

Figure 17.14 **Arthropods (Phylum Arthropoda).**

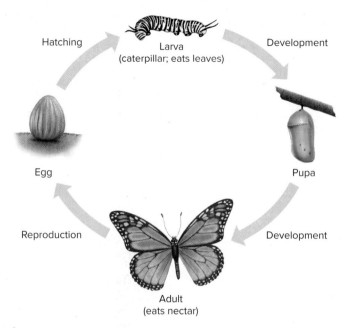

Figure 17.15 **Insect Metamorphosis.** A monarch butterfly's egg hatches into a leaf-eating caterpillar, which eventually develops into a pupa. After metamorphosis is complete, the adult butterfly emerges. Adults reproduce, initiating the next generation.

Support and Movement: The tough exoskeleton protects the animal and gives it its shape. Internal muscles span the joints between body segments and within appendages. This arrangement generates precise, forceful movements as the animal crawls, jumps, swims, or flies.

Reproduction: Most arthropods have separate sexes. In aquatic arthropods, both external and internal fertilization occur. But on land, gametes released in external fertilization would dry out; natural selection therefore has selected for internal fertilization in terrestrial arthropods. The male commonly produces a waterproof packet of sperm. The female takes the sperm packet into her body and typically lays the fertilized eggs, although mites and scorpions bear live young. Ants and bees tend their young, but in most other arthropods, parental care is minimal.

Some insects, such as crickets, have young that resemble the adults; these animals change only gradually from molt to molt. Most insect life cycles, however, include a **metamorphosis,** a developmental process in which the body changes greatly as the animal matures into an adult (figure 17.15). Adult females lay eggs; a hatched egg develops into a larva, which is an immature stage that does not resemble the adult of the species. Caterpillars and maggots are examples of larvae. During metamorphosis, the larvae transform into pupae and then into the corresponding adults (butterflies and houseflies). The larvae often live in different habitats and eat different foods from the adults, an adaptation that may help reduce competition between the generations.

Defense: Besides the protective exoskeleton, many arthropods can bite, sting, pinch, make noises, or emit foul odors or toxins that deter predators. Some have excellent camouflage that enables them to blend into their surroundings. Others have defensive behaviors; they may jump, run, roll into a ball, dig into soil, or fly away when threatened. Some moths unfurl wings with dramatic eyespots that startle or confuse predators.

Effects on Humans: Arthropods intersect with human society in about every way imaginable. Mosquitoes, flies, fleas, and ticks transmit infectious diseases as they consume human blood. Bees and scorpions sting, termites chew wood in our homes, and many insects destroy crops. Yet entire industries rely on arthropods—consider beeswax, honey, silk, and delicacies such as shrimp, crabs, and lobsters. Insects pollinate many plants, and spiders eat crop pests. The fruit fly, *Drosophila melanogaster,* is important in biological research. On a much smaller scale, dust mites eat flakes of skin that we shed as we move about our homes, while follicle mites inhabit our pores (see Why We Care 17.1).

B. Arthropods Are the Most Diverse Animals

Phylum Arthropoda is divided into five subphyla. One contains the 17,000 species of extinct **trilobites** (figure 17.16). Biologists infer the evolutionary relationships among the other four subphyla partly from mouthpart shape (see figure 17.14). Spiders, scorpions, and other **chelicerates** have grasping, claw-like mouthparts called chelicerae. The three subphyla of **mandibulates** have chewing, jawlike mouthparts termed mandibles.

Chelicerates: Spiders and Their Relatives Most chelicerates have two major body regions: an abdomen and a fused head and thorax. They also have chelicerae and four or more pairs of walking legs, but they lack antennae.

The two most familiar groups of chelicerates are horseshoe crabs and arachnids (figure 17.17). **Horseshoe crabs** are primitive-looking arthropods whose name refers to the hard, horseshoe-shaped exoskeleton, which covers a wide abdomen and a long tailpiece. The four species of horseshoe crabs are not true crabs, which are crustaceans. Humans have found an unusual way to

Figure 17.16 **Trilobites.** These extinct marine arthropods have three distinct body regions along the length of the body: a long central lobe plus flanking right and left lobes.

©Francois Gohier/Science Source

Horseshoe crab

a.

Tick

b.

Spider

c.

Scorpion

d.

Figure 17.17 Chelicerate Arthropods. Chelicerates include (a) horseshoe crabs, (b) ticks, (c) spiders, and (d) scorpions.

(a): ©Nature's Images/Science Source; (b): Source: CDC/James Gathany and William Nicholson; (c): ©Diana Lynne/Getty Images RF; (d): ©Digital Vision/PunchStock RF

exploit the horseshoe crab's blood, which contains a unique immune system compound that binds to bacteria. Technicians routinely use this compound to test medical supplies for bacterial contamination.

The more than 100,000 species of **arachnids** are eight-legged arthropods, including mites and ticks; spiders; harvestmen ("daddy longlegs"); and scorpions. The first terrestrial animals to leave fossils resembled scorpions, which explored the land during the Silurian period (about 440 MYA). Spiders make "silk" and use it to produce webs, tunnels, egg cases, and spiderling nurseries. Some spiderlings use silk driftlines to float to a new habitat.

Burning Question 17.1

Are there really only nine kinds of animals?

The nine phyla described in this chapter represent a diverse cross section of the 37 known animal phyla. But it would be a mistake to conclude that these nine groups contain the only important or interesting animals. Here are three additional invertebrate phyla, each representing microscopic animals with unusual features (figure 17.A):

Phylum Placozoa ("flat animals"): This phylum contains just one named species, *Trichoplax adhaerens*. An adult consists of a few thousand cells that are differentiated into just four cell types. The transparent, asymmetrical body resembles a microscopic sandwich, with an upper surface, a lower surface, and a connecting layer in between. Cilia enable the animal to glide or flow along a solid surface. Biologists first discovered placozoans living in aquaria, and the phylogenetic position of these strange animals has been debated ever since. *Trichoplax* has the smallest genome of any animal, and its body plan is even simpler than that of a sponge.

Phylum Rotifera ("wheel bearer"): Like placozoans, the rotifers have tiny, transparent, unsegmented bodies. But the 2000 or so species of rotifers are considerably more complex, with bilateral symmetry and a complete digestive tract. These animals are named for the wheel-like tufts of cilia that sweep particles of decomposing organic matter into the mouth. Rotifers inhabit fresh water and moist soil, and they can survive drying for years by suspending their metabolism. As with the placozoans, the phylogenetic position of rotifers remains controversial.

Phylum Tardigrada ("slow walker"): These charismatic little animals are commonly called "water bears," owing to their aquatic habitat and overall

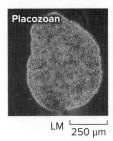
Placozoan
LM ⌐250 μm⌐

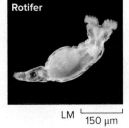
Rotifer
LM ⌐150 μm⌐

Tardigrade
SEM ⌐300 μm⌐
(false color)

Figure 17.A A Placozoan, a Rotifer, and a Tardigrade.

(placozoan): Eitel M, Osigus H-J, DeSalle R, Schierwater B (2013) "Global Diversity of the Placozoa." *PLoS ONE* 8(4): e57131. https://doi.org/10.1371/journal.pone.0057131; (rotifer): ©John Walsh/Science Source; (tardigrade): ©Andrew Syred/Science Source

shape. Their segmented bodies have eight legs, each ending in a claw. Covering the body is a cuticle of chitin, and the animal molts as it grows. These features place the tardigrades in a clade with arthropods and nematodes. More than 1000 species have been collected from habitats all over the world, from the poles to the tropics. Like rotifers, tardigrades can enter suspended animation, a state in which they can survive extreme cold or heat, desiccation, the vacuum of space, high pressure, or radiation doses that would kill a human.

Submit your burning question to
marielle.hoefnagels@mheducation.com

Miniglossary | Arthropod Diversity

Trilobites	Extinct group of arthropods with three-lobed bodies
Chelicerates	Arthropods with clawlike mouthparts (chelicerae) that are typically used for grasping; includes horseshoe crabs and arachnids
Mandibulates	Arthropods with jawlike mouthparts (mandibles) that are typically used for chewing; includes myriapods, crustaceans, and insects
Myriapods	Mandibulates with a head followed by repeating subunits, each with one pair (centipedes) or two pairs (millipedes) of appendages
Crustaceans	Mandibulates with variable body forms and two pairs of antennae; most are aquatic; includes crabs, lobsters, barnacles, and many other animals
Insects	Mandibulates with one pair of antennae, six legs, and a body divided into a head, a thorax, and an abdomen; includes more than 1 million species of beetles, ants, bees, flies, butterflies, and many others

Mandibulates: Centipedes and Millipedes About 13,000 species of **centipedes** and **millipedes** make up a group of terrestrial arthropods called myriapods (figure 17.18a). In these animals, the head features mandibles and one pair of antennae. The rest of the body is divided into repeating subunits, each with one pair (centipedes) or two pairs (millipedes) of appendages. Millipedes eat decaying plants and are generally harmless to humans, but centipedes are predators, and their venomous bite can be very painful.

Mandibulates: Crustaceans The **crustaceans** form a group of about 52,000 species, including crabs, shrimp, and lobsters (figure 17.18b). Smaller aquatic crustaceans include brine shrimp, water fleas *(Daphnia),* copepods, barnacles, and krill. Isopods, commonly known as pill bugs or "roly-polies," are the only terrestrial crustaceans; all other crustaceans live in water. Their bodies are extremely variable, but all have mandibles and two pairs of antennae.

Mandibulates: Insects Scientists know of well over 1 million species of **insects,** with many more awaiting formal description. All of these animals have mandibles; one pair of antennae; a body divided into a head, a thorax, and an abdomen; six legs; and (usually) two pairs of wings.

The ancestors of today's insects colonized land shortly after plants, and they diversified rapidly. Why did this group give rise to so many species? Biologists point to several possible explanations. For example, mutations in homeotic genes can modify the body segments of insects into seemingly unlimited variations; some biologists compare the insect body plan to the versatility of a Swiss army knife. Insect reproductive strategies may also have played a role. Insects have high reproductive rates, and their eggs can survive in dry habitats. As mentioned earlier, metamorphosis can also be adaptive. ⓘ *homeotic genes,* section 13.5

Wings may also partly account for insect success. Although flight later evolved independently in birds and in bats, insects were the first animals to fly. They use their wings to disperse to new habitats, escape predators, court mates, and find food that other animals cannot reach. Many of today's flowering plants evolved in conjunction with flying insects, trading nectar for rapid, efficient pollination services. ⓘ *pollination,* section 22.2C

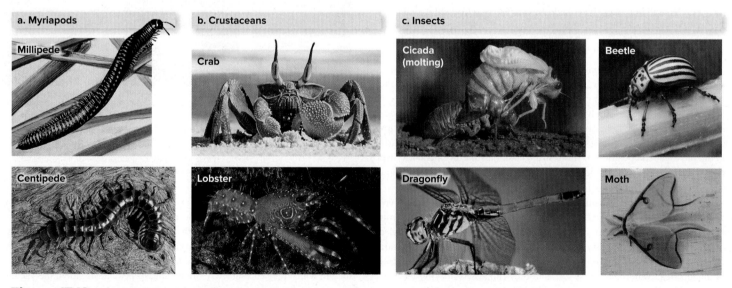

Figure 17.18 Mandibulate Arthropods. (a) Myriapods include millipedes and centipedes. (b) Crustaceans include crabs and lobsters. (c) Insects include cicadas, dragonflies, beetles, and moths.

(a, millipede): ©De Agostini Picture Library/Getty Images; (a, centipede): ©Tom McHugh/Science Source; (b, crab): ©Pete Atkinson/Photographer's Choice/Getty Images RF; (b, lobster): ©Photoshot Holdings Ltd/Alamy Stock Photo; (c, cicada): ©Rob Crandall/Shutterstock RF; (c, dragonfly): ©Thomas Shahan/Getty Images; (c, beetle): Source: USDA/Scott Bauer; (c, moth): ©McGraw-Hill Education/Steven P. Lynch

Animals Live Everywhere   Sponges   Cnidarians   Flatworms   Mollusks   Annelids

TABLE 17.1 Some Major Orders of Insects

Order	Examples
Zygentoma	Silverfish
Ephemeroptera	Mayflies
Odonata	Dragonflies and damselflies
Orthoptera	Crickets and grasshoppers
Phthiraptera	Lice
Hemiptera	Cicadas, aphids
Coleoptera	Beetles
Hymenoptera	Ants, wasps, bees
Lepidoptera	Moths, butterflies
Diptera	True flies
Siphonaptera	Fleas

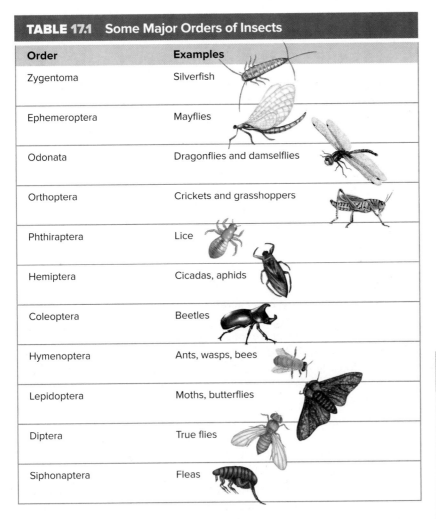

Whatever the explanation for their success, the variety of insect species alive today almost defies description (figure 17.18c and table 17.1). Familiar examples include silverfish, mayflies, dragonflies, roaches, crickets, grasshoppers, lice, cicadas, aphids, beetles, ants, wasps, bees, butterflies, moths, fleas, and flies. Insects range in size from wingless soil-dwellers less than 1 mm long to fist-sized beetles, foot-long walking sticks, and flying insects with 30-cm wingspans. Some extinct dragonflies were even larger—one had a wingspan about as long as an adult human's arm! Most of these species are terrestrial, but some insects live or reproduce in fresh water. The ocean, high altitudes, and extremely cold habitats are about the only places that are nearly devoid of insects.

17.8 Mastering Concepts

1. What features distinguish the arthropods?
2. How do arthropods use their jointed appendages?
3. Describe how arthropods feed, exchange gases, excrete metabolic wastes, sense their environment, move, reproduce, and defend themselves.
4. What are the functions of the exoskeleton?
5. In what ways are arthropods important?
6. How are chelicerates different from mandibulates?

Why We Care 17.1 | Your Tiny Companions

Even when you think you are alone, you aren't; your body may host a diverse assortment of arthropods. Head lice and body lice are biting insects that cause skin irritation. Ticks latch onto the skin and suck your blood, sometimes transmitting the bacteria that cause Lyme disease. The tiny larvae of chigger mites produce saliva that digests small areas of skin tissue, causing intense itching.

Lice, ticks, and chiggers are hard to ignore, but you may never notice one inconspicuous companion: the follicle mite, *Demodex* (figure 17.B). This arachnid, which is less than half a millimeter long, lives in hair follicles and nearby oil glands, where it eats skin secretions and dead skin cells. *Demodex* mites are by no means rare. Nearly everyone has them, and each follicle may house up to 25 of the tiny animals. (If you would like to see your own follicle mites, carefully remove an eyebrow hair or eyelash and examine it with a compound microscope.) Luckily, the infestation is typically symptomless, although occasionally the mites may cause a rash.

Figure 17.B Follicle Mites.
Tiny *Demodex* mites live in skin pores and hair follicles.

(eye): ©Ingram Publishing RF; (mite): ©Andrew Syred/Science Source

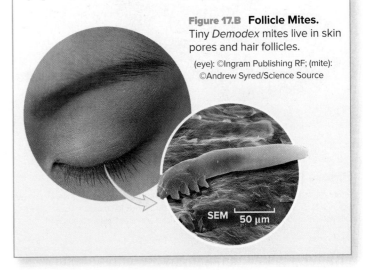

SEM 50 μm

Key features

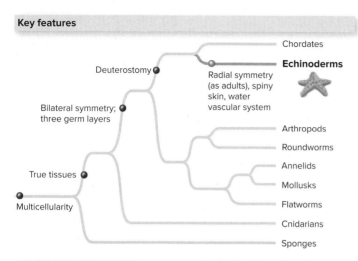

- Chordates
- **Echinoderms**
  - Radial symmetry (as adults), spiny skin, water vascular system
- Deuterostomy
- Bilateral symmetry; three germ layers
  - Arthropods
  - Roundworms
  - Annelids
  - Mollusks
  - Flatworms
- True tissues
  - Cnidarians
- Multicellularity
  - Sponges

Diversity

Sand dollar

Sea cucumber

Sea star   Adult   Larva   LM   500 µm   Sea urchin

Anatomy

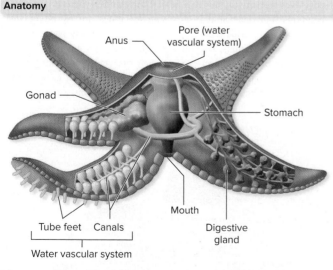

- Anus
- Pore (water vascular system)
- Gonad
- Stomach
- Tube feet   Canals
- Mouth
- Digestive gland
- Water vascular system

Figure 17.19 Echinoderms (Phylum Echinodermata).

Photos: (sand dollar): ©Pat Bonish/Alamy Stock Photo; (sea cucumber): ©Nancy Sefton/Science Source; (sea star adult): ©Comstock/Getty Images RF; (sea star larva): ©FLPA/D P Wilson/age fotostock; (sea urchin): ©Andrew J. Martinez/Science Source

17.9 Echinoderm Adults Have Five-Part, Radial Symmetry

The **echinoderms** (phylum Echinodermata) include some of the most colorful and distinctive sea animals (figure 17.19). Their name means "spiny skin."

Habitat: Marine.

Body Structure: Adult echinoderms have radial symmetry, with the body divided into five parts. Another unique feature of echinoderms is the **water vascular system,** a series of enclosed, water-filled canals that end in hollow tube feet. Coordinated muscle contractions extend and retract each foot, bending it from side to side or creating a suction-cup effect when applied to a hard surface.

Diversity: The most familiar echinoderms are sea stars, sea urchins, sand dollars, and sea cucumbers.

Feeding: Some echinoderms, such as sea stars, are predators. Sea cucumbers eat dead organic matter; sea urchins scrape algae from rocks.

Circulatory, Respiratory, and Excretory Systems: Echinoderms lack complex circulatory, respiratory, and excretory systems, but the versatile water vascular system fulfills many of the same functions. The animal's internal canals can exchange water with the ocean via a specialized pore. As a result, the thin-walled tube feet can function as gills, exchanging gases between ocean water and the internal fluid. Some metabolic wastes can also diffuse out of the tube feet and into the ocean.

Nervous System: Echinoderms lack heads and brains. Nerves extending down the arms or along the body wall connect with a central nerve ring surrounding the gut. Some tube feet detect touch, chemicals, or light.

Support and Movement: Tube feet allow echinoderms to glide slowly while maintaining a firm grip on the substrate. Sand dollars and some sea cucumbers burrow into soft sediments, and some brittle stars can swim by using their appendages as oars.

Reproduction: Echinoderms usually reproduce sexually. The larvae start out with bilateral symmetry, but then a group of cells assembles into a five-sided disc that turns inside out and consumes the remainder of the larva. The animal, now a tiny replica of the adult form, continues to grow to its mature size.

Defense: Besides spines, the skin of echinoderms may also be equipped with small pincers that deter predators. In addition, sea stars have internal skeletal plates; in sea urchins and sand dollars, these plates fuse into a protective shell. The soft-bodied sea cucumbers often produce poisonous chemicals, and many echinoderms can regenerate severed body parts.

Effects on Humans: People harvest some species of sea urchins for their reproductive organs, an edible delicacy called *uni.* Parts of some sea cucumbers are also edible. On the other hand, the crown-of-thorns sea star can cause painful wounds in humans who touch them. These animals eat corals and have wiped out large areas of Australia's Great Barrier Reef.

17.9 Mastering Concepts

1. What characteristics distinguish the echinoderms?
2. What is a water vascular system and why is it adaptive?
3. In what ways are echinoderms important?

17.10 Most Chordates Are Vertebrates

Many people find phylum Chordata to be the most interesting of all, at least in part because it contains humans and many of the animals that we eat, keep as pets, and enjoy observing in zoos and in the wild. The **chordates** are a diverse group of at least 60,000 species. From the tiniest tadpole to fearsome sharks and lumbering elephants, chordates are dazzling in their variety of forms.

The common ancestor of the chordates had several key features; this section concentrates on four of the most important ones. Every chordate has inherited these features and expresses each one at some point during its life (figure 17.20):

1. **Notochord:** The notochord is a flexible rod that extends along the length of a chordate's back. In most vertebrates, the notochord does not persist into adulthood but rather is replaced by the backbone that surrounds the spinal cord.

2. **Dorsal, hollow nerve cord:** The dorsal, hollow nerve cord is parallel to the notochord. In many chordates, the nerve cord develops into the spinal cord and enlarges at the head end, forming a brain.

3. **Pharyngeal slits (or pouches):** In most chordate embryos, slits or pouches form in the pharynx, the muscular tube that begins at the back of the mouth. Invertebrate chordates feed by straining food particles out of water that passes through the slits. In vertebrates, the pouches develop into gills, the middle ear cavity, or other structures.

4. **Postanal tail:** A muscular tail extends past the anus in all chordate embryos. In humans, chimpanzees, and gorillas, the body absorbs most of the tail before birth; only the tailbone remains as a vestige. In fishes, salamanders, lizards, cats, and many other species, adults retain the tail.

Figure 17.21 depicts the evolutionary relationships within phylum Chordata. One of the earliest branching points denotes the evolution of the **cranium,** a bony or cartilage-rich case that surrounds and protects the brain. The next branching point shows the appearance of **vertebrae,** a series of small bone or cartilage structures that make up the backbone. Vertebrae protect the spinal cord and provide attachment points for muscles, giving the animal a greater range of movement. (Vertebrates are chordates that have a backbone.)

Jaws are the bones that frame the entrance to the mouth. The development of hinged jaws from gill supports greatly expanded the ways that vertebrate animals could feed. In many species, the jaw includes teeth or a beak. These features enhance the animal's ability to grasp prey or gather small food items.

Lungs were another important evolutionary milestone. Most fishes have gills that absorb O_2 from water and release CO_2. In contrast, most air-breathing vertebrates have internal saclike **lungs** as the organs of respiration. Lungs are homologous to the swim bladders of bony fishes. These sacs, which allowed fishes to gulp air in shallow water, developed into air-breathing lungs in the ancestors of terrestrial vertebrates.

The evolution of limbs came next. **Tetrapods** are vertebrates with two pairs of limbs that enable the animals to walk on land (*tetrapod* means "four legs"). Amphibians, reptiles (including birds), and mammals are all tetrapods. Some animals classified as tetrapods, however, have fewer than four limbs. Snakes, for example, lack limbs entirely. The limbs of whales, dolphins, and sea lions are either modified into flippers or are too small to project from the body. Anatomical and molecular evidence, however, clearly links all of these animals to tetrapod ancestors (see figure 13.9 and section 13.6).

Key features

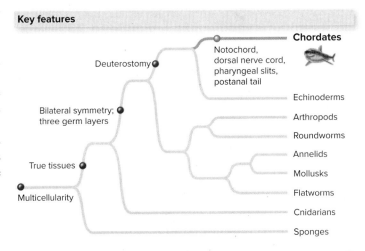

Anatomy

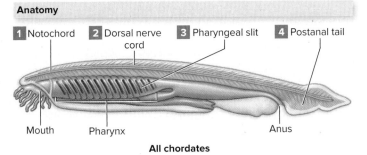

All chordates

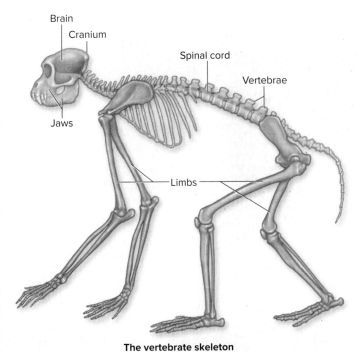

The vertebrate skeleton

Figure 17.20 Chordates (Phylum Chordata).

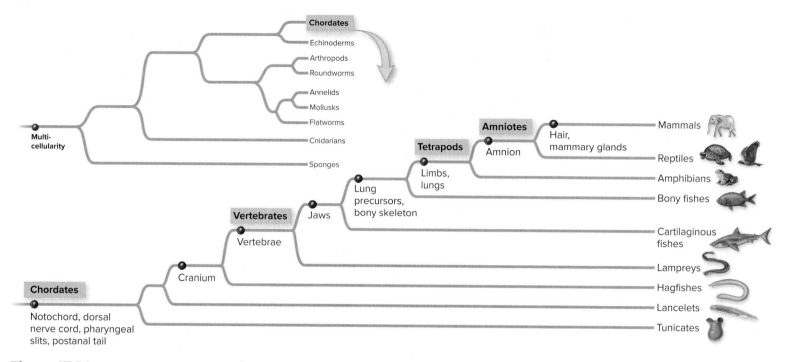

Figure 17.21 **Chordate Diversity.** Chordates include several groups of animals, including the invertebrate tunicates and lancelets and the better-known fishes, amphibians, reptiles (including birds), and mammals.

Loggerhead turtle hatching from egg

Another important event in the evolution of terrestrial chordates was the evolution of eggs that could be laid on dry land. The jellylike eggs of fishes and amphibians do not have a shell; they must remain in water, or the embryos inside will die. In contrast, reptiles and mammals can breed in arid habitats. The egg of a reptile (including a bird) has a leathery or hard outer layer, so the embryo does not dry out and die on land. This **amniotic egg** contains several membranes (the amnion, chorion, and allantois) that cushion the embryo, provide for gas exchange, and store metabolic wastes (figure 17.22). Meanwhile, the egg's yolk nourishes the developing embryo. The internal membranes of the amniotic egg are homologous to the protective structures that surround a developing fetus in the uterus of a female mammal. The **amniote** clade, which consists of reptiles and mammals, reflects this shared evolutionary history.

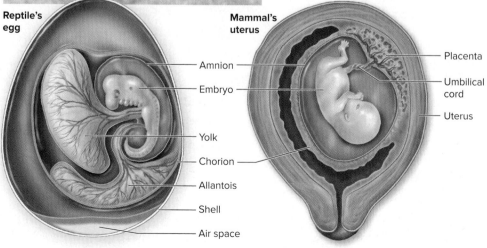

Reptile's egg

Mammal's uterus

Amnion
Embryo
Yolk
Chorion
Allantois
Shell
Air space

Placenta
Umbilical cord
Uterus

Figure 17.22 **The Amnion.** The amnion is a sac that encloses the developing embryo of a reptile or mammal. In an amniotic egg, the embryo is encased in a protective shell, and it is supported internally by three membranes—the amnion, allantois, and chorion. Placental mammals also enclose embryos in an amnion. (The allantois, which is not shown for the mammal, becomes part of the umbilical cord.)

Photo: ©Zankl/Nature Picture Library/Getty Images

Animals Live Everywhere   Sponges   Cnidarians   Flatworms   Mollusks   Annelids

The regulation of body temperature is an additional characteristic that is important in animal biology. Thermoregulation strategies vary widely, but in general, biologists often classify animals as ectotherms or endotherms. The body temperature of an **ectotherm** tends to fluctuate with the environment; these animals lack internal mechanisms that keep their temperature within a narrow range. Invertebrates, fishes, most amphibians, and most nonavian reptiles are ectotherms. Many behaviors, such as basking in the sun or burrowing into the ground, help an ectotherm adjust its temperature. **Endotherms** maintain their body temperature mostly by using heat generated from their own metabolism. Birds and mammals are endotherms, as are a few other types of animals. Endothermy requires an enormous amount of energy, which explains why birds and mammals must eat so much more food than ectotherms of the same size. Fur and feathers help retain heat in these animals.

17.10 Mastering Concepts

1. What are four key defining characteristics of chordates?
2. Which chordates have a cranium, and which are also vertebrates?
3. How did the origin of jaws, lungs, limbs, and the amnion affect the course of vertebrate evolution?
4. What is the difference between an ectotherm and an endotherm?

17.11 Chordate Diversity Extends from Water to Land to Sky

The first part of this chapter explored differences among internal organ systems, highlighting the dramatic transitions between the simplest animals and the most complex. The rest of this chapter takes a slightly different approach. Most chordates have complex respiratory, digestive, excretory, and nervous systems. We therefore focus here on the evolutionary transitions between the main groups of chordates and on the diversity of animals within the phylum.

A. Tunicates and Lancelets Are Invertebrate Chordates

Tunicates and lancelets form two subphyla of invertebrate chordates; their bodies have neither a cranium nor vertebrae (figure 17.23). **Tunicates** are sessile marine animals that resemble a bag with two siphons. Cilia pull water in through one siphon. The animal extracts oxygen and food particles, and the water exits through the other siphon. A covering called a tunic protects the body. Only the free-swimming tunicate larva, which resembles a tadpole, has all four chordate characteristics.

Lancelets resemble small, eyeless fishes with translucent bodies. They live in shallow seas, with the tail buried in sediment and the mouth extending into the water. They feed by filtering food particles out of the water. These animals clearly display all four major chordate characteristics, as well as inklings of the muscular and nervous systems that appear in the vertebrates.

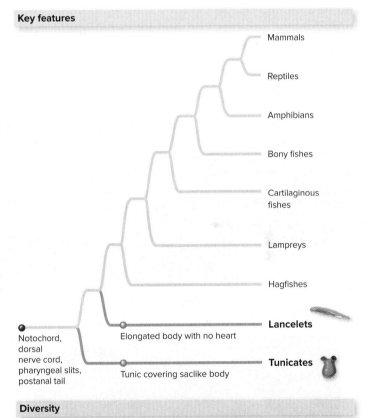

Key features

- Mammals
- Reptiles
- Amphibians
- Bony fishes
- Cartilaginous fishes
- Lampreys
- Hagfishes
- **Lancelets**
- **Tunicates**

Notochord, dorsal nerve cord, pharyngeal slits, postanal tail

Elongated body with no heart

Tunic covering saclike body

Diversity

Tunicates

Lancelet

Figure 17.23 Tunicates and Lancelets.

Photos: (blue tunicate): ©Nancy Sefton/Science Source; (pink tunicates): ©Janna Nichols; (lancelet): ©Natural Visions/Alamy Stock Photo

Key features

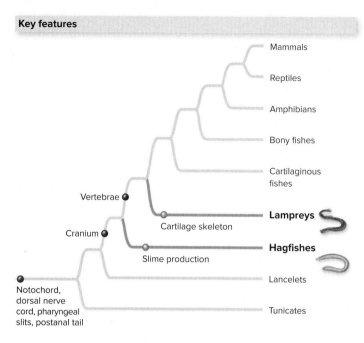

Diversity

Hagfish

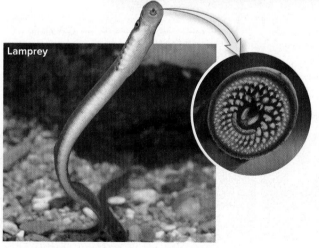

Lamprey

Figure 17.24 **Hagfishes and Lampreys.**

Photos: (hagfish mouth): ©Steven Senne/AP Images; (hagfish): ©Mark Conlin/Alamy Stock Photo; (lamprey): ©David Hosking/Alamy Stock Photo; (lamprey mouth): ©Gena Melendrez/Shutterstock RF

B. Hagfishes and Lampreys Have a Cranium but Lack Jaws

Hagfishes and lampreys are two groups of chordates that share several similarities (figure 17.24): They have long, slender bodies with gills and specialized sense organs clustered near the head end, and their mouths lack jaws. Scientists are still debating the relationship between hagfishes and lampreys; in this section we discuss them together as jawless, fishlike animals.

The long, slender **hagfish** looks something like an eel. Even though hagfish skin is marketed as "eel skin" in boots and wallets, hagfishes are not eels. In a hagfish, cartilage makes up the cranium and supports the tail. But because vertebrae do not surround the nerve cord, hagfishes are not vertebrates.

Hagfishes live in cold ocean waters, eating marine invertebrates such as shrimp and worms or using their raspy tongues to scavenge the soft tissues of dead or near-dead animals. These animals have some unusual abilities. They can slide their flexible bodies in and out of knots to pull on food, escape predation, or clean themselves. Hagfishes are also called "slime hags," in recognition of the glands that release copious amounts of a sticky white slime when the animal is disturbed.

Lampreys are the simplest organisms to have cartilage around the nerve cord, so they are vertebrates. The adults eat small invertebrates, although some species use their suckers to consume the blood of fish. Over the past century, sea lampreys have ventured beyond their natural Lake Ontario range into the other Great Lakes, where they have been largely responsible for the decline in populations of lake trout and whitefish.

C. Fishes Are Aquatic Vertebrates with Jaws, Gills, and Fins

Fishes are the most diverse and abundant of the vertebrates, with more than 30,000 known species that vary greatly in size, shape, and color (figure 17.25). They occupy nearly all types of water, from fresh to salty, from clear to murky, and from frigid to warm, although they cannot tolerate hot springs.

Fishes play important roles in their aquatic habitats. They graze on algae, scavenge dead organic matter, or prey on other animals, eating everything from mosquito larvae and other small invertebrates to one another. Tuna and many other fish species are also an important source of dietary protein for people (and their pets) on every continent. Angling for trout, bass, salmon, and other fishes remains a popular sport. Fishes also inspire a wide range of emotions, from an intense fear of sharks to the peace and tranquility that come from watching tropical fish in a home aquarium.

Fishes originated some 500 MYA from an unknown ancestor with jaws, gills, and paired fins. Several features arose in fishes that would have profound effects on vertebrate evolution. A segmented backbone, with its multiple muscle attachment points, expanded the range of motion. Jaws opened new feeding opportunities, which in turn selected for a more complex brain that could develop a hunting strategy or plan an escape route.

Two of the adaptations that enabled vertebrates to thrive on land originated in fishes: lungs and limbs. Lungs developed in a few species of fishes, and the air-breathing descendants of these animals eventually colonized the land. No fish has true limbs, but some fishes have pectoral fins with stronger bones and more flesh than the delicate, swimming fins of other fishes. These robust fins may have enabled the ancestors of tetrapods to move along the sediments of their shallow-water homes. Whatever their original selective advantage, pectoral and pelvic fins eventually evolved into the limbs that define the tetrapods.

Biologists divide the fishes into two major groups: cartilaginous fishes and bony fishes. The **cartilaginous fishes,** the most ancient group, include sharks, skates, and rays. As the name implies, their skeletons are made of cartilage. Sharks are the most notorious of the cartilaginous fishes. Although some sharks feed on plankton, the carnivorous species are famous for their ability to detect blood in the water. Extending along each side of a fish is a **lateral line,** a sense organ that detects vibration in nearby water and helps the animal to find prey and escape predation. Some species of cartilaginous fishes must swim continuously to keep water flowing over their gills.

The **bony fishes** include 96% of existing fish species. They have skeletons of bony tissue reinforced with mineral deposits of calcium phosphate. Like sharks, bony fishes have a lateral line system. Unlike cartilaginous fishes, however, the bony fishes have a hinged gill covering that can direct water over the gills, eliminating the need for constant swimming. In addition, most bony fishes have a swim bladder that helps the animal to adjust its buoyancy.

Bony fishes are divided into two classes:

- The **ray-finned fishes** include nearly all familiar fishes: eels, minnows, catfish, trout, tuna, salmon, and many others. They have fan-shaped fins, each consisting of a thin sheet of tissue supported by parallel rays made of bone. Their diversity and abundance reflect their superb adaptations to a watery world.

- The **lobe-finned fishes** are the bony fishes most closely related to the tetrapods, based on the anatomical structure of their fleshy paired fins consisting of bone and muscle. This group includes the lungfishes and the coelacanths. **Lungfishes** have lungs that are homologous to those of tetrapods. During droughts, a lungfish burrows into the mud beneath stagnant water, gulping air and temporarily slowing its metabolism. **Coelacanths** are called "living fossils"; they originated during the Devonian period and remain the oldest existing lineage of vertebrates with jaws.

Key features

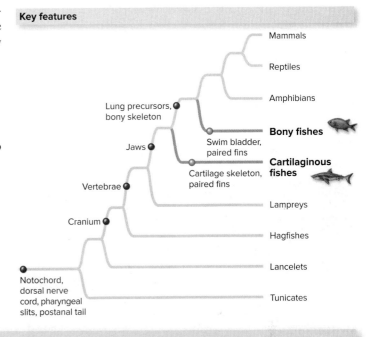

Diversity

Stingray (cartilaginous fish)

Shark (cartilaginous fish)

Ray-finned fish (bony fish)

Lungfish (bony fish)

Coelacanth (bony fish)

Figure 17.25   The Fishes.

Nematodes   Arthropods   Echinoderms   Chordates: Features   Chordates: Diversity   Human Evolution

Key features

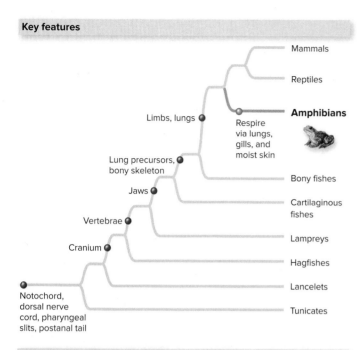

- Mammals
- Reptiles
- **Amphibians** — Respire via lungs, gills, and moist skin
- Limbs, lungs
- Lung precursors, bony skeleton
- Bony fishes
- Jaws
- Cartilaginous fishes
- Vertebrae
- Lampreys
- Cranium
- Hagfishes
- Lancelets
- Notochord, dorsal nerve cord, pharyngeal slits, postanal tail
- Tunicates

Diversity

Frog   Caecilian

Salamander

Figure 17.26 **The Amphibians.**

Photos: (frog) ©Irina Kozorog/Shutterstock RF; (caecilian): ©E.D. Brodie Jr.; (salamander): ©Suzanne L. Collins & Joseph T. Collins/Science Source

D. Amphibians Live on Land and in Water

The word **amphibian** is Greek for "double life," referring to the ability of these tetrapod vertebrates to live in fresh water and on land (figure 17.26). Amphibians are probably the least familiar vertebrates because most people do not eat them or keep them as pets. Nevertheless, they are important in ecosystems, controlling algae and populations of insects that transmit human disease. Scientists are also studying toxins in amphibian skin as possible painkilling drugs.

Amphibians began colonizing the land about 375 MYA. Life on land offered amphibian ancestors space, shelter, food, and plentiful oxygen, compared with the crowded aquatic habitat. But the land also presented new challenges. The animals faced wider swings in temperature, and delicate gills collapsed without the buoyancy of water. The new habitat therefore selected for new adaptations. Lungs improved, and circulatory systems grew more complex and powerful. The skeleton became denser and better able to withstand the force of gravity. Natural selection also favored acute hearing and sight, with tear glands and eyelids keeping eyes moist.

Yet amphibians retain a strong link to the water. Amphibian eggs, which lack protective shells and membranes, will die if they dry out. Also, the larvae respire through external gills, which require water. Although adults typically have lungs, these organs are not very efficient in amphibians. The thin skin provides an additional gas exchange surface and must therefore remain moist.

Today's amphibians include three main lineages: frogs; salamanders and newts; and caecilians. Most amphibian species are **frogs,** a group that includes the smooth-skinned "true frogs" and the warty-skinned toads. In most species, the young start out as legless, aquatic tadpoles that feed on algae. As they mature, tadpoles typically undergo a dramatic change in body form—a metamorphosis. They develop legs and lungs, lose the tail, and acquire carnivorous tastes. **Salamanders** and newts have tails and four legs, so they resemble lizards. Both adults and young are carnivores, eating arthropods, worms, snails, fish, and other salamanders. The limbless **caecilians,** on the other hand, resemble giant earthworms. Most species burrow under the soil in tropical forests, but a few inhabit shallow freshwater ponds. Caecilians are carnivores, eating insects and worms.

E. Reptiles Were the First Vertebrates to Thrive on Dry Land

The changeability of scientific knowledge is evident in any modern discussion of **reptiles** and **birds** (figure 17.27). The word *reptile* traditionally referred only to snakes, lizards, crocodiles, and other amniotes with dry, scaly skin. Birds had feathery body coverings and were considered a separate lineage. But that point of view has changed. We now know that birds form one of several clades of reptiles. As a consequence, modern use of the term *reptile* includes both the nonavian reptiles and the birds.

Reptiles evolved from amphibians about 310 to 320 MYA. They dominated animal life during the Mesozoic era, until their decline beginning 65 MYA. Although many reptile species survived to the present day, many others are known only from fossils. The extinct groups include the marine ichthyosaurs and plesiosaurs, the flying pterosaurs, and the terrestrial dinosaurs (see Burning Question 17.2). We do not know what caused the mass extinction that ended the Mesozoic era, but it coincides with an asteroid impact near the Yucatán peninsula. Biologists estimate that photosynthesis was almost nonexistent for years, as debris from the impact circulated in the atmosphere and blocked the sunlight. With the base of the food chain gone, many animals must have starved.

Unlike their amphibian ancestors, most reptiles have a suite of adaptations that enable them to live and reproduce on dry land. Tough scales reduce water loss from the skin, and the kidneys excrete only small amounts of water. Internal fertilization and amniotic eggs mean that reptiles do not require moist habitats to reproduce (see figure 17.22). Finally, reptiles have greater lung capacity and more efficient circulation than their aquatic ancestors.

Nonavian Reptiles The nonavian reptiles include lizards and snakes; turtles and tortoises; and crocodilians. Like fishes and amphibians, the nonavian reptiles are typically ectothermic.

Almost 95% of nonavian reptile species are **snakes** or **lizards.** These animals periodically shed their scaly skin. They also have strong, flexible jaws that enable them to consume large prey. The main difference between them is that lizards usually have legs and snakes do not.

With a reputation for being "slow and steady," the aquatic **turtles** and their terrestrial relatives (tortoises) may not seem to have a recipe for success. Yet they have persisted in marine, freshwater, and terrestrial habitats since the Triassic period. This group's trademark feature is its shell, made of bony plates and a covering derived from the animal's epidermis. The shell's plates are fused to the animal's vertebrae and ribs, so it forms an integral part of the skeleton.

The **crocodilians** (crocodiles, alligators, and their relatives) are carnivores that live in or near water. Their horizontally held heads have eyes on top and nostrils at the end of the elongated snout. Heavy scales cover their bodies. These reptiles look primitive, yet their behaviors (including nest-guarding and caring for their young) are comparable to those of birds.

Nonavian reptiles are an important link in ecosystems, controlling populations of rodents and insects while providing food for owls and other predatory birds. In some parts of the world, reptiles have a role in the economy; the skins of farm-raised snakes and crocodiles are used in boots, belts, and wallets, and some restaurants serve alligator meat. In addition, some people keep snakes, lizards, or turtles as pets.

Birds The behavioral similarities between crocodilians and birds are unsurprising in light of evolutionary history. Birds, dinosaurs, and crocodilians all belonged to a reptilian group called archosaurs, of which only the birds and crocodilians survive today.

Of course, birds have unique features that set them apart from other reptiles. Most birds can fly, thanks to anatomical adaptations including wings and a tapered body with a streamlined profile. Their lightweight bones are hollow, with internal struts that add support. The powerful heart and unique lungs supply the oxygen that supports the high metabolic demands of flight. In addition, unlike other reptiles, birds are endothermic.

Birds are the only modern animals that have **feathers,** which provide insulation and enable birds to fly. Feathers are also important in mating behavior, as anyone who has watched a peacock show off his plumage can attest. Like a snake's scales, a feather is built of the protein keratin.

Today, birds are a part of everyday human life. People eat hundreds of millions of chickens and turkeys every year, along with countless chicken eggs. We keep caged birds as pets, and we use feathers in everything from hats to blankets. Songbirds enrich the lives of many birdwatchers. Birds are important in ecosystems as well. Some pollinate plants and disperse fruits and seeds, whereas others eat rodents, insects, and other vermin. But birds can also be pests. Starlings and pigeons are a nuisance in cities, fouling buildings and sidewalks with their droppings and speeding the rusting of bridges. Moreover, ducks and other domesticated birds transmit bird flu and other diseases to humans.

Key features

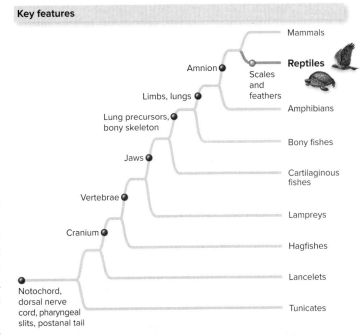

Diversity

Snake   Lizard

Turtle   Alligator   Bird

Figure 17.27 **The Reptiles.**

Photos: (snake): ©Dorling Kindersley/Getty Images; (lizard): ©Tom Horton, Further To Fly Photography/Getty Images RF; (turtle): ©Ed Reschke/Photolibrary/Getty Images; (alligator): ©LaDora Sims/Flickr/Getty Images RF; (bird): ©Image Source RF

Key features

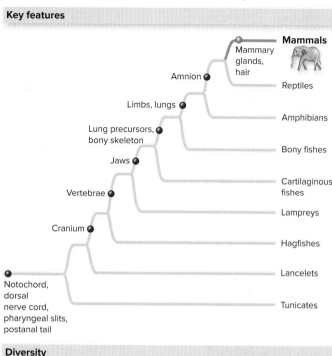

- Mammals — Mammary glands, hair
- Amnion
- Limbs, lungs
- Lung precursors, bony skeleton
- Jaws
- Vertebrae
- Cranium
- Notochord, dorsal nerve cord, pharyngeal slits, postanal tail

Reptiles

Amphibians

Bony fishes

Cartilaginous fishes

Lampreys

Hagfishes

Lancelets

Tunicates

Diversity

Monotremes

Platypus

Echidna

Marsupials

Kangaroo

Opossum

Placental mammals

Human

Dolphin

Fox

Figure 17.28 **The Mammals.**

F. Mammals Are Warm, Furry Milk-Drinkers

Mammals are by far the most familiar vertebrates, not only because we *are* them but also because we surround ourselves with them (figure 17.28). We keep dogs and cats as pets. Farmers raise cows, pigs, and goats for their meat or milk. Lambs and adult sheep provide meat and wool, and leather from cows makes up everything from upholstery to shoes. Horses, oxen, mules, and dogs are important work animals. Many people enjoy hunting deer for food and sport, and trappers kill mink, fox, beaver, and other mammals for their fur.

Mammals also play important roles in ecosystems. Coyotes, wolves, and foxes keep populations of herbivorous deer, rodents, rabbits, and other mammals in check. Some bats eat insects, and others pollinate plants. On the other hand, pests such as rats, mice, and skunks thrive alongside human populations. Some transmit diseases, including hantavirus and rabies.

Mammals arose late in the Triassic period (about 200 MYA). The common ancestor of the mammals had **mammary glands,** structures that secrete milk in the female. (The word *mammal* derives from the Latin *mammae* for "breast.") Infant mammals are nourished by their mother's milk. In addition, mammals produce hair, which is composed of keratin and helps conserve body heat. Even whales and dolphins have hair at birth, but they lose it as they mature into their streamlined shapes.

Most mammals were small until after the mass extinction that occurred 65 MYA. The loss of so many reptiles paved the way for the rapid diversification of many larger species of mammals. At the same time, flowering plants became increasingly prominent, providing new types of food and habitats for mammals. ⓘ *flowering plants,* section 16.5

Biologists divide today's mammals into two subclasses. The **monotremes** are mammals that lay eggs, such as the platypus and the echidna; when a helpless young monotreme hatches, it crawls along its mother's fur until it reaches milk-secreting pores in the skin. The other subclass contains the live-bearing marsupials and placental mammals. **Marsupials,** such as kangaroos and opossums, give birth to tiny, immature young about 4 to 5 weeks after conception. In many marsupials, the babies crawl from the mother's vagina to a pouch, where they suckle milk and continue developing. Some species, however, have poorly developed pouches, and the young drink from exposed nipples.

The **placental mammals** are the most diverse group. In these species, the young develop inside the female's uterus, where a **placenta** connects the maternal and fetal circulatory systems (see figure 17.22). The two largest groups of placental mammals are rodents and bats, but the group also includes carnivores (dogs and cats), hoofed mammals, elephants, and many other familiar animals. Some groups reinvaded the water, including manatees, otters, seals, beavers, hippos, whales, and dolphins. Humans are placental mammals as well. We belong to an order, Primates, that arose some 60 MYA. Section 17.12 describes the evolution of primates.

17.11 Mastering Concepts

1. What is the relationship among tunicates, lancelets, and the vertebrate chordates?
2. Make a table comparing the features of each vertebrate group.
3. Describe the adaptations that mark the transition from fishes to amphibians, reptiles, and mammals.
4. What characteristics place the birds within the reptiles, and how are birds different from other reptiles?

17.12 Fossils and DNA Tell the Human Evolution Story

In many ways, humans are Earth's dominant species. We are not the most numerous—more microbes occupy one person's intestinal tract than there are people on Earth. In the short time of human existence, however, we have colonized most continents, altered Earth's surface, eliminated many species, and changed many others to fit our needs. Where did we come from?

A. Humans Are Primates

If you watch the monkeys or apes in a zoo for a few minutes, it is almost impossible to ignore how similar they seem to humans (figure 17.29). Young ones scramble about and play. They sniff and handle food. Babies cling to their mothers. Adults gather in small groups or sit quietly, snoozing or staring into space.

It is no surprise that we see ourselves reflected in the behaviors of monkeys and apes. All **primates**—including monkeys, apes, and humans—share a suite of physical characteristics. First, primates have grasping hands with opposable thumbs that can bend inward to touch the pads of the fingers. Some primates also have grasping feet with opposable big toes. Second, a primate's fingers and toes have flat nails instead of claws. Third, eyes set in the front of the skull give primates binocular vision with overlapping fields of sight that produce excellent depth perception. Fourth, the primate brain is large in comparison with body size.

Compared with many other groups of mammals, primate anatomy is unusually versatile. For instance, bat wings are useful for flight but not much else; likewise, horse hooves are best for fast running. In contrast, primates have multipurpose fingers and toes that are useful not only for locomotion but also for grasping and manipulating small objects. Primate limbs are similarly versatile.

The primate lineage contains three main groups: prosimians, monkeys, and apes (figure 17.30). *Prosimian* is an informal umbrella term for lemurs, aye-ayes, lorises, tarsiers, and bush babies. Monkeys are divided into two main

Figure 17.29 Two Examples of Primates. The behavioral similarities between baboons and humans are evident in many ways, including parental care.

(baboons): ©Zoonar/Frauke Scholz/age fotostock; (humans): ©Martial Colomb/Getty Images RF

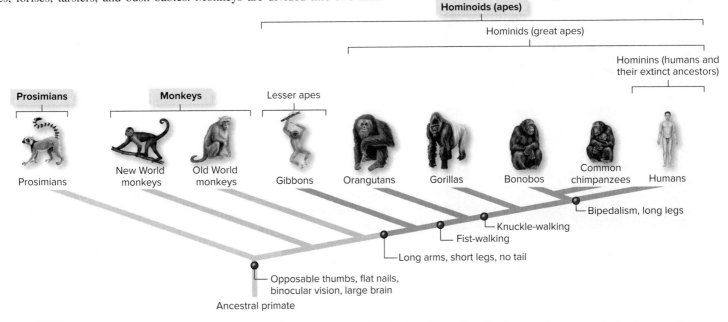

Figure 17.30 Primate Lineages. This evolutionary tree shows the physical traits that differentiate the three main groups of primates: prosimians, monkeys, and hominoids (apes). Molecular data support this hypothesis of the evolutionary relationships among primates.

Figure 17.31 **Clues from Bones.** The skeleton on the left, from a human, shares many similarities with the gorilla skeleton on the right.

©Tom McHugh/Science Source

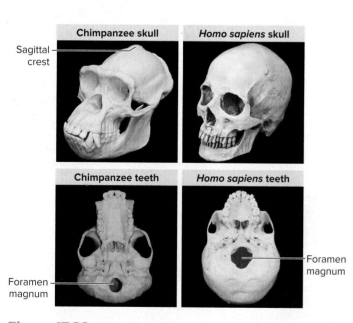

Sagittal crest

Chimpanzee skull

Homo sapiens skull

Chimpanzee teeth

Homo sapiens teeth

Foramen magnum

Foramen magnum

Figure 17.32 **Skulls and Teeth.** The skulls and teeth of a chimpanzee and a human reveal details about posture, diet, and jaw strength. Biologists compare fossils of extinct species to bones of existing primates to learn how our ancestors lived.

(chimp skull and teeth): Skulls Unlimited International, Inc ©David Liebman Pink Guppy; (*Homo sapiens* skull): ©merlinpf/Getty Images RF; (*Homo sapiens* teeth): ©McGraw-Hill Education/Christine Eckel

groups: Old World (native to Africa and Asia) and New World (native to South and Central America). The apes (also called hominoids) are further divided into two groups. One contains the gibbons, or "lesser apes." The other (the hominids) contains all of the "great apes": orangutans, gorillas, chimpanzees (including bonobos), and humans. **Hominins** are humans and their extinct ancestors on the human branch of the primate evolutionary tree.

While studying figure 17.30, keep in mind two important concepts. First, note that humans are not descended from other groups of modern apes (see Burning Question 14.2). Instead, all living humans and chimpanzees share a common ancestor and diverged from that ancestor about 6 MYA. Second, note that gibbons, orangutans, gorillas, and chimpanzees are not "less evolved" than humans. All living species are on an equal evolutionary footing, although some may belong to older lineages.

B. Anatomical and Molecular Evidence Documents Primate Relationships

By examining the physical characteristics of skeletons, paleontologists have learned much about the evolutionary relationships among primates (figure 17.31). Most hominin fossils consist of bones and teeth. Comparing these remains with existing primates reveals surprisingly detailed information about locomotion and diet. For this reason, knowledge of primate skeletal anatomy is essential to interpreting fossils of our extinct ancestors.

Locomotion Adaptations related to locomotion are among the most important characteristics in ape skeletons. Brachiation, for example, is swinging from one arm to the other while the body dangles below. Many apes move through the treetops in this way; in contrast, monkeys run on all fours along the tops of branches. Orangutans spend most of their lives in trees and move by brachiation when they are in treetops. Gibbons, the most superbly acrobatic apes, have long arms and hands. The size and opposability of their thumbs are reduced, but their arms connect to the shoulders by ball-and-socket joints that allow free movement of the arms in 360 degrees. In addition, long collarbones act as braces and keep the shoulders from collapsing toward the chest.

Heavier bodied chimpanzees and gorillas don't brachiate as much as gibbons and orangutans, but like humans, they can do so. Humans seldom brachiate, with the exception of small, light-bodied children playing on schoolyard "monkey bars." Adult human arms are typically too weak to support the heavy torso and legs.

Chimpanzees and gorillas move by knuckle-walking, a behavioral modification that allows an animal to run rapidly on the ground on all fours, with its weight resting on the knuckles. The proportionately longer arms of chimps and gorillas are an adaptation to knuckle-walking.

One important feature distinguishes humans from the other great apes: bipedalism, or the ability to walk upright on two legs. Adaptations to bipedalism include relatively short arms and longer, stronger leg bones. Foot bones form firm supports for walking, with the big toe fixed in place and not opposable. The bowl-shaped pelvis supports most of the weight of the body, and the vertebrae of the lower back are robust enough to bear some body weight.

Bipedalism is also reflected in the bones of the head (figure 17.32). The foramen magnum is the large hole in the skull where the spinal cord leaves the brain. In modern humans, this hole is tucked beneath the skull. In gorillas and chimps, the foramen magnum is located somewhat closer to the rear of the skull; in animals that run on all fours, such as horses and dogs, the hole is at the back of the skull.

Diet Other skeletal characteristics, including the size and shape of the teeth, are related to diet. Upper and lower molar teeth have ridges that fit together, much as the teeth of gears intermesh. Food caught between these surfaces is ground, crushed, and mashed. The size of these teeth is an adaptation that reflects the toughness of the diet.

As you examine the photos in figure 17.32, note that the chimpanzee skull features a sagittal crest. This bony ridge runs lengthwise along the top of the skull and is an attachment point for muscles. A prominent sagittal crest indicates particularly strong jaws, another clue to an animal's diet. Humans lack a sagittal crest.

Other important features in primate skulls include the size of jaw bones, the prominence of the ridge of bone above the eye, the degree to which the jaw protrudes, and the shape of the curve of the tooth row. All of these characteristics allow paleoanthropologists, the scientists who study hominin fossils, to assign each new discovery to a species.

Molecular Data Fossil evidence and anatomical similarities were once the only lines of evidence that paleoanthropologists could analyze in tracing the course of human evolution. Around 1960, however, scientists began to use molecular sequences to investigate relationships among primates. Studies of blood proteins and DNA presented a new picture of primate evolution, as it became clear that humans are a species of great ape. One of the astounding findings of these molecular studies was that the DNA sequences of humans and chimpanzees are nearly identical. The evolutionary tree in figure 17.30 takes into account both the anatomical characteristics and the molecular data.

Other research has further eroded the distinctions between humans and other great apes. Previously, humans had been placed in a separate group, supposedly characterized by upright walking, tool-making, and language. Then, in the 1970s, chimpanzees and wild gorillas were observed to use tools, and captive great apes learned to use sign language to communicate with their trainers. The only characteristic that now remains unique to humans is bipedal locomotion.

C. Human Evolution Is Partially Recorded in Fossils

Even though DNA and proteins provide overwhelming evidence of the relationships between living primates, these molecules deteriorate with time. Scientists therefore cannot usually use molecular data to establish relationships of prehistoric hominins. For this, we must turn to studies of fossilized remains.

To interpret fossils from our family tree, paleoanthropologists compare details of the ancient skeletal features with modern primates and try to reconstruct as much information about diet and lifestyle as they can. So far, hominin fossils have fallen into the following groups (figure 17.33):

- *Ardipithecus.* The oldest representative of the hominin lineage, *Ardipithecus* (or "Ardi" for short), is an extinct species that dates to about 4.4 MYA. Fossils reveal several clues about Ardi's life. The teeth indicate that Ardi was an omnivore. The pelvis supported both upright walking and powerful climbing, as did the feet, which had opposable big toes. The flexible hands had long, grasping fingers with which Ardi could have carried objects while walking upright.

- **Australopiths.** Fossils of several species of extinct small apes have been assigned to the genus *Australopithecus,* meaning "Southern ape-man." The downward position of the foramen magnum indicates that these apes walked upright. *Australopithecus afarensis* (including

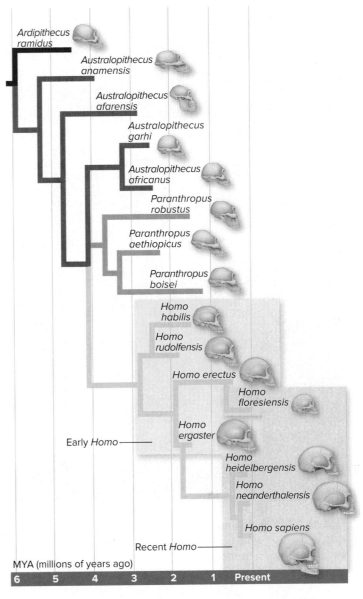

Figure 17.33 **Family Tree.** Fossilized remains place our close relatives into four main groups: *Ardipithecus, Australopithecus, Paranthropus,* and *Homo*. Skull sizes range from about 400 cm³ for *A. afarensis* to about 1450 cm³ for modern humans. The evolutionary relationships in this tree are hypothesized.

Figure It Out

Suppose that a 100-meter track represents Earth's 4.6 billion-year history. Primates originated about 60 million years ago. How close to the end of the track would you mark the origin of primates?

Answer: 1.3 m.

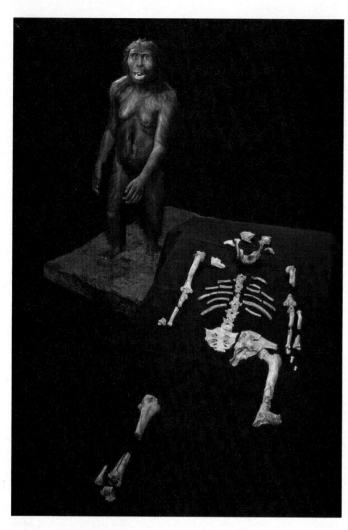

Figure 17.34 **Old Bones.** This reconstruction of the *Australopithecus afarensis* specimen called "Lucy" was based on her 3.2-million-year-old bones.

©Philippe Plailly & Atelier Daynes/Science Source

the famous "Lucy" fossil in figure 17.34) and *A. africanus* are members of this group, which dates from about 4 to 2.5 MYA.

- *Paranthropus.* This extinct group, which dates from about 3 to 1.5 MYA, seems to be an evolutionary dead end that descended from *Australopithecus* but gave rise to no other group. The name literally means "beside humans." These primates had extremely large teeth, protruding jaws, and skulls with a sagittal crest. All of these specializations probably relate to the large jaw muscles needed to crush tough plants or crack nuts.

- *Homo.* All members of genus *Homo* are considered humans, and fossils in this group are associated with stone tools. *Homo* species tend to have larger bodies and larger brains than do australopiths. *Homo habilis, H. ergaster,* and *H. erectus* are among the cluster of extinct species belonging to the group called "early *Homo.*" These species lived from about 2.5 MYA to about 500,000 years ago. New fossil discoveries, including the puzzling *H. naledi* fossils recently uncovered in a South African cave, have called this portion of the family tree into question. Regardless of how these issues are settled, early *Homo* species eventually gave rise to "recent *Homo.*" Recent species of *Homo* have smaller teeth, lighter and less protruding jaws, larger braincases, and lighter brow ridges. Their fossils are associated with evidence of culture (figure 17.35). Recent *Homo* species include *H. heidelbergensis, H. neanderthalensis,* and *H. sapiens.* The only human species alive today is *Homo sapiens.*

One interesting trend in human evolution has been a rapid diversification of species, followed by extinctions. Fossil evidence shows that about 1.8 MYA, as many as five species of humans lived together in Africa. About 200,000 years ago, three species of recent *Homo* coexisted in Europe. Today, however, all except *Homo sapiens* are extinct.

What happened to the other *Homo* species? No one knows, but many anthropologists wonder whether *H. sapiens* contributed to their extinction. Scientists have speculated that Neandertals interbred with *H. sapiens,* effectively causing Neandertals to disappear from the fossil record about 30,000 years ago. Analyses of DNA from Neandertals and *H. sapiens* have supported that hypothesis, although the question remains unsettled.

a.

b.

Figure 17.35 *Homo heidelbergensis.* (a) A skull replica (*left*) is paired with a model showing what *Homo heidelbergensis* might have looked like (*right*). (b) This hand axe, nicknamed "Excalibur," was discovered with a pit of 350,000-year-old *H. heidelbergensis* bones in Spain.

(a): ©Patrick Kovarik/AFP/Getty Images; (b): ©Javier Trueba/Science Source

Animals Live Everywhere   Sponges   Cnidarians   Flatworms   Mollusks   Annelids

D. Environmental Changes Have Spurred Human Evolution

What provoked the first hominins to abandon brachiation in favor of bipedal, upright walking? What allowed the large brains that are characteristic of recent *Homo* to develop? To find the answers, we have to consider a related question: Where did hominins evolve?

Charles Darwin was one of the first to speculate that humans evolved in Africa. About 12 MYA, tectonic movements caused a period of great mountain building. The continental plates beneath India and the Himalayan region collided and ground together, heaving up the Himalayas. The resulting climatic shift had enormous ecological consequences. Cooler temperatures reduced the thick tropical forests that had covered much of Europe, India, the Middle East, and East Africa. Open plains appeared, bringing new opportunities for species that could live there. These included less competition, a new assortment of foods, and a different group of predators. Experts speculate that one type of small ape moved out of the trees and began life on the African savannas.

Perhaps at first this species alternated between running on all fours and bipedal walking. On open plains, however, there are advantages to bipedal walking, especially the elevated vantage point for sensing danger and spotting food and friends. This environment would have selected for apes with the best skeletal adaptations to bipedalism, and the trait would have been preserved and honed in the plains. Bipedalism also freed the hands to carry objects and use the tools that are so characteristic of *Homo* species.

No one knows what might have spurred the evolution of the large brain that characterizes humans. Some experts relate the development of a large brain to tool use; others connect it to language and life in social groups.

E. Migration and Culture Have Changed *Homo sapiens*

DNA sampled from people around the world has revealed a compelling portrait of human migration out of Africa (figure 17.36). Asia, Australia, and Europe all were colonized at least 40,000 years ago, but it took somewhat longer for humans to reach the Americas.

Burning Question 17.2

Did humans and dinosaurs ever coexist?

To answer this question, it is important to first understand what dinosaurs were—and what they were not. As the term is traditionally used, *dinosaurs* were terrestrial reptiles that lived during the Mesozoic era, between 245 and 65 MYA. They ranged in stature from chicken-sized to truly gargantuan—the largest could have peeked into the sixth story of a modern-day building.

Many people mistakenly believe that all reptiles (or even all mammals) that lived during the Mesozoic era were dinosaurs. In reality, lizards, snakes, crocodiles, and other terrestrial reptiles lived alongside the dinosaurs. The Mesozoic also saw marine reptiles such as plesiosaurs and flying reptiles such as pterodactyls, but these animals belonged to clades that were distinct from the dinosaurs. Nor did all dinosaurs live at the same time. As some species appeared, others went extinct throughout the Mesozoic.

Although several comic strips, cartoons, and feature films suggest the contrary, humans never coexisted with *T. rex, Apatosaurus,* or any other nonavian dinosaur. These dinosaurs were gone by the time our own mammalian lineage—the primates—was just getting started. Tens of millions of years later, humans finally roamed the Earth and found the fossils that prove that these huge reptiles once existed, leaving only birds as their modern-day descendants.

Submit your burning question to
marielle.hoefnagels@mheducation.com

(dinosaur): ©Andersen Ross/Getty Images RF

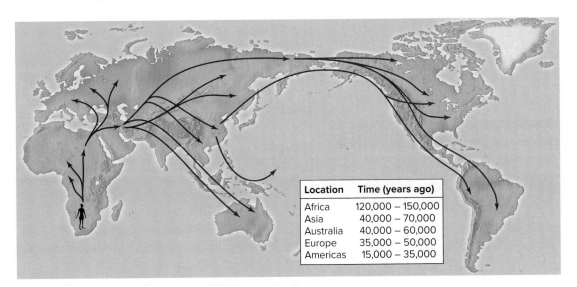

Location	Time (years ago)
Africa	120,000 – 150,000
Asia	40,000 – 70,000
Australia	40,000 – 60,000
Europe	35,000 – 50,000
Americas	15,000 – 35,000

Figure 17.36 Out of Africa. Researchers used mitochondrial DNA sequences to deduce approximately when humans originally settled each continent after migrating out of Africa.

Figure 17.37 Learning from the Past. Thanks to our ability to store and retrieve information, humans can continuously build on the knowledge that previous generations have accumulated.

©Tetra Images/Getty Images RF

As humans spread throughout the world, new habitats selected for different adaptations. Near the equator, for example, sunlight is much more intense than at higher latitudes. One component of sunlight is ultraviolet (UV) radiation, which is both harmful and beneficial. On the one hand, UV radiation damages DNA and causes skin cancer. On the other hand, some UV wavelengths help the skin produce vitamin D, which is essential to bone development and overall health.

These two counteracting selective pressures help explain why skin pigmentation is strongly correlated with the amount of ultraviolet radiation striking the Earth. One pigment that contributes to skin color, melanin, blocks UV radiation. Intense UV radiation selects for alleles that confer abundant melanin. People whose ancestry is near the equator, such as in Africa and Australia, therefore tend to have very dark brown skin. In northern Europe and other areas with weak sunlight, however, heavily pigmented skin would block so much UV light that indigenous people would suffer from vitamin D deficiency. These conditions are correlated with a high frequency of alleles conferring pale, pinkish skin.

No matter where people roamed, one byproduct of the large human brain was **culture:** the knowledge, beliefs, and behaviors that we transmit from generation to generation. Among the earliest signs of culture is cave art from about 14,000 years ago, which indicates that our ancestors had developed fine hand coordination and could use symbols.

By 10,000 years ago, depending on which native plants and animals were available for early farmers to domesticate, agriculture began to replace a hunter–gatherer lifestyle in many places. Agriculture meant increased food production, which profoundly changed societies. Freed from the necessity of producing their own food, specialized groups of political leaders, soldiers, weapon-makers, religious leaders, scientists, engineers, artists, writers, and many other types of workers arose. These new occupations meant improved transportation and communication, better technologies, and the ability to explore the world for new lands and new resources.

Undoubtedly, humans are a special species. We can modify the environment much more than, for example, a slime mold or an earthworm can. We can also alter natural selection, in our own species and in others. Finally, culture allows each generation to build on information accumulated in the past (figure 17.37). The knowledge, beliefs, and behaviors that shape each culture are constantly modified within a person's lifetime, in stark contrast to the millions of years required for biological evolution. Our species is therefore extremely responsive to short-term changes.

Despite our unique set of features, however, we are descended from ancestors with whom we share many characteristics. It is intriguing to think about where the human species is headed, which species will vanish, and how life will continue to diversify in the next 500 million years.

Miniglossary	Human Evolution
Primate	Mammal with grasping hands, opposable thumbs, flat nails, and binocular vision; includes prosimians, monkeys, and apes
Hominoid	An ape; includes lesser apes and great apes
Hominid	A great ape; includes orangutans, gorillas, chimpanzees, and humans
Hominin	Any hominid in genus *Ardipithecus, Australopithecus, Paranthropus,* or *Homo*
Human	A member of genus *Homo,* of which the only living representative is *Homo sapiens*

17.12 Mastering Concepts

1. Name the three groups of contemporary primates. To which group do humans belong?
2. What can skeletal anatomy and DNA sequences in existing primates tell us about human evolution?
3. What are the four groups of species in the hominin family tree, and which still exist today?
4. Which conditions may have contributed to the evolution of humans?

Investigating Life 17.1 | Discovering the "Fishapod"

The idea that early amphibians crawled onto land some 375 MYA is intriguing, especially since the descendants of those early colonists are today's amphibians, reptiles, and mammals. Fossils discovered over the past half-century have clarified the fish–amphibian transition, but some details of this fascinating event remain poorly understood.

Three fossils of an extinct animal unearthed in Arctic Canada, *Tiktaalik roseae*, added new insights (figure 17.C). *Tiktaalik* either crawled or paddled in shallow tropical streams during the late Devonian period, about 380 MYA. (Although today's Canada is anything but tropical, the entire North American continent was near the equator during the Devonian.)

Scientists jokingly call the animal a "fishapod" because of its uncanny mix of fish and tetrapod characteristics. Like a fish, *Tiktaalik* had scales and gills. Like a tetrapod, it had lungs, and its ribs were robust enough to support its body. But the appendages got the most attention. *Tiktaalik* had moveable wrist bones that were sturdy enough to support the animal in shallow water or on short excursions to land. Although the bones were clearly limblike, the "limbs" were fringed with fins, not toes.

The *Tiktaalik* fossils caught the world's eye because they were extraordinarily complete and exquisitely preserved. Scientifically, *Tiktaalik* is important for two reasons. First, it adds to our knowledge about tetrapod evolution. Second, it highlights the predictive power of evolutionary biology. The researchers did not stumble on *Tiktaalik* by accident. Instead, they were looking for a fossil representing the fish–amphibian transition, based on previous knowledge of how and when vertebrates moved onto land. Finding *Tiktaalik* confirmed the prediction.

Interpreting spectacular fossils such as *Tiktaalik* requires countless hours of tedious, labor-intensive work. Researchers scrutinize the limbs, skull, vertebrae, and other parts of each new find to glean every possible piece of information about the animal and how it lived. In this way, fossils contribute immeasurably to our understanding of life's long history.

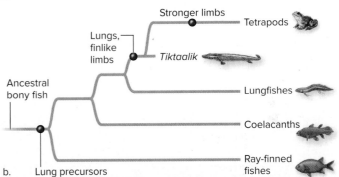

Figure 17.C Fossil "Fishapod." (a) The *Tiktaalik* fossil and a drawing of its limb bones. (b) *Tiktaalik* is a transitional fossil that links fishes and tetrapods.

Photo: ©DEA Picture Library/Photolibrary/Getty Images

Sources: Daeschler, Edward B., Neil H. Shubin, and Farish A. Jenkins, Jr. April 6, 2006. A Devonian tetrapod-like fish and the evolution of the tetrapod body plan. *Nature,* vol. 440, pages 757–763; Shubin, Neil H., Edward B. Daeschler, and Farish A. Jenkins, Jr. April 6, 2006. A pectoral fin of *Tiktaalik roseae* and the origin of the tetrapod limb. *Nature,* vol. 440, pages 764–771.

What's the **Point?** ▼ APPLIED

Most people consume many types of animals. In American grocery stores, the most common animal meats come from chordates (tuna, chickens, turkeys, pigs, and cattle), arthropods (crabs and lobsters), and mollusks (clams and oysters). We also consume many other animal products, including milk, eggs, and honey.

Most people know that meats and milk come from animals, but few may realize that many other foods contain chemicals derived from animal sources. Below are some common (and, in some cases, weird) examples of how the food industry uses animal chemicals.

- Gelatin is a protein derived from the connective tissues of cattle, chickens, pigs, and fish. Food manufacturers use gelatin not only in desserts such as Jell-O but also as a jelling agent in foods such as yogurt, ice cream, and candies. Gelatin also improves the texture of reduced-fat foods.
- Carmine, a bright red pigment produced in certain insects, is often used as a food coloring. Some people are allergic to this buggy ingredient, so it is clearly marked on food labels.

- Shellac is produced by the same insect group that produces carmine. This waxy extract, best known for its role as a wood finish, is sometimes used to coat fruits and preserve their shelf life. It is also used as a glaze on some candies.
- Glycerol, sometimes called glycerin on ingredient labels, is the sugar backbone of a triglyceride. This sugar occurs in candy, marshmallows, and some dairy products, among other foods. Even though this molecule is abundant throughout life, the glycerol used in foods typically comes from animals. ⓘ *triglycerides,* section 2.5E

©Burke/Triolo Productions/ Getty Images RF

- Castoreum, a secretion that beavers use to mark their territory, is sometimes used as a vanilla-flavored food additive. This animal product usually shows up as "natural flavoring" in ingredient lists. But collecting it is difficult, so many food manufacturers use simpler alternatives.
- Cuttlebones that supply calcium to pet birds come from the thick internal shells of cephalopods (mollusks) called cuttlefish.

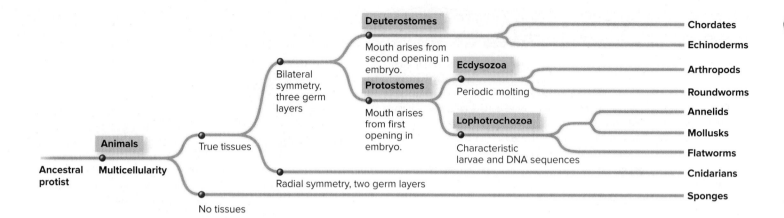

Chordates
- ~60,000 species
- Notochord, dorsal hollow nerve cord, pharyngeal slits, postanal tail
- Fishes, amphibians, reptiles (including birds), mammals

Roundworms
- ~80,000 species
- Unsegmented, cylindrical worms
- Free-living or parasitic
- Pinworms, hookworms, heartworms

Flatworms
- ~25,000 species
- Flat bodies
- Free-living or parasitic
- Marine flatworms, planarians, flukes, tapeworms

Echinoderms
- ~7000 species
- Five-part symmetry
- Water vascular system
- Sea stars, sea urchins, sand dollars

Annelids
- ~15,000 species
- Segmented worms
- Earthworms, polychaetes, leeches

Cnidarians
- ~11,000 species
- Medusa or polyp forms
- Stinging cells
- Hydra, jellyfish, coral

Arthropods
- >1,000,000 species
- Jointed appendages; exoskeleton
- Trilobites, spiders and other arachnids, crustaceans, insects, millipedes, centipedes

Mollusks
- ~112,000 species
- Mantle secretes shell
- Chitons, bivalves, cephalopods, gastropods

Sponges
- ~5000 species
- Porous bodies
- Filter feeders

Phylum	Level of Organization	Symmetry	Cephalization	Coelom	Digestive Tract	Segmentation
Porifera (sponges)	Cellular	Asymmetrical or radial	No	No	Absent	No
Cnidaria	Tissue	Radial	No	No	Incomplete	No
Platyhelminthes (flatworms)	Organ system	Bilateral	Yes	No	Incomplete (when present)	No
Mollusca	Organ system	Bilateral	Yes (usually)	Yes	Complete	No
Annelida	Organ system	Bilateral	Yes	Yes	Complete	Yes
Nematoda (roundworms)	Organ system	Bilateral	Yes	Pseudocoelom	Complete	No
Arthropoda	Organ system	Bilateral	Yes	Yes	Complete	Yes
Echinodermata	Organ system	Bilateral larvae; radial adults	No	Yes	Complete	No
Chordata	Organ system	Bilateral	Yes	Yes	Complete	Yes (usually)

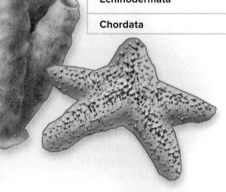

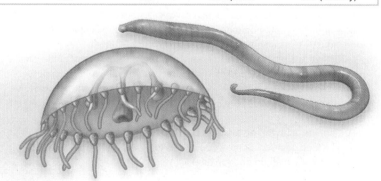

Figure 17.38 Animal Diversity: A Summary.

CHAPTER SUMMARY

17.1 Animals Live Nearly Everywhere

- Of the 1,300,000 species in kingdom **Animalia,** most are **invertebrates;** only one phylum contains **vertebrates,** which have a segmented backbone.
- Figure 17.38 summarizes the characteristics of the nine major animal phyla.

A. What Is an Animal?

- Animals are multicellular, eukaryotic heterotrophs whose cells secrete extracellular matrix but do not have cell walls. Most digest their food internally.
- The **blastula** is a stage in embryonic development that is unique to animals (figure 17.39).

B. Animal Life Began in the Water

- The ancestor of animals probably resembled a choanoflagellate.
- The oldest animal fossils formed about 570 MYA. These animals lived in water, and existing animal diversity reflects this aquatic heritage.

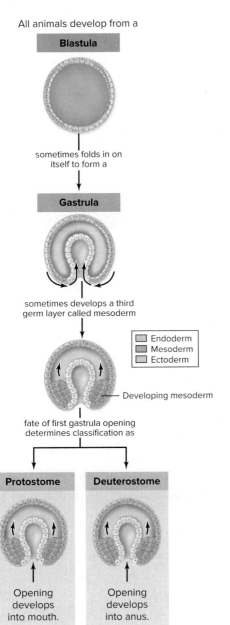

Figure 17.39 Animal Development: A Summary.

- The Ediacarans were soft, flat organisms that lived in the late Precambrian and early Cambrian periods.
- Animals diversified greatly during the Cambrian period.

C. Animal Features Reflect Shared Ancestry

- Differences in development account for major groupings of animal phyla.
- All animals except sponges have true tissues.
- In most phyla, animal bodies have **radial symmetry** or **bilateral symmetry. Cephalization** is correlated with bilateral symmetry.
- An animal zygote divides mitotically to form a blastula and then usually a **gastrula.** In some animals, the gastrula has two tissue layers (**ectoderm** and **endoderm**). In others, a third layer (**mesoderm**) forms between the other two.
- Bilaterally symmetrical animals are **protostomes** if the gastrula's first indentation forms into the mouth. In **deuterostomes,** the first indentation develops into the anus.

D. Biologists Also Consider Additional Characteristics

- Some types of animals have a body cavity (**coelom** or **pseudocoelom**) that can act as a **hydrostatic skeleton.**
- An animal may have an **incomplete digestive tract (gastrovascular cavity)** or a **complete digestive tract.**
- **Segmentation** improves flexibility and increases the potential for specialized body parts.

17.2 Sponges Are Simple Animals That Lack Differentiated Tissues

- **Sponges** are aquatic, sessile animals that are either asymmetrical or radially symmetrical. Their porous bodies filter small food particles out of water. Although they lack tissues, sponges have specialized cell types, including collar cells and amoebocytes.
- A sponge's skeleton consists of spicules or organic fibers (or both).

17.3 Cnidarians Are Radially Symmetrical, Aquatic Animals

- **Cnidarians** are mostly marine animals that capture prey with tentacles and stinging **cnidocytes.** They digest food in a gastrovascular cavity. Cnidarians move by contracting muscle cells that act on a hydrostatic skeleton.
- A cnidarian body form is a sessile **polyp** or a swimming **medusa.**
- Examples of cnidarians include corals, hydras, and jellyfishes.

17.4 Flatworms Have Bilateral Symmetry and Incomplete Digestive Tracts

- **Flatworms** are unsegmented animals that lack a coelom. The flat body shape allows individual cells to exchange gases with their environment.
- These animals include **free-living flatworms** such as planarians; **flukes** and **tapeworms** are parasitic.
- Flatworms lack circulatory and respiratory systems, but specialized structures maintain water balance. They have simple nervous systems and hydrostatic skeletons.

17.5 Mollusks Are Soft, Unsegmented Animals

- **Mollusks** have bilateral symmetry and a complete digestive tract. The main groups of mollusks are **chitons, bivalves, gastropods,** and **cephalopods.**
- The mollusk body includes a **mantle,** a muscular **foot,** and a **visceral mass.** Most mollusks have a shell, and many have a tonguelike **radula.** They are filter feeders, herbivores, or predators.
- Cephalopods such as octopuses have complex sensory and nervous systems.

17.6 Annelids Are Segmented Worms

- **Annelid** bodies consist of repeated segments. Annelids include **earthworms, leeches,** and **polychaetes,** and they feed in diverse ways.

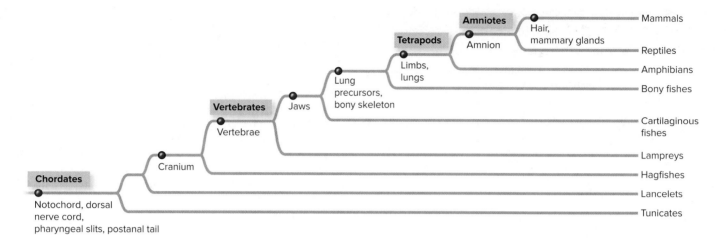

Phylogenetic tree labels:
- **Chordates** — Notochord, dorsal nerve cord, pharyngeal slits, postanal tail
- **Vertebrates** — Vertebrae
- Cranium
- Jaws
- Lung precursors, bony skeleton
- **Tetrapods** — Limbs, lungs
- **Amniotes** — Amnion
- Hair, mammary glands
- Mammals
- Reptiles
- Amphibians
- Bony fishes
- Cartilaginous fishes
- Lampreys
- Hagfishes
- Lancelets
- Tunicates

■ **Mammals**
- ~5800 species
- Hair/fur and mammary glands
- Amnion surrounds developing embryo
- Monotremes, marsupials, placentals

■ **Birds**
- ~10,000 species
- Feathers and hollow bones
- Amniotic eggs

■ **Nonavian reptiles**
- ~8000 species
- Dry, scaly skin; amniotic eggs
- Turtles, lizards, snakes, crocodilians

■ **Amphibians**
- ~6000 species
- Scale-less tetrapods
- Can live on land but reproduce in water
- Frogs, salamanders, caecilians

■ **Fishes**
- ~30,000 species
- Scale-covered bodies with gills and fins
- Cartilaginous fishes (sharks and rays) and bony fishes

■ **Lampreys**
- 38 species
- Fishlike bodies; jawless
- Consume invertebrates or parasitize fish

■ **Hagfishes**
- ~70 species
- Fishlike bodies; jawless
- Marine carnivores
- "Slime hags"

■ **Tunicates and lancelets**
- ~3000 species (tunicates)
- ~30 species (lancelets)
- Invertebrate filter feeders

Group	Cranium	Vertebrae	Jaws	Skeleton	Lungs	Limbs	Amnion	Body Temperature Regulation
Tunicates and lancelets	No	No	No	Absent	No	No	No	Ectotherm
Hagfishes	Yes	No	No	Cartilage	No	No	No	Ectotherm
Lampreys	Yes	Yes	No	Cartilage	No	No	No	Ectotherm
Fishes	Yes	Yes	Yes	Cartilage or bone	No (usually)	No	No	Ectotherm
Amphibians	Yes	Yes	Yes	Bone	Yes	Yes	No	Ectotherm
Nonavian reptiles	Yes	Yes	Yes	Bone	Yes	Yes	Yes	Ectotherm
Birds	Yes	Yes	Yes	Bone	Yes	Yes	Yes	Endotherm
Mammals	Yes	Yes	Yes	Bone	Yes	Yes	Yes	Endotherm

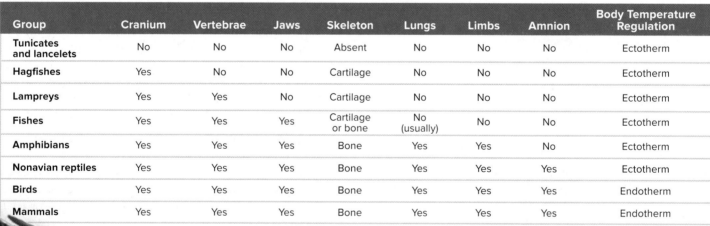
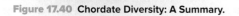

Figure 17.40 Chordate Diversity: A Summary.

- Organ systems include a complete digestive tract, a closed circulatory system, and respiratory, excretory, and nervous systems. The coelom acts as a hydrostatic skeleton.

17.7 Nematodes Are Unsegmented, Cylindrical Worms

- **Roundworms** are unsegmented worms that molt periodically. They include parasitic and free-living species in soil and aquatic sediments.
- Nematodes have diverse diets and complete digestive tracts. The pseudocoelom aids in circulation and is a hydrostatic skeleton.

17.8 Arthropods Have Exoskeletons and Jointed Appendages

- **Arthropods** are segmented animals with jointed appendages and a chitin-rich **exoskeleton.** Like nematodes, arthropods molt periodically.

A. Arthropods Have Complex Organ Systems

- Arthropods exhibit great diversity in feeding, respiratory systems, excretory systems, nervous systems, and reproduction. Life cycles may include **metamorphosis.** They have open circulatory systems.

B. Arthropods Are the Most Diverse Animals

- The extinct **trilobites** were arthropods, as are the **chelicerates** (**horseshoe crabs** and **arachnids**). **Mandibulate** arthropods include **centipedes** and **millipedes; crustaceans;** and **insects.**
- Insects are by far the most diverse arthropods. Segmented bodies, flight, and reproductive strategies all may account for today's enormous number of insect species.

17.9 Echinoderm Adults Have Five-Part, Radial Symmetry

- **Echinoderms** are spiny-skinned marine animals. They are deuterostomes, as are the chordates.
- Adult echinoderms have radial symmetry; larvae are bilaterally symmetrical.
- The **water vascular system** enables echinoderms to move, sense their environment, acquire food, exchange gases, and get rid of metabolic wastes.

17.10 Most Chordates Are Vertebrates

- **Chordates** share four characteristics: a **notochord,** a **dorsal hollow nerve cord, pharyngeal pouches** or **slits** in the pharynx, and a **postanal tail.** Figure 17.40 summarizes diversity within the chordates.
- Most chordates also have a **cranium** that protects the brain.
- Other features that distinguish chordates from one another include **vertebrae, jaws, lungs,** and the presence of limbs in **tetrapods.**
- Reptiles lay **amniotic eggs,** in which the amnion and other membranes protect the developing embryo. A mammal's amnion is homologous to that of the amniotic egg; mammals and reptiles form a clade called **amniotes.**
- **Ectotherms** lack internal temperature control mechanisms, whereas **endotherms** use heat from metabolism to maintain body temperature.
- Figure 17.41 summarizes vertebrate adaptations.

17.11 Chordate Diversity Extends from Water to Land to Sky

A. Tunicates and Lancelets Are Invertebrate Chordates

- **Tunicates** obtain food and oxygen with a siphon system. A tunicate larva has all four chordate characteristics, but adults retain only the pharyngeal slits.
- **Lancelets** resemble eyeless fishes; adults have all four chordate characteristics.

B. Hagfishes and Lampreys Have a Cranium but Lack Jaws

- **Hagfishes** have a cranium of cartilage, but they are not vertebrates. They secrete slime and lack jaws.
- **Lampreys** are vertebrates that resemble fishes, but they are jawless.

Key Vertebrate Adaptations		
Adaptation	**Adaptive Significance**	**Animals with Adaptation**
Vertebrae	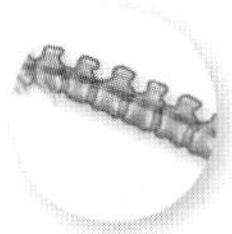 Expand range of motion	Lampreys, fishes, amphibians, reptiles, mammals
Jaws	 Increase feeding versatility	Fishes, amphibians, reptiles, mammals
Lungs	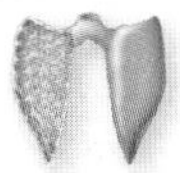 Enable animal to breathe air	Bony fishes (a few species), amphibians, reptiles, mammals
Limbs	 Allow for locomotion on land	Amphibians, reptiles, mammals
Amnion	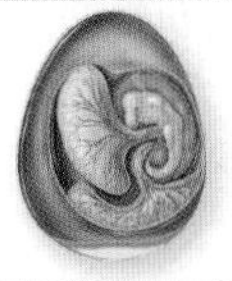 Enables reproduction away from water	Reptiles, mammals

Figure 17.41 Vertebrate Adaptations: A Summary.

C. Fishes Are Aquatic Vertebrates with Jaws, Gills, and Fins

- **Fishes** are abundant and diverse vertebrates.
- Vertebrate adaptations originating in fishes include a vertebral column, jaws, lungs, and paired fins that were later modified as limbs.
- The **cartilaginous fishes** include skates, rays, and sharks. Sharks detect vibrations from prey with a **lateral line** system.
- The **bony fishes** account for 96% of existing fish species. Bony fishes have lateral line systems and swim bladders, which enable them to control their buoyancy. The two groups of bony fishes are the **ray-finned fishes** and **lobe-finned fishes,** which are further subdivided into **lungfishes** and **coelacanths.**

D. Amphibians Live on Land and in Water

- **Amphibians** breed in water and must keep their skin moist to breathe. Adaptations to life on land include a sturdy skeleton, lungs, and limbs.
- Amphibians include **frogs; salamanders** and newts; and **caecilians.**

E. Reptiles Were the First Vertebrates to Thrive on Dry Land

- **Reptiles** (including **birds**) have efficient excretory, respiratory, and circulatory systems. Internal fertilization and amniotic eggs permit reproduction on dry land.
- Nonavian reptiles include **lizards** and **snakes; turtles** and tortoises; and **crocodilians.**
- Birds retain scales and egg-laying from reptilian ancestors. Honeycombed bones, streamlined bodies, and **feathers** are adaptations that enable flight.
- Unlike the other reptiles, birds are endothermic.

F. Mammals Are Warm, Furry Milk-Drinkers

- **Mammals** have fur, secrete milk from **mammary glands,** and have distinctive teeth and highly developed brains.
- **Monotremes** are mammals that hatch from an amniotic egg. The young of **marsupial** mammals are born after a short pregnancy and often develop inside the mother's pouch. **Placental mammals** have longer pregnancies; the young are nourished by the **placenta** in the mother's uterus.

17.12 Fossils and DNA Tell the Human Evolution Story

A. Humans Are Primates

- **Primates** have grasping hands, opposable thumbs, binocular vision, large brains, and flat nails. The three groups are prosimians, monkeys, and apes.
- Hominids are the "great apes," whereas **hominins** are humans and their extinct ancestors on the human branch of the primate evolutionary tree.

B. Anatomical and Molecular Evidence Documents Primate Relationships

- Fossil bones and teeth reveal how extinct species moved and what they ate.
- Protein and DNA analysis has altered how scientists draw the primate family tree.

C. Human Evolution Is Partially Recorded in Fossils

- Early hominins included *Ardipithecus, Australopithecus, Paranthropus,* and *Homo.*

D. Environmental Changes Have Spurred Human Evolution

- Millions of years ago, new mountain ranges arose, causing climate shifts. Savannas replaced tropical forests, and apes—the ancestors of humans— moved from the trees to the savanna.

E. Migration and Culture Have Changed *Homo sapiens*

- After migrating out of Africa, humans encountered new habitats that selected for new allele combinations.
- We humans owe our success to language and **culture.**

MULTIPLE CHOICE QUESTIONS

1. Following gastrulation, the cells that have folded inward develop into
   a. endoderm.
   b. mesoderm.
   c. ectoderm.
   d. All of these are correct.

2. Which of the following groups includes all of the others?
   a. Chordates
   b. Echinoderms
   c. Lizards
   d. Deuterostomes

3. What is a key characteristic of all arthropods?
   a. Six legs
   b. Pseudocoelom
   c. Hydrostatic skeleton
   d. Exoskeleton

4. How is the body structure of an annelid different from that of an arthropod?
   a. Annelids lack jointed appendages.
   b. Annelids have a complete digestive tract.
   c. Annelids have cephalization.
   d. Annelids have bilateral symmetry.

5. When an earthworm is moving through the soil, its muscles are pushing against its
   a. digestive tract.
   b. coelom.
   c. bony skeleton.
   d. cnidocytes.

6. Arthropods molt because
   a. they undergo metamorphosis.
   b. the exoskeleton becomes damaged over time.
   c. the exoskeleton prevents the organism from growing.
   d. they have an open circulatory system.

7. Echinoderms have _____ symmetry as embryos and _____ symmetry as adults.
   a. radial; radial
   b. radial; bilateral
   c. bilateral; radial
   d. bilateral; bilateral

8. Which animal phylum contains the most known species?
   a. Chordata
   b. Nematoda
   c. Arthropoda
   d. Cnidaria

9. Which of the following has pharyngeal slits at some point in its life?
   a. A snake
   b. A sea star
   c. A lobster
   d. All of these are correct.

10. Lobe-finned fishes are important because they
    a. were the first vertebrates.
    b. were the earliest animals.
    c. are closely related to tetrapods.
    d. lack jaws.

11. To which of the following is a salamander most closely related?
    a. A snail
    b. A beetle
    c. A shark
    d. A catfish

12. How do reptiles and mammals differ from amphibians?
    a. Only reptiles and mammals are amniotes.
    b. Only reptiles and mammals are tetrapods.
    c. Only reptiles and mammals have lungs.
    d. All of the above are correct.

13. Which of the following represents the correct order of appearance, from earliest to most recent?
    a. Fishes, reptiles, Ediacarans, primates
    b. Ediacarans, fishes, reptiles, primates
    c. Ediacarans, primates, fishes, reptiles
    d. Primates, reptiles, fishes, Ediacarans

14. Primates share all of the following characteristics except
    a. opposable thumbs.
    b. excellent depth perception.
    c. bipedalism.
    d. flat fingernails.

15. DNA evidence suggests that modern humans
    a. share a common ancestor with *Ardipithecus.*
    b. exist in several species today.
    c. have not evolved in the last 150,000 years.
    d. are evolving from chimpanzees today.

Answers to Multiple Choice questions are in appendix A.

WRITE IT OUT

1. Compare the nine major animal phyla in the order in which the chapter presents them, listing the new features for each group.

2. Suppose you watch a video showing the development of an unknown animal. What clues can the developmental pattern give you about how this organism is classified? Creating a flow chart might be useful.

3. Using the evolutionary trees in this chapter, compare the following groups. For each comparison, what features are similar among the groups? What features are different?
   (a) Compare cnidarians to sponges and to the clade containing flatworms, mollusks, and annelids.
   (b) Compare flatworms to cnidarians, other protostomes, and deuterostomes.
   (c) Compare mollusks to flatworms and annelids.
   (d) Compare nematodes to other protostomes in general and to arthropods in particular.
   (e) Compare echinoderms to protostomes and to chordates.

4. List the criteria used to distinguish: (a) animals from other organisms; (b) vertebrates from invertebrates; (c) protostomes from deuterostomes; (d) ectotherms from endotherms.

5. Compare the digestive tract of an earthworm and a flatworm. How is each worm's digestive system adaptive?

6. Analyze the evolutionary tree in figure 17.3, and then write an argument supporting or refuting this statement: Annelids are more closely related to flatworms than to roundworms.

7. Segmented animals occur in multiple phyla. How might segmentation benefit an animal? If segmentation is adaptive, why do unsegmented animals still exist?

8. Create lists of animal phyla that (a) are cephalized, (b) have an incomplete digestive tract, (c) have segmented bodies, and (d) have a coelom.

9. Both amniotes and insects do very well on land. What characteristics are found in both groups that facilitate their success in terrestrial environments?

10. Draw from memory a phylogenetic tree that traces the evolutionary history of vertebrates. Include the features that mark each branching point in your tree.

11. Tunicates look much different than other chordates. Why are they classified in the same phylum?

12. List the evidence that biologists use to classify earthworms, caecilians, and snakes in different clades despite the superficial similarities between these animals.

13. List three adaptations that enable fishes to live in water, amphibians to live on land, snakes to live in the desert, and birds to fly.

14. How does the changing placement of birds in the vertebrate family tree illustrate the scientific process? Why does this type of research matter?

15. How are fishes, amphibians, nonavian reptiles, birds, and mammals important to humans? How are they important in ecosystems?

16. Explain how a sessile or slow-moving lifestyle, such as that of sponges, sea cucumbers, and tunicates, might select for bright colors and an arsenal of toxic chemicals.

17. What can scientists learn by comparing the fossilized skeletons of extinct primates with the bones of modern species?

18. The foramen magnum in *Australopithecus africanus* is closer to the front of the skull than in gorillas. What does this observation indicate about *A. africanus*?

19. In what ways has culture been an important factor in human evolution?

20. At one time, several species of *Homo* coexisted. Propose at least two hypotheses that might explain why only *Homo sapiens* remains.

21. How do you predict a scientist would respond to a question about whether humans "evolved from monkeys"?

22. Compare the timelines illustrated in figures 16.2 and 17.1. Speculate on some ways that changes in land plants might have driven animal evolution, and vice versa.

SCIENTIFIC LITERACY

1. Invasive animal species are disrupting ecosystems around the world. Search the Internet for a list of invasive animal species in your state or province. Which phyla are represented in the list? What harm do they do? How important do you think it is to try to eradicate invasive species?

2. Review this chapter's *What's the Point? Applied* box, which discusses how chemicals derived from animals are used in processed foods. Use the Internet to compile a list of 10 nonfood products containing chemicals from animals. Once your list is complete, reflect on how difficult it is to be 100% vegan—that is, to avoid using or eating any animal products.

Answers to Mastering Concepts, Write It Out, Scientific Literacy, and Pull It Together questions can be found in the Connect ebook.
connect.mheducation.com

Design element: Burning Question (fire background): ©Ingram Publishing/Super Stock

PULL IT TOGETHER

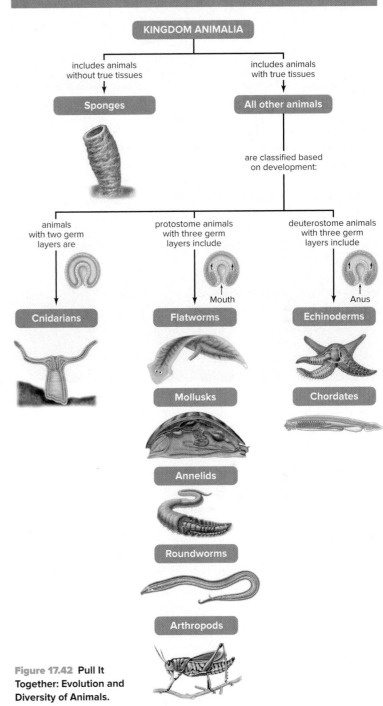

Figure 17.42 Pull It Together: Evolution and Diversity of Animals.

Refer to figure 17.42 and the chapter content to answer the following questions.

1. Review the Survey the Landscape figure in the chapter introduction and figure 17.42. Animal classification is partly based on developmental patterns. Assuming that few embryos left fossils hundreds of millions of years ago, what evidence can biologists use to deduce the early evolutionary history of animals?

2. Write connecting phrases to separate arthropods and roundworms from annelids, mollusks, and flatworms.

3. Draw a concept map that summarizes the chordates, including both invertebrates and vertebrates.

4. Add tunicates, hagfishes, birds, monotremes, vertebrates, marsupials, and placental mammals to this concept map.

5. Other than the ones pictured, give an example of a species in each phylum.

Disaster Relief. Heavy rains caused flooding and landslides in Indonesia in 2016, displacing tens of thousands of people. Crowded conditions at aid stations such as this one promote the spread of disease.

©Donal Husni/NurPhoto via Getty Images

LEARNING OUTLINE

18.1 Ecology Is the Study of Interactions

18.2 A Population's Size and Density Change Over Time

18.3 Births and Deaths Help Determine Population Size

18.4 Natural Selection Influences Life Histories

18.5 Population Growth May Be Exponential or Logistic

18.6 The Human Population Continues to Grow

APPLICATIONS

Burning Question 18.1 *How do biologists count animals in the open ocean?*

Why We Care 18.1 *Controlling Animal Pests*

Investigating Life 18.1 *A Toxic Compromise*

Learn How to Learn
Use Those Office Hours

Most instructors maintain office hours. Do not be afraid to use this valuable resource! Besides offering help with course materials, office hours give you an opportunity to know your professors personally. After all, at some point you may need a letter of recommendation; a letter from a professor who knows you well can carry a lot of weight. If you do decide to visit during office hours, be prepared with specific questions. And if you request a separate appointment, be sure to arrive at the time you have arranged—or let your instructor know you need to cancel.

SURVEY THE LANDSCAPE
Ecology

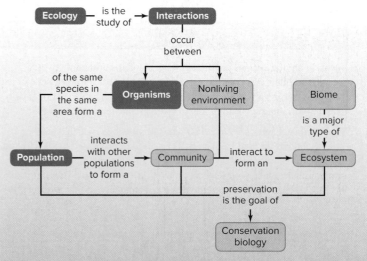

Within a population, individuals interact as they compete for limited nutrients, energy, and mates. Those that acquire these resources most efficiently are most likely to have high reproductive success.

For more details, study the Pull It Together feature in the chapter summary.

What's the **Point?** ▼

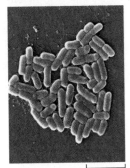

SEM (false color)　3 μm

Source: CDC/Janice
Haney Carr

The photo on the facing page shows the aftermath of flooding and landslides that slammed Indonesia in 2016. Survivors sought shelter in crowded refugee camps, where close quarters and poor sanitation fostered the spread of infectious disease.

Disasters such as floods, landslides, and earthquakes illustrate some of the lessons of population ecology. A population ecologist is a scientist who studies the factors—including disasters and exposure to disease-causing organisms—that determine whether a population grows, shrinks, or stays the same size.

But the study of population ecology extends well beyond the human population. Disease prediction, land management, and the protection of endangered species all rely on population ecology as well. Typical questions that population ecologists might ask include: "Which weather conditions favor reproduction in rodents that transmit human diseases?" or "How many deer should hunters cull to keep the herd healthy?" or "How many humans can Earth support?" Scientists begin to answer these and many other questions by using the principles described in this chapter.

18.1 Ecology Is the Study of Interactions

Take a moment to think of all the ways you have interacted with the world today. You have certainly breathed air, and you have probably sipped a beverage, eaten one or more meals, put on some clothes, and greeted other people. You may have stepped on the grass or driven a car or played with a pet. Such interactions are part of the science of **ecology,** the study of the relationships that organisms have with each other and with the environment.

Ecologists classify these relationships at several levels (figure 18.1). A **population** is a group of interbreeding organisms of one species occupying a location at the same time. Humans form a population, as do the acacia trees in figure 18.1. Population ecology—the study of the factors that influence a population's size over time—is the topic of this chapter.

A **community** includes all of the populations, representing multiple species, that interact in a given area. All of the species in a backyard or park, for example, form a community. Figure 18.1 shows how acacia trees share the landscape with other members of the same community, such as grasses and large grazing animals. Community ecologists study the interactions among these species, such as competition and herbivory. On a still broader scale, an **ecosystem** is a community plus its nonliving environment, including air, water, minerals, and fire. Chapter 19 describes community- and ecosystem-level processes.

ORGANISM
A single living individual
Example: One acacia tree

POPULATION
A group of the same species of organism living in the same place and time
Example: Multiple acacia trees

Figure 18.1 From Organism to Biosphere. Individual organisms make up populations, which in turn make up communities. An ecosystem includes interactions between a community and its nonliving environment, and the biosphere includes all ecosystems on Earth.

Photos: (population): ©Gregory G. Dimijian, M.D./Science Source; (community): ©Daryl Balfour/Gallo Images/Getty Images; (ecosystem): ©Bas Vermolen/Getty Images; (biosphere): ©StockTrek/Getty Images RF

BIOSPHERE
The global ecosystem; the parts of the planet and its atmosphere where life is possible

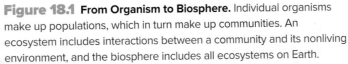

ECOSYSTEM
The living and nonliving components of an area
Example: The savanna

COMMUNITY
All populations that occupy the same region
Example: All populations in a savanna

The science of ecology is global because life exists almost everywhere on Earth, even in places once thought to be much too harsh to support it. Scientists have discovered life in Arctic ice, salt flats, hot springs, hydrothermal vents, and mines that plunge miles below Earth's surface. All of these areas are part of the **biosphere,** the portion of Earth where life exists. Thanks to the global circulation of energy, nutrients, and water, the biosphere forms one huge, interconnected ecosystem.

Within the biosphere are prairies, seashores, deserts, jungles, mountaintops, and a patchwork of other unique landscapes, each with its own set of conditions. Fire regularly ravages the prairie but not the beach; water is scarce in the desert but not in the jungle. The species that are native to each location have adaptations that correspond to these conditions. The same basic evolutionary process—natural selection—has produced unique populations and communities of organisms in nearly every possible habitat. As we shall see in this unit, interactions with both the living and the nonliving world shape the adaptations that contribute to survival and reproductive success.

18.1 Mastering Concepts

1. Distinguish among ecosystems, communities, and populations.

2. Name some living and nonliving parts of your environment.

3. What is the biosphere?

18.2 A Population's Size and Density Change Over Time

To learn more about a population, ecologists begin by describing the population and its habitat. The **habitat** is the physical location where the members of a population normally live. The ocean, desert, and rain forest are typical examples, but an organism's habitat might even be another organism. Your body, for example, is home to billions of microbes. (i) *beneficial microbes,* section 15.2D

Population density is the number of individuals of a species per unit area or unit volume of habitat. Both density and distribution within the habitat vary greatly among species. Figure 18.2, for example, shows a densely packed penguin colony and widely spaced clumps of palm trees. Bacteria often live in extremely dense populations, with billions of bacteria occupying a spoonful of soil.

Hunting, trapping, and fishing regulations are based on population estimates, as are decisions on where to build (or not to build) houses, dams, bridges, and pipelines. But how do researchers know how many individuals of each species inhabit an area?

Simple counts are occasionally possible. For example, aerial photos can reveal the number of caribou in a herd or seals on an island (see figure 18.8). Unless a species is extremely rare or restricted to a limited range, however, it is usually impossible to count each individual. Instead, most population estimates rely on sampling techniques. One common way to estimate the size of a plant population is to count the number of stems in randomly selected locations, such as within a 1-meter square or along a 50-meter line. As another example, a soil ecologist might count the insects or spiders that stumble into a pitfall trap.

A widely used technique to estimate animal populations is called mark–recapture. Suppose researchers want to know how many squirrels inhabit a

a.   b.

Figure 18.2 Portraits of Two Populations. (a) These penguins form a dense breeding colony. (b) Clumps of palms grow sparsely in the sand dunes of Tunisia.

(a): ©Rashman/Shutterstock RF; (b): ©PhotoAlto/PunchStock RF

park. They begin by placing baited nest boxes in trees. After a day or two, the researchers record the weight, sex, age, and health status of each captured squirrel. Each squirrel also receives a unique tattoo or other identifying mark before being released. The proportion of marked squirrels that are subsequently recaptured can help the biologists estimate the population size.

Some organisms, such as those living in the open ocean, are especially hard to count. Burning Question 18.1 describes some of the ways that biologists solve this problem.

Population density measurements provide static "snapshots" of a population at one time. Most populations, however, are not stable. Invasive weeds may take over a habitat as their populations explode; conversely, poaching may devastate a population of rhinos or gorillas. By repeatedly using the same method to measure a population's size over weeks, months, or years, ecologists can determine whether control measures or conservation efforts are working. Collecting long-term data on plant and animal population sizes can also help ecologists measure the effects of catastrophes such as fires or oil spills.

18.2 Mastering Concepts

1 What is population density?

2. What are some ways to measure a population's density?

18.3 Births and Deaths Help Determine Population Size

As we have already seen, population ecologists study the factors that determine whether a population grows, remains stable, or declines. If a population adds more individuals than are subtracted, the population grows. If the opposite happens, the population shrinks (see Why We Care 18.1). The population size remains unchanged if additions exactly balance subtractions.

Table 18.1 summarizes the ways that a population can gain or lose members. Births are the most obvious ways to add individuals. A population's **birth rate** is the number of new individuals produced per individual in a defined time period. For example, the human birth rate worldwide is about 18 births per 1000 people per year.

The number of offspring an individual produces over its lifetime depends on many variables. The number of times it reproduces and the number of offspring per reproductive episode are important, as is the age at first reproduction. All other things being equal, the earlier reproduction begins, the faster the population will grow.

A population's **age structure,** or distribution of age classes, also helps determine its birth rate (figure 18.3). A population with a large fraction of prereproductive individuals will grow. As these individuals enter their reproductive years and produce offspring, the prereproductive age classes swell further, building a foundation that ensures future growth. Conversely, a population that consists mainly of older individuals will be stable or may even decline. This situation can doom a population of endangered plants if, for example, habitat destruction makes it impossible for seedlings to establish themselves. With few young individuals to replace those that die of old age, the population may go extinct.

Death is the most obvious way for a population to lose members. A population's **death rate** is the number of deaths per unit time, scaled by the population size. The causes of death may include accidents, disease, predation, and

Factor	Affected by...
Additions	
Births	Number of reproductive episodes per lifetime
	Number of offspring per reproductive episode
	Age at first reproduction
	Population age structure
Migration into the population	Availability of dispersal mechanism
	Availability of suitable habitat
Subtractions	
Deaths	Accidents
	Disease
	Predation
	Nutrient availability
Migration out of the population	Availability of dispersal mechanism

TABLE 18.1 Factors Affecting Population Growth: A Summary

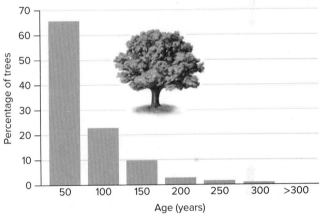

a. White oak

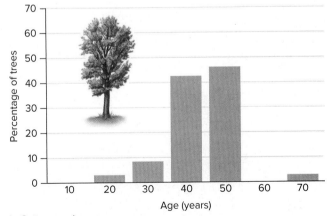

b. Cottonwood

Figure 18.3 Age Structures. (a) This white oak population is dominated by younger individuals, indicating high potential for future reproduction. (b) This population of cottonwoods has few individuals in the youngest age classes. The old trees will likely die soon. Lacking young trees to take their place, the population is probably doomed.

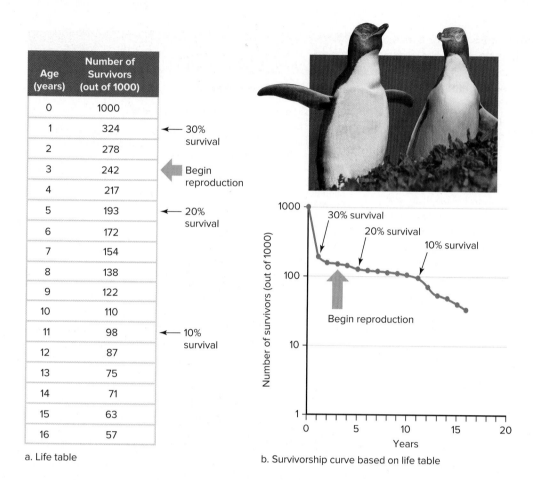

Age (years)	Number of Survivors (out of 1000)
0	1000
1	324
2	278
3	242
4	217
5	193
6	172
7	154
8	138
9	122
10	110
11	98
12	87
13	75
14	71
15	63
16	57

← 30% survival

← Begin reproduction

← 20% survival

← 10% survival

a. Life table

30% survival
20% survival
10% survival
Begin reproduction

b. Survivorship curve based on life table

Figure 18.4 **Penguin Survivorship.** (a) This life table lists the number of survivors remaining each year out of a group of 1000 yellow-eyed penguins that hatch at the same time. (b) A survivorship curve is a graph of the data in a life table. The curve reveals that most penguins die before age 1, after which the death rate levels off.

Photo: ©Kevin Schafer/Corbis/Getty Images

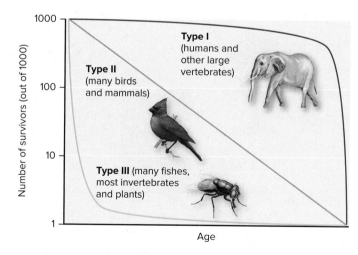

Type I (humans and other large vertebrates)

Type II (many birds and mammals)

Type III (many fishes, most invertebrates and plants)

Age

Figure 18.5 **Three Survivorship Curves.** In type I species, most individuals survive to old age, whereas in type III species, most individuals die young. Type II species are in between, with constant survivorship throughout the life span.

competition for scarce resources. In the human population, for example, the overall death rate is approximately 8 deaths per 1000 people per year; section 18.6 describes human death rates in more detail.

Each individual in a population will eventually die; the only question is when. Some species, such as dandelions, produce many offspring but invest little energy or care in each one. The probability of dying before reaching reproductive age is therefore very high. In other species, such as humans, heavy parental investment in a small number of offspring means that most individuals survive long enough to reproduce.

To help interpret which pattern might apply to a particular species, population biologists developed the **life table,** a chart that shows the probability of surviving to any given age. (Life insurance companies use life tables to compute premiums for clients of different ages.) Figure 18.4a, for example, shows a life table for yellow-eyed penguins. The declining number of survivors in each age class reflects the effects of predation, disease, food scarcity, and all other factors that prevent an individual from reaching its theoretical life span.

The values in a life table are often plotted onto a **survivorship curve,** a graph of the proportion of surviving individuals at each age. Figure 18.4b shows a graph of the values from the penguin life table. Note that the y-axis data are plotted on a logarithmic scale, not a linear one. The log scale makes it easier to see trends along the entire range of values, from 0 to 1000.

The survivorship curves of many species follow one of three general patterns (figure 18.5). Type I species, such as humans and elephants, invest a great deal of energy and time in each offspring. Most individuals live long enough to reproduce, and the death rate is highest as individuals approach the maximum life span. Type II species, including many birds and mammals, may also provide a great deal of parental care. However, the threats of predation and disease are constant throughout life, and these organisms have an equal probability of dying at any age. The type II line is therefore straight. Type III species, such as many fishes and most invertebrates and plants, may produce many offspring but invest little in each one. Most offspring of type III species therefore die at a very young age.

Of course, these generalized examples do not describe all populations. Many species have survivorship curves that fall between two patterns. The yellow-eyed penguins in figure 18.4, for example, have a survivorship curve that combines features of both type II and type III curves.

Birth and death rates are key regulators of a population's size, but migration may also be significant (see table 18.1). Immigration has tremendously increased the human population in the United States over the past 200 or so years. Most Americans are either immigrants or descended from immigrants, and migration is projected to account for most population growth in this country by around the year 2030. Likewise, nonhuman species also disperse to new habitats. They may actively swim, fly, or walk; alternatively, wind or water currents may move individuals into or out of a population.

18.3 Mastering Concepts

1. Under what conditions will a population grow?
2. What factors determine birth and death rates in a population?
3. Describe the three patterns of survivorship curves.

18.4 Natural Selection Influences Life Histories

Although every population's size depends in part on birth and death rates, species differ widely in the timing of these events. Population ecologists therefore find it useful to document a species' **life history,** which includes all events of an organism's life from conception through death. The main focus of life history analysis is the adaptations that influence reproductive success.

The life history of a species includes its typical developmental rate, life span, social behaviors, reproductive timing, mate selection, number and size of the offspring, number of reproductive events, and amount of parental care. This section explores how natural selection shapes some of these life history traits.

A. Organisms Balance Reproduction Against Other Requirements

A species' life history reflects a series of evolutionary trade-offs; after all, supplies of time, energy, and resources are always limited. Just as an investor allocates money among stocks, bonds, and real estate, a juvenile organism divides its efforts among growth, maintenance, and survival. After reaching maturity, another competing demand—reproduction—joins the list.

Reproduction is extremely costly. Many animals, for example, devote time and energy to attracting mates, building and defending nests, and incubating or gestating offspring. Once the young are hatched or born, the parents may feed and protect them. All of these activities limit a parent's ability to feed itself, defend itself, or reproduce again. In plants, the reproductive investment is also substantial. Flowers, fruits, and defensive chemicals cost energy

Burning Question 18.1

How do biologists count animals in the open ocean?

Whales, dolphins, sea turtles, Atlantic bluefin tuna, and many other marine animals are under threat. Scientists will play a critical role in recovery efforts. After all, without solid data on population trends, it is difficult to argue for, or measure the effects of, new regulations intended to protect sensitive populations. How do ecologists track the population sizes of these deep-sea dwellers, which are often hidden beneath an expansive sheet of blue?

Both sight and sound provide critical clues in counting marine mammals. Aerial surveys are sometimes effective; it is easier to see into the water from the air than from a boat. Moreover, whales and other marine mammals have lungs, so they must come to the surface occasionally to breathe. In an aerial survey, a plane flies in a systematic pattern over the region to be surveyed. When a whale surfaces, the population-surveying crew marks the location of the animal on a map. Ecologists also use sound to locate marine mammals. Audio recording devices attached to a boat or a buoy "listen" for whale or dolphin calls.

Sea turtle populations are often monitored using direct counts. Adult females return to the same beach to lay eggs every few years, simplifying data collection. However, many ecologists argue that this small snapshot of the turtle population is not adequate to accurately predict population changes over time. Future studies may focus on survival and reproduction rates, but these data are considerably more difficult to collect.

Scientists must also monitor marine animals that are part of our cuisine. Since the 1950s, Atlantic bluefin tuna have been notoriously overharvested—that is, they are caught faster than they can reproduce. Research crews use a "catch-by-unit-effort" formula to estimate tuna population size. For example, a crew might fish in the same way for the same amount of time each year and record how many tuna are caught. If the population is declining, then the catch each year will also go down. After several such expeditions, the researchers can estimate the population size and rate of decline.

**Submit your burning question to
marielle.hoefnagels@mheducation.com**

(turtle): Source: NOAA/Public and Constituent Affairs/Julie Bedford

to produce and may take away from a plant's photosynthetic area, reducing the ability to capture sunlight.

Besides the total effort allocated to reproduction, the timing is also critical. An organism that delays reproduction for too long may die before producing any offspring at all. On the other hand, reproducing too early diverts energy away from the growth and maintenance that may be crucial to survival. The rest of this section describes how life histories reflect very different solutions to these trade-offs.

B. Opportunistic and Equilibrium Life Histories Reflect the Trade-Off Between Quantity and Quality

Although each species is unique, ecologists have discovered that reproductive strategies fall into patterns shaped by natural selection. One prominent trade-off is the balance between offspring quality and quantity.

At one extreme are species that have an **opportunistic life history** (also called an *r*-selected life history), in which individuals tend to be short-lived, reproduce at an early age, and have many offspring that receive little to no care (figure 18.6a). The population's growth rate can be very high if conditions are optimal. In general, however, each offspring has a very low probability of surviving to reproduce; this pattern is typical of species with type III survivorship curves (see figure 18.5).

Weeds, insects, and many other invertebrates typically have opportunistic life histories. For example, a pigweed plant sheds 100,000 seeds in the one summer of its life; a female winter moth mates once and lays hundreds of fertilized eggs just before she dies. These and many other organisms mature early, produce many offspring in a single reproductive burst, and die. Although their populations can skyrocket when conditions are favorable, competition for resources and stresses such as frost or drought soon limit their growth.

At the other extreme are species with an **equilibrium life history** (also called a *K*-selected life history), in which individuals tend to be long-lived, to be late-maturing, and to produce a small number of offspring that receive extended parental care (figure 18.6b). Coconut trees, many birds, and large mammals (including humans) have equilibrium life histories, maturing late and producing relatively few offspring throughout their long lives. High parental investment in each offspring means that most live long enough to reproduce, so these species have type I survivorship curves.

Just as we saw with survivorship curves, strict adherence to an opportunistic or equilibrium strategy is the exception, not the rule. For example, even populations of large animals with equilibrium life histories fluctuate greatly in response to changes in their environments. Nevertheless, these two strategies illustrate the important point that natural selection shapes a species' life history characteristics. These traits therefore reflect the competing demands of reproduction and survival. Investigating Life 18.1 explores how unique habitats—dark, toxic waters—influence this basic natural selection trade-off.

Opportunistic life history

- High reproduction rate
- Many offspring
- Each offspring receives little parental care
- Low survival rate for juveniles
- Early reproductive maturity
- Type III survivorship curve

Survivors / Age

a.

Equilibrium life history

- Low reproduction rate
- Few offspring
- Each offspring receives extensive parental care
- High survival rate for juveniles
- Late reproductive maturity
- Type I survivorship curve

Survivors / Age

b.

Figure 18.6 Opportunistic and Equilibrium Life Histories. (a) Rice plants and tussock moths produce many small offspring and invest little in each one. (b) Each coconut produced by this tree represents a large energy investment. Likewise, grizzly bears devote extensive time and resources to rearing a small number of young.

Photos: (a): (rice): Source: USDA/Keith Weller; (moth): ©Andrew Darrington/Alamy Stock Photo; (b): (palm): ©Kittichai/Shutterstock RF; (bears): ©DLILLC/Corbis RF

18.4 Mastering Concepts

1. Explain why an organism cannot produce large numbers of offspring that receive extended parental care.
2. Distinguish between opportunistic and equilibrium life histories.

Time (Days)	Population at Start of Day	Number of Individuals Added	Population at End of Day
1	100	22	122
2	122	27	149
3	149	33	182
4	182	40	222
5	222	49	271
6	271	60	331
7	331	73	404
8	404	89	493
9	493	108	601
10	601	132	733
11	733	161	894
12	894	197	1091
13	1091	240	1331
14	1331	293	1624
15	1624	357	1981
16	1981	436	2417
17	2417	532	2949
18	2949	649	3598
19	3598	792	4390
20	4390	966	5356

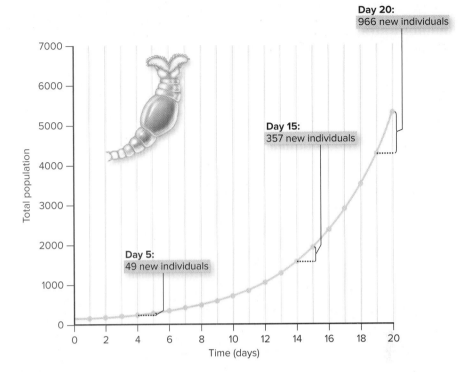

Day 20: 966 new individuals

Day 15: 357 new individuals

Day 5: 49 new individuals

Figure 18.7 Exponential Population Growth. For a population of rotifers with unlimited resources, the number of individuals added each day increases with each generation. Colored rows in the table represent time points that are labeled in the graph.

18.5 Population Growth May Be Exponential or Logistic

Regardless of its life history strategy, any population will grow if the number of individuals added exceeds the number removed. But how fast will it grow, and how large can the population become? Two mathematical models, called exponential and logistic growth, illustrate two simple patterns of population growth.

A. Growth Is Exponential When Resources Are Unlimited

In a population exhibiting **exponential growth,** the number of new individuals is proportional to the population's size; that is, the larger the population, the faster it grows. For example, suppose that 100 aquatic animals called rotifers are placed in a tank under ideal growth conditions. How many new rotifers are produced each day? The answer depends on the size of the population: the more rotifers in the tank, the larger the capacity to add offspring. **Figure 18.7** shows the size of each generation and plots the running total on a graph. A characteristic **J-shaped curve** emerges when exponential growth is plotted over time.

Growth resulting from repeated doubling (1, 2, 4, 8, 16, 32, . . .), such as in bacteria, is exponential. Species introduced to an area where they are not native may also multiply exponentially for a time, since they often have no natural population controls. **Figure 18.8** shows another example. Researchers have counted the grey seal pups born on Sable Island, Nova Scotia, since the 1960s. For four decades, the number of pups was proportional to the total population, implying exponential growth.

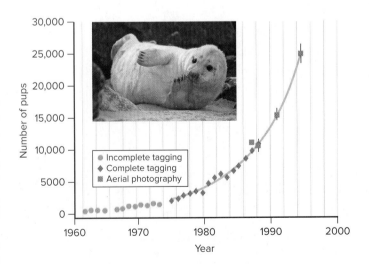

- Incomplete tagging
- Complete tagging
- Aerial photography

Figure 18.8 Seal Pup Population. Tagging and aerial surveys of the grey seal pups on Sable Island, Nova Scotia, revealed 40 years of exponential population growth. (Growth began to slow after the year 2000.)

Photo: ©Ronald Wittek/Getty Images

Time (Days)	Population at Start of Time Interval	Number of Individuals Added	Population at End of Time Interval
1–5	100	153	253
6–10	253	325	577
11–15	577	510	1087
16–20	1087	490	1578
21–25	1578	273	1851
26–30	1851	103	1954
31–35	1954	33	1986
36–40	1986	10	1996

Figure 18.9 **Logistic Population Growth.** When resources are limited, the number of individuals added each day declines. At the habitat's carrying capacity, birth and death rates are equal, and the population no longer grows. Colored rows in the table represent time intervals that are labeled in the graph.

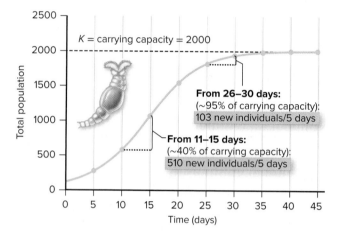

K = carrying capacity = 2000

From 26–30 days:
(~95% of carrying capacity):
103 new individuals/5 days

From 11–15 days:
(~40% of carrying capacity):
510 new individuals/5 days

B. Population Growth Eventually Slows

Exponential growth may occur for a short time, but it cannot continue indefinitely because some resource is eventually depleted. Competition, predation, and anything else that reduces birth rates or increases death rates all can keep a population from reaching its maximum growth rate.

Every habitat has a **carrying capacity,** which is the maximum number of individuals that the ecosystem can support indefinitely. This carrying capacity imposes an upper limit on a population's size. How does this limit affect the population's growth rate? According to the **logistic growth** model, the early growth of a population may be exponential, but growth slows and stops as the population approaches the habitat's carrying capacity (figure 18.9). The resulting **S-shaped curve** shows this change in the growth rate over time.

The carrying capacity of a habitat typically is not fixed. A drought that lasts for a decade may be followed by a year of exceptionally heavy rainfall, causing a flush of new plant growth and a sudden increase in food availability for animals. Alternatively, the food on which a species relies may disappear, or a catastrophic flood can drastically reduce the carrying capacity as habitat is destroyed.

In studying population growth, it is important to understand that some species do not fit neatly into either the exponential or logistic models. In collared lemmings, for example, the population fluctuates on a 4-year cycle (figure 18.10). Predation by stoats (a type of weasel) appears to be one of the main factors regulating the ups and downs of the lemming population.

C. Many Conditions Limit Population Size

A combination of factors determines the size of most populations. Consider the population of songbirds in your town. Some lose their lives to cold weather or food shortages, whereas others succumb to infectious disease or the jaws of a cat. These and other limits on the growth of the bird population fall into two general categories: density-dependent and density-independent (figure 18.11).

Density-dependent factors are conditions whose effects increase as a population grows. Most density-dependent limits are **biotic,** meaning they result from interactions with living organisms. Within a population, competition for space, nutrients, sunlight, food, mates, and breeding sites is density-dependent. When many individuals share limited resources, few may be able to reproduce, and population growth slows or even crashes.

Other species can also exert density-dependent limits on a population's growth. Two or more species may compete fiercely for the same nutrients and space (see figure 19.11). Infectious disease also takes a toll. Viruses, bacteria, and other microbes spread by direct contact between infected individuals and new hosts. The higher the host population density, the more opportunity for these disease-causing organisms to spread. Likewise, a higher population density may lead to a higher probability of death by predation.

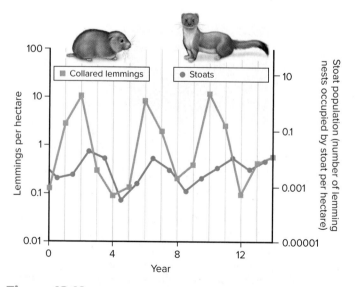

Figure 18.10 **Population Cycle.** The population of collared lemmings fluctuates regularly over a 4-year period. Research suggests that a major influence on the lemming population is the number of stoats, a type of weasel that preys on the lemmings.

Figure 18.11 Factors Limiting Populations. (a) Density-dependent factors become more important as the habitat becomes crowded. This herd of deer may soon overwhelm its habitat. (b) Density-independent factors such as forest fires limit populations at all densities.

Photos: (a): ©Tim Graham/Getty Images; (b): Source: Bureau of Land Management/John McColgan, Alaska Fire Service

Density-independent factors exert effects that are unrelated to population density. Most density-independent limits are **abiotic,** or nonliving. Natural disasters such as fires, earthquakes, floods, volcanic eruptions, and severe weather are typical density-independent factors. The high winds of a hurricane, for example, may destroy 50% of the birds' nests in a forest, without regard to the density of the bird population. Likewise, a lava flow kills everything in its path. The effects of oil spills and other industrial accidents are density-independent, too. In addition, as described in chapter 20, habitat destruction related to human activities is a density-independent factor that is pushing many species to the brink of extinction.

No matter what combination of factors determines a population's size, it is worth remembering the importance of these limits in natural selection and evolution. Many individuals do not survive long enough to reproduce. Those that do manage to breed have the adaptations that allow them to escape the biotic and abiotic challenges that claim the lives of many of their counterparts. As these "fittest" individuals pass their genes on, the next generation contains a higher proportion of offspring with those same adaptations.

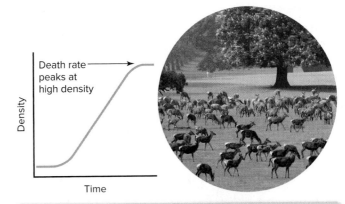

Death rate peaks at high density

Density

Time

Density-dependent limits

- Competition for resources within or among species
- Infectious disease
- Predation

a.

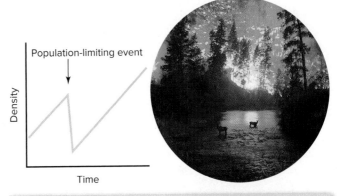

Population-limiting event

Density

Time

Density-independent limits

- Natural disasters
- Industrial accidents
- Habitat destruction

b.

18.5 Mastering Concepts

1. What conditions support exponential population growth?
2. How does logistic growth differ from exponential growth?
3. Distinguish between density-dependent and density-independent factors that limit population size, and give three examples of each.

Why We Care 18.1 | Controlling Animal Pests

Controlling animal pests is a practical application of population ecology. Biologists are tracking many booming populations, such as deer in many parts of the United States, swimming rodents called nutria in Louisiana, brown tree snakes in Guam, and koalas in Australia. In each case, population control means boosting the death rate, cutting the birth rate, or both.

Hunting is a simple way to increase a population's death rate. Nutria are an invasive species in Louisiana, and population control is vital to protecting the native ecosystem. The Coastwide Nutria Control Program offers five dollars for each nutria tail. Thanks to this economic incentive, hundreds of thousands of nutria are harvested each year.

Hunting also reduces populations of deer. In many areas of the United States, humans have eliminated natural predators of deer, and overpopulation is a serious problem. These large animals graze on tender plants and pose a serious risk of collisions

©John White Photos/ Getty Images RF

with vehicles. Regions with too many deer may allow hunting of does (female deer), which reduces the reproductive rate faster than restricting the hunt to bucks (males).

Similarly, efforts to control the brown tree snake population on Guam focus on increasing the death rate. The most common approaches are to trap and kill the snakes or to leave poisoned bait. In the future, introducing parasites that spread among the snakes may be another possibility. Since a population's growth rate reflects both birth and death rates, however, researchers are also investigating ways to prevent snake reproduction.

Reducing birth rates is the primary method for controlling koala populations. Koalas are destroying eucalyptus trees and starving to death in some areas of Australia. Surgically sterilizing male koalas is one method of birth control. Scientists are also giving female koalas birth control implants that block the release of an egg.

Figure 18.12 Historical Growth of the Human Population. The human population has grown exponentially over the past few thousand years, with the most rapid growth occurring within the last 200 years (*inset*).

Data source: U.S. Census Bureau

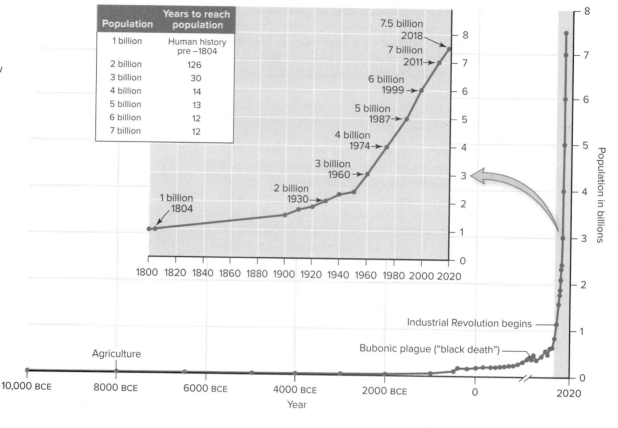

Population	Years to reach population
1 billion	Human history pre –1804
2 billion	126
3 billion	30
4 billion	14
5 billion	13
6 billion	12
7 billion	12

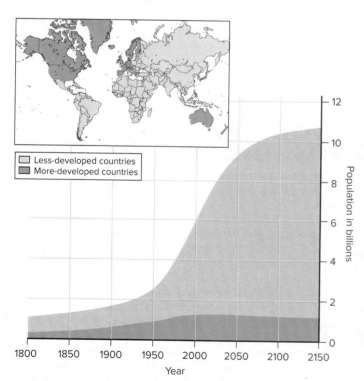

Figure 18.13 Projected Population Growth. Future population growth will continue to be concentrated in less-developed countries.

18.6 The Human Population Continues to Grow

The human population has grown exponentially in the last 2000 years (figure 18.12). So far, we have found ways to escape many of the forces that limit the growth of other animal populations. Yet exponential growth cannot continue indefinitely. This section describes how the principles of population growth apply to the human population.

A. Birth and Death Rates Vary Worldwide

By early 2018, the world's human population was approaching 7.5 billion. Since Earth's land area is about 150 million square kilometers, the average population density is about 50 people per square kilometer of land area, but the distribution is far from random. For the most part, the highest population densities worldwide occur along the coastlines and in the valleys of major rivers; very few people live on high mountains, in the middle of the world's major deserts, or on Antarctica. Two countries—China and India—account for one-third of all humans.

Overall, the human population growth rate is about 1.0% per year and declining. Demographers project that zero population growth may happen during the twenty-second century, but no one is certain. Nor do we know how many people will inhabit Earth when that occurs. Clearly, however, less-developed countries are growing at much faster rates than are more-developed countries (figure 18.13).

What explains the difference in growth rates? Each country's economic development influences its progress along the *demographic transition*, during which birth rates and death rates shift from high to low (figure 18.14). In the first stage of the demographic transition, population growth is minimal because both birth and death rates are high (as they were early in human history). Then, in the next stage of the demographic transition, improved living conditions and disease control lower the death rate, but birth rates remain high. This transitional period therefore sees the rapid population growth typical of the world's less-developed countries. During the third stage of the demographic transition, birth rates fall; the difference between birth and death rates is once again small. The population's growth rate therefore slows once more. The world's more-developed countries have entered this stage, and a few even have declining populations because death rates exceed birth rates.

Factors Affecting Birth Rates As described in section 18.3, a population's age structure helps predict its future birth rate. Figure 18.15 shows the age structures for the world's three most populous countries. In India and many other less-developed countries, a large fraction of the population is entering its reproductive years, suggesting a high potential for future growth (see figure 18.15a). In the United States, as in many other developed countries, the population consists mainly of older individuals (see figure 18.15b). Such populations are stable or declining.

As recently as 1990, China's age structure resembled that of present-day India. But China's population shows an overall decline in the youngest age classes, so its future growth rate should decline (see figure 18.15c). Between 1980 and 2016, the Chinese government controlled runaway population growth by limiting most families to one child. (Couples are now allowed two children.) Although China remains the world's most populous nation, biologists expect India to take the lead by around 2025.

Overall, why do birth rates tend to decline as economic development progresses? One explanation for this trend is the availability of family planning programs, which are relatively inexpensive and have immediate results. Social and economic factors play an important role as well. Educated women are most likely to learn about and use family planning services, have more opportunities outside the home, and may delay marriage and childbearing until after they enter the workforce. Delayed childbearing often means fewer children and therefore slows population growth.

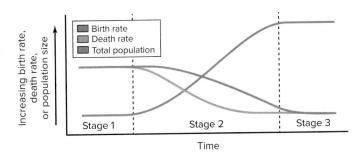

Figure 18.14 **The Demographic Transition.** During stage 1, birth rates and death rates are high, so the population remains small. Population growth is rapid in stage 2, when death rates fall faster than birth rates. In stage 3, birth rates and death rates are both low, and the population stabilizes.

Figure It Out

In Niger, nearly 50% of individuals are younger than 15. Is the population of the Niger likely to grow, remain stable, or decline? Why?

Answer: The population is likely to grow because the proportion of prereproductive individuals is very high.

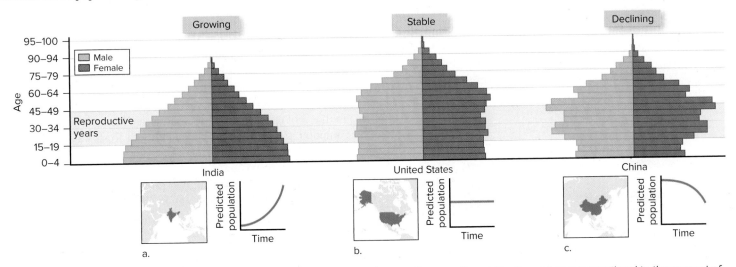

Figure 18.15 **Age Structures for Three Human Populations.** In age structure diagrams, the width of each bar is proportional to the percent of individuals in that age class. (a) India's population is likely to continue to grow because a high proportion of individuals are in the youngest age classes. (b) The population of the United States is stable, with roughly equal numbers of people in each age group. (c) China's future growth rate should decline because most of its members are in older age classes.

Source: Data from U.S. Census Bureau, International Data Base.

TABLE 18.2 Top Five Causes of Death in High- and Low-Income Countries

Rank	High-Income Countries	Low-Income Countries
1	Heart disease (coronary artery blockage)	Lower respiratory infection (pneumonia, acute bronchitis)
2	Stroke	Diarrhea (e.g., cholera, rotavirus)
3	Alzheimer disease and other forms of dementia	Stroke
4	Lung cancer	Heart disease (coronary artery blockage)
5	Chronic obstructive pulmonary disease	HIV/AIDS

Source: World Health Organization Fact Sheet, *Leading Causes of Death by Economy Income Group*

Factors Affecting Death Rates Besides the birth rate, the death rate is the other major factor influencing a population's growth rate. Human life expectancy averages around 69 worldwide; it exceeds 80 in a few developed countries.

Table 18.2 shows the top five causes of death in developed and developing countries. Heart disease, stroke, dementia, and lung cancer top the list in the developed world. Infectious diseases rank relatively low, thanks in large part to sanitation, antibiotics, and vaccines. In contrast, deadly diseases such as respiratory infections, diarrhea, and HIV/AIDS are more prominent in developing countries. Crowded conditions facilitate the spread of cholera and other waterborne diseases, especially in areas with limited access to clean drinking water. AIDS has taken an especially high toll in sub-Saharan Africa, where the epidemic has significantly increased death rates, and life expectancy is below 55 in some countries. ⓘ *HIV*, section 7.8C

B. The Ecological Footprint Is an Estimate of Resource Use

The human population cannot continue to grow exponentially because living space and other resources are finite. Worldwide, increasing numbers of people will mean greater pressure on land, water, air, and fossil fuels as people demand more resources and generate more waste.

Ecologists summarize each country's demands on the planet by calculating an ecological footprint. Just as an actual footprint shows the area that a shoe occupies with each step, an **ecological footprint** measures the amount of land area needed to support a person's or a country's overall lifestyle. The calculation includes, among other measures, energy consumption and the land area used to grow crops for food and fiber, produce timber, and raise cattle and other animals. The land areas occupied by streets, buildings, and landfills are also part of the ecological footprint. Not surprisingly, the world's wealthiest and most populous countries have the largest ecological footprints (figure 18.16).

Energy consumption accounts for about half of the ecological footprint. The wealthiest countries make up less than 20% of the world's population yet consume more than half of the energy. Less-developed countries, however, will take a larger share of energy supplies as their populations grow and their economies become more industrialized. Since the vast majority of the energy comes from fossil fuels, the result will be increased air pollution and acid rain. Moreover, the accumulation of CO_2 and other greenhouse gases is implicated in global climate change.

Food production is another significant element of the ecological footprint. Overall, both the demand for food and agricultural productivity rise each year. To boost food production, people often expand their farms into forests, destroying habitat and threatening biodiversity. ⓘ *biofuels*, section 16.1

Because the ecological footprint focuses on land area, it does not include water consumption. Nevertheless, the availability of fresh water has declined worldwide as people have demanded more water for agriculture, industry, and household use. In many poor countries, less than half the population has access to safe water for drinking and cooking. ⓘ *world water resources*, section 19.3C

Chapter 20 further explores deforestation, species extinctions, increased fuel consumption, global climate change, and other environmental problems related to the expanding human population.

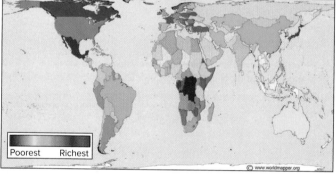

Poorest — Richest

© www.worldmapper.org

a.

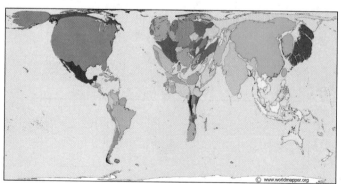

© www.worldmapper.org

b.

Figure 18.16 **Ecological Footprint.** (a) In a traditional map, each country's size is proportional to its land area. (b) Here, each country's size is proportional to its ecological footprint. Note that the mapmakers grouped countries into 12 regions and assigned each region (not each individual country) a color according to its wealth.

18.6 Mastering Concepts

1. What factors affect human birth and death rates worldwide?
2. What are some of the environmental consequences of human population growth?

Investigating Life 18.1 | A Toxic Compromise

Evolution selects for life histories that maximize the number of offspring that live to reproductive age (see section 18.4). Many factors influence life history evolution, including the abundance and predictability of resources in the environment.

Harsh environments can also affect life history traits. Fish called Atlantic mollies are an ideal species for studying the trade-offs of living in a challenging environment. Atlantic mollies live in four distinct habitats in Mexico. The most stressful environments are toxic streams flowing through dark caves (figure 18.A). These streams contain a high concentration of poisonous hydrogen sulfide (commonly associated with its "rotten egg" odor). Other mollies populate nearby caves containing nontoxic water. Atlantic mollies also live in surface waters, which may be either poisonous or nontoxic. Measuring the life history qualities of the mollies in the four habitats could reveal the evolutionary cost of a stressful environment.

To learn more about the life history adaptations of Atlantic mollies, a team of biologists collected pregnant female fish from each habitat. They anesthetized and preserved the animals. Later, at the lab, they measured the number and mass of embryos in each female. They also determined the developmental stage of each embryo and estimated what its mass would have been at birth.

Their results were clear: Females in toxic cave streams produced fewer, larger offspring than females in open, nontoxic streams (figure 18.B). The offspring of females from habitats with only one environmental stress—that is, only toxic water or only darkness—were intermediate in both number and mass. The data suggest that in the harshest environments, females sacrifice offspring quantity in favor of quality.

Why produce fewer, larger babies? The explanation is that food can be scarce in a dark, poisonous habitat. Larger offspring are better swimmers, so they can find more food and are less likely to starve than their smaller counterparts in a dark, toxic stream. These large offspring are most likely to live long enough to reproduce, passing on their "large offspring" genes to the next generation.

Figure 18.A **Dark Water.** Some Atlantic mollies inhabit toxic water in caves.

(both): ©Stephen Alvarez/National Geographic/Getty Images

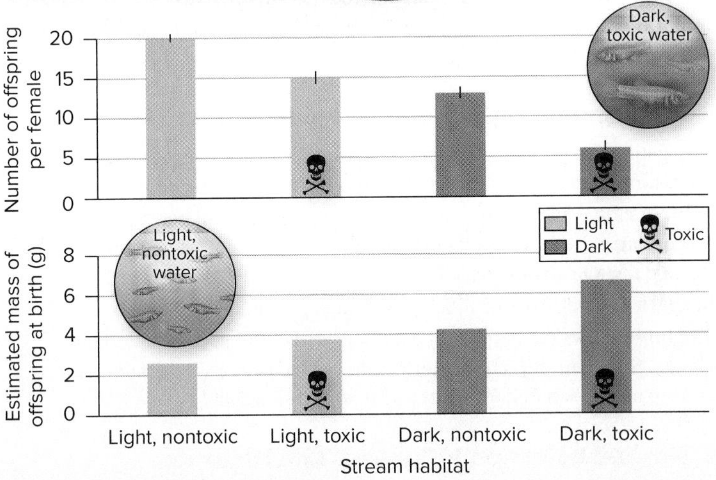

Figure 18.B **Life History Trade-Offs.** Dark, toxic water selected for fewer, larger offspring than did light or nontoxic streams. In the top graph, error bars represent standard errors (see appendix B).

Source: Riesch, Rüdiger, Martin Plath, and Ingo Schlupp. 2010. Toxic hydrogen sulfide and dark caves: life-history adaptations in a livebearing fish (*Poecilia mexicana*, Poeciliidae). *Ecology*, vol. 91, pages 1494–1505.

What's the **Point?** ▼ APPLIED

Ecologists consider many factors—birth rates, death rates, age structure, migration patterns, and more—when monitoring populations. The data are valuable for plant and wildlife conservation because they reveal whether current trends will lead to species extinction or recovery. Similar projections are also important for the human population, as world leaders estimate the future need for water, food, energy, and many other resources.

No one knows exactly how growth rates will change in the future, so it is impossible to predict when or at what level the human population will stabilize. In 2015, the United Nations issued three projections for the world's population, assuming high, medium, and low growth rates. The highest projection says the Earth's population will be around 16.6 billion (and still growing) in 2100. The medium estimate shows the population stabilizing at around 11.2 billion by 2100. The low estimate predicts that the population will peak at just over 8.7 billion in 2050, then decline to about 7.3 billion by 2100.

Almost all of the growth in the coming century will be in less-developed countries. In fact, the UN report suggests that the populations of developed

countries would decrease in the coming century if not for migration from developing countries.

The U.S. Census Bureau maintains population projections for the United States. The U.S. population is expected to grow from approximately 330 million in 2018 to at least 400 million by the year 2060. This growth rate is higher than that projected for many other developed nations, but the United States is also expected to have the highest number of immigrants of any country.

These projections are only as good as their assumptions. For example, researchers assume that the birth rate in developing countries will continue to decline; if it remains at its current level, the population by year 2100 will be nearly 28 billion! But how much will the birth rate decline, and how fast will it happen? The answer to this question will determine the future of Earth's human population.

©Melanie Stetson Freeman/ Christian Science Monitor/ The Image Works

CHAPTER SUMMARY

18.1 Ecology Is the Study of Interactions

- **Ecology** considers interrelationships between organisms and their environment. It includes interactions at the **population, community,** and **ecosystem** levels. The **biosphere,** which includes all areas where life exists, is one interconnected ecosystem.
- Population ecology is the study of the factors that determine a population's size over time.

18.2 A Population's Size and Density Change Over Time

- A **habitat** is the location where an individual normally lives.
- **Population density** is a measure of the number of individuals per unit area of habitat.

18.3 Births and Deaths Help Determine Population Size

- A population grows when more individuals are added through birth or immigration than leave due to death or migration out of the population.
- The **birth rate** is the number of new individuals produced per capita in a defined time period. A population's birth rate depends on many factors, including the **age structure.**
- The **death rate** reflects the number of deaths per unit time.
- A **life table** shows the number of survivors remaining in a population at each age. **Survivorship curves** reflect the balance between the number of offspring and the amount of parental investment in each.

18.4 Natural Selection Influences Life Histories

- The **life history** of a species includes all events from birth to death but typically emphasizes the factors that affect reproduction.

A. Organisms Balance Reproduction Against Other Requirements

- Organisms must allocate limited time, energy, and resources among growth, maintenance, survival, and reproduction.

B. Opportunistic and Equilibrium Life Histories Reflect the Trade-Off Between Quantity and Quality

- Species with an **opportunistic (*r*-selected) life history** produce many offspring but expend little energy on each, whereas species with an **equilibrium (*K*-selected) life history** invest heavily in rearing relatively few young.

18.5 Population Growth May Be Exponential or Logistic

A. Growth Is Exponential When Resources Are Unlimited

- Population growth that is proportional to the size of the population is **exponential** and produces a **J-shaped curve.**

B. Population Growth Eventually Slows

- In response to competition, predation, and other factors that reduce birth rates and increase death rates, the population may stabilize indefinitely at the habitat's **carrying capacity.** A plot of the resulting **logistic growth** produces a characteristic **S-shaped curve.**

C. Many Conditions Limit Population Size

- **Density-dependent factors** such as infectious disease, predation, and competition have the greatest effect on crowded populations. Most such factors are **biotic,** or living.
- **Density-independent factors,** such as natural disasters, kill the same fraction of the population regardless of the population's density. Most such factors are **abiotic** (nonliving).
- **Figure 18.17** summarizes exponential and logistic growth.

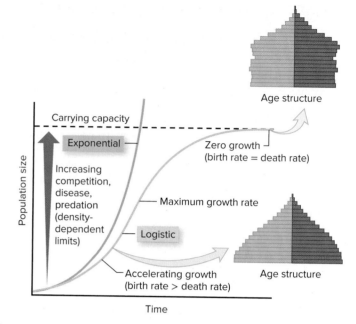

Figure 18.17 Population Growth: A Summary.

18.6 The Human Population Continues to Grow

A. Birth and Death Rates Vary Worldwide

- In less-developed countries, birth rates are high and death rates are low, producing rapid population growth. As economic development increases, birth rates decline and population growth slows.
- Education, access to contraceptives, and government policies affect birth rates.
- The top causes of death vary around the world; they include heart disease, stroke, cancer, and infectious disease.

B. The Ecological Footprint Is an Estimate of Resource Use

- Individuals and countries vary in their **ecological footprint,** a measure of the amount of land required to support their overall lifestyle. Sustained population growth will continue to strain supplies of natural resources such as fossil fuels, farmland, and clean water.

MULTIPLE CHOICE QUESTIONS

1. Which of the following is arranged from least inclusive to most inclusive?
   a. Community < biosphere < population < ecosystem
   b. Ecosystem < population < biosphere < community
   c. Biosphere < ecosystem < community < population
   d. Population < community < ecosystem < biosphere

2. Population size increases when
   a. the sum of birth rate and death rate exceeds the sum of migration into the population and migration out of the population.
   b. the sum of birth rate and migration into the population exceeds the sum of death rate and migration out of the population.
   c. the sum of birth rate and migration out of the population exceeds the sum of death rate and migration into the population.
   d. the sum of death rate and migration into the population exceeds the sum of birth rate and migration out of the population.

3. As a habitat becomes more crowded, which of the following is likely to increase?
   a. Predation   c. Competition for food
   b. Disease   d. All of these are correct.

4. An opportunistic life history emphasizes
   a. large offspring size.
   b. high offspring quantity.
   c. late reproduction.
   d. type I survivorship.

5. As a population's size increases toward the ecosystem's carrying capacity,
   a. the carrying capacity decreases.
   b. the rate of population growth slows.
   c. logistic growth shifts toward exponential growth.
   d. All of the above are correct.

6. Which of the following is typical of developed countries such as the United States?
   a. Smaller ecological footprint than less-developed countries
   b. Sustained exponential growth
   c. Low birth rates and low death rates
   d. A shift from opportunistic to equilibrium life history

Answers to Multiple Choice questions are in appendix A.

WRITE IT OUT

1. List some of the ways you have interacted with your surroundings today. Categorize each item on your list as a population-, community-, or ecosystem-level interaction.

2. Why might an ecologist be interested in studying population dynamics?

3. Refer to the rotifer population in figure 18.7. After 21 days, will the population be greater than, less than, or equal to 6300 individuals? Explain your answer.

4. Domesticated animals, such as house cats, may disturb ecosystems. Are outdoor cats a density-independent or density-dependent factor limiting the populations of birds? Explain your answer. Then, brainstorm as many ways as you can think of for how home bird feeders might raise or lower bird population sizes.

5. White nose syndrome is a disease in bats caused by an infectious fungus (see figure 15.38). The mortality rate of bats that contract the infection is nearly 90%. Is white nose syndrome a density-dependent or a density-independent factor limiting the size of bat populations? As white nose syndrome becomes more common, how does the carrying capacity for bats change?

6. What would the graph in figure 18.17 look like for a declining population? How would the birth rate compare to the death rate? How would the age structure diagram look?

7. A species with an opportunistic life history occupies a habitat where conditions fluctuate in two-year cycles; that is, years with optimal conditions for population growth alternate with suboptimal years. Graph the population size of the species over 6 years, then indicate on the same graph how the population of a species with an equilibrium life history would change over the same period.

SCIENTIFIC LITERACY

What is your ecological footprint? To find out, search for the Personal Footprint Calculator on the Global Footprint Network website (or find a similar calculator on the Internet). What can you do to reduce your ecological footprint? Do you have an ethical obligation to live sustainably? Why or why not?

PULL IT TOGETHER

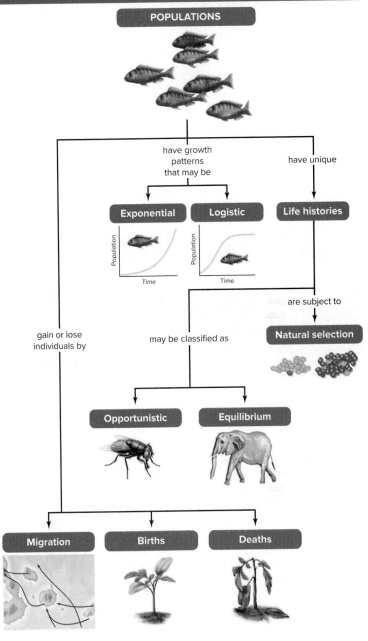

Figure 18.18 **Pull It Together: Populations.**

Refer to figure 18.18 and the chapter content to answer the following questions.

1. Review the Survey the Landscape figure in the chapter introduction, and then add *ecosystems, nonliving environment,* and *communities* to the Pull It Together concept map.

2. Add *growth rate* to the concept map. Explain how a population's growth rate changes throughout exponential and logistic growth.

3. Add the following terms to this concept map: *density-dependent, density-independent, carrying capacity,* and *age structure.*

Design element: Burning Question (fire background): ©Ingram Publishing/Super Stock

LEARNING OUTLINE

19.1 Organisms Interact Within Communities and Ecosystems

19.2 Earth Has Diverse Climates

19.3 Biomes Are Ecosystems with Distinctive Communities of Life

19.4 Community Interactions Occur Within Each Biome

19.5 Succession Is a Gradual Change in a Community

19.6 Ecosystems Require Continuous Energy Input

19.7 Chemicals Cycle Within Ecosystems

APPLICATIONS

Burning Question 19.1 *Why is there a "tree line" above which trees won't grow?*

Why We Care 19.1 *What Happens After You Flush*

Why We Care 19.2 *Mercury on the Wing*

Burning Question 19.2 *Is bottled water safer than tap water?*

Why We Care 19.3 *The Nitrogen Cycle in Your Fish Tank*

Investigating Life 19.1 *Winged Migrants Sidestep Parasites*

Unbalanced Ecosystem. Chickens and other farm animals may have little interaction with other species or with the outdoors.

©Pedro Ladeira/SambaPhoto/Getty Images

Learn How to Learn
Don't Throw That Exam Away!

Whether or not you were satisfied with your last exam, take the time to learn from your mistakes. Mark the questions that you missed and the ones that you got right but were unsure about. Then figure out what went wrong for each question. For example, did you neglect to study the information, thinking it wouldn't be on the test? Did you memorize a term's definition without understanding how it fits with other material? Did you misread the question? After you have finished your analysis, look for patterns and think about what you could have done differently. Then revise your study plan so that you can avoid making the same mistakes in the future.

SURVEY THE LANDSCAPE
Ecology

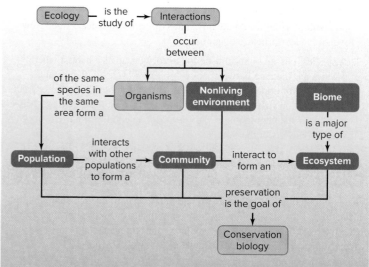

From the tropics to the poles, multiple populations form complex webs of interactions. Along the way, each individual exchanges energy and materials with its environment.

For more details, study the Pull It Together feature in the chapter summary.

©Ingram Publishing/SuperStock RF

The chapter opening photo shows a concentrated animal feeding operation, or CAFO. Much of the chicken, turkey, pork, and beef consumed in the United States comes from operations such as this one.

What does a CAFO have to do with a chapter on communities and ecosystems? As you will see, ecosystems rely on interactions among multiple species (the community) and the nonliving environment. A typical ecosystem on land includes plants that absorb inorganic nutrients and harvest energy from the sun. Animals eat the plants, and decomposers return inorganic nutrients to the soil. As the photo shows, however, a CAFO has an extremely high density of animals, and plants are nowhere in sight.

The CAFO ecosystem is therefore "subsidized" with food brought from outside. Moreover, the animals release immense volumes of organic wastes—too much for the local decomposers to handle. Spreading the wastes on nearby land can cause nutrient overloads, throwing those ecosystems out of balance as well. Solving these and other environmental problems requires an understanding of both community and ecosystem ecology.

19.1 Organisms Interact Within Communities and Ecosystems

Chapter 18 explored the ecology of **populations,** which consist of members of the same species that inhabit the same area. Douglas fir trees, for example, form a population in the forests of the Pacific Northwest. But no population lives in isolation; this chapter therefore extends the study of ecology to communities and ecosystems.

A **community** is a group of interacting populations. Figure 19.1 illustrates three distinctly different examples: the reef shark and coral that live in the ocean; the trees and mushrooms of a coniferous forest; and the cacti and other plants of a scorching desert. **Biotic** interactions are those that occur among the living species in a community. Moreover, each community exists within the context of its physical and chemical surroundings. The **ecosystem** includes the community plus the **abiotic,** or nonliving, environment.

Each species in a community has a characteristic home and way of life. Recall from chapter 18 that a **habitat** is the place where members of a population typically live, such as a forest canopy or the bottom of a river. The habitat is one part of the **niche,** which is the total of all the resources a species requires for its survival, growth, and reproduction. In addition to the habitat, the niche also includes the temperature, light, water availability, salinity, fire, and other abiotic conditions where the species lives. Biotic interactions, such as an organism's place in the food chain, are part of the niche as well.

This chapter describes the interactions that characterize both communities and ecosystems. We begin at the broadest possible scale, by explaining why climates differ across the planet. Next, the chapter describes the major ecosystems on Earth, with a focus on how climate and other features of the physical environment select for a unique community of organisms. The species that occupy each habitat are interconnected, setting the stage for a discussion of community-level interactions. Finally, the chapter returns its focus to ecosystems and the ways that communities exchange materials and energy with the physical environment.

a. b. c.

Figure 19.1 Distinctive Communities. (a) A reef shark swims near a sea fan, a type of coral. (b) The forest floor community includes plants, fungi, and many unseen animals and microbes. (c) A saguaro cactus is one of many plant species in this Arizona desert community.

(a): ©David Nardini/Taxi/Getty Images; (b): ©IT Stock Free/Alamy Stock Photo RF; (c): ©Kristy-Anne Glubish/Design Pics/Corbis RF

In studying these topics, keep in mind that population-, community-, and ecosystem-level interactions are the selective forces that shape the evolution of each species. Plants, animals, and all other organisms must be able to defend themselves and to acquire resources for growth, maintenance, and reproduction. The adaptations that characterize each species—the ability to produce thorns or live in salt water or catch a gazelle—ultimately trace their origins to genetic mutations. But these features persist over multiple generations because they have enhanced reproductive success in a dangerous and competitive world.

19.1 Mastering Concepts

1. Distinguish between ecosystems, communities, and populations.
2. Which abiotic conditions influence the distribution of species in the biosphere?
3. What is the relationship between an organism's habitat and its niche?

19.2 Earth Has Diverse Climates

Earth has a wide variety of climates, from the year-round warmth and moisture of the tropics to the perpetually chilly poles. Why does each part of the planet have a different climate? The answer relates to the curvature of the planet and to the tilt of its axis (figure 19.2).

At the equator, the sun is overhead (or nearly so) all year; equatorial regions therefore receive the most intense sunlight, and the temperature is warm year-round. Thanks to Earth's curvature, however, the sun's rays hit other parts of the surface at a slant. Because the same amount of sunlight is distributed over a larger area, the average temperature falls with distance from the equator. Figure 19.3 shows the resulting broad temperature bands from the equator to the poles.

The tilt of Earth's axis accounts for seasonal temperature changes in non-equatorial regions. From March through September, the northern hemisphere tilts toward the sun and experiences the warm temperatures of spring and summer. During the rest of the year, cooler temperatures prevail as the northern hemisphere tilts away from the sun. The seasons are the opposite in the southern hemisphere.

Equatorial regions receive not only the most light but also the most precipitation. When sunlight heats the air over the equator, the air rises, expands, and cools. Because cool air cannot hold as much moisture as warm air, the excess water vapor condenses, forming the clouds that pour rain over the tropics.

Air that rises near the equator also travels north and south (figure 19.4). As the air cools at higher latitudes, its density increases, and it sinks back down to Earth at about 30° North and South latitude. Here the warming air absorbs moisture from the land, creating the vast deserts of Asia, Africa, the Americas, and Australia. Some of the air continues toward the poles, rising and cooling at about 60° North and South latitude, bringing the rains that support temperate (midlatitude) forests in these areas. The air rises, and some again continues toward the poles, where precipitation is quite low. The rest returns to the equator, where the air heats up again, and the cycle begins anew.

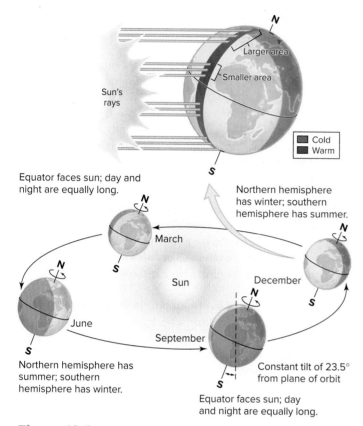

Equator faces sun; day and night are equally long.

Northern hemisphere has winter; southern hemisphere has summer.

March

Sun

December

June

September

Northern hemisphere has summer; southern hemisphere has winter.

Constant tilt of 23.5° from plane of orbit

Equator faces sun; day and night are equally long.

Sun's rays

Larger area

Smaller area

Cold
Warm

Figure 19.2 Earth's Seasons. The tilt of Earth's axis produces distinct seasons in the northern and southern hemispheres as Earth travels around the sun. The inset shows the uneven distribution of sunlight across Earth's surface, which explains why temperature falls with distance from the equator.

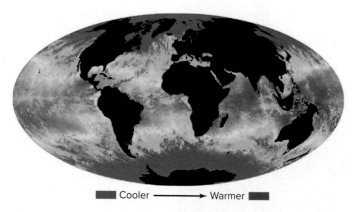

Cooler ⟶ Warmer

Figure 19.3 From Warm to Cold. The colored bands on this map show that Earth is warmest at the equator and coolest at the poles.
Source: MODIS Oceans Group, NASA Goddard Space Flight Center

A cycle of heating and cooling, rising and falling air is called a convection cell. The planet has six such convection cells (three north of the equator and three south). Earth's major winds correspond to these convection cells. (From Earth, the winds appear deflected toward the east and west because the planet rotates beneath them.) Together, these winds power major ocean currents, including the Gulf Stream along the east coast of North America (figure 19.5).

Ocean currents, in turn, influence coastal climates, in part because they transport cold and warm water around the globe. For example, bands of cold water flow along the west coast of North America, while currents along the east coast are relatively warm. In addition, large water bodies heat up and cool down much more slowly than does the land. Coastal regions therefore often have milder climates than do inland areas at the same latitude. On a hot summer day, beachgoers notice this effect as they enjoy cooling breezes from the sea. Conversely, during the winter, the ocean releases stored heat.

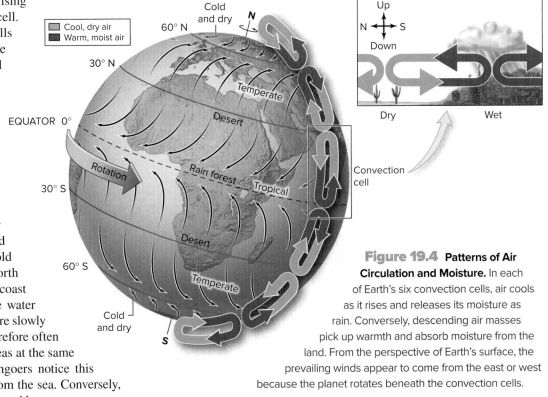

Figure 19.4 Patterns of Air Circulation and Moisture. In each of Earth's six convection cells, air cools as it rises and releases its moisture as rain. Conversely, descending air masses pick up warmth and absorb moisture from the land. From the perspective of Earth's surface, the prevailing winds appear to come from the east or west because the planet rotates beneath the convection cells.

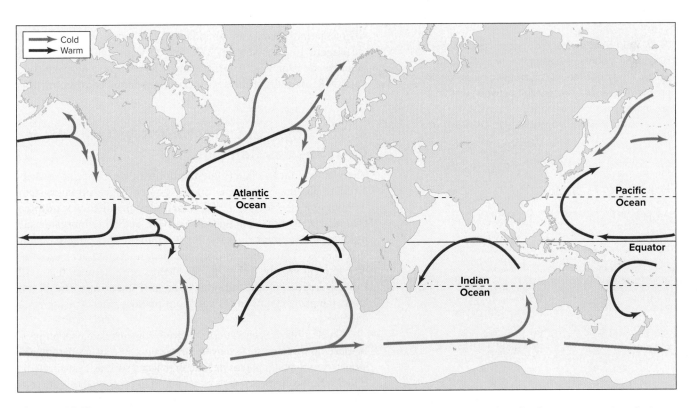

Figure 19.5 Ocean Currents. Earth's prevailing winds produce the major ocean currents, which redistribute water and nutrients throughout the oceans.

Succession Is a Gradual Change in a Community   Ecosystems Require Continuous Energy Input   Chemicals Cycle Within Ecosystems

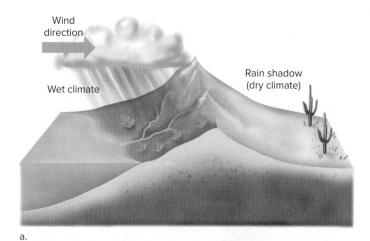

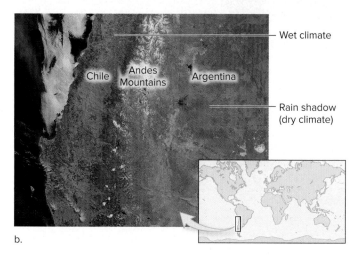

Figure 19.6 Rain Shadow. (a) Precipitation falls on the windward side of the mountain, leaving the other side with a dry climate. (b) The Andes Mountains create a massive rain shadow; note the difference in vegetation between wet Chile and dry Argentina in this satellite image.

Photo: (b): Source: Jacques Descloitres, MODIS Land Rapid Response Team, NASA/GSFC

Mountain ranges also influence climate, in two ways. First, the top of a mountain is generally cooler than its base. Second, mountains often block wind and moisture-laden clouds on their upwind side. The **rain shadow** on the downwind side of the mountain has a much drier climate (figure 19.6).

> ### 19.2 Mastering Concepts
>
> 1. Explain this statement: "If Earth's axis were not tilted, there would be no seasons."
> 2. How do prevailing winds, ocean currents, and mountain ranges affect climate?

19.3 Biomes Are Ecosystems with Distinctive Communities of Life

Ecologists divide the biosphere into **biomes,** which are the major types of ecosystems. Forests, deserts, and grasslands are examples of terrestrial biomes. Lakes, streams, and oceans are water-based ecosystems. Each biome is characterized by a distinctive group of species. Although it is convenient to classify each ecosystem as belonging to one biome or another, keep in mind that no ecosystem exists in isolation. Water, air, sediments, and organisms can travel freely from one part of the biosphere to another.

This section explains the distribution of the main biomes on land and in water. As you read this material, remember that the biomes we see today have not been in place forever. Over hundreds of millions of years, the continents have moved, and sea levels have risen and fallen. The central United States, for example, was once under the sea, which explains why fossils of marine animals are abundant in landlocked states such as Oklahoma (see figure 13.3). Likewise, 375 million years ago, the landmass that now includes the islands of Arctic Canada was once very near the equator. Long-buried fossils tell the tales of these tremendous ecosystem shifts. ⓘ *plate tectonics,* section 13.3A

A. The Physical Environment Dictates Where Each Species Can Live

Many abiotic factors determine the limits of each species' distribution. The ultimate abiotic factor is an energy source, since no ecosystem can exist without one (see section 19.6). A **primary producer,** or **autotroph** ("self-feeder"), is any organism that can use CO_2 and other inorganic substances to produce all the organic molecules it requires. This process requires energy. A few ecosystems, such as deep-sea hydrothermal vents, are based entirely on chemical energy. There, the producers are microbes that extract energy from hydrogen sulfide and other inorganic chemicals. But in most ecosystems, the primary producers carry out photosynthesis, using sunlight as the energy source.

On land, plants such as trees and grasses are the dominant primary producers. In water, however, most photosynthesis occurs courtesy of **phytoplankton:** microscopic, free-floating, photosynthetic organisms such as cyanobacteria and algae. Either way, the organic molecules produced in photosynthesis are eaten by a host of grazers, which are in turn consumed by predators. These herbivores and carnivores are **consumers,** or **heterotrophs** ("other-feeders"), which obtain carbon by eating the organic molecules that make up another organism.

Besides sunlight, the major abiotic factors that determine the numbers and types of plants on land are temperature and moisture. All organisms are adapted to a limited temperature range; trees, for example, cannot live where the temperature is too low (see Burning Question 19.1). In addition, all life requires water. The plants that grow where water is abundant, such as the tropical rain forest, have very different adaptations from the vegetation that characterizes a desert ecosystem.

Nutrient availability is another crucial abiotic factor that often determines an ecosystem's productivity. On land, soil provides essential mineral elements such as nitrogen and phosphorus. In aquatic ecosystems, both nutrients and sunlight are often scarce, especially with increasing depth and distance from the shoreline.

Fire is an essential abiotic condition in some terrestrial biomes. In grasslands, for example, periodic fires kill trees that might otherwise take over. In coniferous forests, many adult trees die in fires, but their cones open and release their seeds only after prolonged exposure to heat. The seeds germinate after the fire, and the young trees thrive with little competition for sunlight or nutrients.

Other abiotic factors may also be important in some locations. In aquatic ecosystems, one notable example is the amount of dissolved oxygen, which influences the types of microbes and animals that can live in the water. Likewise, many organisms are adapted to seawater or salty soils, but others are not.

B. Terrestrial Biomes Range from the Lush Tropics to the Frozen Poles

Earth's climatic zones give rise to huge bands of characteristic types of vegetation, which correspond to the terrestrial biomes. Temperature and moisture are the main factors that determine the dominant plants in each location (figure 19.7). The overall pattern of vegetation, in turn, influences which microorganisms and animals can live in a biome.

Soils form the framework of terrestrial biomes because they directly support plant life. Although soil may seem like "just dirt," it is actually a complex mixture of rock fragments, organic matter, and microbes. Climate influences soil development in many ways. Heavy rain may leach nutrients from surface layers and deposit them in deeper layers, or it may remove them entirely from the soil. In addition, in a warm, moist climate, rapid decomposition may leave little organic material in the soil. In cold, damp areas, on the other hand, undecomposed peat may accumulate in the soil.

Figure 19.8 describes 10 major terrestrial biomes. The map at the center of the figure shows the original range of each biome. It is important to remember, however, that humans have drastically reduced many natural biomes, replacing them with farmland, suburban housing, and cities. In addition, as described in chapter 20, human activities threaten much of the native habitat that remains.

Burning Question 19.1

Why is there a "tree line" above which trees won't grow?

The tree line, or timberline, is an edge beyond which trees cannot survive. The alpine tree line is the highest elevation at which they can grow; the Arctic and Antarctic tree lines are the farthest points north and south, respectively, that trees can live. In each case, the tree line usually defines the point at which the environment simply becomes too cold to support trees.

Most species near the tree line are evergreen conifers such as pine, spruce, larch, and fir. Their needles have a waxy coating and an arrangement of stomata that minimizes water loss in the thin, dry air. Eventually, however, chilly temperatures and biting winds get the best of even these hardy plants. At the timberline, the trees shorten to low, stunted bushes, and beyond the tree line, it is too cold for seeds to germinate. ⓘ *stomata,* section 21.3B

Wind, salt, and a dry climate can also produce other types of tree lines. For example, along coasts, a tree line can result from high winds and salt spray that make life impossible for trees. Beyond the desert tree line, rainfall is insufficient to support trees.

Submit your burning question to marielle.hoefnagels@mheducation.com

(mountain): ©Christopher Boswell/Shutterstock RF

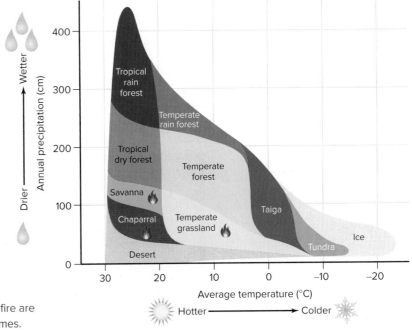

Figure 19.7 Biome Classification. Temperature, precipitation, and fire are important factors influencing the distribution of the major terrestrial biomes.

Figure 19.8 **Earth's Major Terrestrial Biomes.**

Taiga (boreal forest)
Climate: Relatively dry, with short summers and long, cold winters. Moisture can be scarce in winter, when water may remain frozen for months. *Soil:* Cold, damp, acidic, and nutrient-poor. *Plants:* Plants include conifers such as spruce, fir, and pine with evergreen needles that resist water loss. *Animals:* Many migratory birds visit during the summer, whereas the mammals and birds that live in the taiga year-round have thick layers of insulation that retain body heat. Herbivorous mammals, such as caribou, moose, and snowshoe hares, eat whatever vegetation they can find during the cold winters. Others, such as the black bear, hibernate for the winter. Predatory mammals include lynx, gray wolves, and wolverines.

Temperate coniferous forest (also called temperate rain forest)
Climate: Mild winters, cool summers, and abundant rain and fog. Water is plentiful. *Soil:* Deep, well-drained, somewhat acidic, and high in organic matter. *Plants:* Most trees are evergreens such as spruce, cedar, Douglas fir, and hemlock, all of which have waxy, needlelike leaves adapted to year-round photosynthesis. Moisture-loving mosses, ferns, and lichens cover nearly every surface. *Animals:* Many fishes, amphibians, birds, small mammals, and deer consume the plants and invertebrates in the forest and its streams. Owls, black bears, and cougars hunt small animals.

Temperate deciduous forest
Climate: Summers are warm, winters are cold, and rainfall is approximately constant throughout the year. *Soil:* Nutrient-rich, owing to abundant leaf litter. *Plants:* The cold winters select for trees such as oak, hickory, and maple that shed their foliage in autumn. Shade-tolerant shrubs grow beneath the trees. Below them, small flowering plants grow in early spring, when light penetrates the leafless tree canopy. *Animals:* Many of the animals adjust to the seasons by putting on fat in the summer and hibernating in the winter, or they may store seeds and nuts that sustain them when food is scarce. Still others migrate to warmer areas. Herbivores include seed- and nut-eating mice and birds, whitetail deer, and gray squirrels. Red foxes and snakes are common predators.

Temperate grassland
Climate: Moderately moist, with hot summers and cold winters. *Soil:* Deep, dark, and fertile. *Plants:* Wind-pollinated grasses dominate, although other flowering plants are also common. Summertime drought, along with fire and grazing, suppress tree growth; grasses survive because their perennial buds lie protected below the soil surface. *Animals:* Bison, elk, and pronghorn antelope, whose teeth and digestive systems are adapted to a grassy diet, were originally the large, grazing herbivores of North America. Other herbivores include prairie chickens, insects, and rodents such as prairie dogs and mice. Some of these small animals burrow into the soil to hide from predators, whereas others have camouflage. Coyotes, bobcats, snakes, and birds of prey feed on the herbivores.

Tropical rain forest
Climate: Warm and wet year-round. *Soil:* Rapid decomposition and leaching means soils are nutrient-poor. *Plants:* Since climate does not limit plant growth, the main selective forces are competition for light and nutrients. Broadleaf evergreen trees with tall, straight trunks form the forest canopy. Vegetation in the shade beneath the canopy includes climbing vines and small plants that grow on the surfaces of other plants. In the deep shade of the forest floor, plants often have large leaves that maximize the capture of scarce light. *Animals:* These forests house an incredible diversity of arthropods, fishes, amphibians, reptiles, and mammals that eat plants or other animals. Many of these animals have similar adaptations to life in the trees: bright colors that maximize visibility, coupled with loud calls that are audible throughout the forest canopy.

Tundra
Climate: Snow covers the Arctic and Antarctic tundra during the bitterly cold and dark winter. Temperatures venture above freezing for a few months each year. *Soil:* Poorly drained and rich in organic matter. Below the surface, a layer of permafrost remains frozen year-round. *Plants:* Permafrost limits rooting depth and prevents the establishment of large plants, but reindeer lichens, mosses, dwarf shrubs, and low-growing perennial plants such as sedges, grasses, and broad-leafed herbs are common. *Animals:* In the summer, migratory birds raise their young and feed on the insects that flourish in the tundra. Year-round inhabitants of the Arctic tundra include caribou, lemmings, hares, foxes, and wolverines, all of which have thick layers of fat and warm fur. Animal life in the Antarctic tundra is much less diverse.

Mediterranean shrubland (chaparral)
Climate: Summers are hot and dry; winters are mild and moist. *Soil:* Sandy soils retain little water. *Plants:* Shrubs and trees have thick bark and small, leathery, evergreen leaves that slow moisture loss during the dry summers. This biome is especially susceptible to fires because the vegetation dries out during the summer. Fire-adapted plants resprout from underground parts or produce seeds that germinate only after the heat of a fire. *Animals:* Herbivores include jackrabbits, deer, and birds and rodents that forage for seeds under the shrub canopy; some of their predators include coyotes, foxes, snakes, and hawks.

Desert
Climate: Dry with less than 20 cm of rainfall per year. The temperature can vary dramatically. In a hot desert such as the Sonoran, which spans parts of Arizona and Mexico, the days can be scorchingly hot. In China's cold Gobi Desert the average temperature is below freezing. *Soil:* Dry soils have little to no organic matter. *Plants:* Desert plants often have long taproots, quick life cycles that exploit the brief rainy periods, fleshy stems or leaves that store water, and spines or toxins that guard against thirsty herbivores. *Animals:* Most burrow or seek shelter during the day, then become active at night. Standing water is scarce, so most water comes from the animal's food. Herbivores include jackrabbits and kangaroo rats, which eat seeds and leaves. Snakes and cougars hunt the herbivores.

Tropical savanna (grassland with scattered shrub or trees)
Climate: Warm year-round, with distinct wet and dry seasons. *Soil:* Well-drained, with a thin layer of organic matter. *Plants:* Perennial grasses dominate along with scattered patches of drought- and fire-resistant trees and shrubs such as palms, acacias, and baobab trees. These plants have deep roots, thick bark, and trunks that store water. *Animals:* During the dry season, great herds of animals migrate enormous distances in search of water. The Australian savanna is home to many birds and kangaroos, whereas the savanna in Africa features herds of zebra, giraffes, wildebeests, gazelles, and elephants. Lions, cheetahs, wild dogs, birds of prey, and hyenas prey on the herbivores, and vultures and other scavengers eat the leftovers.

Polar ice
Climate: Both Antarctica and the Arctic ice cap are extremely cold, dry, and windy year-round. *Soil:* None. *Plants:* None. The primary producers in ice and the surrounding ocean are phytoplankton. The light passing through the ice is dim, even in the summer. *Animals:* The phytoplankton support a unique food web consisting of worms, crustaceans, and icefishes. All of these organisms have antifreeze chemicals that prevent deadly ice crystals from forming in their cells. Larger animals that exploit the Arctic food web have insulating fat and either fur or feathers. Examples include polar bears, seals, whales, and birds. On Antarctica, vertebrates include penguins and seals. Whales and squid inhabit the waters surrounding Antarctica as well.

Photos: (taiga): ©Kari Niemeläinen/Alamy Stock Photo RF; (coniferous forest): ©Taylor S. Kennedy/National Geographic/Getty Images RF; (deciduous forest): ©Digital Archive Japan/Alamy Stock Photo RF; (grassland): ©Corbis Premium RF/Alamy Stock Photo RF; (rain forest): ©Muzhik/Shutterstock RF; (tundra): ©Michael DeYoung/Perspectives/Getty Images; (chaparral): ©Emma Lee/Life File/Getty Images RF; (desert): ©Westend61/Getty Images RF; (savanna): ©Arthur Morris/Corbis/Getty Images; (polar ice): ©Kelly Cheng/Moment/Getty Images RF

Succession Is a Gradual Change in a Community   Ecosystems Require Continuous Energy Input   Chemicals Cycle Within Ecosystems

Freshwater Biomes

Standing water: lakes and ponds

Light penetrates the regions of a lake to differing degrees. The shallow, nutrient-rich shoreline is part of the photic zone, where light is sufficient for photosynthesis. Rooted plants and phytoplankton thrive along the shore, providing food and shelter for invertebrates, fishes, amphibians, and other animals. In open water, phytoplankton are the dominant producers of the photic zone; zooplankton and fishes are typical consumers. But light does not penetrate the deeper water, where scavengers and decomposers such as insect larvae and bacteria rely on organic material from above to supply energy and nutrients.

Running water: rivers

A river carries water and sediment from land toward the ocean, providing moisture and habitat to aquatic and terrestrial organisms. At the headwaters, the water is clear, the stream channel is narrow, and the current may be swift. Turbulence mixes air with water, so the water is rich in oxygen. Algae, mosses, and insects cling to any available surface, such as rocks and logs. The river widens as small streams drain additional land areas and contribute water, sediments, and nutrients. As the land flattens, the current slows. The river is now murky, restricting photosynthesis to the banks and water surface. As a result, the oxygen content is relatively low. Typical animals in a slow-moving river include crayfish, snails, bass, and catfish; worms burrow in the muddy bottom.

Marine Biomes

Open ocean

The oceans, which cover most of Earth's surface, contain 97% of the planet's water. Both light and nutrients are abundant in the shallow waters above the continental shelf, supporting high primary productivity and extensive marine food webs such as the great kelp forests that fringe many cool-water coastal areas. Beyond the continental shelf, the open ocean's photic zone houses phytoplankton and the zooplankton that feed on them; fishes and whales, in turn, scoop up vast quantities of krill and zooplankton. Below the photic zone, light is too dim for photosynthesis, but a continual rain of organic matter supports great numbers of jellyfishes, fishes, whales, dolphins, mollusks, echinoderms, and crustaceans. The communities that occupy hydrothermal vents add biodiversity to the ocean floor.

Estuary

An estuary is an area where the fresh water of a river meets the salty ocean. When the tide is out, the water may not be much saltier than water in the river. The returning tide, however, may make the water nearly as salty as the sea. Organisms that can withstand these extremes receive nutrients from both the river and the tides. Estuaries therefore house some of the world's most productive ecosystems. In the open water, phytoplankton account for most of the productivity, whereas salt-tolerant plants dominate the salt marshes that often occur along the fringes of an estuary. Together, these producers support many species of fish, shellfish, and migratory birds.

Intertidal zone

Along coastlines, the intertidal zone is the area between the high tide and low tide marks. This region is alternately exposed and covered with water as the tide rises and falls. A sandy beach is one familiar example. Constantly shifting sands mean that few producers can take root on the beach, but ocean water delivers a constant supply of organic matter that feeds crabs and shorebirds. In a rocky intertidal zone, seaweeds and mussels attach to rocks, whereas sea anemones, sea urchins, sea stars, and snails occupy the small tide pools.

Coral reef

Coral reefs border tropical coastlines where the water is clear and sediment-free. These vast underwater structures of calcium carbonate are built by coral animals. The tissues of the coral animals house algae that are essential for the coral's—and the ecosystem's—survival. Sunlight penetrates the clear, shallow water, allowing photosynthesis to occur, and constant wave action brings in additional nutrients. The nooks and crannies of the reef provide food and habitat for a huge variety of algae, fishes, sponges, snails, sea stars, sea urchins, sea turtles, and countless microorganisms.

Figure 19.9 **Aquatic Biomes.**

(oligotrophic lake): ©Ray Bouknight/Moment/Getty Images RF; (eutrophic lake): ©McGraw-Hill Education/Pat Watson; (stream): ©dvande/Shutterstock RF; (river): ©Westend61/Getty Images RF; (ocean): ©ESB Professional/Shutterstock RF; (estuary): ©John Warburton-Lee Photography/Alamy Stock Photo; (intertidal): ©Craig Tuttle/Design Pics/Getty Images RF; (reef): ©Digital Vision/Getty Images RF

C. Aquatic Biomes Include Fresh Water and the Oceans

Although terrestrial biomes are most familiar to us, the aquatic ecosystems illustrated in figure 19.9 occupy much more space. Water moves continuously among the ocean, atmosphere, land surface, and groundwater, providing vital connections among all biomes. Section 19.7 describes the water cycle in detail.

Biologists divide aquatic biomes into two main categories: freshwater and marine. Lakes and rivers, such as those in figure 19.9, contain only about 0.3% of the freshwater supply; the rest is in groundwater or locked in ice caps and glaciers (figure 19.10). Nevertheless, that tiny sliver of the global water "pie" is vital to humans for drinking water and irrigation. Most other terrestrial species rely on this fresh water as well.

At the opposite extreme, the oceans cover 70% of Earth's surface and contain 97% of the planet's water; they therefore form the world's largest biome. Most photosynthesis on Earth occurs in the vast oceans, contributing enormous amounts of oxygen to the atmosphere. Moreover, oceans absorb so much heat from the sun that they help stabilize Earth's climate.

All aquatic communities need sunlight and nutrients. The **photic zone** is the layer of water with sufficient light for photosynthesis, and nutrients are most abundant near land. Ecosystem productivity is therefore highest in the shallow, well-lit waters near the shore. In deep water away from the shore, however, both energy and nutrients can be scarce. Figure 19.9 explains how each combination of conditions selects for a unique community of life.

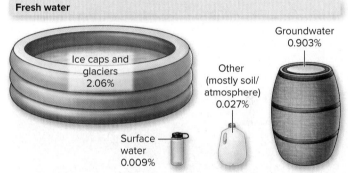

Figure 19.10 World Water Resources. This analogy depicts Earth's water supply in a series of containers, ranging from a large aboveground swimming pool (the oceans) to a small bottle (surface water).

19.3 Mastering Concepts

1. How do climate and soil composition determine the characteristics of terrestrial biomes?
2. Infer one adaptation of plants and one adaptation of animals to the abiotic conditions in any four biomes.
3. Describe the types of organisms that live in each zone of a lake.
4. How does a river change from its headwaters to its mouth?
5. Describe some of the adaptations that characterize organisms in the open ocean, estuaries, intertidal zones, and coral reefs.

19.4 Community Interactions Occur Within Each Biome

Communities usually include many species. Some are easily visible, whereas others are microscopic. One on one, their interactions may seem simple—a whale eats an otter, or a wasp kills a caterpillar. But an attempt to map all interactions within a community quickly becomes complicated. Individuals of different species compete for limited resources, live in or on one another, eat one another, and try to avoid being eaten. Table 19.1 summarizes these interactions.

TABLE 19.1   Species Interactions: A Summary

Interaction	Effects on Species 1	Effects on Species 2
Competition	–	–
Symbiosis		
Mutualism	+	+
Commensalism	+	0
Parasitism	+	–
Herbivory	+	–
Predation	+	–

Figure 19.11 Competition. A blackbird (*left*) fights over apples with a fieldfare (*right*) in England.

©FLPA/Alamy Stock Photo

Figure 19.12 Zebra Mussels. Tiny, invasive zebra mussels coat the shell of a painter's mussel.

©blickwinkel/Alamy Stock Photo

A. Many Species Compete for the Same Resources

Competition occurs when two or more organisms vie for the same limited resource, such as shelter, nutrients, water, light, or food (figure 19.11). Since neither participant obtains all of the resource that it needs, the effects of competition are negative for both.

Competition can help shape the species composition of a community. Consider the **competitive exclusion principle,** which states that two species cannot coexist indefinitely in the same niche. The two species will compete for the limited resources that they both require, such as food, nesting sites, or soil nutrients. According to the competitive exclusion principle, the species that acquires more of the resources will eventually "win." The less successful species dies out.

Introduced species sometimes displace native species by competitive exclusion. Zebra mussels, for example, are native to the Caspian Sea in Asia. These mollusks were accidentally introduced into the Great Lakes in the 1980s and have since spread to many waterways in the United States and Canada. The tiny filter feeders reproduce rapidly and have crowded out native mussel species, with which they compete for food and oxygen (figure 19.12). The effects of the zebra mussel invasion have rippled through the rest of the lake community as well. Zebra mussels have greatly increased water clarity, which has changed the amount of light available to aquatic plant communities. In turn, the altered plant species composition has triggered changes in the community of fishes. ⓘ *invasive species,* section 20.5A

Competitive exclusion, however, is not inevitable; coexistence in overlapping niches is also possible. After all, if competition between species reduces fitness, then natural selection should favor organisms that avoid competition. Therefore, another possible outcome of competition is **resource partitioning,** in which multiple species use the same resource in a slightly different way or at a different time (figure 19.13). For example, multiple species of rockhopper penguins live on islands in the southern Indian Ocean, occupying similar niches: The birds all appear similar, and they all eat similar foods. When researchers tracked penguin movements, however, they found that the populations feed in different places and at different times. Their feeding locations and times reduce competition and therefore improve the reproductive success of all populations.

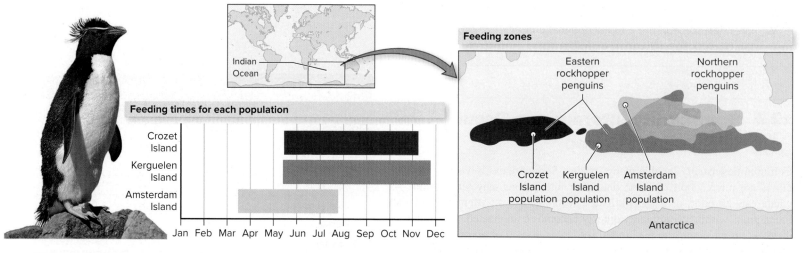

Figure 19.13 Resource Partitioning. Two species of rockhopper penguins form three populations. All have similar diets, but each feeds at a different time or place. Photo: ©Enrique R Aguirre Aves/Oxford Scientific/Getty Images

Introduction: Communities and Ecosystems   Earth Has Diverse Climates   Biomes: Ecosystems with Distinctive Communities   Community Interactions Occur Within Each Biome

B. Symbiotic Interactions Can Benefit or Harm a Species

In a **symbiosis** (literally, "living together"), two species share a close (and often lifelong) relationship in which one typically lives in or on the other. The relationship between symbiotic species may take several forms, defined by the effect on each participant.

Mutualistic relationships are symbioses that improve the fitness of both partners (figure 19.14a). Sea anemones and their clownfish companions are a classic example, as are mycorrhizal fungi. In the latter case, the fungus acquires nutrients and water that it shares with its host plant; the plant feeds sugars to its live-in partner. Many of the bacteria in our intestines are also mutualistic; these microbes consume nutrients from our food but also produce vitamins and defend us against disease. ⓘ *mycorrhizae,* section 15.5C; *beneficial microbes,* section 15.2D

Commensalism is a type of symbiosis in which one species benefits, but the other is not significantly affected. Most humans, for example, never notice the tiny mites that live, eat, and breed in our hair follicles (see Why We Care 17.1). Similarly, the reproductive success of a tree is neither helped nor harmed by the moss plants and lichens that grow on its trunk and branches (figure 19.14b).

In a symbiotic relationship called **parasitism,** one species acquires resources at the expense of a living host. The most familiar parasites are disease-causing bacteria, protists, fungi, and worms. Some of these organisms can suppress their host's immune system, an adaptation that favors long-term colonization of the body (see Investigating Life 29.1). Plants may also be parasites. Mistletoe, for example, is a parasitic plant that taps into the water- and nutrient-conducting "pipes" of a host plant. ⓘ *parasitic plants,* section 21.6C

C. Herbivory and Predation Link Species in Feeding Relationships

All animals must obtain energy and nutrients by eating other organisms, living or dead. An **herbivore** is an animal that consumes plants; a **carnivore** eats meat. A **predator** is a carnivore that kills and eats other animals, called **prey.** As in parasitism, the fitness of the herbivore or predator increases at the expense of the organism being consumed. In some cases, predator–prey interactions are directly responsible for fluctuations in an animal's population size (see section 18.5).

Natural selection favors plant defenses against herbivores, which may eat leaves, roots, stems, flowers, fruits, or seeds (figure 19.15). The loss of leaf and root tissue reduces the plant's ability to carry out photosynthesis; consumption of flowers or immature fruits and seeds compromises the plant's reproductive success. Natural selection therefore favors plant defenses against herbivory. Some plant species deter herbivores with thorns, a milky sap, or distasteful or poisonous chemicals. The spicy hot chemicals in chili peppers, for example, discourage attack by both fungi and small mammals (see Investigating Life 22.1). At the same time, many herbivores have adaptations that correspond to the plant's defenses. The caterpillars of monarch butterflies, for example, tolerate the noxious chemicals in milkweed plants.

Likewise, predation exerts strong selective pressure on prey animals, which often have adaptations that help them avoid being eaten. They may protect themselves with weapons and structural defenses such as hard shells, spikes, pincers, stingers, and other deterrents (some of which are also useful in capturing their own prey). Prey animals also display a repertoire of defensive behaviors, including fleeing, fighting, releasing noxious chemicals, or forming a tight group.

a.

b.

Figure 19.14 Symbiosis Examples. (a) A clownfish in Indonesia is safe from predators among the tentacles of a sea anemone; the fish chases away animals that would otherwise nibble on its host. This relationship is therefore mutualistic. (b) Moss plants and lichens live on the branches and trunks of trees, enhancing their own sun exposure without harming their hosts. This is a commensal relationship.

(a): ©Reinhard Dirscherl/WaterFrame/Getty Images; (b): ©Anton Foltin/Shutterstock RF

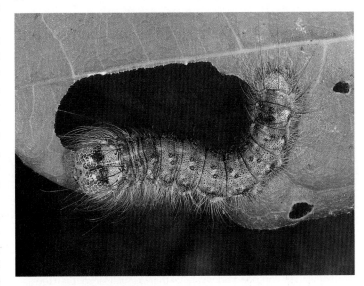

Figure 19.15 Herbivory. A greater oak dagger moth caterpillar munches on an oak leaf. ©Steven P. Lynch RF

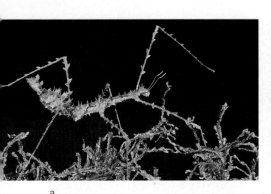

a.   b.   c.

Figure 19.16 **Prey Defenses.** (a) Camouflage helps prey species hide. This insect from Madagascar resembles the leaves in its habitat. (b) Warning coloration advertises a poison dart frog's defenses. (c) These jumping spiders look like ants. Many predators avoid ants, which are aggressive and unpalatable.

(a): ©Kevin Schafer/The Image Bank/Getty Images; (b): ©MedioImages/SuperStock RF; (c): ©Simon D. Pollard/Science Source

Camouflage and warning coloration are two other examples of prey defenses (figure 19.16). An interesting variation on the theme of warning coloration is mimicry, in which different species develop similar appearances. For example, a harmless species of fly may have yellow and black stripes similar to those of a bee. The stripes deter predators even though the fly cannot sting.

Only those predators that can defeat prey defenses will live long enough to reproduce and care for their young. Acute senses, agility, sharp teeth, and claws are common among predators. Camouflage is adaptive in predators as well as prey. Tigers and other big cats, for example, have markings that hide their shape against their surroundings, which helps them sneak up on their prey. Hunting in groups is a behavioral adaptation that helps predators capture large prey (figure 19.17).

D. Closely Interacting Species May Coevolve

Some connections between species are so strong that the species directly influence one another's evolution. In **coevolution,** a genetic change in one species selects for subsequent changes in the genome of another species. Of course, all interacting species in one community have the potential to influence one another, and they are all "evolving together." These genetic changes are considered coevolution only if scientists can demonstrate that adaptations specifically result from the interactions between the species.

One example of coevolution is the relationship between lodgepole pines and birds called crossbills that eat the trees' seeds (figure 19.18). In areas with crossbills, the pine trees produce large seed cones with thick, protective scales. The birds, however, have a corresponding adaptation: Their bills are largest in forest regions where pines have thick cones.

Another example of an evolutionary "arms race" links predators and prey animals. Amphibians called rough-skinned newts produce exceedingly high concentrations of a toxin that binds to sodium channels in a predator's muscles, usually causing paralysis and death. Garter snakes, however, routinely eat the newts. Their muscle cells have modified sodium channels that are resistant to the poison. Snakes with resistant sodium channels have greater reproductive success than those with typical channels; on the other hand, newts are vulnerable to resistant snakes. Natural selection therefore simultaneously favors more potent newt toxins and more resistant snakes.

Figure 19.17 **Predator Cooperation.** By working together, a pack of wolves can bring down an elk that is much larger than the wolves themselves. Sharp teeth and claws aid in prey capture as well.
Source: U.S. National Park Service/Doug Smith

Flowering plants and insects have also coevolved. As described in section 22.2C, a plant may rely on one insect species for pollination, and the insect may eat nectar from only that plant. The plant–insect relationship may also select for specialized structures and behaviors, such as exclusive plant "homes" for insects that defend their hosts (see Investigating Life 21.1).

E. A Keystone Species Has a Pivotal Role in the Community

Sometimes, many species in a community depend on one **keystone species,** a type of organism that makes up a small portion of the community yet has a large influence on community diversity. Note that *keystone* does not simply mean *essential*. For example, the grasses that underlie the prairie biome are obviously essential to the other species in the grassland, but they are not considered keystone species because they make up the bulk of the community.

One example of a keystone species is the sea otter, which plays a critical role in the vast underwater kelp forests that fringe the Pacific Northwest coast (figure 19.19). Otters eat sea urchins, which devour kelp. In the absence of otters, sea urchin populations explode. The loss of kelp, in turn, eliminates the habitat for many species of marine shrimps, fishes, sea stars, and snails. The otter is a keystone species because it keeps other predators in check.

Many keystone species, including sea otters, are versatile predators. But mutualists may also be keystone species. For example, mycorrhizal fungi help coniferous trees acquire nutrients from soil, and they produce underground fruiting bodies that small rodents eat. Owls and other predators hunt these small mammals. The fungi are considered keystone species because their small biomass is disproportionate to their enormous influence on community structure.

Figure 19.18 Coevolution. Crossbills eat the seeds of lodgepole pines, selecting for pine cones with thick, protective scales. In turn, these cones select for birds with larger, stronger bills.
©Craig W. Benkman

19.4 Mastering Concepts

1. What is the competitive exclusion principle?
2. Give examples of three types of symbiotic relationships.
3. Describe some adaptations that protect against herbivory and predation.
4. Define *coevolution* and describe an example.
5. Woodpeckers dig nesting cavities in tree trunks; many other bird species subsequently use abandoned woodpecker cavities. In these forest communities, what term might be used to describe woodpeckers?

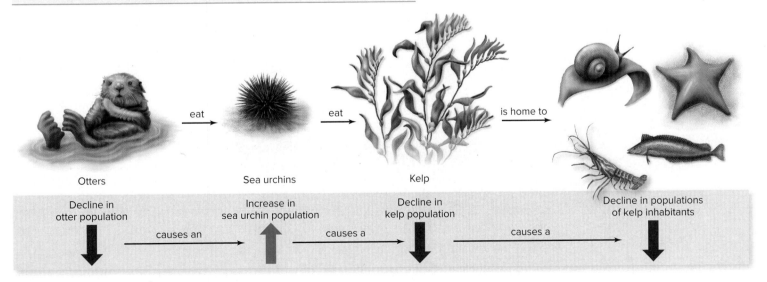

Otters	Sea urchins	Kelp	
Decline in otter population	Increase in sea urchin population	Decline in kelp population	Decline in populations of kelp inhabitants

eat → eat → is home to

causes an → causes a → causes a

Figure 19.19 Keystone Species. Sea otters are a keystone species in the Pacific Northwest, thanks to their taste for kelp-eating sea urchins.

Succession Is a Gradual Change in a Community   Ecosystems Require Continuous Energy Input   Chemicals Cycle Within Ecosystems

Species richness

Higher (# species = 5)   Lower (# species = 3)

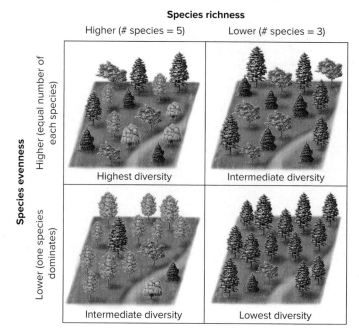

Species evenness

Higher (equal number of each species)

Highest diversity   Intermediate diversity

Lower (one species dominates)

Intermediate diversity   Lowest diversity

Figure 19.20 **Measures of Community Diversity.** Species diversity combines two components: species richness and species evenness.

19.5 Succession Is a Gradual Change in a Community

From oceans to mountaintops, many species share each habitat, but these communities vary greatly in diversity. Ecologists consider diversity to be a function of two measures, called species richness and species evenness (figure 19.20). One way to measure **species richness** is simply to count the species occupying a habitat. A patch of prairie, for instance, may contain about 100 plant species, whereas an equal-sized area of desert might house only six types of plants. In this example, the prairie has greater species richness than the desert.

But two communities with the same species richness may not be equally diverse. **Species evenness,** or relative abundance, describes the proportion of the community that each species occupies. In our patch of prairie, for example, suppose that one type of plant accounts for 90% of the individuals in the community, with 99 species making up the remaining 10%. Because one species has such high relative abundance, that community is less diverse than one in which, say, each of the 100 species makes up 1% of the community.

The numbers and types of species that form each community may seem constant, but that is only because we usually observe them over a relatively short period. **Succession** is a gradual change in a community's species composition. Ecologists define two major types of succession: primary and secondary.

Primary succession occurs in an area where no community previously existed. When a volcano erupts, for example, lava may obliterate existing life, a little like suddenly replacing an intricate painting with a blank canvas. Road cuts and glaciers that scour the landscape also expose virtually lifeless areas on which new communities eventually arise.

Figure 19.21 illustrates primary succession in New England, a region typically occupied by deciduous forest. The process begins with a patch of bare rock. Hardy **pioneer species** such as lichens are first to colonize the area. Lichens

Figure 19.21

Primary Succession.

It takes centuries for a mature forest community to develop on a patch of bare rock. The example shown here includes plant species typical of New England; a region with another soil type and climate would have a different mix of species.

Bare rock | Lichens | Mosses | Herbs, weeds | Grasses | Shrubs | Pines, hickories, immature oaks | Oaks, hickories, black walnuts, maples, tulip poplars, beeches

Time (hundreds of years)

release organic acids that erode the rock, producing crevices where sand and dust accumulate. Decomposing lichens add organic material, eventually forming a thin covering of soil. Then rooted plants such as herbs and grasses invade. Soil continues to form, and larger plants such as shrubs appear. As these new plants take root, a changing variety of birds, mammals, and other vertebrates joins the community as well. Next come the young trees. Finally, hundreds of years after lichens first arrived on the bare rock, the soil becomes rich enough to support a stable, mature forest community. ⓘ *lichens,* section 15.5C

In contrast to primary succession, **secondary succession** occurs where a community is disturbed but not destroyed. Because some soil and life remain, secondary succession occurs faster than primary succession. Fires, hurricanes, and agriculture commonly trigger secondary succession (figure 19.22).

Primary and secondary succession share a common set of processes. The first plants to arrive are usually opportunistic, with rapid reproduction and efficient dispersal. These early colonists often alter the physical conditions in ways that enable other species to become established. The new arrivals, in turn, continue to change the environment. Some early colonists do not survive the new challenges, further altering the community. When pine trees invade a site, for example, they simultaneously shade out lower-growing plants while attracting species that grow or feed on pines. Later in succession, the dominant species are usually long-lived, late-maturing, equilibrium species that are strong competitors in a stable environment. ⓘ *opportunistic and equilibrium life histories,* section 18.4B

A century ago, ecologists hypothesized that primary and secondary succession would eventually lead to a so-called **climax community,** which is a community that remains fairly constant. We now know, however, that few (if any) communities ever reach true climax conditions. In the Pacific Northwest, for example, old-growth forests are 500 to 1000 years old, yet they are still changing in their structure and composition. Major disturbances such as fire, disease, and severe storms can leave a mark that lasts for centuries. On a smaller scale, pockets of local disturbance, such as the area affected when a large tree blows over, create a patchy distribution of successional stages across a landscape.

Succession is not limited to land; it occurs in aquatic communities as well. A young lake, for example, is too low in nutrients to support abundant phytoplankton. These lakes are therefore clear and sparkling blue. As a lake ages, however, nutrients accumulate from decaying organisms and sediment. Algae thrive, turning the water green and murky. In time, a lake continues to fill with sediments and transforms into a freshwater wetland, where the soil is permanently or seasonally saturated with water. Wetlands often host spectacularly diverse assemblages of plants and animals that rely on the interface between land and water. Eventually, the wetland fills in completely and becomes dry land.

Figure 19.22 Secondary Succession. A forest fire can devastate an existing community. Soon, however, seedlings sprout and absorb the nutrients in the ashes. The forest eventually regrows, with fresh green foliage obscuring the burned stumps.
©Jerry Dodrill/Aurora/Getty Images

19.5 Mastering Concepts

1. How do ecologists measure species diversity in a community?
2. How is natural selection apparent in ecological succession?
3. Distinguish between primary and secondary succession.
4. What processes and events contribute to primary and secondary succession?
5. How do disturbances prevent true climax communities from developing?

Miniglossary | Diversity and Succession

Species richness	Measure of diversity based on the number of species in an area
Species evenness	Measure of diversity based on the relative abundance of each species in an area
Primary succession	Change in species composition of an area that previously had no community
Secondary succession	Change in species composition of an area where the existing community was disturbed

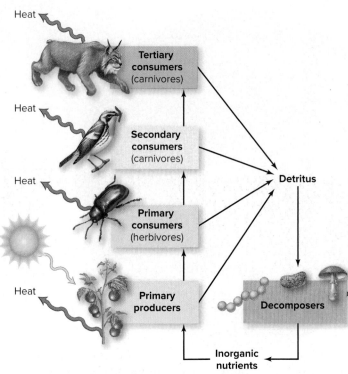

Figure 19.23 **Trophic Levels.** Primary producers are at the base of the food chain; consumers occupy the other trophic levels. All organisms contribute detritus (wastes and dead bodies) to the ecosystem. Decomposers are microbes that return the nutrients to their inorganic form, which producers absorb.

Why We Care 19.1 │ What Happens After You Flush

Community wastewater treatment plants harness the power of microorganisms to consume the organic matter in sewage before it enters waterways. In trickling filters, for example, sewage-eating bacteria and archaea are given "dream homes"—all the organic matter they can eat, along with plenty of moisture and O_2 (see figure 15.17c). After the microbes have done their job, the treated water contains a very low concentration of organic matter.

The presence of these microscopic workers explains why communities prohibit dumping used motor oil or organic solvents down the drain. Toxic chemicals can poison the bacteria and archaea that degrade sewage, making water treatment impossible.

The most commonly used forms of sewage treatment fail to remove a wide array of household chemicals and pharmaceutical drugs from the waste stream. The antibiotics in soaps and hand sanitizers are especially common, as are hormones excreted by women taking birth control pills. Ecologists are still studying the effects of these chemicals on wildlife. In the meantime, experts recommend against flushing medications—or anything other than human waste—down the toilet.

Photo: ©C Squared Studios/Getty Images RF

19.6 Ecosystems Require Continuous Energy Input

Sections 19.4 and 19.5 described the biotic interactions among members of a community. We now turn to the ecosystem-level processes by which communities interact with the nonliving environment.

To understand the interactions in an ecosystem, it is useful to recall from unit 1 that all organisms consist of both matter and energy. One way to remember this is to picture a candy bar's nutrition label, which lists the nutrient and calorie (energy) content of the snack. The food label mirrors the contents of a living cell. Inside each cell are organic molecules such as fats, sugars, and proteins. These molecules consist of carbon, hydrogen, oxygen, nitrogen, and other elements. Moreover, the covalent bonds of organic molecules store potential energy that cells can use to do work. ⓘ *organic molecules,* section 2.5

Both energy and nutrients are critical to the two properties shared by all ecosystems on Earth. First, energy flows through ecosystems in one direction only. All ecosystems therefore rely on a continuous supply of energy from some outside source, usually the sun. Second, the atoms that make up every object in an ecosystem are constantly recycled. This section and the next describe these two properties and their consequences for ecosystem function.

A. Food Webs Depict the Transfer of Energy and Atoms

Many energy and nutrient transfers occur in the context of food chains and food webs. A **food chain** is a linear sequence of feeding relationships: A beetle eats a plant, a bird eats the beetle, and so on (figure 19.23). Each organism's **trophic level** describes its position in the food chain.

Trophic levels are defined relative to the ecosystem's energy source, which is sunlight in figure 19.23. The first trophic level in any food chain is a primary producer, and all of the other trophic levels consist of consumers. For example, the primary consumers in figure 19.23 are herbivores, which eat the primary producers. Secondary consumers are carnivores (meat-eaters) that eat primary consumers, and tertiary consumers eat secondary consumers.

All organisms leave behind **detritus** consisting of dead tissue and organic wastes such as feces. Scavengers are animals that eat this material. Vultures, crows, raccoons, flies, earthworms, and many other animals are scavengers. **Decomposers,** such as many fungi and bacteria, are microbes that complete the recycling process; they secrete enzymes that digest the remaining organic molecules in detritus. As they do so, they fuel their own growth and reproduction, but they also return carbon, nitrogen, phosphorus, and other inorganic nutrients to the environment. Without these crucial microbes, dead bodies and organic wastes would tie up all useful nutrients, and ecosystems would grind to a halt. Why We Care 19.1 describes how we employ decomposers in community wastewater treatment facilities.

Autotrophs and decomposers have opposite roles in ecosystems. Whereas autotrophs absorb inorganic nutrients and produce organic molecules, decomposers return the elements in those organic molecules to their inorganic form. Both roles are critical to ecosystem function.

Of course, feeding relationships in an ecosystem are more complex than a simple food chain might suggest. A **food web** is a network of interconnected food chains, such as the Antarctic web in figure 19.24. Keep in mind that this

Figure 19.24 **The Antarctic Web of Life.** The interactions among Antarctic residents form a complex network. Note that producers, consumers, and scavengers are all present, and that decomposers release inorganic nutrients that producers can use. For simplicity, heat energy released by the food web is not illustrated.

Succession Is a Gradual Change in a Community   Ecosystems Require Continuous Energy Input   Chemicals Cycle Within Ecosystems

Why We Care 19.2 | Mercury on the Wing

Water pollution can have surprising effects on terrestrial ecosystems. For example, researchers have discovered that mercury in a contaminated Virginia river has reduced the reproductive success of nearby birds.

Carolina wrens eat insects, spiders, and small vertebrates in the areas surrounding rivers. Plants in the river and on the shore accumulate mercury, which moves up the food chain to the wrens. Concentrations in some wrens were as high as 3 parts per million.

Carolina wren

The mercury lowers wren fitness; birds with the highest mercury concentrations in their blood might produce 60% fewer offspring than they otherwise would have. Other studies suggest that mercury also distorts song performance, possibly further reducing reproductive success.

The human diet can contain mercury, too. Tuna and other predatory fish contain high concentrations of mercury, another example of waterborne toxins returning to the land.

Photo: ©William Leaman/Alamy Stock Photo RF

Figure It Out

According to figure 19.24, baleen whales mainly eat herbivorous zooplankton, and sperm whales mainly eat squid. About how much more energy is available in the ecosystem for baleen whales than for sperm whales?

Answer: About 10 times more energy is available for baleen whales.

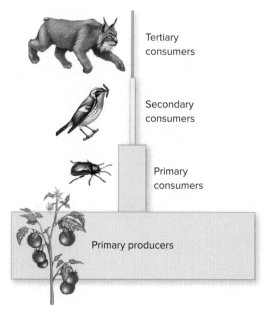

Tertiary consumers

Secondary consumers

Primary consumers

Primary producers

Figure 19.25 **Pyramid of Energy.** Each block depicts the amount of energy stored in a trophic level. This example assumes that about 10% of the energy in any level is available to the next.

---

diagram is highly simplified. Not shown, for example, are the worms, hagfishes, sharks, and other organisms that feed for months or years on the carcasses of dead whales that sink to the seafloor. Nor does the diagram depict the many transfers between aquatic ecosystems and the land.

Careful examination of a food web diagram reveals that even the fiercest top predator, such as a killer whale, relies on other organisms, many of them microscopic. The same principle applies to the food web in which humans participate. Think of everything you have eaten today: In all likelihood, your meals and snacks have included both plant and animal products. Humans have developed an extremely complex global food chain that includes organisms harvested from land and water all over the world.

Some of the connections are surprising. For example, pigs and chickens raised on commercial farms around the world eat millions of tons of fish harvested near South America each year. Conversely, commercially raised catfish eat soybean meal, corn, and rice—all of which grow on land.

B. Heat Energy Leaves Each Food Web

The total amount of energy that is trapped, or "fixed," by all autotrophs in an ecosystem is called gross primary production. Autotrophs use much of this energy in respiration, generating ATP for their own growth, maintenance, and reproduction (see chapter 6). As they do so, they release heat energy (section 4.1B explains why). No organism can store heat or use it as an energy source, so it is exchanged with the environment and leaves the food web. The remaining energy in the producer level is called **net primary production;** it is the amount of energy available for consumers to eat.

Primary production varies widely across ecosystems on land, depending largely on temperature and moisture. The warm, wet tropical rain forests therefore have among the highest rates of net primary production per square meter; deserts have the lowest. In aquatic ecosystems, the availability of inorganic nutrients such as phosphorus is more important. Nutrient-polluted waters therefore often become overgrown with algae (see section 19.7E).

Consumers also produce heat energy in every metabolic reaction. Because of these inefficiencies, only a small fraction of the potential energy in one trophic level fuels the growth and reproduction of organisms at the next trophic level. On average, about one tenth of the energy at one trophic level is available to the next rank in the food chain (figure 19.25).

The "10% rule" provides a convenient estimate, but it ignores the fact that food quality varies widely. The transfer efficiency from one trophic level to the next actually ranges from about 2% to 30%. Primary consumers that eat hard-to-digest plants convert only a small percentage of the energy available to them into animal tissue, whereas meat is easy to digest. In addition, ectothermic animals such as insects, fishes, and lizards use energy much more efficiently than do endothermic mammals and birds. A trophic level consisting of lizards therefore consumes much less energy than a trophic level consisting of an equal weight of birds. ⓘ *endotherms and ectotherms,* section 23.5

Eventually, as organic molecules pass from trophic level to trophic level, all of the stored energy leaves the food web in the form of heat. Thus, energy flows through an ecosystem in one direction: from source (usually the sun), through organisms, to heat. For the ecosystem to persist, it must have a continual supply of energy. If the energy source goes away, so does the ecosystem.

A **pyramid of energy** represents each trophic level as a block whose size is directly proportional to the energy stored in that level (see figure 19.25). Because every organism loses heat to the environment, the energy pyramid explains why food chains rarely extend beyond four trophic levels. An organism in a still higher trophic level would have to expend tremendous effort just to find the small amount of food available, and that small amount would not be enough to make all that effort pay off.

The loss of energy at each trophic level suggests a way to maximize the benefit we get from crops we grow for food. The most energy available in an ecosystem is at the producer level. Therefore, the lower we eat on the food chain, the more people we can feed. A person can do this by getting protein from beans, grains, and nuts instead of from meat and dairy (figure 19.26).

C. Harmful Chemicals May Accumulate in the Highest Trophic Levels

The shape of the energy pyramid has another consequence for ecosystems. In **biomagnification,** a chemical becomes most concentrated in organisms at the highest trophic levels. Biomagnification happens for pollutants and other chemicals that share two characteristics. First, they dissolve in fat. This characteristic is important because animals eliminate water-soluble chemicals in their urine but retain fat-soluble chemicals in fatty tissues. Second, chemicals that biomagnify are not readily degraded. A highly degradable chemical would not persist long enough in the environment to ascend food chains.

Mercury is a persistent pollutant that illustrates biomagnification; see figure 19.27 and Why We Care 19.2. Coal-fired power plants and mines release this element into the air and water. Moreover, many household products contain mercury that enters water, air, or soil via sewage treatment plants and incinerators. Imagine mercury entering water from a nearby power plant. Bacteria in the sediments soon convert the mercury into a fat-soluble form called methylmercury. Its concentration in the water is initially low. But methylmercury that enters an organism's body is not eliminated in urine or other watery wastes. As one animal eats another, all of the methylmercury stored in the prey ends up in the predator. Each predator eats many prey, so the mercury accumulates in the predator's tissues. In organisms at the fourth level of the food chain, mercury concentrations may be 100,000 times greater than at the base of the food web. High levels of mercury have prompted health professionals to recommend that people limit their consumption of long-lived, carnivorous fish such as shark and tuna.

19.6 Mastering Concepts

1. Identify the trophic levels in a food chain.
2. What roles do producers and decomposers play in ecosystems?
3. How efficient is energy transfer between trophic levels?
4. Draw an energy pyramid for an ecosystem with three levels of consumers.
5. Explain how biomagnification disproportionately affects organisms at the top of a food chain.

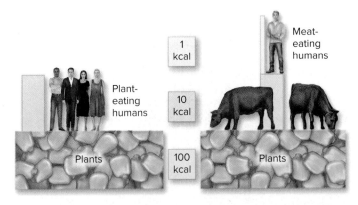

Figure 19.26 The Energetic Cost of Meat. These simple pyramids show that the energy available in plants can support many more vegetarians than it can meat-eaters.

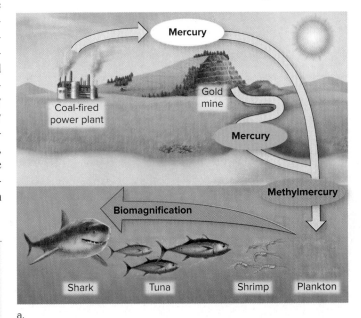

a.

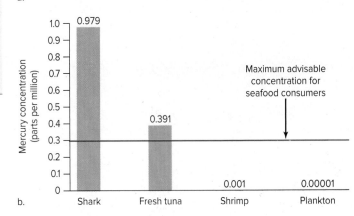

b.

Figure 19.27 Biomagnification. (a) Power plants and gold mines are among the main sources of mercury pollution. Mercury is converted to methylmercury in sediments; after entering the food chain, the methylmercury concentration increases with each successive trophic level. (b) Concentrations of mercury in shark meat, tuna, and shrimp vividly illustrate biomagnification.

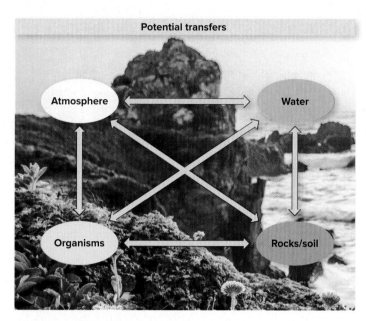

Potential transfers

Figure 19.28 Biogeochemical Cycle. Water and inorganic nutrients cycle among four storage reservoirs. A coastal ecosystem illustrates all four.

Photo: ©Adam Hester/Blend Images LLC RF

19.7 Chemicals Cycle Within Ecosystems

Energy flows in one direction, but all life must use the elements that were present when Earth formed. In **biogeochemical cycles,** interactions of organisms and their environment continuously recycle these elements. If not for this worldwide recycling program, supplies of essential elements would have been depleted as they became bound in the bodies of organisms that lived eons ago.

Whatever the element, all biogeochemical cycles have features in common. Each element is distributed among four major storage reservoirs: organisms, the atmosphere, water, and rocks and soil (figure 19.28). Depending on the reservoir, the element may combine with other elements and form a solid, liquid, or gas.

Transfers among the reservoirs form the basis of each biogeochemical cycle (figure 19.29). Along the way, the elements undergo chemical changes. For example, plants and other autotrophs take up the inorganic forms of elements from abiotic reservoirs and incorporate them into organic molecules such as lipids, carbohydrates, proteins, and nucleic acids. If an animal eats the plant, the elements may become part of animal tissue. If another animal eats the herbivore, the elements may be incorporated into the predator's body, and so on. Eventually, decomposers consume the dead bodies and organic wastes, releasing inorganic forms of the elements back into the environment.

This section describes the water, carbon, nitrogen, and phosphorus cycles. All of these substances are essential to life and abundant in cells. As you will see, however, the water cycle is somewhat different from the other three cycles. Most processes in the water cycle are physical, not biological. Its status as a *bio*geochemical cycle is therefore tenuous. Nevertheless, carbon, nitrogen, and phosphorus compounds can dissolve in water, and water movement is important in transporting these elements among the storage pools. Knowledge of the water cycle is therefore essential to understanding the three nutrient cycles.

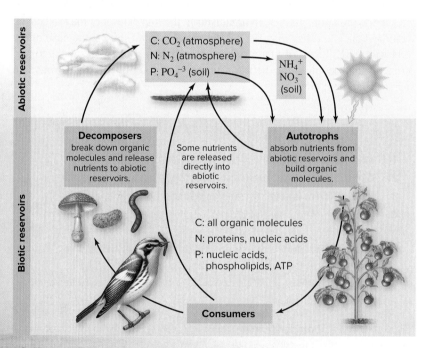

Figure 19.29 Abiotic and Biotic Reservoirs. Inorganic forms of carbon, nitrogen, and phosphorus come from soil or the atmosphere. The elements enter the food web upon absorption by autotrophs. Organisms eventually release the elements back into abiotic reservoirs, continuing the cycle.

(grassland): ©Brand X Pictures/PunchStock RF

A. Water Circulates Between the Land and the Atmosphere

Water covers much of Earth's surface, primarily as oceans but also as lakes, rivers, streams, ponds, swamps, snow, and ice (see figure 19.9 and figure 19.30). Water also occurs below the land surface as groundwater.

The main processes that transfer water among these major storage compartments are evaporation, precipitation, runoff, and percolation. The sun's heat evaporates water from land and water surfaces (figure 19.30, step 1). Water vapor rises on warm air currents, then cools and forms clouds (step 2). If air currents carry this moisture higher or over cold water, more cooling occurs, and the vapor condenses into water droplets that fall as rain, snow, or other precipitation (step 3). Some of this precipitation falls on land, where it may run along the surface. Streams unite into rivers that lead back to the ocean (step 4), where the sun's energy again heats the surface, continuing the cycle.

Rain and melted snow may also soak (percolate) into the ground, restoring soil moisture and groundwater (step 5). This underground water feeds the springs that support many species. Spring water evaporates or flows into streams, linking groundwater to the overall water cycle.

Although most processes in the water cycle are physical, organisms do participate (step 6); after all, water is essential to life (see Burning Question 19.2). Plant roots absorb water from soil and release much of it from their leaves in transpiration. The lush plant life of the tropical rain forests draws huge amounts of water from soil and returns it to the atmosphere. In addition, animals drink water and consume it with their food, returning it to the environment through evaporation and urination. ⓘ *essential water*, section 2.3; *transpiration*, section 21.6A

Burning Question 19.2

Is bottled water safer than tap water?

A liter of bottled water may cost thousands of times more than the same volume of tap water. Many people are willing to pay the extra cost because they think bottled water is the safest option. For some residents of North America, the public water supply is indeed contaminated with lead or other toxins. But for most people, perfectly safe water flows from the kitchen faucet. What's more, a lot of bottled water is actually nothing more than tap water embellished with a fancy label.

©McGraw-Hill Education

Bottled water has hidden environmental costs as well. It takes energy to transport and refrigerate bottled water, and bottles that are not recycled end up in landfills. Moreover, chemicals from the bottle can leach into the water inside, and pores in the plastic can harbor bacteria once the bottle is opened. These risks intensify if a disposable plastic bottle is reused.

Instead of buying bottled water, consider filling your own reusable bottle with tap water. If necessary, a high-quality filter can remove most contaminants and odors.

Submit your burning question to
marielle.hoefnagels@mheducation.com

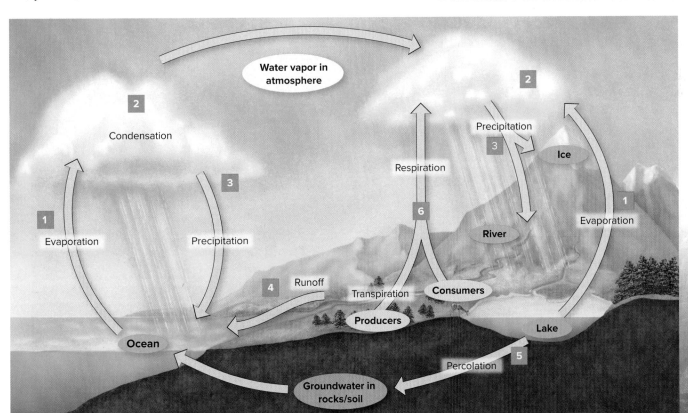

Figure 19.30 **The Water Cycle.** Water falls to Earth as precipitation. Organisms use some water, and the remainder evaporates, runs off into streams and the ocean, or enters the ground. Transpiration and respiration return water to the environment.

(stream): ©Natural Selection John Reddy/Design Pics RF

(sky): ©Ingram Publishing/SuperStock RF

B. Autotrophs Obtain Carbon as CO_2

Carbon is a part of all organic molecules, and organisms continually exchange it with the atmosphere (figure 19.31). Autotrophs absorb atmospheric CO_2 and use photosynthesis to produce organic compounds, which they incorporate into their tissues (step 1). Cellular respiration releases carbon back to the atmosphere as CO_2 (step 2). Dead organisms and wastes contribute organic carbon to soil or water (step 3). Bacteria and fungi decompose these organic compounds and release CO_2 to the soil, air, and water as they respire (step 4).

Some types of archaea also participate in the carbon cycle by metabolizing organic compounds and releasing methane (CH_4) into the atmosphere. Most of these microbes live in anaerobic habitats such as wetlands, marine sediments, and the intestines of humans, cattle, and other animals.

The exchange of carbon between organisms and the environment is relatively rapid, but more stable pools of carbon also exist. Fossil fuels such as coal and oil formed long ago from the remains of dead organisms (step 5). When these fuels burn (step 6), carbon returns to the atmosphere as CO_2. Decades of accumulation of CO_2 and other greenhouse gases in the atmosphere are likely responsible for Earth's gradually warming climate. Chapter 20 describes this topic in more detail.

One of the largest reservoirs of carbon is the ocean. CO_2 from the atmosphere dissolves in ocean water. Most of the dissolved gas reacts with the water to form carbonic acid (H_2CO_3). Some of this carbon reacts with calcium to

Figure 19.31 **The Carbon Cycle.** Carbon dioxide (CO_2) in the air and water enters ecosystems through photosynthesis and then passes along food chains. Respiration and combustion return carbon to the abiotic environment. Carbon can be retained for long periods in carbonate rocks and fossil fuels. The archaea that produce methane (CH_4) are omitted from the cycle for simplicity.

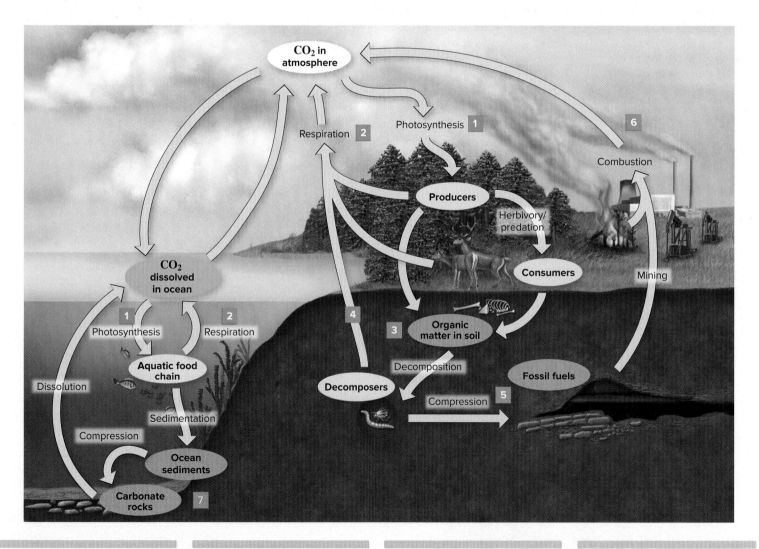

CHAPTER SUMMARY

19.1 Organisms Interact Within Communities and Ecosystems

- **Communities** consist of coexisting **populations** of multiple species.
- An **ecosystem** includes a **biotic** community plus its **abiotic** environment.
- Each species in a community has a place where it normally lives **(habitat)** and a set of resources necessary for its life **(niche).**

19.2 Earth Has Diverse Climates

- Solar radiation is most intense at the equator and least intense at the poles (figure 19.35). The resulting uneven heating creates the patterns of precipitation, prevailing winds, and ocean currents that influence climate.
- A mountain range can influence local climate by producing a **rain shadow.**

19.3 Biomes Are Ecosystems with Distinctive Communities of Life

- **Biomes** are major types of ecosystems that occupy large geographic areas and share a characteristic climate and group of species.

A. The Physical Environment Dictates Where Each Species Can Live

- **Autotrophs (primary producers)** directly or indirectly support the **heterotrophs (consumers)** in each biome. On land, the most important primary producers are plants; in water, **phytoplankton** are most important.
- The major abiotic factors that limit a species' distribution on land include sunlight, temperature, moisture, salinity, and fire.

B. Terrestrial Biomes Range from the Lush Tropics to the Frozen Poles

- The forest biomes include the **northern coniferous forest, temperate coniferous forest, temperate deciduous forest,** and **tropical rain forest.**
- Grasslands include the tropical **savannas** and the **temperate grasslands.** Fire, grazing, and seasonal drought keep trees from dominating these biomes.
- **Desert** plants have adaptations that help them obtain and store water.
- **Mediterranean shrublands** have dry summers and fire-adapted plants.
- **Tundras** have very cold, long winters. A layer of permafrost prevents the growth of trees.
- **Polar ice** forms at the poles, where the climate is extremely cold and dry.

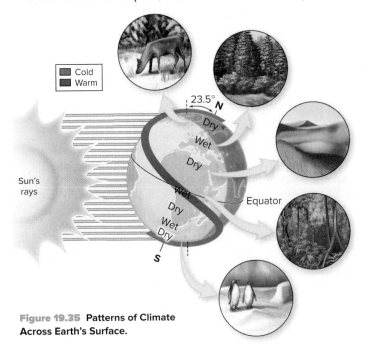

Figure 19.35 Patterns of Climate Across Earth's Surface.

C. Aquatic Biomes Include Fresh Water and the Oceans

- In a lake's **photic zone,** light is sufficient for photosynthesis. Rooted plants fringe the shallow shoreline, whereas phytoplankton are the primary producers in the upper layer of open water. Nutrients from the upper layers support life in the deeper zones, where light does not penetrate.
- Near the headwaters of a river, the channel is narrow, and the current is swift. As the river accumulates water and sediments, the current slows, and the channel widens.
- The ocean has a productive, nutrient-rich shoreline. Photosynthesis occurs in the photic zone, but life in the deep ocean relies on organic matter from above.
- An **estuary** is a highly productive area where rivers empty into oceans, and life is adapted to fluctuating salinity. Residents of the rocky **intertidal zone** are adapted to stay in place as the tide ebbs and flows. In tropical regions, **coral reefs** support many thousands of species.

19.4 Community Interactions Occur Within Each Biome

- Within a community, species interact in many ways.

A. Many Species Compete for the Same Resources

- Populations that share a habitat often compete for limited resources. **Competition** reduces the fitness of both species.
- According to the **competitive exclusion principle,** two species cannot indefinitely occupy exactly the same niche.
- In **resource partitioning,** competition between multiple species with similar niches restricts each species to a subset of available resources.

B. Symbiotic Interactions Can Benefit or Harm a Species

- **Symbiotic** relationships include **mutualism** (both species benefit), **commensalism** (one species benefits, whereas the other is unaffected), and **parasitism** (one species benefits, but the host is harmed).

C. Herbivory and Predation Link Species in Feeding Relationships

- **Herbivory** is an interaction in which a consumer eats a plant; a **predator** is a **carnivore** that kills and eats another animal (its **prey**).
- Plants and prey animals have defenses against herbivores and predators; in addition, animals have adaptations that help them capture food.

D. Closely Interacting Species May Coevolve

- In **coevolution,** the interaction between species is so strong that genetic changes in one population select for genetic changes in the other.

E. A Keystone Species Has a Pivotal Role in the Community

- A **keystone species** makes up a small proportion of a community's biomass but has a large influence on the community's composition.

19.5 Succession is a Gradual Change in a Community

- The diversity of a community depends on both **species richness** (the number of species) and **species evenness** (relative abundance).
- As species interact with one another and their physical habitats, they change the composition of the community. This process is called **succession.**
- **Primary succession** occurs in a previously unoccupied area, beginning with **pioneer species** that allow soil to develop, paving the way for additional organisms to thrive.
- **Secondary succession** is more rapid than primary succession because soil does not have to build anew.
- Succession may lead toward a stable **climax community,** but true long-term stability is rare. Pockets of local disturbance mean that most communities are a patchwork of successional stages.

- Lakes also undergo succession as they age. Young lakes contain few nutrients to support phytoplankton. As nutrients gradually accumulate, algae tint the water green. A continuous influx of sediments from the land eventually transforms the lake into a wetland.

19.6 Ecosystems Require Continuous Energy Input

A. Food Webs Depict the Transfer of Energy and Atoms

- An organism's **trophic level** depends on the number of steps between it and the ultimate source of energy in the ecosystem.
- At the base of each **food chain** are primary producers that harness energy from the sun or inorganic chemicals.
- Consumers make up the next trophic levels. Primary consumers (herbivores) eat the primary producers. A secondary consumer may eat the primary consumer, and a tertiary consumer may eat the secondary consumer. **Decomposers** break down **detritus** (nonliving organic material) into inorganic nutrients.
- Interconnected food chains form **food webs.**

B. Heat Energy Leaves Each Food Web

- The total amount of energy that producers convert to chemical energy in an ecosystem is gross primary production. After subtracting energy for maintenance and growth, the energy remaining in producers is **net primary production.**
- Food chains rarely extend beyond four trophic levels because only a small percentage of the energy in one trophic level transfers to the next level.
- A **pyramid of energy** is a diagram that depicts the amount of energy at each trophic level.

C. Harmful Chemicals May Accumulate in the Highest Trophic Levels

- **Biomagnification** concentrates nondegradable, fat-soluble chemicals in the highest trophic levels because the chemical passes to the next consumer rather than being metabolized or eliminated in urine.

19.7 Chemicals Cycle Within Ecosystems

- **Biogeochemical cycles** are geological and chemical processes that recycle chemicals essential to life, including transfers among organisms, the atmosphere, water, and rocks and soil.

A. Water Circulates Between the Land and the Atmosphere

- Water is transferred from the atmosphere to the land or to bodies of water as precipitation. Organisms release water in transpiration, evaporation, or urination.

B. Autotrophs Obtain Carbon as CO_2

- Autotrophs use carbon in CO_2 to produce organic molecules. Cellular respiration and burning fossil fuels release CO_2. Decomposers release carbon from once-living material.

C. The Nitrogen Cycle Relies on Bacteria

- **Nitrogen-fixing** microbes convert atmospheric nitrogen to ammonium, which plants can incorporate into their tissues. Decomposers convert the nitrogen in dead organisms back to ammonium. **Nitrification** converts the ammonium to nitrates, whereas **denitrification** converts these compounds to nitrogen gas.

D. The Phosphorus Cycle Begins with the Weathering of Rocks

- Rocks release phosphorus as they erode. Autotrophs incorporate the phosphates into organic molecules; decomposers return inorganic phosphates to the environment.

E. Excess Nitrogen and Phosphorus Cause Problems in Water

- Excess nutrients enter water in **eutrophication,** triggering algal blooms. Microbes decompose dead algae, depleting oxygen in the water.

MULTIPLE CHOICE QUESTIONS

1. Which of the following is an example of an ecological community?
   a. The many types of microbes living in a human intestine
   b. An ant colony
   c. The people living in your neighborhood
   d. The cells of a platypus

2. Why are the poles colder than the equator?
   a. Because the poles receive solar radiation that is less intense
   b. Because the poles receive less rainfall than the equator
   c. Because ocean currents bring warm water to the equator
   d. Because the tilt of the Earth points the poles away from the sun

3. A biome with high average temperature and moderate annual rainfall is a
   a. tropical savanna.   c. temperate deciduous forest.
   b. tropical rain forest.   d. taiga.

4. Why do estuaries have such high primary production?
   a. The changing salinity of the water
   b. High nutrient input from the land and tides
   c. The absence of consumers that eat producers
   d. All of the above are correct.

5. Two species of photosynthetic bacteria live in ocean water but use different wavelengths of light in photosynthesis. This example illustrates
   a. competitive exclusion.
   b. commensalism.
   c. resource partitioning.
   d. biomagnification.

6. Some types of crabs live in clumps of coral. The crab defends its home, protecting the coral from sea stars and other predators. The interaction between crabs and corals is an example of
   a. resource partitioning.   c. commensalism.
   b. competitive exclusion.   d. mutualism.

7. A large tree falls over in an old-growth forest, allowing light to reach a formerly shaded area. A few weeks later, what types of rooted plants will dominate the open patch?
   a. Lichens and mosses   c. Large shrubs
   b. Herbs and weeds   d. Oak trees

8. When you eat a carrot, you are acting as a(n)
   a. heterotroph and primary consumer.
   b. autotroph and herbivore.
   c. heterotroph and secondary consumer.
   d. autotroph and primary producer.

9. Which statement best explains why persistent, fat-soluble chemicals such as methylmercury accumulate in the highest trophic levels?
   a. Animals at the highest trophic levels eat the lowest-quality food.
   b. The amount of biomass increases in each trophic level, and large organisms accumulate the most methylmercury.
   c. Large organisms are often herbivores, so they consume methylmercury directly.
   d. Methylmercury does not leave an animal's body; each predator therefore consumes the mercury contained in many prey.

10. Which cycle relies the least on decomposers?
    a. Carbon cycle   c. Phosphorus cycle
    b. Water cycle   d. Nitrogen cycle

Answers to Multiple Choice questions are in appendix A.

WRITE IT OUT

1. How does a community differ from an ecosystem?
2. How does the fact that Earth is a sphere tilted on its axis influence the distribution of life?
3. How can the poor soil of the tropical rain forest support such diverse and abundant life?
4. List adaptations of desert and polar animals to the climate of their respective biomes.
5. Use the clues provided to determine which biome houses each of the following four fish. Yellowfin tuna require salt water; young sea bass occupy areas where fresh and salt water mix; brook trout require clear, cool, oxygen-rich fresh water; and catfish prefer warm, quiet fresh water with a slow current.
6. Three Galapagos finch species have different beak sizes and specialize in different types of food. Explain how these three species can share the same habitat without driving each other to extinction.
7. List examples of adaptations that enable an organism to compete with other species, live inside another species, find food, and avoid herbivory or predation. How does each adaptation contribute to the organism's fitness?
8. Mountain yellow-legged frogs live in the Sierra Nevada range. Their tadpoles mainly eat algae. One predator of adult frogs is a garter snake, which is eaten by bullfrogs. Recently, a chytrid fungus has killed many adult mountain yellow-legged frogs. How might this change affect the algae, garter snakes, and bullfrogs?
9. Suppose a plot of forest is cleared of trees in anticipation of a new shopping mall. However, after the bulldozers are gone, the company runs out of money, and the land sits undisturbed for many years. Describe the events that may occur in the years following the damage to the forest. What are these community changes called?
10. Imagine that you could build a covered enclosure around a small ecosystem, blocking out all light and preventing gas exchange with the environment. How would the total amount of organic material, available energy, and nutrients in the ecosystem change over time?
11. Alaska salmon spend much of their adult lives in the ocean, eating crustaceans and small fish. When it is time to mate, they swim upstream to spawn in the rivers where they hatched. Along the way, bears and other predators catch and eat many salmon. What does this story tell us about the source and fate of nutrients in ecosystems?

SCIENTIFIC LITERACY

Search the Internet to find a definition for *anthropogenic biome* or *anthrome*. In which anthrome do you live? Considering the effects of humans on ecosystems, do you think biome classifications are still relevant, or should biologists only discuss anthromes?

Answers to Mastering Concepts, Write It Out, Scientific Literacy, and Pull It Together questions can be found in the Connect ebook.
connect.mheducation.com

PULL IT TOGETHER

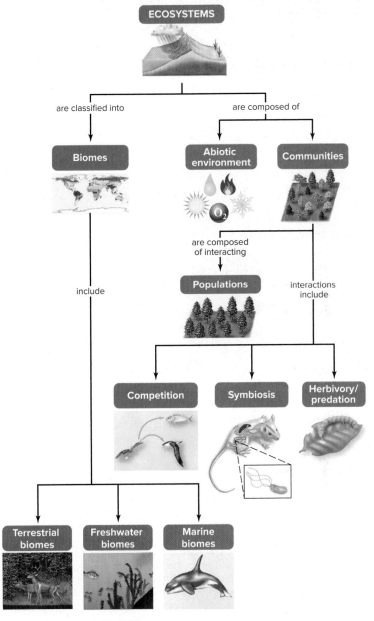

Figure 19.36 Pull It Together: Communities and Ecosystems.

Refer to figure 19.36 and the chapter content to answer the following questions.

1. Review the definitions of *ecology, populations, communities,* and *ecosystems* using the Survey the Landscape figure in the chapter introduction and the Pull It Together figure. Then, classify the three biologists' questions below as relating to *population ecology, community ecology,* or *ecosystem ecology*: (a) How do gophers and moles compete for resources? (b) How do monkeys establish dominance within a group? (c) How does pollution affect crop plants?
2. Add *mutualism, commensalism,* and *parasitism* to this concept map.
3. Make another concept map that shows nutrient cycling on land. Include *producers, consumers, decomposers, carbon, nitrogen, phosphorus, atmosphere,* and *soil;* you may also add other concepts.

Preserving Biodiversity

Death by Plastic. This young albatross died after eating plastic garbage in the ocean.

Source: CINMS/NOAA/Claire Fackler

LEARNING OUTLINE

20.1 Earth's Biodiversity Is Dwindling

20.2 Many Human Activities Destroy Habitats

20.3 Pollution Degrades Habitats

20.4 Global Climate Change Alters and Shifts Habitats

20.5 Exotic Invaders and Overexploitation Devastate Many Species

20.6 Some Biodiversity May Be Recoverable

APPLICATIONS

Burning Question 20.1 *What are the best ways to reverse habitat destruction?*

Burning Question 20.2 *How can people reduce their contribution to water pollution?*

Burning Question 20.3 *What does the ozone hole have to do with global climate change?*

Burning Question 20.4 *How can small lifestyle changes reduce air pollution and global climate change?*

Burning Question 20.5 *How can people help slow the spread of invasive species?*

Burning Question 20.6 *Can everyday buying decisions help protect overharvested species?*

Why We Care 20.1 *Environmental Legislation*

Investigating Life 20.1 *Up, Up, and Away*

Learn How to Learn

How to Use a Study Guide

Some professors provide study guides to help students prepare for exams. Used wisely, a study guide can be a valuable tool. One good way to use a study guide is AFTER you have studied the material, when you can go through the guide and make sure you haven't overlooked any important topics. Alternatively, you can cross off topics as you encounter them while you study. No matter which technique you choose, don't just memorize isolated facts. Instead, try to understand how the items on the study guide relate to one another. If you're unclear on the relationships, be sure to ask your instructor.

SURVEY THE LANDSCAPE
Ecology

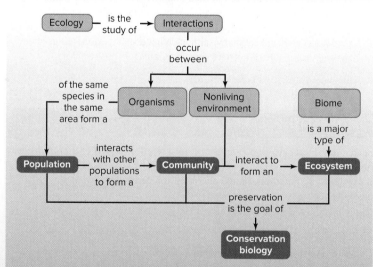

Human activities are associated with a dramatic reduction in biodiversity, but conservation efforts at the population, community, and ecosystem levels can help reverse the decline.

For more details, study the Pull It Together feature in the chapter summary.

Do you think that humans are at the "top of the heap"? That we can go it alone or with just a few carefully selected plant and animal companions? Think again. The truth is that we rely on other species for a huge variety of eco-system services, many of which are not obvious until the species that provide them are gone.

Source: USFWS/USGS/ John Klavitter

Chapter 19 described the interactions that unite species in ecosystems; we now turn to the many strands in the web of life that are being torn. You may find this chapter depressing—the list of problems goes on and on because human activities threaten other species in so many ways. The albatross in this chapter's opening photo is a particularly striking example. It died after consuming plastic garbage that its mother mistook for food floating in the ocean. A photographer captured the evidence in this case, but countless additional animals, plants, and other organisms are dying as well.

Given our reliance on other species, the loss of biodiversity is not only sad but also dangerous. But as you will see, humans can undo some of the damage if we are willing to commit the time, effort, and money to preserve the other species that share our planet.

20.1 Earth's Biodiversity Is Dwindling

For more than 3 billion years, evolution has produced an extraordinary diversity of life, both obvious and unseen; unit 3 provided an overview of Earth's inhabitants. Humans simply cannot live without these other species. We use a wide variety of other organisms for food, shelter, energy, clothing, and drugs. Microbes carry out indispensable tasks, including digesting food in our intestines, decaying organic matter in soil, fixing nitrogen, and producing oxygen (O_2). Plants and microbes absorb carbon dioxide (CO_2) and purify the air, soil, and water. Wetland plants reduce the severity of floods. Insects pollinate our crops. The remains of species that lived millions of years ago provide the fossil fuels that sustain our economies. The list goes on and on.

Clearly, our existence as a species depends on **biodiversity**—the variety of life on Earth. Biologists measure biodiversity at three levels: genetic, species, and ecosystem. Genetic diversity is the amount of variation that exists within a species. This aspect of biodiversity is essential for populations to adapt to changing conditions. The next level, species diversity, accounts for the number of species that occupy the biosphere. Finally, ecosystem diversity means the variety of ecosystems on Earth, such as deserts, rain forests, grasslands, and mountaintops. ⓘ *biomes,* section 19.3; *species richness,* section 19.5

One way to monitor species biodiversity is to count how many species are at risk of extinction. **Extinction** means that the last individual of a species has perished. (Note that conservation biologists sometimes distinguish between species that are totally extinct and those that are extinct in the wild. The dodo is extinct, whereas a bird called the Guam rail exists in captivity but is extinct in its natural habitat.) An **endangered species** has a high risk of extinction in the near future, and a **vulnerable species** is likely to become extinct in the more distant future. The International Union for Conservation of Nature (IUCN) combines endangered and vulnerable species into one umbrella category ("threatened").

The data suggest that Earth is in the midst of a biodiversity crisis (figure 20.1). The current extinction rate of vertebrates is some 100 to 1000 times the "background" species extinction rate, which estimates how quickly

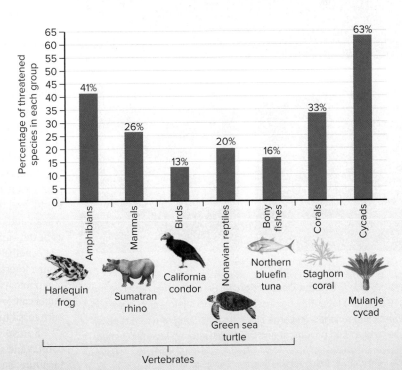

Figure 20.1 Biodiversity Crisis. Species throughout the tree of life are in peril, but some are harder hit than others. Each bar represents the percentage of known species that are considered threatened; one example from each group is shown.

Source: Data from IUCN Red List

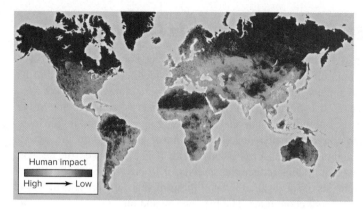

Figure 20.2 **Big Footprint.** Yellow, orange, and red areas on the map depict areas where human impacts on the biosphere are most intense.

Source: Venter O, Sanderson EW, Magrach A, et al. Sixteen years of change in the global terrestrial human footprint and implications for biodiversity conservation. *Nature Communications.* 2016;7:12558. doi:10.1038/ncomms12558.

a.

b.

c.

Figure 20.3 **Habitat Destruction.** (a) Farmland has replaced the native prairie in Iowa. (b) Logging destroys the rainforest to make room for farms and ranches. (c) Large cities such as San Francisco replace native habitat with pavement and buildings, altering land and water.

(a): Source: USDA Natural Resources Conservation Service/Lynn Betts; (b): ©Marcelo Horn/E+/Getty Images RF; (c): ©Jenny Solomon/Shutterstock RF

species disappeared before human intervention. According to the IUCN, threatened groups include amphibians, mammals, reptiles (including birds), bony fishes, corals, plants, and many other types of organisms. Overall, the current biodiversity crisis is comparable to the five mass extinctions that have occurred in the past 500 million years. ⓘ *extinction,* section 14.5

Conservation biologists study the preservation of biodiversity at all levels. These scientists try to determine why species disappear, and they develop strategies for maintaining diversity. This chapter focuses first on the main causes of the loss of biodiversity, which are sometimes abbreviated as HIPPO: *h*abitat destruction and degradation; *i*nvasive species; human *p*opulation size; *p*ollution; and *o*verexploitation. In reality, however, the human population is not a separate issue. Instead, it is intimately related to the other four main problems, each of which is likely to become worse as our population continues to grow. ⓘ *human population,* section 18.6

Figure 20.2 illustrates where humanity's effects on the biosphere are most intense. The scale of our influence is so profound that scientists have proposed adding a new epoch—the Anthropocene—to the geologic timescale (see figure 13.2). Nevertheless, the chapter ends on a hopeful note, with some ways people can counteract the biodiversity crisis. (See also the Burning Question boxes throughout the chapter to learn what you can do to help.)

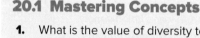

20.1 Mastering Concepts

1. What is the value of diversity to humans and to ecosystems as a whole?
2. Describe the relationships among the three levels of biodiversity. Why is each level important?
3. Differentiate among extinct, endangered, and vulnerable species.
4. What is conservation biology?

20.2 Many Human Activities Destroy Habitats

Habitat destruction is the primary threat to biodiversity (figure 20.3). Humans have altered more than 50% of the land, replacing prairies, wetlands, and forests with farms, rangeland, and cities. The link to biodiversity is obvious: Destroying a habitat makes it difficult or impossible for its occupants to survive and reproduce.

Prairies and other temperate grasslands have disappeared as the human population has expanded. Grassland soils are among the most fertile in the world, and their rolling hills are ideal for cultivation. Fields of corn, soybeans, wheat, and other crops have replaced diverse grasses, leaving only small patches of the vast grasslands that once occupied three continents (see figure 20.3a).

Another form of habitat destruction is deforestation, the removal of all trees from a forested area. Like all plants, trees absorb CO_2 from the atmosphere and use it in photosynthesis, a process that also releases O_2. The wood of a living tree is an important long-term "carbon sink" that helps offset CO_2 released when humans burn fossil fuels. ⓘ *carbon cycle,* section 19.7B

Forests harbor a tremendous diversity of organisms that provide wildlife habitat and supply many of the resources that we use every day: lumber, paper, furniture, and foods such as wild mushrooms and nuts. As plants near extinction, we lose potential medicines along with the ancestors of many of our most important domesticated plants. These older varieties are a valuable resource to plant breeders, who are always searching for disease-resistance genes to breed into modern crop plants.

Trees also play a critical role in the water cycle, returning water to the atmosphere via transpiration. Removing trees may therefore trigger changes in the global water cycle and, by extension, in the global climate. In addition, in an intact forest, rain and melting snow seep slowly through decaying plant litter and into the soil, helping to recharge groundwater. Forest destruction removes this uppermost layer of organic matter, so rainwater enters streams rather than soaking into the soil. As the soil erodes away, waterways become choked with nutrients and sediments. ⓘ *water cycle,* section 19.7A

All of the world's forest biomes are under threat. Logging, mining, and exploration for oil and gas are rapidly depleting the northern coniferous forest. In addition, nearly all of the world's native temperate forests are already gone. In North America, for example, people have cleared the land to create farmland, obtain fuel, and make room for cities. Today, less than 1% of the original temperate forest survives.

Closer to the equator, people are logging and burning tropical rain forests to make room for crops and domesticated animals (see figure 20.3b). Networks of roads, originally built for logging, oil and gas exploration, or mining, are opening up more and more areas of previously untouched forest to farming and ranching. When trees burn, they release stored carbon into the atmosphere, contributing to the greenhouse effect (see section 20.4). Ironically, the same soils that support the lush rain forest produce poor crop yields. The tropical climate explains this apparent contradiction. Warm temperatures promote rapid decomposition of organic matter, and heavy rains deplete soil nutrients. Once native plants give way to crops or grazing animals, the nutrient-poor soils harden into a cementlike crust. Species disappear and food webs topple, threatening biodiversity in the entire region.

Grasslands and forests are shrinking worldwide, but deserts are expanding as unsustainable agriculture eats away at forests and savannas. For example, widespread drought and overgrazing by domesticated animals threaten to turn large areas of tropical savanna to desert. But human activities can also destroy native deserts (figure 20.4). Many people are drawn to the warm climate of the southwestern United States. Desert cities such as Phoenix, Las Vegas, and Palm Springs demand huge amounts of water for household use, irrigation, and recreation. The water comes from faraway rivers, changing the desert ecosystem and reducing the river's flow.

Freshwater habitats in general are vulnerable to destruction. Damming, for example, completely alters river ecosystems (figure 20.5). Worldwide, the number of large dams (over 15 meters high) is estimated at more than 48,000. How do dams reduce biodiversity? Deep reservoirs replace waterfalls, rapids, and wetlands, where birds and many other species breed. Areas that were once seasonally flooded become dry. Water temperature, oxygen content, and nutrient levels all change, triggering shifts in species diversity and food webs both above and below the dam. Dams also disrupt the migration of fishes and other aquatic animals.

Another threat to freshwater biodiversity is alterations to a river's path. Along the banks of the Mississippi River, for example, levees built to prevent flooding alter the pattern of sediment deposition. The levees confine the river to its channel, so the water's flow rate increases. Fast-moving water, in turn, erodes nutrient-rich sediments. These sediments, which once spread over floodplains during periodic floods, now remain in the river. Once the nutrients arrive at the Gulf of Mexico, they stimulate the growth of algae, causing the additional problems described in section 20.3A.

Coastlines are also suffering from habitat destruction (see figure 20.3c). The world's coasts are a vital source of food, transportation, recreation, and

Figure 20.4 **Thirsty Ecosystem.** Water diverted from the Colorado River helps keep this Palm Springs golf course lush.
©David Falk/Getty Images RF

Figure 20.5 **Big Dam.** The Stave Falls Dam provides electricity to British Columbia in Canada. Dams provide many benefits, but they also eliminate streamside habitat and disrupt the migration of fishes and other animals.
©Ellen Atkin/Design Pics Inc/Alamy Stock Photo

Exotic Invaders and Overexploitation   Some Biodiversity May Be Recoverable

Figure 20.6 Water Pollution. A child wades in garbage-strewn water in Borneo. Heavily polluted waters are likely to be fouled with many contaminants at the same time.

Photo: ©Digital Vision/PunchStock RF

Examples of chemical water pollutants	
Organic	**Inorganic**
Sewage	Chloride ions
Pharmaceutical drugs	Heavy metals (mercury, lead,
Cosmetics	chromium, zinc, nickel, copper,
Antibacterial soaps	cadmium)
Detergents	Nitrogen from fertilizer
Pesticides	Phosphorus from fertilizer
Petroleum	and sewage
Persistent organic pollutants	Cyanide
(PCBs, PAHs)	Selenium

waste disposal. Many fishes and invertebrates spend part of their lives in estuaries, and diverse algae and flowering plants support the food web. Yet humans have drained and filled estuaries for urbanization, housing, tourism, dredging, mining, and agriculture. These activities affect life in the oceans, too. The loss of coastal habitats can threaten populations of commercially important species of marine animals such as bluefin tuna, grouper, and cod. These problems are only expected to get worse. Most of the world's largest cities are located along coastlines, and human populations in those cities will probably continue to rise in the future. ⓘ *estuaries,* section 19.3C

20.2 Mastering Concepts

1. Which human activities account for most of the loss of terrestrial habitat?
2. How do dams and levees alter river ecosystems?
3. Why is damage to estuaries and coastlines especially devastating?

20.3 Pollution Degrades Habitats

Pollution is any chemical, physical, or biological change in the environment that harms living organisms. Pollution degrades the quality of air, water, and land, threatening biodiversity worldwide.

A. Water Pollution Threatens Aquatic Life

A diverse array of pollutants affects rivers, lakes, and groundwater (figure 20.6). For example, mining operations often release inorganic pollutants such as heavy metals or cyanide into the water, whereas shipping accidents and leaking oil wells add petroleum. Toxins also leach into water from landfills.

Raw sewage can also be a major pollutant. In addition to carrying disease-causing organisms, sewage also contains organic matter and nutrients such as nitrogen and phosphorus. When released into waterways, organic matter fuels the growth of bacteria, whose respiration depletes the water of oxygen. Fish and other organisms die. Meanwhile, in a process called eutrophication, nutrients from the sewage fertilize phytoplankton in the water (see figure 19.34). The resulting algal blooms are unsightly. Moreover, when the algae die, the microbes that decompose their dead bodies further deplete dissolved oxygen in the water. Starved of oxygen, many aquatic organisms die.

Fertilizer and animal wastes that enter waterways also cause eutrophication, sometimes on a massive scale. For example, nutrients from agricultural and urban lands drained by the Mississippi River find their way into the Gulf of Mexico. There, the nutrients fuel algal blooms. On a local scale, toxins from the algae may kill fishes, manatees, and other sea life. A larger-scale problem, however, is the oxygen-depleted zone that forms each summer near the seafloor off the coast of Louisiana (figure 20.7). The seasonal lack of oxygen in this "dead zone" kills many animals, disrupting not only the Gulf's food web but also its economy: Commercially important fish and shrimp cannot live in oxygen-depleted water. ⓘ *harmful algal bloom,* section 15.4B

Some pollutants seem deceptively harmless. Sediments, for example, reduce photosynthesis by blocking light penetration into water. Even heat can be a pollutant. Hot water discharged from power plants reduces the ability of a river to carry dissolved oxygen, harming fishes and other aquatic organisms.

Toxic chemicals and trash also pollute the open ocean. For example, ocean currents have concentrated millions of tons of plastic into two huge, interconnected "garbage patches" in the northern Pacific. Floating just below the water surface are tiny pellets, called "nurdles," that form when plastic debris disintegrates. Fishes, sea turtles, and sea birds mistake the plastic for their natural food. The pellets can lodge in intestines and kill the animals outright, as the albatross in the chapter-opening photo illustrates. Alternatively, chemicals from the plastic may accumulate in animal tissues.

The chemicals leached from plastic are a small subset of human-made **persistent organic pollutants,** carbon-containing molecules that do not degrade (or that degrade extremely slowly). Many pesticides fall into this category, as do some solvents and pharmaceutical drugs. These substances contaminate ecosystems over long periods. Some of these compounds cause cancer; others are hormone mimics that disrupt reproduction (see chapter 25). Because they do not biodegrade, these fat-soluble chemicals become more concentrated as they ascend the food chain. This process, called biomagnification, accounts for the high concentrations of toxic chemicals in the fatty tissues of tunas, polar bears, and other top predators. ⓘ *biomagnification,* section 19.6C

Organisms living in polluted areas are often exposed to several pollutants at once. A waterway may contain not only garbage but also raw sewage, crude oil, gasoline, lead and other heavy metals, pesticides, and countless other toxic substances spilled from nearby factories. The toxic soup threatens not only human health but also the entire aquatic food chain.

B. Air Pollution Causes Many Types of Damage

Smog is a type of air pollution that forms a visible haze in the lower atmosphere (figure 20.8). Industrial smog occurs in urban and industrial regions where power plants, factories, and households burn coal and oil. The resulting smoke and sulfur dioxide (SO_2) may form a dark haze. Photochemical smog forms when nitrogen oxides and emissions from vehicle tailpipes undergo chemical reactions in the presence of light, producing ozone (O_3) and other harmful chemicals that injure plants and cause severe respiratory problems in humans. Warm, sunny areas with heavy automobile traffic have the most photochemical smog, but winds may carry the pollutants to sparsely populated areas.

Air also carries suspended **particulates,** tiny bits of matter that float in the air. Examples include road dust, volcanic ash, soot from partially burned fossil fuels, mold spores, pollen, and acidic particles. The damage they cause extends beyond the occasional need to dust off bookshelves and window sills. Most harmful are particles that are 2.5 μm in diameter or smaller. Not only do they become trapped deep within the lungs, but the heavy metals and toxic organic compounds in these particles make them especially likely to trigger inflammation, shortness of breath, asthma, or even cancer.

Chemicals that destroy ozone can also be extremely harmful air pollutants. Ozone is an atmospheric molecule with two faces. As we have already seen, ozone in photochemical smog at

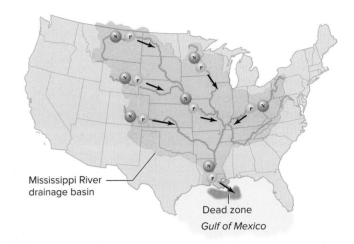

Mississippi River drainage basin

Dead zone

Gulf of Mexico

Figure 20.7 **Dead Zone.** Nutrient-rich runoff enters the major river systems draining into the Mississippi River. The combined waters then pour into the Gulf of Mexico. The result is a zone of seasonal oxygen depletion off the coast of Louisiana.

Examples of air pollutants	
Particulates	Nitrogen oxides (NO, NO_2)
Ozone (ground level)	Lead
Sulfur oxides (SO_2)	Chlorofluorocarbons (CFCs)

Figure 20.8 **Smog.** Air pollution from cars, power plants, and industrial sources plagues many cities. This is Santiago, Chile.

Photo: ©Martin Bernetti/AFP/Getty Images

Figure 20.9 **Acid Deposition.**
(a) Sulfur and nitrogen oxides react with water in the atmosphere, forming acids. These acids return to Earth as dry particles or dissolved in precipitation.
(b) Rainfall with a pH of about 4 severely damaged this fir forest in the Czech Republic.

(b): ©Oliver Strewe/Stone/Getty Images

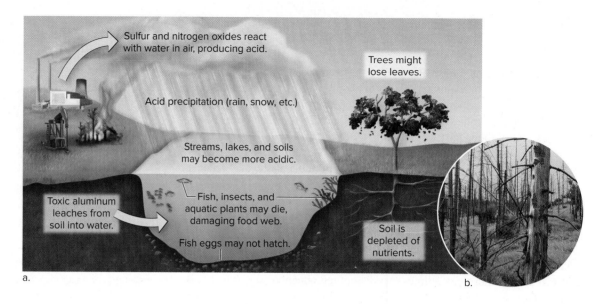

a.

b.

Earth's surface is harmful. In the upper atmosphere, however, the stratospheric **ozone layer** blocks damaging ultraviolet (UV) radiation from the sun.

In the past several decades, the ozone layer has thinned over parts of Asia, Europe, North America, Australia, and New Zealand, and a "hole" has formed over Antarctica (see Burning Question 20.3). As the ozone layer thins, UV radiation increases at Earth's surface. In humans, exposure to UV radiation can cause skin cancer or cataracts. Ozone depletion may also indirectly contribute to species extinctions. For example, increasing UV radiation may be one of many factors causing amphibian populations to plummet.

What is damaging the ozone layer? The main culprits are persistent, human-made chlorofluorocarbon (CFC) gases. These compounds were once used in refrigerants such as Freon, as propellants in aerosol cans, and to produce foamed plastics. They can persist for decades in the upper atmosphere, catalyzing chemical reactions that break down ozone. An international treaty signed in 1987, the Montreal Protocol, banned the use of CFCs. Experts estimate that at mid-latitudes, the ozone layer should recover by 2050; healing the hole over Antarctica might take some 25 years longer.

Some types of air pollutants produce **acid deposition:** acidic rain, snow, fog, dew, or dry particles (figure 20.9). Burning fossil fuels release sulfur and nitrogen oxides (SO_2, NO, and NO_2) into the atmosphere. These compounds react with water, forming sulfuric acid and nitric acid. The acids return to the Earth as acid deposition. ⓘ *pH scale,* section 2.4

Coal-burning power plants release the most sulfur and nitrogen oxides, although emissions of these pollutants have declined since the 1980s. In the United States, winds carry airborne acids hundreds of miles east and northeast of the power plants in the Midwest. As a result, rainfall in the eastern United States has an average pH of about 4.6, compared to a normal pH of 5.6. Acid deposition also affects the Pacific Northwest, the Rockies, Canada, Europe, East Asia, and the former Soviet Union.

Most lakes have a pH between 6 and 8; acid deposition can lower it to 5 or less. The acid leaches toxic metals such as aluminum and mercury from soils and sediments, causing fish eggs to die or yield deformed offspring. Lake-clogging algae replace aquatic flowering plants. Organisms that consume the doomed species must seek alternative food sources or starve, which disrupts or topples food webs. Eventually, lake life dwindles to a few species that can tolerate increasingly acidic conditions.

Burning Question 20.3

What does the ozone hole have to do with global climate change?

These two satellite images show the change in the Antarctic ozone hole over the last four decades. As this hole grows, scientists are recording rising global temperatures. But are these two observations related?

Many people confuse global climate change and the ozone hole. These two problems are largely separate, but they do share two common threads. First, the chlorofluorocarbon gases that deplete the ozone layer are also greenhouse gases, contributing to a warmer atmosphere. Second, the greenhouse effect may cause the hole in the ozone layer to grow. A thick, heat-trapping "blanket" of greenhouse gases in the lowest part of the atmosphere means less heat reaches the stratosphere, where the ozone layer is. A cooler stratosphere, in turn, extends the time that stratospheric clouds blanket the polar regions in winter. These clouds of ice and nitric acid speed the chemical reactions that deplete stratospheric ozone.

Submit your burning question to
marielle.hoefnagels@mheducation.com

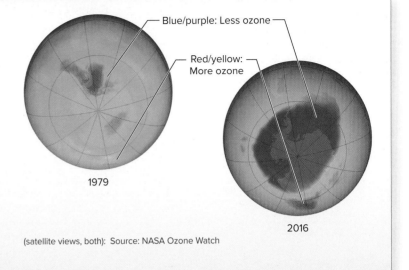

Blue/purple: Less ozone

Red/yellow: More ozone

1979

2016

(satellite views, both): Source: NASA Ozone Watch

Acid deposition alters forests, too. As soil pH drops, aluminum ions released from soil enter roots and stunt tree growth. Affected trees become less able to resist infection or to survive harsh weather. As a result, acid deposition is thinning high-elevation forests throughout Europe and on the U.S. East Coast from New England to South Carolina.

20.3 Mastering Concepts

1. How do toxic chemicals, nutrients, sediments, and heat affect aquatic ecosystems?
2. What are major sources of industrial smog, photochemical smog, particulates, and acid deposition?
3. What effects do smog, particulates, the thinning ozone layer, and acid deposition have on life?

20.4 Global Climate Change Alters and Shifts Habitats

We now turn to air pollutants with the potential to do the most harm of all: greenhouse gases. In the past, scientists debated whether human activities could actually change something as complex as Earth's overall climate. Now, the scientific consensus is clear: We can and do.

A. Greenhouse Gases Warm Earth's Surface

CO_2 is a colorless, odorless gas present in the atmosphere at a concentration of about 400 parts per million. Although it makes up a tiny fraction of the atmosphere, CO_2 is one of several gases that contribute to the **greenhouse effect,** an increase in surface temperature caused by heat-trapping gases in Earth's atmosphere. As illustrated in figure 20.10, sunlight passes through the atmosphere and reaches Earth's surface. Some of the energy is reflected, but some is absorbed and reradiated as heat. Greenhouse gases block the escape

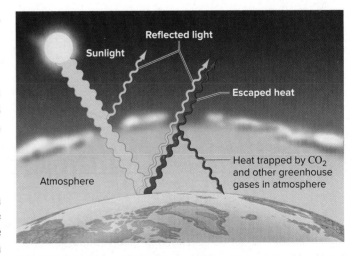

Reflected light

Sunlight

Escaped heat

Heat trapped by CO_2 and other greenhouse gases in atmosphere

Atmosphere

Figure 20.10 **The Greenhouse Effect.** Solar radiation heats Earth's surface. Some of this heat energy is trapped near the surface by CO_2 and other greenhouse gases.

of this heat from the atmosphere, just as transparent panes of glass trap heat inside a greenhouse.

Carbon dioxide is one greenhouse gas; others include methane, nitrous oxide (N_2O), and CFCs. These other gases actually trap heat much more efficiently than does CO_2, but because they are less abundant, they contribute only half as much to the greenhouse effect.

In a sense, the greenhouse effect supports life, because Earth's average temperature would be much lower without its blanket of greenhouse gases. And greenhouse gases have many sources—including volcanoes—that have nothing to do with our actions.

But CO_2 has been steadily accumulating in the atmosphere for the past century (figure 20.11). The primary culprit is the use of fossil fuels such as coal, oil, and gas. Tropical deforestation and other combustion activities also add a share. All together, human activities release some 38 billion metric tons of CO_2 into the atmosphere each year. Photosynthesis temporarily removes some of this carbon from the atmosphere and incorporates it into plant tissues. Overall, however, more CO_2 is added than is removed. ⓘ *carbon cycle,* section 19.7B

This accumulation of CO_2—along with climbing levels of other greenhouse gases—was accompanied by an increase in average global temperatures in the twentieth century (see figure 20.11). Climate models predict that these trends will continue unless we take immediate action. The United Nations Climate Change Conference in 2015 laid the groundwork for change, with 196 nations agreeing to hold the increase in global temperature below 2°C and to balance greenhouse gas emissions with removals before 2100.

Figure It Out

Using the graph in figure 20.11, calculate the percent increase in atmospheric CO_2 concentration between 1960 and today.

Answer: 405/317 = 1.28; the CO_2 concentration is 28% higher today.

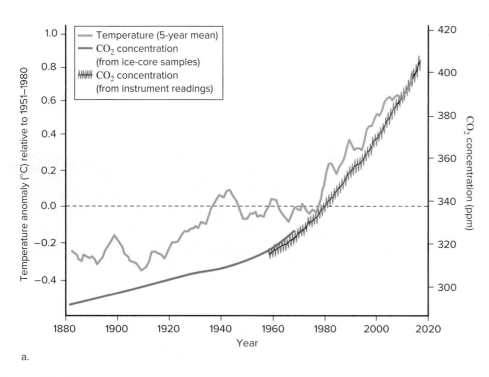

a.

b.

Figure 20.11 CO_2 and Global Average Temperature. (a) CO_2 continues to accumulate in the atmosphere. The low point of each year's CO_2 concentration reflects peak photosynthesis in the northern hemisphere. As CO_2 accumulates, the average global temperature is also increasing. (b) Coal-fired power plants are among the main culprits in the rising levels of CO_2.

(a): Data sources: NOAA Earth System Research Laboratory; NASA Goddard Institute for Space Studies Surface Temperature Analysis; (b): ©Corbis RF

Earth's Biodiversity Is Dwindling   Many Human Activities Destroy Habitats   Pollution Degrades Habitats   Global Climate Change Alters and Shifts Habitats

Earth's average temperature is rising overall, but some areas will become warmer and others will become cooler. The commonly used phrase "global warming" is therefore somewhat misleading. **Global climate change** is a more accurate term for past and future changes in Earth's weather patterns.

B. Global Climate Change Has Severe Consequences

The gradual warming that has already occurred has been associated with shrinking alpine glaciers and polar ice sheets (figure 20.12). This process is self-reinforcing. Melting sea ice exposes seawater, which absorbs more heat than ice. The water becomes slightly warmer, which further accelerates the melt, exposing more seawater, and so on.

When floating sea ice melts, sea level does not rise (just as ice melting in a full glass of lemonade does not cause the liquid to overflow). However, the loss of ice over land—such as the ice covering Greenland and Antarctica—has contributed to a rise in sea level. This change could eventually alter the ocean ecosystem and flood coastal cities.

Weather conditions in tropical and temperate regions are also changing, thanks in part to a rise in sea surface temperature. In the tropics, warmer water means higher winds and more evaporation, which increases the amount of rainfall in a severe storm. In dry areas, on the other hand, global climate change may mean more intense and longer droughts, fewer cold snaps, and more heat waves.

These changes kill some organisms outright, whereas others become stressed and vulnerable to disease. Still others may migrate to higher latitudes or higher elevations (see Investigating Life 20.1). At the southern ends of their ranges, where temperatures are rising, some species have become locally extinct. In addition, scientists are tracking events known to occur at the same time each year. In the United Kingdom, butterflies are emerging and amphibians are mating a few days earlier than usual; in North America, many plants are flowering and birds are migrating earlier.

Continued climate change will affect not only wild organisms but also agriculture and public health. Growing seasons in temperate areas are lengthening, and the southern United States may become too dry to sustain many traditional crops. Drought-related water shortages may affect more than a billion people worldwide. Tropical diseases such as malaria, African sleeping sickness, dengue fever, and river blindness may move into temperate areas.

Ocean life is also vulnerable to CO_2 accumulation. One problem is that CO_2 buildup causes ocean water to become more acidic. The lowered pH causes the calcium carbonate shells of oysters, clams, and other mollusks to dissolve, along with the exoskeletons of coral animals. The loss of these organisms disrupts the entire ocean food web.

Burning Question 20.4

How can small lifestyle changes reduce air pollution and global climate change?

The same strategies that reduce your "carbon footprint" also apply to air pollution in general:

- Eat less meat, dairy, and eggs. Farm animals and their manure emit copious greenhouse gases—especially methane—into the atmosphere. Moreover, eating lower on the food chain is energy-efficient (see section 19.6).

- Conserve energy. Use energy-efficient lightbulbs, recycle, carpool, drive a fuel-efficient car, ride a bicycle, or turn down the thermostat. Pouring filtered tap water into a reusable bottle rather than buying bottled water not only saves energy but also reduces landfill waste.

- Check with your electric company to see whether you can select renewable energy sources, such as wind or solar energy.

- Use less stuff. Manufacturing, packaging, transporting, and storing consumer goods use energy and raw materials and produce waste. The less you buy, the fewer resources you consume and the less waste you discard.

- Reduce the transportation and storage costs of food by shopping close to home and by selecting fruits and vegetables that are in season and locally grown.

Submit your burning question to
marielle.hoefnagels@mheducation.com

20.4 Mastering Concepts

1. Why is CO_2 accumulating in Earth's atmosphere?

2. Describe how and why Earth's climate has changed during the past century.

3. How does global climate change threaten biodiversity?

Figure 20.12   Melting Ice. The Arctic polar ice cap is shrinking, owing to higher temperatures at the North Pole.

Data source: http://climate.nasa.gov/vital-signs/arctic-sea-ice/; (bear): Source: USGS/Canadian Coast Guard/Patrick Kelley

Burning Question 20.5

How can people help slow the spread of invasive species?

Individual choices can help control the spread of invasive species:

- If you garden, avoid invasive species. Instead, choose native plant species that attract wildlife.
- Don't buy exotic fish, amphibians, reptiles, or mammals as pets.
- Never release unwanted pets into the wild.
- If you have a boat, take care not to spread invasive aquatic plants and animals between waterways.
- If you fish, don't release live bait into the water; better yet, use artificial lures. Disinfect those lures (and your waders) after you fish so they don't spread harmful microbes to new bodies of water.
- If you camp, buy firewood locally to avoid spreading invasive insects and other pests between campgrounds.

Submit your burning question to
marielle.hoefnagels@mheducation.com

Figure 20.13 **Invasive Species.** (a) European starlings form enormous, destructive flocks in North America. (b) An invasive population of kudzu in Virginia blankets every surface in sight.

(a): ©Alex Saberi/National Geographic/Getty Images; (b): ©Steve St. John/Getty Images RF

20.5 Exotic Invaders and Overexploitation Devastate Many Species

In addition to habitat destruction and pollution, two other important threats to biodiversity are invasive species and overexploitation.

A. Invasive Species Displace Native Organisms

An introduced species (also called a nonnative, alien, or exotic species) is one that humans bring to an area where it did not previously occur. When people move from one location to another, we often bring along our pets, crops, livestock, and ornamental plants. We also unintentionally carry microbes, parasites, and stowaways such as rodents and insects on ships, cars, and planes.

This transport may seem harmless at first, and many introduced species die. Even if they survive in their new homes, they may not cause problems. For example, the house sparrow was introduced to the United States from Europe in the 1850s. Although it has spread throughout the North American continent, it has not caused obvious ecological problems. In addition, at least 5000 nonnative plant species live in U.S. ecosystems, introduced from agriculture and urbanization. Most have done no apparent harm.

If a nonnative species becomes invasive, however, it can cause immense destruction (figure 20.13). To be considered an **invasive species,** an introduced species must begin breeding in its new location and spread widely from the original point of introduction. In addition, according to some definitions, the species must harm the environment, human health, or the economy. Of every 100 species introduced, only one persists to take over a niche. Nonetheless, the Global Invasive Species Database lists 498 types of invasive plants, animals, and microorganisms in the United States alone; many more invasive species occur worldwide. ⓘ *zebra mussels,* section 19.4A

Examples of invasive species in North America include the following:

- European starlings are birds that were released in New York City's Central Park in 1890. Huge flocks of starlings now reside all across North America, fouling cities with their droppings and devouring crops (see figure 20.13a).
- The marine toad is a voracious omnivore that competes with and preys on native amphibians in Florida.
- Asian carp are large, fast-breeding fish that were originally introduced to catfish ponds in the South but have since spread along the Mississippi River toward the Great Lakes. They gobble up huge amounts of algae, outcompeting other plankton-eaters for food.
- Hungry caterpillars of the gypsy moth devour the foliage of hundreds of species of hardwood trees in North America.
- Kudzu is a fast-growing plant that smothers native plants in the southeastern United States (see figure 20.13b).
- Hydrilla is an aquatic plant that forms dense mats on water surfaces. These mats block light, alter food webs, and reduce the recreational use of lakes and rivers.
- Fungi have all but eradicated American chestnut and American elm trees.

When an invasion does occur, the harm may be ecological and economic. A nonnative species not only changes the composition of a community but also may carry diseases that spread to native species. The economic costs include everything from the purchase of herbicides that kill invasive weeds, to the loss of grain eaten by hungry birds and rodents, to declining tax revenue when invasive aquatic plants interfere with boating and recreation.

B. Overexploitation Can Drive Species to Extinction

Another cause of species extinction is **overexploitation:** harvesting the members of a species faster than they can reproduce (figure 20.14). The market for exotic pets, for example, is harming populations of many species of mammals, birds, snakes, lizards, amphibians, and fishes.

Many of the most famous examples of species extinctions result from overhunting of terrestrial animals. The dodo, for example, was a flightless bird that once lived on the Indian Ocean island of Mauritius. In the late seventeenth century, humans hunted the dodo for food while introducing other species to the island. The dodo soon went extinct. In the United States, commercial-scale hunting nearly drove the American bison to extinction in the 1800s. The passenger pigeon and Carolina parakeet did go extinct in the 1900s, victims of overhunting and habitat destruction.

The best illustration of widespread overexploitation is the recent collapse of ocean fisheries. Since the 1950s, some 90% of the world's large, predatory ocean fishes have disappeared, including tuna, flounder, halibut, swordfish, and cod. Superefficient fishing boats harvest the adults faster than the fishes can reproduce. Moreover, fishing pressure has shifted to other species as predatory fishes have vanished. Fishing equipment scrapes and scours the seafloor, destroying habitat for many other species. In addition, many marine mammals, seabirds, sea turtles, and nontarget fish species are killed accidentally when they are caught up as bycatch in the nets set for the target species. Even farmed seafood may contribute to the problem: Ocean fishes are fed to farmed shrimp and salmon, depleting marine food webs.

Improved management practices can help an overharvested population recover its numbers. Consider, for example, the Chesapeake Bay blue crab. Due to overharvesting, pollution, and habitat loss, the federal government declared the crab fishery a disaster in 2008. The state of Virginia had already stopped issuing new commercial crabbing licenses in the 1990s, but in 2009, the state began buying back existing licenses as well. The overall goal of this approach, combined with other regulations, was to reduce pressure on the dwindling blue crab population. Officials are carefully monitoring crab abundance and are optimistic that the population will rebound.

a.

b.

Figure 20.14 **Overexploitation.** (a) Officials in Lhasa, Tibet, inspect confiscated tiger, leopard, and otter skins. Poaching is one of the main threats to mammal populations in Asia. (b) This young sea turtle was accidentally caught in a fishing net.

(a): ©China Photos/Getty Images News; (b): ©Jeffrey Rotman/Science Source

20.5 Mastering Concepts

1. What features characterize an invasive species?
2. How do invasive species disrupt ecosystems?
3. List examples of species declines caused by overexploitation.

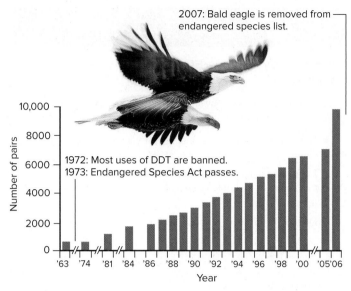

2007: Bald eagle is removed from endangered species list.

10,000

8000

6000

4000

2000

0

1972: Most uses of DDT are banned.
1973: Endangered Species Act passes.

Number of pairs

'63 '74 '81 '84 '86 '88 '90 '92 '94 '96 '98 '00 '05 '06

Year

Figure 20.15 Good News for Bald Eagles. Bald eagles were near extinction in the 1960s, due to a combination of habitat loss, hunting, and exposure to the pesticide DDT. In the 1970s, however, new laws stimulated a steady recovery for the species. Numbers in this graph represent breeding pairs in the contiguous United States.

Data source: U.S. Fish and Wildlife Service

Figure 20.16 Taking the High Road. Fish use "ladders" such as this one to cross large dams in a series of small steps.

©David R. Frazier Photolibrary, Inc.

20.6 Some Biodiversity May Be Recoverable

As the human population continues to grow, pressure on natural resources will only increase. One key to reversing environmental decline will therefore be to slow the growth of the human population. In addition, although some species are gone forever, humans may have the power to undo some of our past mistakes (see Why We Care 20.1). For example, thanks in part to the Endangered Species Act of 1973, some species that faced extinction, such as the bald eagle, have recovered (figure 20.15).

A. Protecting and Restoring Habitat Saves Many Species at Once

One important conservation tool is to set aside parks, wildlife refuges, and other natural areas and to protect them from destruction, invasive species, and hunting. Preserving critical habitat is a good conservation tool because it saves not just one endangered species but also the many other species that share its habitat. For example, the red-cockaded woodpecker is native to the southeastern United States, where it builds its nest in a cavity that it excavates high on the trunk of a mature pine tree. Although most of its preferred habitat has been logged, private property owners and the government have cooperated to save some of the remaining habitat. Other animals that use the nesting cavities, including birds, mammals, snakes, amphibians, and insects, also benefit from the woodpecker recovery plan.

Reversing habitat destruction is a second conservation tool. Major restoration projects show that recovery, although costly and difficult, may yet reverse some species declines. For example, human activities have damaged Florida's Everglades region for more than a century. Over many decades, engineers built a massive system of canals, levees, and pumps that direct water to agricultural and urban areas. These changes have altered the natural course of the rivers and starved the wetlands of fresh water. Dozens of species have become endangered. But an ambitious restoration project is underway. Native vegetation, waterfowl, aquatic invertebrates, and fishes have already returned to areas where the project is complete.

On a smaller scale, we can also help species bypass degraded habitats by supplying wildlife corridors through housing developments or building "fish ladders" over dams (figure 20.16). Bioremediation, a strategy that uses plants and microbes to clean up polluted soils, may also play a role in some cases.

B. Some Conservation Tools Target Individual Species

One straightforward way to save a species is to protect its individuals. For example, it is illegal to collect endangered carnivorous plants such as Venus flytraps and pitcher plants in the wild. In the ocean, northern and southern right whales were nearly hunted to extinction for their blubber in the 1800s. They remain endangered, but it is now illegal to kill them. Likewise, the catastrophic decline of Atlantic cod in the past few decades prompted the closure of some fisheries off Newfoundland's coast, along with strict quotas for the overall catch. Whether conservation efforts are in time to save the cod fishery remains to be seen.

Predator control programs can also help. For example, rats, weasels, dogs, and other predators brought by European settlers endanger the great spotted kiwi, a flightless bird native to New Zealand. Removing these predators from nature preserves has made life much easier for the endangered kiwis.

In areas where wildlife poaching is a problem, changing the local economic incentives can be a powerful conservation tool. Poaching is profitable if a dead animal is worth more money than a live one in its natural habitat. But ecotourism can turn this economic calculation on its head. By attracting visitors who pay to see endangered animals, conservation may bring in more money than poaching. A black rhino conservation program in Namibia, for example, hires former poachers as armed guards who protect the animals from hunters. Meanwhile, guides lead tourists hoping to spot a rhino in the wild (figure 20.17).

Protective laws and predator control programs may help halt the decline of a threatened species, but sometimes it is possible to boost reproduction as well. A captive breeding program can be very useful. Biologists can capture adults from the wild, allow them to reproduce in captivity, then nurture and protect the young until they are old enough to return to their native habitat. The California condor, red wolf, and black-footed ferret are notable examples. But this solution does not work for species whose habitat is gone (submerged after dam construction, for example) or still under the same pressures that threatened the species in the first place.

The biotechnology revolution plays a role in yet another approach to species conservation. In one project, researchers are using DNA to identify bison whose genes are uncontaminated with those of domesticated cattle. The "purest" bison are set aside as the best candidates for reestablishing wild bison herds. In the future, it may even be possible to recover extinct species using DNA extracted from preserved specimens. Scientists who are sequencing DNA from a frozen baby mammoth, for example, may one day be able to produce a live mammoth by using a cloning technique similar to the one used to make Dolly the sheep. One possible approach would be to replace the DNA in a fertilized elephant egg with mammoth DNA and then implant the resulting embryo into an elephant's uterus. After gestation, a woolly mammoth would be born—some 10,000 years after its species went extinct. ⓘ *cloning,* section 11.3B

C. Conserving Biodiversity Involves Scientists and Ordinary Citizens

Regardless of which tools are used to preserve biodiversity, all conservation efforts require a scientific approach. To get a true measure of Earth's biodiversity, taxonomists must catalog all organisms, not just vertebrates and plants. Evolutionary biologists must continue to analyze the relationships among all species. Preserving biodiversity also requires an understanding of which species need help, whether current conservation efforts are working, and the consequences to ecosystems as species disappear.

But not every important question has a scientific answer. Are the only species worth saving the photogenic ones, such as giant pandas? Or do we also commit to saving the worms, algae, bacteria, and fungi so essential to global ecology? How much money should we spend on conservation? Should developed countries help poor nations with their efforts? How do we balance the need for conservation with the need for economic growth? Which of the tangled threads that tie all life together should we sacrifice to other interests?

It would be very difficult to halt life on Earth completely, short of a global catastrophe such as a meteor collision or a nuclear holocaust. Just the presence of life, however, does not guarantee that the surviving species will have the

Figure 20.17 **Black Rhino.** Ecotourism may provide some hope for the survival of the endangered black rhinoceros.
©Michele Burgess/Alamy Stock Photo

Why We Care 20.1 | Environmental Legislation

Since the Environmental Protection Agency was established in 1970, Congress has passed laws to combat some of the worst environmental problems in the United States. Listed here are a few major pieces of environmental legislation:

- The **Endangered Species Act** of 1973 requires that the U.S. Secretary of the Interior identify threatened and endangered species. The overall goals are to prevent extinction and to help endangered species recover their numbers.

- The **Clean Air Act,** passed in 1970 and amended several times since, sets minimum air quality standards for many types of air pollutants. Since 1970, emissions of nitrogen and sulfur oxides, lead, carbon monoxide, and other pollutants have declined, leading to significant improvements in air quality.

- Among other provisions, the **Clean Water Act** of 1972 required nearly every city to build and maintain a sewage treatment plant, drastically reducing discharge of raw sewage into rivers and lakes. The 1987 **Water Quality Act** regulates water pollution from industry, agricultural runoff, overflow from sewage treatment plants during storms, and runoff from city streets.

©Brand X Pictures/PunchStock RF

diversity that humans value. It is safest to try to protect the remaining resources for the future, while maintaining a reasonable standard of living for all people. Scientists and politicians, as well as ordinary citizens, share this heavy burden. Part of the solution lies within you and how you choose to live (see the Burning Questions scattered throughout the chapter). Do whatever you can to preserve the diversity of life, for in diversity lies resiliency and the future of life on Earth.

20.6 Mastering Concepts

1. What is the relationship between human population growth and conservation biology?

2. List and describe the tools that conservation biologists use to preserve biodiversity.

3. How can scientists, governments, and ordinary citizens work together for conservation?

Investigating Life 20.1 | Up, Up, and Away

Animal and plant specimens deposited years ago in museum collections can offer glimpses into past biodiversity. If these relics still contain DNA, researchers can directly compare the genes of past populations with those of today.

Biologists at the University of California, Berkeley, seized one such opportunity to compare past and present. A century ago, naturalists from the university's Museum of Vertebrate Zoology visited Yosemite National Park. They trapped many small mammals, noted their locations, and preserved them in the museum's collections. In the 100 years since those surveys occurred, the average temperature at Yosemite has climbed by about 3°C. How might the elevated temperatures affect park wildlife? A research team used the museum's preserved alpine chipmunks to investigate this question.

The researchers knew that alpine chipmunks now live about 500 meters higher than they did a century ago. Perhaps the chipmunks responded to the gradually warming climate by moving to higher elevations, where temperatures are cooler. Populations that had once overlapped were therefore increasingly fragmented into subpopulations, each occupying its own mountaintop. With fewer opportunities to mix the gene pool, the subpopulations should become genetically isolated (figure 20.A). To test this prediction, the team compared DNA from museum specimens with modern chipmunk DNA.

The results strongly supported the authors' prediction. That is, modern populations were much more different from each other than were historical populations (figure 20.B). The researchers also determined that increasing isolation among modern alpine chipmunk populations has led to a significant loss of genetic diversity overall. If genetic diversity continues to erode as alpine chipmunks retreat to still higher elevations, then the future of the species may be in question. After all, a population with low genetic variation has less potential to adapt to a changing environment.

The consequence of climate change on alpine chipmunks is therefore two-pronged: Increasing temperatures cause lower-elevation populations to disappear, which increases the genetic isolation of higher-elevation populations and boosts their chance of extinction. As the climate continues to shift, the same trend is likely to endanger many other species, especially those in mountain habitats.

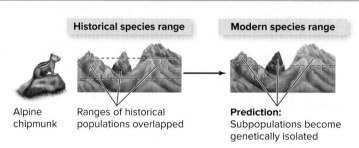

Figure 20.A Moving Up. Alpine chipmunks have moved to higher elevations over the past century, as temperatures in Yosemite have climbed. Researchers predicted that habitat fragmentation would increase genetic isolation among the chipmunks.

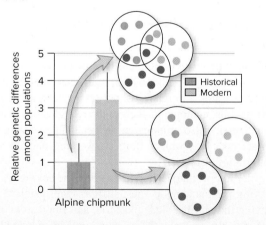

Figure 20.B Alpine Isolation. As predicted, populations of alpine chipmunks have diverged genetically relative to historical populations. Numbers on the Y axis represent differences among modern chipmunk populations, relative to the differences in the historical samples. Inset circles illustrate the degree of overlap among subpopulations. Bars represent 95% confidence intervals; see appendix B.

Source: Rubidge, Emily M. and five coauthors, including Craig Moritz. 2012. Climate-induced range contraction drives genetic erosion in an alpine mammal. *Nature Climate Change*, vol. 2, pages 285–288.

What's the **Point?** ▼ APPLIED

Humans affect the global ecosystem in many ways that cause or speed up species extinctions. Some argue that preserving species is important because nature has intrinsic value. But species diversity also has practical value. We need a diverse array of species to maintain ecosystem processes that are vital to our survival.

One way to learn more about our need for biodiversity is to set up a self-contained ecosystem, such as the large glass compound called Biosphere 2 in southern Arizona. Within its 12,000 square meters, the enclosure houses five mini-biomes: an ocean, a rain forest, a desert, a wetland, and a savanna, as well as an area for agriculture. Each biome contains diverse plant, animal, and microbial life. For two years in the early 1990s, eight researchers lived inside Biosphere 2. The goal was to isolate the entire system from the outside world (except for sunlight), so scientists could study and manipulate ecological interactions.

The researchers in Biosphere 2 maintained good health; some even reported health improvements, perhaps due to a calorie-restricted and nutrient-rich diet. On the other hand, the ecosystems were unstable during the experiment. Some species thrived while others crashed. Plants had weak stems and sparse leaves. Atmospheric O_2

Biosphere 2 in Oracle, Arizona

©Camerique/ClassicStock/Getty Images

dropped, and some researchers reported fatigue and sleep apnea. The facility was therefore infused with O_2 from the outside. Meanwhile, soil bacteria produced more CO_2 than anticipated. Many insects, including vital pollinators, died.

On a more positive note, humans played a sustainable and nurturing role to the ecosystems in Biosphere 2. Nothing was wasted. The eight researchers maintained a productive farm without the added chemicals of conventional agriculture. The ecosystem also recycled human wastewater, which was pumped into the wetland ecosystem. Anaerobic microbes decomposed the waste, and aquatic plants saw a boost in growth from the nutrient input. The researchers then fed the plants to their livestock.

Even if you're not a farmer, the Biosphere 2 experiments teach important lessons about sustainable living. We are part of Biosphere 1—the Earth—and we cannot treat the global ecosystem as an experiment. If Earth's ecosystems crash or if atmospheric conditions fluctuate wildly, we will not be able to rescue our biosphere from the outside. We must act from within to keep it hospitable. Recycling, composting, and using minimal resources are just some simple strategies you can use to live a more sustainable life.

CHAPTER SUMMARY

20.1 Earth's Biodiversity Is Dwindling

- **Biodiversity** means the variety of life on Earth.
- Increasing numbers of species are threatened with **extinction** or are **endangered** or **vulnerable. Conservation biologists** study and attempt to preserve biodiversity.
- Figure 20.18 summarizes the major causes of species extinctions.

20.2 Many Human Activities Destroy Habitats

- Prairies and other grasslands have been destroyed, usually to make room for crops. Agriculture, logging, and urbanization contribute to deforestation.
- Drought and overgrazing are expanding Earth's deserts. At the same time, native desert habitats are being lost to urbanization.
- Dams and levees alter the species that live in and near rivers.
- Preserving estuaries is important because they are breeding grounds for many species.

20.3 Pollution Degrades Habitats

- **Pollution** is any environmental change that harms living organisms.

A. Water Pollution Threatens Aquatic Life

- Excessive nutrient levels can cause an algal bloom, altering a water body's ecosystem. Sediments and heat also pollute aquatic ecosystems.
- Water and sediments can be contaminated by a mixture of toxic substances such as **persistent organic pollutants,** heavy metals, spilled oil, and plastics.

B. Air Pollution Causes Many Types of Damage

- Air pollutants include heavy metals, particulates, and emissions from fossil fuel combustion. Some of these pollutants react in light to form photochemical **smog.**

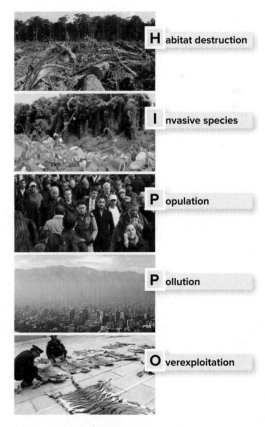

Figure 20.18 Threats to Biodiversity: A Summary.

(Photos): (destruction): ©Marcelo Horn/E+/Getty Images RF; (invasive): ©Steve St. John/Getty Images RF; (population): ©Melanie Stetson Freeman/Christian Science Monitor/The Image Works; (pollution): ©Martin Bernetti/AFP/Getty Images; (exploitation): ©China Photos/Getty Images News

- **Particulates** are bits of matter suspended in the air. When inhaled, the tiniest particles can cause lung problems.
- Use of chlorofluorocarbon compounds (CFCs) has thinned the stratospheric **ozone layer,** which protects life from damaging ultraviolet radiation.
- **Acid deposition** forms when nitrogen and sulfur oxides react with water in the atmosphere to form nitric and sulfuric acids. These acids return to Earth as dry particles or in precipitation.

20.4 Global Climate Change Alters and Shifts Habitats

A. Greenhouse Gases Warm Earth's Surface

- Human activities produce CO_2 and other gases that trap heat near Earth's surface, producing the **greenhouse effect.**
- This accumulation of greenhouse gases is causing Earth's average temperature to rise. The result is **global climate change.**

B. Global Climate Change Has Severe Consequences

- As the temperature increases, polar ice is melting, sea level is rising, coral reefs are declining, and species ranges are shifting.

20.5 Exotic Invaders and Overexploitation Devastate Many Species

A. Invasive Species Displace Native Organisms

- **Invasive species** consume or outcompete native organisms.

B. Overexploitation Can Drive Species to Extinction

- **Overexploitation** means individuals are harvested faster than they can reproduce. High fishing pressure endangers global fisheries.

20.6 Some Biodiversity May Be Recoverable

A. Protecting and Restoring Habitat Saves Many Species at Once

- Protected reserves, habitat restoration, and wildlife corridors are conservation tools that protect multiple species simultaneously.

B. Some Conservation Tools Target Individual Species

- Harvest management, predator exclusion, economic incentives, captive breeding programs, and biotechnology can help save one species at a time.

C. Conserving Biodiversity Involves Scientists and Ordinary Citizens

- Conservation biologists can monitor biodiversity trends and recommend strategies for saving threatened species, but everyone can choose actions that preserve or deplete biodiversity.

MULTIPLE CHOICE QUESTIONS

1. Which of the following is *not* one of the main causes of today's biodiversity crisis?
   a. Natural disasters, such as earthquakes
   b. Habitat destruction and degradation
   c. Overexploitation
   d. Introduction of nonnative species

2. What activity is the main cause of expanding deserts?
   a. Housing   c. Pollution
   b. Agriculture   d. Tourism

3. What is the connection between agriculture in the midwestern United States and the Gulf of Mexico's "dead zone"?
   a. Pesticides from farmlands are killing ocean life.
   b. Nutrient enrichment causes oxygen depletion in the waters of the Gulf.
   c. River sediments block out the light needed for photosynthesis in the Gulf.
   d. Farmlands use up all of the nitrogen in the water, so the Gulf waters are starved for nutrients.

4. What is the greenhouse effect?
   a. The filtering of specific wavelengths of light by Earth's atmosphere
   b. The increase in global plant growth due to enhanced photosynthesis
   c. The trapping of heat by gases in the atmosphere
   d. The reduction in the amount of CO_2 in Earth's atmosphere

5. Why is deforestation associated with global climate change?
   a. Because the loss of forest animals reduces the CO_2 released into the atmosphere
   b. Because the loss of trees reduces the amount of photosynthesis occurring on the planet
   c. Because the loss of forest makes more land available for agriculture
   d. Because the loss of tree root systems leads to erosion

6. An invasive species is one that
   a. is introduced into a new habitat.
   b. causes species extinctions wherever it exists.
   c. establishes a breeding population in a new habitat.
   d. Both a and c are correct.

7. What might limit the effectiveness of a captive breeding program for the restoration of an extremely rare endangered species?
   a. A very rare species has limited genetic diversity.
   b. Captive breeding programs do not preserve the habitat of an organism.
   c. Determining the conditions required to nurture the offspring to reproductive age may be difficult.
   d. All of the above are correct.

8. Which strategy would be most likely to restore overexploited ocean species with the least potential for harm to the ecosystem?
   a. Eating only farmed seafood
   b. Adding nutrients to the ocean to increase primary production
   c. Passing regulations that limit harvests
   d. Adding invasive species to increase competition in aquatic food webs

Answers to Multiple Choice questions are in appendix A.

WRITE IT OUT

1. List the main threats to biodiversity worldwide.

2. Considering the many types of community and ecosystem interactions (see sections 19.4 and 19.6), explain why the loss of one species is likely to lead to declines in other populations.

3. As part of an effort to combat habitat loss, some conservationists suggest "assisted migration," in which scientists move endangered or threatened organisms to new habitats. Discuss the possible advantages, disadvantages, and challenges of this strategy.

4. When trees are removed from an area, patches or strips of untouched trees often intersperse the deforested land. How is the abiotic environment on the edge of these strips or patches different from before the area was disturbed? What changes in vegetation would you expect to see in the next few years? How might animals be affected by forest fragmentation?

5. Nanoparticles are tiny bits of metal that are used in sunscreens, as a wastewater treatment, and for many other purposes. Recent evidence suggests that nanoparticles are toxic to phytoplankton, the primary producers at the base of many aquatic food chains. Phytoplankton use the energy in sunlight to produce organic matter, and they consume CO_2 and release O_2. Predict some possible consequences to biodiversity if nanoparticles become a more common pollutant.

6. How does the Gulf of Mexico's "dead zone" demonstrate the connections among land and water ecosystems? How would you expect bird populations in the dead zone to be affected?

7. Suppose you throw a small piece of plastic in the garbage. List three places where the plastic might be found months later.

8. Use the Internet to learn how pharmaceutical drugs end up in the water supply. What is the proper way to dispose of pharmaceuticals in your area?

9. Use the Internet to research ways to make homes more energy-efficient. How does reducing your monthly energy bill relate to the conservation of biodiversity?

10. How does the combustion of fossil fuels influence such different phenomena as acid deposition and global climate change?

11. In what ways is the greenhouse effect both beneficial and harmful?

12. Explain the logic behind planting trees to reduce global climate change.

13. One way to combat invasive species is to kill the invaders. In Hawaii, officials shoot feral cats, goats, and pigs. In Australia, the government fought zebra mussels by adding chlorine and copper to a bay, killing everything living in the water. Do you think that these approaches are reasonable? Suggest alternative strategies.

14. List three ways you can alter your lifestyle to promote conservation.

15. Search the Internet for information on the Convention on Biological Diversity and on the international agreement called CITES. How does each approach tackle the biodiversity crisis on a global scale?

16. Refer back to section 12.7, which describes the bottleneck effect. With this information in mind, why might recovery be difficult for species, such as cheetahs, that are nearly extinct?

17. Phytoremediation is the use of plants to treat environmental problems. Search the Internet for applications of phytoremediation. What are the benefits of phytoremediation? If you were trying to discover plants suitable for use in phytoremediation, what qualities would you look for? Can you foresee any problems with phytoremediation?

18. Use the Internet to learn how aluminum is mined. How can recycling aluminum help save rain forests?

SCIENTIFIC LITERACY

1. Explain why preserving biodiversity is important (or is not important) to you. Use the Internet to explore ethical arguments about conservation.

2. DNA evidence recently confirmed the existence of a "pizzly bear," the offspring of a polar bear and a grizzly bear. Scientists hypothesize that some polar bears are staying on the mainland because of the warming climate, so polar bears are encountering grizzlies more often than in the past. Pizzly bears may be less fit than either polar bears or grizzlies, so some people are advocating that they be killed. Make an argument for or against this strategy. If human actions are contributing to the breeding behavior, do we have an ethical obligation to intervene? Do you think polar bears have a better chance at avoiding extinction if humans eliminate their hybrid offspring? Would evidence that polar bears and grizzlies had interbred in the past change your answers?

Answers to Mastering Concepts, Write It Out, Scientific Literacy, and Pull It Together questions can be found in the Connect ebook.
connect.mheducation.com

Design element: Burning Question (fire background): ©Ingram Publishing/Super Stock

PULL IT TOGETHER

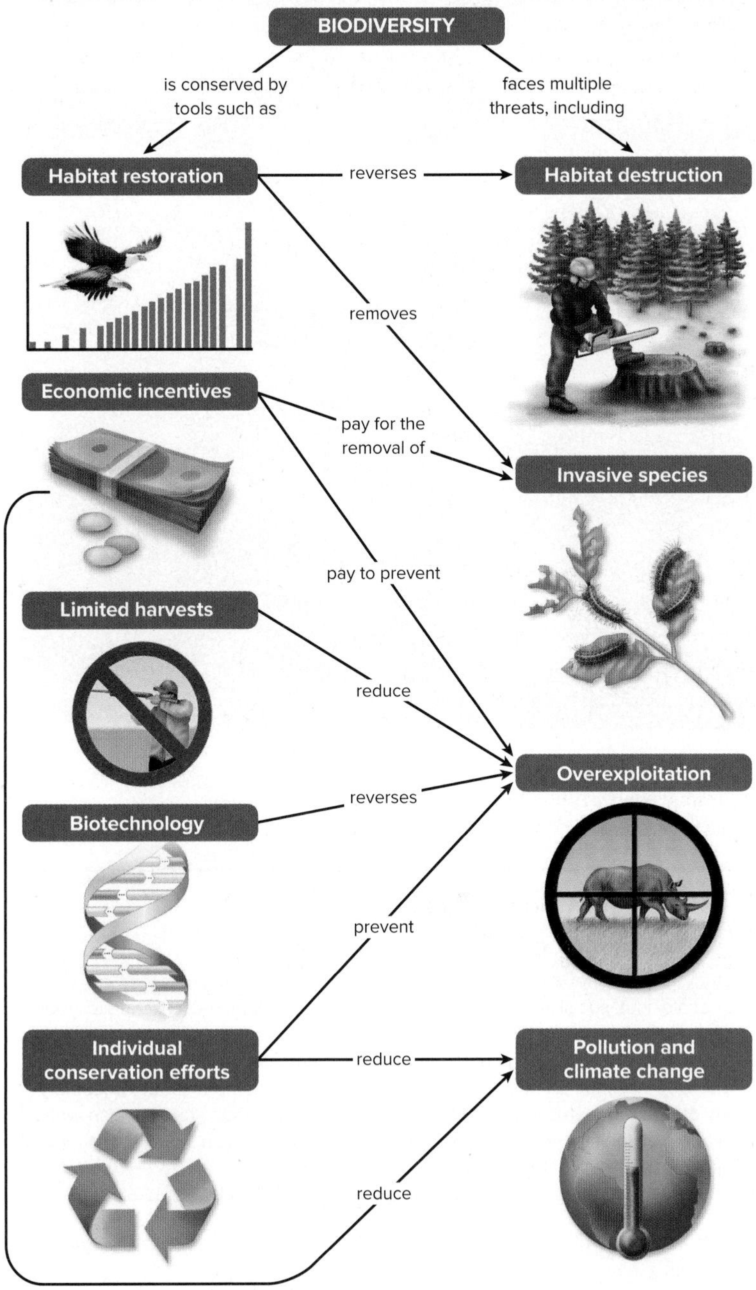

Figure 20.19 **Pull It Together: Preserving Biodiversity.**

Refer to figure 20.19 and the chapter content to answer the following questions.

1. Review the Survey the Landscape figure in the chapter introduction. Add the terms *population, community,* and *ecosystem* along the right edge of figure 20.19. For each threat to biodiversity, connect to one of those three terms. Each new connecting phrase should explain how the threat affects biodiversity at the selected level of biological organization. For example, one connecting phrase might explain how invasive species affect communities.

2. Figure 20.19 has arrows connecting conservation strategies with threats to biodiversity. For each arrow, list a specific way to implement the strategy.

LEARNING OUTLINE

21.1 Vegetative Plant Parts Include Stems, Leaves, and Roots

21.2 Soil and Air Provide Water and Nutrients

21.3 Plant Cells Build Tissues

21.4 Tissues Build Stems, Leaves, and Roots

21.5 Plants Have Flexible Growth Patterns, Thanks to Meristems

21.6 Vascular Tissue Transports Water, Minerals, and Sugar

APPLICATIONS

Burning Question 21.1 *What's the difference between fruits and vegetables?*

Why We Care 21.1 *Boost Plant Growth with Fertilizer*

Burning Question 21.2 *What are controlled burns?*

Burning Question 21.3 *Where does maple syrup come from?*

Investigating Life 21.1 *An Army of Tiny Watchdogs*

A Buggy Meal. Sticky secretions on the leaves of this carnivorous sundew plant have trapped a blue damselfly.

©Bob Gibbons/Alamy Stock Photo

Learn How to Learn
Studying in Groups

Study groups offer a great way to learn from other students, but they can also dissolve into social events that accomplish little real work. Of course, your choice of study partners makes a huge difference, so try to pick classmates who are at least as serious as you are about learning. To stay focused, plan activities that work well for groups. For example, you can agree on a list of vocabulary words and take turns adding them to a group concept map. You can also write exam questions for your study partners to answer, or you can simply explain the material to each other in your own words.

SURVEY THE LANDSCAPE
Plant Anatomy and Physiology

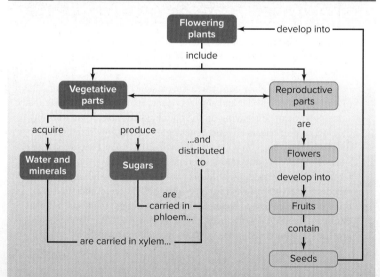

A variety of plant cells make up tissues, which in turn compose the familiar organs of the vegetative plant body: stems, leaves, and roots. A transportation network composed of specialized "pipes" carries water, minerals, and sugars throughout the plant.

For more details, study the Pull It Together feature in the chapter summary.

What's the Point? ▼

The insect-eating sundew in the chapter opening photo and the cactus shown here may look unusual, but they are typical plants in most respects. They are composed of cells that are assembled into tissues, which form organs such as leaves, flowers, stems, and roots. They require the same essential elements. Their roots extract water and nutrients from soil, and their chloroplasts carry out photosynthesis. They produce flowers, fruits, and seeds.

©Brand X Pictures/ PunchStock RF

Why do they look so different? Natural selection provides the answer. Nitrogen-poor soil selects for the carnivorous lifestyle, whereas the sharp spines and fleshy stem of the cactus are water-saving adaptations.

Plants provide us with food, textiles, building materials, fuel, paper, pharmaceutical drugs, and much more. All of these useful items derive their properties from the cells and tissues illustrated in this chapter, which describes the form and function of a plant's nonreproductive parts.

21.1 Vegetative Plant Parts Include Stems, Leaves, and Roots

Imagine a rose bush that produces a bounty of sweet-smelling flowers. Biologists would divide your plant into two sets of parts. The gorgeous flowers, which will eventually give rise to fruits called rose hips, are the reproductive parts. Chapter 22 describes flowers and fruits in detail. The rest of the plant consists of **vegetative** parts that do not participate directly in reproduction.

This chapter focuses on the **anatomy** (form) and **physiology** (function) of a plant's vegetative organs—its stems, leaves, and roots. As you will see, a plant's **organs** are composed of multiple interacting tissues. A **tissue,** in turn, is a group of cells that interact to provide a specific function. This section begins our tour of the plant body with an overview of its major organs.

A plant's vegetative organs work together (figure 21.1). The **shoot** is the aboveground part of the plant; a **terminal bud** contains undeveloped tissue at the shoot tip. The shoot's **stem** supports the **leaves,** which produce carbohydrates such as sucrose by photosynthesis. A large portion of this sugar moves down the stem and nourishes the **roots,** which are usually belowground. Root cells depend completely on the shoots to provide energy for their metabolism. At the same time, however, roots anchor the plant and absorb water and minerals that move via the stem to the leaves.

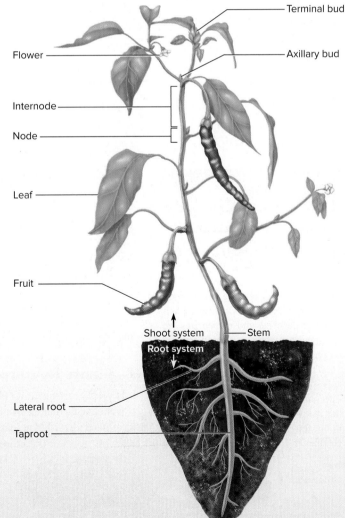

Specialized stems

Climbing (grape tendrils)

Specialized leaves

Nutrient storage (onion)

Specialized roots

Nutrient storage (carrot)

Figure 21.1 Parts of a Flowering Plant. A plant consists of a shoot system and a root system. Stems, leaves, and roots are vegetative organs; flowers and fruits are reproductive structures (see chapter 22). Tendrils, onion bulbs, and carrots are specialized stems, leaves, and roots.

Photos: (grape): ©Franz Krenn/Science Source; (onion): ©YAY Media AS/Alamy Stock Photo RF; (carrots): ©Huw Jones/ Photolibrary/Getty Images

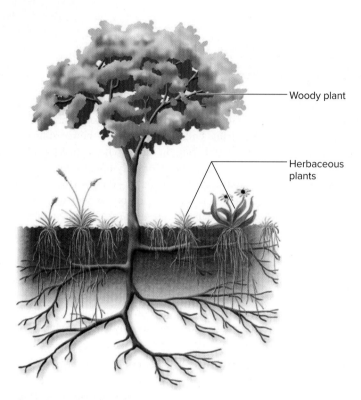

Woody plant

Herbaceous plants

Figure 21.2 Herbaceous and Woody Plants. Herbaceous plants, such as grasses and daisies, have soft green stems. Trees are woody plants; they have stems and roots made of wood.

A close look at a stem reveals that it consists of alternating nodes and internodes (see figure 21.1). A **node** is a point at which one or more leaves attach to the stem. **Internodes** are the stem areas between the nodes. Along with at least one leaf, each node features an **axillary bud** (also called a **lateral bud**), an undeveloped shoot that forms in the angle between the stem and leaf stalk. Although some buds elongate to form a branch or flower, many remain dormant.

Informally, biologists divide plants into two categories based on the characteristics of the stem (figure 21.2). A **herbaceous plant** has a soft, green stem at maturity. The chili plant in figure 21.1 is herbaceous, as are grasses, daisies, dandelions, and radishes. The stems and roots of **woody plants** such as elm and cedar trees are made of tough wood covered with bark.

Natural selection ensures that stems, leaves, and roots do not always look exactly like those in figures 21.1 and 21.2. Depending on the habitat, plants may contend with everything from hungry animals to extreme drought to continuous flooding to frozen winters. These selective forces have sculpted the vegetative plant body into a tremendous diversity of forms.

Humans have used artificial selection to modify stems, leaves, roots, flowers, and fruits to suit our own needs (see figure 21.1). We cultivate onions for their fleshy leaves and carrot plants for their nutritious roots. We have also selected Brussels sprouts for their axillary buds, spinach plants for their leaves, and tomato plants for their fruits. Most people divide these foods into two categories: fruits and vegetables. Burning Question 21.1 explains more about these overlapping terms.

21.1 Mastering Concepts

1. How do stems, leaves, and roots support one another?

2. What are the other major parts of the plant body?

Burning Question 21.1

What's the difference between fruits and vegetables?

Have you ever heard the rumor that tomatoes are really fruits? If so, you may also have wondered about a larger question: the difference between fruits and vegetables.

Botanically speaking, tomatoes are fruits because they contain seeds. Any seed-bearing structure produced by a flowering plant is a fruit. Apples, cherries, oranges, peaches, and raspberries are fruits. So are foods that we think of as vegetables, such as zucchinis, green beans, bell peppers, and pumpkins. But the term *vegetable* also includes plant parts that don't contain seeds, such as spinach leaves or carrot roots or Brussels sprouts.

Unlike fruits, vegetables do not have a botanical definition—just a culinary one. That is, foods that we consider vegetables may come from any part of a plant, but they are typically used in salty or savory dishes, not sweet ones.

Submit your burning question to marielle.hoefnagels@mheducation.com

(vegetables): ©Getty Images RF

21.2 Soil and Air Provide Water and Nutrients

To stay healthy, a person needs water and the right dietary mix of fat, protein, carbohydrates, vitamins, and minerals (see section 28.3). Plants have similar needs, but they do not acquire these raw materials by eating and drinking. Instead, they are autotrophs that use photosynthesis to assemble organic molecules from elements absorbed from their surroundings (see chapter 5).

This section describes the sources of the elements that plants require. The rest of the chapter focuses on how plants absorb and transport water, minerals, and organic substances such as sugar within their tissues.

A. Plants Require 16 Essential Elements

Like every organism, a plant requires certain **essential nutrients,** which are chemicals that are vital for metabolism, growth, and reproduction. Biologists have identified at least 16 elements essential to all plants (figure 21.3). Nine are **macronutrients,** meaning that they are needed in fairly large amounts. The macronutrients are carbon (C), oxygen (O), hydrogen (H), nitrogen (N), potassium (K), calcium (Ca), magnesium (Mg), phosphorus (P), and sulfur (S). The others are **micronutrients,** which are required in much smaller amounts.

Macronutrients	Form Taken Up by Plants	Percent Dry Weight	Selected Functions
Carbon (C)	CO_2	45	Part of organic compounds
Oxygen (O)	H_2O, O_2, CO_2	45	Part of organic compounds
Hydrogen (H)	H_2O	6	Part of organic compounds
Nitrogen (N)	NO_3^-, NH_4^+	1.5	Part of nucleic acids, amino acids, chlorophyll, ATP
Potassium (K)	K^+	1.0	Controls opening and closing of stomata, activates enzymes
Calcium (Ca)	Ca^{2+}	0.5	Cell wall component, activates enzymes, second messenger in signal transduction, maintains membranes
Magnesium (Mg)	Mg^{2+}	0.2	Part of chlorophyll, activates enzymes, participates in protein synthesis
Phosphorus (P)	$H_2PO_4^-$, HPO_4^{2-}	0.2	Part of nucleic acids, sugar phosphates, ATP, phospholipids
Sulfur (S)	SO_4^{2-}	0.1	Part of cysteine and methionine (amino acids), coenzyme A

Micronutrients	Form Taken Up by Plants	Percent Dry Weight	Selected Functions
Chlorine (Cl)	Cl^-	0.01	Water balance
Iron (Fe)	Fe^{3+}, Fe^{2+}	0.01	Chlorophyll synthesis, part of electron carriers
Boron (B)	BO_3^-, $B_4O_7^{2-}$	0.002	Growth of pollen tubes, sugar transport, regulates certain enzymes
Zinc (Zn)	Zn^{2+}	0.002	Hormone synthesis, activates enzymes, stabilizes ribosomes
Manganese (Mn)	Mn^{2+}	0.005	Activates enzymes, electron transfer, photosynthesis
Copper (Cu)	Cu^{2+}	0.0006	Part of plastid pigments, lignin synthesis, activates enzymes
Molybdenum (Mo)	MoO_4^{2-}	0.00001	Nitrate reduction

Carbon, oxygen, and hydrogen (96% of dry weight)

Other macronutrients (~3.5%)

Micronutrients (~0.5%)

Figure 21.3 Essential Nutrients for Plants. The nine most abundant elements in plants are macronutrients; the seven micronutrients occur in much lower concentrations.

Among the essential elements, C, H, and O are by far the most abundant, together accounting for about 96% of the dry weight of a plant. The six other macronutrients account for another 3.5%. Of these, N, P, and K are the most common ingredients in commercial fertilizers (see Why We Care 21.1). Gardeners and farmers use fertilizers to prevent or treat nutrient deficiencies such as those shown in figure 21.4.

B. Leaves and Roots Absorb Essential Elements

Plants obtain their three most abundant elements (C, H, and O) from water and the atmosphere. Water (H_2O) enters the plant through the roots, as described in section 21.6. Carbon and oxygen atoms come from the atmosphere in the form of CO_2 gas, which diffuses into the leaf or stem through pores called stomata. ⓘ *diffusion,* section 4.5A

As roots absorb water, they also take up all of the other mineral elements listed in figure 21.3. These nutrients dissolve in the soil's water when rock particles disintegrate or when microbes decompose dead organisms.

Plants often have help from soil organisms in obtaining water and nutrients. Recall from chapter 15 that the roots of most land plants are colonized with mycorrhizal fungi. In exchange for sugars produced in photosynthesis, the fungal filaments explore the soil and absorb water and minerals that the roots could not otherwise reach. In particular, phosphorus is poorly soluble in water and does not move easily to roots. Mycorrhizae therefore especially boost phosphorus absorption. ⓘ *mycorrhizae,* section 15.5C

a.

b.

Figure 21.4 Nutrient Deficiencies. (a) Iron deficiency causes yellowed leaves, but the veins remain green. (b) Phosphorus deficiency causes plants to develop purplish leaves.

Photos: (a, b): ©Nigel Cattlin/Science Source

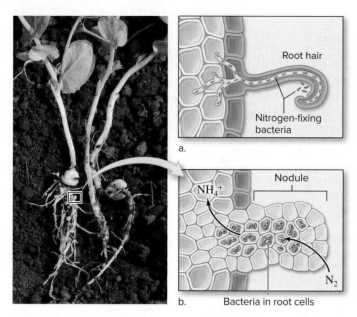

Figure 21.5 Root Nodules. (a) Nitrogen-fixing bacteria enter through root hairs and trigger nodule development. (b) In a mature nodule, the bacteria live within the root's cells. The plant provides the energy the bacteria need to fix nitrogen.

Photo: ©WILDLIFE GmbH/Alamy Stock Photo

Why We Care 21.1 | Boost Plant Growth with Fertilizer

Farmers and gardeners often amend soil with commercial synthetic fertilizers or with nutrient-rich organic matter, such as manure or compost. Plant growth surges if the added material provides a nutrient that was previously scarce.

Commercial fertilizer labels prominently display three numbers that indicate the content of nitrogen, phosphorus, and potassium (figure 21.A). These are the three elements that are most commonly deficient in soils. The label also lists other macro- and micronutrients in the fertilizer.

Chemically, nutrients from a synthetic fertilizer are equivalent to those from manure or compost. So why should a gardener bother adding organic matter to soil? The answer is that organic matter not only contains nutrients but also aerates the soil, increases soil's ability to hold water, and provides food for beneficial microbes and animals. Manure and compost also release nutrients slowly, providing long-lasting results.

6-12-6 LIQUID PLANT FOOD PLUS
Contains Plant Food Supplements PLUS Humate Humic Acid and Medina Soil Activator

GUARANTEED ANALYSIS

Total Nitrogen (N)	6.00%
Available Phosphoric Acid (P_2O_5)	12.00%
Soluble Potash (K_2O)	6.00%

Figure 21.A Read the Label. Fertilizer labels list the concentrations of each nutrient. ©David Tietz/Editorial Image, LLC

The source of nitrogen also deserves special mention. As described in chapter 2, nitrogen atoms occur in proteins, nucleic acids, and chlorophyll. Nitrogen gas (N_2) makes up 78% of the atmosphere, but plants cannot use it. Fortunately for plants (and ultimately animals), several types of bacteria use **nitrogen-fixing** enzymes to convert N_2 to NH_4^+, a form of nitrogen that plants can use. Many nitrogen-fixing bacteria are free-living; others live in **nodules,** small swellings on the roots of some types of plants (figures 21.5 and 15.15). The bacteria consume sugars that the host plant produces by photosynthesis. The plant, in turn, incorporates the nitrogen atoms from NH_4^+ into its own tissues. When the plant dies, decomposers make the nitrogen available to other organisms. The first step—nitrogen fixation—is therefore the key to the entire nitrogen cycle, bringing otherwise inaccessible nitrogen to plants, microbes, and all other life (see figure 19.32 for a detailed look at the nitrogen cycle).

The most famous nitrogen-fixing bacteria, in genus *Rhizobium,* stimulate nodule formation on the roots of plants called legumes (clover, beans, peas, peanuts, soybeans, and alfalfa). A farming technique called crop rotation alternates legume crops with nitrogen-hungry plants such as corn or cotton. The legume replenishes the soil's nitrogen, reducing the need for fertilizer.

A few plants acquire nutrients from more unusual sources. For example, carnivorous plants often live in waterlogged, acidic soils where organic matter decays slowly and nutrients are scarce. They obtain nitrogen and phosphorus from their prey (see the chapter-opening photo).

21.2 Mastering Concepts

1. Which macro- and micronutrients do all plants require?
2. How do plants acquire C, H, O, N, and P?
3. How would planting nodule-producing legumes be useful in crop rotations?

21.3 Plant Cells Build Tissues

A cactus, an elm tree, and a dandelion have distinctly different stems, leaves, roots, and growth patterns. They may seem to have little in common, but a closer look reveals that all consist of the same types of cells and tissues. Before examining these building blocks, it may be helpful to review the structure of a plant cell in figure 3.7 and of the cell wall in figure 3.24. Note that in many plant cells, the wall has two layers: a thin, flexible primary one and a thick, rigid secondary one that forms after the cell is fully grown.

A. Plants Have Several Cell Types

Plants consist of several cell types. This section lists the most common ones, which are illustrated in figure 21.6.

Ground Tissue Cells **Ground tissue** makes up the majority of the body of a herbaceous plant. It consists of three main cell types: parenchyma, collenchyma, and sclerenchyma.

Parenchyma cells are the most abundant. These cells are alive at maturity, have thin primary cell walls, and retain the ability to divide, which enables them to differentiate in response to injury or a changing environment. It is parenchyma cells, for example, that divide to produce the roots that emerge from a houseplant cutting placed in water. Parenchyma cells have vital functions, including photosynthesis, respiration, gas exchange, and the storage of starch and other materials.

Collenchyma cells are elongated living cells with unevenly thickened primary walls that can stretch as the cells grow. These cells provide elastic support without interfering with the growth of young stems or expanding leaves. Collections of collenchyma cells make up the tough, flexible "strings" in celery stalks.

Sclerenchyma cells provide inelastic support to parts of a plant that are no longer growing. These cells, which are dead at maturity, have thick, rigid secondary cell walls that occupy most of the cell's volume. The secondary walls typically contain **lignin,** a tough, complex molecule that adds great strength to the cell walls. Some sclerenchyma cells are elongated fibers; linen, for example, comes from the soft fibers of the stems of the flax plant. Other sclerenchyma cells form hard layers or clusters, such as the small groups of cells that create a pear's gritty texture.

Conducting Cells in Xylem and Phloem

Vascular tissues transport water, minerals, carbohydrates, and other dissolved compounds throughout the plant. Two types of vascular tissue are xylem and phloem.

Xylem is a tissue that transports water and dissolved minerals from the roots to all parts of the plant. The water-conducting cells of xylem are elongated and have thick, lignin-rich secondary walls. They are dead at maturity, which means that no cytoplasm blocks water flow.

The two kinds of water-conducting cells in xylem are tracheids and vessel elements. **Tracheids** are long, narrow cells that overlap at their tapered ends. Water moves from tracheid to tracheid through pits, which are thin areas in the cell wall. **Vessel elements** are short, wide, barrel-shaped conducting cells that stack end to end, forming long, continuous tubes. Their side walls have pits, but their end walls are either perforated or absent. Water moves much faster in vessels than in tracheids, both because of their greater diameter and because water can pass easily through each vessel element's end wall. On the other hand, the narrower tracheids are less vulnerable to air bubble formation.

Phloem tissue transports dissolved organic compounds, primarily sugars produced in photosynthesis. The main conducting cells are **sieve tube elements,** which align end to end to make a single functional unit called a sieve tube. The sieve tube elements are alive, but they lack a nucleus and have little cytoplasm. Adjacent to each sieve tube element is at least one **companion cell,** a specialized parenchyma cell that retains all of its organelles. Companion cells transfer carbohydrates into and out of the sieve tube elements, and they provide energy and proteins to the conducting cells.

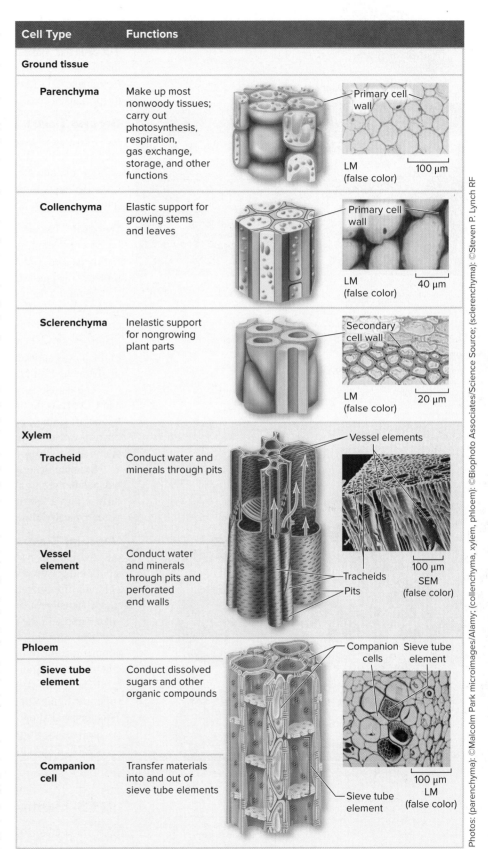

Cell Type	Functions
Ground tissue	
Parenchyma	Make up most nonwoody tissues; carry out photosynthesis, respiration, gas exchange, storage, and other functions
Collenchyma	Elastic support for growing stems and leaves
Sclerenchyma	Inelastic support for nongrowing plant parts
Xylem	
Tracheid	Conduct water and minerals through pits
Vessel element	Conduct water and minerals through pits and perforated end walls
Phloem	
Sieve tube element	Conduct dissolved sugars and other organic compounds
Companion cell	Transfer materials into and out of sieve tube elements

Figure 21.6 Plant Cell Types: A Summary. Parenchyma, collenchyma, and sclerenchyma cells compose the majority of a herbaceous plant's body; other specialized cell types make up the xylem and phloem that transport materials within the plant.

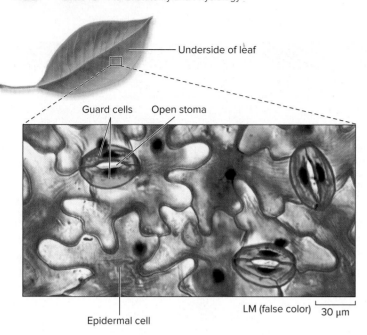

Figure 21.7 **Stomata.** Plants exchange gases with the atmosphere through open stomata. Guard cells control whether each stoma is open or closed.

Photo: ©Dr. Keith Wheeler/Science Source

B. Plant Cells Form Three Main Tissue Systems

The cells that make up a plant form three main tissue systems: ground tissue, dermal tissue, and vascular tissue. For comparison, animals have four types of tissues (see chapter 23). Each tissue type derives its properties from a unique combination of specialized cells. Together, these cells carry out all of the plant's functions.

Ground Tissue   Ground tissue often fills the spaces between more specialized cell types inside roots, stems, leaves, fruits, and seeds. For example, the pulp of an apple, the photosynthetic area inside a leaf, and the starch-containing cells of a potato all consist of ground tissue composed mainly of parenchyma cells. Although most ground tissue cells are structurally unspecialized, they are important sites of photosynthesis, respiration, and storage.

Dermal Tissue   Dermal tissue covers the leaves, stems, and roots. In a herbaceous plant, dermal tissue consists of the **epidermis,** a single layer of tightly packed, flat, transparent parenchyma cells. In woody plants, tough bark replaces the epidermis in stems and roots (see section 21.5D).

  Natural selection has shaped dermal tissue over hundreds of millions of years. When plants first moved onto land, they were at risk for drying out. This condition selected for new water-conserving adaptations. For instance, in land plants, epidermal cells secrete a **cuticle,** a waxy layer that coats the epidermis of the leaves and stem. The cuticle conserves water and protects the plant from predators and fungi.

  The cuticle is impermeable not only to water but also to gases such as O_2 and CO_2. How do these gases pass through the cuticle? The answer is **stomata** (singular: stoma), pores through which leaves and stems exchange gases with the atmosphere (figure 21.7). A pair of specialized **guard cells** surrounds each stoma and controls its opening and closing.

  Stomata occupy about 1% to 2% of the leaf surface area, allowing for a balance between gas exchange and water conservation. Open stomata let CO_2 diffuse into a leaf for photosynthesis but also allow water to diffuse out; the pores typically close when conditions are too dry.

Vascular Tissue   Vascular tissues—xylem and phloem—form a continuous distribution system embedded in the ground tissue of shoots and roots. In stems and leaves, a **vascular bundle** is a strand of tissue containing xylem and phloem, often together with collenchyma tissue or sclerenchyma fibers. The tough fibers protect against animals that might otherwise tap into the rich sugar supply in the phloem.

  The transition from water to land selected for vascular tissues, which accommodate the division of labor between roots and shoots. Roots absorb water and minerals; shoots produce food. Xylem and phloem shuttle these materials throughout the plant's body. But vascular tissue also has another function: support. Lignin strengthens the walls of xylem cells and sclerenchyma fibers. This physical support enables vascular plants to tower over their nonvascular counterparts, an important adaptation in the intense competition for sunlight.

  Figure 21.8 summarizes the locations of the tissue systems in plants.

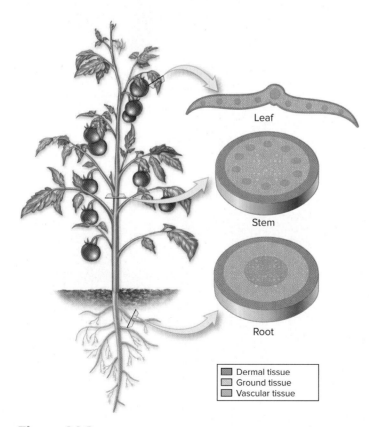

Figure 21.8 **Three Tissue Types.** Dermal, ground, and vascular tissues make up the leaves, stem, and roots of a plant.

21.3 Mastering Concepts

1.   Describe the cell types that make up a plant body.

2.   Where in the plant does ground tissue occur?

3.   Compare and contrast tracheids, vessel elements, and sieve tube elements.

21.4 Tissues Build Stems, Leaves, and Roots

The tissues described in section 21.3 make up the stems, leaves, and roots of vascular plants. We now return to these organs to examine their structures more closely. Note that flowering plants (angiosperms) differ in the internal anatomy of their main vegetative organs. As described in section 16.5, two groups—eudicots and monocots—account for 97% of all angiosperms. Eudicots, the largest group, include everything from chili peppers to sunflowers to elm trees. Monocots include orchids, grasses, corn, rice, and wheat. This section describes the structural similarities and differences between these two groups of plants.

A. Stems Support Leaves

Ground tissue occupies most of the volume of the stem of a herbaceous plant. The ground tissue, which consists mostly of parenchyma cells, stores water and starch. The cells are often loosely packed, allowing for gas exchange between the stem interior and the atmosphere.

Vascular bundles are embedded in the stem's ground tissue. The vascular bundles, which typically have phloem to the outside and xylem toward the inside, are arranged differently in monocots and eudicots (figure 21.9). In most monocot stems, vascular bundles are scattered throughout the ground tissue. Most eudicot stems, in contrast, have a single ring of vascular bundles. Ground tissue occupies most of the rest of the eudicot stem: The **cortex** fills the area between the epidermis and vascular tissue, and **pith** occupies the center.

Some stems have specialized functions. The stems of climbing plants may form tendrils that coil around objects, maximizing exposure of the leaves to the sun (see figure 21.1). The succulent, fleshy stems of cacti stockpile water. Rhizomes are thickened underground stems that produce both shoots and roots; tubers, such as potatoes, are swollen regions of rhizomes that store starch. Still other stems are protective; the thorns on hawthorn plants, for example, are modified branches.

B. Leaves Are the Primary Organs of Photosynthesis

Most leaves have two main parts: The stalklike **petiole** attaches to the stem and supports the broad, flat **blade.** The leaf blade maximizes the surface area available to capture light. For example, a large maple tree has about 100,000 leaves, with a total surface area that would cover six basketball courts.

Biologists categorize leaves according to their basic forms (figure 21.10).

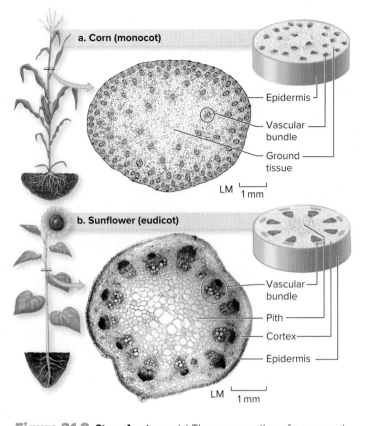

Figure 21.9 Stem Anatomy. (a) The cross section of a monocot stem features vascular bundles scattered in ground tissue. (b) A eudicot stem has a ring of vascular bundles surrounding a central pith.

Photos: (a): ©Steven P. Lynch RF; (b): ©Jupiterimages/Getty Images RF

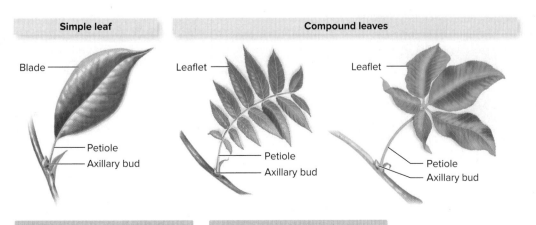

Figure 21.10 Leaf Types. A simple leaf has an undivided blade, whereas compound leaves consist of multiple leaflets. An axillary bud defines the base of each leaf.

Meristems Enable Flexible Plant Growth   Vascular Tissue Transports Water, Minerals, Sugar

A **simple leaf** has an undivided blade, whereas **compound leaves** are divided into leaflets. How can you tell the difference between a simple leaf and one leaflet of a compound leaf? A leaf has an axillary bud at its base, whereas an individual leaflet does not (see figure 21.10).

Veins are vascular bundles inside leaves, and they are often a leaf's most prominent external feature. Networks of veins occur in two main patterns (figure 21.11). Many monocots have parallel veins, with several major longitudinal veins connected by smaller minor veins. Most eudicots have netted veins, with minor veins branching in all directions from large, prominent midveins.

The ground tissue inside a leaf is called **mesophyll,** and it is composed mostly of parenchyma cells (figure 21.12). Most mesophyll cells have abundant chloroplasts and produce sugars by photosynthesis. When the stomata are open, mesophyll cells exchange CO_2 and O_2 directly with the atmosphere.

Mesophyll cells also exchange materials with vascular tissue. Xylem in the tiniest leaf veins delivers water and minerals to nearby mesophyll cells. Meanwhile, sugars produced in photosynthesis move from the mesophyll cells to the phloem's companion cells and then to the sieve tube elements. The sugars, along with other organic compounds, travel within the phloem to the roots and other nonphotosynthetic plant parts.

In addition to carrying out photosynthesis, leaves can store nutrients, provide protection, and even trap animals. Onion bulbs, for example, are collections of the fleshy bases of leaves that store nutrients. Cactus spines are modified leaves that deter predators. Some flower parts, such as petals, are modified leaves. And in carnivorous plants, leaves attract, capture, and digest prey, as shown in the chapter-opening photo.

Figure 21.11 Leaf Veins. (a) Leaves of many monocots, such as this dracaena, have prominent parallel veins. (b) Leaves of eudicots, including this pumpkin plant, have a netlike pattern of veins.

(a): ©Steven P. Lynch RF; (b): ©Easton Manley/age fotostock/SuperStock

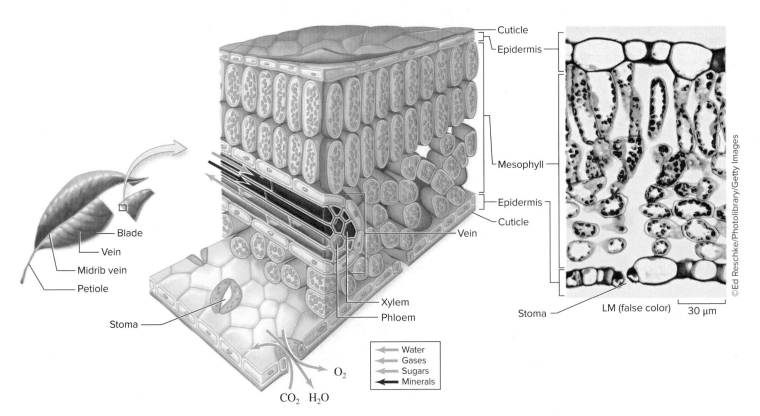

Figure 21.12 Leaf Anatomy. Leaf mesophyll consists of cells that carry out photosynthesis. Stomata are often most abundant on the lower leaf surface. Leaf veins deliver water and minerals, and they carry off the products of photosynthesis.

C. Roots Absorb Water and Minerals, and Anchor the Plant

Roots grow in two main patterns (figure 21.13). A **fibrous root system** consists of a widespread network of slender roots arising from the plant's stem. Grasses and other monocots usually have fibrous root systems. Because they are relatively shallow, these roots rapidly absorb minerals and water near the soil surface and prevent erosion. A **taproot system,** on the other hand, features a thick main root from which lateral branches emerge. Taproots grow fast and deep, maximizing support and enabling a plant to use minerals and water deep in the soil. Most eudicots develop taproot systems.

Figure 21.13 also reveals that the vascular cylinders of monocot and eudicot roots have different arrangements. In most monocot roots, a ring of vascular tissue surrounds a central core (pith) of parenchyma cells. In most eudicots, the vascular cylinder consists of a solid core of xylem, with ridges that project toward the root's exterior. Phloem strands are generally located between the "arms" of the xylem core.

In both fibrous and taproot systems, countless root tips explore the soil for water and nutrients. The **root cap** protects the growing tip from abrasion. Root cap cells, which slough off and are continually replaced, secrete a slimy substance that lubricates the root as it pushes through the soil. The cells of the root cap also play a role in sensing gravity. ⓘ *gravitropism,* section 22.6

Miniglossary	Plant Anatomy
Cortex	Ground tissue between the epidermis and the vascular tissue in some stems and roots
Pith	Ground tissue in the center of some stems and roots
Simple leaf	An undivided blade connected to the stem by a petiole
Compound leaf	A blade divided into leaflets, all connected to the stem by the same petiole
Mesophyll	Ground tissue inside a leaf; mesophyll cells usually contain abundant chloroplasts
Fibrous root system	Widespread network of slender roots arising from a stem; typical of monocots
Taproot system	Root system consisting of a large central root and its lateral branches; typical of eudicots

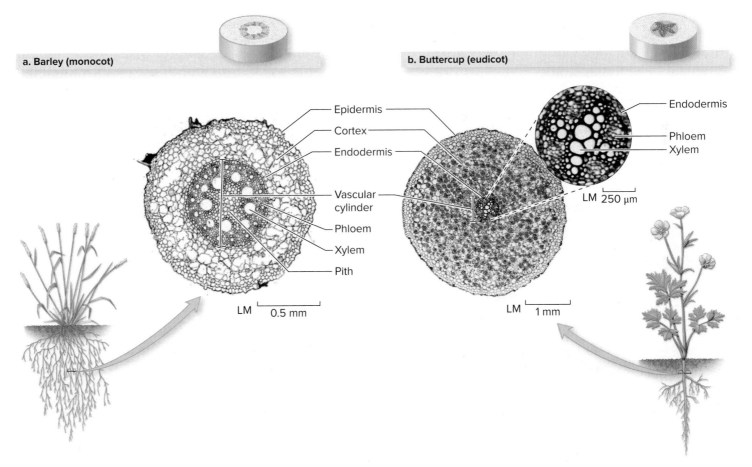

a. Barley (monocot)

b. Buttercup (eudicot)

Epidermis
Cortex
Endodermis
Vascular cylinder
Phloem
Xylem
Pith

Endodermis
Phloem
Xylem

LM 250 μm

LM 0.5 mm

LM 1 mm

Figure 21.13 Root Anatomy. (a) A monocot's fibrous root system features numerous slender roots, each with a ring of vascular tissue surrounding a central pith. (b) A eudicot's taproot system features a central, thick root with large lateral branches. The cortex surrounds a central cylinder of vascular tissue.

Photos: (a): ©McGraw-Hill Education/Al Telser; (b), both): ©Ed Reschke/Photolibrary/Getty Images

Meristems Enable Flexible Plant Growth   Vascular Tissue Transports Water, Minerals, Sugar

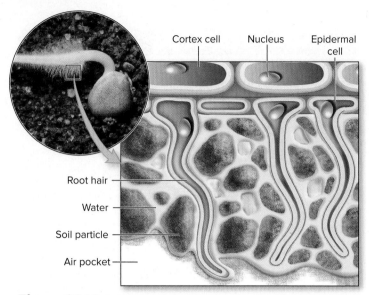

Cortex cell   Nucleus   Epidermal cell

Root hair

Water

Soil particle

Air pocket

Figure 21.14 Root Hairs. These epidermal cell outgrowths greatly increase the surface area of this radish seedling's root.

Photo: ©Dr. Jeremy Burgess/Science Source

TABLE 21.1	**Apical and Lateral Meristems: A Summary**	
Type	Locations	Function
Apical	Terminal and axillary buds of shoots; root tips	Produces tissues that lengthen the tips of shoots and roots
Lateral	Internal cylinder along the length of roots and stems of woody plants	Thickens roots and stems

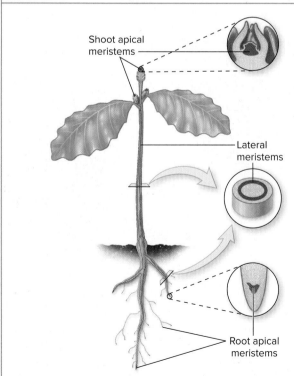

Shoot apical meristems

Lateral meristems

Root apical meristems

The epidermis surrounds the entire root except the root cap. Root hairs are extensions of epidermal cells, maximizing the surface area for absorption of water and minerals (figure 21.14). Just internal to the epidermis is the cortex, which consists of loosely packed cells that may store starch or other materials. The spaces between the cells allow for both gas exchange and the free movement of water.

The **endodermis** is the innermost cell layer of the cortex. The walls of its tightly packed cells contain a ribbon of waxy, waterproof material. These deposits form a barrier that blocks the passive diffusion of water and dissolved substances into the xylem. The endodermis therefore acts as a filter, enabling the plant to exclude toxins and control the concentrations of some minerals.

Some roots have functions other than absorption, including storage and gas exchange. Beet and carrot roots stockpile starch, for example, and desert plant roots may store water. In oxygen-poor habitats such as swamps, specialized roots grow up into the air, allowing oxygen to diffuse in.

21.4 Mastering Concepts

1. Name the cell layers that occur in the stem of a monocot and a eudicot, moving from the epidermis to the innermost tissues.
2. List the parts of a simple and a compound leaf.
3. Describe the internal anatomy of a leaf.
4. Corn is a monocot and sunflower is a eudicot. Make a chart that compares the stems, leaves, and roots of these plants.

21.5 Plants Have Flexible Growth Patterns, Thanks to Meristems

Consider the plight of a green plant. Rooted in place, it seems vulnerable and defenseless against drought, flood, wind, fire, and hungry herbivores. Yet plants dominate nearly every habitat on land. How do they do it? In many ways, plants owe their success to modular growth.

A. Plants Grow by Adding New Modules

To understand modular growth, imagine a landowner who plans to build a motel. Money is tight at first, so she starts with just a few rooms. As her business grows, however, she adds more units to the basic plan. Plant growth is similar. Shoots become larger by adding units ("modules") consisting of repeated nodes and internodes; roots can branch repeatedly as they explore the soil.

Some plants, such as dandelions, have **determinate growth,** meaning that they stop growing after they reach their mature size. Most plants, however, can grow indefinitely by adding module after module—think of ivy climbing up a building. Such **indeterminate growth** can persist as long as environmental conditions allow it.

Modular growth enhances a plant's ability to respond to the environment. For example, plants produce the most root tips in pockets of soil with the richest nutrients. Likewise, a shrub growing in partial shade can add new branches where it receives the most sunlight, while the shaded limbs remain unchanged. The motel owner mentioned earlier might use a similar strategy. That is, if clients will pay more for an ocean view than for a city view, she would be wise to build additional rooms overlooking the sea.

In addition, modular growth means that the loss of a branch or root does not harm a plant as much as, say, the loss of a leg affects a cat. Neighboring branches can add modules to compensate for a broken tree limb, but the cat's body cannot regenerate a leg. Modular growth is one key feature that distinguishes plants from animals.

B. Plant Growth Occurs at Meristems

All of a plant's new cells come from **meristems,** regions that undergo active mitotic cell division (see chapter 8). Meristems are patches of "immortality" that allow a plant to grow, replace damaged parts, and respond to environmental change.

Most plants have two main types of meristems (table 21.1). **Apical meristems** are small patches of actively dividing cells near the tips of roots and shoots. When cells in the apical meristem divide, they give rise to new cells that differentiate into all of the tissue types described in section 21.3.

Woody plants also have **lateral meristems,** which produce cells that thicken a stem or root. A lateral meristem is usually an internal cylinder of cells extending along most of the length of the plant. When the cells divide, they typically produce tissues to both the inside and the outside of the meristem.

C. In Primary Growth, Apical Meristems Lengthen Stems and Roots

Primary growth lengthens the shoot or root tip by adding cells produced by the apical meristems. Figure 21.15 shows how a stem grows and differentiates at its tip. New cells originate at the apical meristem. The daughter cells eventually give rise to ground tissue, the epidermis, and vascular tissue. The stem elongates as the vacuoles of the new cells absorb water, pushing the apical meristem upward. Meanwhile, new leaves originate on the flanks of the meristem.

Remnants of the apical meristem remain in the axillary buds that form at stem nodes. These buds may either remain dormant or "awaken" to form side branches. When a shoot loses its terminal bud, cells in one or more dormant axillary buds begin to divide. The result is a bushy growth form. Gardeners exploit this phenomenon by pinching off the tips of young tomato or basil stems, a practice that promotes the growth of side branches and therefore greatly increases yields. ⓘ *apical dominance,* section 22.4A

Roots also grow at their tips. Just behind the tip of each actively growing root is an apical meristem (figure 21.16). Some of the cells produced at this meristem differentiate into the root cap. Other cells elongate by absorbing water into their vacuoles. As the cells become larger, the root grows farther into the soil. Beyond this zone of elongation is a zone of maturation, in which cells complete their differentiation into the functional ground, dermal, and vascular tissues that make up the root.

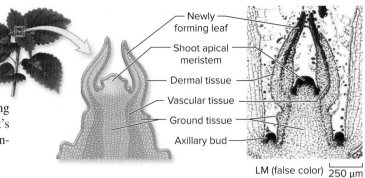

LM (false color) 250 µm

Figure 21.15 **Shoot Apical Meristem.** The apical meristem at the tip of a growing shoot gives rise to the tissues that make up the aboveground parts of the plant.

Photo: ©Steven P. Lynch/McGraw-Hill Education

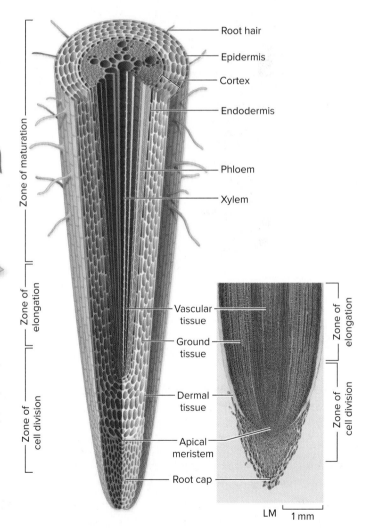

Figure 21.16 **Root Apical Meristem.** The apical meristem at the tip of a growing root produces root cap cells, ground tissue, vascular tissue, and the epidermis.

Photo: ©Oxford Scientific/Getty Images

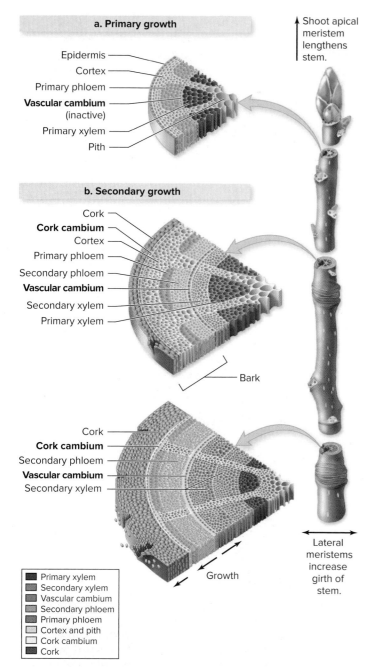

a. Primary growth

- Epidermis
- Cortex
- Primary phloem
- **Vascular cambium** (inactive)
- Primary xylem
- Pith

Shoot apical meristem lengthens stem.

b. Secondary growth

- Cork
- **Cork cambium**
- Cortex
- Primary phloem
- Secondary phloem
- **Vascular cambium**
- Secondary xylem
- Primary xylem

Bark

- Cork
- **Cork cambium**
- Secondary phloem
- **Vascular cambium**
- Secondary xylem

Growth

Lateral meristems increase girth of stem.

- Primary xylem
- Secondary xylem
- Vascular cambium
- Secondary phloem
- Primary phloem
- Cortex and pith
- Cork cambium
- Cork

Figure 21.17 Secondary Growth. (a) In a primary shoot, the microscopic vascular cambium has not yet started producing secondary vascular tissue. (b) In secondary growth, the vascular cambium produces secondary xylem toward the inside of the stem and secondary phloem toward the outside. Cork cambium, meanwhile, produces cork cells to the outside and (in some species) parenchyma to the inside. These parenchyma cells are not shown here.

D. In Secondary Growth, Lateral Meristems Thicken Stems and Roots

In many habitats, plants compete for sunlight. The tallest plants reach the most light, so selection for height has been a powerful force in the evolutionary history of plants. But primary tissue is not strong enough to support a very tall plant. The increasing competition for light therefore selected for additional support, in the form of **secondary growth** that increases the girth of stems and roots in woody plants.

Wood and Bark Formation Wood and bark are tissues that arise from secondary growth originating at two types of lateral meristems: vascular cambium and cork cambium (figure 21.17). These meristems occur in gymnosperms (conifers and their relatives) and many eudicots, but not in monocots.

The **vascular cambium** is an internal cylinder of meristem tissue that produces most of the diameter of a woody root or stem. This lateral meristem forms a thin layer between the primary xylem and phloem (see figure 21.17a). When a cell in the vascular cambium divides, it produces two daughter cells. One of the two cells remains a meristem cell. If the other cell matures to the inside of the cambium, it becomes secondary xylem; if it matures to the outside, it becomes secondary phloem (see figure 21.17b).

A stem or root's **wood** consists of secondary xylem, and it can accumulate to massive proportions. For example, a giant sequoia tree in California is 100 meters tall and more than 7 meters in diameter. Secondary phloem occupies much less volume. This tissue forms the live, innermost layer of **bark,** a collective term for all tissues to the outside of the vascular cambium. Bark protects against water loss and many other dangers, including fire; see Burning Question 21.2.

The **cork cambium** is a lateral meristem that gives rise to cork cells to the outside (see figure 21.17b) and, in some plants, parenchyma cells to the inside. Cork consists of layers of densely packed, waxy cells on the surfaces of mature stems and roots. The cells are dead at maturity and form waterproof, insulating layers. The cork used in wine bottles comes from oak trees that grow in the Mediterranean region. Every 10 years, harvesters remove much of the cork cambium and the thick cork layer, which grows back (see chapter 16's opening photo). Cork is also important in the history of biology; in 1665, Robert Hooke became the first person to see cells when he used a primitive microscope to gaze at cork.

Wood Is Durable and Useful Few plant products are as versatile or economically important as wood. Lumber forms the internal frame that supports many buildings. Firewood provides heat and cooking fuel. Most paper comes from wood. Throughout history, humans have fashioned wood into furniture, pencils, cabinets, boats, baseball bats, serving bowls, roofing shingles, jewelry, picture frames, and countless other items.

Figure 21.18a illustrates the internal anatomy of a tree trunk, including the bark, vascular cambium, and secondary xylem. Nearly all of the trunk consists of secondary xylem, or wood. The cross section in figure 21.18b reveals that the **heartwood,** or innermost wood, is darker than the **sapwood,** or outer portion. This color difference arises as the tree ages. The lighter-colored sapwood, located nearest the vascular cambium, transports water and dissolved minerals. Meanwhile, as the years pass, the oldest secondary xylem at the center of the trunk—the heartwood—gradually becomes unable to conduct water. As its function declines, dark-colored chemicals accumulate in the heartwood.

Another feature of a trunk's cross section is tree rings. In temperate climates, cells in the vascular cambium are dormant in winter, but they divide to

Figure 21.18 Anatomy of a Woody Stem. (a) Wood is secondary xylem, and bark is all of the tissue outside the vascular cambium. (b) At the center of the stem, the darker-colored heartwood is nonfunctional secondary xylem. (c) Wood that forms in the spring has larger cells than wood that forms in the summer, thanks to differences in soil moisture. This size difference is visible as tree rings.

Photos: (b): ©Siede Preis/Getty Images RF; (c): ©Herve Conge/Medical Images

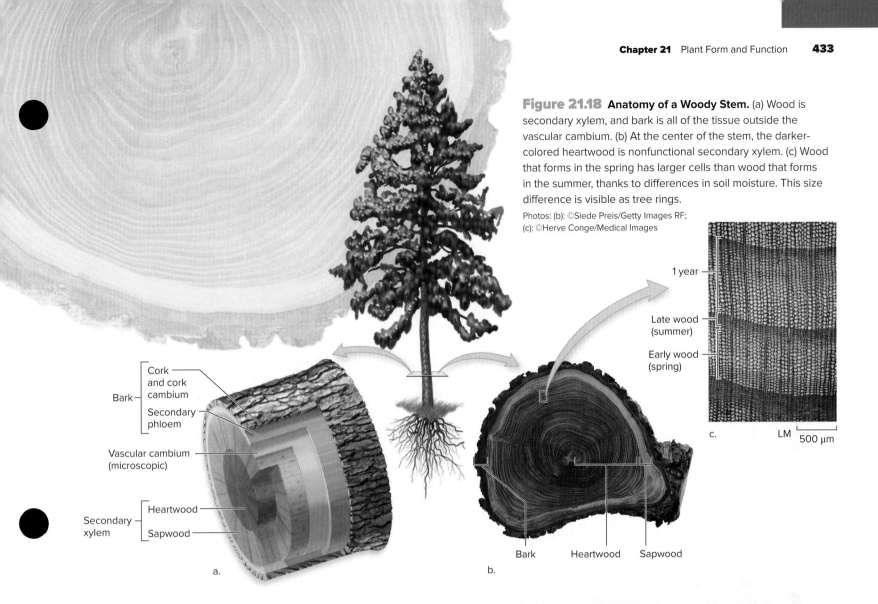

produce wood during the spring and summer. During the moist days of spring, the vascular cambium produces large, water-conducting cells. During the drier days of summer, new wood has smaller cells. The contrast between the summer wood of one year and the spring wood of the next highlights each annual tree ring (figure 21.18c).

Fortunately, it is not necessary to cut down a tree to see its growth rings; a slender core extracted from the trunk reveals the pattern without harming the tree. By counting the rings, a forester can estimate a tree's age. Growth rings also provide clues about climate and significant events throughout a tree's life. A thick ring indicates plentiful rainfall and good growing conditions. Narrow tree rings may reflect stress from herbivory, disease, or fierce competition for light or water. A fire leaves behind a charred "burn scar."

21.5 Mastering Concepts

1. What is the difference between determinate and indeterminate growth?
2. What are the locations and functions of meristems?
3. What are the functions of wood and bark?
4. Explain the origin of tree rings.

(background image of tree rings): ©Siede Preis/Getty Images RF

Burning Question 21.2

What are controlled burns?

Many people live near forests, where wildfires were historically common. For much of the twentieth century, firefighters protected life and property by quickly extinguishing any forest fire. Suppressing every blaze meant that dead wood and leaves accumulated on forest floors. Over the years, this stockpiled fuel boosted the risk of uncontrollable, catastrophic wildfires.

The U.S. Forest Service therefore altered its policy over the past couple of decades. Land managers now intentionally set fires in some forests, but only when conditions make an accidental wildfire unlikely. These small, relatively low-temperature fires not only reduce the risk of wildfires but also improve the health of the forest ecosystem. Controlled fires kill young competitors of the mature forest trees. Invasive plants, which may not be fire-adapted, may also die. Meanwhile, some native trees survive. Giant sequoias, for example, have thick, fibrous, protective bark. Lodgepole pines have thinner bark, but heat triggers their cones to release seeds. They quickly germinate in the nutrient-rich, post-fire soil.

Submit your burning question to marielle.hoefnagels@mheducation.com

21.6 Vascular Tissue Transports Water, Minerals, and Sugar

We turn next to a closer look at the function of vascular tissue, the transportation system that connects the plant's roots, stems, leaves, flowers, and fruits (figure 21.19). This section describes how xylem and phloem interact to move substances throughout the plant.

A. Water and Minerals Are Pulled Up to Leaves in Xylem

Xylem is the vascular tissue that transports **xylem sap,** a dilute solution consisting mostly of water and dissolved minerals absorbed from soil. How do these substances move within the plant?

Water Evaporates from Leaves The easiest way to visualize how plants acquire and transport water is to begin at the end. Plants lose water through **transpiration,** the evaporation of water from a leaf. Heat from the sun

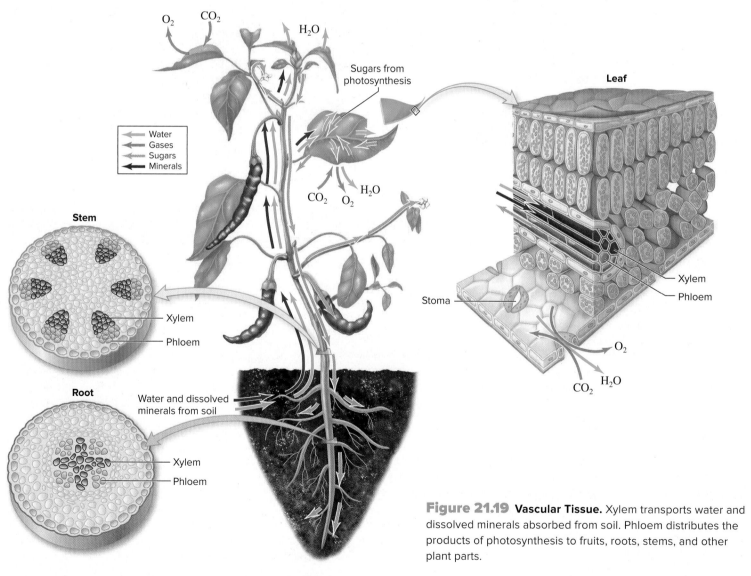

Figure 21.19 Vascular Tissue. Xylem transports water and dissolved minerals absorbed from soil. Phloem distributes the products of photosynthesis to fruits, roots, stems, and other plant parts.

causes water in the cell walls to evaporate into the spaces between the leaf's cells. This evaporation establishes a gradient. That is, the concentration of water molecules inside the leaf is higher than in the air surrounding the leaf. Water vapor therefore diffuses from inside the leaf to the outside air. Most transpiration occurs through open stomata (see figure 21.7).

The bigger the concentration gradient between the leaf interior and the surrounding air, the faster the transpiration rate. Low humidity, wind, and high temperatures therefore all cause the transpiration rate to soar. If the air is too hot or too dry, however, the plant's stomata close. Transpiration slows when the stomata close, but so does photosynthesis; after all, a plant that closes its stomata cuts off its supply of CO_2. But because plants store extra food as starch, a temporary drop in sugar production causes far less harm than does a rapid loss of water.

Most plants conserve water by closing their stomata after dark, when photosynthesis cannot occur anyway. Many plants in dry habitats also use water-saving variations on the photosynthetic pathway. ① C_4 *and CAM plants,* section 5.7

Xylem Transport Relies on Cohesion

The plant must replace the moisture lost in transpiration by transporting water upward from the roots. The functional cells of the xylem, however, are dead at maturity (see figure 21.6). Metabolic activity in xylem cells therefore cannot drive water transport in a plant. So how does water in the xylem get from the roots to the leaves?

The **cohesion–tension theory** explains how xylem sap moves within a plant (figure 21.20). As its name implies, the cohesion–tension theory hinges on the cohesive properties of water—the tendency for water molecules to form hydrogen bonds and "cling" together. As water molecules evaporate through the stomata, additional water diffuses out of leaf veins and into the mesophyll (figure 21.20, step 1). Water molecules leaving the vein attract molecules adjacent to them in the xylem, pulling them toward the vein ending. Each water molecule tugs on the one behind it (step 2). ① *cohesion,* section 2.3A

As evaporation from leaf surfaces pulls water up the stem, additional water enters roots from the soil (step 3). Just behind each growing root tip, the epidermis is fringed with root hairs (see figure 21.14). The plant's millions of root hairs, coupled with filaments of mycorrhizal fungi, add up to an enormous surface area for water and mineral absorption.

The solution flows among and within the cells that make up the outer portion of the root until it reaches the endodermis. At that point, a waxy barrier forces the materials that had gone around cells to now enter the cells of the endodermis. Water enters by osmosis, because the concentration of solutes in cells is generally higher than in the soil. ① *osmosis,* section 4.5A

Materials that cross the endodermis continue into the xylem, enter the transpiration stream, and move up the plant. As we have already seen, the water eventually returns to the atmosphere through the open stomata in the leaves and stem; the dissolved nutrients are incorporated into the plant's tissues.

As long as sufficient moisture is available in the soil, the cohesion between water molecules is enough to move continuous, narrow columns of xylem sap upward against the force of gravity. This mechanism exploits the physical properties of water and requires no energy input from the plant. If the soil is too dry to replace water lost in transpiration, however, water movement stops. The plant's cells quickly lose turgor, and the leaves wilt. This interruption in xylem flow explains the drooping foliage of a neglected houseplant (figure 21.21).

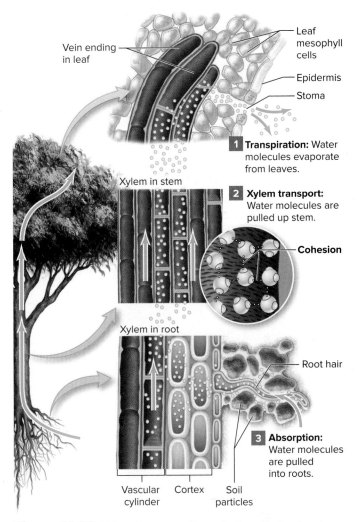

Figure 21.20 Xylem Transport. Transpiration of water from leaves pulls water up a plant's stem from the roots. The cohesiveness of water makes xylem transport possible.

Figure 21.21 Wilted. Drooping leaves indicate that a houseplant needs water.

©Image Source/Getty Images RF

Miniglossary | Plant Transport

Xylem sap	Solution of water and dissolved minerals in xylem
Cohesion–tension theory	Explanation of xylem sap movement; proposes that xylem sap ascends in a plant when water evaporating from leaves pulls water upward from the roots
Transpiration	Evaporation of water from a leaf
Phloem sap	Solution of water, sugars, minerals, and other substances in phloem
Pressure flow theory	Explanation of phloem sap movement; proposes that phloem sap moves under positive pressure from sugar sources (such as leaves) to sinks (such as flowers, fruits, and roots)

B. Sugars Are Pushed to Nonphotosynthetic Cells in Phloem

With sufficient light, water, and nutrients, a photosynthetic cell will produce sugars that can be transported in phloem to the plant's nonphotosynthetic cells, which cannot produce food on their own. The major transport structures of phloem are the microscopic sieve tubes (see figure 21.6). Unlike cells of xylem, the cells that make up sieve tubes are alive.

The organic compounds carried in phloem are dissolved in the **phloem sap,** a solution that also includes water and minerals from the xylem. The carbohydrates in phloem sap are mostly dissolved sugars such as sucrose (see figure 2.19). Phloem sap also contains amino acids, hormones, enzymes, and messenger RNA molecules. (Although phloem sap is the most common vehicle for sugar transport, it is not the only one, as Burning Question 21.3 explains.)

The explanation of phloem transport is called the **pressure flow theory,** which suggests that phloem sap moves under positive pressure from "sources" to "sinks." A **source** is any plant part that produces or releases sugars; a **sink** is any plant part that receives these sugars. Examples of sinks include flowers, fruits, shoot apical meristems, roots, and storage organs. If these cells do not receive enough sugar to generate the ATP they require, the plant may die or fail to reproduce.

Figure 21.22 illustrates the pressure flow theory. Inside a leaf or other sugar source, companion cells load sucrose into sieve tube elements by active transport (step 1). Because sucrose becomes so much more concentrated in the sieve tubes than in the adjacent xylem, water moves by osmosis out of the xylem and into the phloem sap (step 2). The resulting increased pressure drives phloem sap through the sieve tubes (step 3). ⓘ *active transport,* section 4.5B

Figure 21.22 **Phloem Transport.** Sugar produced in green "source" organs such as leaves moves under positive pressure to roots, fruits, and other nonphotosynthetic "sinks."

Photo: ©Ingram Publishing RF

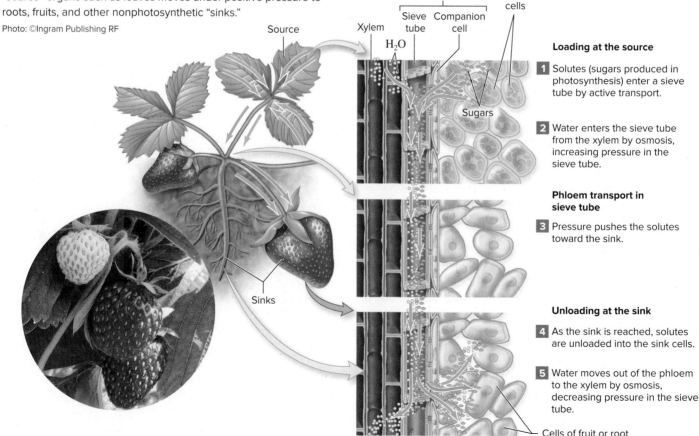

Loading at the source

1 Solutes (sugars produced in photosynthesis) enter a sieve tube by active transport.

2 Water enters the sieve tube from the xylem by osmosis, increasing pressure in the sieve tube.

Phloem transport in sieve tube

3 Pressure pushes the solutes toward the sink.

Unloading at the sink

4 As the sink is reached, solutes are unloaded into the sink cells.

5 Water moves out of the phloem to the xylem by osmosis, decreasing pressure in the sieve tube.

At a root, flower, fruit, or other sink, cells take up the sucrose (and other compounds in the phloem sap) through facilitated diffusion or active transport (step 4). As the sucrose is unloaded from the sieve tubes, the concentration of solutes in the phloem sap declines. Water therefore moves by osmosis from the sieve tube to the surrounding tissue, which is often xylem (step 5). Movement of water out of the sieve tube relieves the pressure, so the phloem sap in the sieve tube continues to flow toward the sink.

A given organ may act as either a sink or a source. For example, a developing potato tuber is a sink, storing the plant's sugars in the form of starch. Later, when the plant uses those stored reserves to fuel the growth of new tissues, that same tuber becomes a source. The starch in the tuber breaks down into simple sugars, which are loaded into phloem sap for transport to other plant parts.

A similar mobilization occurs each spring when a deciduous tree produces new stems and leaves, using carbohydrates stored in roots. Later in the growing season, leaves approach their mature size and produce sugar of their own. The leaves are then sources, and the roots are again sinks.

C. Parasitic Plants Tap into Another Plant's Vascular Tissue

Of the hundreds of thousands of plant species, most are self-sufficient. They produce their own food by photosynthesis, and they absorb their own nutrients and water from soil (often with the help of mycorrhizal fungi). Some plant species, however, are parasites that exploit the hard-won resources of other plants.

Parasitic plants acquire water, minerals, and food by tapping into the vascular tissues of their hosts. The story begins with the parasite's seeds, which are carried to the host by birds or released explosively from seed pods. Either way, a seed germinates, and the seedling secretes an adhesive that sticks the young plant to its host. The seedling's root pushes through the host's epidermis and connects the parasite's vascular tissues to those of the host.

The most common parasitic plants are the many species of mistletoe. These dark green shrubs live in the branches of host trees throughout the United States (figure 21.23). Most plants infected with mistletoe are weakened but do not die, an observation that makes sense from an evolutionary perspective. The most successful parasites extract enough resources to survive and reproduce—but not so much that the host dies. After all, a dead plant is of no use to a parasite that requires a living host.

Mistletoe plants

Figure 21.23 Mistletoe. The branches of this apple tree are heavily infested with parasitic mistletoe plants. ©Mark Boulton/Alamy Stock Photo

21.6 Mastering Concepts

1. What are the components of xylem sap?
2. Trace the path of water and dissolved minerals from soil, into the root's xylem, and up to the leaves.
3. Explain the pressure flow theory of phloem transport.
4. Distinguish between a source and a sink. How can the same plant part act as both a source and a sink?
5. How does a parasitic plant infect and harm a host?

Burning Question 21.3

Where does maple syrup come from?

The source of maple syrup is the xylem sap of the sugar maple tree. During the winter and early spring, these trees produce copious amounts of xylem sap. To harvest the sap, collectors drill a hole and insert a spout through the tree's bark and into the xylem. The xylem sap drips off the end of the spout and into a container (figure 21.B). Each tap produces about 40 liters of xylem sap each year, which boils down to about 1 liter of finished syrup.

Figure 21.B Sap Flow. The sugar maple tree produces sweet xylem sap, the precursor to maple syrup.
©Kurt Werby/All Canada Photos/Getty Images

The upward flow of this sweet fluid was once a bit of a puzzle. After all, during the sap flow period, sugar maples lack leaves, which are required for transpiration. The cohesion–tension theory, which explains most water movement in the xylem, therefore cannot apply.

Instead, the sap flow apparently results from alternating freezing and thawing of the xylem tissues. During the day, respiring cells in the stem produce CO_2. At night, compressed CO_2 bubbles are trapped in ice that forms in the xylem. When the xylem thaws during the day, the gases expand once more, pushing the sap up the tree.

Submit your burning question to marielle.hoefnagels@mheducation.com

Investigating Life 21.1 | An Army of Tiny Watchdogs

For many people, a dog is both a friend and a protector, warning of burglars and other intruders. In exchange, we provide our canine companions with shelter and food. Likewise, some plants welcome ants and other invertebrates with open arms—or, more precisely, with open stems.

In the tropics, hundreds of tree species provide ants with room and board. Some tree stems have hollow areas, called domatia, where ants live. Moreover, the young leaves of these trees secrete sugary nectar or other food that their guests eat.

Supplying nectar to ants costs energy, which a plant could otherwise put toward its own reproduction. How can natural selection allow plants to extend such hospitality to ants? Biologists in France and India teamed up to find out. The researchers monitored five trees patrolled exclusively by the tramp ant and 15 trees patrolled by other ant species. To exclude ants, they dabbed glue at the base of some leaves of each tree; control leaves remained untreated.

After 10 days, the researchers scored each leaflet as either intact or damaged. As illustrated in figure 21.C, trees patrolled by the tramp ant suffered less herbivory overall than trees harboring other ant species. In addition, leaves from which the tramp ants were excluded had more damaged leaflets.

Overall, the research team concluded that domatia improve the plants' reproductive success by housing protective ants. Defending against hungry caterpillars translates into more leaf area for photosynthesis; more food could, in turn, promote reproductive output. Natural selection has shaped the anatomy of the stem and every other part of the plant body. In this case, the unique hollowed-out stems create a home for ants—an army of tiny watchdogs that protect the plant by chasing off intruders.

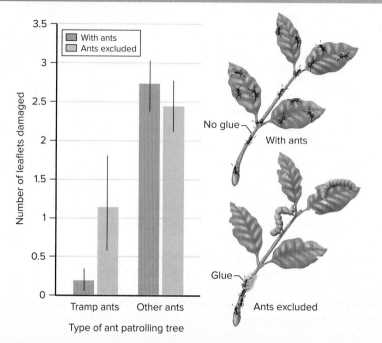

Figure 21.C **Ants on Patrol.** When tramp ants were excluded, leaves suffered extra damage from herbivores. Excluding other types of ants did not have an effect. (Error bars represent standard errors; see appendix B.)

Source: Gaume, Laurence, Merry Zacharias, Vladimir Grosbois, and Renee M. Borges. 2005. The fitness consequences of bearing domatia and having the right ant partner: experiments with protective and non-protective ants in a semi-myrmecophyte. *Oecologia,* vol. 145, pp. 76–86.

What's the **Point?** ▼ APPLIED

Plants absorb essential nutrients and assemble them into an astonishing variety of sugars, proteins, nucleic acids, fats, vitamins, and other organic molecules. Humans can therefore obtain all components of a healthy diet by eating roots, stems, leaves, flowers, and fruits. A carrot root, for example, is rich in vitamin A and potassium. An asparagus stem provides vitamin K and calcium. Dark green lettuce contains B vitamins, iron, and manganese. A handful of cabbage flower heads (broccoli) or a large orange nearly meets our daily requirements for vitamin C.

How can we maximize the nutritional content of the plants we eat? In general, fruits and vegetables are at their peak nutrient level immediately after harvest, regardless of whether the plant was organically or conventionally grown. That's because once a plant is removed from the ground, vitamins begin to break down upon exposure to heat, light, oxygen, and other stressors.

Buying local produce decreases the time between harvest and consumption. As a result, local fruits and vegetables typically have higher nutrient content than do those shipped from distant sources. Canning and freezing also deplete many vitamins, but once processing is complete, the fruits and vegetables show

©Roy Hsu/Getty Images RF

little change in nutrient content. Therefore, by the time it is consumed, frozen or canned produce may have vitamin levels similar to those in unprocessed fruits and vegetables picked weeks ago.

Food storage and preparation techniques may also affect nutrient content. Gentle handling, cool temperatures, and high humidity typically delay vitamin breakdown in harvested plants. Cutting the produce into small pieces exposes more surfaces to the vitamin-damaging conditions, so chopping should be delayed until just before food is prepared. Finally, boiling vegetables leaches water-soluble vitamins into the hot water; steaming the food minimizes this loss.

In making everyday food choices, it is tempting to ignore fruits and vegetables in favor of cheaper, more convenient processed foods. Eating fresh foods, however, is correlated with good health. Fortunately, healthy eating does not require us to obsess about complicated rules, "forbidden foods," "miracle foods," and calories. Instead, food writer Michael Pollan recommends this simple strategy for healthy eating: "Eat food, not too much, mostly plants." Regardless of how you select or prepare them, a diet rich in fruits and vegetables is one simple key to good health.

CHAPTER SUMMARY

21.1 Vegetative Plant Parts Include Stems, Leaves, and Roots

- **Anatomy** is the study of an organism's form, and **physiology** is the study of its function. A plant's body consists of **organs** composed of **tissues** with specific functions.
- The **vegetative** plant body includes a **shoot** and **roots** that depend on each other. Photosynthesis occurs in the shoot, whereas roots absorb water and dissolved minerals.
- The **stem** is the central axis of a shoot, which ends in a **terminal bud.** A stem consists of **nodes,** where **leaves** attach, and **internodes** between leaves. An **axillary bud** (also called a **lateral bud**) is located at each node.
- **Herbaceous plants** typically have soft, green stems; **woody plants** have stems and roots strengthened with wood.

21.2 Soil and Air Provide Water and Nutrients

A. Plants Require 16 Essential Elements

- Like all organisms, plants require water and **essential nutrients.**
- In all plants, the nine essential **macronutrients** are C, H, O, P, K, N, S, Ca, and Mg. The seven **micronutrients** are Cl, Fe, B, Mn, Zn, Cu, and Mo.

B. Leaves and Roots Absorb Essential Elements

- Plants obtain CO_2 and O_2 from the atmosphere, and they acquire hydrogen and oxygen atoms from H_2O in soil. The other elements also come from soil.
- Mycorrhizal fungi add to the root's surface area for absorbing water and nutrients, especially phosphorus. Several types of **nitrogen-fixing** bacteria live in root **nodules,** converting atmospheric nitrogen into forms that plants can use.

21.3 Plant Cells Build Tissues

A. Plants Have Several Cell Types

- **Parenchyma** cells are alive at maturity. They are relatively unspecialized and often function in metabolism or storage.

- **Collenchyma** cells are also alive. Their thick primary cell walls provide elastic support to growing shoots.
- **Sclerenchyma** cells are dead at maturity. Their thick secondary cell walls contain **lignin,** supporting plant parts that are no longer growing.
- Water-conducting cells in **xylem** include long, narrow **tracheids** and barrel-shaped **vessel elements.** Both cell types have thick walls and are dead when functioning. Water moves through pits in tracheids and through the end walls of vessel elements.
- Sugar-conducting cells in **phloem** include **sieve tube elements. Companion cells** help transfer carbohydrates into sieve tubes.

B. Plant Cells Form Three Main Tissue Systems

- The most abundant tissue in a herbaceous plant is **ground tissue,** which consists of relatively unspecialized parenchyma cells that fill the space between dermal and vascular tissues.
- **Dermal tissue** includes the **epidermis,** a single cell layer covering the plant. The epidermis secretes a waxy **cuticle** that coats the shoot. Gas exchange in the shoot occurs through **stomata** bounded by **guard cells.**
- **Vascular tissue** is conducting tissue. Xylem transports water and dissolved minerals from roots upward. Phloem transports dissolved carbohydrates and other substances throughout a plant. Xylem and phloem occur together with other tissues to form **vascular bundles.**

21.4 Tissues Build Stems, Leaves, and Roots

- Figure 21.24 summarizes the structural differences between monocots and eudicots.

A. Stems Support Leaves

- Vascular bundles are scattered in the ground tissue of monocot stems but form a ring of bundles in eudicot stems. Between a eudicot stem's epidermis and vascular tissue lies the **cortex,** made of ground tissue. **Pith** is ground tissue in the center of a stem.

B. Leaves Are the Primary Organs of Photosynthesis

- A stalklike **petiole** supports each leaf **blade.** A **simple leaf** has one undivided blade, and a **compound leaf** forms leaflets. **Veins** are vascular bundles in leaves; they may be in either netted or parallel formation.

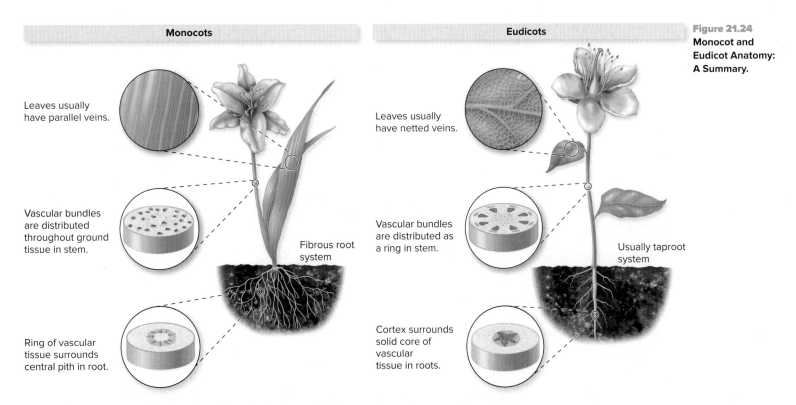

Monocots

Leaves usually have parallel veins.

Vascular bundles are distributed throughout ground tissue in stem.

Fibrous root system

Ring of vascular tissue surrounds central pith in root.

Eudicots

Leaves usually have netted veins.

Vascular bundles are distributed as a ring in stem.

Usually taproot system

Cortex surrounds solid core of vascular tissue in roots.

Figure 21.24
Monocot and Eudicot Anatomy: A Summary.

- The cells that make up a leaf's epidermis are tightly packed, transparent, and mostly nonphotosynthetic. Leaf ground tissue includes **mesophyll** cells that carry out photosynthesis. Stomata enable gas exchange.

C. Roots Absorb Water and Minerals, and Anchor the Plant

- **Fibrous root systems** consist of shallow, branched, relatively fine roots, whereas **taproot systems** have a large, persistent major root with small branches emerging from it.
- A **root cap** protects the tip of a growing root. The root cortex consists of storage parenchyma and the **endodermis.** A waxy strip surrounding the cells of the endodermis ensures that the solution entering the root's xylem first passes through the living cells of the endodermis.

21.5 Plants Have Flexible Growth Patterns, Thanks to Meristems

A. Plants Grow by Adding New Modules

- Plants with **determinate growth** stop growing when they reach their mature size; plants with **indeterminate growth** grow indefinitely.

B. Plant Growth Occurs at Meristems

- **Meristems** are localized collections of cells that retain the ability to divide throughout the life of the plant. **Apical meristems** are at the tips of shoots and roots; **lateral meristems** are cylinders of cells at the periphery of a woody stem or root.

C. In Primary Growth, Apical Meristems Lengthen Stems and Roots

- Apical meristems at the root and shoot tips provide **primary growth.** As the tip lengthens, cells arising from the apical meristems differentiate into the three tissue types.

D. In Secondary Growth, Lateral Meristems Thicken Stems and Roots

- **Secondary growth** increases the girth of the stem or root. The **vascular cambium** is a lateral meristem that produces **wood** (secondary xylem) and secondary phloem.
- Another type of lateral meristem, the **cork cambium,** produces mainly cork. The cork makes up the majority of a woody plant's **bark.**
- **Heartwood** is the central, dark-colored, nonfunctional wood in a tree. The light-colored **sapwood** transports water and minerals within a tree.
- Tree rings result from seasonal differences in the size of wood's xylem cells.

21.6 Vascular Tissue Transports Water, Minerals, and Sugar

- Vascular tissue, which consists of xylem and phloem, transports materials within plants (figure 21.25).

A. Water and Minerals Are Pulled Up to Leaves in Xylem

- **Xylem sap** consists mostly of water and dissolved minerals.
- Leaves lose water by **transpiration** through open stomata.
- According to the **cohesion–tension theory,** water molecules evaporating from leaves are replaced by those pulled up from below, a consequence of the cohesive properties of water.
- Water enters roots by osmosis and moves through the root's epidermis and cortex. The endodermis controls which substances enter the xylem.

B. Sugars Are Pushed to Nonphotosynthetic Cells in Phloem

- **Phloem sap** includes sugars, hormones, and other organic molecules, along with water and minerals from xylem.
- According to the **pressure flow theory,** phloem sap flows under positive pressure through sieve tubes from a **source** (such as a leaf) to a **sink.**
- Sources are photosynthetic or sugar-storing parts that load carbohydrates into phloem. Water follows by osmosis. The increase in pressure drives phloem sap to nonphotosynthetic sinks such as roots, flowers, and fruits.

C. Parasitic Plants Tap into Another Plant's Vascular Tissue

- Mistletoe and other parasitic plants absorb water, minerals, and sugar from a host plant's xylem and phloem.

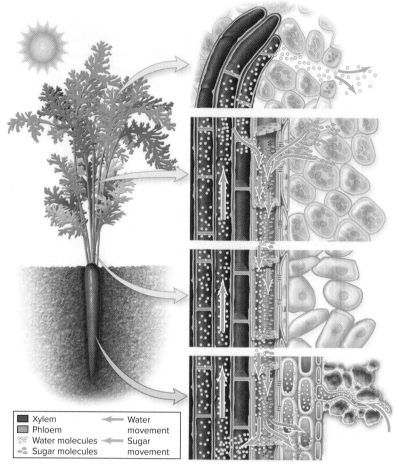

Xylem	← Water
Phloem	movement
Water molecules	← Sugar
Sugar molecules	movement

Figure 21.25 **Plant Transport: A Summary.**

MULTIPLE CHOICE QUESTIONS

1. Which of the following is NOT a vegetative organ in a plant?
   a. Stem   b. Leaf   c. Flower   d. Root

2. Which of the following is a supportive cell type with a thick secondary cell wall?
   a. Collenchyma cell   c. Guard cell
   b. Sclerenchyma cell   d. Vascular cambium cell

3. Which structures provide abundant surface area for water and mineral absorption?
   a. Leaves   c. Veins
   b. Meristems   d. Root hairs

4. The ability of a sunflower plant to become taller is directly due to its
   a. apical meristem.   c. mesophyll cells.
   b. lateral meristems.   d. taproot system.

5. Which tissue type occupies most of the volume of a tree trunk?
   a. Secondary xylem   c. Bark
   b. Secondary phloem   d. Vascular cambium

6. Which of the following structures is necessary for transpiration to occur?
   a. Sieve tubes   b. Leaves   c. Fruits   d. Roots

7. Where do the simple sugars in phloem sap originate?
   a. Soil   c. Photosynthetic tissue
   b. Fruits   d. Flowers

Answers to Multiple Choice questions are in appendix A.

WRITE IT OUT

1. When Chris mows the grass, she faces a choice between discarding the clippings and leaving them on the lawn. How would each choice influence the nutrient content of the soil? Explain your answer.

2. If left in the same pot for multiple years, a houseplant may become "root-bound," meaning that the roots grow in circles along the inner surface of the pot. Why do root-bound plants eventually show signs of nutrient deprivation?

3. Many biology labs use slides of root tips to demonstrate the stages of mitosis. Why is a root tip a better choice than a mature leaf?

4. Thorns, spines, and tendrils are so highly modified that it can be difficult to tell whether they derive from leaves or stems. How could a biologist use his or her knowledge of plant anatomy to determine their origin?

5. Explain why it might be more adaptive for a tree to produce thousands of small leaves rather than one huge leaf.

6. Mammals exchange gases in the alveoli of the lungs. How do the structures and functions of leaf mesophyll compare with those of alveoli?

7. Girdling is cutting away or severing the living bark in a ring around a tree's trunk. Which part of a girdled tree do you expect to die first, the roots or the shoot? Why? Would the tree be harmed as much by a vertical gash? Why or why not?

8. Suppose you drive a metal spike from the outermost bark layer to the center of a tree's trunk. Which tissues does your spike encounter as it moves through the stem? What type of meristem produced each type?

9. Suppose you use a rubber band to secure a clear plastic bag around a few leaves on a live plant. What do you think will happen?

10. Explain the role of cohesion in xylem transport.

11. How can phloem transport occur either with or against gravity?

12. Basil is common in vegetable gardens. Many gardeners grow this plant for its leaves, which provide flavor in sauces and other dishes. Leaf production is higher when young flowers are pruned off the plant before they have a chance to develop. Explain this observation.

SCIENTIFIC LITERACY

1. The Cork Forest Conservation Alliance is a nonprofit environmental organization dedicated to preserving the cork forests of the Mediterranean. Visit its website and then explain which is more environmentally friendly: buying wine closed with natural cork or buying wine closed with a screw cap.

2. Review Burning Question 21.2. While controlled burns may benefit a forest ecosystem, some residents living near burn areas oppose the practice. Use the Internet to research potential drawbacks of controlled burns. As a forest manager, how would you weigh the pros and cons?

Answers to Mastering Concepts, Write It Out, Scientific Literacy, and Pull It Together questions can be found in the Connect ebook.
connect.mheducation.com

Design element: Burning Question (fire background): ©Ingram Publishing/Super Stock

PULL IT TOGETHER

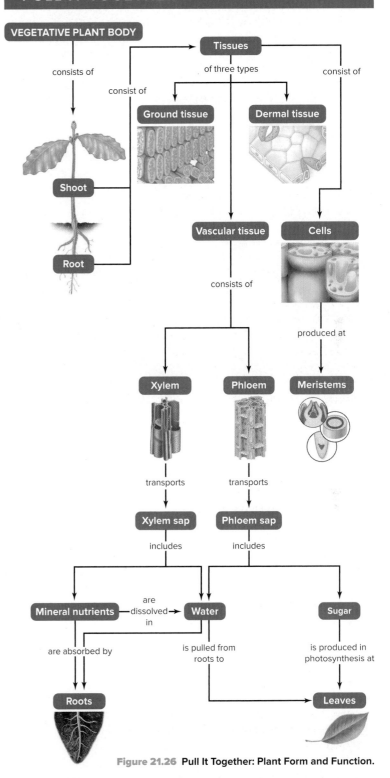

Figure 21.26 **Pull It Together: Plant Form and Function.**

Refer to figure 21.26 and the chapter content to answer the following questions.

1. Review the Survey the Landscape figure in the chapter introduction, and then add *vegetative parts*, *reproductive parts*, and *flowers* to the Pull It Together concept map. Connect *vegetative parts* and *reproductive parts* to the concept map in at least two ways each.

2. Write a phrase connecting *water* to *sugar*.

3. Add *soil*, *source*, *sink*, *pressure flow*, and *transpiration* to this map.

22 Reproduction and Development of Flowering Plants

LEARNING OUTLINE

22.1 Angiosperms Reproduce Sexually and Asexually

22.2 The Angiosperm Life Cycle Includes Flowers, Fruits, and Seeds

22.3 Plant Growth Begins with Seed Germination

22.4 Hormones Regulate Plant Growth and Development

22.5 Light Is a Powerful Influence on Plant Life

22.6 Plants Respond to Gravity and Touch

APPLICATIONS

Burning Question 22.1 *How can a fruit be seedless?*
Why We Care 22.1 *Talking Plants*
Investigating Life 22.1 *A Red Hot Chili Pepper Paradox*

Seed Eater. A bright yellow American goldfinch devours a sunflower seed. Many plant species rely on animals for seed dispersal.

©Gay Bumgarner/Alamy Stock Photo

Learn How to Learn
How to Use a Tutor

Your school may provide tutoring sessions for your class, or perhaps you have hired a private tutor. How can you make the most of this resource? First, meet regularly with your tutor for an hour or two each week; don't wait until just before an exam. Second, if possible, tell your tutor what you want to work on before each session, so he or she can prepare. Third, bring your textbook, class notes, and questions to your tutoring session. Fourth, be realistic. Your tutor can discuss difficult concepts and help you practice with the material, but don't don't expect him or her to simply give you the answers to your homework.

SURVEY THE LANDSCAPE
Plant Anatomy and Physiology

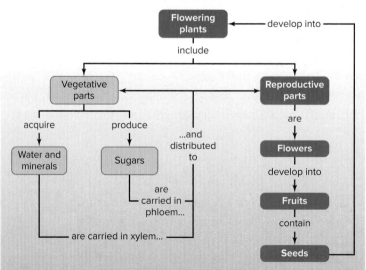

Have you ever wondered how an apple's seeds get inside the fruit? The answer traces to sexual reproduction in flowering plants. After pollination, the seeds develop inside the apple flower, while surrounding flower parts develop into the fruit.

For more details, study the Pull It Together feature in the chapter summary.

©Martin Ruegner/Photodisc/Getty Images RF

Flowers, fruits, and seeds enrich all of our lives. Humans depend on plants, so we have a strong interest in understanding plant reproduction and development.

Other animals have a stake in plant reproduction, too. The bee shown here is gathering pollen and nectar; the goldfinch in the chapter-opening photo is eating a sunflower seed. Each animal is serving its own interests, using plants for food. But at the same time, each is also contributing to a plant's reproductive success. After all, the pollen contains the plant's sperm, which the bee may deliver to another plant's egg. And the bird will knock many uneaten seeds to the ground. If the seeds are scattered any distance from the flower, the finch has helped to reduce competition between the members of the next generation of sunflower plants.

The connections between flowering plants and animals are fascinating and intricate. So are the interactions among genes, hormones, and environmental cues that enable a seed to germinate and grow. This chapter assembles some of the pieces of the complex puzzle of plant reproduction and development.

22.1 Angiosperms Reproduce Sexually and Asexually

Flowering plants (angiosperms) dominate many terrestrial landscapes. From grasslands to deciduous forests and from garden plots to large-scale agriculture, angiosperms have been extremely successful. These plants first evolved only about 144 million years ago, yet they subsequently branched into more than 250,000 species and now occupy nearly every habitat on land.

Angiosperms share the land with three other groups of plants: the mosses and their relatives; the ferns and their relatives; and the conifers and other gymnosperms (see chapter 16). But the flowering plants are by far the most diverse and widespread. Angiosperms owe their success to three adaptations. First, pollen enables sperm to fertilize an egg in the absence of free water. In contrast, the sperm cells of mosses and ferns must swim to the egg, so these plants can reproduce sexually only in moist habitats. Second, the seed protects the embryo during dormancy and nourishes the developing seedling. Third, flowers not only promote pollination but also develop into fruits that help disperse the seeds far from the parent plant.

Most angiosperms reproduce sexually. **Sexual reproduction** yields genetically variable offspring with a mix of traits derived from two parents. Scrambling genes in this way is adaptive in a changing environment. After all, a gene combination that is successful today might not work in the future if selective pressures change. Producing variable offspring improves reproductive success in an uncertain world. ⓘ *why sex?*, section 9.1

Many plants also reproduce asexually, forming new individuals by mitotic cell division. In **asexual reproduction,** a parent organism produces offspring that are genetically identical to it and to each other—they are clones. Asexual reproduction, also called vegetative reproduction, is advantageous when conditions are stable and plants are well-adapted to their surroundings; the clones will be equally suited to the same environment.

Plants often reproduce asexually by forming new plants from portions of their roots, stems, or leaves (figure 22.1). The roots of quaking aspen trees, for example, can sprout identical aerial shoots called "suckers." Suckers can also grow upward from buds on the roots of cherry, pear, apple, and black locust trees. If these shoots are cut or broken away from the parent plant, they can become new individuals.

The rest of this chapter begins with a look at sexual reproduction in flowering plants. As you will see, the plant packages its offspring inside a seed. When the seed germinates, the young plant faces a host of challenges. The second half of this chapter explores the hormonal signals and environmental factors that influence the development of the angiosperm throughout its life.

Leaf   Plantlets

a.   b.   5 mm

Figure 22.1 **Asexual Reproduction.** (a) Quaking aspen trees are clones connected by a common root system. (b) The leaf of this kalanchöe plant is producing genetically identical plantlets.
(a, b): ©Steven P. Lynch RF

22.1 Mastering Concepts

1. What adaptations contribute to the reproductive success of angiosperms?

2. When are sexual and asexual reproduction each adaptive?

3. What are some examples of asexual reproduction in plants?

22.2 The Angiosperm Life Cycle Includes Flowers, Fruits, and Seeds

When humans reproduce, sexual intercourse brings sperm to an egg cell, and the embryo develops into a fetus (see chapter 30). Childbirth separates woman from baby. Clearly, flowering plant sexual reproduction is different from our own. How does the sperm get to the egg in angiosperms? How does the angiosperm embryo develop (along with surrounding tissues) into a seed? How do the seeds separate from the "mother" plant?

This section explains sexual reproduction in flowering plants. Figure 22.2 summarizes the life cycle; you may find it helpful to refer to this illustration frequently as you study this section. As you orient yourself to the figure, look first for the two adaptations that are unique to the angiosperms: flowers and fruits. **Flowers** are reproductive organs where eggs and sperm unite. Parts of the flower develop into a **fruit,** which protects and disperses the seeds.

A closer look reveals that the angiosperm life cycle—like that of all other plants—includes an **alternation of generations** with multicellular diploid and

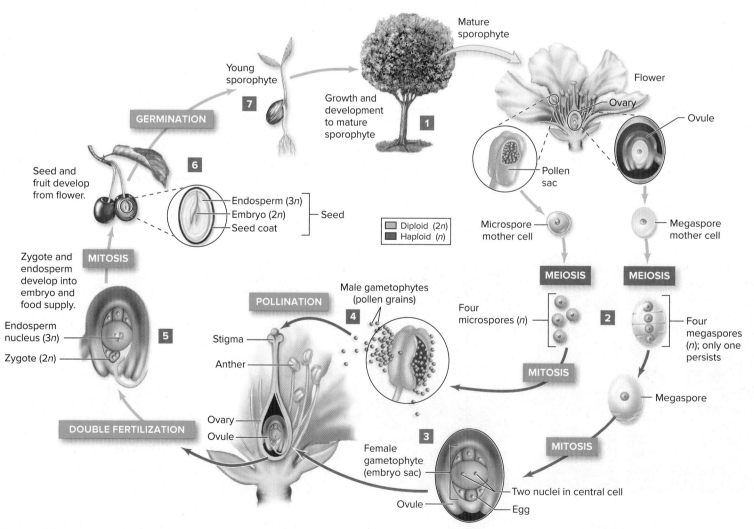

Figure 22.2 Flowering Plant Life Cycle. (1) Cells in a flower undergo (2) meiosis to produce microspores and megaspores, which (3, 4) develop into the gametophytes that produce sperm and egg cells. (5) Fertilization yields the diploid zygote and triploid endosperm. (6) The resulting seeds are enclosed in fruits. (7) Seed germination reveals the young sporophyte.

haploid stages (see figure 16.5). The **sporophyte,** or diploid generation, produces haploid **spores** by meiosis; in figure 22.2, the sporophyte is the tree at top center. Inside the tree's flowers, each spore divides mitotically to produce a multicellular haploid **gametophyte,** which undergoes mitosis to generate haploid **gametes** (egg cells or sperm). These tiny gametophytes appear at the bottom of figure 22.2. In **fertilization,** gametes fuse to form a diploid **zygote.** The zygote develops into an embryo, which is packaged inside a seed (which is itself within a fruit). With additional growth, the embryo becomes a mature sporophyte, and the cycle begins anew.

A. Flowers Are Reproductive Organs

We begin our exploration of angiosperm reproduction with the formation of flowers on the mature sporophyte (see figure 22.2, step 1). Figure 22.3 shows the anatomy of a typical flower—in this case, a cherry blossom. A part of the floral stalk called the **receptacle** is the attachment point for four types of structures, all of which are modified leaves. The outermost whorl (ring of parts) consists of **sepals,** which are leaflike structures that enclose and protect the inner floral parts. Next is a whorl of **petals,** which often (but not always) have bright colors that attract pollinators. The sepals and petals do not play a direct role in sexual reproduction.

The two innermost whorls of a flower, however, are essential for sexual reproduction. The male flower parts consist of **stamens,** which are filaments that bear pollen-producing bodies called **anthers** at their tips. The female part at the center of a flower is composed of one or more **carpels.** The base of each carpel is called an **ovary,** and it encloses one or more egg-bearing **ovules.** The upper part of each carpel is a stalklike **style** that bears a structure called a **stigma** at its tip. Stigmas receive pollen.

The flower in figure 22.3 is "complete" because it includes all four whorls, including both male and female parts. In some species, however, each flower has either male or female parts but not both. An individual plant may have both types of single-sex flowers, or there may be separate male and female plants. A holly plant, for example, is either male or female; only the female plants produce the distinctive red berries.

Recall from chapters 16 and 21 that the two largest clades of angiosperms are monocots and eudicots. Flower structure is one feature that distinguishes the two groups. Most monocots, such as lilies and tulips, have petals, stamens, and other flower parts in multiples of three. Most eudicots, on the other hand, have flower parts in multiples of four or five. Buttercups and geraniums are examples of eudicots with five prominent petals on each flower.

B. The Pollen Grain and Embryo Sac Are Gametophytes

Once the flowers have formed, the next step is to produce the microscopic male and female gametophytes (see figure 22.2, steps 2–4). Inside the anther's pollen sacs, diploid cells divide by meiosis to produce four haploid **microspores.** Each microspore then divides mitotically and produces a two-celled, thick-walled structure called a **pollen** grain, which is the young male gametophyte. One of the haploid cells inside the pollen grain divides by mitosis to form two sperm nuclei.

Meanwhile, meiosis also occurs in the female flower parts. The ovary may contain one or more ovules, each containing a diploid cell that divides by meiosis to produce four haploid **megaspores.** In many species, three of these cells quickly disintegrate, leaving one large megaspore. The megaspore undergoes three mitotic divisions to form the **embryo sac,** which is a female

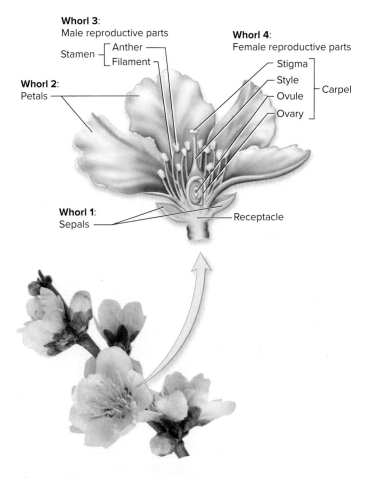

Figure 22.3 Parts of a Flower. A cross section of a complete flower reveals four whorls: sepals, petals, stamens (male reproductive parts), and one or more carpels (female reproductive parts).

Photo: ©Burke/Triolo Productions/Getty Images RF

Miniglossary	Alternation of Generations in Angiosperms
Alternation of generations	Life cycle featuring multicellular diploid (sporophyte) and haploid (gametophyte) stages
Sporophyte	Diploid stage of the plant life cycle; in angiosperms, the sporophyte produces flowers
Spores	Haploid cells produced by meiotic cell division within a flower; microspores (male) and megaspores (female) eventually develop into gametophytes
Gametophyte	Haploid stage of the plant life cycle, originating from a single spore; the male gametophyte is a pollen grain, and the female gametophyte is an embryo sac
Gametes	Sperm or egg cells produced by mitotic cell division in the male or female gametophyte
Zygote	Product of fertilization; first cell of the sporophyte generation

a.

b.

c.

d.

e.

f.

Figure 22.4 **Pollination.** Animal pollinators include
(a) hummingbirds, (b) butterflies, and (c) bats. (d) These yellow petals
appear uniform to our eyes, but (e) the flower actually has distinctive
markings that attract insects and are visible only in ultraviolet light.
(f) Birch trees are wind-pollinated.

(a): ©Angela Arenal/iStockphoto RF; (b): ©MedioImages/Getty Images RF; (c): ©Merlin
D. Tuttle/Bat Conservation International/Science Source; (d, e): ©Leonard Lessin/
Science Source; (f): ©Dr. Jeremy Burgess/Science Source

gametophyte. At first, the embryo sac contains eight haploid nuclei. Cell walls soon form, dividing the female gametophyte into seven cells. One of these cells is the egg. In addition, a large, central cell contains a pair of nuclei; as we shall see, both the egg and the central cell's two nuclei participate in fertilization.

C. Pollination Brings Pollen to the Stigma

Eventually, the pollen sac opens and releases millions of pollen grains. The next step is **pollination:** the transfer of pollen from an anther to a receptive stigma. Usually, either animals or wind carry the pollen (figure 22.4).

Flower color, shape, and odor attract animal pollinators. For example, hummingbirds are attracted to red, tubular flowers. Beetles visit dull-colored flowers with spicy scents, whereas blue or yellow sweet-smelling blooms attract bees. Bee-pollinated flowers often have markings that are visible only at ultraviolet wavelengths of light, which bees can perceive. Butterflies prefer red or purple flowers with wide landing pads. Moths and bats pollinate white or yellow, heavily scented flowers, which are easy to locate at night.

Many animals benefit from their association with plants: They obtain food in the form of pollen or sugary nectar, seek shelter among the petals, or use the flower for a mating ground (see Investigating Life 6.1). Sometimes, the connection between plant and pollinator is so strong that a genetic change in one partner directly alters the selective pressure on the other, and vice versa; this reciprocal selective pressure is called **coevolution.** For example, some hummingbirds have long, curved bills that fit precisely into the tubular flowers from which they sip nectar. Any modification in flower shape selects for corresponding changes in birds; likewise, altered bill shapes select for changes in flowers. ⓘ *coevolution,* section 19.4D

About 10% of angiosperms (and most gymnosperms) use wind, not animals, to carry pollen. Wind-pollinated flowers are small and odorless, and their petals are typically reduced or absent; perfume, nectar, and showy flowers are not necessary for wind to disperse pollen. One advantage of wind pollination is that the plant does not spend energy on nectar or other lures. On the other hand, an animal delivers pollen directly to another plant, whereas the wind is "wasteful." That is, wind-blown pollen may land on the ground, on water, or on the wrong plant species. Wind-pollinated plants therefore manufacture abundant pollen, an adaptation that compensates for this inefficiency. The large quantities of pollen produced by oaks, cottonwoods, ragweed, and grasses provoke allergies in many people. ⓘ *allergies,* section 29.5C

D. Double Fertilization Yields Zygote and Endosperm

After a pollen grain lands on a stigma of the correct species, a pollen tube emerges (figure 22.5, step 1). The pollen grain's two haploid sperm nuclei enter the pollen tube as it grows through the tissue of the style toward the ovary (figure 22.5, step 2). When the pollen tube reaches an ovule, it discharges its two sperm nuclei into the embryo sac.

Then, in **double fertilization,** the sperm nuclei fertilize the egg and the central cell's two nuclei (figure 22.5, step 3; see also figure 22.2, step 5). That is, one sperm nucleus fuses with the haploid egg nucleus and forms a diploid zygote, which will develop into the embryo. The second sperm nucleus fuses with the haploid nuclei in the central cell. The resulting triploid nucleus divides to form a tissue called **endosperm,** which is composed of parenchyma cells that

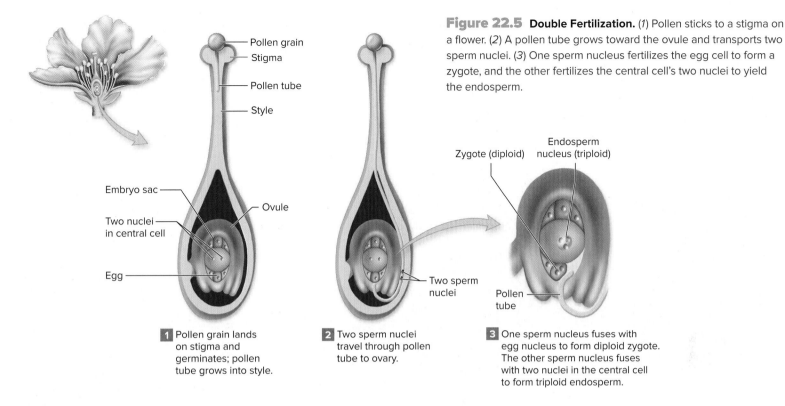

Figure 22.5 Double Fertilization. (*1*) Pollen sticks to a stigma on a flower. (*2*) A pollen tube grows toward the ovule and transports two sperm nuclei. (*3*) One sperm nucleus fertilizes the egg cell to form a zygote, and the other fertilizes the central cell's two nuclei to yield the endosperm.

1 Pollen grain lands on stigma and germinates; pollen tube grows into style.

2 Two sperm nuclei travel through pollen tube to ovary.

3 One sperm nucleus fuses with egg nucleus to form diploid zygote. The other sperm nucleus fuses with two nuclei in the central cell to form triploid endosperm.

store food for the developing embryo. Familiar endosperms are the "milk" and "meat" of a coconut and the starchy part of a rice grain. The starchy endosperm of corn is an important source of not only food but also ethanol, a biofuel (see Burning Question 16.2).

Double fertilization reduces the energetic cost of reproduction. In gymnosperms, which lack this adaptation, the female gametophyte stockpiles food for the embryo in advance of fertilization. The investment is wasted if no zygote ever forms. In contrast, double fertilization saves energy because the angiosperm produces food for the embryo only if a sperm nucleus actually fertilizes the egg.

E. A Seed Is an Embryo and Its Food Supply Inside a Seed Coat

Immediately after fertilization, the ovule contains an embryo sac with a diploid zygote and a triploid endosperm. The ovule eventually develops into a **seed:** a plant embryo together with its stored food, surrounded by a seed coat (see figure 22.2, step 6). Where do these parts come from?

The zygote divides to form the embryo (figure 22.6). Among the first features of the developing embryo are the **cotyledons,** or seed leaves. (The cotyledons are called "seed leaves" because in many species they emerge from the soil with the seedling and carry out photosynthesis for a short time. But they are not true leaves.) Soon, the shoot and root apical meristems form at opposite ends of the embryo.

What about the embryo's food supply? Inside the developing seed, the endosperm cells divide more rapidly than the zygote and thus form a large multicellular mass. This endosperm supplies nutrients to the developing embryo, although the timing depends on the type of plant. In many eudicots, the paired cotyledons absorb the endosperm during seed development; when the seed begins to germinate, the cotyledons transfer the

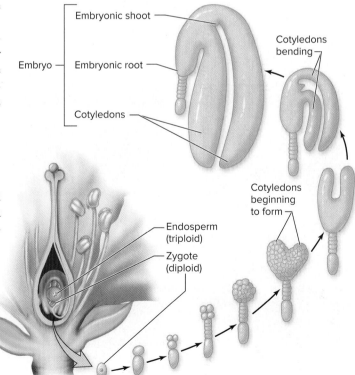

Figure 22.6 From Zygote to Embryo. As a seed develops inside a flower, the zygote divides repeatedly to form the tiny embryonic plant.

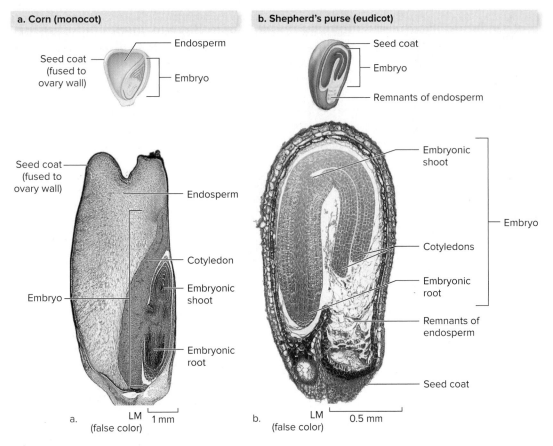

a. Corn (monocot)

Endosperm

Seed coat (fused to ovary wall)

Embryo

Seed coat (fused to ovary wall)

Endosperm

Embryo

Cotyledon

Embryonic shoot

Embryonic root

a. LM 1 mm (false color)

b. Shepherd's purse (eudicot)

Seed coat

Embryo

Remnants of endosperm

Embryonic shoot

Embryo

Cotyledons

Embryonic root

Remnants of endosperm

Seed coat

b. LM 0.5 mm (false color)

Figure 22.7 **Mature Seeds.** (a) Corn is a monocot. A grain of corn is a fruit containing one seed, including a starchy endosperm and an embryo with one cotyledon. (b) In a eudicot, the two cotyledons may absorb much of the endosperm as the seed develops.

Photos: (a, b): ©Steven P. Lynch RF

Pollination occurs.

Petals are shed.

Fruit protects and disperses seeds.

Ovary and receptacle swell as seeds develop.

Figure 22.8 **Development of a Fruit.** After pollination and fertilization, the apple tree's flower begins to develop into a fruit.

(flowers): ©Nigel Cattlin/Science Source; (petal drop): ©Harry Rogers/Science Source; (swelling): ©Bhandol/Alamy Stock Photo; (apples): ©moonlightbgd/Shutterstock RF

stored food to the rest of the embryo. In monocots, the mature seed retains the endosperm. During germination, the stored nutrients pass from the endosperm through the cotyledon's vascular tissue to the embryonic shoot and root. Figure 22.7 illustrates these differences between the seeds of monocots and eudicots.

As the embryo and endosperm develop, the seed coat begins to form. In many species, the **seed coat** is a tough outer layer that protects both the embryo and its food supply from damage and hungry animals. In other plants, including corn, the thin seed coat fuses with the ovary wall as the seed matures.

At some point in seed development, hormonal signals "tell" cells in the embryo and endosperm to stop dividing, and the seed gradually loses moisture and enters dormancy. The ripe, mature seeds are firm, dry, and ready for dispersal. Depending on the species, the dormant period can last for days, weeks, months, years, decades, or even centuries.

Seed dormancy is a crucial adaptation because it ensures that seeds have time to disperse away from the parent plant before germinating. Moreover, dormancy enables seeds to postpone development if the environment is unfavorable, such as during a drought or frost. Favorable conditions trigger embryo growth to resume when young plants are more likely to survive.

Producing seeds is costly: The parent plant uses precious sugars, lipids, and other organic molecules to produce both the embryo and its stored food supply. The seed continues to consume its parent's resources until it enters dormancy. From that time until the young seedling begins carrying out photosynthesis on its own, however, the only energy available to the embryo is the fuel stored inside the seed.

F. The Fruit Develops from the Ovary

The rest of the flower also changes as the seeds develop (see figure 22.2, step 6). When a pollen tube begins growing, the stigma produces ethylene, a hormone that stimulates the stamens and petals to wither and drop; these parts are no longer needed. Developing seeds also produce another hormone, auxin, which triggers fruit formation.

In many angiosperms, the ovary grows rapidly to form the fruit, which may contain one or more seeds. (Burning Question 22.1 explores how seedless fruits develop.) In some species, additional plant parts also contribute to fruit development. The pulp of an apple, for example, derives from a cup-shaped region of the receptacle. The apple's core is derived from the carpel walls, which enclose the seeds. Figure 22.8 shows how the parts of an apple flower give rise to the fruit.

Fruits come in many forms (table 22.1). A simple fruit develops from a flower with one ovary. The fruit may have one seed, as in a cherry, or many seeds, as in a tomato. An aggregate fruit develops from one flower with many carpels. Strawberries and raspberries are examples of aggregate fruits. A multiple fruit develops from clusters of flowers that fuse into a single fruit as they mature. Pineapples and figs are multiple fruits.

Flowers never form on roots, so it may seem surprising that some fruits develop underground. The yellow flowers of peanut plants, for example, form on the shoot. After fertilization, the petals wither, and the young fruit produces a peg that turns downward and buries itself in the soil. Three to five months later, farmers dig up the plants to harvest the mature fruits. Each fruit consists of a fibrous shell enclosing one to three peanuts—the seeds.

G. Fruits Protect and Disperse Seeds

Fruits have two main functions, one of which is the protection of the seeds. Many developing fruits contain chemicals that animals find distasteful. Anyone who has ever bitten into an unripe apple, plum, or tomato can testify to the hard texture and intensely sour flavor. Animals avoid these unappealing, unripe fruits, so the immature seeds inside remain safe from herbivory. Once the seeds are mature, however, the fruit changes: It becomes soft, sweet, and appealing to animals.

These changes relate to the second function of fruits: to promote seed dispersal by animals, wind, and water (figure 22.9). Regardless of the transportation mode, the seeds are often deposited far from the parent plant, promoting reproductive success by minimizing competition between parent and offspring.

Many animals disperse fruits and seeds. Colored berries attract birds and other animals that carry the ingested seeds to new locations, only to release them in their droppings. Birds and mammals spread seeds when spiny fruits attach to their feathers or fur. Squirrels and other nut-hoarding animals also disperse seeds. Although they later eat many of the seeds they hide, they also forget some of their cache locations. The uneaten seeds may germinate.

Wind and water can also distribute seeds. Wind-dispersed fruits, such as those of dandelions and maples, have wings or other structures that catch air currents. Water-dispersed fruits include gourds and coconuts, which may drift across the ocean before colonizing distant lands.

22.2 Mastering Concepts

1. Label as many flower parts as possible in the photo of cherry blossoms in figure 22.3.
2. How does pollen move from one flower to another, and why is this process essential for sexual reproduction?
3. Describe the difference between pollination and seed dispersal.
4. How does endosperm form, and how is it important to plants?
5. What are the components of a seed?
6. Which flower parts develop into a fruit?
7. What are the two main functions of fruits?

TABLE 22.1	**Types of Fruits: A Summary**	
Fruit Type	**Characteristics**	**Examples**
Simple	Derived from one flower with one ovary	Olive, cherry, peach, plum, coconut, grape, tomato, pepper, eggplant, apple, pear
Aggregate	Derived from one flower with many separate ovaries	Blackberry, strawberry, raspberry, magnolia
Multiple	Derived from tightly clustered flowers whose ovaries fuse as fruit develops	Pineapples, figs

(cherry and pineapple): ©Ingram Publishing/Alamy Stock Photo RF; (strawberry): ©Corbis RF

a.   b.

c.   d.

Figure 22.9 **Seed Dispersal.** (a) A mockingbird helps disperse the seeds of possumhaw, a type of holly. (b) Hooks on the surface of the burdock fruit easily attach to the fur of a passing animal. (c) Dandelion fruits have fluff that enables them to float on a breeze. (d) Coconuts are water-dispersed fruits.

(a): ©Bill Draker/Getty Images RF; (b): ©Scott Camazine/Science Source; (c): ©Dimitri Vervitsiotis/Getty Images RF; (d): ©Ethan Daniels/Shutterstock RF

22.3 Plant Growth Begins with Seed Germination

If a seed arrives at a suitable habitat, it may germinate, beginning the life of an independent young plant (see figure 22.2, step 7). **Germination** is the resumption of growth and development after a period of seed dormancy. It usually requires water, oxygen, and a favorable temperature. First, the seed absorbs water. The incoming water swells the seed, rupturing the seed coat and exposing the plant embryo to oxygen. Water also may cause the embryo to release hormones that stimulate the production of starch-digesting enzymes (see section 22.4B). The stored starch in the seed breaks down to sugars, providing energy for the now-growing embryo.

Growth and development continue after the seed coat ruptures (figure 22.10). The root emerges first from the germinating seed, and then the shoot begins to elongate. Rapidly dividing cells in apical meristems continually add length to both roots and shoots (see section 21.5). The new cells differentiate into the ground tissue, vascular tissue, and dermal tissue that make up the plant body. ⓘ *plant tissue types,* section 21.3B

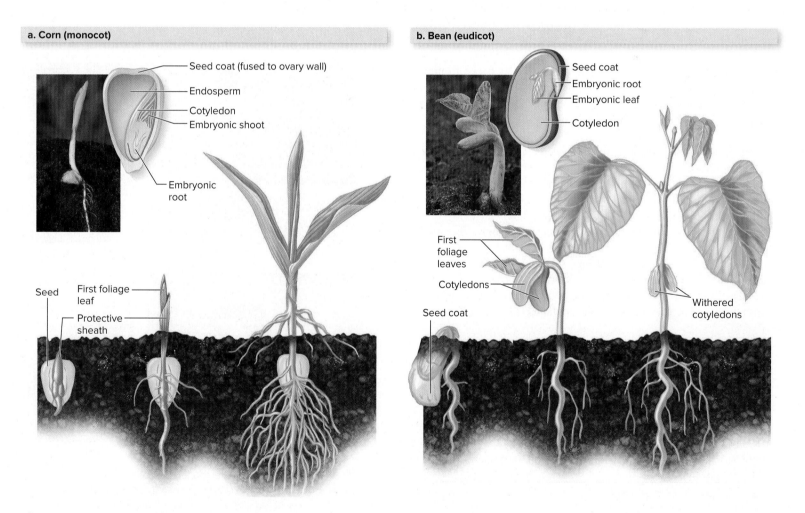

a. Corn (monocot)

Seed coat (fused to ovary wall)
Endosperm
Cotyledon
Embryonic shoot
Embryonic root

Seed
First foliage leaf
Protective sheath

b. Bean (eudicot)

Seed coat
Embryonic root
Embryonic leaf
Cotyledon

First foliage leaves
Cotyledons
Seed coat
Withered cotyledons

Figure 22.10 **Seed Germination and Early Seedling Development.** (a) In a monocot such as corn, a protective sheath covers the shoot until the first foliage leaves develop. The cotyledon remains belowground. (b) In green beans and some other eudicots, the cotyledons emerge from the ground. They carry out photosynthesis until the first foliage leaves develop from the embryonic leaves.

Photos: (a, b): ©Ed Reschke/Photolibrary/Getty Images

The seedling soon begins to take on its mature form. Young roots grow downward in response to gravity, anchoring the plant in the soil and absorbing water and minerals. The shoot produces leaves as it grows upward toward the light. Initially, the energy source for the seedling's growth is stored food inside the seed. By the time the seedling has depleted these reserves, the new green leaves should begin producing food by photosynthesis. But if a seed is buried too deep in the soil, the young plant will never emerge—in effect, it will starve to death before reaching the light.

The size of a plant's seeds therefore reflects an evolutionary trade-off. Large, heavy seeds contain ample nutrient reserves to fuel seedling growth but may not travel far. Small seeds, on the other hand, store limited nutrients but tend to disperse far and wide. Interestingly, the crops that humans cultivate typically have larger seeds than do their wild ancestors. We gather and plant the seeds ourselves, removing the selection pressure favoring small seed size.

Depending on the species, a plant may keep growing for weeks, months, years, decades, or even centuries. When the plant reaches reproductive maturity, it too will develop flowers, seeds, and fruits, continuing the life cycle.

22.3 Mastering Concepts

1. Why must seeds absorb water before germinating?
2. What are the events of early seedling development?
3. How does natural selection influence seed size?

22.4 Hormones Regulate Plant Growth and Development

A plant's responses to environmental stimuli usually seem much more subtle than those of an animal. Plants cannot hide, bite, or flee; instead they must adjust their growth and physiology to external conditions. The rest of this chapter explores some of the ways in which a plant responds to the changing environment as it grows and develops.

The environment affects plant growth in many ways. Shoots grow up, toward light and against gravity; roots grow down. Many plants leaf out in the spring, produce flowers and fruits, then enter dormancy in autumn—all in response to seasonal changes. Other responses are immediate. When the weather is hot, plants reduce transpiration by closing their stomata. A Venus flytrap snaps shut when a fly wanders across a leaf. Plants may even send signals to one another, warning of such dangers as attacks by insects and diseases (see Why We Care 22.1).

Chemicals called hormones regulate many aspects of plant growth, flower and fruit development, and responses to environmental change. A **hormone** is a biochemical produced in one part of an organism and transported to another location, where it triggers a response from target cells (see also chapter 25). A plant's hormones may either diffuse from cell to cell or move larger distances by entering xylem or phloem. Either way, when a hormone reaches a target cell, it binds to a receptor protein. This interaction begins a cascade of chemical reactions that ultimately change the expression of genes in target cells.

The "classic five" plant hormones are auxins, cytokinins, gibberellins, ethylene, and abscisic acid (table 22.2). A plant must produce auxins and cytokinins if it is to develop at all. Both of these hormones occur in all major

Burning Question 22.1

How can a fruit be seedless?

©Deborah Jaffe/Getty Images RF

Given the role of seeds and fruits in plant reproduction, seedless fruits might seem puzzling. After all, from a plant's point of view, what's the point of producing a fruit without seeds inside? Natural selection clearly would not favor such a trait in the wild! But humans often consider seeds a nuisance and have selected for seedless varieties.

Seedless fruits can form in two ways. Often, the fruit develops in the absence of fertilization. Seedless oranges and watermelons are two examples. Alternatively, if fertilization does occur, the embryos may die during development. Tiny, immature seeds remain inside the fruit, but they do not interfere with eating. Seedless grapes and bananas illustrate this second path to seedlessness.

Since seedless fruits lack seeds, how do farmers grow more of them? Most are produced asexually by grafting or taking cuttings. But, surprisingly, seedless watermelons do come from seeds! To make "seedless watermelon seeds," plant breeders first cross a regular diploid watermelon plant with a tetraploid plant (with four sets of chromosomes). Farmers then sow the triploid seeds arising from this union. The new triploid plants produce flowers, which the grower pollinates with diploid pollen. Pollination stimulates fruit production, but, like a mule, the triploid plant is sterile. The "seeds" that do develop inside seedless watermelons are actually empty, soft hulls that are easy to chew and swallow.

**Submit your burning question to
marielle.hoefnagels@mheducation.com**

TABLE 22.2 Five Plant Hormones: A Summary

Class	Selected Actions
Auxins	• Stimulate elongation of cells in stems and fruits • Control phototropism, gravitropism, thigmotropism • Stimulate growth of roots from stem cuttings • Suppress growth of lateral buds in shoots (apical dominance)
Cytokinins	• Stimulate cell division in seeds, roots, young leaves, fruits • Delay shedding of leaves • Stimulate growth of lateral buds
Gibberellins	• Stimulate cell division and elongation in roots, shoots, young leaves • Break seed dormancy
Ethylene	• Hastens fruit ripening • Stimulates shedding of leaves, flowers, and fruits • Participates in thigmotropism
Abscisic acid	• Inhibits shoot growth and maintains bud dormancy • Induces and maintains seed dormancy • Stimulates closure of stomata

Why We Care 22.1 | Talking Plants

If you have ever tried to grow a garden, you know that ravenous insect larvae—including caterpillars—can quickly devour a plant's leaves. But plants may not be as passive as they seem. Their hidden defenses include a biochemical arsenal that punishes the offending caterpillar, warns the neighboring plants, and summons the insect's enemies. The plants "talk" by using chemical chatter, not sounds.

Imagine a caterpillar munching on a leaf of a tomato plant. The plant's cells release multiple hormones in response to the injury. One such hormone, systemin, travels in phloem throughout the plant and stimulates the release of another hormone, jasmonic acid. Among other actions,

Painted lady caterpillar
©Steven P. Lynch RF

this hormone triggers the synthesis of chemicals that destroy a caterpillar's digestive enzymes. Plus, jasmonic acid forms a gas that prompts the injured plant's neighbors to strengthen their own defenses against herbivores. The jasmonic acid message lures caterpillar-killing wasps to the scene, too.

Understanding the language of plants is especially important to agriculture. Hungry insects constantly attack the plants we grow for food. Learning to boost a plant's own defenses may mean less reliance on pesticides in the future.

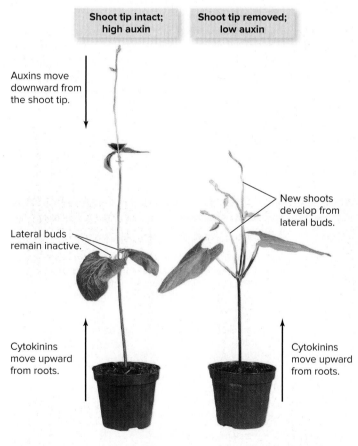

Shoot tip intact; high auxin

Shoot tip removed; low auxin

Auxins move downward from the shoot tip.

New shoots develop from lateral buds.

Lateral buds remain inactive.

Cytokinins move upward from roots.

Cytokinins move upward from roots.

Figure 22.11 Apical Dominance. The plant on the right had its shoot tip removed, promoting the growth of lateral buds.

Photo: ©Nigel Cattlin/Alamy Stock Photo

organs of all plants at all times, and biologists have never found a mutant plant lacking either one. Other hormones are required for normal development, but plants can complete their life cycles without them.

A. Auxins and Cytokinins Are Essential for Plant Growth

Auxins (from the Greek meaning "to grow") are hormones that promote cell elongation in stems and fruits but have the opposite effect in roots. Auxins also control plant responses to light and gravity (see sections 22.5 and 22.6).

Auxins have commercial uses. These hormones stimulate the growth of roots from cuttings, which is important in the asexual production of plants. A synthetic auxin called 2,4-D (2,4-dichlorophenoxyacetic acid) is a widely used herbicide, although the mechanism by which this compound kills plants is unclear. (Learn about other weed killers in Why We Care 5.1.)

Cytokinins earned their name because they stimulate cytokinesis, or cell division. In flowering plants, most cytokinins affect roots and developing organs such as seeds, fruits, and young leaves. Cytokinins also slow the aging of mature leaves, so these hormones are used to extend the shelf lives of leafy vegetables.

The actions of cytokinins and auxins compete with each other. Cytokinins are more concentrated in the roots, whereas auxins are more concentrated in shoot tips. Cytokinins move upward within the xylem and stimulate lateral bud sprouting. In a counteracting effect called **apical dominance,** the terminal bud of a plant secretes auxins that move downward and suppress the growth of lateral buds. If the shoot tip is removed, the concentration of auxins in lateral buds decreases. Meristem cells in the buds then begin dividing, thanks to the ever-present cytokinins. Apical dominance explains why gardeners can promote bushier growth by pinching off a plant's shoot tip (figure 22.11).

B. Gibberellins, Ethylene, and Abscisic Acid Influence Plant Development in Many Ways

Gibberellins are another class of plant hormone that causes shoot elongation. Young shoots produce gibberellins, which stimulate both cell division and cell elongation (figure 22.12). Farmers therefore use these hormones in agriculture to stimulate stem elongation and fruit growth in seedless grapes. But gibberellins also have other functions. For example, they trigger seed germination by inducing the production of enzymes that digest starch in the seed.

Ethylene is a gaseous hormone that ripens fruit in many species. Ethylene released from one overripe apple can hasten the ripening, and eventual spoiling, of others nearby, leading to the expression "one bad apple spoils the bushel." Exposure to ethylene also ripens immature fruits after harvest. For example, shipping can damage soft, vine-ripened tomatoes. Farmers therefore pick the fruit while it is still hard and green. Ethylene treatment just before distribution to supermarkets yields ripe-looking (if not good-tasting) tomatoes.

All parts of flowering plants synthesize ethylene, especially the shoot apical meristem, nodes, flowers, and ripening fruits. Like other hormones, ethylene has several effects. In most species, it causes petals and leaves to fade and wither. This effect was noticed in Germany in 1864, when ethylene in a mixture of gases in street lamps caused nearby trees to lose their leaves. In addition, a damaged plant part produces ethylene, which hastens aging; the plant then sheds the affected part before the problem spreads.

A fifth plant hormone, **abscisic acid** (abbreviated ABA), counters the growth-stimulating effects of many other hormones. Stresses such as frost and drought stimulate the production of ABA. One role of this hormone is to induce bud dormancy during winter (figure 22.13). When moisture is scarce, another effect of ABA is to trigger stomata to close, which helps plants conserve water. ABA also inhibits seed germination, opposing the effects of gibberellins. Commercial growers apply ABA to inhibit the growth of nursery plants so that shipping is less likely to damage them.

Researchers once recognized only the five types of plant hormones listed in table 22.2. We now know that plants produce several additional hormones. One example is florigen, a protein that induces a shoot apical meristem to produce flowers. Another is salicylic acid. This molecule, familiar to most people as aspirin, helps a plant defend against viruses and other disease-causing agents. When a plant's cells detect an attack, they release salicylic acid, which induces the surrounding cells to die. This response keeps the pathogen from spreading to additional tissue, and the accumulation of salicylic acid makes the entire plant resistant to a wide variety of pathogens.

22.4 Mastering Concepts

1. What is a hormone?

2. How does a plant hormone exert its effects?

3. List the major classes of plant hormones, and name some of their functions.

4. Give an example of how plant hormones interact.

Figure 22.12 Gibberellins and Shoot Elongation. When gibberellins were applied to these cabbage plants, elongated stems developed in place of compact heads. Untreated cabbage plants are on the ground next to the ladder.
©Al Fenn/The Life Picture Collection/Getty Images

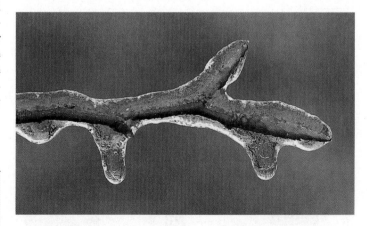

Figure 22.13 Dormancy. As winter approaches, shoot tips produce ABA, preparing the plant for dormancy during the cold season. Among many other effects, ABA triggers the production of the protective, light brown scales covering the buds at this branch tip.
©L. West/Science Source

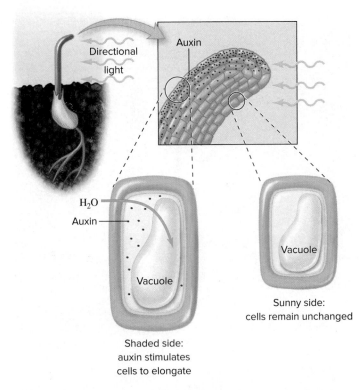

Figure 22.14 Auxin and Cell Elongation. Phototropism occurs because auxins (*red dots*) move to the shaded side of a shoot, stimulating elongation of the affected cells.

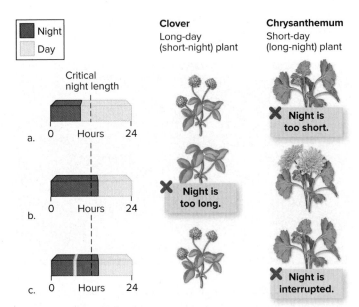

Figure 22.15 Night Length Matters. (a) When nights are shorter than a critical length, long-day (short-night) plants produce flowers. (b) When nights exceed a critical length, short-day (long-night) plants flower. (c) If a flash of light interrupts a long night, the short-night plants flower, but the long-night plants do not.

22.5 Light Is a Powerful Influence on Plant Life

Plants are exquisitely attuned to light. Their lives depend on it because light is their sole energy source for photosynthesis (see chapter 5). This section illustrates some of the ways that light influences a plant's life.

A response to light requires **photoreceptors,** molecules that detect the wavelength and intensity of light. In general, light absorption triggers a change in the photoreceptor's shape, ultimately leading to a growth response in the plant. This response may involve changes in the production, release, or transport of one or more hormones.

Consider, for example, a **tropism:** the orientation of a plant part toward or away from a stimulus such as light, gravity, or touch. All tropisms result from differential growth, in which one side of the responding organ grows faster than the other. In **phototropism,** a plant grows toward or away from light. This response is adaptive because it bends stems toward light and roots toward darkness.

Phototropism occurs in stems when cells on the shaded side elongate more than cells on the opposite side. Photoreceptors and auxins participate in the response. Photoreceptors absorb light, which somehow causes auxins to migrate to the shaded side of the stem. Water follows the auxins into the cells, and the resulting increase in turgor pressure causes the cells to elongate. The stem therefore bends toward the light (figure 22.14).

The plant commonly sold as "lucky bamboo" often has a curled stem, illustrating the effects of phototropism. Farmers grow the plants for a year or more in greenhouses, exposing only one side to light. Periodically rotating each plant directs the stem's growth into a twist.

The timing of light exposure also influences many aspects of a plant's life. For example, plants respond in several ways to the **photoperiod,** or day length. Consider the changes that occur in a deciduous forest throughout the year. The short days that accompany the approach of winter are associated with the formation of buds, the loss of leaves, and dormancy. In the spring, as days grow longer, buds resume growth and rapidly transform a barren deciduous forest into a leafy canopy. These seasonal changes illustrate the interactions among photoreceptors, environmental signals, hormones, and the plant's genes.

Photoperiod regulates flowering in some species. Traditionally, biologists used the term "long-day plants" for plants that flower when days are longer than a critical length, usually 9 to 16 hours. These plants typically bloom in the spring or early summer and include lettuce, spinach, beets, clover, and irises. Likewise, "short-day plants" produce flowers when days are shorter than some critical length, usually in late summer or fall. Asters, strawberries, poinsettias, potatoes, soybeans, ragweed, and goldenrods are short-day plants. "Day-neutral plants" such as tomatoes flower at maturity, regardless of day length.

However, experiments eventually confirmed that flowering actually requires a defined period of uninterrupted darkness, rather than a certain day length (figure 22.15). Thus, short-day plants are really long-night plants, because they flower only if their uninterrupted dark period exceeds a critical length. Similarly, long-day plants are really short-night plants.

22.5 Mastering Concepts

1. What is auxin's role in phototropism?
2. How does light help regulate flowering time?

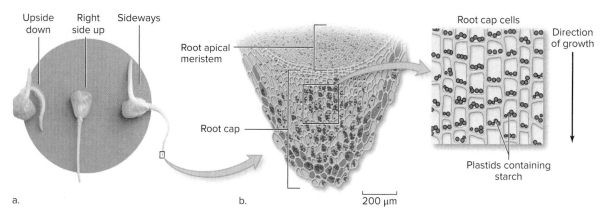

Figure 22.16 **Gravitropism.** (a) Shoots grow up and roots grow down, whether a seed is oriented upside down, right side up, or sideways. (b) Starch-filled organelles in the root cap may help the plant detect gravity.

(a): ©Martin Shields/Alamy Stock Photo

22.6 Plants Respond to Gravity and Touch

Besides light, a developing plant also responds to countless other environmental cues. For example, the more CO_2 in the atmosphere, the lower the density of stomata on leaves. Likewise, a plant in soil with abundant nitrogen produces fewer lateral roots than it would in nutrient-poor soil. Plants can also sense temperature; many require a prolonged cold spell before producing buds or flowers. A warm period in December, before temperatures have really plummeted, does not stimulate apple and cherry trees' buds to "break," but a similar warm-up in late February does induce growth.

Gravity is another important environmental cue. **Gravitropism** is directional growth in response to gravity (figure 22.16). As a seed germinates, its shoot points upward toward light, and its roots grow downward into the soil. Turn the plant sideways, and the stem and roots bend according to the new direction of gravity.

No one knows exactly how gravitropism works, although it is clear that the root cap must be present for roots to respond to gravity. One hypothesis centers on starch-rich plastids inside the root cap cells. These organelles function as gravity detectors, sinking to the bottoms of the cells (see figure 22.16). Somehow the position of the plastids tells the cells which direction is down. Turning a root sideways causes the plastids to move, redistributing calcium ions and auxins in a way that bends the root downward. ⓘ *plastids,* section 3.4D

Besides ever-present gravity, a plant also encounters a changing variety of mechanical stimuli, including contact with wind, rain, animals, and other plants. The coiling tendrils of twining plants exhibit **thigmotropism,** a directional response to touch (figure 22.17). Specialized epidermal cells detect contact with an object, which induces the tendril to bend. In only 5 to 10 minutes, the tendril completely encircles the object. Auxins and ethylene apparently control thigmotropism.

22.6 Mastering Concepts

1. How do plastids and auxins participate in gravitropism?

2. How does thigmotropism help plants climb?

Figure 22.17 **Thigmotropism.** A tendril's epidermis is sensitive to touch. This tendril of a white bryony vine wraps around a fence.

©blickwinkel/Alamy Stock Photo

Investigating Life 22.1 | A Red Hot Chili Pepper Paradox

Chili peppers are famous for their hot, spicy taste. Chilies get their kick from a unique chemical compound called capsaicin. The sensation ranges from the pleasantly mild pimento to the outright painful habanero.

The pungency of chilies is a bit of a paradox. After all, the main functions of fruits are protection and seed dispersal. Many plants produce chemicals that make their fruits unpalatable until the seeds are mature. Once they are ripe, however, most fruits lose their defensive chemicals. But chilies retain capsaicin even after the fruit is fully developed. Why?

One clue is that some fruit-eating animals destroy seeds as they chew their food, whereas others disperse seeds intact. Perhaps hot chilies deter the seed-destroying animals without affecting the beneficial dispersers.

Researchers used field observations and laboratory feeding studies to test this hypothesis. First, they wanted to learn which animals consumed the fruits. After videotaping wild chiltepin chili plants in the desert, they found that birds called thrashers ate most of the fruits. On the other hand, small mammals seemed to avoid the chilies.

Later, in the laboratory, the team studied the feeding preferences of cactus mice, pack rats, and thrashers. Each animal was offered three types of fruit: hot chilies from the field sites, nonpungent mutant chilies, and desert hackberries. The birds ate all three fruits; the mice and pack rats consumed the hackberries but avoided the chilies (figure 22.A).

Finally, the researchers measured the germination rates of seeds that had passed through bird and rodent digestive systems; control seeds remained uneaten. Seeds eaten by thrashers remained viable, whereas those consumed by mice and pack rats failed to germinate. The mammals' molars apparently crushed and destroyed the chili seeds.

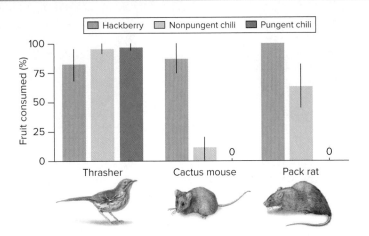

Figure 22.A Some Like It Hot. Thrashers readily ate pungent chilies, but mice and pack rats rejected the spicy fruits. (See appendix B for an explanation of error bars.)

These experiments help explain the evolutionary forces that select for capsaicin production in mature chili peppers. Thrashers and other birds seem unaffected by capsaicin, while most mammals avoid it. The real paradox is how humans transformed this innate mammalian aversion into a worldwide love affair with the chili pepper.

Source: Tewksbury, Joshua J., and Gary P. Nabhan. 2001. Directed deterrence by capsaicin in chillies. *Nature*, vol. 412, pages 403–404.

What's the Point? ▼ APPLIED

After pollination, a flower develops into a fruit. That is, each of the flower's ovules becomes a seed that contains one plant embryo. A fruit surrounds the seeds, protecting the offspring and aiding in their dispersal from the parent plant.

The botanical definition of *fruit*, then, is a seed-containing structure that develops from a pollinated flower. In contrast, the culinary definition is narrower, typically referring to a sweet, fleshy, and possibly acidic plant part containing seeds. A lemon is therefore a fruit in both the botanical and the culinary sense, whereas an eggplant—which has seeds but is not sweet—is a fruit only in the botanical sense.

Applying this definition of fruits to the human diet reveals how important these plant reproductive structures are. The most commonly recognized fruits, such as apples, kiwis, and strawberries, are sweet. On the other hand, tomatoes, bell peppers, and pumpkins are fruits that are slightly less sweet and that are usually considered vegetables. Fatty fruits include olives, avocados, and nuts. The starchy fruits and seeds of grain plants offer many useful products, including wheat, oats, and corn. The latter grain is a major food source for animals raised for milk and meat.

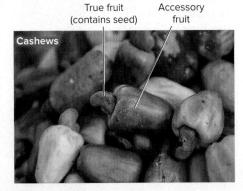

True fruit (contains seed)   Accessory fruit

Cashews

©McGraw-Hill Education/Barry Barker

Some bizarre fruits are greatly altered by the time they reach supermarket shelves. Consider, for example, the fruit of the cashew tree. The cashews that we eat as snacks are barely recognizable as the brownish "stems" of the brightly colored "fruits" shown here. In reality, each "stem" is actually a fruit with a toxic exterior protecting the kidney-shaped seed inside. This seed and its surrounding fruit are attached to the top of a brightly colored "accessory fruit." Unlike the true fruit, the accessory fruit is fleshy and appealing to animals. A bird or bat that collects the fruit is therefore likely to discard the seed after consuming the sweet reward. (Lucky for us, cashew harvesters remove both the toxic outer shell of the true fruit and the accessory fruit before the seed enters the market.)

It is hard to imagine how humans could survive without flowering plants. Their reproductive structures provide all fruits and grains, many vegetables (such as eggplant and squash), and nuts. Vegetarians consume nuts, beans, and grains for protein; meat, eggs, and dairy products come from farm animals that may have eaten soybeans or grains. Directly or indirectly, flowers and fruits are therefore unmatched in their contribution to our diet.

CHAPTER SUMMARY

22.1 Angiosperms Reproduce Sexually and Asexually

- **Sexual reproduction** produces genetically variable offspring, increasing reproductive success in a changing environment.
- In **asexual reproduction,** identical clones develop from the roots, stems, or leaves of a parent plant. Asexual reproduction is advantageous in a stable environment where plants are well-adapted to their surroundings.

22.2 The Angiosperm Life Cycle Includes Flowers, Fruits, and Seeds

- Plant life cycles include an **alternation of generations** (figure 22.18). The diploid **sporophyte** undergoes meiosis and produces haploid **spores,** which give rise to the haploid **gametophyte** generation. The gametophytes, in turn, produce **gametes.** These haploid cells fuse at **fertilization** to produce the diploid **zygote,** which develops into a new sporophyte.
- Two features unique to the angiosperm life cycle are **flowers** and **fruits.**

A. Flowers Are Reproductive Organs

- Flowers are reproductive structures built of whorls of parts attached to a **receptacle.** The **sepals** and **petals** are accessory parts. The **stamens** have pollen-containing **anthers** at their tips. **Carpels** consist of the **ovary** (which contains one or more **ovules**) and the **style,** topped by the **stigma.**

B. The Pollen Grain and Embryo Sac Are Gametophytes

- In the anther, specialized cells divide meiotically, each producing four haploid **microspores.** The microspores divide mitotically to yield haploid cells. The **pollen** grain is the male gametophyte; it produces two identical sperm nuclei.

- In ovules, specialized cells divide meiotically to yield four haploid cells, one of which persists as a haploid **megaspore** that divides mitotically three times. The resulting female gametophyte, or **embryo sac,** contains seven cells. One is the egg, and another is the central cell (which has two nuclei).

C. Pollination Brings Pollen to the Stigma

- Animals and wind carry pollen from anthers to a stigma. Flowers are usually adapted to either animal or wind **pollination.** Animal pollinators and flowers select for changes in one another in **coevolution.**

D. Double Fertilization Yields Zygote and Endosperm

- Once on a stigma, a pollen grain grows a pollen tube, and its two sperm nuclei move through the tube toward the ovary.
- In **double fertilization,** one sperm nucleus fertilizes the egg to form the diploid zygote; the other sperm nucleus fertilizes the central cell's two nuclei, forming the triploid **endosperm.**

E. A Seed Is an Embryo and Its Food Supply Inside a Seed Coat

- A **seed** is an embryo, endosperm, and **seed coat.** The endosperm nourishes the developing embryo as cells in apical meristems divide to produce the embryonic shoot and root. As the embryo grows, one or two **cotyledons** develop.
- Seeds enter a dormancy period in which the embryo postpones development.

F. The Fruit Develops from the Ovary

- After fertilization, nonessential flower parts fall off, and hormones influence the ovary (and sometimes other plant parts) to develop into a fruit.

G. Fruits Protect and Disperse Seeds

- The fruit protects the seeds and aids in dispersal.
- Animals, wind, and water disperse seeds to new habitats, reducing competition between parent plants and their offspring.

22.3 Plant Growth Begins with Seed Germination

- Seed **germination** requires oxygen, water, and a favorable temperature. The embryo bursts from the seed coat, and its development resumes.

22.4 Hormones Regulate Plant Growth and Development

- Plants respond to the environment with changes in growth, mediated by the action of **hormones** (figure 22.19).

A. Auxins and Cytokinins Are Essential for Plant Growth

- **Auxins** stimulate cell elongation in shoot tips, embryos, young leaves, flowers, fruits, and pollen. Auxins are most concentrated at the main shoot tip, which blocks the growth of lateral buds (**apical dominance).**
- **Cytokinins** stimulate cell division in actively developing plant parts, including lateral buds.

B. Gibberellins, Ethylene, and Abscisic Acid Influence Plant Development in Many Ways

- **Gibberellins** stimulate cell division and elongation, and they help break seed dormancy.
- **Ethylene** is a gas that speeds ripening, aging, and the loss of leaves and petals.
- **Abscisic acid** counters the growth-inducing effects of other hormones by inducing dormancy and inhibiting shoot growth.

22.5 Light Is a Powerful Influence on Plant Life

- **Photoreceptors** absorb light energy and influence a plant's growth, development, or other response to the environment.
- A **tropism** is a growth response toward or away from an environmental stimulus. In **phototropism,** light stimulates auxin to move to the shaded side of the stem, which therefore bends toward the light.
- Some plants use **photoperiod** (day length) as a cue to produce flowers. Short-day (long-night) plants flower only when the duration of uninterrupted darkness is greater than a critical length. Long-day (short-night) plants require a dark period shorter than a critical length.

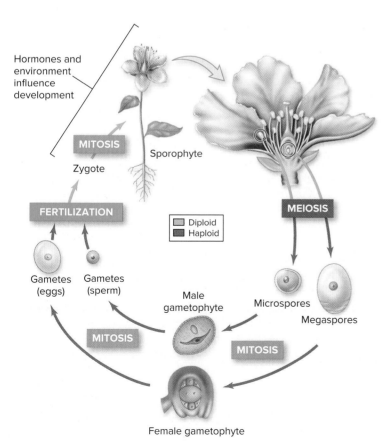

Hormones and environment influence development

MITOSIS

Sporophyte

Zygote

FERTILIZATION

☐ Diploid
☐ Haploid

MEIOSIS

Gametes (eggs)   Gametes (sperm)

Male gametophyte

Microspores

Megaspores

MITOSIS

MITOSIS

Female gametophyte

Figure 22.18 Angiosperm Reproduction: A Summary.

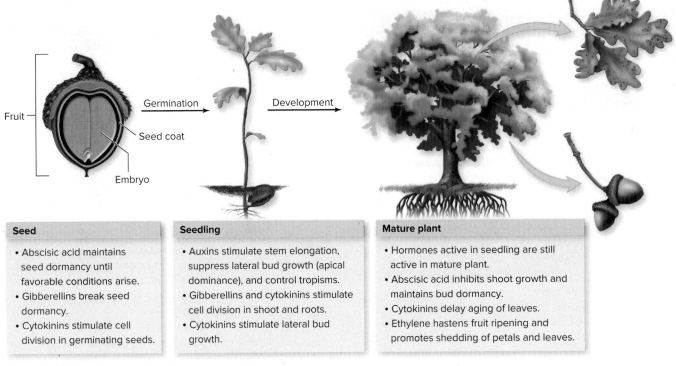

Figure 22.19 **Hormones and Plant Development: A Summary.**

Seed
- Abscisic acid maintains seed dormancy until favorable conditions arise.
- Gibberellins break seed dormancy.
- Cytokinins stimulate cell division in germinating seeds.

Seedling
- Auxins stimulate stem elongation, suppress lateral bud growth (apical dominance), and control tropisms.
- Gibberellins and cytokinins stimulate cell division in shoot and roots.
- Cytokinins stimulate lateral bud growth.

Mature plant
- Hormones active in seedling are still active in mature plant.
- Abscisic acid inhibits shoot growth and maintains bud dormancy.
- Cytokinins delay aging of leaves.
- Ethylene hastens fruit ripening and promotes shedding of petals and leaves.

22.6 Plants Respond to Gravity and Touch

- **Gravitropism** is growth toward or away from the direction of gravity. Examples are the upward growth of shoots and the downward growth of roots. The positions of starch-rich plastids in cells apparently help plants detect gravity.
- **Thigmotropism** is growth directed toward or away from a mechanical stimulus such as wind or touch.
- Table 22.3 summarizes three types of tropisms.

TABLE 22.3 Plant Tropisms: A Summary

Type	Stimulus	Response	Example
Phototropism	Light	Photoreceptors absorb light energy; auxins move to shaded side of stem and stimulate cell elongation.	Stem grows toward window.
Gravitropism	Gravity	Starch-rich organelles "sink" within root cap cells; auxins cause root to bend downward (hypothesized mechanism).	Roots grow downward into soil.
Thigmotropism	Touch	Unknown	Tendril coils around trellis.

MULTIPLE CHOICE QUESTIONS

1. The new gene combinations associated with sexual reproduction in plants are the result of
   a. mitosis.
   b. meiosis.
   c. cloning.
   d. Both b and c are correct.

2. Assuming that every egg in a flower is fertilized, the number of ovules in the flower is typically _____ the number of seeds in the resulting fruit.
   a. less than
   b. equal to
   c. greater than
   d. unrelated to

3. Where would you find a male gametophyte?
   a. Inside a ripe pollen sac
   b. Inside an ovule
   c. Inside the embryo sac
   d. Both b and c are correct.

4. How does the endosperm develop inside the seed?
   a. It receives nutrients from the embryo.
   b. It receives nutrients from the parent plant.
   c. It produces its own nutrients as it divides.
   d. All of the above are correct.

5. A fungus causes "foolish seedling" disease in rice. The stems of infected plants grow so rapidly that they fall over from their own weight. Which plant hormone is the fungus secreting?
   a. Abscisic acid
   b. Gibberellin
   c. Ethylene
   d. Cytokinin

6. An overripe orange releases _____, promoting ripening of other nearby oranges.
   a. gibberellins
   b. abscisic acid
   c. ethylene
   d. cytokinins

7. Chrysanthemums are long-night plants that normally flower in the fall. If you could manipulate photoperiod, what would be the simplest way to prevent mums from blooming (without killing the plants)?
   a. Never expose the mums to light at all.
   b. Interrupt each night with a flash of light.
   c. Interrupt each day with a brief period of darkness.
   d. Make sure each uninterrupted night lasts longer than the critical period.

8. What type of tropism is adaptive to a developing plant?
   a. Phototropism   c. Thigmotropism
   b. Gravitropism   d. All are adaptive.

Answers to Multiple Choice questions are in appendix A.

WRITE IT OUT

1. Give an example of asexual reproduction in a plant.
2. Explain how flowers, fruits, and seeds contribute to the reproductive success of angiosperms.
3. Describe the male and female gametophytes of flowering plants.
4. What is double fertilization? How does it increase the reproductive success of the parent plant?
5. How does an exclusive relationship between a plant and its pollinator benefit each partner? What are the risks of exclusivity?
6. If fruit production is a measure of fitness, why wouldn't a plant spend all of its energy producing fruits instead of roots and leaves? Why do you think some annual plants lose their foliage as they produce fruits?
7. An oak tree may produce thousands of acorns, which squirrels bury or eat. Why does the tree make so many acorns? Why might a tree whose seeds disperse far from the parent have better reproductive success than one whose seeds fall at the base of the parent plant?
8. How do fruits and seeds disperse to new habitats?
9. How does seed dormancy promote reproductive success?
10. List the major plant hormones and describe some of their actions.
11. Pruning stimulates the growth of new branches. What hormonal changes occur in the lateral buds after the shoot tip is pruned?
12. What is the function of photoreceptors?
13. Describe three tropisms. Which hormone is common to all three?
14. Explain how plants grow toward light. Why is this response adaptive?
15. Develop a hypothesis that explains why it might be adaptive for a plant to flower in response to photoperiod rather than temperature.
16. How is gravitropism adaptive?

SCIENTIFIC LITERACY

Review Burning Question 22.1, which explains how humans select for seedless watermelons and other fruits. Some people worry that producing seedless fruits reduces genetic diversity. How might that occur? From a farmer's standpoint, what are the risks of reduced genetic diversity?

Answers to Mastering Concepts, Write It Out, Scientific Literacy, and Pull It Together questions can be found in the Connect ebook.
connect.mheducation.com

Design element: Burning Question (fire background): ©Ingram Publishing/Super Stock

PULL IT TOGETHER

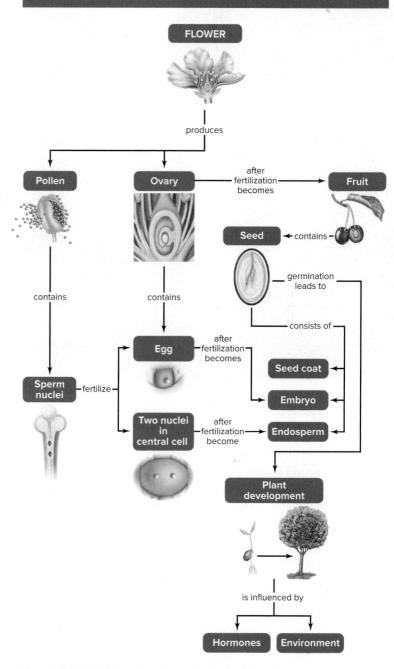

Figure 22.20 Pull It Together: Reproduction and Development of Flowering Plants.

Refer to figure 22.20 and the chapter content to answer the following questions.

1. Review the Survey the Landscape figure in the chapter introduction, and then add *flowering plants, leaves,* and *sugars* to the Pull It Together concept map.
2. Add the following terms to the concept map: *stamen, anther, carpel, ovule, stigma.*
3. Name structures in the concept map that are haploid, diploid, and triploid.
4. Add to the concept map at least three conditions implied by *environment,* and explain how each affects plant development.

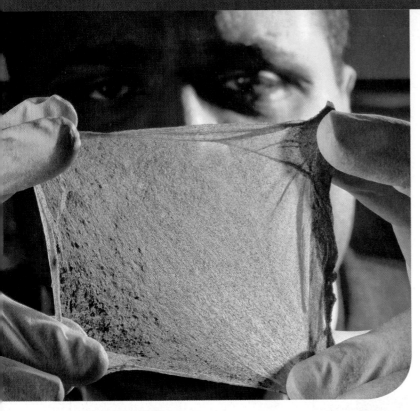

LEARNING OUTLINE

23.1 Specialized Cells Build Animal Bodies

23.2 Animals Consist of Four Tissue Types

23.3 Organ Systems Are Interconnected

23.4 Organ System Interactions Promote Homeostasis

23.5 Animals Regulate Body Temperature

APPLICATIONS

Why We Care 23.1 *Two Faces of Plastic Surgery*
Burning Question 23.1 *How does the body react to food poisoning?*
Burning Question 23.2 *Can biologists build artificial organs?*
Investigating Life 23.1 *Sniffing Out the Origin of Feathers*

Artificial Tissue. A researcher holds a thin sheet of living cells embedded in an artificial matrix. Produced by a bioprinter, tissues such as this one may someday be incorporated into artificial organs.
©James King-Holmes/Science Source

Learn How to Learn
Pay Attention in Class

It happens to everyone occasionally: Your mind begins to wander while you are sitting in class, so you check your phone or doze off. How can you keep from wasting your class time this way? One strategy is to get plenty of sleep and eat well, so your mind stays active. Another is to prepare for class in advance, since getting lost can be an excuse for drifting off. When you get to class, sit near the front, listen carefully, and take good notes. Finally, a friendly reminder can't hurt; make a small PAY ATTENTION sign to put on your desk where you can always see it.

SURVEY THE LANDSCAPE
Animal Anatomy and Physiology

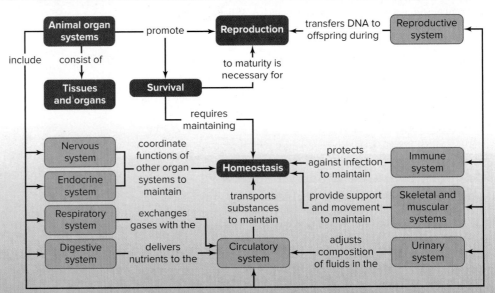

An animal's parts include familiar organs such as skin, bones, muscles, and the brain. These organs are composed of tissues, which are themselves made of specialized cell types. All of these parts interact to maintain homeostasis in the animal body.

For more details, study the Pull It Together feature in the chapter summary.

What's the Point? ▼

©DLILLC/Corbis RF

The two photos that introduce this chapter may seem unrelated, but each illustrates an important point about animal life. One photo shows a scientist holding a sheet of artificial tissue, a medical advance made possible by basic research into the biology of cells, tissues, and organs. This chapter introduces these levels of organization in the animal body; subsequent chapters in this unit explore each organ system in detail.

The other photo shows a penguin, a bird that spends much of its life in cold water. The penguin's organ systems work together to keep its body warm despite the frigid surroundings. This state of internal constancy, which applies not only to temperature but also to other vital conditions in the body, is called homeostasis. We introduce the idea of homeostasis in this chapter, but every other chapter in this unit reinforces the concept with additional examples.

23.1 Specialized Cells Build Animal Bodies

Everywhere we look, form and function are entwined. The broad, flat surface of a plant's leaf maximizes its exposure to light. A neuron's many branches permit the cell-to-cell connections essential to communication in the nervous system. In birds, fluffy down feathers trap pockets of air and conserve warmth.

Anatomy, the study of an organism's structure, describes the parts that compose the body—that is, its form. **Physiology** is a related discipline that considers how those parts work—their function. Unit 5 described the anatomy and physiology of plants; unit 6 turns to animals.

Biologists describe the animal body in terms of an organizational hierarchy (figure 23.1). Most animals have **tissues,** which are groups of specialized cells that interact and provide a specific function. The inner lining of the stomach, for example, is a tissue that secretes stomach acid. An **organ** consists of two or more interacting tissues that function as a unit. The stomach is an organ that consists of muscle, blood, and nerves in addition to the tissue of its inner lining. Still farther up the organizational hierarchy are **organ systems,** which consist of two or more organs that are physically or functionally joined. The human digestive system includes not only the stomach but also the small intestine, large intestine, and other organs.

Everyone is familiar with the overall form of the human body. Other animal bodies have wildly different shapes, from the flattened tapeworm to the squishy squid to the armored lobster to the scaly snake. But all of these animals have organs that carry out the same basic functions as our own: They sense their environment, acquire food and oxygen, eliminate wastes, protect themselves from injury and disease, and reproduce. Although this unit describes some notable adaptations in other animals, the focus is mainly on humans.

This chapter introduces the basic parts that build animal bodies. Chapters 24 through 30 then consider the organ systems one at a time.

23.1 Mastering Concepts

1. What is the difference between anatomy and physiology?

2. How are cells, tissues, organs, and organ systems related?

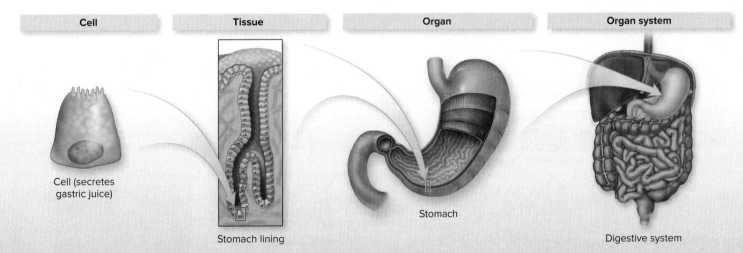

Figure 23.1 Organizational Hierarchy Within the Body. Cells make up tissues, which build the organs that form organ systems. In this example, a cell that secretes gastric juice is one specialized cell type in the tissue that lines the stomach. This organ, in turn, is one of many that make up the digestive system.

Simple squamous epithelial tissue

Composition: Single layer of flattened cells

Functions: Allows substances to pass by diffusion and osmosis

Locations: Lining of blood vessels; alveoli of lungs

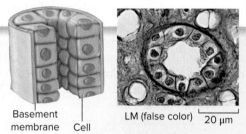

Basement membrane   Cell   LM (false color)   30 μm

23.2 Animals Consist of Four Tissue Types

Animal bodies contain a spectacular diversity of specialized cells, all of which ultimately descend from stem cells in the embryo (see figure 3.6 to review the structure of a basic animal cell). The vertebrate body, including that of humans, has at least 260 cell types. But these cells do not act alone. Instead, some cell types tend to occur together and to share common functions. Together, these interacting cells form tissues that fall into four broad categories: epithelial, connective, muscle, and nervous. This section summarizes their characteristics, functions, and locations; Why We Care 23.1 describes how plastic surgeons reposition these tissues to improve a person's appearance or repair damage. ⓘ *stem cells,* section 11.3A

All animal tissues have a feature in common: The cells are embedded in a nonliving **extracellular matrix,** which is a mixture of water, carbohydrates, lipids, and (usually) protein fibers such as collagen and elastin. Interestingly, most normal body cells cannot survive or replicate when removed from the extracellular matrix. Somehow, cancer cells escape this "anchorage dependence," breaking away from the extracellular matrix yet retaining the ability to divide. These abnormal cells also often secrete enzymes that destroy the fibers of the extracellular matrix, clearing the way for cells from a cancerous tumor to invade adjacent tissues. ⓘ *cancer,* section 8.6

A. Epithelial Tissue Covers Surfaces

Epithelial tissues coat the body's internal and external surfaces with one or more layers of tightly packed cells (figure 23.2). They cover organs and line the inside of hollow organs and body cavities. The diverse functions of epithelial tissues include protection, nutrient absorption along the intestinal tract, and gas diffusion in the lungs. These tissues also form **glands,** organs that secrete substances into ducts or into the bloodstream. Glands release breast milk, sweat, saliva, tears, mucus, hormones, enzymes, and many other important secretions.

Simple cuboidal epithelial tissue

Composition: Single layer of cube-shaped cells

Functions: Secretes and absorbs substances

Locations: Glands; lining of kidney tubules

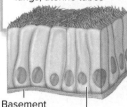

Basement membrane   Cell   LM (false color)   20 μm

Simple columnar epithelial tissue

Composition: Single layer of column-shaped cells (may be ciliated)

Functions: Secretes and absorbs substances; sweeps egg/embryo along uterine tube

Locations: Lining of digestive tract; bronchi of lungs; uterine tubes

Basement membrane   Cell   LM (false color)   10 μm

Stratified squamous epithelial tissue

Composition: Multiple layers of flattened cells

Functions: Protects areas subject to abrasion; prevents water loss and infection

Location: Outer layer of skin; lining of body openings

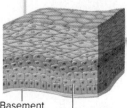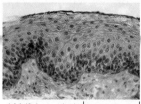

Figure 23.2 Epithelial Tissues.

Photos: (simple squamous, simple cuboidal): ©Ed Reschke/Photolibrary/Getty Images; (simple columnar, stratified squamous): ©Victor P. Eroschenko RF

Basement membrane   Cell   LM (false color)   120 μm

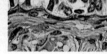

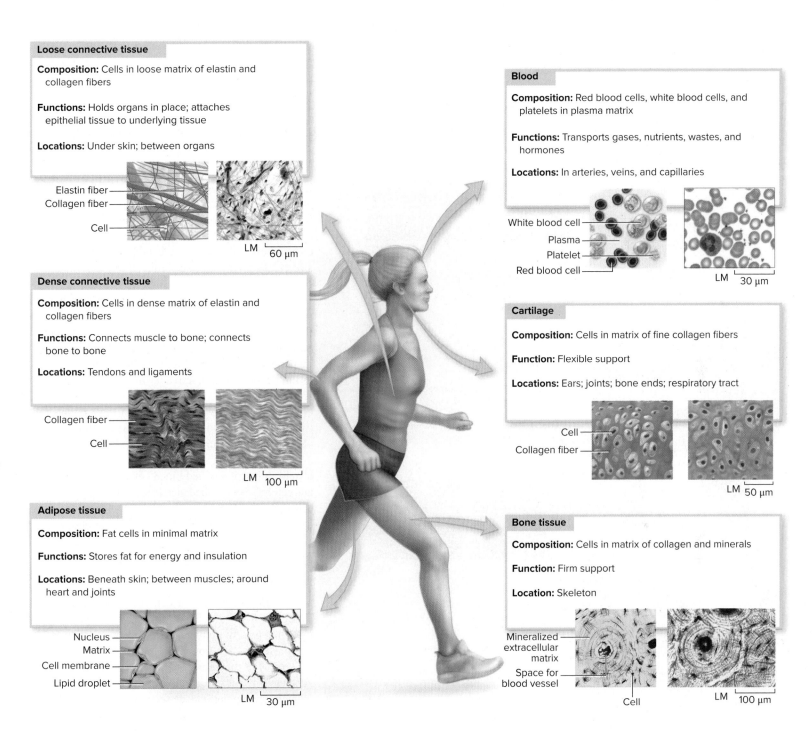

Loose connective tissue

Composition: Cells in loose matrix of elastin and collagen fibers

Functions: Holds organs in place; attaches epithelial tissue to underlying tissue

Locations: Under skin; between organs

Elastin fiber
Collagen fiber
Cell

LM 60 μm

Dense connective tissue

Composition: Cells in dense matrix of elastin and collagen fibers

Functions: Connects muscle to bone; connects bone to bone

Locations: Tendons and ligaments

Collagen fiber
Cell

LM 100 μm

Adipose tissue

Composition: Fat cells in minimal matrix

Functions: Stores fat for energy and insulation

Locations: Beneath skin; between muscles; around heart and joints

Nucleus
Matrix
Cell membrane
Lipid droplet

LM 30 μm

Blood

Composition: Red blood cells, white blood cells, and platelets in plasma matrix

Functions: Transports gases, nutrients, wastes, and hormones

Locations: In arteries, veins, and capillaries

White blood cell
Plasma
Platelet
Red blood cell

LM 30 μm

Cartilage

Composition: Cells in matrix of fine collagen fibers

Function: Flexible support

Locations: Ears; joints; bone ends; respiratory tract

Cell
Collagen fiber

LM 50 μm

Bone tissue

Composition: Cells in matrix of collagen and minerals

Function: Firm support

Location: Skeleton

Mineralized extracellular matrix
Space for blood vessel
Cell

LM 100 μm

Epithelial tissues always have a "free" surface that is exposed either to the outside or to a space within the body. On the opposite side, epithelium is anchored to underlying tissues by a layer of extracellular matrix called the basement membrane. The epithelial cells are often connected to one another, forming leak-proof sheets. The tightly knit structure of epithelial tissue is closely tied to its function as a border between the body's tissues and an open space.

Epithelial tissues are classified partly by the shapes of their cells: squamous (flattened), cuboidal (cube-shaped), or columnar (tall and thin). The number of cell layers is also important. Simple epithelial tissues consist of a single layer of cells, whereas stratified epithelial tissues are made of multiple cell layers.

Figure 23.3 Connective Tissues.

Photos: (loose connective tissue, blood): ©McGraw-Hill Education/Al Telser; (dense connective tissue, adipose, bone): ©McGraw-Hill Education/Dennis Strete; (cartilage): ©Chuck Brown/Science Source

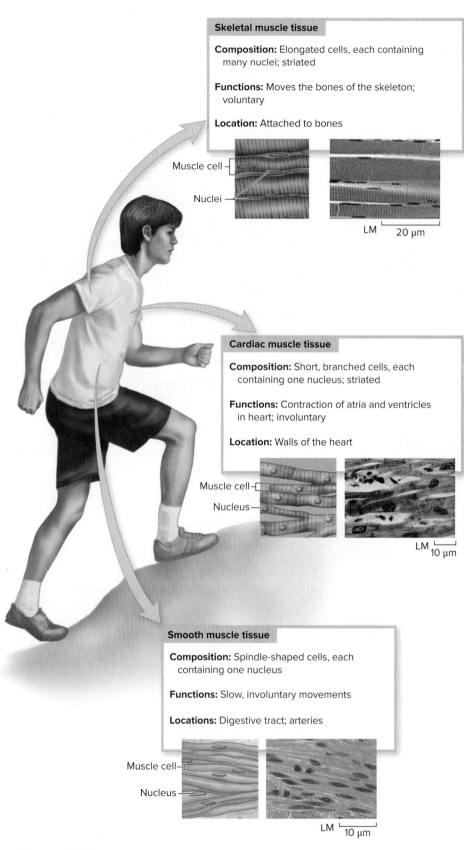

Skeletal muscle tissue

Composition: Elongated cells, each containing many nuclei; striated

Functions: Moves the bones of the skeleton; voluntary

Location: Attached to bones

Muscle cell

Nuclei

LM   20 µm

Cardiac muscle tissue

Composition: Short, branched cells, each containing one nucleus; striated

Functions: Contraction of atria and ventricles in heart; involuntary

Location: Walls of the heart

Muscle cell

Nucleus

LM   10 µm

Smooth muscle tissue

Composition: Spindle-shaped cells, each containing one nucleus

Functions: Slow, involuntary movements

Locations: Digestive tract; arteries

Muscle cell

Nucleus

LM   10 µm

Figure 23.4 Muscle Tissues.

Photos: (skeletal): ©Corbis RF; (cardiac): ©Fuse/Getty Images RF; (smooth): ©McGraw-Hill Education/Dennis Strete

About 90% of human cancers arise in epithelial tissues. Such a cancer is called a carcinoma. The most common carcinomas include cancers of the skin, breast, lung, prostate, and colon.

B. Most Connective Tissues Bind Other Tissues Together

The most widespread tissue type in a vertebrate's body is **connective tissue,** which consists of cells that are embedded within the extracellular matrix rather than being attached to one another. Connective tissues fill spaces, attach epithelium to other tissues, protect and cushion organs, and provide both flexible and firm structural support. Unlike epithelial tissues, connective tissues never coat any body surface.

Connective tissues are extremely variable in both structure and function (figure 23.3). Loose connective tissue binds other tissues together and fills the space between organs; dense connective tissue builds ligaments and tendons; and adipose tissue stores energy as fat. Blood is a connective tissue, as are the cartilage and bone that make up the vertebrate skeleton. A close look at figure 23.3 reveals that in all connective tissues except adipose tissue, the extracellular matrix occupies more volume than do the cells.

Blood is an unusual tissue because it is a liquid, not a solid. It consists of red blood cells, white blood cells, and cell fragments called platelets, all traveling in a liquid called plasma. The extracellular matrix of blood is unique in two ways: It is a fluid and it lacks protein fibers.

C. Muscle Tissue Provides Movement

Muscle tissue consists of cells that contract (become shorter) when stimulated. Contraction occurs when long, thin protein filaments slide past one another inside the muscle cells. Abundant mitochondria in muscle cells provide the energy for contraction, and the heat generated by muscle contraction is important in body temperature regulation (see section 23.5).

The most familiar function of muscle tissue, however, is to move other tissues and organs. Muscle cells attach to soft tissue or bone; when the cells contract, the body part moves. Digestion, the elimination of wastes, blood circulation, and the motion of the limbs all rely on muscle contraction.

Animal bodies contain three types of muscle tissue (figure 23.4). Skeletal muscle tissue consists of long cells. When viewed with a microscope, this tissue appears striped, or striated, because the protein filaments that fill each cell align in a repeated pattern. Most skeletal muscle attaches to bone and provides voluntary movements that a person can consciously control.

Why We Care 23.1 | Two Faces of Plastic Surgery

The "plastic" in "plastic surgery" has nothing to do with the substance that makes up water bottles. Instead, the word derives from the Greek word *plastikos,* which means "to mold." Plastic surgeons "mold" a person's appearance by moving and reshaping tissues such as fat, bone, and cartilage.

Cosmetic surgery is one field of plastic surgery. Among the most common procedures is a "nose job," in which a surgeon removes or repositions some of the cartilage and bone of the nose to create a new shape. Liposuction is also popular (figure 23.A). The surgeon makes small incisions and removes excess fat from the thighs, buttocks, arms, neck, or stomach. Face lifts are common as well. In a typical face lift, a surgeon makes a long incision at the hairline, lifts the skin of the face, and repositions the muscle and connective tissue under the skin. He or she then tightens the skin and trims the

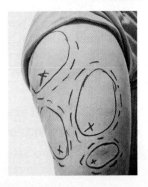

Figure 23.A
Liposuction Markings.
©image100/Corbis RF

excess before reattaching the skin to the face. Other popular procedures include breast augmentation, breast reduction, breast lifts, buttock lifts, tummy tucks, laser skin resurfacing, hair transplants, and collagen injections.

Cosmetic surgery may enhance a healthy person's appearance, but reconstructive plastic surgery has a different goal: to restore the function of damaged body parts. For example, some children are born with a cleft palate (a gap in the bones between the nose and mouth). To repair a cleft palate, a plastic surgeon rearranges the epithelium, muscles, and bones of the roof of the mouth. Reconstructive surgery may also include skin grafts (for burn patients) or breast reconstruction (for women who have lost one or both breasts to cancer). Far from frivolous, these procedures can dramatically improve a patient's ability to function.

Cardiac muscle tissue, which occurs only in the heart, is also striated, but the cells are shorter, and their control is involuntary. Cardiac muscle cells are electrically coupled with one another, so they contract simultaneously to produce the heartbeat. Smooth muscle tissue is not striated, and its contraction is involuntary. This type of muscle pushes food along the intestinal tract, regulates the diameter of blood vessels, and controls the size of the pupil of the eye.

D. Nervous Tissue Forms a Rapid Communication Network

Nervous tissue uses electrochemical signals to convey information rapidly within the body. Sensory cells detect stimuli such as the scent of a rose or a prick of its thorn. Other cells then transmit that information along nerves to the central nervous system (brain and spinal cord), which helps you interpret what you experience.

Two main cell types occur in nervous tissue: neurons and neuroglia (figure 23.5). Neurons form communication networks that receive, process, and transmit information. The cell may connect to another neuron at a junction called a synapse, or it may stimulate a muscle or gland. Neuroglia are cells that support neurons and assist in their functioning. The drawing in figure 23.5 shows the neuroglia that form insulating sheaths of myelin around parts of a neuron. As explained in chapter 24, the myelin sheath speeds the conduction of electrical impulses. Other types of neuroglia surround and support the neurons in the photo in figure 23.5.

23.2 Mastering Concepts

1. If you were given a microscope slide with a slice of tissue on it, how would you classify it into one of the four main tissue types?

2. Suppose you determine that a tissue sample consists of unconnected cells embedded in an extracellular matrix. What criteria would you use to classify it into a specific subtype of tissue?

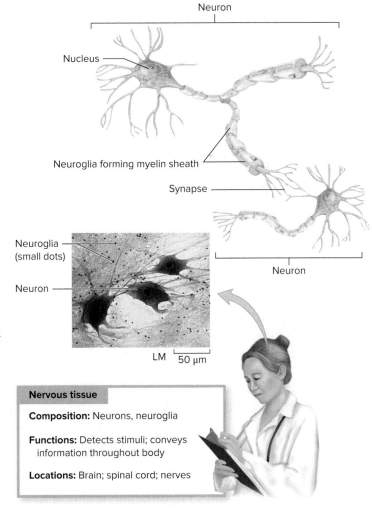

Nervous tissue

Composition: Neurons, neuroglia

Functions: Detects stimuli; conveys information throughout body

Locations: Brain; spinal cord; nerves

Figure 23.5 Nervous Tissue.

Photo: ©Ed Reschke/Photolibrary/Getty Images

Burning Question 23.1

How does the body react to food poisoning?

©Alix Minde/PhotoAlto/ Getty Images RF

In a healthy body, the organ systems operate so seamlessly that you are unlikely to notice them. But suppose you eat food that is tainted with bacteria, viruses, mold, or other contaminants. Within a few hours to a few days, your body's reactions are hard to miss.

First, receptors in the digestive system signal the brain that toxins are in the gut. The brain responds by triggering vomiting, which ejects partially digested food from the stomach. Meanwhile, water moves from the circulatory system into the intestines. This fluid contributes to diarrhea that flushes toxins out of the body.

Vomiting and diarrhea dehydrate the body. In response, endocrine glands release a hormone called ADH (antidiuretic hormone) into the blood. ADH travels in blood vessels and binds to receptors in the kidneys. The kidney's cells respond by saving water, returning it to the blood instead of eliminating it in urine.

Some forms of food poisoning are accompanied by fever, shivers, and fatigue. These are responses of the immune system as it fights invaders. A feverish body is inhospitable to some bacteria. The shivers—a rapid series of muscle contractions—help raise the body's temperature. And what about fatigue? The body uses a lot of energy to maintain a fever and to produce immune cells. Ordinarily, we eat to replenish our reserves. But if food and liquids won't stay down, the digestive system cannot absorb nutrients or water. Both nutrient depletion and dehydration contribute to low energy.

Submit your burning question to marielle.hoefnagels@mheducation.com

23.3 Organ Systems Are Interconnected

The tissues described in section 23.2 build organs, which form organ systems. This section provides a brief overview of how each organ system contributes to the body's function (figure 23.6). Each system may seem distinct, but the function of each one relies on extensive interactions with the others (see Burning Question 23.1).

A. The Nervous and Endocrine Systems Coordinate Communication

The human **nervous system,** which consists of the brain, spinal cord, and nerves, specializes in rapid communication. Some neurons are sensory receptors that detect stimuli; others relay the sensory input to the spinal cord and brain. Still other neurons carry impulses from the brain or spinal cord to muscles or glands, which contract or secrete products in response.

The **endocrine system** includes glands that secrete hormones, which are communication molecules that affect development, reproduction, mental health, metabolism, and many other functions. Hormones travel within the circulatory system and stimulate a characteristic response in target organs. Hormones act relatively slowly, but their effects last longer than nerve impulses.

B. The Skeletal and Muscular Systems Support and Move the Body

The **skeletal system** consists of bones, cartilage, ligaments, and tendons. Bones protect underlying soft tissues and serve as attachment points for muscles. The marrow within some bones produces the components of blood; bones also store minerals such as calcium.

Individual skeletal muscles are the organs that make up the **muscular system.** When a skeletal muscle contracts, it moves another body part or helps support a person's posture. As noted in section 23.2, the heat released by contracting skeletal muscles also helps maintain body temperature.

Figure 23.6 Human Organ Systems.

Communication		Support and movement		Acquiring energy
Nervous system	Endocrine system	Skeletal system	Muscular system	Digestive system
Detects, interprets, and responds to stimuli from outside and within the body. With endocrine system, coordinates all organ functions.	Produces hormones and works with the nervous system to control many body functions, including reproduction, response to stress, and metabolism.	Provides framework for muscles to attach, making movement possible. Houses bone marrow. Protects soft organs. Stores minerals.	Supports posture and enables body to move. Helps maintain body temperature.	Breaks down nutrients into chemical components that are small enough to enter the blood. Eliminates undigested food.

Male

Female

C. The Digestive, Circulatory, and Respiratory Systems Work Together to Acquire Energy

The organs of the **digestive system** dismantle food, absorb the small molecules, and eliminate indigestible wastes. All cells of the body use the digested food molecules, either to generate energy in cellular respiration or as raw materials in maintenance and growth.

The **circulatory system** transports these food molecules (and many other substances) throughout the body. Nutrients absorbed by the digestive system enter blood at the intestines. The heart pumps the nutrient-laden blood through blood vessels that extend to all of the body's cells.

The **respiratory system** exchanges gases with the atmosphere. Cellular respiration requires not only food but also oxygen gas (O_2), which diffuses into blood at the lungs. The circulatory system delivers the O_2 throughout the body. Blood also carries carbon dioxide gas (CO_2), a waste product of cellular respiration, to the lungs to be exhaled.

D. The Urinary, Integumentary, Immune, and Lymphatic Systems Protect the Body

Cell metabolism generates many waste products in addition to CO_2. These wastes enter the blood, which circulates through the kidneys. These organs are part of the **urinary system,** the organs that remove water-soluble nitrogenous wastes and other toxins from blood and eliminate them in urine. The kidneys also have other protective functions; they adjust the concentrations of many ions, balance the blood's pH, and regulate blood pressure.

One line of physical protection is the **integumentary system,** which consists of skin, associated glands, hair, and nails. Skin is a waterproof barrier that helps keep the underlying tissues moist, blocks the entry of microorganisms, senses the environment, and helps maintain body temperature.

The body also fights infection and cancer. The **immune system** is a huge army of specialized cells, organs, and transport vessels. This complex system attacks cancer cells, viruses, microbes, and other foreign substances. Moreover, the immune system has a "memory" of previous infections. Vaccines build upon

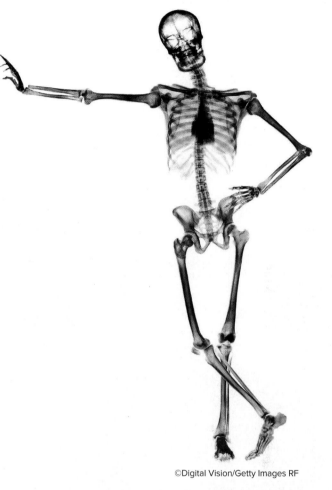

©Digital Vision/Getty Images RF

Acquiring energy		Protection			Reproduction
Circulatory system	**Respiratory system**	**Urinary system**	**Integumentary system**	**Immune and lymphatic systems**	**Reproductive system**
	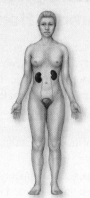	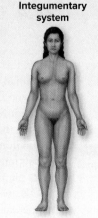	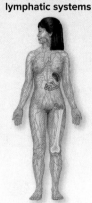		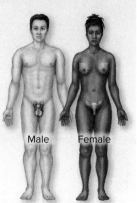
					Male Female
Vessels carry blood throughout the body, nourishing cells, delivering oxygen, and removing wastes.	Delivers oxygen to blood and removes carbon dioxide.	Excretes nitrogenous wastes and maintains volume and composition of body fluids.	Protects the body, controls temperature, and conserves water.	Protect the body from infection, injury, and cancer.	Manufactures gametes and enables the female to carry and give birth to offspring.

Animals Regulate Body Temperature

this memory by "teaching" the immune system about disease-causing agents the body has never actually encountered.

The **lymphatic system** is a bridge between the immune system and the circulatory system. Lymph originates as fluid that leaks out of blood capillaries and fills the spaces around the body's cells. Lymph capillaries absorb the excess fluid and pass it through the lymph nodes, where immune system cells destroy foreign substances. The fluid then returns to the circulatory system.

E. The Reproductive System Produces the Next Generation

The **reproductive system** consists of organs that produce and transport sperm and egg cells (gametes). Examples include the testes and penis in males and the ovaries and vagina in the female. The female body also can nurture developing offspring in the uterus. Moreover, hormones from the testes and ovaries promote the development of secondary sex characteristics in adults, including the facial hair of a man and the breasts and wide hips of a woman.

The reproductive system illustrates how the organ systems are, in a sense, not separate at all. Consider the uterus, the pear-shaped sac that houses the embryo and fetus. This organ consists mainly of muscle. It also contains nervous tissue, which is why a woman feels cramps when it contracts. Hormones from the endocrine system stimulate these contractions. The entire system is richly supplied with the circulatory system's blood vessels, which also deliver the cells and chemicals of the immune system.

Note that the organs and organ systems introduced in section 23.3 each sustain life in different ways. Organ failure can therefore be deadly. Burning Question 23.2 addresses one possible solution: artificial organs.

23.3 Mastering Concepts

1. Which organ systems contribute to each of the five general functions of life?
2. What are three examples of interactions between organ systems?

23.4 Organ System Interactions Promote Homeostasis

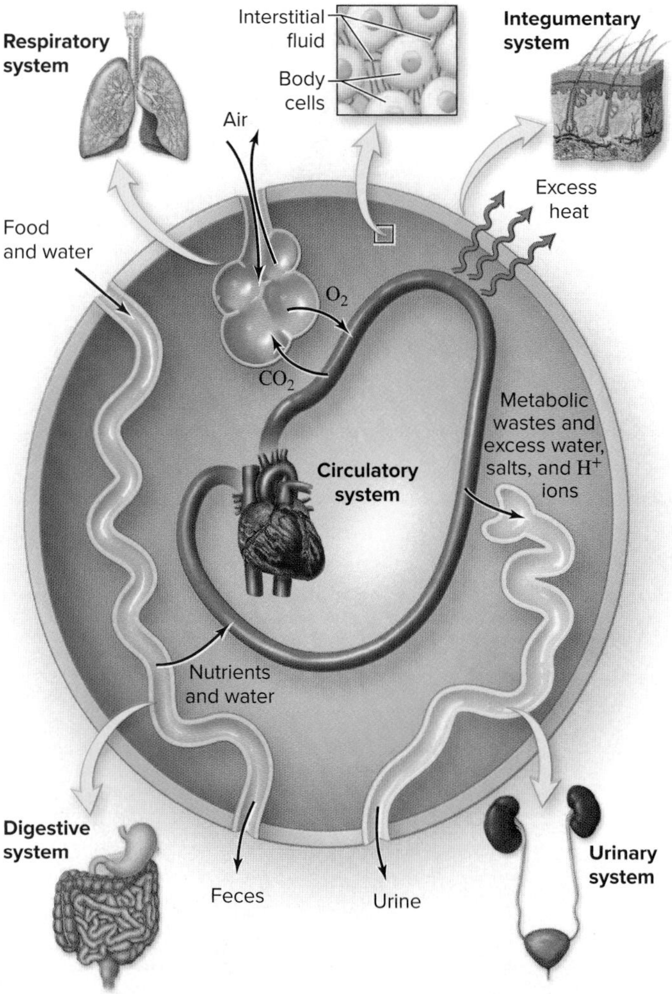

So far, this chapter has emphasized cells, tissues, organs, and organ systems. An animal's body, however, consists mostly of water. Some of this moisture makes up the cytoplasm inside every cell. The rest of it forms blood plasma and the **interstitial fluid** that bathes the body's cells. Because interstitial fluid is inside the body but outside the cells, biologists consider it part of the "internal environment." Many organ systems interact to help maintain the correct concentrations of nutrients, salts, hydrogen ions, and dissolved gases in body fluids (figure 23.7).

Yet the external environment, which surrounds the body, changes constantly. Temperatures rise and fall; food may be abundant or scarce; water comes and goes. In the midst of this variability, an animal's body must maintain its internal temperature, its blood pressure, and the chemical composition of its fluids within certain limits. **Homeostasis** is this state of internal stability.

All organisms maintain homeostasis. Without it, a body system may stop functioning, and the organism may die. As just one example, consider what

Figure 23.7 Organ System Interactions. This diagram of a generic animal illustrates how organ systems work together to maintain a constant body temperature and optimal concentrations of O_2, CO_2, nutrients, and other substances in the body's fluids.

happens if the lungs fill with water. The body can no longer acquire O_2 or dispose of CO_2, yet cells continue to respire. Soon, all available O_2 in the blood is consumed, and CO_2 accumulates to toxic levels. Cells begin to die, and the person will drown unless rescuers arrive quickly.

Many physiological mechanisms that maintain homeostasis use **negative feedback,** in which a change in a condition triggers action that reverses the change. Figure 23.8 illustrates negative feedback in a familiar situation: maintaining room temperature. When the room gets too warm, the heater turns off. When the temperature is too low, the thermostat signals the heater to switch on.

In all negative feedback systems, a sensor detects a stimulus such as body temperature or blood pH. If the value is too high or too low, a control center activates one or more effectors. The effector responds by counteracting the original change. In the body, the control center that coordinates much of the action is an almond-sized part of the brain called the hypothalamus. If blood pressure rises too high, for example, sensors in the walls of blood vessels signal the hypothalamus to slow the contraction of the heart. The pressure drops. If blood pressure falls too low, the hypothalamus signals the heart to speed up, sending out more blood.

As you will see throughout this unit, the hypothalamus participates in many negative feedback loops. If oxygen is scarce, the hypothalamus stimulates faster breathing. If the body's temperature deviates from normal, the hypothalamus initiates mechanisms that help the body release or conserve heat. If the blood's salt concentration is too high, hormones from the hypothalamus signal the kidney to release more salt into urine, and so on.

Only a few biological functions demonstrate **positive feedback,** in which the body reacts to a change by amplifying it. Blood clotting and childbirth are examples of positive feedback—once started, they perpetuate their activity. By itself, positive feedback therefore does not maintain homeostasis. Ultimately, however, other controls cut off the positive feedback loop and restore equilibrium.

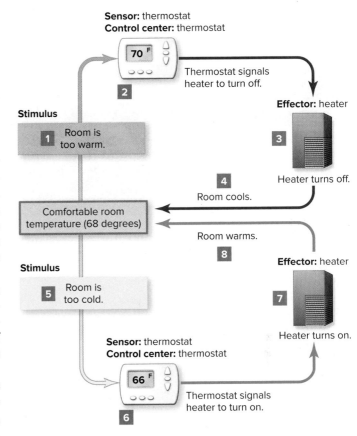

Figure 23.8 **Negative Feedback: An Example.** A negative feedback system maintains room temperature within comfortable limits. In this example, the room temperature is the stimulus, the thermostat is both the sensor and the control center, and the heater is the effector. (*1*) If the room is too warm, (*2*) the thermostat sends a signal to the heater. (*3*) The heater shuts off, and (*4*) the room cools. (*5*) If the room is too cold, (*6*) the thermostat sends a different signal to the heater, (*7*) which switches on. (*8*) This action warms the room.

23.4 Mastering Concepts

1. Use figure 23.7 to explain which materials enter and exit the body and how they do so.
2. Distinguish between negative and positive feedback.

23.5 Animals Regulate Body Temperature

Whether an animal lives in Antarctica or the Amazonian rain forest, its body temperature must remain within certain limits. Part of the reason is that extreme temperatures alter biological molecules. Excessive heat can ruin a protein's three-dimensional shape and disrupt its function. Extreme cold makes membranes less fluid, inhibiting their function. Overall, if cells become too warm or too cool, enzymes function less efficiently, and vital chemical reactions slow down or even stop. ⓘ *protein shape,* section 2.5C; *membranes,* section 3.3; *enzymes,* section 4.4

Thermoregulation is the control of body temperature, and it requires the ability to balance heat gained from and lost to the environment. An animal can also maintain homeostasis by controlling how much heat it produces. When cells generate ATP in aerobic respiration (see chapter 6), they also produce metabolic heat. The more active the animal, the higher its metabolic rate and the more heat it produces.

Miniglossary	Negative Feedback
Stimulus	Condition being monitored and adjusted (e.g., temperature, pH, blood glucose)
Sensor	Receptor that detects the level of the stimulus
Control center	Structure that receives input from a sensor and sends signals to an effector
Effector	Muscle or gland that counteracts the change that the sensor detected

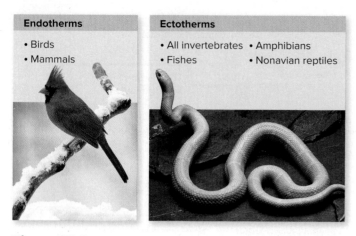

Endotherms	Ectotherms	
• Birds	• All invertebrates	• Amphibians
• Mammals	• Fishes	• Nonavian reptiles

Figure 23.9 Endotherm and Ectotherm. A bird is an endotherm; its metabolism generates most of its body heat. In contrast, an ectotherm such as a snake alters its behavior to manage the exchange of heat with the environment.

Photos: (bird): ©Daniel Dempster Photography/Alamy Stock Photo RF; (snake): ©IT Stock/PunchStock RF

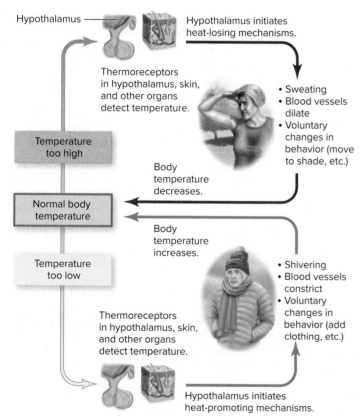

Figure 23.10 Thermoregulation in Humans. Thermoreceptors signal the hypothalamus to trigger the responses that maintain a stable body temperature.

The main source of an animal's body heat may be internal or external (figure 23.9). An **endotherm** regulates its body temperature internally. Most endotherms maintain a relatively constant body temperature by balancing heat generated in metabolism (especially in the muscles) with heat lost to the environment. Mammals and birds are endotherms; insulation in the form of fat, feathers, or fur helps retain their body heat.

Other adaptations also help endotherms maintain a constant body temperature. The hypothalamus detects blood temperature, receives information from sensory receptors in the skin and other organs, and controls many of the negative feedback loops that maintain homeostasis (figure 23.10). In cold weather, the animal may shiver; the contraction of skeletal muscle generates heat. Blood vessels in the limbs constrict, retaining more blood in the warmer core of the body. At the same time, muscles in the skin cause feathers or fur to "stand on end," trapping an insulating air layer next to the skin. (In humans, this hair-raising response is useless because we have so little body hair. Nevertheless, goose bumps form when the hair muscles contract.) Animals may also huddle together to conserve heat (figure 23.11a), or they may migrate to warmer climates during winter. ⓘ *vestigial structures,* section 13.4B

An **ectotherm** relies mostly on external conditions to thermoregulate. It moves to areas where it can gain or lose heat, so its temperature varies with the environment. The vast majority of animals are ectotherms, including all invertebrates, plus fishes, amphibians, and nonavian reptiles (see figure 23.9). Like endotherms, ectotherms have a repertoire of heat-conserving behaviors. They may seek sunlight, sprawl on warm rocks or roadways, build insulated burrows, and tuck wings and legs near their bodies.

So far, the focus has been on conserving heat, but animals also must maintain homeostasis when the environment is too hot (figure 23.11b). Evaporative cooling from the skin or respiratory surfaces is one way to lower body temperature. For example, humans sweat to cool off, whereas a panting dog allows water to evaporate from the moist lining of its mouth. Likewise, an owl flutters loose skin under its throat to move air over moist surfaces in the mouth.

In addition, when the environment is warm, blood vessels in the limbs dilate and allow more blood to approach the relatively cool body surface. Tiny veins in the face and scalp also reroute blood cooled near the body's surface toward the brain. This adaptation explains why vigorous exercise causes the face of a light-skinned person to turn red.

Behavioral strategies can also help both ectotherms and endotherms to cool off. Many animals escape the sun's heat by swimming, covering themselves with cool mud, or retreating to the shade. Some burrow underground and emerge only at night; others extend their wings to promote cooling. Humans shed extra layers of clothing, and we consume cold food and drinks. We also swim or fan ourselves to increase heat loss.

Both ectothermy and endothermy have advantages and disadvantages. The ectotherm uses much less energy, and therefore requires less food and O_2, than an endotherm. However, an ectothermic animal must be able to seek or escape environmental heat. An injured snake that could not squeeze into a crevice to avoid the broiling sun would cook to death. Ectotherms also become sluggish when the temperature is low, which can make it hard for them to escape from predators.

On the other hand, an endotherm typically maintains its body temperature even in cold weather or in the middle of the night. But this internal constancy comes at a cost. The metabolic rate of an endotherm is generally five times that of an ectotherm of similar size and body temperature. Endotherms therefore require much more food than do ectotherms. However, during cold periods, some birds and mammals cut the demand for food by

a.

b.

Figure 23.11 **Too Cold or Too Hot.** (a) Snow monkeys are adapted to cold winters. Their thick fur retains body heat, as does their huddling behavior. (b) This panting dog loses excess heat through its mouth, while the man is sweating. The man has also employed a behavioral strategy that helps him cool off: He has shed his excess clothing. (a): ©Akira Kaede/Digital Vision/Getty Images RF; (b) ©Dynamic Graphics/PictureQuest RF

temporarily allowing their body temperature and metabolic rate to fall (a condition called torpor). Hummingbirds and bats undergo this process each night; in contrast, bears are famous for entering long periods of torpor—commonly known as hibernation—during the winter.

23.5 Mastering Concepts

1. Describe the difference between endotherms and ectotherms.
2. What are the advantages and disadvantages of endothermy and ectothermy?

Figure It Out

As the sun sets and the external temperature gets colder, will a frog's body temperature go down, stay the same, or go up?

Answer: It will go down.

Burning Question 23.2

Can biologists build artificial organs?

Medical technology can help replace body parts damaged by disease or injury. Transplantable organs—taken from living donors or cadavers—include corneas, pancreases, kidneys, skin, livers, lungs, bone marrow, parts of the digestive tract, and hearts. Surgeons have even transplanted entire hands and faces.

Unfortunately, the demand for transplantable organs far exceeds the supply. Artificial organs, grown in the lab, may one day offer an attractive option (see the chapter-opening photo). One example is a new technique that uses three-dimensional (3-D) printing to create an exact replica of a human ear. A 3-D bioprinter uses a laser scan of an ear to create a collagen mold. Technicians inject the mold with human cartilage cells that grow within the collagen matrix, just as they would in a genuine ear. Using a similar technique, researchers in China and San Diego have "printed" functional mini-livers—complete with blood

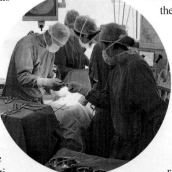

©Chris Ryan/age fotostock RF

vessels and other tissues. So far, however, neither the printed ears nor the mini-livers have been connected to a human.

Whether an organ comes from a person or a lab bench, one challenge in transplant medicine is preventing the recipient's immune system from rejecting the foreign tissues. Transplant surgeons minimize rejection by carefully matching organ donors with recipients, but the immune system attacks nonliving materials, too. Biologists are therefore trying to build artificial organs out of materials that are "invisible" to the body's defenses. In addition, physicians can prescribe drugs that suppress the immune system, but this strategy increases the risk of deadly infection.

Submit your burning question to
marielle.hoefnagels@mheducation.com

Investigating Life 23.1 | Sniffing Out the Origin of Feathers

Have you wondered about the origin of a bird's beautiful plumage? Evolutionary biologists working to answer this question think metabolism could be part of the answer. Existing nonavian reptiles such as lizards and crocodiles are ectotherms, whereas birds and mammals are endotherms. Yet fossil and DNA evidence clearly indicate that birds arose from an ancient lineage of nonavian reptiles. Feathers provide insulation that helps maintain a bird's body temperature. Did the elevated metabolic rates associated with endothermy evolve at the same time as insulating feathers?

To test this hypothesis of simultaneous evolution, scientists need to trace animal ancestry back in time. Feathers occasionally show up in fossils, but how can we know whether an extinct animal was an endotherm or an ectotherm? Surprisingly, the ideal indicator of endothermy may be the inside of the nose. Endotherms have a high metabolic rate, which in turn means a huge demand for O_2 to fuel respiration. This elevated O_2 demand requires a high breathing rate; in general, endotherms should therefore have broader nasal cavities than ectotherms.

Researchers tested this prediction by measuring the cross-sectional areas of the nasal cavities of 21 living species of birds, mammals, and nonavian reptiles. In every case, the cross-sectional area of the nasal cavity was larger for an endotherm than for an ectotherm of equal size (figure 23.B). The team then used modern imaging technologies to measure nasal cavities inside the fossilized skulls of three dinosaur species that lived about 70 million years ago (mya) and are closely related to modern birds. The results indicate that the reptilian ancestors to birds probably were ectotherms. Yet fossil evidence suggests that dinosaurs with feathers already existed by 150 mya—more than 80 million years earlier.

Feathers are adaptations that help birds stay warm, so it is easy to assume they evolved hand-in-hand with endothermy. By pairing old-fashioned comparative anatomy with modern technology, however, researchers have turned this assumption on its head.

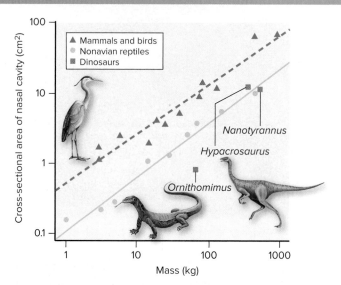

Figure 23.B **The Nose Knows.** The nasal cavities of existing endotherms (ranging in size from herons to African cape buffaloes) have a higher cross-sectional area than those of ectotherms of equal size (including iguanas, monitor lizards, and crocodiles). Dinosaurs had nasal cavities similar in size to those of existing ectotherms.

Sources: Ruben, John A., Willem J. Hillenius, Nicholas R. Geist, et al. 1996. The metabolic status of some late Cretaceous dinosaurs. *Science,* vol. 273, pages 1204–1207.

Ruben, John A., and Terry D. Jones. 2000. Selective factors associated with the origin of fur and feathers. *American Zoologist,* vol. 40, pages 585–596.

What's the **Point?** ▼ APPLIED

Running a marathon is an impressive athletic endeavor that places extreme stress on many organ systems. How does the body maintain homeostasis before, during, and after the race?

Imagine that you have registered to run a marathon, and the race is tomorrow morning. To prepare, you've eaten lots of starchy foods today. Your digestive system physically and chemically broke down the meals; sugars and other nutrients then moved from your small intestine into your circulatory system. Your blood delivered the nutrients to your body cells, so your energy reserves are high.

But your preparation actually began long ago. Months of endurance training increased your heart's ability to pump blood throughout your body. You have also felt the discomfort of physical exhaustion many times. You know how to relax when you are tired, minimizing wasted energy. You are ready.

Fast-forward to the morning of the race. The starter's gun goes off, and you start to run. Every step requires countless interactions between your nervous system and your muscular system. Neurons send signals to muscle cells, telling them it is time to contract. Each muscle, in turn, pulls on a bone. The brain seamlessly orchestrates these individual movements so your overall stride is smooth and balanced.

©Erik Isakson/Blend Images LLC RF

Communication systems also coordinate changes in your metabolism, heartbeat, and breathing rate. The nervous system senses that your muscle cells have ramped up their respiration rate. In response, neurons signal your breathing rate to increase, allowing more oxygen into your body and more carbon dioxide to escape. Blood shuttles these gases between body cells and the lungs at an increasing speed as your pulse rate climbs. As your blood glucose level drops, hormones stimulate your liver and fat cells to release stored nutrients into your blood, which delivers these molecules to the energy-hungry muscle cells. Meanwhile, signals from the nervous system shunt blood away from the digestive system, saving energy for the processes that your body needs to run. Sweat pours out of your skin and blood vessels near the body surface dilate, releasing excess body heat.

After you cross the finish line, your pulse, metabolism, breathing rate, and core body temperature slowly return to their resting levels. Your digestive system receives blood once again, preparing your body for its post-marathon meal—a welcome reward after a job well done.

CHAPTER SUMMARY

23.1 Specialized Cells Build Animal Bodies

- **Anatomy** and **physiology** are interacting studies of the structure and function of organisms.
- Specialized cells function together to form **tissues.** Tissues build **organs,** and interacting organs form **organ systems.**

23.2 Animals Consist of Four Tissue Types

- Animal tissues consist of cells within an **extracellular matrix** of water, dissolved substances, and (usually) protein fibers.

A. Epithelial Tissue Covers Surfaces

- **Epithelial tissue** lines and covers organs; it also forms **glands.** This tissue protects the underlying tissue, senses stimuli, and secretes substances.
- Epithelium may be simple (one layer) or stratified (more than one layer), and the cells may be flat, cube-shaped, or columnar.

B. Most Connective Tissues Bind Other Tissues Together

- **Connective tissues** have diverse structures and functions. Most consist of scattered cells and a prominent extracellular matrix.
- The six major types of connective tissues are loose connective tissue, dense connective tissue, adipose tissue, cartilage, bone, and blood.

C. Muscle Tissue Provides Movement

- **Muscle tissue** consists of cells that contract when protein filaments slide past one another.
- Three types of muscle tissue are skeletal, cardiac, and smooth muscle.

D. Nervous Tissue Forms a Rapid Communication Network

- Neurons and neuroglia make up **nervous tissue.**
- A neuron functions in rapid communication; neuroglia support neurons.
- Table 23.1 summarizes the four types of tissue in an animal's body.

23.3 Organ Systems Are Interconnected

A. The Nervous and Endocrine Systems Coordinate Communication

- The **nervous system** and **endocrine system** coordinate all other organ systems.
- Neurons form networks of cells that communicate rapidly, whereas hormones produced by the endocrine system act more slowly.

B. The Skeletal and Muscular Systems Support and Move the Body

- The bones of the **skeletal system** protect and support the body. Bones also store calcium and other minerals.
- The **muscular system** enables body parts to move and generates body heat.

TABLE 23.1 Animal Tissue Types: A Summary		
Tissue Type	**Description**	**Functions**
Epithelial	Single or multiple layer of flattened, cube-shaped, or columnar cells	Cover interior and exterior surfaces of organs; protection; secretion; absorption
Connective	Cells scattered in prominent extracellular matrix	Support, adhesion, insulation, attachment, and transportation
Muscle	Elongated cells that contract when stimulated	Movement
Nervous	Cells that transmit electrochemical impulses	Rapid communication among cells

C. The Digestive, Circulatory, and Respiratory Systems Work Together to Acquire Energy

- The **digestive system** provides nutrients. The **respiratory system** obtains O_2, and the **circulatory system** delivers nutrients and O_2 to tissues.
- The body's cells use O_2 to extract energy from food molecules. The circulatory and respiratory systems eliminate the waste CO_2.

D. The Urinary, Integumentary, Immune, and Lymphatic Systems Protect the Body

- The **urinary system** removes metabolic wastes from the blood and reabsorbs useful substances.
- The **integumentary system** provides a physical barrier between the body and its surroundings.
- The **immune system** protects against infection, injury, and cancer.
- The **lymphatic system** connects the circulatory and immune systems, passing the body's fluids through the lymph nodes.

E. The Reproductive System Produces the Next Generation

- The male and female **reproductive systems** are essential for the production of offspring.
- Figure 23.12 summarizes the human body's organ systems.

23.4 Organ System Interactions Promote Homeostasis

- **Homeostasis** is stability in the internal environment. Animals may maintain homeostasis in body temperature and in the chemical composition of the blood plasma and the **interstitial fluid.**
- **In negative feedback,** sensors detect a change in the internal environment, and a control center activates effectors that counteract the change. The overall effect is to restore the parameter to its normal range.
- **Positive feedback** reinforces the effect of a change.

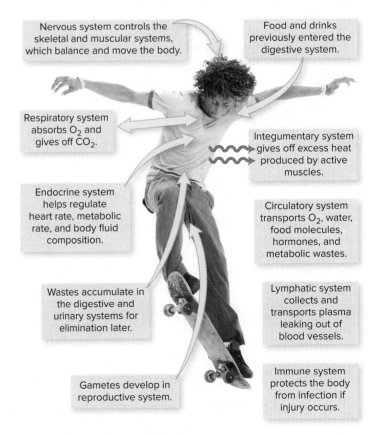

Nervous system controls the skeletal and muscular systems, which balance and move the body.

Food and drinks previously entered the digestive system.

Respiratory system absorbs O_2 and gives off CO_2.

Integumentary system gives off excess heat produced by active muscles.

Endocrine system helps regulate heart rate, metabolic rate, and body fluid composition.

Circulatory system transports O_2, water, food molecules, hormones, and metabolic wastes.

Wastes accumulate in the digestive and urinary systems for elimination later.

Lymphatic system collects and transports plasma leaking out of blood vessels.

Gametes develop in reproductive system.

Immune system protects the body from infection if injury occurs.

Figure 23.12 Organ System Functions: A Summary.
Photo: ©James Woodson/Getty Images RF

23.5 Animals Regulate Body Temperature

- Animals regulate their body temperatures (**thermoregulate**) with physiological and behavioral adaptations.
- **Endotherms** use internal metabolism to thermoregulate. **Ectotherms** use the environment to regulate body temperature.
- Adaptations to cold include insulation, the constriction of blood vessels near the body's surface, and shivering.
- Adaptations to heat include evaporative cooling, routing blood toward the body surface, and behaviors that accelerate heat loss.

MULTIPLE CHOICE QUESTIONS

1. Which of the following represents the correct order of organization of an animal's body?
    a. Cells; organs; organ systems; tissues
    b. Cells; tissues; organ systems; organs
    c. Tissues; cells; organs; organ systems
    d. Cells; tissues; organs; organ systems

2. Epithelial tissue consisting of multiple layers of flattened cells is called _____.
    a. simple cuboidal
    b. simple squamous
    c. stratified cuboidal
    d. stratified squamous

3. Which of the following features do all types of connective tissue share?
    a. Cells are connected firmly to each other.
    b. Each cell contacts an extracellular matrix.
    c. Abundant protein fibers surround cells.
    d. The extracellular matrix is fluid.

4. Blood differs from other types of connective tissue in that it
    a. lacks fibers in its extracellular matrix.
    b. has cells.
    c. binds epithelium to underlying tissues.
    d. contains minerals.

5. Smooth muscle is *different* from skeletal muscle because
    a. smooth muscle contraction is involuntary.
    b. skeletal muscle is striated (striped).
    c. smooth muscle connects to bones.
    d. Both a and b are correct.

6. Ovaries produce gametes and hormones; these organs therefore belong to the _____ systems.
    a. immune and integumentary
    b. reproductive and endocrine
    c. circulatory and nervous
    d. urinary and lymphatic

7. Which of the following scenarios does NOT illustrate negative feedback?
    a. In childbirth, contractions stimulate the release of oxytocin, which provokes more contractions.
    b. Body temperature climbs so high that a person begins to sweat, which cools the body.
    c. The salt concentration in blood is too high, so the kidneys eliminate salt in urine.
    d. Eating a meal causes a rise in blood sugar, which stimulates body cells to absorb sugar from blood.

8. Why do endotherms require more energy than ectotherms?
    a. Because they need more energy to move between hot and cold environments
    b. Because they use metabolic energy to maintain an internal temperature
    c. Because they lack insulation that retains body heat
    d. Both b and c are correct.

9. Which animal's body temperature would drop the fastest in a cold environment?
    a. A whale
    b. An Arctic bird
    c. A tropical bird
    d. A lizard

Answers to Multiple Choice questions are in appendix A.

WRITE IT OUT

1. Distinguish between the following pairs of terms.
    a. Organs and organ systems
    b. Simple squamous and stratified squamous epithelial tissue
    c. Loose and dense connective tissue
    d. Skeletal and cardiac muscle tissue
    e. Neurons and neuroglia
    f. Negative and positive feedback

2. Where do epithelial tissues occur, and how are they named?

3. List and describe six types of connective tissue.

4. Explain the similarities and differences among the three types of muscle tissue.

5. What are the two main cell types in nervous tissue?

6. Use the Internet to research cosmetic surgery. Write a paragraph explaining how a plastic surgeon might manipulate each of the four main tissue types.

7. Explain how you use your organ systems (except your reproductive system) while you are exercising at the gym, and provide three hypotheses that might explain why identical workouts might feel easier on some days than others.

8. Describe how the circulatory system connects the digestive system with the urinary system.

9. Make a chart that compares and contrasts the organization of the animal body with that of a plant (see chapter 21).

10. What is homeostasis, and how is it important?

11. Provide nonbiological examples of negative and positive feedback, other than those mentioned in the chapter.

12. When a person gets cold, he or she may begin to shiver. If the weather is too hot, the heart rate increases and blood vessels dilate, sending more blood to the skin. How does each scenario illustrate homeostasis?

13. Observe what happens to the size of your eye's pupil when you leave a dark room and enter the sunshine. What happens in the opposite situation, when you enter a dark room? How do the opposing reactions of your eye illustrate negative feedback?

14. Birds and insects frequently collect nectar from plants. Birds are endothermic, and insects are ectothermic. Do you think a greater mass of insects or of birds can be supported on 100 g of nectar? Explain your answer.

15. Would an alligator require more, less, or the same amount of food as a horse of the same size? Explain.

16. Explorers of Antarctica must eat thousands more Calories per day than people exploring the tropics. Explain this observation.

17. Woolly mammoths are extinct relatives of modern-day elephants. The mammoths were heavier and shaggier than elephants, their ears were smaller, and they had a thick fat layer under their skin. Explain each of these differences in light of the fact that today's elephants originate in Asia and Africa, whereas mammoths lived on the tundra.

SCIENTIFIC LITERACY

Review Burning Question 23.2 and then search the Internet for "pig–human liver chimeras." Scientists are attempting to use human stem cells to produce human organs within developing pigs, with the goal of reducing the transplantable organ shortage. Make an ethical argument both for and against this practice.

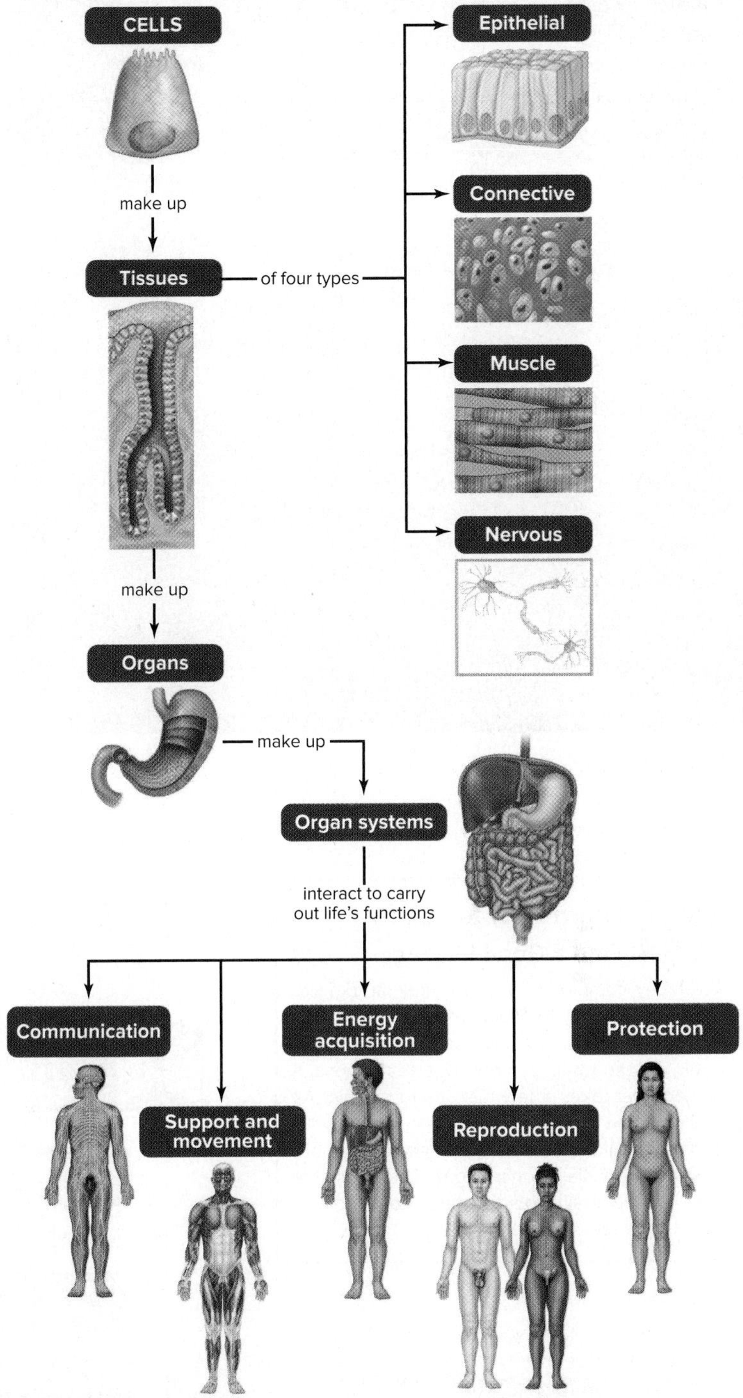

Figure 23.13 Pull It Together: Animal Tissues and Organ Systems.

Refer to figure 23.13 and the chapter content to answer the following questions.

1. Review the Survey the Landscape figure in the chapter introduction, and then connect *homeostasis* to figure 23.13 in at least two ways. One connection to *homeostasis* should include a detailed description of how one "life function" (e.g., *energy acquisition*) maintains homeostasis.

2. Add the names of the 11 organ systems to this concept map.

3. Connect *thermoregulation* to the concept map in three ways.

Answers to Mastering Concepts, Write It Out, Scientific Literacy, and Pull It Together questions can be found in the Connect ebook.
connect.mheducation.com

Design element: Burning Question (fire background): ©Ingram Publishing/Super Stock

Mapping Brain Activity. Electrodes detect charge fluctuations in this man's neurons as he meditates.

©Cary Wolinsky/Getty Images

LEARNING OUTLINE

24.1 The Nervous System Forms a Rapid Communication Network

24.2 Neurons Are the Functional Units of a Nervous System

24.3 Action Potentials Convey Messages

24.4 Neurotransmitters Pass the Message from Cell to Cell

24.5 The Peripheral Nervous System Consists of Nerve Cells Outside the Central Nervous System

24.6 The Central Nervous System Consists of the Spinal Cord and Brain

24.7 The Senses Connect the Nervous System with the Outside World

24.8 The General Senses Detect Touch, Temperature, and Pain

24.9 The Senses of Smell and Taste Detect Chemicals

24.10 Vision Depends on Light-Sensitive Cells

24.11 The Sense of Hearing Begins in the Ears

APPLICATIONS

Burning Question 24.1 *Do neurons communicate at the speed of light?*
Why We Care 24.1 *Drugs and Neurotransmitters*
Burning Question 24.2 *Do I really use only 10% of my brain?*
Burning Question 24.3 *Do humans have pheromones?*
Why We Care 24.2 *Correcting Vision*
Burning Question 24.4 *What is an ear infection?*
Investigating Life 24.1 *Scorpion Stings Don't Faze Grasshopper Mice*

Learn How to Learn
Find a Good Listener

For many complex topics, you may struggle to know how well you really understand what is going on. One tip is to try explaining what you think you know to somebody else. Choose a subject that takes a few minutes to explain. As you describe the topic in your own words, your partner should ask follow-up questions and note where your explanation is vague. Those insights should help draw your attention to important details that you have overlooked.

In the nervous system, neurons are the cells that communicate with one another and with cells in other organ systems. Sensory structures pass along information about the outside world to the body's internal communication networks, triggering responses that maintain homeostasis.

For more details, study the Pull It Together feature in the chapter summary.

SURVEY THE LANDSCAPE
Animal Anatomy and Physiology

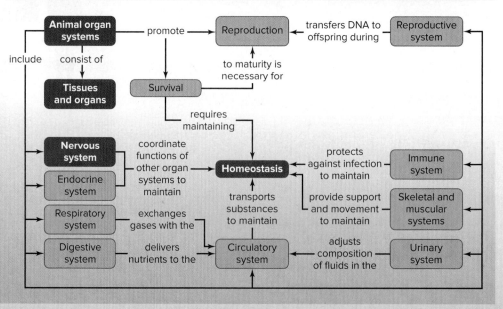

©Digital Vision RF

It is almost impossible to conceive of the nervous system's complexity. Intricate connections among 100 billion neurons simultaneously influence everything from wakefulness to muscle contraction to memory.

As complex as the nervous system is, however, the neurons all work on the same basic principles: Electrochemical signals zip from cell to cell, producing instant responses. Disease or trauma may damage or break these vital communication lines. In that case, messages are lost in transit; memory, sleep, speech, sensation, movement, and many other functions may be affected.

This chapter explores the basic structure and function of the nervous system before focusing on one familiar component: the senses. The eyes, ears, skin, nose, and mouth are our gateways to the outside world. Thanks to the blend of our senses, we experience the world as an exciting, multidimensional place.

24.1 The Nervous System Forms a Rapid Communication Network

Love, happiness, tranquility, sadness, jealousy, rage, fear, and excitement—all of these emotions spring from the cells of the nervous system. So do language, the sensation of warmth, memories of your childhood, and your perception of pain, color, sound, smell, and taste. The muscles that move when you chew, blink, or breathe all are controlled by the nervous system, as is the unseen motion that propels food along your digestive tract.

Among the nervous system's most critical functions are the "behind the scenes" activities that keep the body's temperature, ion balance, and other conditions within optimal levels. The negative feedback loops that maintain homeostasis require communication between the sensors that detect each condition, the control center, and the effectors (muscles and glands) that make adjustments. Together, the **nervous system** and the **endocrine system** provide this communication. A major difference between these two organ systems is the speed with which they act. The nervous system's electrochemical impulses travel so rapidly that their effects are essentially instantaneous. The endocrine system, the subject of chapter 25, acts much more slowly. Endocrine glands secrete chemical messages called hormones that can take minutes or hours to take effect, but these signals last much longer. ⓘ *negative feedback,* section 23.4

Nervous tissue includes two basic cell types: interconnected neurons and their associated neuroglia. The **neurons** are the cells that communicate with one another (and with muscles and glands). The more numerous **neuroglia** are cells that provide physical support, help maintain homeostasis in the fluid surrounding the neurons, guide neuron growth, and play many other roles that researchers are just beginning to discover. ⓘ *nervous tissue,* section 23.2D

These basic building blocks form a wide variety of animal nervous systems, which range from diffuse networks of neurons in jellyfishes to the highly centralized nervous systems of vertebrates. As nervous systems increased in complexity, so too did animals' abilities to detect stimuli, coordinate responses, form memories, solve problems, and communicate. These capabilities are most highly developed in vertebrates.

The vertebrate nervous system has two main divisions: the central and peripheral nervous systems. The **central nervous system** consists of the **brain** (inside the skull) and **spinal cord.** The main function of these two organs is to integrate sensory information and coordinate the body's response. The **peripheral nervous system** carries information between the central nervous system and the rest of the body.

To understand how the nervous system regulates virtually all other organ systems, imagine a lynx hunting a hare **(figure 24.1)**. Sensory

Sensory input (ears, eyes, and nose detect prey)

Sensory integration (brain and spinal cord interpret sensory input)

Motor response (muscles and glands react)

Figure 24.1 Roles of the Nervous System. Sensory organs such as the eyes, ears, and nose receive sensory input. The central nervous system integrates the information and sends signals that initiate appropriate motor responses.

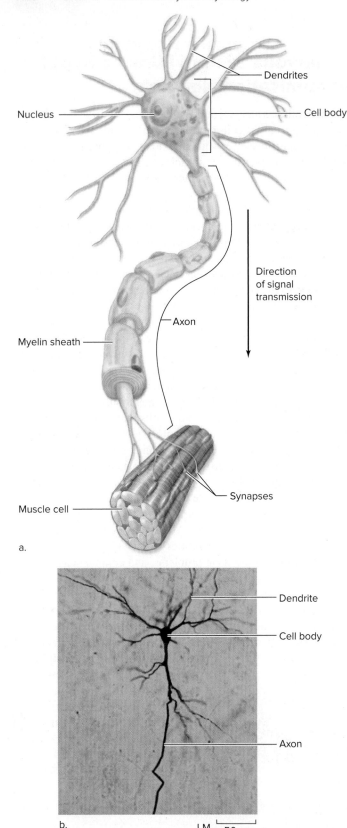

a.

b.   LM ⊢—⊣
                         50 μm

Figure 24.2 **Parts of a Neuron.** (a) A neuron consists of a cell body, one or more dendrites that transmit information to the cell body, and an axon. In many neurons, the axon is encased in a myelin sheath. (b) The neuron's parts are clearly visible in this micrograph.

(b): ©Biophoto Associates/Science Source

neurons in the peripheral nervous system enable the cat to hear, see, and smell its prey. The lynx's central nervous system interprets this sensory input and decides how to act, and then motor neurons coordinate the skeletal muscles that move the lynx into position to catch the hare. Meanwhile, the cat's heart pumps blood, and its lungs inhale and exhale—all under the control of the central and peripheral nervous systems.

24.1 Mastering Concepts

1. How are nervous systems adaptive to animals?
2. What are the roles of neurons and neuroglia?
3. Distinguish between the central and peripheral nervous systems.

24.2 Neurons Are the Functional Units of a Nervous System

The nervous system's function is rapid communication by electrical and chemical signals. Neurons are the cells that do the communicating, either with one another or with muscles and glands. To understand how neurons carry out their function, it helps to first learn about their structure.

A. A Typical Neuron Consists of a Cell Body, Dendrites, and an Axon

All neurons have the same basic parts (figure 24.2). The enlarged, rounded **cell body** contains the nucleus, mitochondria that supply ATP, ribosomes that manufacture proteins, and other organelles. **Dendrites** are short, branched extensions that transmit information toward the cell body. The number of dendrites may range from one to thousands, and each can receive input from many other neurons. ⓘ *animal cell structure,* section 3.2B

The **axon,** also called the nerve fiber, conducts nerve impulses away from the cell body. An axon is typically a single long extension that is finely branched at its tip. Each tiny terminal extension communicates with another cell at a junction called a synapse. In many neurons, a **myelin sheath** composed of fatty material coats sections of the axon, speeding nerve impulse conduction (see section 24.3C).

To picture the relative sizes of a typical neuron's parts, imagine its cell body is the size of a tennis ball. The axon might then be up to 1.5 kilometers long—the length of about 63 tennis courts—but only a few centimeters thick. The mass of dendrites extending from the cell body would fill an average-size living room.

B. The Nervous System Includes Three Classes of Neurons

Biologists divide neurons into three categories, based on general function (figure 24.3).

- A **sensory neuron** brings information to the central nervous system from the rest of the body. Sensory neurons respond to light, pressure from sound waves, heat, touch, pain, and chemicals detected as odors or taste. The dendrites, cell body, and most of the axon of each sensory neuron lie in the peripheral nervous system, whereas the axon's endings reside in the central nervous system.

- About 90% of all neurons are **interneurons,** which connect one neuron to another within the spinal cord and brain. Interneurons receive information from sensory neurons, process this information, and generate the messages that the motor neurons carry to muscles and glands.

- A **motor neuron** conducts its message from the central nervous system toward a muscle or gland cell. A motor neuron's cell body and dendrites reside in the central nervous system, but its axon extends into the peripheral nervous system. Thus, motor neurons stimulate muscle cells to contract and stimulate glands to secrete their products into the bloodstream or into a duct. (They are called motor neurons because most lead to muscle cells, not glands.)

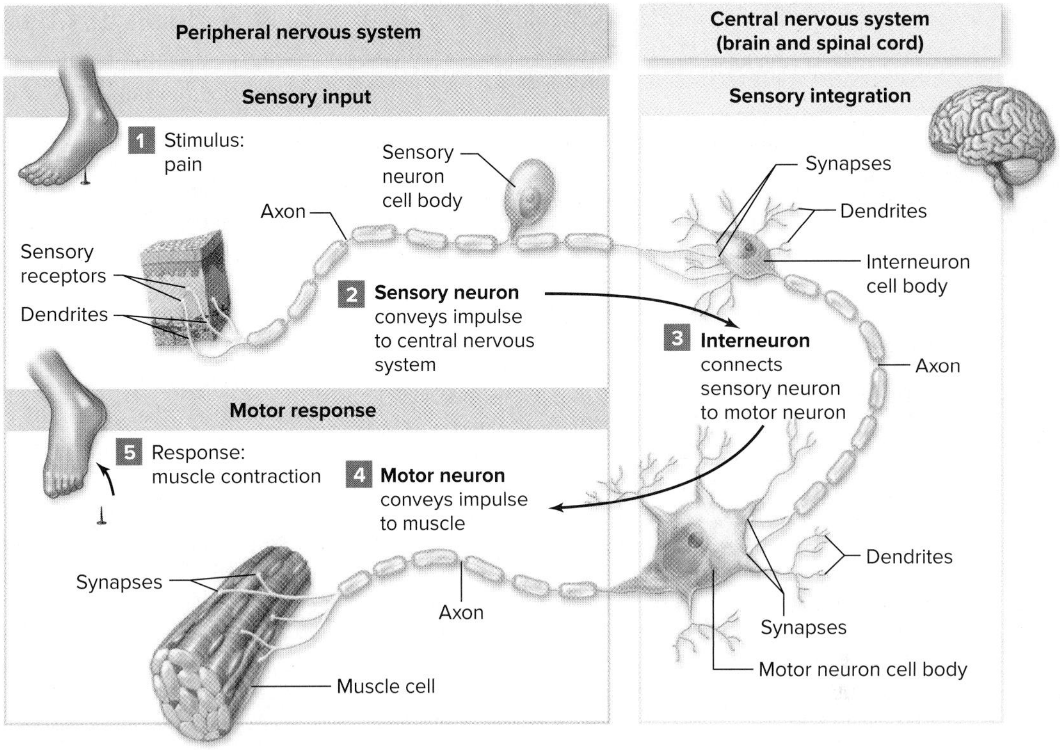

Figure 24.3 shows a simplified example of how the three types of neurons work together to coordinate the body's reaction to a painful stimulus. The process begins when a person steps on a tack. Sensory neurons whose dendrites are in the skin of the foot convey the information to the spinal cord. There, the sensory neuron transmits the signal to an interneuron, which in turn synapses on a motor neuron that stimulates muscle contraction in the leg and foot. The interneuron in the spinal cord also sends action potentials to an area of the brain that interprets the sensation as pain. The signals move so quickly that the foot withdraws at about the same time as the brain perceives the pain. (Burning Question 24.1 explores the nervous system's speed.) ⓘ *motor neurons and muscle contraction,* section 26.4C

Figure 24.3 Categories of Neurons. Sensory neurons transmit information from sensory receptors to the central nervous system. Interneurons connect sensory neurons to motor neurons, which send information from the central nervous system to muscles or glands.

24.2 Mastering Concepts

1. Describe the parts of a typical neuron.
2. In what direction does a message move within a neuron?
3. Describe the three categories of neurons.

24.3 Action Potentials Convey Messages

Neurons send messages by conveying action potentials; a neural impulse is the propagation of action potentials like a wave along an axon. As you will see, action potentials result from the movement of charged particles (ions) across a neuron's cell membrane. This section describes the distribution of ions in neurons, both when the cell is "at rest" and when it is transmitting a neural impulse. ⓘ *ions,* section 2.1B; *cell membrane,* section 3.3

Miniglossary | Neuron Anatomy

Cell body	Enlarged portion of neuron; contains most of the cell's organelles
Dendrite	Short, branched extension that receives impulses from other neurons and transmits the signals toward the cell body
Axon	Long extension that transmits action potentials away from the cell body; portions of the axon may have a myelin sheath
Synapse	Specialized junction at which an impulse from one neuron is transmitted to another cell

Do neurons communicate at the speed of light?

Many people assume that neurons work like the metal wiring in a home, conducting electricity at nearly the speed of light (about 300 million meters per second). In fact, biologists once thought that nerve impulses were instantaneous. But experiments eventually demonstrated that a mammal's neurons conduct impulses at a top speed of about 100 meters per second (224 miles per hour). That speed is impressive, and it allows athletes and ordinary people to react quickly to incoming stimuli, but it is just a tiny fraction of the speed of light.

A simple exercise can help you estimate the speed of neural communication in your body. You'll need a partner, a ruler, paper and pencil, and a calculator. Have your partner hold the ruler vertically by the 30 cm mark, and cup your own hand around the 0 cm mark. Your partner should then drop the ruler without saying anything. Grab the ruler as quickly as possible after your partner drops it, and note the centimeter mark where you caught it. Repeat three times; use the average of the three results as your distance measurement (*D*). Then use the following equation to convert *D* into a time measurement (*T*):

$$T = \sqrt{(2 \times D)/980}$$

T represents the time (in seconds) it took for receptors in your eye to send neural impulses to the brain, for the brain to process the information and send a neural impulse along a motor neuron to your arm, and for your muscles to close your hand around the ruler. Keep your reaction time in mind as you learn how cells in the nervous system communicate.

Submit your burning question to
marielle.hoefnagels@mheducation.com

(batter): ©Akihiro Suhimoto/age fotostock RF

A. A Neuron at Rest Has a Negative Charge

To understand how ions move in a neural impulse, it helps to be familiar with the **membrane potential,** which is the difference in electrical charge between the inside and outside of a neuron. The membrane potential can change, depending on whether or not the neuron has received a stimulus.

The neuron's **resting potential** is its membrane potential when it is not conducting a neural impulse. At rest, the inside of a neuron carries a negative electrical charge relative to the outside because it maintains an unequal distribution of ions across its membrane (figure 24.4). In particular, the concentration of potassium (K^+) is much higher inside the cell than outside, while the reverse is true for sodium (Na^+). One membrane protein that helps maintain this gradient is the sodium–potassium pump, which pumps three Na^+ out of the cell for every two K^+ that enter; the energy cost is one ATP per cycle. The sodium–potassium pump operates continuously. ⓘ *active transport,* section 4.5B

The neuron's resting potential reflects a balance of forces on K^+. On one hand, the sodium–potassium pump concentrates K^+ inside the cell. Membrane proteins called "leakage channels" allow some of this K^+ to diffuse back out of the cell along its concentration gradient. On the other hand, positively charged Na^+ ions outside the cell repel K^+, while large, negatively charged proteins (and other negative ions) inside the cell attract K^+. When the opposing forces—the concentration gradient and charge interactions—are equal, the membrane has a net positive charge on the outside and a net negative charge on the inside. This difference in charge is the resting potential. ⓘ *facilitated diffusion,* section 4.5A

The term *resting potential* is a bit misleading because the neuron consumes a tremendous amount of energy while "at rest." In fact, the nervous system devotes about three quarters of its total energy budget to maintaining the ion gradients that characterize the resting potential. The resulting state of readiness allows each neuron to respond quickly when a stimulus does arrive. The resting potential is therefore like holding back the string on a bow to be constantly ready to shoot an arrow.

B. A Neuron's Membrane Potential Reverses During an Action Potential

A change in pH, a touch, or a signal from another neuron may trigger a neuron to "fire," meaning that action potentials occur along the neuron's axon. An **action potential** is a brief reversal in membrane potential that propagates like a wave along the membrane of the axon; figure 24.5 summarizes how this occurs.

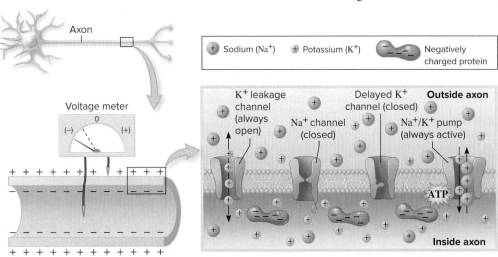

Figure 24.4 **The Resting Potential.** Multiple ion channels and the sodium–potassium pump generate the neuron's resting potential. At rest, the cell's interior is negatively charged relative to the outside (see the voltage meter). Note that the concentration of K^+ is higher inside the cell than it is outside the cell; conversely, the concentration of Na^+ is higher outside the cell than inside.

As described in section 24.3A, a neuron's resting potential keeps it primed to convey messages at any moment (figure 24.5, step 1). If a stimulus does arrive, some sodium channels in a neuron's membrane open and then immediately close, usually at the dendrites or cell body. A small amount of Na⁺ leaks into the cell through the open channels, and the cell's interior therefore becomes slightly less negative. This initial change does not, however, constitute the action potential. Instead, an action potential will occur only if the cell's **threshold potential** is reached (figure 24.5, step 2). Once the action potential does begin, additional sodium channels open, and more Na⁺ pours into the cell. The axon's membrane now has a positive charge at its interior side.

This reversal of the membrane potential lasts only for an instant. Near the peak of the action potential, sodium channels close, again preventing Na⁺ from entering the cell (figure 24.5, step 3). However, the delayed K⁺ channels visible in figure 24.4 are now open. K⁺ diffuses out of the cell, again making the inside of the axon negative relative to the outside. Meanwhile, the sodium–potassium pump continues to run. Resting potential is restored (figure 24.5, step 4). The entire process, from the initial influx of Na⁺ to the restoration of the resting potential, takes only 1 to 5 milliseconds to complete.

Figure 24.5 shows how an action potential occurs at one small patch of a neuron's membrane. To transmit a neural impulse, however, the signal must move from near the cell body to the end of the axon. How does this occur? During an action potential, some of the Na⁺ ions that rush into the cell diffuse along the interior of the membrane. As a result of this local influx of Na⁺, the neighboring patch of membrane reaches its threshold potential as well, triggering a new influx of Na⁺. The resulting chain reaction carries the impulse forward. The impulse does not spread "backward" because the membrane must reestablish its resting potential before another action potential can occur.

A neural impulse is similar to people "doing the wave" in a stadium. Although the participants do not change their locations, the wave travels around the stadium as successive groups of spectators stand and then quickly sit.

Figure 24.5 **The Action Potential.** An action potential is a brief reversal in the membrane potential at a patch of membrane in an axon. (*1*) At rest, the interior of an axon's membrane carries a negative charge relative to the outside. (*2*) If a stimulus arrives and the cell's threshold potential is reached, an action potential begins. Na⁺ ions pour into the axon, and the membrane's interior becomes positively charged relative to the outside. (*3*) The membrane potential is restored as K⁺ ions diffuse out of the cell; the action potential moves on to the next patch of membrane. (*4*) Resting potential is reestablished.

Movement of Na⁺

Movement of K⁺

1 Resting potential

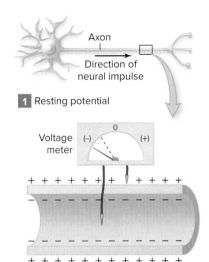

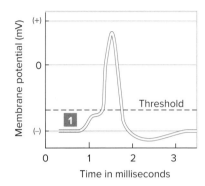

2 Threshold potential reached; action potential begins

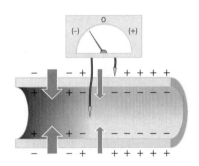

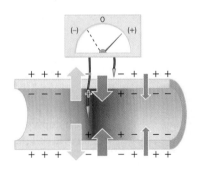

3 Peak of action potential; recovery begins

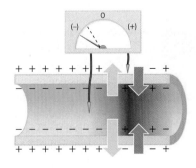

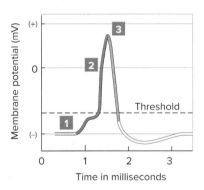

4 Return to resting potential

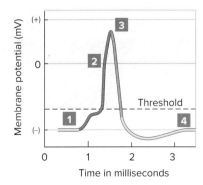

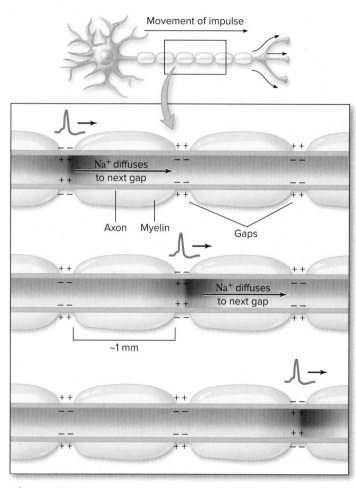

Figure 24.6 **The Role of the Myelin Sheath.** In myelinated axons, Na⁺ channels occur only at the gaps in the myelin sheath. Action potentials appear to "jump" between the gaps, which speeds impulse transmission along the axon.

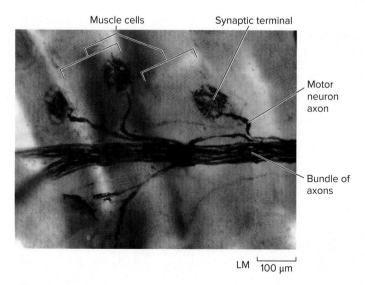

LM ⊢ 100 μm ⊣

Figure 24.7 **Talking to a Muscle.** The axons of motor neurons convey impulses to the muscle cells they control.

©McGraw-Hill Education/Al Telser

C. The Myelin Sheath Speeds Communication

The greater the diameter of an axon, the faster it conducts an impulse. A squid's "giant axons" are up to 1 millimeter in diameter. (Much of what biologists know about action potentials comes from studies on these large-diameter nerve fibers.) Axons from vertebrates are a hundredth to a thousandth the diameter of the squid's. Yet even thin vertebrate axons can conduct impulses very rapidly when they are coated with a myelin sheath (figure 24.6).

Myelin prevents ion flow across the membrane. At first glance, it might therefore seem that myelin should *prevent* the spread of action potentials. But the entire axon is not coated with myelin. Instead, ions can move across the membrane at the gaps in the sheath. When an action potential happens at one gap, Na⁺ entering the axon diffuses to the next gap. The incoming Na⁺ makes the membrane potential more positive and stimulates the sodium channels to open at the second gap, triggering an action potential there. In this way, when a neural impulse travels along the axon, it appears to "jump" from gap to gap.

The neural impulse moves up to 100 times faster when it leaps between gaps in the myelin sheath than when it spreads along an unmyelinated axon. Not surprisingly, myelinated axons occur in neural pathways where speed is essential, such as those that transmit motor commands to skeletal muscles. Thanks to myelin, a sensory message travels from the toe to the spinal cord in less than 1/100 of a second (about one third the speed of sound). Unmyelinated axons occur in pathways where speed is less important, such as in the neurons that trigger the secretion of stomach acid.

24.3 Mastering Concepts

1. What is the difference between the resting potential, the threshold potential, and an action potential?
2. How does an axon generate and transmit a neural impulse?
3. How does myelin speed neural impulse transmission?

24.4 Neurotransmitters Pass the Message from Cell to Cell

To form a communication network, a neuron conducting action potentials must convey the impulse to another cell—another neuron, a gland cell, or a muscle cell (figure 24.7). Most neurons do not touch each other, so the electrical impulse cannot travel directly from cell to cell. Instead, an action potential that reaches the tip of an axon causes the release of a **neurotransmitter,** a chemical signal that travels from a "sending" cell to a "receiving" cell across a tiny space.

A **synapse** is a specialized junction at which the axon of a neuron communicates with another cell. Figure 24.8 shows a synapse between two neurons. Note that every synapse has three components: the neuron sending the message, the cell receiving the message, and the **synaptic cleft,** which is the space between the two cells.

As illustrated in figure 24.7 and the upper portion of figure 24.8, the axon of the sending neuron enlarges at its tip to form a knob-shaped **synaptic terminal.** Each tiny knob contains many small sacs, or vesicles, that hold neurotransmitter molecules. The lower half of figure 24.8 shows the receiving cell's membrane. Immediately opposite the synaptic terminal are receptor proteins that can bind to the neurotransmitters.

The steps in figure 24.8 show how two neurons communicate at a synapse. A neural impulse travels along the membrane of the sending neuron until it reaches the synaptic terminal (step 1). There, the neural impulses stimulate vesicles inside the synaptic terminal to fuse with the cell membrane (step 2) and dump neurotransmitters into the synaptic cleft (step 3). The neurotransmitter molecules diffuse across the synaptic cleft and attach to ion channel proteins on the membrane of the receiving cell (step 4). The ion channels open, meaning that the cell has received the signal from the neuron that sent it.

The interaction between a neurotransmitter and the ion channel may be excitatory; that is, the membrane of the receiving cell may become more positive, increasing the probability of an action potential. In figure 24.8, the Na^+ ions entering the receiving cell are having this effect. Conversely, the interaction may be inhibitory. Opening membrane channels that admit chlorine ions (Cl^-), for example, makes the interior of the cell more negative and reduces the likelihood of an action potential.

What happens to a neurotransmitter after it has done its job? It might diffuse away from the synaptic cleft, be destroyed by an enzyme, or be taken back into the sending axon, an event called reuptake.

The human brain uses at least 100 neurotransmitters. The two most common are the amino acids glutamate and GABA (gamma aminobutyric acid). Other neurotransmitters occur at fewer synapses but still are vital. Serotonin, dopamine, epinephrine, norepinephrine, and acetylcholine are examples.

Too much or too little of a neurotransmitter can cause serious illness; table 24.1 lists a few examples of disorders that are at least partly associated with neurotransmitter imbalances. Moreover, some drugs can alter the functioning of the nervous system by either halting or enhancing the activity of a neurotransmitter (see Why We Care 24.1).

Notice that a synapse is asymmetrical; that is, nerve impulses travel from sending neuron to receiving cell and not in the opposite direction. This one-way traffic of information was a key adaptation that permitted the evolution of dedicated circuits in which one set of neurons communicated with a limited set of receiving cells. Over time, some circuits became associated with specific functions, controlling complex behaviors and forming the specialized sense organs typical of many animals (including vertebrates).

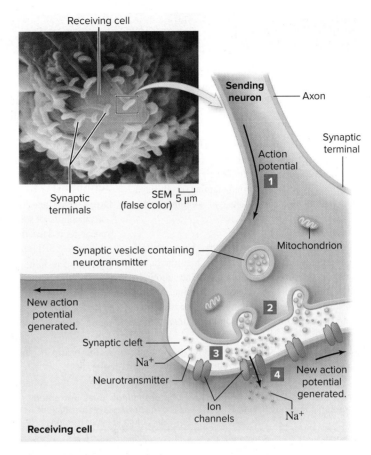

Figure 24.8 **The Synapse.** An action potential reaches the synaptic terminal (*1*) and triggers the release of neurotransmitters (*2*), which diffuse across the synaptic cleft (*3*) and bind with ion channel proteins in the receiving cell membrane. The channels open (*4*), changing the likelihood of an action potential in the receiving cell. The inset shows synaptic terminals from many neurons converging on the cell body of a neuron. Photo: ©ER Lewis, YY Zeevi and TE Everhart

1 Action potential arrives at synaptic terminal.

2 Vesicle loaded with neurotransmitter fuses with sending neuron's membrane.

3 Neurotransmitters are released into synaptic cleft.

4 Neurotransmitters bind to ion channels in receiving cell's membrane, stimulating the channels to open.

24.4 Mastering Concepts

1. Describe the structure of a synapse.
2. What event stimulates a neuron to release neurotransmitters?
3. What happens to a neurotransmitter after its release?

TABLE 24.1	**Disorders Associated with Neurotransmitter Imbalances**	
Condition	**Neurotransmitter Imbalance**	**Symptoms**
Alzheimer disease	Deficient acetylcholine	Memory loss, depression, disorientation, dementia, hallucinations, death
Epilepsy	Excess GABA, norepinephrine, and dopamine	Seizures, loss of consciousness
Huntington disease	Deficient GABA	Uncontrollable movements, dementia, behavioral and personality changes, death
Parkinson disease	Deficient dopamine	Tremors of hands, slowed movements, muscle rigidity
Schizophrenia	Deficient GABA, excess dopamine	Inappropriate emotional responses, hallucinations

Why We Care 24.1 | Drugs and Neurotransmitters

Understanding how neurotransmitters work helps explain the action of some mind-altering illicit and pharmaceutical drugs. The following are some examples, organized by the neurotransmitter affected.

Norepinephrine

Amphetamine drugs are chemically similar to norepinephrine; they bind to norepinephrine receptors and trigger the same changes in the receiving cell's membrane. The resulting enhanced norepinephrine activity heightens alertness and mood. Cocaine, which is chemically related to amphetamine, produces a short-lived feeling of euphoria, in part by blocking reuptake of norepinephrine.

Acetylcholine

Nicotine crosses the blood–brain barrier and reaches the brain within seconds of inhaling from a cigarette. An acetylcholine mimic, nicotine binds to acetylcholine receptor proteins in neuron cell membranes. The nicotine-stimulated neurons signal other brain cells to release dopamine, which provides the pleasurable feelings associated with smoking. Nicotine addiction stems from two sources: seeking the dopamine release and avoiding painful withdrawal symptoms.

Excess acetylcholine accounts for the deadly effects of poisonous nerve gases and some insecticides. These toxic chemicals prevent acetylcholine from breaking down in the synaptic cleft. The resulting buildup of acetylcholine overstimulates skeletal muscles, causing them to contract continuously. The twitching legs of a cockroach sprayed with insecticide demonstrate the effects.

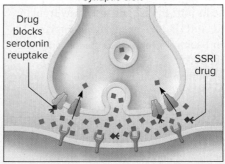
©Comstock Images/Alamy Stock Photo RF

Serotonin

Drugs that increase the amount of norepinephrine or serotonin in a synapse appear to reduce the symptoms of depression. Selective serotonin reuptake inhibitors (SSRIs) block the reuptake of serotonin at the synaptic cleft, causing the neurotransmitter to accumulate (figure 24.A).

Endorphins

Humans produce several types of endorphins, molecules that influence mood and perception of pain. Opiate drugs such as morphine, heroin, codeine, and opium are potent painkillers that bind endorphin receptors in the brain. In doing so, they elevate mood and make the pain easier to tolerate.

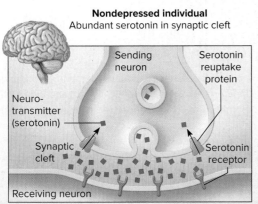
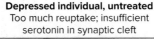
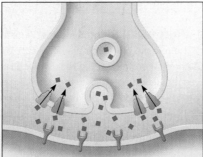

Nondepressed individual
Abundant serotonin in synaptic cleft

Depressed individual, untreated
Too much reuptake; insufficient serotonin in synaptic cleft

Depressed individual, treated with SSRI
Reuptake blocked; abundant serotonin in synaptic cleft

Sending neuron

Serotonin reuptake protein

Neuro-transmitter (serotonin)

Synaptic cleft

Serotonin receptor

Receiving neuron

Drug blocks serotonin reuptake

SSRI drug

Figure 24.A Anatomy of an Antidepressant. Selective serotonin reuptake inhibitors (SSRIs) block the reuptake of serotonin, making more of the neurotransmitter available in the synaptic cleft. The precise mechanism by which SSRIs relieve depression is not well understood, but some research suggests that they indirectly restore neuronal connections in dysfunctional areas of the brain.

24.5 The Peripheral Nervous System Consists of Nerve Cells Outside the Central Nervous System

The neurons of the brain and spinal cord interact constantly with those of the peripheral nervous system—the nerve cells outside the central nervous system (figure 24.9a). The peripheral nervous system consists mainly of **nerves,** which are bundles of axons encased in connective tissue. The nerves, in turn, are classified based on where they originate. Cranial nerves emerge directly from the brain; examples include the nerves that transmit information from the eyes and ears to the brain. Spinal nerves emerge from the spinal cord and control most functions from the neck down.

The peripheral nervous system is functionally divided into sensory and motor divisions (figure 24.9b). Sensory pathways carry signals to the central nervous system from sensory receptors in the skin, skeleton, muscles, and other organs. Motor pathways, on the other hand, convey information from the central nervous system to muscles and glands. In most nerves, the sensory and motor nerve fibers form a single cable.

The motor pathways of the peripheral nervous system include the somatic (voluntary) nervous system and the autonomic (involuntary) nervous system. The **somatic nervous system** carries signals from the brain to voluntary skeletal muscles, such as those that enable you to ride a bicycle, shake hands, or talk. The **autonomic nervous system** transmits impulses from the brain to smooth muscle, cardiac muscle, and glands, enabling internal organs to function without conscious awareness.

The autonomic nervous system is further subdivided into the sympathetic and parasympathetic nervous systems. The **sympathetic nervous system** dominates under stress, including emergencies. When you are startled, you can immediately feel your sympathetic nervous system leap into action as your heart pounds and your breathing rate increases. Neurons of the sympathetic nervous system also slow digestion and boost blood flow toward vital organs such as the heart, the brain, and the muscles necessary for "fight or flight." Still others trigger the adrenal glands to secrete hormones that prolong these effects. The **parasympathetic nervous system** returns body systems to normal during relaxed times ("rest and repose"); heart rate and respiration slow, and digestion resumes. Despite the "fight or flight" and "rest and repose" nicknames, the autonomic nervous system is always active; the two subdivisions maintain homeostasis without conscious thought. ⓘ *adrenal hormones,* section 25.4C

Some illnesses interfere with the function of the peripheral nervous system. In Guillain–Barré syndrome, for example, the immune system attacks and destroys the nerves of the peripheral nervous system. The disease can be life-threatening if it causes paralysis, breathing difficulty, and heart problems. In a person with Bell's palsy, another peripheral nervous system disorder, the cranial nerve that controls the muscles on one side of the face is damaged. The facial paralysis typically strikes suddenly and may either resolve on its own or be permanent. The cause is unknown.

24.5 Mastering Concepts

1. Which structures make up the peripheral nervous system?
2. How do the sensory and motor pathways of the peripheral nervous system differ?
3. Describe the relationships among the motor, somatic, autonomic, sympathetic, and parasympathetic nervous systems.
4. Cite an example of the opposing actions of the sympathetic and parasympathetic nervous systems.

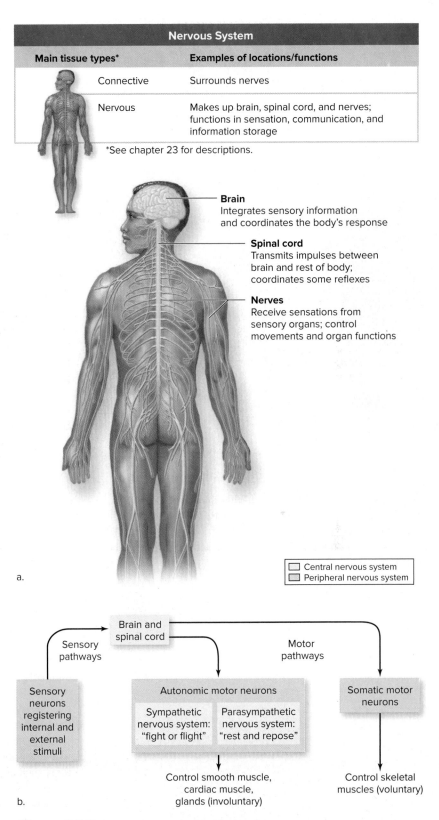

Nervous System	
Main tissue types*	**Examples of locations/functions**
Connective	Surrounds nerves
Nervous	Makes up brain, spinal cord, and nerves; functions in sensation, communication, and information storage

*See chapter 23 for descriptions.

Brain
Integrates sensory information and coordinates the body's response

Spinal cord
Transmits impulses between brain and rest of body; coordinates some reflexes

Nerves
Receive sensations from sensory organs; control movements and organ functions

☐ Central nervous system
☐ Peripheral nervous system

a.

Brain and spinal cord

Sensory pathways   Motor pathways

Sensory neurons registering internal and external stimuli

Autonomic motor neurons

Sympathetic nervous system: "fight or flight"

Parasympathetic nervous system: "rest and repose"

Somatic motor neurons

Control smooth muscle, cardiac muscle, glands (involuntary)

Control skeletal muscles (voluntary)

b.

Figure 24.9 The Human Nervous System. (a) The brain and spinal cord make up the central nervous system, and nerves compose the peripheral nervous system. (b) Sensory pathways of the peripheral nervous system provide input to the central nervous system. The brain and spinal cord, in turn, regulate the motor pathways of the peripheral nervous system.

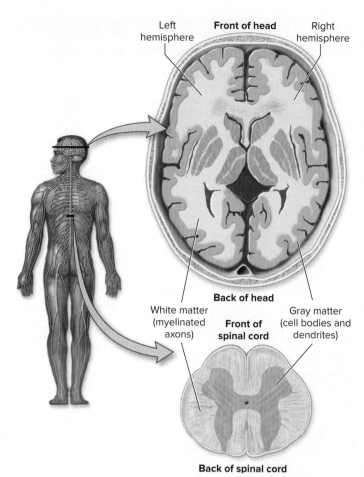

Left hemisphere

Front of head

Right hemisphere

Back of head

White matter (myelinated axons)

Front of spinal cord

Gray matter (cell bodies and dendrites)

Back of spinal cord

Figure 24.10 **Gray Matter and White Matter.** Gray matter makes up the exterior of the brain and some internal structures. It also makes up the central core of the spinal cord. Myelin-rich white matter is at the periphery of the spinal cord and forms most of the brain's interior.

24.6 The Central Nervous System Consists of the Spinal Cord and Brain

The nerves of the peripheral nervous system spread across the body, but the brain and spinal cord form the largest part of the nervous system. Together, these two organs make up the central nervous system.

Two types of nervous tissue occur in the central nervous system (figure 24.10). **Gray matter** consists of neuron cell bodies and dendrites, along with the synapses by which they communicate with other cells. Information processing occurs in the gray matter. **White matter** consists of myelinated axons transmitting information throughout the central nervous system.

A. The Spinal Cord Transmits Information Between Body and Brain

The spinal cord is a tube of neural tissue that emerges from the base of the brain and extends along the back of the body. This critical component of the central nervous system is encased in the bony armor of the vertebral column, or backbone. The backbone protects the delicate nervous tissue and provides points of attachment for muscles.

The spinal cord handles reflexes without communicating with the brain. A **reflex** is a rapid, involuntary response to a stimulus. For example, if a flying insect or a splash of water hits your face, your eyes close immediately, without you being conscious of the need to do so. This response is a reflex because it does not require input from the brain. (Nevertheless, impulses must be relayed to the brain for awareness to occur.)

B. The Brain Is Divided into Several Regions

The human brain weighs, on average, about 1.4 to 1.6 kilograms; it looks and feels like grayish pudding. The brain requires a large and constant energy supply to oversee organ systems and to provide the qualities of "mind"—learning, reasoning, and memory. At any time, brain activity consumes 20% of the body's oxygen and 15% of its blood glucose (see Burning Question 24.2). Permanent brain damage occurs after just 5 minutes without oxygen.

Anatomically, the brain has three main subdivisions: the hindbrain, the midbrain, and the forebrain (figure 24.11). The **hindbrain** is located toward the lower back of the skull. The **midbrain** is a narrow region that connects the hindbrain with the forebrain. The **forebrain** is the front of the brain. All three subdivisions are obvious early in embryonic development, but the forebrain's rapid growth soon obscures the midbrain and much of the hindbrain.

The midbrain and parts of the hindbrain make up the **brainstem,** the stalk-like lower portion of the brain. The brainstem regulates essential survival functions such as breathing and heartbeat. In addition, most of the cranial nerves emerge from the brainstem. Among other functions, these nerves control movements of the eyes, face, neck, and mouth along with the senses of taste and hearing.

The brainstem includes two parts of the hindbrain: the medulla oblongata and the pons. The **medulla oblongata** is a continuation of the spinal cord; this region not only regulates breathing, blood pressure, and heart rate, but it also contains reflex centers for vomiting, coughing, sneezing, defecating, swallowing, and hiccupping. The **pons,** which means "bridge," is the area above the medulla. White matter in this oval mass connects the forebrain to the medulla and to another part of the hindbrain, the cerebellum.

The midbrain is also part of the brainstem. Portions of the midbrain help control consciousness and participate in hearing and eye reflexes. In addition, nerve fibers that control voluntary motor function pass from the forebrain through the brainstem; the death of certain neurons in the midbrain results in the uncontrollable movements of Parkinson disease.

Behind the brainstem is the **cerebellum,** the largest part of the hindbrain. The neurons of the cerebellum refine motor messages and coordinate muscle movements subconsciously. Thanks to the cerebellum, you can complete complex physical skills—such as tying your shoes or brushing your teeth—smoothly and rapidly.

By far the largest part of the human brain is the forebrain, which contains structures that participate in complex functions such as learning, memory, language, motivation, and emotion. Three major parts of the forebrain are the thalamus, hypothalamus, and cerebrum.

The **thalamus** is a mass of gray matter located between the midbrain and the cerebrum. This central relay station processes sensory input and sends it to the appropriate part of the cerebrum. The almond-sized **hypothalamus,** which lies below the thalamus, occupies less than 1% of the brain volume, but it plays a vital role in maintaining homeostasis by linking the nervous and endocrine systems. Cells in the hypothalamus are sensitive not only to neural input arriving via the brainstem but also to hormones circulating in the bloodstream. The autonomic nervous system, in turn, relays neural signals from the hypothalamus to involuntary muscles and glands. Moreover, hormones produced in the hypothalamus coordinate the production and release of many other hormones (see chapter 25). All together, neural and hormonal signals from the hypothalamus regulate body temperature, heartbeat, water balance, and blood pressure, along with hunger, thirst, sleep, and sexual arousal.

The other major region of the forebrain is the **cerebrum,** which controls the qualities of what we consider the "mind"—that is, personality, intelligence, learning, perception, and emotion. In humans, the cerebrum occupies 83% of the brain's volume. It is divided into two **hemispheres,** which gather and process information simultaneously. The cerebral hemispheres work together, interconnected by a thick band of nerve fibers called the corpus callosum.

Although each side of the brain participates in most brain functions, some specialization does occur. In most people, for example, parts of the left hemisphere are associated with speech, language skills, mathematical ability, and reasoning, whereas the right hemisphere specializes in spatial, intuitive, musical, and artistic abilities. In addition, each hemisphere controls the opposite side of the body.

The cerebrum consists mostly of white matter—myelinated axons that transmit information within the cerebrum and between the cerebrum and other parts of the brain. But the outer layer of the cerebrum, the **cerebral cortex,** consists of gray matter that processes information (see figure 24.10). The human cerebral cortex is only a few millimeters thick, but it boasts about 10 billion neurons that form some 60 trillion synapses. In humans and other large mammals, deep folds enhance the surface area of the cerebral cortex.

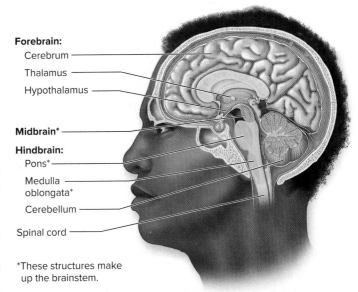

Forebrain:
 Cerebrum
 Thalamus
 Hypothalamus

Midbrain*

Hindbrain:
 Pons*
 Medulla oblongata*
 Cerebellum

Spinal cord

*These structures make up the brainstem.

Structure	Selected functions
Hindbrain	
Medulla oblongata	Regulates essential physiological processes such as blood pressure, heartbeat, and breathing
Pons	Connects forebrain with medulla and cerebellum
Cerebellum	Controls posture and balance; coordinates subconscious muscular movements
Midbrain	Relays information about voluntary movements from forebrain to spinal cord
Forebrain	
Thalamus	Processes information and relays it to the cerebrum
Hypothalamus	Homeostatic control of most organs
Cerebrum	
White matter	Transmits information within brain
Gray matter (cerebral cortex)	Sensory, motor, and association areas

Figure 24.11 The Human Brain. The three major areas of the vertebrate brain are the hindbrain, the midbrain, and the forebrain.

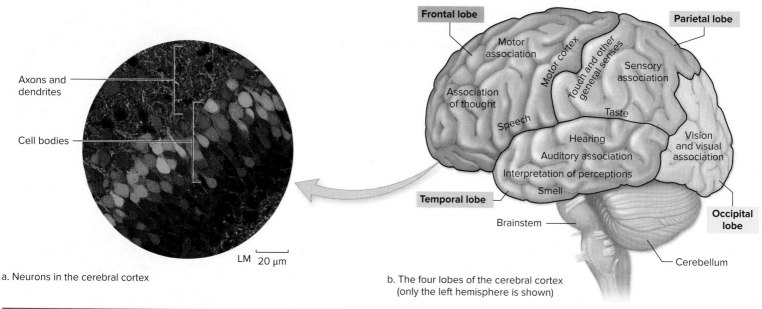

a. Neurons in the cerebral cortex

LM 20 μm

b. The four lobes of the cerebral cortex
(only the left hemisphere is shown)

Division	Function(s)	Brain region(s)
Sensory	General senses and the senses of vision, hearing, smell, and taste	Parietal, occipital, and temporal lobes
Motor	Voluntary movements	Frontal lobe
Association	Judgment, analysis, learning, creativity	Frontal lobe and parts of the parietal, occipital, and temporal lobes

c. Functional divisions of the cerebral cortex

Figure 24.12 The Cerebral Cortex. (a) Special labeling techniques reveal the individual neurons that make up the intricate circuits of the cerebral cortex. (b) Each hemisphere of the cerebrum is divided into four lobes. (c) The functional divisions of the brain (sensory, motor, and association) have been mapped to these lobes.

(a): ©AFP/Getty Images

Burning Question 24.2

Do I really use only 10% of my brain?

The notion that humans only use 10% of their brains is a myth. After all, damage to even a tiny part of the brain can be devastating. Moreover, the brain demands a huge amount of energy and oxygen; it does not make sense that the nervous system would waste valuable resources that the body could use in productive ways.

The myth originated in the late 1800s with psychologists William James and William Sidis, who theorized that humans carry out physical and mental tasks well below their capabilities. Years later, writers began assigning percentages to our supposed intellectual shortfall, and the 10% estimate stuck. However, brain scans now show that a human uses the entire brain, and many parts are active all the time.

**Submit your burning question to
marielle.hoefnagels@mheducation.com**

Anatomically, the cerebral cortex of each hemisphere is divided into four main parts (figure 24.12): the frontal, parietal, temporal, and occipital lobes. The functions of the cerebral cortex, however, overlap across these lobes. Sensory areas receive and interpret messages from sense organs. Motor areas send impulses to skeletal muscles, which produce voluntary movements. Association areas analyze, integrate, and interpret information from many brain areas. These are the seats of judgment, problem solving, learning, abstract thought, language, and creativity.

The cerebrum also houses most of the **limbic system,** a loosely defined collection of brain structures that is sometimes called the emotional center of the brain. The thalamus and hypothalamus are part of the limbic system, as are two nearby parts of each temporal lobe: the hippocampus and the amygdala. The hippocampus participates in long-term memory formation, whereas the amygdala is a center for emotions such as pleasure or fear. The amygdala sends signals to the hypothalamus, which activates the autonomic nervous system and coordinates the physical sensations that accompany strong emotions.

C. Many Brain Regions Participate in Memory

Why is it that you can't remember the name of someone you met a few minutes ago, but you can easily picture your first-grade teacher or your best friend from childhood? The answer relates to the difference between short-term and long-term memories. In response to the new acquaintance's name, your brain apparently created a **short-term memory** that remained available only for a few moments. You remember the teacher, however, because you interacted with that person every day for months at a time. This repeated reinforcement allowed your brain to produce a **long-term memory,** which can last a lifetime.

Much of what scientists know about memory comes from research on people with damage to specific parts of the brain. One famous example is a man called Henry Molaison, known in the medical literature by the initials H. M. until his death in 2008. Surgeons removed portions of his temporal lobes and hippocampus in 1953 in an effort to alleviate his severe epilepsy. The surgery accomplished its goal but had an unintended consequence: H. M. was unable to form new memories. Although he could recall events that occurred before the surgery, he could not remember what he had eaten for breakfast. Clues from H. M. and other patients suggest that the hippocampus is essential in the formation of long-term memories, but memories are not actually stored there.

No one knows exactly what happens to the brain's neurons and synapses when a new memory forms, but researchers are actively trying to learn more. Practical applications could include drugs that enhance memory in patients with disorders that cause memory loss, including Alzheimer disease. Conversely, pharmaceuticals that selectively erase memories could help people who are struggling in the aftermath of traumatic experiences.

D. Damage to the Central Nervous System Can Be Devastating

The central nervous system is generally well protected from physical injury. The bones of the skull and vertebral column shield nervous tissue from bumps and blows. **Meninges** are layered membranes that jacket the central nervous system. **Cerebrospinal fluid** bathes and cushions the brain and spinal cord; this fluid further insulates the central nervous system from injury. And the **blood–brain barrier,** formed by specialized neuroglia lining the brain's capillaries, helps protect the brain from harmful chemicals (figure 24.13).

Nevertheless, trauma and illness can harm the central nervous system (figure 24.14). If the spinal cord is severed, motor impulses from the brain cannot descend below the site of the injury. The result may be paralysis of the arms, torso, and legs. Damage to the brain also has serious consequences and can be caused by infectious agents, degenerative diseases, or a more subtle killer: stroke. In a stroke, a burst or blocked blood vessel can interrupt the flow of blood to part of the brain. Deprived of oxygen, some brain cells die, often so many that the stroke is fatal. The brain is also vulnerable to hard blows to the head. A concussion occurs when the neck jerks forward or backward so forcefully that the brain crashes into the interior of the skull. Neurons stretch and tear and may require days to recover.

Whatever its cause, part of the difficulty in reversing damage to the central nervous system is that mature neurons typically do not divide. The brain and spinal cord therefore cannot simply heal themselves by producing new cells, as your skin does after a minor cut. The neurons that survive the damage can, however, form some new connections that compensate for the loss. Therapy can therefore help restore some function to injured tissues. Moreover, stem cells and gene therapy may one day improve the outlook for patients with brain damage or disease. ⓘ *stem cells,* section 11.3A; *gene therapy,* section 11.4D

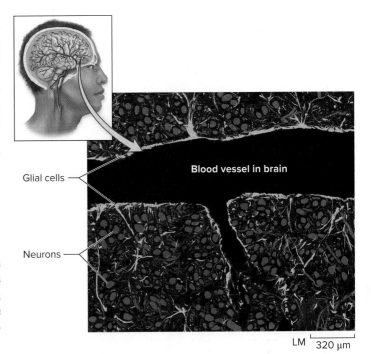

Figure 24.13   **Blood–Brain Barrier.** Neuroglia (*green*) guide the development of the blood–brain barrier, which protects the brain's neurons (*red*) from harmful chemicals.

Photo: ©C.J. Guerin, PhD, MRC Toxicology Unit/Science Source

Figure 24.14   **Nervous System Damage.** (a) Mark Zupan broke his neck in a car accident and later became famous as a wheelchair rugby player. (b) A series of mini-strokes left former President George H.W. Bush confined to a wheelchair. (c) A hard blow to the head—such as from a field hockey stick—can cause a concussion.

(a): ©Entertainment Pictures/Moviestills/age fotostock; (b): ©Mark Wilson/Getty Images News; (c): ©Mark Kolbe/Getty Images Sport

---

24.6 Mastering Concepts

1. What is the relationship between gray matter and white matter in the spinal cord?
2. What are the major structures in the hindbrain, midbrain, and forebrain, and what are their functions?
3. How do short- and long-term memories differ?
4. What are some examples of disorders that affect the central nervous system?
5. To what extent can the nervous system heal itself?

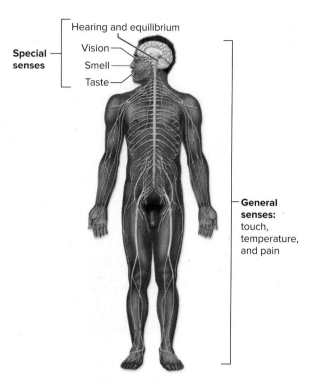

Special senses
- Hearing and equilibrium
- Vision
- Smell
- Taste

General senses: touch, temperature, and pain

Figure 24.15 **Overview of the Human Senses.** The general senses include touch, pain, and other senses with receptors located throughout the body. Receptors for the special senses, such as vision and hearing, are limited to the head.

Figure 24.16 **Sensory Blend.** This woman is experiencing the senses of touch, sight, and sound, among others.
©RubberBall Productions/Getty Images RF

24.7 The Senses Connect the Nervous System with the Outside World

The senses are an integral part of the nervous system (figure 24.15). A **sensation** is the raw input from the peripheral nervous system that arrives at the central nervous system. For example, your eyes and hands may inform your brain that a particular object is small, round, red, and smooth. The brain integrates all of this sensory input and consults memories to form a **perception,** or interpretation of the sensations—in this case, of a tomato.

The human senses paint a complex portrait of our surroundings. Consider, for example, the woman in figure 24.16. The tips of her fingers feel the banjo strings, her eyes see the instrument, and her ears hear the music. Her skin senses the warmth of the sun. She also maintains her balance, thanks to both her ability to feel the position of her limbs and her inner ear's sense of equilibrium. When she pauses for a snack, she will be able to smell and taste her food.

As rich as our own senses are, other animals can detect stimuli that are imperceptible to us. Dogs, for example, have an extremely well-developed sense of smell, which explains why these animals are so useful in sniffing out illicit drugs and other contraband. Bats have an entirely different ability, called echolocation. As a bat flies, it emits high-frequency pulses of sound. The animal's large ears pick up the sound waves that bounce off of prey and other objects, and its brain analyzes these echoes to "picture" the surroundings.

The senses convey vital information about food, danger, mates, and other stimuli. Sensory information also helps animals maintain homeostasis. Many of the negative feedback loops that stabilize the body's internal conditions operate without our awareness; for example, we can't directly "feel" our blood pH or hormone concentrations. But we are aware of sights, sounds, smells, and many other stimuli. The central nervous system responds to many types of sensory input by coordinating the actions of muscles and glands, which make adjustments as necessary to maintain homeostasis.

A. Sensory Receptors Respond to Stimuli by Generating Action Potentials

All sense organs ultimately derive their information from **sensory receptor** cells that detect stimuli. The human body includes several types of sensory receptors. **Mechanoreceptors** respond to physical stimuli such as sound or touch. **Thermoreceptors** respond to temperature. **Pain receptors** detect tissue damage, extreme heat and cold, and chemicals released from damaged cells. **Photoreceptors** respond to light, and **chemoreceptors** detect chemicals. (Table 24.2, in section 24.11, summarizes the sensory receptors.)

Each of these cell types "translates" sensory information into the language of the nervous system. **Transduction** is the process by which a sensory receptor converts energy from a stimulus into action potentials. Generally, a stimulus alters the shape of a protein embedded in a sensory receptor's cell membrane, causing the membrane's permeability to ions to change. The resulting movement of ions across the membrane triggers a **receptor potential,** which is a change in the membrane potential of a sensory receptor cell (figure 24.17). Three of the green lines in figure 24.17 depict receptor potentials that are below the cell's threshold and therefore do not trigger action potentials; the stimulus remains undetected. But if the receptor potential does exceed the threshold potential, as in the uppermost green line, an action potential occurs in the sensory receptor (red line in the figure). The frequency of action potentials arriving at the brain from specific groups of receptors conveys information about the type and intensity of the stimulus.

B. Continuous Stimulation May Cause Sensory Adaptation

You may have noticed that your perceptions of some stimuli can change over time. Your first thought when you roll out of bed may be "I smell coffee." But by the time you stand up, pull your clothes on, and wander to the kitchen, you hardly notice the coffee odor anymore. Likewise, the steaming water in a bathtub may seem too hot at first, but it soon becomes tolerable, even pleasant.

These examples illustrate **sensory adaptation,** a phenomenon in which sensations become less noticeable with prolonged exposure to the stimulus. The explanation is that sensory receptors generate fewer action potentials under constant stimulation. Generally, the response returns only if the intensity of the stimulus changes.

Without sensory adaptation, our nervous system would constantly react to old information, and detecting new stimuli would be challenging. Many receptors adapt quickly. Pain receptors, however, are very slow to adapt. The constant awareness of pain is uncomfortable, but it also alerts us to tissue damage and prompts us to address the source of the pain.

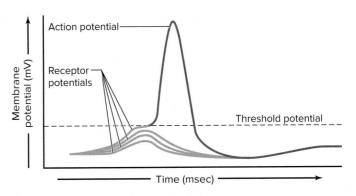

Figure 24.17 Receptor Potentials. Green lines in this figure show receptor potentials, three of which do not exceed the threshold potential and therefore do not trigger action potentials. The largest receptor potential, however, exceeds the threshold potential, stimulating an action potential (*red line*). The central nervous system detects only stimuli that provoke action potentials.

24.7 Mastering Concepts

1. Distinguish between sensation and perception.
2. What role do the senses play in maintaining homeostasis?
3. What are the major types of sensory receptors?
4. What is a receptor potential?
5. What is sensory adaptation, and how is it beneficial?

24.8 The General Senses Detect Touch, Temperature, and Pain

The general senses allow you to detect touch, temperature, or pain with any part of your skin. Each of these senses uses its own types of receptors.

The sense of touch comes from several types of mechanoreceptors in the skin (figure 24.18). The receptors all work in essentially the same way: Pressure pushes the flexible sides of the receptor cell inward, generating an action potential in the nerve fiber. Some detect light touch, whereas others respond to deeper pressure. In addition, the dendrites of some mechanoreceptors snake around each hair follicle and sense when the hair bends. Thanks to these sensitive hair follicle receptors, you can tell when a mosquito or other tiny insect lands on your arm.

The density of touch receptors varies across the body. As a result, the fingertips and tongue are much more sensitive to touch than, say, the skin of the lower back. This observation explains why medical professionals generally administer injections in the buttocks, shoulders, and thighs. These parts of the body have relatively few nerve endings and therefore are the least sensitive to needles and other painful stimuli.

Sensory receptors also enable the skin to sense temperature and pain. The brain integrates input from many cold and heat thermoreceptors to determine whether a stimulus is cool, hot, or somewhere in between. Pain receptors detect tissue damage. These neurons respond to the mechanical damage that follows a sharp blow, a cut, or a scrape. Pain receptors also detect extreme heat, extreme cold, and chemicals released from damaged cells.

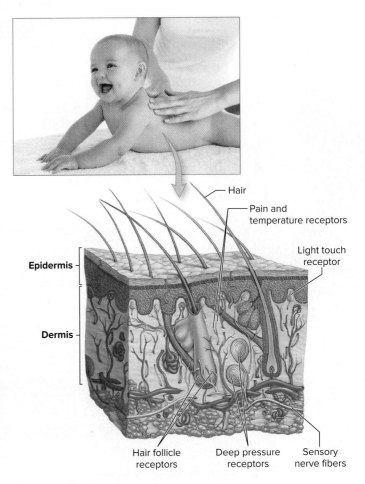

Figure 24.18 Skin Senses Many Stimuli. Sensory receptors in the skin respond to touch, temperature, and pain.

Photo: ©Monkey Business Images/Shutterstock RF

Miniglossary	The Senses
Sensation	Raw input from sensory neurons that arrives at the central nervous system
Perception	Central nervous system's interpretation of a sensation
Sensory receptor	Cell that detects stimuli
Transduction	Process by which a sensory receptor converts energy from a stimulus into action potentials
Receptor potential	Change in membrane potential in a sensory receptor cell

Pain is an unpleasant but important response; people who are unable to perceive pain can unknowingly injure themselves. Nevertheless, temporarily suspending the body's pain response with drugs called anesthetics can make some medical treatments tolerable. These drugs work in multiple ways. Local anesthetics such as a dentist's procaine (Novocain) stop pain-sensitive neurons from transmitting action potentials in a limited area of the body, such as one side of the mouth. General anesthetics cause a loss of consciousness that prevents the brain from perceiving any pain.

Ultimately, a sensory receptor cell that is stimulated by touch, temperature, or pain generates action potentials. These signals travel along spinal or cranial nerves to the central nervous system, which integrates the information. We can therefore tell where on the body a sensation is originating and identify its characteristics. For example, the brain's cerebral cortex can rapidly process multiple sensory signals to perceive that the right hand is touching the hot, smooth hood of a car.

24.8 Mastering Concepts

1. Which structures provide the senses of touch, temperature, and pain?

2. How does the brain participate in the general senses?

24.9 The Senses of Smell and Taste Detect Chemicals

Chemoreception is probably the most ancient sense. Bacteria and protists use chemical cues to approach food or move away from danger, so the ability to detect external chemicals must have arisen long before animals evolved.

The senses of smell and taste both depend on the body's ability to detect chemicals. Not surprisingly, these two senses have properties in common. In each case, the stimulus molecule must dissolve in a watery solution, such as saliva or the moist lining of a nasal passage. In addition, the molecule must interact with a chemoreceptor on a sensory cell's membrane.

The sense of smell begins at the **nose,** which forms the entrance to the nasal cavity inside the head. Specialized olfactory receptor neurons are located high in the nasal cavity (figure 24.19). Each olfactory neuron expresses one type of receptor protein on its cell membrane; each receptor protein, in turn, can bind to a limited set of odorants. A molecule that enters the nose in inhaled air binds to a receptor protein, and the cell then transduces this chemical signal into receptor potentials. Each olfactory receptor cell synapses with neurons in the brain's olfactory bulb. The brain interprets the information from multiple receptors and identifies the odor.

The nose detects odor molecules in inhaled air, so we can perceive scents originating from near or distant objects. Chemoreceptors in the **mouth,**

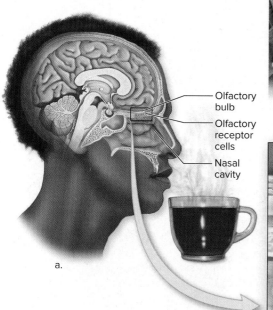

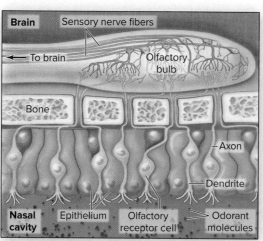

Figure 24.19 The Sense of Smell. (a) Chemoreceptors in the nose detect the odor of coffee. (b) An olfactory receptor cell binds an odorant molecule and transmits neural impulses to cells in the olfactory bulb.

Photo: ©Kelly Redinger/Design Pics RF

Labels in figure: Olfactory bulb; Olfactory receptor cells; Nasal cavity; a.; Brain; Sensory nerve fibers; To brain; Olfactory bulb; Bone; Axon; Dendrite; Nasal cavity; Epithelium; Olfactory receptor cell; Odorant molecules; b.

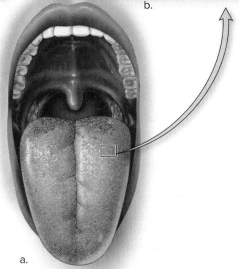

Papillae

b.

Epithelium

Sensory nerve fibers

Taste receptor cells

To brain

Taste bud

c.

a.

Figure 24.20

The Sense of Taste. (a) The tongue's surface is covered with (b) circular papillae. Taste buds within the papillae contain the taste receptor cells. (c) The receptor cells that make up a taste bud synapse on sensory neurons, which convey the information to the brain.

Photo: ©White Rock/Getty Images RF

however, can taste items only at very close range. The tongue's surface is covered with papillae; these bumps house **taste buds,** the organs associated with the sense of taste (figure 24.20). Each of the mouth's 10,000 taste buds contains 50 to 150 chemoreceptor cells that generate action potentials when dissolved food molecules bind to them. At a taste bud's base, the receptor cells form synapses with sensory neurons that lead to the brain. Our sense of taste depends on the pattern and intensity of activity across all taste neurons.

Many arthropods use chemicals in communication. **Pheromones** are chemical substances that elicit specific responses in other members of the same species. For example, female silk moths release pheromones that attract males, who "smell" the chemical signal from up to several kilometers away. Female scorpions also attract mates with chemicals: Males "taste" female pheromones deposited on sand. The role of pheromones in humans remains an open question, as described in Burning Question 24.3.

24.9 Mastering Concepts

1. How does the brain detect and identify odors?
2. How does a taste bud function?
3. What are pheromones?

Burning Question 24.3

Do humans have pheromones?

Advertisements for "human pheromone" colognes appeal to the desire to attract the opposite sex. Dab some on, they say, and watch your love life blossom. But are there really human pheromones?

This question is surprisingly difficult to answer, in part because human behavior is so complex; it is hard to find chemicals that elicit predictable responses. Nevertheless, at least some mammals do produce pheromones. A male hamster smeared with vaginal secretions from a female will stimulate sexual advances from another male—but only if the responding male has an intact vomeronasal organ, a tiny offshoot of the olfactory system. This structure is apparently the pheromone detector.

Studies have also demonstrated that pheromones from human females influence the menstrual cycles of other women. However, researchers still know little about how humans detect pheromones. We do have a vomeronasal organ, but no one has ever shown that it is functional. We therefore do not know whether the vomeronasal organ plays a role in human life or is just a vestige of our evolutionary history.

Submit your burning question to marielle.hoefnagels@mheducation.com

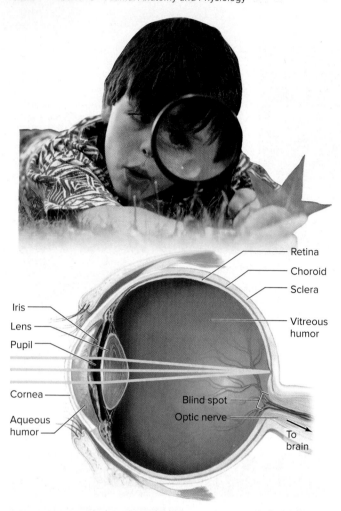

Figure 24.21 labels: Iris, Lens, Pupil, Cornea, Aqueous humor, Retina, Choroid, Sclera, Vitreous humor, Blind spot, Optic nerve, To brain

Figure 24.21 The Vertebrate Eye. Light passes through several layers before striking the retina, where sensory cells transmit information to the optic nerve.

Photo: ©Comstock/Getty Images RF

24.10 Vision Depends on Light-Sensitive Cells

An **eye** is an organ that produces the sense of sight. **Figure 24.21** depicts the vertebrate eye, which is composed of several layers. The **sclera** is the white, outermost layer that protects the inner structures of the eye. Toward the front of the eye, the sclera is modified into the **cornea;** this transparent, curved window bends incoming light rays. The **choroid** is the layer internal to the sclera. Behind the cornea, the choroid becomes the **iris,** which is the colored part of the eye. The iris regulates the size of the **pupil,** the hole in the middle of the iris. In bright light, the pupil is tiny, shielding the eye from excess stimulation. The pupil grows larger as light becomes dimmer.

A portion of the choroid also thickens into a structure that holds the flexible **lens,** which further bends the incoming light. Muscles regulate the curvature of the lens to focus on objects at any distance. When a person gazes at a faraway object, the lens is flattened and relaxed. To examine an article closely, however, muscles must pull the lens into a more curved shape.

Blood vessels in the choroid supply nutrients and oxygen to the **retina,** a sheet of photoreceptors that forms the innermost layer of the eye. The **optic nerve** is a cranial nerve that connects each retina to the brain. The point where the optic nerve exits the retina is called the blind spot because it lacks photoreceptors and therefore cannot sense light.

Each eyeball also contains fluid that helps bend light rays and focus them on the retina. The watery aqueous humor lies between the cornea and the lens. This fluid cleanses and nourishes the cornea and the lens and maintains the shape of the eyeball. Behind the lens is the vitreous humor, a jellylike substance that fills most of the eyeball's volume. Light rays pass through the cornea, lens, and humors of the eye and are focused on the retina. (Glasses and surgery can improve poor eyesight by redirecting light before it enters the eye; see Why We Care 24.2.)

Oddly, light has to pass through several layers of cells before reaching the photoreceptors at the back of the retina **(figure 24.22).** The photoreceptors are

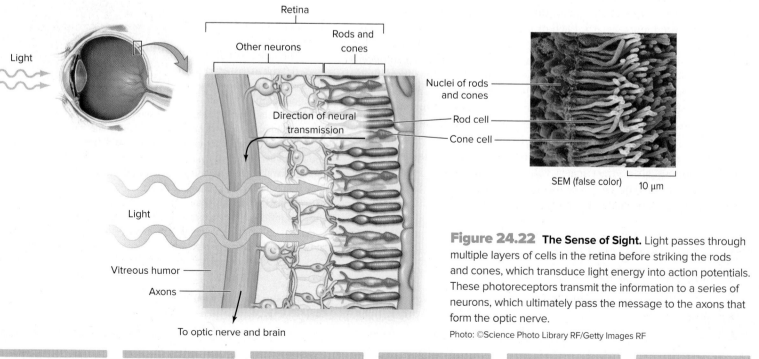

Figure 24.22 labels: Light, Retina, Other neurons, Rods and cones, Nuclei of rods and cones, Direction of neural transmission, Rod cell, Cone cell, SEM (false color), 10 µm, Light, Vitreous humor, Axons, To optic nerve and brain

Figure 24.22 The Sense of Sight. Light passes through multiple layers of cells in the retina before striking the rods and cones, which transduce light energy into action potentials. These photoreceptors transmit the information to a series of neurons, which ultimately pass the message to the axons that form the optic nerve.

Photo: ©Science Photo Library RF/Getty Images RF

neurons called rods and cones. **Rod cells,** which are concentrated around the edges of the retina, provide black-and-white vision in dim light and enable us to see at night. **Cone cells** require more light, but they can detect color; they are concentrated toward the center of the retina. The human eye contains about 125 million rod cells and 7 million cone cells.

The membranes of rods and cones are studded with pigment molecules that absorb light of different wavelengths. A pigment molecule that absorbs light energy changes shape, ultimately triggering receptor potentials that stimulate other neurons in the retina. Eventually, the action potentials are relayed through the optic nerve to the brain's visual cortex for processing and interpretation.

Humans have three cone types: "blue" cones absorb shorter wavelengths of light, "green" cones absorb medium wavelengths, and "red" cones absorb long wavelengths. People who lack a cone type entirely, due to a genetic mutation, are color-blind. Because the genes encoding these pigments are on the X chromosome, red–green color blindness is more common in males than females. (i) *X-linked disorders,* section 10.7A

Miniglossary | Vision

Step	Description
Pupil	Hole in the iris through which light enters the eye
Lens	Eye structure that focuses light on the retina
Retina	Sheet of light-absorbing photoreceptors (rod and cone cells) at the back of the eye
Rod cell	Photoreceptor that provides black-and-white vision in dim light
Cone cell	Photoreceptor that provides color vision in bright light
Optic nerve	Bundle of ganglion cell axons; conveys visual information from the retina to the thalamus
Visual cortex	Brain structure that processes and interprets visual information

24.10 Mastering Concepts

1. What are the parts of the vertebrate eye?
2. What are the roles of photoreceptors and pigments in vision?
3. Trace the pathway of information flow from the retina to the visual cortex of the brain.

Why We Care 24.2 | Correcting Vision

Perfect vision requires that the cornea, lens, and eyeball are a certain shape, so that light rays focus precisely on the retina. For those of us whose eyes are not perfectly formed, corrective lenses (eyeglasses and contact lenses) can treat blurry vision by altering the path of light (figure 24.B).

A more recent technology for correcting vision problems is laser eye surgery, which vaporizes tiny parts of the cornea, changing the path of light to the retina.

Sometimes, the cornea becomes clouded or misshapen. Surgeons can replace the defective cornea with one taken from a cadaver. Corneal transplant surgery carries a low risk of immune system rejection because, unlike other transplantable organs, the cornea lacks blood vessels. Another common eye disorder is a cataract, in which the lens of the eye becomes opaque. Cataract surgery is a simple procedure that replaces the clouded lens with a plastic implant.

Even people with perfectly shaped eyeballs and corneas usually need reading glasses after the age of about 40. To focus on a very close object, a muscle inside the eye must curve the lens so that it can bend incoming light rays at sharper angles. As we age, the lens becomes less flexible. It therefore becomes difficult for the muscles in the eye to bend the lens enough to clearly focus on nearby objects or printed words. Glasses can help, but laser surgery cannot correct this age-related decline in eyesight.

Figure 24.B Correcting Vision. If the cornea, lens, and eyeball are perfectly shaped, light rays focus precisely on the retina. Eyeglasses alter the path of light, correcting the blurry vision that occurs when rays focus elsewhere.

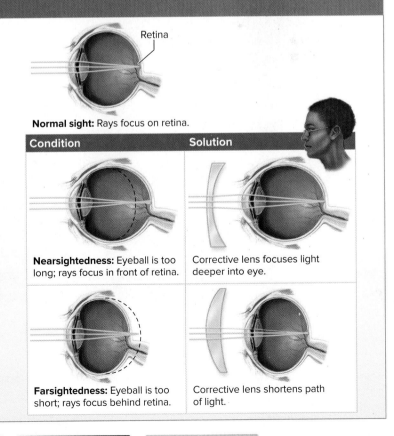

Normal sight: Rays focus on retina.

Condition	Solution
Nearsightedness: Eyeball is too long; rays focus in front of retina.	Corrective lens focuses light deeper into eye.
Farsightedness: Eyeball is too short; rays focus behind retina.	Corrective lens shortens path of light.

Burning Question 24.4

What is an ear infection?

Normally, the bones of the middle ear operate in air. But when we have a cold, sore throat, or other illness, fluid leaks into the middle ear from surrounding tissues. This fluid should drain out of the middle ear through the auditory (Eustachian) tube, which empties into the back of the nasal cavity (and from there to the throat).

If the auditory tube swells or becomes plugged with mucus, fluid accumulates in the middle ear. Bacteria may colonize this fluid, causing an ear infection. The fluid can also put pressure on the inside of the eardrum, causing pain and possibly making hearing difficult.

Babies and toddlers have narrow auditory tubes, which do not drain fluid as well as their adult counterparts. Partly because of this anatomical difference, young children get ear infections more often than adults. Fortunately, an ear infection usually only lasts for a few days, and the immune system can often fight the bacteria without the help of antibiotics.

Submit your burning question to
marielle.hoefnagels@mheducation.com

(girl): ©Pixtal/age fotostock RF

24.11 The Sense of Hearing Begins in the Ears

The clatter of a train, the notes of a symphony, a child's wail—what do they have in common? All are sounds that originate when something vibrates and creates repeating pressure waves in the surrounding air.

In humans, the sense of hearing begins with the fleshy outer part of the **ear,** which traps sound waves and funnels them down the **auditory canal** to the **eardrum** (figure 24.23). Sound pressure waves in air vibrate the eardrum, which moves three small bones in the middle ear. (Burning Question 24.4 describes infections that affect this area.) These bones, called the hammer, anvil, and stirrup, transmit and amplify the incoming sound. When the stirrup moves, it pushes on the **oval window,** a membrane that connects the middle ear with the inner ear. The oval window transfers the vibration to the snail-shaped **cochlea,** where sound is transduced into neural impulses.

The walls of the cochlea, which are made of bone, enclose three fluid-filled ducts (figure 24.24). One of these ducts contains the mechanoreceptors that transduce sound to action potentials. These mechanoreceptors, called **hair cells,** initiate the transduction of mechanical energy to receptor potentials. When the oval window vibrates, the fluid inside the cochlea moves, causing cilia on the hair cells to move relative to an overlying membrane. As the cilia bend, the hair cells initiate action potentials in the **auditory nerve.** The information then passes to the brain's auditory cortex for interpretation.

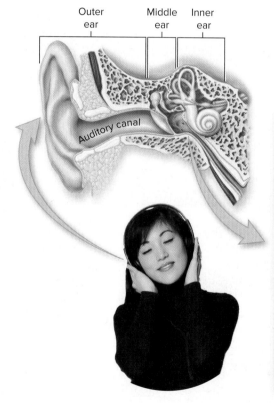

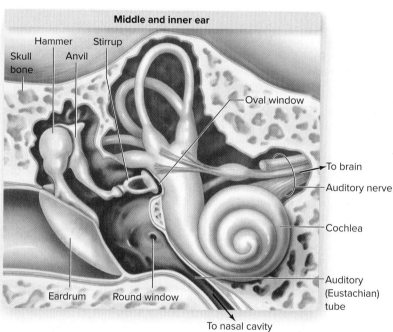

Figure 24.23 **The Human Ear.** Sound enters the outer ear and vibrates the three bones of the middle ear. The bones, in turn, cause vibrations in the fluid inside the pea-sized cochlea in the inner ear.

Photo: ©Rubberball Productions RF

The Nervous System   Neurons Are Functional Units   Action Potentials   Neurotransmitters   The Peripheral Nervous System   The Central Nervous System

Each sound stimulates a different region of the cochlea. The high-pitched tinkle of a dinner bell vibrates the wide, rigid region at the base of the cochlea; the low-pitched tones of a tuba stimulate the cochlea's tip, deep inside the spiral. The brain interprets the input from different regions of the cochlea as sounds of different pitches. Sound intensity is important as well. Louder sounds stimulate more hair cells, each of which fires rapid bursts of action potentials. The brain interprets the resulting increase in the rate and number of neurons firing as an increase in loudness.

The sense of hearing requires the interaction of many parts of the ear and nervous system. Deafness can occur if any of those components fails to function correctly. For example, the middle ear may not transmit sounds to the inner ear, or the auditory nerve may not function, or the brain may not respond to input from the nerve.

Depending on the type of deafness, hearing aids may offer relief. A hearing aid amplifies sound waves so they move the eardrum more than they otherwise would, helping the person hear more clearly. But if the middle ear cannot transmit sound, a conventional hearing aid is useless. One alternative is a cochlear implant. A surgeon places the device under the skin behind the ear. A microphone in the implant picks up sound; a processor then decomposes it into separate frequency components. Electrodes placed directly in the cochlea stimulate the parts of the auditory nerve corresponding to each frequency. By sending signals directly to the nervous system, cochlear implants compensate for nonfunctioning parts of the middle and inner ear.

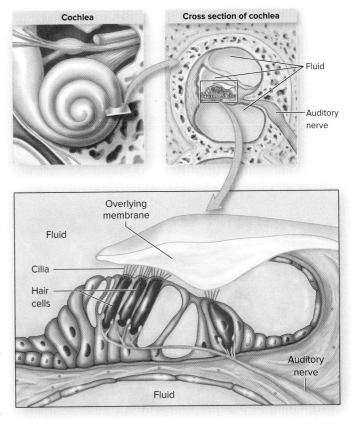

Figure 24.24 The Sense of Hearing. When the fluid inside the cochlea moves, cilia on the hair cells bend relative to the overlying membrane. These hair cells transduce sound waves into action potentials, which travel along the auditory nerve to the brain.

24.11 Mastering Concepts

1. What is the role of mechanoreceptors in the sense of hearing?
2. What are the parts of the ear, and how do they transmit sound?

TABLE 24.2 Sense Organs and Receptors: A Summary

Sense	Sense Organ(s)	Stimulus	Type of Receptor
General senses			
Touch	Skin	Pressure, vibration	Mechanoreceptor
Temperature	Skin	Heat, cold	Thermoreceptor
Pain	Everywhere except the brain	Damage to body tissues	Pain receptor
Special senses			
Smell	Nasal cavity	Airborne molecules	Chemoreceptor
Taste	Mouth and tongue	Dissolved molecules	Chemoreceptor
Vision	Eyes	Light	Photoreceptor
Hearing	Ears	Air pressure waves	Mechanoreceptor
Equilibrium	Ears	Motion of fluid in inner ear	Mechanoreceptor

Investigating Life 24.1 | Scorpion Stings Don't Faze Grasshopper Mice

Of the many types of arthropods that can sting, scorpions are distinctive. The scorpion uses the stinger on its tail to inject venom when threatened. Proteins called neurotoxins (poisons that act on the nervous system) cause most of the sting's worst effects. Some scorpion neurotoxins cause an axon's sodium channels to become stuck in the "open" position. The resulting flood of ions triggers a continuous barrage of action potentials, accompanied by simultaneous sensations of pain, heat, cold, and touch.

Scorpion venom causes intense pain in many mammals, including humans and rodents. However, grasshopper mice are an exception. These small rodents prey on bark scorpions in the Arizona desert. When it encounters a scorpion, a grasshopper mouse attempts to bite the scorpion's head. The scorpion fights back, stinging the mouse many times. Oddly, the mouse barely responds, as if it hardly feels the sting.

To learn more, researchers injected scorpion venom or a control solution into the hind paws of typical lab mice and of grasshopper mice. The lab mice licked a venom-injected paw far more vigorously than they did a control paw, an indication of pain. But venom did not induce the paw-licking behavior in grasshopper mice.

Clearly, the nervous systems of these two types of mice respond differently to scorpion venom, but how? A research team focused on pain receptors in the rodents. Specifically, they used electrodes to measure the number of action potentials in pain receptors as the animals were injected with a control solution, a low dose of scorpion venom, or a high dose of venom.

The results were clear (figure 24.C). In lab mice, a higher venom dose translated into more action potentials in pain receptors. In grasshopper mice, scorpion venom actually *inhibited* action potentials in pain receptors. Apparently, a diet rich in scorpions has selected for specialized pain receptors in grasshopper mice. Many generations from now, random changes in

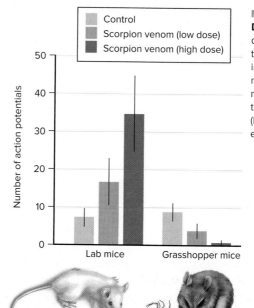

Figure 24.C **Venom Disarmed.** The higher the dose of scorpion venom, the more action potentials in a laboratory mouse's pain receptors. In grasshopper mice, the venom reduced the activity of pain receptors. (Error bars reflect standard errors; see appendix B.)

the scorpion toxin may again make grasshopper mice vulnerable to the sting. But in this arms race between predator and prey, the predator has taken the lead for now.

Source: Rowe, Ashlee H., and four coauthors. 2013. Voltage-gated sodium channel in grasshopper mice defends against bark scorpion toxin. *Science,* vol. 342, pages 441–446.

What's the **Point?** ▼ APPLIED

Imagine that you're cheering at a crowded concert. Sensory cells in your eyes, ears, nose, and skin are simultaneously sending many signals to your brain. But we do not perceive all of these stimuli. Instead, the nervous system filters out much of the incoming sensory information. This process begins at the level of the sensory receptor, which may adapt and stop responding to stimuli that persist for a long period. Sensory filtering also occurs in the brain, which often processes only some of the signals received from sensory neurons.

Thieves capitalize on our inability to notice everything in our surroundings. For example, when a pickpocket brusquely bumps your shoulder in a crowd, you likely won't realize that he also reached into your pocket. Your brain perceives the more obvious sensory stimulus at the expense of the less noticeable one.

An online video featuring a crowd of basketball players also illustrates the selective attention of our sensory system. In the video, two teams—one wearing white uniforms and the other in black—each has a basketball. The viewer is asked to count how many times the players in white pass their ball, cuing viewers to focus on white. Their distracted brains therefore do not perceive the man in the black bear suit, moonwalking across the scene in plain view.

Scientists debate about when the brain "decides" which stimuli to perceive and which to ignore. Some argue that the brain makes this decision early in processing. Others say that the brain fully processes all stimuli, but only some are consciously perceived. Recent research points toward the latter hypothesis, suggesting the intriguing idea that a viewer's brain may "see" the moonwalking bear but choose to ignore it.

Our selective attention can mean the difference between life and death. Most people know, for instance, that texting and driving is dangerous; it is impossible to simultaneously watch the road and focus on a cell phone's screen. Even eyewitness testimony, often considered the most persuasive evidence in a criminal trial, is error-prone. The human brain does not record events as a video camera does. Instead, we fail to notice key events surrounding a crime as it occurs, leaving a spotty memory that is vulnerable to suggestion by police or prosecutors eager to close the case or win a conviction. Innocent people, wrongly convicted of the crime, pay a high price for these unintentional errors of perception.

©momcilog/Getty Images RF

CHAPTER SUMMARY

24.1 The Nervous System Forms a Rapid Communication Network

- The **nervous system** and **endocrine system** work together to coordinate the feedback systems that maintain homeostasis (table 24.3).
- Nervous tissue consists of **neurons** and **neuroglia.**
- The vertebrate **central nervous system** consists of the **brain** and **spinal cord.** The **peripheral nervous system** conveys information between the central nervous system and the rest of the body.
- Overall, the nervous system receives sensory information, integrates it, and coordinates a response.

24.2 Neurons Are the Functional Units of a Nervous System

A. A Typical Neuron Consists of a Cell Body, Dendrites, and an Axon

- A neuron has a **cell body, dendrites** that receive impulses and transmit them toward the cell body, and an **axon** that conducts impulses away from the cell body. Fatty neuroglia wrap around portions of some axons to form the **myelin sheath.**

B. The Nervous System Includes Three Classes of Neurons

- A **sensory neuron** carries information toward the central nervous system; an **interneuron** conducts information between two neurons and coordinates responses; a **motor neuron** carries information away from the central nervous system and stimulates a muscle or gland.

24.3 Action Potentials Convey Messages

A. A Neuron at Rest Has a Negative Charge

- A neuron has a **membrane potential** that changes depending on the cell's activity.
- When not conducting a neural impulse, a neuron has a **resting potential** that is slightly negative. The sodium–potassium pump uses ATP to maintain a chemical gradient in which the K^+ concentration is much greater inside the cell than outside and the Na^+ concentration is greater outside than inside.

B. A Neuron's Membrane Potential Reverses During an Action Potential

- If a stimulus causes enough Na^+ to enter a cell, the membrane may reach its **threshold potential.** An electrical change called an **action potential** begins. Na^+ and K^+ quickly redistribute across a small patch of the axon's membrane, creating a series of changes that propagate like a wave along the axon.

C. The Myelin Sheath Speeds Communication

- The myelin sheath increases the speed of neural impulse transmission. The neural impulse rapidly "jumps" between gaps in the myelin sheath.

24.4 Neurotransmitters Pass the Message from Cell to Cell

- A **synapse** is a junction between a neuron and another cell.
- An action potential reaching the end of an axon causes **synaptic terminals** to release **neurotransmitters** into the **synaptic cleft.** These chemicals bind to ion channels on the membrane of the cell receiving the message.
- A neurotransmitter may have an excitatory effect, making an action potential more probable in the receiving cell; an inhibitory interaction has the opposite effect.
- Used neurotransmitter molecules diffuse away from the synaptic cleft, are destroyed, or are reabsorbed into the sending cell.

24.5 The Peripheral Nervous System Consists of Nerve Cells Outside the Central Nervous System

- **Nerves** are bundles of axons in the peripheral nervous system.
- The peripheral nervous system is divided into the sensory and motor pathways, and it includes all of the nerves that transmit sensations from sensory receptors and stimulate muscles and glands.
- The motor pathways of the peripheral nervous system consist of the **somatic** (voluntary) division and the **autonomic** (involuntary) division. The autonomic nervous system receives sensory information and conveys impulses to smooth muscle, cardiac muscle, and glands.
- Within the autonomic nervous system, the **sympathetic nervous system** controls physical responses to stressful events, and the **parasympathetic nervous system** restores a restful state. Both systems are always active in maintaining homeostasis.

24.6 The Central Nervous System Consists of the Spinal Cord and Brain

A. The Spinal Cord Transmits Information Between Body and Brain

- **White matter** on the periphery of the spinal cord conducts impulses to and from the brain; the central **gray matter** processes information.
- A **reflex** is a quick, automatic, protective response.

B. The Brain Is Divided into Several Regions

- The **brainstem** consists of the midbrain and portions of the hindbrain.
- The **hindbrain** includes three main subdivisions: the **medulla oblongata,** which controls many vital functions; the **cerebellum,** which coordinates unconscious movements; and the **pons,** which bridges the medulla and higher brain regions and connects the cerebellum to the cerebrum.
- The **midbrain** connects the hindbrain and forebrain.
- The major parts of the **forebrain** are the **thalamus,** a relay station between brain regions; the **hypothalamus,** which regulates vital physiological processes; and the **cerebrum.** The **limbic system,** including the amygdala and hippocampus, also resides in the gray matter of the forebrain.
- The outer layer of the cerebrum is the **cerebral cortex,** where information is processed and integrated. Each cerebral **hemisphere** receives sensory input from and directs motor responses to the opposite side of the body.

C. Many Brain Regions Participate in Memory

- Biologists have much to learn about memory, but it appears that the brain stores **short-term memories** and **long-term memories** in different ways.
- The formation of long-term memories requires an intact hippocampus.

D. Damage to the Central Nervous System Can Be Devastating

- The skull, vertebrae, **meninges, cerebrospinal fluid,** and **blood–brain barrier** protect the central nervous system.
- Trauma, infectious agents, degenerative diseases, and strokes all can damage the nervous system.

TABLE 24.3	Nervous and Endocrine Systems Compared	
Property	**Nervous System**	**Endocrine System**
Transmission method	Action potentials and neurotransmitters	Hormones
Speed	Nearly instantaneous	Relatively slow, but lasting
Range	Signal molecules affect a limited number of nearby cells	Signal molecules affect many cells throughout the body

24.7 The Senses Connect the Nervous System with the Outside World

- Sense organs send information about internal and external stimuli to the central nervous system. A **sensation** is the raw input received by the central nervous system; a **perception** is the brain's interpretation of the sensation.

A. Sensory Receptors Respond to Stimuli by Generating Action Potentials

- **Sensory receptors** are cells that detect stimuli. Types of sensory receptors include **mechanoreceptors, thermoreceptors, pain receptors, photoreceptors,** and **chemoreceptors** (figure 24.25).
- A sensory receptor selectively responds to a single form of energy and **transduces** it to **receptor potentials,** which change membrane potential in proportion to stimulus strength. If a receptor potential exceeds the cell's threshold, the cell generates action potentials.

B. Continuous Stimulation May Cause Sensory Adaptation

- In **sensory adaptation,** sensory receptors cease to respond to a constant stimulus.

24.8 The General Senses Detect Touch, Temperature, and Pain

- The skin's mechanoreceptors respond to touch; other sensory receptors include thermoreceptors and pain receptors.

24.9 The Senses of Smell and Taste Detect Chemicals

- The senses of smell and taste detect chemicals dissolved in watery solutions.
- Odorant molecules bind to receptors in the **nose.**
- Humans perceive taste when chemicals in the **mouth** stimulate receptors within **taste buds** on the tongue.
- **Pheromones** are chemicals that many animals use to communicate with others of the same species.

24.10 Vision Depends on Light-Sensitive Cells

- Photoreceptors in the **eye** contain light-sensitive pigments associated with membranes.
- The human eye's outer layer, the **sclera,** forms the transparent **cornea** in the front of the eyeball.

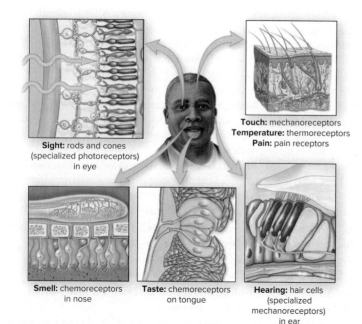

Sight: rods and cones (specialized photoreceptors) in eye

Touch: mechanoreceptors
Temperature: thermoreceptors
Pain: pain receptors

Smell: chemoreceptors in nose

Taste: chemoreceptors on tongue

Hearing: hair cells (specialized mechanoreceptors) in ear

Figure 24.25 Sensory Receptors: A Summary.

- The next layer, the **choroid,** supplies nutrients to the **retina.** At the front of the eye, the choroid holds the muscle that controls the shape of the **lens,** which focuses light on the photoreceptors. The **iris** adjusts the amount of light entering the eye by constricting or dilating the **pupil.**
- The innermost eye layer is the retina, and the **optic nerve** connects the retina with the brain.
- The retina's photoreceptors are **rod cells,** which provide black-and-white vision in dim light, and **cone cells,** which provide color vision in brighter light.
- Light stimulation alters the pigments embedded in the membranes of rod and cone cells. The resulting change in the charge across the membrane may generate an action potential.
- Photoreceptor cells synapse with multiple layers of neurons in the retina. Eventually, axons of some of these neurons leave the retina as the optic nerve, which carries information to the visual cortex.

24.11 The Sense of Hearing Begins in the Ears

- Mechanoreceptors bend in response to the motion of fluid in the inner **ear.**
- Sound enters the **auditory canal,** vibrating the **eardrum.** Three bones in the middle ear amplify these vibrations. The movements of these bones are transmitted through the **oval window,** changing the pressure in fluid within the **cochlea.** At the base of the cochlea, vibration moves cilia on **hair cells.** The **auditory nerve** transmits the impulses to the brain.
- The brain perceives the pitch of a sound based on the location of the moving hair cells in the cochlea. Louder sounds generate more action potentials than softer ones.

MULTIPLE CHOICE QUESTIONS

1. The peripheral nervous system _____, whereas the central nervous system _____.
   a. receives sensory stimuli; integrates sensory information
   b. produces sensory information; produces movements
   c. integrates sensory information; receives sensory stimuli
   d. coordinates voluntary actions; coordinates involuntary actions

2. What event triggers an action potential?
   a. Opening of sodium channels
   b. Opening of delayed potassium channels
   c. High concentration of negative ions outside the cell
   d. Activation of the sodium–potassium pump

3. Which division of the nervous system is responsible for a rapid heartbeat?
   a. Autonomic
   b. Sympathetic
   c. Parasympathetic
   d. Both a and b are correct.

4. Underdevelopment of the medulla oblongata could lead to
   a. poor processing of sensory information.
   b. limited language skills.
   c. erratic changes in blood pressure.
   d. unconscious muscle movements.

5. As you snuggle into bed, you feel the weight of the blankets on your body, but you soon become unaware of the covers. What has happened?
   a. Your skin's touch receptors became unable to receive information about new stimuli.
   b. Your skin's touch receptors adapted to the feeling of the blankets.
   c. All of your body's sensory receptors became unable to receive information about new stimuli.
   d. All of your body's sensory receptors adapted to the feeling of the blankets.

6. In what way are the senses of smell and taste different?

    a. Chemoreceptors detect smell, whereas mechanoreceptors detect taste.

    b. Olfactory receptors bind to chemicals dissolved in gas, whereas taste receptors bind to chemicals dissolved in water.

    c. Olfactory receptors generate action potentials, whereas taste receptors generate only receptor potentials.

    d. Smell can detect chemicals from distant objects, whereas taste is limited to chemicals at close range.

Answers to Multiple Choice questions are in appendix A.

WRITE IT OUT

1. How do the nervous and endocrine systems differ in how they communicate?

2. Explain how sensory neurons, interneurons, and motor neurons work together as an insect moves toward a chemical stimulus.

3. Describe the distribution of charges in the membrane of a resting neuron.

4. Ordinarily, the beginning of a neuron's axon (near the cell body) is activated before any other part of the axon, so a wave of action potentials occurs as sodium channels open in the direction of the synaptic terminal. What would happen if you artificially stimulated an axon to reach threshold potential midway along its length, rather than at the connection with the cell body?

5. How does myelin alter the conduction of a neural impulse along an axon? What would happen to neural impulse transmission in an axon without gaps in the myelin sheath? Explain.

6. Speculate about why synapses and neurotransmitters are beneficial adaptations to animal nervous systems, compared to direct connections between neurons.

7. Neuroglia outnumber neurons by about 10 to 1. In addition, neuroglia retain the ability to divide, unlike neurons. How do these two observations relate to the fact that most brain cancers begin in neuroglia?

8. Suppose you put on eyeglasses belonging to someone who is more farsighted than you. Draw how light passes through the glasses and into your eyes. Why will the glasses blur your vision?

9. What are the roles of rods and cones in the sense of sight?

10. In a rare condition called synesthesia, stimulation of one sense triggers the perception of another sense. For example, people with synesthesia have reported seeing bursts of color when stimulated with loud noises. Would you expect synesthesia to be a problem with sensory receptors, peripheral nerves, or the central nervous system? Explain.

SCIENTIFIC LITERACY

Acupuncture is the insertion of thin needles into the skin, often with the goal of providing pain relief. Use the Internet to learn more about acupuncture. Can you find scientific evidence that acupuncture is effective? If so, how does acupuncture help the patient, and how do you know that the findings are scientific? If you cannot find scientific evidence, then propose an explanation for why the practice is common.

Answers to Mastering Concepts, Write It Out, Scientific Literacy, and Pull It Together questions can be found in the Connect ebook.
connect.mheducation.com

PULL IT TOGETHER

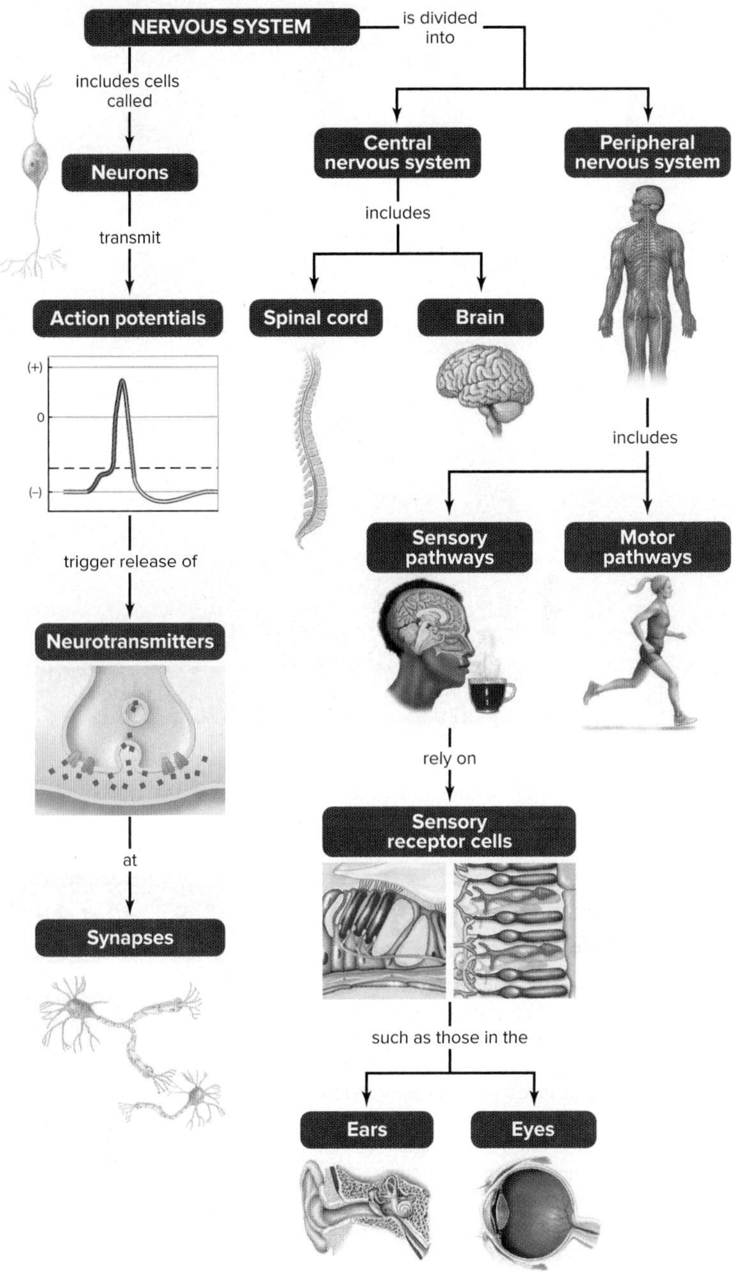

Figure 24.26 Pull It Together: The Nervous System and the Senses.

Refer to figure 24.26 and the chapter content to answer the following questions.

1. Review the Survey the Landscape figure in the chapter introduction, and then add *homeostasis* to the Pull It Together concept map.

2. Add *axons* and *myelin* to this concept map.

3. Add three types of sensory receptor cells to this concept map.

LEARNING OUTLINE

25.1 The Endocrine System Uses Hormones to Communicate

25.2 Hormones Stimulate Responses in Target Cells

25.3 The Hypothalamus and Pituitary Gland Oversee Endocrine Control

25.4 Hormones from Many Glands Regulate Metabolism

25.5 Hormones from the Ovaries and Testes Control Reproduction

APPLICATIONS

Burning Question 25.1 *What are endocrine disruptors?*
Why We Care 25.1 *Anabolic Steroids in Sports*
Investigating Life 25.1 *Addicted to Affection*

Yoga on the Beach. Hormones control the body's stress response, which begins at the brain. Gentle exercises such as yoga relieve tension and calm the mind. ©Purestock/SuperStock RF

Learn How to Learn
Don't Waste Old Exams

If you are lucky, your instructor may make old exams available to your class. If so, it is usually a bad idea to simply look up and memorize the answer to each question. Instead, use the old exam as a chance to test yourself before it really counts. Put away your notes and textbook, and set up a mock exam. Answer each question without "cheating," then check how many you got right. Use the questions you got wrong—or that you guessed right—as a guide to what you should study more.

SURVEY THE LANDSCAPE
Animal Anatomy and Physiology

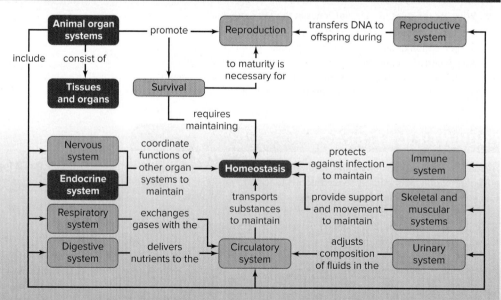

Hormones are communication molecules produced by the organs of the endocrine system. These molecules circulate throughout the body, altering the activities of the body's other organs. These adjustments help the animal body maintain homeostasis.

For more details, study the Pull It Together feature in the chapter summary.

©Getty Images RF

The endocrine system is one of the body's two communication networks. Hormones, the endocrine system's communication molecules, take a relatively leisurely route—the bloodstream—through the body. The responses may be slower than those coordinated by the nervous system, but they are just as important in maintaining homeostasis. Without the nervous and endocrine systems calling the shots, the body's other organ systems could not do their jobs.

One way to appreciate the importance of hormones is to consider what happens when they are produced in quantities that are too small or too large. A patient who fails to produce a hormone called insulin has type 1 diabetes. If human growth hormone is lacking, the bones do not grow properly, resulting in a type of dwarfism. Too much growth hormone, on the other hand, produces a giant.

Understanding the endocrine system has led to many practical applications. Insulin injections have saved the lives of many diabetic patients, and the synthetic hormones in birth control pills have prevented countless unwanted pregnancies. This chapter introduces these hormones and many more.

25.1 The Endocrine System Uses Hormones to Communicate

Animals communicate with one another in many ways, including color displays, sounds, body language, and scents. Likewise, the cells that make up a multicellular organism's body send and receive signals; these cell-to-cell messages coordinate the actions of the body's organ systems.

An animal's body has two main communication systems. The **nervous system,** described in chapter 24, is a network of cells that specialize in sending speedy signals that vanish as quickly as they arrive. The **endocrine system** is the other main communication system. As you will see, the endocrine system does not act with the speed of neural impulses, but its chemical messages have something else: staying power.

The endocrine system has two main components: glands and hormones. An **endocrine gland** consists of cells that produce and secrete hormones into the bloodstream, which carries the secretions throughout the body. A **hormone** is a biochemical that travels in the bloodstream and alters the metabolism of one or more cells. ⓘ *circulatory system,* section 27.2

The endocrine system would be ineffective if every hormone acted on every cell in the body. Instead, a limited selection of **target cells** respond to each hormone. Inside or on the surface of each target cell is a receptor protein, which binds to the hormone and initiates the cell's response.

Hormones are analogous to the radio signals that multiple stations simultaneously broadcast into the atmosphere. The receptor proteins, then, are like individual radios. Even when dozens of signals are present, each radio is tuned to one frequency and therefore picks up the signal of just one station. Likewise, each receptor binds to one of the many hormones that may be circulating in the bloodstream. Moreover, just as one house may contain many radios, each tuned to a different station, one target cell may also have receptors for many hormones, each of which initiates a unique response.

To illustrate the power of the endocrine system, consider one stage of life that famously involves hormones: puberty. During this period, hormones transform a child's body into that of an adult. Females develop enlarged breasts and wider hips, males acquire a deeper voice and more muscular physique, and new body hair sprouts in both sexes. The same hormones also affect mood, emotions, and feelings of sexual attraction.

Hormones figure prominently in the lives of other animals, too. For example, a caterpillar undergoes a dramatic metamorphosis as it develops into a butterfly, as does a tadpole transforming into an adult frog (figure 25.1). The endocrine system's effects are not always so extreme, but they are nonetheless present throughout life.

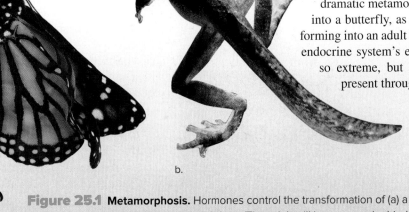
a.   b.

Figure 25.1 **Metamorphosis.** Hormones control the transformation of (a) a caterpillar into a butterfly and (b) a tadpole into an adult frog. (The adult will have muscular hind legs and no tail.)
(a, both): ©McGraw-Hill Education/Ken Cavanagh; (b): ©Robert Clay/Alamy Stock Photo RF

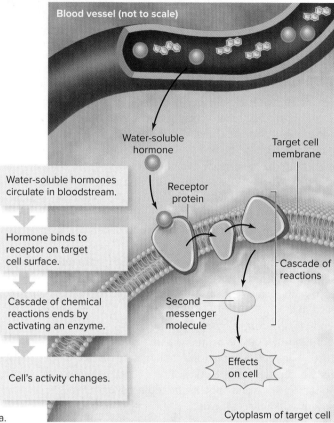

Blood vessel (not to scale)

Water-soluble hormone

Target cell membrane

Water-soluble hormones circulate in bloodstream.

Receptor protein

Hormone binds to receptor on target cell surface.

Cascade of reactions

Cascade of chemical reactions ends by activating an enzyme.

Second messenger molecule

Cell's activity changes.

Effects on cell

Cytoplasm of target cell

a.

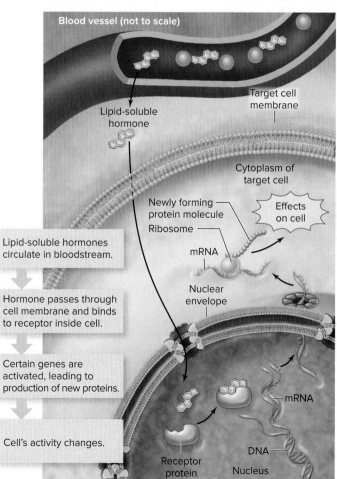

Blood vessel (not to scale)

Lipid-soluble hormone

Target cell membrane

Cytoplasm of target cell

Lipid-soluble hormones circulate in bloodstream.

Newly forming protein molecule

Effects on cell

Ribosome

mRNA

Hormone passes through cell membrane and binds to receptor inside cell.

Nuclear envelope

Certain genes are activated, leading to production of new proteins.

mRNA

Cell's activity changes.

DNA

Receptor protein

Nucleus

b.

Because hormones are so powerful, an animal's body strictly regulates the levels of these molecules in the bloodstream. This tight control often occurs by negative feedback. Recall from chapter 23 that in negative feedback, a change in a condition triggers action that reverses the change. Feedback loops ensure that endocrine glands adjust the secretion of all hormones as required to maintain homeostasis. (Some everyday chemicals may disturb this delicate balance; see Burning Question 25.1.) (i) *negative feedback*, section 25.4

25.1 Mastering Concepts

1. What is the overall function of the endocrine system?
2. Describe the relationships among endocrine glands, hormones, and target cells.
3. Compare and contrast the nervous and endocrine systems.

25.2 Hormones Stimulate Responses in Target Cells

Just as a key fits a lock, each hormone affects only target cells bearing specific receptor molecules. The term *target cells* is a little misleading, because it implies that hormones somehow travel straight from their source to a limited set of cells. In reality, hormones circulate throughout the entire body in blood. Each hormone's target cells are simply those with the corresponding receptors.

This section describes how the interaction between a hormone and its receptor initiates the target cell's response. In general, receptors for water-soluble hormones are on the surface of the target cell. In contrast, lipid-soluble hormones typically interact with receptors inside cells, either in the cytoplasm or in the nucleus.

A. Water-Soluble Hormones Trigger Second Messenger Systems

Most water-soluble hormones are either proteins or "peptide hormones" (short chains of amino acids). Because these hormones are water-soluble, they cannot pass readily through the cell membrane. Instead, they bind to receptors on the surface of target cells (figure 25.2a). (i) *cell membrane*, section 3.3

The hormone-receptor interaction triggers a cascade of chemical reactions within the cell. The product of this chain reaction is a **second messenger,** and it is the molecule that actually provokes the cell's response. The second messenger typically activates the enzymes that produce the hormone's effects.

Whatever the outcome, the entire cascade of reactions converts the external "message"—the arrival of the hormone at the outer membrane—into a signal that acts inside the cell. In general, water-soluble hormones act rapidly, within minutes of their release. Target cells respond quickly because all of the participating biochemicals are already in place when the hormone binds to the receptor.

Figure 25.2 **Target Cell Responses to Hormones.** (a) Water-soluble hormones bind to receptors on the surface of target cells. A series of chemical reactions initiates the target cell's response. (b) Lipid-soluble hormones pass through cell membranes and bind to receptors in the cytoplasm or nucleus. The target cell responds by altering the expression of one or more genes.

B. Lipid-Soluble Hormones Directly Alter Gene Expression

Some hormones are lipid-soluble. The most familiar are the **steroid hormones,** such as testosterone and estrogen. The body synthesizes these and other steroid hormones from cholesterol, which is one reason humans need at least some cholesterol in their diets. Two other lipid-soluble hormones, the thyroid hormones, are derived from a single amino acid. ⓘ *lipids,* section 2.5E

Lipid-soluble hormones easily cross the cell membrane (figure 25.2b); no second messenger is involved. Once inside the cell, the hormone may enter the nucleus and bind to a receptor associated with DNA, stimulating or inhibiting the production of particular proteins. Alternatively, the hormone may bind to a receptor in the cytoplasm, and the two molecules may travel together to the nucleus. Either way, response time is much slower than for water-soluble hormones, because the cell must produce new proteins (or stop producing them) before the hormone takes effect. ⓘ *DNA function,* section 7.2

25.2 Mastering Concepts

1. How does a hormone affect some cells but not others?
2. What is the role of second messengers in hormone action?
3. Describe the locations of the receptors that bind to water- and lipid-soluble hormones.

25.3 The Hypothalamus and Pituitary Gland Oversee Endocrine Control

Many organs produce hormones. The main endocrine organs in vertebrates are the hypothalamus, pituitary gland, pineal gland, thyroid gland, parathyroid glands, adrenal glands, pancreas, ovaries, and testes (figure 25.3). The heart, kidneys, liver, stomach, small

Endocrine System	
Main tissue types*	**Examples of locations/functions**
Epithelial	Makes up the bulk of most glands and secretes many types of hormones
Connective	Blood circulates hormones throughout the body
Nervous	Parts of the brain secrete some hormones and control release of others; some neurons secrete hormones

*See chapter 23 for descriptions.

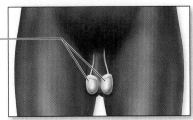

Testes (in male)
Produce testosterone, which promotes sperm maturation and secondary sex characteristics

Miniglossary | Hormones and Responses

Water-soluble hormone	Protein or short chain of amino acids (peptide hormone); biochemical that travels in blood and binds to a receptor on a target cell's surface, starting reactions that alter cell metabolism
Lipid-soluble hormone	Steroid hormone; biochemical that travels in blood and passes through a target cell's membrane, binds to a receptor in the cell, and turns particular genes on or off
Target cell	Cell with receptor proteins that bind to a particular hormone

Figure 25.3 Human Endocrine Glands. The endocrine system includes several glands containing specialized cells that secrete hormones. The hormones circulate throughout the body in blood vessels, which are not shown in this figure.

Hypothalamus
(shown in green) Produces hormones that stimulate or inhibit the release of hormones from the pituitary gland

Pituitary gland
(shown in orange) Produces numerous hormones that affect target tissues directly or stimulate other endocrine glands

Pineal gland
(shown in blue) Produces melatonin, which helps regulate sleep–wake cycles

Thyroid gland
Releases thyroid hormones, which regulate metabolism

Parathyroid glands
(behind thyroid) Secrete parathyroid hormone, which helps regulate blood calcium

Adrenal glands
Produce hormones that regulate kidney function and contribute to the body's stress response

Pancreas
Releases hormones that regulate blood glucose levels

Ovaries (in female)
Produce estrogen and progesterone, which mediate monthly changes in the uterine lining and promote secondary sex characteristics

The Ovaries and Testes Control Reproduction

Figure 25.4 **Hormones of the Hypothalamus and Pituitary.** The hypothalamus and the pituitary gland secrete hormones that coordinate the actions of the other endocrine glands.

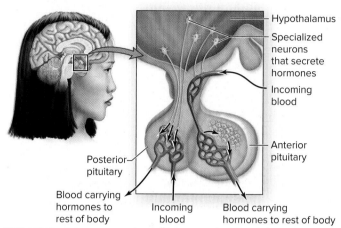

intestine, and placenta also contain scattered hormone-secreting cells. Together, these organs release dozens of hormones that simultaneously regulate every aspect of our lives, from conception through death.

Two of these structures, the hypothalamus and the pituitary gland, coordinate much of the endocrine system's action. The almond-sized **hypothalamus** is a part of the forebrain, and the **pituitary gland** is a pea-sized structure attached to a stalk extending from the hypothalamus. The pituitary is really two glands in one (figure 25.4): the larger **anterior pituitary** (toward the front) and the smaller **posterior pituitary** (toward the back). Anatomically, the posterior pituitary is a continuation of the hypothalamus, whereas the anterior pituitary consists of endocrine cells.

Information about body temperature, body fluid composition, and many other stimuli travels from the body's sensory neurons to the hypothalamus. In response, specialized neurons extending from the hypothalamus secrete hormones into both parts of the pituitary gland, forming a direct link between the nervous and endocrine systems (see the anatomical diagram in figure 25.4). As you will see, the overall effect is to maintain homeostasis.

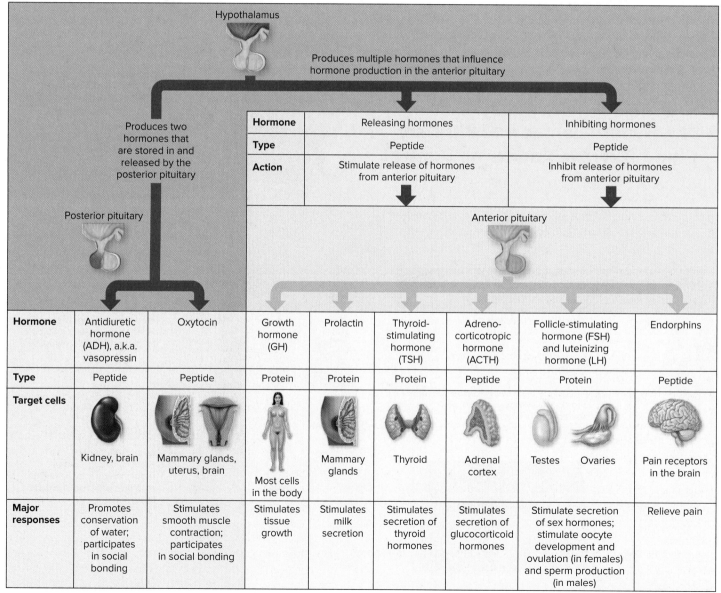

Hormone		Releasing hormones		Inhibiting hormones	
Type		Peptide		Peptide	
Action		Stimulate release of hormones from anterior pituitary		Inhibit release of hormones from anterior pituitary	

Hormone	Antidiuretic hormone (ADH), a.k.a. vasopressin	Oxytocin	Growth hormone (GH)	Prolactin	Thyroid-stimulating hormone (TSH)	Adreno-corticotropic hormone (ACTH)	Follicle-stimulating hormone (FSH) and luteinizing hormone (LH)	Endorphins
Type	Peptide	Peptide	Protein	Protein	Protein	Peptide	Protein	Peptide
Target cells	Kidney, brain	Mammary glands, uterus, brain	Most cells in the body	Mammary glands	Thyroid	Adrenal cortex	Testes Ovaries	Pain receptors in the brain
Major responses	Promotes conservation of water; participates in social bonding	Stimulates smooth muscle contraction; participates in social bonding	Stimulates tissue growth	Stimulates milk secretion	Stimulates secretion of thyroid hormones	Stimulates secretion of glucocorticoid hormones	Stimulate secretion of sex hormones; stimulate oocyte development and ovulation (in females) and sperm production (in males)	Relieve pain

The table in figure 25.4 summarizes the roles of the hypothalamus, posterior pituitary, and anterior pituitary. Note that the posterior pituitary does not synthesize any hormones, but it does store and release two hormones that the hypothalamus produces. The hypothalamus controls the anterior pituitary too, but in a different way—by secreting hormones that reach the anterior pituitary through a specialized system of blood vessels. These hormones, in turn, regulate the release of the anterior pituitary's hormones.

A. The Posterior Pituitary Stores and Releases Two Hormones

One of the two hormones produced by the hypothalamus and released by the posterior pituitary is **antidiuretic hormone (ADH),** also called vasopressin. If cells in the hypothalamus detect that the body's fluids are too concentrated, the posterior pituitary releases more ADH. This hormone stimulates kidney cells to return water to the blood (rather than eliminating the water in urine). The body's fluids become more dilute. Once balance is restored, ADH production slows. Figure 28.25 shows this feedback loop. ⓘ *negative feedback,* section 23.4

Oxytocin is the other posterior pituitary hormone. When a baby suckles, sensory neurons in the mother's nipple relay the information to the brain, which stimulates the release of oxytocin. The hormone causes cells in the breast to contract, squeezing the milk through ducts leading to the nipple. Oxytocin also triggers muscle contraction in the uterus, which pushes a baby out during labor. Physicians use synthetic oxytocin to induce labor or accelerate contractions in a woman who is giving birth. ⓘ *childbirth,* section 30.5E

Both ADH and oxytocin also act on the brain, playing a role in bonding, affection, and social recognition in at least some species (see Investigating Life 25.1). These hormones may also participate in human social attachment and in disorders such as autism.

B. The Anterior Pituitary Produces and Secretes Six Hormones

One of the six hormones that the anterior pituitary gland produces is **growth hormone (GH).** This hormone promotes growth and development in all tissues by increasing protein synthesis and cell division rates. Levels of GH peak in the preteen years and help spark adolescent growth spurts. A severe deficiency of GH during childhood leads to pituitary dwarfism (extremely short stature); at the other extreme, a child with too much GH becomes a pituitary giant (figure 25.5). In an adult, GH does not affect height because the long bones of the body are no longer growing. However, excess GH can cause acromegaly, a thickening of the bones in the hands and face.

Prolactin is an anterior pituitary hormone that stimulates milk production in a woman's breasts after she gives birth. In males and in women who are not nursing an infant, a hormone from the hypothalamus suppresses prolactin synthesis. In nursing mothers, however, suckling by an infant triggers nerve impulses that overcome this inhibition.

Four other anterior pituitary hormones all influence hormone secretion by other endocrine glands. **Thyroid-stimulating hormone (TSH)** prompts the thyroid gland to release hormones, whereas **adrenocorticotropic hormone (ACTH)** stimulates hormone release from parts of the adrenal glands. The remaining two anterior pituitary hormones stimulate hormone release from the ovaries and testes: **follicle-stimulating hormone (FSH)** and **luteinizing hormone (LH).** Sections 25.4 and 25.5 describe the roles of these four hormones in more detail.

a.

b.

Figure 25.5 Growth Hormone Abnormalities. (a) Lavinia Warren, who was born in 1841, was just 81 centimeters (32 inches) tall. She had a pituitary disorder that caused her short stature and is pictured here at about age 20. (b) A pituitary giant, Robert Wadlow, poses with his father and young brother. At about 2.7 meters (just under 9 feet) tall, Wadlow is thought to have been the tallest person in history. He died in 1940 at age 22.

The anterior pituitary also produces **endorphins,** which are natural painkillers that bind to receptors on target cells in the brain. Usually, however, endorphins are not detectable in the blood, so their status as hormones is questionable.

25.3 Mastering Concepts

1. How does the hypothalamus interact with the posterior and anterior pituitary glands?

2. List the names and functions of the hormones released by the posterior and anterior pituitary glands.

25.4 Hormones from Many Glands Regulate Metabolism

The thyroid gland, parathyroid glands, adrenal glands, and pancreas secrete hormones that influence metabolism (figure 25.6). Hormones from the anterior pituitary control many, but not all, of the activities of these glands (see figure 25.4).

A. The Thyroid Gland Sets the Metabolic Pace

The **thyroid gland** is a two-lobed structure in the neck. The lobes secrete two thyroid hormones, **thyroxine** and **triiodothyronine,** that increase the rate of metabolism in target cells. Under thyroid stimulation, the lungs exchange gases faster, the small intestine absorbs nutrients more readily, and fat levels in cells and in blood plasma decline.

Source	Thyroid		Parathyroid	Adrenal medulla	Adrenal cortex		Pancreas		Pineal gland
Hormone	Thyroxine, triiodothyronine	Calcitonin	Parathyroid hormone (PTH)	Epinephrine, norepinephrine	Mineralo-corticoids	Gluco-corticoids	Insulin	Glucagon	Melatonin
Type	Amine	Peptide	Protein	Amine	Steroid	Steroid	Protein	Peptide	Amine
Target cells	All tissues	Bone	Bone, digestive organs, kidneys	Blood vessels	Kidney	All tissues	All tissues	Liver, adipose tissue	Other endocrine glands
Major responses	Increase metabolic rate	Increases rate of calcium deposition	Releases calcium from bone, increases calcium absorption in digestive organs and kidneys	Raise blood pressure, constrict blood vessels, slow digestion	Maintain blood volume and electrolyte balance	Increase glucose levels in blood and brain	Increases uptake of glucose	Stimulates breakdown of glycogen into glucose and of fats into fatty acids	Regulates effects of light–dark cycles

Figure 25.6 Hormones That Regulate Metabolism. Hormones from several endocrine glands simultaneously influence many metabolic processes. (Note that amines are derived from amino acids. With the exception of thyroid hormones, which are lipid-soluble, amines are water-soluble.)

The thyroid hormones illustrate how the hypothalamus and pituitary interact in negative feedback loops (figure 25.7). When blood levels of thyroid hormones are low, the hypothalamus secretes thyrotropin-releasing hormone (TRH), which stimulates the anterior pituitary to increase production of thyroid-stimulating hormone (TSH). In response, cells in the thyroid secrete thyroxine and triiodothyronine. In the opposite situation, TRH secretion slows, so the thyroid glands reduce their production of hormones.

One disorder that affects the thyroid gland is hypothyroidism, a condition in which the thyroid does not release enough hormones. The metabolic rate slows, and weight increases. Synthetic hormones can treat many cases of hypothyroidism. In the past, the most common cause of hypothyroidism was iodine deficiency. Both thyroid hormones contain iodine; a deficiency of this essential element causes a goiter, or swollen thyroid gland. Today, iodine-deficient goiter is rare in nations where iodine is added to table salt.

An overactive thyroid causes hyperthyroidism. This disorder is associated with hyperactivity, an elevated heart rate, a high metabolic rate, and rapid weight loss. Graves disease is the most common type of hyperthyroidism.

Scattered cells throughout the thyroid gland produce a third hormone, **calcitonin,** which decreases blood calcium level by increasing the deposition of calcium in bone. The overall physiological importance of calcitonin in adult humans, however, is usually minimal.

B. The Parathyroid Glands Control Calcium Level

The **parathyroid glands** are four small groups of cells embedded in the back of the thyroid gland. **Parathyroid hormone (PTH)** increases calcium levels in blood and tissue fluid by releasing calcium from bones and by enhancing calcium absorption at the digestive tract and kidneys. PTH action therefore opposes that of calcitonin.

Calcium is vital to muscle contraction, blood clotting, bone formation, and the activities of many enzymes. Underactivity of the parathyroids can therefore be fatal. Excess PTH can also be harmful if calcium leaves bones faster than it accumulates. This condition, called osteoporosis, is most common in women who have reached menopause (cessation of menstrual periods). The estrogen decrease that accompanies menopause makes bone-forming cells more sensitive to PTH, which depletes bone mass. ⓘ *osteoporosis, section 26.3D*

C. The Adrenal Glands Coordinate the Body's Stress Responses

The paired, walnut-sized **adrenal glands** sit on top of the kidneys (*ad-* means "near," *renal* means "kidney"). Each adrenal gland has two regions, which are controlled in different ways and secrete different hormones (figure 25.8). The **adrenal medulla,** the inner portion, releases its hormones when stimulated by the sympathetic nervous system. The **adrenal cortex** is the outer portion, and it is under endocrine control. ⓘ *sympathetic nervous system, section 24.5*

The adrenal medulla's hormones, **epinephrine** (adrenaline) and **norepinephrine** (noradrenaline),

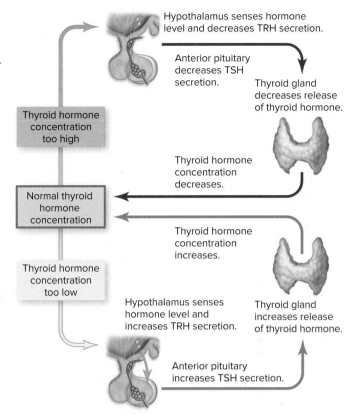

Figure 25.7 Thyroid Hormone Regulation. A negative feedback loop maintains the proper concentration of thyroid hormones in blood.

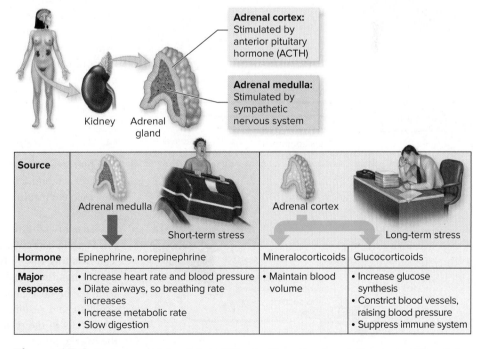

Figure 25.8 Hormones of the Adrenal Glands. The adrenal medulla secretes epinephrine and norepinephrine, which help the body respond to short-term stresses. Mineralocorticoids and glucocorticoids from the adrenal cortex enable the body to survive prolonged stress. The adrenal cortex also secretes small amounts of sex hormones (not shown).

Figure It Out

If the anterior pituitary doesn't produce enough ACTH, will the level of cortisol in the blood rise, fall, or stay the same? [*Hint:* Consult figure 25.4.]

Answer: It will fall.

help the body respond to exercise, trauma, fear, excitement, and other short-term "fight-or-flight" stresses. These water-soluble hormones cause heart rate and blood pressure to climb. In addition, the airway increases in diameter, making breathing easier. The metabolic rate increases, while digestion and other "nonessential" processes slow.

Epinephrine can save the lives of people with severe allergic reactions to bee stings or specific foods. Moments after contacting the allergen, a massive immune system reaction causes the airway to constrict. People with known allergies may therefore carry a self-injectable dose of epinephrine. The epinephrine temporarily reverses the allergic reaction, but symptoms may recur. Anyone experiencing a life-threatening allergic reaction should therefore seek emergency medical help. (i) *allergies,* section 29.5C

Unlike the adrenal medulla, the adrenal cortex secretes steroid hormones, including mineralocorticoids, glucocorticoids, and even a small amount of testosterone. The **mineralocorticoids** maintain blood volume and salt balance. One example, aldosterone, stimulates the kidneys to return sodium ions and water to the blood while excreting potassium ions. This action conserves water and increases blood pressure, which is especially important in compensating for fluid loss from severe bleeding. (i) *aldosterone,* section 28.9C

Glucocorticoids are essential in the body's response to prolonged stress. Cortisol is the most important glucocorticoid. This hormone mobilizes energy reserves by stimulating the production of glucose from amino acids. Glucocorticoids also indirectly constrict blood vessels, which slows blood loss and prevents inflammation after an injury. These same effects, however, also account for the unhealthy consequences of chronic stress. Narrowed blood vessels can lead to heart attacks, and the suppressed immune system leaves a person vulnerable to illness.

Prednisone, like other synthetic glucocorticoids, is an anti-inflammatory drug that mimics cortisol's effects. This drug can treat arthritis, allergic reactions, and asthma, but it also suppresses the immune system. In addition, with long-term use of the drug, the adrenal cortex may stop producing its own glucocorticoids. Abruptly stopping treatment may therefore cause a "steroid withdrawal" condition, with symptoms including fatigue, low blood pressure, and nausea. In severe cases, the patient may go into shock, which can be fatal.

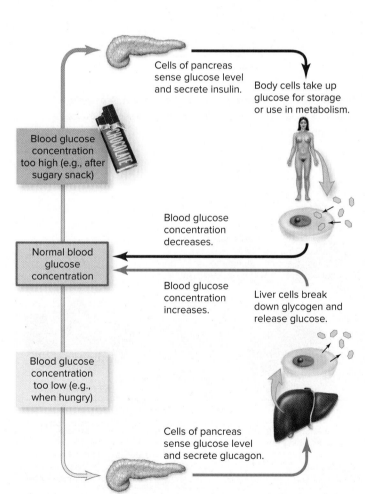

Figure 25.9 **Blood Glucose Regulation.** In a negative feedback loop, insulin and glucagon control the concentration of glucose in the blood.

D. The Pancreas Regulates Blood Glucose

The **pancreas** is an elongated gland, about the size of a hand, located beneath the stomach and attached to the small intestine. Clusters of cells in the pancreas secrete insulin and glucagon, two hormones that regulate the body's use of glucose.

Insulin and glucagon regulate blood glucose levels (figure 25.9). After a meal rich in carbohydrates, glucose enters the circulation at the small intestine. The resulting rise in blood sugar triggers specialized cells in the pancreas to secrete **insulin,** which stimulates cells throughout the body to absorb glucose from the bloodstream. As cells take up sugar, the blood glucose concentration declines, and insulin secretion slows. If blood sugar dips too low, however, other cells in the pancreas secrete **glucagon,** which stimulates target cells in the liver to release stored glucose into the bloodstream.

Too Much Glucose in Blood: Diabetes Failure to regulate blood sugar can be deadly. In **diabetes,** glucose accumulates to dangerously high levels in the bloodstream. Diabetes is a paradox: Sugar pours out of the body in urine, yet the body's cells starve for lack of glucose. (Centuries ago, before lab tests for blood sugar were available, physicians diagnosed diabetes from the sweet taste of a patient's urine.)

Symptoms of diabetes include frequent urination, excessive thirst, extreme hunger, blurred vision, weakness, fatigue, irritability, nausea, and weight loss. If the illness remains untreated, complications may include kidney failure, blindness, or a loss of sensation in the hands and feet. The nerve damage, in turn, can contribute to poor healing of wounds, as undetected cuts and scrapes become infected with bacteria and fungi. Severe diabetes can eventually result in coma and death.

The accumulation of blood sugar can occur for multiple reasons, but two forms of diabetes are most common. In type 1 diabetes, the pancreas fails to produce insulin, so the body's cells never receive the signal to "open the door" and admit glucose. In type 2 diabetes, the body's cells fail to absorb glucose even when insulin is present; this condition is called insulin resistance. In type 2 diabetes, then, insulin "rings the doorbell," but the cell never opens the door.

Fifteen percent of affected individuals have type 1 diabetes, which usually begins in childhood or early adulthood. Typically, the underlying cause is an autoimmune attack on cells of the pancreas, which therefore cannot produce insulin. Type 1 diabetes is sometimes also called insulin-dependent diabetes because insulin injections can replace the missing hormone (figure 25.10). ⓘ *autoimmune disorders,* section 29.5A

Type 2 diabetes is much more common. Although it usually begins in adulthood, the incidence of type 2 diabetes among both adults and children is rising in developed countries (including the United States). This disease is strongly associated with obesity (figure 25.11); nearly all type 2 diabetes patients are overweight. The cause–effect relationship between obesity and type 2 diabetes, however, is unclear.

Medicines can help lower blood glucose levels, but the best strategies to prevent and treat type 2 diabetes are to be physically active, reduce calorie intake, and choose healthy foods. For obese patients, gastric bypass surgery may also help. This procedure helps people cut calories by reducing the size of the stomach. But as a bonus, many gastric bypass patients enjoy an immediate reduction in diabetic symptoms, perhaps due to shifting hormones. ⓘ *healthy diet,* section 28.3

Not Enough Glucose in Blood: Hypoglycemia The opposite of diabetes is **hypoglycemia,** in which excess insulin production or insufficient carbohydrate intake causes low blood sugar. A person with this condition feels weak, sweaty, anxious, and shaky; in severe cases, hypoglycemia can cause seizures or loss of consciousness. Frequent, small meals low in sugar and high in protein and complex carbohydrates can prevent insulin surges and help relieve symptoms of hypoglycemia.

E. The Pineal Gland Secretes Melatonin

The **pineal gland,** a small structure located deep within the brain, produces the hormone **melatonin.** Darkness stimulates melatonin synthesis, whereas exposing the eye to light inhibits melatonin production. The amount of melatonin in blood therefore "tells" the other cells of the body how much light the eyes are receiving. This interaction, in turn, sets the stage for the regulation of sleep–wake cycles and other circadian rhythms.

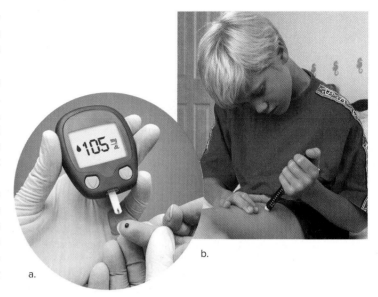

Figure 25.10 **Type 1 Diabetes.** (a) A blood glucose meter is a device that measures blood sugar. (b) A diabetic boy injects himself with insulin.

(a): ©Piotr Adamowicz/Getty Images RF; (b): ©Saturn Stills/Science Source

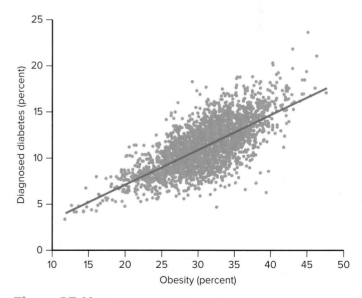

Figure 25.11 **Diabetes and Obesity.** Data from the Centers for Disease Control and Prevention show that the prevalence of type 2 diabetes is strongly correlated with obesity. Each dot represents one county in the United States in 2013 (the latest year for which data were available).

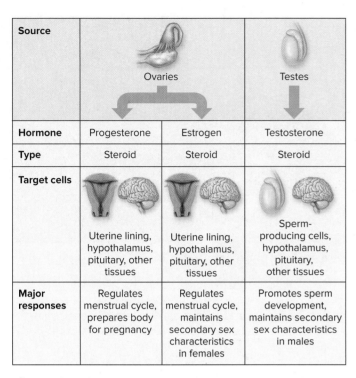

Source	Ovaries		Testes
Hormone	Progesterone	Estrogen	Testosterone
Type	Steroid	Steroid	Steroid
Target cells	Uterine lining, hypothalamus, pituitary, other tissues	Uterine lining, hypothalamus, pituitary, other tissues	Sperm-producing cells, hypothalamus, pituitary, other tissues
Major responses	Regulates menstrual cycle, prepares body for pregnancy	Regulates menstrual cycle, maintains secondary sex characteristics in females	Promotes sperm development, maintains secondary sex characteristics in males

Figure 25.12 Hormones of the Ovaries and Testes. Hormones produced in the ovaries and testes coordinate reproduction and the development of secondary sex characteristics.

Why We Care 25.1 | Anabolic Steroids in Sports

The anabolic steroids that regularly make headlines in the sporting news are synthetic forms of testosterone. Despite their notoriety, these drugs have a legitimate place in medicine. A physician might prescribe steroids for a person who produces too little testosterone, for example, or for someone with an illness that causes muscles to waste away.

Although anabolic steroids are legal only by prescription, some athletes abuse these drugs as a shortcut to greater muscle mass. Steroid users may improve strength and performance in the short term, but the drugs are harmful in the long run. In males, the body mistakes synthetic steroids for the natural hormone and lowers its own production of testosterone, causing infertility once use of the drug stops. Impotence, shrunken testicles, and the growth of breast tissue are other possible side effects. Females who abuse steroids may develop a masculine physique, a deeper voice, and facial or body hair. In adolescents, steroids hasten adulthood, stunting growth and causing early hair loss. Finally, research suggests that high doses of steroids may cause psychological side effects such as aggression, mood swings, and irritability. For all of these reasons, health professionals strongly advise against the use of illegal steroids.

A form of depression called seasonal affective disorder may be linked to abnormal melatonin secretion. Exposure to additional daylight (or full-spectrum lightbulbs) can elevate mood. Because melatonin levels fall as we age, some people believe that taking extra melatonin might delay aging. However, additional evidence is needed to support this claim.

25.4 Mastering Concepts

1. Describe the functions of each of the thyroid's hormones.
2. What is the function of parathyroid hormone (PTH)?
3. Compare the hormones of the adrenal cortex and medulla.
4. Describe the opposing roles of insulin and glucagon.
5. How do darkness and light affect melatonin secretion?

25.5 Hormones from the Ovaries and Testes Control Reproduction

The reproductive organs include the **ovaries** in females and the **testes** in males. These egg- and sperm-producing organs secrete the steroid hormones that enable these gametes to mature (figure 25.12). Hormones from the ovaries and testes also promote the development of secondary sex characteristics, which are features that differentiate the sexes but do not participate directly in reproduction. This section briefly introduces the sex hormones; chapter 30 explains their role in reproduction in more detail.

In a woman of reproductive age, the levels of several sex hormones cycle approximately every 28 days. The hypothalamus produces a hormone that stimulates the anterior pituitary to release FSH and LH into the bloodstream. At target cells in the ovary, these two hormones trigger the events that lead to the release of an egg cell. Meanwhile, cells surrounding the egg produce the sex hormones **estrogen** and **progesterone,** which exert negative feedback control on both the hypothalamus and the pituitary. Estrogen also promotes development of the female secondary sex characteristics, such as breasts and wider hips, whereas progesterone helps prepare the uterus for pregnancy.

In males, FSH stimulates the early stages of sperm formation in the testes. Sperm production is completed under the influence of LH, which also prompts cells in the testes to release the sex hormone **testosterone.** This hormone stimulates the formation of male structures in the embryo and promotes later development of male secondary sex characteristics, including facial hair, deepening of the voice, and increased muscle growth (see Why We Care 25.1).

25.5 Mastering Concepts

1. Which organs contain target cells for FSH and LH?
2. What are the functions of estrogen, progesterone, and testosterone?

Investigating Life 25.1 | Addicted to Affection

One of the rarest and least understood animal behaviors is monogamy—that is, exclusive mating with one partner for life. Unlike 97% of mammals, small rodents called prairie voles are monogamous. What makes these animals different from their more promiscuous counterparts?

Antidiuretic hormone (ADH) apparently plays a role in the vole's social attachments (see section 25.3A). ADH receptors in prairie voles occur in the area of the brain where addictive drugs act; as a result, males seem to derive feelings of reward from being with their mates and young. Researchers wondered whether adding extra ADH receptors into the reward area might induce males to form social attachments even more readily.

To find out, the researchers inserted the gene encoding the ADH receptor into a virus, then injected the modified virus into the reward areas of vole brains. Males in a control group received the ADH receptor gene in a different brain area. Afterward, each test animal spent 17 hours in a cage with an adult female. Each male was then placed into a choice chamber, where he was free to spend as much time as he wanted with his previous female companion or with a different female that was a stranger to him. The voles with the extra ADH receptors in the reward region of the brain spent much more time in contact with their previous companions than did control voles (figure 25.A).

The extra ADH receptors evidently made the male prairie voles especially likely to form pair bonds. Of course, it is important to remember that the explanation of monogamy is more complex than this quirk in a vole's brain chemistry might suggest. Nevertheless, this study is interesting because it explicitly links genes, brain chemistry, and social behavior. Moreover, the location of ADH receptors in the brain may also play a role in human social behaviors; examples include not only sexual fidelity but also autism, a disorder in which individuals have difficulty forming social attachments.

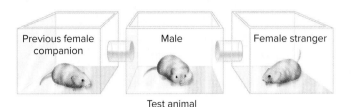

a. The male test animal can choose to spend time with a previous companion or a stranger.

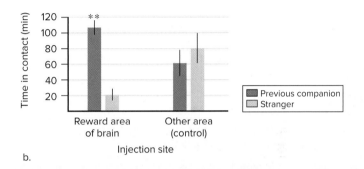

b.

Figure 25.A More Receptors, More Bonding. Prairie voles injected with ADH receptor genes in the reward center of the brain spent significantly more time with a female companion than with a stranger. Control voles were less likely than treated voles to bond with their partners. (Asterisks indicate a statistically significant difference within a group; see appendix B.)

Source: Pitkow, Lauren, and five coauthors (including Larry Young). 2001. Facilitation of affiliation and pair-bond formation by vasopressin receptor gene transfer into the ventral forebrain of a monogamous vole. *Journal of Neuroscience,* vol. 21, no. 18, pages 7392–7396.

What's the **Point?** ▼ A P P L I E D

Maybe midterms are around the corner or your personal relationships are on the rocks. Whatever the cause, stress can promote weight gain, suppress your immune system, and have other serious side effects. How can a mental state—chronic stress—alter your physical health?

Hormones play a big role in this process. When you are frightened, the adrenal medulla releases epinephrine (adrenaline), a hormone that makes your heart race. Long-term stress, however, stimulates the adrenal cortex to release different hormones; the most important of these is cortisol. The body responds in multiple ways. For example, cortisol initiates reactions that prompt liver and fat cells to release stored glucose and fatty acids into the blood. Having these energy sources readily available is useful during stress, when metabolic demands might be higher. The effects of cortisol are therefore adaptive when stress is infrequent and temporary.

Chronic stress, however, maintains high levels of cortisol. As cells release glucose and fatty acids, appetite increases. Food helps replenish depleted energy reserves, but overeating causes body weight to climb.

©Juanmonino/Getty Images RF

Cortisol triggers the redistribution of fat in the body, increasing the likelihood that fat is stored in the abdomen. Excess "belly fat" is therefore often associated with a high-stress lifestyle. Moreover, the perpetually high blood glucose associated with chronic stress may, over time, lead to diabetes.

Cortisol also suppresses the immune system, decreasing swelling in tissues damaged by physical stress. But lowered immune function may become harmful over the long term, as the body's ability to fight infection dwindles. Not surprisingly, illness often follows a stressful week.

Additional possible long-term effects of chronic stress include high blood pressure and other cardiovascular conditions. Males under stress may have low sperm production and erectile dysfunction; females may have irregular menstrual cycles and decreased sexual desire.

Exercise can enhance or suppress the cortisol response. Intense exercise puts physical stress on the body and temporarily increases cortisol levels. Yoga and other gentle forms of exercise, in contrast, relieve stress, as do deep breathing and stretching.

CHAPTER SUMMARY

25.1 The Endocrine System Uses Hormones to Communicate

- The **nervous system** and **endocrine system** specialize in the intercellular communication needed to maintain homeostasis in an animal body.
- The nervous system acts faster and more locally than the endocrine system.
- The endocrine system includes several **endocrine glands** and scattered cells, plus the **hormones** they secrete into the bloodstream. Hormones interact with **target cells** to exert their effects (figure 25.13).
- Negative feedback loops regulate hormone levels.

25.2 Hormones Stimulate Responses in Target Cells

A. Water-Soluble Hormones Trigger Second Messenger Systems

- Water-soluble hormones such as peptides and proteins bind to the surface receptors of target cells. A **second messenger** triggers the hormone's effect.
- Water-soluble hormones act relatively rapidly.

B. Lipid-Soluble Hormones Directly Alter Gene Expression

- Most **steroid hormones** cross cell membranes and bind to receptors inside target cells, activating or silencing particular genes.
- Because protein production takes time, lipid-soluble hormones act relatively slowly.

25.3 The Hypothalamus and Pituitary Gland Oversee Endocrine Control

- Neurons from the **hypothalamus** influence the release of hormones from the **pituitary gland,** which consists of two parts: the **posterior pituitary** and **anterior pituitary** glands. These hormones regulate many other processes in the body.

A. The Posterior Pituitary Stores and Releases Two Hormones

- Two hormones produced by the hypothalamus are released from the posterior pituitary. **Antidiuretic hormone (ADH)** regulates body fluid composition, and **oxytocin** stimulates muscle contraction in the uterus and milk ducts.

B. The Anterior Pituitary Produces and Secretes Six Hormones

- Hormones from the hypothalamus regulate the production and release of six hormones from the anterior pituitary.
- **Growth hormone (GH)** stimulates cell division, protein synthesis, and growth throughout the body. **Prolactin** stimulates milk production. **Thyroid-stimulating hormone (TSH)** prompts the thyroid gland to release hormones. **Adrenocorticotropic hormone (ACTH)** stimulates the adrenal cortex to release hormones. **Follicle-stimulating hormone (FSH)** and **luteinizing hormone (LH)** stimulate hormone release from the ovaries and testes.
- **Endorphins** are natural painkillers with target cells in the brain.

25.4 Hormones from Many Glands Regulate Metabolism

A. The Thyroid Gland Sets the Metabolic Pace

- **Thyroxine** and **triiodothyronine** from the **thyroid gland** speed metabolism. **Calcitonin** lowers the level of calcium in the blood.

B. The Parathyroid Glands Control Calcium Level

- The **parathyroid glands** secrete **parathyroid hormone (PTH),** which increases the blood calcium level.

C. The Adrenal Glands Coordinate the Body's Stress Responses

- The **adrenal gland** has an inner portion, the **adrenal medulla,** which secretes **epinephrine** and **norepinephrine.** These hormones ready the body to cope with a short-term emergency. The outer portion, or **adrenal cortex,** secretes **mineralocorticoids** and **glucocorticoids,** which mobilize energy reserves during stress and maintain blood volume and blood composition.

D. The Pancreas Regulates Blood Glucose

- Cells in the **pancreas** secrete **insulin,** which stimulates cells to take up glucose. **Glucagon** increases blood glucose levels.

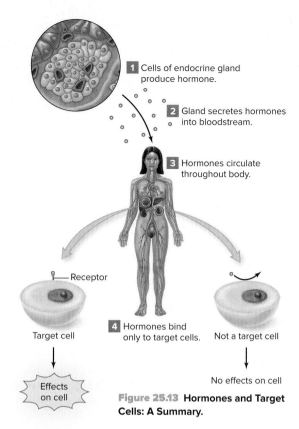

Figure 25.13 Hormones and Target Cells: A Summary.

1. Cells of endocrine gland produce hormone.
2. Gland secretes hormones into bloodstream.
3. Hormones circulate throughout body.
4. Hormones bind only to target cells.

Receptor
Target cell
Effects on cell

Not a target cell
No effects on cell

- In the most common types of **diabetes,** blood sugar concentrations rise to dangerous levels. Type 1 diabetes occurs when the pancreas fails to produce insulin; in type 2 diabetes, the body's cells do not respond to insulin. Type 2 diabetes is associated with obesity.
- **Hypoglycemia** is low blood sugar.

E. The Pineal Gland Secretes Melatonin

- The **pineal gland** secretes a hormone, **melatonin,** that may regulate how other glands respond to light–dark cycles.

25.5 Hormones from the Ovaries and Testes Control Reproduction

- In females, FSH and LH stimulate the **ovaries** to secrete **estrogen** and **progesterone,** hormones that stimulate the development of female secondary sex characteristics and control the menstrual cycle.
- In males, FSH and LH stimulate the **testes** to secrete **testosterone,** which stimulates sperm cell production and the development of secondary sex characteristics.

MULTIPLE CHOICE QUESTIONS

1. The effect of a water-soluble peptide hormone such as insulin is generally quicker than that of a steroid hormone such as estrogen because
   a. peptide hormones are synthesized within target cells, whereas steroid hormones must travel to target cells.
   b. steroid hormones stimulate or inhibit protein synthesis, which takes time.
   c. steroid hormones cannot pass through the cell membrane.
   d. All of these are correct.

2. Which of the following glands releases hormones when the thyroid hormone concentration in the blood is too low?
   a. Hypothalamus   b. Pancreas   c. Pineal gland   d. Adrenal gland

3. The parathyroid gland releases hormones when
   a. blood sugar is too high.   c. blood calcium levels are too low.
   b. bone growth is too slow.   d. urine is too dilute.

4. Which hormone is lipid-soluble and helps conserve water?
   a. ADH   b. Aldosterone   c. Estrogen   d. All of these

5. The body receives a series of stress-inducing stimuli throughout the day. In response, glucocorticoids are released from the
   a. adrenal cortex.   c. hypothalamus.
   b. anterior pituitary.   d. All of these are correct.

6. Secretion of melatonin is regulated by
   a. light.   b. temperature.   c. stress.   d. glucose.

7. To increase male fertility, it would be logical to develop a drug that boosts hormone synthesis at any of the following structures *except* the
   a. hypothalamus.   b. thyroid.   c. anterior pituitary.   d. testes.

Answers to Multiple Choice questions are in appendix A.

WRITE IT OUT

1. How does the endocrine system interact with the circulatory system?

2. Write a paragraph describing the events that occur from the time an endocrine gland releases a steroid hormone to the time the hormone exerts its effects on a target cell.

3. Describe an example of negative feedback on a hormone released from the anterior pituitary.

4. Give two examples of hormones counteracting the effects of one another.

5. Which hormone(s) match each of the following descriptions?
   a. Produced by a woman who is breast feeding
   b. Causes fatigue if too little is present
   c. Causes increase in blood calcium level
   d. Causes decrease in blood glucose level
   e. Synthetic steroids mimic the muscle-building effects of this hormone.

6. Alcohol and caffeine inhibit the effects of antidiuretic hormone. Explain why drinking beer or coffee increases the frequency of urination.

7. Some major league baseball players use human growth hormone (a banned substance) to aid in fast recovery after difficult workouts. How would GH help speed a player's recovery? Is GH use the same as anabolic steroid use?

8. Why can a stressful lifestyle lead to heart attacks? Which hormone is released in response to long-term stress?

9. Would an insulin injection help a person with type 2 diabetes? Why or why not?

10. Identify the target cells and effects of FSH, LH, estrogen, progesterone, and testosterone.

SCIENTIFIC LITERACY

1. Review Burning Question 25.1, which defines endocrine disruptors as molecules that either mimic or block the activity of a hormone. Propose a way to test the hypothesis that microwaving foods in plastic containers releases endocrine disruptors.

2. Review Why We Care 25.1, which explains how synthetic steroids affect the body. Find a website that supports the use of "testosterone boosters" and a website that says that they are dangerous. Which individual or organization produced each website? Which website might be more reliable, and why do you think so?

Answers to Mastering Concepts, Write It Out, Scientific Literacy, and Pull It Together questions can be found in the Connect ebook.
connect.mheducation.com

PULL IT TOGETHER

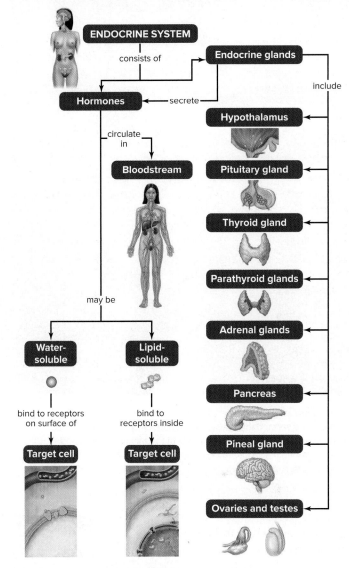

Figure 25.14 Pull It Together: The Endocrine System.

Refer to figure 25.14 and the chapter content to answer the following questions.

1. Review the Survey the Landscape figure in the chapter introduction, and then add at least two other organ systems to the Pull It Together concept map. Connect an endocrine gland to each new term, and explain how hormones released from the gland affect the organ system in a way that maintains homeostasis.

2. Connect each hormone discussed in this chapter to the gland that produces it and to either the "Water-soluble" or "Lipid-soluble" box.

3. Label the endocrine glands that are primarily regulated by signals from the nervous system; by hormones from the hypothalamus; by hormones from the anterior pituitary; and by the concentration of calcium or sugar in the blood.

26

The Skeletal and Muscular Systems

Helping Hand. Patrick Kane, who lost the fingers of his left hand to illness when he was a baby, shows off his battery-operated prosthetic hand. All five fingers bend independently at the joints; a mobile app lets him program his hand with a variety of grip positions.

©Jeff J Mitchell/Getty Images

LEARNING OUTLINE

26.1 Skeletons Take Many Forms

26.2 The Vertebrate Skeleton Features a Central Backbone

26.3 Bones Provide Support, Protect Internal Organs, and Supply Calcium

26.4 Muscle Movement Requires Contractile Proteins and ATP

26.5 Muscle Cells Generate ATP in Multiple Ways

26.6 Muscle Fiber Types Influence Athletic Performance

APPLICATIONS

Why We Care 26.1 *Bony Evidence of Murder, Illness, and Evolution*

Burning Question 26.1 *Is creatine a useful dietary supplement?*

Burning Question 26.2 *Why does heat soothe sore muscles and joints?*

Investigating Life 26.1 *Did a Myosin Gene Mutation Make Humans Brainier?*

 Learn How to Learn

Study as You Go

Last-minute cramming for exams may be a classic college ritual, but it is not usually the best strategy. If you try to memorize everything right before an exam, you may become overwhelmed and find yourself distracted by worries that you'll never learn it all. Instead, work on learning the material as the course goes along. Then, on the night before the exam, get plenty of rest, and don't forget to eat on exam day. If you are too tired or too hungry to think, you won't be able to give the exam your best shot.

Most animals can move. They may approach food, mates, or a source of warmth; they may flee from danger. These movements require the coordinated actions of the skeletal and muscular systems.

For more details, study the Pull It Together feature in the chapter summary.

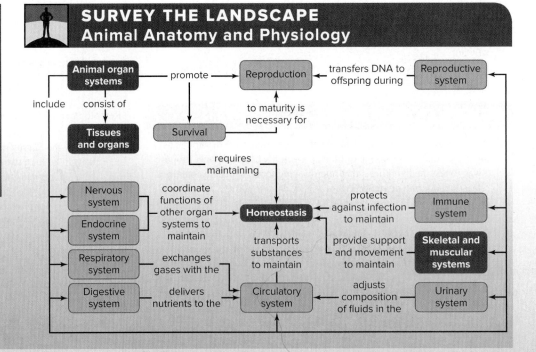

SURVEY THE LANDSCAPE
Animal Anatomy and Physiology

What's the **Point?** ▼

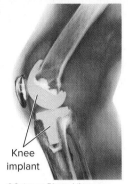

Knee implant

©Science Photo Library - Dr. P. Marazzi/Getty Images RF

From Halloween skeletons to a body builder's exaggerated brawn, bones and muscles are familiar parts of an animal's body. What many people do not realize, however, is that skeletons and muscles are intimately related. As you will learn in this chapter, most muscles attach to the ends of bones. Hundreds of finely controlled muscle movements produce the range of motion possible in the human body. The nervous system coordinates the action.

The intricate relationships among nerves, bones, and muscles make it all the more amazing that surgeons can replace a missing limb or a damaged joint with a functional, if not a biological, alternative. The man pictured in the chapter opening photo has an artificial hand, and a replacement knee joint is shown here. These high-tech materials integrate with a patient's own flesh and bone, restoring mobility. This astonishing feat begins with a basic understanding of the skeletal and muscular systems.

26.1 Skeletons Take Many Forms

Ask a child what sets animals apart from other organisms, and he or she will likely answer "movement." This response is technically wrong—many bacteria, archaea, protists, fungi, and plants have swimming, creeping, or gliding cells. Yet the child correctly recognizes that animal movements are unmatched in their drama, versatility, and power.

The ability to hop, dig, fly, slither, scuttle, or swim comes from two closely allied, interacting organ systems: the **muscular system** and the **skeletal system,** which together function under the direction of the nervous system. In most animals, organs called **muscles** provide motion; the **skeleton** adds a firm supporting structure that muscles pull against. The skeleton also gives shape to an animal's body and protects the internal organs.

Figure 26.1 shows three types of skeletons. The simplest is a **hydrostatic skeleton** (*hydro-* means water), which consists of fluid constrained within a layer of flexible tissue. Many of the invertebrate animals described in chapter 17 have hydrostatic skeletons. The bell of a jellyfish, for example, consists mostly of a gelatinous substance constrained between two tissue layers. To swim, the animal rhythmically contracts the muscles acting on this hydrostatic skeleton, forcibly ejecting water from its body. Snails, squids, flatworms, earthworms, and nematodes also use hydrostatic skeletons in locomotion.

The most common type of skeleton is an **exoskeleton** (*exo-* means outside), which acts as a "suit of armor" that protects the animal from the outside. Internal muscles pull against the exoskeleton, enabling the animal to move. Animals with exoskeletons include arthropods such as crabs, lobsters, and insects. Mollusks, including clams and snails, also produce exoskeletons.

Figure 26.1 **Types of Skeletons.**
(a) Hydrostatic skeleton. (b) Exoskeleton.
(c) Endoskeleton.

(a): ©Gabriel Bouys/AFP/Getty Images;
(b): ©Iconotec/Alamy Stock Photo RF;
(c): ©Sebastian Alberto Greco/Getty Images RF

a. Jellyfish

b. Crab

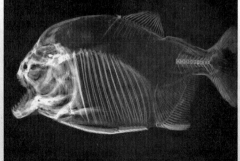

c. Bony fish (piranha)

517

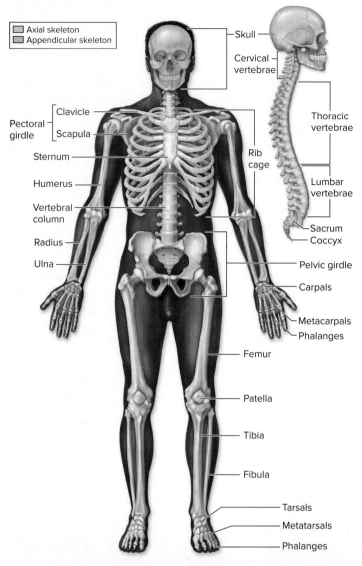

Axial skeleton
Appendicular skeleton

Skull

Cervical vertebrae

Clavicle

Pectoral girdle

Scapula

Sternum

Humerus

Vertebral column

Radius

Ulna

Thoracic vertebrae

Rib cage

Lumbar vertebrae

Sacrum

Coccyx

Pelvic girdle

Carpals

Metacarpals

Phalanges

Femur

Patella

Tibia

Fibula

Tarsals

Metatarsals

Phalanges

Skeletal System	
Main tissue types*	Examples of locations/functions
Connective	Makes up bone, cartilage, tendons, ligaments, marrow of vertebrate skeleton
Muscle	Skeletal muscle connects to movable bones, enabling voluntary movements
Nervous	Senses body position and controls muscles

*See chapter 23 for descriptions.

Figure 26.2 The Human Skeleton. The axial skeleton in humans includes the bones of the head, vertebral column, and rib cage. The bones that compose and support the limbs constitute the appendicular skeleton.

Exoskeletons have advantages and disadvantages. The hard covering protects soft internal organs and provides excellent leverage for muscles. On the other hand, a growing arthropod must periodically molt; until its new skeleton has hardened, the animal is vulnerable to predators.

An **endoskeleton** (*endo-* means inner) is an internal support structure. Sea stars and other echinoderms, for example, produce rigid, calcium-rich spines and internal plates. Vertebrate animals have endoskeletons made of cartilage or bone. Sharks and rays are fishes with cartilage skeletons. Other fishes, and all land vertebrates, have skeletons composed primarily of bone.

Like exoskeletons, endoskeletons represent an evolutionary trade-off. On the plus side, an internal skeleton grows with the animal, eliminating the problems associated with molting. Also, an endoskeleton consumes less of an organism's total body mass, so it can support animals as large as a whale. One disadvantage, however, is that the endoskeleton does not protect soft tissues at the body surface.

The vertebrate skeleton's capacity to change over evolutionary time is striking. Each vertebrate species has a distinctive skeleton, yet all are composed of the same types of cells and have similar arrangements. The characteristics of each species' muscles and skeleton reflect common ancestry and the selective forces in its environment. (i) *natural selection,* section 12.3; *homologous structures,* section 13.4A

26.1 Mastering Concepts

1. How do the skeletal and muscular systems interact?
2. Describe similarities and differences among the three main types of skeletons.
3. How do vertebrate skeletons reveal common ancestry?

26.2 The Vertebrate Skeleton Features a Central Backbone

Bones, the organs that compose the vertebrate skeleton, are grouped into two categories (figure 26.2). The **axial skeleton,** so named because it is located along the central axis of the body, consists of the bones of the head, vertebral column, and rib cage. The **appendicular skeleton** consists of the appendages (limbs) and the bones that support them.

The axial skeleton shields soft body parts. The skull, which protects the brain and many of the sense organs, consists of hard, dense bones that fit together like puzzle pieces. All of the head bones are attached with immovable joints, except for the lower jaw and the middle ear. These movable jaw and ear bones enable us to chew food, speak, and hear. (i) *hearing,* section 24.11

The **vertebral column** supports and protects the spinal cord. A human vertebral column typically consists of 33 vertebrae, separated by cartilage disks that cushion shocks and enhance flexibility. A "slipped," or herniated, disk occurs when these pads tear or rupture, causing a bulge that presses painfully on a nearby nerve. Scoliosis, in which the vertebral column curves to the side, is also a disorder of the axial skeleton (figure 26.3).

Attached to the human vertebral column are 12 pairs of ribs, which protect the heart and lungs. Flexible cartilage in the rib cage allows the chest to expand during breathing.

In the appendicular skeleton, the **pectoral girdle** connects the forelimbs to the axial skeleton; it includes the collarbones (clavicles) and shoulder blades (scapulas). Likewise, the **pelvic girdle** attaches the hindlimb bones to the axial skeleton. The hipbones join the backbone in the rear and meet each other in front, creating a bowl-like pelvic cavity. (The term *pelvis* is Latin for "basin.") The bony pelvis protects the lower digestive organs, the bladder, and some reproductive structures (especially in the female).

Why We Care 26.1 illustrates how bones reveal clues that are useful to people in several professions.

26.2 Mastering Concepts

1. What are the two subdivisions of the human skeleton?

2. What are the locations of the pectoral and pelvic girdles?

26.3 Bones Provide Support, Protect Internal Organs, and Supply Calcium

The skeleton not only supports and protects the body, but it also has several other functions that may at first glance seem unrelated to one another (table 26.1). Bones connected to muscles provide movement, and bone minerals supply calcium and phosphorus to the rest of the body. Blood cells also form at the marrow inside bones.

A. Bones Consist Mostly of Bone Tissue and Cartilage

A glance back at figure 26.2 reveals that bones take many shapes. Long bones make up the arms and legs, whereas the wrists and ankles consist mainly of short bones. Flat bones include the ribs and skull. Vertebrae are irregularly shaped.

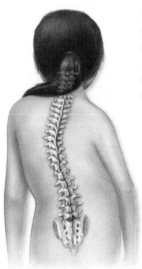

Figure 26.3 Scoliosis. This young girl's spine curves to the side. To straighten the spine, physicians may recommend a specially designed back brace or spinal surgery.

TABLE 26.1	Functions of the Vertebrate Endoskeleton: A Summary
Function	**Explanation**
Support	The skeleton supports an animal's body against gravity. It largely determines the body's shape.
Movement	The vertebrate skeleton is a system of muscle-operated levers. Typically, the two ends of a skeletal (voluntary) muscle attach to different bones that connect in a structure called a joint. When the muscle contracts, one bone moves.
Protection of internal structures	The backbone surrounds and shields the spinal cord, the skull protects the brain, and ribs protect the heart and lungs.
Production of blood cells	The long bones of the arms and legs contain red marrow, a tissue that produces red blood cells, white blood cells, and platelets.
Mineral storage	The skeleton stores calcium and phosphorus.

Why We Care 26.1 | Bony Evidence of Murder, Illness, and Evolution

Skeletons sometimes provide useful clues to past events. Hard, mineral-rich bones and teeth remain intact long after a corpse's soft body parts decay (figure 26.A). These durable remains can help solve crimes, lend insight into human history, and shed light on evolution.

Detectives can use bones to identify the sex of a decomposed murder victim. This technique relies on the differences between male and female skeletons. Most obviously, the average male is larger than the average female. In addition, the front of the female pelvis is broader and larger than the male's, and it has a wider bottom opening that accommodates the birth of a baby. These same features allow anthropologists to determine the sex of ancient human fossils.

Bones can also reveal events and illnesses unique to each person's life. Healed breaks may indicate

Figure 26.A Old Bones. An archaeologist inspects a grave from the ninth century.

©Viktor Chlad/isifa/Getty Images

accidents or abuse. Egypt's King Tut, for example, suffered a severe leg break shortly before he died. Crooked joints may be evidence of arthritis, and patterns of bone thickenings tell whether a person spent his or her life in hard physical labor.

The shapes and sizes of fossilized bones also reveal some of the details of human evolution. Section 17.12 explains how the skeletons and teeth of primate fossils provide clues to brain size, diet, and posture in our ancestors. Animal skeletons also tell the larger story of vertebrate evolution. For example, paleontologists can examine skeletal features to determine whether an extinct animal was terrestrial or aquatic. Air is much less supportive than water, so land dwellers tend to have sturdier skeletons than their aquatic relatives. Radiometric dating reveals when the new adaptations arose. ⓘ *radiometric dating*, section 13.2B

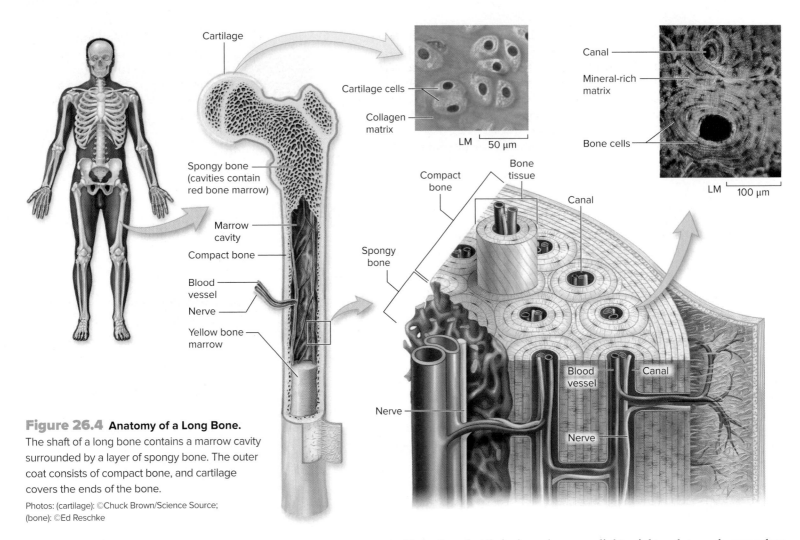

Figure 26.4 Anatomy of a Long Bone.
The shaft of a long bone contains a marrow cavity surrounded by a layer of spongy bone. The outer coat consists of compact bone, and cartilage covers the ends of the bone.

Photos: (cartilage): ©Chuck Brown/Science Source; (bone): ©Ed Reschke

Miniglossary	Skeletal System
Bone	An organ of the skeletal system
Red bone marrow	Soft, spongy tissue that is a nursery for blood cells and platelets
Yellow bone marrow	Fatty tissue that occupies the marrow cavity of adult bones
Compact bone tissue	Hard, dense bone tissue, with canals for blood vessels and nerves
Spongy bone tissue	Light, porous bone tissue with spaces filled with marrow, blood vessels, and nerves
Cartilage	Smooth, rubbery connective tissue that absorbs shocks and reduces friction at joints
Tendon	Connective tissue that attaches a bone to a muscle, stabilizing a movable joint
Ligament	Connective tissue that attaches a bone to another bone, stabilizing a movable joint

No matter what their shape, bones are lightweight and strong because they are porous (figure 26.4). The weight of long bones is further reduced by the **marrow cavity,** a space occupying the center of the shaft. Bones contain two types of marrow. **Red bone marrow** is a nursery for blood cells and platelets; in adults, **yellow bone marrow** replaces the red marrow in the marrow cavity. The fatty yellow marrow does not produce blood; however, if blood cells are in short supply, yellow marrow can revert to red marrow.

Besides marrow, bones also contain nerves and blood vessels. But the majority of the vertebrate skeleton consists of two types of connective tissue: bone and cartilage. Figures 23.3 and 26.4 offer a closer look at both of these tissues.

Bone tissue consists of multiple cell types suspended in a hard extracellular matrix. Some bone cells secrete the matrix, which consists mainly of collagen and minerals. Collagen is a protein that gives the bone flexibility, elasticity, and strength. The hardness and rigidity of bone come from the minerals, primarily calcium and phosphate, that coat the collagen fibers. Other bone cells degrade the matrix at the bone surface, releasing calcium and phosphorus into the blood as needed to maintain homeostasis.

Bones include both compact and spongy bone tissue. **Compact bone tissue** is hard and dense, with canals that house blood vessels and nervous tissue. **Spongy bone tissue** is much lighter than compact bone tissue, thanks to a web of hard, bony struts enclosing large spaces filled with red marrow. The shaft of a long bone consists mostly of compact bone overlying a layer of spongy bone. The bulbous tips also contain spongy bone.

Besides bone, **cartilage** is the other main connective tissue in the skeleton. This rubbery material, which covers the ends of bones, consists mostly of tough, elastic proteins. Cartilage therefore resists breakage and stretching, even when bearing great weight. Moreover, the protein network in cartilage holds a great deal of water, making it an excellent shock absorber. But cartilage lacks a blood supply. As the body moves, water within cartilage cleanses the tissue and bathes it with dissolved nutrients from nearby blood vessels. Nevertheless, the absence of a dedicated blood supply means that injured cartilage is slow to heal.

B. Bone Meets Bone at a Joint

A **joint** is an area where two bones meet. Many joints are freely movable, such as those of the knees, hips, elbows, fingers, and toes (figure 26.5). These joints consist of movable bones joined by a fluid-filled capsule of fibrous connective tissue. Together, the lubricating fluid and slippery cartilage allow bones to move against each other in a nearly friction-free environment.

Tendons and ligaments help stabilize movable joints. **Tendons** are tough bands of connective tissue that attach bone to muscle; **ligaments** are similar structures that attach bone to bone. A strain is an injury to a muscle or tendon, whereas a sprain is a stretched or torn ligament. A torn anterior cruciate ligament (ACL) is a common type of knee sprain, especially in sports such as basketball and volleyball. The ACL is one of two ligaments that crisscross at the knee, connecting the thighbone to the shinbone. Surgical reconstruction of the ACL enables many injured athletes to return to their sports.

Arthritis is a common disorder of joints. A very severe form, rheumatoid arthritis, is an inflammation of the joint membranes, usually in the hands and feet. In the more common osteoarthritis, joint cartilage wears away. As the bone is exposed, small bumps of new bone begin to form, and the joints become stiff and painful. Osteoarthritis usually appears after age 40.

C. Bones Are Constantly Built and Degraded

The bones of a developing embryo originate as cartilage "models" (figure 26.6). As the fetus grows, each model's matrix hardens with calcium salts. After birth, bone growth becomes concentrated near the ends of the long bones in thin disks of cartilage called "growth plates." The bones continue to elongate

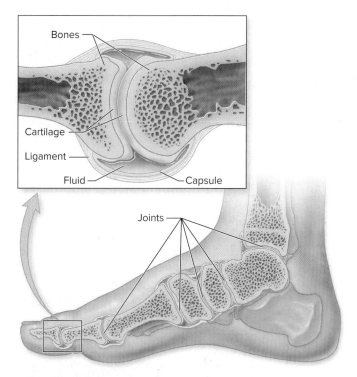

Figure 26.5 Movable Joints. The bones of the foot are connected by joints that enable the foot and toes to move. The inset shows a fluid-filled capsule of fibrous connective tissue surrounding a movable joint in the big toe.

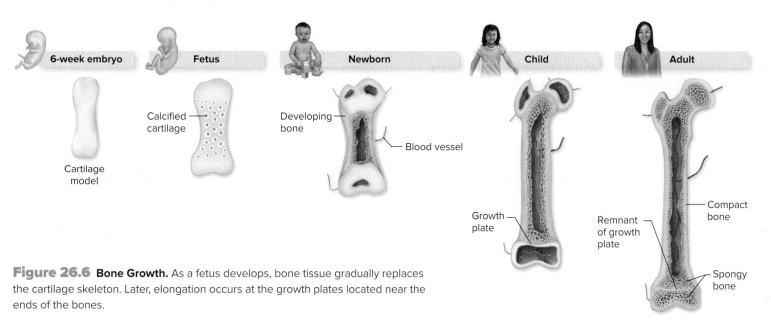

Figure 26.6 Bone Growth. As a fetus develops, bone tissue gradually replaces the cartilage skeleton. Later, elongation occurs at the growth plates located near the ends of the bones.

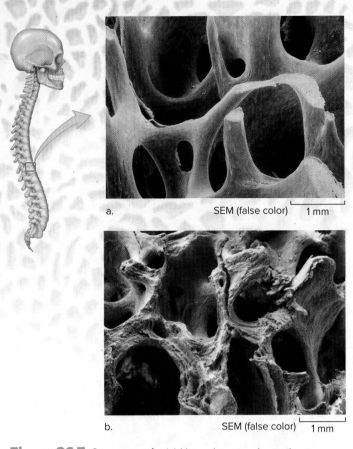

a.   SEM (false color)   1 mm

b.   SEM (false color)   1 mm

Figure 26.7 **Osteoporosis.** (a) Normal spongy bone tissue.
(b) Calcium loss from this spongy bone tissue has led to osteoporosis.

Photos: (a): ©Prof. P.M. Motta/Univ. "La Sapienza," Rome/Science Source; (b): ©Dee
Breger/Science Source

until the late teens, when bone tissue begins to replace the cartilage growth
plates. By the early twenties, bone growth is complete.

Even after a person stops growing, bone is continually being remodeled.
Bones become thicker and stronger with strenuous exercise such as weight
lifting. On the other hand, less-used bones lose mass. For example, astronauts lose bone density if they are in a prolonged weightless environment
because their bodies don't have to work as hard as they do against Earth's
gravity.

Moreover, broken bones can repair themselves. Bone cells near the site
of the fracture produce new bone tissue, so that after several weeks, the
injury is all but healed.

D. Bones Help Regulate Calcium Homeostasis

Throughout life, bones are a reservoir for calcium. This mineral is vital for muscle
contraction, blood clotting, the activity of some enzymes, and many other essential
functions. The body therefore maintains calcium homeostasis by constantly shut-
tling calcium between blood and bone. Hormones from the parathyroid and thyroid
glands control this exchange in a negative feedback loop. ⓘ *negative feedback
loop,* section 23.4; *parathyroid glands,* section 25.4B

Bones sometimes lose more calcium than they add, leading to **osteoporosis,**
a condition in which bones become less dense (figure 26.7). An astronaut's
"disuse osteoporosis" is one example. Much more familiar, however, is the
age-related osteoporosis that causes shrinking stature, back pain, and frequent
fractures in the elderly.

Both men and women can suffer from osteoporosis, but the disorder is most
common in females. To prevent bone loss, doctors therefore advise all women
to exercise regularly and to take calcium supplements daily. Several drugs can
also slow or reverse bone loss.

26.3 Mastering Concepts

1. What are the main parts of a long bone?
2. Describe the structures and functions of bone tissue and
   cartilage.
3. What are the relationships among joints, tendons, and ligaments?
4. How are bones remodeled and repaired throughout life?
5. How do bones participate in calcium homeostasis?

Function	Explanation
TABLE 26.2	**Functions of Skeletal Muscles: A Summary**
Voluntary movement	Contraction of muscles attached to bones produces movements under voluntary control.
Control of body openings	Skeletal muscles provide voluntary control of the eyelids, mouth, and anus.
Maintain posture and joint stability	Muscles attached to bones keep the body upright and stabilize joints.
Communication	Skeletal muscle movements enable facial expressions, speech, writing, and gesturing.
Maintain body temperature	Metabolic activity in skeletal muscle generates abundant heat.

26.4 Muscle Movement Requires Contractile Proteins and ATP

As we have already seen, movement relies on the interaction between bones
and muscles. The human muscular system includes more than 600 **skeletal
muscles,** which generate voluntary movements. (This number does not include
smooth muscle and cardiac muscle, which are involuntary and are not typically
considered part of the muscular system.) Table 26.2 lists some functions of
skeletal muscles, and figure 26.8 identifies a few of the major skeletal muscles
in a human.

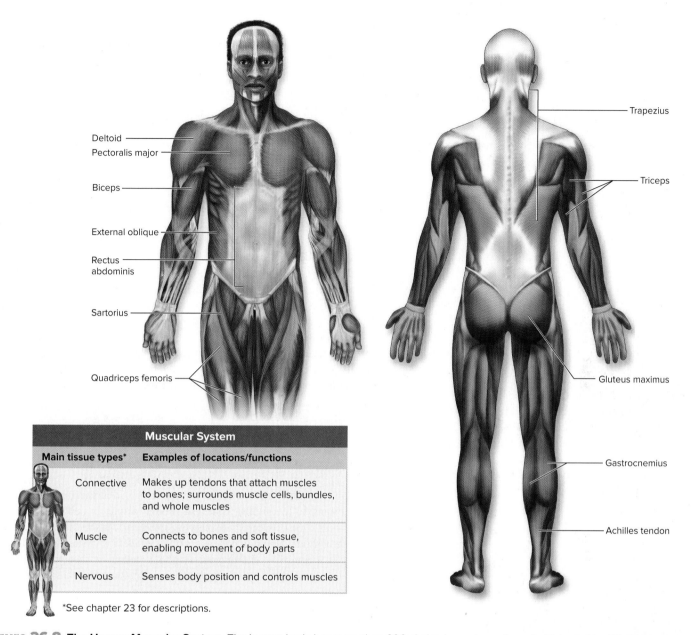

Muscular System	
Main tissue types*	**Examples of locations/functions**
Connective	Makes up tendons that attach muscles to bones; surrounds muscle cells, bundles, and whole muscles
Muscle	Connects to bones and soft tissue, enabling movement of body parts
Nervous	Senses body position and controls muscles

*See chapter 23 for descriptions.

Figure 26.8 The Human Muscular System. The human body has more than 600 skeletal muscles, a few of which are identified here.

Pairs of muscles often work together to generate body movements. Figure 26.9, for example, shows the biceps and triceps muscles. Tendons attach both of these muscles to the bones of the shoulder and lower arm. Each contracting muscle can pull a bone in one direction but cannot push the bone the opposite way. The elbow can bend and straighten because the biceps and triceps operate in opposite directions. That is, when you contract the biceps (the bulge that appears when you "make a muscle"), the triceps relaxes and the arm bends at the elbow joint. Conversely, the arm extends when the triceps contracts and the biceps relaxes. Many other skeletal muscles occur in similar antagonistic pairs that permit back-and-forth movements.

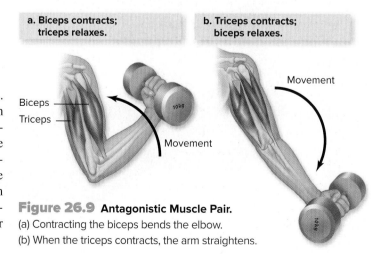

a. Biceps contracts; triceps relaxes.

b. Triceps contracts; biceps relaxes.

Figure 26.9 Antagonistic Muscle Pair.
(a) Contracting the biceps bends the elbow.
(b) When the triceps contracts, the arm straightens.

A. Actin and Myosin Filaments Fill Muscle Cells

Picture a softball player swinging a bat, an action that requires the controlled contraction of many skeletal muscles in the legs, arms, and torso. Each muscle moves a different body part, yet all are organized in essentially the same way. Figure 26.10 illustrates several levels of muscle anatomy, zooming from the whole organ to the microscopic scale.

The left side of figure 26.10 shows a whole muscle, an organ that consists of multiple tissue types. Connective tissue, for example, makes up tendons and the sheath that wraps around each muscle like a banana peel. Blood vessels in the muscle deliver nutrients and oxygen while removing wastes. Nerves transmit information to and from the central nervous system (see chapter 24).

The bulk of the muscle, however, consists of skeletal muscle tissue. The cross sections in the center of figure 26.10 show that muscle tissue is composed of parallel bundles of **muscle fibers,** which are individual muscle cells ranging in length from 1 to 40 mm. The right half of the figure focuses on one muscle cell. Each cell contains multiple nuclei and other organelles. But most of the cell's volume is occupied by hundreds of thousands of cylindrical **myofibrils,** bundles of parallel protein filaments running the length of the cell.

The inset in figure 26.10 depicts the proteins that compose each myofibril. A **thick filament** is made of a protein called **myosin.** A **thin filament** consists primarily of two entwined strands of another protein, **actin.** Interactions between thick and thin protein filaments are the basis of muscle contraction.

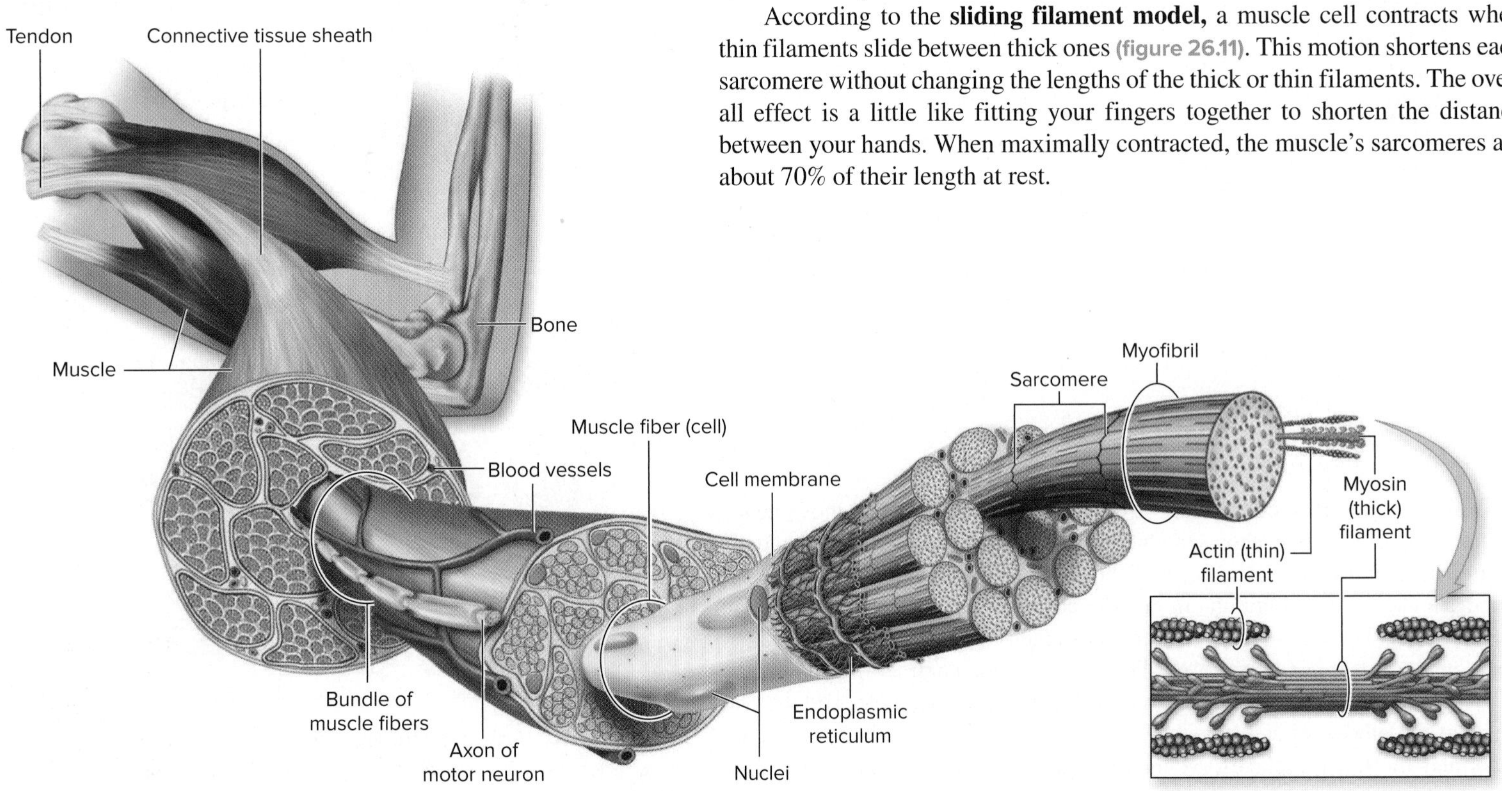

B. Sliding Filaments Are the Basis of Muscle Cell Contraction

Skeletal muscle tissue appears striped, or striated, because of the alternating arrangement of thick and thin filaments (see figure 26.10). These striations divide each myofibril into many functional units, called **sarcomeres.**

According to the **sliding filament model,** a muscle cell contracts when thin filaments slide between thick ones (figure 26.11). This motion shortens each sarcomere without changing the lengths of the thick or thin filaments. The overall effect is a little like fitting your fingers together to shorten the distance between your hands. When maximally contracted, the muscle's sarcomeres are about 70% of their length at rest.

Figure 26.10 Skeletal Muscle Organization. A muscle is an organ enclosed in connective tissue, nourished by blood vessels, and controlled by nerves. Tendons attach skeletal muscles to bones. Bundles of muscle fibers make up most of the muscle's volume. Each muscle fiber is a single cell with many nuclei. Most of the cell's volume is occupied by myofibrils, which are, in turn, composed of filaments of the proteins actin and myosin.

For thick and thin filaments to move past each other and contract a muscle cell, actin and myosin must touch. The physical connection between the two types of filaments is the pivoting club-shaped "head" portion of each myosin molecule. A myosin head forms a cross bridge when it swings out to contact an actin molecule (figure 26.12).

As detailed in figure 26.12, the sliding interaction between actin and myosin requires energy in the form of ATP. In step 1 of the figure, the myosin heads are not yet connected to actin. Soon, however, a myosin head forms a cross bridge by attaching to an exposed actin subunit on a thin filament (step 2). The cross bridge bends, pulling on actin and causing it to slide past myosin in the same way that an oar's motion moves a boat (step 3). The myosin head then binds a molecule of ATP and releases the actin (step 4). ATP splits into ADP and a phosphate group, prompting the myosin head to swivel back to its original position (step 5). The myosin head is now ready to contact another actin subunit farther down the thin filament. ⓘ *ATP*, section 4.3

Figure 26.12 **ATP's Role in Muscle Contraction.** ATP provides the energy required for myosin filaments to "ratchet" past actin filaments as the muscle contracts.

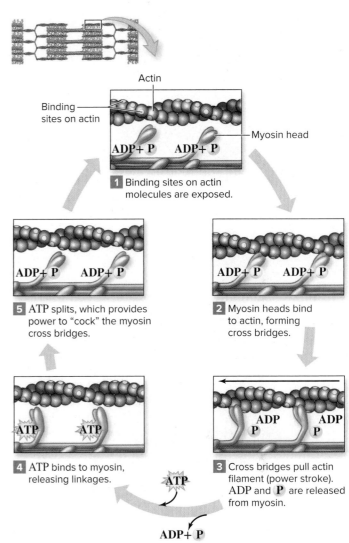

Actin

Binding sites on actin

Myosin head

ADP + P   ADP + P

1 Binding sites on actin molecules are exposed.

ADP + P   ADP + P

5 ATP splits, which provides power to "cock" the myosin cross bridges.

ADP + P   ADP + P

2 Myosin heads bind to actin, forming cross bridges.

ATP   ATP

4 ATP binds to myosin, releasing linkages.

ATP

ADP + P

ADP P   ADP P

3 Cross bridges pull actin filament (power stroke). ADP and P are released from myosin.

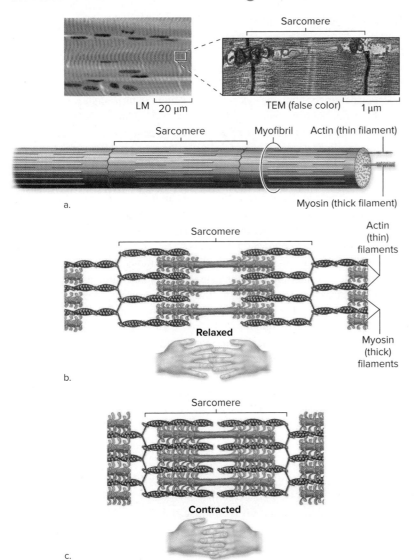

Sarcomere

LM 20 µm

TEM (false color) 1 µm

Sarcomere   Myofibril   Actin (thin filament)

Myosin (thick filament)

a.

Sarcomere

Actin (thin) filaments

Relaxed

Myosin (thick) filaments

b.

Sarcomere

Contracted

c.

Figure 26.11 **The Sliding Filament Model of Muscle Contraction.**
(a) Myofibrils are divided into units called sarcomeres. (b–c) During muscle cell contraction, thin filaments slide past thick filaments, decreasing the length of each sarcomere.

Photos: (LM): ©McGraw-Hill Education/Al Telser; (TEM): ©Biology Pics/Science Source

Muscle Cells Generate ATP in Multiple Ways   Muscle Fiber Types Influence Athletic Performance

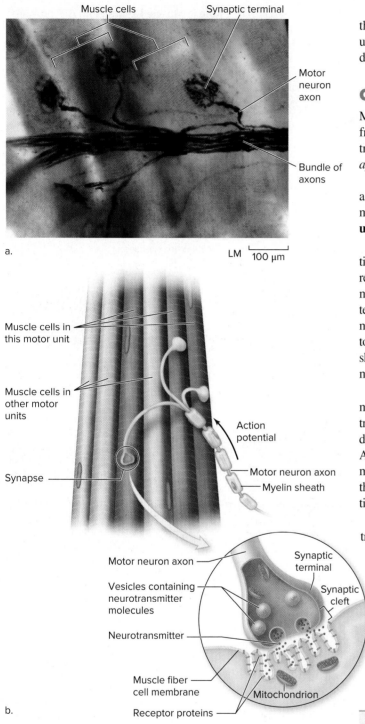

a.

LM ⊢ 100 μm ⊣

Muscle cells in this motor unit

Muscle cells in other motor units

Synapse

Action potential

Motor neuron axon

Myelin sheath

Motor neuron axon

Vesicles containing neurotransmitter molecules

Neurotransmitter

Synaptic terminal

Synaptic cleft

Muscle fiber cell membrane

Mitochondrion

Receptor proteins

b.

Figure 26.13 How Neurons Control Muscle Cells. (a) Motor neurons communicate with muscle cells at synapses; several synapses appear in this micrograph. (b) This motor unit is composed of one motor neuron axon and three muscle cells. (The remaining muscle cells belong to other motor units.) When action potentials in the axon stimulate the release of a neurotransmitter at the synapses, the three muscle cells contract.

(a): ©McGraw-Hill Education/Al Telser

This sliding action repeats about a hundred times per second on each of the hundreds of myosin molecules of a thick filament. Although each individual movement is minuscule, a skeletal muscle contracts quickly and forcefully due to the efforts of many thousands of "rowers" in each sarcomere.

C. Motor Neurons Stimulate Muscle Contraction

Muscles do not contract at random; rather, they wait for electrical stimulation from the central nervous system. A **motor neuron** delivers the signal to contract at a specialized synapse between the neuron and a muscle cell. ⓘ *synapse,* section 24.4

Each motor neuron's axon branches at its tip, with each branch leading to a different muscle cell (figure 26.13). One motor neuron can therefore control multiple cells; together, a motor neuron and its muscle cells make up a **motor unit.** The motor unit in figure 26.13b, for example, includes three muscle cells.

When the central nervous system sends the signal to contract, action potentials are conveyed along the motor neuron's axon. These signals stimulate the release of a chemical message called a **neurotransmitter** at each synapse with a muscle cell (see figure 26.13b, inset). The neurotransmitters bind to receptor proteins on the cell surface, causing an electrical wave to race along the muscle cell membrane. This electrical signal causes the muscle cell's endoplasmic reticulum to release calcium ions into the cytosol. The calcium, in turn, slightly changes the shape of the thin filaments, allowing the myosin heads to bind to actin. The sarcomeres in the muscle cell are now free to contract. ⓘ *action potential,* section 24.3

Within one muscle, motor units vary in size from tens to hundreds of cells per motor neuron. When a neural impulse arrives, all of the cells in a motor unit contract at the same time. A motor neuron that controls only a few muscle cells produces fine, small-scale responses, such as the eye movements required for reading. A motor unit consisting of hundreds of muscle cells produces large, coarse movements, such as those required for throwing a ball. The more motor units activated, the stronger the force of contraction. In this way, by recruiting different combinations of motor units, the same hand can both grip a hammer and pick up a tiny nail.

Some diseases interfere with the neural signals that stimulate muscle contraction. The virus that causes polio, for example, causes paralysis by destroying motor neurons. Another disease-causing organism is the bacterium that produces botulinum toxin (commonly known as Botox). This poison blocks the release of neurotransmitters from motor neurons. Affected muscles therefore never receive the signal to contract. Ingesting botulinum in tainted foods can cause paralysis, which can be fatal. Injecting tiny amounts of Botox around the eyes and forehead temporarily paralyzes some of the muscles of the face. This treatment reduces the appearance of wrinkles but also causes parts of the face to become expressionless and "frozen."

26.4 Mastering Concepts

1. What is an antagonistic pair of muscles?
2. Describe the levels of organization of a muscle.
3. Describe how sliding filaments shorten a sarcomere.
4. How do ATP, motor neurons, and calcium ions participate in muscle contraction?
5. How can the same muscle generate both small and large movements?

26.5 Muscle Cells Generate ATP in Multiple Ways

Skeletal muscle contraction requires huge amounts of ATP to break the connection between actin and myosin. Muscle cells have several ways to produce this ATP. A resting muscle cell uses O_2 to generate ATP in aerobic respiration, but it stores only small amounts of ATP (see chapter 6). When muscle activity begins, this stored ATP is depleted within a second or two. However, **creatine phosphate** molecules rapidly donate high-energy phosphates to ADP, temporarily restoring the ATP supply. (Burning Question 26.1 discusses the value of creatine phosphate as a dietary supplement.)

Less than a minute after intense exercise starts, the supply of creatine phosphate is gone. At that point, aerobic respiration can continue to produce ATP for as long as the lungs and blood deliver sufficient O_2. Once the muscle's demand for O_2 exceeds its supply, muscle cells switch to fermentation. This metabolic route does not require O_2, but it has a drawback: It generates far less ATP than aerobic respiration. At one time, researchers blamed lactic acid produced in fermentation for the pain associated with muscle fatigue; that conclusion is now being challenged. ⓘ *fermentation,* section 6.8

Intense exercise may lead to a period of **oxygen debt,** during which the body requires extra O_2 to restore resting levels of ATP and to recharge the proteins that carry oxygen in blood and muscle. Heavy breathing for several minutes after intense muscle activity is a sign of oxygen debt.

Shortly after death, muscles can no longer generate any ATP at all. One consequence is rigor mortis, the stiffening of muscles that occurs within a few hours after a person dies. Without ATP, the myosin cross bridges cannot release from actin. The muscles remain in a stiff position for the next couple of days, until the protein filaments begin to decay.

26.5 Mastering Concepts

1. Describe the role of creatine phosphate in muscle metabolism.
2. What happens when a muscle cell cannot generate ATP by aerobic respiration?

©Jupiterimages/DigitalVision/Getty Images RF

Burning Question 26.1

Is creatine a useful dietary supplement?

You may have seen jars of creatine powder on nutrition store shelves, marketed as a muscle-building aid.

In theory, an increase in creatine phosphate levels should help skeletal muscle cells generate ATP, providing an energy boost during brief, intense bouts of exercise. After all, creatine phosphate in muscle cells donates its phosphate to ADP, quickly regenerating ATP soon after muscle activity starts (see section 26.5).

But does this supplement really work? Although the idea seems logical, people differ in their response to creatine. For some, taking creatine powder does increase the amount of creatine phosphate inside skeletal muscle cells.

©Alan Mather/Alamy Stock Photo

But not everyone's athletic performance improves. Results vary from no effect to small gains in sprints and other short-term intense exercises. Endurance athletes show no gains from creatine supplements, which makes sense: Creatine phosphate plays its short-lived role in muscle metabolism soon after exercise begins.

As a note of caution, the long-term consumption of creatine powder may be harmful. Much of the extra creatine ends up in urine, indicating stress on the kidneys. The possibility of kidney toxicity requires further study.

Submit your burning question to
marielle.hoefnagels@mheducation.com

26.6 Muscle Fiber Types Influence Athletic Performance

Most skeletal muscles contain fibers of two main types, distinguished by the duration of each twitch (figure 26.14). A "twitch" is a cycle of contraction and relaxation in one muscle cell. **Slow-twitch fibers** have a relatively small diameter and produce twitches of relatively long duration. Abundant capillaries deliver oxygen-rich blood, and the muscle cells have a high content of a red pigment that stores oxygen. The O_2, in turn, supports the aerobic respiration that regenerates ATP in the fibers' plentiful mitochondria. High-endurance, slow-twitch muscle fibers predominate in body parts that are active for extended periods, such as the flight muscles ("dark meat") of ducks and geese or the back muscles that maintain our upright posture.

Fast-twitch fibers, in contrast, are larger-diameter cells that split ATP quickly in short-duration twitches. Short bouts of rapid, powerful contraction are characteristic of fast-twitch fibers. Anaerobic pathways generate the ATP in these cells, which tire quickly. Muscles dominated by fast-twitch fibers appear white because they have few capillaries and a lower content of the red oxygen-storing pigment. The white breast muscle of a domesticated chicken, for example, can power barnyard flapping for a short time but cannot support sustained, long-distance flight.

The proportion of slow-twitch to fast-twitch fibers affects athletic performance. People with a high proportion of slow-twitch fibers excel at endurance sports, such as long-distance biking, running, and swimming. Athletes who have a higher proportion of fast-twitch fibers perform best at short, fast events, such as weight lifting, hurling the shot put, and sprinting. Genetics largely determines the balance between slow- and fast-twitch fibers in each person's muscles, although intensive training can alter this proportion in some people.

Regardless of the mix of slow- and fast-twitch fibers, regular exercise strengthens the muscular system. During the few months after a runner begins

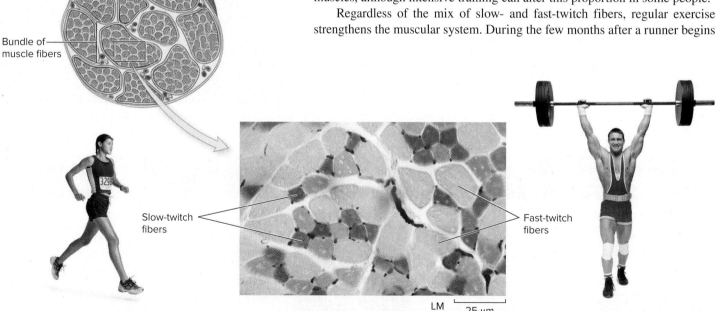

Muscle

Bundle of muscle fibers

Slow-twitch fibers

Fast-twitch fibers

LM   25 μm

Figure 26.14 **Slow-Twitch and Fast-Twitch Fibers.** Each muscle contains a mix of fibers. Special stains reveal mitochondrial activity in each cell type. The red slow-twitch fibers in the photo contain more mitochondria and therefore sustain ATP production for a longer duration than do the yellow fast-twitch fibers.

Photos: (runner): ©RubberBall/Getty Images RF; (fibers): ©Biophoto Associates/Science Source; (weight lifter): ©Jack Mann/Photodisc/Getty Images RF

Characteristic	Slow-Twitch Fibers	Fast-Twitch Fibers
Metabolism	Aerobic	Anaerobic
Energy use	Slow, steady	Quick, explosive
Endurance	High	Low

training, for example, leg muscles noticeably enlarge. This increase in muscle mass comes from the growth of individual muscle cells rather than from an increase in their number. Exercise-induced muscle growth is more pronounced in a weight lifter, because the resistance of the weights greatly stresses the muscles. Anabolic steroids boost muscle growth by activating the genes encoding muscle proteins, but the potential side effects of illicit steroid use are serious (see Why We Care 25.1).

A trained athlete's muscle fibers also use energy more efficiently than those of an inactive person. The athlete's cells contain more enzymes and mitochondria, so his or her muscles can withstand more exertion before fermentation begins. The athlete's muscles also receive more blood and store more glycogen than those of an untrained person. Muscle cells can break down the glycogen into glucose molecules when energy demands increase.
ⓘ *glycogen,* section 2.5B

Like bones, muscles can degenerate from lack of use. After just two days of inactivity, mitochondrial enzyme activity drops in skeletal muscle cells. After a week without exercise, aerobic respiration efficiency falls by 50%. The number of small blood vessels surrounding muscle fibers declines, lowering the body's ability to deliver O_2 to the muscle. Glycogen reserves fall, and the breakdown of lactic acid occurs less efficiently. After a few months of inactivity, the benefits of regular exercise all but disappear.

Athletic ability aside, exercise is often followed by muscle soreness and joint pain. A soak in a hot tub can offer some relief; Burning Question 26.2 explains why this remedy works.

©technotr/Getty Images RF

26.6 Mastering Concepts

1. Why do endurance sports require a high proportion of slow-twitch muscle fibers, whereas power sports require more fast-twitch muscle fibers?

2. How does exercise strengthen muscles?

Burning Question 26.2

Why does heat soothe sore muscles and joints?

Few experiences are as relaxing as slipping into a bubbling hot tub. Muscle tension subsides, and aches seem to melt away. Why does heat have this effect?

Conventional wisdom says that heat causes muscles to relax, but this answer is misleading. Technically, a muscle relaxes when actin and myosin filaments slide past one another in a way that makes the sarcomeres longer. Heat does not make muscles relax in this sense.

Instead, most of heat's soothing effects on muscles and joints are indirect. First, heat causes the proteins that make up tendons and ligaments to become more fluid and stretchier. Loosening up this connective tissue not only relieves tension on the muscles but also makes stiff joints more mobile. Second, warmth helps to relieve pain in sore, overworked muscles and

©liquidlibrary/PictureQuest RF

joints. Heat stimulates thermoreceptors in the skin, which may override impulses from pain receptors. Third, heat may reduce the activity of motor neurons that trigger painful muscle spasms. And finally, heat increases blood flow near the body's surface. This blood speeds healing of damaged muscles in two ways: by delivering nutrients and oxygen that cells require to make repairs, and by removing the remains of damaged cells.

Sports balms or creams may mimic the sensation of heat, but they do not actually raise the temperature of the muscles. These products therefore may relieve pain, but they do not offer the other benefits of a well-placed heating pad or a soak in a warm bath.

Submit your burning question to
marielle.hoefnagels@mheducation.com

Investigating Life 26.1 | Did a Myosin Gene Mutation Make Humans Brainier?

Human chewing muscles are considerably smaller than those of most primates. Fossilized remains suggest that primates had large skull bones and robust jaw muscles until about 2 million years ago (MYA). These large complexes occurred in the earliest human ancestors—*Australopithecus* and *Paranthropus*—and in contemporary primates such as macaques and gorillas. But more delicate chewing muscles appeared in early humans (*Homo erectus/ergaster*). ⓘ *human evolution*, section 17.12

These differences have spurred biologists to ask this question: How have jaw muscles evolved in humans and other primates? An important clue came from an unexpected source: researchers trying to learn more about muscular dystrophy. The team was trying to catalog every myosin gene in the human genome. Recall that myosin makes up the thick filaments inside muscle cells. Myosin, however, is not just one protein; it is a family of proteins encoded by at least 40 closely related genes. During a search of chromosome 7, the researchers stumbled upon an inactive gene that encoded a nonfunctional myosin protein.

The team compared the gene to homologous DNA in seven species of nonhuman primates. The results were clear: The researchers had discovered a human myosin gene containing mutations that are not present in nonhuman primates (figure 26.B). Moreover, the researchers discovered that the muscles that express the gene participate in the up-and-down jaw movements required for chewing.

The ancient mutations in the human version of the gene may have made our chewing muscles small and weak, at least compared with those of our closest relatives. Interestingly, the timing of the mutation coincides with a significant trend in human evolution: increasing brain size. Enhanced brain power may have eventually led to the development of culture, profoundly changing the course of human evolution. Perhaps the myosin mutations changed the chewing muscles in a way that released a constraint on the size of the brain.

Of course, one mutated myosin gene is, by itself, not likely to have set in motion the entire course of human history. Other changes in the skeletal and nervous systems must also have occurred to spur the evolution of the human brain's unique capabilities. But these results give us something to chew on as we contemplate the evolutionary history of our species.

	30	40	50
Human	CCCTCCATAGC--CGCACCCCATTTTGTC		
Nonhuman			
Woolly monkey	CCCTCCACAGCACTGTACCCCATTTTGTC		
Pigtail macaque	CCCTCCACAGCACTGTACCCCATTTTGTC		
Rhesus macaque	CCCTCCACAGCACTGTACCCCATTTTGTC		
Orangutan	CCCTCCACAGCACTGTACCCCATTTTGTC		
Gorilla	CCCTCCACAGCACTGTACCCCATTTTGTC		
Bonobo	CCCTCCACAGCACTGTACCCCATTTTGTC		
Chimpanzee	CCCTCCACAGCACTGTACCCCATTTTGTC		

Figure 26.B Myosin Mutation. The DNA sequences for a small portion of the myosin gene are shown for humans and a selection of nonhuman primates. The fragments start at nucleotide 26 in the gene. Differences between the genes are highlighted with shaded boxes. The skulls show the attachment area (*red*) for one of the main chewing muscles in humans, macaques, and gorillas. In humans, this area is relatively small, resulting in a much weaker jaw.

Source: Stedman, Hansell, Benjamin W. Kozyak, Anthony Nelson, et al. 2004. Myosin gene mutation correlates with anatomical changes in the human lineage. *Nature,* vol. 428, pages 415–418.

What's the Point? ▼ APPLIED

We use our arms and legs so often that it's hard to imagine life without them. Luckily, people with missing limbs have options. Careful studies of the structure and function of the skeletal and muscular systems, coupled with advances in materials science, have enabled bioengineers to develop sophisticated limb replacements.

Modern prosthetic limbs are much more realistic than the old "wooden legs" of the past. Engineers can shape strong, lightweight plastics and carbon fiber into prosthetics that absorb weight stresses and mimic the other functions of bones and joints. In fact, some artificial limbs might even work better than their natural counterparts, so prosthetics must be evaluated before amputee athletes are allowed to compete in the Olympics.

In the past, a person could control a prosthetic limb by operating cables or switches accessible by a healthy limb. Being able to move an artificial limb just by thinking about it would be a huge improvement. When an amputee thinks about

©Michael Svoboda/Vetta/Getty Images RF

moving a missing limb, however, the nervous system's signals reach dead ends. A prosthetic that could "listen" to signals in the brain would simulate the natural communication between nerves and muscles.

One step in that direction is a surgical procedure in which the nerves that previously stimulated an amputated muscle are relocated to an intact muscle in the chest. Afterward, when the patient thinks about moving the amputated arm, the chest muscle contracts instead. Electrodes attached to the chest muscle transmit the electrical signal to a prosthetic arm, causing it to move.

One drawback of this approach is a limited range of motion; another is that the chest muscles also contract each time the patient moves the prosthetic arm. Prosthetics are therefore being developed that will respond directly to electrical signals in the brain. In laboratory studies, monkeys have already used thought-powered artificial limbs, and human trials are underway.

CHAPTER SUMMARY

26.1 Skeletons Take Many Forms

- The **muscular system** and the **skeletal system** enable an animal to move.
- An animal's **skeleton** supports its body and protects soft tissues. **Muscles** act on the skeleton to provide motion (figure 26.15).
- A **hydrostatic skeleton** requires constrained fluid. An **exoskeleton** is on the organism's exterior, and an **endoskeleton** forms inside the body.

26.2 The Vertebrate Skeleton Features a Central Backbone

- The **axial skeleton** consists of the **bones** of the head, **vertebral column,** and rib cage.
- The **appendicular skeleton** includes the limbs and the bones (**pectoral girdle** and **pelvic girdle**) that attach them to the axial skeleton.

26.3 Bones Provide Support, Protect Internal Organs, and Supply Calcium

- Bones are strong and lightweight because they are porous. Long bones have a **marrow cavity** that contains **red bone marrow** or **yellow bone marrow.** As we age, the proportion of yellow marrow increases.

A. Bones Consist Mostly of Bone Tissue and Cartilage

- Bone tissue derives its strength from collagen and its hardness from minerals. **Compact bone tissue** is hard and dense. **Spongy bone tissue** has many spaces separated by a web of bony supports.

- **Cartilage** is a connective tissue that entraps a great deal of water, which makes it an excellent shock absorber.

B. Bone Meets Bone at a Joint

- **Joints** attach bones to each other. Some joints are immovable.
- Freely moving joints consist of cartilage and a connective tissue capsule that contains lubricating fluid. **Ligaments** connect bone to bone, whereas **tendons** connect bones to muscles.

C. Bones Are Constantly Built and Degraded

- Even after growth stops, bone continually degenerates and renews itself.
- Exercise strengthens bones; conversely, bones weaken with disuse.

D. Bones Help Regulate Calcium Homeostasis

- Hormones control the exchange of calcium between blood and bones, maintaining homeostasis. **Osteoporosis** results when bones lose more calcium than they replace.

26.4 Muscle Movement Requires Contractile Proteins and ATP

- Many **skeletal muscles** form antagonistic pairs, which enable bones to move in two directions.

A. Actin and Myosin Filaments Fill Muscle Cells

- Each skeletal **muscle fiber** is a cylindrical cell that contains **myofibrils** composed of two types of protein filaments. The **thick filaments** are **myosin,** and the **thin filaments** are composed primarily of **actin.**

B. Sliding Filaments Are the Basis of Muscle Cell Contraction

- A myofibril is a chain of contractile units called **sarcomeres.**
- According to the **sliding filament model,** muscle contraction occurs when thick and thin filaments move past one another.
- Muscle contraction requires ATP. A myosin head forms a cross bridge when it touches actin. When the cross bridge bends, the actin filament slides past the myosin filament. An ATP molecule then binds to the myosin head, and the link to actin breaks. The myosin head returns to its original position, and a new cross bridge forms farther along the filament.

C. Motor Neurons Stimulate Muscle Contraction

- A **motor neuron** and all of the muscle cells it touches form a **motor unit.** When a motor neuron receives a signal from the central nervous system, it releases a **neurotransmitter** at a specialized synapse. Electrical waves then spread along the muscle cell membrane, releasing calcium ions into the cytosol. Calcium prompts actin filaments to change shape in a way that allows myosin to bind to it, and the muscle contracts.
- The more motor units stimulated, the greater the contraction of the muscle.

26.5 Muscle Cells Generate ATP in Multiple Ways

- The energy that powers muscle contraction comes first from stored ATP, then from **creatine phosphate** stored in muscle cells, then from aerobic respiration, and finally from fermentation.
- **Oxygen debt** is a temporary deficiency of O_2 after intense exercise.

26.6 Muscle Fiber Types Influence Athletic Performance

- **Slow-twitch fibers** use ATP slowly and regenerate it by aerobic respiration. **Fast-twitch fibers** use ATP quickly and use mostly anaerobic pathways (fermentation) to replenish it.
- People vary in their proportion of fast- and slow-twitch muscle fibers.
- A muscle that is exercised regularly increases in size because each muscle cell thickens. An unused muscle shrinks. Regular exercise causes changes in muscle cells that enable them to use energy more efficiently.

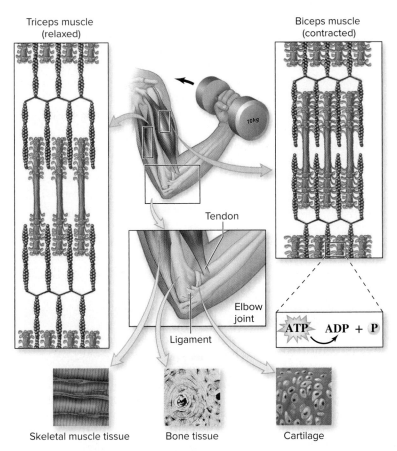

Triceps muscle (relaxed)

Biceps muscle (contracted)

Tendon

Elbow joint

Ligament

ATP → ADP + P

Skeletal muscle tissue   Bone tissue   Cartilage

Figure 26.15 Muscles Move Bones: A Summary.

Miniglossary | Muscular System

Skeletal muscle	An organ of the muscular system; generates voluntary movements
Muscle fiber	An individual muscle cell; consists of hundreds or thousands of myofibrils
Myofibril	A bundle of parallel proteins within a muscle fiber; consists of hundreds or thousands of sarcomeres
Sarcomere	The functional unit of a myofibril; each sarcomere shortens as a muscle contracts
Thick filament	A filament of myosin proteins in a myofibril
Thin filament	A filament of actin proteins in a myofibril
Sliding filament model	An explanation of muscle contraction; proposes that sarcomeres shorten as thin filaments slide between thick filaments
Motor neuron	A neuron that transmits a message from the central nervous system toward a muscle or gland
Motor unit	A motor neuron and all the muscle fibers it contacts; all of the fibers in a motor unit contract when the motor neuron releases neurotransmitters

MULTIPLE CHOICE QUESTIONS

1. Exoskeletons differ from endoskeletons in
   a. their ability to protect an animal from the outside.
   b. their ability to grow along with an organism.
   c. their function as a framework for muscle attachment.
   d. Both a and b are correct.

2. The axial skeleton is to the appendicular skeleton as
   a. a tree's branches are to its trunk.
   b. a car's body is to its wheels.
   c. a cell's cytoplasm is to its nucleus.
   d. a finger is to a toe.

3. Bone matrix is composed of _____, which give bones flexibility and strength, and _____, which give bones rigidity.
   a. cartilage cells; bone cells
   b. minerals; collagen proteins
   c. bone cells; cartilage cells
   d. collagen proteins; minerals

4. The function of a ligament is to connect
   a. cartilage to bone.   c. bone to muscle.
   b. bone to bone.   d. muscle to muscle.

5. Which of the following is arranged in order from smallest to largest?
   a. Motor unit < sarcomere < muscle cell < actin subunit
   b. Muscle cell < actin subunit < sarcomere < motor unit
   c. Actin subunit < sarcomere < muscle cell < motor unit
   d. Sarcomere < motor unit < actin subunit < muscle cell

6. Which of the following is NOT part of the sliding filament model?
   a. Sarcomeres become shorter as muscle contracts.
   b. ATP splits when the myosin head returns to its original position.
   c. Actin and myosin filaments slide past one another as muscle contracts.
   d. Actin filaments shorten as muscle contracts.

7. Within the first few seconds of a 5-minute race, stored ATP is quickly depleted. How do muscles obtain energy for the rest of the race?
   a. High-energy phosphates are transferred from glycogen to ADP.
   b. Creatine phosphate restores the ATP supply at first; then aerobic respiration generates ATP as long as O_2 is available.
   c. Actin filaments donate the phosphates that myosin filaments require.
   d. Fermentation begins as soon as stored ATP is depleted.

8. Slow-twitch muscle cells have a high concentration of red pigment, which
   a. provides oxygen to the mitochondria to promote ATP synthesis.
   b. helps promote the growth of the muscle cells in response to exertion.
   c. prevents coordinated movements.
   d. slows the rate of ATP hydrolysis by myosin.

9. Which of the following is NOT a consequence of exercise?
   a. Increased bone density
   b. Enhanced muscle cell metabolism
   c. Stronger cross bridges between myosin and actin
   d. Increased size of the muscle cells

Answers to Multiple Choice questions are in appendix A.

WRITE IT OUT

1. Distinguish among a hydrostatic skeleton, an exoskeleton, and an endoskeleton. What are the advantages and disadvantages of each type of skeleton? Give an example of an animal with each type.

2. Explain the observation that animals with exoskeletons and endoskeletons are better represented in the fossil record than are animals with hydrostatic skeletons. How might this difference affect scientific interpretations of the fossil record?

3. Use the Internet to research bone marrow transplants. What do patients who receive these transplants typically have in common?

4. What are the differences between spongy bone and compact bone?

5. Suppose a young boy severely fractures his femur. Describe how the injury could affect his growth.

6. Bones typically become stronger with exercise. However, some athletes develop stress fractures from overexercising. Why might light exercise strengthen bones but intense exercise cause fractures?

7. How can an imbalance in calcium homeostasis lead to osteoporosis?

8. How do antagonistic muscle pairs move bones? Give an example of such a pair.

9. Write the sequence of events that leads to muscle contraction, starting with "An action potential travels along the axon of a motor neuron."

10. How might your muscles lengthen when you stretch? Use *sarcomere, myosin, actin,* and *tendon* in your answer.

11. How do the effects of exercise (or lack thereof) illustrate homeostasis in bones and muscles?

12. How does the muscular system interact with the nervous system? The skeletal system? The respiratory system? The circulatory system?

13. Search the Internet for disorders of the skeletal or muscular system. Choose one such illness to research in more detail. Describe how the disorder interferes with bone or muscle function. What causes the disorder? Is a treatment or cure available?

14. What is the role of calcium in bones? In muscle contraction?

15. The following table shows recent men's world-record times for various running events. Graph the distance traveled against the average running speed, in meters per second. How does the production of ATP by muscles over time explain the graph?

Distance (m)	Time	Average m/sec
100	9.58 sec	10.44
200	19.19 sec	10.05
400	43.03 sec	9.30
800	1 min, 40.91 sec	7.93
1500	3 min, 26.00 sec	7.28
5000	12 min, 37.35 sec	6.60

SCIENTIFIC LITERACY

Design an experiment to test whether changes in the atmosphere (such as an incoming thunderstorm) cause joint pain. Then, use the Internet to learn whether researchers have found evidence to support a connection between weather and joint pain. Is the evidence that you found online convincing?

PULL IT TOGETHER

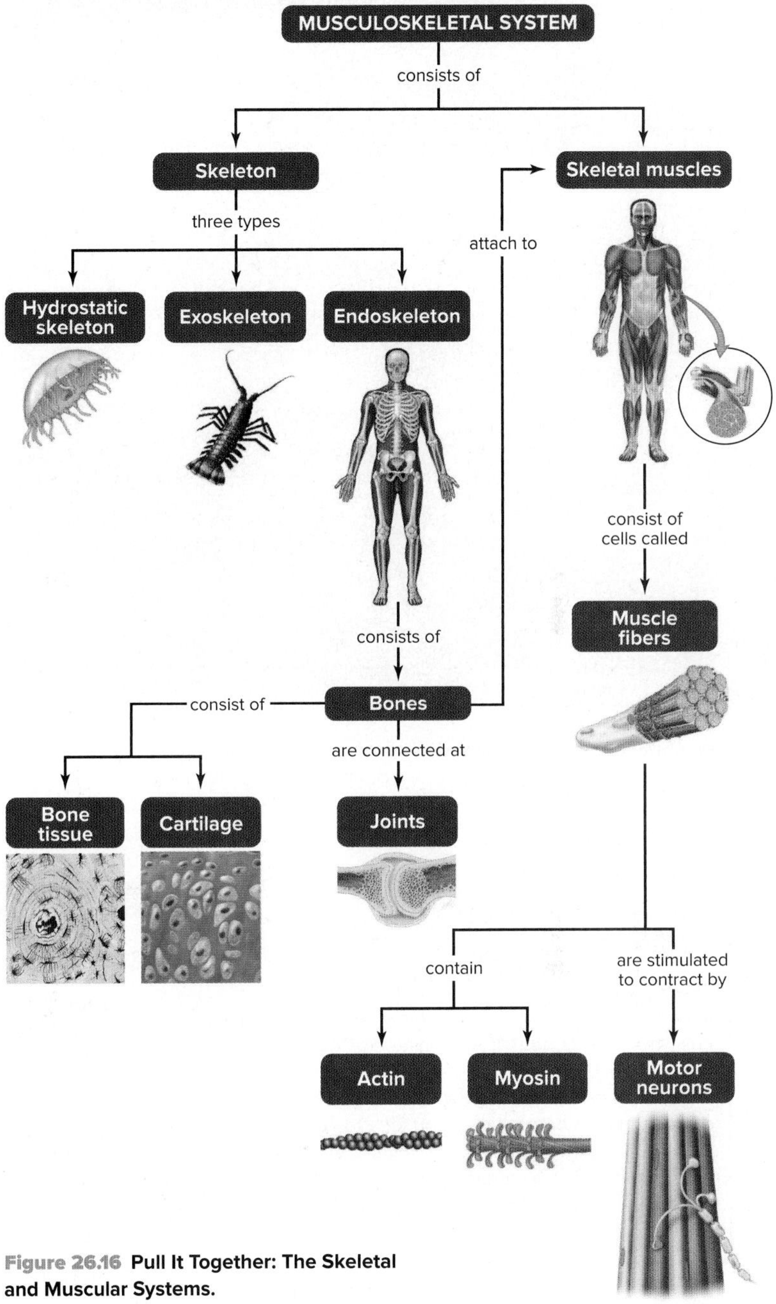

Figure 26.16 **Pull It Together: The Skeletal and Muscular Systems.**

Refer to figure 26.16 and the chapter content to answer the following questions.

1. Using the Survey the Landscape figure in the chapter introduction and the Pull It Together concept map, explain some ways that the musculoskeletal system maintains homeostasis.
2. Connect *exercise* to this concept map in at least three different places.
3. Add *neurotransmitters* and *ATP* to this concept map.
4. Add *fast-* and *slow-twitch muscle fibers* to this concept map. How do these two cell types differ?

Answers to Mastering Concepts, Write It Out, Scientific Literacy, and Pull It Together questions can be found in the Connect ebook.
connect.mheducation.com

Design element: Burning Question (fire background): ©Ingram Publishing/Super Stock

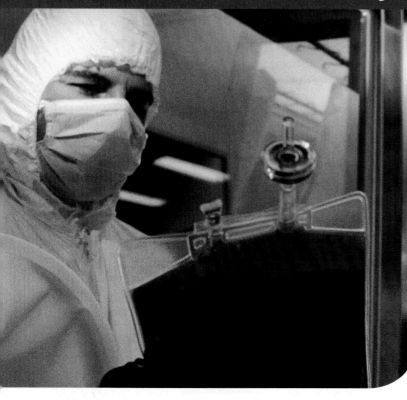

LEARNING OUTLINE

27.1 Blood Plays a Central Role in Maintaining Homeostasis

27.2 Animal Circulatory Systems Range from Simple to Complex

27.3 Blood Circulates Through the Heart and Blood Vessels

27.4 The Human Heart Is a Muscular Pump

27.5 Blood Vessels Form the Circulation Pathway

27.6 The Human Respiratory System Delivers Air to the Lungs

27.7 Breathing Requires Pressure Changes in the Lungs

27.8 Red Blood Cells Carry Most Oxygen and Carbon Dioxide

APPLICATIONS

Burning Question 27.1 *What is the difference between donating whole blood and donating plasma?*

Burning Question 27.2 *What causes bruises?*

Burning Question 27.3 *If some exercise is good, is more exercise better?*

Why We Care 27.1 *Unhealthy Circulatory and Respiratory Systems*

Investigating Life 27.1 *In (Extremely) Cold Blood*

Raw Material for Artificial Blood? A technician handles a bag of hemoglobin, an oxygen-toting protein purified from human blood. The hemoglobin is being tested for use in a blood substitute.

©Philippe Plailly/Science Source

Learn How to Learn

Skipping Class?

Attending lectures is important, but you may need to skip class once in a while. How will you find out what you missed? If your instructor does not provide complete lecture notes, you may be able to copy them from a friend. Whenever you borrow someone else's notes, it's a good idea to compare them with the assigned reading to make sure they are complete and accurate. You might also want to check with the instructor if you have lingering questions about what you missed.

 SURVEY THE LANDSCAPE
Animal Anatomy and Physiology

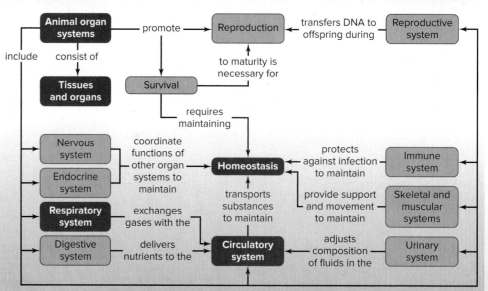

Blood vessels form the body's highways, carrying hormones, nutrients, wastes, and dissolved gases throughout the body. The lungs exchange oxygen and carbon dioxide between the blood and the external environment.

For more details, study the Pull It Together feature in the chapter summary.

©Don Hammond/Design Pics RF

The sight of blood is alarming to many people, and for good reason: This bright red fluid is essential to life. Powered by the heart, the bloodstream is a river that carries hormones, immune system cells, water, nutrients, and oxygen throughout the body. The same transportation network also removes waste products that cells produce. It is no wonder that a damaged heart or the loss of blood can threaten life.

Likewise, we risk death if we stop breathing for even a few minutes. Blood and breathing are closely connected. As we inhale, our lungs acquire fresh oxygen. This gas diffuses into the bloodstream at millions of tiny air sacs, each enclosed in a basket of blood vessels. At the same time, waste carbon dioxide gas diffuses out of the blood and into the lungs. Exhaled air carries this gas out of the body. This chapter describes the intimate relationship between the circulatory and respiratory systems.

27.1 Blood Plays a Central Role in Maintaining Homeostasis

Watch a crime drama on TV, and it won't be long until a gunshot wound leaves someone lying in a pool of blood. Unless help arrives immediately, life quickly fades—a vivid reminder of blood's importance.

This bright red fluid is the most visible and familiar part of the **circulatory system,** which consists of blood (or a comparable fluid), a network of vessels that contain the blood, and a heart. The **heart** is a pump that keeps the blood moving through these vessels; animals may have one or more hearts. Overall, the circulatory system's function is to transport materials throughout the body.

Blood is the fluid of the circulatory system. It carries many substances along its journey, among them glucose and oxygen gas (O_2). Without these resources, the body's cells could not carry out **aerobic respiration,** which generates the ATP required for life (see chapter 6). Blood delivers these and other raw materials, and it carries off wastes such as carbon dioxide (CO_2). ⓘ *ATP,* section 4.3

The circulatory system has extensive connections with organ systems that exchange materials with the environment (see figure 23.7). For example, blood vessels acquire O_2 and unload CO_2 at gills, lungs, or other organs of the respiratory system. Nutrients enter the circulatory system at blood vessels near the intestines, which form part of the digestive system. Blood also circulates through the kidneys, which eliminate many water-soluble metabolic wastes (see chapter 28). Moreover, blood carries hormones from the endocrine system (see chapter 25), participates in immune reactions (see chapter 29), and helps maintain homeostasis in several ways.

Blood is a connective tissue consisting of cells and cell fragments (platelets) suspended in a liquid extracellular matrix called plasma (figure 27.1). The cell types are diverse: A milliliter of blood normally contains about 5 million red blood cells, 7000 white blood cells, and 250,000 platelets. This section describes the functions of each component of blood; Burning Question 27.1 explains how donations of plasma and whole blood can save lives. ⓘ *connective tissue,* section 23.2B

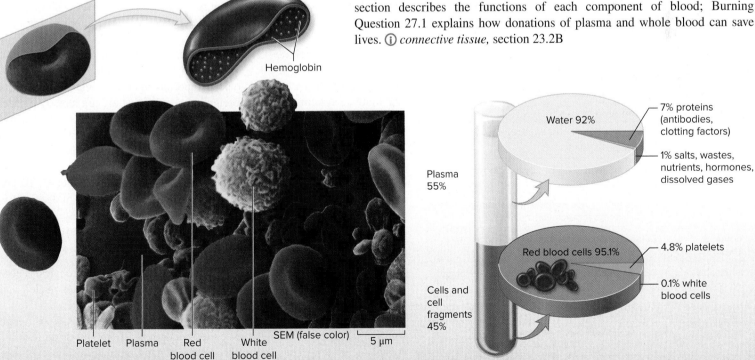

Hemoglobin

Water 92%

7% proteins (antibodies, clotting factors)

1% salts, wastes, nutrients, hormones, dissolved gases

Plasma 55%

Red blood cells 95.1%

4.8% platelets

0.1% white blood cells

Cells and cell fragments 45%

Platelet   Plasma   Red blood cell   White blood cell   SEM (false color)   5 μm

Figure 27.1 Blood Composition. Human blood is a mixture of red blood cells, white blood cells, and platelets suspended in a liquid called plasma.

Photo: ©National Cancer Institute/SPL/Getty Images

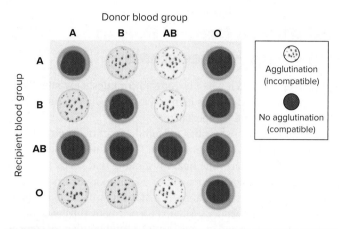

Donor blood group

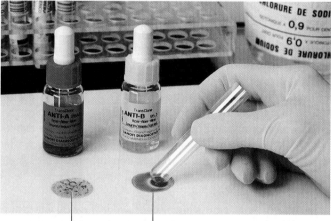

Agglutination   No agglutination

Figure 27.2 ABO Blood Groups. In this chart, the agglutination (clumping) reactions reveal which blood types are incompatible with one another. The blood sample being tested here contains molecule A (*left*) but not molecule B (*right*).

Photo: ©Jean Claude Revy - ISM/Medical Images

TABLE 27.1	Functions of Blood: A Summary
Function	**Explanation**
Gas exchange	Carries O_2 from lungs to tissues; carries CO_2 to the lungs to be exhaled
Nutrient transport	Carries nutrients absorbed by the digestive system throughout the body
Waste transport	Carries urea (a waste product of protein metabolism) to the kidneys for excretion in urine
Hormone transport	Carries hormones secreted by endocrine glands
Formation of interstitial fluid	Blood plasma leaking out of capillaries becomes interstitial fluid that surrounds cells
Maintenance of homeostasis (temperature, water, pH)	Absorbs heat and dissipates it at the body's surface; regulates cells' water content; buffers in blood help maintain pH of interstitial fluid
Protection	Blood clots plug damaged vessels; white blood cells destroy foreign particles and participate in inflammation

A. Plasma Carries Many Dissolved Substances

Plasma is the liquid matrix of blood. This fluid, which makes up more than half of the blood's volume, is 90% to 92% water. The main function of plasma is to exchange water and dissolved substances with the fluid that surrounds the body's cells. Cells depend upon this exchange to maintain homeostasis in their levels of water, salts, and other nutrients.

Besides water, more than 70 types of dissolved proteins make up the largest component of plasma. These proteins have many functions. For example, antibodies participate in the body's immune response; lipoproteins transport cholesterol; and clotting factors help stop bleeding following an injury (see section 27.1D). ⓘ *antibodies,* section 29.3B; *cholesterol,* section 2.5E

About 1% of plasma consists of dissolved salts, hormones, metabolic wastes, CO_2, nutrients, and vitamins. The concentrations of these dissolved molecules are low, but they are critical. For example, blood usually contains about 0.1% glucose; if the concentration falls to 0.06%, the body begins to convulse.

B. Red Blood Cells Transport Oxygen

Red blood cells are saucer-shaped disks that participate in the exchange of O_2 and CO_2. As they fill with the pigment **hemoglobin**—the protein that carries O_2—the red blood cells of humans and most other mammals lose their nuclei, ribosomes, and mitochondria. This adaptation maximizes the space available for hemoglobin but also means that the cells cannot divide.

Red blood cells originate from stem cells in red bone marrow at a rate of 2 million to 3 million per second. Mature red blood cells leave the bone marrow and enter the circulation. During its life of about 120 days, each red blood cell pounds against artery walls and squeezes through tiny capillaries. Eventually, the spleen destroys the cell and recycles most of its components. ⓘ *bone marrow,* section 26.3A

A person's blood type derives from carbohydrates and other molecules embedded in the outer membranes of red blood cells. In the ABO blood group system, external carbohydrate molecules called A and B determine an individual's blood type: A, B, AB, or O. That is, a person's cells may express only molecule A (type A blood), only molecule B (type B), both A and B (type AB), or neither A nor B (type O). ⓘ *ABO blood type,* section 10.6A

Knowing a person's blood type is important in blood transfusions because the immune system reacts to "foreign" molecules that are not already present in the blood. Antibodies produced against incompatible blood types cause **agglutination,** a reaction in which the cells clump together (figure 27.2). Agglutination following a transfusion of incompatible blood can be fatal. For this reason, a person with type A blood cannot receive a transfusion of type B or type AB blood; his or her antibodies will react against molecule B. For people with blood type AB, however, neither molecule A nor molecule B is "foreign." Type O blood reacts against all blood types except O.

C. White Blood Cells Fight Infection

Blood also contains five types of **white blood cells,** or leukocytes. These immune system cells are larger than red blood cells, retain their nuclei, and lack hemoglobin.

White blood cells originate from stem cells in red bone marrow. Although some enter the bloodstream, most either wander in body tissues or settle in the lymphatic system. These cells participate in many immune responses. Some secrete signaling molecules that provoke inflammation, whereas others destroy microbes or produce antibodies. Chapter 29 explains the lymphatic system and the interactions of white blood cells in more detail.

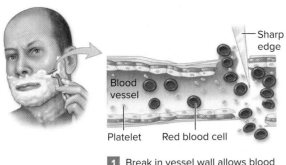
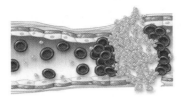
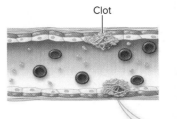
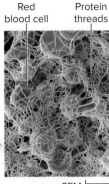

1 Break in vessel wall allows blood to escape; vessel constricts.

2 Platelets adhere to each other and to end of broken vessel. Platelet plug temporarily helps control blood loss.

3 Exposure of blood to surrounding tissue activates clotting factors. The resulting protein threads trap red blood cells, forming a clot.

SEM 5 μm (false color)

White blood cell numbers that are too high or too low can indicate illness. For example, **leukemias** are cancers in which bone marrow overproduces white blood cells. The abnormal white cells form at the expense of red blood cells, so when the patient's "white cell count" rises, the "red cell count" falls. Thus, leukemia also causes anemia. Having too few white blood cells, on the other hand, leaves the body vulnerable to deadly infections. Viruses such as HIV destroy white blood cells. Likewise, exposure to radiation or toxic chemicals can severely damage bone marrow, killing many white blood cells. ⓘ *cancer,* section 8.6; *HIV,* section 7.8C

Figure 27.3 Blood Clotting. (*1*) A cut blood vessel immediately constricts. (*2*) Platelets aggregate at the injured site. (*3*) Proteins called clotting factors participate in a cascade of reactions, producing a meshwork of protein threads. The inset photo shows blood cells trapped by the protein threads in a clot.

Photo: ©Steve Gschmeissner/SPL/Getty Images RF

D. Blood Clotting Requires Platelets and Plasma Proteins

Platelets are small, colorless cell fragments that participate in blood clotting. A platelet originates as part of a huge cell containing rows of vesicles that divide the cytoplasm into distinct regions, like a sheet of stamps. The vesicles enlarge and join together, "shedding" fragments that become platelets.

In a healthy circulatory system, platelets travel freely within the vessels. Sometimes, however, a wound nicks a blood vessel, or the blood vessel's inner lining becomes obstructed. Platelets "catch" on the obstacle and form a clump that temporarily plugs the leak. The platelets attract plasma proteins called clotting factors, which participate in reactions that ultimately produce a web of protein threads. These threads trap red blood cells and platelets, forming a **blood clot**—a plug of solidified blood (figure 27.3).

Blood that clots too slowly can lead to severe blood loss. Hemophilias, for example, are inherited bleeding disorders caused by absent or abnormal clotting factors. Deficiencies of vitamin C or K can also slow clotting and wound healing. Blood that clots too readily is also extremely dangerous. For example, platelets may snag on rough spots in blood vessel linings, producing a clot that may stay in place or travel in the bloodstream to another location. The obstruction may be deadly if it blocks blood flow to the lungs, brain, or heart.

Table 27.1 summarizes the functions of blood.

27.1 Mastering Concepts

1. What are the components of blood?
2. What are the functions of white blood cells?
3. Where do red and white blood cells originate?
4. Describe the process of blood clotting.

Burning Question 27.1

What is the difference between donating whole blood and donating plasma?

A person who "gives blood" donates 450 to 500 milliliters (about a pint) of blood to a nonprofit blood bank. After being screened for disease-causing agents, the blood may go to patients who need transfusions following trauma or surgery. More commonly, however, the blood is separated into its components, such as red blood cells, platelets, or clotting proteins. In this way, a single blood donation can help several different patients.

Plasma donation is another option. In this process, a machine separates out the plasma from a donor's blood. The red blood cells and other components are returned to the donor. The plasma center sells the fluid to pharmaceutical companies, which use it to manufacture treatments for hemophilia, hepatitis, and other diseases.

In the United States, it is illegal to pay a donor for whole blood. This law promotes a safe blood supply because donors have no incentive to lie about illnesses that might disqualify them from donating. Plasma donors, however, can receive money. The companies that process the plasma purify each fraction separately, removing viruses and other harmful components.

Submit your burning question to marielle.hoefnagels@mheducation.com

(donor): ©BSIP/Universal Images Group/Getty Images

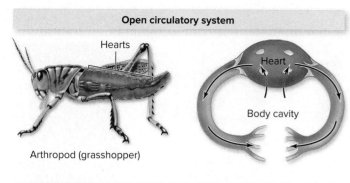

Open circulatory system

Hearts

Heart

Body cavity

Arthropod (grasshopper)

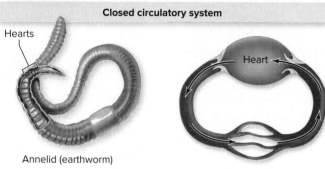

Closed circulatory system

Hearts

Heart

Annelid (earthworm)

Figure 27.4 **Open and Closed Circulatory Systems.** In an open circulatory system, fluid leaves vessels and bathes cells directly. In a closed circulatory system, blood is confined within vessels.

27.2 Animal Circulatory Systems Range from Simple to Complex

Some types of animals lack blood and a dedicated circulatory system. Flatworms, for example, use their incomplete digestive tracts not only to absorb nutrients but also for gas exchange. ⓘ *flatworms,* section 17.4

Most animals, however, have a circulatory system, and it may be open or closed (figure 27.4). In an **open circulatory system,** fluid is pumped through short vessels that lead to open spaces in the body cavity. There, the fluid can exchange materials with the body's cells before flowing back into the heart through pores. Animals with open circulatory systems include arthropods and most mollusks. ⓘ *mollusks,* section 17.5; *arthropods,* section 17.8

In a **closed circulatory system,** blood remains within vessels that exchange materials with the fluid surrounding the body's cells. Examples of animals with closed circulatory systems include vertebrates, annelids, and some mollusks such as squids and octopuses. ⓘ *annelids,* section 17.6

Both types of circulatory systems have advantages. Open circulatory systems require fewer vessels, and the blood moves under low pressure. The energetic costs of circulation are therefore relatively low. But closed circulatory systems tend to be more efficient than open systems. In a closed circulatory system, blood flows at higher pressure, so nutrient delivery and waste removal can occur more rapidly. Moreover, the vessels of a closed system can direct blood flow toward or away from specific areas of the body, depending on metabolic demands.

All vertebrates have closed circulatory systems, but these systems do not all look alike (figure 27.5). Among the vertebrates, fishes and tadpoles have the simplest circulatory systems. A fish's heart has just two chambers: an **atrium** where blood enters, and a **ventricle** from which blood exits. The heart pumps blood through the gills to pick up O_2 and unload CO_2. The blood then circulates to the rest of the body before returning to the heart.

Other vertebrates divide the circulatory system into two interrelated circuits. In the **pulmonary circulation,** blood absorbs O_2 and releases CO_2 at the lungs and returns to the heart; in the **systemic circulation,** blood circulates throughout the rest of the body to unload O_2 and pick up CO_2 before returning to the heart.

The heart of a bird or mammal has four chambers: two atria and two ventricles. In contrast, the heart of an adult amphibian and of most nonavian reptiles has only three chambers: two atria and one ventricle. The three-chambered heart is less efficient because oxygenated blood from the pulmonary circuit mixes inside the ventricle with oxygen-poor blood returning from the systemic circuit.

The rest of this chapter focuses on the human circulatory system and its interactions with the respiratory system.

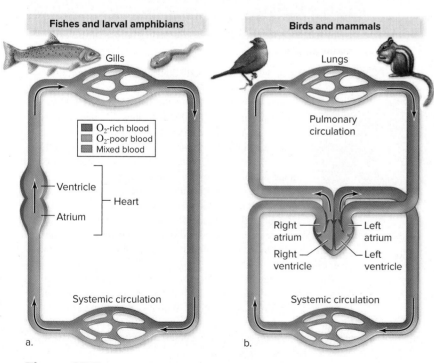

Fishes and larval amphibians

Gills

O₂-rich blood
O₂-poor blood
Mixed blood

Ventricle

Heart

Atrium

Systemic circulation

a.

Birds and mammals

Lungs

Pulmonary circulation

Right atrium — Left atrium
Right ventricle — Left ventricle

Systemic circulation

b.

Figure 27.5 **Vertebrate Circulatory Systems.** (a) A fish's two-chambered heart pumps blood in a single circuit around the body. (b) A bird or mammal has a four-chambered heart, maximizing the separation of the pulmonary and systemic circuits.

27.2 Mastering Concepts

1. Compare open and closed circulatory systems.
2. What is the difference between pulmonary and systemic circulation?

27.3 Blood Circulates Through the Heart and Blood Vessels

The plasma, cells, and platelets that make up blood circulate throughout the body in an elaborate system of blood vessels, thanks to the relentless pumping of the heart. The **cardiovascular system** is this entire transportation network; *cardio-* refers to the heart, *vascular* to the vessels. (Burning Question 27.2 explains what happens when blood vessels under the skin break.)

Figure 27.6 shows the largest of the body's blood vessels, which are classified by size and the direction of blood flow. **Arteries** are large vessels that conduct blood away from the heart; the left half of the figure lists some of the body's major arteries. These branch into **arterioles,** smaller vessels that then diverge into a network of **capillaries,** the body's tiniest blood vessels.

Water and dissolved substances diffuse between each capillary and the **interstitial fluid,** the liquid that bathes the body's cells. The interstitial fluid, in turn, exchanges materials with the tissue cells.

To complete the circuit, capillaries empty into slightly larger vessels, called **venules,** which unite to form the **veins** that carry blood back to the heart. The right half of figure 27.6 lists some major veins.

Burning Question 27.2

What causes bruises?

Anyone who has bumped into a piece of heavy furniture knows that minor injuries often leave bruises. The collision breaks blood vessels, which leak blood into the surrounding tissues. Blood that collects near the skin surface produces the familiar discoloration of a bruise.

©Ingram Publishing RF

The color of a bruise indicates its progression through the healing process. Recent bruises are reddish-blue. As the hemoglobin (the oxygen-containing molecules in blood) breaks down, the bruise turns bluish-purple. Meanwhile, white blood cells remove decayed blood products. A yellow-brown component of blood, called bilirubin, is usually last to disappear.

Submit your burning question to
marielle.hoefnagels@mheducation.com

Cardiovascular System	
Main tissue types*	**Examples of locations/functions**
Epithelial	Forms inner lining of heart wall; lines veins and arteries; makes up capillary walls
Connective	Surrounds heart; forms outer layers of veins and arteries; blood is a connective tissue
Nervous	Regulates heart rate and blood pressure
Muscle	Heart wall is mostly cardiac muscle; smooth muscle forms middle layer of arteries and veins; skeletal muscle propels blood in veins

*See chapter 23 for descriptions.

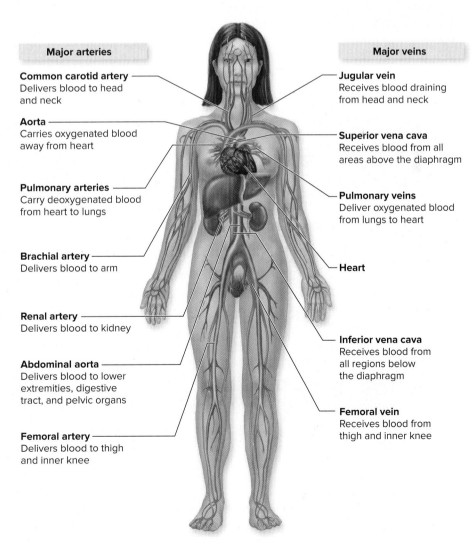

Major arteries

Common carotid artery
Delivers blood to head and neck

Aorta
Carries oxygenated blood away from heart

Pulmonary arteries
Carry deoxygenated blood from heart to lungs

Brachial artery
Delivers blood to arm

Renal artery
Delivers blood to kidney

Abdominal aorta
Delivers blood to lower extremities, digestive tract, and pelvic organs

Femoral artery
Delivers blood to thigh and inner knee

Major veins

Jugular vein
Receives blood draining from head and neck

Superior vena cava
Receives blood from all areas above the diaphragm

Pulmonary veins
Deliver oxygenated blood from lungs to heart

Heart

Inferior vena cava
Receives blood from all regions below the diaphragm

Femoral vein
Receives blood from thigh and inner knee

Figure 27.6 Human Circulatory System.

Notice in figure 27.6 that vessels carrying oxygen-rich blood are red and those carrying oxygen-poor blood are blue. This convention, coupled with the bluish appearance of blood vessels under lightly pigmented skin, has led to the misconception that the blood in veins is actually blue. In fact, blood is always red, whether it is fully oxygenated or not. Veins only appear blue because of the way that light of various wavelengths interacts with the skin. As for arteries, these blood vessels tend to be located in deeper tissues, far from the skin's surface. If arteries were visible through skin, they would appear blue, too.

27.3 Mastering Concepts

1. What is the cardiovascular system?
2. Describe the relationship among arteries, arterioles, capillaries, venules, and veins.

27.4 The Human Heart Is a Muscular Pump

Each day, the human heart sends a volume equal to more than 7000 liters of blood through the body, and it beats more than 2.5 billion times in a lifetime. The heart of a 70-year-old person has therefore pumped enough blood to fill about 70 Olympic-sized swimming pools. This section explores the structure and function of the hard-working human heart.

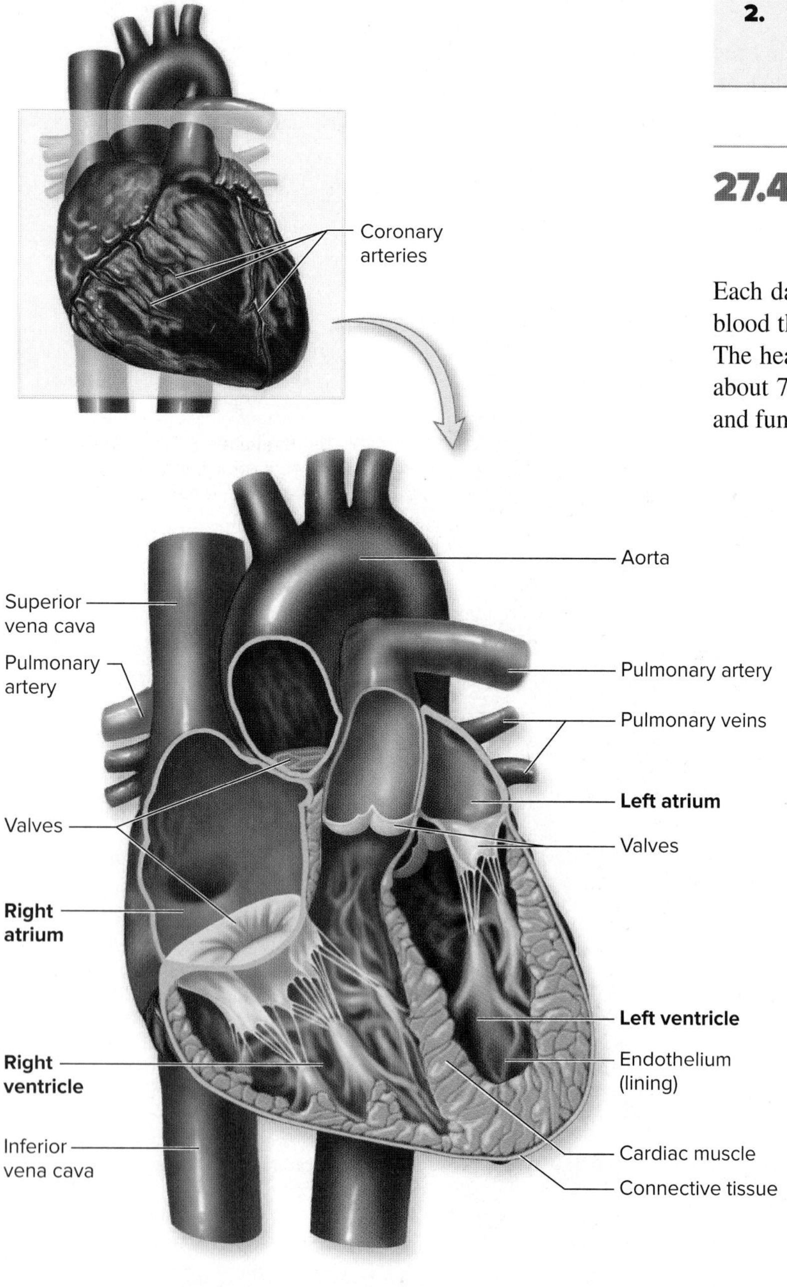

A. The Heart Has Four Chambers

Figure 27.7 illustrates the fist-sized human heart. A tough connective tissue sac encloses the heart and anchors it to surrounding tissues. This protective structure consists of two tissue layers. Thanks to lubricating fluid between the two layers, the heart is free to move, even during vigorous beating.

The heart's wall consists mostly of a thick layer of muscle. Contraction of the **cardiac muscle tissue** in the wall of the heart provides the force that propels blood through arteries and arterioles. The innermost lining of the heart (and of all blood vessels) consists of **endothelium,** a one-cell-thick layer of simple squamous epithelium. ⓘ *epithelial tissue,* section 23.2A; *cardiac muscle tissue,* section 23.2C

The human heart has four chambers: two upper atria and two lower ventricles. The atria are "primer pumps" that send blood to the ventricles, which pump the blood to the lungs or the rest of the body. Four heart valves ensure that blood moves in one direction. Two of the valves keep blood from moving back into an atrium when a ventricle contracts, and two prevent backflow into the ventricles from the arteries leaving the heart.

B. The Right and Left Halves of the Heart Deliver Blood Along Different Paths

A schematic view of the circulatory system shows the pathway of blood as it travels to and from the heart (figure 27.8). The two

Figure 27.7 **A Human Heart.** This illustration depicts the four chambers, the valves, and the major blood vessels of the human heart.

Labels: Coronary arteries, Aorta, Superior vena cava, Pulmonary artery, Pulmonary artery, Pulmonary veins, Left atrium, Valves, Valves, Right atrium, Left ventricle, Endothelium (lining), Right ventricle, Inferior vena cava, Cardiac muscle, Connective tissue

largest veins in the body, the superior vena cava and the inferior vena cava, deliver blood from the systemic circulation to the right atrium. From there, blood passes into the right ventricle and through the **pulmonary arteries** to the lungs, where blood picks up O_2 and unloads CO_2. The **pulmonary veins** carry oxygen-rich blood from the lungs to the left atrium of the heart, completing the pulmonary circuit.

The blood then flows from the left atrium into the left ventricle, the most powerful heart chamber. The massive force of contraction of the left ventricle sends blood into the **aorta,** the largest artery in the body. The blood then circulates throughout the body before returning to the veins that deliver blood to the right side of the heart. The systemic circuit is complete.

How does the heart muscle receive its blood supply? Blood does not seep from the heart's chambers directly to the cardiac muscle. Instead, the **coronary arteries**—two vessels that branch off from the aorta—supply blood to the heart muscle (see figure 27.7). A vein entering the right atrium returns blood that has been circulating within the walls of the heart. Blockage of a coronary artery is the most common cause of a heart attack.

C. Cardiac Muscle Cells Produce the Heartbeat

A **cardiac cycle,** or a single beat of the heart, consists of the events that occur with each contraction and relaxation of the heart muscle.

Each heartbeat requires the forceful contraction of cardiac muscle in the wall of the heart. The sliding filament model of muscle contraction described in chapter 26 applies to cardiac muscle, just as it does to skeletal muscle. Unlike skeletal muscle, however, cardiac muscle does not require stimulation from motor neurons to contract.

Instead, cardiac muscle is "self-excitable"; many cardiac muscle cells contract in unison without input from the central nervous system. Cardiac muscle cells are interconnected, forming an almost netlike pattern (see figure 23.4). Synchronized waves of action potentials therefore spread from cell to cell.

The signal to contract begins at the **sinoatrial (SA) node** (commonly called the **pacemaker**), a region of specialized cardiac muscle cells in the upper wall of the right atrium (figure 27.9). The pacemaker sets the tempo of the beat (normally about 75 beats per minute). Each time the cells of the pacemaker fire, they stimulate the cardiac cells of the atria to contract. After a brief delay, which gives the ventricles time to fill, a "relay station" called the

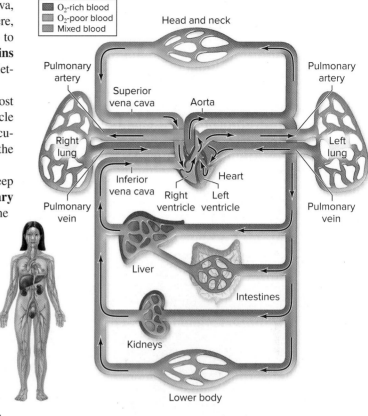

Figure 27.8 Blood's Journey in the Circulatory System. Oxygen-depleted blood leaving the right side of the heart goes to the lungs to pick up O_2. The oxygenated blood enters the left side of the heart, which pumps the blood throughout the body.

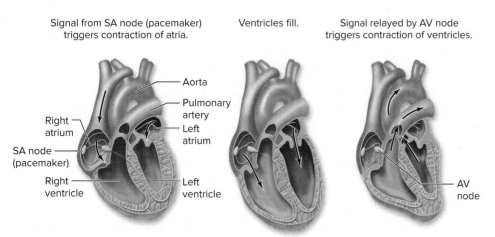

Figure 27.9 Heartbeat. Electrical signals that trigger the heartbeat start in the SA node (pacemaker), travel through the wall of the right atrium to the AV node, and pass to the ventricle walls.

TABLE 27.2   Target Heart Rates by Age

Age	Theoretical Maximum Heart Rate (beats per minute)	Target Heart Rate During Exercise (beats per minute)
20	200	140–170
25	195	137–166
30	190	133–162
40	180	126–153
50	170	119–145
60	160	112–136
70	150	105–128

©Duncan Smith/Getty Images RF

Burning Question 27.3

If some exercise is good, is more exercise better?

Any amount of exercise is beneficial to the body. Even gentle exercise such as slow walking lowers blood pressure and relieves stress. However, a person with a healthy heart should also participate regularly in moderate to vigorous exercise, which reduces the risk of cardiovascular disease, diabetes, and many other conditions. The World Health Organization recommends 150 minutes of moderate exercise a week, and the U.S. Centers for Disease Control and Prevention says "more time equals more health benefits."

Too much exercise, however, may actually harm the heart. Extreme endurance athletes who exercise between 2 and 5 hours a *day* may have an increased risk of stiffened cardiac muscles and artery walls, boosting the chance of arrhythmias and other heart conditions. However, this connection remains tentative, and extreme endurance athletes are generally at low risk for poor cardiovascular health. For now, the benefits of endurance training—even for long races—seem to outweigh the costs.

Submit your burning question to
marielle.hoefnagels@mheducation.com

atrioventricular (AV) node conducts the electrical stimulation throughout the ventricle walls. The cardiac cells of the ventricles then contract in unison.

The familiar "lub-dup" sound of the beating heart comes from the two sets of heart valves closing, preventing blood from flowing backward during each contraction. A heart murmur is a variation on the normal "lub-dup" sound, and it often reflects abnormally functioning valves.

D. Exercise Strengthens the Heart

After you circle the bases in a softball game, you may notice that your heart is beating faster than normal. The explanation for your elevated heart rate relates to the activity of your skeletal muscles, which require lots of ATP. Regenerating that ATP by aerobic respiration requires O_2; as we have already seen, one function of blood is to deliver this essential gas to the body's cells.

As you exercise, your heart meets the increased demand for O_2 by increasing its **cardiac output,** a measure of the volume of blood that the heart pumps each minute. Cardiac output is a function of the heart rate and the volume of blood pumped per stroke. Elevating the heart rate therefore quickly boosts cardiac output during an exercise session. With regular exercise, however, the stroke volume will also increase. An active person can therefore pump the same amount of blood at 50 beats per minute as a sedentary person's heart pumps at 75 beats per minute.

Exercise provides several other cardiovascular benefits as well (see Burning Question 27.3). The number of red blood cells increases in response to regular exercise, and these cells are packed with more hemoglobin, delivering more O_2 to tissues. Exercise can also lower blood pressure and reduce the amount of cholesterol in blood. Moreover, regular activity spurs the development of extra blood vessels within the walls of the heart, which may help prevent a heart attack by providing alternative pathways for blood to flow to the heart muscle.

To achieve the most benefit from exercise, the heart rate must be elevated to 70% to 85% of its "theoretical maximum" for at least half an hour three times a week. One way to calculate your theoretical maximum is to subtract your age from 220 (table 27.2). If you are 18 years old, your theoretical maximum is 202 beats per minute; 70% to 85% of this value is 141 to 172 beats per minute. Tennis, skating, skiing, racquetball, vigorous dancing, hockey, basketball, biking, or brisk walking can elevate your heart rate to this level.

27.4 Mastering Concepts

1. Why is the heart sometimes called "two hearts that beat in unison"?
2. Trace the pathway of an O_2 molecule from the lungs to a respiring cell at the tip of your finger.
3. How does a heartbeat originate and spread?
4. How does exercise affect the circulatory system?

27.5 Blood Vessels Form the Circulation Pathway

As the heart's ventricles contract, they push blood to the lungs and the rest of the body. This section describes the system of vessels through which blood travels as it delivers nutrients and removes wastes.

A. Arteries, Capillaries, and Veins Have Different Structures

Arteries carry blood away from the heart, whereas veins return blood to the heart. Despite these opposite functions, the walls of arteries and veins share some similarities (figure 27.10). The outer layer is a sheath of connective tissue. The middle layer is made mostly of **smooth muscle tissue,** and endothelium forms the inner layer. ⓘ *smooth muscle tissue,* section 23.2C

One feature that characterizes arteries is the thick layer of smooth muscle. The muscular walls of major arteries can withstand the high-pressure surges of blood leaving the heart. Farther from the heart, as arteries branch into arterioles, their walls become thinner, and the outermost layer of connective tissue may taper away. Arterioles do retain a layer of smooth muscle that helps regulate blood pressure; section 27.5B describes how this occurs.

Arterioles branch into **capillary beds,** networks of tiny blood vessels that connect an arteriole and a venule (see the lower half of figure 27.10). Capillaries are tiny but very numerous, providing extensive surface area where materials are exchanged with the interstitial fluid. Because their walls consist of a single layer of endothelial cells, nutrients and gases easily diffuse into and out of capillaries. ⓘ *diffusion,* section 4.5A

From the capillary beds, blood flows into venules, which converge into veins. These vessels receive blood at low pressure. The smooth muscle layer in their walls is much reduced or even absent (see figure 27.10, upper left portion); in fact, unlike an artery, a vein collapses when empty.

If pressure in veins is so low, what propels blood back to the heart, against the force of gravity? In many veins, valves keep blood flowing in one direction (figure 27.11). These valves are especially numerous in the legs. As skeletal

Figure 27.10 **Types of Blood Vessels.** The walls of arteries and veins consist of connective tissue, smooth muscle, and endothelium. Arteries, which are subject to high blood pressure, are much more muscular than veins. In some veins, valves keep blood moving toward the heart. A capillary bed is a network of tiny vessels that lies between an arteriole and a venule. The capillary wall consists only of endothelium through which nutrients, wastes, and gases pass.

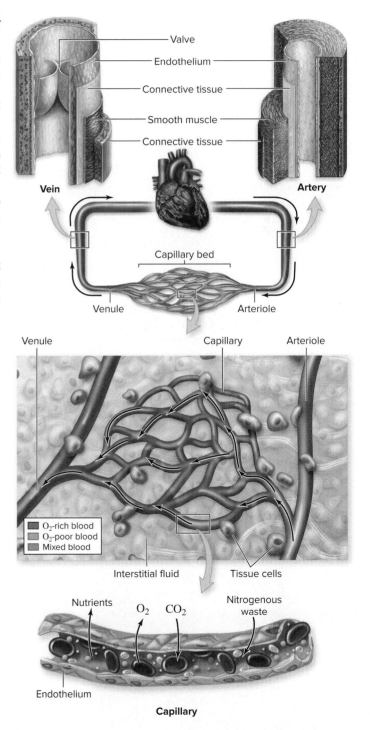

Figure 27.11 **Valves in Veins.** When skeletal muscles are relaxed, valves prevent blood from flowing backward in veins. Contracted skeletal muscles squeeze the veins, propelling blood through the open valves. In this illustration, the veins appear much larger than they are relative to real muscles.

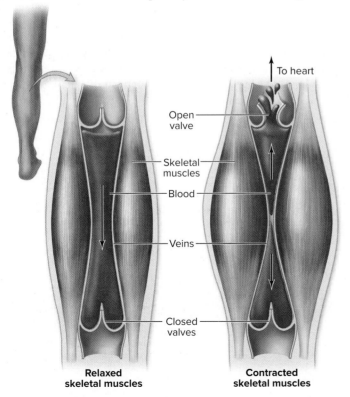

Miniglossary | Circulation

Artery	Large vessel that conducts blood away from the heart; blood pressure is highest in these vessels
Arteriole	Small vessel that carries blood away from the heart; arterioles have thinner walls than arteries
Capillary	Tiny blood vessel that exchanges nutrients and gases with interstitial fluid surrounding body cells
Venule	Small vessel that receives blood from capillaries; venules converge into veins
Vein	Vessel that carries blood back to the heart; blood pressure is lowest in these vessels
Blood pressure	The force that blood exerts on vessel walls; medical devices measure blood pressure at an artery in the upper arm
Vasoconstriction	Narrowing of blood vessels, increasing blood pressure
Vasodilation	Widening of blood vessels, decreasing blood pressure

Figure 27.12 Measuring Blood Pressure. (a) The cuff of a sphygmomanometer is inflated until circulation to the lower arm is cut off. Then, as the cuff slowly deflates, the stethoscope detects the sound of returning blood flow. (b) The value on the gauge when thumping is first audible is the systolic blood pressure; this sound is the blood rushing through the arteries past the deflating cuff. The pressure reading when the sound ends is the diastolic blood pressure.

(a): ©Blend Images LLC RF

muscles in the leg contract, they squeeze veins and propel blood through the open valves in the only direction it can move: toward the heart. Varicose veins result in part from faulty valves that allow blood to pool in the veins of the lower legs. The walls of these distended blood vessels form prominent bulges under the skin.

B. Blood Pressure and Velocity Differ Among Vessel Types

A routine doctor's office visit always includes a blood pressure reading, a good indication of overall cardiovascular health. **Blood pressure** is the force that blood exerts on vessel walls.

A device called a sphygmomanometer measures blood pressure in an artery in the upper arm, close to the heart (figure 27.12). The **systolic pressure,** or upper number in a blood pressure reading, reflects the powerful contraction of the ventricles. The **diastolic pressure,** or low point, occurs when the ventricles relax.

Blood pressure readings are in units of "millimeters of mercury," abbreviated "mm Hg," because older devices measured the distance over which blood pressure could push a column of mercury. A typical blood pressure reading for a young adult is 110 mm Hg for the systolic pressure and 70 mm Hg for the diastolic pressure, expressed as "110 over 70" (written 110/70). "Normal" blood pressure, however, varies with age, sex, race, and other factors.

Blood pressure decreases with distance from the heart; that is, blood in arteries has the highest pressure, followed by capillaries and then veins. Blood velocity, however, is lowest in the capillaries. The reason is that the total cross-sectional area of capillaries is much greater than that of the arteries or veins. Just as the velocity of a river slows as the water spreads out over a delta, so does the flow of blood slow as it is divided among countless tiny capillaries. This leisurely flow of blood allows adequate time for nutrients and wastes to diffuse across the capillary walls.

Past the capillaries, venules converge into veins. The total cross-sectional area of these blood vessels is again smaller than that of the capillaries. The resulting reduction in cross-sectional area helps speed blood flow back to the heart. To understand why, picture water flowing out of a hose. If you put your thumb over the nozzle, you reduce the area of the opening. What happens? The velocity of water through the nozzle increases.

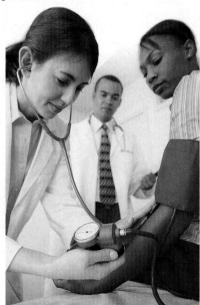

a.

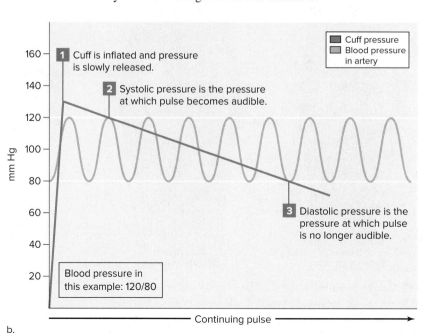

b.

160 — 1 Cuff is inflated and pressure is slowly released.

140 — 2 Systolic pressure is the pressure at which pulse becomes audible.

120 —

100 —

mm Hg

80 —

60 — 3 Diastolic pressure is the pressure at which pulse is no longer audible.

40 —

20 — Blood pressure in this example: 120/80

■ Cuff pressure
■ Blood pressure in artery

← Continuing pulse →

Overall, a person's blood pressure reflects many factors, including blood vessel diameter, heart rate, and blood volume. The body regulates blood pressure over the long term by raising or lowering the volume of blood. Chapter 28 describes how the kidneys adjust the blood's volume by controlling the amount of fluid excreted in urine.

In the short term, blood vessel diameter and heart rate are under constant regulation by negative feedback (figure 27.13). Pressure receptors within the walls of major arteries detect blood pressure and pass that information to the medulla, in the brainstem. The medulla, via the autonomic nervous system, adjusts both heart rate and the diameter of arterioles to maintain homeostasis. ⓘ *negative feedback,* section 23.4; *autonomic nervous system,* section 24.5

The role of the arterioles deserves special mention. **Vasoconstriction** is the narrowing of blood vessels that occurs when smooth muscle in arteriole walls contracts. When arteriole diameter decreases, blood pressure rises. The opposite effect, **vasodilation,** is the widening of blood vessels that occurs when the same muscles relax. Altering arteriole diameter allows the body to increase blood delivery to regions that need it most. During physical activity, for example, skeletal muscles receive additional blood at the expense of organs not in immediate use, such as those in the digestive tract.

Blood pressure that is too low or too high can cause health problems. Hypotension, which is blood pressure that is significantly lower than normal, may cause fainting. At the opposite end of the spectrum, consistently elevated blood pressure, or hypertension, may severely damage the circulatory system and other organs. High blood pressure affects 15% to 20% of adults residing in industrialized nations, including the United States. The exact cause is usually unknown, but poor diet, smoking, and stress all increase a person's risk of developing hypertension.

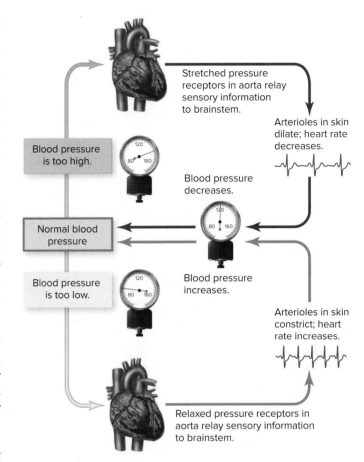

Figure 27.13 Regulation of Blood Pressure. A negative feedback loop regulates blood pressure in the short term. Stretch-sensitive neurons detect pressure in major arteries. If blood pressure climbs too high, signals from the central nervous system cause the blood vessels to dilate and the heart rate to slow. If blood pressure is too low, vessels constrict and the heart beats faster.

27.5 Mastering Concepts

1. Compare and contrast the structures of arteries, capillaries, and veins.

2. Explain why blood pressure is highest in the arteries and lowest in the veins.

3. How is the regulation of blood pressure an example of negative feedback?

27.6 The Human Respiratory System Delivers Air to the Lungs

As we have already seen, one function of blood is to deliver O_2 to cells and to collect the CO_2 waste that cells produce. In most animals, the **respiratory system** exchanges these gases with the environment. The functions of the circulatory and respiratory systems are therefore closely connected.

Each of us breathes some 20,000 times a day. Most of the time, you inhale and exhale without thinking—unless you happen to be using your breath to fog a mirror, spin a pinwheel, or play the trumpet. Fun aside, breathing is obviously a vital function; if a person is deprived of air, death can occur within minutes.

Breathing is so automatic that it is easy to forget why we do it. Cells use ATP to power protein synthesis, movement, DNA replication, cell division, growth, reproduction, and countless other activities that require energy. Animal cells generate ATP in aerobic respiration, which consumes O_2 and generates CO_2 as a waste product. Without gas exchange, cells die.

Figure It Out

If you breathe 20,000 times a day, about how many breaths do you take each minute?

Answer: About 14.

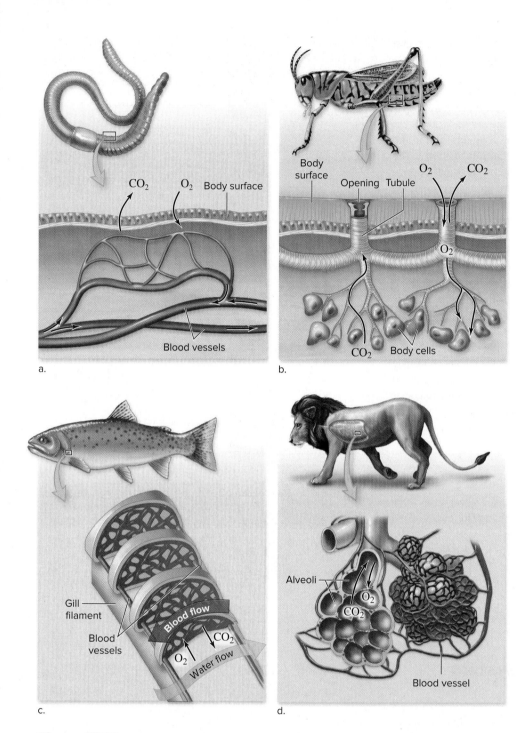

Figure 27.14 Respiratory Surfaces. Gas exchange may occur (a) across the body surface, (b) via tubules that connect the body's cells to the atmosphere, (c) at gills, or (d) in lungs.

Besides its use in aerobic cellular respiration, the term *respiration* also has two additional meanings. One is breathing (ventilation), the physical movement of air into and out of the body. Respiration can also mean the act of exchanging gases. External respiration is gas exchange between an animal's body and its environment; internal respiration is gas exchange between tissue cells and the bloodstream.

An animal's **respiratory surface** is the area of its body where external respiration occurs. In humans and other terrestrial vertebrates, the respiratory surface is in the **lungs,** but other animals use different surfaces (figure 27.14). All of these structures share three characteristics. First, the respiratory surface must come into contact with air or water. Second, each respiratory surface consists of a moist membrane across which O_2 and CO_2 diffuse. Third, its surface area must be relatively large. Not only must the respiratory surface meet the animal's demand for O_2, but it must also eliminate CO_2 fast enough to keep this waste gas from accumulating to toxic levels.

As you examine figure 27.14, note that either air or water can be a source of O_2 and a "dumping ground" for CO_2. In general, air offers two advantages over water. Air has a higher concentration of O_2 than does water; in addition, air is lighter than water. Less energy is therefore required to move air across a respiratory surface than to move an equal volume of water. However, the requirement that the respiratory surface remain moist puts air-breathing animals at a disadvantage: A respiratory surface exposed to air may dry out, rendering it useless.

The human respiratory system is a continuous network of tubules that delivers O_2 to the circulatory system and unloads CO_2 into the lungs. Table 27.3 summarizes the functions of the respiratory system, and figure 27.15 presents an overview of its anatomy. You may find it helpful to refer to this figure as you read the rest of this section.

A. The Nose, Pharynx, and Larynx Form the Upper Respiratory Tract

The **nose,** which forms the external entrance to the nasal cavity, functions in breathing, immunity, and the sense of smell. Stiff hairs at the entrance of each nostril keep dust and other large particles out. If a large particle is inhaled, a sensory cell in the nose may signal the brain to orchestrate a sneeze, which forcefully ejects the object. ⓘ *sense of smell,* section 24.9

Epithelial tissue in the nose secretes a sticky mucus that catches most airborne bacteria and dust particles that get past the hairs. Enzymes in the mucus destroy some of the would-be invaders, and immune system cells under the epithelial layer await any disease-causing organisms that penetrate the mucus.

The nasal cavity also adjusts the temperature and humidity of incoming air. Blood vessels lining the nasal cavity release heat, and mucus contributes moisture to the air. This function ensures that the respiratory surface of the lungs remains moist.

The back of the nose and mouth leads into the **pharynx,** or throat. Like the mouth, the pharynx is part of the digestive and respiratory systems, since both swallowed food and inhaled air pass through the pharynx.

Just below and in front of the pharynx is the **larynx,** or Adam's apple, a boxlike structure that produces the voice. Stretched over the larynx are the two **vocal cords,** elastic bands of tissue that vibrate as air from the lungs passes through a slitlike opening called the **glottis.** Vibrations of the vocal cords produce the sounds of speech. A male's voice becomes deeper during puberty because the vocal cords grow longer and thicker. The cords therefore vibrate more slowly during speech, producing lower-frequency sounds that we perceive as a deeper voice.

Another function of the larynx is to direct chewed food toward the esophagus—the tube leading from the mouth to the stomach—and away from the respiratory system. During swallowing, a cartilage flap called the **epiglottis** covers the glottis so that food enters the esophagus, not the lungs (see chapter 28).

The entire upper respiratory tract is lined with epithelium that secretes mucus. Dust and other inhaled particles trapped in the mucus are swept out by waving cilia. Coughing brings the mucus up, to be either spit out or swallowed.

TABLE 27.3	Functions of the Human Respiratory System: A Summary
Function	**Explanation**
Gas exchange	Lungs exchange O_2 and CO_2 with blood.
Sense of smell	Breathing moves air to the odor receptor cells in the nose.
Production of sounds, including speech	Movement of air across the vocal cords in the larynx produces sounds.
Maintenance of homeostasis (blood pH)	Breathing volume and rate determine the concentration of CO_2 in blood, which affects blood pH.

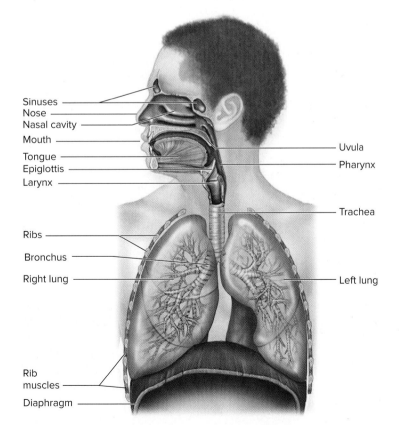

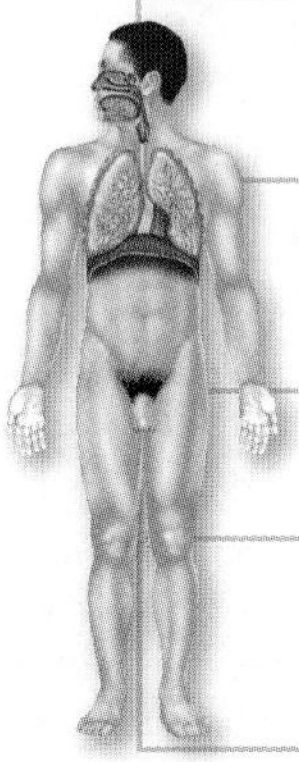

Respiratory System		
Main tissue types*	**Examples of locations/functions**	
Epithelial	Enables diffusion across walls of alveoli and capillaries; secretes mucus along respiratory tract.	
Connective	Blood (a connective tissue) exchanges gases with lungs; cartilage makes up part of the nose, trachea, bronchi, and larynx.	
Nervous	Autonomic nervous system controls smooth muscle in bronchi.	
Muscle	Smooth muscle in lungs regulates airflow to alveoli; skeletal muscle in diaphragm expands lungs.	

*See chapter 23 for descriptions.

Labels on figure:
Sinuses
Nose
Nasal cavity
Mouth
Tongue
Epiglottis
Larynx
Ribs
Bronchus
Right lung
Rib muscles
Diaphragm
Uvula
Pharynx
Trachea
Left lung

Figure 27.15 **The Human Respiratory System.** Inhaled air passes through the mouth and trachea, which divides into two bronchi. The airways branch into increasingly narrow tubes ending in tiny alveoli, where gas exchange occurs (see figure 27.17).

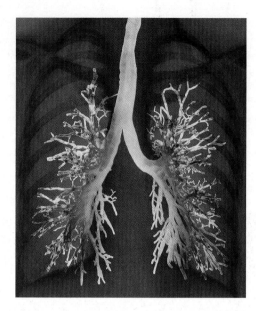

Figure 27.16 **The Bronchial Tree.** The respiratory passages of the lungs form a complex branching pattern. This is an X-ray image called a bronchogram.

©Innerspace Imaging/Science Source

B. The Lower Respiratory Tract Consists of the Trachea and Lungs

The **trachea,** or windpipe, is a tube just beneath the larynx. C-shaped rings of cartilage hold the trachea open and accommodate the expansion of the esophagus during swallowing. You can feel these rings in the front of your neck, near the lower portion of your throat. Cilia and mucus coat the trachea's inside surface, trapping debris and moistening the incoming air.

The trachea branches into two **bronchi,** one leading to each lung. The bronchi branch repeatedly, each branch decreasing in diameter and wall thickness (figure 27.16). **Bronchioles** ("little bronchi") are the finest branches. The bronchioles have no cartilage, but their walls contain smooth muscle. The autonomic nervous system controls the contraction of these muscles, adjusting airflow in response to metabolic demands.

Each bronchiole narrows into several alveolar ducts, and each duct opens into a grapelike cluster of alveoli, where gas exchange occurs (figure 27.17). Each **alveolus** is a tiny sac with a wall of epithelial tissue that is one cell layer thick. A vast network of capillaries surrounds each cluster of alveoli. Oxygen and CO_2 diffuse through the thin walls of the alveoli and the neighboring capillaries. The interface between the alveoli and the capillaries is the respiratory surface in humans, and it is enormous: The total surface area of the alveoli in a pair of lungs is about 150 square meters, or more than half the area of a tennis court!

27.6 Mastering Concepts

1. What is the main function of the respiratory system?
2. What is the relationship between the circulatory and respiratory systems?
3. List the parts of the upper and lower respiratory tracts.
4. Describe the relationships among the trachea, bronchi, bronchioles, and alveoli.

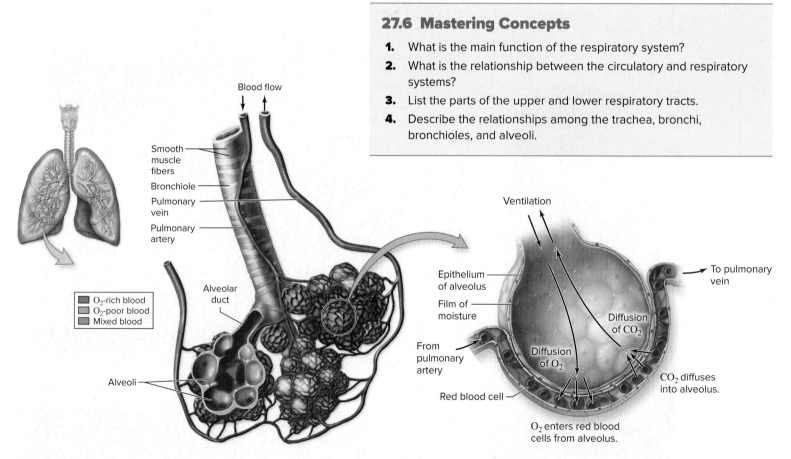

Figure 27.17 **Alveoli.** Each human lung contains some 300 million alveoli, which are tiny air sacs that give the lungs a spongy texture. Gas exchange occurs at the lush capillary network that surrounds each cluster of alveoli.

Why We Care 27.1 | Unhealthy Circulatory and Respiratory Systems

Heart disease, lung cancer, and respiratory problems top the list of the causes of death in the United States and other high-income countries (see table 18.2). Here is an overview of these and other problems that can affect the circulatory and respiratory systems.

Atherosclerosis

Fatty deposits inside the walls of arteries reduce blood flow to the brain or heart muscle. A diet high in fat and cholesterol is associated with this "hardening of the arteries," also called atherosclerosis (*athero-* is from the Greek word for "paste," and *sclerosis* means "hardness"). Atherosclerosis can cause several ailments, including strokes (blocked blood flow to the brain), heart attacks, and aneurysms. ⓘ *stroke,* section 24.6D

Heart Attack

Blocked blood flow in a coronary artery prevents oxygen delivery to part of the heart muscle (see figure 27.7). Starved for oxygen, muscle cells die. This is a heart attack, and it may come on suddenly. A common treatment for a blocked coronary artery is a bypass operation. A surgeon creates a bridge around the blockage by sewing pieces of blood vessel taken from the patient's chest or leg onto the blocked artery. The procedure's name often includes the number of arteries repaired. A quadruple bypass operation, for example, bridges four obstructed arteries.

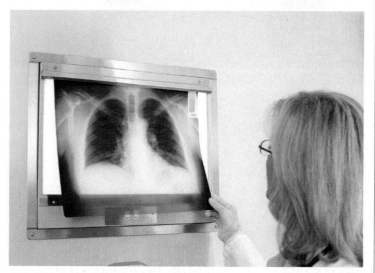

©Echo/Getty Images RF

Aneurysm

The wall of an artery can weaken and bulge, forming a pulsating, enlarging sac called an aneurysm. Aneurysms are dangerous because they may rupture without warning. Depending on the location and volume of bleeding, a burst aneurysm can be fatal; common locations include the aorta and the arteries of the brain. Aneurysms may result from atherosclerosis, a congenitally weakened area of an arterial wall, trauma, infection, persistently high blood pressure, or an inherited disorder such as Marfan syndrome.

Anemias

Anemias are a collection of more than 400 disorders resulting from a decrease in the oxygen-carrying capacity of blood. One symptom is fatigue, reflecting a shortage of O_2 at the body's cells. Some types of anemia are inherited; a mutation in a gene encoding hemoglobin, for example, can cause an inherited form of anemia called sickle cell disease. Other types of anemia are related to diet, especially an iron deficiency, which can prevent cells from producing hemoglobin. Iron-deficiency anemia is most common in women because of blood loss in menstruation. In still other forms of anemia, red blood cells may be too small, be manufactured too slowly, or die too quickly. ⓘ *sickle cell disease,* section 7.6A

COPD (Emphysema and Chronic Bronchitis)

Emphysema, a term derived from the Greek word for "inflate," is an abnormal accumulation of air in the lungs. Long-term exposure to cigarette smoke and other irritants causes a loss of elasticity in lung tissues. As they become less elastic, they can no longer absorb the pressure changes that accompany coughing. The delicate walls of the alveoli tear, impeding airflow and reducing the surface area for gas exchange. The patient experiences shortness of breath,

fatigue, wheezing, an expanded chest, and hyperventilation. Emphysema is often accompanied by chronic bronchitis—that is, inflammation of the bronchi. Together, these two illnesses are called chronic obstructive pulmonary disease (COPD).

Pneumonia

Pneumonia is an inflammation of the alveoli, usually resulting from an infection with bacteria, viruses, or fungi. Mucus and white blood cells accumulate in inflamed alveoli, impeding gas exchange. Symptoms include coughing, often accompanied by green or yellow sputum; fever; chest pain; and shortness of breath.

Tuberculosis

Infection with the bacterium *Mycobacterium tuberculosis* causes tuberculosis (also called TB). Symptoms of advanced TB include painful coughs, bloody sputum, fever, and weight loss. When a patient with an active TB infection coughs or sneezes, he or she expels bacteria-laden droplets that nearby people can inhale. Once in the lungs, the bacteria replicate inside immune system cells. Other immune system cells clump around the infected cells. This response keeps the bacteria from spreading, but these clusters also block airways and provide a place for the bacteria to become dormant. The resulting latent infection may later reemerge as full-blown (active) TB. Tuberculosis is especially important today because of the emergence of *Mycobacterium* strains that resist antibiotic treatment.

Lung Cancer

Lung cancer is the most common cancer worldwide, and 85% of cases occur in smokers. Cigarette smoke, asbestos, and radioactive substances contain chemicals that mutate DNA in lung cells; these altered cells may develop into tumors that obstruct the airway. Patients experience chest pain, chronic coughing, and shortness of breath. Moreover, cancerous cells may break away from the tumor and spread throughout the body. ⓘ *cancer,* section 8.6

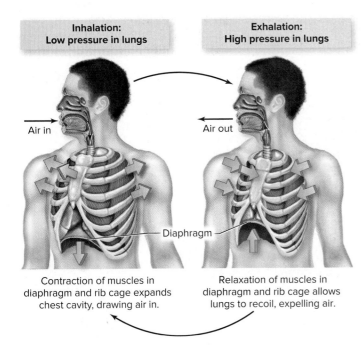

Inhalation:
Low pressure in lungs

Air in

Diaphragm

Contraction of muscles in diaphragm and rib cage expands chest cavity, drawing air in.

Exhalation:
High pressure in lungs

Air out

Relaxation of muscles in diaphragm and rib cage allows lungs to recoil, expelling air.

Figure 27.18 How We Breathe. When we inhale, the muscles of the diaphragm and rib cage contract, expanding the chest cavity. Because the expanded chest cavity has lower air pressure than the atmosphere, air moves into the lungs. When we exhale, the diaphragm relaxes and the rib cage lowers, reversing the process and pushing air out of the lungs.

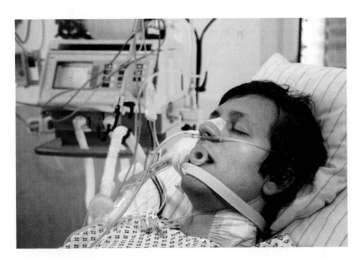

Figure 27.19 Ventilator. A patient with impaired lung function may need a ventilator, a machine that blows air into the respiratory system through a tube.

©Jochen Tack/imageBROKER/Getty Images

27.7 Breathing Requires Pressure Changes in the Lungs

Pay attention to your own breathing for a moment. Each **respiratory cycle** consists of one inhalation and one exhalation (figure 27.18). Each time you **inhale,** air moves into the lungs; when you **exhale,** air flows out of the lungs.

What drives this back-and-forth motion? The answer is that air flows from areas of high pressure to areas of low pressure. Therefore, air enters the body when the pressure inside the lungs is lower than the pressure outside the body. Conversely, air moves out when the pressure in the lungs is greater than the atmospheric pressure.

The body generates these pressure changes by altering the volume of the chest cavity. As a person inhales, skeletal muscles of the rib cage and diaphragm contract, expanding and elongating the chest cavity. The resulting increase in volume lowers the air pressure within the space between the lungs and the outer wall of the chest. The lungs therefore expand, lowering pressure in the alveoli. Air rushes in. Inhalation requires energy because muscle contraction uses ATP (see section 26.4).

The muscles of the rib cage and the diaphragm then relax. As a result, the rib cage falls to its former position, the diaphragm rests up in the chest cavity again, and the elastic tissues of the lung recoil. The pressure in the lungs now exceeds atmospheric pressure, so air flows out. At rest, exhalation is passive—that is, it requires only muscle relaxation, not contraction—and therefore does not require ATP.

Hiccups briefly interrupt the respiratory cycle. In a hiccup, the diaphragm contracts unexpectedly, causing a sharp intake of air; the "hic" sound occurs as the epiglottis closes. Why We Care 13.1 explores the evolutionary origin of hiccups, which have no known function.

Illness or injury may prevent contraction of the diaphragm and rib muscles, so breathing stops. Mechanical ventilators compensate for this loss of function (figure 27.19). A ventilator is a machine that blows air into a patient's respiratory tract. The air reaches the lungs through a tube inserted into the patient's mouth, nose, or trachea (through an incision). In most cases, the tube disrupts the larynx, so a patient on a ventilator cannot speak.

Medical professionals can test lung function by having a patient blow into a device called a spirometer. One measure of lung capacity is the **tidal volume,** or the amount of air inhaled or exhaled during a quiet breath taken at rest. In a young adult male, the tidal volume is about 500 mL. **Vital capacity,** on the other hand, is the total amount of air that a person can exhale after taking the deepest possible breath, about 4700 mL in a young man.

As we age, vital capacity declines. Illnesses can also interfere with a person's lung function by reducing the elasticity of the lungs, obstructing the airways, or weakening the muscles of the chest. Why We Care 27.1 describes a few disorders that affect the respiratory and circulatory systems.

27.7 Mastering Concepts

1. What is the relationship between the volume of the chest cavity and the air pressure in the lungs?

2. Describe the events of one respiratory cycle.

3. What is the difference between tidal volume and vital capacity? What do these measurements indicate about lung function?

27.8 Red Blood Cells Carry Most Oxygen and Carbon Dioxide

Gas exchange in the alveoli and at the body's other tissues relies on simple diffusion (figure 27.20). In external respiration, which occurs at the lungs, O_2 diffuses down its concentration gradient from the alveoli into the blood. At the same time, CO_2 diffuses from the blood to the air in the lungs. The heart then pumps the freshly oxygenated blood to the rest of the body. In internal respiration, O_2 diffuses from blood to the tissue fluid and then into the body's respiring cells, which have the lowest O_2 level. CO_2 diffuses in the opposite direction.

Blood carries O_2 and CO_2 in different ways. Red blood cells transport at least 99% of O_2 in blood; the rest is dissolved in plasma. Hemoglobin attracts O_2 in red blood cells. Tucked into each hemoglobin protein are four iron atoms, each of which can combine with one O_2 molecule picked up in the lungs.

Not surprisingly, illness or death results when hemoglobin cannot bind O_2. Carbon monoxide (CO), for example, is a colorless, odorless gas in cigarette smoke and in exhaust from car engines, kerosene heaters, wood stoves, and home furnaces. CO binds to hemoglobin more readily than O_2 does. When 30% of the hemoglobin molecules carry CO instead of O_2, a person loses consciousness and may go into a coma or even die.

Hemoglobin also carries some CO_2, but it does not bind this gas as readily as it binds to O_2. Instead, an enzyme in red blood cells converts most CO_2 to bicarbonate ions (HCO_3^-) that subsequently diffuse into plasma. The following sequence of chemical reactions produces bicarbonate (the two double arrows indicate that the reactions are reversible):

$$CO_2 + H_2O \rightleftharpoons H_2CO_3 \text{ (carbonic acid)}$$

$$H_2CO_3 \rightleftharpoons H^+ \text{ (hydrogen ions)} + HCO_3^- \text{ (bicarbonate)}$$

Notice that the second reaction produces hydrogen ions (H^+), a byproduct that allows the brain to regulate the depth and rate of breathing (figure 27.21). During intense exercise, for example, actively respiring muscle cells release abundant CO_2 into the blood. When CO_2 levels rise, blood becomes more acidic. Receptors in the arteries and in the medulla of the brain detect this change in blood pH. In response, the brain stimulates an increase in the breathing rate. The body acquires additional O_2 and releases the excess CO_2, maintaining homeostasis in blood gas concentrations.

Perhaps surprisingly, O_2 is less important than CO_2 in regulating breathing. In fact, the concentration of O_2 in blood affects breathing rate only if it falls dangerously low. Sometimes this O_2-regulating system fails, especially in the very young. Sudden infant death syndrome, in which a baby dies while asleep, may occur when receptors fail to detect low oxygen levels in arterial blood.

27.8 Mastering Concepts

1. Describe the O_2 and CO_2 diffusion gradients in the lungs and in the rest of the body.
2. In what forms does blood transport O_2 and CO_2?
3. How does the brain regulate breathing rate?

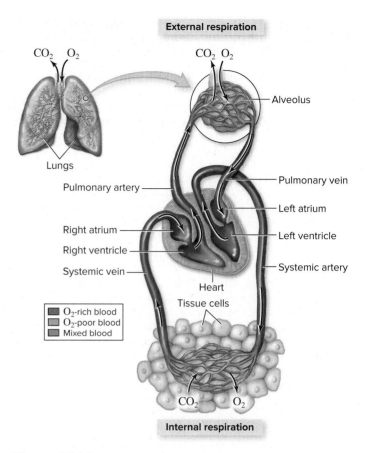

Figure 27.20 Gas Exchange at the Lungs and Tissues. CO_2 diffuses out of blood and into the alveoli of the lungs. O_2 moves in the opposite direction. In the rest of the body, O_2 diffuses out of the blood and into the tissues, while CO_2 moves into the bloodstream.

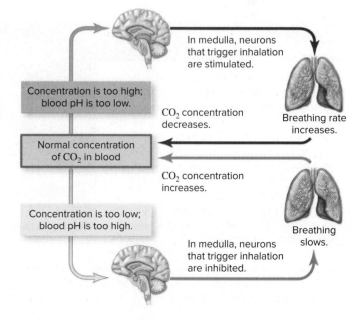

Figure 27.21 Breathing Control. The control of breathing illustrates negative feedback. If blood contains too much CO_2, the pH drops, and the medulla stimulates faster breathing; in the opposite situation, breathing slows.

Investigating Life 27.1 | In (Extremely) Cold Blood

Sometimes, an obscure discovery can turn into a big story. So it was with a report that appeared in the journal *Nature* in 1954, when a researcher named J. T. Ruud confirmed the existence of a fish with colorless blood. The animal in Ruud's report was an icefish that lives deep in the extremely cold ocean waters surrounding Antarctica. The blood of the icefish is a ghostly white. Microscopic examination of the blood revealed white blood cells, but few if any red blood cells and never any hemoglobin. The story represented a biological curiosity because it contradicted the commonly held idea that all vertebrate life requires hemoglobin.

More than 50 years later, biologists used the tools of biotechnology to learn more about icefish blood. They already knew that two genes encode the chains that make up the hemoglobin protein. The researchers collected specimens representing all 16 species of Antarctic icefishes, plus 17 species representing the icefishes' red-blooded relatives. They extracted DNA from the fishes' cells and sequenced the regions where the hemoglobin genes are normally located. ⓘ *DNA sequencing*, section 11.2B

The gene sequences revealed a clear pattern (figure 27.A). In red-blooded fishes, both hemoglobin genes are functional. In contrast, the icefishes have two nonfunctional genes. In fact, 15 of the 16 icefish species lack one of the genes entirely and retain just a small fragment of the other one. The remaining icefish species has two mutated genes, plus a nonfunctional remnant of an additional hemoglobin gene. ⓘ *mutations*, section 7.6

The genetic analysis explained why icefishes have colorless blood: Unlike their red-blooded relatives, icefishes simply cannot produce hemoglobin. Evidently, in the very cold waters of the Antarctic, metabolic rates and oxygen demand are so low that blood plasma alone—without red blood cells and hemoglobin—meets the icefish's gas exchange requirements.

Source: Near, Thomas J., Sandra K. Parker, and H. William Detrich III. 2006. A genomic fossil reveals key steps in hemoglobin loss by the Antarctic icefishes. *Molecular Biology and Evolution*, vol. 23, no. 11, pages 2008–2016.

Red-blooded fishes

Functional genes (hemoglobin present)

Gene 1 — Regulatory DNA — Gene 2

3 2 1   1 2 3

Icefishes

Gene 1 fragment, gene 2 absent (no hemoglobin)

Gene 1 fragment

3

Typical icefishes (15 of 16 species)

Both genes present but mutated (no hemoglobin)

Gene 1   Gene 2

3 2 1   1 2 3

One icefish species   Additional hemoglobin gene remnant

Figure 27.A Shattered Genes. Analysis of DNA sequences revealed that red-blooded fishes have intact hemoglobin genes. No icefish, however, produces hemoglobin. Most icefishes have just a fragment of one gene and lack the other one entirely, but one type of icefish has nonfunctional versions of both genes. (Numbers below each gene represent identifiable gene fragments.)

What's the **Point?** ▼ APPLIED

©incamerastock/Alamy Stock Photo

The circulatory system transports O_2 from the respiratory system to the body's tissues. Air enters the lungs when we inhale, and O_2 diffuses into the blood vessels surrounding the alveoli. Red blood cells carry O_2 to body cells, which use it along with glucose to produce ATP in aerobic respiration. Meanwhile, respiring cells release CO_2 into the blood. At the alveoli, CO_2 diffuses into the lungs and is exhaled.

If either the circulatory or the respiratory system is not functioning as usual, then aerobic respiration slows and cells starve for ATP. For example, damage to the heart muscle during a heart attack impairs delivery of O_2 to body tissues. Likewise, a lung damaged by a gunshot or stab wound has a reduced ability to exchange gases with the air. With limited oxygen, patients feel shortness of breath, rapid heart rate, and fatigue.

Some people with well-functioning circulatory and respiratory systems limit their O_2 intake for fun. Competitive breath-holding tests the limits of human oxygen requirements. Intense training of the body and the mind results in extraordinary accomplishments—the world record for underwater breath-holding is an astonishing 21 minutes! Many of us can't even last a minute without taking a breath.

How do they do it? A deep understanding of both circulation and respiration helps elite breath-holders tip the competitive balance in their favor. First, they breathe pure O_2 for several minutes before submerging, packing their blood with O_2 reserves. Second, the competitions occur in cold water. The chilly surroundings initiate an oxygen-conserving response called the "diving reflex." The heart rate slows, and blood is shunted toward the head and brain.

However, breath-holding is not just fun and games. Even during the diving reflex, muscle cells in the extremities continue to produce ATP, releasing carbon dioxide that is carried away only slowly by the sluggish blood flow. CO_2 accumulation at these tissues produces painful muscle cramps and acidifies the blood. No one knows whether competitive breath-holders face long-term health consequences. But it is clear that attempting underwater breath-holding without proper monitoring can result in fatal blackouts—a stark reminder of our unyielding need for oxygen.

CHAPTER SUMMARY

27.1 Blood Plays a Central Role in Maintaining Homeostasis

- A **circulatory system** consists of **blood** or a similar fluid, a network of vessels, and a **heart** that pumps the fluid throughout the body. The blood delivers nutrients and oxygen (O_2), removes metabolic wastes such as carbon dioxide (CO_2), and transports other substances.
- Human blood is a mixture of water, proteins, and other dissolved substances, plus various cells and cell fragments (table 27.4).

A. Plasma Carries Many Dissolved Substances

- **Plasma** is the fluid component of blood; it transports all other blood components.

B. Red Blood Cells Transport Oxygen

- **Red blood cells** contain abundant **hemoglobin,** a pigment that binds O_2 molecules. Like other blood cells, these cells originate in red bone marrow.
- Surface molecules on red blood cells react with antibodies in **agglutination** reactions that reveal a person's blood type.

C. White Blood Cells Fight Infection

- **White blood cells** provoke inflammation, destroy infectious organisms, and secrete antibodies. **Leukemia** is a type of cancer in which red bone marrow produces too many white blood cells.

D. Blood Clotting Requires Platelets and Plasma Proteins

- **Platelets** are cell fragments that collect near a wound. Damaged tissue activates plasma proteins that trigger the formation of a network of fibers, trapping additional platelets and perpetuating **blood clot** formation.

27.2 Animal Circulatory Systems Range from Simple to Complex

- In an **open circulatory system,** blood bathes tissues directly in open spaces before returning to the heart.
- In a **closed circulatory system,** such as that of vertebrates, the heart pumps blood through a continuous system of vessels.
- A fish has a two-chambered heart, with an **atrium** that receives blood and a **ventricle** that pumps blood out.
- In other vertebrates, the **pulmonary circulation** delivers oxygen-depleted blood to the lungs, and the **systemic circulation** brings freshly oxygenated blood to the rest of the body.
- Most land vertebrates have a three- or four-chambered heart.

27.3 Blood Circulates Through the Heart and Blood Vessels

- The heart is the muscular pump that propels blood through the vessels of the human **cardiovascular system.**
- **Arteries** are blood vessels that carry blood away from the heart. Arteries branch into smaller **arterioles,** which lead to tiny capillaries.

TABLE 27.4	Components of Blood: A Summary
Component	**Function**
Plasma	Liquid component of blood; exchanges water and many dissolved substances with the interstitial fluid surrounding body cells
Red blood cells	Carry O_2
White blood cells	Destroy foreign substances, initiate inflammation
Platelets	Initiate clotting

- **Capillaries** exchange materials with the **interstitial fluid** surrounding the body's cells.
- Capillaries empty into **venules,** which converge into the **veins** that return blood to the heart.

27.4 The Human Heart Is a Muscular Pump

A. The Heart Has Four Chambers

- A sac of connective tissue surrounds the heart. **Cardiac muscle tissue** makes up most of the heart wall. **Endothelium** lines the inside of the heart and all of the body's blood vessels.
- The heart has two atria that receive blood and two ventricles that propel blood throughout the body. The heart's four valves ensure one-way blood flow.

B. The Right and Left Halves of the Heart Deliver Blood Along Different Paths

- **Pulmonary arteries** and **pulmonary veins** transport blood between the right side of the heart and the lungs.
- Blood exits the left side of the heart at the **aorta,** the artery that carries blood toward the rest of the body.
- **Coronary arteries** supply blood to the heart muscle itself.

C. Cardiac Muscle Cells Produce the Heartbeat

- A **cardiac cycle** consists of a single contraction and relaxation of the heart muscle.
- The **pacemaker,** or **sinoatrial (SA) node,** is a collection of specialized cardiac muscle cells in the wall of the right atrium. The SA node sets the heart rate. From there, the heartbeat spreads to the **atrioventricular (AV) node** and then through the ventricles.

D. Exercise Strengthens the Heart

- Exercise increases the heart's **cardiac output** and lowers blood pressure.

27.5 Blood Vessels Form the Circulation Pathway

A. Arteries, Capillaries, and Veins Have Different Structures

- The walls of arteries and veins consist of an inner layer of endothelium, a middle layer of **smooth muscle tissue,** and an outer layer of connective tissue. Arteries have thicker, more elastic walls than veins.
- Nutrient and waste exchange occur at the **capillary beds,** where blood vessels consist of a single layer of endothelium.

B. Blood Pressure and Velocity Differ Among Vessel Types

- The pumping of the heart and the diameter of the blood vessels determine **blood pressure. Systolic pressure** reflects the force exerted on artery walls when the ventricles contract. The low point of a blood pressure reading, **diastolic pressure,** occurs when the ventricles relax.
- Blood pressure is highest in the arteries and lowest in the veins. Because of their high total cross-sectional area, the capillaries have the lowest blood velocity.
- The autonomic nervous system speeds or slows the heart rate. **Vasoconstriction** and **vasodilation** in the arterioles adjust the blood pressure.

27.6 The Human Respiratory System Delivers Air to the Lungs

- Cells use O_2 in **aerobic respiration** to release the energy in food and store the energy in ATP. CO_2 forms as a byproduct of respiration and must be eliminated from the body.
- **Respiratory systems** exchange O_2 and CO_2 with air or water, often in conjunction with a circulatory system that transports gases within the body.
- O_2 and CO_2 are exchanged by diffusion across a moist **respiratory surface** such as the body surface, gills, or **lungs.**

A. The Nose, Pharynx, and Larynx Form the Upper Respiratory Tract

- The **nose** purifies, warms, and moisturizes inhaled air. The air then flows through the **pharynx** and **larynx.**
- **Vocal cords** stretched over the larynx produce the voice as air passes through the **glottis.** The **epiglottis** prevents food from entering the trachea through the glottis.

B. The Lower Respiratory Tract Consists of the Trachea and Lungs

- Cartilage rings hold open the **trachea,** which branches into **bronchi** that deliver air to the lungs. The bronchi branch extensively and form smaller air tubules, **bronchioles,** which end in clusters of tiny, thin-walled, saclike **alveoli.**
- Many capillaries surround each alveolus. O_2 diffuses into the blood from the alveolar air, while CO_2 diffuses from the blood into the alveoli.
- Figure 27.22 summarizes gas exchange.

27.7 Breathing Requires Pressure Changes in the Lungs

- A **respiratory cycle** consists of one **inhalation** and one **exhalation.**
- When the diaphragm and rib cage muscles contract, the chest cavity expands. This reduces air pressure in the lungs, drawing air in. When

these muscles relax and the chest cavity shrinks, the pressure in the lungs increases and pushes air out.
- Measurements of lung function include **tidal volume** and **vital capacity.**

27.8 Red Blood Cells Carry Most Oxygen and Carbon Dioxide

- Almost all oxygen transported to cells is bound to hemoglobin in red blood cells. Carbon monoxide (CO) poisoning occurs when CO prevents O_2 from binding hemoglobin.
- Some CO_2 in the blood is bound to hemoglobin or dissolved in plasma. Most CO_2, however, is converted to bicarbonate ions, generated from carbonic acid that forms when CO_2 reacts with water.
- The brain typically uses blood pH (an indirect measure of CO_2 concentration) to adjust the depth and rate of breathing.

MULTIPLE CHOICE QUESTIONS

1. What component of blood is matched correctly with one of its functions?
   a. Red blood cell: initiates clotting
   b. White blood cell: produces antibodies
   c. Platelet: carries oxygen and carbon dioxide
   d. Plasma: engulfs and destroys foreign substances

2. The protein that carries O_2 in the circulatory system is _____, which is contained in _____.
   a. a plasma protein; platelets
   b. lymph; plasma
   c. an antibody; red blood cells
   d. hemoglobin; red blood cells

3. What is the advantage of a four-chambered heart?
   a. It beats four times more often than a one-chambered heart.
   b. It maximizes the amount of O_2 reaching tissues.
   c. It enhances the mixing of blood from the pulmonary and systemic circulation.
   d. Both a and c are correct.

4. Rank the following blood vessels from highest blood pressure to lowest blood pressure. Which type of vessel is *third* in your list?
   a. A capillary   c. A vein
   b. An artery   d. An arteriole

5. Air is pulled into the human respiratory tract mainly because of volume changes in the
   a. nose.   c. chest cavity.
   b. pharynx.   d. trachea.

6. How does the concentration of CO_2 in an expanding bubblegum bubble compare to the concentration of CO_2 in the surrounding air? (*Hint:* Refer to figure 27.22.)
   a. The concentration of CO_2 is higher in the bubble.
   b. The concentration of CO_2 is higher in the surrounding air.
   c. The concentration of CO_2 is equal in the bubble and the surrounding air.
   d. More information is necessary because the concentration of CO_2 depends on the concentration of exhaled O_2.

7. Breathing rate is mostly determined by
   a. the concentration of O_2 in the air.
   b. the concentration of O_2 in the blood.
   c. the concentration of CO_2 in the blood.
   d. the amount of hemoglobin in the blood.

Answers to Multiple Choice questions are in appendix A.

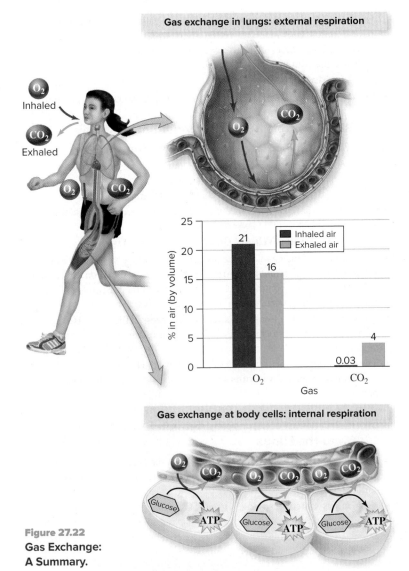

Gas exchange in lungs: external respiration

O_2 Inhaled

CO_2 Exhaled

% in air (by volume)

| | Inhaled air | Exhaled air |

21 — 16 — 0.03 — 4

O_2 CO_2

Gas

Gas exchange at body cells: internal respiration

Glucose ATP   Glucose ATP   Glucose ATP

Figure 27.22
Gas Exchange: A Summary.

WRITE IT OUT

1. Some athletes turn to blood doping to gain an unfair competitive advantage. For example, they may take supplements of erythropoietin (EPO), a hormone that stimulates red blood cell production. Why would increasing the number of circulating red blood cells help an athlete? What might be the dangers of having too many red blood cells?

2. Referring to figure 27.2, speculate about which blood type is considered the "universal donor" and which is called the "universal recipient." Explain your answer.

3. One effect of aspirin is to prevent platelets from sticking together. Why do some people take low doses of aspirin to help prevent a heart attack?

4. Describe the events that occur during one cardiac cycle.

5. What is the function of heart valves? Speculate about why a person with a leaky heart valve may feel short of breath.

6. Make a chart that compares systemic arteries, capillaries, and systemic veins. Consider the following properties: structure; amount of smooth muscle; presence of valves; cross-sectional area; blood pressure; blood velocity; direction of blood flow relative to the heart; O_2 content of blood.

7. Explain how a sphygmomanometer (blood pressure gauge) measures systolic and diastolic pressure.

8. Describe the interactions between the circulatory system and the respiratory, immune, digestive, and endocrine systems.

9. What is the function of breathing?

10. Differentiate between aerobic cellular respiration, internal respiration, and external respiration.

11. Explain how an animal's environment influences the structure and function of its respiratory surface.

12. People who suffer from claustrophobia are afraid of being enclosed in small areas. Some claustrophobes fear that they will "use all of the air" in the space and suffocate. Why is it impossible to use all of the air in a space? What does happen to the air in an enclosed space as you respire? Are the changes in the air dangerous?

13. A track athlete runs 100 meters while holding his breath. Predict the gas composition of the breath he exhales as he crosses the finish line. How might it compare to the gas composition of a normally exhaled breath?

14. The concentration of O_2 in the atmosphere declines with increasing elevation. Why do you think the times of endurance events at the 1968 Olympics, held in Mexico City (elevation: 2200 m), were relatively slow?

SCIENTIFIC LITERACY

1. Use the Internet to learn more about disorders of the cardiovascular or respiratory system. Choose one to investigate in more detail. What causes the disease you chose? Who is affected, and what are the consequences? Are there ways to prevent, treat, or cure the disease?

2. Review Burning Question 27.1. The World Health Organization urges all countries to collect blood only from unpaid, volunteer donors. In some countries, however, blood donors receive incentives such as a day off work or priority transfusions during shortages. What are the advantages and disadvantages of these nonmonetary incentives?

Answers to Mastering Concepts, Write It Out, Scientific Literacy, and Pull It Together questions can be found in the Connect ebook.
connect.mheducation.com

PULL IT TOGETHER

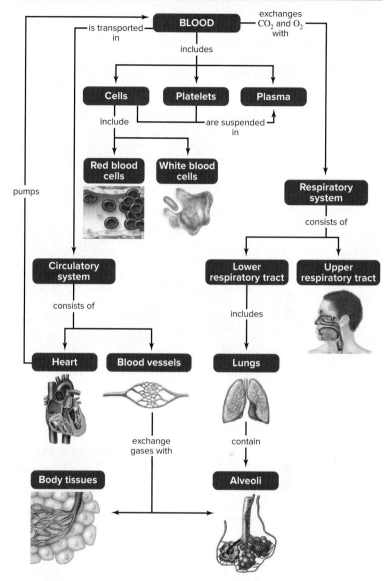

Figure 27.23 Pull It Together: The Circulatory and Respiratory Systems.

Refer to figure 27.23 and the chapter content to answer the following questions.

1. Review the Survey the Landscape figure in the chapter introduction. Then add O_2 and CO_2 to the Pull It Together concept map. Connect these terms with *body tissues, blood,* and *alveoli* to reveal how the respiratory system and circulatory system work together to maintain homeostasis.

2. How do the pulmonary and systemic circulatory pathways fit into this concept map?

3. Write the relative percentages (by volume) of cells, platelets, and plasma into the boxes for these terms.

4. Add terms to this concept map to explain the pressure changes that occur during inhalation and exhalation.

Food and Drink. Humans and many other types of animals need to eat and drink several times a day to maintain homeostasis.

©Ingram Publishing RF

LEARNING OUTLINE

28.1 Animals Maintain Nutrient, Water, and Ion Balance

28.2 Digestive Systems Derive Energy and Raw Materials from Food

28.3 A Varied Diet Is Essential to Good Health

28.4 Body Weight Reflects Food Intake and Activity Level

28.5 Most Animals Have a Specialized Digestive Tract

28.6 The Human Digestive System Consists of Several Organs

28.7 Animals Eliminate Nitrogenous Wastes and Regulate Water and Ions

28.8 The Urinary System Produces, Stores, and Eliminates Urine

28.9 Nephrons Remove Wastes and Adjust the Composition of Blood

APPLICATIONS

Burning Question 28.1 *Which diets lead to the most weight loss?*
Burning Question 28.2 *What is lactose intolerance?*
Why We Care 28.1 *The Unhealthy Digestive System*
Why We Care 28.2 *Urinary Incontinence*
Burning Question 28.3 *What can urine reveal about health and diet?*
Why We Care 28.3 *Kidney Failure, Dialysis, and Transplants*
Investigating Life 28.1 *The Cost of a Sweet Tooth*

Learn How to Learn
Avoid Distractions

Despite your best intentions, constant distractions may take you away from your studies. Friends, music, TV, social media, text messages, video games, and online shopping all offer attractive diversions. How can you stay focused? One answer is to find your own place to study where no one can find you. Turn your phone off for a few hours; the world will get along without you while you study. And if you must use your computer, create a separate user account with settings that prevent you from visiting favorite websites during study time.

SURVEY THE LANDSCAPE
Animal Anatomy and Physiology

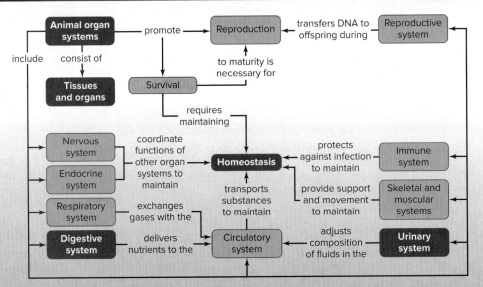

The digestive system dismantles food into small molecules and delivers them to the blood for distribution throughout the body. Cells absorb nutrients and release wastes. Meanwhile, the urinary system adjusts the composition of the blood to maintain homeostasis.

For more details, study the Pull It Together feature in the chapter summary.

©Mandy Godbehear/Alamy Stock Photo RF

The animal body's inputs and outputs are familiar: Food and drink go in, and wastes come out. But what happens inside the body in the meantime?

It all begins with the digestive system, which dismantles the nutrients in food and drink. Blood vessels snaking around the intestines absorb the nutrients and distribute them throughout the body. Meanwhile, the body eliminates the indigestible food as feces.

Urine has a different source: the kidneys. These organs cleanse the blood and release urine as a byproduct. This surprisingly complex fluid contains water, metabolic wastes, and many other substances. The kidneys adjust urine composition to maintain the body's delicate balance between water and dissolved salts (ions).

These two organ systems play critical roles in maintaining homeostasis in the animal body. The digestive system provides a steady supply of carbohydrates, proteins, fats, water, minerals, and vitamins to replace energy and materials that leave the body. Meanwhile, the urinary system eliminates water-soluble wastes and fine-tunes the blood's volume and chemical composition.

Figure 28.1 Different Habitats. Penguins thrive on Antarctic ice, parrots inhabit the tropics, and roadrunners live in the desert. The adaptations of each bird reflect its diet, the availability of water and essential ions, and many other selective forces.

(penguins): ©Paul Souders/Getty Images RF; (parrots): ©Art Wolfe/Mint Images/ Getty Images RF; (roadrunner): Source: U.S. Fish & Wildlife Service/Gary Karamer

28.1 Animals Maintain Nutrient, Water, and Ion Balance

Animals live nearly everywhere on Earth (figure 28.1). The salty Antarctic Ocean contrasts sharply with the perpetual humidity of a tropical rain forest and the dry, scorching home of a roadrunner. Most animals live in more moderate environments, but each species has adaptations that enable it to regulate its nutrient levels and the composition of its body fluids.

Like all animals, the birds in figure 28.1 constantly lose heat energy, water, dead cells, and other substances to their environment. To sustain growth and the everyday operation and maintenance of its body, each bird's food and drinks must supply all of its energy and essential nutrients. Their diets vary widely. Parrots live on a carbohydrate-rich, low-protein diet of fruits, seeds, and other plant parts. Penguins and roadrunners, on the other hand, mostly eat other animals, a diet that is relatively high in protein and fat but low in carbohydrates. The digestive systems of all of these birds reflect their varied diets.

An animal also maintains homeostasis in the chemical composition of its blood and other body fluids. Cells must retain water even when conditions are dry, yet too much water is damaging. Sodium, chlorine, hydrogen, and other ions are vital to life, but not in excess. The roadrunner's dry habitat demands water conservation, or the bird's cells will not function. In the jungle, abundant rainfall ensures that parrots have plenty of fresh water. But the same water also carries away scarce ions. Penguins have the opposite problem. They are surrounded by ocean water, and they eat a salty diet; together, these salts dissolve into ions that pull moisture out of the bird's cells. An animal in this habitat retains water only if it continually pumps excess ions out of its body. ⓘ *ions,* section 2.1B; *water,* section 2.3

This chapter begins by describing nutrition and the digestive system; it concludes with the structure and function of the urinary system, which helps maintain the composition of body fluids.

28.1 Mastering Concepts

1. Why do digestive systems differ widely among animals?
2. What challenges do animals face in balancing the ion content of body fluids?

Figure 28.2 Delicious. Eating provides the raw materials and energy required for life.
©Stockbyte/PunchStock RF

Figure 28.3 Ravenous. These chicks are begging for the food that will fuel their rapid growth.
©Digital Vision/PunchStock RF

28.2 Digestive Systems Derive Energy and Raw Materials from Food

Is it true that "you are what you eat"? In some ways, the answer is yes. After all, the atoms and molecules that make up your body came from food that you ate. But in other ways, the answer is no. The woman in figure 28.2 may enjoy eating fruit, yet she looks nothing like a mango. Clearly, food is not incorporated whole into her body, even though atoms and molecules derived from food compose her and every other animal.

The explanation for this apparent contradiction lies in the **digestive system,** the organs that ingest food, break it down, absorb the small molecules, and eliminate undigested wastes. As this section describes, some of the molecules absorbed from food do become part of the animal body, but others are used to generate the energy needed for life.

You may recall that an **autotroph** uses inorganic raw materials such as water and carbon dioxide to make its own food; the energy source is often sunlight. Plants and algae are examples of autotrophs. Unlike these organisms, animals need to eat; that is, they are heterotrophs. A **heterotroph** is an organism that must consume food—organic molecules—to obtain carbon and energy. Fungi and many other microbes are heterotrophs, too.

An animal's food is its source of **nutrients,** which are substances required for metabolism, growth, maintenance, and repair. The six main types of nutrients in food are carbohydrates, proteins, lipids, water, vitamins, and minerals.

These nutrients contain two important resources. One is the potential energy stored in the chemical bonds of carbohydrates, proteins, and lipids. As described in chapter 6, cells use each of these fuels in respiration to generate ATP, the molecule that powers most cellular activities. ⓘ *ATP, section 4.3*

The second resource in food is the chemical building blocks that make up the animal's body. Simple sugars, fatty acids, amino acids, nucleotides, water, vitamins, and minerals are the raw materials that build, repair, and maintain all parts of the body. To the extent that your body incorporates these materials, you really are what you eat.

An animal's metabolic rate largely determines its need for food. Animals that maintain a constant body temperature, such as birds and mammals, typically have the highest metabolic rates. Body size also influences the need for food. In general, the larger the animal, the more it needs to eat. When corrected for body size, however, the smallest animals typically have the highest metabolic rates. A hummingbird, for example, has a much higher surface area relative to its body mass than does an elephant. Because the tiny bird rapidly loses heat to the environment, it must consume a lot of food to maintain a constant body temperature. A hummingbird therefore eats its own weight in food every day; an elephant takes 3 months to do the same.

Another factor that affects metabolic rate is an animal's physiological state. Growth and reproduction require more energy and nutrients than simply maintaining the adult body. Baby animals therefore often have ravenous appetites, and a new parent may spend much of its time finding food to fuel its offsprings' rapid growth (figure 28.3).

28.2 Mastering Concepts

1. What are two reasons that animals must eat?
2. Explain the factors that affect an animal's metabolic rate.

28.3 A Varied Diet Is Essential to Good Health

Nutrients fall into two categories. **Macronutrients** are required in large amounts. Water is a macronutrient; all living cells require water as a solvent and as a participant in many reactions. Organisms use three other macronutrients—carbohydrates, proteins, and lipids—to build cells and to generate ATP. Despite the many nonfat foods on grocery store shelves, the diet must include *all* of these nutrients, including moderate amounts of fat.

Unlike macronutrients, **micronutrients** are required in very small amounts. Two examples are vitamins and minerals, neither of which are used as fuel. Instead, these micronutrients participate in many aspects of cell metabolism. Table 28.1 provides details on the sources and functions of several vitamins and minerals. Many processed foods are fortified with vitamins and minerals, so deficiencies in developed countries are rare.

The best way to acquire all required nutrients is to eat a varied diet. The U.S. government's food guidelines emphasize grains, fresh vegetables, and low-fat dairy products, along with fruits and limited amounts of meat and fat.

©McGraw-Hill Education/Ken Karp

TABLE 28.1   Selected Vitamins and Minerals in the Human Diet

	Food Sources	Function(s)	Selected Deficiency Symptoms
Vitamins			
Water-Soluble Vitamins			
B complex vitamins			
Niacin	Liver, meat, peas, beans, whole grains, fish	Growth, energy use	Pellagra (diarrhea, dementia, dermatitis)
Folic acid	Liver, navy beans, dark green vegetables	Manufacture of red blood cells, metabolism	Weakness, fatigue, diarrhea, neural tube defects in fetus
Vitamin C	Citrus fruits, tomatoes, peppers, strawberries, cabbage	Antioxidant, production of connective tissue and neurotransmitters	Scurvy (weakness, gum bleeding, weight loss)
Fat-Soluble Vitamins			
Vitamin A	Liver, dairy products, egg yolk, vegetables, fruit	Night vision, new cell growth	Blindness, impaired immune function
Vitamin D	Fish liver oil, milk, egg yolk	Bone formation	Skeletal deformation (rickets)
Minerals			
Calcium	Milk products, green leafy vegetables	Electrolyte*, bone and tooth structure, blood clotting, hormone release, nerve transmission, muscle contraction	Muscle cramps and twitches; weakened bones, heart malfunction
Iron	Meat, liver, fish, shellfish, egg yolk, peas, beans, dried fruit, whole grains	Part of hemoglobin and some enzymes	Anemia, learning deficits in children
Phosphorus	Meat, fish, eggs, poultry, whole grains	Bone and tooth structure; part of DNA, ATP, and cell membranes	Weakness, mineral loss from bones
Potassium	Fruits, potatoes, meat, fish, eggs, poultry, milk	Electrolyte, nerve transmission, muscle contraction, nucleic acid synthesis	Weakness, loss of appetite, muscle cramps, confusion, heart arrhythmia
Sodium	Table salt, meat, fish, eggs, poultry, milk	Electrolyte, nerve transmission, muscle contraction	Muscle cramps, nausea, weakness

*An electrolyte is an ion that helps regulate osmotic balance.

Figure 28.4 Nutrition Information. The packaging of processed foods includes a standard nutrition label that indicates the Calorie and nutrient content in each serving.

©McGraw-Hill Education. Tara McDermott, photographer

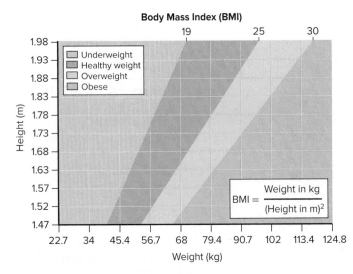

Figure 28.5 Body Mass Index. This chart is a quick substitute for a BMI calculation; the intersection of a person's height (in meters) and weight (in kilograms) indicates the BMI range.

Source: U.S. Department of Agriculture: Dietary Guidelines for Americans

The Harvard School of Public Health suggests a somewhat different diet that minimizes dairy products, red meat, and starchy processed grains. Whole grains and vegetable oils, along with abundant vegetables, make up the base of this pyramid. (Burning Question 28.1 explores other diets.)

The indigestible components of food help maintain good health, too. Dietary fiber, for example, is composed of cellulose from plant cell walls. Humans do not produce cellulose-digesting enzymes, so fiber contributes only bulk—not nutrients—to food. This increased mass eases movement of the food through the digestive tract, so cancer-causing ingredients in food contact the walls of the intestines for a shorter period. People who consume abundant fiber in their food therefore have a lower incidence of colorectal cancer. A high-fiber diet also reduces blood cholesterol (by mechanisms that remain unclear) and slows sugar absorption, minimizing blood sugar spikes. ⓘ *cellulose,* section 2.5B

A balanced diet delivers many long-term health benefits, including a reduced risk of type 2 diabetes, cancer, osteoporosis, high blood pressure, and heart disease. Fortunately, the labels on packaged foods list not only the ingredients but also the nutrient content in each serving (figure 28.4).

28.3 Mastering Concepts

1. Which nutrients are macronutrients and which are micronutrients?

2. How does indigestible fiber contribute to a healthy diet?

28.4 Body Weight Reflects Food Intake and Activity Level

Healthy eating has two main components. First, the diet must include all of the nutrients necessary to sustain life. A second consideration is the Calorie content of food, which must balance a person's metabolic rate and activity level.

As you can see in figure 28.4, nutrition labels list a food's energy content in units called **Calories.** Most young adults require 2000 to 2400 Calories per day, depending on sex and level of physical activity. A gram of carbohydrate or protein yields 4 Calories, whereas 1 gram of fat yields 9 Calories. These values help explain the link between a fatty diet and weight gain; gram for gram, fats supply over twice as much energy as any other type of food.

A. Body Mass Index Can Identify Weight Problems

No matter what we eat or drink, taking in more Calories than we expend causes weight gain; those who consume fewer Calories than they expend lose weight and may even starve. But what constitutes a "healthful" weight? The most common measure is the body mass index, or BMI (figure 28.5). To calculate BMI, divide a person's weight (in kilograms) by his or her squared height (in meters): $BMI = weight/(height)^2$. Alternatively, multiply weight (in pounds) by 704.5, and divide by the square of the height (in inches).

Many health professionals consider a person whose BMI is less than 19 to be underweight. A BMI between 19 and 25 is healthy; an overweight person has a BMI greater than 25; and a BMI greater than 30 denotes **obesity.** Morbid obesity is defined as a BMI greater than 40. One limitation of BMI is that it cannot account for many of the details that affect health. An extremely muscular person, for example, will have a high BMI because muscle is denser than fat, yet he or she would not be considered overweight.

Due in part to these limitations, many health professionals rely on other indicators of healthy weight, including the circumference of the abdomen. People whose fat accumulates at the waistline ("apples") are more susceptible to health problems such as insulin resistance than are "pears," who are bigger around the hips.

B. Starvation: Too Few Calories to Meet the Body's Needs

A diet is inadequate if it either contains too few Calories to sustain life or fails to provide an essential nutrient (figure 28.6). In some areas of the world, famine is a constant threat, and millions of people starve to death every year. Hunger strikes, inhumane treatment of prisoners, and eating disorders can also cause starvation.

A healthy human can survive for 50 to 70 days without food—much longer than without air or water. The exact timeline varies from person to person, but the starving human body essentially digests itself. After only a day without food, reserves of sugar and glycogen are dwindling, and the body begins extracting energy from stored fat and muscle protein. Gradually, metabolism slows, blood pressure drops, the pulse slows, and chills set in. Skin becomes dry and hair falls out as the proteins that form these structures are digested. When the body dismantles the immune system's antibody proteins, protection against infection declines. Mouth sores and anemia develop, the heart beats irregularly, and bones begin to degenerate. Near the end, the starving human is blind, deaf, and emaciated.

Anorexia nervosa, or self-imposed starvation, is refusal to maintain normal body weight. The condition affects about 1 in 250 adolescents, more than 90% of whom are female. The sufferer's body image is distorted; although she may be extremely thin, she perceives herself as overweight. A person with anorexia eats barely enough to survive, losing as much as 25% of her original body weight. She may further lose weight by vomiting, taking laxatives and diuretics, or exercising intensely. About 15% to 21% of people with anorexia die from the disease.

Bulimia is another eating disorder that mainly affects females. Rather than avoiding food, a person with **bulimia** eats large quantities and then intentionally vomits or uses laxatives shortly afterward, a pattern called "binge and purge." A person with bulimia may or may not be underweight.

Fortunately, eating disorders are often treatable under the combined care of a physician, dietician, and psychologist. These health professionals can help the patient return to a normal body weight, develop healthy eating habits, and address underlying psychological problems.

C. Obesity: More Calories Than the Body Needs

Obesity is increasingly common in the United States, and the health consequences can be serious. People who accumulate fat around their waists are susceptible to type 2 diabetes, high blood pressure, and atherosclerosis. High body weight is also correlated with heart disease, acid reflux, urinary incontinence, low back pain, stroke, sleep disorders, and other health problems. In addition, obese people also face a higher risk of cancers of the colon, breast, and uterus. ⓘ *diabetes,* section 25.4D

Excess body weight accumulates when a person consumes more Calories than he or she expends. For most people, the main culprits are an inactive

Figure 28.6 Malnutrition and Starvation. The malnourished child on the left has a swollen belly, a sign of a severe protein deficiency. The girl on the right suffers from anorexia nervosa, or self-imposed starvation.

(malnutrition): ©Brennan Linsley/AP Images; (anorexia): ©Bubbles Photolibrary/Alamy Stock Photo

Burning Question 28.1

Which diets lead to the most weight loss?

Every year or so, a new fad diet promises quick, easy weight loss. Some dieters report amazing results after eating nothing but "miracle" foods such as grapefruits or cabbage. Other diet plans exclude entire categories of food; the Paleolithic diet, for example, restricts intake of grains,  dairy products, and other "modern" foods. Still another strategy is to eat or avoid certain food combinations, based on anything from the dieter's blood type to the time of day. Scientific studies have yielded limited or no evidence that these strategies increase dieting success, and some can actually be dangerous.

©Image Source/Glow Images RF

Fortunately, you do not need to be a dieting expert or a biologist to lose weight. A successful diet is one that you can healthfully maintain for the rest of your life. Being physically active and using food labels to count Calories are simple, sustainable ways to lose weight. (See *What's the Point? Applied* in chapter 4 to learn more about boosting metabolism.)

Submit your burning question to marielle.hoefnagels@mheducation.com

Figure 28.7 **Hormone Deficiency.** In a normal mouse, leptin secreted by fat cells affects target cells in the hypothalamus, helping the brain regulate appetite and metabolism. The obese mouse on the left cannot produce leptin and therefore has a ravenous appetite.

©Chicago Tribune/KRT/Newscom

Figure It Out

Consider the following nutritional facts. Bacon cheeseburger: 23 g fat, 25 g protein, 2 g carbohydrates; large fries: 25 g fat, 6 g protein, 63 g carbohydrates; large soda: 86 g carbohydrates. How many Calories are in this meal?

Answer: 1160 Calories.

a.   b.

c.   d.

Figure 28.8 **A Selection of Animal Diets.** (a) A giant panda munches on bamboo. (b) This leopard is eating its kill. (c) A mosquito takes a blood meal. (d) This basking shark filters food from water.

(a): ©MelindaChan/Moment/Getty Images RF; (b): ©Alain Pons/PhotoAlto sas/Alamy Stock Photo RF; (c): Source: CDC; (d): ©image100/PunchStock RF

lifestyle coupled with a diet loaded with sugar and fat. Nevertheless, many genes contribute to appetite, digestion, and metabolic rate, so a person's family history plays at least a small part in the risk for obesity (see Why We Care 10.1).

Some of these genes encode hormones and hormone receptor proteins, which interact in complex and poorly understood ways to maintain the balance between food intake and energy expenditure. Hormones from the thyroid gland, for example, help regulate the metabolic rate. Another weight-related hormone is leptin. Stored fat releases leptin into the bloodstream, ultimately inhibiting food intake and increasing metabolic activity. This action explains why leptin-deficient mice become extremely obese (figure 28.7). Leptin deficiency, however, is very rare in humans. ⓘ *thyroid hormones,* section 25.4A

Scientific studies of leptin and other appetite-related hormones may someday yield new treatments for obesity. In the meantime, the concern over expanding waistlines has fueled demand for low-Calorie artificial sweeteners and fats (see chapter 2). Although fad diets remain popular, the most healthful way to lose weight is to exercise and reduce Calorie intake while maintaining a balanced diet. For people who have difficulty losing weight in this way, stomach-reduction surgery and drugs that either reduce appetite or block fat absorption offer other options.

28.4 Mastering Concepts

1. Describe the relationship of body weight to Calorie intake and energy expenditure.
2. What is body mass index?
3. Describe the events of starvation.
4. What are some of the causes and effects of obesity?

28.5 Most Animals Have a Specialized Digestive Tract

Biologists divide animals into categories based on what they eat and how they eat it (figure 28.8). **Herbivores,** such as cows and rabbits, eat only plants. Eagles, cats, wolves, and other **carnivores** are predators or scavengers that eat the flesh of other animals. **Detritivores** consume decomposing organic matter; dung beetles and earthworms illustrate this diet. Finally, **omnivores** eat a broad variety of foods, including plants and animals. Humans are omnivores, as are raccoons, pigs, chickens, and many other animals.

Many animals rely on one or a few kinds of food. Some animals, such as anteaters, flycatchers, praying mantises, and most spiders, eat only insects. Other animals eat only fish or fruits. The giant panda is a leaf-eater. Because it eats only bamboo, a wild panda can survive only where that plant thrives. Animals with more flexible diets, such as raccoons, can live in a broader range of habitats.

A. Acquiring Nutrients Requires Several Steps

Although diets differ, all animals have the same four-step process of obtaining and using food (figure 28.9). First, **ingestion** is the entrance of food into the digestive tract. The second stage, **digestion,** is the physical and chemical breakdown of food. In mammals, this process begins with chewing, which tears food into small pieces mixed with saliva. Chewing therefore softens food and increases the surface area exposed to digestive enzymes. In chemical digestion, enzymes split large nutrient molecules into their smaller components. In the third stage, **absorption,** the nutrients enter the cells lining the digestive tract and move into the bloodstream to be transported throughout the body (see chapter 27). Fourth, in **elimination,** the animal's body expels undigested food. **Feces** are the solid wastes that leave the digestive tract. ⓘ *enzymes,* section 4.4

One potential source of confusion is the distinction between feces and urine, both of which are animal waste products. Feces are composed partly of undigested food that never enters the body's cells. Urine, on the other hand, is a watery fluid containing dissolved nitrogen and other metabolic wastes produced by the body's cells. Section 28.9 describes how the kidneys produce urine.

B. Digestive Tracts May Be Incomplete or Complete

Because heterotrophs and their food consist of the same types of chemicals, digestive enzymes could just as easily attack an animal's body as its food. Digestion therefore occurs within specialized compartments that are protected from enzyme action.

These compartments may be located inside or outside of cells (figure 28.10). Intracellular digestion occurs entirely inside a cell; sponges are the only animals that rely solely on intracellular digestion. Collar cells lining the body wall of a sponge take in nutrients and enclose the food in a food vacuole. A loaded food vacuole fuses with another sac containing digestive enzymes that break down nutrient molecules. ⓘ *sponges,* section 17.2

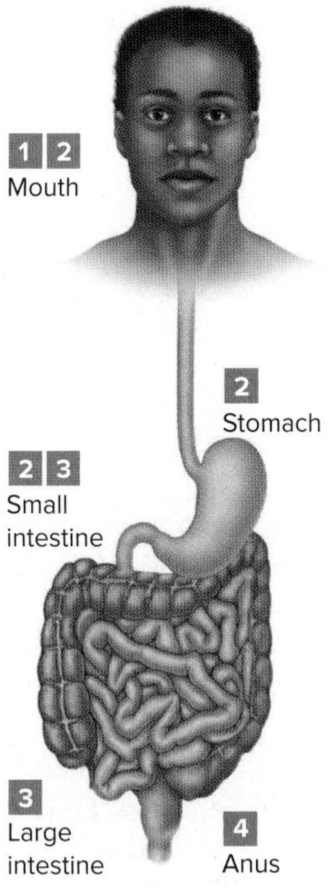

1 2
Mouth

2
Stomach

2 3
Small intestine

3
Large intestine

4
Anus

Stage	Description	Location(s) in Human Body
1 Ingestion	Food enters digestive tract.	Mouth
2 Digestion		
• Mechanical	Food is physically broken down into small particles.	Mouth, stomach
• Chemical	Digestive enzymes break food molecules into small subunits.	Mouth, stomach, small intestine
3 Absorption	Water and digested food enter bloodstream from digestive tract.	Small intestine (food and water), large intestine (water)
4 Elimination	Undigested food exits digestive tract with feces.	Anus

Figure 28.9 Acquiring and Using Food. The four stages by which animals acquire and use food are ingestion, digestion, absorption, and elimination.

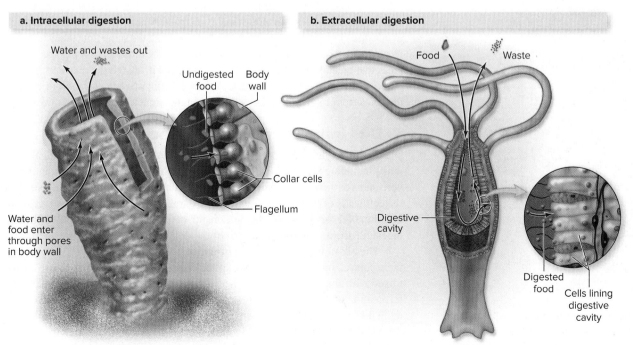

a. Intracellular digestion

Water and wastes out

Undigested food

Body wall

Collar cells

Flagellum

Water and food enter through pores in body wall

b. Extracellular digestion

Food

Waste

Digestive cavity

Digested food

Cells lining digestive cavity

Figure 28.10

Intracellular and Extracellular Digestion.

(a) A sponge uses intracellular digestion, which occurs entirely inside cells.
(b) Extracellular digestion occurs in a digestive cavity containing enzymes that break down food. Cells lining the digestive cavity absorb the nutrients.

a. Incomplete digestive tract: one opening

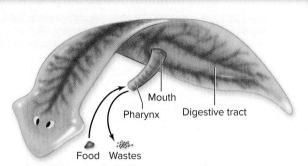

Mouth

Pharynx

Digestive tract

Food   Wastes

b. Complete digestive tract: two openings

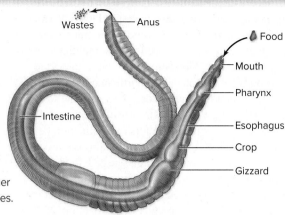

Wastes — Anus

Food

Mouth

Pharynx

Intestine

Esophagus

Crop

Gizzard

Figure 28.11 Incomplete and Complete Digestive Tracts. (a) In flatworms and other animals with an incomplete digestive tract, the mouth acquires food and eliminates wastes. (b) A complete digestive tract has two openings: the mouth and anus.

a. Ruminant herbivore

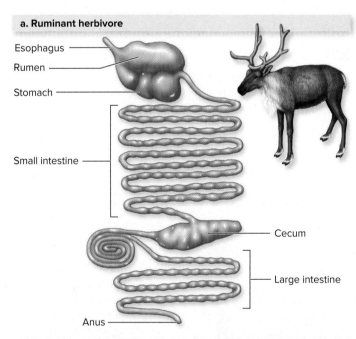

Esophagus

Rumen

Stomach

Small intestine

Cecum

Large intestine

Anus

b. Carnivore

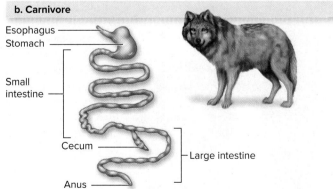

Esophagus

Stomach

Small intestine

Cecum

Large intestine

Anus

Figure 28.12 Digestive System Adaptations. (a) Ruminant herbivores have a rumen and a long digestive tract, an adaptation to a grassy diet. (b) The protein-rich diet of carnivores is easy to digest, so the digestive tract is much shorter and has a reduced cecum.

More complex animals use extracellular digestion, releasing enzymes into a digestive cavity connected with the outside world. The enzymes dismantle large food particles; cells lining the cavity then absorb the products of digestion. Food remains outside the body's cells until it is digested and absorbed. Extracellular digestion eases waste removal because indigestible components of food never enter the cells. Instead, the digestive tract simply ejects the waste.

The cavity in which extracellular digestion occurs may have one or two openings (figure 28.11). An **incomplete digestive tract** has only one opening: a mouth that both ingests food and ejects wastes. The animal must digest food and eliminate the residue before the next meal can begin. This two-way traffic limits the potential for specialized compartments that might store, digest, or absorb nutrients. Cnidarians such as jellyfish and *Hydra* have incomplete digestive tracts, as do flatworms. In these organisms, the digestive tract is also called a **gastrovascular cavity** because it doubles as a circulatory system that distributes nutrients to the body cells. ⓘ *cnidarians,* section 17.3; *flatworms,* section 17.4

Most animals have a **complete digestive tract** with two openings; the mouth is the entrance, and the **anus** is the exit. This tubelike digestive cavity is called the **alimentary canal** or **gastrointestinal (GI) tract.** Notice that food passes through in one direction, so the animal can digest one meal at the same time that it acquires its next one. Another advantage is that regions of the tube can develop specialized areas that break food into smaller particles, digest it, absorb the nutrients into the bloodstream, and eliminate wastes. A complete digestive tract therefore extracts nutrients from food more efficiently than an incomplete digestive tract does.

C. Diet Influences Digestive Tract Structure

Diet and lifestyle differences select for digestive system adaptations in all animals, including mammals. Figure 28.12 shows the digestive tracts of an herbivore and a carnivore. An herbivore's diet is rich in hard-to-digest cellulose from the cell walls of plants. The long digestive tract allows extra time for digestion. The diet of a carnivore, on the other hand, consists mostly or entirely of highly digestible meat. The overall length of the digestive tract is therefore short.

The elk in figure 28.12 is a **ruminant,** which is an herbivore with a complex, four-chambered organ that specializes in the digestion of grass. Saliva mixed with chewed grass enters the first and largest chamber, the rumen, where fermenting microorganisms break the plant matter down into balls of cud. The animal regurgitates the cud into its mouth; chewing the cud breaks the food down further. When the animal swallows again, the food bypasses the rumen, continuing digestion in the remaining chambers. Cows, sheep, deer, and goats are familiar ruminants. ⓘ *fermentation,* section 6.8

Figure 28.12 also depicts another structure whose size varies with diet: the pouchlike **cecum,** which forms the entrance to the large intestine. In herbivores, the cecum is large, and it houses bacteria that ferment plant matter. Carnivores have a small or absent cecum. The cecum is medium-sized in omnivores, reflecting a diet based partly on plants.

28.5 Mastering Concepts

1. Define the terms *herbivore, carnivore, detritivore,* and *omnivore.*
2. What four processes does food undergo when an animal eats?
3. Distinguish between intracellular and extracellular digestion.
4. How do incomplete and complete digestive tracts differ?
5. Compare and contrast the digestive systems of an elk and a wolf.

28.6 The Human Digestive System Consists of Several Organs

The human digestive system consists of the gastrointestinal tract and accessory structures (figure 28.13). The salivary glands and pancreas are accessory structures that produce digestive enzymes; the liver and gallbladder produce and store bile, which assists in fat digestion. The teeth and tongue are also accessory organs.

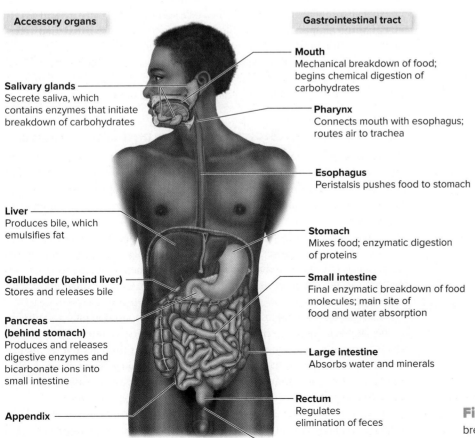

Accessory organs

Salivary glands
Secrete saliva, which contains enzymes that initiate breakdown of carbohydrates

Liver
Produces bile, which emulsifies fat

Gallbladder (behind liver)
Stores and releases bile

Pancreas (behind stomach)
Produces and releases digestive enzymes and bicarbonate ions into small intestine

Appendix

Gastrointestinal tract

Mouth
Mechanical breakdown of food; begins chemical digestion of carbohydrates

Pharynx
Connects mouth with esophagus; routes air to trachea

Esophagus
Peristalsis pushes food to stomach

Stomach
Mixes food; enzymatic digestion of proteins

Small intestine
Final enzymatic breakdown of food molecules; main site of food and water absorption

Large intestine
Absorbs water and minerals

Rectum
Regulates elimination of feces

Anus

Digestive System	
Main tissue types*	**Examples of locations/functions**
Epithelial	Secretes hormones, enzymes, and mucus into digestive tract; absorbs products of digestion; protects mouth, esophagus, and anal canal from pathogens and abrasion.
Connective	Blood (a connective tissue) transports nutrients from the digestive system to all parts of the body; supports esophagus, liver, and digestive lining.
Muscle	Smooth muscle moves food along digestive tract and aids in mechanical digestion; skeletal and smooth muscle control the mouth, tongue, esophagus, and anal canal.
Nervous	Stretch receptors signal presence of food in stomach; nerves regulate activity of digestive organs.

*See chapter 23 for descriptions.

Figure 28.13 The Human Digestive System. Food breaks down as it moves along the digestive tract. Accessory organs aid in digestion.

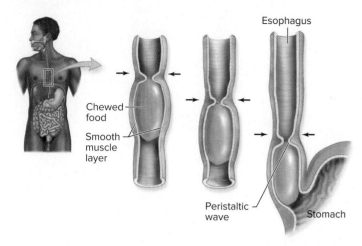

Figure 28.14 Peristalsis. Waves of smooth muscle contraction push food along the esophagus and the rest of the digestive tract.

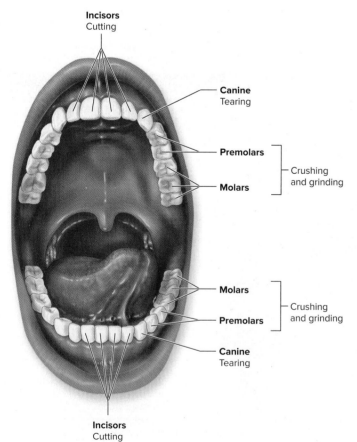

Figure 28.15 The Mouth. Mechanical digestion begins with the teeth, which grasp, cut, crush, and grind food into pieces small enough to be swallowed. Salivary glands (such as those under the tongue) contribute fluid and some enzymes.

A. Muscles Underlie the Digestive Tract

How does food make its way along an animal's digestive tract? Layers of smooth muscle underlying the entire digestive tract undergo **peristalsis,** or rhythmic waves of contraction that propel food in one direction (figure 28.14). These contractions also churn the food, mixing it with enzymes to form a liquid. Unlike skeletal muscle, smooth muscle contraction is involuntary: It does not require input from motor neurons. Instead, the autonomic nervous system stimulates smooth muscle to contract. ⓘ *smooth muscle tissue,* section 23.2C; *autonomic nervous system,* section 24.5

Muscles also control the openings between digestive organs. **Sphincters** are muscular rings that can contract to block the passage of materials. The sphincters at the mouth and anus are composed of skeletal muscle and are under voluntary control, so we can decide when to open our mouth or eliminate feces. The remaining sphincters within the digestive tract, however, are composed of involuntary smooth muscle.

B. Digestion Begins in the Mouth

A tour of the digestive system begins at the **mouth.** The taste of food triggers salivary glands in the mouth to secrete saliva. This fluid contains an enzyme that starts to break down starch (a polysaccharide) into sugars. Meanwhile, the **teeth**—mineral-hardened structures embedded in the jaws—grasp and chew the food (figure 28.15). Chewing is a form of mechanical digestion: Water and mucus in saliva aid the teeth as they tear food into small pieces, increasing the surface area available for chemical digestion. The muscular **tongue** at the floor of the mouth mixes the food with saliva and pushes it to the back of the mouth to be swallowed.

The chewed mass of food passes first through the **pharynx,** or throat, the tube that also conducts air to the trachea (see section 27.6). During swallowing, the **epiglottis** temporarily covers the opening to the trachea so that food enters the digestive tract instead of the lungs. From the pharynx, swallowed food and liquids pass to the **esophagus,** a muscular tube leading to the stomach. Food does not merely slide down the esophagus under the influence of gravity; instead, contracting muscles push it along in a wave of peristalsis.

C. The Stomach Stores, Digests, and Churns Food

The **stomach** is a muscular bag that receives food from the esophagus (figure 28.16). The stomach is about the size of a large sausage when empty, but when very full, it can expand to hold as much as 3 or 4 liters of food. Ridges in the stomach's lining can unfold like the pleats of an accordion to accommodate a large meal.

The stomach's main function is to continue the mechanical and chemical digestion of food. Swallowed chunks of food break into smaller pieces as waves of peristalsis churn the stomach's contents. At the same time, the stomach lining produces **gastric juice,** a mixture of water, mucus, ions, hydrochloric acid, and enzymes.

This gastric juice comes from specialized cells housed in pits in the lining of the stomach. Some cells produce mucus; others secrete a protein that becomes **pepsin,** an enzyme that digests proteins. Still others release hydrochloric acid. The pH of the gastric juice is therefore low, about 1.5 or 2. The acidity denatures the proteins in food, kills most disease-causing organisms, and activates pepsin. ⓘ *epithelial tissue,* section 23.2A; *pH,* section 2.4

If gastric juice breaks down protein in food, how does the stomach keep from digesting itself? First, the stomach produces little gastric juice until food is present. Second, mucus coats and protects the stomach lining. Tight junctions between cells in the stomach lining also prevent gastric juice from seeping through to the tissues below. ⓘ *animal cell junctions,* section 3.6

Although digestion begins in the stomach, this organ absorbs very few nutrients. It can absorb some water and ions, a few drugs (e.g., aspirin), and, like the rest of the digestive tract, alcohol. As a result, we feel alcohol's intoxicating effects quickly.

Chyme is the semifluid mixture of food and gastric juice in the stomach. At regular intervals, small amounts of chyme squirt through the sphincter that links the stomach and the upper part of the small intestine.

D. The Small Intestine Digests and Absorbs Nutrients

The **small intestine** is a tubular organ that completes digestion and absorbs nutrients and water. Although narrow in comparison with the large intestine, the small intestine's 3- to 7-meter length makes it the longest organ in the digestive system.

Anatomy of the Small Intestine   The duodenum makes up the first 25 centimeters of the small intestine. Glands in the wall of the duodenum secrete mucus that protects and lubricates the small intestine. In addition, ducts from the pancreas and liver open into the duodenum.

The rest of the small intestine absorbs nutrients from digested food and passes them to the bloodstream. Close examination of the hills and valleys along the small intestine's lining reveals millions of **intestinal villi,** tiny finger-like projections that absorb nutrients (figure 28.17). The epithelial cells on the

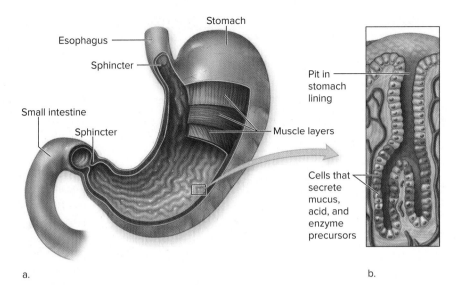

a.   b.

Figure 28.16 **The Stomach.** (a) The stomach receives food from the esophagus, mixes it with gastric juice, and passes it to the small intestine. (b) The stomach lining contains cells that secrete mucus and gastric juice.

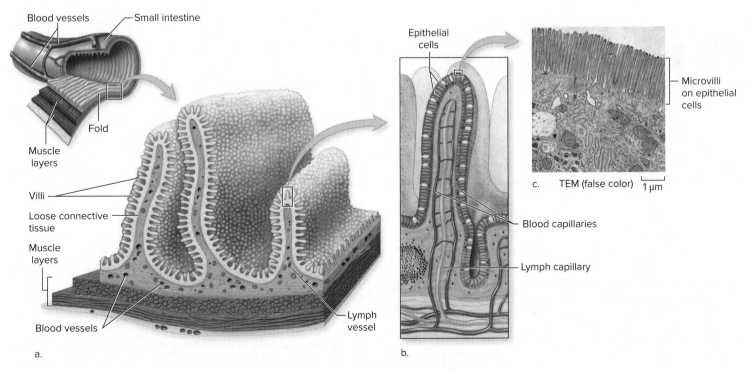

a.   b.

Figure 28.17 **The Lining of the Small Intestine.** (a) Fingerlike villi project from each ridge of the small intestine's lining. (b) Within each villus, blood-filled capillaries absorb digested carbohydrates and proteins, and lymph capillaries absorb digested fats. (c) Microvilli extending from the epithelial cells add tremendous surface area for absorption.

(c): Courtesy of David H. Alpers, M.D.

Miniglossary | Digestive Fluids

Saliva	Fluid secreted by glands in the mouth; contains starch-digesting enzymes
Gastric juice	Mixture of water, mucus, ions, hydrochloric acid, and enzymes that the stomach lining secretes
Pepsin	Protein-digesting enzyme in the stomach
Pancreatic juice	Mixture of enzymes that the pancreas delivers to the small intestine; includes trypsin and chymotrypsin (protein digestion), amylase (starch digestion), lipase (fat digestion), and nuclease (nucleotide digestion)
Bile	Chemical that suspends fats in water to speed digestion by lipase; produced in the liver but stored and released by the gallbladder

Burning Question 28.2

What is lactose intolerance?

Lactose, or milk sugar, is a disaccharide in milk. In an infant's small intestine, an enzyme called lactase breaks down lactose. Most people stop producing lactase after infancy; after all, milk is not typically part of an adult mammal's diet. If a person without lactase consumes milk, bacteria in the large intestine ferment the undigested sugar. Their byproducts create the symptoms of lactose intolerance: abdominal pain, gas, diarrhea, bloating, and cramps.

People with lactose intolerance can prevent these problems simply by avoiding fresh milk. Instead, they can choose fermented dairy products such as yogurt, buttermilk, and cheese; bacteria have already broken down the lactose in those foods. Taking lactase tablets can also prevent symptoms.

Interestingly, people with roots in northern Europe and a few other locations continue to produce lactase as adults, thanks to a long-ago genetic mutation that was adaptive in dairy-herding regions of the world.

Submit your burning question to
marielle.hoefnagels@mheducation.com

(milk): ©D. Hurst/Alamy Stock Photo RF

surface of each villus bristle with hundreds of **microvilli,** extensions of the cell membrane. Villi and microvilli increase the surface area of the small intestine at least 600 times, allowing for the efficient extraction of nutrients from food.

The capillaries that snake throughout each villus take up the newly absorbed nutrients and water, then empty into veins that carry the nutrient-laden blood to the liver. Also inside each villus is a lymph capillary that receives digested fats. Cells throughout the body use all of these nutrients to generate energy and to build new proteins, carbohydrates, lipids, and nucleic acids.

The Role of the Pancreas, Liver, and Gallbladder The small intestine absorbs water, minerals, free amino acids, cholesterol, and vitamins without further digestion. Most molecules in food, however, require additional processing.

Digestive enzymes released by cells lining the small intestine act on short polysaccharides and disaccharides to release simple sugars, which the small intestine immediately absorbs. These enzymes are called carbohydrases. People who lack one such enzyme, lactase, cannot digest milk sugar; Burning Question 28.2 discusses this condition.

Most of the digestive enzymes in the small intestine, however, come from the **pancreas** (see figure 28.13). This accessory organ sends about a liter of pancreatic juice to the duodenum each day. The fluid from the pancreas contains many enzymes. Trypsin and chymotrypsin break polypeptides into amino acids; pancreatic amylase digests starch; pancreatic lipase breaks down fats; and nucleases split nucleic acids such as DNA into nucleotides. In addition to these enzymes, pancreatic juice also contains alkaline sodium bicarbonate, which neutralizes the acid from the stomach.

Fats present an interesting challenge to the digestive system. Lipase enzymes are water soluble, but fats are not. Therefore, lipase can act only at the surface of a fat droplet, where it contacts water. **Bile** is a greenish-yellow biochemical that disperses the fat into tiny globules suspended in water. The resulting mixture, called an emulsion, increases the surface area exposed to lipase.

Bile comes from the **liver,** a large accessory organ with more than 200 functions. The liver's only *direct* contribution to digestion is the production of bile. The **gallbladder** is an accessory organ that stores this bile until chyme triggers its release into the small intestine. The cholesterol in bile can crystallize, forming gallstones that partially or completely block the duct to the small intestine. Gallstones are very painful and may require removal of the gallbladder. A person can survive without this organ because surgeons simply redirect the flow of bile from the liver to the small intestine. ⓘ *cholesterol,* section 2.5E

The liver's other functions include detoxifying alcohol and other harmful substances in the blood, storing glycogen and fat-soluble vitamins, and producing blood-clotting proteins. Nutrient-laden blood arrives from the intestines and passes through the liver's extensive capillary beds, which remove bacteria and toxins. The liver gets "first dibs" on the nutrients in the blood before it is pumped to the rest of the body. (In a condition called cirrhosis, scar tissue blocks this vital blood flow through the liver.)

Like the stomach, the small intestine protects itself against self-digestion by producing digestive biochemicals only when food is present. In addition, mucus protects the intestinal wall from digestive juices and neutralizes stomach acid. Nevertheless, many cells of the intestinal lining die in the caustic soup. Rapid division of the small intestine's epithelial cells compensates for the loss, replacing the lining every 36 hours.

Figure 28.18 Overview of Chemical Digestion. Although digestion begins in the mouth and stomach, the small intestine digests and absorbs most molecules.

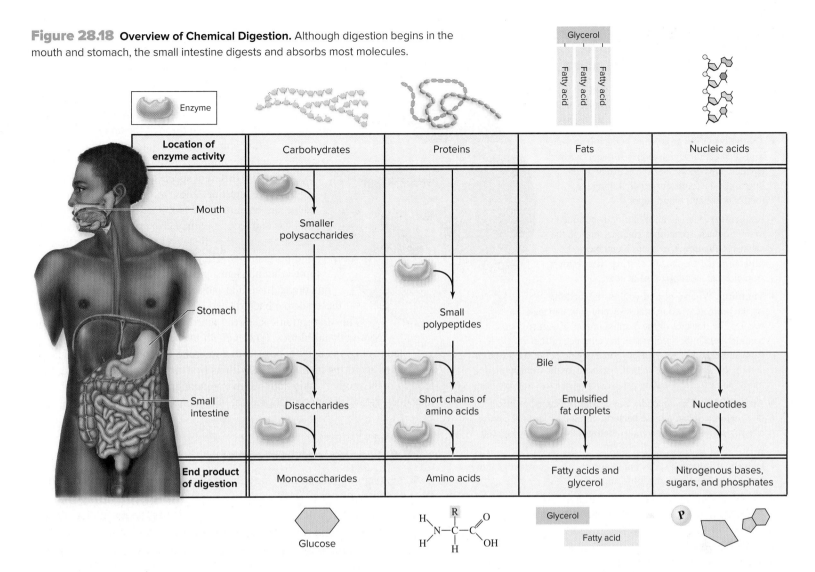

Location of enzyme activity	Carbohydrates	Proteins	Fats	Nucleic acids
Mouth	Smaller polysaccharides			
Stomach		Small polypeptides		
Small intestine	Disaccharides	Short chains of amino acids	Bile → Emulsified fat droplets	Nucleotides
End product of digestion	Monosaccharides	Amino acids	Fatty acids and glycerol	Nitrogenous bases, sugars, and phosphates

Glucose

Glycerol / Fatty acid

Figure 28.18 summarizes the locations of the major digestive enzymes, along with their products. Ultimately, as we have seen, the small intestine absorbs these nutrients and passes them to the bloodstream. Cells throughout the body then use the nutrients to generate energy and to build new proteins, carbohydrates, lipids, and nucleic acids (see chapter 2).

E. The Large Intestine Completes Nutrient and Water Absorption

The material remaining in the small intestine moves next into the **large intestine,** which extends to the anus while forming a "frame" around the small intestine (figure 28.19). At 1.5 meters long, the large intestine is much shorter than the small intestine, but its diameter is greater (about 6.5 centimeters). Its main functions are to absorb water and ions and to eliminate the remaining wastes as feces.

The start of the large intestine is the pouchlike cecum. Dangling from the cecum is the appendix, a thin, worm-shaped tube. Trapped bacteria or undigested food can cause the appendix to become irritated, inflamed, and infected, producing severe pain. A burst appendix can spill its contents into the abdominal cavity and spread the infection. (Why We Care 28.1 describes some other examples of illnesses affecting the gastrointestinal tract.)

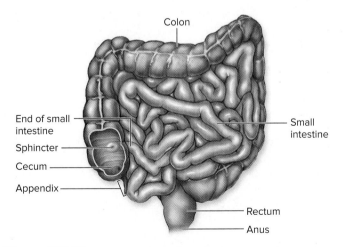

Figure 28.19 The Large Intestine. The large intestine receives chyme from the small intestine. The lining of the colon absorbs water, ions, and minerals; whatever is left is eliminated as feces.

Why We Care 28.1 | The Unhealthy Digestive System

The entire length of the human digestive tract is subject to numerous disorders. Some are a nuisance or easily treated, whereas others can be deadly. A few are listed here:

- **Tooth decay:** Bacteria living in the mouth secrete acids that eat through a tooth's surface, causing cavities. This decayed area can extend to the interior of the tooth, eventually killing the tooth's nerve and blood supply.

- **Acid reflux:** Gastric juice, normally confined to the stomach, sometimes emerges through the sphincter guarding the stomach entrance. The acid burns the esophagus, causing "heartburn." Antacids can neutralize the acidity.

- **Vomiting:** When a person vomits, the medulla (in the brainstem) coordinates several simultaneous events. The muscles of the diaphragm and abdomen contract, while the sphincter at the entrance of the stomach relaxes. These actions force chyme out of the stomach and up the esophagus. Alcohol, bacterial toxins from spoiled foods, and excessive eating can trigger queasiness and vomiting.

- **Appendicitis:** The appendix can become inflamed or infected. If it ruptures, it can release bacteria into the abdominal cavity, causing serious infection and sometimes even death when left untreated.

- **Diarrhea:** If the intestines fail to absorb as much water as they should, the feces become loose and watery.

- **Constipation:** An obstructed large intestine, loss of peristalsis, dehydration, starvation, or anxiety might cause hard, dry feces that are difficult to eliminate.

- **Colon (colorectal) cancer:** Cancerous tumors may arise in the rectum, colon, or appendix. Colon cancer is a leading cause of death worldwide; a diet high in fiber aids in prevention.

Photo: ©SchulteProductions/iStock/Getty Images RF

The colon forms the majority of the large intestine. Here, the large intestine absorbs most of the remaining water, ions, and minerals from chyme. Veins carry blood from vessels surrounding the large intestine to the liver.

The remnants of digestion consist mostly of bacteria, undigested fiber, and intestinal cells. These materials, plus smaller amounts of other substances, collect in the rectum as feces. When the rectum is full, the feces are eliminated through the anus.

Our partnership with our intestinal microbes deserves special mention. Trillions of bacteria, representing about 500 different species, are normal inhabitants of the large intestine. These microscopic residents produce the characteristic foul-smelling odors of intestinal gas and feces, but they also provide many benefits. Most notably, they help prevent infection by harmful microorganisms. They also decompose cellulose and some other nutrients, produce B vitamins and vitamin K, and break down bile and some drugs. Antibiotics often kill these normal bacteria and allow other microbes to take over. This disruption causes the diarrhea that may accompany treatment with antibiotics. ⓘ *beneficial microbes,* section 15.2D

Interestingly, a baby is born with a microbe-free digestive tract. The infant begins acquiring its microbiota with its first meal of milk or formula. Bacteria from the mother's skin and the environment gradually colonize the baby's intestines, establishing populations that will persist throughout life.

28.6 Mastering Concepts

1. Explain the action and importance of peristalsis and sphincters in digestion.
2. Describe the functions of saliva, teeth, and the tongue in digestion.
3. How does food move from the mouth to the stomach?
4. Describe the mechanical and chemical digestion that occurs in the stomach.
5. What is the structure and function of the small intestine?
6. How do the pancreas, liver, and gallbladder aid digestion?
7. Describe the events that occur as food passes through the large intestine.
8. How does undigested food leave the body?

28.7 Animals Eliminate Nitrogenous Wastes and Regulate Water and Ions

As we have already seen, the digestive system provides energy and raw materials required in metabolism. As cells carry out their metabolic reactions, they also release wastes into the bloodstream. **Excretion** is the elimination of these metabolic wastes.

For example, as described in chapter 27, the respiratory system excretes CO_2, a waste produced in aerobic cellular respiration. A less familiar byproduct of metabolism is the nitrogen-containing (nitrogenous) wastes that cells produce when they break down proteins (figure 28.20). Proteins are composed of amino acids, which can enter the energy-generating pathways described in chapter 6. During this process, amino groups (–NH$_2$) are stripped from amino acids. In mammals, liver cells incorporate the nitrogen into a waste molecule called **urea.** (Other animals eliminate the nitrogen in other forms.) This nitrogenous waste then moves to the bloodstream and is excreted in urine.

Besides eliminating urea, urine also participates in the regulation of the balance between ions and water inside the body. In most habitats, organisms must **osmoregulate;** that is, they control the concentration of ions in their body fluids as the environment changes. Osmoregulation requires managing the gain and loss of water, ions, or both.

A brief review of how water and ions move across membranes will help explain how osmoregulation works. Osmosis is the diffusion of water across a semipermeable membrane; the net direction of water movement is toward the side with the higher concentration of dissolved solutes. If a cell's environment is saltier than the cell itself, water moves out of the cell. In the opposite situation, water moves into the cell. ⓘ *membrane transport,* section 4.5

Osmoregulation often requires cells to move ions against their concentration gradient; that is, from where they are less concentrated to where they are more concentrated. This process, called active transport, requires energy in the form of ATP. Water may follow the ions by osmosis; our kidneys exploit this mechanism to conserve water during the production of urine.

Land animals use a combination of strategies to obtain and conserve water (figure 28.21). Whereas humans ingest most of their water in food and drink, desert kangaroo rats derive most of their water as a byproduct of metabolism, especially cellular respiration. Animals lose water through evaporation from lungs and the skin surface, in feces, and in urine. The rest of this chapter describes how the human urinary system carries out its functions.

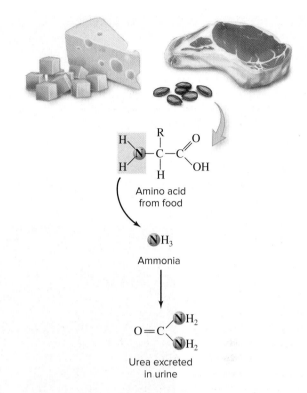

Figure 28.20 **Nitrogenous Waste.** Protein breakdown yields ammonia, which the liver converts to urea for elimination in urine.

28.7 Mastering Concepts

1. What is the main nitrogenous waste in a mammal's urine?

2. How do land animals gain and lose water?

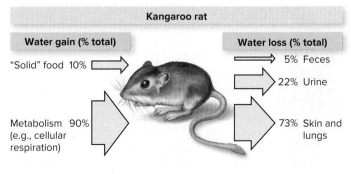
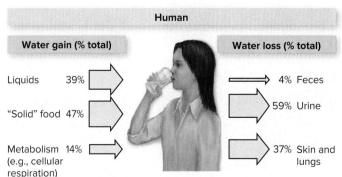

Figure 28.21 **Water Gain and Loss.** The desert-dwelling kangaroo rat gets all of its water from its food and its metabolism; it loses little in feces and urine. In contrast, a human gets most water from food and drink, and loses most of it in urine.

Why We Care 28.2 | Urinary Incontinence

The ability to choose where and when we urinate requires functioning sphincter muscles, bladder muscles, and nerves. If one or more of these elements fails, a person can lose control of the bladder—a condition called urinary incontinence.

For some people, incontinence is caused by a temporary condition such as a urinary tract infection. For others, it is a persistent problem, with underlying causes ranging from pregnancy to prostate enlargement to spinal cord injuries. Depending on the cause, treatment may include strengthening the muscles that control urination. Behavioral changes, alone or coupled with medication, may also help. Surgical procedures to treat incontinence include the surgical implantation of an artificial sphincter.

Many adults, however, manage incontinence with low-tech solutions. Some wear adult-sized diapers or absorbent undergarments. Others rely on a catheter, which is a flexible tube that drains urine directly from the bladder into a portable collection bag.

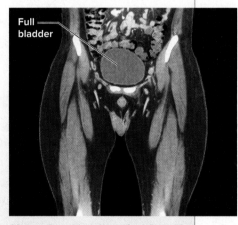

Full bladder

©Science Photo Library/Alamy Stock Photo RF

28.8 The Urinary System Produces, Stores, and Eliminates Urine

The human **urinary system** filters blood, eliminates nitrogenous wastes, and helps maintain the ion concentration of body fluids. The paired **kidneys** are the major excretory organs in the urinary system (figure 28.22). Located near the rear wall of the abdomen, each kidney is about the size of an adult fist and weighs about 230 grams.

As the kidneys cleanse blood, a liquid waste called **urine** forms; section 28.9 describes this process in detail. Besides eliminating urea and other toxic substances, kidneys have other functions as well. These organs conserve water, ions, glucose, amino acids, and other valuable nutrients. They also help regulate blood pH.

One other function of the kidneys is to regulate the volume of blood. Look back at the water balance in figure 28.21. What happens if we drink too many fluids or if we lose too much moisture in sweat and breath? Rather than swelling like a balloon or drying up like a leaf, the body adjusts: The kidneys maintain the volume of blood by controlling the amount of water lost in urine. This function is important because blood volume is one factor that influences blood pressure. ⓘ *blood pressure,* section 27.5B

The urine from each kidney drains into a **ureter,** a narrow muscular tube about 28 centimeters long. Waves of peristalsis squeeze the fluid along the two ureters and squirt it into the **urinary bladder,** a saclike muscular organ that collects urine. Two sphincters at the bladder's exit remain closed, except during urination.

The **urethra** is the tube that connects the bladder with the outside of the body. In females, the urethra opens between the clitoris and vagina. In males, the urethra extends the length of the **penis.** The urethra also carries semen in males (see chapter 30). The term *urogenital tract* reflects the intimate connection between the urinary and reproductive systems.

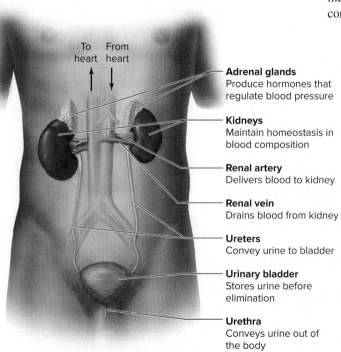

To heart   From heart

Adrenal glands
Produce hormones that regulate blood pressure

Kidneys
Maintain homeostasis in blood composition

Renal artery
Delivers blood to kidney

Renal vein
Drains blood from kidney

Ureters
Convey urine to bladder

Urinary bladder
Stores urine before elimination

Urethra
Conveys urine out of the body

Urinary System	
Main tissue types*	**Examples of locations/functions**
Epithelial	Enables diffusion between nephron and blood; also lines ureters and bladder.
Connective	Blood (which kidneys filter) is a connective tissue.
Muscle	Smooth muscle controls flow of blood to and from nephrons; smooth and skeletal muscle sphincters control urine release.
Nervous	Sensory cells in hypothalamus coordinate negative feedback loops that maintain homeostasis.

*See chapter 23 for descriptions.

Figure 28.22 The Human Urinary System. The human urinary system includes the kidneys, ureters, urinary bladder, and urethra. This generalized depiction omits the differences between male and female organs. Figures 30.4 and 30.8 show the sex differences in more detail.

The adult bladder can hold about 600 milliliters of urine. The accumulation of urine stimulates stretch receptors in the bladder. These sensory neurons send impulses to the spinal cord and brain, generating a strong urge to urinate. We can suppress this urge for a short time. Eventually, however, the brain's cerebral cortex directs the sphincters at the bladder's exit to relax, and bladder muscle contractions force the urine out of the body. (For a variety of reasons, some adults lose control of the bladder; Why We Care 28.2 describes this condition.)

28.8 Mastering Concepts

1. List the organs that make up the human urinary system.
2. What are the functions of the kidneys?

28.9 Nephrons Remove Wastes and Adjust the Composition of Blood

The body's entire blood supply courses through the kidney's blood vessels every 5 minutes. At that rate, the equivalent of 1600 to 2000 liters of blood passes through the kidneys each day. Most of the fluid that the nephrons process, however, is reabsorbed into the blood and not released in urine. As a result, a person produces only about 1.5 liters of urine daily.

A. Nephrons Interact Closely with Blood Vessels

Each kidney contains 1.3 million tubular **nephrons,** the functional units of the kidney. As illustrated in figure 28.23, each kidney receives blood via a renal artery, which branches into capillaries surrounding each nephron. The kidney's capillaries eventually converge into the renal vein, which carries cleansed blood out of the kidney and (ultimately) to the heart.

Burning Question 28.3

What can urine reveal about health and diet?

Urinalysis is a routine part of many medical examinations. Laboratory tests of the chemical components of urine can reveal many health problems:

- More than a trace of glucose may be a sign of diabetes, a high-carbohydrate diet, or stress. Stress causes the adrenal glands to release excess epinephrine, which stimulates the liver to break down more glycogen into glucose. (i) *diabetes,* section 25.4D

- Albumin may be a sign of damaged nephrons, since this plasma protein does not normally fit through the pores of the filter at the head of the nephron.

- Together, pus and an absence of glucose indicate a urinary tract infection. The pus consists of infection-fighting white blood cells along with bacteria, which consume the glucose.

- Traces of marijuana, cocaine, and other substances may appear in the urine. Athletes and employees of some organizations routinely submit urine samples for drug testing.

Urine is normally pale yellow, thanks to a pigment that the liver produces as it breaks down dead blood cells. Colorless urine usually indicates excessive water intake or the ingestion of diuretics such as coffee or beer. A reddish tinge may suggest anything from bleeding in the urinary tract to beet consumption to mercury poisoning. Either vitamin C or carrot consumption can color urine orange.

Submit your burning question to marielle.hoefnagels@mheducation.com

(specimen cup): ©JodiJacobson/E+/Getty Images RF

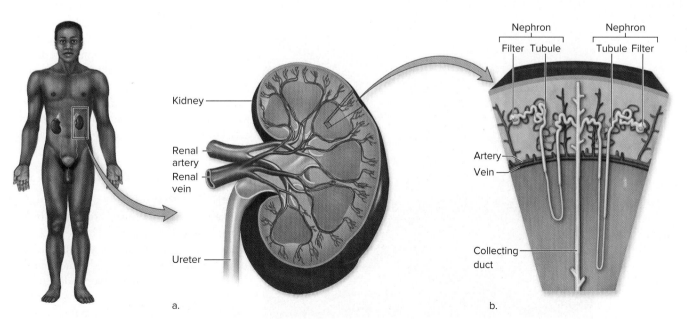

a.   b.

Figure 28.23 Anatomy of a Kidney. (a) Blood arrives at each kidney via the renal artery and leaves via the renal vein. (b) The nephron is the functional unit of the kidney. The urine produced by each nephron enters a collecting duct, which delivers the fluid to the ureter.

Examine the anatomy of the nephrons in figure 28.23b. Each nephron consists of two main parts: a filter and a tubule. The filter receives fluid from the blood. From there, the solution travels along the tubule, which is a winding passageway that dips toward the kidney's center before returning to the outer portion of the kidney. The entire nephron, when stretched out, is about 3 to 4 centimeters long.

A **collecting duct** receives the fluid from several nephrons. Urine from many collecting ducts empties into the funnel-like upper portion of the ureter. The fluid flows down the ureter to the urinary bladder and eventually moves out of the body through the urethra (see section 28.8).

B. Urine Formation Includes Filtration, Reabsorption, and Secretion

The chemical composition of urine reflects three processes (figure 28.24):

1. **Filtration:** Water and dissolved substances are filtered out of the blood at the entrance to the nephron. Blood pressure drives substances across the capillary walls into the nephron. Pores in the filter allow water, urea, glucose, ions, and amino acids to pass, but large structures such as plasma proteins, blood cells, and platelets remain in the bloodstream.
2. **Reabsorption:** Useful materials such as ions, water, glucose, and amino acids return from the nephron and collecting duct to the blood.
3. **Secretion:** Toxic substances and drug residues are secreted into the nephron to be eliminated in urine, as are hydrogen ions (H^+) and other surplus ions.

After reabsorption and secretion, the filtrate is urine. Urine contains water, urea, and several types of ions. Nearly all of the glucose and amino acids present in the original filtrate, however, return to the blood. Nevertheless, the exact chemical composition of urine can vary, as described in Burning Question 28.3.

When the nephrons fail to do their job, nitrogenous wastes and other toxins accumulate in the blood to harmful levels; in addition, a person may lose too much water and become dehydrated. Without treatment, kidney failure can be fatal (see Why We Care 28.3).

C. Hormones Regulate Kidney Function

Kidney function adjusts continuously to maintain homeostasis. For example, if water is scarce, we produce small amounts of concentrated urine. When we drink too much water, urine is abundant and dilute.

Hormones help make these adjustments in kidney function (figure 28.25). When we are dehydrated, receptor cells in the hypothalamus send impulses to the

Figure 28.24 **Overview of Urine Formation.** Three processes contribute to urine production: filtration, reabsorption, and secretion. The nephron exchanges materials with blood along its entire length.

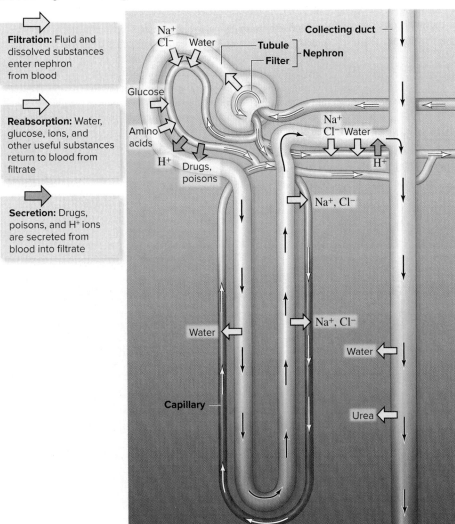

Filtration: Fluid and dissolved substances enter nephron from blood

Reabsorption: Water, glucose, ions, and other useful substances return to blood from filtrate

Secretion: Drugs, poisons, and H^+ ions are secreted from blood into filtrate

Why We Care 28.3 | Kidney Failure, Dialysis, and Transplants

Kidneys can fail for many reasons. The most common causes are diabetes and high blood pressure, which together account for about two thirds of all cases. Other causes are chronic inflammation that causes progressive loss of nephrons, polycystic kidney disease (the accumulation of cysts in the kidneys), scarring from untreated kidney or urinary tract infections, and obstructed urine flow.

A person with diabetes has too much sugar in his or her blood. Over time, high blood sugar damages small blood vessels in the kidneys (and throughout the body). As a result, diabetes reduces kidney function. Likewise, high blood pressure also damages the kidneys' blood vessels. Kidney damage, in turn, may cause the blood to retain excess fluid, driving blood pressure even higher and setting into motion a dangerous cycle.

One treatment option for kidney patients is dialysis (figure 28.A). The dialysis machine pumps blood out of the patient's body and past a semipermeable membrane. Wastes and toxins, along with water, diffuse across the membrane, but blood cells do not. The cleaned blood then circulates back to the patient's body. The procedure requires hours a day, several times a week.

Dialysis membranes cannot replace all of the kidney's functions. For example, nephrons selectively recycle useful components such as glucose

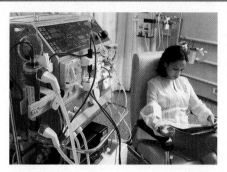

Figure 28.A Kidney Dialysis. A dialysis machine can take over some, but not all, of a healthy kidney's functions.
©AJPhoto/Science Source

and ions to the blood. The dialysis machine cannot do this, although a technician can adjust the concentrations of dissolved compounds removed from the blood.

Transplantation is a second option. The transplanted kidney may come from a cadaver, or a living person may donate one healthy kidney to a recipient. Surgeons connect the new kidney to the recipient's blood supply and attach its ureter to the bladder, usually leaving the old kidneys in place.

posterior pituitary gland, which secretes a peptide hormone called **antidiuretic hormone (ADH),** or vasopressin. ADH triggers the reabsorption of more water at the nephron and collecting duct, so the urine becomes more concentrated. Conversely, if blood plasma is too dilute, ADH production stops, and more water is eliminated in the urine. ⓘ *posterior pituitary,* section 25.3A

A diuretic is a substance that increases the volume of urine. The ethanol in alcoholic beverages is one example. Alcohol stimulates urine production partly by reducing ADH secretion, thereby decreasing water reabsorption. By increasing water loss to urine, alcoholic beverages actually intensify thirst; dehydration also causes the discomfort of a hangover.

Hormones from the adrenal glands also act on the kidneys to regulate blood pressure (see figure 25.8). One factor that determines blood pressure is the volume of blood in the body. When blood pressure dips too low, the adrenal cortex releases the steroid hormone **aldosterone.** This hormone stimulates the production of sodium channels in the nephron. Na^+ ions move from the nephron into the blood, and water follows by osmosis. The volume of blood goes up, and blood pressure rises. ⓘ *adrenal glands,* section 25.4C

28.9 Mastering Concepts

1. Trace the path of blood as it moves through a kidney.
2. What is a nephron?
3. What three processes occur in urine formation?
4. What is the function of the collecting duct?
5. Describe the roles of antidiuretic hormone and aldosterone in regulating kidney function.

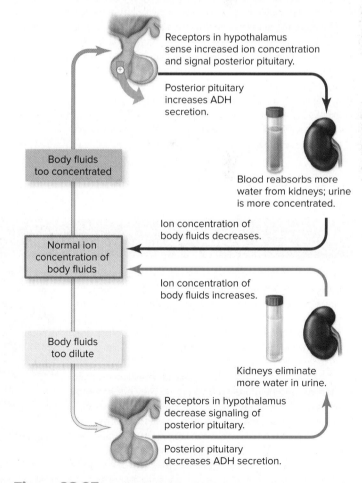

Receptors in hypothalamus sense increased ion concentration and signal posterior pituitary.

Posterior pituitary increases ADH secretion.

Body fluids too concentrated

Blood reabsorbs more water from kidneys; urine is more concentrated.

Ion concentration of body fluids decreases.

Normal ion concentration of body fluids

Ion concentration of body fluids increases.

Body fluids too dilute

Kidneys eliminate more water in urine.

Receptors in hypothalamus decrease signaling of posterior pituitary.

Posterior pituitary decreases ADH secretion.

Figure 28.25 Osmoregulation in Humans. A feedback loop regulates the amount of water that blood reabsorbs from the kidneys.

Investigating Life 28.1 | The Cost of a Sweet Tooth

Most children know that eating candy causes tooth decay. Sweets fuel the growth of oral bacteria, which release tooth-eroding acids as they metabolize sugar. The result? Cavities. Bacteria on teeth also form plaque, which hardens into tartar that inflames the gums and can ultimately lead to tooth loss.

The human diet has not always been as sugary as it is now. Thousands of years ago, in Mesolithic (mid-Stone Age) cultures, people were hunters and gatherers. Sweets would have been a rare luxury. Neolithic people were the first farmers. They domesticated wheat, barley, and other grains, triggering a major dietary shift toward starch. Human diets remained stable until the Industrial Revolution (about 150 years ago), when refined sugars became commonplace.

How did the bacteria colonizing the mouth change over this time? To answer this question, researchers scraped tartar from the teeth of 34 skeletons from Mesolithic, Neolithic, Bronze Age, and Medieval times; they also collected tartar from living people. The researchers then sequenced the DNA from bacteria in the tartar and searched for DNA specific to two species known to cause tooth decay. ⓘ *DNA sequencing,* section 11.2B

The diversity of bacteria turned out to be lower in today's mouths than it was in earlier times. And the species associated with tooth decay? One of them has remained relatively steady since Mesolithic times, but the other has exploded along with the modern availability of sugar (figure 28.B).

This research documents an ecological shift in the bacterial communities occupying the human mouth. But it also raises evolutionary questions. Why hasn't natural selection optimized the human oral ecosystem? That is, why can't our bodies fight off the bacteria that cause tooth decay? One explanation is that our diet has shifted faster than our defenses have evolved. Similarly, the super-clean, germ-free homes of the developed world may have reduced our exposure to pathogens faster than our immune systems could adjust, leading to today's high rates of asthma and allergies (see Investigating Life 29.1). In both cases, it seems that our rapid cultural shifts have outstripped the pace of evolution. ⓘ *ecological succession,* section 19.5

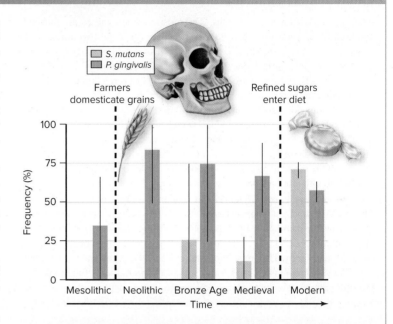

Figure 28.B **The Changing Oral Ecosystem.** Researchers collected tartar from the teeth of modern humans and of skeletons ranging from 7500 years old (Mesolithic) to 400 years old (Medieval). The samples revealed that *Streptococcus mutans,* a bacterium associated with tooth decay, has become more common as the human diet has become more sugary. The incidence of another bacterium, *Porphyromonas gingivalis,* has remained stable. (Error bars are a measure of uncertainty in the data; see appendix B.)

Source: Adler, Christina J. and 11 coauthors, including Alan Cooper. 2013. Sequencing ancient calcified dental plaque shows changes in oral microbiota with dietary shifts of the Neolithic and Industrial revolutions. *Nature Genetics,* vol. 45, pages 450–455.

What's the Point? ▼ APPLIED

Food and water are among the familiar necessities of life. Digestive organs break down foods and absorb nutrients, which the body uses to generate energy and build its tissues. The small and large intestines absorb most of the nutrients and water in food. The urinary system eliminates excess water and ions, along with metabolic wastes produced when cells use food.

Trillions of microbes participate in digestion, produce vitamins, and help break down drugs. They are essential for survival throughout life. As mentioned in section 28.6, babies are born with microbe-free intestines, but bacteria soon move in. An infant's weak immune system may be an adaptation that helps speed this crucial colonization process.

When it comes to the gut, not just any microbes will do; the optimal mix of species is vital. In fact, certain digestive problems can be treated by simply changing the intestinal community. For example, yogurt and other foods may include probiotics, which are live microbes that may benefit the digestive tract. And for patients with stubborn intestinal problems, physicians sometimes order a "fecal transplant." That is, they transplant feces from a

Probiotic bacteria from yogurt

SEM (false color) ⊢—⊣ 4 μm

©Steve Gschmeissner/Science Photo Library/Corbis RF

healthy donor into the patient's intestines. This simple procedure can save lives by restoring a healthy microbial community.

The gut's microbes may be linked to disorders, too. For example, autistic children harbor types of bacteria that are not present in non-autistic children; it is not yet known whether autism promotes a shift in intestinal microbes or if the microbes cause autism. Moreover, some microscopic gut inhabitants may increase the risk of obesity. Intestinal bacteria may even form a link between red meat and heart attack risk in humans. The microbes inside frequent beef-eaters—but not vegetarians—react to a meaty diet by releasing chemicals that can cause heart disease.

Unlike the digestive system, the urinary tract is normally microbe free. The bacteria that cause tuberculosis in cattle, however, may be used to treat certain forms of bladder cancer. A solution containing live bacteria is placed into the bladder. No one knows how it works, but in many patients, the microbes somehow stimulate the immune system to eliminate the cancer cells from the bladder lining.

CHAPTER SUMMARY

28.1 Animals Maintain Nutrient, Water, and Ion Balance

• The digestive system helps an animal maintain homeostasis by providing energy and essential nutrients; the urinary system disposes of metabolic wastes and helps maintain the composition of body fluids. An animal's habitat selects for variations on these adaptations.

28.2 Digestive Systems Derive Energy and Raw Materials from Food

• The **digestive system** acquires and breaks down food.
• Plants and algae produce their own food and are therefore **autotrophs,** whereas animals are **heterotrophs** that consume food. **Nutrients** in food provide energy and raw materials needed for growth and maintenance.
• An animal's metabolic rate determines its need for food. Metabolic rate, in turn, reflects the animal's body temperature regulation, body size, and physiological state.

28.3 A Varied Diet Is Essential to Good Health

• Metabolism, growth, maintenance, and repair of body tissues all require nutrients from food. **Macronutrients** include carbohydrates, proteins, fats, and water, whereas **micronutrients** include vitamins and minerals.

28.4 Body Weight Reflects Food Intake and Activity Level

A. Body Mass Index Can Identify Weight Problems

• **Calories** measure the energy stored in food. Fat has more Calories per gram than either carbohydrate or protein.

B. Starvation: Too Few Calories to Meet the Body's Needs

• If a person does not eat enough over a long period, the body uses reserves of fat and protein to fuel essential processes. **Anorexia nervosa** and **bulimia** are eating disorders that may reduce Calorie intake to dangerously low levels.

C. Obesity: More Calories Than the Body Needs

• A person who eats more Calories than he or she expends will gain weight. **Obesity** is associated with many health problems.

28.5 Most Animals Have a Specialized Digestive Tract

• **Herbivores** eat plants, **carnivores** eat meat, **detritivores** consume decaying organic matter, and **omnivores** have a varied diet.

A. Acquiring Nutrients Requires Several Steps

• Food is **ingested, digested,** and **absorbed** into the bloodstream and then indigestible wastes are **eliminated** as **feces** (figure 28.26).
• In chemical digestion, enzymes dismantle large molecules into their smaller subunits.

B. Digestive Tracts May Be Incomplete or Complete

• Sponges have intracellular digestion. Their cells engulf food and digest it in food vacuoles.
• Other animals use extracellular digestion, which occurs in a cavity within the body. An **incomplete digestive tract** (also called a **gastrovascular cavity**) has one opening. A **complete digestive tract,** or **alimentary canal (gastrointestinal tract),** has two openings. Food enters through the **mouth** and is digested and absorbed; undigested material leaves through the **anus.**

C. Diet Influences Digestive Tract Structure

• The length of the digestive tract and size of the **cecum** are adaptations to specific diets. In **ruminants,** bacteria inhabiting the rumen help break down hard-to-digest plant matter.

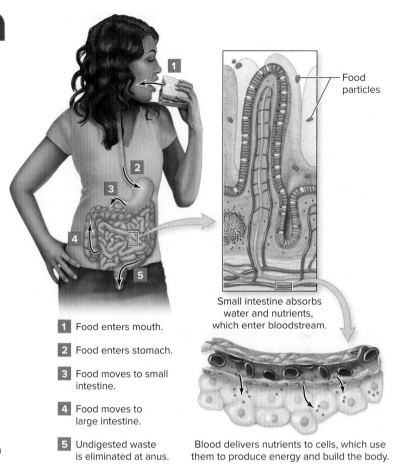

1 Food enters mouth.

2 Food enters stomach.

3 Food moves to small intestine.

4 Food moves to large intestine.

5 Undigested waste is eliminated at anus.

Small intestine absorbs water and nutrients, which enter bloodstream.

Food particles

Blood delivers nutrients to cells, which use them to produce energy and build the body.

Figure 28.26 The Digestive System: A Summary.

28.6 The Human Digestive System Consists of Several Organs

A. Muscles Underlie the Digestive Tract

• Waves of contraction called **peristalsis** move food along the digestive tract. Muscular **sphincters** control movement from one compartment to another.

B. Digestion Begins in the Mouth

• In the mouth, **teeth** break food into small pieces. Salivary glands produce saliva, which moistens food and begins starch digestion.
• With the help of the **tongue,** swallowed food moves past the **pharynx** and through the **esophagus** to the stomach. The **epiglottis** prevents food from entering the trachea.

C. The Stomach Stores, Digests, and Churns Food

• The **stomach** stores food, mixes it with **gastric juice,** and churns it into liquefied **chyme.** Hydrochloric acid in the gastric juice kills most microorganisms and denatures proteins. The protein-splitting enzyme **pepsin** begins protein digestion.

D. The Small Intestine Digests and Absorbs Nutrients

• The **small intestine** is the main site of digestion and nutrient absorption. **Intestinal villi** absorb the products of digestion; **microvilli** on each villus provide tremendous surface area.
• The **pancreas** supplies bicarbonate to the small intestine, along with digestive enzymes that break down carbohydrates, polypeptides, lipids, and nucleic acids.
• The **liver** produces **bile,** which emulsifies fat; the **gallbladder** stores the bile and releases it to the small intestine.

E. The Large Intestine Completes Nutrient and Water Absorption

- Material remaining after absorption in the small intestine passes to the **large intestine,** which absorbs water, minerals, and ions. Bacteria digest the remaining nutrients and produce useful vitamins that are then absorbed. Feces exit the body through the anus.

28.7 Animals Eliminate Nitrogenous Wastes and Regulate Water and Ions

- Animals **excrete** metabolic wastes. In mammals, the nitrogenous waste called **urea** is a byproduct of protein breakdown.
- **Osmoregulation** is the control of ion concentrations in body fluids. Depending on its habitat, an animal may need to conserve or eliminate water and ions.

28.8 The Urinary System Produces, Stores, and Eliminates Urine

- The **urinary system** excretes nitrogenous wastes (mostly urea) and regulates water and electrolyte levels.
- The **kidneys** produce **urine.** Each kidney drains into a **ureter,** which delivers urine to the **urinary bladder.** The **urethra** opens near the vagina (females) or extends along the **penis** (males).

28.9 Nephrons Remove Wastes and Adjust the Composition of Blood

A. Nephrons Interact Closely with Blood Vessels

- **Nephrons** are the functional units of the kidney; they eliminate wastes while returning valuable substances to blood.
- Each tubular nephron receives blood from capillaries originating at the renal artery. The renal vein carries cleansed blood away from the kidney.

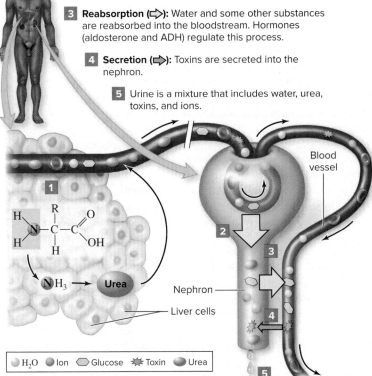

1. Liver cells metabolize amino acids and release urea into the bloodstream.

2. **Filtration (⇨):** Urea, glucose, ions, and other solutes are filtered from the blood at the nephrons.

3. **Reabsorption (⇨):** Water and some other substances are reabsorbed into the bloodstream. Hormones (aldosterone and ADH) regulate this process.

4. **Secretion (⇨):** Toxins are secreted into the nephron.

5. Urine is a mixture that includes water, urea, toxins, and ions.

Blood vessel

Nephron

Liver cells

● H_2O ● Ion ⬡ Glucose ✳ Toxin ● Urea

Figure 28.27 Nephron Function: A Summary.

- The two main regions of a nephron are the filter and a winding tubule. Waste fluid moves from the nephron into a **collecting duct.**

B. Urine Formation Includes Filtration, Reabsorption, and Secretion

- **Figure 28.27** summarizes the function of a nephron.
- Blood **filtration** occurs at the entrance to a nephron. Blood pressure forces some components of blood through the filter into the nephron.
- Along a nephron's tubule, capillaries **reabsorb** useful components such as glucose, amino acids, ions, and water. Other adjustments maintain blood pH.
- Additional wastes are **secreted** from blood into the filtrate in each nephron.
- Urine is the fluid that nephrons release into the kidney's collecting ducts.

C. Hormones Regulate Kidney Function

- **Antidiuretic hormone (ADH),** secreted by the posterior pituitary gland, increases the amount of water reabsorbed into the bloodstream and therefore causes urine to be more concentrated.
- The adrenal cortex releases **aldosterone** in response to low blood pressure. Aldosterone causes additional Na^+ to be reabsorbed into the blood; water follows by osmosis, raising blood volume (and blood pressure).

MULTIPLE CHOICE QUESTIONS

1. Which of the following animals uses the most energy per gram of body weight?
   a. Horse   b. Eagle   c. Mouse   d. Python

2. A person's body mass index is calculated based on
   a. caloric intake.   c. weight.
   b. height.   d. Both b and c are correct.

3. At which stage do nutrients enter an organism's bloodstream?
   a. Ingestion   b. Digestion   c. Absorption   d. Elimination

4. Which digestive organ uses both mechanical and chemical digestion?
   a. Mouth   b. Stomach   c. Intestines   d. Both a and b

5. If a person is hanging upside down, food can still move along the esophagus to the stomach, thanks to
   a. microvilli.   b. chyme.   c. the epiglottis.   d. peristalsis.

6. The protein you eat is mostly
   a. incorporated into your body without modification.
   b. eliminated in feces.
   c. converted to fat before digestion.
   d. dismantled into individual amino acids.

7. Complete this analogy: Urine is to _____ as feces is to undigested food.
   a. unabsorbed drinks   c. old blood cells
   b. metabolic wastes   d. All of the choices are correct.

8. Which of the following functions of the kidney is correctly paired with its description?
   a. Filtration: Fluid containing solutes moves into the nephron.
   b. Reabsorption: Water returns to the blood.
   c. Secretion: Hydrogen ions move into the nephron.
   d. All of these are correctly paired.

9. Which of the following is NOT reabsorbed into the bloodstream from the nephron?
   a. Glucose   b. Urea   c. Water   d. Amino acids

10. You may feel the urge to urinate after drinking an alcoholic beverage because
    a. alcohol travels quickly through the digestive tract.
    b. the digestive tract does not absorb alcohol.
    c. alcohol is easily reabsorbed at nephrons.
    d. alcohol decreases the reabsorption of water at nephrons.

Answers to Multiple Choice questions are in appendix A.

WRITE IT OUT

1. What are the two main functions of food in an animal's body?

2. Write down the foods you ate today, and use nutrition labels to determine how many Calories you consumed. Then, use calculators on the Internet to estimate how many Calories you require each day based on your age, weight, sex, and activity level. Did you consume more or fewer Calories than you needed? If you repeated these eating habits for several weeks, would you gain or lose weight?

3. Compare and contrast the digestive systems of a whale and a sponge.

4. Identify a part of the digestive system that includes the following: duodenum; cecum, appendix, rectum, and anus; villi and microvilli.

5. Trace the movement of food in the digestive tract from mouth to anus.

6. What are the digestive products of carbohydrates, proteins, and fats?

7. Many children believe that a piece of swallowed chewing gum will remain in the body for 7 years. Chewing gum is made of an indigestible polymer that does not dissolve in water. Since the gum cannot be digested, what happens to it after it is swallowed? Given your answer, does the 7-year timescale make sense? Propose an alternative prediction for how long it might take instead, and explain your reasoning.

8. Imagine you are adrift at sea. If you drink seawater, you will dehydrate much faster than if you have access to fresh water. Explain.

9. Shortly after you drink a large glass of water, you will feel the urge to urinate. Explain this observation. Begin by tracing the path of the water, starting at the stomach and ending with the arrival of urine in the bladder.

10. Diuretic drugs prevent ions from being reabsorbed from the nephron. Explain why a diuretic drug might lower blood pressure. Why do some patients have low levels of potassium and sodium in their blood while taking a diuretic?

11. Review the action of steroid and peptide hormones in chapter 25. Which hormone should act faster, ADH or aldosterone? Why?

12. In a disease called diabetes insipidus, ADH activity is insufficient. Would a person with this disease produce more or less urine than normal? Explain.

13. In the project described in Investigating Life 28.1, the researchers sequenced bacterial DNA in hardened plaque on the teeth of decaying skeletons. They also analyzed bacterial DNA from *inside* the teeth and from soil and water near the lab where the work was carried out. Why were these additional DNA samples important in interpreting the results?

SCIENTIFIC LITERACY

1. Fructose and glucose are both monosaccharides, but these sugars affect the body differently. For example, glucose stimulates insulin release from the pancreas (see chapter 25); fructose does not. Moreover, insulin stimulates leptin release. Use this information to propose an explanation for the correlation between the skyrocketing consumption of high fructose corn syrup since 1970 and the rise in obesity during the same period.

2. Review Burning Question 28.1. Use the Internet to learn more about the term *fad diet*. How can you distinguish a fad diet from a healthy, lifelong change in your approach to food?

Answers to Mastering Concepts, Write It Out, Scientific Literacy, and Pull It Together questions can be found in the Connect ebook.
connect.mheducation.com

PULL IT TOGETHER

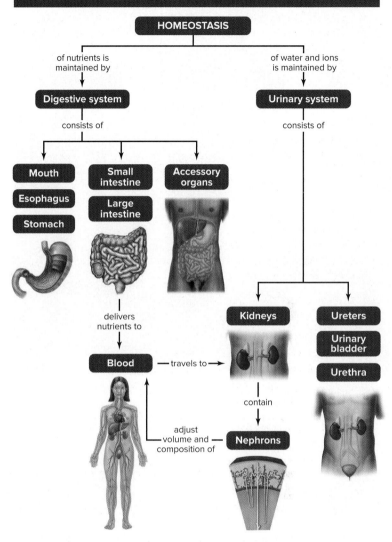

Figure 28.28 Pull It Together: The Digestive and Urinary Systems.

Refer to figure 28.28 and the chapter content to answer the following questions.

1. Review the Survey the Landscape figure in the chapter introduction and the Pull It Together concept map. How do the digestive system and the urinary system rely on the circulatory system to maintain homeostasis?

2. Add the terms *ingestion, digestion, absorption, elimination, chyme, bacteria,* and *peristalsis* to this concept map.

3. What are the accessory organs required for digestion? Add them to this concept map. What is the function of each?

4. Write a phrase that connects *liver* to *gallbladder* in this concept map. How would removing the gallbladder affect the digestion of fats?

5. Add *water, ions, toxins,* and *urine* to this concept map.

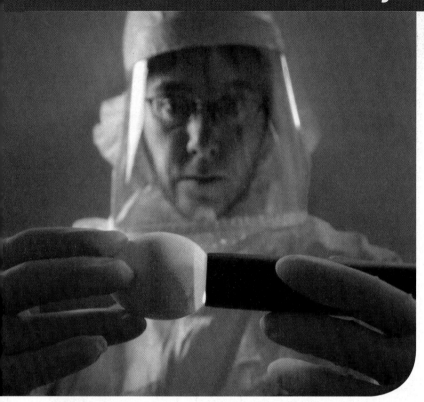

LEARNING OUTLINE

29.1 Many Cells, Tissues, and Organs Defend the Body

29.2 Innate Defenses Are Nonspecific and Act Early

29.3 Adaptive Immunity Defends Against Specific Pathogens

29.4 Vaccines Jump-Start Immunity

29.5 Several Disorders Affect the Immune System

APPLICATIONS

Why We Care 29.1 *Severe Burns*

Why We Care 29.2 *Protecting a Fetus from Immune Attack*

Burning Question 29.1 *Why do we need multiple doses of some vaccines?*

Burning Question 29.2 *Can people be allergic to meat?*

Investigating Life 29.1 *The Hidden Cost of Hygiene*

Vaccines from Eggs. Influenza viruses replicate in fertilized chicken eggs. This microbiologist is using a bright light to check an egg in a laboratory that produces the flu vaccine. The vaccine prevents illness by "teaching" the immune system to recognize several flu viruses.

Source: CDC/James Gathany

Learn How to Learn
Practice Your Recall

Here's an old-fashioned study tip that still works. When you finish reading a passage, close the book and write what you remember—in your own words. In this chapter, for example, you will learn about the parts of the human immune system. After you read about them, can you list and describe them without peeking at your book? Try it and find out!

SURVEY THE LANDSCAPE
Animal Anatomy and Physiology

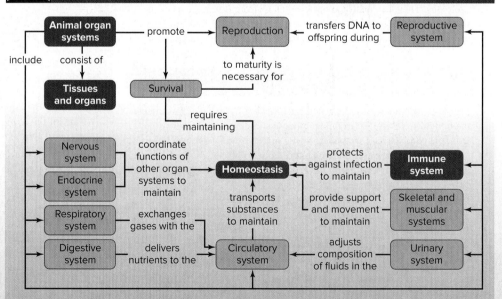

The environment teems with microbes and other organisms that compete for access to the resources inside an animal's body. The body's defenses, including the immune system, fight invading pathogens in many ways.

For more details, study the Pull It Together feature in the chapter summary.

What's the **Point?** ▼

©Image Source RF

The immune system is unusual. Unlike most other body systems, it does not feature prominent organs such as the heart, muscles, or brain. Instead, it consists of a widespread collection of cells and chemicals, and we are usually unaware of its activities.

The immune system's main job is to prevent illness. Its cells quietly patrol your body's tissues, attacking virus-infected cells, cancerous cells, and anything that does not belong to your body. What's more, the immune system "remembers" the invaders that it encountered in the past.

Most of the time, we only become aware of the immune system when something goes wrong. Its cells and chemicals may overreact and attack the body's own tissues; they may launch an allergic reaction; or they may not function at all. This chapter describes the activities and disorders of the human immune system.

29.1 Many Cells, Tissues, and Organs Defend the Body

Disease-causing agents—**pathogens**—are nearly everywhere. Viruses, bacteria, protists, fungi, and worms are in water, food, soil, and air. These pathogens can enter our bodies whenever we eat, drink, breathe, or interact with people and other animals. Yet we are not constantly sick. The explanation is that the **immune system** enables the body to recognize its own cells and to defend against infections, cancer, and foreign substances. The vertebrate immune system consists of organs, scattered cells, defensive chemicals, and fluids that permeate the body; this section introduces them.

A. White Blood Cells Play Major Roles in the Immune System

Blood is critical to immune function. Plasma, the liquid matrix of blood, carries defensive proteins called antibodies. In addition, infection-fighting **white blood cells** are suspended in blood plasma and occupy the interstitial fluid between cells. Stem cells in **red bone marrow,** the spongy tissue inside bones, give rise to white blood cells. ⓘ *bone marrow,* section 26.3A; *blood,* section 27.1

White blood cells play many roles in the body's defenses (figure 29.1). About 75% of white blood cells function primarily as **phagocytes,** which are scavenger cells that engulf and destroy bacteria and debris. **Macrophages** form one important class of phagocytes. Some types of macrophages wander throughout the body; others remain in just one tissue. As described in sections 29.2 and 29.3, macrophages that consume foreign particles play important roles in initiating the body's defenses. ⓘ *phagocytosis,* section 4.5C

The remaining white blood cells are mostly **lymphocytes,** which include several cell types. **B cells** are lymphocytes that mature in red bone marrow, then migrate to the blood and other tissues. Lymphocytes called **T cells** also originate in red bone marrow but mature in the **thymus,** a small immune organ in the chest ("T" is for thymus). From there, T cells migrate throughout the body.

Figure 29.1 **White Blood Cells.** (a) Human blood contains three main classes of white blood cells. (b) A phagocyte (purple) engulfs a yeast cell (red).
(b): ©SPL/Science Source

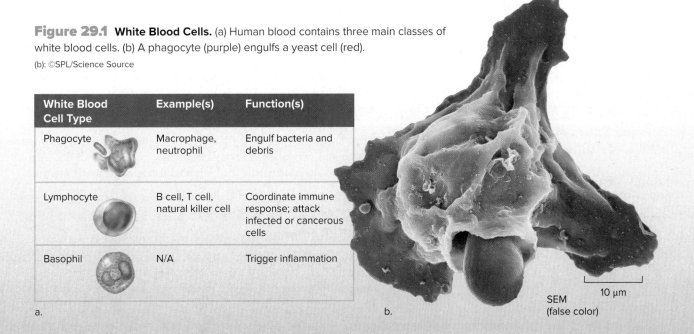

White Blood Cell Type	Example(s)	Function(s)
Phagocyte	Macrophage, neutrophil	Engulf bacteria and debris
Lymphocyte	B cell, T cell, natural killer cell	Coordinate immune response; attack infected or cancerous cells
Basophil	N/A	Trigger inflammation

a.

b.

SEM
(false color)

10 μm

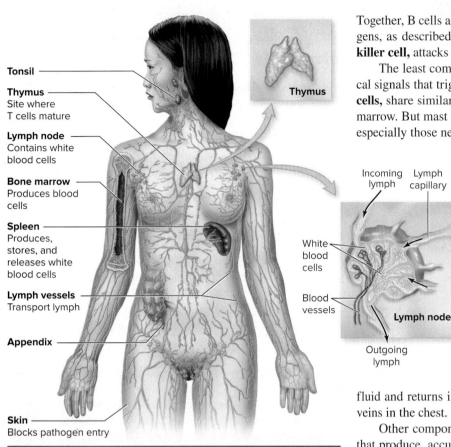

Tonsil

Thymus
Site where
T cells mature

Lymph node
Contains white
blood cells

Bone marrow
Produces blood
cells

Spleen
Produces,
stores, and
releases white
blood cells

Lymph vessels
Transport lymph

Appendix

Skin
Blocks pathogen entry

Thymus

Incoming   Lymph
lymph   capillary

White
blood
cells

Blood
vessels

Lymph node

Outgoing
lymph

Lymphatic and Immune Systems

Main tissue types*	Examples of locations/functions
Epithelial	Thymus, spleen, and tonsils consist partly of epithelial tissue; lines lymphatic and blood vessels; lymphoid tissue lies beneath epithelial tissues of the digestive, respiratory, and urinary tracts, guarding potential points of entry for pathogens.
Connective	Immune system cells and chemicals circulate in blood and lymph, which are connective tissues; bone marrow is connective tissue that produces lymphocytes; thymus, spleen, and lymph nodes consist partly of connective tissue.

*See chapter 23 for descriptions.

Figure 29.2   The Human Lymphatic and Immune Systems.

Together, B cells and T cells coordinate the body's responses to specific pathogens, as described in section 29.3. Another type of lymphocyte, the **natural killer cell,** attacks cancerous or virus-infected cells.

The least common white blood cells are **basophils,** which release chemical signals that trigger inflammation and allergies. Their close relatives, **mast cells,** share similar functions. Like basophils, mast cells originate in red bone marrow. But mast cells do not circulate in blood. Rather, they settle in tissues, especially those near the skin, digestive tract, and respiratory system.

B. The Lymphatic System Produces and Transports Many Immune System Cells

The **lymphatic system** is the part of the immune system that collects fluid that leaks from blood vessels, removes bacteria, debris, and cancer cells, and returns the liquid to the blood (figure 29.2). **Lymph,** the colorless fluid of the lymphatic system, originates as plasma that seeps out of blood vessels and into the interstitial fluid. Lymph capillaries throughout the body absorb this fluid, along with bacteria, viruses, white blood cells, cancer cells, and other large particles. The lymphatic system "recycles" this fluid and returns it to the bloodstream at large lymph vessels that empty into veins in the chest.

Other components of the lymphatic system include the lymphoid organs that produce, accumulate, or aid in the circulation of lymphocytes. Red bone marrow and the thymus are two examples of lymphoid organs. Another is the **spleen,** a large organ containing masses of lymphocytes and macrophages that destroy pathogens in the blood.

A **lymph node** is one of the many small, bean-shaped lymphoid organs located along the lymph vessels. Inside each lymph node, millions of white blood cells engulf dead cells and pathogens circulating in lymph. Lymph nodes also release B and T cells to lymph, which carries them to the blood. When you have an infection, lymph nodes in the neck, armpits, or groin may become swollen and tender as they accumulate extra white blood cells. Many people call these enlarged lymph nodes "swollen glands."

The lymphatic system can also carry cancer cells that break off of tumors, seeding new tumors elsewhere in the body. A biopsy of cancerous tissue therefore often includes a sample of nearby lymph nodes. If these lymph nodes are cancer-free, abnormal cells may not have begun to invade other tissues, improving the chance of successful treatment. ⓘ *cancer,* section 8.6

In addition to the major lymphoid organs already described, scattered concentrations of lymphoid tissues also guard the mucous membranes where pathogens may enter the body. Examples include the tonsils (near the throat), the appendix, and patches of lymphoid tissue in the small intestine.

C. The Immune System Has Two Main Subdivisions

Biologists divide the human immune system into two parts: innate defenses and adaptive immunity (figure 29.3). Together, innate defenses and adaptive immunity interact in highly coordinated ways to defend the body against pathogens.

Innate defenses provide a broad defense against any infectious agent. *Innate* refers to the fact that these defenses are always present and ready to function, as described in section 29.2. In **adaptive immunity,** the body's immune cells not only recognize specific parts of a pathogen, but they also "remember" previous encounters. Section 29.3 describes adaptive immunity in detail.

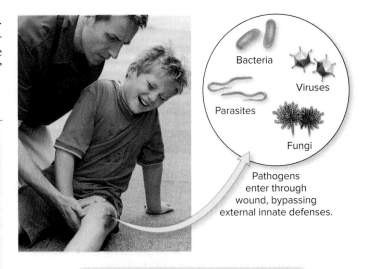

Pathogens enter through wound, bypassing external innate defenses.

29.1 Mastering Concepts

1. List the cell types that participate in the body's defenses, along with some of their functions.
2. How does the immune system interact with the circulatory system?
3. List and describe the components of the lymphatic system.
4. What are the two subdivisions of the immune system?

29.2 Innate Defenses Are Nonspecific and Act Early

The body's innate defenses include many external and internal components. This arm of the immune system is called "nonspecific" because it acts against any type of invader.

A. External Barriers Form the First Line of Defense

Physical barriers block pathogens and foreign substances from entering the body. Unpunctured skin is the most extensive and obvious wall (see Why We Care 29.1). Human skin has two major layers (figure 29.4): the epidermis and the dermis. The **epidermis** is the outermost layer, and it consists partly of dry, dead cells that help keep invaders outside the body. Below the epidermis is the **dermis,** which houses nerve endings, sweat and oil glands, and the blood vessels that nourish both skin layers.

Skin has many protective functions. For example, calluses are thick, scaly accumulations that protect skin from disease-causing organisms and abrasions. The skin's color derives from melanin, a pigment that absorbs ultraviolet radiation and therefore protects against some types of skin cancer.

The body's other physical and chemical barriers include mucus that traps inhaled dust particles in the nose; wax in the ears; tears that wash irritants from the eyes and contain antimicrobial substances; and cilia that sweep bacteria out of the respiratory system. In addition, a bath of strong acid destroys most microbes that reach the stomach. ⓘ *stomach, section 28.6C*

The immune system: two main components

Innate defenses (external and internal)	Adaptive immune response
• Prevent entry of pathogens (external) or attack pathogens that enter the body (internal) • Always active; immediate response • Nonspecific	• Delayed response • Response is strongest and fastest for previously encountered pathogens • Specific to particular pathogens

Figure 29.3 Overview of Body Defenses. The immune system's two arms protect the body against disease. Innate defenses are nonspecific but always active, whereas the adaptive immune response creates a "memory" of specific pathogens.

Photo: ©Digital Vision RF

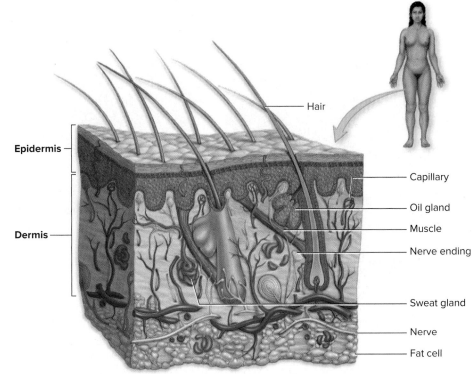

Figure 29.4 Human Skin. Flattened, dead cells make up the outer layer of the epidermis, the main physical barrier against pathogens.

Why We Care 29.1 | Severe Burns

Intact skin is the body's first line of defense against infection. Severe burns from fires, hot liquids, or corrosive chemicals, however, can damage the dermis, giving bacteria and other pathogens direct access to the underlying tissues. As a result, burn patients face a high risk of infection. Aggressive treatment with antibiotics can help prevent this life-threatening outcome.

©Chris Cheadle/All Canada Photos/ Getty Images

Burn patients also confront a second risk: dehydration. Intact skin is waterproof and therefore helps keep underlying tissues moist. If burns are extensive, body fluids evaporate rapidly through the wounds. Patients with large burns therefore receive intravenous fluids that compensate for the loss and help maintain blood volume.

An often underappreciated component of this first line of defense is the body's normal microbiota. Resident microbes on the skin, in the gut, and elsewhere help prevent colonization by pathogens. ⓘ *beneficial microbes,* section 15.2D

B. Internal Innate Defenses Destroy Invaders

A large collection of defensive cells and molecules awaits any microbe that manages to breach the body's external barriers.

White Blood Cells White blood cells play many roles in the body's innate defenses, consuming bacteria by phagocytosis or secreting substances that destroy pathogens. Basophils and mast cells provoke inflammation, attracting additional white blood cells. Natural killer cells destroy cancerous or virus-infected cells. Meanwhile, macrophages and other phagocytes consume pathogens and promote fever. And, as described in section 29.3, macrophages also play a critical role in activating the body's adaptive immune response.

Inflammation Inflammation is an immediate, localized reaction to an injury or to any pathogen that breaches the body's barriers. The area surrounding the wound or infection site becomes red, warm, swollen, and painful. Overall, this nonspecific defense recruits immune system cells, helps clear debris, and creates an environment hostile to microorganisms.

Figure 29.5 illustrates the events of inflammation in response to a minor injury from a splinter. Damaged cells release substances that provoke basophils and mast cells in the skin's dermis to release **histamine,** a chemical that dilates (widens) blood vessels and causes them to become "leakier"—that is, more permeable to fluids and white blood cells.

As capillaries near the injury become dilated, additional blood arrives, turning the area warm and red. Blood plasma, which carries antimicrobial substances, leaks out of the blood vessels. This fluid dilutes the toxins secreted by bacteria and causes localized swelling. Pressure on the swollen tissues, coupled with chemical signals released from the injured cells, stimulates pain receptors in the skin.

Figure 29.5 Inflammation. Immediately after a splinter pierces the skin, chemicals released from damaged cells trigger the inflammatory response.

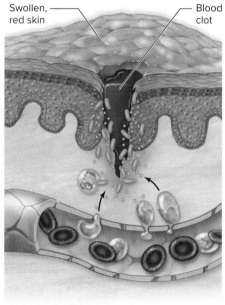

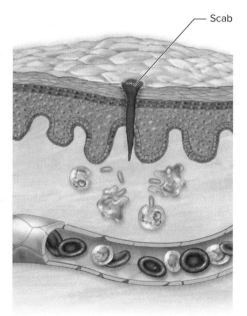

1 Damaged cells trigger mast cells and basophils to release histamine.

2 Histamine causes blood vessels to dilate and become more permeable. White blood cells move into the damaged area.

3 White blood cells engulf and destroy bacteria and damaged cells.

Meanwhile, macrophages and other phagocytes squeeze through the blood vessel walls and move into the area, engulfing and destroying bacteria and damaged cells. Pus may accumulate; this whitish fluid contains white blood cells, bacteria, and debris from dead cells.

In medical terminology, the suffix *-itis* indicates inflammation. For example, dermatitis (a rash) signifies inflamed skin, often resulting from contact with an irritant such as poison ivy. Appendicitis is inflammation of the appendix, usually caused by bacterial infection. Aspirin and ibuprofen reduce pain and swelling by blocking the enzymes required for inflammation to occur.

Inflammation may be acute or chronic. Acute inflammation is an adaptation that prevents infection after an injury. The effects usually last only a few days or less, as illustrated by the short-term discomfort of a minor burn or "paper cut." Chronic inflammation, on the other hand, is a prolonged response that may last for months or years. The persistent presence of pathogens or toxins can cause any tissue in the body to become chronically inflamed; genetic mutations may also play a role. Medical problems associated with chronic inflammation include rheumatoid arthritis, gum disease, diabetes, Alzheimer disease, celiac disease (gluten intolerance), and many other serious illnesses.

Complement Proteins and Cytokines Many antimicrobial biochemicals participate in the innate defenses. For example, **complement** proteins help to destroy pathogens in the body. When activated, some trigger a chain reaction that punctures bacterial cell membranes. Others cause mast cells to release histamine, and still others attract phagocytes.

Other chemical defenses include **cytokines,** messenger proteins that bind to immune cells and promote cell division, activate defenses, or otherwise alter their activity. For example, cells infected by viruses release interferons, which are cytokines that "sound an alarm" to alert other components of the immune system to the infection. White blood cells release interleukins, the largest group of cytokines. Their name comes from their role in communicating (*inter-*) between leukocytes, or white blood cells (*-leukins*).

Fever Cytokines travel throughout the body in the bloodstream. At the hypothalamus, they can trigger a temporary increase in the set point of the body's thermostat. **Fever,** a rise in the body's temperature, is therefore a common reaction to infection. Although the shivering and chills that accompany fever feel uncomfortable, a mild fever can help fight infection in several ways. A higher body temperature directly inhibits some bacteria and viruses. Fever also counters microbial growth indirectly because an elevated body temperature reduces the iron level in the blood. Bacteria and fungi require more iron as the temperature rises, so a fever stops the replication of these pathogens. Phagocytes also attack more vigorously when the temperature climbs. ⓘ *hypothalamus,* section 25.3

Figure 29.6 summarizes the innate defenses.

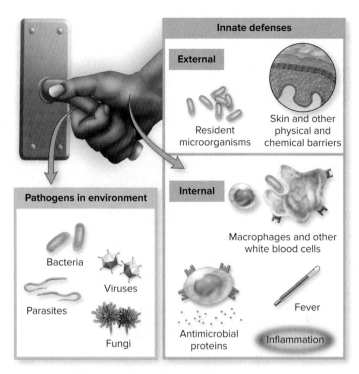

Figure 29.6 Overview of Innate Defenses. These nonspecific defenses prevent bacteria, viruses, and other pathogens from entering the body, or they attack those that do breach the physical and chemical barriers.

29.2 Mastering Concepts

1. List the main categories of innate defenses.
2. Describe the external barriers to infection.
3. Which white blood cells participate in innate defenses?
4. How can inflammation be both helpful and harmful?
5. How is fever protective?

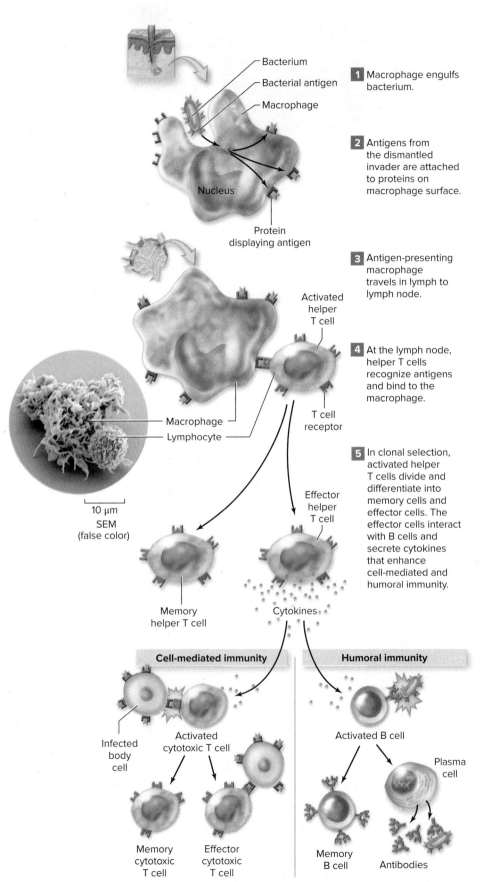

Bacterium

Bacterial antigen

Macrophage

1 Macrophage engulfs bacterium.

Nucleus

2 Antigens from the dismantled invader are attached to proteins on macrophage surface.

Protein displaying antigen

3 Antigen-presenting macrophage travels in lymph to lymph node.

Activated helper T cell

4 At the lymph node, helper T cells recognize antigens and bind to the macrophage.

Macrophage
Lymphocyte

T cell receptor

5 In clonal selection, activated helper T cells divide and differentiate into memory cells and effector cells. The effector cells interact with B cells and secrete cytokines that enhance cell-mediated and humoral immunity.

10 μm
SEM
(false color)

Effector helper T cell

Memory helper T cell

Cytokines

Cell-mediated immunity

Humoral immunity

Activated cytotoxic T cell

Activated B cell

Infected body cell

Plasma cell

Memory cytotoxic T cell

Effector cytotoxic T cell

Memory B cell

Antibodies

29.3 Adaptive Immunity Defends Against Specific Pathogens

The innate defenses described in section 29.2 are broad-spectrum weapons. Adaptive immune responses, on the other hand, act against individual targets. Two classes of lymphocytes, B cells and T cells, provide the ammunition in these precision defenses.

The target in an adaptive immune response is an **antigen,** which is any molecule that stimulates an immune reaction by B and T cells. Most antigens are carbohydrates or proteins. Examples include parts of a bacterial cell wall or virus, proteins on the surface of a mold spore or pollen grain, and unique molecules on the surface of a cancer cell.

The word *antigen* (short for *anti*body-*gen*erating) reflects a crucial part of adaptive immunity: the production of **antibodies,** which are Y-shaped proteins that recognize specific antigens. As described later in this section, each B and T cell is genetically programmed to produce receptors that recognize and bind to only one target antigen. But because every foreign particle contains dozens of molecules that can act as antigens, many sets of B cells and T cells respond to invasion by one pathogen.

Note that the cells of the immune system can respond to antigens from pathogens that are "floating" in the body's fluids, but they cannot see inside any cell. They can, however, respond to molecules on the surface of a cell. As you will soon see, the immune system therefore relies on specialized proteins that display antigens on cell surfaces.

A. Helper T Cells Play a Central Role in Adaptive Immunity

One of the first cell types to respond to infection is the macrophage, which both participates in innate defenses and triggers adaptive immunity. If a macrophage encounters a bacterium or other foreign substance in the body, it engulfs the invader, dismantles it, and links each antigen to a protein on the macrophage surface (figure 29.7, steps 1 and 2).

A macrophage displaying an antigen on its surface travels in lymph to a lymph node, where the cell encounters collections of T and B cells (figure 29.7, step 3). **Helper T cells** are "master cells" of the immune system because they initiate and coordinate the adaptive immune response.

Figure 29.7 Adaptive Immunity: A Summary.
A macrophage displays bacterial antigens on its surface. At a lymph node, a helper T cell binds to the macrophage. The activated helper T cell divides, producing effector T cells that help activate and enhance the cell-mediated and humoral immune responses.
Photo: ©Dr. Olivier Schwartz, Institute Pasteur/SPL/Science Source

When an antigen-presenting macrophage meets a helper T cell with receptors specific to the antigen it is displaying, the two cells bind (figure 29.7, step 4). In a process called **clonal selection,** the activated helper T cell immediately divides into many identical copies. Some of the copies differentiate into **memory cells,** which remain in the body long after the initial infection subsides. (As you will see, memory cells may be either T cells or B cells.) In general, memory cells launch a quick immune response upon subsequent exposure to the antigen; that is, they "remember" antigens the immune system has already encountered. The other cell copies are "effectors" that act immediately, initiating the cell-mediated and humoral arms of the adaptive immune system (step 5).

In **cell-mediated immunity,** defensive cells kill body cells that are defective or have already been infected by a pathogen. **Humoral immunity,** on the other hand, relies primarily on secreted antibodies (the term *humoral* refers to substances that circulate in body fluids). The rest of this section describes these defenses in more detail.

B. Cytotoxic T Cells Provide Cell-Mediated Immunity

Cytotoxic T cells provide cell-mediated immunity by physically binding to and destroying "suspicious" cells—that is, those that are cancerous, damaged, foreign to the body, or infected with viruses or bacteria. Activation of cell-mediated immunity requires a cytotoxic T cell to bind to a cell presenting an antigen (this requirement is one difference between cytotoxic T cells and the natural killer cells participating in innate immunity). Once activated, a cytotoxic T cell divides, and the resulting cells differentiate into memory cells and effector cells. Cytokines from helper T cells enhance the rate of cell division.

Figure 29.8 shows how an effector cytotoxic T cell kills a cancer cell. After binding to an antigen on the surface of the cancer cell, the cytotoxic T cell releases proteins that poke holes in the cancer cell's membrane. The cancer cell

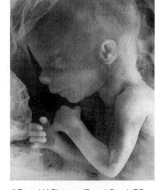

Why We Care 29.2 | Protecting a Fetus from Immune Attack

Since the immune system rejects "foreign" cells, it may seem surprising that a woman's body does not destroy her fetus. After all, the developing child is not genetically identical to its mother. In general, the female immune response dampens during pregnancy so that it doesn't reject the embryo and fetus. Full immune function returns after the woman gives birth.

One possible source of problems, however, traces to an antigen called the Rhesus (Rh) factor. Some people produce this ©Brand X Pictures/PunchStock RF protein on the surfaces of their red blood cells. A person can be Rh-positive (Rh$^+$) or Rh-negative (Rh$^-$). If your blood type is positive (such as "A positive"), the Rh antigen is present; if your blood is Rh-negative (such as "O negative"), your cells lack the Rh antigen. ⓘ *blood types,* section 10.6A

Suppose an Rh$^-$ woman becomes pregnant with an Rh$^+$ baby. When the baby is born, some of its cells may enter the mother's bloodstream. Her immune system therefore produces antibodies to the Rh antigen. In all subsequent Rh$^+$ pregnancies, these antibodies can cross the placenta and destroy the fetus's blood. A transfusion of Rh$^-$ blood at birth can save the newborn's life, but this is rarely necessary. Instead, women receive an injection of a drug, Rho(D) immune globulin, which prevents the immune response to the Rh antigen.

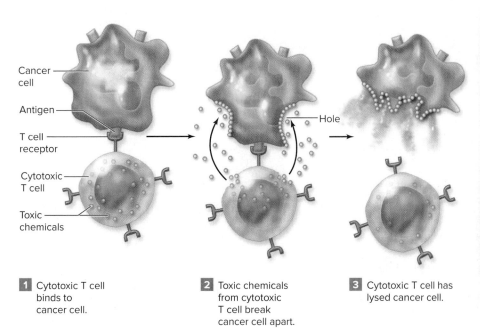

Cancer cell

Antigen

T cell receptor

Cytotoxic T cell

Toxic chemicals

Hole

1 Cytotoxic T cell binds to cancer cell.

2 Toxic chemicals from cytotoxic T cell break cancer cell apart.

3 Cytotoxic T cell has lysed cancer cell.

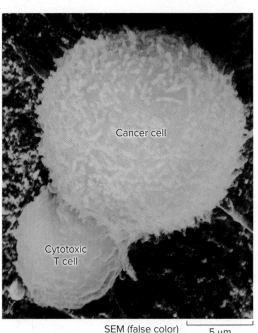

Cancer cell

Cytotoxic T cell

SEM (false color) 5 μm

©Dr. Andrejs Liepins/Science Source

Figure 29.8 Cytotoxic T Cells. (*1*) An effector cytotoxic T cell binds to a cancer cell and (*2*) secretes proteins that form holes in the cell membrane. (*3*) The cancer cell dies as its membrane disintegrates. The photo shows a small cytotoxic T cell attacking a large cancer cell.

Figure 29.9 **Antibody Structure.** The simplest antibody molecules consist of four polypeptide chains, two long ("heavy") and two short ("light"). Each chain has constant and variable regions; the variable portions form the antigen-binding sites.

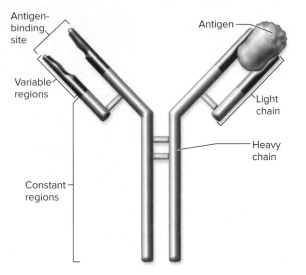

soon dies. By destroying the cell before it can replicate, cytotoxic T cells can stop a potential problem in its tracks. Moreover, the memory cells, which linger long after the infection subsides, differentiate immediately into cell-killing "machines" if the body encounters the same problem again.

Cell-mediated immunity is one factor that complicates organ transplants. The body perceives any foreign object, including a donated kidney, heart, or skin graft, as something to be destroyed. Cytotoxic T cells bind to and destroy the transplanted cells, provoking tissue rejection (see Burning Question 23.1). Immune-suppressing drugs can reduce the risk of rejection, but the cost is increased vulnerability to cancer and infectious disease (see section 29.5).

C. B Cells Direct the Humoral Immune Response

The humoral immune response includes millions of different B cells, each producing a unique antibody. Before learning how B cells operate, it is important to understand what antibodies are.

Antibodies Are Defensive Proteins Antibodies are the main weapons of humoral immunity. These large proteins circulate freely in blood plasma, lymph, and interstitial fluid. Their function is to attack pathogens in the body's fluids, not inside infected cells.

The simplest antibody molecule consists of four polypeptides: two identical light chains and two identical heavy chains (figure 29.9). Together, the four

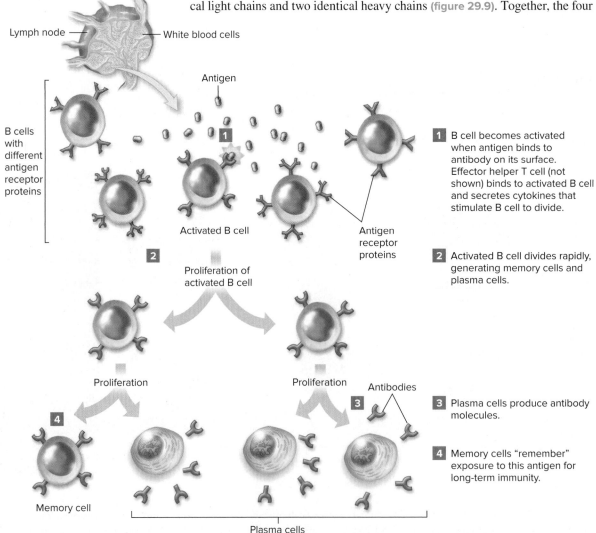

Figure 29.10 **Clonal Selection in Humoral Immunity.** The body produces B cells with many different receptors, but only the cell that binds an antigen proliferates; its descendants develop into memory cells or plasma cells.

1 B cell becomes activated when antigen binds to antibody on its surface. Effector helper T cell (not shown) binds to activated B cell and secretes cytokines that stimulate B cell to divide.

2 Activated B cell divides rapidly, generating memory cells and plasma cells.

3 Plasma cells produce antibody molecules.

4 Memory cells "remember" exposure to this antigen for long-term immunity.

chains form a shape like the letter Y. Each chain has constant and variable regions. The **constant regions** have amino acid sequences that are very similar in all antibody molecules, but the **variable regions** differ a great deal among antibodies. These variable regions determine the specific target antigen to which an antibody binds. ⓘ *protein shape,* section 2.5C

Antibodies are potent weapons that attack pathogens in many ways. The binding of an antibody to an antigen can inactivate a microbe or neutralize its toxins. Antibodies can cause pathogens to clump, making them more apparent to macrophages. They can coat viruses, preventing them from contacting target cells. Antibodies also activate complement proteins, which destroy microorganisms.

Activated B Cells Produce a Surge of Identical Antibodies
On the surface of each B cell is a receptor, which is a version of the antibody that the cell will produce. Until a B cell encounters the antigen it is genetically programmed to recognize, it remains dormant. But when an antigen binds to this surface antibody, the B cell begins to activate. An effector helper T cell binds to the B cell and secretes cytokines that stimulate cell division, completing the activation.

In another example of clonal selection, an activated B cell divides rapidly, generating an army of memory cells and plasma cells that are clones of the original B cell (figure 29.10). The **plasma cells** immediately secrete huge numbers of antibodies—thousands of molecules each second. Although the plasma cells vanish after the infection is over, the memory cells "remember" exposure to the antigen for years or decades.

Humoral Immunity Is Active or Passive
The humoral immune response is divided into two categories: active and passive (table 29.1).

Active immunity results from the body's own production of antibodies after exposure to antigens in the environment. A person who is bitten by a tick, for example, may begin producing antibodies against the bacteria that cause Lyme disease. Immunity following an illness is called "natural" active immunity. Vaccines stimulate "artificial" active immunity because they induce memory cells to form in the absence of illness.

In passive immunity, a person receives intact antibodies from another individual. For example, an infant acquires antibodies from its mother in breast milk. Administering antivenom to a victim of a snakebite also illustrates passive immunity.

Genetic Recombination Yields a Huge Variety of Antibodies and Antigen Receptors
Of the human genome's 25,000 or so genes, fewer than 250 encode proteins that specifically bind to antigens. How can one person's lymphocytes produce enough unique antibody proteins and antigen receptor proteins to defeat millions of potential pathogens? As it turns out, generating these diverse molecules is a little like using the limited number of words in a language to compose an infinite variety of stories.

The genes that encode antibodies contain hundreds of small DNA segments that are rearranged in developing lymphocytes. The result: countless lineages of cells that each produce a unique antigen receptor and antibody. Most lymphocytes will never encounter a pathogen with the corresponding antigen, but a few will. Thanks to genetic recombination, the immune system can respond to even newly emerging pathogens.

Producing an enormous assortment of antigen receptors poses a problem: Some of them will no doubt correspond to the body's own molecules. In a process called **clonal deletion,** lymphocytes that recognize the body's own cell

Miniglossary	Adaptive Immunity
Antigen	Molecule (such as a protein or carbohydrate on a pathogen's surface) that stimulates an immune system reaction
Antibody	Protein that binds to an antigen; an antibody binding to an invader may neutralize the pathogen in several ways
Helper T cell	Immune cell that coordinates and enhances the activity of B cells and cytotoxic T cells
Cell-mediated immunity	Portion of the adaptive immune system in which defensive cells destroy damaged, cancerous, foreign, or infected cells by direct cell-to-cell contact
Cytotoxic T cell	Immune cell that binds to and destroys cells that are defective or are infected by a pathogen
Humoral immunity	Portion of the adaptive immune system in which defensive cells secrete antibodies
Plasma cell	B cell that secretes antibodies specific to an ongoing infection
Effector cell	Activated cell that carries out an immune response (e.g., by secreting antibodies or killing infected cells)
Memory cell	Inactive cell that lingers after an immune response is complete and launches a new immune response if the same pathogen is encountered again

TABLE 29.1 Ways to Acquire Immunity: A Summary

Type	Description	Examples
Active immunity	Individual produces antibodies to an antigen; long-lasting (memory cells are produced)	• Having chickenpox confers future immunity to that disease ("natural" active immunity). • Influenza vaccine triggers production of memory cells specific to antigens in the vaccine ("artificial" active immunity).
Passive immunity	One individual acquires antibodies from another individual; temporary (no memory cells are produced)	• Fetus acquires antibodies from mother via placenta or milk. • Dog bite victim receives injections of antibodies to rabies virus. • Snakebite victim receives intravenous antivenom (antibodies to snake venom).

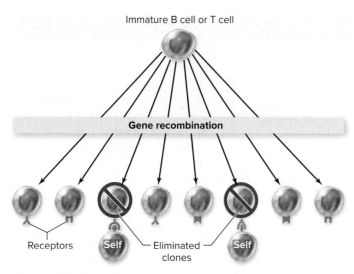

Immature B cell or T cell

Gene recombination

Receptors   Self   Eliminated   Self
                    clones

Figure 29.11 Clonal Deletion. As lymphocytes develop in a fetus, they are tested against proteins and polysaccharides on the body's own cell surfaces. Clones that match self antigens are eliminated.

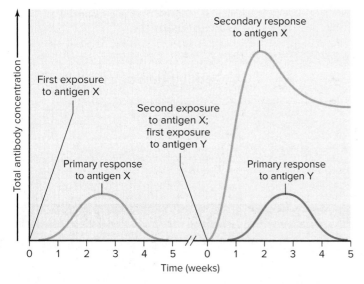

Total antibody concentration

Secondary response
to antigen X

First exposure
to antigen X

Second exposure
to antigen X;
first exposure
to antigen Y

Primary response
to antigen X

Primary response
to antigen Y

0   1   2   3   4   5   0   1   2   3   4   5
Time (weeks)

Figure 29.12 Primary and Secondary Immune Responses. The primary immune response leaves memory cells that stimulate a stronger immune response on subsequent exposure to the same antigen.

surfaces and molecules are "weeded out" by apoptosis, or programmed cell death (figure 29.11). This process, which begins before birth, helps prevent self-immunity. The developing immune system somehow also learns not to attack antigens in food. ⓘ *apoptosis,* section 8.1B

D. The Secondary Immune Response Is Stronger Than the Primary Response

The **primary immune response** is the adaptive immune system's first reaction to a foreign antigen. Because the clonal selection process takes time, days or even weeks may elapse before antibody concentrations reach their peak. During this time, the pathogen can cause severe damage or death. If a person survives, however, the memory B cells and memory T cells leave a lasting impression—that is, immunological memory.

Thanks to memory cells, the **secondary immune response**—the immune system's reaction the next time it detects the same foreign antigen—is much stronger than the primary response (figure 29.12). Memory B cells transform into rapidly dividing plasma cells. Within hours, billions of antigen-specific antibodies are circulating throughout the host body, destroying the pathogen before it takes hold. Usually, there is no hint that a second infection ever occurred. As described in section 29.4, vaccines create this immunological memory without risking an initial infection.

It is important to remember that turning off an immune response once an infection has been halted is as important as turning it on. After all, powerful immune biochemicals can attack not only pathogens but also the body's healthy tissues. Immunologists continue to learn more about the precise signals that cause the immune system to "back down" after a threat is removed.

29.3 Mastering Concepts

1. What is the relationship between antigens and antibodies?
2. What is the role of macrophages in adaptive immunity?
3. In your own words, write a paragraph describing the events of adaptive immunity, beginning with a pathogen entering the body and ending with the production of memory cells.
4. Describe the structure and function of an antibody.
5. Explain the difference between the primary and secondary immune responses.

29.4 Vaccines Jump-Start Immunity

The immune system is remarkably effective at keeping pathogens and cancer cells from taking over our bodies. Nevertheless, humans do suffer from many incurable infectious diseases caused by viruses and other pathogens. Besides sanitation, the best way to prevent many of these illnesses is vaccination.

A **vaccine** is a substance that stimulates active immunity against a pathogen without actually causing illness. The vaccine consists partly of antigens that "teach" the recipient's immune system to recognize a pathogen. Once taken into the body, the antigens stimulate a primary immune response. Memory cells linger after this initial exposure, ensuring that a subsequent encounter with the real pathogen triggers the rapid secondary immune response (see Burning Question 29.1). Vaccination programs therefore reduce both the incidence and the spread of infectious disease.

The use of vaccines in medicine dates to the late eighteenth century, when an incurable viral disease called smallpox was ravaging the human population. The fatality rate among infected people was about 25%, and about two thirds of the survivors were left horribly scarred and sometimes blind. But in 1796, a British country physician named Edward Jenner invented a vaccine against smallpox. The preparation contained vaccinia viruses, which cause a mild infection called cowpox. The two viruses—smallpox and vaccinia—are related to each other, and they share some antigens. Exposure to the vaccinia virus therefore confers immunity to smallpox.

This vaccine became the centerpiece of a worldwide smallpox eradication campaign, which began in 1967. By 1980, the World Health Organization had declared that "smallpox is dead." The declaration heralded a milestone in medicine—the eradication of a disease. Today, all known stocks of smallpox virus reside in two labs, one in the United States and the other in Russia. Only the threat of bioterrorism maintains the demand for the smallpox vaccine among some military personnel and "first responders."

Scientists have developed vaccines against many other pathogens as well. The antigens in the vaccines take several different forms (table 29.2). Measles and mumps vaccines, for example, contain weakened viruses. Others, such as the vaccine against hepatitis A, contain inactivated viruses that cannot cause an infection. Diphtheria and tetanus vaccines incorporate an inactivated toxin that bacterial pathogens produce. Still others, such as the hepatitis B vaccine, incorporate only a part of the pathogen's surface. As you examine table 29.2, note that most vaccines do not contain live pathogens and therefore cannot cause the disease they are designed to prevent.

Although vaccines have saved countless lives since Jenner's time, they cannot prevent all infectious diseases. For example, researchers have been unable to develop a vaccine against HIV, the rapidly evolving virus that causes AIDS. Influenza viruses also mutate rapidly, so each vaccine is effective for only one flu season. And it has so far proved impossible to develop one vaccine that will prevent infection by the many viruses that cause the common cold.

So far, only a handful of vaccines are used to prevent cancer. One example is the "cervical cancer vaccine," which prevents infection with the human papillomavirus. Cancer vaccines are part of a larger field of research called immunotherapy. The overall goal is to fight cancer using weapons from both cell-mediated and humoral immunity. For example, a physician may harvest T cells specific to antigens on a cancer patient's tumor. The T cells replicate outside of the body and are then re-injected into the tumor to act as a powerful, customized anticancer "drug." In other forms of immunotherapy, a patient may receive injections of antibodies that target antigens on his cancer cells, or he may receive cytokines that signal his immune system to produce more cancer-fighting cells. As research progresses, new cancer vaccines—and other forms of immunotherapy—will become increasingly common.

29.4 Mastering Concepts

1. What is a vaccine?
2. List the main types of vaccine formulations.
3. Why haven't scientists been able to develop vaccines against HIV and the common cold?

TABLE 29.2   Types of Vaccines

Vaccine Formulation	Examples
Live, weakened (attenuated) pathogen	Polio (oral vaccine), influenza (nasal spray), measles, mumps, rubella, chickenpox
Inactivated pathogen	Polio (injectable vaccine), influenza (injectable vaccine), hepatitis A
Inactivated toxins	Tetanus, diphtheria
Subunits of pathogens	Whooping cough (pertussis), hepatitis B, human papillomavirus, Lyme disease (experimental)

Burning Question 29.1

Why do we need multiple doses of some vaccines?

Vaccines stimulate the immune system to produce memory cells that "remember" their exposure to the antigens in the vaccine. These memory cells can last for decades, triggering the production of antibodies when the real pathogen comes along. Yet most childhood vaccines require a series of shots, spaced out over months or years. And the tetanus and diphtheria vaccines require booster shots at least every 10 years. Why isn't one dose enough? The answer has to do with the number of memory cells that the body produces after exposure to the vaccine.

The vaccine's formulation helps determine the need for booster shots. A vaccine that consists of active viruses causes a mild infection that lasts for about a week. This lengthy exposure to viral antigens stimulates a robust immune response, so booster shots are not usually necessary. But when a vaccine contains toxins or inactivated viruses, no infection occurs. The body's exposure to the antigens is therefore relatively brief, and the number of memory cells may be too small to launch an effective immune response against the pathogen. In that case, repeated shots help boost the number of memory cells over time.

Submit your burning question to marielle.hoefnagels@mheducation.com

(syringe): ©Yuri Kevhiev/Alamy Stock Photo RF

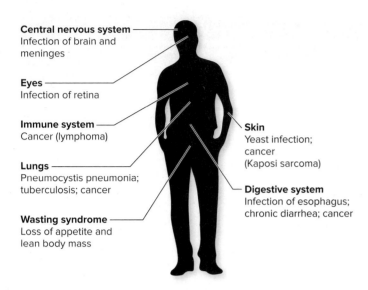

Central nervous system
Infection of brain and
meninges

Eyes
Infection of retina

Immune system
Cancer (lymphoma)

Lungs
Pneumocystis pneumonia;
tuberculosis; cancer

Wasting syndrome
Loss of appetite and
lean body mass

Skin
Yeast infection;
cancer
(Kaposi sarcoma)

Digestive system
Infection of esophagus;
chronic diarrhea; cancer

Figure 29.13 Opportunistic Illnesses. People with immunodeficiencies are vulnerable to a wide variety of illnesses that are rare in people with intact immune systems.

Burning Question 29.2

Can people be allergic to meat?

Imagine eating a hamburger and then a few hours later, breaking out in itchy hives. For thousands of people in the United States, this scenario became a reality after they were bitten by a Lone Star tick. When a tick bites a person, its saliva enters the victim's bloodstream. In response, the person's immune system develops antibodies against the many chemicals contained in the saliva. One of those chemicals is a sugar called alpha-gal, which also occurs in meat. When the person subsequently eats meat, antibodies bind to alpha-gal sugars and initiate an allergic response. In rare cases, the immune reaction is strong enough to put the victim's life at risk.

After developing the allergy, people must avoid meat. Researchers advise using caution when exploring wooded areas in the Lone Star tick's range. To help prevent tick bites, wear clothing that covers the body and use bug spray containing DEET.

Submit your burning question to
marielle.hoefnagels@mheducation.com

29.5 Several Disorders Affect the Immune System

The immune system may turn against the body's own cells, or it may fail to respond to disease-causing organisms. In addition, harmless substances sometimes trigger an immune response. (A woman's immune system can even attack her own fetus, as described in Why We Care 29.2.)

A. Autoimmune Disorders Are Devastating and Mysterious

Ideally, the immune system does not attack the body's own cells; as a person develops, lymphocytes corresponding to molecules already present in the body should be eliminated by clonal deletion. In an **autoimmune disorder,** however, the immune system attacks the body's self antigens.

The resulting damage to tissues and organs may be severe. In type 1 diabetes, for example, antibodies attack the insulin-producing cells of the pancreas. Without insulin, the body's cells starve because they cannot absorb glucose. Another disease, rheumatoid arthritis, arises from an autoimmune attack on cells lining the skeleton's joints. Pain and joint deformity are the result.

B. Immunodeficiencies Lead to Opportunistic Infections

An **immunodeficiency** is a condition in which the immune system lacks one or more essential components. A weakened immune system leaves a person vulnerable to **opportunistic pathogens** and cancers that do not normally affect people with healthy immune systems (figure 29.13). Viruses such as HIV can weaken the immune system, as can some inherited diseases and pharmaceutical drugs such as prednisone. ⓘ *prednisone,* section 25.4C

Human immunodeficiency virus (HIV) kills immune cells, causing acquired immune deficiency syndrome (AIDS). A person can acquire HIV by sexual contact or by using contaminated needles when injecting drugs. A mother can also transmit HIV to her baby, either during delivery or in breast milk.

Helper T cells are HIV's main target (see figure 7.20). Infected helper T cells die as they assemble and release new viruses, which then infect additional helper T cells. For months to a decade or more, however, no AIDS symptoms appear, because the body can produce enough new T cells to compensate for the loss. During this latent period, B cells manufacture antibodies to the virus; rapid tests for HIV exposure detect these proteins. Unfortunately, the antibodies do not prevent new viruses from forming. As helper T cell counts decline, the immune system's ability to fight the virus also weakens. Eventually, the immune system fails entirely, and the opportunistic infections and cancers of AIDS begin.

AIDS is a consequence of a viral infection, but immune deficiency can also be inherited. Each year, a few children are born defenseless against infection due to **severe combined immunodeficiency (SCID),** a disorder in which neither T cells nor B cells function. Decades ago, the parents of a child with SCID had one option: try to isolate the youngster from all possible infectious diseases. Today, SCID has a cure. Most children born with SCID receive bone marrow transplants before they are 3 months old, replacing their defective cells with marrow from a healthy donor. In addition, gene therapy has been used to replace faulty genes in some SCID patients. ⓘ *gene therapy,* section 11.4D

Immunodeficiency is also common in organ transplant recipients. To avoid rejection of a donated organ, transplant recipients must take immune-suppressing drugs for the rest of their lives. Like other people with immunodeficiencies, these patients are vulnerable to opportunistic infections.

C. Allergies Misdirect the Immune Response

In an **allergy,** the immune system is overly sensitive, launching an exaggerated attack on a harmless substance (see figure 29.14 and Burning Question 29.2). Common **allergens,** or antigens that trigger an allergy, include foods, dust mites, pollen, fur, and oils in the leaves of plants such as poison ivy. The allergens activate B cells to produce antibodies. A first exposure to the allergen initiates a step called sensitization, in which antibodies bind to mast cells and basophils. On subsequent exposure, the allergens bind to the molecules attached to the cells, causing them to explosively release histamine and other allergy mediators.

The symptoms of an allergic response depend on where in the body the cells release mediators. Many mast cells are in the skin, respiratory passages, and digestive tract, so allergies often affect these organs. The result: hives, runny nose, watery eyes, asthma, nausea, vomiting, and diarrhea. Antihistamine drugs relieve these symptoms by preventing the release of histamine or by keeping it from binding to target cells.

Some individuals react to allergens with **anaphylactic shock,** a rapid, widespread, and potentially life-threatening reaction in which mast cells and basophils release allergy mediators throughout the body. The person may at first feel an inexplicable apprehension. Then, suddenly, the entire body itches and erupts in hives. Histamine causes blood vessels to dilate, lowering blood pressure. As blood rushes to the skin, not enough of it reaches the brain, and the person becomes dizzy and may lose consciousness. Breathing becomes difficult as the airways in the lungs become constricted. Meanwhile, the face, tongue, and larynx begin to swell. Unless the person receives an injection of epinephrine and sometimes an incision into the trachea to restore breathing, death can come within minutes.

Anaphylactic shock most often results from an allergy to penicillin, insect stings, or foods. Peanut allergy, for example, affects 6% of the U.S. population and is on the rise. The fact that the initial allergic reaction to peanuts occurs at an average age of 14 months, typically after eating peanut butter, suggests that sensitization occurs even earlier, during breast feeding or before birth. Fortunately, the immune system can sometimes be "retrained." That is, small children can be given foods containing peanuts in gradually increasing doses (under a physician's supervision) until the allergy subsides.

Early exposure to microorganisms and viruses may be crucial to the development of the immune system. A growing body of evidence has led to the "hygiene hypothesis," which suggests that excessive cleanliness has contributed to recent increases in the incidence of asthma and some allergies (see Investigating Life 29.1). Apparently, ultraclean surroundings decrease stimulation of the immune system early in life.

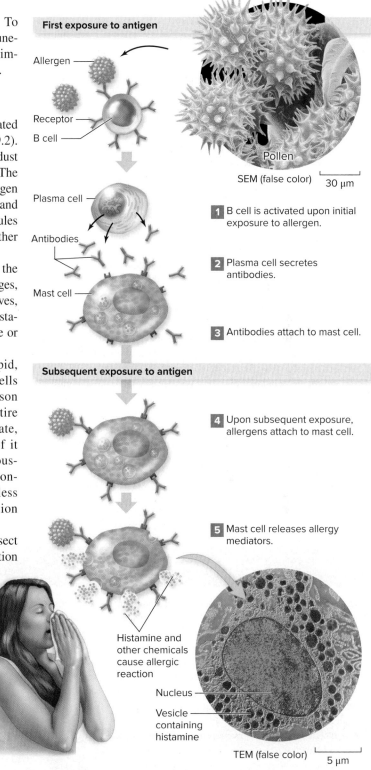

First exposure to antigen

Allergen

Receptor

B cell

Pollen

SEM (false color) 30 μm

Plasma cell

Antibodies

Mast cell

1 B cell is activated upon initial exposure to allergen.

2 Plasma cell secretes antibodies.

3 Antibodies attach to mast cell.

Subsequent exposure to antigen

4 Upon subsequent exposure, allergens attach to mast cell.

5 Mast cell releases allergy mediators.

Histamine and other chemicals cause allergic reaction

Nucleus

Vesicle containing histamine

TEM (false color) 5 μm

Figure 29.14 Allergy. The first exposure to an allergen such as pollen sensitizes mast cells and basophils (not shown here). Subsequent encounters with the same type of pollen trigger the allergic response.

Photos: (pollen): ©Susumu Nishinaga/Science Source; (mast cell): ©Institut Pasteur/Medical Images

29.5 Mastering Concepts

1. How might autoimmunity arise?
2. How does HIV harm the immune system?
3. Which cells and biochemicals participate in an allergic reaction?

Investigating Life 29.1 | The Hidden Cost of Hygiene

Healthy immune function requires a delicate balance. On the one hand, the immune system must be strong enough to protect the body from dangerous pathogens. But if the immune response is too strong, it may overreact to harmless substances—as in allergies—or launch an autoimmune attack against the body's own cells.

People who live in developed countries have a high incidence of allergies, asthma, and autoimmune disorders when compared to their counterparts in developing countries. According to the hygiene hypothesis, the difference may stem from the sanitation, vaccines, antibiotics, and ultraclean surroundings that are typical of wealthy countries.

For millions of years, the human immune system has coevolved with countless bacteria, viruses, and parasitic worms. Many of these hidden residents produce substances that suppress our immune systems, an adaptation that allows them to "fly under the radar" and maintain long-term, chronic infections. Perhaps constant exposure to these pathogens keeps the human immune system in check.

To test this hypothesis, researchers studied the incidence of allergies and chronic worm infection among schoolchildren in the developing country of Gabon. The team scratched potential allergens into each child's forearm and tested for an allergic reaction. They also checked for infection with several types of parasitic worms, including a liver fluke. ⓘ *flatworms*, section 17.4

Of the 513 children, 57 tested positive for dust-mite allergies. Forty-six of these allergic children were worm-free; only 11 were infected with flukes (figure 29.A). Statistical analysis of the data indicated that harboring worms significantly lowers the risk for allergies. Worms apparently induce white blood cells to release an interleukin that dampens the overall immune response and suppresses allergies. This adaptation allows the parasites to "fly under the radar" and maintain long-term, chronic infections.

This study suggests the intriguing possibility that parasites might have a role to play in the fight against allergies. The long-term outcome may be

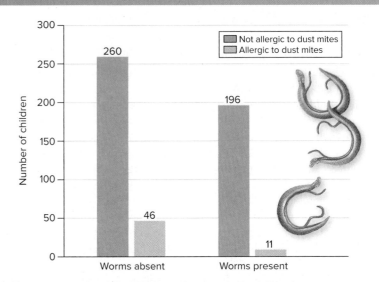

Figure 29.A **Protective Worms?** Of the 306 children without parasitic worms, dust-mite allergies appeared in 46 (about 15%). In contrast, of the 207 schoolchildren infected with worms, only 11 (about 5%) also had allergies.

new drugs to treat immune disorders. Of course, no one is suggesting that we return to the days before sanitation and clean water or that we deny those resources to people who lack them now. At the same time, it is interesting to know that allergies and autoimmune diseases may be the price we pay to live a cleaner life.

Source: Anita H. J. van den Biggelaar and six coauthors, including Maria Yazdanbakhsh. 2000. Decreased atopy in children infected with *Schistosoma haematobium*: a role for parasite-induced interleukin-10. *Lancet*, vol. 356, pages 1723–1727.

What's the **Point?** ▼ APPLIED

The immune system protects the body against infection. Vaccines represent a tangible benefit of scientific research on this complex network of cells, tissues, and organs. After immunization, the immune system creates memory cells that fight off a subsequent infection before symptoms even begin.

Vaccines save millions of lives and prevent enormous suffering. For example, the Centers for Disease Control and Prevention estimates that smallpox, diphtheria, measles, polio, and rubella once killed nearly 650,000 people a year in the United States. Ever since routine vaccinations for these five diseases started, however, mortality has plunged below 100 per year.

Children have immature immune systems, so they are typically more susceptible to disease than are adults. Most doctors therefore strongly recommend vaccines and booster shots for children. Nevertheless, some people refuse to vaccinate their children. They cite a variety of arguments. One stems from the suspicion that vaccines cause autism. This argument comes primarily from parents who noticed signs of autism shortly after a child was vaccinated. The events may occur at the same time, but that does not

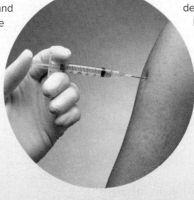

©Jeffrey Hamilton/Getty Images RF

mean that one causes the other. In fact, numerous epidemiological studies have found no connection between vaccines and autism.

A second objection is that vaccines can have harmful side effects. It is true that no vaccine is 100% safe, and some children do have medical conditions that preclude vaccines. But for a healthy child, the risk of a deadly reaction to a vaccine is much smaller than the risk of death from an infectious disease. Before refusing a vaccine, learn more about it on your own, and ask your doctor whether it endangers your child.

When people decide not to vaccinate their children, they increase the chance that many now-rare diseases will flare up again. For example, although measles has been largely eradicated from the Americas, international travel occasionally leads to outbreaks in the United States. Another example is pertussis, which has recently become more common (partly because doctors switched to a weaker vaccine with fewer side effects). It is hard to anticipate what other diseases may be on the rise in coming years. Declining a vaccination puts your child—and other unvaccinated children—at risk.

CHAPTER SUMMARY

29.1 Many Cells, Tissues, and Organs Defend the Body

- The **immune system** protects the body against **pathogens** and cancer cells.

A. White Blood Cells Play Major Roles in the Immune System

- Several types of **white blood cells** participate in immune responses.
- **Macrophages** and some other white blood cells are **phagocytes,** cells that engulf and destroy bacteria and debris.
- **B cells** and **T cells** are **lymphocytes** that mature in the **red bone marrow** and the **thymus,** respectively. **Natural killer cells** are also lymphocytes.
- **Basophils** are white blood cells that participate in inflammation; they are closely related to noncirculating **mast cells.**

B. The Lymphatic System Produces and Transports Many Immune System Cells

- The **lymphatic system** plays a crucial role in the immune response. The vessels of the lymphatic system distribute a fluid called **lymph.**
- Besides the thymus, other lymphoid organs include the **spleen** and **lymph nodes.** Immune cells are also concentrated in the tonsils, appendix, and digestive tract.

C. The Immune System Has Two Main Subdivisions

- **Innate defenses** provide broad protection against all pathogens, whereas **adaptive immunity** is directed against specific pathogens (figure 29.15). Only adaptive immunity produces an immunological "memory" that protects against future exposure to a previously encountered pathogen.

29.2 Innate Defenses Are Nonspecific and Act Early

A. External Barriers Form the First Line of Defense

- Intact skin consists of two layers (the outer **epidermis** and inner **dermis**) that block pathogens. Mucous membranes, tears, earwax, cilia, and beneficial microbes are other examples of external barriers to infection.

B. Internal Innate Defenses Destroy Invaders

- White blood cells destroy bacteria and promote inflammation; natural killer cells destroy cancerous or virus-infected cells; macrophages consume pathogens, promote fever, and activate the immune response.
- Basophils and mast cells trigger **inflammation,** which is an immediate reaction to injury. These cells release **histamine,** a biochemical that causes blood vessels to dilate.
- Redness, warmth, swelling, and pain are associated with inflammation.
- **Complement** proteins interact in a cascade that ends with the destruction of bacterial cells.
- **Cytokines** are antimicrobial molecules that communicate with immune system cells and stimulate the development of a fever. Interferons and interleukins are examples of cytokines.
- The elevated body temperature of a mild **fever** helps discourage microbial replication.

29.3 Adaptive Immunity Defends Against Specific Pathogens

- Adaptive immunity is directed against specific **antigens.**

A. Helper T Cells Play a Central Role in Adaptive Immunity

- A macrophage that engulfs a pathogen links antigens from the microbe to specialized proteins on its cell surface.
- A **helper T cell** binding to the antigen-presenting cell initiates the **cell-mediated** and **humoral** components of the adaptive immune response. In **clonal selection,** activated helper T cells divide and differentiate into **memory cells** and into effector cells that help activate cytotoxic T cells and B cells.

B. Cytotoxic T Cells Provide Cell-Mediated Immunity

- **Cytotoxic T cells** kill cells that are cancerous, damaged, or infected with viruses or bacteria. Memory cytotoxic T cells contribute to long-term immunity.

C. B Cells Direct the Humoral Immune Response

- An **antibody** is a Y-shaped protein composed of two heavy and two light polypeptide chains. Each chain has a **constant region** and a **variable region.** Antibodies bind antigens and form complexes that attract other immune system components.
- An activated B cell multiplies rapidly (clonal selection), generating an army of identical **plasma cells** that all churn out the same antibody. Some also differentiate into memory cells.
- In **active immunity,** a person makes his or her own antibodies. In **passive immunity,** a person receives antibodies from someone else.

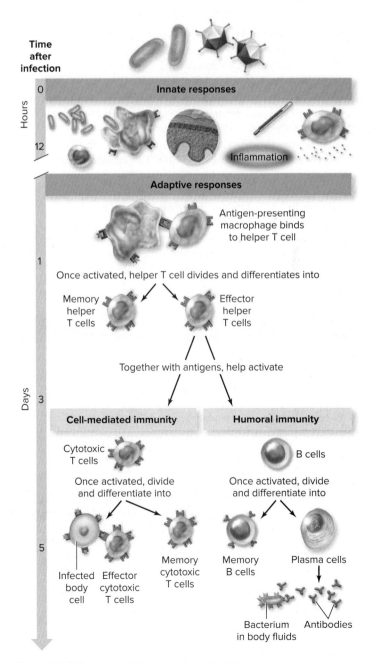

Figure 29.15 Innate and Adaptive Immunity: A Summary.

- Antibodies and antigen receptors are incredibly diverse because DNA segments shuffle during early lymphocyte development. **Clonal deletion** subsequently eliminates lymphocytes corresponding to self antigens.

D. The Secondary Immune Response Is Stronger Than the Primary Response

- The first encounter with an antigen provokes the **primary immune response,** which is relatively slow. Its legacy is memory cells that greatly speed the **secondary immune response** on subsequent exposure to the same antigen.
- Tissue damage can occur if the immune response does not turn off after eliminating a pathogen.

29.4 Vaccines Jump-Start Immunity

- A **vaccine** "teaches" the immune system to recognize specific components of a pathogen, bypassing the primary immune response.

29.5 Several Disorders Affect the Immune System

A. Autoimmune Disorders Are Devastating and Mysterious

- An **autoimmune disorder** occurs when the immune system produces antibodies that attack the body's own tissues.

B. Immunodeficiencies Lead to Opportunistic Infections

- An **immunodeficiency** is the absence of one or more essential elements of the immune system. These disorders leave patients vulnerable to cancer and **opportunistic pathogens.**
- The **human immunodeficiency virus (HIV)** kills helper T cells, causing AIDS.
- **Severe combined immunodeficiency (SCID)** is an inherited disease in which the adaptive immune response is defective.
- Drugs that prevent organ transplant rejection also weaken the immune system.

C. Allergies Misdirect the Immune Response

- An **allergy** is an immune reaction to a harmless substance. An **allergen** triggers the production of antibodies, which bind mast cells and basophils. On subsequent exposure, these cells release allergy mediators such as histamine.
- **Anaphylactic shock** is a life-threatening allergic reaction.

MULTIPLE CHOICE QUESTIONS

1. What is lymph?
   a. Blood that contains abundant pathogens
   b. Plasma that leaks out of blood vessels
   c. Any part of the body that contains immune cells
   d. Pathogen-catching fluid produced at the spleen

2. Histamine acts on the _____ , causing redness and swelling.
   a. white blood cells
   b. cells lining blood vessels
   c. smooth muscle cells
   d. red blood cells

3. Ibuprofen dampens the immune system's inflammation response. What might be a short-term consequence of taking ibuprofen?
   a. Blood vessels may become less permeable.
   b. Antibodies may become less active.
   c. Cytotoxic T cells may begin to divide.
   d. All of the above are correct.

4. The innate immune response is characterized by its
   a. rapid response to invading pathogens.
   b. ability to "remember" pathogens it has already encountered.
   c. ability to produce antibodies.
   d. Both b and c are correct.

5. Which of the following ranks the immune reactions in the order they respond to a pathogen?
   a. Innate response; primary response; secondary response
   b. Secondary response; primary response; innate response
   c. Primary response; secondary response; innate response
   d. Innate response; secondary response; primary response

6. Antibody function requires that the shape of the _____ corresponds to the shape of the antigen.
   a. constant region
   b. stem
   c. variable region
   d. amino acid

7. During the humoral immune response,
   a. B cells divide and produce thousands of different types of antibodies, a few of which will bind to antigens on the pathogen.
   b. only B cells that produce antigen-specific antibodies divide.
   c. clonal deletion removes B cells that do not bind to the antigens of the invader.
   d. histamine recruits B cells to engulf pathogens.

8. Why is the secondary immune response so much stronger than the primary response?
   a. Because high concentrations of antibodies are already present
   b. Because the phagocytes present the antigens more rapidly
   c. Because memory cells can rapidly convert to plasma cells
   d. Because protein synthesis occurs more quickly in memory cells

9. How do vaccines prevent infectious disease?
   a. By killing bacteria and viruses
   b. By boosting overall immune function
   c. By stimulating a primary immune response
   d. By passive immunity

10. HIV causes immunodeficiency by attacking
    a. B memory cells.
    b. helper T cells.
    c. plasma cells.
    d. cytotoxic T cells.

Answers to Multiple Choice questions are in appendix A.

WRITE IT OUT

1. Explain the observation that lymphoid tissues are scattered in the skin, lungs, stomach, and intestines.

2. Explain why a scraped knee increases the chance that pathogens will trigger an adaptive immune response.

3. In an effort to reduce allergic responses, a drug company wishes to develop a medication that binds to and kills mast cells. What would be a side effect of this drug?

4. Dead phagocytes are one component of pus. Why is pus a sure sign of infection?

5. Since fever has protective effects, should we avoid taking fever-reducing medications when ill? Use the Internet to research the consequences of overmedicating a fever and the risk of allowing a fever to rise too high.

6. Briefly explain the function of each innate and adaptive defense shown in figure 29.15.

7. What do a plasma cell and a memory cell descended from the same B cell have in common, and how do they differ?

8. To what must a B cell bind to become activated? To what must the activated B cell be exposed before it proliferates?

9. As a treatment for bladder cancer, physicians may inject bovine tuberculosis bacteria into the patient's bladder. The bacteria bind to the bladder wall. Scientists do not fully understand why this treatment causes the patient's immune system to launch an attack against the cancer cells. Using your knowledge of how the immune system is activated, write a prediction for how an injection of bovine tuberculosis bacteria might help a patient's immune system fight cancer.

10. How does a cytotoxic T cell kill an infected body cell?

11. Influenza viruses mutate rapidly, whereas the chickenpox virus does not. Why are people encouraged to receive vaccinations against influenza every year, whereas immunity to chickenpox lasts for decades?

12. Explain the difference between: clonal deletion and clonal selection; a natural killer cell and a cytotoxic T cell; antibodies and antigens; cell-mediated and humoral immunity; an autoimmune disorder and an immunodeficiency.

13. Search the Internet for information about immune system disorders. Choose one illness to study in more detail. What are the characteristics of the disorder? Who is primarily affected? What causes the illness, and is there a treatment or cure?

14. Humans (and all other organisms) are in an evolutionary battle with a wide variety of pathogens. How does natural selection favor (a) an immune system that adjusts to a changing variety of pathogens and (b) pathogens that evade the immune system?

SCIENTIFIC LITERACY

Review the *What's The Point? Applied* box in this chapter, then consider each of the following assertions:

(a) Cancer-causing ingredients in vaccines harm children.

(b) A parent's choice to not vaccinate a child will not affect unrelated individuals.

(c) Memory cells derived from an actual illness are more powerful than memory cells derived from a vaccine (i.e., "natural" immunity is more effective than vaccine-induced immunity).

For each assertion, find a website that supports the statement and another that refutes it. Which of the two websites makes a more compelling case? In your opinion, which websites base their arguments on scientific evidence, and which base their arguments on opinions or anecdotal evidence? How can you tell?

Answers to Mastering Concepts, Write It Out, Scientific Literacy, and Pull It Together questions can be found in the Connect ebook.
connect.mheducation.com

Design element: Burning Question (fire background): ©Ingram Publishing/Super Stock

PULL IT TOGETHER

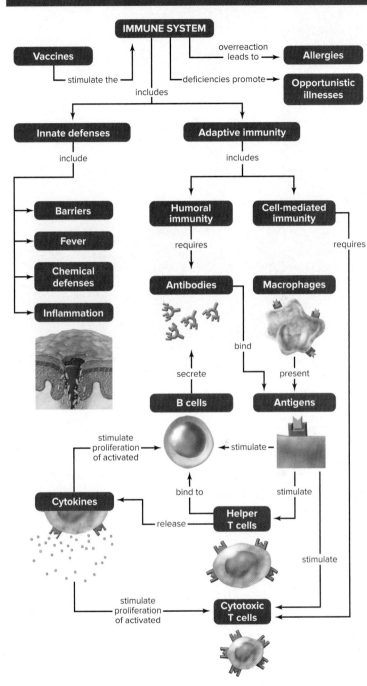

Figure 29.16 Pull It Together: The Immune System.

Refer to figure 29.16 and the chapter content to answer the following questions.

1. Review the Survey the Landscape figure in the chapter introduction and the Pull It Together concept map. How do the actions of the immune system help an animal maintain homeostasis?

2. Add *memory B cells, plasma cells, memory cytotoxic T cells, primary immune response, secondary immune response,* and *autoimmune disorders* to this concept map.

3. Connect *vaccines* to other parts of this concept map, including those that you added for question 2.

4. How do lymph and lymph nodes fit into this concept map?

LEARNING OUTLINE

30.1 Animal Development Begins with Reproduction

30.2 Males Produce Sperm Cells

30.3 Females Produce Egg Cells

30.4 Reproductive Health Considers Contraception and Disease

30.5 The Human Infant Begins Life as a Zygote

APPLICATIONS

Burning Question 30.1 *When can conception occur?*
Why We Care 30.1 *Substances That Cause Birth Defects*
Investigating Life 30.1 *Playing "Dress Up" on the Reef*

Quite a Mouthful. In some animal species, males have unusual reproductive roles. This male yellowhead jawfish protects the eggs he has fertilized by keeping them in his mouth until they hatch. Until that happens, the father must go without food.

©UIG via Getty Images

Learn How to Learn
Make Your Own Review Sheet

If you are facing a big exam, how can you make sense of everything you have learned? One way is to make your own review sheet. The best strategy will depend on what your instructor expects you to know, but here are a few ideas to try: Make lists; draw concept maps that link ideas within and between chapters; draw diagrams that illustrate important processes; and write mini-essays that explain the main points in each chapter's learning outline.

SURVEY THE LANDSCAPE
Animal Anatomy and Physiology

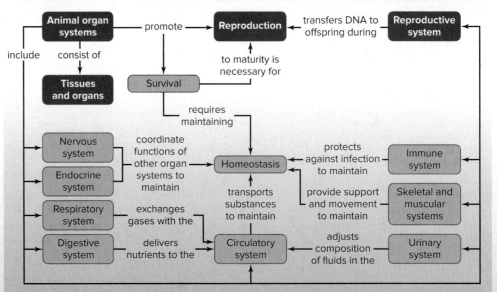

Reproductive success is the cornerstone of natural selection. The reproductive system produces the sperm and egg cells that give rise to the next generation.

For more details, study the Pull It Together feature in the chapter summary.

Reproductive success is the key to natural selection, and the animal kingdom is full of intriguing strategies. The male jawfish in the chapter opening photo is one example, but there is no shortage of others. Some animals, like the aphids on garden plants, can reproduce without a mate. Others, like butterflies and moths, go through a dramatic transformation as they develop from a crawling caterpillar into a flying adult.

©Samuel Borges
Photography/Shutterstock RF

Humans are no less amazing. Sperm and egg cells come together in a woman's body. The resulting cell begins to divide, first into two cells, then four, then eight, and so on. The resulting ball of cells soon hollows out and develops an outside and an inside. Slowly, as cells continue to divide and specialize, organs develop and begin to work together. After 9 months, a new baby emerges into the world.

This chapter explores these two interrelated topics—reproduction and development—with a focus on our own species. Along the way we will explain contraception, list and describe sexually transmitted infections, and introduce a few of the other reproductive perils we face.

a.

Figure 30.1 External and Internal Fertilization. (a) A sea urchin releases sperm cells into the water. Meanwhile, females release eggs; fertilization is external. (b) A male black-winged stilt mates with a female. Their offspring will develop inside the female's body until she lays three or four hard-shelled eggs in a nest.

(a): ©Andrew J. Martinez/Science Source; (b): ©sysasya photography/Shutterstock RF

30.1 Animal Development Begins with Reproduction

A monarch butterfly emerges from its chrysalis; a baby bird hatches from an egg; a kitten becomes a full-grown cat. All of these familiar examples illustrate growth and development. Together, reproduction and development are shared features of all multicellular life. Chapter 22 described how flowering plants reproduce and grow; this chapter picks up the topic for animals.

A. Reproduction Is Asexual or Sexual

Like plants, animals may reproduce asexually or sexually (or both). In **asexual reproduction,** the offspring contain genetic information from only one parent. Aside from mutations that occur during replication, all offspring are identical to the parent and to one another. Asexually reproducing animals include sponges, sea anemones, aphids, and some types of lizards. In general, asexual reproduction is advantageous in environments that do not change much over time.

Sexual reproduction requires two parents, each of which contributes half the DNA in each offspring. In many species, sexual reproduction entails high energy costs for attracting mates, copulating, and defending against rivals (see Investigating Life 30.1). Nevertheless, the benefits of genetic diversity apparently outweigh these costs, especially in a changing environment. Sexual reproduction is extremely common among animals. ⓘ *why sex?,* section 9.1

In organisms that reproduce sexually, haploid **gametes** are the sex cells that carry the genetic information from each parent. The gametes—sperm cells from males and egg cells from females—are the products of meiosis, a specialized type of cell division. In meiosis, a diploid cell containing two sets of chromosomes divides into four haploid cells, each containing just one chromosome set. **Fertilization** is the union of two gametes; the product of fertilization is the **zygote,** the diploid first cell of the new offspring.

Sperm and egg come together in a variety of ways. In **external fertilization,** males and females release gametes into the same environment, and fertilization occurs outside the body (figure 30.1a). This strategy is especially common in aquatic animals. Salmon, for example, spawn in streams. Females lay eggs in gravelly nests, and then males shed sperm over them. Other animals with external fertilization include sponges, corals, sea urchins, some crustaceans (such as American lobsters), and some amphibians. Unique "recognition" proteins on the surfaces of the gametes ensure that sperm cells fertilize egg cells of the correct species.

b.

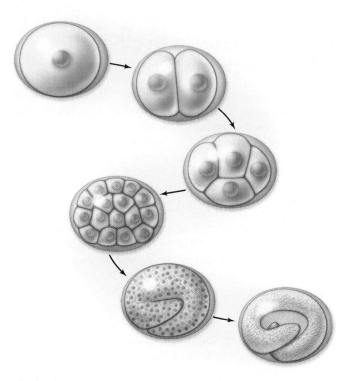

Figure 30.2 Tiny Worm. The life of a nematode worm begins as a single fertilized egg cell. Researchers understand nematode development in detail, describing and mapping the fates of all cells produced from the original zygote.

In **internal fertilization,** a male deposits sperm inside a female's body, where fertilization occurs (see figure 30.1b). Land animals, including mammals, nonavian reptiles, and birds, commonly use internal fertilization. After copulation, the female may lay hard-shelled, fertilized eggs that provide both nutrition and protection to developing offspring. A chicken egg is a familiar example. Alternatively, the female may bear live young, as humans and other mammals do.

B. Development Is Indirect or Direct

No matter what the reproductive strategy, development of sexually reproducing animals begins with the zygote. That first cell begins to divide soon after fertilization is complete. As the embryonic animal grows, cells divide and die in coordinated ways to shape the body's distinctive form and function. Developmental biologists study the stages of an animal's growth as cells specialize and interact to form tissues, organs, and organ systems (figure 30.2). The similarities and differences among developing animals can yield important clues to evolution, as illustrated in figure 13.12.

One key to animal development is **differentiation,** the process by which cells acquire their specialized functions by activating unique combinations of genes in the skin, brain, eyes, and other organs. Another essential process is **pattern formation,** in which genes determine the overall shape and structure of the animal's body, such as the number of segments or the placement of limbs. (i) *control of gene expression,* section 7.5

Differentiation and pattern formation involve complex interactions between the DNA inside cells and external signals such as hormones. These interactions ultimately regulate the formation of each structure. Section 7.6 described the importance of homeotic genes in establishing the correct placement of a developing animal's parts. Scientists first discovered homeotic genes by studying mutant flies with legs growing out of their heads. Since that time, additional studies have verified that homeotic genes orchestrate development in humans and all other animal species.

Although many details of animal development remain undiscovered, clear patterns do emerge on a broader scale. For example, biologists distinguish between indirect and direct development. An animal that undergoes **indirect development** spends the early part of its life as a larva, an immature stage that looks different from the adult. A caterpillar, for example, is a larva that looks nothing like its butterfly parents; likewise, a tadpole resembles a fish, not the adult frog or salamander that it will grow up to become. Caterpillars, tadpoles, and other larvae often spend most of their time eating and growing. Then, during a process called metamorphosis, the larva matures into an adult.

Humans and many other familiar animals undergo **direct development:** An infant resembles a smaller version of its parents. The hatching turtle in figure 30.3, for example, is a miniature version of the adult.

This chapter combines reproduction and development, starting with the reproductive anatomy of human males and females. Before you begin, you may find it helpful to review mitosis (chapter 8), meiosis (chapter 9), and the basics of hormone function (chapter 25). The second half of the chapter describes where babies come from—that is, how a fertilized egg grows into a fully formed infant.

Direct development

Figure 30.3 From Hatchling to Adult. Like other animals that undergo direct development, a newly hatched tortoise is a miniature version of the adult.

(hatchling): ©Daniel Heuclin/Science Source; (adult): ©Juniors Bildarchiv GmbH/Alamy Stock Photo RF

30.1 Mastering Concepts

1. What is the difference between asexual and sexual reproduction?
2. How is internal fertilization different from external fertilization?
3. How do genes participate in differentiation and pattern formation?
4. Differentiate between indirect and direct development.

30.2 Males Produce Sperm Cells

Both the male and female **reproductive systems** consist of the organs that produce and transport gametes. Each system includes paired **gonads** (testes or ovaries), which contain the **germ cells** that give rise to gametes. Both reproductive systems also include tubes that transport the gametes.

In both sexes, hormones control reproduction and the development of **secondary sex characteristics,** features that distinguish the sexes but do not participate directly in reproduction. Examples include enlarged breasts and menstruation in adult females and facial hair and deep voices in adult males.

Although the male and female reproductive systems share some similarities, there are also obvious differences. This section details the features and processes that are unique to males.

A. Male Reproductive Organs Are Inside and Outside the Body

Figure 30.4 illustrates the human male reproductive system. The paired **testes** (singular: testis) are the male gonads. The testes lie in a sac called the **scrotum.** Their location outside of the abdominal cavity allows the testes to maintain a

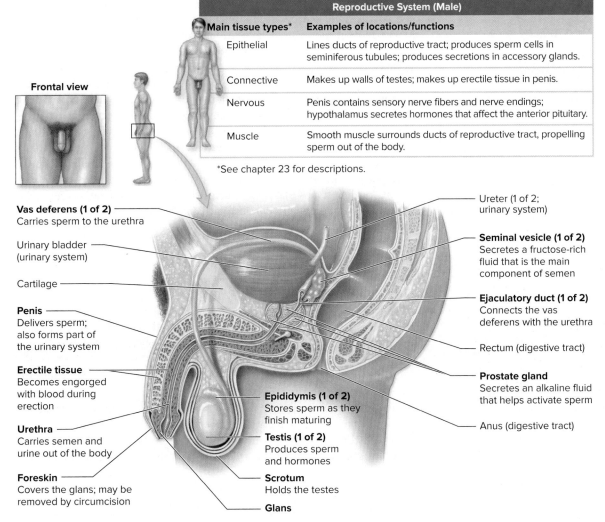

Reproductive System (Male)	
Main tissue types*	**Examples of locations/functions**
Epithelial	Lines ducts of reproductive tract; produces sperm cells in seminiferous tubules; produces secretions in accessory glands.
Connective	Makes up walls of testes; makes up erectile tissue in penis.
Nervous	Penis contains sensory nerve fibers and nerve endings; hypothalamus secretes hormones that affect the anterior pituitary.
Muscle	Smooth muscle surrounds ducts of reproductive tract, propelling sperm out of the body.

*See chapter 23 for descriptions.

Frontal view

Vas deferens (1 of 2)
Carries sperm to the urethra

Urinary bladder
(urinary system)

Cartilage

Penis
Delivers sperm;
also forms part of
the urinary system

Erectile tissue
Becomes engorged
with blood during
erection

Urethra
Carries semen and
urine out of the body

Foreskin
Covers the glans; may be
removed by circumcision

Epididymis (1 of 2)
Stores sperm as they
finish maturing

Testis (1 of 2)
Produces sperm
and hormones

Scrotum
Holds the testes

Glans

Ureter (1 of 2;
urinary system)

Seminal vesicle (1 of 2)
Secretes a fructose-rich
fluid that is the main
component of semen

Ejaculatory duct (1 of 2)
Connects the vas
deferens with the urethra

Rectum (digestive tract)

Prostate gland
Secretes an alkaline fluid
that helps activate sperm

Anus (digestive tract)

Figure 30.4 The Human Male Reproductive System. The paired testes manufacture sperm cells, which travel through a series of ducts before exiting the body via the urethra in the penis.

The Human Infant Begins Life as a Zygote

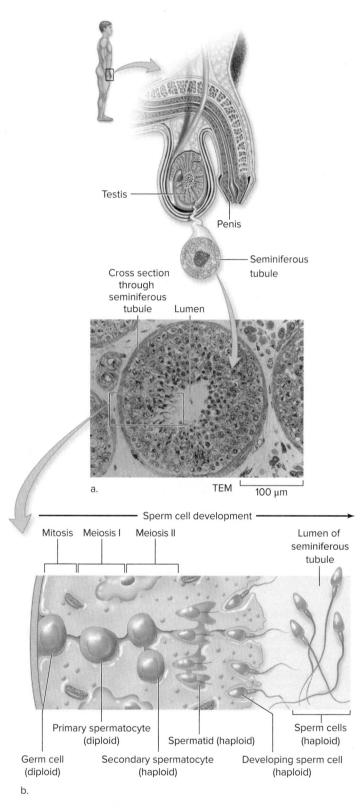

Testis

Penis

Cross section through seminiferous tubule

Seminiferous tubule

Lumen

a.   TEM   100 μm

Sperm cell development

Mitosis   Meiosis I   Meiosis II

Lumen of seminiferous tubule

Germ cell (diploid)

Primary spermatocyte (diploid)

Secondary spermatocyte (haploid)

Spermatid (haploid)

Developing sperm cell (haploid)

Sperm cells (haploid)

b.

Figure 30.5 Sperm Production. (a) Anatomy of a testis. (b) In the walls of the seminiferous tubules, diploid germ cells divide mitotically. Some of the daughter cells undergo meiosis, giving rise to four haploid cells that mature into sperm cells.

Photo: (a): © Larry Johnson/Dept. of Veterinary Anatomy and Public Health

temperature about 3°C cooler than the rest of the body, which is necessary for sperm to develop properly. Muscles surrounding each testis can bring the scrotum closer to the body, conserving warmth when the temperature is too cold. If conditions are too warm, the scrotum descends away from the body.

A maze of small ducts carries developing sperm to the left or right **epididymis,** a tightly coiled tube that receives and stores sperm from one testis. Each epididymis opens into a **vas deferens,** a duct that travels upward out of the scrotum, bends behind the bladder, and connects with the left or right **ejaculatory duct.** The two ejaculatory ducts empty into the **urethra,** the tube that extends the length of the cylindrical **penis** and carries both urine and semen out of the body. Although these two fluids share the urethra, a healthy male cannot urinate when sexually aroused because a ring of smooth muscle temporarily seals the exit from the urinary bladder.

Semen, the fluid that carries sperm cells, includes secretions from several accessory glands. The two **seminal vesicles,** one of which opens into each vas deferens, secrete most of the fluid in semen. The secretions include fructose (a sugar that supplies energy) and prostaglandins. Prostaglandins are hormone-like lipids that may stimulate contractions in the female reproductive tract, helping to propel sperm. In addition, the single, walnut-sized **prostate gland** wraps around part of the urethra and contributes a thin, milky, alkaline fluid that activates the sperm to swim.

During sexual arousal, the penis becomes erect, enabling it to penetrate the vagina and deposit semen in the female reproductive tract. At the peak of sexual stimulation, a pleasurable sensation called **orgasm** occurs, accompanied by rhythmic muscular contractions that eject the semen through the urethra and out the penis. **Ejaculation** is the discharge of semen from the penis. One human ejaculation typically delivers more than 100 million sperm cells.

The organs of the male reproductive system may become cancerous. Prostate cancer is the second most common type of cancer in men (behind lung cancer). The resulting prostate enlargement constricts the urethra and may interfere with urination and ejaculation. Prostate cancer usually strikes men older than 50; noncancerous (benign) prostate enlargement affects many older men as well. Testicular cancer, which usually occurs in men younger than 40, is much rarer than prostate cancer. Mutated cells in the testes divide out of control, forming lumps that may be detected in a self-examination. Testicular cancer has a very high cure rate, if detected before it spreads to other parts of the body. ⓘ *cancer*, section 8.6

B. Spermatogenesis Yields Sperm Cells

Spermatogenesis, the production of sperm, begins when a male reaches puberty and continues throughout life.

Figure 30.5 illustrates the internal anatomy of a testis. Each testis contains about 200 tightly coiled, 50-centimeter-long **seminiferous tubules,** which produce the sperm cells. Endocrine cells fill the spaces between the seminiferous tubules and secrete male sex hormones.

Sperm production begins with diploid germ cells that reside within the wall of a seminiferous tubule (see figure 30.5b). The germ cells have nuclei containing 46 chromosomes. When a germ cell divides mitotically, one daughter cell remains in the tubule wall and acts as a stem cell, continually giving rise to cells that become sperm. The other cell becomes a diploid **primary spermatocyte** that accumulates cytoplasm and moves closer to the tubule's lumen (central cavity).

In the wall of the seminiferous tubule, the primary spermatocyte undergoes meiosis I, yielding two haploid **secondary spermatocytes.** These cells undergo meiosis II, forming four round, haploid cells called **spermatids** that each contain 23 chromosomes.

As the spermatids move into the lumen of the seminiferous tubule, they complete their differentiation into mature sperm cells. They separate into individual cells and develop flagella. They also lose much of their cytoplasm, acquire a streamlined shape, and package their DNA into a distinct head (figure 30.6). Mitochondria just below the head region generate the ATP the sperm needs to move toward an egg cell. The caplike **acrosome** covers the head and releases enzymes that will help the sperm penetrate the egg cell. The entire process, from germ cell to mature sperm cell, takes 74 days in humans. ⓘ *ATP,* section 4.3

C. Hormones Influence Male Reproductive Function

Hormones play a critical role in male reproduction (figure 30.7). In the brain, the hypothalamus secretes **gonadotropin-releasing hormone (GnRH).** This water-soluble hormone travels in the bloodstream to the anterior pituitary, where it stimulates the release of two other water-soluble hormones: **follicle-stimulating hormone (FSH)** and **luteinizing hormone (LH).** Blood carries FSH and LH throughout the body. ⓘ *water-soluble hormones,* section 25.2A

LH induces endocrine cells in the testes to release the steroid hormone **testosterone** and other male sex hormones (androgens). In the presence of FSH, testosterone affects the body in multiple ways. In adolescents, the hormone stimulates the development of secondary sex characteristics. The testes and penis begin to enlarge at puberty, and hair grows on the face, in the armpits, and at the groin. Testosterone also stimulates the secretion of growth hormone, causing a growth spurt that increases height, boosts muscle mass, and deepens the voice. In adults, testosterone stimulates sperm production, sustains the libido, and controls the activity of the prostate gland. ⓘ *steroid hormones,* section 25.2B

Negative feedback loops maintain homeostasis in the concentrations of these hormones. Negative feedback also explains one well-known consequence of abusing anabolic steroids: infertility or low sperm counts (see Why We Care 25.1). The body mistakes the synthetic steroids for testosterone, causing the testes to produce less of the real sex hormone. Without testosterone, sperm do not form. ⓘ *negative feedback,* section 23.4

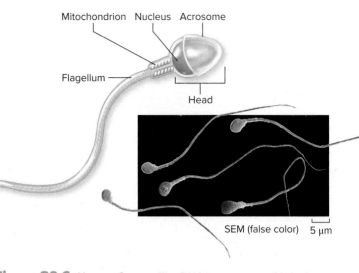

Figure 30.6 Human Sperm. The DNA in a sperm cell is in the nucleus, which enters the egg cell. The long flagellum propels the sperm.

Photo: ©Eye of Science/Science Source

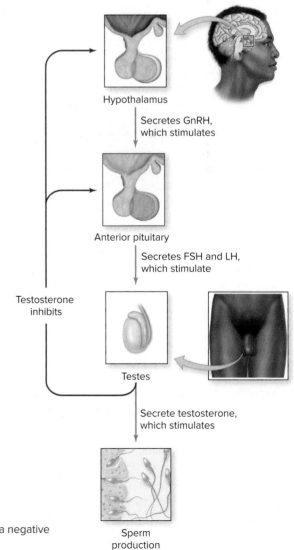

Figure 30.7 Male Reproductive Hormones. GnRH, FSH, LH, and testosterone interact in a negative feedback loop to regulate male reproductive function.

30.2 Mastering Concepts

1. What are the relationships among gonads, germ cells, gametes, and the zygote?
2. Describe the role of each part of the male reproductive system.
3. Where in the testes do sperm develop?
4. What are the stages of spermatogenesis?
5. What are the parts of a mature sperm cell?
6. How do hormones regulate sperm production?

30.3 Females Produce Egg Cells

Egg cell production in females is somewhat more complicated than is sperm formation in males, in at least two ways. First, in females, meiosis begins before birth, pauses, and resumes at sexual maturity. Meiosis does not complete until after a sperm cell fertilizes the egg cell. Second, egg cell production is cyclical, under the control of several interacting hormones whose levels fluctuate monthly during a woman's reproductive years. Keep these differences in mind while reading this section.

A. Female Reproductive Organs Are Inside the Body

Female sex cells develop within the **ovaries,** which are paired gonads in the abdomen (figure 30.8). Ovaries produce both egg cells and sex hormones. They do not contain ducts comparable to the seminiferous tubules of the male's testes. Instead, within each ovary of a newborn female are about a million oocytes, the cells that give rise to mature egg cells. Nourishing **follicle cells** surround each oocyte.

Approximately once a month, beginning at puberty, one ovary releases the single most mature oocyte. Beating cilia sweep the mature oocyte into the fingerlike projections of one of the two **uterine tubes** (also called Fallopian tubes or oviducts). If sperm are present, fertilization occurs in a uterine tube. The tube carries the oocyte or zygote into a muscular saclike organ, the **uterus.** During

Frontal view

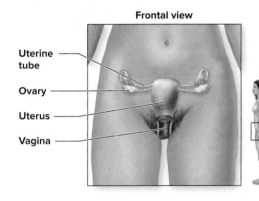

- Uterine tube
- Ovary
- Uterus
- Vagina

Reproductive System (Female)	
Main tissue types*	**Examples of locations/functions**
Epithelial	Lines uterus, uterine tubes, and vagina; produces oocytes in ovaries; forms external surface of umbilical cord.
Connective	Makes up walls of ovaries, uterus, and vagina.
Nervous	Clitoris contains sensory nerve fibers and nerve endings; hypothalamus secretes hormones that affect the anterior pituitary.
Muscle	Smooth muscle surrounds uterine tubes, uterus, and vagina.

*See chapter 23 for descriptions.

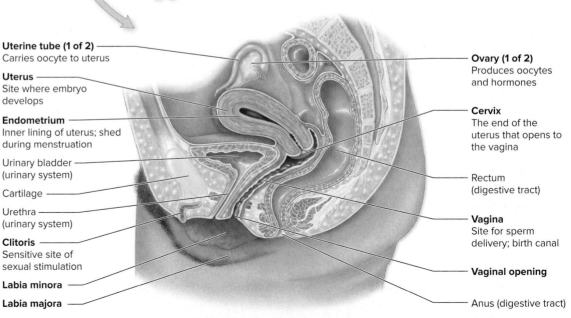

Uterine tube (1 of 2)
Carries oocyte to uterus

Uterus
Site where embryo develops

Endometrium
Inner lining of uterus; shed during menstruation

Urinary bladder
(urinary system)

Cartilage

Urethra
(urinary system)

Clitoris
Sensitive site of sexual stimulation

Labia minora

Labia majora

Ovary (1 of 2)
Produces oocytes and hormones

Cervix
The end of the uterus that opens to the vagina

Rectum
(digestive tract)

Vagina
Site for sperm delivery; birth canal

Vaginal opening

Anus (digestive tract)

Figure 30.8 The Human Female Reproductive System. One of the two ovaries periodically releases an oocyte. This egg cell enters a nearby uterine tube. If a sperm cell fertilizes the oocyte, the offspring develops in the uterus and is delivered through the vagina.

pregnancy, the fetus develops inside the uterus, also called the womb. The **endometrium,** or inner lining of the uterus, has a rich blood supply that is important in both menstruation and pregnancy.

The **cervix** is the necklike narrowing at the lower end of the uterus. The cervix opens into the **vagina,** the tube that leads outside the body. The vagina receives the penis during intercourse, and it is the birth canal. Like many other areas of the body, the vagina harbors a community of resident microorganisms. These bacteria lower the pH of the vagina, which helps prevent colonization by harmful bacteria and the yeast *Candida albicans*. Taking antibiotics can disrupt this microbial community and create an opportunity for *Candida* to overgrow, causing a vaginal yeast infection. ⓘ *beneficial microbes,* section 15.2D

Two pairs of fleshy folds protect the vaginal opening on the outside: the labia majora (major lips) and the thinner, underlying flaps of tissue they protect, called the labia minora (minor lips). The **clitoris** is a 2-centimeter-long structure at the upper junction of both pairs of labia. Rubbing the clitoris stimulates females to experience orgasm. Together, the labia, clitoris, and vaginal opening constitute the **vulva,** or external female genitalia.

The female secondary sex characteristics include the wider and shallower shape of the pelvis, the accumulation of fat around the hips, a higher pitched voice than that of the male, and the breasts. The **breasts** produce milk that nourishes a nursing infant; each breast has fatty tissue, collagen, milk ducts, and a nipple.

Cancers of the female reproductive system often develop in the breasts, cervix, or ovaries. Breast cancer is the most common cancer type in women. The abnormally dividing cells may originate in the breast's milk-forming tissues or in the milk ducts. Some, but not all, forms of breast cancer have a strong heritable component. A family history is also the leading risk factor for ovarian cancer, which usually starts in the outer lining of the ovary. In contrast, nearly all cases of cervical cancer are associated with the sexually transmitted human papillomavirus (see section 30.4). The Pap test, in which a medical professional uses a microscope to look for abnormal cervical cells, is an important early-detection tool for cervical cancer.

B. Oogenesis Yields Egg Cells

The making of an egg cell—**oogenesis**—begins with a diploid germ cell containing 46 chromosomes (figure 30.9). Each germ cell grows, accumulates cytoplasm, replicates its DNA, and divides mitotically, becoming two **primary oocytes.** The subsequent divisions of meiosis partition the cytoplasm unequally, so that oogenesis (unlike spermatogenesis) produces cells of different sizes. By the end of meiosis I, the primary oocyte has divided into a small, haploid **polar body** and a larger, haploid **secondary oocyte,** each containing 23 chromosomes.

Ovulation is the release of a secondary oocyte from its follicle. As the follicle ruptures, fingerlike projections of the uterine tube move across the ovary and usher the egg into the tube. Following ovulation, the now-ruptured follicle transforms into a gland called a **corpus luteum.** Meanwhile, meiosis halts at metaphase II and does not resume unless a sperm contacts the secondary oocyte. In that case, the secondary oocyte again divides unequally to produce a small additional polar body and the mature egg cell (or ovum), which contains 23 chromosomes and a large amount of cytoplasm.

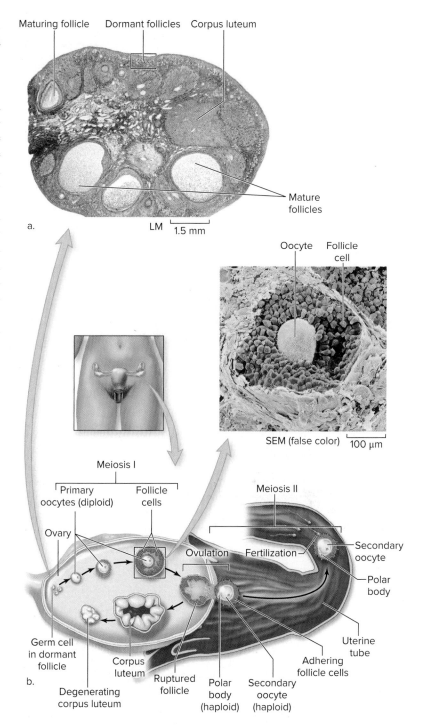

Figure 30.9 **Egg Cell Production.** (a) A cross section of a cat's ovary shows follicles at several stages of development. (b) This illustration traces the stages of egg cell development in a human ovary. Follicles contain diploid germ cells, which divide mitotically to produce primary oocytes. These cells undergo meiosis. Every month between puberty and menopause, the most mature follicle ruptures and a secondary oocyte bursts out of the ovary, an event called ovulation. The secondary oocyte completes meiosis II only if fertilized by a sperm cell.

Photos: (a): ©Victor P. Eroschenko RF; (b): ©Prof. P.M. Motta, G. Macchiarelli, S.A. Nottola/Science Source

TABLE 30.1 **Spermatogenesis and Oogenesis Compared**

Process	Diploid Starting Cell	Product of Mitosis (Diploid)	Products of Meiosis I (Haploid)	Products of Meiosis II (Haploid)
Spermatogenesis	Germ cell in seminiferous tubule	Primary spermatocyte	Two secondary spermatocytes	Four equal-sized spermatids
Oogenesis	Germ cell in ovary	Primary oocyte	One large secondary oocyte + one small polar body	One large egg cell + three small polar bodies

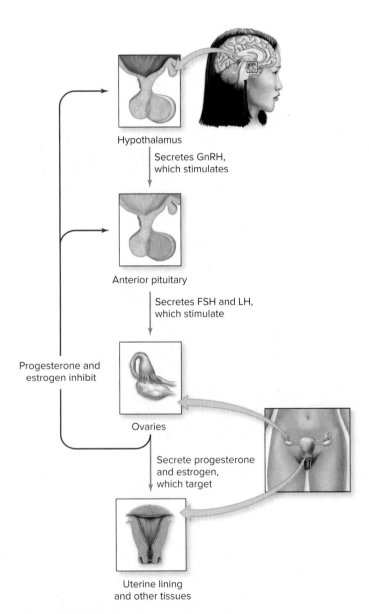

Figure 30.10 **Female Reproductive Hormones.** Hormones from the hypothalamus regulate the activities of the ovaries and the uterine lining. Progesterone and estrogen, in turn, regulate hormone release from the hypothalamus and anterior pituitary in a negative feedback loop.

The polar body produced in meiosis I may divide into two additional polar bodies, or it may decompose.

The egg cell, in receiving most of the cytoplasm, contains all of the biochemicals and organelles that the zygote will use until its own DNA begins to function. The polar bodies normally play no further role in development. Rarely, however, sperm can fertilize polar bodies, and a mass of tissue that does not resemble an embryo grows until the woman's body rejects it. A fertilized polar body accounts for about 1 in 100 miscarriages.

From puberty to menopause (when menstruation stops entirely), monthly hormonal cues prompt an ovary to release one secondary oocyte into a uterine tube. If a sperm penetrates the oocyte membrane, meiosis in the oocyte completes, and the nuclei from the two parents combine to form the diploid zygote. If the secondary oocyte is not fertilized, it leaves the body with the endometrium in the menstrual flow.

In some ways, oogenesis is similar to spermatogenesis (table 30.1). Each process starts with a diploid germ cell that eventually gives rise to the haploid gametes. Also, both testes and ovaries contain gametes in various stages of development. Of course, the two processes also differ. For example, spermatogenesis gives rise to four equal-sized sperm cells, whereas in females, one germ cell yields one functional egg cell and three smaller polar bodies.

Also, the timetable for oogenesis differs greatly from that of spermatogenesis. A male takes about 74 days to produce a sperm cell. In contrast, oogenesis stretches from before birth until after puberty. The ovaries of a 3-month-old female fetus contain 2 million or more primary oocytes. From then on, the oocytes slowly degenerate. At birth, a million primary oocytes are present, their development arrested in prophase I. Only about 400,000 remain by the time of puberty, after which one or a few oocytes complete meiosis I each month. These secondary oocytes stop meiosis again, this time at metaphase II. Meiosis is completed only if fertilization occurs.

C. Hormones Influence Female Reproductive Function

The male and female reproductive systems rely on many of the same hormones (figure 30.10). Females, however, produce these hormones in different quantities and on a different schedule.

In females, hormonal fluctuations produce two interrelated cycles. The **ovarian cycle** controls the timing of oocyte maturation in the ovaries, and the **menstrual cycle** prepares the uterus for pregnancy. Figure 30.11 tracks changes in the follicle, the uterine lining, and the levels of four hormones during the ovarian and menstrual cycles.

Menstruation begins on the first day of the menstrual cycle. Low blood levels of two sex hormones, **estrogen** and **progesterone,** signal the hypothalamus to secrete GnRH. This hormone prompts the anterior pituitary to release FSH and LH into the bloodstream. In the ovaries, receptors on the surfaces of follicle cells bind to FSH, stimulating follicles to mature and to release estrogen.

Estrogen has two seemingly contradictory roles in the ovarian cycle. At low concentrations, estrogen inhibits the release of both LH and FSH. But at around the midpoint of the cycle, a spike in estrogen accumulation triggers the release of additional LH and FSH from the anterior pituitary. This LH surge in the bloodstream triggers ovulation and transforms the ruptured follicle into a corpus luteum.

The corpus luteum, in turn, secretes progesterone and estrogen, which have multiple effects. These two hormones act together to promote the thickening of the endometrium, preparing the uterus for possible pregnancy. Progesterone and estrogen also act on target cells in the hypothalamus, inhibiting the production of GnRH, LH, and FSH.

If pregnancy does not occur, the corpus luteum degenerates into an inactive scar. Over the next several days, levels of progesterone and estrogen gradually decline (section 30.5 describes what happens if pregnancy does occur). The reduced levels of these hormones no longer maintain the endometrium, which then exits the body through the cervix and vagina as menstrual flow. Lowered progesterone and estrogen levels also release their inhibition of GnRH, LH, and FSH in the brain, and the cycle begins anew.

D. Hormonal Fluctuations Can Cause Discomfort

Fluctuating concentrations of hormones trigger a variety of conditions that are unique to women. One example is premenstrual syndrome (PMS), a collection of symptoms that appears in the second half of the menstrual cycle. In the days before her menstrual period begins, a woman may experience headache, breast tenderness, cramping, depression, irritability, or dozens of other signs of PMS. The cause of each symptom is unknown, but the correlation with hormonal changes is clear. Over-the-counter drugs provide relief from many symptoms.

Menstrual cramps, which may occur with or without PMS, arise from smooth muscle contractions in the uterus. The function of these painful contractions is to help menstrual fluid exit the uterus; heavy bleeding is correlated with severe cramps.

As a woman nears the end of her reproductive years, her hormonal fluctuations cease. Ironically, the estrogen withdrawal that accompanies menopause can cause symptoms as well. For example, hot flashes affect most women going through menopause. A hot flash, which typically lasts a few minutes, is a temporary sensation of heat in the face and upper body, coupled with flushed skin and sweating. The exact cause is unknown, but researchers speculate that the withdrawal of estrogen somehow affects the temperature set point of the hypothalamus.

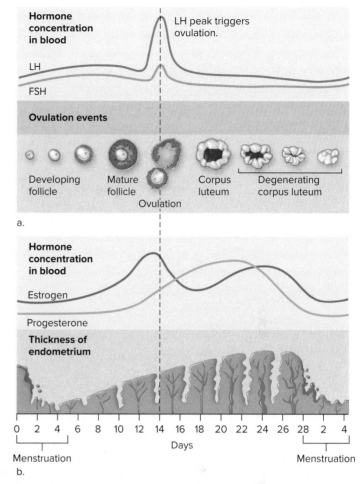

Figure 30.11 The Ovarian and Menstrual Cycles. (a) In the ovarian cycle, LH and FSH coordinate follicle maturation and oocyte release from the ovary. (b) Estrogen and progesterone signal the endometrium to thicken, then disintegrate, during each menstrual cycle. Note that this diagram shows a 28-day cycle; in reality, a woman's cycle may last anywhere from 21 to 35 days or longer, and it may not be the same length each time.

30.3 Mastering Concepts

1. What is the role of each part of the female reproductive system?
2. Where do egg cells develop?
3. What are the stages of oogenesis?
4. What are polar bodies, and where do they come from?
5. How do hormones regulate the ovarian and menstrual cycles?
6. List some side effects of hormone fluctuations in females.

30.4 Reproductive Health Considers Contraception and Disease

Birth control, or **contraception,** is the use of devices or practices that work "against conception"; that is, they prevent the union of sperm and egg. Table 30.2 summarizes several methods and estimates the pregnancy rate based on typical (not perfect) use. For comparison, 85 out of 100 fertile women will become pregnant within one year if they are sexually active and use no birth control at all. (Burning Question 30.1 explains how to know when conception is likely.)

Besides abstinence, the most effective contraceptives are surgical; next are those that adjust hormone concentrations in the woman's body. Birth control pills, patches, vaginal rings, injections, and implants all contain a synthetic form of progesterone. If used correctly, each of these methods prevents ovulation and therefore precludes fertilization. Other methods kill sperm, block the meeting of sperm and oocyte, or prevent a developing embryo from implanting in the lining of the uterus.

Only latex and polyurethane condoms, however, simultaneously prevent pregnancy and protect against **sexually transmitted infections (STIs),** which spread to new hosts during sexual contact. Vaginal intercourse, oral sex, and anal sex all can provide direct, person-to-person transmission for a wide variety of infectious agents. Interestingly, humans are not the only ones to suffer from STIs. Other animal species have STIs of their own; so do plants, which can pass viroids in pollen. ⓘ *viroids,* section 7.10

Burning Question 30.1

When can conception occur?

The short answer to this question is that conception can occur as soon as an ovary releases an egg. Predicting ovulation and avoiding sex near that time can therefore be an effective method of birth control.

After ovulation, sperm may fertilize an egg for about 24 hours. If unfertilized, the egg disintegrates or the body reabsorbs it. However, a woman is fertile for longer than one day each cycle. During some cycles, the ovaries may release one or more additional eggs over a couple of days after initial ovulation. Also, sperm may survive inside a female for up to 5 days. Therefore, the fertile period begins approximately 5 days before ovulation and lasts about 3 or 4 days after ovulation.

Monitoring body clues can help a woman determine when an egg is likely to be released. For example, body temperature (measured upon waking) is elevated shortly after ovulation (figure 30.A). Other clues, such as the quantity and quality of cervical fluid, as well as the position of the cervix, may also help pinpoint ovulation. To track their fertility, many women record their temperature and other measurements in charts each day.

Careful, daily record-keeping is essential to monitoring fertility. Although it may seem logical to predict ovulation by simply counting days after the start of menstruation, this method is less effective. Many healthy women have cycles that average shorter or longer than 28 days, so ovulation will not always occur on the fourteenth day of the cycle. Further, any given cycle may be longer or shorter than usual, depending on a woman's diet, stress level,

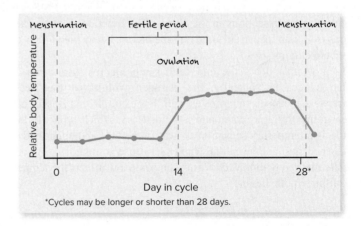

Figure 30.A Thermal Shift. Body temperature—measured upon waking each morning—increases after an egg is released. However, the fertile period begins a few days before ovulation.

and other factors. Averaging the duration of previous cycles is therefore not a reliable predictor for when the next egg will be released.

Submit your burning question to
marielle.hoefnagels@mheducation.com

		TABLE 30.2	**Birth Control Methods**		

Method	Area(s) Targeted	Mechanism	Advantages	Disadvantages	Pregnancies per 100 Women per Year**
Abstinence	N/A	No intercourse; sperm never encounter egg cell	No cost	Difficult to do	0
Vasectomy*	1	Cuts each vas deferens, so sperm cells never reach urethra	Permanent; does not interrupt spontaneity	Requires minor surgery; difficult to reverse	< 1
Withdrawal*	2	Removal of penis from vagina before ejaculation	No cost	Not recommended for sexually inexperienced men; requires great self-control; sperm may leak from penis before withdrawal; sperm spilled on vulva may cause pregnancy	18
Latex or polyurethane condom	2	Worn over penis or inserted into vagina; keeps sperm out of vagina	Protects against sexually transmitted infections	Disrupts spontaneity; reduces sensation	15
Cervical cap used with spermicide*	3	Kills sperm and blocks cervix	Inexpensive; can be kept in for 24 hours	May slip out of place; must be fitted; less effective for women who have already given birth vaginally	20
Intrauterine device (IUD)*	4	Prevents implantation of preembryo	Does not interrupt spontaneity	Severe menstrual cramps; risk of infection	< 1
Tubal ligation*	5	Cuts uterine tubes, so oocytes never reach uterus	Permanent; does not interrupt spontaneity	Requires surgery; risk of infection; difficult to reverse	< 1
Fertility awareness method*	6	No intercourse during fertile times, as inferred from body temperature and other clues (see Burning Question 30.1)	No cost	Requires careful record-keeping	20
Hormonal contraception (estrogen and/or progesterone)*	4, 6	Prevents ovulation and implantation	Does not interrupt spontaneity; easy to use	Menstrual changes; weight gain; headaches	2–9

*Does not protect against sexually transmitted infections (STIs).

**For some methods, the number of pregnancies may be much lower if the method is used perfectly every time. Using no birth control results in 85 pregnancies per 100 fertile women per year.

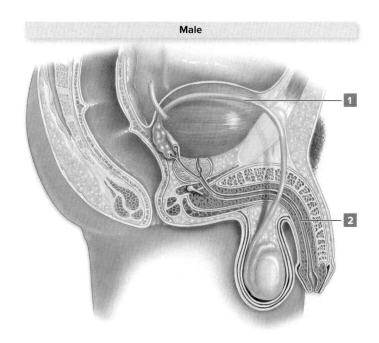

Male

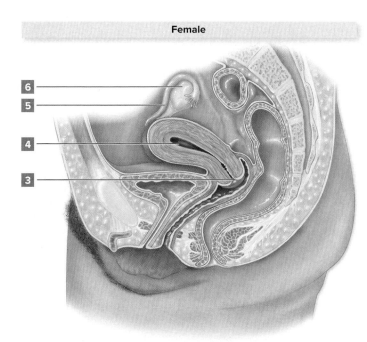

Female

TABLE 30.3 Examples of Sexually Transmitted Diseases

Disease	Agent	Treatment
Viruses		
HIV/AIDS	Human immunodeficiency virus (HIV)	Combination of drugs that reduce viral replication
Genital warts	Human papillomavirus (HPV)	Removal of warts
Cervical cancer	Human papillomavirus (HPV)	Surgery, radiation, chemotherapy
Genital herpes	Herpes simplex virus	Medications that reduce outbreak frequency and duration
Hepatitis B	Hepatitis B virus	None
Bacteria		
Chlamydia	*Chlamydia trachomatis*	Antibiotics
Gonorrhea	*Neisseria gonorrhoeae*	Antibiotics
Syphilis	*Treponema pallidum*	Antibiotics
Protists		
Trichomoniasis	*Trichomonas vaginalis*	Antiprotozoan drugs
Fungi		
Yeast infection	*Candida albicans*	Antifungal drugs

Viruses, bacteria, protists, and fungi all can cause STIs; table 30.3 lists a sampling of some of the most common ones, along with their treatments. In the United States, the most common sexually transmitted disease is genital warts, which can be caused by more than 40 strains of the human papillomavirus (HPV). Many infections remain symptomless, but others trigger the growth of visible bumps on the male or female genitals. Some strains of HPV that do not cause warts can create a much bigger problem: cancer. Nearly all cervical cancer tumors test positive for DNA from HPV, and the same virus has also been linked to cancers of the mouth and throat in males. If administered before exposure to the virus, the HPV vaccine (commonly called the "cervical cancer vaccine") can help prevent these cancers in both sexes.

Another disease that affects millions of people each year is trichomoniasis, caused by a protist called *Trichomonas vaginalis* (see figure 15.29b). Males can transmit this organism, usually without visible symptoms. Infected women may notice unusual vaginal discharge along with pain, irritation, and itching.

A third common disease, chlamydia, is caused by the bacterium *Chlamydia trachomatis*. Typical symptoms include discharge from the penis or vagina; an infected person may also experience pain or a burning sensation when urinating. Chlamydia, however, is sometimes called a "silent disease" because some 75% of infected women and 50% of infected men remain symptomless.

Infections with HPV, trichomoniasis, chlamydia, and other agents listed in table 30.3 often remain invisible, but that does not mean they are harmless. For example, in addition to the risk of cancer associated with HPV, untreated chlamydia infections in women can spread to the uterine tubes. Pelvic inflammatory disease, one possible complication, can cause pain and infertility. Moreover, syphilis, gonorrhea, and chlamydia can pass to infants during childbirth.

Most STIs are also associated with an elevated risk for acquiring and transmitting HIV. One explanation is that many STIs cause sores through which viral particles can enter or leave the body. A second reason is that any infection can cause inflammation and other defensive reactions. As described in chapter 29, these responses attract the types of white blood cells that HIV infects. Efforts to prevent STIs in general can therefore also help reduce infections by HIV in particular.

The best way to prevent STIs is to abstain from sex entirely. The second best way is to develop a long-term, monogamous relationship with a partner who has recently been tested for STIs and is therefore known to be infection-free. The third best option is to properly use a latex or polyurethane condom throughout a sexual encounter. In addition, vaccines can protect against some sexually transmitted viruses, including HPV and hepatitis B.

30.4 Mastering Concepts

1. Describe how three birth control methods work.
2. List and describe three common STIs.
3. Describe two reasons that a symptomless STI can be harmful.

30.5 The Human Infant Begins Life as a Zygote

So far, this chapter has described gamete production in the male and female reproductive systems. This section now turns to the development of a baby. As you will see, fertilization produces the zygote; this single cell divides many times as it develops into a preembryo, an embryo, a fetus, and a newborn baby. Why We Care 30.1 explains how problems during development can cause birth defects.

A. Fertilization Initiates Pregnancy

After intercourse, sperm cells swim toward the oocyte. Of the 100 million sperm that begin the journey in the vagina, only about 200 approach the egg cell in a uterine tube.

Those that make it must penetrate two layers to contact the ovum. An outer layer of follicle cells surrounds a thin, clear "jelly layer" of proteins and carbohydrates that encases the oocyte (figure 30.12). On contact with the follicle cells surrounding the oocyte, each sperm's acrosome bursts, spilling enzymes that digest both outer layers.

Fertilization begins when the outer membranes of one sperm cell and the secondary oocyte touch. At that time, physical and chemical changes across the oocyte surface produce a "fertilization envelope" that prevents other sperm from entering the same egg cell. The sperm's head releases its DNA as it enters the secondary oocyte. Meanwhile, the female cell completes meiosis. The sperm's DNA and egg's DNA combine, completing zygote formation. This diploid cell has 46 chromosomes, in 23 pairs; one chromosome of each pair comes from each parent. (i) *homologous chromosomes,* section 9.2

Occasionally a woman ovulates two or more egg cells at once, and a different sperm fertilizes each one. If all the zygotes complete development, the result is twins, triplets, quadruplets, or even higher-order multiple births (see Why We Care 9.1). Because each zygote is derived from a separate sperm and egg cell, the siblings will not be genetically identical, and they may be of different sexes. At the opposite end of the spectrum are couples that have trouble conceiving any child at all; table 30.4 explains how technology can help.

1. Sperm squeezes between follicle cells adhering to egg cell.

2. Sperm's acrosome bursts, releasing enzymes that digest follicle cells and jelly layer surrounding egg cell.

3. Sperm cell membrane fuses with egg cell membrane.

4. Sperm cell nucleus enters egg cell and fuses with its nucleus.

Figure 30.12 Fertilization. A sperm cell releases enzymes that help its nucleus enter the oocyte. Meanwhile, changes to the egg cell's surface produce a fertilization envelope, which ensures that only one sperm fertilizes the egg.

TABLE 30.4	Assisted Reproductive Technologies
Technology	**Description**
Artificial insemination	Donated sperm is placed in a woman's reproductive tract.
In vitro fertilization (IVF)	Sperm fertilizes an oocyte in a laboratory dish; one or more preembryos are placed in a woman's uterus.
Oocyte donation	Variant of *in vitro* fertilization in which the oocytes come from a donor instead of from the woman who is trying to become pregnant.
Intracytoplasmic sperm injection (ICSI)	Variant of *in vitro* fertilization in which a sperm cell is injected directly into an egg.
Gamete intrafallopian transfer (GIFT)	Oocytes are collected and placed with sperm into a woman's uterine tube; fertilization occurs in the woman's body.
Zygote intrafallopian transfer (ZIFT)	Sperm fertilizes an oocyte in a laboratory dish; the zygote is placed in a woman's uterine tube 24 hours later.

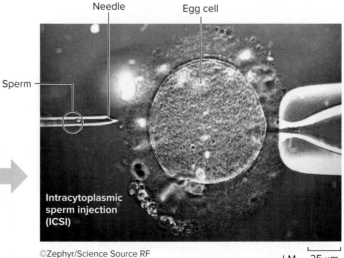

Intracytoplasmic sperm injection (ICSI)

©Zephyr/Science Source RF

LM   25 μm

Figure 30.13 From Ovulation to Implantation. Fertilization occurs in the uterine tube, producing the zygote. The first mitotic divisions occur while the preembryo moves toward the uterus. By day 7, the preembryo has begun to implant in the uterine lining.

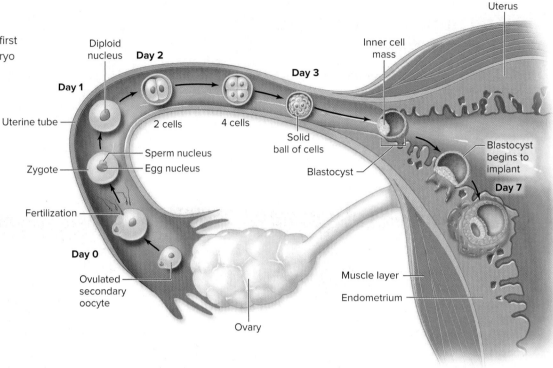

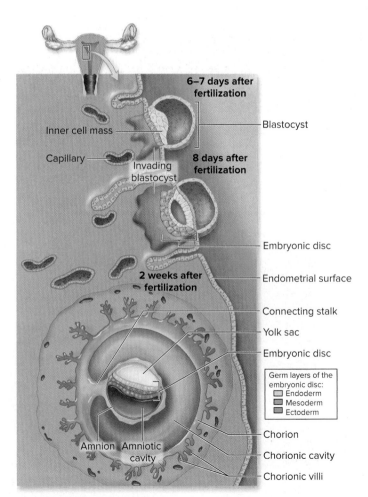

B. The Preembryonic Stage Ends When Implantation Is Complete

The first 2 weeks of prenatal development have a variety of names, but we will call this period the **preembryonic stage.**

About 24 hours after fertilization, the zygote divides for the first time (figure 30.13). This event signals the beginning of a period of rapid mitotic cell division called **cleavage.** The result is a solid ball of 16 or more cells, and it reaches the uterus 3 to 6 days after fertilization. The preembryo then hollows out, its center filling with fluid that seeps in from the uterus. This fluid-filled ball of cells is called a **blastocyst.** The few cells inside the blastocyst form the **inner cell mass,** the cells that will develop into the embryo itself. (The inner cell mass is also the source of embryonic stem cells.) ⓘ *stem cells,* section 11.3A

In a process called **implantation,** the blastocyst becomes embedded in the lining of the uterus (figure 30.14). Implantation begins within a week of fertilization. The outer layer of the blastocyst not only secretes digestive enzymes that eat through the outer layer of the uterine lining, but it also sends projections into the uterine lining. Until the placenta forms, the preembryo obtains nutrients from these digested endometrial cells.

The blastocyst now secretes **human chorionic gonadotropin (hCG),** the hormone that is detected by pregnancy tests. For a while, hCG maintains progesterone production at the corpus luteum; progesterone prevents menstruation and further ovulation. In this way, the blastocyst helps to ensure its own survival; if the uterine lining were shed, the blastocyst would leave the woman's body, too. The placenta eventually takes over hCG production, which peaks about 10 weeks after fertilization.

Figure 30.14 Implantation and Gastrulation. The preembryo implants into the endometrium and develops three germ layers by 2 weeks after fertilization.

Why We Care 30.1 | Substances That Cause Birth Defects

About 97% of newborns in the United States are apparently normal, but birth defects occasionally occur. A birth defect is any abnormality that occurs during development and causes death or disability in the child. Genetic mutations, chromosomal abnormalities, and certain substances in the mother's body all may cause birth defects; in many cases, however, the cause remains unknown.

This box describes some of the substances and conditions associated with birth defects. Avoiding them can help boost the chance of delivering a healthy baby.

- **Alcohol:** A child with fetal alcohol syndrome has impaired intellect, ranging from minor to severe learning disabilities. Physicians therefore advise all pregnant women to avoid alcohol.

- **Caffeine:** The caffeine in coffee, tea, soft drinks, and chocolate constricts blood vessels in the placenta, reducing blood flow to the fetus. Caffeine also crosses the placenta and accumulates in the fetal brain. Heavy caffeine consumption during pregnancy is associated with low birth weight and a small head size in the newborn.

- **Cigarettes:** Smoking during pregnancy increases the risk of miscarriage, stillbirth, and prematurity. Carbon monoxide from cigarette smoke crosses the placenta and robs the fetus of oxygen.

- **Illicit drugs:** A pregnant woman's use of cocaine, heroin, methamphetamine, and other illicit drugs is associated with low birth weight.

- **Prescription drugs:** Some prescription drugs cause birth defects. During the late 1950s, for example, physicians prescribed thalidomide as a treatment for morning sickness in pregnant women. The drug caused severe limb shortening in the children of these women (figure 30.B).

- **Excess or insufficient vitamins:** Too much vitamin A can cause miscarriages and defects of the heart, nervous system, and face. Vitamin deficiencies can also cause birth defects. Spina bifida, for example, is associated with insufficient folic acid in the mother's diet. (i) *vitamins,* section 28.3

- **Starvation:** Inadequate nutrition during pregnancy increases the incidence of miscarriage and damages the placenta, causing low birth weight, short stature, tooth decay, delayed sexual development, and possibly intellectual disabilities. Starvation while in the uterus greatly raises the risk of developing obesity, type 2 diabetes, and other conditions later in life. (i) *starvation,* section 28.4B

- **Viral infection:** The virus that causes rubella (German measles) causes deafness, cataracts, and heart disease; herpes simplex viruses can infect the fetal nervous system and cause neurological problems such as intellectual disabilities or cerebral palsy. HIV, the virus that causes AIDS, infects 15% to 30% of infants born to HIV-positive women. Fetuses infected with HIV are at risk for low birth weight, prematurity, and stillbirth. The chance of having an HIV-positive baby drops sharply, however, if an infected woman takes antiviral drugs while pregnant.

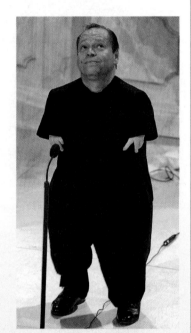

Figure 30.B Effects of Thalidomide. Grammy award-winning baritone Thomas Quasthoff has unusually short arms and legs, and his hands resemble flippers. His mother took thalidomide to combat morning sickness early in pregnancy.

©Peer Grimm, Pool/AP Images

During the second week of development, the blastocyst completes implantation. A space called the amniotic cavity forms within a sac called the amnion (see figure 30.14). The inner cell mass flattens and forms the **embryonic disc,** which will develop into the embryo.

As the preembryo continues to develop, one layer of the embryonic disc becomes **ectoderm,** and another layer becomes **endoderm.** Soon, a middle **mesoderm** layer forms from the ectoderm. (As you will see, each of these layers develops into different organs during the embryonic stage.) The **gastrula** is the resulting three-layered structure. The formation of the gastrula is called gastrulation, and it begins at about 2 weeks postfertilization (see figure 30.14). Although the woman has not yet missed her menstrual period, she might notice effects of her shifting hormones, such as swollen, tender breasts and fatigue. By now, her urine contains enough hCG for an at-home pregnancy test to detect. (i) *germ layers,* section 17.1C

The preembryo may split during the first 2 weeks of development, forming identical twins. Depending on when the split occurs, the twins may or may not share the same amnion and placenta; the later the split, the more structures the twins will share. If a preembryo splits after day 12 of pregnancy, the twins are unlikely to separate completely, and they may be conjoined.

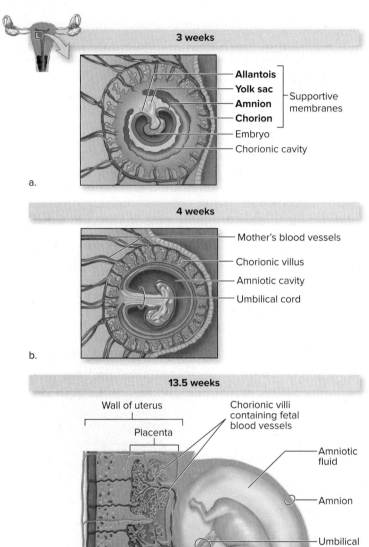

3 weeks

- Allantois ⎫
- Yolk sac ⎬ Supportive
- Amnion ⎪ membranes
- Chorion ⎭
- Embryo
- Chorionic cavity

a.

4 weeks

- Mother's blood vessels
- Chorionic villus
- Amniotic cavity
- Umbilical cord

b.

13.5 weeks

- Wall of uterus
- Placenta
- Chorionic villi containing fetal blood vessels
- Amniotic fluid
- Amnion
- Umbilical cord
- Mother's blood
- Mother's blood vessels

c.

Figure 30.15 **Protective Membranes and the Placenta.**
(a) The four membranes that protect the embryo are present by
3 weeks after fertilization. (b) Chorionic villi initiate placenta formation.
(c) In the fully developed placenta, materials pass between fetal
blood vessels and pools of maternal blood.

C. Organs Take Shape During the Embryonic Stage

Once implantation is complete, the **embryonic stage** begins; it lasts until the end of the eighth week. During this stage of development, cells of the three layers continue to divide and differentiate, forming all of the body's organs. Structures that support the embryo also develop during this period.

Four Membranes, the Placenta, and the Umbilical Cord Support the Embryo Four thin layers of tissue, called membranes, support, protect, and nourish the embryo. Their presence is an important clue to our evolutionary history because each occurs not only in humans but also in other mammals, in birds, and in nonavian reptiles. One membrane, the **yolk sac,** manufactures blood cells until about the sixth week; in addition, parts of the yolk sac develop into the intestines and germ cells. By the third week, an outpouching of the yolk sac forms the **allantois,** another membrane. It, too, manufactures blood cells, and it gives rise to blood vessels in the umbilical cord. The **amnion** is the transparent sac that contains the amniotic fluid. This fluid cushions the embryo, maintains a constant temperature and pressure, and protects the embryo if the woman falls. Fetal cells collected from amniotic fluid are used in many prenatal medical tests. ⓘ *amnion,* section 17.10

The **chorion** is the outermost membrane. **Chorionic villi** are fingerlike projections from the chorion that extend into the uterine lining. These structures establish the beginnings of the **placenta,** which will connect the developing embryo with its mother's uterus (figure 30.15). One side of the placenta comes from the embryo; the other side consists of endometrial tissue and blood from the pregnant woman's circulation.

The placenta begins to form in the embryonic stage, but it is not completely functional until about 10 weeks after fertilization (early in the fetal stage). Arteries and veins in the **umbilical cord** connect the fetus to the placenta, delivering waste-laden fetal blood to the placenta and returning oxygen- and nutrient-rich blood to the developing baby. Note that the mother's blood does not actually mix with the embryo's. Instead, the two bloodstreams pass very close to each other and exchange materials at the placenta.

The fully developed placenta consists of pools of the mother's blood surrounding the chorionic villi, which contain fetal blood vessels. Many substances are exchanged at the placenta, although the cells of the chorionic villi block most large proteins and red blood cells. Carbon dioxide, urea, and other wastes from the fetus pass to the mother's blood. Glucose, amino acids, oxygen, ions, and some antibodies from the mother move in the opposite direction, nourishing and protecting the offspring. At the same time, harmful substances can also cross from mother to fetus; examples include some disease-causing agents and drugs such as cocaine, alcohol, and nicotine.

Organ Formation Begins in the Third Week of Development
As the placenta and umbilical cord develop, so does the embryo itself. Beginning during the third week of prenatal development, distinct organs form. Over several months, ectoderm cells develop into the nervous system, sense organs, outer skin layers, hair, nails, and skin glands. Splits in the mesoderm form the coelom, which becomes the chest and abdominal cavities. Mesoderm cells develop into bone, muscle, blood, the inner skin layer, and the reproductive organs. Endoderm cells form the organs of the digestive and respiratory systems. ⓘ *coelom,* section 17.1D

Between the third and the fifth week after fertilization, many changes shape the organ systems of the developing embryo. The ectoderm differentiates into the central nervous system. At about the same time, a reddish bulge containing the heart appears; it begins to beat around day 22. Blood cells begin to form and to fill developing blood vessels. Immature lungs and kidneys appear. Small buds appear that will develop into arms and legs. The embryo also develops a distinct head and jaw and early evidence of eyes, ears, and nose. A woman carrying this embryo, which is now only about 6 millimeters long, may suspect that she is pregnant because her menstrual period is about 2 weeks late.

All early embryos have unspecialized reproductive structures. At week 7, however, a sex-determining gene on the Y chromosome is activated in male embryos. Hormones then begin to stimulate development of male reproductive organs and glands. If there is no Y chromosome (or if the sex-determining gene never switches on), female reproductive structures develop (figure 30.16).

During weeks 7 and 8, a cartilage skeleton appears. The placenta is now almost fully formed and functional, secreting estrogen and progesterone that maintain the blood-rich uterine lining. The embryo is about the size and weight of a paper clip. By the end of the eighth week, all organ systems are in place, and the embryonic stage of prenatal development is complete. The offspring is now a fetus.

D. Organ Systems Become Functional in the Fetal Stage

The third stage of prenatal development is the **fetal stage,** which lasts from the beginning of the ninth week through the full 38 weeks of development. The fetal stage therefore begins about two thirds of the way through the first trimester (3-month period) of pregnancy and ends 9 months after fertilization. During this time, the fetus grows considerably (figure 30.17). Organs begin to function and interact, forming organ systems.

As the first trimester progresses, the body proportions of the fetus begin to appear more like those of a newborn. Bone begins to form and will eventually replace most of the cartilage, which is softer. Soon, as the nerves and muscles begin to coordinate their actions, the fetus will move its arms and legs. Physical differences between the sexes are usually detectable by ultrasound after the twelfth week. From this time, the fetus sucks its thumb, kicks, and makes fists and faces, and baby teeth begin to form in the gums. More than half of all pregnant women experience the nausea and vomiting of morning sickness in their first trimester.

During the second trimester, body proportions become even more like those of a newborn. By the fourth month, the fetus has hair, eyebrows, eyelashes, nipples, and nails. Bone continues to replace the cartilage skeleton. The fetus's muscle movements become stronger, and the woman may begin to feel a slight fluttering in her abdomen. By the end of the fifth month, the fetus curls into the classic head-to-knees "fetal position." As the second trimester ends, the woman feels distinct kicks and jabs and may even detect a fetal hiccup. The fetus is now about 30 centimeters long.

In the final trimester, fetal brain cells rapidly connect into networks, and organs differentiate further and grow. A layer of fat develops beneath the skin. The digestive and respiratory systems mature last, which is why infants born prematurely often have difficulty digesting milk and breathing. The fetus may move vigorously, causing back pain and frequent urination in the woman as it presses against her bladder. About 266 days (38 weeks) after a single sperm burrowed into an oocyte, a baby is ready to be born.

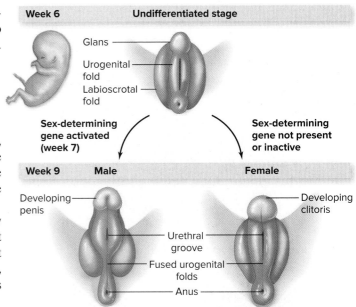

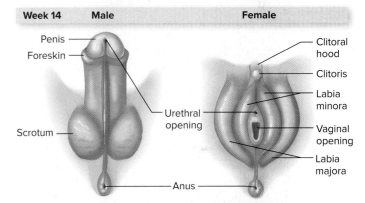

Figure 30.16 **Development of Male and Female Genitals.** Until about week 7 of development, embryos have undifferentiated genitals. If the sex-determining gene on the Y chromosome is activated, development continues as a male. Otherwise, female genitals develop.

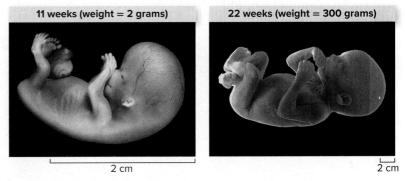

Figure 30.17 **The Fetal Stage.** These photographs show the development of a fetus at 11 weeks and 22 weeks after fertilization.
(11 weeks): ©Science Source; (22 weeks): ©James Stevenson/Science Source

a. First stage: Amniotic sac breaks and cervix dilates

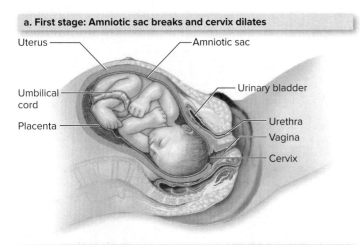

Uterus

Amniotic sac

Umbilical cord

Urinary bladder

Placenta

Urethra

Vagina

Cervix

b. Second stage: Delivery of baby

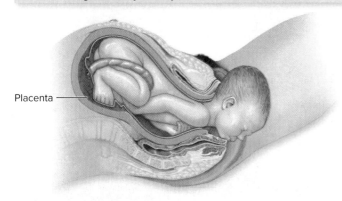

Placenta

c. Third stage: Delivery of placenta

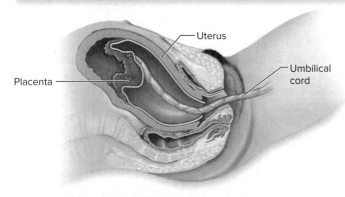

Uterus

Umbilical cord

Placenta

Figure 30.18   Childbirth. (a) In the first stage of labor, the amniotic sac breaks and the cervix dilates. (b) In the second stage, the baby is pushed out of the birth canal. (c) The final stage is the delivery of the placenta.

E. Muscle Contractions in the Uterus Drive Childbirth

Labor refers to the strenuous work that a woman performs as she gives birth. The process typically occurs in three stages (figure 30.18). During the first stage of labor, the fetus presses down and ruptures the amniotic sac, causing an abrupt leakage of amniotic fluid ("water breaking"). Labor may also begin with a discharge of blood and mucus from the vagina, or a woman may feel mild contractions in her lower abdomen about every 20 minutes.

As labor proceeds, hormones prompt the smooth muscle that makes up the wall of the uterus to contract with increasing frequency and intensity. The cervix dilates (opens) a little more each time the baby's head presses against it. By the end of the first stage of labor, the cervix has stretched open to about 10 centimeters.

During the second stage of labor, the baby is typically delivered head-first through the cervix and vagina. In the third (and last) stage of labor, the uterus expels the placenta.

The events of childbirth illustrate **positive feedback** because the process reinforces itself (as opposed to negative feedback, in which a process counteracts an existing condition). At the onset of labor, stretched sensory receptors at the cervix relay the message to the hypothalamus, which triggers release of the hormone oxytocin from the posterior pituitary. Oxytocin travels in the bloodstream and stimulates muscles in the uterus to contract, pushing the baby out (for this reason, physicians sometimes use synthetic oxytocin to induce labor). As the baby emerges, the cervix stretches farther and stimulates more oxytocin production, which intensifies the contractions, and so on. After the baby is born, other hormonal changes stop the cycle.

Not all births go according to plan. For example, about 3% of babies have a "breech presentation," in which the baby's feet, knees, or buttocks appear first instead of the head. Breech deliveries are more difficult than head-first deliveries and account for some cesarean sections ("C-sections"). In this procedure, a surgeon removes the child from the uterus through an incision in the abdomen. Nearly one third of all babies in the United States are delivered by cesarean section, either by choice or because the life of the mother or the baby is at risk.

Sometimes, a pregnancy ends with a premature birth. An infant is premature if it is born before completing 35 weeks (about 8 months) of gestation. A fetus born before 22 weeks of gestation is not viable, but babies born after that time may survive. Premature infants are at risk for many health problems, so they typically spend their first weeks or months in incubators at a neonatal intensive care unit. The incubators control the temperature and protect the infants from infection.

30.5 Mastering Concepts

1. What are the events of fertilization?
2. What are the relationships among the zygote, blastocyst, inner cell mass, and gastrula?
3. What is implantation, and when does it occur?
4. Which supportive structures develop during the embryonic period? What are their functions?
5. When do sex differences appear, and what triggers them?
6. What are the events of the second and third trimesters?
7. Which events make up the three stages of labor?

Investigating Life 30.1 | Playing "Dress Up" on the Reef

Across the animal kingdom, many paths lead to reproductive success. In some animals, head-to-head battles determine which male wins the exclusive "right" to mate. In other species, it pays to avoid conflict.

Consider, for example, the unusual behavioral adaptations in a squidlike mollusk called the Australian giant cuttlefish. During the winter mating season, these animals congregate by the hundreds of thousands on the reefs of southern Australia. With a sex ratio of at least four males to every female, the competition for mates is fierce. (i) *mollusks,* section 17.5

When a female accepts a male's mating attempt, he inserts a sperm packet into her body. At the same time, he also tries to flush out the sperm that other males have deposited. The largest males fight off rivals after mating to prevent subsequent insemination of their mates.

Nevertheless, smaller males do get to mate. How do they gain access to females? A team of biologists noticed that some males are female impersonators. A small male can disguise himself as a female by hiding the long arms that characterize male cuttlefish and changing his skin color to a pattern that is typical of females. The ploy is so realistic that other males, including other female impersonators, often try to mate with the mimic.

The researchers wondered how often these mimics successfully fertilized eggs. Video evidence revealed that three of 62 disguised males actually managed to mate with females. Moreover, DNA analysis showed that two of these three males fertilized the first egg laid after mating (figure 30.C).

Even within the same species, there may be more than one path to reproductive success. The brawniest males can fight off rivals, but the fierce competition and constant distractions often leave females with additional mating opportunities. A small male has little chance of winning a head-to-head competition with a larger rival, but he does have another tool—deception.

Figure 30.C
The Disguise Works. Small cuttlefish males that impersonated females occasionally mated and sired offspring.

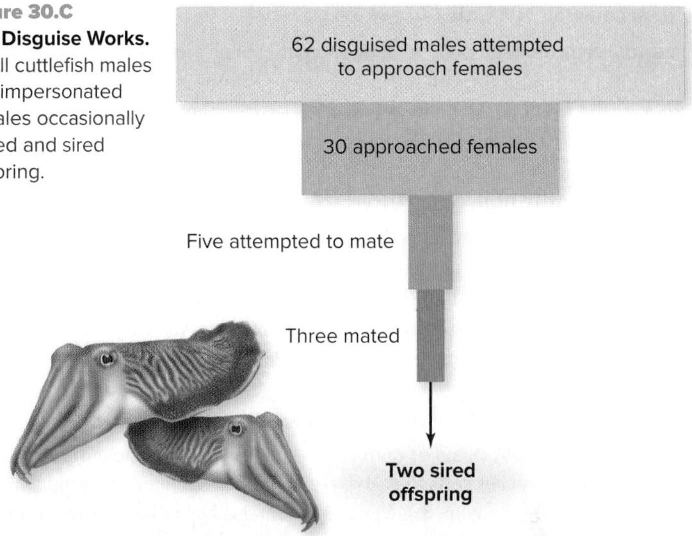

Sources: Hanlon, Roger T., Marié-Jose Naud, Paul W. Shaw, and Jonathan N. Havenhand. 2005. Transient sexual mimicry leads to fertilization. *Nature,* vol. 433, page 212.

Naud, Marié-Jose, Roger T. Hanlon, Karina C. Hall, et al. 2004. Behavioural and genetic assessment of reproductive success in a spawning aggregation of the Australian giant cuttlefish, *Sepia apama. Animal Behaviour,* vol. 67, pages 1043–1050.

Norman, Mark D., Julian Finn, and Tom Tregenza. 1999. Female impersonation as an alternative reproductive strategy in giant cuttlefish. *Proceedings of the Royal Society of London B.,* vol. 266, pages 1347–1349.

What's the **Point?** ▼ APPLIED

Many people believe that there are just two kinds of people: males and females. But the truth is considerably more interesting. After all, the development of the genitals depends on interactions among multiple genes, hormones, and embryonic tissues. Any deviation from this complex sequence of events can lead to unexpected results.

Some children are born with an intersex condition, in which the anatomy does not match the "standard" male or female. Intersex conditions may affect people who are genetically male (XY) or female (XX) or who have sex chromosome abnormalities. Experts disagree on the exact incidence because it is difficult to determine the criteria that define an intersex individual; estimates range between 1 in 1500 and 1 in 5000 births. (i) *sex chromosome abnormalities,* section 9.7B

How does biology explain intersex conditions? One cause is androgen insensitivity. In this case, the cells of a child conceived as a boy cannot bind to masculinizing hormones, such as testosterone and other androgens. The genitals may then be ambiguous or appear female. Klinefelter syndrome (XXY), Turner syndrome (XO), and 5-alpha reductase deficiency are also associated with intersex conditions.

Caster Semenya
©Gallo Images/Alamy Stock Photo

Sex determination has implications in sports. In 2009, the International Association of Athletics Federations (IAAF) questioned Caster Semenya's sex due to her "masculine" build and her easy victory in the women's 800-meter run at the World Athletic Championships. She underwent multiple complicated tests to reveal her sex. Months later, the IAAF agreed to let Semenya keep her medal without announcing the results of the tests. However, she subsequently competed in other athletic events for women and won a gold medal in the 2016 Olympics.

Other athletes did not have the same outcome. Following the 2006 Asian Games, runner Santhi Soundarajan was required to return her silver medal after a test revealed that she was genetically male. She was unaware of her condition prior to competing. The public exposure of the devastating news was psychologically taxing on Soundarajan, who reportedly attempted suicide. Stories such as Soundarajan's raise concerns about whether sex testing is fair to women, especially those whose development does not conform exactly to the female norm. They also stimulate conversations about how much our genes define us. Clearly, gender cannot be considered as only a matter of X and Y.

CHAPTER SUMMARY

30.1 Animal Development Begins with Reproduction

A. Reproduction Is Asexual or Sexual

- **Asexual reproduction** does not require a partner and yields identical offspring.
- In **sexual reproduction,** two haploid **gametes** unite and form a **zygote,** the first cell of a new offspring. **Fertilization** may occur in the environment (**external fertilization**) or inside an animal's body (**internal fertilization**).

B. Development Is Indirect or Direct

- Development requires **differentiation,** the formation of specialized cells. In **pattern formation,** the animal takes on its overall shape and structure.
- Animals that undergo **indirect development** have a larva stage that does not resemble the adult. Metamorphosis transforms the larva into an adult.
- In **direct development,** young animals resemble miniature adults.

30.2 Males Produce Sperm Cells

- The **reproductive systems** of both males and females include **gonads,** which house the **germ cells** that give rise to gametes. Other sex organs deliver or nurture the gametes. Males and females have different **secondary sex characteristics,** traits that do not directly participate in reproduction.

A. Male Reproductive Organs Are Inside and Outside the Body

- Developing sperm originate in the **seminiferous tubules** within the paired **testes.** Also inside the testes are endocrine cells that secrete hormones. A pouch called the **scrotum** contains the testes.
- Sperm travel through the **epididymis, vas deferens,** and **ejaculatory duct,** and they exit the body with **semen** through the **urethra** (within the **penis**) during **orgasm** and **ejaculation.**
- The **prostate gland** and **seminal vesicles** add secretions to semen.

B. Spermatogenesis Yields Sperm Cells

- **Spermatogenesis** begins with a diploid germ cell, which divides mitotically to yield a stem cell and a **primary spermatocyte.** The first meiotic division produces two haploid **secondary spermatocytes.** In meiosis II, the secondary spermatocytes divide, yielding four haploid **spermatids.** The spermatids develop into mature sperm cells.
- A mature sperm cell has a flagellum and a head that contains the chromosomes. A caplike **acrosome** covers the head.

C. Hormones Influence Male Reproductive Function

- **Gonadotropin-releasing hormone (GnRH)** from the hypothalamus stimulates the anterior pituitary gland to release **follicle-stimulating hormone (FSH)** and **luteinizing hormone (LH).** In males, these hormones affect the testes, triggering the release of **testosterone** necessary for sperm formation and the development of secondary sex characteristics.

30.3 Females Produce Egg Cells

A. Female Reproductive Organs Are Inside the Body

- Egg cells (and their nourishing **follicle cells**) originate in the **ovaries.** Approximately once a month after puberty, one ovary releases an egg cell into a **uterine tube,** which leads to the **uterus.** A blood-rich **endometrium** lines the inside of the uterus. The **cervix** leads to the **vagina,** which connects the uterus with the outside of the body.
- The **vulva,** or external genitalia, consists of the labia, **clitoris,** and vaginal opening. **Breasts** deliver milk to infants.

B. Oogenesis Yields Egg Cells

- In **oogenesis,** germ cells divide mitotically to form two **primary oocytes.** In meiosis I, the cytoplasm of the primary oocyte divides unevenly as it splits into one large haploid **secondary oocyte** and a much smaller haploid **polar body.** In meiosis II, the secondary oocyte again divides unequally, yielding the large haploid ovum and another small polar body.
- Meiosis in the female begins before birth and completes at fertilization.
- The steps of sperm and oocyte formation are similar, and the products of each process are haploid. But the male and female gametes look different, and they form according to different timetables (figure 30.19).

C. Hormones Influence Female Reproductive Function

- The ovaries secrete **estrogen** and **progesterone,** hormones that stimulate development of female sexual characteristics. GnRH, FSH, LH, estrogen, and progesterone control the **ovarian cycle** and the **menstrual cycle.**

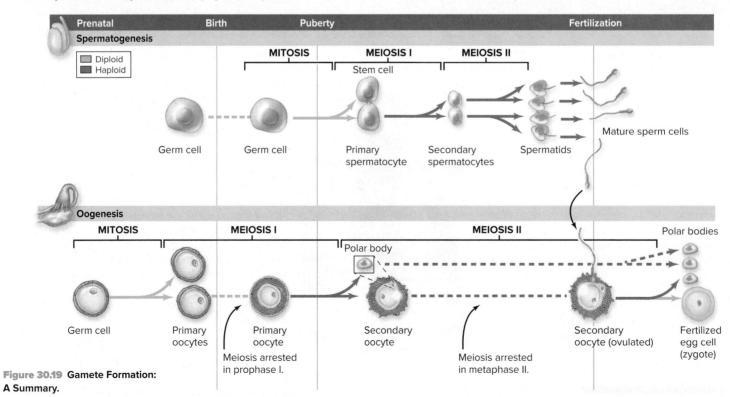

Figure 30.19 Gamete Formation: A Summary.

TABLE 30.5	The Roles of GnRH, LH, and FSH in Human Males and Females			
Hormone	**Source**	**Target Cells**	**Function(s) in Males**	**Function(s) in Females**
GnRH (gonadotropin-releasing hormone)	Hypothalamus	Anterior pituitary	Stimulates release of LH and FSH (continuously)	Stimulates release of LH and FSH (cyclically)
LH (luteinizing hormone)	Anterior pituitary	Testes, ovaries	Stimulates production of testosterone	Stimulates production of estrogen; triggers ovulation midway through menstrual cycle; stimulates development of corpus luteum, which secretes progesterone
FSH (follicle-stimulating hormone)	Anterior pituitary	Testes, ovaries	Enhances function of cells that support spermatogenesis	Stimulates follicle maturation

- After **ovulation,** the ruptured follicle develops into the **corpus luteum,** which secretes hormones that prepare the body for pregnancy. If pregnancy does not occur, the endometrium is shed during **menstruation.**
- Table 30.5 summarizes the hormones required for reproduction in males and females.

D. Hormonal Fluctuations Can Cause Discomfort

- Premenstrual syndrome, menstrual cramps, and hot flashes are associated with changing hormone concentrations.

30.4 Reproductive Health Considers Contraception and Disease

- **Contraception** is the use of behaviors, barriers, hormones, spermicidal chemicals, or other devices that prevent pregnancy.
- **Sexually transmitted infections (STIs)** spread via sexual contact. Viruses and bacteria cause most STIs, which may result in infertility.

30.5 The Human Infant Begins Life as a Zygote

A. Fertilization Initiates Pregnancy

- A male produces sperm cells and delivers them to the female's body, where fertilization occurs (figure 30.20).
- In fertilization, a sperm cell burrows through the two layers surrounding a secondary oocyte. The two united cells constitute the diploid zygote.

B. The Preembryonic Stage Ends When Implantation Is Complete

- The **preembryonic stage** lasts from fertilization until the end of the second week of development.
- After fertilization, **cleavage** divisions produce a ball of cells. Between days 3 and 6, the ball arrives at the uterus and hollows out, forming a **blastocyst.** An outer layer of cells and an **inner cell mass** form. The blastocyst (and

later, the placenta) secretes **human chorionic gonadotropin (hCG),** which prevents menstruation. **Implantation** occurs between days 6 and 14.
- During the second week, the amniotic cavity forms as the inner cell mass flattens, forming the **embryonic disc. Ectoderm** and **endoderm** form, and then **mesoderm** appears, establishing the three primary germ layers of the **gastrula.**

C. Organs Take Shape During the Embryonic Stage

- The **embryonic stage** lasts from the second week through the eighth week of development.
- **Chorionic villi** extending from the **chorion** start to develop into the **placenta.** The **yolk sac, allantois,** and **umbilical cord** form as the **amnion** swells with fluid.
- By the end of the embryonic period, organ systems are in place but are not yet functional.

D. Organ Systems Become Functional in the Fetal Stage

- Structures continue to elaborate during the **fetal stage,** which lasts from the ninth week of development until the baby is ready for birth.
- Figure 30.21 summarizes the stages of development.

E. Muscle Contractions in the Uterus Drive Childbirth

- **Labor** begins as the fetus presses against the cervix. In a **positive feedback** loop, cervical stretching triggers the release of oxytocin, which stimulates uterine contractions that push the baby farther out, causing even more stretching. Eventually, the mother expels the baby and the placenta.

Name		Duration	Description
Zygote		About 24 hours after fertilization	Fertilized egg cell
Preembryonic stage		From first cell division until implantation (about 2 weeks)	Solid ball of cells differentiates into blastocyst and then gastrula
Embryonic stage		From implantation until about 8 weeks after fertilization	Germ layers differentiate into organ systems
Fetal stage		From end of eighth week until birth (end of first trimester, plus second and third trimesters)	Organ systems become functional

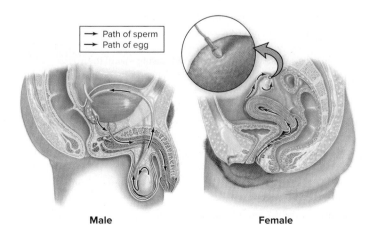

Path of sperm
Path of egg

Male   Female

Figure 30.20 Path to Fertilization.

Placenta
Implantation   develops   Birth

2wk   1mo   2mo   3mo   4mo   5mo   6mo   7mo   8mo   9mo

Fertilization Gastrulation   Organs develop

Figure 30.21 Stages in Development from Fertilization to Birth: A Summary.

MULTIPLE CHOICE QUESTIONS

1. Unchanging environments are advantageous to organisms
   a. that reproduce asexually.
   b. that undergo indirect development.
   c. in which fertilization occurs internally.
   d. that undergo pattern formation.

2. Which hormones play central roles in both male and female reproductive function?
   a. Estrogen and testosterone
   b. Progesterone and LH
   c. LH and FSH
   d. Testosterone and FSH

3. After ovulation, the follicle becomes a
   a. dormant follicle.
   b. corpus luteum.
   c. polar body.
   d. germ cell.

4. Whereas a human sperm cell has _____, an egg cell has _____.
   a. a flagellum; a large amount of cytoplasm
   b. 23 chromosomes; 46 chromosomes
   c. nutrients; mobility
   d. a cell membrane; a cell wall

5. Why can't a fertilized polar body develop into a fetus?
   a. Because it has too few chromosomes
   b. Because it has too many chromosomes
   c. Because it lacks sufficient cytoplasm and organelles
   d. Because it carries excess flagella

6. Which of the following is a birth control method that protects against STIs?
   a. Vasectomy
   b. Condom
   c. Withdrawal
   d. HPV vaccine

7. The acrosome's enzymes allow a sperm cell to penetrate the
   a. proteins and carbohydrates encasing the oocyte.
   b. mucus coating the cervix.
   c. plasma membrane of the secondary oocyte.
   d. Both a and b are correct.

8. What is the correct order of structures to develop after fertilization?
   a. Zygote, blastocyst, gastrula, embryo, fetus
   b. Zygote, fetus, gastrula, embryo, blastocyst
   c. Fetus, zygote, embryo, blastocyst, gastrula
   d. Zygote, embryo, blastocyst, gastrula, fetus

9. The umbilical cord's function is to
   a. transport food to the fetal digestive system.
   b. connect each ovary to the other reproductive organs.
   c. connect the developing fetus to the placenta.
   d. transfer nutrients from uterine tube to the uterus.

10. The hormones that determine the development of male or female reproductive structures first become active during
    a. fertilization.
    b. the preembryonic stage.
    c. the embryonic stage.
    d. the fetal stage.

Answers to Multiple Choice questions are in appendix A.

WRITE IT OUT

1. What are the advantages and disadvantages of asexual and sexual reproduction? Of internal and external fertilization?

2. How are the human male and female reproductive tracts similar, and how are they different? How are the functions of the testis and ovary similar and different?

3. Briefly explain the contradictory roles of estrogen in the human female reproductive cycle.

4. How are the timetables different for oogenesis and spermatogenesis?

5. Suppose that a single ejaculate contains 240 million sperm cells. How many primary spermatocytes must enter meiosis to produce this many sperm cells, and how many secondary spermatocytes are produced along the way?

6. In many species, a female may mate with multiple males, increasing her chance of conception. Speculate about how this behavior may select for males with faster sperm or more numerous sperm per ejaculate.

7. Is each of the following cell types haploid or diploid? How does each cell type relate to the others?
   a. A germ cell
   b. A primary spermatocyte
   c. A spermatid
   d. A mature egg cell
   e. A polar body produced at the end of meiosis II

8. Write a paragraph describing the path of sperm to egg, starting with sperm in the epididymis.

9. Describe some of the most common contraceptives.

10. After ovulation, LH transforms the ruptured follicle into the corpus luteum. As LH levels decline, the corpus luteum degenerates. However, if fertilization occurs, the preembryo starts producing a hormone called hCG, which is similar in structure to LH and prevents the corpus luteum from degenerating. Why is it adaptive for the preembryo to maintain the corpus luteum?

11. Use the Internet to learn more about sexually transmitted infections. Choose one to study in detail. Is the pathogen a virus, bacterium, protozoan, or fungus? What are the symptoms and long-term consequences of infection? Can a person be infected without showing symptoms? Is a treatment or cure available? Who is most affected?

12. For each of the following pairs of phrases, indicate if the first is greater than the second, if the second is greater than the first, or if the two are equal.
    a. The number of cells in the inner cell mass; the number of cells in a gastrula
    b. The thickness of the endometrium after ovulation; the thickness of the endometrium after menstruation
    c. The number of eggs released from an ovary during each ovarian cycle; the number of sperm released during ejaculation
    d. The number of chromosomes in an egg cell; the number of chromosomes in a sperm cell

13. A zygote is a single cell whose weight is immeasurably small, yet after countless cell divisions, it develops into a newborn weighing approximately 3.5 kilograms. While in the uterus, the developing infant obtains the atoms that make up its ever-increasing mass from its mother. Where does the mother obtain these atoms?

14. Provide a general explanation for why men have nipples. *Hint:* Figure 30.16 may help.

15. List some of the causes of birth defects.

16. What are the events of childbirth?

17. During labor, a baby may remain in the birth canal for hours. How does the baby not suffocate during this time?

18. Some newborn mammals can walk and carry out other life functions independently. Human babies, on the other hand, are born helpless. Speculate about the trade-offs in these two reproductive strategies. What selective forces might limit the stage of development at which humans are born?

19. Use this textbook or the Internet to learn more about human cloning and stem cell therapies. How does each topic relate to human reproduction and development? Why are these techniques controversial?

SCIENTIFIC LITERACY

1. Most physicians clamp and cut the umbilical cord immediately after delivering a baby, separating the infant from the nutrients and respiratory gases it receives from the placenta; other physicians wait to cut the cord for several minutes after birth, allowing the remaining cord blood to enter the baby's body. Design an experiment to test whether immediate clamping or delayed clamping of the umbilical cord is a better practice for healthy babies.

2. Review Why We Care 30.1. What kinds of studies and information would be necessary to determine whether exposing a man or woman to a particular chemical can cause birth defects in children born a year later?

Answers to Mastering Concepts, Write It Out, Scientific Literacy, and Pull It Together questions can be found in the Connect ebook.
connect.mheducation.com

Design element: Burning Question (fire background): ©Ingram Publishing/Super Stock

PULL IT TOGETHER

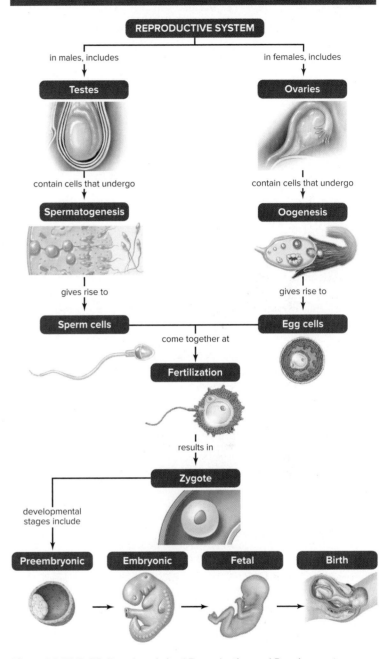

Figure 30.22 Pull It Together: Animal Reproduction and Development.

Refer to figure 30.22 and the chapter content to answer the following questions.

1. Review the Survey the Landscape figure in the chapter introduction and the Pull It Together concept map. Explain why an evolutionary biologist might say that the reproductive system is "selfish" compared to the body's other organ systems.

2. Add *ovulation* to this concept map. What hormonal changes typically accompany ovulation? Add those hormones to the concept map as well.

3. Add the terms *placenta, follicle, polar body, gonad, gametes, meiosis,* and *mitosis* to this concept map.

Appendix A · Answers to Multiple Choice Questions

Answers to *Mastering Concepts, Write It Out, Scientific Literacy,* and *Pull It Together* questions can be found in the Connect eBook at: www.mcgrawhillconnect.com.

Chapter 1
1. b 2. d 3. a 4. a 5. d 6. b 7. b 8. b 9. c 10. d

Chapter 2
1. a 2. c 3. a 4. c 5. c 6. a 7. b 8. a 9. d 10. a

Chapter 3
1. d 2. a 3. c 4. b 5. c

Chapter 4
1. b 2. a 3. a 4. c 5. a 6. d 7. b 8. b

Chapter 5
1. c 2. d 3. d 4. c 5. d 6. d

Chapter 6
1. a 2. d 3. b 4. a 5. c 6. c 7. b 8. a 9. a

Chapter 7
1. a 2. c 3. a 4. a 5. a 6. b 7. a 8. c

Chapter 8
1. c 2. d 3. b 4. a 5. b 6. a 7. c

Chapter 9
1. c 2. a 3. b 4. d 5. c 6. d 7. c

Chapter 10
1. c 2. d 3. b 4. b 5. b 6. b

Chapter 11
1. c 2. a 3. c 4. b 5. b 6. c 7. c 8. a 9. d 10. b

Chapter 12
1. c 2. b 3. b 4. c 5. a 6. b 7. c 8. c 9. a 10. b

Chapter 13
1. d 2. b 3. d 4. c 5. a 6. b 7. a 8. c 9. a 10. a

Chapter 14
1. a 2. c 3. d 4. c 5. b 6. d 7. b 8. c 9. b 10. b

Chapter 15
1. d 2. d 3. d 4. c 5. c

Chapter 16
1. d 2. a 3. d 4. d 5. d 6. b 7. b 8. c 9. b 10. a

Chapter 17
1. a 2. d 3. d 4. a 5. b 6. c 7. c 8. c 9. a 10. c 11. d 12. a 13. b 14. c 15. a

Chapter 18
1. d 2. b 3. d 4. b 5. b 6. c

Chapter 19
1. a 2. a 3. a 4. b 5. c 6. d 7. b 8. a 9. d 10. b

Chapter 20
1. a 2. b 3. b 4. c 5. b 6. d 7. d 8. c

Chapter 21
1. c 2. b 3. d 4. a 5. a 6. b 7. c

Chapter 22
1. b 2. b 3. a 4. b 5. b 6. c 7. b 8. d

Chapter 23
1. d 2. d 3. b 4. a 5. d 6. b 7. a 8. b 9. d

Chapter 24
1. a 2. a 3. d 4. c 5. b 6. d

Chapter 25
1. b 2. a 3. c 4. b 5. a 6. a 7. b

Chapter 26
1. d 2. b 3. d 4. b 5. c 6. d 7. b 8. a 9. c

Chapter 27
1. b 2. d 3. b 4. a 5. c 6. a 7. c

Chapter 28
1. c 2. d 3. c 4. d 5. d 6. d 7. b 8. d 9. b 10. d

Chapter 29
1. b 2. b 3. a 4. a 5. a 6. c 7. b 8. c 9. c 10. b

Chapter 30
1. a 2. c 3. b 4. a 5. c 6. b 7. a 8. a 9. c 10. c

Appendix B · Brief Guide to Statistical Significance

Experiments often yield numerical data, such as the height of a plant or the incidence of illness in vaccinated children (see, for example, figure 1.14). But how are we to know whether an observed difference between two samples is "real"? For example, if we do find that 100 vaccinated children become sick slightly less often than 100 unvaccinated ones, how can we make sure that this outcome does not simply reflect random variation between samples of 100 children?

A statistical analysis can help. The dictionary definition of *statistics* is "the science that deals with the collection, analysis, and interpretation of numerical data, often using probability theory." Note that the analysis is grounded in probability theory, a branch of mathematics that deals with random events. A statistical test is therefore a mathematical tool that assesses variation, with the goal of determining whether any observed differences between treatments can be explained by the variation that random events produce.

Researchers use many types of statistical tests, depending on the type of data collected and the design of the experiment. A description of these tests is beyond the scope of this appendix. For now, it is enough to understand that in each statistical test, the researcher computes a value (the "test statistic") that takes into account the sample size and the variability in the data. The researcher then determines the likelihood that the observed test statistic could be explained by chance alone.

An imaginary experiment will help you understand the role of variability in accepting or rejecting a hypothesis. Suppose that you have two friends, Pat and Kris, both of whom play softball. Pat claims to be able to hit a ball farther than Kris, but Kris disagrees. You therefore set up a test of the null hypothesis, which is that Pat and Kris can hit the ball equally far. You ask both of your friends to hit the ball one time, and Pat's ball does go farther. But Kris wants to redo the test. This time, Kris's ball goes farther. Evidently, two hits apiece is not sufficient for you to settle the matter.

You therefore decide to improve the experiment (figure B.1). This time, each player gets to hit 10 balls, and you use a tape measure to determine how far each ball traveled from home plate. Figure B.2 shows two possible outcomes of the contest. In each scenario, Pat's average distance is 63 meters, compared with 49 meters for Kris. Pat therefore appears to be the better hitter.

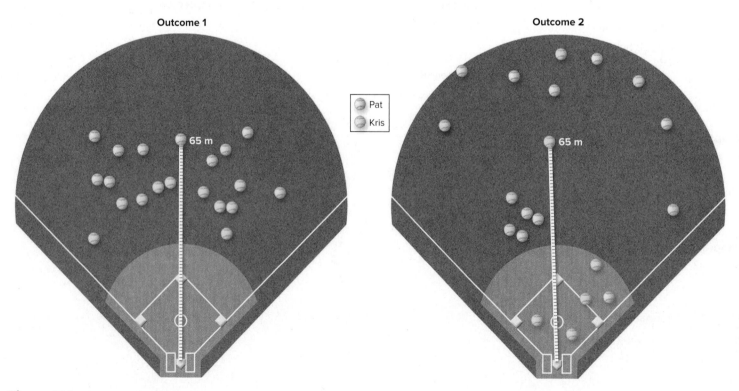

Figure B.1 Hitting Competition. These illustrations show two possible outcomes in a hitting contest between Pat and Kris. Note that the batting distances in Outcome 1 were much less variable than they were in Outcome 2.

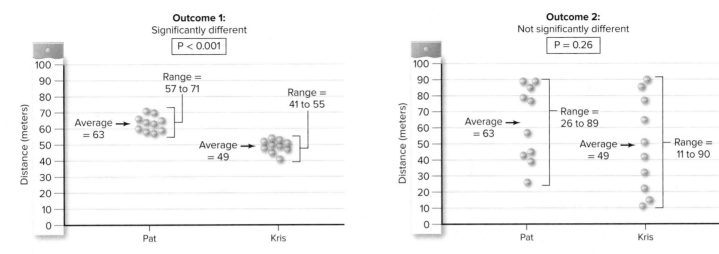

Figure B.2 **Statistical Significance.** In Outcome 1, the difference between Pat and Kris is considered highly significant at P < 0.001; in Outcome 2, however, the variability in hitting distances means that the difference between the two batters is not statistically significant.

But look more closely at the data in figure B.2. In Outcome 1, the hitting distances are much less variable than they are in Outcome 2. How does this variability influence our conclusions about Pat and Kris?

The answer lies in statistical tests. The goal of each statistical test is to calculate the probability that an observed outcome would arise *if the null hypothesis were true*. This probability is called the P value. The lower the P value, the greater the chance that the difference between two treatments is "real."

Generally, biologists accept a P value of 0.05 or smaller as being statistically significant. In other words, a difference between two treatments is statistically significant when the probability that we would observe that result by chance alone is 5% or less. Moreover, a P value of 0.01 or smaller is considered highly significant. Note that the meaning of the word *significant* is different from its use in everyday language. Ordinarily, "significant" means "important." Not so in statistics, where "significant" simply means "likely to be true," and "highly significant" means "very likely to be true."

Returning to our experiment, the null hypothesis is that Pat and Kris are equally good hitters. The amount of variability in Outcome 1 was small, and the calculated value of our test statistic suggests that we can reject this null hypothesis with a P value of < 0.001. In other words, there is a 99.999% chance that Pat really is a better hitter than Kris and that our results are not simply due to chance (see figure B.2). In Outcome 2, however, the hitting distances were much more variable, and the calculated P value is 0.26. We therefore cannot reject the null hypothesis with confidence. After all, given these data, the chance that Pat really is the better batter is only 74%. In science, a 26% chance of incorrectly rejecting the null hypothesis is unacceptably high.

Scientists often include information about statistical analyses along with their data (figure B.3). The most common technique is to graph the average for each treatment and then add error bars that reflect the amount of variability in the data. The longer the error bar, the greater the amount of variability and the less confident we are in the accuracy of our result.

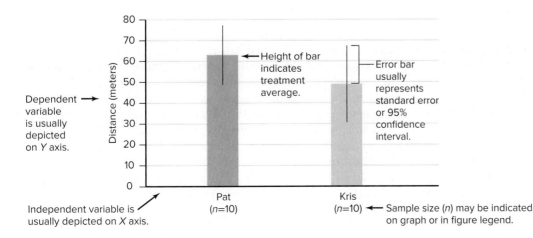

Figure B.3 **Representing Statistics on a Graph.** The basic parts of a bar graph include the axes, treatment averages, and error bars.

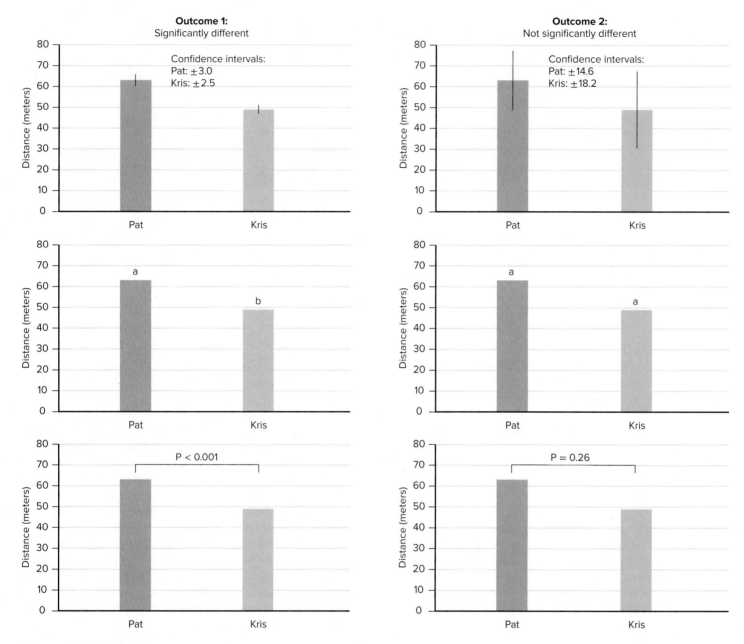

Figure B.4 **Three Ways to Represent Statistics.** Error bars, letters, and P values are all ways to indicate whether the difference between two treatments is statistically significant.

An error bar usually indicates either the standard error or the 95% confidence interval; you may wish to consult a statistics reference to learn more about each. Because these two measures have slightly different meanings, researchers should always note the type of error bar depicted on a graph.

Figure B.4 shows three common ways to indicate whether an observed difference is statistically significant. Examine the error bars on the top pair of graphs in figure B.4. These bars indicate the 95% confidence intervals that were calculated for the two possible outcomes in our hitting competition. Because the data in Outcome 1 are much less variable than for Outcome 2, the error bars for Outcome 1 are smaller than for

Outcome 2. The overlapping error bars for Outcome 2 indicate that the difference between Pat and Kris is not statistically significant.

Figure B.4 also shows that researchers sometimes use other methods to illustrate which treatment averages are significantly different from one another. The middle pair of graphs in figure B.4 shows one approach. In Outcome 1 of our batting experiment, Pat was a significantly better hitter than Kris, so the bars depicting their averages are topped with different letters (*a* and *b*). In Outcome 2, however, Pat and Kris were not significantly different, so their bars have the same letter, *a*. A third possibility is to simply report the P values between pairs of treatments; the last pair of graphs in figure B.4 illustrates this strategy.

Appendix C · Units of Measure

Metric Prefixes

Symbol	Prefix	Multiplier
G	giga	One billion = 1,000,000,000 = 10^9
M	mega	One million = 1,000,000 = 10^6
k	kilo	One thousand = 1,000 = 10^3
h	hecto	One hundred = 100 = 10^2
da	deka (or deca)	Ten = 10 = 10^1
d	deci	One-tenth = 0.1 = 10^{-1}
c	centi	One-hundredth = 0.01 = 10^{-2}
m	milli	One-thousandth = 0.001 = 10^{-3}
μ	micro	One-millionth = 0.000001 = 10^{-6}
n	nano	One-billionth = 0.000000001 = 10^{-9}

Metric Units and Conversions

Measurement Type	Metric Unit	Metric to English Conversion	English to Metric Conversion
Length	1 meter (m)	1 km = 0.62 mile 1 m = 1.09 yards = 39.37 inches 1 cm = 0.394 inch 1 mm = 0.039 inch	1 mile = 1.609 km 1 yard = 0.914 m 1 foot = 0.305 m = 30.5 cm 1 inch = 2.54 cm
Mass	1 gram (g) 1 metric ton (t) = 1,000,000 g = 1,000 kg	1 t = 1.102 tons (U.S.) 1 kg = 2.205 pounds 1 g = 0.0353 ounce	1 ton (U.S.) = 0.907 t 1 pound = 0.4536 kg 1 ounce = 28.35 g
Volume (liquids)	1 liter (l)	1 l = 1.06 quarts 1 mL = 0.034 fluid ounce	1 gallon = 3.79 l 1 quart = 0.95 l 1 pint = 0.47 l 1 fluid ounce = 29.57 mL
Temperature	Degrees Celsius (°C)	°C = (°F − 32)/1.8	°F = (°C × 1.8) + 32
Energy and Power	1 joule (J)	1 J = 0.239 calorie 1 kJ = 0.239 kilocalorie ("food Calorie")	1 calorie = 4.186 J 1 kilocalorie ("food Calorie") = 4186 J
Time	1 second (sec)		

Appendix D · Periodic Table of the Elements

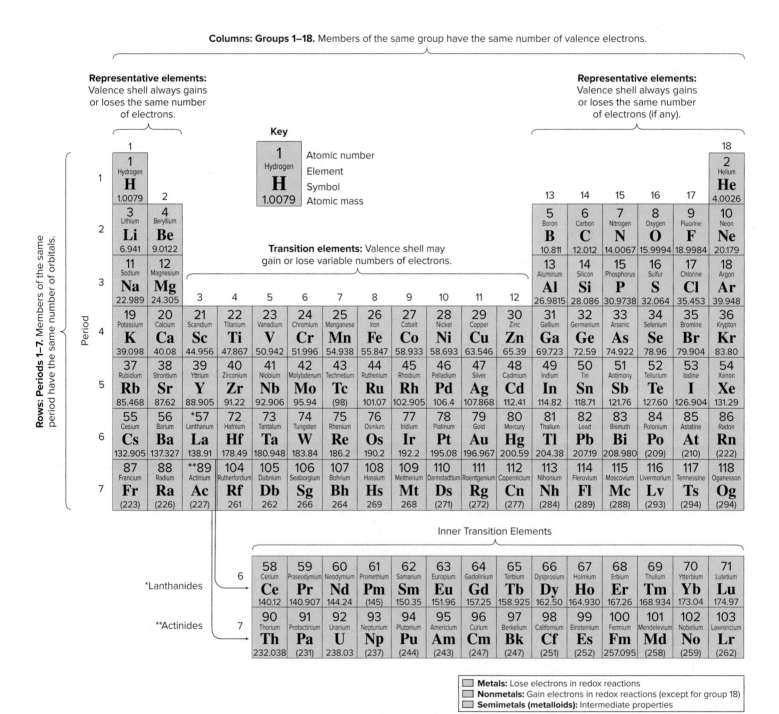

Columns: Groups 1–18. Members of the same group have the same number of valence electrons.

Representative elements: Valence shell always gains or loses the same number of electrons.

Representative elements: Valence shell always gains or loses the same number of electrons (if any).

Key

1
Hydrogen
H
1.0079

Atomic number
Element
Symbol
Atomic mass

Transition elements: Valence shell may gain or lose variable numbers of electrons.

Rows: Periods 1–7. Members of the same period have the same number of orbitals.

Period

1	2	3	4	5	6	7	8	9	10	11	12	13	14	15	16	17	18
1 Hydrogen **H** 1.0079																	2 Helium **He** 4.0026
3 Lithium **Li** 6.941	4 Beryllium **Be** 9.0122											5 Boron **B** 10.811	6 Carbon **C** 12.012	7 Nitrogen **N** 14.0067	8 Oxygen **O** 15.9994	9 Fluorine **F** 18.9984	10 Neon **Ne** 20.179
11 Sodium **Na** 22.989	12 Magnesium **Mg** 24.305	3	4	5	6	7	8	9	10	11	12	13 Aluminum **Al** 26.9815	14 Silicon **Si** 28.086	15 Phosphorus **P** 30.9738	16 Sulfur **S** 32.064	17 Chlorine **Cl** 35.453	18 Argon **Ar** 39.948
19 Potassium **K** 39.098	20 Calcium **Ca** 40.08	21 Scandium **Sc** 44.956	22 Titanium **Ti** 47.867	23 Vanadium **V** 50.942	24 Chromium **Cr** 51.996	25 Manganese **Mn** 54.938	26 Iron **Fe** 55.847	27 Cobalt **Co** 58.933	28 Nickel **Ni** 58.693	29 Copper **Cu** 63.546	30 Zinc **Zn** 65.39	31 Gallium **Ga** 69.723	32 Germanium **Ge** 72.59	33 Arsenic **As** 74.922	34 Selenium **Se** 78.96	35 Bromine **Br** 79.904	36 Krypton **Kr** 83.80
37 Rubidium **Rb** 85.468	38 Strontium **Sr** 87.62	39 Yttrium **Y** 88.905	40 Zirconium **Zr** 91.22	41 Niobium **Nb** 92.906	42 Molybdenum **Mo** 95.94	43 Technetium **Tc** (98)	44 Ruthenium **Ru** 101.07	45 Rhodium **Rh** 102.905	46 Palladium **Pd** 106.4	47 Silver **Ag** 107.868	48 Cadmium **Cd** 112.41	49 Indium **In** 114.82	50 Tin **Sn** 118.71	51 Antimony **Sb** 121.76	52 Tellurium **Te** 127.60	53 Iodine **I** 126.904	54 Xenon **Xe** 131.29
55 Cesium **Cs** 132.905	56 Barium **Ba** 137.327	*57 Lanthanum **La** 138.91	72 Hafnium **Hf** 178.49	73 Tantalum **Ta** 180.948	74 Tungsten **W** 183.84	75 Rhenium **Re** 186.2	76 Osmium **Os** 190.2	77 Iridium **Ir** 192.2	78 Platinum **Pt** 195.08	79 Gold **Au** 196.967	80 Mercury **Hg** 200.59	81 Thalium **Tl** 204.38	82 Lead **Pb** 207.19	83 Bismuth **Bi** 208.980	84 Polonium **Po** (209)	85 Astatine **At** (210)	86 Radon **Rn** (222)
87 Francium **Fr** (223)	88 Radium **Ra** (226)	**89 Actinium **Ac** (227)	104 Rutherfordium **Rf** 261	105 Dubnium **Db** 262	106 Seaborgium **Sg** 266	107 Bohrium **Bh** 264	108 Hassium **Hs** 269	109 Meitnerium **Mt** 268	110 Darmstadtium **Ds** (271)	111 Roentgenium **Rg** (272)	112 Copernicium **Cn** (277)	113 Nihonium **Nh** (284)	114 Flerovium **Fl** (289)	115 Moscovium **Mc** (288)	116 Livermorium **Lv** (293)	117 Tennessine **Ts** (294)	118 Oganesson **Og** (294)

Inner Transition Elements

*Lanthanides 6

58 Cerium **Ce** 140.12	59 Praseodymium **Pr** 140.907	60 Neodymium **Nd** 144.24	61 Promethium **Pm** (145)	62 Samarium **Sm** 150.35	63 Europium **Eu** 151.96	64 Gadolinium **Gd** 157.25	65 Terbium **Tb** 158.925	66 Dysprosium **Dy** 162.50	67 Holmium **Ho** 164.930	68 Erbium **Er** 167.26	69 Thulium **Tm** 168.934	70 Ytterbium **Yb** 173.04	71 Lutetium **Lu** 174.97

**Actinides 7

90 Thorium **Th** 232.038	91 Protactinium **Pa** (231)	92 Uranium **U** 238.03	93 Neptunium **Np** (237)	94 Plutonium **Pu** (244)	95 Americium **Am** (243)	96 Curium **Cm** (247)	97 Berkelium **Bk** (247)	98 Californium **Cf** (251)	99 Einsteinium **Es** (252)	100 Fermium **Fm** 257.095	101 Mendelevium **Md** (258)	102 Nobelium **No** (259)	103 Lawrencium **Lr** (262)

Metals: Lose electrons in redox reactions
Nonmetals: Gain electrons in redox reactions (except for group 18)
Semimetals (metalloids): Intermediate properties

Appendix E · Amino Acid Structures

Nonpolar

Alanine
(Ala; A)

Glycine
(Gly; G)

Proline
(Pro; P)

Valine
(Val; V)

Leucine
(Leu; L)

Phenylalanine
(Phe; F)

Isoleucine
(Ile; I)

Methionine
(Met; M)

Tryptophan
(Trp; W)

Polar

Cysteine
(Cys; C)

Serine
(Ser; S)

Threonine
(Thr; T)

Asparagine
(Asn; N)

Tyrosine
(Tyr; Y)

Glutamine
(Gln; Q)

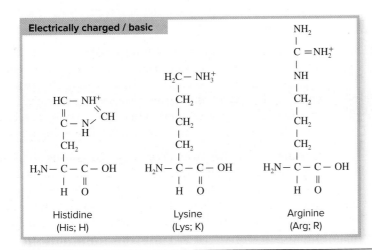

Electrically charged / basic

Histidine
(His; H)

Lysine
(Lys; K)

Arginine
(Arg; R)

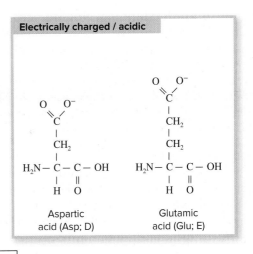

Electrically charged / acidic

Aspartic
acid (Asp; D)

Glutamic
acid (Glu; E)

Note: Amino acids are arranged according to each R group's chemical properties.

Appendix F · Learn How to Learn

Chapter 1 | Real Learning Takes Time

You got good at basketball, running, dancing, art, music, or video games by putting in lots of practice. Likewise, you will need to commit time to your biology course if you hope to do well. To get started, look for the Learn How to Learn tip in each chapter of this textbook. Each hint is designed to help you use your study time productively. With practice, you'll discover that all concepts in biology are connected. The Survey the Landscape figure in every chapter highlights each chapter's place in the "landscape" of the entire unit. Use it, along with the more detailed Pull It Together concept map in the chapter summary, to see how each chapter's content fits into the unit's big picture.

Chapter 2 | Organize Your Time, and Don't Try to Cram

Get a calendar and study the syllabus for every class you are taking. Write each due date in your calendar. Include homework assignments, quizzes, and exams, and add new dates as you learn them. Then, block out time well before each due date to work on each task. Success comes much more easily if you take a steady pace instead of waiting until the last minute.

Chapter 3 | Interpreting Images from Microscopes

Any photo taken through a microscope should include information that can help you interpret what you see. First, read the caption and labels so you know what you are looking at—usually an organ, a tissue, or an individual cell. Then study the scale bar and estimate the size of the image. (For a review of metric units, consult appendix C.) Finally, check whether a light microscope (LM), scanning electron microscope (SEM), or transmission electron microscope (TEM) was used to create the image. Note that stains and false colors are often added to emphasize the most important features.

Chapter 4 | Focus on Understanding, Not Memorizing

When you are learning the language of biology, be sure to concentrate on how each new term fits with the others. Are you studying multiple components of a complex system? Different steps in a process? The levels of a hierarchy? As you study, always make sure you understand how each part relates to the whole. For example, you might jot down brief summaries in the margins of your notes, or you could use lists of boldfaced terms in a chapter to make your own concept map.

Chapter 5 | A Quick Once-Over

Unless your instructor requires you to read your textbook in detail before class, try a quick preview. Start by reviewing the Survey the Landscape figure at the start of each chapter to see how the material fits with the rest of the unit, then read the chapter outline to identify the main ideas. It is also a good idea to look at the figures and the key terms in the narrative. Previewing a chapter should help you follow the lecture because you will already know the main ideas. In addition, note-taking will be easier if you recognize new vocabulary words from your quick once-over. Return to your book for an in-depth reading after class to help nail down the details.

Chapter 6 | Don't Skip the Figures

As you read the narrative in the text, pay attention to the figures; they are there to help you learn. Some figures summarize the narrative, making it easier for you to see the "big picture." Other illustrations show the parts of a structure or the steps in a process. Still others summarize a technique or help you classify information. Flip through this book and see if you can find examples of each type.

Chapter 7 | Explain It, Right or Wrong

As you work through any multiple choice question, such as those at the end of each chapter or on an exam, make sure you can explain why each correct choice is right. You can also test your understanding by taking the time to explain why each of the other choices is wrong.

Chapter 8 | Write It Out—Really!

Get out a pen and a piece of scratch paper, and answer the open-ended "Write It Out" questions at the end of each chapter. This tip applies even if the exams in your class are multiple choice. Putting pen to paper (as opposed to just saying the answer in your head) forces you to organize your thoughts and helps you discover the difference between what you know and what you only THINK you know.

Chapter 9 | Don't Neglect the Boxes

You may be tempted to skip the boxed readings in a chapter because they're not "required." Read them anyway. The contents should help you remember and visualize the material you are trying to learn. And who knows? You may even find them interesting.

Chapter 10 | Be a Good Problem Solver

This chapter is about the principles of inheritance, and it contains many practice problems. Need help? The **How to Solve a Genetics Problem** guide at the end of this chapter shows a systematic, step-by-step approach to solving three of the most common types of problems. Keep using the guide until you feel comfortable solving any problem type.

Chapter 11 | Write Your Own Test Questions

Have you ever tried putting yourself in your instructor's place by writing your own multiple choice test questions? It's a great way to pull the pieces of a chapter together. The easiest questions to write are based on definitions and vocabulary, but those will not always be the most useful. Try to think of questions that integrate multiple ideas or that apply the concepts in a chapter. Write 10 questions, and then let a classmate answer them. You'll probably both learn something new.

UNIT 3 Evolution and Diversity

Chapter 12 | Take the Best Possible Notes

Some students take notes only on what they consider "important" during a lecture. Others write down words but not diagrams. Both strategies risk losing vital information and connections between ideas that could help in later learning. Instead, write down all you can during lecture, including examples, analogies, and sketches of the diagrams the instructor uses. It will be much easier to study later if you have a complete picture of what happened in every class.

Chapter 13 | Why Rewrite Your Notes?

Your notes are your record of what happened in class, so why should you rewrite them after a lecture is over? One excellent reason is that the abbreviations and shorthand that make perfect sense while you take notes will become increasingly mysterious as the days or weeks go by. Rewriting the information in complete sentences not only reinforces learning but also makes your notes much easier to study before an exam.

Chapter 14 | Take Notes on Your Reading

Taking notes as you read should help you not only to retain information but also to identify what you don't understand. Before you take notes, skim through the assigned pages once; otherwise, you may have trouble distinguishing between main points and minor details. Then read them again. This time, pause after each section and write the most important ideas in your own words. What if you can't remember or don't understand well enough to summarize the passage? Read it again, and if that doesn't work, ask for help with whatever isn't clear.

Chapter 15 | Use All Your Resources

Online quizzes, animations, and other resources can help you learn biology. As you study, take regular breaks from reading and explore the digital resources associated with your book. For example, practice questions can help you learn the material and test your understanding. Also, check for animations that take you through complex processes one step at a time. Sometimes the motion of an animation can help you understand what's happening more easily than studying a static image.

Chapter 16 | Think While You Search the Internet

Your class assignments may require you to use the Internet. But the Internet is full of misinformation, so you must evaluate every site you visit. Collaborative sites such as Wikipedia may be unreliable because anyone can change any article. For other sites, ask the following questions: Are you looking at someone's personal page? Is there an educational, governmental, nonprofit, or commercial sponsor? Is the author reputable? Does the page contain facts or opinions? Are the facts backed up with documentation? Taking the time to find the answers to these questions will help ensure that the sites you use are credible.

Chapter 17 | Flashcard Excellence

While making flashcards, you may be tempted to focus on definitions. For example, after reading this chapter, you might make a flashcard with "amnion" on one side and "membrane surrounding an embryo" on the other. This description is correct, but it won't help you understand the bigger picture. Instead, your flashcards should include realistic questions that cover both the big picture and the small details. Try making flashcards that pose a question, such as "Which animals are amniotes?" or "How are amniotes adapted for life on land?" Write the full answer on the other side, then practice writing the answers on scratch paper until you are sure you have them right.

UNIT 4 Ecology

Chapter 18 | Use Those Office Hours

Most instructors maintain office hours. Do not be afraid to use this valuable resource! Besides offering help with course materials, office hours give you an opportunity to know your professors personally. After all, at some point you may need a letter of recommendation; a letter from a professor who knows you well can carry a lot of weight. If you do decide to visit during office hours, be prepared with specific questions. And if you request a separate appointment, be sure to arrive at the time you have arranged—or let your instructor know you need to cancel.

Chapter 19 | Don't Throw That Exam Away!

Whether or not you were satisfied with your last exam, take the time to learn from your mistakes. Mark the questions that you missed and the ones that you got right but were unsure about. Then figure out what went wrong for each question. For example, did you neglect to study the information, thinking it wouldn't be on the test? Did you memorize a term's definition without understanding how it fits with other material? Did you misread the question? After you have finished your analysis, look for patterns and think about what you could have done differently. Then revise your study plan so that you can avoid making the same mistakes in the future.

Chapter 20 | How to Use a Study Guide

Some professors provide study guides to help students prepare for exams. Used wisely, a study guide can be a valuable tool. One good way to use a study guide is AFTER you have studied the material, when you can go through the guide and make sure you haven't overlooked any important topics. Alternatively, you can cross off topics as you encounter them while you study. No matter which technique you choose, don't just memorize isolated facts. Instead, try to understand how the items on the study guide relate to one another. If you're unclear on the relationships, be sure to ask your instructor.

UNIT 5 Plant Anatomy and Physiology

Chapter 21 | Studying in Groups

Study groups offer a great way to learn from other students, but they can also dissolve into social events that accomplish little real work. Of course, your choice of study partners makes a huge difference, so try to pick classmates who are at least as serious as you are about learning. To stay focused, plan activities that are well-suited for groups. For example, you can agree on a list of vocabulary words and take turns adding them to a group concept map. You can also write exam questions for your study partners to answer, or you can simply explain the material to each other in your own words.

Chapter 22 | How to Use a Tutor

Your school may provide tutoring sessions for your class, or perhaps you have hired a private tutor. How can you make the most of this resource? First, meet regularly with your tutor for an hour or two each week; don't wait until just before an exam. Second, if possible, tell your tutor what you want to work on before each session, so he or she can prepare. Third, bring your textbook, class notes, and questions to your tutoring session. Fourth, be realistic. Your tutor can discuss difficult concepts and help you practice with the material, but don't don't expect him or her to simply give you the answers to your homework.

UNIT 6 Animal Anatomy and Physiology

Chapter 23 | Pay Attention in Class

It happens to everyone occasionally: Your mind begins to wander while you are sitting in class, so you check your phone or doze off. How can you keep from wasting your class time this way? One strategy is to get plenty of sleep and eat well, so your mind stays active. Another is to prepare for class in advance, since getting lost can be an excuse for drifting off. When you get to class, sit near the front, listen carefully, and take good notes. Finally, a friendly reminder can't hurt; make a small PAY ATTENTION sign to put on your desk where you can always see it.

Chapter 24 | Find a Good Listener

For many complex topics, you may struggle to know how well you really understand what is going on. One tip is to try explaining what you think you know to somebody else. Choose a subject that takes a few minutes to explain. As you describe the topic in your own words, your partner should ask follow-up questions and note where your explanation is vague. Those insights should help draw your attention to important details that you have overlooked.

Chapter 25 | Don't Waste Old Exams

If you are lucky, your instructor may make old exams available to your class. If so, it is usually a bad idea to simply look up and memorize the answer to each question. Instead, use the old exam as a chance to test yourself before it really counts. Put away your notes and textbook, and set up a mock exam. Answer each question without "cheating," then check how many you got right. Use the questions you got wrong—or that you guessed right—as a guide to what you should study more.

Chapter 26 | Study as You Go

Last-minute cramming for exams may be a classic college ritual, but it is not usually the best strategy. If you try to memorize everything right before an exam, you may become overwhelmed and find yourself distracted by worries that you'll never learn it all. Instead, work on learning the material as the course goes along. Then, on the night before the exam, get plenty of rest, and don't forget to eat on exam day. If you are too tired or too hungry to think, you won't be able to give the exam your best shot.

Chapter 27 | Skipping Class?

Attending lectures is important, but you may need to skip class once in a while. How will you find out what you missed? If your instructor does not provide complete lecture notes, you may be able to copy them from a friend. Whenever you borrow someone else's notes, it's a good idea to compare them with the assigned reading to make sure they are complete and accurate. You might also want to check with the instructor if you have lingering questions about what you missed.

Chapter 28 | Avoid Distractions

Despite your best intentions, constant distractions may take you away from your studies. Friends, music, TV, social media, text messages, video games, and online shopping all offer attractive diversions. How can you stay focused? One answer is to find your own place to study where no one can find you. Turn your phone off for a few hours; the world will get along without you while you study. And if you must use your computer, create a separate user account with settings that prevent you from visiting favorite websites during study time.

Chapter 29 | Practice Your Recall

Here's an old-fashioned study tip that still works. When you finish reading a passage, close the book and write what you remember—in your own words. In this chapter, for example, you will learn about the parts of the human immune system. After you read about them, can you list and describe them without peeking at your book? Try it and find out!

Chapter 30 | Make Your Own Review Sheet

If you are facing a big exam, how can you make sense of everything you have learned? One way is to make your own review sheet. The best strategy will depend on what your instructor expects you to know, but here are a few ideas to try: Make lists; draw concept maps that link ideas within and between chapters; draw diagrams that illustrate important processes; and write mini-essays that explain the main points in each chapter's learning outline.

Glossary

A

abiotic: nonliving

abscisic acid: plant hormone that inhibits shoot growth, maintains bud dormancy, induces seed dormancy, and stimulates stomatal closing

absolute dating: determining the age of a fossil in years

absorption: the process of taking in and incorporating nutrients or energy

accessory pigment: photosynthetic pigment other than chlorophyll *a* that extends the range of light wavelengths useful in photosynthesis

acetyl CoA: molecule that enters the Krebs cycle in cellular respiration; product of partial oxidation of pyruvate

acid: a molecule that releases hydrogen ions into a solution

acid deposition: low pH precipitation or dry particles that form when air pollutants react with water in the upper atmosphere

acrosome: structure covering the head of a sperm cell; contains enzymes that enable the sperm to penetrate the layers surrounding an oocyte

actin: protein that forms thin filaments in muscle cells; also forms part of the cytoskeleton

action potential: brief reversal in membrane potential that propagates along the cell membrane of a neuron

activation energy: energy required for a chemical reaction to begin

active immunity: immunity generated when an individual produces his or her own antibodies

active site: the part of an enzyme to which substrates bind

active transport: movement of a substance across a membrane against its concentration gradient, using a carrier protein and energy from ATP

adaptation: inherited trait that permits an organism to survive and reproduce

adaptive immunity: defense system in vertebrates that recognizes and remembers specific antigens

adenosine triphosphate: See **ATP**

adhesion: the attraction of water molecules to substances other than water

adrenal cortex: outer portion of an adrenal gland; secretes steroid hormones such as mineralocorticoids and glucocorticoids in response to long-term stress

adrenal gland: one of two endocrine glands atop the kidneys; secretes stress-related hormones

adrenal medulla: inner portion of an adrenal gland; secretes epinephrine and norepinephrine in response to short-term stress

adrenocorticotropic hormone (ACTH): hormone produced in the anterior pituitary; stimulates secretion of hormones from the adrenal cortex

adult stem cell: cell that can give rise to a limited subset of cells in the body

aerobic respiration: complete oxidation of glucose to CO_2 in the presence of O_2, producing ATP

age structure: distribution of age classes in a population

agglutination: clumping together of cells

alcoholic fermentation: metabolic pathway in which NADH reduces the pyruvate from glycolysis, producing ethanol and CO_2

aldosterone: mineralocorticoid hormone produced in the adrenal cortex; stimulates water conservation at the kidneys

alga (pl. algae): aquatic, photosynthetic protist

alimentary canal: two-opening (complete) digestive tract; also called gastrointestinal (GI) tract

alkaline: having a pH greater than 7

allantois: protective membrane that manufactures blood cells and gives rise to the umbilical cord's blood vessels

allele: one of two or more alternative forms of a gene

allele frequency: number of copies of one allele, divided by the number of alleles in a population

allergen: antigen that triggers an allergic reaction

allergy: exaggerated immune response to a harmless substance

allopatric speciation: formation of new species after a physical barrier separates a population into groups that cannot interbreed

alternation of generations: the sexual life cycle of plants and many green algae, which alternates between a diploid sporophyte stage and a haploid gametophyte stage

alveolus (pl. alveoli): microscopic air sac where gas exchange occurs in mammalian lungs

amino acid: an organic molecule consisting of a central carbon atom bonded to a hydrogen atom, an amino group, a carboxyl group, and an R group

amino group: a nitrogen atom single-bonded to two hydrogen atoms

amnion: protective membrane that contains amniotic fluid surrounding an embryo or fetus

amniote: a vertebrate in which protective membranes surround the embryo (amnion, chorion, and allantois); reptiles and mammals

amniotic egg: reptile or monotreme egg containing fluid and nutrients within membranes that protect the embryo

amoeboid protozoan: unicellular protist that produces pseudopodia

amphibian: type of tetrapod vertebrate that may live on land but reproduces in water

analogous: similar because of convergent evolution, not common ancestry

anaphase: stage of mitosis in which the spindle pulls sister chromatids toward opposite poles of the cell

anaphase I: anaphase of meiosis I, when spindle fibers pull homologous chromosomes toward opposite poles of the cell

anaphase II: anaphase of meiosis II, when centromeres split and spindle fibers pull sister chromatids toward opposite poles of the cell

anaphylactic shock: rapid, widespread, severe allergic reaction

anatomy: the study of an organism's structure

ancestral character: characteristic already present in the ancestor of a group being studied

anchoring (or adhering) junction: connection between two adjacent animal cells; anchors intermediate filaments in a single spot on the cell membrane

angiosperm: a seed plant that produces flowers and fruits; includes monocots and eudicots

Animalia: kingdom containing multicellular eukaryotes that are heterotrophs by ingestion

annelid: segmented worm; phylum Annelida

anorexia nervosa: eating disorder characterized by refusal to maintain normal body weight

antenna pigment: photosynthetic pigment that passes photon energy to the reaction center of a photosystem

anterior pituitary: the front part of the pituitary gland

anther: pollen-producing structure at tip of stamen

antibody: defensive Y-shaped protein that binds to a specific antigen

anticodon: a three-base portion of a tRNA molecule; the anticodon is complementary to one codon

antidiuretic hormone (ADH): hormone released from the posterior pituitary; promotes water conservation; also called vasopressin

antigen: molecule that elicits an immune reaction by B and T cells

anus: exit from a complete digestive tract

aorta: the largest artery leaving the heart

apical dominance: the suppression of the growth of lateral buds by the intact terminal bud of a plant

apical meristem: meristem at the tip of a root or shoot

apicomplexan: non-motile protist with cell containing an apical complex; obligate animal parasite

apoptosis: programmed cell death that is a normal part of development

appendicular skeleton: the limb bones and the bones that support them in the vertebrate skeleton

arachnid: type of chelicerate arthropod with eight legs; spiders, ticks, mites, and scorpions

Archaea: one of two domains of prokaryotes

arteriole: small artery

artery: vessel that carries blood away from the heart

arthropod: segmented animal with an exoskeleton and jointed appendages; phylum Arthropoda

artificial selection: selective breeding strategy in which a human allows only organisms with desired traits to reproduce

ascomycete: fungus that produces sexual spores in a specialized sac

asexual reproduction: form of reproduction in which offspring arise from only one parent

atom: a particle of matter; composed of protons, neutrons, and electrons

atomic number: the number of protons in an atom's nucleus

atomic weight: the average mass of all isotopes of an element

ATP (adenosine triphosphate): a molecule whose high-energy phosphate bonds power many biological processes

ATP synthase: enzyme complex that admits protons through a membrane, triggering the production of ATP

atrioventricular (AV) node: specialized cardiac muscle cells that delay the heartbeat, giving the ventricles time to fill

atrium: heart chamber that receives blood

auditory canal: ear canal; funnels sounds from the outer ear to the eardrum

auditory nerve: bundle of nerve fibers connecting the cochlea to the brain

autoimmune disorder: immune reaction to antigens associated with the body's own cells

autonomic nervous system: in the peripheral nervous system, motor pathways that lead to smooth muscle, cardiac muscle, and glands

autosomal dominant: inheritance pattern of a dominant allele on an autosome

autosomal recessive: inheritance pattern of a recessive allele on an autosome

autosome: a nonsex chromosome

autotroph: organism that produces organic molecules by acquiring carbon from inorganic sources; primary producer

auxin: plant hormone that promotes cell elongation in stems and fruits, controls tropisms, stimulates growth of roots from stem cuttings, and suppresses growth of lateral buds

axial skeleton: the central axis of a vertebrate skeleton; consists of the bones of the head, vertebral column, and rib cage

axillary bud: undeveloped shoot in the angle between stem and petiole; a lateral bud

axon (nerve fiber): a neuron extension that transmits messages away from the cell body and toward another cell

B

B cell: type of lymphocyte that matures in red bone marrow; if activated, B cells secrete antibodies

bacillus (pl. bacilli): rod-shaped prokaryote

background extinction rate: steady, gradual loss of species through natural competition or loss of genetic diversity

Bacteria: one of two domains of prokaryotes

bacteriophage (phage): a virus that infects bacteria

bark: tissues outside the vascular cambium

base: a molecule that either releases hydroxide ions into a solution or removes hydrogen ions from it

basidiomycete: fungus that releases sexual spores from club-shaped structures

basophil: type of white blood cell that triggers inflammation and allergy

benign tumor: mass of abnormal cells that does not have the potential to spread

bilateral symmetry: body form in which only one plane divides the animal into mirror image halves

bile: digestive biochemical that emulsifies fats

binary fission: type of asexual reproduction in which a prokaryotic cell divides into two identical cells

biodiversity: the variety of life on Earth

biogeochemical cycle: geological and biological processes that recycle elements vital to life

biogeography: the study of the distribution patterns of species across the planet

biological species concept: the idea that a species is a population, or group of populations, whose members can interbreed and produce fertile offspring

biomagnification: tendency of certain chemicals to accumulate in the highest trophic levels

biome: one of the world's several major types of ecosystems, characterized by a distinctive climate and group of species

biosphere: part of Earth where life can exist

biotic: living

bird: type of reptile with feathers and wings

birth rate: the number of new individuals produced per individual per unit time

bivalve: type of mollusk with a two-part, hinged shell

blade: flattened part of a leaf

blastocyst: preembryonic stage consisting of a fluid-filled ball of cells

blastula: stage of early animal embryonic development; a sphere of cells surrounding a fluid-filled cavity

blood: type of connective tissue consisting of cells and platelets suspended in a liquid matrix

blood–brain barrier: close-knit cells that form capillaries in the brain, limiting the substances that can enter

blood clot: plug of solidified blood

blood pressure: force that blood exerts against artery walls

bone: organ consisting of bone tissue, cartilage, and other tissues

bone tissue: connective tissue consisting of cells embedded in a mineralized matrix

bony fish: type of vertebrate chordate with a bony skeleton and a swim bladder

bottleneck: sudden reduction in the size of a population

brain: a distinct concentration of nervous tissue, often encased within a skull, at the anterior end of an animal

brainstem: continuation of the spinal cord into the vertebrate hindbrain; includes the midbrain, pons, and medulla oblongata

breast: milk-producing organ in female mammals

bronchiole: small branched airway that connects bronchi to alveoli

bronchus (pl. bronchi): one of two large tubes that branch from the trachea

brown alga: multicellular photosynthetic aquatic protist with swimming spores and brownish accessory pigments

bryophyte: plant that lacks vascular tissue; includes liverworts, hornworts, and mosses

buffer: weak acid/base pair that resists changes in pH

bulimia: eating disorder in which a person eats large quantities and then intentionally vomits or uses laxatives shortly afterward

bulk element: an element that an organism requires in large amounts

bundle-sheath cell: thick-walled plant cell surrounding veins; site of Calvin cycle in C_4 plants

C

C_3 pathway: carbon fixation pathway in which CO_2 is incorporated into a three-carbon carbohydrate; the Calvin cycle

C_4 pathway: carbon fixation pathway in which CO_2 combines with a three-carbon molecule to form a four-carbon compound

caecilian: type of amphibian that lacks limbs

calcitonin: thyroid hormone that decreases blood calcium levels

Calorie: a measure of the energy content in food; equal to 1000 calories or 1 kilocalorie

calorie: the energy required to raise the temperature of 1 gram of water by 1°C under standard conditions

Calvin cycle: in photosynthesis, carbon fixation pathway that uses NADPH and ATP to incorporate CO_2 molecules into three-carbon carbohydrates; C_3 pathway

CAM pathway: carbon fixation that occurs at night; CO_2 is later released for use in the Calvin cycle during the day

cancer: class of diseases characterized by uncontrolled division of cells that invade or spread to other tissues

capillary: tiny blood vessel that connects an arteriole with a venule

capillary bed: network of capillaries where the circulatory system and body tissues exchange materials

carbohydrate: compound containing carbon, hydrogen, and oxygen in a ratio 1:2:1

carbon fixation: the initial incorporation of carbon from CO_2 into an organic compound

carbon reactions: the reactions of photosynthesis that use ATP and NADPH to synthesize carbohydrates from CO_2

cardiac cycle: sequence of contraction and relaxation that makes up the heartbeat

cardiac muscle tissue: involuntary muscle tissue composed of branched, striated, single-nucleated contractile cells

cardiac output: the volume of blood that the heart pumps each minute

cardiovascular system: circulatory system

carnivore: animal that eats the flesh of other animals

carpel: female part at a flower's center; consists of an ovary, a style, and a stigma

carrying capacity: maximum number of individuals that a habitat can support indefinitely

cartilage: type of connective tissue consisting of cells surrounded by a rubbery collagen matrix

cartilaginous fish: type of vertebrate chordate with a cartilage skeleton

cecum: the entrance to the large intestine

cell: smallest unit of life that can function independently

cell body: enlarged portion of a neuron that contains most of the organelles

cell cycle: sequence of events that occur in an actively dividing cell

cell membrane: the boundary of a cell, consisting of proteins embedded in a phospholipid bilayer; also called the plasma membrane

cell plate: in plants, the materials that begin to form the wall that divides two cells

cell theory: the ideas that all living matter consists of cells, cells are the structural and functional units of life, and all cells come from preexisting cells

cell wall: a rigid boundary surrounding cells of many prokaryotes, protists, plants, and fungi

cell-mediated immunity: branch of adaptive immune system in which defensive cells kill invaders by direct cell–cell contact

cellular slime mold: protist in which feeding stage consists of individual cells that come together as a multicellular "slug" when food runs out

centipede: type of mandibulate arthropod with one pair of legs per repeating subunit

central nervous system: the brain and spinal cord

centromere: small section of a chromosome where sister chromatids attach to each other

centrosome: part of the cell that organizes microtubules

cephalization: development of sensory structures and a brain at the head end of an animal

cephalopod: type of marine mollusk with arms connected to its head

cerebellum: area of the hindbrain that coordinates subconscious muscular responses

cerebral cortex: outer layer of the cerebrum; participates in perception, voluntary movement, language, thought, and other functions

cerebrospinal fluid: fluid that bathes and cushions the central nervous system

cerebrum: region of the forebrain that controls intelligence, learning, perception, and emotion

cervix: lower, narrow part of the uterus

chaparral: See **Mediterranean shrubland**

charophyte: type of green alga thought to be most closely related to terrestrial plants

chelicerate: arthropod with clawlike mouthparts (chelicerae); horseshoe crabs and arachnids

chemical bond: attractive force that holds atoms together

chemical reaction: interaction in which bonds break and new bonds form

chemoreceptor: sensory receptor that responds to chemicals

chemotroph: organism that derives energy by oxidizing inorganic or organic chemicals

chiton: type of mollusk with eight flat, overlapping shells

chlorophyll *a*: green pigment that plants, algae, and cyanobacteria use to harness the energy in sunlight

chloroplast: organelle housing the reactions of photosynthesis in eukaryotes

chordate: animal that at some time during its development has a notochord, hollow nerve cord, pharyngeal pouches or slits, and a postanal tail; phylum Chordata

chorion: outermost membrane surrounding an embryo; helps establish the placenta

chorionic villus: fingerlike projection extending from the chorion to the uterine lining

choroid: middle layer of the eyeball, between the sclera and the retina

chromatid: one of two identical DNA molecules that make up a replicated chromosome

chromatin: collective term for all of the DNA and its associated proteins in the nucleus of a eukaryotic cell

chromosome: a continuous molecule of DNA wrapped around protein in the nucleus of a eukaryotic cell; also, the genetic material of a prokaryotic cell

chyme: semifluid mass of food and gastric juice that moves from the stomach to the small intestine

chytridiomycete (chytrid): microscopic fungus that produces spores with flagella

ciliate: protist with cilia-covered cell surface

cilium (pl. cilia): one of many short, movable protein projections extending from a cell

circulatory system: organ system consisting of the heart and vessels that distribute blood (or a comparable fluid) throughout the body

clade: group of organisms consisting of a common ancestor and all of its descendants

cladistics: phylogenetic system that defines groups by distinguishing between ancestral and derived characters

cladogram: treelike diagram built using shared derived characteristics

cleavage: period of rapid cell division following fertilization

cleavage furrow: in dividing animal cells, the indentation that begins the process of cytokinesis

climax community: community that persists indefinitely if left undisturbed

clitoris: small, highly sensitive female sexual organ at the junction of the labia

clonal deletion: elimination of lymphocytes with receptors for self antigens

clonal selection: rapid division of a stimulated lymphocyte, generating memory cells and effector cells that are clones of the original lymphocyte

closed circulatory system: circulatory system in which blood remains confined to vessels

club moss: type of seedless vascular plant with scales or needles and club-shaped reproductive structures

cnidarian: animal with radial symmetry, two germ layers, a jellylike interior, and cnidocytes; phylum Cnidaria

cnidocyte: cell in cnidarians that can fire a toxic barb in predation or defense

coccus (pl. cocci): spherical prokaryote

cochlea: spiral-shaped part of the inner ear, where vibrations are translated into nerve impulses

codominance: mode of inheritance in which two alleles are fully expressed in a heterozygote

codon: a triplet of mRNA bases that specifies a particular amino acid

coelacanth: rare type of lobe-finned fish; oldest existing lineage of jawed vertebrates

coelom: fluid-filled animal body cavity that forms completely within mesoderm

coevolution: reciprocal selective pressure in which a genetic change in one species selects for subsequent change in another species and vice versa

cohesion: the attraction of water molecules to one another

cohesion–tension theory: theory that explains how water moves under tension in xylem

collecting duct: tubule in the kidney into which nephrons drain urine; site of final adjustments to urine composition

collenchyma: elongated living plant cells with thick, elastic cell walls

commensalism: type of symbiosis in which one member increases its fitness without affecting the other member

community: group of interacting populations that inhabit the same region

compact bone tissue: hard, dense bone tissue, with canals for blood vessels and nerves

companion cell: in phloem, a specialized parenchyma cell that transfers materials into and out of a sieve tube element

competition: struggle between organisms for the same limited resource

competitive exclusion principle: the idea that two or more species cannot indefinitely occupy the same niche

competitive inhibition: change in an enzyme's activity occurring when an inhibitor binds to the active site, competing with the enzyme's normal substrate

complement: group of proteins that help destroy pathogens

complementary: in DNA and RNA, the precise pairing of nucleotide bases (A with T or U; C with G)

complete digestive tract: digestive tract through which food passes in one direction from mouth to anus; also called alimentary canal or gastrointestinal (GI) tract

compound: a molecule including different elements

compound leaf: leaf that is divided into leaflets

concentration gradient: difference in solute concentrations between two adjacent regions

cone: a pollen- or ovule-bearing structure in many gymnosperms

cone cell: photoreceptor cell in the retina that detects colors

conifer: type of gymnosperm with cones and evergreen leaves that are needlelike or scalelike

connective tissue: animal tissue consisting of widely spaced cells in a distinctive extracellular matrix

conservation biologist: scientist who studies the preservation of biodiversity

constant region: amino acid sequence that is the same for all antibodies

consumer: organism that eats other organisms; heterotroph

contraception: use of devices or practices that prevent pregnancy

control: untreated group used as a basis for comparison with a treated group in an experiment

convergent evolution: the evolution of similar adaptations in organisms that do not share the same evolutionary lineage

coral reef: underwater deposit of calcium carbonate formed by colonies of coral animals

cork cambium: lateral meristem that produces cork cells and parenchyma in woody plant

cornea: in the eye, a modified portion of the sclera that forms a transparent curved window that admits light

coronary artery: artery that provides blood to the heart muscle

corpus luteum: gland formed from a ruptured ovarian follicle that has recently released an oocyte

cortex: ground tissue between epidermis and vascular tissue in roots and stems

cotyledon: seed leaf in angiosperms

covalent bond: type of chemical bond in which two atoms share electrons

cranium: part of the skull that encloses the brain

creatine phosphate: molecule stored in muscle fibers; donates its high-energy phosphate to ADP, regenerating ATP

crista (pl. cristae): fold of the inner mitochondrial membrane along which many of the reactions of cellular respiration occur

crocodilian: type of reptile that lives in or near water; its elongated snout has eyes on top

crossing over: exchange of genetic material between homologous chromosomes during prophase I of meiosis

crustacean: type of mandibulate arthropod with two pairs of antennae

culture: the knowledge, beliefs, and behaviors that humans transmit from generation to generation

cuticle: waterproof layer covering the aerial epidermis of a plant

cycad: type of gymnosperm with palmlike leaves and large cones

cytokine: messenger protein synthesized in immune cells that influences the activity of other immune cells

cytokinesis: distribution of cytoplasm into daughter cells in cell division

cytokinin: plant hormone that stimulates cell division, delays leaf shedding, and promotes growth of lateral buds

cytoplasm: the watery mixture that occupies much of a cell's volume; in eukaryotic cells, it consists of all materials, including organelles, between the nuclear envelope and the cell membrane

cytoskeleton: framework of protein rods and tubules in eukaryotic cells

cytosol: the fluid portion of the cytoplasm

cytotoxic T cell: lymphocyte that participates in adaptive immunity by binding to and killing cancerous, damaged, foreign, or infected cells

D

death rate: the number of deaths per individual per unit time

decomposer: organism that consumes wastes and dead organic matter, returning inorganic nutrients to the ecosystem

dehydration synthesis: formation of a covalent bond between two molecules by the loss of water

deletion mutation: removal of one or more nucleotides from a gene

denature: to modify a protein's shape so that its function is destroyed

dendrite: thin neuron branch that receives neural messages and transmits information to the cell body

denitrification: conversion of nitrate to N_2

density-dependent factor: population-limiting condition whose effects increase when a population's density increases

density-independent factor: population-limiting condition that acts irrespective of population density

deoxyribonucleic acid: See **DNA**

dependent variable: response that may be under the influence of an independent variable

derived character: characteristic not found in the ancestor of a group being studied

dermal tissue: tissue covering a plant's surface

dermis: layer of connective tissue that lies beneath the epidermis in vertebrate skin

desert: type of terrestrial biome with very low precipitation and variable temperature; plants are adapted to dry conditions

determinate growth: growth that halts at maturity

detritivore: animal that eats decomposing organic matter

detritus: feces and dead organic matter

deuterostome: clade of bilaterally symmetrical animals in which the first opening in the gastrula develops into the anus

diabetes: disease associated with elevated blood sugar levels

diastolic pressure: lower number in a blood pressure reading; reflects relaxation of the ventricles

diatom: photosynthetic aquatic protist with a two-part silica wall

differentiation: process by which cells acquire specialized functions

diffusion: movement of a substance from a region where it is highly concentrated to an area where it is less concentrated

digestion: the physical and chemical breakdown of food

digestive system: organ system consisting of the intestines and other organs that ingest and dismantle food, absorb nutrient molecules, and eliminate wastes

dihybrid cross: mating between two individuals that are heterozygous for two genes

dinoflagellate: unicellular aquatic protist with two flagella of unequal length; many have cellulose plates

diploid cell: cell containing two full sets of chromosomes, one from each parent; also called $2n$

direct development: development pattern in which a juvenile animal resembles an adult

directional selection: form of natural selection in which one extreme phenotype is fittest, and the environment selects against the others

disaccharide: a simple sugar that consists of two bonded monosaccharides

disruptive selection: form of natural selection in which the two extreme phenotypes are fittest

DNA (deoxyribonucleic acid): genetic material consisting of a double helix of nucleotides, each containing the sugar deoxyribose

DNA polymerase: enzyme that adds new DNA nucleotides and corrects mismatched base pairs in DNA replication

DNA probe: labeled, single-stranded fragment of DNA used to reveal the presence of a complementary DNA sequence

DNA profiling: use of variable parts of the genome to detect genetic differences between individuals

DNA technology: practical application of knowledge about DNA

domain: broadest (most inclusive) taxonomic category

dominant: describes an allele that is expressed whenever it is present

dorsal, hollow nerve cord: tubular nerve cord that forms dorsal to the notochord; one of the four characteristics of chordates

double fertilization: in angiosperms, process by which one sperm nucleus fertilizes the egg and another fertilizes the two haploid nuclei in the female gametophyte's central cell

E

ear: sense organ of hearing and equilibrium

eardrum: structure that transmits sound from air to the middle ear

earthworm: type of annelid that ingests soil

echinoderm: unsegmented deuterostome with a five-part body plan, radial symmetry in adults, and a spiny outer covering; phylum Echinodermata

ecological footprint: measure of the land area needed to support an individual or a population

ecology: study of relationships among organisms and the environment

ecosystem: a community and its nonliving environment

ectoderm: outermost germ layer in an animal embryo; develops into skin and nervous system

ectotherm: animal that lacks an internal mechanism that keeps its temperature within a narrow range; invertebrates, fishes, amphibians, and nonavian reptiles

ejaculation: discharge of semen through the penis

ejaculatory duct: tube that deposits sperm into the urethra

electromagnetic spectrum: range of naturally occurring radiation

electron: a negatively charged particle that orbits the atom's nucleus

electron transport chain: membrane-bound molecular complex that shuttles electrons to slowly extract their energy

electronegativity: an atom's tendency to attract electrons

electrophoresis: technique that uses an electric field to sort DNA fragments by size

element: a pure substance consisting of atoms containing a characteristic number of protons

elimination: the expulsion of waste from the body

embryo sac: female gametophyte in angiosperms

embryonic disc: in the preembryo, a flattened, two-layered mass of cells that develops into the embryo

embryonic stage: stage of human development lasting from the end of the second week until the end of the eighth week of gestation

embryonic stem cell: stem cell that can give rise to all types of cells in the body

emergent property: quality that results from interactions of a system's components

endangered species: a species facing a high risk of extinction in the near future

endocrine gland: concentration of hormone-producing cells in an animal

endocrine system: organ system consisting of glands and cells that secrete hormones

endocytosis: form of transport in which a membrane engulfs substances to bring them into a cell

endoderm: innermost germ layer in an animal embryo; develops into digestive tract and other internal organs

endodermis: the innermost cell layer of root cortex

endomembrane system: eukaryotic organelles that exchange materials in transport vesicles

endometrium: inner uterine lining that is shed during menstruation and supports an embryo during pregnancy

endophyte: fungus that colonizes a plant without triggering disease symptoms

endoplasmic reticulum (ER): interconnected membranous tubules and sacs in a eukaryotic cell

endorphin: pain-killing protein produced in the anterior pituitary

endoskeleton: skeleton on the inside of an animal

endosperm: triploid tissue that stores food for the embryo in an angiosperm seed

endospore: dormant, thick-walled structure that enables some bacteria to survive harsh conditions

endosymbiont theory: the idea that mitochondria and chloroplasts originated as free-living bacteria engulfed by other cells

endothelium: layer of epithelial tissue that lines blood vessels and the heart

endotherm: animal that maintains its body temperature by using heat generated from its own metabolism; birds and mammals

energy: the ability to do work

energy shell: group of electron orbitals that share the same energy level

entropy: randomness or disorder

envelope: the lipid- and protein-rich layer surrounding the protein coat of some viruses

enzyme: an organic molecule that catalyzes a chemical reaction without being consumed

epidermis (animal): the outermost layer of skin

epidermis (plant): single layer of cells covering the leaves, stem, and roots

epididymis: tube that receives and stores sperm from one testis

epigenetics: the study of changes in gene expression that do not involve changes in the DNA sequence

epiglottis: cartilage that covers the glottis, routing food to the digestive tract during swallowing

epinephrine (adrenaline): hormone secreted by the adrenal medulla; readies the body for short-term stress; also can act as a neurotransmitter

epithelial tissue (epithelium): animal tissue consisting of tightly packed cells that form linings, coverings, and glands

equilibrium (*K*-selected) life history: reproductive strategy characterized by long-lived, late-maturing individuals that produce few offspring, with each receiving heavy parental investment

esophagus: muscular tube that leads from the pharynx to the stomach

essential nutrient: substance vital for an organism's metabolism, growth, and reproduction

estrogen: steroid sex hormone produced in ovaries of female vertebrates

estuary: area where fresh water in a river meets the salty water of an ocean

ethylene: volatile plant hormone that ripens fruit, stimulates leaf and fruit shedding, and participates in thigmotropism

eudicot: type of angiosperm; embryo has two cotyledons, and pollen grains have three or more pores

Eukarya: domain containing all eukaryotes

eukaryote: organism composed of one or more cells containing a nucleus and other membrane-bounded organelles

eutrophication: addition of nutrients to a body of water

evaporation: the conversion of a liquid to a vapor (gas)

evolution: descent with modification; change in allele frequencies in a population over time

excretion: elimination of metabolic wastes, especially by the urinary system

exhalation: movement of air out of the lungs

exocytosis: form of transport in which vesicles fuse with the cell membrane to carry materials out of a cell

exon: portion of an mRNA that is translated after introns are removed

exoskeleton: skeleton on the outside of an animal

experiment: a test of a hypothesis under controlled conditions

exponential growth: population growth pattern in which the number of new individuals is proportional to the size of the population

external fertilization: release of gametes by males and females into the same environment; fertilization occurs outside the body

extinction: death of all individuals of a species

extracellular matrix: nonliving substances that surround animal cells; includes water and several types of organic molecules

eye: organ that detects light and produces the sense of sight

F

F_1 generation: the offspring of the P generation in a genetic cross

F_2 generation: the offspring of the F_1 generation in a genetic cross

facilitated diffusion: form of passive transport in which a substance moves down its concentration gradient with the aid of a transport protein

facultative anaerobe: organism that can live with or without O_2

$FADH_2$: electron carrier molecule used in respiration

fast-twitch fiber: large-diameter muscle cell that produces twitches of short duration

fatty acid: long-chain hydrocarbon terminating with a carboxyl group

feather: in birds, an epidermal outgrowth composed of keratin

feces: solid waste that leaves the digestive tract

fermentation: metabolic pathway in the cytoplasm in which NADH from glycolysis reduces pyruvate

fertilization: the union of two gametes

fetal stage: stage of human development lasting from the beginning of the ninth week of gestation through birth

fever: rise in the body's temperature

fibrous root system: branching root system arising from a plant's stem

filtration: removal of water and solutes from the blood, as occurs at the glomerulus

fish: vertebrate animal with fins and external gills

fitness: an organism's contribution to the next generation's gene pool

flagellated protozoan: unicellular heterotrophic protist with one or more flagella

flagellum (pl. flagella): a long whiplike appendage that a cell uses for motility

flatworm: unsegmented worm lacking a coelom; phylum Platyhelminthes

flower: reproductive structure in angiosperms; produces pollen and eggs

fluid mosaic: two-dimensional structure of movable phospholipids and proteins that form biological membranes

fluke: type of parasitic flatworm

follicle cell: nourishing cell surrounding an oocyte

follicle-stimulating hormone (FSH): reproductive hormone produced in the anterior pituitary; stimulates secretion of sex hormones in both sexes

food chain: linear series of organisms in which those in one level eat those at a lower level

food web: network of interconnecting food chains

foot (mollusk): ventral muscular structure that provides movement

foraminiferan: amoeboid protozoan with a calcium carbonate shell

forebrain: front part of the vertebrate brain

fossil: any evidence of an organism from more than 10,000 years ago

founder effect: genetic drift that occurs when a small, nonrepresentative group of individuals leaves their ancestral population and begins a new settlement

frameshift mutation: type of mutation in which nucleotides are added or deleted by any number other than a multiple of three, altering the reading frame

free-living flatworm: planarian or marine flatworm

frog: type of amphibian in which adults have legs and lungs but lack a tail

fruit: seed-containing structure in angiosperms

fruiting body: organ that produces sexual spores in a fungus

Fungi: kingdom containing mostly multicellular eukaryotes that are heterotrophs by external digestion

G

G₁ phase: gap stage of interphase in which the cell grows and carries out its basic functions

G₂ phase: gap stage of interphase in which the cell makes its final preparations for division

gallbladder: organ that stores bile from the liver and releases it into the small intestine

gamete: a sex cell; sperm or egg cell

gametophyte: haploid, gamete-producing stage of the plant life cycle

gap junction: connection between two adjacent animal cells that allows cytoplasm to flow between them

gastric juice: mixture of water, mucus, salts, hydrochloric acid, and enzymes produced at the stomach lining

gastrointestinal (GI) tract: two-opening (complete) digestive tract; also called alimentary canal

gastropod: type of mollusk with a broad, flat "stomach-foot"

gastrovascular cavity: digestive chamber with one opening; incomplete digestive tract

gastrula: early animal embryo consisting of two (cnidarians) or three (other animals) tissue layers

gene: sequence of DNA that encodes a specific protein or RNA molecule

gene pool: all of the genes and their alleles in a population

gene therapy: treatment that supplements a faulty gene in a cell with a functioning version of the gene

genetic code: correspondence between specific nucleotide sequences and amino acids

genetic drift: change in allele frequencies that occurs purely by chance

genome: all the genetic material in an organism

genotype: an individual's combination of alleles for a particular gene

genotype frequency: number of individuals of one genotype, divided by the number of individuals in the population

genus: taxonomic category that groups closely related species

geologic timescale: a division of Earth's history into eons, eras, periods, and epochs defined by major geological or biological events

germ cell: specialized cell that gives rise to gametes

germination: resumption of growth after seed dormancy is broken

gibberellin: plant hormone that breaks seed dormancy and stimulates cell division in roots, shoots, and leaves

ginkgo: type of gymnosperm with fan-shaped leaves

gland: organ that secretes substances into the bloodstream or into a duct

global climate change: long-term changes in Earth's weather patterns

glomeromycete: mycorrhiza-forming fungus lacking sexual spores

glottis: slitlike opening between the vocal cords

glucagon: pancreatic hormone that raises blood sugar level by stimulating liver cells to break down glycogen into glucose

glucocorticoid: hormone secreted by the adrenal cortex; among other functions, boosts blood glucose levels in response to long-term stress

glycerol: a three-carbon molecule that forms the backbone of triglycerides and phospholipids

glycolysis: a metabolic pathway occurring in the cytoplasm of all cells; one molecule of glucose splits into two molecules of pyruvate

gnetophyte: type of gymnosperm with some characteristics resembling flowering plants

Golgi apparatus: a system of flat, stacked, membrane-bounded sacs that packages cell products for export

gonad: gland that manufactures hormones and gametes in animals; ovary or testis

gonadotropin-releasing hormone (GnRH): reproductive hormone produced in the hypothalamus in both sexes; stimulates the release of FSH and LH from the anterior pituitary

gradualism: theory that proposes that evolutionary change occurs gradually, in a series of small steps

granum (pl. grana): a stack of flattened thylakoid discs in a chloroplast

gravitropism: directional growth response to gravity

gray matter: nervous tissue in the central nervous system; consists mostly of neuron cell bodies, dendrites, and synapses

green alga: photosynthetic protist that has pigments, starch, and cell walls similar to those of land plants

greenhouse effect: increase in surface temperature caused by CO_2 and other atmospheric gases

ground tissue: plant tissue that makes up most of the body of a herbaceous plant; composed mostly of parenchyma cells

growth hormone (GH): hormone produced in the anterior pituitary; stimulates tissue growth

guard cells: pair of cells flanking a stoma

gymnosperm: a plant with seeds that are not enclosed in a fruit; includes conifers, ginkgo, gnetophytes, and cycads

H

habitat: physical place where an organism normally lives

hagfish: type of invertebrate chordate with a cranium but no jaws

hair cell: mechanoreceptor that initiates sound transduction in the cochlea

half-life: the time it takes for half the atoms in a sample of a radioactive substance to decay

haploid cell: cell containing one set of chromosomes; also called *n*

Hardy–Weinberg equilibrium: situation in which allele frequencies and genotype frequencies do not change from one generation to the next

heart: muscular organ that pumps blood (or a comparable fluid) throughout the body

heartwood: dark-colored, nonfunctioning secondary xylem in woody plant

helper T cell: lymphocyte that coordinates activities of other immune system cells

hemisphere: one of two halves of the cerebrum

hemoglobin: pigment that carries O_2 in red blood cells

herbaceous plant: plant with a green, nonwoody stem at maturity

herbivore: animal that eats plants

heterotroph: organism that obtains carbon and energy by eating another organism; consumer

heterozygous: possessing two different alleles for a particular gene

hindbrain: lower, posterior portion of the vertebrate brain

histamine: biochemical that dilates blood vessels and increases their permeability; involved in inflammation and allergies

HIV: See **human immunodeficiency virus**

homeostasis: state of internal constancy in the presence of changing external conditions

homeotic gene: any gene that, when mutated, leads to organisms with structures in the wrong places

hominins: humans and their extinct ancestors on the human branch of the primate evolutionary tree

homologous: similar in structure or position because of common ancestry

homologous pair: two chromosomes that look alike and have the same sequence of genes

homozygous: possessing two identical alleles for a particular gene

hormone: biochemical synthesized in small quantities in one place and transported to another

hornwort: type of bryophyte with a tapered, hornlike sporophyte

horseshoe crab: type of chelicerate arthropod with a hard, horseshoe-shaped exoskeleton

horsetail: type of seedless vascular plant with abrasive stems and leaves

human chorionic gonadotropin (hCG): hormone secreted by the blastocyst and placenta; indirectly prevents menstruation

human immunodeficiency virus (HIV): virus that causes acquired immune deficiency syndrome (AIDS)

humoral immunity: branch of adaptive immune system in which B cells secrete antibodies in response to a foreign antigen

hybrid (genetics): producing a mix of offspring for one or more traits; heterozygous

hybrid (speciation): the offspring of individuals from two different species

hydrogen bond: weak chemical bond between opposite partial charges on two molecules or within one large molecule

hydrolysis: splitting a molecule by adding water

hydrophilic: attracted to water

hydrophobic: repelled by water

hydrostatic skeleton: skeleton consisting of constrained fluid in a closed body compartment

hypertonic: describes a solution in which the solute concentration is higher than on the other side of a semipermeable membrane

hypha (pl. hyphae): a fungal filament; the basic structural unit of a multicellular fungus

hypoglycemia: low blood sugar caused by excess insulin or insufficient carbohydrate intake

hypothalamus: small forebrain structure beneath the thalamus that controls homeostasis and links the nervous and endocrine systems

hypothesis: a testable, tentative explanation based on prior knowledge

hypotonic: describes a solution in which the solute concentration is less than on the other side of a semipermeable membrane

I

immune system: organ system consisting of cells and organs that defend the body against infections, cancer, and foreign substances

immunodeficiency: condition in which the immune system lacks one or more components

impact theory: idea that mass extinctions were caused by impacts of extraterrestrial origin

implantation: embedding of the blastocyst into the uterine lining

incomplete digestive tract: digestive tract with one opening that takes in food and ejects wastes; gastrovascular cavity

incomplete dominance: mode of inheritance in which a heterozygote's phenotype is intermediate between the phenotypes of the two homozygotes

independent variable: a factor that is hypothesized to influence a dependent variable

indeterminate growth: growth that persists indefinitely

indirect development: development of a juvenile animal into an adult while passing through intervening larval stages

inflammation: immediate, localized reaction to an injury or any pathogen that breaches the body's barriers

ingestion: the act of taking food into the digestive tract

inhalation: movement of air into the lungs

innate defense: cell or substance that provides generalized protection against all infectious agents

inner cell mass: cells in the blastocyst that develop into the embryo

insect: type of mandibulate arthropod with one pair of antennae, a three-part body, six legs, and wings

insertion mutation: addition of one or more nucleotides to a gene

insulin: pancreatic hormone that lowers blood sugar level by stimulating body cells to take up glucose from the blood

integumentary system: organ system consisting of skin and its outgrowths, which sense the external environment and protect the body

intermediate filament: component of the cytoskeleton; intermediate in size between a microtubule and a microfilament

intermembrane compartment: the space between a mitochondrion's two membranes

internal fertilization: use of a copulatory organ to deposit sperm inside a female's body

interneuron: neuron that connects one neuron to another in the central nervous system

internode: stem area between two points of leaf attachment

interphase: stage preceding mitosis or meiosis, when the cell carries out its functions, replicates its DNA, and grows

interstitial fluid: liquid that bathes cells in a vertebrate's body

intertidal zone: region along a coastline between the high and low tide marks

intestinal villus (pl. villi): tiny projection on the inner lining of the small intestine

intron: portion of an mRNA molecule that is removed before translation

invasive species: introduced species that establishes a breeding population in a new location and spreads widely from the original point of introduction

invertebrate: animal without a backbone

ion: an atom or group of atoms that has lost or gained electrons, giving it an electrical charge

ionic bond: attraction between oppositely charged ions

iris: colored part of the eye; regulates the size of the pupil

isotonic: condition in which a solute concentration is the same on both sides of a semipermeable membrane

isotope: any of the forms of an element, each having a different number of neutrons in the nucleus

J

jaws: bones that frame the entrance to the mouth

joint: area where two bones meet

J-shaped curve: plot of exponential growth over time

K

karyotype: a size-ordered chart of the chromosomes in a cell

keystone species: a species whose effect on community structure is disproportionate to its biomass

kidney: excretory organ in the vertebrate urinary system

kilocalorie (kcal): a measure of the energy content in food; equal to 1000 calories or one food Calorie

kinetic energy: energy being used to do work; energy of motion

kingdom: taxonomic category below domain

Krebs cycle: stage in cellular respiration that completely oxidizes the products of glycolysis

L

labor: the process of childbirth

lac **operon:** in *E. coli,* three lactose-degrading genes plus the promoter and operator that control their transcription

lactic acid fermentation: metabolic pathway in which NADH from glycolysis reduces pyruvate, producing lactic acid

lamprey: type of vertebrate chordate with a cartilage skeleton and cranium but no jaws

lancelet: type of invertebrate chordate with an elongated body

large intestine: part of the digestive tract that connects the small intestine to the anus; absorbs water and ions and eliminates wastes

larynx: boxlike structure in front of the pharynx; produces the voice

latent infection: infection in which viral genetic material in a host cell is not expressed

lateral bud: undeveloped shoot on the side of a stem; an axillary bud

lateral line: network of canals that extends along the sides of fishes and detects vibrations

lateral meristem: meristem whose daughter cells thicken a root or stem

law of independent assortment: Mendel's law stating that during gamete formation, the segregation of the alleles for one gene does not influence the segregation of the alleles for another gene

law of segregation: Mendel's law stating that the two alleles of each gene are packaged into separate gametes

leaf: flattened organ that carries out photosynthesis

leech: type of annelid; most live in fresh water and suck blood or eat other animals

lens: structure in the eye that bends incoming light

leukemia: cancer in which bone marrow overproduces white blood cells

lichen: association of a fungus and a green alga or cyanobacterium

life history: the events of an organism's life, especially those that are related to reproduction

life table: chart that shows the probability of surviving to any given age

ligament: band of fibrous connective tissue that connects bone to bone across a joint

ligase: enzyme that catalyzes formation of covalent bonds in the DNA sugar-phosphate backbone

light reactions: photosynthetic reactions that harvest light energy and store it in molecules of ATP or NADPH

lignin: tough, complex molecule that strengthens the walls of some plant cells

limbic system: collection of forebrain structures involved in emotion and memory

linkage group: group of genes that tend to be inherited together because they are on the same chromosome

linkage map: diagram of gene order and spacing on a chromosome, based on crossover frequencies

linked genes: genes on the same chromosome

lipid: hydrophobic organic molecule consisting mostly of carbon and hydrogen

liver: organ that produces bile and has many other functions

liverwort: type of bryophyte with upright or flattened gametophytes

lizard: type of reptile that typically has legs and periodically sheds its skin

lobe-finned fish: type of bony fish with fleshy fins; closely related to tetrapods

logistic growth: population growth pattern in which the growth rate slows as the population's size approaches the habitat's carrying capacity

long-term memory: memory that can last from hours to a lifetime

lung: saclike structure where gas exchange occurs in air-breathing vertebrates

lungfish: type of lobe-finned fish with lungs

luteinizing hormone (LH): reproductive hormone produced in the anterior pituitary; stimulates secretion of sex hormones in both sexes

lymph: colorless fluid that originates when interstitial fluid enters lymph capillaries

lymph node: one of many lymphoid organs located along lymph capillaries; contains white blood cells and fights infection

lymphatic system: organ system consisting of lymphoid organs and lymph vessels that recover excess tissue fluid and aid in immunity

lymphocyte: type of white blood cell; T cell, B cell, or natural killer cell

lysogenic infection: type of viral infection in which the genetic material of a bacteriophage is replicated along with the host cell's chromosome

lysosome: organelle in a eukaryotic cell that buds from the Golgi apparatus and enzymatically dismantles molecules, bacteria, and worn-out cell parts

lytic infection: type of viral infection in which a bacteriophage enters a cell, replicates, and causes the host cell to burst (lyse) as it releases the new viruses

M

macroevolution: large-scale evolutionary change

macronutrient: nutrient required in large amounts

macrophage: type of phagocyte that also helps initiate the adaptive immune response

malignant tumor: mass of abnormal cells that has the potential to invade adjacent tissues and spread throughout the body

mammal: type of tetrapod vertebrate with hair and mammary glands; embryo is enclosed in an amnion

mammary gland: milk-producing gland in mammals

mandibulate: arthropod with jawlike mouthparts (mandibles); crustaceans, insects, centipedes, and millipedes

mantle: dorsal fold of tissue that secretes a shell in most mollusks

marrow cavity: space in a bone shaft that contains marrow

marsupial: type of mammal that bears live young, which typically complete development in the mother's pouch

mass extinction: the disappearance of many species over relatively short expanses of time

mass number: the total number of protons and neutrons in an atom's nucleus

mast cell: immune system cell that triggers inflammation and allergy; settles in tissues rather than circulating in the blood

matrix: the inner compartment of a mitochondrion

matter: substance that takes up space and is made of atoms

mechanoreceptor: sensory receptor sensitive to physical deflection

Mediterranean shrubland (chaparral): type of terrestrial biome with rainy, mild winters, and dry hot summers; dominant plants have thick bark and leathery leaves

medulla oblongata: part of the brainstem that controls many involuntary vital functions

medusa: free-swimming form of a cnidarian

megaspore: in seed plants, spore that gives rise to female gametophyte

meiosis: division of genetic material that halves the chromosome number and yields genetically variable gametes

melatonin: hormone produced in the pineal gland; regulates effects of light–dark cycles

membrane potential: difference in electrical charge across a neuron's cell membrane

memory cell: lymphocyte produced in an initial infection; launches a rapid immune response upon subsequent exposure to an antigen

meninges: membranes that cover and protect the central nervous system

menstrual cycle: hormonal cycle that prepares the uterus for pregnancy

menstruation: discharge of the endometrium through the vagina during a menstrual period

meristem: localized region of active cell division in a plant

mesoderm: embryonic germ layer between ectoderm and endoderm in an animal embryo; develops into muscles, bones, the circulatory system, and several other structures

mesophyll: photosynthetic tissue in a leaf's interior

messenger RNA (mRNA): a molecule of RNA that encodes a protein

metabolism: the biochemical reactions of a cell

metamorphosis: developmental process in which an animal changes drastically in body form during the transition between juvenile and adult

metaphase: stage of mitosis in which chromosomes are aligned down the center of a cell

metaphase I: metaphase of meiosis I, when homologous chromosome pairs are arranged down the center of a cell

metaphase II: metaphase of meiosis II, when chromosomes are arranged down the center of a cell

metastasis: spreading of cancer

microevolution: relatively short-term changes in allele frequencies within a population or species

microfilament: component of the cytoskeleton; made of the protein actin

micronutrient: nutrient required in small amounts

microspore: in seed plants, a spore that gives rise to a male gametophyte

microtubule: component of the cytoskeleton; made of subunits of the protein tubulin

microvillus (pl. microvilli): extension of the cell membrane; increases the surface area of an epithelial cell of an intestinal villus

midbrain: part of the brain between the forebrain and hindbrain

millipede: type of mandibulate arthropod with two pairs of legs per repeating subunit

mineralocorticoid: hormone secreted by the adrenal cortex; maintains blood volume and salt balance

mitochondrion (pl. mitochondria): organelle that houses the reactions of cellular respiration in eukaryotes

mitosis: division of genetic material that yields two genetically identical nuclei

modern evolutionary synthesis: the idea that genetic mutations create the variation upon which natural selection acts

molecular clock: application of the rate at which DNA mutates to estimate when two types of organisms diverged from a shared ancestor

molecule: two or more atoms joined by chemical bonds

mollusk: unsegmented animal with a soft body, mantle, muscular foot, and visceral mass; phylum Mollusca

monocot: type of angiosperm; embryo has one cotyledon, and pollen grains have one pore

monohybrid cross: mating between two individuals that are heterozygous for the same gene

monomer: a single unit of a polymer

monosaccharide: a sugar molecule that contains five or six carbon atoms

monotreme: egg-laying mammal

moss: type of bryophyte with leaflike gametophytes topped with stalklike sporophytes

motor neuron: neuron that transmits a message from the central nervous system toward a muscle or gland

motor unit: a motor neuron and all of the muscle fibers it contacts

mouth: entrance to the digestive system; functions in feeding, breathing, and vocalizing

muscle: organ that powers movements in animals by contracting; consists of muscle tissue and other tissue types

muscle fiber: muscle cell

muscle tissue: animal tissue consisting of contractile cells that provide motion

muscular system: organ system consisting of skeletal muscles whose contractions form the basis of movement and posture

mutagen: any external agent that causes a mutation

mutant: a genotype, phenotype, or allele that is not the most common in a population or that

has been altered from the "typical" (wild type) condition

mutation: a change in a DNA sequence

mutualism: type of symbiosis that improves the fitness of both partners

mycelium: assemblage of hyphae that forms an individual fungus

mycorrhiza: mutually beneficial association of a fungus and the roots of a plant

myelin sheath: fatty material that insulates some nerve fibers in vertebrates, speeding nerve impulse transmission

myofibril: cylindrical subunit of a muscle fiber, consisting of parallel protein filaments

myosin: protein that forms thick filaments in muscle cells

N

NADH: molecule that carries electrons in glycolysis and respiration

NADPH: molecule that carries electrons in photosynthesis

natural killer cell: type of lymphocyte that participates in innate defenses by destroying cancerous or virus-infected cells

natural selection: differential reproduction of organisms based on inherited traits

negative feedback: regulatory mechanism in which a change in a condition triggers action that reverses the change

nephron: functional unit of the kidney; adjusts composition of blood and produces urine

nerve: bundle of nerve fibers (axons) bound together in a sheath of connective tissue

nervous system: organ system consisting of nerves, the brain, and other structures that specialize in rapid communication

nervous tissue: tissue type whose cells (neurons and neuroglia) form a communication network

net primary production: energy available to consumers in a food chain, after cellular respiration and heat loss by producers

neuroglia: type of cell in nervous tissue; supports and assists neurons

neuron: type of cell in nervous tissue; receives, processes, and transmits information

neurotransmitter: chemical passed from a neuron to receptors on another neuron or on a muscle or gland cell

neutral (solution): neither acidic nor basic

neutron: a particle in an atom's nucleus that is electrically neutral

niche: all resources a species uses for survival, growth, and reproduction

nitrification: conversion of ammonia to nitrate

nitrogen fixation: conversion of nitrogen gas (N_2) to ammonium (NH_4^+), a form of nitrogen that plants can use

nitrogenous base: a nitrogen-containing compound that forms part of a nucleotide

node: point at which leaves attach to a stem

nodule: root growth housing nitrogen-fixing bacteria

noncompetitive inhibition: change in an enzyme's shape occurring when an inhibitor binds to a site other than the active site

nondisjunction: failure of chromosomes to separate at anaphase I or anaphase II of meiosis

nonpolar covalent bond: a covalent bond in which atoms share electrons equally

norepinephrine (noradrenaline): hormone secreted by the adrenal medulla; readies the body for short-term stress; also can act as a neurotransmitter

nose: organ that forms the entrance to the nasal cavity inside the head; functions in breathing and the sense of smell

notochord: flexible rod extending along a chordate's back; one of the four characteristics of chordates

nuclear envelope: the two membranes bounding a cell's nucleus

nuclear pore: a hole in the nuclear envelope

nucleic acid: a long polymer of nucleotides; DNA or RNA

nucleoid: the part of a prokaryotic cell where the DNA is located

nucleolus: a structure within the nucleus where components of ribosomes are assembled

nucleosome: the basic unit of chromatin; consists of DNA wrapped around eight histone proteins

nucleotide: building block of a nucleic acid; consists of a phosphate group, a nitrogenous base, and a five-carbon sugar

nucleus (atom): central part of an atom

nucleus (cell): the membrane-bounded sac that contains DNA in a eukaryotic cell

nutrient: any substance that an organism uses for metabolism, growth, maintenance, and repair of its tissues

O

obesity: condition characterized by an unhealthy amount of body fat; body mass index greater than 30

obligate aerobe: organism that requires O_2 for generating ATP

obligate anaerobe: organism that must live in the absence of O_2

omnivore: animal that eats many types of food, including plants and animals

oogenesis: the production of egg cells

open circulatory system: circulatory system in which fluid circulates freely through the body cavity (is not confined to vessels)

operator: in an operon, the DNA sequence between the promoter and the protein-encoding regions

operon: group of related bacterial genes plus a promoter and operator that control the transcription of the entire group at once

opportunistic pathogen: infectious agent that normally does not cause disease in a person with a healthy immune system

opportunistic (*r*-selected) life history: reproductive strategy characterized by short-lived, early-maturing individuals that have many offspring, with each receiving little parental investment

optic nerve: nerve fibers that connect the retina to the brain

orbital: volume of space where a particular electron is likely to be

organ: two or more tissues that interact and function as an integrated unit

organ system: two or more physically or functionally linked organs

organelle: compartment of a eukaryotic cell that performs a specialized function

organic molecule: compound containing both carbon and hydrogen

organism: a single living individual

orgasm: pleasurable sensation, accompanied by involuntary muscle contractions, associated with sexual activity

osmoregulation: control of an animal's ion concentration

osmosis: simple diffusion of water through a semipermeable membrane

osteoporosis: condition in which bones become less dense

outgroup: basis for comparison in a cladistics analysis

oval window: membrane between the middle ear and the inner ear

ovarian cycle: hormonal cycle that controls the timing of oocyte maturation in the ovaries

ovary (animal): female gonad; organ that produces egg cells and hormones

ovary (plant): base of a carpel in a flower; encloses one or more ovules

overexploitation: harvesting members of a species faster than they can reproduce

ovulation: release of an oocyte from an ovarian follicle

ovule: egg-bearing structure that develops into a seed in gymnosperms and angiosperms

oxidation: the loss of one or more electrons by a participant in a chemical reaction

oxidation–reduction (redox) reaction: chemical reaction in which one reactant is oxidized and another is reduced

oxygen debt: after vigorous exercise, a period in which the body requires extra O_2 to restore ATP and creatine phosphate to muscle and to recharge O_2-carrying proteins

oxytocin: hormone released from the posterior pituitary; stimulates muscle contraction in breasts and uterus

ozone layer: atmospheric zone rich in ozone gas (O_3), which absorbs the sun's ultraviolet radiation

P

P (parental) generation: the first generation (true-breeding) in a genetic cross

pacemaker: specialized cardiac muscle cells that set the tempo of the heartbeat; the sinoatrial (SA) node

pain receptor: sensory receptor that detects mechanical damage, temperature extremes, or chemicals released from damaged cells

paleontology: the study of fossil remains or other clues to past life

pancreas: gland between the spleen and the small intestine; produces hormones, digestive enzymes, and bicarbonate

parasitism: type of symbiosis in which one member increases its fitness at the expense of a living host

parasympathetic nervous system: part of the autonomic nervous system; opposes the sympathetic nervous system; dominates during relaxed times

parathyroid gland: one of four small groups of cells behind the thyroid gland; secretes parathyroid hormone

parathyroid hormone (PTH): hormone produced in the parathyroid gland; stimulates increase in blood calcium level

parenchyma: unspecialized plant cells that make up majority of ground tissue

parental chromatid: chromatid containing genetic information from only one parent

particulate: small piece of matter suspended in air

passive immunity: immunity generated when one individual receives antibodies from another individual

passive transport: movement of a solute across a membrane without the direct expenditure of energy

pathogen: disease-causing agent

pattern formation: developmental process that establishes the body's overall shape and structure

PCR: See **polymerase chain reaction**

pectoral girdle: bones that connect the forelimbs to the axial skeleton

pedigree: chart showing family relationships and phenotypes

peer review: evaluation of scientific results by experts before publication in a journal

pelvic girdle: bones that connect the hindlimbs to the axial skeleton

penis: cylindrical male organ of copulation and urination

pepsin: enzyme that begins the digestion of proteins in the stomach

peptide bond: a covalent bond between adjacent amino acids; results from dehydration synthesis

peptidoglycan: material in bacterial cell wall

perception: the brain's interpretation of a sensation

periodic table: chart that lists elements according to their properties

peripheral nervous system: portion of the nervous system that transmits information to and from the central nervous system

peristalsis: waves of muscle contraction that propel food along the digestive tract

peroxisome: membrane-bounded sac that houses enzymes that break down fatty acids and dispose of toxic chemicals

persistent organic pollutant: carbon-based chemical pollutants that remain in ecosystems for long periods

petal: flower part interior to sepals

petiole: stalk that supports a leaf blade

PGD: See **preimplantation genetic diagnosis**

pH scale: a measurement of how acidic or basic a solution is

phagocyte: cell that engulfs and digests foreign material and cell debris

phagocytosis: form of endocytosis in which the cell engulfs a large particle

pharyngeal slit (or pouch): opening in the pharynx of a chordate embryo; one of the four characteristics of chordates

pharynx: tube just behind the oral and nasal cavities; the throat

phenotype: observable characteristic of an organism

pheromone: volatile chemical an organism releases that elicits a response in another member of the species

phloem: vascular tissue that transports sugars and other dissolved organic substances in plants

phloem sap: solution of water, minerals, sucrose, and other biochemicals in phloem

phospholipid: molecule consisting of glycerol attached to two hydrophobic fatty acids and a hydrophilic phosphate group

phospholipid bilayer: double layer of phospholipids that forms in water; forms the majority of a cell's membranes

photic zone: region in a water body where light is sufficient for photosynthesis

photon: a packet of light or other electromagnetic radiation

photoperiod: day length

photoreceptor: molecule or cell that detects quality and quantity of light

photorespiration: a metabolic pathway in which rubisco reacts with O_2 instead of CO_2, counteracting photosynthesis

photosynthesis: biochemical reactions that enable organisms to harness sunlight energy to manufacture organic molecules

photosystem: cluster of pigment molecules and proteins in a chloroplast's thylakoid membrane

phototroph: organism that derives energy from sunlight

phototropism: directional growth response to unidirectional light

phylogenetics: field of study that attempts to explain the evolutionary relationships among species

physiology: the study of the functions of organisms and their parts

phytoplankton: microscopic photosynthetic organisms that drift in water

pilus (pl. pili): short projection made of protein on a prokaryotic cell

pineal gland: small gland in the brain that secretes melatonin

pioneer species: the first species to colonize an area devoid of life

pith: ground tissue inside a ring of vascular bundles in roots and stems

pituitary gland: pea-sized endocrine gland attached to the hypothalamus; secretes hormones that influence other endocrine glands

placebo: inert substance used as an experimental control

placenta: structure that connects the developing fetus to the maternal circulation in placental mammals

placental mammal: mammal in which the developing fetus is nourished by a placenta

Plantae: kingdom consisting of multicellular, eukaryotic autotrophs

plasma: watery, protein-rich fluid that forms the matrix of blood

plasma cell: B cell that secretes large quantities of one antibody

plasmid: small circle of double-stranded DNA separate from a cell's chromosome

plasmodesma (pl. plasmodesmata): connection that allows cytoplasm to flow between adjacent plant cells

plasmodial slime mold: protist in which feeding stage consists of a huge cell containing many nuclei

plate tectonics: theory that Earth's surface consists of several plates that move in response to forces acting deep within the planet

platelet: cell fragment that orchestrates clotting in blood

pleiotropy: multiple seemingly unrelated phenotypic effects of one genotype

polar body: small, haploid byproduct of female meiosis; typically cannot be fertilized

polar covalent bond: a covalent bond in which electrons are attracted more to one atom's nucleus than to the other

polar ice: type of terrestrial biome that is cold, dry, and windy year-round; phytoplankton in ice and surrounding water are the primary producers

pollen: male gametophyte in seed plants (gymnosperms and angiosperms)

pollination: transfer of pollen to female reproductive part

pollution: physical, chemical or biological change in the environment that harms organisms

polychaete: type of marine annelid with paddle-shaped appendages

polygenic: caused by more than one gene; polygenic traits are typically expressed as a continuum of possible phenotypes

polymer: a long molecule composed of similar subunits (monomers)

polymerase chain reaction (PCR): biotechnology tool that rapidly produces millions of copies of a DNA sequence of interest

polyp: sessile form of a cnidarian

polypeptide: a long polymer of amino acids; it is called a protein once it folds into its functional shape

polyploid: containing extra chromosome sets

polysaccharide: carbohydrate consisting of hundreds of monosaccharides

pons: oval mass in the brainstem where white matter connects the forebrain to the medulla and cerebellum

population: interbreeding members of the same species occupying the same region

population density: number of individuals of a species per unit area or volume of habitat

positive feedback: a process that reinforces an existing condition

postanal tail: muscular tail that extends past the anus; one of the four characteristics of chordates

posterior pituitary: the back part of the pituitary gland

postzygotic reproductive isolation: separation of species due to selection against hybrid offspring

potential energy: stored energy available to do work

predator: animal that kills and eats other animals

prediction: anticipated outcome of the test of a hypothesis

preembryonic stage: first two weeks of human development

preimplantation genetic diagnosis (PGD): use of DNA probes to detect genetic illness in an embryo before implanting it into a uterus

pressure flow theory: theory that explains how phloem sap moves under pressure from source to sink

prey: animal that a predator kills and eats

prezygotic reproductive isolation: separation of species due to factors that prevent the formation of a zygote

primary growth: cell division in apical meristems; lengthens shoot and root tips

primary immune response: immune system's response to its first encounter with a foreign antigen

primary oocyte: in oogenesis, a diploid cell that undergoes meiosis I and yields a haploid polar body and a haploid secondary oocyte

primary producer: species forming the base of a food web by extracting energy and nutrients from nonliving sources; autotroph

primary spermatocyte: a diploid cell that undergoes meiosis I and yields two haploid secondary spermatocytes

primary structure: the amino acid sequence of a protein

primary succession: series of changes in a community's species composition in an area previously devoid of life

primate: mammal with opposable thumbs, eyes in front of the skull, a relatively large brain, and flat nails instead of claws; includes prosimians, monkeys, and apes

prion: infectious protein particle

product: the result of a chemical reaction

product rule: the chance of two independent events occurring equals the product of the individual chances of each event

progesterone: steroid hormone produced in ovaries of female vertebrates

prokaryote: a cell that lacks a nucleus and other membrane-bounded organelles; bacteria and archaea

prolactin: hormone produced in the anterior pituitary; stimulates milk secretion

promoter: a control sequence at the start of a gene; attracts RNA polymerase and (in eukaryotes) transcription factors

prophage: DNA of a lysogenic bacteriophage that is inserted into a host cell's chromosome

prophase: stage of mitosis when chromosomes condense and the mitotic spindle begins to form

prophase I: prophase of meiosis I, when chromosomes condense and become visible, and crossing over occurs

prophase II: prophase of meiosis II, when chromosomes condense and become visible

prostate gland: male structure between the bladder and the penis; produces a milky, alkaline fluid that activates sperm

protein: a polymer consisting of amino acids and folded into its functional three-dimensional shape

protein coat: structural component that surrounds the genetic material of a virus

protist: eukaryotic organism that is not a plant, fungus, or animal

proton: a particle in an atom's nucleus carrying a positive charge

protostome: clade of bilaterally symmetrical animals in which the first opening in the gastrula develops into the mouth

protozoan: unicellular protist that is heterotrophic and (usually) motile

pseudocoelom: fluid-filled animal body cavity lined by endoderm and mesoderm

pulmonary artery: artery that leads from the right ventricle to the lungs

pulmonary circulation: blood circulation between the heart and lungs

pulmonary vein: vein that leads from the lungs to the left atrium

punctuated equilibrium: theory that life's history has been characterized by bursts of rapid evolution interrupting long periods of little change

Punnett square: diagram that uses the genotypes of the parents to reveal the possible results of a genetic cross

pupil: opening in the iris that admits light into the eye

pyramid of energy: diagram depicting energy stored at each trophic level at a given time

pyruvate: the three-carbon product of glycolysis

Q

quaternary structure: the shape arising from interactions between multiple polypeptide subunits of the same protein

R

R group: an amino acid side chain

radial symmetry: body form in which parts are arranged around a central axis

radioactive isotope: atom that emits particles or rays as its nucleus disintegrates

radiometric dating: type of absolute dating that uses known rates of radioactive decay to date fossils

radula: a chitin-rich, tonguelike strap in many mollusks

rain shadow: downwind side of a mountain, with a drier climate than the upwind side

ray-finned fish: type of bony fish with fan-shaped fins

reabsorption: return of useful substances to the blood at the nephron and collecting duct

reactant: a starting material in a chemical reaction

reaction center: a molecule of chlorophyll *a* (and associated proteins) that participates in the light reactions of photosynthesis

receptacle: attachment point for flower parts

receptor potential: localized change in membrane potential in a sensory receptor

recessive: describes an allele whose expression is masked if a dominant allele is present

recombinant chromatid: chromatid containing genetic information from both parents as a result of crossing over

recombinant DNA: genetic material spliced together from multiple sources

red alga: multicellular, photosynthetic, marine protist with red or blue accessory pigments

red bone marrow: marrow that gives rise to blood cells and platelets

red blood cell: disk-shaped blood cell that contains hemoglobin

reduction: the gain of one or more electrons by a participant in a chemical reaction

reflex: instantaneous, automatic response to a stimulus

relative dating: placing a fossil into a sequence of events without assigning it a specific age

repressor: in an operon, a protein that binds to the operator and prevents transcription

reproductive system: organ system consisting of organs that produce and transport gametes and that may nurture developing offspring

reptile: tetrapod vertebrate with scaly or feathery skin; embryo is enclosed in an amniotic egg

resource partitioning: use of the same resource in different ways or at different times by multiple species

respiratory cycle: one inhalation followed by one exhalation

respiratory surface: part of an animal's body that exchanges gases with the environment

respiratory system: organ system consisting of the lungs or comparable organs that acquire O_2 and release CO_2

resting potential: membrane potential for a neuron not conducting a nerve impulse

restriction enzyme: enzyme that cuts double-stranded DNA at a specific base sequence

retina: sheet of photoreceptors that forms the innermost layer of the eye

ribonucleic acid: See **RNA**

ribosomal RNA (rRNA): a molecule of RNA that, along with proteins, forms a ribosome

ribosome: a structure built of RNA and protein where mRNA anchors during protein synthesis

ribulose bisphosphate (RuBP): the five-carbon molecule that reacts with CO_2 in the Calvin cycle

RNA (ribonucleic acid): nucleic acid typically consisting of a single strand of nucleotides, each containing the sugar ribose

RNA polymerase: enzyme that uses a DNA template to produce a molecule of RNA

RNA world: the idea that independently replicating RNA molecules were precursors to life

rod cell: photoreceptor in the retina that provides black-and-white vision

root: belowground part of most plants

root cap: cells that protect the root apical meristem from abrasion

rough endoplasmic reticulum: ribosome-studded portion of the ER where secreted proteins are synthesized

roundworm: unsegmented worm with a pseudocoelom; phylum Nematoda

rubisco: enzyme that adds CO_2 to ribulose bisphosphate in the carbon reactions of photosynthesis

ruminant: herbivore with a four-chambered organ specialized for grass digestion

S

S phase: the synthesis phase of interphase, when DNA replicates

S-shaped curve: plot of logistic growth over time

salamander: type of amphibian that superficially resembles a lizard

sample size: number of subjects in each experimental group

sapwood: light-colored, functioning secondary xylem in woody plant

sarcomere: one of many repeated units in a myofibril of a muscle cell

saturated fatty acid: a fatty acid with single bonds between all carbon atoms

savanna: type of terrestrial biome with wet and dry seasons and warm weather year-round; grassland with scattered trees

SCID: See **severe combined immunodeficiency**

scientific method: a systematic approach to understanding the natural world based on evidence and testable hypotheses

sclera: the outermost layer of the eye; the white of the eye

sclerenchyma: rigid plant cells that support mature plant parts

scrotum: the sac containing the testes

second messenger: molecule that translates a stimulus at the cell's exterior into an effect inside the cell

secondary growth: cell division in lateral meristems; increases girth of shoots and roots

secondary immune response: immune system's response to subsequent encounters with a foreign antigen

secondary oocyte: haploid cell that undergoes meiosis II and yields a haploid polar body and a haploid egg cell

secondary sex characteristic: trait that distinguishes the sexes but does not participate directly in reproduction

secondary spermatocyte: haploid cell that undergoes meiosis II and yields two haploid spermatids

secondary structure: a "substructure" within a protein, resulting from hydrogen bonds between parts of the peptide backbone

secondary succession: change in a community's species composition following a disturbance

secretion: addition of substances to the fluid in a nephron

seed: in gymnosperms and angiosperms, a plant embryo packaged with a food supply inside a tough outer coat

seed coat: protective outer layer of seed

seedless vascular plant: plant with vascular tissue but not seeds; includes true ferns, club mosses, whisk ferns, and horsetails

segmentation: division of an animal body into repeated subunits

selective permeability: the property that enables a membrane to admit some substances and exclude others

semen: fluid that carries sperm cells out of the body

seminal vesicle: structure that contributes fluid, fructose, and prostaglandins to semen

seminiferous tubule: tubule within a testis where sperm form and mature

sensation: raw sensory input that reaches the central nervous system

sensory adaptation: lessening of sensation with prolonged exposure to a stimulus

sensory neuron: neuron that transmits information from a stimulated body part to the central nervous system

sensory receptor: cell that detects stimulus information

sepal: part of the outermost whorl of a flower

severe combined immunodeficiency (SCID): inherited immune system disorder in which neither T cells nor B cells function

sex chromosome: a chromosome that carries genes that determine sex

sex-linked: describes genes or traits on the X or Y chromosome

sexual dimorphism: difference in appearance between males and females

sexual reproduction: the combination of genetic material from two individuals to create a third individual

sexual selection: type of natural selection resulting from variation in the ability to obtain mates

sexually transmitted infection (STI): infection caused by a virus, bacterium, protist, or fungus that spreads during sexual contact

shoot: aboveground part of a plant

short tandem repeat (STR): short DNA sequence that is repeated; people vary in the number of repeats they inherit

short-term memory: memory available for only a few moments

sieve tube element: conducting cell that makes up a sieve tube in phloem

simple diffusion: form of passive transport in which a substance moves down its concentration gradient without the use of a transport protein

simple leaf: leaf with an undivided blade

sink: plant part that receives sugars in phloem

sinoatrial (SA) node: specialized cardiac muscle cells that set the pace of the heartbeat; the pacemaker

skeletal muscle: organ composed of bundles of skeletal muscle cells and other tissue types that generates voluntary movements between pairs of bones

skeletal muscle tissue: voluntary muscle tissue consisting of long, unbranched, striated cells containing multiple nuclei

skeletal system: organ system consisting of bones, cartilage, ligaments, and tendons; functions include support, movement, and protection

skeleton: structure that supports an animal's body

sliding filament model: sliding of actin and myosin past each other to shorten a muscle cell

slow-twitch fiber: small-diameter muscle fiber that produces twitches of long duration

small intestine: the part of the digestive tract that connects the stomach with the large intestine; site of most chemical digestion and absorption

smog: type of air pollution that forms a visible haze in the lower atmosphere

smooth endoplasmic reticulum: portion of the ER that produces lipids and detoxifies poisons

smooth muscle tissue: involuntary muscle tissue consisting of nonstriated, spindle-shaped cells

snake: type of reptile that lacks legs and periodically sheds its skin

sodium–potassium pump: protein that uses energy from ATP to transport Na⁺ out of cells and K⁺ into cells

solute: a chemical that dissolves in a solvent, forming a solution

solution: a mixture of a solute dissolved in a solvent

solvent: a chemical in which other substances dissolve, forming a solution

somatic cell: body cell that does not give rise to gametes

somatic cell nuclear transfer: technique used to clone a mammal from an adult cell

somatic nervous system: in the peripheral nervous system, motor pathways carrying signals to skeletal (voluntary) muscles

source: plant part that produces or releases sugar

speciation: formation of new species

species: a distinct type of organism

species evenness: measure of biodiversity; the relative abundance of the species in a community

species richness: measure of biodiversity; the number of species in a community

spermatid: one of four haploid cells produced in meiosis II of spermatogenesis

spermatogenesis: the production of sperm

sphincter: muscular ring that contracts to close an opening

spinal cord: tube of nervous tissue that extends through the vertebral column

spindle: a structure of microtubules that aligns and separates chromosomes during cell division in eukaryotes

spirillum (pl. spirilla): spiral-shaped prokaryote

spleen: abdominal organ that produces and stores lymphocytes and destroys worn-out red blood cells

sponge: simple animal lacking true tissues and gastrulation; phylum Porifera

spongy bone tissue: bone tissue with large spaces between a web of bony struts

spore (fungus): microscopic reproductive cell, produced sexually or asexually

spore (plant): haploid product of meiosis in the sporophyte; develops mitotically into the gametophyte

sporophyte: diploid, spore-producing stage of the plant life cycle

stabilizing selection: form of natural selection in which extreme phenotypes are less fit than the optimal intermediate phenotype

stamen: male flower part interior to petals; consists of a filament and an anther

standardized variable: any factor held constant for all subjects in an experiment

statistically significant: unlikely to be attributed to chance

stem: part of a plant that supports leaves

stem cell: undifferentiated cell that divides to give rise to additional stem cells and cells that specialize

steroid: lipid consisting of four interconnected carbon rings

steroid hormone: a lipid-soluble hormone that can freely diffuse through a cell membrane and bind to a receptor inside the cell

STI: See **sexually transmitted infection**

stigma: in angiosperms, pollen-receiving tip of style

stoma (pl. stomata): pore in a plant's epidermis through which gases are exchanged with the atmosphere

stomach: J-shaped compartment that mechanically and chemically breaks down food received from the esophagus

STR: See **short tandem repeat**

stroma: the fluid inner region of the chloroplast

style: in angiosperms, the stalklike upper part of a carpel

substitution mutation: replacement of one nucleotide in a gene with another

succession: change in the species composition of a community over time

survivorship curve: graph of the proportion of individuals that survive to a particular age

symbiosis: close and often lifelong ecological relationship in which one species typically lives in or on another

sympathetic nervous system: part of the autonomic nervous system; opposes the parasympathetic nervous system; dominates under stress

sympatric speciation: formation of a new species within the habitat boundaries of a parent species

synapse: junction at which a neuron communicates with another cell

synaptic cleft: space into which neurotransmitters are released between two cells at a synapse

synaptic terminal: enlarged tip of an axon; contains synaptic vesicles filled with neurotransmitters

systematics: field of study that includes taxonomy and phylogenetics

systemic circulation: blood circulation between the heart and the rest of the body, except the lungs

systolic pressure: upper number in a blood pressure reading; reflects contraction of the ventricles

T

T cell: type of lymphocyte that matures in the thymus; may become a helper T cell or a cytotoxic T cell

taiga (boreal forest): type of terrestrial biome with long, dry winters and cool summers with moderate precipitation; coniferous trees dominate

tapeworm: type of parasitic flatworm

taproot system: large central root and its lateral branches

target cell: cell that expresses receptors for a particular hormone

taste bud: cluster of cells that detect chemicals in food

taxon: a group of organisms at any rank in the taxonomic hierarchy

taxonomy: the science of describing, naming, and classifying organisms

technology: the practical application of scientific knowledge

telophase: stage of mitosis in which chromosomes arrive at opposite poles and nuclear envelopes form

telophase I: telophase of meiosis I, when homologs arrive at opposite poles

telophase II: telophase of meiosis II, when chromosomes arrive at opposite poles and nuclear envelopes form

temperate coniferous forest: type of terrestrial biome with cool summers, mild winters, and moderate to high precipitation all year; coniferous trees dominate

temperate deciduous forest: type of terrestrial biome with warm summers, cold winters, and moderate precipitation all year; deciduous trees dominate

temperate grassland: type of terrestrial biome with hot summers, cold winters, and low to moderate precipitation; grasses dominate; grazing, fire, and drought restrict tree growth

template strand: in a DNA double helix, the strand that is transcribed

tendon: band of fibrous connective tissue that attaches a muscle to a bone

terminal bud: undeveloped shoot at a stem's tip

terminator: sequence in DNA that signals where the gene's coding region ends

tertiary structure: the overall shape of a polypeptide, resulting mostly from interactions between amino acid R groups and water

test cross: a mating of an individual of unknown genotype to a homozygous recessive individual to reveal the unknown genotype

testis (pl. testes): male gonad; organ that produces sperm and hormones

testosterone: steroid hormone produced primarily in the testes of male vertebrates

tetrapod: vertebrate with four limbs

thalamus: forebrain structure that relays sensory input to the cerebrum

theory: well-supported scientific explanation

thermoreceptor: sensory receptor that responds to temperature

thermoregulation: control of an animal's body temperature

thick filament: in muscle cells, a filament composed of myosin

thigmotropism: directional growth response to touch

thin filament: in muscle cells, a filament composed of actin

threshold potential: membrane potential at which an action potential is triggered

thylakoid: pancake-shaped structure that makes up the inner membrane of a chloroplast

thylakoid space: the inner compartment of the thylakoid

thymus: lymphoid organ in the upper chest where T cells learn to distinguish foreign antigens from self antigens

thyroid gland: gland in the neck that secretes two thyroid hormones (thyroxine and triiodothyronine) and calcitonin

thyroid-stimulating hormone (TSH): hormone produced in the anterior pituitary; stimulates secretion of thyroid hormones

thyroxine: one of two thyroid hormones; increases the rate of cellular metabolism

tidal volume: volume of air inhaled or exhaled during a normal breath

tight junction: connection between two adjacent animal cells that prevents fluid from flowing past the cells

tissue: group of cells that interact and provide a specific function

tongue: muscular structure on the floor of the mouth; mixes food and aids in swallowing

tooth: mineral-hardened structure embedded in the jaw; grasps and chews food

trace element: an element that an organism requires in small amounts

trachea: the respiratory tube that connects the larynx to the bronchi; the "windpipe"

tracheid: long, narrow conducting cell in xylem

transcription: production of RNA using DNA as a template

transcription factor: in a eukaryotic cell, a protein that binds a gene's promoter and regulates transcription

transduction: process by which a sensory cell converts energy from one form to another

trans fat: unsaturated fat with straight fatty acid tails

transfer RNA (tRNA): a molecule of RNA that binds an amino acid at one site and an mRNA codon at its anticodon site

transgenic: containing DNA from multiple species

translation: assembly of an amino acid chain according to the sequence of nucleotides in mRNA

transpiration: evaporation of water from a leaf

triglyceride: lipid consisting of one glycerol bonded to three fatty acids

triiodothyronine: one of two thyroid hormones; increases the rate of cellular metabolism

trilobite: extinct type of arthropod with three body lobes

trophic level: an organism's position along a food chain

tropical rain forest: type of terrestrial biome with year-round high temperatures and precipitation; abundant, diverse plant species include large trees

tropism: orientation toward or away from a stimulus

true fern: type of seedless vascular plant with fronds bearing sporangia on their undersides

true-breeding: always producing offspring identical to the parent for one or more traits; homozygous

tumor: abnormal mass of tissue resulting from cells dividing out of control

tundra: type of terrestrial biome with low temperature and short growing season; lichens, mosses, grasses, and shrubs dominate

tunicate: type of invertebrate chordate with a tunic covering its saclike body

turgor pressure: the force of water pressing against the cell wall

turtle: type of reptile with a shell made of bony plates

U

umbilical cord: ropelike structure containing blood vessels that connect an embryo or fetus with the placenta

unsaturated fatty acid: a fatty acid with at least one double bond between carbon atoms

urea: nitrogenous waste derived from ammonia

ureter: muscular tube that transports urine from the kidney to the bladder

urethra: tube that transports urine (and semen in males) out of the body

urinary bladder: muscular sac where urine collects

urinary system: organ system consisting of the kidneys and other organs that maintain body fluid composition

urine: liquid waste produced by kidneys

uterine tube: tube that conducts an oocyte from an ovary to the uterus

uterus: muscular, saclike organ where embryo and fetus develop

V

vaccine: substance that initiates a primary immune response to a pathogen without actually causing a disease

vacuole: membrane-bounded storage sac in a cell, especially the large central vacuole in a plant cell

vagina: conduit from the uterus to the outside of the body; receives the penis during intercourse; the birth canal

valence shell: outermost occupied energy shell of an atom

variable: any changeable element in an experiment

variable region: amino acid sequence that is different for every antibody

vas deferens: tube that transports sperm from an epididymis to an ejaculatory duct

vascular bundle: a strand of xylem, phloem, and other tissues in a stem or leaf

vascular cambium: lateral meristem that produces secondary xylem and secondary phloem

vascular tissue: conducting tissue for water, minerals, and organic substances in plants

vasoconstriction: decrease in the diameter of a blood vessel

vasodilation: increase in the diameter of a blood vessel

vegetative plant parts: nonreproductive parts (roots, stems, and leaves)

vein (animal): vessel that returns blood to the heart

vein (plant): vascular bundle inside leaf

ventricle: heart chamber that pumps blood out of the heart

venule: small vein

vertebra: one unit of the vertebral column; composed of bone or cartilage that supports and protects the spinal cord

vertebral column: bone or cartilage that supports and protects the spinal cord

vertebrate: animal with a backbone

vesicle: a membrane-bounded sac that transports materials within a cell

vessel element: short, wide conducting cell in xylem

vestigial: having no apparent function in one organism, but homologous to a functional structure in another species

viroid: infectious RNA molecule

virus: infectious agent that consists of genetic information enclosed in a protein coat

visceral mass: part of a mollusk that contains the digestive, circulatory, excretory, and reproductive systems

vital capacity: maximum volume of air that can be forced out of the lungs during one breath

vocal cord: elastic tissue band that covers the larynx and vibrates as air passes, producing sound

vulnerable species: species facing a high risk of extinction in the distant future

vulva: external female genitalia; the labia, clitoris, and vaginal opening

W

water mold: filamentous, heterotrophic protist with swimming spores

water vascular system: system of canals in echinoderms; functions in locomotion, feeding, sensation, gas exchange, and excretion

wavelength: the distance a photon moves during a complete vibration

whisk fern: type of seedless vascular plant with branched stems but no obvious leaves

white blood cell: one of five types of blood cells that participate in the immune response

white matter: nervous tissue in the central nervous system; consists of myelinated axons

wild-type: the most common allele, genotype, or phenotype

wood: secondary xylem

woody plant: plant with stems and roots made of wood and bark

X

X inactivation: turning off all but one X chromosome in each cell of a mammal (usually female) early in development

X-linked: describes traits controlled by genes on the X chromosome

xylem: vascular tissue that transports water, dissolved minerals, and some hormones in plants

xylem sap: solution of water, dissolved minerals, and hormones in xylem

Y

yellow bone marrow: fatty marrow that replaces red bone marrow as bones age

yolk sac: membrane that manufactures blood cells that support an early embryo

Z

zygomycete: fungus that produces zygospores

zygote: fused egg and sperm; develops into a diploid individual

Index

Note: Page numbers followed by an f indicate figures; numbers followed by a t indicate tables.

A

Abdominal aorta, 539f
Abiotic, 365, 373, 376–377
Abiotic reservoirs, 392f
ABO blood groups, 34–35, 182, 183f, 536, 536f
Abortion, spontaneous, 166
Abscisic acid, 451t, 453, 453f
Absolute dating, 242–243
Absorption, in digestion, 563
Abstinence, 609t, 610
Acacia tree, 4f, 357f
Accessory pigments, 86
Acetylcholine, 483, 484
Acetyl CoA, 103, 103f
Achilles tendon, 523f
Achondroplasia, 151, 174, 186f, 186t
Acid, 32, 32f, 34, 60
Acid deposition, 34, 396, 408–409, 408f
Acidophiles, 281
Acid reflux, 570
Acrosome, 603, 603f, 611f
Actin filaments, 38f, 63, 63f, 64, 524, 524f, 525, 525f
Actinobacteria, 281
Action potentials, 479–482, 480f–482f, 490
Activated B cells, 586f, 588f, 589
Activation energy, 74, 74f
Active immunity, 589, 589t
Active site, 74, 74f
Active transport, 76t, 78–79, 79f, 436, 480, 571
Activity level, and body weight, 560–562, 560f–562f
Adam's apple, 547
Adaptations
   in amphibians, 340, 349, 349f
   in carnivores, 564–565, 564f
   in convergent evolution, 246, 246f
   definition of, 7, 223
   in herbivores, 383, 383f, 564–565, 564f
   and natural selection, 7, 8, 223, 223f
   of plants, to life on land, 302f–305f, 303–305, 426
   in prey animals, 383–384, 384f
   in primates, 344–345
   in reptiles, 341

   sensory, 491
   and speciation, 259–260, 259f, 262
Adaptive immunity, 583, 583f, 586–590, 586f–588f, 589t, 590f
Adaptive therapy, 150, 150f
Adenine, 39, 39f, 72f, 113, 114, 114f, 115f
Adenosine diphosphate (ADP). See ADP
Adenosine triphosphate (ATP). See ATP
Adhering junction, 64, 64f
Adhesion, in water, 29, 29f, 31t
Adhesion proteins, 55
Adipose tissue, 43, 463f, 464
ADP, ATP and, 72
Adrenal cortex, 508f, 509, 509f, 510, 513
Adrenal glands, 505f, 509–510, 509f, 572, 575
Adrenaline. See Epinephrine
Adrenal medulla, 508f, 509–510, 509f, 513
Adrenocorticotropic hormone (ACTH), 506f, 507
Adrenoleukodystrophy, 60
Adult stem cells, 205, 205f, 206
Aerobe, 280, 280t
Aerobic respiration, 99, 99f, 100f, 103–106, 103f, 104f, 527, 528, 535, 546
Africa, 347
African sleeping sickness, 290f, 291
Agar, 289
Age structure, 359, 359f, 367, 367f
Agglutination, 536
Aggregate fruit, 449, 449t
Agrobacterium tumefaciens, 199–200, 200f
AIDS, 592, 610t. See also Human immunodeficiency virus
Air
   mass of, 91
   plants obtaining resources from, 422–424
Air pollution, 407–409, 407f, 408f
Alanine, 118f
Albatross, 402f, 403
Albinism, 186f, 186t
Albumin, 116, 573
Alcohol
   as diuretic, 575
   in hand sanitizers, 75
   prenatal exposure to, 188, 613
Alcoholic fermentation, 108, 108f
Aldosterone, 510, 575
Algae, 287–289
   as autotrophs, 280
   bioluminescent, 297, 297f

   definition of, 287
   photosynthetic, 85, 95, 287–289, 301
   plants compared to, 301–302
Algal bloom, 285, 287–288, 397, 406
Alimentary canal, 564
Alkaline, 32
Allantois, 614, 614f
Alleles. See also Gene(s)
   crossing over and, 160–161, 160f
   definition of, 156, 171, 175t
   dominant, 172f–175f, 173–175, 175t
   frequencies of, 217, 217f, 218t, 226–227, 227f
   in gametes, 175–177, 176f, 177f
   genetic drift and, 231–233, 231f
   in homologous chromosomes, 156, 156f
   mutant, 175, 175t
   recessive, 172f–175f, 173–175, 175t
   wild-type, 175, 175t
Allergens, 593, 593f
Allergies, 295, 446, 510, 592, 593, 593f, 594
Alligators, 341, 341f
Allopatric speciation, 259–260, 259f, 260f, 262, 263t
Alpine chipmunks, 416, 416f
Alternation of generations, 304, 304f, 305t, 444–445, 445t
Alveolar duct, 548, 548f
Alveoli, 548, 548f
Alzheimer disease, 120, 483t, 489, 585
Amanatin, 120
Amino acids, 34t
   bonding of, 36, 36f
   as energy source, 107, 107f
   in origin of life, 274–275, 274f
   and protein folding, 36–38, 37f, 120
   structure of, 36, 36f
Amino group, 36, 36f
Amish people, 232, 232f
Ammonia, 274, 275, 571, 571f
Ammonites, 241, 241f
Ammonium, 282, 395
Amnion, 336f, 613, 614, 614f
Amniote, 336
Amniotic egg, 336, 336f
Amniotic fluid, 614, 614f, 616
Amniotic sac, 616, 616f
Amoeba, 51f, 60, 291
Amoeboid protozoa, 291, 291f

Amphetamine, 484
Amphibians, 241, 340, 340f, 349, 349f, 599
Amygdala, 488
Amylase, 568
Anabolic steroids, 512, 529, 603
Anaerobe, 280, 280t
Analogous structures, 246, 246f
Analysis
    in scientific method, 11f, 12
    statistical, 13, 13f, 203
Anaphase
    in meiosis, 158f, 159, 159f
    in mitosis, 147, 147f
Anaphylactic shock, 593
Anatomical comparisons, 245–246,
        245f, 246f
Anatomy, definition of, 421, 461
Ancestral characters, 266, 266f, 266t
Anchoring junction, 64, 64f
Androgen insensitivity, 617
Androgens, 603
Anemias, 22, 537, 549
Anesthetics, 492
Aneurysm, 549
Angiosperms (flowering plants), 302f, 312–313,
        312f, 442–459
    coevolving with insects, 384–385, 446
    evolution of, 262, 443
    gravitropism in, 455, 455f
    growth of, 450–451, 450f
    gymnosperms versus, 447
    hormones in, 451–453, 451t, 452f, 453f
    leaves of, 427–428, 427f, 428f
    life cycle of, 312, 313f, 444–449, 444f–449f
    light and, 454, 454f
    pollination in, 312, 444f, 446, 446f
    reproduction in, 304, 305, 305f, 312–313,
        443, 443f
    roots of, 429–430, 429f, 430f
    stem of, 427, 427f
    thigmotropism in, 455, 455f
Animal(s). See also specific animals
    asexual reproduction in, 207
    body cavity in, 322, 322f
    cell organization in, 321
    characteristics of, 320–322, 321f, 322f
    definition of, 319
    digestive tract in, 322, 322f
    diversity of, 319, 320, 320f
    embryonic development in, 319,
        321, 321f
    evolution of, 319–320, 319f
    fungal diseases in, 294–295, 295f
    fungi compared to, 292
    as heterotrophs, 85, 280, 319
    importance of, 319
    plants versus, 431
    segmentation in, 322
    tissue organization in, 321

    tissue types in, 462–465, 462f–465f
    transgenic, 200
Animal cells
    active transport in, 78–79
    anatomy of, 53, 53f
    cell division in, 145, 146, 146f–147f, 148, 148f
    connections of, 64, 64f
Animalia, 9f, 10, 319
Annelids, 327, 327f, 538, 538f
Anorexia nervosa, 561, 561f
Antacids, 34
Antagonistic pairs, 523, 523f
Antarctic web of life, 388–390, 389f
Antenna pigments, 88, 91
Anterior cruciate ligament (ACL), 521
Anterior pituitary, 506–508, 506f, 507f, 603,
        603f, 606–607, 606f
Anthers, 445, 445f, 446
Anthocyanins, 89
Anthrax, 281
Antibiotic resistance
    binary fission and, 143
    causes of, 132
    on farms, 234, 234f
    mutations and, 125
    natural selection and, 8, 8f, 223f, 228, 229,
        231, 234
    and new treatment options, 65
    plasmids and, 278
    and "superbugs," 229
Antibiotics
    development of, 65
    fungi secreting, 294
    and intestinal microbes, 570
    mechanisms of action of, 52, 65, 278, 282
    and ribosomes, 120
    in sewage, 388
    and viruses, 132
    and yeast infections, 605
Antibodies, 536, 581, 586, 588–590, 588f,
        593, 593f
Anticodon, 118, 119f
Antidepressants, 55
Antidiuretic hormone (ADH), 466, 506f, 507,
        513, 513f, 575
Antigen, 586, 586f
Antiviral drugs, 132
Ants, 44, 44f, 268, 268f, 330, 438, 438f
Anus, 564, 565f, 569f, 570
Anvil, 496, 496f
Aorta, 539f, 540f, 541
Apes, 343f, 344
Aphids, 599
Apical dominance, 431, 452, 452f
Apical meristems, 430t, 431, 431f
Apicomplexans, 291–292
Apoptosis, 140, 140f, 590
Appendicitis, 570, 585
Appendicular skeleton, 518, 518f, 519

Appendix, 565f, 569, 569f, 570, 582f
Apples, 448, 448f
Aquarium, 396
Aquatic biomes, 376, 377, 380f, 381, 381f, 390,
        406–407
Aquatic plants, 303
Aqueous humor, 494, 494f
Aqueous solution, 30
Arachnids, 331, 331f
Archaea
    bacteria compared to, 278f
    binary fission in, 142–143, 142f
    in carbon cycle, 394
    cells in, 52f, 53
    as domain, 9, 9f, 277, 277f, 280t, 281, 281f
    human uses of, 284
    importance of, 281–284
    mRNA in, 117
    in nitrification, 395f, 396
    reproduction in, 139, 142–143, 142f
    structure of, 278, 279
Archean eon, 240f
Archeology, 519, 519f
Ardipithecus, 345, 345f
Arginine, 118f
Aristotle, 218f, 220
Arsenic, 103
Arteries, 463f, 464f, 539, 539f, 540, 543,
        543f, 544
Arterioles, 539, 543, 543f, 544t, 545
Arthritis, 521
Arthropods, 329–333, 329f–332f, 333t, 493,
        517, 538, 538f
Artificial blood, 534f
Artificial cells, 55
Artificial hand, 516f
Artificial insemination, 611t
Artificial limbs, 530
Artificial organs, 471
Artificial selection
    in dogs, 220
    and natural selection, 221, 221f, 225
    nonrandom mating in, 233
    in plants, 198, 198f, 221, 221f, 422
Artificial sweeteners, 15, 42, 42f
Ascomycetes, 292f, 293, 294, 295, 296
Asexual reproduction, 139, 155, 155f, 599
    in animals, 207
    in archaea, 139
    in bacteria, 6, 139, 142–143, 142f
    cloning as, 206–207
    in cnidarians, 324
    definition of, 6, 7f, 443
    in flatworms, 325
    in fungi, 6, 293
    mutations in, 143, 599
    in plants, 6, 7f, 443, 443f
    in protists, 140f
    in protozoa, 290f

sexual reproduction versus, 155, 163t
in sponges, 6, 323, 599
Asian carp, 412
Asparagine, 118f
Aspartame, 42, 42f, 174
Aspartic acid, 118f
Aspergillus flavus, 297
Aspirin, 75
Assembly, viral, 127f, 128
Assisted reproductive technologies, 611t
Astronauts, 522
Astronomy, 273
Atherosclerosis, 549, 561
Athletic performance, 528–529, 528f
Atlantic cod, 414
Atlantic mollies, 369, 369f
Atmosphere
    in biogeochemical cycle, 392, 392f
    early Earth, 274, 276, 276f, 301
    greenhouse effect in, 409, 409f
Atom(s), 21–23
    in chemical bonds, 24–28, 24f–27f
    definition of, 23t
    energy shells of, 24f, 25
    mass number of, 23
    nucleus of, 22, 22f
    in organizational hierarchy of life, 4f, 5
    structure of, 22, 22f
    types of particles in, 22, 22f, 22t
Atomic number, 22, 23t
Atomic weight, 23, 23t
ATP (adenosine triphosphate)
    ADP and, 72
    in aerobic respiration, 99, 99f, 100f, 103–106, 103f, 104f
    in cells, 72–73, 72f, 73f, 99, 99f
    in cellular respiration, 100, 100f
    in DNA replication, 142
    in electron transport chain, 100, 100f, 105, 105f, 106, 106f
    as energy, in cell, 72–73, 72f, 73f
    in energy storage, 73
    in fermentation, 108, 108f
    function of, 40
    in glycolysis, 100, 100f, 102–103, 102f, 106, 106f
    in Krebs cycle, 100, 100f, 104, 104f, 106, 106f
    mitochondria and, 101, 101f
    muscle cell generation of, 527
    in muscle movement, 522–526, 525f, 526f
    in photosynthesis, 88–93, 89f, 90f
    structure of, 72, 72f
    yield of, from glucose, 106, 106f
ATP synthase, 90f, 90t, 91, 100, 104t
Atrioventricular (AV) node, 541f, 542
Atrium, 538, 538f, 540f, 541
Attachment, viral, 127, 127f
Attention, selective, 498
Auditory canal, 496, 496f

Auditory nerve, 496, 496f, 497f
Auditory tube, 496, 496f
Australian giant cuttlefish, 617, 617f
Australian redback spider, 225
Australopithecus, 530
Australopithecus afaransis, 345–346, 345f, 346f
Australopithecus africanus, 345f, 346
Australopithecus anamensis, 345f
Australopithecus garhi, 345f
Australopiths, 345–346, 345f
Autism, 507, 513, 576, 594
Autoimmune disorders, 511, 592
Autonomic nervous system, 485, 545, 548, 566
Autosomal dominant disorders, 186, 186f, 186t
Autosomal recessive disorders, 186, 186f, 186t
Autosomes, 156, 171–172
Autotrophs (primary producers), 10, 33, 85, 280
    in biogeochemical cycles, 392f
    in carbon cycle, 394, 394f
    in ecosystem, 281, 376
    energy fixed by, 390
    energy source of, 6, 6f, 301, 558
    in nitrogen cycle, 395, 395f, 396
    in phosphorus cycle, 396f, 397
    as trophic level, 388, 388f, 390, 390f
    in water cycle, 393, 393f
Auxins, 448, 451t, 452, 452f, 454, 454f
Axial skeleton, 518, 518f
Axillary bud, 421f, 422, 427f, 428, 431, 431f, 452
Axon, 478, 478f, 479t

B

Bacillus, 279, 279f
Bacillus anthracis, 281
Bacillus thuringiensis, 200
Background extinction rate, 263, 263t, 403–404
Bacon, 41
Bacteria
    acid-producing, 34
    antibiotic resistance in, 8, 8f, 125
    archaea compared to, 278f
    as autotrophs, 281
    in biofilms, 297, 297f
    cells in, 52–53, 52f
    as chemoheterotrophs, 280
    as decomposers, 6, 394
    definition of, 8
    DNA in, 52, 52f, 125
    as domain, 9, 9f, 277, 277f, 280–281, 280t, 281f
    endospores in, 279–280, 279f, 279t
    evolution in, 8, 8f, 234
    hand sanitizers killing, 75
    in human body, 57
    human uses of, 283–284, 283f
    importance of, 281–284, 282f, 283f
    infections caused by, 282–283
    intestinal, 570, 576, 584
    magnetic fields and, 65, 65f

in mouth, 570, 576
    movement of, 53
    mRNA in, 117
    in nitrification, 395f, 396
    nitrogen-fixing, 282, 282f, 395, 424, 424f
    observation of, 14
    photosynthesis in, 85, 281
    as primary producers, 6, 6f
    reproduction in, 6, 139, 142–143, 142f
    sexually transmitted infections caused by, 610, 610t
    structure of, 52–53, 52f, 278, 278f, 279, 279f
    transgenic, 199, 199f, 284
    viruses attacking, 127, 128–129, 129f
    viruses versus, 8, 126, 126f
Bacteriophages, 126f, 127, 128–129, 129f, 130t
Bald eagle, 414, 414f
Baldness, 183, 186
Bark, 432, 432f
Barley, 429f
Base, 32, 32f, 34
Basement membrane, 462f, 463
Basidiomycetes, 292f, 293, 293f, 294, 295, 296
Basophils, 581–582, 581f, 582, 584, 584f
Bats, 295, 446, 446f, 471, 490
B cells, 581–582, 586, 586f, 588–590, 588f
Beagle (ship), 219–220
Bees, 330, 446
Beetles, 318f, 332f, 446
Behavioral isolation, 257f, 258
Bell's palsy, 485
Benign tumor, 149, 149f
Beringia, 314, 314f
Biceps, 523, 523f
Bighorn rams, 230, 230f
Bilateral symmetry, 321, 321f, 325
Bile, 32f, 568, 568t
Bilirubin, 539
Binary fission, 142–143, 142f, 142t
Biodiesel, 304
Biodiversity, 402–419. *See also* Diversity
    climate change and, 409–411, 409f–411f
    conservation of, 415–416
    crisis of, 403–404, 403f
    decline in, 403–404
    definition of, 403
    in forests, 404–405
    habitat destruction and, 404–406, 404f, 405f
    invasive species and, 412–413, 412f
    overexploitation and, 413, 413f
    pollution and, 406–409, 406f–408f
    recovery of, 414–416, 414f, 415f
Biofilms, 297, 297f, 396
Biofuels, 304, 368
Biogeochemical cycles, 392–397, 392f–397f
Biogeography, 243–244
Biological species concept, 255–256, 256t
Biology, definition of, 3
Bioluminescence, 297, 297f

Biomagnification, 391, 391f, 407
Biomes
    aquatic, 376, 377, 380f, 381, 381f, 390,
        406–407
    classification of, 377, 377f
    community interactions in, 381–385, 381t,
        382f–385f
    definition of, 376
    as ecosystems, 376–381
    freshwater, 405
    terrestrial, 376, 377, 377f–379f
Bioprinter, 471
Bioremediation, 284, 414
Biosphere
    definition of, 357f, 358
    in organizational hierarchy of life, 4f, 5
Biosphere 2, 417
Biotechnology, 283, 415
Biotic, 364, 373
Biotic reservoirs, 392f
Bipedalism, 344, 347
Bird(s), 341
    circulatory system of, 538, 538f
    counting, 12
    evolution of, 238f, 241, 246, 472, 472f
    habitats of, 557, 557f
    as reptiles, 340, 341f
Bird-of-paradise, 230f
Birth control, 608, 609t
Birth defects, 613
Birth rate, 359, 365, 367, 367f, 369
Births, and population size, 359, 359f
Bison, 232, 413, 415
Bisphenol A (BPA), 508
Bivalves, 326, 326f
Blackcap warblers, 268
Black mamo, 264t
Black rhino, 415, 415f
Black-winged stilts, 599f
Bladder, urinary, 572–573, 572f
Bladder cancer, 15, 576
Blade, in leaves, 427, 427f
Blastocyst, 612, 612f, 613
Blastula, 319, 321, 321f
Blind spot, 494
Blood
    artificial, 534f
    circulation of, 539–540, 539f
    clotting, 537, 537f
    colorless, 552
    composition of, 535, 535f
    as connective tissue, 463f, 464, 535
    definition of, 535
    donation, 537
    embryonic development of, 614, 615
    functions of, 535, 536–537, 536t, 581
    glucose in, 510–511, 510f, 575
    in homeostasis, 535–537, 535f–537f
    kidneys regulating volume of, 572

    in respiration, 551, 551f
    sugar in, 36
    transfusions, 536
Blood-brain barrier, 489, 489f
Blood pressure, 469, 510, 544–545, 544f, 544t,
        561, 572, 575
Blood types, 34–35, 182, 183f, 536, 536f, 587
Blood vessels, 539, 539f, 542–545
    blood pressure in, 544–545, 544f
    in bones, 520, 520f
    functions of, 540
    in muscles, 524, 524f
    in pregnancy, 614, 614f, 615
    structure of, 543–544, 543f
Blue crabs, 413
Boa constrictor, 245, 245f
Body cavity, 322, 322f
Body mass index (BMI), 560–561, 560f
Body symmetry, 321, 321f
Body temperature
    fever and, 466, 585, 585f
    in hibernation, 107, 471
    maintaining constant, 468f, 522t, 558
    ovulation and, 608, 608f
    regulation of, 6, 6f, 337, 469–471, 470f, 471f
Body weight, 560–562, 560f–562f
Bollworms, 190, 190f
Bonds. See Chemical bonds
Bone(s). See also Skeletal system
    in archaeology, 519, 519f
    calcium in, 522
    definition of, 518, 520t
    development of, 614, 615
    in forensics, 519
    functions of, 519–522, 519t
    in joints, 521, 521f
    modeling of, 521–522, 521f
    in osteoporosis, 509, 522, 522f
    in skeleton, 518, 518f
    tissue in, 463f, 464, 519–521, 520f
Bone marrow, 520t, 582f. See also Red bone
        marrow; Yellow bone marrow
Bone marrow transplant, 592
Bony fishes, 339, 339f
Boreal forest, 378f
Boron, 423f
Bottleneck effect, 232–233, 232f
Botulinum toxin, 526
Botulism, 279, 283
Brachial artery, 539f
Brachiation, 344
Brain
    damage to, 5, 489, 489f
    emergent property in, 5, 5f
    evolution of, 530, 530f
    fetal, 615
    functions of, 485f, 486
    gray matter in, 486, 486f, 487
    human, 347, 348, 486–489, 486f–488f

    and memory, 5, 5f, 488–489
    in nervous system, 477
    nervous tissue in, 465f
    neurotransmitters in, 483
    regions of, 486–488, 487f
    white matter in, 486, 486f, 487
Brain cancer, 145
Brainstem, 486–487, 487f
BRCA1 gene, 209
Breast cancer, 149, 209, 605
Breast milk, production and secretion of, 56f,
        57–59, 80, 116, 342, 507
Breasts, 605
Breath-holding, 552
Breathing, 546. *See also* Respiration;
        Respiratory system
    carbon dioxide in, 545–546, 546f
    cellular respiration and, 99, 99f
    control of, 551, 551f
    in exercise, 551
    in hibernation, 107
    mechanically assisted, 550, 550f
    pressure changes in, 550, 550f
Breech delivery, 616
Breeding
    captive, 415
    selective (*See* Artificial selection)
Bronchi, 547f, 548
Bronchial tree, 548f
Bronchioles, 548, 548f
Bronchitis, 549
Bronchogram, 548f
Brown algae, 288–289, 288f
Brown tree snake, 365
Bruises, 539
Bryophytes, 302f, 304, 305f–307f,
        306–307
Bt toxin, 190, 200, 224
Buffers, 32
Buffon, Georges-Louis, 218f, 219
Bulimia, 561
Bulk elements, 22
Bundle-sheath cells, 93
Burgess Shale, 320, 320f
Burkitt lymphoma, 131
Burns, 584
Butter, 41, 41f
Buttercup, 429f
Butterflies, 446, 446f, 503, 503f, 600
Bypass operation, 549

C

C_3 pathway, 93, 93f, 94f
C_4 pathway, 93, 93f, 94f
Cactus, 94, 94f, 246f, 427, 428
Caecilians, 340, 340f
Caffeine, 81, 613
Calcitonin, 508f, 509

Calcium
  in bones, 522
  as bulk element, 22
  in diet, 559t
  as essential plant nutrient, 422, 423f
  in muscle contraction, 526
  parathyroid hormone and, 509
Calico cats, 185, 185f
Calorie, 69–70, 560
Calvin cycle, 92, 92f, 93, 94f
"Cambrian explosion," 241, 320
Cambrian period, 240f, 320
Camouflage, 384, 384f
CAM pathway, 93–94, 94f
Cancer. *See also specific cancers*
  cell division in, 145, 148–150, 149f
  cytotoxic T cells and, 587–588, 587f
  definition of, 149
  environment and, 150
  and extracellular matrix, 462
  genes and, 150
  lymphatic system and, 582
  malignant cells and, 149, 149f
  metastasis of, 149, 149f
  mutagens and, 125
  risk reduction in, 149t, 150
  transcription factors and, 122
  treatment of, 149, 150, 150f, 234
  vaccines in prevention of, 591
Candida albicans, 295, 605, 610t
Canine teeth, 567f
Capillaries, 463f, 539, 543f,
      544, 544t
Capillary beds, 543, 543f
Capsaicin, 456, 456f
Capsule, 52, 52f
Captive breeding, 415
Carbohydrate(s)
  complex, 34t, 35
  definition of, 34
  digestion of, 569f
  energy extraction from,
      107, 107f
  functions of, 34, 34t, 35
  as organic molecules, 34–35, 34t
  simple, 34–35, 34t, 35f
  structure of, 34t, 35f
Carbohydrate-based fat substitutes, 42
Carbon
  as bulk element, 22
  covalent bonds of, 27f, 28
  energy shells of, 24f, 25
  as essential plant nutrient, 422,
      423, 423f
  in hydrocarbons, 33
  isotopes of, 23, 23f
  in origin of life, 276
  in periodic table, 21f
  versatility of, 33

Carbon cycle, 282, 394–395, 394f
Carbon dioxide
  blood carrying, 541, 551, 551f
  in carbon cycle, 394–395, 394f
  in chemical reactions, 31
  in early atmosphere, 275, 276, 301
  fossil fuels and, 92
  as greenhouse gas, 409–411, 409f, 410f
  in photosynthesis, 85, 85f
  in respiration, 546, 546f
Carbon fixation, 92, 92f
Carbon-14 dating, 242–243, 242f
Carbonic acid, 394–395
Carboniferous period, 240f
Carbon monoxide, 24, 103, 551
Carbon reactions, 89, 89f, 90f, 92, 92f
Carboxyl group, 36, 36f
Carcinoma, 464
Cardiac cycle, 541–542, 541f
Cardiac muscle cells, 541–542, 541f
Cardiac muscle tissue, 464f, 465, 540, 540f
Cardiac output, 542
Cardiovascular system, 539
Carmine, 349
Carnivores, 376, 383, 388, 388f, 562, 564–565, 564f
Carnivorous plants, 420f, 421, 424, 428
Carolina parakeet, 413
Carolina wrens, 390
Carotenoids, 86, 89
Carpals, 518f
Carpels, in flowers, 445, 445f
Carrageenan, 289
Carrots, 421f, 430, 438
Carrying capacity, 364, 364f
Cartilage
  in bone modeling, 521, 521f
  as connective tissue, 463f, 464, 520
  definition of, 520t
  in embryonic development, 615
  functions of, 521
  in long bone, 520f
Cartilaginous fishes, 339, 339f
Cashews, 456
Cas9 (enzyme), 210
Caspian tiger, 264t
Castoreum, 349
Cataracts, 408, 495
Caterpillars, 190, 330, 330f, 383, 383f, 412, 452,
      503, 503f, 600
Catheter, 572
Cats, 185, 185f, 188f
Cave crayfish, 246f
Cecum, 564f, 565, 569, 569f
Cell(s), 48–65. *See also* specific cell types
  anchoring junction between, 64, 64f
  animal, 53, 53f, 64, 78–79, 321
  artificial, 55
  ATP in, 72–73, 72f, 73f, 99, 99f
  in bacteria, 52–53, 52f

  in cancer, 145, 148–150, 149f
  common features of, 51, 51f
  as component of life, 3, 4f
  death of, 59, 128, 129, 140, 140f, 590
  definition of, 49
  division of, 139–140, 140f, 142–148, 142f–144f
  DNA in, 3, 3f
  energy conversion by, 70, 70f
  in eukaryotes (*See* Eukaryotic cells)
  in extracellular matrix, 462
  gap junction between, 64, 64f
  homeostasis maintained by, 6
  human, 57
  ice crystals formed in, 31
  life cycle of, 139–140, 139f, 140f, 145, 145f
  microscopic viewing of, 49–50, 49f, 50f, 278
  movement of, 63
  nucleus in, 9, 9f, 52, 53, 53f, 57, 57f
  organic molecules in, 33–43
  in organizational hierarchy, 4f, 5, 461f
  origin of, 274–276, 274f
  outer membrane of, 3
  pH in, 32, 32f, 59
  plant (*See* Plant cells)
  in prokaryotes, 52–53, 52f, 53f
  proteins produced by, 36, 51
  size of, 51
  specialized, 62, 62f
  surface area of, 51, 51f
  tight junction between, 64, 64f
  as units of life, 49–51, 49f–51f
  viruses compared to, 126–127
Cell biology, 49
Cell body, in neurons, 478, 478f, 479t
Cell-mediated immunity, 586f, 587–588, 587f
Cell membrane, 51, 52, 52f, 53f, 54–55, 54f, 55f
Cell plate, 148, 148f
Cell theory, 49, 276
Cellular respiration
  ATP in, 100, 100f
  glucose in, 70, 71
  in mitochondria, 60, 61f
Cellular slime molds, 286, 290, 290f
Cellulose, 35, 64, 64f, 294, 301, 304, 560, 564
Cell wall, 52, 52f, 64, 64f, 278f, 279, 279t
Cenozoic era, 240f
Centipedes, 322, 332, 332f
Central nervous system, 477, 479f, 485f–489f,
      486–489
Centriole, 53f, 146f
Centromere, 144, 144f, 144t, 147
Centrosome, 53f, 63, 145, 146f
Cephalization, 321, 321f
Cephalopods, 326, 326f
Cerebellum, 487, 487f
Cerebral cortex, 487–488, 488f
Cerebrospinal fluid, 489
Cerebrum, 487, 487f
Cervical cancer, 131, 591, 605, 610, 610t

Cervical cap, 609t
Cervical vertebrae, 518f
Cervix, 604f, 605
Cesarean section, 616
C₄ pathway, 93, 93f, 94f
CFTR gene, 209f
Chagas disease, 291
Chaparral, 379f
Chargaff, Erwin, 113, 113f
Charophytes, 301, 301f
Cheetahs, 232, 232f
Chelicerates, 330–331, 331f, 332t
Chemical bonds
    in amino acids, 36, 36f
    atoms in, 24–28, 24f–27f
    covalent, 25t, 26–28, 27f
    definition of, 25
    hydrogen, 25t, 28, 28f
    ionic, 25–26, 25t, 26f
Chemical cycles, in ecosystems, 392–397,
        392f–397f
Chemical digestion, 563, 566, 569f
Chemical pollutants, 391, 391f, 406–407, 406f
Chemical reactions
    definition of, 31
    energy in, 71, 71f
    enzymes in, 74–75, 74f, 75f
    networks of, 71–72
    in photosynthesis, 71, 85
    water and, 31, 31t
Chemistry, 21
Chemoreceptors, 490, 492–493
Chemotherapy, 149, 150, 234
Chemotrophs, 280, 280t
Chesapeake Bay blue crabs, 413
Chest cavity
    in breathing, 550, 550f
    embryonic development of, 614
Chewing, 562–563, 566
Chicken, 140f, 372f
Childbirth, 507, 616, 616f
Chili plant, 421f, 422, 456, 456f
Chimera, 167
Chimpanzees, 250, 345
China, 367
Chinese paddlefish, 264t
Chipmunks, 416, 416f
Chitin, 35, 292, 329
Chitons, 326, 326f
Chlamydia, 610, 610t
Chlamydomonas, 286f, 289
Chloramphenicol, 120, 282
Chloride, 29–30, 30f
Chlorine, 24, 25–26, 26f, 423f
Chlorofluorocarbons (CFCs), 408, 409, 410
Chlorophyll, 61, 86, 88–89, 89f, 90f, 91
Chloroplasts
    endosymbiosis and, 284–286, 285f, 286f
    mitochondria and, 101, 101f

photosynthesis in, 60–62, 61f, 62f, 62t,
        87–88, 87f, 88f, 91, 428
    in plant cell anatomy, 53, 53f
Choanoflagellates, 319
Cholera, 80, 279
Cholesterol, 43, 43f, 54, 55f, 186t, 505, 568
Chordates
    common features of, 335, 335f
    as deuterostomes, 321
    diversity of, 335, 336f, 336f–342f,
        337–342
    embryonic development in, 336, 336f
    thermoregulation in, 337
    as vertebrates, 335–337
Chorion, 614, 614f
Chorionic villi, 614, 614f
Choroid, 494, 494f
Christmas Bird Count, 12
Chromatids, 144, 144f, 144t, 160f, 161, 161t,
        180, 181f, 181t
Chromatin, 143–144, 144f, 144t, 158
Chromosome(s)
    definition of, 114t, 144t, 156, 175t
    DNA in, 114, 141, 141f, 143, 143f, 171
    homologous pairs of, 156, 156f, 160–162,
        160f, 161t, 171f, 172, 180, 181f, 611
    human, 156, 156f, 171–172, 171f
    independent inheritance of genes on,
        178–179, 178f, 179f
    linked genes on, 180–181, 180f, 181f
    in meiosis, 158–159, 158f–159f
    in mitosis, 146–147, 146f–147f
    nondisjunction and, 164–166, 165f
    number of, 157
    polyploidy and, 164
    in prokaryotes, 143
    structure of, 143–144, 144f
    testing of, in pregnancy, 190
Chromosome maps, 181, 181f
Chyme, 567, 568
Chymosin, 199
Chymotrypsin, 568
Chytridiomycetes, 292f, 293
Cicadas, 332f
Cichlids, 260–261, 260f
Cigarettes, 149t, 549, 613
Cilia, 63, 63f, 291, 548
Ciliates, 63, 291, 291f
Circulatory system. *See also* Blood
    in annelids, 327, 538, 538f
    in arthropods, 329, 538, 538f
    closed, 538, 538f
    diseases of, 549
    in echinoderms, 334
    in flatworms, 325
    functions of, 467, 467f
    human, 539–545, 539f, 541f
    in mollusks, 326, 538
    in nematodes, 328

    open, 538, 538f
    in pregnancy, 614, 614f
Clade, 266t, 267
Cladistics, 265–267, 266f, 266t
Cladograms, 266–267, 266t, 267f
Clams, 326, 517
Classes, 265
Classification systems, 265–267
    biological species concept in, 255–256
    cladistics in, 265–267, 266f
    cladograms in, 266–267, 267f
    Linnaean, 255, 265
    taxonomic hierarchy in, 9–10, 10f, 265, 265f
Clavicle, 518f, 519
Clay, 275, 275f
Clean Air Act, 415
Clean Water Act, 415
Cleavage, 612
Cleavage furrow, 148, 148f
Cleft palate, 465
Climate(s), 374–376, 374f–376f
    convection cells and, 375, 375f
    mountain ranges and, 376, 376f
    ocean currents and, 375, 375f
    and photosynthesis, 93–94, 94f
    water and, 30
Climate change, 409–411
    and biodiversity, 409–411, 409f–411f
    consequences of, 411, 411f
    definition of, 411
    and extinction, 263
    fossil fuels and, 92, 304
    and human evolution, 347
    ozone layer and, 409
Climax community, 387
Clindamycin, 120
Clitoris, 604f, 605, 615f
Clock, molecular, 240f, 249
Clonal deletion, 589–590, 590f
Clonal selection, 586f, 587, 588f, 589
Cloning, 206–207, 206f, 207f
Closed circulatory system, 538, 538f
Clostridium botulinum, 279, 283
Clostridium tetanum, 280, 281
Clotting factors, 536
Club mosses, 308, 308f
Cnidarians, 322, 324, 324f, 564
Cnidocytes, 324, 324f
Coastlines, 405–406
Coccus, 279, 279f
Coccyx, 518f
Cochlea, 496, 496f, 497, 497f
Cochlear implant, 497
Cod, 414
Codominance, 182, 183f, 183t
Codon, 115, 119f
Coelacanths, 339, 339f
Coelom, 322, 322f, 327, 327f
Coevolution, 384–385, 385f, 446

Cohesion, in water, 29, 29f, 31t, 435
Cohesion-tension theory, 435, 435f, 436t
Colds, 128, 132
Cold sores, 129
Coleoptera, 333t
Collagen, 38f, 520, 520f
Collar cells, 323, 323f
Collecting ducts, 573f, 574, 574f
Collenchyma, 425, 425f
Colon, 569f, 570
Colon cancer, 35, 570
Color blindness, 185t, 186f, 495
Colorectal cancer, 560, 570
Commensalism, 381t, 383, 383f
Common carotid artery, 539f
Communication
    endocrine system in, 466, 466f, 477,
        503–504, 503f
    among eukaryotic cells, 64, 64f
    human, origin of, 133, 133f
    nervous system in, 466, 466f, 477–478, 477f,
        482–483, 482f, 483f, 503
    pheromones in, 493
Community
    climax, 387
    definition of, 357, 357f, 373
    diversity of, 386, 386f
    examples of, 373f
    interactions of, in biomes, 381–385, 381t,
        382f–385f
    succession in, 386, 386f, 387f
Community ecology, 357
Compact bone, 520, 520f, 520t
Companion cells, 425, 425f
Competition, for resources, 381t, 382, 382f
Competitive exclusion principle, 382
Competitive inhibition, 74, 75f
Complementary base pairs, 114, 114f, 115f
Complement proteins, 585
Complete digestive tract, 322, 564, 564f
Complete dominance, 183t
Compound, definition of, 24
Compound leaves, 427f, 428, 429t
Compound microscope, 50, 50f
Concentrated animal feeding operation (CAFO),
        372f, 373, 398
Concentration gradient, 78, 79, 435
Conception, 608
Concussion, 489, 489f
Condensation reaction. *See* Dehydration
        synthesis
Condoms, 608, 609t, 610
Conducting cells, in plants, 29, 425, 425f
Cone cells, 494f, 495, 495t
Cones, 310, 311, 311f
Confocal microscope, 50, 50f
Coniferous forest, 378f, 405
Conifers, 310, 310f, 314, 377
Conjoined twins, 167, 613

Connective tissue, 463f, 464, 520–521, 524,
        524f, 535
Conservation biologists, 404
Constant regions, 588f, 589
Constipation, 570
Consumers (heterotrophs)
    in biogeochemical cycles, 392f
    in ecosystem, 376
    energy source of, 6, 6f, 33, 85, 280, 319, 376,
        558
    fungi as, 292
    in nitrogen cycle, 395f, 396
    in phosphorus cycle, 396f, 397
    protists as, 289–292, 289f–291f
    as trophic level, 388, 388f, 390f
    in water cycle, 393, 393f
Continental drift, 243–244, 243f, 244f, 263, 376
Contraception, 608, 609t
Contractions
    in muscles, 464, 524–526, 525f, 526f, 541,
        548, 566, 566f
    uterine, 468, 507, 607, 616
Control center, 469t
Controlled experiments, 12, 12f
Controls, 13, 13t
Convection cell, 375, 375f
Convergent evolution, 246, 246f
COPD (chronic obstructive pulmonary disease),
        549
Copepods, 297, 297f
Copper, 423f
Coral reef, 324f, 380f, 395
Cordyceps, 294
Cork cambium, 432, 432f
Corn, 94f
Cornea, 494, 494f, 495
Corneal transplant surgery, 495
Coronary arteries, 540f, 541
Corpus callosum, 487
Corpus luteum, 605, 605f, 607, 607f
Cortex, in plants, 427, 427f, 429f, 429t, 430, 430f
Cortisol, 510, 513
Cortisone, 43
Cosmetic surgery, 465
Cotton, 190, 261, 261f
Cotyledons, 312, 447–448, 447f
Coughing, 547
Covalent bonds, 25t, 26–28, 27f
Cowpox, 591
Crabs, 332f, 413, 517
Cranial nerves, 484, 486, 494
Cranium, 335, 335f
Crayfish, 246f
Crazy ants, 44, 44f
Creatine phosphate, 527
Cretaceous period, 240f
Crick, Francis, 113, 113f, 115, 141
Criminal justice, 203, 204, 204f, 212
CRISPR-Cas9, 210–211, 210f

Cristae, 60, 61f, 101
Crocodilians, 341, 341f
Crossbills, 384, 385f
Cross-fertilization, 173, 173f
Crossing over, 160–161, 160f, 161t, 180, 181f
Crustaceans, 35, 332, 332f, 332t
Cryptosporidium, 291
C_3 pathway, 93, 93f, 94f
Cucumber mosaic virus, 131f
Culture, 348
Cuticle, 303, 303f, 426
Cuttlebones, 349
Cuttlefish, 617, 617f
Cyanide, 103, 406
Cyanobacteria, 280, 281, 281f, 296, 301
Cycads, 310, 310f
Cyclosporine, 294, 294f
Cysteine, 118f
Cystic fibrosis, 80, 80f, 120, 124, 134, 171, 177,
        177f, 184, 186t, 189, 208–209, 208f
Cytochrome *c*, 248, 249f
Cytokines, 585, 586f
Cytokinesis, 142t, 145, 148, 148f, 158f, 159,
        162–163, 452
Cytokinins, 451t, 452, 452f
Cytoplasm, 51, 52f, 53, 53f, 54–55, 148, 148f, 606
Cytosine, 39, 39f, 113, 114, 114f, 115f
Cytoskeleton, 53f, 62–64, 63f, 64
Cytosol, 51, 52f, 53f, 58, 58f, 60f, 62, 101f, 102
Cytotoxic T cells, 587–588, 587f

D

Damming, 405, 405f
Dandelions, 430, 449, 449f
Darwin, Charles, 8, 10–11, 14, 14f, 16, 218f,
        219–221, 219f, 255, 347
Data collection, in scientific method, 11f, 12
"Dead zone," in Gulf of Mexico, 406, 407f
Deafness, 497
Death, cell, 59, 128, 129, 140, 140f, 590
Death rate, 359–360, 365, 367, 367f, 368, 368t
Deaths, and population size, 359, 359f
Deciduous forest, 378f, 454
Decomposers, 6, 6f, 294, 294f
    in biogeochemical cycles, 392f
    in carbon cycle, 394, 394f
    in nitrogen cycle, 395f, 396
    in phosphorus cycle, 396f, 397
    as trophic level, 388
Deer, 365
Deforestation, 404, 404f, 405
Dehydration synthesis, 33, 33f, 34, 39, 39f, 40
Deletion mutation, 124, 124t
Deltoid, 523f
Demographic transition, 366–367, 367f
Denaturation, of proteins, 38, 38f, 75
Dendrites, 478, 478f, 479t
Denitrification, 395f, 396

Dense connective tissue, 463f, 464
Density, population, 358–359, 358f
Density-dependent factors, 364–365, 365f
Density-independent factors, 365, 365f
Deoxyribonucleic acid (DNA). *See* DNA
Deoxyribose, 39, 114, 114f, 115, 115f
Dependent variables, 12, 13t
Depression, 484, 484f, 512
Derived characters, 266, 266f, 266t
Dermal tissue, in plants, 426, 426f, 431f
Dermatitis, 585
Dermis, 583, 583f
Desert
    as biome, 379f
    net primary production in, 390
    plants in, 246, 246f, 303, 310, 430
Determinate growth, in plants, 430
Detritivores, 562
Detritus, 388
Deuterostomes, 321
Development. *See also* Embryonic development
    of bones, 521, 521f
    as characteristic of life, 7
    direct, 600, 600f
    of human infant, 611–616, 611f, 612f,
        614f–616f
    indirect, 600
    in plants, 451–453, 451t, 452f, 453f
    reproduction and, 599–600, 599f
Devil's Hole pupfish, 260, 260f
Devonian period, 240f, 349
Diabetes, 36, 510–511, 575, 585
    Type 1, 503, 511, 511f, 592
    Type 2, 511, 511f, 561
Dialysis, 575, 575f
Diaphragm, 550, 550f
Diarrhea, 466, 570
Diastolic pressure, 544, 544f
Diatoms, 288, 288f
Dickinsonia, 320f
Diet. *See also* Food
    artificial sweeteners in, 15, 42, 42f
    and body weight, 560–562, 560f–562f
    cancer and, 149t
    and digestive tract structure, 564–565, 564f
    essential elements in, 22
    fads in, 561
    fat in, 40–41, 42, 43
    habitat and, 557, 557f
    healthful, 43, 438, 511
    high-fiber, 35, 560
    junk food in, 43
    minerals in, 559, 559t
    and tooth decay, 576, 576f
    varied, 559–560, 559t, 560f
    vitamins in, 559, 559t
Dietary supplements, 527
Diet pills, 105
Diet soda, 174

Differentiation, 600
Diffusion, 76–78, 76f, 76t, 77f
    facilitated, 76t, 78, 78f, 480
    in gas exchange, 551
    simple, 76t, 77
Digestion, 562–563
    acids in, 34
    chemical, 563, 566, 569f
    extracellular, 563f, 564
    intracellular, 563, 563f
    mechanical, 567
    muscle contraction in, 464
Digestive enzymes, 36, 38, 38f, 107, 563–564,
        567–569, 569f
Digestive system
    in animals, 322
    definition of, 558
    embryonic development of, 614
    food poisoning and, 466
    functions of, 466f, 467, 558
    in homeostasis, 468f
    human, 565–570, 565f
    unhealthy, 570
Digestive tract
    complete, 322, 564, 564f
    diet and structure of, 564–565, 564f
    incomplete, 322, 322f, 564, 564f
    muscles in, 566, 566f
    specialized, 562–565, 562f–564f
    tissues in, 462f, 464f
Dihybrid cross, 177t, 178, 178f
Dihydrotestosterone, 186
Dikaryotic mycelium, 293f, 294, 294f
Dimorphism, sexual, 230
Dinoflagellates, 287–288, 287f, 297, 297f
Dinosaurs, 340, 347
Diphtheria toxin, 120
Diphtheria vaccine, 594
Diploid cells, 156, 156f, 171, 293f, 294, 294f
Diptera, 333t
Direct development, 600, 600f
Directional selection, 228, 228f
Disaccharides, 34, 34t, 35f
Discovery science, 12
Disruptive selection, 228, 228f
Dissolution, in water, 29–30, 30f, 31t, 32
Diuretics, 575
Diversity, 9f. *See also* Biodiversity
    of amphibians, 340, 340f
    of animals, 319, 320, 320f
    of annelids, 327, 327f
    of arthropods, 329f, 330–333
    of chordates, 335, 336f, 336f–342f, 337–342
    of cnidarians, 324
    of community, 386, 386f
    of echinoderms, 334
    of flatworms, 325
    of fungi, 292f, 293–294
    of mammals, 342, 342f

    of mollusks, 326, 326f
    of nematodes, 328, 328f
    of plants, 302f, 303
    of prokaryotes, 278–280
    of protists, 287, 287f
    of reptiles, 340, 341f
Diving reflex, 552
Dizygotic (fraternal) twins, 162, 162f, 611
DNA. *See also* Gene(s)
    availability, 122, 122f
    in bacteria, 52, 52f, 125
    in chloroplasts, 62, 62t
    in chromosome, 114, 141, 141f, 143, 143f, 171
    in classification systems, 256
    cloning, 206–207, 206f, 207f
    discovery of, 113, 113f
    as double helix, 40, 40f, 113–114, 113f, 114f,
        116, 116f, 141, 141f
    in egg cell, 605, 606, 611
    evolutionary evidence in, 40, 248
    function of, 40, 116, 197, 197f
    in gene doping, 211
    in gene therapy, 209
    as informational molecule of life, 3, 3f
    in meiosis, 158–159
    mitochondrial, 60, 62, 203–204
    in molecular clock, 249, 249f
    mutations in (*See* Mutations)
    as nucleic acid, 39
    in plastids, 62
    in primate evolution, 345
    of prokaryotes, 9, 278
    recombinant, 198–199
    replication, 40, 124, 141–142, 141f, 145, 158,
        201, 202, 202f
    RNA versus, 115, 115f
    sequencing, 200–202, 201f, 248, 256
    in sperm, 603, 603f, 611
    structure of, 39, 39f, 113, 114, 114f, 115f,
        141, 197, 197f
    testing, 204
    transcription, 115–117, 115f–117f
    translation, 115, 115f, 118–120, 118f–120f
    viral, 126t, 127f, 128
DNA polymerase, 141, 141f, 202
DNA probes, 208, 208f
DNA profiling, 202–203, 203f, 204f
DNA technology, 196–215
    cloning in, 206–207, 206f, 207f
    definition of, 197
    ethics of, 200, 206, 207, 211
    genes in, 198–204, 199f–204f
    importance of, 197–198
    in medicine, 198, 199, 208–211, 208f–210f
    in species conservation, 415
    stem cells in, 205–206, 205f
DNP, 103
Dodo, 264t, 403, 413
Dogs, 220, 490

Dolly (cloned sheep), 207
Dolphins, 246
Domains, 9–10, 52–53, 52f, 53f, 265
Domatia, 438
Dominance
    complete, 183t
    incomplete, 182, 183f, 183t
Dominant alleles, 172f–175f, 173–175, 175t
Dopamine, 483
Dormancy, in plants, 448, 453, 453f
Dorsal nerve cord, 335, 335f
Double fertilization, 312, 444f, 446–447, 447f
Double helix, 40, 40f, 113–114, 113f, 114f, 116,
    116f, 141, 141f
Down syndrome, 166, 166f
Dragonflies, 332f
Drugs
    and birth defects, 613
    and neurotransmitters, 484
Drug testing, 573
Dry rot, 294
Duchenne muscular dystrophy, 185t
Duck, 140, 140f
Duodenum, 567, 568
Dwarfism, 151, 174, 186f, 186t, 507, 507f

E

Eagle, bald, 414, 414f
Ear
    artificial, 471
    embryonic development of, 615
    in hearing, 496–497, 496f, 497f
Eardrum, 496, 496f
Ear infection, 496
Earth
    age of, 239
    climates of, 374–376, 374f–376f
    convection cells of, 375, 375f
    origin of life on, 273–276, 273f–276f
    seasons of, 374, 374f
    temperature of, 410, 410f, 411
    water supply of, 381, 381f
Earthworms, 322, 327, 327f, 517
Eating disorders, 561, 561f
Ecdysozoans, 321
Echidna, 342, 342f
Echinoderms, 321, 334, 334f, 518
Echolocation, 490
E. coli, 108, 121, 198, 199f, 280, 281, 283, 283f
Ecological footprint, 368, 368f
Ecology, 357–358, 357f
Ecosystems
    autotrophs in, 85
    biomes as, 376–381
    carrying capacity of, 364, 364f
    chemical cycles in, 392–397, 392f–397f
    definition of, 357, 357f, 373
    diversity of, 403

energy in, 388–391
    in organizational hierarchy of life, 4f, 5
    prokaryotes in, 282, 282f
    self-contained, 417
    unbalanced, 372f, 398
Ectoderm, 321, 321f, 613, 614–615
Ectomycorrhizae, 295–296
Ectotherms, 337, 390, 470, 470f
Ediacaran organisms, 319–320, 320f
Effector, in negative feedback, 469t
Effector helper T cells, 586f, 588f, 589
Egg
    cooked, 38, 38f
    evolution of, 336
Egg cells, 604–607
    in angiosperm life cycle, 444f, 445,
        446–447
    in fertilization, 155, 611, 611f
    as gamete, 139, 139f, 157, 157f, 599
    production of, 512, 605–606, 605f, 606t
Ejaculation, 602
Ejaculatory duct, 601f, 602
Electromagnetic spectrum, 86, 86f
Electron(s)
    in atoms, 22, 22f, 22t
    in bonding, 25, 25t
    in covalent bonds, 25t, 26–28, 27f
    energy carried by, 71–72, 72f
    in hydrogen bonds, 25t, 28, 28f
    in ionic bonds, 25–26, 25t, 26f
    in ions, 22
    in isotopes, 23, 23f
    in orbitals, 25
Electronegativity, 25, 25f, 26, 27, 28
Electron microscopes, 49f, 50, 50f
Electron transport chain, 72, 72f, 90, 90f, 100,
    100f, 104t, 105, 105f, 106, 106f
Electrophoresis, 201, 201f
Elements
    atoms in, 22
    bulk, 22
    definition of, 21, 23t
    as fundamental types of matter, 21–22, 21f
    isotopes of, 23, 23f
    molecules in, 24
    in periodic table, 21, 21f, 22
    trace, 22
Elimination, 563, 570–573
Ellis-van Creveld syndrome, 232, 232f
Elongation
    of DNA transcription, 116, 116f
    of DNA translation, 119–120, 119f
Embryo
    bones of, 521, 521f
    seed as, 447–448, 447f
Embryonic development
    in animals, 319, 321, 321f
    in chordates, 336, 336f
    evolutionary clues in, 246–247, 246f, 247f

Embryonic disc, 612f, 613
Embryonic stage, 614–615, 614f, 615f
Embryonic stem cells, 205, 205f, 206
Embryo sac, 445–446, 447
Emergent properties, 5, 5f
Emphysema, 549
Endangered species, 403
Endangered Species Act, 415
Endocrine disruptors, 508
Endocrine glands, 503, 505f
Endocrine system, 502–515. See also Hormones
    communication function of, 466, 466f, 477,
        503–504, 503f
    control in, 505–508, 505f, 506f
    stress and, 509–510, 509f, 513
Endocytosis, 76t, 79–80, 79f, 128
Endoderm, 321, 321f, 613
Endodermis, in plants, 429f, 430, 435
Endomembrane system, 57
Endometrium, 604f, 605, 606, 607, 607f
Endophytes, 295
Endoplasmic reticulum (ER), 53f, 56f, 58,
    58f, 59
Endorphins, 484, 506f, 508
Endoskeleton, 517f, 518
Endosperm, 446–448, 447f
Endospores, 279–280, 279f, 279t
Endosymbiosis, 62, 284–286, 285f, 286f
Endothelium, 540, 540f
Endotherms, 337, 390, 470, 470f
Endotoxins, 283
Energy
    activation, 74, 74f
    ATP as, in cell, 72–73, 72f, 73f
    calorie as unit of, 69–70, 560
    in chemical reactions, 71, 71f
    conversion of, by cells, 70, 70f
    definition of, 21, 69
    digestive system and, 558
    and ecological footprint, 368
    in ecosystems, 388–391
    entropy and, 70, 70f
    in food, 69–70, 99, 99f, 107, 107f, 558
    in food webs, 388–390, 388f–390f
    kinetic, 69, 69f
    lipids as source of, 43
    in membrane transport, 75–80, 76f–80f, 76t
    mitochondria and, 60
    organ systems acquiring, 466f–467f, 467
    potential, 69, 69f
    pyramid of, 390–391, 390f
    as requirement of life, 5–6, 6f
    sexual reproduction costing, 599
    solar, converted to chemical (See
        Photosynthesis)
    storage of, 34t, 35, 73
    sugars providing, 34, 34t
    use of, by cells, 99, 99f
    yield of, from glucose, 106, 106f

Energy shell, 24f, 25
Entropy, 70, 70f
Envelope (viral), 126–127
Environment
    and cancer risk, 150
    and human evolution, 347
    and natural selection, 224, 224f
    and phenotype, 188–189, 188f
    plants responding to, 430–431, 451, 455, 455f
    and species' distribution, 376–377
Enzymes
    activity of, 74–75, 74f, 75f
    in bacteria, 121, 121f
    and cell membrane, 55
    definition of, 74
    in dehydration synthesis, 33
    digestive, 36, 38, 38f, 107, 563–564,
        567–569, 569f
    and DNA availability, 122
    in DNA replication, 141, 141f
    functions of, 113, 120
    in hydrolysis, 34
    inhibition of, 74, 75f
    in Krebs cycle, 100
    nitrogen-fixing, 424
    in peroxisomes, 60
    in photosynthesis, 92, 93
    in reactions, 74–75, 74f, 75f
    in respiration, 547
    in saliva, 566
    in transgenic technology, 199, 199f
    viruses using, 129
Eocene epoch, 240f
Ephemeroptera, 333t
Epidermis. See also Skin
    in immune system, 583, 583f
    in plants, 426, 426f, 429f, 430, 431, 435, 435f
Epididymis, 601f, 602
Epigenetics, 122
Epiglottis, 248, 248f, 547, 547f, 566
Epilepsy, 483t
Epinephrine, 483, 508f, 509–510, 509f, 513, 593
Epithelial tissue, 462–464, 462f, 546
Epstein-Barr virus, 131
Equator, 374, 405
Equilibrium, 497t
    Hardy-Weinberg, 226–227, 227f
    punctuated, 262, 262f, 263t
    sense of, 497t
Equilibrium life history, 362, 362f
Erectile tissue, 601f
Erosion, 405, 429
Erythromycin, 282
Erythropoietin, 211
Esophagus, 565f, 566
Essay on the Principle of Population (Malthus), 220
Essential elements, 422–423, 423f
Estrogen, 43, 512, 512f, 606, 606f, 607,
    607f, 615

Estuary, 380f, 406
Ethanol, 108, 304, 575
Ethics
    of cloning, 207
    of gene therapy, 211
    of genetic testing, 211
    of in vitro fertilization, 211
    of stem cells, 206
    of transgenic technology, 200
Ethylene, 448, 451t, 453
Eudicots, 312
    flower structure in, 445
    leaf veins of, 428, 428f
    roots of, 429, 429f
    seeds of, 447–448, 448f
    stem anatomy of, 427, 427f
Eukarya, 9, 9f
Eukaryotes
    evolution of, 284–286, 284f–286f
    prokaryotes versus, 9, 277, 277f
    protists as, 53, 287–292
Eukaryotic cells, 52, 52f, 53
    ATP production in, 101, 101f
    cell division in (See Meiosis; Mitosis)
    cell membrane of, 54–55, 54f, 55f
    communication among, 64
    connections of, 64, 64f
    cytoskeleton supporting, 62–64, 63f
    gene expression in, 121–123, 122f
    mRNA in, 117, 117f
    organelles in, 53, 56–62, 56f–62f, 62t
    origin of, 284–286, 284f–286f
    size of, 53
    structure of, 53, 53f
European starlings, 268, 412, 412f
Eustachian tube, 496, 496f
Eutrophication, 285, 397, 397f, 406
Evaporation, 30, 30f, 393, 393f, 434–435, 470
Evolution. See also Natural selection; Speciation
    adaptation and, 7, 223, 223f
    of amphibians, 241, 340, 349, 349f
    anatomical comparisons in, 245–246,
        245f, 246f
    of animals, 319–320, 319f
    "backwards," 250, 250f
    in bacteria, 8, 8f, 234
    of brain, 530, 530f
    in bursts, 262, 262f
    as characteristic of life, 7–8
    clues to, 239–241, 239f, 240f
    coevolution, 384–385, 385f, 446
    convergent, 246, 246f
    definition of, 217
    DNA in evidence for, 40, 248
    of eggs, 336
    embryonic development revealing clues to,
        246–247, 246f, 247f
    of endoskeleton, 518
    of eukaryotes, 284–286, 285f, 286f

    as fact, 14
    of feathers, 472, 472f
    fossils as evidence of, 239, 239f, 241–243,
        241f, 242f, 262
    genetic drift and, 231–233, 231f
    gradual, 262, 262f
    of hiccups, 248, 248f
    of humans, 343–348, 343f–347f, 519, 530, 576
    of language, 133, 133f
    of limbs, 338
    of lungs, 338
    macroevolution, 218t, 255
    of mammals, 342
    microevolution, 218, 218t, 256
    misconceptions about, 222t, 224, 226, 231
    molecular clock studies of, 249, 249f
    mutations and, 231
    observing, 258
    of photosynthesis, 276, 301
    "pinnacle" of, 226
    of plants, 262, 301, 302f, 305, 443
    and populations, 217, 217f, 226–229, 227f–229f
    predictions in theory of, 14, 14f, 16, 349
    protein sequences in evidence for, 248, 249f
    simultaneous, 472
    species distribution and, 244, 244f
    theory of, 8, 14, 218–222, 218f, 219f, 221f,
        222f, 222t
    validity of, 239
    of vertebrates, 335, 338
    in viruses, 234
Evolutionary developmental biology, 247
Evolutionary tree, 9f
Excretion, 570–571
Excretory system
    in annelids, 327
    in arthropods, 329
    in echinoderms, 334
    in flatworms, 325
    in mollusks, 326
    in nematodes, 328
Exercise
    amount of, 542
    breathing in, 551
    and cortisol, 513
    and heart rate, 542, 542t
    and lactic acid fermentation, 108, 108f
    and metabolism, 81
    muscle fiber types and, 528–529, 528f
Exhalation, 550, 550f. See also Breathing
Exocytosis, 76t, 80, 80f, 128
Exons, 117, 117f
Exoskeleton, 35, 329, 330, 517–518, 517f
Exotoxins, 283
Experiments
    controlled, 12, 12f
    definition of, 12
    design of, 12–13, 13f
    limitations of, 14–15

Exponential population growth, 363, 363f, 366, 366f
External fertilization, 599, 599f
External oblique, 523f
External respiration, 546, 551
Extinction, 224, 225f
   age structure of population and, 359
   background rate of, 263, 263t, 403–404
   current rate of, 268, 403–404, 403f
   definition of, 256t, 263, 403
   human actions and, 263, 264, 264t, 268
   mass, 262, 262f, 263, 263t, 340, 404
   overexploitation and, 413, 413f
   recent, 264, 264t
   risk of, 403
Extracellular digestion, 563f, 564
Extracellular matrix, 462
Extrafloral nectaries, 268, 268f
Extremophiles, 281, 281f
Eye
   of cave animals, 246, 246f
   embryonic development of, 615
   vestigial, 245, 245f
   in vision, 494, 494f, 495f
Eye color, 189

F

F_1 generation, 175, 175t, 176, 178
F_2 generation, 175, 175t, 176, 178
Face lift, 465
Facilitated diffusion, 76t, 78, 78f, 480
Fact, definition of, 14
Facultative anaerobes, 280
$FADH_2$, 100, 100f, 104, 105
Familial hypercholesterolemia, 186t
Families, 265
Farsightedness, 495, 495f
Fast-twitch fibers, 528, 528f
Fat. *See* Lipids
Fat-based fat substitute, 42, 42f
Fat-soluble vitamins, 559t
Fat substitutes, 42, 42f
Fatty acids, 34t, 40, 41, 41f
Feathers, 341, 472, 472f
Fecal transplant, 576
Feces, 563, 569, 570
Feedback, 469
   negative (*See* Negative feedback)
   positive, 469, 616
Feeding
   in annelids, 327
   in arthropods, 329
   in cnidarians, 324
   in echinoderms, 334
   in flatworms, 325
   in mollusks, 326
   in nematodes, 328

relationships in, 383–384, 383f, 384f
   in sponges, 323
Female reproductive organs, 604–605, 604f, 615, 615f
Female reproductive system, 604–607, 604f–607f, 606t
Femoral artery, 539f
Femoral vein, 539f
Femur, 518f
Fermentation, 108, 108f, 199, 283, 283f, 294, 527, 564
Ferns, 304, 308, 308f, 309f
Fertility awareness method, as birth control, 609t
Fertility treatments, 164, 206, 208–209, 211, 611t
Fertilization
   definition of, 139, 172, 599
   double, 312, 444f, 446–447, 447f
   external, 599, 599f
   internal, 599f, 600
   in life cycle, 139, 139f, 157, 157f
   monitoring fertility and, 608
   in plants, 444f, 445
   and pregnancy, 611, 611f, 612f
   in uterine tube, 604, 611
Fertilizers, 395, 395f, 396f, 397, 398, 406, 423, 424
Fetal alcohol syndrome, 188
Fetal stage, 615, 615f
Fetus, immune system and, 587
Fever, 466, 585, 585f
Fiber, dietary, 35, 560
Fibrous root system, 429, 429f, 429t
Fibula, 518f
"Fight or flight," 485, 510
Filter, in urinary system, 573f, 574, 574f
Filtration, in urine formation, 574, 574f
Finches, 220, 221
Fins, 338
Fire, 377, 433
Fire ants, 44, 44f
Fish aquarium, 396
Fisheries, 413, 414
Fishes, 338–339, 339f
   bony, 339, 339f
   cartilaginous, 339, 339f
   circulatory system in, 538, 538f
   lobe-finned, 339, 339f
   ray-finned, 339, 339f
   skeleton of, 517f, 518
Fishing, 413, 414
Fitness, natural selection and, 225, 225f
Flagellated protozoa, 290–291, 290f
Flagellum, 52f, 53, 63, 278f, 279, 279t, 287, 287f, 603, 603f
Flatworms, 322, 325, 325f, 517, 538, 564, 564f, 594
Flight, 246, 332, 341
Florigen, 453
Flowering plants. *See* Angiosperms

Flowers, 312f
   in angiosperm life cycle, 312–313, 313f, 444f, 445
   definition of, 305t, 444
   functions of, 303f, 305
   heat production in, 109, 109f
   in mutant plants, 125
   photoperiod and, 454, 454f
   in plant anatomy, 421f, 445, 445f
   viral infection and, 131, 131f
Fluid mosaic, 54, 55f
Flukes, 325, 325f
Fluoxetine, 55
Flu vaccine, 591
Flu virus, 127, 128, 129, 132, 234
Folic acid, 559t, 613
Follicle cells, 604, 605, 605f, 611, 611f
Follicle mites, 333
Follicle-stimulating hormone (FSH), 506f, 507, 512, 603, 603f, 606f, 607
F_1 generation, 175, 175t, 176, 178
Food. *See also* Diet
   E. coli in, 283, 283f
   and ecological footprint, 368
   energy in, 69–70, 99, 99f, 107, 107f, 558
   fungi and, 294, 294f
   genetically modified, 198
   junk, 43
   labeling, 21, 44, 44f, 560, 560f
   metabolic rate and, 558
   mold on, 294, 294f
   natural, 35
   nutrients in, 558
   organic, 35
   plants as, 438
   preserving, 296
   raw materials in, 558
   steps in acquisition of, 562–563, 563f
   survival without, 109
Food and Drug Administration (FDA), 15, 105, 314
Food chain, 388
Food poisoning, 466
Food safety, 204
Food webs, 6, 6f, 388–390, 388f–390f
Foot, in mollusks, 326, 326f
Foramen magnum, 344, 344f
Foraminiferans, 291, 291f
Forebrain, 486, 487, 487f
Forensics, 519
Foreskin, 601f
Forests. *See also* Rain forest
   acid rain and, 408f, 409
   biodiversity in, 404–405
   boreal, 378f
   coniferous, 378f, 405
   deciduous, 378f
   deforestation, 404, 404f, 405
Formic acid, 44

Fossil(s), 219, 220
  age of, 242–243, 242f
  and background extinction rate, 263
  bones, 519
  Darwin on, 220, 221
  definition of, 241
  early theories on, 219
  as evolutionary evidence, 239, 239f,
      241–243, 241f, 242f, 262
  feathers in, 472
  gradualism in, 262
  human evolution in, 345–346, 345f, 346f
  as incomplete record, 241
  of marsupials, 244
  of plants, 304
  of prokaryotes, 277
  punctuated equilibrium in, 262
  of reptiles, 340
  Tiktaalik, 241, 349, 349f
  transitional forms in, 241
Fossil fuels, 92, 304, 394, 408, 410
Founder effect, 232, 232f
FOXP2 gene, 133, 133f
Fragile X syndrome, 185t
Frameshift mutation, 124, 124t
Franklin, Rosalind, 113, 113f
Fraternal (dizygotic) twins, 162, 162f, 611
Free-living flatworms, 325
Freezing, of water, 30–31, 31f, 31t
Freshwater biomes, 380f, 381, 381f, 405
Frogs, 267, 340, 340f, 503, 503f, 600
Frontal lobe, 488, 488f
Fructose, 34, 35f
Fruit(s)
  as adaptation, 305
  definition of, 305t, 444, 456
  development of, 312, 313f, 448–449,
      448f, 453
  in diet, 438
  functions of, 303f, 449, 449f
  in plant anatomy, 421f
  seedless, 451
  types of, 313, 449, 449t
  vegetables versus, 422, 456
Fruit fly, 123f, 175, 180, 180f, 181, 181f, 330
Fruiting body, 293, 293f
F$_2$ generation, 175, 175t, 176, 178
Fungi, 292–296
  anatomy of, 292–294, 293f
  animals compared to, 292
  characteristics of, 292
  chitin in, 35, 292
  classification of, 292f, 293–294
  as decomposers, 6, 6f, 294, 294f, 394
  diseases caused by, 294–295, 295f, 297
  endophytes as, 295
  in food, 294, 294f
  as heterotrophs, 280, 292
  as invasive species, 412
  kingdom of, 9f, 10
  in lichens, 296, 296f
  in medicine, 294, 294f
  mycorrhizae as, 295–296, 295f, 397, 423
  plants compared to, 292
  reproduction in, 6, 7, 293–294
  sexually transmitted infections caused by,
      610, 610t
  transgenic, 199

G

G$_1$ phase, 145, 145f, 147f
G$_2$ phase, 145, 145f, 146f
GABA (gamma aminobutyric acid), 483
Galápagos Islands, 219f, 220, 221, 259–260, 259f
Gallbladder, 565f, 568
Gallstones, 568
Gamete(s)
  definition of, 305t
  in meiosis, 139, 139f, 157, 172, 599
  in plants, 304, 304f, 445, 445t
  tracking genes in, 175–177, 176f, 177f
Gamete intrafallopian transfer (GIFT), 611t
Gametic isolation, 257f, 258
Gametophytes
  in angiosperms, 305f, 312, 313f, 444f,
      445–446
  in bryophytes, 305f, 306, 306f, 307f
  definition of, 305t, 445t
  gametes produced by, 304, 304f
  in gymnosperms, 305f, 310, 311f
  in seedless vascular plants, 305f, 308, 309f
  size of, 305f
Gap junction, 64, 64f
Garter snakes, 384
Gas exchange, 551, 551f. *See also* Breathing;
      Respiratory system
Gastric bypass surgery, 511
Gastric juice, 567, 568t, 570
Gastrocnemius, 523f
Gastrointestinal (GI) tract, 564, 565f
Gastropods, 326, 326f
Gastrovascular cavity, 564
Gastrula, 321, 321f, 613
Gastrulation, 612f, 613
Gatenby, Robert, 150
Gelatin, 349
Gelsinger, Jesse, 210
Gene(s). *See also* Alleles; DNA; Inheritance
  bottleneck effect on, 232–233, 232f
  and cancer risk, 150
  definition of, 114t, 117, 171, 175t
  in DNA technology, 198–204, 199f–204f
  founder effect on, 232, 232f
  in gametes, 175–177, 176f, 177f
  homeotic, 247, 247f, 600
  independent inheritance of, 178–179,
      178f, 179f
  linked, 180–181, 180f, 181f, 181t
  number of, in humans, 114, 171, 201, 589
  and obesity, 188
  sex-linked, 184–185, 184f, 185f, 185t
  variety of, meiosis and, 160–162
  in viruses, 126, 126t
Gene doping, 211
Gene editor, 210, 210f
Gene expression, 117, 121–123
Gene pool, 217, 218t, 256, 256t
Gene therapy, 209–210, 209f, 211, 592
Genetic code, 118, 118f
Genetic diagnosis, preimplantation, 208–209, 208f
Genetic diversity, 6–7, 125, 143, 155, 403,
      416, 599
Genetic drift, 231–233, 231f, 231t
Genetic engineering, 198
Genetic mutations. *See* Mutations
Genetic recombination, and immune function,
      589–590, 590f
Genetic testing, 209, 211
Genital herpes, 610t
Genital warts, 610, 610t
Genome, 114, 114t, 141
Genotype, 174, 175f, 175t, 183–184
Genotype frequencies, 226–227, 227f
Gentamicin, 120
Genus, 9, 265
Geographic barriers, speciation and, 259–260,
      259f, 260f
Geologic timescale, 239, 240f, 273, 273f
Germ(s), 8
Germ cells, 157, 601
Germination, 450–451, 450f, 453, 455, 455f
Germ layers, 612f, 613
Germ theory, 13–14
Giant, pituitary, 507, 507f
Giardia, 291
Gibberellins, 451t, 453, 453f
Gigantism, 151
Gills, 338
Ginkgo, 310, 310f
Glands
  definition of, 462, 503
  hormone secretion in, 466, 503, 505f
  tissues in, 462, 462f
Glans, 601f
Global warming. *See* Climate change
Glomeromycetes, 292f, 293, 295
Glottis, 547, 547f
Glucagon, 508f, 510, 510f
Glucocorticoids, 508f, 509f, 510
Glucose, 35
  in blood, 510–511, 510f, 575
  blood carrying, 535
  energy yield from, 106, 106f

in glycolysis, 100, 100f, 102–103, 102f, 106, 106f
in Krebs cycle, 100, 100f, 104, 104f, 106, 106f
in photosynthesis, 85, 85f, 92
structure of, 35f
in urine, 573
Glutamate, 483
Glutamic acid, 118f
Glutamine, 118f
Gluten, 313
Gluteus maximus, 523f
Glycerol, 34t, 40, 41f, 349
Glycine, 36, 118f
Glycogen, 35, 292, 529
Glycolysis, 100, 100f, 102–103, 102f, 104t, 106, 106f
Glyphosate, 212, 212f
Gnetophytes, 310, 310f
Goiter, 22
Goldfinch, 442f
Golgi apparatus, 53f, 56f, 58–59, 59f
Gonadotropin-releasing hormone (GnRH), 603, 603f, 606–607, 606f
Gonads, 601
G_1 phase, 145, 145f, 147f
Gonorrhea, 610t
Gould, John, 220
Gradualism, 262, 262f, 263t
Granum, 61, 61f, 87, 87f
Grasses, 312, 385, 429, 564
Grasshopper mice, 498, 498f
Grassland
    grasses in, 385
    human activity and, 404, 405
    temperate, 378f
    tropical, 379f
Graves disease, 509
Gravitropism, 429, 455, 455f
Gray matter, 486, 486f, 487
Great apes, 343f, 344
Great auk, 264t
Green algae, 289, 289f, 296, 301–302, 301f
Greenhouse effect, 394, 405, 409–410, 409f
Greenhouse gases, 409–411, 409f
Grey seals, 363, 363f
Ground tissue, in plants, 424–428, 425f, 427f, 431, 431f
Groundwater, 393, 393f, 406
Growth. See also Development
    as characteristic of life, 7
    modular, in plants, 430–433, 430t, 431f–433f, 452
Growth hormone (GH), 503, 506f, 507, 507f, 603
G_2 phase, 145, 145f, 146f
Guanine, 39, 39f, 113, 114, 114f, 115f
Guard cells, 426, 426f
Guillain-Barré syndrome, 485

Gulf of Mexico, dead zone in, 406, 407f
Gymnosperms, 302f, 304, 305, 305f, 310–311, 310f, 311f, 447
Gypsy moths, 412

H

Habitats
    of annelids, 327
    of arthropods, 329
    carrying capacity of, 364, 364f
    climate change and, 409–411, 409f–411f
    of cnidarians, 324
    definition of, 358, 373
    and diet, 557, 557f
    of echinoderms, 334
    of flatworms, 325
    human destruction of, 404–406, 404f, 405f
    isolation of, 257f, 258, 416, 416f
    of mollusks, 326
    of nematodes, 328
    pollution of, 406–409, 406f–408f
    protection of, 406, 414, 414f
    shared, and speciation, 260–261, 260f
    of sponges, 323
Hadean eon, 240f
Hagfishes, 338, 338f
Hair, 342, 614
Hair cells, in ear, 496, 497f
Half-life, 23, 242
Hammer, 496, 496f
Hamsters, 493
Hand, prosthetic, 516f
Hand sanitizer, 75
Haploid cells, 157, 157f, 172, 293f, 294, 294f
Hardy-Weinberg equilibrium, 226–227, 227f
Hearing, 496–497, 496f, 497f, 497t
Hearing aids, 497
Heart, 535, 538, 538f–541f, 540–542, 542t, 615. See also under Cardiac
Heart attack, 549
Heartbeat, 541–542, 541f
Heartburn, 55, 570
Heart muscle cells, 62, 62f
Heart rate, 542, 542t
Heartwood, 432, 433f
Heat, effect of
    on muscles, 529
    on proteins, 38, 38f
Heat energy
    in energy conversions, 70
    and evaporation, 30
    in food webs, 6, 6f, 390–391
Heavy metals, 406
Helium, 25
Helper T cells, 586–587, 586f, 592
Hemiptera, 333t
Hemispheres, cerebral, 487

Hemoglobin, 38, 113, 120, 534f, 536, 539, 549, 551, 552
Hemophilia, 184, 184f, 185t, 537
Hepatitis A, 591
Hepatitis B, 591, 610t
Herbaceous plants, 422, 422f, 426, 427
Herbal remedies, 314
Herbicides, 93, 200
Herbivores, 376, 383, 562, 564–565, 564f
Herbivory, 381t, 383, 383f
Hermaphrodites, 167, 167f, 323, 327
Herpes, genital, 610t
Herpes simplex virus, 126f, 128, 129, 610t, 613
Heterotrophs. See Consumers
Heterozygous genotype, 174, 175f, 176, 176f, 228–229
Hibernation, 107, 471
Hiccups, 248, 248f, 550
Hindbrain, 486–488, 487f
Hippocampus, 488, 489
Histamine, 584, 584f, 585, 593
Histidine, 118f
Histones, 144, 144f
HIV. See Human immunodeficiency virus
Holocene epoch, 240f
Homeostasis
    blood in, 535–537, 535f–537f
    breathing, 551, 551f
    calcium, 522, 522f
    as characteristic of life, 5t, 6
    definition of, 6, 468
    hormones in, 504
    kidney function, 574–575, 575f
    negative feedback in (See Negative feedback)
    organ system interactions in, 468–469, 468f
    positive feedback in, 469, 616
    temperature, 6, 6f
    water balance, 571, 571f
Homeotic genes, 125, 247, 247f, 600
Hominids, 343f, 344, 348t
Hominins, 343f, 344, 348t
Hominoids, 343f, 344, 348t
Homo erectus, 345f, 346, 530
Homo ergaster, 345f, 346, 530
Homo floresiensis, 345f
Homo habilis, 345f, 346
Homo heidelbergensis, 345f, 346
Homologous pair, 156, 156f, 160–162, 160f, 161t, 171f, 172, 180, 181f, 611
Homologous structures, 245, 245f
Homo naledi, 346
Homo neanderthalensis, 345f, 346
Homo rudolfensis, 345f
Homo sapiens, 9, 10f, 265, 345f, 346
Homosexuality, 125
Homozygous dominant genotype, 174, 175f, 176, 176f

Homozygous recessive genotype, 174, 175f, 176, 176f, 179
Hooke, Robert, 49
Hormones. *See also* Endocrine system
   in birth control, 609t
   and body weight, 562
   and breast milk production, 507
   definition of, 451, 503
   and female reproductive function, 606–607, 606f, 607f
   importance of, 503
   and kidney function, 507, 509f, 510, 574–575, 575f
   lipid-soluble, 504f, 505, 505t
   and male reproductive function, 603, 603f
   in metabolism, 508–512, 508f–511f
   in metamorphosis, 503, 503f
   negative feedback regulation of, 504, 510f, 512
   in plants, 451–453, 451t, 452f, 453f
   in pregnancy, 612, 613
   in puberty, 503, 603
   synthetic, 509, 510
   target cells and, 503, 504–505, 504f, 505t, 506f
   water-soluble, 504, 504f, 505t, 510
Hornworts, 306, 306f
Horseshoe crabs, 330–331, 331f
Horsetails, 308, 308f
Hot flashes, 607
Human(s)
   autosomal disorders in, 186t
   bacterial infections in, 282–283
   brain of, 347, 348, 486–489, 486f–488f
   cells of, 57, 145
   chimpanzee genome compared to genome of, 250
   chromosomes of, 156, 156f, 171–172, 171f
   circulatory system of, 539–545, 539f, 541f
   cloning, 207
   culture of, 348
   definition of, 348t
   development of, 611–616, 611f, 612f, 614f–616f
   digestive system of, 565–570, 565f
   endocrine system of, 505–512, 505f, 506f
   evolution of, 343–348, 343f–347f, 519, 530, 576
   extinction caused by, 263, 264, 264t, 268
   eye color of, 189
   fungal diseases in, 295, 297
   habitat destruction by, 404–406, 404f, 405f
   height of, 151, 189
   immune system of, 580–597
   life cycle of, 139, 139f, 157, 157f
   microbiota in, 282
   migration of, 347–348, 347f
   muscular system of, 523f
   nervous system of, 485f
   number of genes in, 114, 171, 201, 589
   origin of language of, 133, 133f
   osmoregulation in, 575f
   pheromones of, 493
   as "pinnacle of evolution," 226
   as primates, 343–344, 343f
   reproductive system of, 601–607, 601f–607f, 606t
   respiratory system of, 545–551, 547f
   scientific name for, 9, 10f, 265
   skeleton of, 518f
   skin color of, 189f, 348
   thermoregulation in, 470, 470f
   urinary system of, 572–575, 572f, 573f
   vestigial organs in, 246
   viruses infecting, 126t
   X-linked disorders in, 185t
Human chorionic gonadotropin (hCG), 612, 613
Human Genome Project, 201
Human immunodeficiency virus (HIV), 127, 128, 130, 234, 537, 591, 592, 610, 610t, 613
Human papilloma virus (HPV), 130–131, 591, 605, 610, 610t
Human population, 366–368
   age structure in, 367, 367f
   allele frequencies in, 217, 217f, 227
   birth rates in, 367, 367f, 369
   death rates in, 360, 367, 367f, 368, 368t
   ecological footprint of, 368, 368f
   exponential growth of, 366, 366f
   immigration and, 361
   Malthus on, 220
   projected growth of, 366, 366f, 369
Humerus, 518f
Hummingbirds, 446, 446f, 471, 558
Humoral immunity, 586f, 587
Hunting, 365, 413, 413f
Huntington disease, 174, 184, 186t, 483t
Huxley, Thomas Henry, 221
Hybrid breakdown, 257f, 258
Hybrid infertility, 257f, 258
Hybrid inviability, 257f, 258
Hybrids, 173, 173f, 175, 175t, 257
Hydra, 324, 324f, 564
Hydrilla, 412
Hydrocarbons, 33
Hydrochloric acid, 32f, 34, 567
Hydrogen
   in amino acids, 36, 36f
   atomic number of, 22
   as bulk element, 22
   energy shell of, 24f, 25
   as essential plant nutrient, 422, 423, 423f
   forms of, 22, 22f
   in hydrocarbons, 33
   in origin of life, 274
   in periodic table, 21f
   structure of, 24f
   in water molecule, 28, 28f
Hydrogen bonds, 25t, 28, 28f, 29, 30, 40, 40f
Hydrogen cyanide, 275
Hydrogen ions, 22, 22f, 32, 32f, 78
Hydrogen sulfide, 276, 369
Hydrolysis, 33–34, 41f, 73, 73f
Hydrophilic substances, 30, 54, 54f, 78
Hydrophobic substances, 30, 40, 54, 54f
Hydrostatic skeleton, 322, 324, 328, 517, 517f
Hydroxide, 22, 32
Hygiene, and immune function, 593, 594
Hymenoptera, 333t
Hypertension, 545, 561
Hyperthyroidism, 509
Hypertonic solution, 77f, 78
Hyphae, 292, 293f, 294, 295
Hypoglycemia, 511
Hypophosphatemic rickets, 185t
Hypotension, 545
Hypothalamus, 487, 487f
   in endocrine system, 505–508, 505f, 506f
   in kidney function, 574–575, 575f
   in limbic system, 488
   in negative feedback loop, 469, 470, 512
   in reproduction, 512, 603, 603f, 606, 606f
   in thermoregulation, 470, 585
Hypotheses
   definition of, 11
   evidence supporting, 11, 12
   in scientific method, 11, 11f
   testing, 12–13
   theory versus, 13–14
Hypothyroidism, 509
Hypotonic solution, 77f, 78

I

Ice, 30–31, 31f, 31t
Icefish, 31, 552, 552f
Ichthyophthirius multifilis, 291
Identical (monozygotic) twins, 162, 162f, 167, 613
Immune system, 580–597
   adaptive immunity in, 583, 583f, 586–590, 586f–588f, 589t, 590f
   allergies and, 592, 593, 593f, 594
   clonal deletion in, 589–590, 590f
   clonal selection in, 586f, 587, 588f, 589
   components of, 581–583, 582f
   disorders of, 592–593, 592f, 593f
   and fever, 466, 585, 585f
   functions of, 467–468, 467f, 581
   gene doping and, 211
   gene therapy and, 210
   genetic recombination in, 589–590, 590f
   hygiene and, 593, 594
   innate defenses in, 583–585, 583f–585f
   lymphatic system in, 582, 582f
   organ transplantation and, 588
   pathogens and, 581, 585f
   during pregnancy, 587

primary immune response in, 590, 590f
secondary immune response in, 590, 590f, 591
stress and, 513
vaccines and, 590–591, 591t, 594
viral infections and, 128, 129
white blood cells in, 536–537, 581–582, 581f, 584–585, 584f
Immunodeficiencies, 592–593, 592f
Impact theory, 263, 263f
Implantation, 612–613, 612f
Incisor teeth, 567f
Incomplete digestive tract, 322, 322f, 564, 564f
Incomplete dominance, 182, 183f, 183t
Incontinence, urinary, 572
Independent assortment, law of, 178, 179f
Independent variables, 12, 13t
Indeterminate growth, in plants, 430
Indirect development, 600
Infant, development of, 611–616, 611f, 612f, 614f–616f
Infections
bacterial, 282–283, 283f
and birth defects, 613
in burn patients, 584
ear, 496
latent, 129–130, 130t
lysogenic, 128–129, 129f, 130t
lytic, 128, 129f, 130t
opportunistic, 592–593, 592f
viral, 128–131, 129f, 130f
white blood cells fighting, 536–537, 584–585, 584f
Inferior vena cava, 539f, 540f, 541, 541f
Infertility treatments, 164, 206, 208–209, 211
Inflammation, 584–585, 584f, 585f
Influenza vaccine, 591
Influenza virus, 127, 128, 129, 132, 234
Infrared radiation, 86
Ingestion, 562
Inhalation, 550, 550f. *See also* Breathing
Inheritance, 170–195
complex patterns of, 182–183, 182f, 183f
environment and, 188–189, 188f
linked genes in, 180–181, 180f, 181f
Mendel's experiments on, 172–175, 172f–175f
modes of, 186–187, 187f
one-gene, 175–177, 176f, 189
polygenic traits in, 189, 189f
of sex, 184, 184f
sex-linked, 184–185, 184f, 185f, 185t
two-gene, 178–179, 178f, 179f
Inhibition, in enzyme activity, 74, 75f
Initiation
of DNA transcription, 116, 116f
of DNA translation, 119, 119f
Innate defenses, in immune system, 583–585, 583f–585f
Inner cell mass, 612, 612f
Inner mitochondrial membrane, 101, 101f, 101t

Insecticides, 200
Insects, 35, 190, 246, 322, 332–333, 332f, 333t, 384–385, 446, 517
Insertion mutation, 124, 124t
Insulin, 36, 120, 123, 503, 508f, 510, 510f, 511, 592
Integumentary system. *See also* Skin
functions of, 467, 467f
in homeostasis, 468f
Interbreeding, species and, 255–256
Interleukins, 585
Intermediate filaments, 63, 63f
Intermembrane compartment, 101, 101f, 101t
Internal fertilization, 599f, 600
Internal respiration, 546, 551
International Union for Conservation of Nature (IUCN), 403–404
Interneurons, 479, 479f
Internodes, 421f, 422
Interphase
in meiosis, 158, 158f
in mitosis, 145, 145f, 146f, 147f
Intersex condition, 617
Interstitial fluid, 468, 468f, 539
Intertidal zone, 380f
Intestine. *See* Large intestine; Small intestine
Intracellular digestion, 563, 563f
Intracytoplasmic sperm injection (ICSI), 611t
Intrauterine device (IUD), 609t
Introns, 117, 117f, 201
Invasive species, 382, 382f, 412–413, 412f
Invertebrates, 319
In vitro fertilization (IVF), 164, 206, 208–209, 211, 611t
Iodine, 22, 509
Ionic bonds, 25–26, 25t, 26f
Ions
definition of, 22
as hydrophilic substances, 30
Iris, 494, 494f
Iron, 22, 23, 423f, 551, 585
Iron-deficiency anemia, 549
Iron pyrite, 275, 275f
Isolation, reproductive, 256–258, 257f, 258f, 416, 416f
Isoleucine, 118f
Isopods, 332
Isotonic solution, 77–78, 77f
Isotopes, 23, 23f, 23t, 242

J

Jacobs syndrome, 166t
Jasmonic acid, 452
Javan tiger, 264t
Jaws, 335, 335f, 338, 345, 615
Jellyfish, 324, 324f, 517, 517f, 564
Jenner, Edward, 591
Joints, 521, 521f
J-shaped curve, 363, 363f

Jugular vein, 539f
Junk food, 43
Jurassic period, 240f

K

Kangaroo rats, 571, 571f
Kangaroos, 342, 342f
Karyotype, 156, 156f
Kelp, 288–289, 288f, 385, 385f
Keratin, 38
Keystone species, 385, 385f
Kidney(s)
functions of, 572
hormones and, 507, 509f, 510, 574–575, 575f
nephrons in, 573–575, 573f
tissues in, 462f
in urinary system, 572, 572f
Kidney cells, 60, 78
Kidney dialysis, 575, 575f
Kidney failure, 575
Kidney transplants, 575
Kidney tubules, 573f, 574, 574f
Kilocalories, 69–70
Kinetic energy, 69, 69f
Kingdoms, 265
Klinefelter syndrome, 166t, 617
Knuckle-walking, 344
Koala, 365
Krebs cycle, 100, 100f, 104, 104f, 104t, 106, 106f
Kudzu, 412, 412f

L

Labels, nutrition, 21, 44, 44f, 560, 560f
Labia majora, 604f, 605, 615f
Labia minora, 604f, 605, 615f
Labor, 616, 616f
lac operon, 121, 121f
Lactase, 74
Lactic acid fermentation, 108, 108f, 527
Lactobacillus, 108
Lactose, 34, 74, 568
Lactose intolerance, 74, 568
Lakes, 380f, 387, 406, 408
Lamarck, Jean-Baptiste, 218f, 219
Lampreys, 338, 338f
Lancelets, 337, 337f
Land plants, 302f–305f, 303–305, 426
Language, origin of, 133, 133f
Large intestine, 565f, 569–570, 569f
Larva, 330, 330f, 600
Larynx, 547, 547f
Laser eye surgery, 495
Las Vegas dace, 264t
Latent infection, 129–130, 130t
Lateral bud. *See* Axillary bud
Lateral line, 339
Lateral meristems, 430t, 431, 431f, 432–433, 432f

Lateral root, 421f
Law of independent assortment, 178, 179f
Law of segregation, 176–177, 177f
Laysan honeycreeper, 264t
Lazarus Project, 268
Leaves
    anatomy of, 427–428, 428f
    cells of, 62, 62f
    color change by, 89
    compound, 427f, 428, 429t
    essential elements absorbed by, 423–424
    of gymnosperms, 310
    nutrient deficiencies in, 423f
    photosynthesis in, 87, 87f, 427–428
    pigments in, 86, 86f
    as plant adaptation, 303, 303f
    in plant anatomy, 421, 421f
    simple, 427f, 428, 429t
    stems supporting, 427, 427f
    tissues in, 426
    water in, 434–435
Leeches, 327, 327f
Lemmings, 364, 364f
Lemon juice, 32, 32f, 60
Lens, 494, 494f, 495, 495t
Lepidoptera, 333t
Leptin, 188, 562, 562f
Lesser apes, 343f, 344
Leucine, 118f
Leukemia, 149, 537
Leukocytes. *See* White blood cells
Lice, 333
Lichens, 296, 296f, 386–387, 386f
Life
    cells as units of, 49–51, 49f–51f
    characteristics of, 5–8, 5t
    components of, 3, 4f
    development as characteristic of, 7
    energy as requirement of, 5–6, 6f
    evolution as characteristic of, 7–8
    growth as characteristic of, 7
    informational molecule of, 3, 3f
    internal constancy in, 6, 6f
    meaning of, 16
    organizational hierarchy of, 4f, 5
    origin of, 273–276, 273f–276f
    reproduction as characteristic of, 6–7, 7f
    tree of, 9, 9f
Life histories, 361–362, 362f, 369, 369f
Life table, 360, 360f
Ligaments, 463f, 520t, 521
Ligase, 141
Light
    electromagnetic spectrum of, 86, 86f
    in greenhouse effect, 409, 409f
    in photosynthesis, 85, 85f, 86, 88–89, 89f, 454
    plants influenced by, 454, 454f

Light microscope, 49f, 50, 50f, 278
Light reactions, 88–91, 89f, 90f, 90t
Lignin, 294, 303f, 304, 425, 426
Limb(s)
    in chordates, 335, 335f
    embryonic development of, 615
    evolution of, 338
    homeotic genes and, 247, 247f
    homologous, 245, 245f
    of primates, 343, 344
    prosthetic, 530
    in snake evolution, 245, 245f, 250, 250f
    vestigial, 245, 245f
Limbic system, 488
Linkage groups, 180, 180f
Linkage maps, 181, 181f
Linked genes, 180–181, 180f, 181f, 181t
Linnaeus, Carolus, 255, 265
Lipase, 568
Lipids, 40–43
    definition of, 40
    digestion of, 568, 569f
    energy extraction from, 107, 107f
    functions of, 34t, 43
    as organic molecules, 34t, 40–43
    structure of, 34t, 40, 41, 41f
    types of, 40
Lipid-soluble hormones, 504f, 505, 505t
Lipoproteins, 536
Liposuction, 465, 465f
Liver
    artificial, 471
    in digestive system, 565f, 568
    functions of, 568
Liver cells, 59, 60
Liverworts, 306, 306f
Lizards, 341, 341f, 599
Lobe-finned fishes, 339, 339f
Lobsters, 332f, 517
Locomotion
    in annelids, 327
    in arthropods, 330
    in cnidarians, 324
    in echinoderms, 334
    in flatworms, 325
    in mollusks, 326
    muscles in, 522t, 523
    in nematodes, 328
    organ systems supporting, 466, 466f
    in primates, 344
    vertebrate skeleton in, 519t
Locus, 175t
Lodgepole pines, 384, 385f, 433
Logistic population growth, 364, 364f
Lone Star ticks, 592
Long-term memory, 488
Loose connective tissue, 463f, 464
Lophotrochozoans, 321

Lumbar vertebrae, 518f
Lung(s), 335
    evolution of, 338
    gas exchange in, 551, 551f
    measuring function of, 550
    respiratory surfaces in, 546, 546f
    tissues in, 462f
Lung cancer, 149, 549
Lungfishes, 339, 339f
Luteinizing hormone (LH), 506f, 507, 512, 603, 603f, 606f, 607
Lyell, Charles, 218f, 219
Lyme disease, 281, 333
Lymph, 468, 582, 586
Lymphatic system
    functions of, 467f, 468, 582
    in immune system, 582, 582f
Lymph capillaries, 582
Lymph nodes, 582, 582f, 586
Lymphocytes, 581–582, 581f, 586, 586f, 589–590
Lymph vessels, 582, 582f
Lynx, 477–478, 477f
Lysine, 118f
Lysogenic infection, 128–129, 129f, 130t
Lysosome, 53f, 59, 59f, 79
Lytic infection, 128, 129f, 130t

M

Macroevolution, 218t, 255
Macronutrients, 422–423, 423f, 559
Macrophages, 581, 584, 585, 586–587, 586f
Mad cow disease, 120, 132f, 133
Magnesium, 22, 422, 423f
Magnetic fields, bacteria and, 65, 65f
Magnetosomes, 65
Malaria, 229, 229f, 291–292
Male pattern baldness, 183, 186
Male reproductive organs, 601–602, 601f, 615, 615f
Male reproductive system, 601–603, 601f–603f
Malignant tumor, 149, 149f
Malnutrition, 561, 561f
Malthus, Thomas, 220
Maltose, 34
Mammals, 244, 262, 342, 342f, 538, 538f
Mammary glands, 342
Mammoths, 415
Mandibulates, 330, 332–333, 332f, 332t
Manganese, 423f
Mantle, 326, 326f
Maple syrup, 437
Marfan syndrome, 183, 183f, 186t
Marine biomes, 380f, 381, 381f
Marine toads, 412
Mark-recapture, in population estimations, 358–359
Marrella, 320f
Marrow cavity, 520, 520f
Marsupials, 244, 342, 342f

Martinique lizard, 264t
Mass extinction, 239, 262, 262f, 263, 263t, 340, 404
Mass number, 23, 23t
Mast cells, 582, 584, 584f, 593, 593f
Mating. *See also* Reproduction
    nonrandom, 231t, 233, 233f
Matrix
    extracellular, 462
    mitochondrial, 101, 101f, 101t
Matter
    definition of, 21
    elements as fundamental types of, 21–22, 21f
Measles, 591, 594
Meat
    allergic reaction to, 592
    energetic cost of, 391, 391f
Mechanical digestion, 567
Mechanical isolation, 257f, 258
Mechanoreceptors, 490, 491, 496
Medicine
    DNA technology in, 198, 199, 208–211,
        208f–210f
    fungi used in, 294, 294f
    stem cells in, 205–206
Mediterranean shrubland, 379f
Medulla oblongata, 486–487, 487f, 545, 551, 551f
Medusa, 324
Megaspores, 444f, 445
Meiosis, 139, 139f, 154–169, 599
    definition of, 139, 142t, 157, 172
    DNA in, 158–159
    in egg cell production, 604, 605f
    errors in, 164–166, 165f, 166f, 166t
    and genetic variety, 160–162
    importance of, 157
    law of independent assortment and, 178, 179f
    law of segregation and, 176–177, 177f
    mitosis versus, 158, 159, 162–163, 163f
    in sperm production, 602–603, 602f
    stages of, 158–159, 158f–159f
Melanin, 348, 583
Melatonin, 508f, 511–512
Membrane, origin of, 276, 284, 284f
Membrane potential, 480
Membrane transport, 75–80, 76f–80f, 76t
Memory
    brain and, 5, 5f, 488–489
    as emergent property, 5, 5f
    immunological, 590, 590f, 591
    long-term, 488
    short-term, 488
Memory cells, 586f, 587, 588f, 590, 590f, 591
Mendel, Gregor
    and Darwin, 222
    experiments on inheritance by, 171, 172–175,
        172f–175f
    on law of independent assortment, 178
    on law of segregation, 176–177, 177f

    on one-gene inheritance, 175–176, 176f
    on two-gene inheritance, 178
Mendeleyev, Dmitry, 21
Meninges, 489
Menopause, 607
Menstrual cramps, 607
Menstrual cycle, 606, 607f
Menstruation, 606, 607f, 608f
Mercury, 103, 390, 391, 391f
Meristems, 430–433, 430t, 431f–433f, 452
Mesoderm, 321, 321f, 613, 614
Mesophyll, 87, 87f, 93, 93f, 94f, 428, 428f, 429t
Mesozoic era, 240f, 340, 347
Messenger RNA. *See* mRNA
Metabolic rate, 509, 510, 558
Metabolism. *See also* Cellular respiration
    body temperature and, 470
    definition of, 69, 71, 81
    exercise and, 81
    hormones in, 508–512, 508f–511f
    in prokaryotes, 280, 280f
    in starvation, 109
Metacarpals, 518f
Metamorphosis, 330, 330f, 503, 503f, 600
Metaphase
    in meiosis, 158f, 159, 159f, 161
    in mitosis, 146f, 147
Metastasis, 149, 149f
Metatarsals, 518f
Methane, 24, 26, 27f, 31, 33, 274, 275, 410
Methanogens, 281
Methylmercury, 391, 391f
Microbes. *See also* Archaea; Bacteria; Protists
    beneficial, 570, 576, 584, 605
    classification of, 278
    human uses of, 283–284, 283f
    importance of, 273, 277, 282
    intestinal, 570, 576
    pathogenic, 282–283
    photosynthesis in, 276
Microevolution, 218, 218t, 256
Microfilament, 63, 63f
Micronutrients, 422, 423f, 559
Microscopes, 49–50, 49f, 50f, 278
Microspores, 444f, 445
Microtubule, 63, 63f, 145
Microvilli, 567f, 568
Midbrain, 486, 487, 487f
Migration
    definition of, 231t, 233
    and evolution, 233, 233f
    human, 347–348, 347f
    and population size, 361
    as tool against parasites, 398, 398f
Milk, production and secretion of, 56f, 57–59,
        80, 116, 342, 507
Milkweed, 398
Miller, Stanley, 274–275

Miller experiment, 274–275, 274f
Millipedes, 322, 332, 332f
Mineralocorticoids, 508f, 509f, 510
Minerals, dietary, 22, 559, 559t
Miocene epoch, 240f
Miscarriage, 166, 606, 613
Mistletoe, 437, 437f
Mites, follicle, 333
Mitochondria
    ATP produced by, 101, 101f
    in cell anatomy, 53f
    definition of, 101, 101t
    energy extraction by, 60, 61f
    origin of, 284–286, 285f, 286f
    in sperm, 603f
Mitochondrial DNA, 60, 62, 203–204
Mitosis, 139–140, 139f, 145–148, 157
    definition of, 139, 142t
    in egg cell production, 605
    meiosis versus, 158, 159, 162–163, 163f
    in sperm production, 602, 602f
    stages of, 146–147, 146f–147f
Moa, 264t
Modern evolutionary synthesis, 222, 222f
Modular growth, in plants, 430–433, 430t, 431f–433f
Molaison, Henry, 488–489
Molars, 567f
Mold
    and allergies, 295
    on food, 294
    slime, 286, 290, 290f
    water, 289, 289f
Mole, 245, 245f
Molecular clock, 249, 249f
Molecular formula, 24
Molecules
    definition of, 24
    organic (*See* Organic molecules)
    in organizational hierarchy of life, 4f, 5
Mollusks, 326, 326f, 517, 538, 617, 617f
Molting, 329, 518
Molybdenum, 423f
Monarch butterflies, 398, 398f
Monkeys, 343–344, 343f
Monocots, 312
    flower structure in, 445
    leaf veins of, 428, 428f
    roots of, 429, 429f
    seeds of, 448, 448f
    stem anatomy of, 427, 427f
Monogamy, 513, 513f
Monohybrid cross, 176, 176f, 177t
Monomers, 33, 35, 36, 275, 275f
Monosaccharides, 34, 34t, 35f
Monosodium glutamate (MSG), 43
Monotremes, 342, 342f
Monozygotic (identical) twins, 162, 162f, 167, 613
Morgan, Thomas Hunt, 178, 180

Morning sickness, 613, 615

Mosquitoes, 229, 229f

Mosses, 306, 306f, 307f

Moths, 14, 14f, 16, 332f, 446, 493

Motor neurons, 479, 479f, 526, 526f

Motor response, 477f, 478, 479f

Motor unit, 526, 526f

Mountains, climate of, 376, 376f

Mouth, 492–493, 493f, 547, 547f, 564, 565f, 566, 566f, 569f

Movement. *See* Locomotion

mRNA
  degradation of, 122, 122f
  DNA transcription and, 116f, 117, 117f
  DNA translation and, 118–120, 118f, 119f
  function of, 57, 115
  in milk production, 56f, 57
  processing, 117, 117f, 122, 122f

Mucus
  in digestive tract, 566
  in respiratory tract, 546–547, 548

Mules, 160, 258, 258f

Müller, Herman J., 139

Multicellularity, 286, 286f. *See also* Eukaryotes; Eukaryotic cells

Multiple births, 611

Multiple fruit, 449, 449t

Mumps, 591

Muscle(s)
  definition of, 517
  in digestive tract, 566, 566f
  embryonic development of, 614
  skeletal (*See* Skeletal muscles)

Muscle cells, 108, 108f

Muscle contraction, 63, 524–526, 525f, 526f, 541, 548, 566, 566f

Muscle fibers, 524, 524f, 528–529, 528f

Muscle movement, 522–526, 525f, 526f

Muscle pairs, 523, 523f

Muscle tissue, 464–465, 464f

Muscular system, 466, 466f, 517, 522–529

Mussels, 382, 382f

Mutagen, 124–125

Mutant allele, 175, 175t

Mutations, 123–125, 123f–125f
  in asexual reproduction, 143, 599
  causes of, 124–125
  definition of, 123, 141, 231t
  and evolution, 231
  and genetic diversity, 7, 125, 143
  in homeotic genes, 247, 247f
  and phenotype, 183
  radiation-induced, 23
  researchers inducing, 198
  types of, 123–124, 124f, 124t
  in viruses, 8, 132

Mutualism, 381t, 383, 383f

Mycelium, 292, 293f

Mycobacterium tuberculosis, 549

Mycorrhizae, 295–296, 295f, 397, 423

Myelin sheath, 478, 478f, 482, 482f

Myofibrils, 524, 524f

Myosin filaments, 38f, 524, 524f, 525, 525f

Myosin gene, 530, 530f

Myriapods, 332f, 332t

N

NADH, 100, 100f, 103, 104, 105, 108, 108f

NADPH, 88–91, 89f, 90f

Najash, 250, 250f

Names, scientific, 9, 265

Nasal cavity, 547, 547f

National Audubon Society, 12

Native plants, 398

Natural food, 35

Natural killer cells, 582, 584, 587

Natural selection. *See also* Evolution
  adaptation and, 7, 8, 223, 223f
  and antibiotic resistance, 8, 8f, 223f, 228, 229, 231, 234
  and cystic fibrosis, 80, 80f
  Darwin on, 220–221, 221t
  definition of, 7, 220, 231t
  directional, 228, 228f
  disruptive, 228, 228f
  environment and, 224, 224f
  fitness and, 225, 225f
  mutations and, 7, 125
  as not having goal, 224–225
  and plant anatomy, 422
  populations shaped by, 8, 228–229, 228f, 361–362, 362f
  sexual selection in, 230, 230f
  stabilizing, 228, 228f
  in *Staphylococcus aureus,* 8f, 223f
  in viruses, 8

Nearsightedness, 495, 495f

Negative feedback, 469, 469f
  in blood pressure regulation, 545
  in breathing control, 551, 551f
  definition of, 74
  endocrine system and, 477, 504, 509, 509f, 510f, 512
  nervous system and, 477
  in reproduction, 512, 603, 603f, 606f
  in thermoregulation, 470

Neisseria gonorrhoeae, 610t

Nematodes, 328, 328f, 517, 600

Neogene period, 240f

Neon, 24f, 25

Nephrons, 573–575, 573f

Nerves
  in bones, 520, 520f
  definition of, 484
  functions of, 485f
  in muscles, 524, 524f, 526, 526f

Nervous system
  in annelids, 327
  in arthropods, 329
  autonomic, 485, 545, 548, 566
  central, 477, 479f, 485f–489f, 486–489
  communication function of, 466, 466f, 477–478, 477f, 482–483, 482f, 483f, 503
  disorders of, 483, 483t, 485
  in echinoderms, 334
  embryonic development of, 614, 615
  in flatworms, 325
  human, 485f
  in mollusks, 326
  in nematodes, 328
  neurons in, 477, 478–479, 478f, 479f
  parasympathetic, 485
  peripheral, 477, 479f, 484–485, 485f
  roles of, 477–478, 477f
  and senses, 490–497
  somatic, 485
  speed of, 479, 480, 482
  sympathetic, 485

Nervous tissue, 465, 465f, 477. *See also* Neurons

Net primary production, 390

Neuroglia, 465, 465f, 477

Neurons, 48f
  action potentials in, 479–482, 480f–482f
  anatomy of, 478, 478f
  division in, 145
  exocytosis in, 80
  function of, 62, 62f
  and memory, 5, 5f
  in muscle contraction, 526, 526f
  in nervous system, 477, 478–479, 478f, 479f
  as nervous tissue, 465, 465f
  types of, 478–479, 479f

Neurotransmitters, 482–484, 482f–484f, 526, 526f

Neutral solution, 32, 32f

Neutrons
  in atoms, 22, 22f, 22t
  in isotopes, 23, 23f

Newts, 384

Niacin, 559t

Niche, 373

Nicotine, 484

Night length, and plants, 454, 454f

Nitric acid, 408

Nitrification, 395f, 396

Nitrogen
  in acid deposition, 408, 408f
  as bulk element, 22
  energy shells of, 24f
  as essential plant nutrient, 422, 423, 423f, 424
  and eutrophication, 397, 397f
  in periodic table, 21f
  in soil, 377
  structure of, 24

Nitrogen cycle, 282, 395–396, 395f

Nitrogen fixation, 282, 282f, 395, 424, 424f

Nitrogen-fixing enzymes, 424
Nitrogenous base, 39, 39f
Nitrogenous waste, 570–571, 571f
Nitrogen oxides, 408, 408f
Nitrous oxide, 410
Node, 421f, 422
Nodules, 424, 424f
Nonavian reptiles, 340, 341, 341f
Noncompetitive inhibition, 74, 75f
Nondisjunction, 164–166, 165f
Nonpolar covalent bond, 27–28
Nonrandom mating, 231t, 233, 233f
Non-vascular plants, 304, 306
Noradrenaline. *See* Norepinephrine
Norepinephrine, 483, 484, 508f, 509–510, 509f
Nose
    in breathing, 546–547, 547f
    embryonic development of, 615
    in sense of smell, 492, 492f
"Nose job," 465
Notochord, 335, 335f
Nuclear envelope, 56f, 57, 57f, 146f
Nuclear pores, 57, 57f
Nucleic acids. *See also* DNA; RNA
    definition of, 38–39
    digestion of, 569f
    functions of, 34t, 40
    as organic molecules, 34t, 38–40
    structure of, 34t, 39, 39f, 40f
Nucleoid, 52, 52f, 278, 278f, 279t
Nucleolus, 57, 57f, 146f
Nucleosomes, 144, 144f
Nucleotide, 34t, 39–40, 39f, 113, 114, 114f
Nucleus
    of atom, 22, 22f
    cells containing, 9, 9f, 52, 53, 53f, 57, 57f
    in meiosis, 139
    in mitosis, 139
    in neuron, 478f
Nutrients. *See also* Macronutrients;
        Micronutrients
    as abiotic factor, 377
    absorption of, 568
    definition of, 558
    essential, for plants, 422–423, 423f
    steps in acquisition of, 562–563, 563f
    in varied diet, 559–560, 559t, 560f
Nutrition labels, 21, 44, 44f, 560, 560f

O

Obesity, 149t, 188, 511, 511f, 560, 561–562
Obligate aerobes, 280
Obligate anaerobes, 280
Observation, in scientific method, 10–11, 11f, 14
Occipital lobe, 488, 488f
Ocean(s)
    acidification of, 395, 411
    as biome, 380f

counting animals in, 361
    currents in, 375, 375f, 407
    pollution of, 407
Ocean fisheries, 413
Octopuses, 326, 326f, 538
Odonata, 333t
Olestra, 42, 42f
Olfactory bulb, 492, 492f
Olfactory receptors, 492, 492f
Oligocene epoch, 240f
Olive oil, 41, 41f
Omega-3 fatty acids, 43
Omeprazole, 55
Omnivores, 562, 565
Onions, 62, 62f, 421f, 428
On the Origin of Species (Darwin),
        220–221
Oocyte(s), 604–606, 605f, 611, 611f. *See also*
        Egg cells
Oocyte donation, 611t
Oogenesis, 605–606, 605f, 606t
Oparin, Alex I., 274
Open circulatory system, 538, 538f
Operator, 121
Operon, 121
Opossums, 342, 342f
Opportunistic infections, 592–593, 592f
Opportunistic life history, 362, 362f
Optic nerve, 494, 494f, 495t
Orbitals, 25
Orchids, 14, 14f, 16, 224, 224f
Orders, 265
Ordovician period, 240f
Organ(s). *See also specific organs*
    artificial, 471
    definition of, 461
    development of, 614–615, 615f
    in organizational hierarchy, 4f, 5, 461f
    of plants, 421–422, 421f, 422f
Organelles
    in eukaryotes, 53, 56–62, 56f–62f, 62t
    in organizational hierarchy of life, 4f, 5
Organic food, 35
Organic molecules, 33–43
    carbohydrates as, 34–35, 34t
    definition of, 33
    lipids as, 34t, 40–43
    nucleic acids as, 34t, 38–40
    origin of, 274–275, 274f
    proteins as, 34t, 36–38
    subunits of, 33–34, 33f
Organisms
    in biogeochemical cycle, 392, 392f
    as component of life, 3, 4f
    definition of, 357f
    in organizational hierarchy of life,
        4f, 5
    source of energy for, 6, 6f
    transgenic, 198

Organizational hierarchy
    in body, 461f
    of life, 4f, 5
Organ systems
    definition of, 461
    in homeostasis, 468–469, 468f
    interconnectedness of, 466–468, 466f–467f
    in organizational hierarchy, 4f, 5, 461f
Organ transplantation
    and immunodeficiency, 593
    organs for, 471, 575
    and tissue rejection, 588
Orgasm, 602, 605
Origin of life, 273–276, 273f–276f
Origin of Life, The (Oparin), 274
Orlistat, 105
Orthoptera, 333t
Osmoregulation, 571, 574–575, 575f
Osmosis, 76t, 77–78, 77f, 78f, 435–437, 571
Osteoarthritis, 521
Osteoporosis, 509, 522, 522f
Outgroup, 266, 266t
Ovalbumin, 38f
Oval window, 496, 496f
Ovarian cancer, 605
Ovarian cycle, 606, 607f
Ovary(ies)
    in female reproductive system, 604,
        604f, 606f
    as human endocrine glands, 505f, 512, 512f
    in plants, 444f, 445, 445f
Overexploitation, 413, 413f
Ovulation, 605, 605f, 607f, 608, 608f, 612f
Ovules, 310, 311f, 444f, 445, 445f
Ox beetles, 230f
Oxidation, 71, 72f, 89
Oxidation-reduction reactions, 71–72, 71f,
        89, 100
Oxygen
    absence of, in early atmosphere, 274,
        276, 301
    in aquatic ecosystems, 377
    blood carrying, 535, 536, 541, 551, 551f
    as bulk element, 22
    in carbon dioxide, 31
    electronegativity of, 25
    energy shells of, 24f, 25
    as essential plant nutrient, 422, 423, 423f
    molecule of, 26, 27, 27f, 28
    in periodic table, 21f
    in photosynthesis, 85, 85f, 301
    prokaryotes and, 280, 280f
    in respiration, 546, 546f
    structure of, 24
    in water molecule, 28, 28f
Oxygen debt, 527
Oxytocin, 506f, 507, 616
Ozone, 95, 407
Ozone layer, 408, 409, 409f

P

Pacemaker, 541, 541f
Pain receptors, 490, 491
Pain sensation, 491–492, 497t
Paleocene epoch, 240f
Paleogene period, 240f
Paleolithic diet, 561
Paleontology, 239, 273
Paleozoic era, 240f
Palestinian painted frog, 264t
Pancreas, 505f, 508f, 510–511, 510f, 565f, 568, 592
Pancreatic juice, 32f, 568, 568t
Papillae, on tongue, 493f
Pap test, 605
Paralysis, 489, 489f, 526
Paramecium, 60, 291, 291f
Paranthropus, 345f, 346, 530
Paranthropus aethiopicus, 345f
Paranthropus boisei, 345f
Paranthropus robustus, 345f
Parasites, 291–292, 328, 398, 398f, 594
Parasitic plants, 437, 437f
Parasitism, 381t, 383, 383f
Parasympathetic nervous system, 485
Parathyroid glands, 505f, 508f, 509
Parathyroid hormone (PTH), 508f, 509
Parenchyma, 424–425, 425f
Parental chromatid, 160f, 161, 161t, 180, 181f, 181t
Parietal lobe, 488, 488f
Parkinson disease, 483t
Parrots, 557, 557f
Particles, in atom, 22, 22f, 22t
Particulates, 407
Passenger pigeons, 264t, 413
Passive immunity, 589, 589t
Passive transport, 76–78, 76f–78f, 76t
Patella, 518f
Pathogens, 581, 585f
Pattern formation, 600
Peanut allergy, 593
Peanut plants, 449
Pea plants, 172f–175f, 173–175, 303f
Peat moss, 307, 314
Pectoral girdle, 518f, 519
Pectoralis major, 523f
Pedigrees, 186–187, 186t, 187f
Peer review, in scientific method, 11f, 12
Pelvic girdle, 518f, 519
Penetration, viral, 127f, 128
Penguins, 360, 360f, 361, 382, 382f, 461, 557, 557f
Penicillin, 52, 282, 294, 593
Penicillium, 294
Penis, 572, 601f, 602, 615f
Pepsin, 567, 568t
Peptide bond, 36
Perception, 490, 492t, 498
Percolation, 393, 393f
Periodic table, 21, 21f, 22

Peripheral nervous system, 477, 479f, 484–485, 485f
Peristalsis, 566, 566f
Permafrost, 314, 314f
Permian period, 240f
Peroxisome, 60, 60f
Persistent organic pollutants, 407
Pertussis, 594
Pest control, 365
Pesticides, 407
Petals, 428, 445, 445f, 448, 453
Petiole, 427, 427f
Pflesteria, 297
P generation, 175, 175t, 178
pH
    in acid rain, 408, 408f, 409
    in cells, 32, 32f, 59
    and enzyme activity, 74
    of gastric juice, 567
    lysosomes and, 59
    and proteins, 38
    scale, 32, 32f
    vacuoles and, 60
    of vagina, 605
    of water, 32, 32f
Phagocytes, 581, 581f, 584, 585
Phagocytosis, 79, 584
Phalanges, 518f
Phanerozoic eon, 240f
Pharyngeal slits, 335, 335f
Pharynx, 547, 547f, 565f, 566
Phenotype, 175, 175f
    codominance and, 182, 183f
    definition of, 175, 175t
    environment and, 188–189, 188f, 224, 224f
    incomplete dominance and, 182, 182f
    pedigree charts of, 186–187, 187f
    pleiotropy and, 183, 183f
    sex-linked, 184–185
Phentermine, 105
Phenylalanine, 74, 118, 118f, 174
Phenylketonuria (PKU), 74, 174, 186t, 209
Pheromones, 493
Philodendron, 109, 109f
Phloem, 303f
    conducting cells in, 425f, 425f
    functions of, 304, 425, 426, 436–437, 436f
    in leaves, 428, 428f
    in roots, 429
    secondary, 432, 432f, 433, 433f
    in seedless vascular plants, 308
Phloem sap, 436, 436t
Phloem transport, 436–437, 436f
Phosphate, 114, 114f
Phospholipid(s), 34t, 40, 54, 54f, 276
Phospholipid bilayer, 54, 54f, 75
Phosphorus
    as bulk element, 22
    in diet, 559t
    energy shells of, 24f

    as essential plant nutrient, 422, 423f
    and eutrophication, 397, 397f
    in soil, 377
Phosphorus cycle, 396f, 397
Photic zone, 381
Photons, 86
Photoperiod, 454, 454f
Photoreceptors, 454, 490, 494–495, 494f
Photorespiration, 93
Photosynthesis, 84–97
    in algae, 85, 95, 287–289, 301
    ATP in, 88–93, 89f, 90f
    carbon reactions in, 89, 89f, 90f, 92, 92f
    chemical reactions in, 71, 85
    in chloroplasts, 60–62, 61f, 62f, 62t, 87–88, 87f, 88f, 91, 428
    in cyanobacteria, 281, 301
    definition of, 85
    in ecosystem, 376
    energy conversion in, 70
    evolution of, 276, 301
    importance of, 85
    in leaves, 87, 87f, 427–428
    light in, 85, 85f, 86, 88–89, 89f, 454
    light reactions in, 88–91, 89f, 90f, 90t
    in microbes, 276
    pathways of, 93–94, 93f, 94f
    pigments in, 86, 86f
    plant adaptations for, 303–304, 303f
    in protists, 287–289
    stages of, 88–89, 89f
    water in, 31, 31t
Photosystems, 88, 88f, 90–91, 90f, 90t
Phototroph, 280, 280t
Phototropism, 454, 454f
Phrenic nerve, 248, 248f
Phthiraptera, 333t
Phylogenetics, 265
Phylum, 10, 265
Physiology, definition of, 421, 461
Phytophthora infestans, 289, 289f
Phytoplankton, 376, 406
Pigments
    in photosynthesis, 86, 86f
    in vision, 495
Pigs, 234, 234f
Pili, 278f, 279, 279t
Pine, 304, 310–311, 310f, 311f, 384, 385f
Pineal gland, 505f, 508f, 511–512
Pineapple, 74, 94
Pinocytosis, 79
Pioneer species, 386–387, 386f
Pith, in plants, 427, 427f, 429f, 429t
Pituitary dwarfism, 507, 507f
Pituitary giant, 507, 507f
Pituitary gland, 505–508, 505f, 506f, 512, 575. *See also* Anterior pituitary; Posterior pituitary
Placebo, 13
Placenta, 342, 614, 614f

Placental mammals, 244, 342, 342f
Placozoa, 331, 331f
Planarians, 325f
Plant(s). *See also* Flowers; Fruit; Leaves;
    Photosynthesis; Root; Seed; Stem
    adaptations in, to life on land, 302f–305f,
        303–305, 426
    adhesion in, 29
    animals versus, 431
    aquatic, 303
    artificial selection in, 198, 198f, 221, 221f, 422
    asexual reproduction in, 6, 7f
    as autotrophs, 6, 6f, 10, 33, 85, 280, 301
    as biofuels, 304
    carnivorous, 420f, 421, 424, 428
    cellulose in, 35
    classification of, 302f, 303
    cloning, 206f, 207
    coevolution in, 384–385, 385f, 446
    cohesion in, 29
    defenses of, against herbivores, 383, 383f, 452
    in desert, 246, 246f, 303, 310, 430
    dormancy in, 448, 453, 453f
    endophytes and, 295
    essential elements required by, 422–423, 423f
    evolution of, 262, 301, 302f, 305, 443
    extrafloral nectaries in, 268, 268f
    fertilization in, 444f, 445
    fitness in, 225, 225f
    flowering (*See* Angiosperms)
    as food, 438
    fossilized, 239f
    fungal diseases in, 294, 295f
    fungi compared to, 292
    germination in, 450–451, 450f, 453, 455, 455f
    gravitropism in, 429, 455, 455f
    growth of, 60, 430–433, 430t, 431f–433f,
        450–451, 450f, 452
    heat production in, 109, 109f
    herbaceous, 422, 422f, 426, 427
    hormones in, 451–453, 451t, 452f, 453f
    human uses of, 314
    importance of, 301
    light and, 454, 454f
    Mendel's experiments on, 172–175
    mutant, 125
    mycorrhizae and, 295–296, 423
    native, 398
    non-vascular, 304, 306
    organizational hierarchy of, 4f
    parasitic, 437, 437f
    as phototrophs, 280
    polygenic traits in, 189
    polyploidy in, 164, 261, 261f
    reproduction in, 6, 7, 7f, 155, 304–305, 304f,
        305f, 443
    reproductive barriers in, 258
    speciation in, 261, 261f, 262
    starch in, 35

    thigmotropism in, 455, 455f
    at timberline, 377
    tissues in, 303f, 304, 421, 426–431, 426f,
        431f, 434–437
    transgenic, 199–200, 200f, 212, 212f
    vascular, 304, 305f, 308–313
    vegetative parts of, 421–422, 421f, 422f
    viral infections of, 131, 131f, 453
    water absorbed by, 423, 429–431, 429f, 435, 435f
    woody, 422, 422f, 426, 430t, 431, 432–433, 433f
Plantae, 9f, 10
Plant cells
    anatomy of, 53, 53f
    cell division in, 145, 146, 146f–147f, 148, 148f
    chloroplasts in, 60–62, 61f
    types of, 424–425, 425f
    vacuoles in, 59–60, 60f, 431
    walls of, 64, 64f
Plasma, 535, 535f, 536, 537, 537f, 581, 584
Plasma cells, 588f, 589, 593, 593f
Plasma membrane. *See* Cell membrane
Plasmid, 199, 199f, 278, 278f
Plasmodesmata, 64, 64f, 131
Plasmodial slime mold, 290
Plasmodium, 229, 229f, 291–292
Plastic surgery, 465, 465f
Plastids, 61–62
Platelets, 535, 535f, 537, 537f
Plate tectonics, 243–244, 243f, 244f, 263, 376
Platyhelminthes, 325, 325f
Platypus, 342, 342f
Pleiotropy, 183, 183f, 183t
Pleistocene epoch, 240f
Pliocene epoch, 240f
Pneumonia, 549
Poaching, 415
Poisons, 103, 120
Polar body, 605, 605f, 606
Polar covalent bond, 27, 28
Polar ice, 379f, 411, 411f
Polio, 526, 594
Pollen, 303f, 304, 305, 312, 313f
Pollen grain, 304, 305t, 310, 312, 445, 446
Pollen sac, 444f, 445, 446
Pollen tube, 310–311, 446, 448
Pollination, 304, 310–311, 312, 384–385, 444f,
    446, 446f
Pollinators, 446, 446f
Pollution, 406–409, 406f–408f
    air, 407–409, 407f, 408f
    and biomagnification, 391, 391f, 407
    reducing, 408, 411
    water, 390, 391, 391f, 406–408, 406f, 407f
Polychaetes, 327, 327f
Polycystic kidney disease, 575
Polygenic traits, 189, 189f
Polymerase chain reaction (PCR), 202, 202f
Polymers, 33, 35, 275, 275f
Polymyxin, 282

Polyp, 324
Polypeptides, 36, 37, 37f, 38
Polyploidy, 261, 261f
Polysaccharides, 34t, 35, 35f
Polyunsaturated omega-3 fatty acids, 43
Ponds, 380f
Pons, 487, 487f
Population(s), 356–371
    adaptations in, 223, 223f
    age structure of, 359, 359f
    bottleneck, 232
    definition of, 217, 218t, 357, 357f, 373
    density of, 358–359, 358f
    estimation of, 358–359, 361
    evolution and, 217, 217f, 226–229, 227f–229f
    exponential growth of, 363, 363f
    human (*See* Human population)
    logistic growth of, 364, 364f
    Malthus on, 220
    natural selection shaping, 8, 228–229, 228f,
        361–362, 362f
    in organizational hierarchy of life, 4f, 5
    sampling error in, 231–232, 231f
    size of, 359–361, 359f, 359t, 360f, 364–365, 365f
Population ecology, 357, 359
Porifera, 323, 323f
Positive feedback, 469, 616
Postanal tail, 335, 335f
Posterior pituitary, 506, 506f, 507, 575, 616
Postzygotic reproductive isolation, 257, 257f, 258
Potassium
    active transport of, 78–79, 79f
    as bulk element, 22
    in diet, 559t
    as essential plant nutrient, 422, 423, 423f
    and neuron action potential, 480, 480f,
        481, 481f
Potassium-argon dating, 243
Potato blight, 289, 289f
Potential energy, 69, 69f
Prairie chickens, 232
Prairie voles, 513, 513f
Praying mantis, 225
Precambrian supereon, 240f, 320
Precipitation, 393, 393f
Predation, 381t, 383–384, 384f
Predator control, 415
Predators, 383
Prediction
    in evolutionary theory, 14, 14f, 16, 349
    in scientific method, 11, 11f, 14
Prednisone, 510, 592
Preembryonic stage, 612–613, 612f
Pregnancy, 605, 611–616, 611f, 612f, 614f–616f
    alcohol consumption in, 188, 613
    and childbirth, 616, 616f
    chromosome testing in, 190
    fertilization and, 611, 611f, 612f
    immune system during, 587

miscarriage in, 166, 606, 613
    preimplantation diagnosis in, 208–209, 208f
    Toxoplasma and, 291
Preimplantation genetic diagnosis (PGD),
    208–209, 208f
Premature birth, 616
Premenstrual syndrome (PMS), 607
Premolars, 567f
Pressure
    blood, 469, 510, 544–545, 544f, 544t
    changes, in breathing, 550, 550f
    turgor, 60, 78, 78f, 435
Pressure flow theory, 436–437, 436t
Prey, 383
Prezygotic reproductive isolation, 257, 257f, 258
Primary consumers, 388, 388f, 390f
Primary growth, in plants, 431, 431f, 432f
Primary immune response, 590, 590f
Primary oocytes, 605, 605f, 606
Primary producers. *See* Autotrophs
Primary spermatocyte, 602
Primary structure (protein), 37, 37f
Primary succession, 386–387, 386f, 387t
Primates
    definition of, 348t
    lineages of, 343–344, 343f
    skeleton of, 344–345, 344f
Primers, 201, 201f
Prions, 132–133, 132f
Probiotics, 576
Producers. *See* Autotrophs
Product rule, 178–179, 179f
Products, in chemical reaction, 31, 31t
Progesterone, 512, 512f, 606, 606f, 607, 607f, 615
Proinsulin, 120
Prokaryotes. *See also* Archaea; Bacteria
    cells in, 52–53, 52f, 53f
    classification of, 278–280
    definition of, 9, 277, 280t
    domains of, 280–281
    eukaryotes versus, 9, 277, 277f
    human uses of, 283–284, 283f
    importance of, 281–284, 282f, 283f
    metabolism in, 280, 280f
    reproduction in, 142–143, 142f
    structure of, 277, 277f, 278–280, 278f, 279f
    success of, 277–284
Prolactin, 506f, 507
Proline, 118f
Promoter, 116, 116f, 121, 122, 123f
Prophage, 129, 129f
Prophase
    in meiosis, 158, 158f, 159, 159f, 160
    in mitosis, 146, 146f
Prosimians, 343, 343f
Prostaglandins, 602
Prostate cancer, 602
Prostate gland, 601f, 602
Prosthetic hand, 516f

Prosthetic limbs, 530
Protein(s)
    adhesion, 55
    antibodies as, 588–589, 588f
    in blood plasma, 536
    cells producing, 36, 51
    complement, 585
    in cytoskeleton, 63, 63f
    definition of, 36
    degradation of, 123
    denaturation of, 38, 38f, 75
    digestion of, 569f
    diversity of, 36, 38, 38f
    DNA translation and, 118–120, 118f–120f
    energy extraction from, 107, 107f
    evolutionary evidence in, 248, 249f
    in extracellular matrix, 462
    in facilitated diffusion, 78, 78f
    folding of, 36–38, 36f, 37f, 120, 132
    functions of, 34t, 38, 113, 134
    as organic molecules, 34t, 36–38
    processing, 123
    receptor, 55
    recognition, 55
    regulation in synthesis of, 121–123, 121f–123f
    structure of, 34t, 36–38, 36f, 37f
    transport, 54, 78
    in viral structure, 126–128, 126f, 127f
Protein-based fat substitutes, 42
Protein coat, viral, 126, 126f
Proteobacteria, 281, 281f
Proterozoic eon, 240f
Protista, 9f, 10, 287
Protists, 287–292, 287f–292f
    cilia in, 63
    classification of, 287, 287f
    definition of, 287
    as eukaryotes, 53, 287–292
    as heterotrophs, 289–292, 289f–291f
    photosynthesis in, 287–289
    reproduction in, 140f
    vacuoles in, 60
Protons
    in atoms, 22, 22f, 22t
    in isotopes, 23, 23f
Protostomes, 321
Protozoa, 290–292, 290f, 291f
Pseudocoelom, 322, 322f, 328, 328f
Pseudogenes, 201
Pseudopodia, 291
Puberty, 503, 603, 606
Pulmonary arteries, 539f, 540f, 541, 541f
Pulmonary circulation, 538, 538f, 541
Pulmonary veins, 539f, 540f, 541, 541f
Punctuated equilibrium, 262, 262f, 263t
Punnett square, 175–176, 176f, 177t, 178, 178f,
    179, 179f, 180f
Pupa, 330, 330f
Pupfish, 260, 260f

Pupil, 494, 494f, 495t
Pus, 585
Pygmy seahorse, 7f
Pyramid of energy, 390–391, 390f
Pyrenean ibex, 264f, 264t
Pyrite, 275, 275f
Pyruvate, 100–103, 100f, 102f, 103f, 108, 108f

Q

Quadriceps femoris, 523f
Quadruplets, 164, 611
Quagga, 264t
Quaternary period, 240f
Quaternary structure (protein), 37f, 38
Questions, in scientific method, 10–11, 11f
Quintuplets, 164

R

Rabbits, 233, 233f, 267, 477–478, 477f
Radial symmetry, 321, 321f, 324, 334
Radiation sickness, 23
Radiation therapy, 149
Radioactive isotopes, 23, 242
Radiolarians, 291f
Radiometric dating, 242–243
Radius, 518f
Radula, in mollusks, 326, 326f
Rain forest
    temperate, 378f
    tropical, 378f, 390, 405
Rain shadow, 376, 376f
Random orientation, 161, 161t
Rats, 15
Ray-finned fishes, 339, 339f
Reabsorption, in urine formation, 574, 574f
Reactants, 31, 31t, 74
Reaction center, 88, 90f
Receptacle, in flowers, 445, 445f
Receptor potential, 490, 491f, 492t
Receptor proteins, 55
Recessive alleles, 172f–175f, 173–175, 175t
Recognition proteins, 55
Recombinant chromatid, 160f, 161, 161t, 180,
    181f, 181t
Recombinant DNA, 198–199
Recombinant offspring, 180, 181t
Reconstructive plastic surgery, 465
Rectum, 565f, 569f, 570
Rectus abdominis, 523f
Red algae, 288f, 289
Red blood cells
    in blood composition, 535, 535f
    decline in number of, 22
    functions of, 536, 551
    osmosis in, 77–78, 77f
Red bone marrow, 520, 520f, 520t, 536, 581, 582
Red-green color blindness, 185t, 186f, 495

Redox reactions, 71–72, 71f, 89, 100
Red tide, 287–288
Reduction, 71, 72f, 89
Reflex, 486
Relative dating, 242
Release, viral, 127f, 128
Renal artery, 539f, 572f, 573, 573f
Renal vein, 572f, 573, 573f
Replication
    DNA, 40, 124, 141–142, 141f, 145, 158, 201,
        202, 202f
    viral, 127–128, 127f
Repressor, 121
Reproduction, 598–621
    asexual (See Asexual reproduction)
    as characteristic of life, 6–7, 7f
    development and, 599–600, 599f
    fitness and, 225–226, 225f
    in fungi, 7, 293–294
    other requirements balanced against,
        361–362, 369, 369f
    in plants, 6, 7, 7f, 155, 304–305, 304f,
        305f, 443
    sexual (See Sexual reproduction)
    unequal success in, 221t
Reproductive barrier, 256–258, 256t, 257f, 258f
Reproductive organs
    female, 604–605, 604f, 615, 615f
    male, 601–602, 601f, 615, 615f
Reproductive system
    cancers of, 602, 605
    embryonic development of, 614, 615, 615f
    female, 604–607, 604f–607f, 606t
    functions of, 467f, 468
    health of, 608–610, 609t, 610t
    hormones in, 512, 512f
    male, 601–603, 601f–603f
Reptiles, 340–341, 341f
Resource partitioning, 382, 382f
Resources
    competition for, 382, 382f
    and evolution, 221t
    plant acquisition of, 303–304
Respiration. See also Breathing
    aerobic, 99, 99f, 100f, 103–106, 103f, 104f,
        527, 528, 535, 546
    cellular (See Cellular respiration)
    external, 546, 551
    internal, 546, 551
Respiratory cycle, 550, 550f
Respiratory surfaces, 546, 546f
Respiratory system. See also Breathing
    in amphibians, 340
    in annelids, 327
    in arthropods, 329
    blood in, 551, 551f
    in chordates, 335
    definition of, 545
    diseases of, 549

    in echinoderms, 334
    embryonic development of, 614
    in flatworms, 325
    functions of, 467, 467f, 546, 547, 547t
    in homeostasis, 468f
    human, 545–551, 547f
    in mollusks, 326
    in nematodes, 328
Respiratory tract
    cilia in, 63, 63f, 548
    lower, 547f, 548
    upper, 546–547, 547f
Resting potential, 480, 480f, 481
"Rest or repose," 485
Restriction enzymes, 199, 199f
Retina, 494, 494f, 495t
Retinitis pigmentosa, 185t
Retrovirus, 130
Rett syndrome, 185, 185t
R group, 36, 36f, 38
Rhesus (Rh) factor, 587
Rheumatoid arthritis, 521, 585, 592
Rhinoceros, 415, 415f
Rhizobium, 282, 282f, 395, 424
Rhizoids, 306
Rhizome, 308, 427
Rib cage, 518, 518f, 550, 550f
Ribonucleic acid (RNA). See RNA
Ribose, 35f, 39, 72f, 115, 115f
Ribosomal RNA. See rRNA
Ribosomes, 278, 278f
    in animal cells, 53f
    antibiotics and, 120
    in bacteria, 52f
    definition of, 115
    in DNA translation, 119, 120, 120f
    in milk production, 56f
    in plant cells, 53f, 61f
    proteins produced by, 51, 58, 58f
    in translation, 118–119, 120f
Ribulose bisphosphate (RuBP), 92, 92f
Rice, 362f
Rickets, 185t
Rifamycin, 282
Rigor mortis, 527
Rivers, 380f, 405, 406
RNA, 39
    in bacteria, 52
    definition of, 115
    DNA transcription and, 115–117, 115f, 116f
    DNA versus, 115, 115f
    function of, 40
    messenger (See mRNA)
    origin of, 275
    ribosomal, 115, 117, 201
    structure of, 39, 39f, 40, 115f
    transfer, 115
RNA polymerase, 116, 116f, 122, 123f
RNA world, 275

Roadrunners, 557, 557f
Rockhopper penguins, 382, 382f
Rocks, 392, 392f, 396f, 397
Rod cells, 494f, 495, 495t
Root(s), 421f
    anatomy of, 429, 429f
    cohesion in, 29
    functions of, 421, 423–424, 424f, 429–430
    growth of, 431, 431f, 452
    hypertonic to soil, 78
    as plant adaptation, 303, 303f
    in seed germination, 450, 450f, 455, 455f
    tissues in, 426
Root caps, 429, 431, 431f, 455, 455f
Root hairs, 424f, 430, 430f, 435, 435f
Root nodules, 424, 424f
Roquefort cheese, 294, 294f
Rotavirus, 126f
Rotavirus vaccine, 12–13, 13f
Rotifera, 331, 331f
Rough endoplasmic reticulum, 53f, 56f, 58, 58f, 59
Rough-skinned newts, 384
Roundworms, 167, 167f, 328, 328f
rRNA, 115, 117, 201
Rubella, 594, 613
Rubisco enzyme, 92, 92f, 93
Rumen, 564, 564f
Ruminant, 564, 564f
Runoff, 393, 393f
Ruud, J. T., 552

S

Saccharin, 15, 42, 42f
Sacrum, 518f
Sagittal crest, 344f, 345
Salamanders, 95, 95f, 140f, 246f, 340, 340f, 600
Salicylic acid, 453
Saliva, 566, 568t
Salivary glands, 565f
Salmon, 599
Salmonella, 281, 283
Salt, 24, 26, 26f, 29–30, 30f, 38
Sample size, 12, 13t
Sampling error, 231–232, 231f
Sand dollars, 334, 334f
Sap
    phloem, 436, 436t
    xylem, 434, 436t, 437
Sapwood, 432, 433f
Sarcomeres, 524, 524f, 525f
Sartorius, 523f
Saturated fatty acid, 41, 41f
Savanna, 379f
Scallops, 326, 326f
Scanning electron microscope (SEM), 50, 50f
Scapula, 518f, 519
Scavengers, 388
Schizophrenia, 483t

Schleiden, Mathias J., 49
Schwann, Theodor, 49
Scientific inquiry, 11f
    limitations of, 14–15
    scientific method in, 11f
    technology and, 16
Scientific method, 10–16, 11f
Scientific names, 9, 265
Sclera, 494, 494f
Sclerenchyma, 425, 425f, 426
Scoliosis, 518, 519f
Scorpions, 331, 331f, 493, 498, 498f
Scrotum, 601–602, 601f, 615f
Sea cucumbers, 334, 334f
Seahorse, pygmy, 7f
Sea otter, 385, 385f
Seasonal affective disorder, 512
Seasons, 374, 374f
Sea stars, 334, 334f, 518
Sea urchins, 258, 334, 334f, 385, 385f, 599, 599f
Secondary consumers, 388, 388f, 390f
Secondary growth, 432–433, 432f, 433f
Secondary immune response, 590, 590f, 591
Secondary oocytes, 605, 605f, 611
Secondary sex characteristics, 601, 603, 605
Secondary spermatocyte, 602–603
Secondary structure (protein), 37, 37f
Secondary succession, 387, 387f, 387t
Second messenger, 504, 504f
Secretion, in urine formation, 574, 574f
Sediment, as pollutant, 406
Seed(s)
    cost of producing, 448
    definition of, 305, 305t
    as embryo, 447–448, 447f, 448f
    in fruits, 312–313, 313f, 422, 449, 449f
    functions of, 303f, 305
    "naked," 310–311, 310f
    size of, 451
Seed coat, 448, 448f, 450, 450f
Seed germination, 450–451, 450f, 453, 455, 455f
Seedless fruits, 451
Seedless vascular plants, 302f, 305f, 308, 308f, 309f
Seedling, 450–451, 450f
Seed plants, 310–313
Segmentation, 322, 327, 329
Segregation, law of, 176–177, 177f
Selective attention, 498
Selective breeding. See Artificial selection
Selective permeability, 75, 77–78, 77f, 78f
Self-fertilization, 173, 173f
Semen, 602
Semenya, Caster, 617, 617f
Seminal vesicle, 601f, 602
Seminiferous tubules, 602, 602f, 603
Sensation, definition of, 490, 492t
Senses
    embryonic development of, 614
    of equilibrium, 497t

general, 490f, 491–492, 491f, 497t
    of hearing, 496–497, 496f, 497f, 497t
    nervous system and, 490–497
    of pain, 491–492, 497t
    of sight, 494–495, 494f, 495f, 497t
    of smell, 492, 492f, 497t
    special, 490f, 492–497, 497t
    of taste, 492–493, 493f, 497t
    of temperature, 491–492, 497t
    of touch, 491–492, 491f, 497t
Sensor, 469t
Sensory adaptation, 491
Sensory input, 477f, 478, 479f
Sensory integration, 477f, 478, 479f
Sensory neurons, 478, 479, 479f
Sensory receptors, 490, 492t
Sepals, in flowers, 445, 445f
Sequencing, DNA, 200–202, 201f, 248, 256
Serine, 118f
Serotonin, 55, 483, 484
Severe combined immunodeficiency (SCID), 592
Sewage, 406
Sewage treatment, 284, 388
Sex, inheritance of, 184, 184f
Sex characteristics, secondary, 601, 603, 605
Sex chromosomes, 156, 156f, 172, 184, 184f
Sex-linked genes, 184–185, 184f, 185f, 185t
Sex pilus, 143, 143f
Sexual dimorphism, 230
Sexually transmitted infections (STIs), 608–610, 610t
Sexual orientation, 125
Sexual reproduction, 139, 139f, 155, 155f, 599
    in amphibians, 340, 599
    in angiosperms, 304–305, 305f, 312–313, 313f
    in annelids, 327
    in arthropods, 330
    asexual reproduction versus, 155, 163t
    bacteria and, 167, 167f
    in bryophytes, 304, 305f, 306–307, 307f
    chromosomes in, 157
    in cnidarians, 324
    definition of, 6, 7f, 443
    in echinoderms, 334
    energy costs of, 599
    in flatworms, 325
    in fungi, 7, 293–294, 293f
    and genetic diversity, 6–7, 155, 167, 599
    in gymnosperms, 304–305, 305f, 310–311, 311f
    in insects, 332
    in mollusks, 326, 617, 617f
    in nematodes, 328
    in plants, 7, 155, 304–305, 304f, 305f, 443
    in seedless vascular plants, 304, 305f, 308, 309f
    in sponges, 323, 599
Sexual selection, 230, 230f
Sharks, 246, 339f, 518
Shellac, 349
Shellfish poisoning, 288, 297

Shivering, 6, 6f
Shock, anaphylactic, 593
Shoots, 421, 421f, 426, 431, 450, 452, 452f, 453, 453f
Short tandem repeats (STRs), 203, 203f
Short-term memory, 488
Siamese cats, 188f
Siberian permafrost, 314, 314f
Sickle cell mutation, 124, 124f, 229, 229f, 292
Sieve tube elements, 425, 425f, 436–437
Sight, 494–495, 494f, 495f, 497t
Silk moths, 493
Silurian period, 240f, 331
Simple columnar epithelial tissue, 462f
Simple cuboidal epithelial tissue, 462f
Simple diffusion, 76t, 77
Simple fruit, 449, 449t
Simple leaves, 427f, 428, 429t
Simple squamous epithelial tissue, 462f
Simultaneous evolution, 472
Sink, in plants, 436–437, 436f
Sinoatrial (SA) node, 541, 541f
Sinuses, 547f
Siphonaptera, 333t
Skeletal muscles, 522–529
    athletic performance of, 528–529, 528f
    ATP generated in, 527
    contractions of, 524–526, 525f, 526f
    functions of, 522–523, 522t
    heat and, 529
    number of, 522
    organization of, 524, 524f
Skeletal muscle tissue, 464, 464f, 524
Skeletal system, 466, 466f, 517. See also Bone(s)
Skeleton
    definition of, 517
    development of, 615
    functions of, 519–522, 519t
    of primates, 344–345, 344f
    types of, 517–518, 517f
    vertebrate, 518–522, 518f
Skin. See also Integumentary system
    anatomy of, 583, 583f
    in body temperature regulation, 470
    color of, 189f, 348, 583
    embryonic development of, 614
    functions of, 583
    in immune system, 583–584, 584f
    sensory receptors in, 491–492, 491f
    tissues in, 462f
Skin cancer, 149, 150, 408, 583
Skin cells, 139
Skull, 246, 246f, 344, 344f, 345, 518, 518f
Sliding filament model, 524–526, 525f, 541
Slime capsules, 278f, 279
Slime molds, 286, 290, 290f
Slow-twitch fibers, 528, 528f
Slugs, 326, 326f
Small intestine, 565f, 567–569, 567f, 569f

Smallpox, 132, 591, 594

Smell, 492, 492f, 497t

Smog, 407, 407f

Smoking, 149t, 549, 613

Smooth endoplasmic reticulum, 53f, 56f, 58, 58f

Smooth muscle, 548, 566

Smooth muscle tissue, 464f, 465, 543

Snails, 326, 326f, 517

Snakes, 245, 245f, 250, 250f, 341, 341f

Snapdragons, 182, 182f

Sodium
  active transport of, 78–79, 79f
  as bulk element, 22
  characteristics of, 24
  in diet, 559t
  dissolution of, 29–30, 30f
  ionic bonding of, 25–26, 26f
  and neuron action potential, 480–483, 480f, 481f

Sodium bicarbonate, 568

Sodium chloride, 24, 26, 26f, 29–30, 30f

Sodium ions, 22

Sodium-potassium pump, 79, 79f, 480, 480f, 481

Soil, 377, 387, 392, 392f, 396f, 397, 405, 409,
    422–424, 435

Solutes, 30, 30f

Solution, 30, 30f

Solvent, water as, 30, 30f

Somatic cell nuclear transfer, 207, 207f

Somatic cells, 157, 207, 207f

Somatic nervous system, 485

Soundarajan, Santhi, 617

Source, in plants, 436

Southern day frog, 264t

Speciation. *See also* Evolution
  allopatric, 259–260, 259f, 260f, 262, 263t
  in bursts, 262, 262f
  definition of, 255–256, 256t
  geographic barriers and, 259–260, 259f, 260f
  gradual, 262, 262f
  after mass extinction, 262, 262f
  observing, 258
  reproductive barriers and, 256–258, 257f, 258f
  sympatric, 259f–261f, 260–261, 263t

Species
  biological, 255–256, 256t
  classification of (*See* Classification systems)
  coevolving, 384–385, 385f
  competition between, 382, 382f
  conservation of, 414–416, 415f
  definition of, 9, 255–256
  distribution of, 244, 244f, 376–377
  divergence of, 256–258, 257f
  diversity of, 403 (*See also* Diversity)
  endangered, 403
  evenness of, 386, 386f, 387t
  interactions of, 381–385, 381t, 382f–385f
  and interbreeding, 255–256
  invasive, 382, 382f, 412–413, 412f
  keystone, 385, 385f

life histories of, 361–362, 362f
  naming, 9
  overexploitation of, 413, 413f
  pioneer, 386–387, 386f
  richness of, 386, 386f, 387t
  symbiotic, 381t, 383, 383f
  threatened, 403–404, 403f
  vulnerable, 403

Spectrum, electromagnetic, 86, 86f

Speech, 547

Spermatids, 602f, 603

Spermatocytes, 602–603

Spermatogenesis, 602–603, 602f, 606, 606t

Sperm cells, 601–603
  anatomy of, 603, 603f
  in angiosperm life cycle, 444f, 445, 446–447
  in fertilization, 155, 611, 611f
  flagella of, 63, 63f, 603, 603f
  as gamete, 139, 139f, 157, 157f, 599
  production of, 512, 602–603, 602f, 606, 606t

Spermicide, 609t

S phase, 145, 145f

Sphincters, 566, 567f

Sphinx moth, 16

Sphygmomanometer, 544, 544f

Spiders, 35, 330–331, 331f

Spike mosses, 308, 308f

Spina bifida, 613

Spinal cord, 518
  damage to, 489, 489f
  functions of, 485f, 486
  gray matter in, 486, 486f
  in nervous system, 477
  nervous tissue in, 465f
  white matter in, 486, 486f

Spinal nerves, 484

Spindle, 145, 146f, 147, 158, 158f, 159

Spirillum, 279, 279f

Spirochaetes, 281

Spirometer, 550

Spleen, 536, 582, 582f

Sponges, 6, 321–323, 323f, 563, 563f, 599

Spongy bone, 520, 520f, 520t, 522f

Spontaneous abortion, 166

Sporangium, 306, 307f, 309f, 310

Spores
  fungal, 292–293, 294
  in plants, 304, 304f, 305t, 445, 445t

Sporophytes, 304, 304f
  in angiosperms, 305f, 312, 313f, 444f, 445
  in bryophytes, 305f, 306, 307f
  definition of, 305t, 445t
  in gymnosperms, 305f, 310, 311f
  in seedless vascular plants, 305f, 308, 309f
  size of, 305f

Sports, anabolic steroids in, 512, 529

Sports drinks, 30

Squids, 326, 326f, 517, 538

S-shaped curve, 364, 364f

Stabilizing selection, 228, 228f

Stamens, 445, 445f, 448

Standardized variables, 12–13, 13t

Staphylococcus, 279, 281

Staphylococcus aureus, 8f, 223f, 283

Starch, 35, 107, 430, 437, 450, 576

Starlings, 268, 412, 412f

Starvation, 109, 561, 561f, 613

Statistical analysis, 13, 13f, 203

Statistical significance, 13

Steller's sea cow, 264t

Stem(s), 421–422, 421f
  functions of, 303, 421, 427
  growth of, 431, 431f
  tissues in, 426, 427, 427f

Stem cells, 205–206, 205f, 612

Sterility, 257f, 258

Sterilization, 284

Sternum, 518f

Steroid(s), 43
  anabolic, 512, 529, 603
  in cell membrane, 54, 55f
  functions of, 34t
  structure of, 34t, 43f

Steroid hormones, 43, 505, 510, 512, 512f, 603

Stickleback fish, 297, 297f

Stigma, 445, 445f, 446, 448

Stimulus, 469t

Stingrays, 339f

Stirrup, 496, 496f

Stoats, 364, 364f

Stomach, 565f, 566–567, 567f, 569f

Stomach acid, 32f, 34, 64

Stomach lining, 566, 567f

Stomata, 87, 87f, 93, 303, 303f, 304, 426, 426f,
    428, 428f, 434f, 435, 435f

Stratified squamous epithelial tissue, 462f

Strawberry, 6, 7f

Streptococcus, 281

Streptomycin, 278, 281, 282

Stress, 485, 502f, 509–510, 509f, 513

Stroke, 489

Stroma, 61, 61f, 62t, 87, 87f, 90f

Style, in flowers, 445, 445f

Substitution mutation, 124, 124t

Succession, 386–387, 386f, 387f

Sucralose, 42, 42f

Sucrose, 34, 35f, 436–437

Sudden infant death syndrome, 551

Sugar(s), 34–35, 34t, 35f, 436–437

Sugar maple tree, 437

Sugar substitutes, 15, 42, 42f

Sulfanilamide, 282

Sulfur, 22, 276, 408, 408f, 422, 423f

Sulfur dioxide, 407

Sulfuric acid, 408

Sundew plant, 420f, 421

Sunlight. *See* Light

"Superbugs." *See* Antibiotic resistance

Superior vena cava, 539f, 540f, 541, 541f
Supplements
  dietary, 527
  herbal, 314
Surface area, of cells, 51, 51f
Surface tension, 29
"Survival of the fittest," 225, 225f
Survivorship curves, 360–361, 360f
Swallowing, 547, 548, 566
Sweating, 6, 6f, 30, 30f, 470
Sweeteners, artificial, 15, 42, 42f
Sycamore, 94f
Symbiosis, 381t, 383, 383f
Symmetry, 321, 321f
Sympathetic nervous system, 485
Sympatric speciation, 259f–261f, 260–261, 263t
Synapse, 465, 465f, 478, 478f, 479t, 482–483, 483f
Synaptic cleft, 482, 483, 483f
Synaptic terminal, 482, 483, 483f
Synthesis
  ATP, 73, 73f
  viral, 127f, 128
Synthetic cells, 55
Syphilis, 610t
Systematics, 265, 266t
Systemic circulation, 538, 538f
Systemin, 452
Systolic pressure, 544, 544f

T

Tadpoles, 248, 503, 503f, 538, 600
Taiga, 378f
Tail, postanal, 335, 335f
Tapeworms, 325, 325f
Taproot system, 421f, 429, 429f, 429t
Tardigrada, 331, 331f
Target cells, 503, 504–505, 504f, 505t, 506f
Tarsals, 518f
Taste, 492–493, 493f, 497t
Taste buds, 493, 493f
Tawny crazy ants, 44, 44f
Taxon, 265
Taxonomic hierarchy, 9–10, 10f
Taxonomy, 9, 265
Tay-Sachs disease, 59, 186t
T cells, 130, 581–582, 586, 586f, 587–588,
    587f, 592
Tea bag, 76–77, 76f
Technology, 16
Tectonic plates, 243–244, 243f, 244f, 263, 376
Teeth, 344f, 345, 566, 566f, 570, 576
Telophase
  in meiosis, 158f, 159, 159f
  in mitosis, 147, 147f
Temperate coniferous forest, 378f
Temperate deciduous forest, 378f
Temperate grassland, 378f

Temperature
  as abiotic factor, 377
  and enzyme activity, 74–75, 75f
  greenhouse effect and, 409–410, 409f, 410f
Temperature homeostasis, 6, 6f
Temperature regulation
  in body, 6, 6f, 337, 469–471, 470f, 471f
  water in, 30, 30f, 31t
Temperature sensation, 491–492, 497t
Template strand, 116, 116f
Temporal isolation, 257f, 258
Temporal lobe, 488, 488f
Tendons, 463f, 520t, 521
Tendrils, 421f, 427, 455, 455f
Tentacles, 324
Terminal bud, 421, 421f, 431
Termination
  of DNA transcription, 116–117, 116f
  of DNA translation, 119f, 120
Terrestrial biomes, 376, 377, 377f–379f
Tertiary consumers, 388, 388f, 390f
Tertiary structure (protein), 37f, 38
Test cross, 176, 176f, 177t
Testes, 505f, 512, 512f, 601–602, 601f–603f
Testicular cancer, 602
Testosterone, 43, 43f, 186, 510, 512, 512f, 603, 603f
Tetanus, 280, 281
Tetracyclines, 120
Tetrapods, 335
Thalamus, 487, 487f, 488
Thalidomide, 613, 613f
Theory, 13–14
Thermal cycler, 202
Thermophiles, 281
Thermoreceptors, 490, 491
Thermoregulation, 6, 6f, 337, 469–471, 470f, 471f
Thick filament. *See* Myosin filament
Thigmotropism, 455, 455f
Thin filament. *See* Actin filament
Thoracic vertebrae, 518f
Thrashers, 456, 456f
Threatened species, 403–404, 403f
Threonine, 118f
Threshold potential, 481, 481f
Throat. *See* Pharynx
Thylakoid, 61, 61f, 62t, 87, 87f, 88f
Thylakoid membrane, 87–88, 87f, 88f, 101
Thylakoid space, 87, 87f, 90f
Thymine, 39, 39f, 113, 114, 114f, 115f
Thymus, 581, 582f
Thyroid gland, 22, 505f, 508–509, 508f, 509f, 562
Thyroid hormones, 508–509, 508f, 509f, 562
Thyroid-stimulating hormone (TSH), 506f, 507, 509
Thyrotropin-releasing hormone (TRH), 509
Thyroxine, 508, 508f
Tibia, 518f
Ticks, 331, 331f, 333
Tidal volume, 550

Tight junction, 64, 64f
Tiktaalik, 241, 349, 349f
Timberline, 377
Timescale, geologic, 239, 240f, 273, 273f
Tissue(s)
  adipose, 43, 463f, 464
  animal, 321
  bone, 463f, 464, 519–521, 520f
  cardiac muscle, 464f, 465, 540, 540f
  connective, 463f, 464, 520–521, 524, 524f, 535
  definition of, 461
  epithelial, 462–464, 462f, 546
  muscle, 464–465, 464f
  nervous, 465, 465f, 477
  in organizational hierarchy of life, 4f, 5
  plant, 303f, 304, 421, 426–431, 426f, 431f,
      434–437
  types of, 462–465, 462f–465f
Tissue rejection, 588
Tobacco (plant), 200, 200f
Tobacco mosaic virus, 126f, 131
Tobacco use, 149t, 549, 613
Tongue, 493, 493f, 547f, 566
Tonsils, 582f
Tooth decay, 570, 576
Torpor, 471
Tortoises, 259–260, 259f, 600f
Touch, 491–492, 491f, 497t
Toxic shock syndrome, 283
Toxins, 103, 120, 406, 407, 498
Toxoplasma gondii, 291
Trace elements, 22
Trachea, 329, 547f, 548
Tracheids, 425, 425f
Traits
  in cladistics, 265–266
  environment and, 188
  polygenic, 189, 189f
Transcription, DNA, 115–117, 115f–117f
Transcription factors, 122, 123f
Transduction, 490, 492t
Trans fat, 41f, 43
Transfer RNA. *See* tRNA
Transgenic animals, 200
Transgenic bacteria, 199, 199f, 284
Transgenic plants, 199–200, 200f, 212, 212f
Transgenic technology, 198–200, 199f, 200f
Transitional forms, 241
Transition step, 103, 103f, 104t
Translation, DNA, 115, 115f, 118–120,
    118f–120f
Transmission electron microscope (TEM), 50, 50f
Transpiration, in leaves, 434–435, 435f, 436t
Transplantation. *See* Organ transplantation
Transport, membrane, 75–80, 76f–80f, 76t
Transport proteins, 54, 78
Transport vesicle, 56f
Transposons, 201

Trapezius, 523f
Tree(s), 404–405, 408f, 409, 422, 432–433, 433f, 438
Tree line, 377
Tree rings, 432–433, 433f
Treponema pallidum, 610t
Triassic period, 240f, 342
Triceps, 523, 523f
Trichinella, 328
Trichomonas vaginalis, 290f, 291, 610, 610t
Trichomoniasis, 610, 610t
Trichonympha, 290f
Trichoplax adhaerens, 331, 331f
Triglycerides, 34t, 40–43, 41f, 54
Triiodothyronine, 508, 508f
Trilobites, 330, 330f, 332t
Triphosphate, 72f
Triplets, 164, 611
Triploidy, 164
Triplo-X, 166, 166t
Trisomy 13, 166
Trisomy 18, 166
Trisomy 21, 166, 166f
tRNA, 115, 117–120, 119f, 201
Trophic level, 388, 388f, 390–391, 390f
Tropical rain forest, 378f, 390, 405
Tropical savanna, 379f
Tropism, 454
True-breeding, 173, 173f, 175, 175t
True ferns, 308, 308f
Trypanosoma brucei, 290f, 291
Trypsin, 568
Tryptophan, 36, 118f
Tubal ligation, 609t
Tube feet, 334
Tuberculosis, 549
Tubeworms, 327
Tubules
    in respiratory systems, 546f
    in urinary system, 573f, 574, 574f
Tubulin, 63
Tumor, 149, 149f, 150. *See also* Cancer
Tundra, 379f
Tunicates, 337, 337f
Turgor pressure, 60, 78, 78f, 435
Turner syndrome, 166t, 617
Turtles, 341, 341f, 361, 413f, 600, 600f
Twins, 162, 162f, 164, 167, 188, 611, 613
Tyrosine, 118f

U

Ulna, 518f
Ultraviolet (UV) radiation, 86, 149t, 279, 348, 408
Umbilical cord, 614, 614f
Unsaturated fatty acid, 41, 41f
Uracil, 39, 39f, 115, 115f

Uranium, 22
Urea, 571, 571f, 574
Ureter, 572, 572f
Urethra, 572, 572f, 601f, 602, 615f
Urey, Harold, 274
Urinalysis, 573
Urinary bladder, 572–573, 572f
Urinary incontinence, 572
Urinary system
    definition of, 572
    functions of, 467, 467f
    in homeostasis, 468f
    human, 572–575, 572f, 573f
Urinary tract infection, 572, 573
Urine, 32f, 563, 571–574, 571f
Urine formation, 572–574, 572f, 574f
Urogenital tract, 572
Uterine contractions, 468, 507, 607, 616
Uterine lining, 606f, 612, 612f, 615
Uterine tubes, 462f, 604, 604f, 611
Uterus, 468, 604, 604f, 607
Uvula, 547f

V

Vaccines, 594
    definition of, 131–132
    and immune system, 590–591, 591t
    rotavirus, 12–13, 13f
    against viral infections, 131–132, 131f
Vacuoles, 59–60, 60f, 431
Vagina, 604f, 605
Vaginal opening, 604f, 615f
Vaginal yeast infection, 605, 610t
Valence shell, 25, 28
Valine, 118f
Valves
    cardiac, 540f
    in veins, 543–544, 543f
Variable regions, 588f, 589
Variables, in experimental design, 12–13, 13t
Varicose veins, 544
Vascular bundle, in plants, 426, 427, 427f
Vascular cambium, 432, 432f, 433, 433f
Vascular cylinders, 429, 429f
Vascular plants, 302f, 304, 305f, 308–313
Vascular tissue, in plants, 303f, 304, 425, 426, 426f, 428, 429, 431, 431f, 434–437
Vas deferens, 601f, 602
Vasectomy, 609t
Vasoconstriction, 544t, 545
Vasodilation, 544t, 545
Vasopressin. *See* Antidiuretic hormone
Vegetables, 422, 438, 456
Vegetative plant parts, 421–422, 421f, 422f
Vegetative reproduction. *See* Asexual reproduction

Veins
    in human body, 463f, 539, 539f, 540, 543–544, 543f, 544, 544t
    in leaves, 428, 428f
Venom, 44, 44f
Ventilation. *See* Breathing
Ventilator, 550, 550f
Ventricle, 538, 538f, 540f, 541
Venules, 539, 543, 543f, 544t
Vertebrae, 335, 335f
Vertebral column, 518, 518f
Vertebrates, 319
    chordates as, 335–337
    circulatory system in, 538, 538f
    evolution of, 335, 338
    segmentation in, 322
    skeleton in, 518–522, 518f
Vesicles, 57, 58, 58f, 76t, 80f
Vessel elements, 425, 425f
Vestigial structures, 245–246, 245f
Villi, 567, 567f
Vinegar, 32, 32f
Virchow, Rudolf, 49
Viroids, 132, 132f, 608
Virus(es), 126–128
    as alive, 8
    bacteria versus, 8, 126, 126f
    cells versus, 126–127
    definition of, 8, 126, 130t
    drugs against, 131, 132
    evolution in, 234
    examples of, 126t
    features of, 126–127, 126f
    infection by, 128–131, 129f, 130f
    mutations in, 8, 132
    in plants, 131, 131f, 453
    in pregnancy, 613
    replication of, 127–128, 127f
    sexually transmitted infections caused by, 610, 610t
    structure of, 126–127, 126f
    vaccination against, 131–132, 131f
Visceral mass, in mollusks, 326, 326f
Vision, 494–495, 494f, 495f, 497t
Vision correction, 495
Visual cortex, 495t
Vital capacity, 550
Vitamin A, 559t, 613
Vitamin C, 559t
Vitamin D, 559t
Vitamins, 559, 559t, 613
Vitreous humor, 494, 494f
Vocal cords, 547
Volvox, 286f, 289
Vomeronasal organ, 493
Vomiting, 466, 561, 570
Vulnerable species, 403
Vulva, 605

W

Wadlow, Robert, 151, 507f
Wallace, Alfred Russel, 8, 14, 16, 218f, 221, 244
Wallace's line, 244, 244f
Warblers, 268
Warning coloration, 384, 384f
Warren, Lavinia, 151, 507f
Warts, genital, 610, 610t
Wasps, 224, 224f, 268
Waste management, 284, 388
Water
    as abiotic factor, 377
    adhesion in, 29, 29f, 31t
    in biogeochemical cycle, 392, 392f
    in blood, 536
    in body, regulation of, 570–571, 571f, 572
    bottled versus tap, 393
    and chemical reactions, 31, 31t
    cohesion in, 29, 29f, 31t, 435
    diffusion of, 77–78, 77f, 78f
    dissolution in, 29–30, 30f, 31t, 32
    eutrophication in, 397, 397f, 406
    expansion of, in freezing, 30–31, 31f, 31t
    hydrogen bonds in, 28, 28f, 29, 30
    large intestine absorbing, 569
    in leaves, 434–435
    molecule of, 24, 26, 27f, 28, 28f
    in origin of life, 274
    pH of, 32, 32f
    phospholipids in, 54, 54f
    in photosynthesis, 85, 85f
    plants absorbing, 423, 429–431, 429f, 435, 435f
    properties of, 29–31, 31t
    resources of, 368
    in seed germination, 450
    small intestine absorbing, 568
    and temperature regulation, 30, 30f, 31t
    in urine, 574
Water cycle, 393, 393f, 405
Water-dispersed fruits, 449, 449f
Water mold, 289, 289f
Water pollution, 390, 391, 391f, 406–407, 406f, 407f, 408
Water Quality Act, 415

Water-soluble hormones, 504, 504f, 505t, 510
Water-soluble substances, 30
Water-soluble vitamins, 559t
Water strider, 29f
Water treatment, 284, 388
Water vascular system, in echinoderms, 334
Watson, James, 115, 141
Wavelength, of photons, 86, 86f
Weather, 411. See also Climate(s)
Weaver birds, 230f
Webbed feet, 140, 140f
Weed killers, 93
Wetlands, 387
Whales, 245, 245f, 361
Wheat, 312, 313
Whisk ferns, 308, 308f
White blood cells
    in blood composition, 535, 535f
    endocytosis in, 79, 79f
    functions of, 536–537
    in immune system, 581–582, 584–585, 584f
    lysosomes in, 59
    types of, 581–582, 581f
White matter, 486, 486f, 487
White nose syndrome, 295
"Widow's peak" hairline, 183, 186
Wildfires, 433
Wild-type allele, 175, 175t
Wilkins, Maurice, 113, 113f
Wilmut, Ian, 207
Wind-dispersed fruits, 449, 449f
Windpipe. See Trachea
Wind-pollinated flowers, 446
Wings, 246, 332, 341
Withdrawal, as birth control, 609t
Woese, Carl, 52
Wood, 432–433, 432f, 433f
Woody plants, 422, 422f, 426, 430t, 431–433, 433f
Woolly mammoths, 415
Wrens, 390

X

X chromosome, 156, 156f, 172, 184, 184f
X inactivation, 184–185, 185f

X-linked dominant disorders, 185, 185t
X-linked recessive disorders, 184, 184f, 185t
X-ray diffraction, 113, 113f
XXY syndrome, 166t, 617
Xylem, 303f
    conducting cells in, 425, 425f
    functions of, 304, 425, 426, 434–435, 434f, 435f
    in leaves, 428, 428f
    in roots, 429
    secondary, 432, 432f, 433f
    in seedless vascular plants, 308
Xylem sap, 434, 436t, 437
Xylem transport, 435, 435f
XYY syndrome, 166t

Y

Yangtze River dolphin, 264t
Y chromosome, 156, 156f, 172, 184, 184f, 615
Yeast(s)
    characteristics of, 292
    in food, 294
    transgenic, 199
Yeast infection, 605, 610t
Yellow bone marrow, 520, 520f, 520t
Yellow-eyed penguins, 360, 360f, 361
Yellowhead jawfish, 598f
Yolk sac, 614, 614f
Yunnan box turtle, 264t

Z

Zebra mussels, 382, 382f
Zinc, 22, 423f
Zygentoma, 333t
Zygomycetes, 292f, 293, 294
Zygote
    in angiosperms, 444f, 445, 446–447, 447f
    definition of, 445t
    human, 611, 612, 612f
    as product of fertilization, 139, 139f, 157, 157f, 599, 600, 600f
Zygote intrafallopian transfer (ZIFT), 611t